Microgrids

Microgrids

Theory and Practice

Edited by

Peng Zhang
Stony Brook University, New York, USA

IEEE Press Series on Power and Energy Systems
Ganesh Kumar Venayagamoorthy, Series Editor

Published by John Wiley & Sons, Inc., Hoboken, New Jersey.
Published simultaneously in Canada.

For general information on our other products and services or for technical support, please contact our Customer Care Department within the United States at (800) 762-2974, outside the United States at (317) 572-3993 or fax (317) 572-4002.

Wiley also publishes its books in a variety of electronic formats. Some content that appears in print may not be available in electronic formats. For more information about Wiley products, visit our web site at www.wiley.com.

Library of Congress Cataloging-in-Publication Data Applied for:

Hardback ISBN: 9781119890850

Cover Design: Wiley
Cover Image: © PopTika/Shutterstock

Set in 9.5/12.5pt STIXTwoText by Straive, Chennai, India

Contents

23   A Time of Energy Transition at Princeton University   *555*

Edward T. Borer, Jr.

24   Considerations for Digital Real-Time Simulation, Control-HIL, and Power-HIL in Microgrids/DER Studies   *579*

Juan F. Patarroyo, Joel Pfannschmidt, K. S. Amitkumar, Jean-Nicolas Paquin, and Wei Li

About the Editor

Peng Zhang is a Professor of Electrical and Computer Engineering and a SUNY Empire Innovation Professor at Stony Brook University, NY, USA. He has made major contributions to strategic areas including AI-enabled smart grids, quantum-engineered resilient power grids (Quantum Grids), programmable microgrids, cybersecurity, and formal methods and reachability analysis. Prof. Zhang published over 200 journal and conference papers, technical reports and patents. He is the author of Networked Microgrids published by Cambridge University Press.

Prof. Zhang pioneered a series of quantum computing, quantum security, quantum networking and quantum machine learning algorithms that have been successfully implemented on today's noisy intermediate-scale quantum computers to solve challenging power system problems. Prof. Zhang and his team have been building an AI-operated programmable platform that integrates various deep neural network frameworks, reachability analysis, formal control, and runtime assurance technologies to enable scalable, self-protecting, autonomic and provably resilient power grids and microgrids. He actively participates in CIGRE Working Groups and has made various contributions. When he was a Performance Planning Engineer at BC Hydro in Canada, he planned, designed, and commissioned British Columbia's first large wind farm Bear Mountain Wind Project and provided planning services to multiple large wind generation projects with a total capacity of two gigawatts.

List of Contributors

Tuncay Altun
Department of Electrical and Electronics
Engineering
Yozgat Bozok University
Yozgat
Turkey

Amitkumar K. S.
Opal-RT Technologies Inc.
Montreal
Canada

William W. Anderson, Jr.
Naval Facilities Engineering Systems
Command (NAVFAC)
Engineering and Expeditionary Warfare Center
(EXWC)
Port Hueneme, CA
USA

Pouya Babahajiani
Department of Electrical and Computer
Engineering
Stony Brook University
Stony Brook, NY
USA

Rômulo G. Bainy
Department of Electrical and Computer
Engineering
University of Idaho
Moscow, ID
USA

Phil Barker
Nova Energy Specialists
Schenectady, NY
USA

Yiheng Bian
Xi'an Jiaotong University
State Key Laboratory of Electrical Insulation
and Power Equipment
School of Electrical Engineering
Xi'an
China

Zhaohong Bie
Xi'an Jiaotong University
State Key Laboratory of Electrical Insulation
and Power Equipment
School of Electrical Engineering
Xi'an
China

Edward T. Borer, Jr.
Facilities Engineering Department
Princeton University
Princeton, NJ
USA

Mikhail A. Bragin
Department of Electrical and Computer
Engineering
University of California, Riverside
Riverside, CA
USA

Clayton Burns
National Grid
Syracuse, NY
USA

Jing Dai
Energy Internet Research Institute
Tsinghua University
Beijing
China

Amol Damare
Department of Computer Science
Stony Brook University
Stony Brook, NY
USA

Ali Davoudi
Department of Electrical Engineering
The University of Texas at Arlington
Arlington, TX
USA

Hui Ding
RTDS Technologies Inc.
Winnipeg
Canada

Shamsun Nahar Edib
Department of Electrical and Computer
Engineering
University of Massachusetts Lowell
Lowell, MA
USA

Fei Feng
Department of Electrical and Computer
Engineering
Stony Brook University
Stony Brook, NY
USA

Feng Gao
Department of Industrial Engineering and
Management
College of Engineering
Peking University
Beijing
China

Michael Gouzman
Department of Electrical and Computer
Engineering
Stony Brook University
Stony Brook, NY
USA

Jacky Xiangyu Han
Data Science, Fulcrum3D
Artarmon, NSW
Australia

Guannan He
Department of Industrial Engineering and
Management
College of Engineering
Peking University
Beijing
China

Heqing Huang
Department of Electrical and Computer
Engineering
New York University
Brooklyn, NY
USA

Tong Huang
Department of Electrical and Computer
Engineering
San Diego State University
San Diego, CA
USA

Milad Izadi
Department of Electrical and Computer
Engineering
University of California
Riverside, CA
USA

Christian Jegues
RTDS Technologies Inc.
Winnipeg
Canada

Soumitri Jena
Department of Electrical and Computer
Engineering
Stony Brook University
Stony Brook, NY
USA

Brian K. Johnson
Department of Electrical and Computer
Engineering
University of Idaho
Moscow, ID
USA

Aysegul Kahraman
Department of Wind and Energy Systems
Technical University of Denmark (DTU)
Kgs. Lyngby
Denmark

Akash Kumar
Department of Electrical and Microelectronic
Engineering
Rochester Institute of Technology
Rochester, NY
USA

Frank L. Lewis
Department of Electrical Engineering
The University of Texas at Arlington
Arlington, TX
USA

Wei Li
Opal-RT Technologies Inc.
Montreal
Canada

Zhixin Li
Marketing Service Center of State Grid Jiangsu
Electric Power Company
Nanjing
China

Xuheng Lin
Binghamton University
Electrical and Computer Engineering
Vestal
USA

Yuzhang Lin
Department of Electrical and Computer
Engineering
New York University
Brooklyn, NY
USA

Zongli Lin
Charles L. Brown Department of Electrical and
Computer Engineering
University of Virginia
Charlottesville, VA
USA

Tianyun Ling
Center of Excellence in Wireless and
Information Technology
Stony Brook University
Stony Brook, NY
USA

Qi Liu
College of Electrical Engineering and
Automation
Shandong University of Science and
Technology
Qingdao
China

Yu Liu
Center for Intelligent Power and Energy
Systems (CiPES)
School of Information Science and Technology
ShanghaiTech University
Shanghai
China

Xiaonan Lu
School of Engineering Technology
Purdue University
West Lafayette, IN
USA

Serge Luryi
Department of Electrical and Computer
Engineering
Stony Brook University
Stony Brook, NY
USA

Claran Martis
Department of Electrical and Computer
Engineering
Stony Brook University
Stony Brook, NY
USA

Tingyang Meng
Charles L. Brown Department of Electrical and
Computer Engineering
University of Virginia
Charlottesville, VA
USA

Priyanka Mishra
Department of Electrical and Computer
Engineering
Stony Brook University
Stony Brook, NY
USA

Sheik M. Mohiuddin
Energy and Environment Directorate
Pacific Northwest National Laboratory
Richland, WA
USA

Hamed Mohsenian-Rad
Department of Electrical and Computer
Engineering
University of California
Riverside, CA
USA

Tom Ortmeyer
Clarkson University
Potsdam, NY
USA

Jean-Nicolas Paquin
Opal-RT Technologies Inc.
Montreal
Canada

Juan F. Patarroyo
Computer Science and Engineering
Department
University of Puerto Rico, Mayaguez Campus
Mayaguez, PR
USA

Joel Pfannschmidt
Opal-RT Technologies Inc.
Montreal
Canada

Junjian Qi
Department of Electrical Engineering and
Computer Science
South Dakota State University
Brookings, SD
USA

Yi Qi
RTDS Technologies Inc.
Winnipeg
Canada

Shouvik Roy
Department of Computer Science
Stony Brook University
Stony Brook, NY
USA

Yacov A. Shamash
Department of Electrical and Computer
Engineering
Stony Brook University
Stony Brook, NY
USA

Alex Shevchenko
Department of Electrical and Computer
Engineering
Stony Brook University
Stony Brook, NY
USA

Xianghua Shi
RTDS Technologies Inc.
Winnipeg
Canada

Scott A. Smolka
Department of Computer Science
Stony Brook University
Stony Brook, NY
USA

Jie Song
Department of Industrial Engineering and
Management
College of Engineering
Peking University
Beijing
China

Scott D. Stoller
Department of Computer Science
Stony Brook University
Stony Brook, NY
USA

Qiuye Sun
The State Key Laboratory of Synthetical
Automation for Processes Industries and the
School of Information Science and Engineering
Northeastern University
Shenyang
China

Zefan Tang
Department of Electrical and Computer
Engineering
Stony Brook University
Stony Brook, NY
USA

Douglas L. Van Bossuyt
Department of Systems Engineering
Naval Postgraduate School
Monterey, CA
USA

Vinod M. Vokkarane
Department of Electrical and Computer
Engineering
University of Massachusetts Lowell
Lowell, MA
USA

Yan Wan
Department of Electrical Engineering
University of Texas at Arlington
Fort Worth, TX
USA

Jianxiao Wang
National Engineering Laboratory for Big Data
Analysis and Applications
Peking University
Beijing
China

Lizhi Wang
Department of Electrical and Computer
Engineering
Stony Brook University
Stony Brook, NY
USA

Shouxiang Wang
School of Electrical and Information
Engineering
Tianjin University
Tianjin
China

Xin Wang
Department of Electrical and Computer
Engineering
Stony Brook University
Stony Brook, NY
USA

Xuan Wang
School of Electrical and Information
Engineering
Tianjin University
Tianjin
China

Qing Xia
State Key Laboratory of Power Systems and
Generation Equipment
Department of Electrical Engineering
Tsinghua University
Beijing
China

Weidong Xiao
School of Electrical and Computer Engineering
University of Sydney
Darlington, NSW
Australia

Yucheng Xing
Department of Electrical and Computer
Engineering
Stony Brook University
Stony Brook, NY
USA

Bing Yan
Department of Electrical and Microelectronic
Engineering
Rochester Institute of Technology
Rochester, NY
USA

Guangya Yang
Department of Wind and Energy Systems
Technical University of Denmark (DTU)
Kgs. Lyngby
Denmark

Lingxiao Yang
The School of Artificial Intelligence
Anhui University
Hefei
China

Nanpeng Yu
Department of Electrical and Computer
Engineering
University of California, Riverside
Riverside, CA
USA

Md. Zahidul Islam
Department of Electrical and Computer
Engineering
New York University, New York
USA

Ning Zhang
The School of Electrical Engineering and
Automation
Anhui University
Hefei
China

Peng Zhang
Department of Electrical and Computer
Engineering
Stony Brook University
Stony Brook, NY
USA

Tiance Zhang
State Key Laboratory of Alternate Electrical
Power System with Renewable Energy Sources
School of Electrical and Electronic Engineering
North China Electric Power University
Beijing
China

Xuan Zhang
School of Electrical and Information
Engineering
Tianjin University
Tianjin
China

Yi Zhang
RTDS Technologies Inc.
Winnipeg
Canada

Ziang Zhang
Binghamton University
Electrical and Computer Engineering
Vestal
USA

Qianyu Zhao
School of Electrical and Information
Engineering
Tianjin University
Tianjin
China

Rong Zhao
Center of Excellence in Wireless and
Information Technology
Stony Brook University
Stony Brook, NY
USA

Haiwang Zhong
State Key Laboratory of Power Systems and
Generation Equipment
Department of Electrical Engineering
Tsinghua University
Beijing
China

Yifan Zhou
Department of Electrical and Computer
Engineering
Stony Brook University
Stony Brook, NY
USA

Shan Zuo
Department of Electrical and Computer
Engineering
University of Connecticut
Storrs, CT
USA

Preface

This book is a collection of cutting-edge research in microgrids as well as industry insights in designing, operating, and maintaining microgrids.

In the past decades, the United States has been suffering from many power blackouts annually due to extreme events such as winter storms and wildfires. Meanwhile, we still have under-served communities where their power infrastructures lag behind and their residents suffer from poor electricity reliability and resilience. Microgrids could be a promising solution for the aforementioned problems. By coordinating local loads, distributed energy resources, and storage, a microgrid provides continuous energy supply even when the utility grid is unreliable. Even better, if local microgrids can be networked and coordinated, it is believed that they could provide more resilience benefits for customers and utilities. As a result, the number of microgrids in operation has been increasing in recent years. Yet there are various remaining challenges that lead to high capital expenditure and operational expenditure in microgrid applications. As well, cyber-physical attacks are being emerging challenges for microgrid adoptions. Those challenges have inspired many research groups around the world to continue developing new techniques and methods toward flexible, affordable, and resilient microgrids that can benefit more communities. Meanwhile, the industry and academia have gained more first-hand experience in designing, planning, and operating microgrids. For readers who would like to understand microgrid technologies, therefore, it would be helpful to have a reference book that provides them the start-of-the-art in the microgrid research the state-of-the-practice in the microgrid development.

Bearing this in mind, in December 2021, I tried to reach out to a few eminent power engineers and researchers who are actively working on microgrids, asking if they would like to work together to produce a two-volume reference book for both theory and practice of microgrids. To my pleasant surprise, all of them immediately agreed to work on this initiative. One year later, all the 37 chapters were finished on time and were integrated into this book as we planned.

All chapters have been written as self-contained as possible to make the book suitable for self-study purposes. Further, readers will find exercises at the end of most chapters. Thus, the book is not only a reference for professionals but hopefully a textbook suitable for classroom use.

Stony Brook, New York, USA   *Peng Zhang*
January, 2024

Acknowledgments

During my academic journey, many people have taught, influenced, and collaborated with me. I deeply appreciate all of them for their help and support.

I would like to thank my department and Stony Brook University. Many thanks to Petar M. Djurić, Yacov A. Shamash, Mónica Bugallo, Susan Nastro, and all my colleagues who created a warm, family-like environment for everyone in our department. I am privileged to work with Yacov A. Shamash, Ann-Marie Scheidt, Scott Smolka, Xin Wang, Scott Stoller, Yifan Zhou, Rong Zhao, Ji Liu, Yue Zhao, and many colleagues at Stony Brook, and the discussions with them have been inspiring in various aspects. For this book, Yifan Zhou has spent tremendous efforts to integrate the contributions from all the teams.

I would like to express my sincere thanks to our sponsors, the National Science Foundation, the Office of Naval Research, the Department of Energy, and various other agencies, for their strong support that allow us to present the research results summarized in this book.

We would like to thank Wiley's editors: Mary Hatcher, Victoria Bradshaw, Teresa Netzler, Indirakumari S., and their colleagues, who have been consistently helpful and supportive.

Surely my deep appreciation goes to all the contributors of this book. Together with you, the writing of this book became so enjoyable. Further, I am wholeheartedly grateful to be blessed with my family who loves and supports me. Thank You!

Peng Zhang

1

Introduction

Peng Zhang

1.1 Background

A basic definition of microgrid documented by the US Department of Energy Microgrid Exchange Group is that a microgrid is "a group of interconnected loads and distributed energy resources within clearly defined electrical boundaries that acts as a single controllable entity with respect to the grid." Normally, microgrid refers to a localized autonomous network designed to supply electrical and heat loads for a local community (e.g. a university campus, a commercial building or a residential area). It can be connected with the main grid (grid-connected mode) or isolated during main grid emergencies (islanded mode). Occasionally, intentionally islanded transmission or distribution grids during grid emergencies are also called "dynamic" microgrids, where short-term microgrid formation can serve as a means of grid restoration.

Increased adoption of microgrids is largely driven by the fast-growing market for renewable energy resources. As integrators and coordinators of distributed energy resources, microgrids offer higher electricity resilience benefits, as renewable energy resources can hardly ride through sustained grid contingencies by themselves. Given the rapid development of power electronics technologies as well as primary, secondary, and tertiary control techniques in recent years, their resilience and economic benefits have been more promising nowadays. Many prior works are summarized in *Microgrids: Architecture and Control*, edited by Dr. Nikos Hatziargyriou and in existing literature.

Although microgrids are effective and promising, transforming community power infrastructures into microgrids remains difficult, thus hindering broader adoption. Multiple issues remain: (i) Microgrids are still expensive solutions. Forming a microgrid requires expensive generation, protection, automation, and control facilities. Retrofitting and redesigning of existing infrastructures, designated space for microgrid hardware, and long-term operations and maintenance are all expensive. New technologies that reduce capital expenditure and operational expenditure are highly desirable. (ii) There is a need for high-speed communication infrastructure and scalable microgrid analytics. The frequent changing of states, ubiquitous uncertainties, fast ramping, and non-synchronism that characterize low-inertia community microgrids create significant challenges to optimize operations and reliably assess and enhance stability and resilience – key to microgrids serving as dependable resilience resources. (iii) Cyber-physical security and privacy are of growing concern to the communities. In addition, there are also social acceptance issues, and how to help underserved communities afford microgrids and allow them to become energy "prosumers" remains an open question.

Microgrids: Theory and Practice, First Edition. Edited by Peng Zhang.
© 2024 The Institute of Electrical and Electronics Engineers, Inc. Published 2024 by John Wiley & Sons, Inc.

This book aims to summarize some of the ongoing cutting-edge research for tackling the remaining challenges aforementioned. Further, we will also introduce selected industrial experiences in designing, operating, and maintaining microgrids. A salient feature is the incorporation of new cyber-physical system technologies for enabling microgrids as resiliency resources, as well as hard-to-find information, including in-depth theories and hands-on microgrid design and operation experiences. Therefore, this book is expected to serve as a good starting point for researchers who plan to carry out microgrid research. Meanwhile, the practical subjects covered in this book might be useful for readers in industry as a guidance to analyze, design, and operate microgrids and networked microgrid systems.

1.2 Reader's Manual

This book will include up-to-date knowledge of the design and operations of resilient and secure microgrids. Overall, the book adopts a cyber-physical systems approach, presenting a unification of the underpinning theories and the existing practices of microgrids.

1.2.1 Volume I: Theory

1.2.1.1 Platform
In Chapter 2, we introduce AI-Grid: AI-enabled, smart programmable microgrids. AI-Grid is a programmable platform that integrates deep learning, reachability analysis, formal control and runtime assurance technologies to enable scalable, self-protecting, autonomic, and resilient networked microgrids. A prototype of AI-Grid developed at Stony Brook is presented with details.

1.2.1.2 Steady-State Analysis
In Chapter 3, we present power flow algorithms for steady-state analysis of microgrids. It focuses on a distributed algorithm to provide fast power flow analytics for highly meshed, hierarchically controlled networked microgrids. In addition, an augmented continuation power flow is devised to identify voltage insecurities in networked microgrids.

In Chapter 4, we discuss state and parameter estimation for inferring unknown states and models of microgrids from measurements. State estimation is a fundamental tool for power flow and other situational awareness functions. This chapter provides mathematical formulation and solutions to the state estimation of inverter-based-resources-based microgrids.

1.2.1.3 Dynamics and Stability
In Chapter 5, a delayed eigenanalysis tool based on ordinary-differential-equation-based solution operator discretization (ODE-SOD) is introduced to accurately estimate the eigenspectra and delay margins (ability margins formed by critical delays) for networked microgrids. This allows us to incorporate communication delay effects in the small signal stability of networked microgrids.

In Chapter 6, we introduce learning-based dynamic model discovery. This new method can construct a nonlinear ODE model from measurement data, which can preserve dynamics of networked microgrids without assuming a priori any specific dynamic modes. This modeling approach can effectively address the data-rich, information-poor problem existing in today's microgrids.

In Chapter 7, we describe transient stability analysis for microgrids under large disturbances. This chapter focuses on inverter control models in the transient stability analysis and provides discussions on the effects of grid-following and grid-forming inverters on the transient stability performance.

In Chapter 8, we present a learning-based transient stability assessment which constructs a neural network structured Lyapunov function for certifying the transient stability of networked microgrids. The function learned leads to a region of attraction that can quantify the disturbances' magnitudes that the networked microgrids can tolerate.

In Chapter 9, we discuss challenges microgrid operations pose for protection schemes commonly applied to power distribution and subtransmission systems. We present protective relaying solutions applied for different types of microgrids in practice and finish with a discussion of emerging research directions to meet the needs of future microgrids.

1.2.1.4 Resilience

In Chapter 10, we leverage microgrids to enhance electricity resilience against extreme events. Those techniques include siting and sizing of distributed energy resources as well as infrastructure hardening during the planning stage, mobile energy resources prepositioning and proactive management of microgrids during the preparation stage, and coordinated networked microgrid operations and microgrids formation during the restoration stage.

In Chapter 11, we introduce in situ resilience quantification to provide a signal temporal logic-based formal method for quantitatively monitoring the microgrid states and resilience under varying operating conditions by only using locally available data.

1.2.1.5 Control and Optimization

In Chapter 12, we introduce distributed voltage regulation for DC microgrids. It provides insights into designing controls to regulate microgrid voltages to possibly time-varying reference voltages, regardless of the influence of local time-varying current loads and the coupling among the units.

In Chapter 13, droop-free distributed control for AC microgrids is discussed, including two realizations: one achieves average voltage regulation and perfect power sharing, and the other achieves average voltage regulation, voltage variance regulation, and relaxed reactive power sharing.

In Chapter 14, we introduce an optimal distributed voltage control in AC microgrids with an objective function that makes a trade-off between voltage regulation and reactive power sharing. A distributed primal-dual gradient-based algorithm is presented to solve the formulated optimization problem.

In Chapter 15, we focus on cyber-resilient microgrid control. A continuous-time push-sum-based synchronization mechanism is devised to enable resilient distributed control with impaired communications. Further, quantum synchronization schemes are developed to achieve inherent cybersecurity.

In Chapter 16, crypto-control for microgrids is introduced. An enhanced partial homomorphic encryption and a secret sharing scheme are integrated to provably preserve the privacy of distributed energy systems while ensuring fast, flexible distributed control in networked microgrids.

In Chapter 17, we introduce AI-enabled cooperative control and optimization of microgrids in the context of energy hubs. Distributed algorithms are introduced to coordinate microgrid components for stable operation. Hybrid-reinforcement-learning-enabled optimized scheduling is developed to realize safe and economical operations of microgrids.

In Chapter 18, we use deep neural networks (DNNs) to solve the transactive energy management problem and determine a suitable charging schedule for electric vehicles, while ensuring loss reduction and network safety in microgrids. A price coordinator promotes interactions between distribution system operators and aggregators.

1.2.1.6 Cyber Infrastructure and Cybersecurity

In Chapter 19, we review commonly used sensing and communication technologies for microgrid monitoring, and discuss the planning and operation solutions for achieving cost-effective observability, managed end-to-end data transfer delay, and multi-domain resilience against component failures.

In Chapter 20, distributed attack-resilient control frameworks are introduced to defend unbounded attacks on the input channels of the secondary control loops of AC and DC microgrids. New adaptive control techniques are developed to preserve uniformly ultimately bounded stability.

In Chapter 21, quantum security for microgrids is discussed. We introduce the design of a quantum-key-distribution-based cybersecurity scheme, including how to integrate quantum security into a single microgrid and networked microgrids and how to make quantum security practical in microgrids.

1.2.2 Volume II: Practice

1.2.2.1 Community Microgrids

In Chapter 22, we discuss the design process of community microgrids, focusing on several key dynamic and power quality aspects such as voltage flicker, temporary overvoltage, harmonic distortion, frequency regulation, and black start capability. The team provides case study details and their insights in designing the Potsdam Microgrid in New York.

In Chapter 23, the process of energy transition at Princeton University is presented. It offers insights on developing campus microgrids based on co-generation and multiple forms of energy storage. Rich data and first-hand experiences are provided by the author who led this energy transition.

In Chapter 24, we summarize Opal-RT's experiences in the control-hardware-in-the-loop and power-hardware-in-the-loop testing for microgrids. Smart inverter modeling and digital simulations, digital twins, cyber network emulation and a grid hardware open-source testbed are also introduced.

In Chapter 25, RTDS Technologies provides experiences in real-time digital simulation studies of microgrids. A universal converter model for microgrids is presented. Two practical microgrid cases, including an aircraft electrical system and Banshee microgrid, are introduced.

1.2.2.2 Control, Protection, and Analytics

In Chapter 26, we present practical coordinated control of DC microgrids. A piecewise linear formation for droop gains, a distributed average voltage sharing scheme, and a method of DC bus signaling, together with various practical tips, allow for efficient coordinated operations of DC microgrids.

In Chapter 27, we establish the foundations of microgrid resilience. Invulnerability and recoverability are introduced to calculate microgrid resilience. Further, we suggest analyzing climate resilience primarily through the amount of CO_2 a microgrid produces. Efficacy of the resilience measures is demonstrated on a representative Navy microgrid.

In Chapter 28, we summarize AC–DC microgrid planning and design experiences. Reliability evaluation is performed to support the selection of AC–DC configurations. A coordinated optimization strategy is used to improve the voltage profile and stability, and to achieve an optimized power flow distribution. Both strategies are employed in a microgrid demonstration project.

In Chapter 29, we describe a self-organizing system of sensors for real-time monitoring and diagnostics of a microgrid. The core technology is a network topology identification algorithm that uses GPS coordinates, current magnitudes, and phases from energy flow sensors.

In Chapter 30, we present microgrid situational awareness using waveform measurement units, a new class of smart grid sensors. This chapter demonstrates the effectiveness of using synchro-waveform measurements for event detection, event classification, and event location identification.

In Chapter 31, we describe a traveling wave-based protection scheme to identify and locate faults in microgrids. The new features include an improved filtering mechanism to avoid traveling wave superimposition, an accurate fault location algorithm, and a current-only protection scheme. Tests on a real-world microgrid validate its ultra-high speed and high accuracy in determining fault locations.

In Chapter 32, we introduce neuro-dynamic state estimation, a learning-based dynamic state estimation algorithm for networked microgrids under unknown subsystems. Extensive case studies demonstrate the efficacy of dynamic state estimation and its variants under different noise levels, control modes, power sources, observabilities, and model knowledge, respectively.

1.2.2.3 Microgrid as a Service

In Chapter 33, hydrogen-supported microgrids are exploited to facilitate a low-carbon energy transition. By using excessive wind and solar energy in a microgrid for hydrogen production, renewable energy consumption on-site can be promoted. Further, utilizing hydrogen-based resources including electric vehicles, the impact from microgrids on the main grid can be alleviated.

In Chapter 34, sharing economy in microgrids is discussed. A microgrid serves as an aggregator, which organizes a number of electricity users to cooperate as a single interest entity. Then, an incentive mechanism is implemented to allocate the benefits to users that incentivizes users' participation in energy sharing without users' bidding on electricity price.

In Chapter 35, using microgrids is suggested as a pathway to mitigate greenhouse impact of rural electrification. A few policy-level suggestions for China and other countries are given to facilitate the accommodation of renewable energy and to improve electric heating efficiency in microgrids.

In Chapter 36, we further discuss the operations of microgrids with meshed topology under uncertainty. Stochastic modeling through Markov chains is integrated with the mix-integer programming solvers where AC power flow, droop effects and tap changers are considered in the microgrid modeling.

In Chapter 37, we discuss practical operation optimization of microgrids with renewables that commit and dispatch distributed energy devices with renewable generation to minimize the total energy and emission cost while meeting the forecast electrical and thermal demands. Tests are performed on a planned microgrid of Kings Plaza in Brooklyn, New York.

The book is suitable for classroom use or as a reference for professionals. It consists of not only modeling details for microgrid components, but also various microgrid test cases. The book can be used for two semester-long courses (one for theory, one for practice and implementations) aimed at senior undergraduates in electrical engineering, computer engineering or computer science. It can also be suited for one semester-long course for Master's and PhD students in power engineering or relevant majors. Instructors are suggested to select a subset of the chapters (e.g. 6–10 chapters) for one semester's use. Power engineering graduate certificate programs, training courses or short courses can adopt some chapters as training materials.

2

AI-Grid: AI-Enabled, Smart Programmable Microgrids

Peng Zhang, Yifan Zhou, Scott A. Smolka, Scott D. Stoller, Xin Wang, Rong Zhao, Tianyun Ling, Yucheng Xing, Shouvik Roy, and Amol Damare

2.1 Introduction

In this chapter, we introduce artificial intelligence (AI)-Grid, a software-defined, AI-enabled microgrid (MG)/networked microgrids (NMs) platform, which aims at significantly lower costs and improve energy resilience. It has been developed at Stony Brook University in collaboration with communities and utilities.

Rolling blackouts in Texas and California in the past few years signaled that our power infrastructures are vulnerable to weather events and difficult to meet ever-expanding energy demands from our communities. In fact, based on the Federal Government's statistics, "U.S. power customers experienced an average of nearly 5 hours of interruptions in 2019," with the top-five impacted states ranging from "almost 7 hours in Mississippi to more than 15 hours in Maine." Cyber-attacks and surging online activities due to the pandemic have also stressed power grids and compromise grid resiliency. Currently, distributed energy resources (DERs) are increasingly interconnected to the power grids. However, today's inverter-based resources have sensitive trip-off or deactivation settings for grid contingencies, meaning a minor or remote fault can lead to a sudden reduction of power generation from DERs. This implies that DERs are often not reliable resilience resources for customers and the grids. A typical example is the Odessa Disturbance that occurred in Texas on 9 May 2021 – voltage sags initiated by a single-line-to-ground fault at a combined-cycle power plant caused more than 1.1 GW reduction of solar photovoltaic (PV) and wind generation up to 200 miles away from the fault location.

Meanwhile, microgrids have become a promising new paradigm for electricity resiliency. A microgrid is an autonomous local power grid formed by a collection of power loads, distributed energy resources, and coordinated control. It can flexibly host renewables and energy storages and operates continuously even without the utility supply. Naturally, microgrids can serve as aggregators and integrators of PV, wind, and various grid-edge facilities, such as smart buildings, electric vehicle charging infrastructures and Power-to-X technologies, to provide grid supports and resilience benefits for customers.

Further, if multiple microgrids are interconnected and well controlled, those NMs have the potential to offer much more resilience benefit, and can even help utilities to quickly restore power. Several main challenges, however, have prohibited the wide adoption of NMs: (1) Lack of understanding and control of NMs dynamics; (2) Limited and unscalable analytics to support real-time decision-making and control; and (3) NMs are vulnerable to cyberattacks. Moreover, it is oftentimes prohibitively expensive for a community to build and operate a microgrid. Forming a microgrid

Microgrids: Theory and Practice, First Edition. Edited by Peng Zhang.

requires protection, automation, and control (PAC) embedded in expensive hardware facilities. Coping with challenges of renewables and storms demands frequent, expensive retrofitting and forced re-designs of PAC systems.

To resolve the aforementioned challenges, our team has been working on AI-enabled, provably resilient NMs (AI-Grid). Our key contribution is a programmable platform that integrates reliable modeling and prediction of system states under uncertainty, reachability analysis, formal control, and cybersecurity technologies to enable scalable, self-protecting, autonomous and resilient microgrids and NMs capable of coordinating many distributed energy systems and serving as backbone infrastructures of smart communities. Microgrids operated by AI-Grid platform will achieve decoupled cyber and physical layers, making microgrid control and management software-defined and hardware-independent and thus achieving more affordable, scalable, and configurable microgrids.

2.2 AI-Grid Platform

AI-Grid is a new microgrid management platform that features software-defined-network-based architectures and techniques to enable secure, reliable, and fault-tolerant algorithms for resilient networked systems. With the availability of data from smart sensors (sometimes under data-rich, information-poor situations), AI-Grid can perform various AI-based microgrid operations and monitoring. Our team has worked with end-users to understand, their concerns about extreme renewable integration issues, communication delays, data deficiencies, and broadened cyber-attack surfaces. To address the end-user demands, we have been integrating a series of new solutions into the AI-Grid technology, including new features such as neuro-reachability, programmable active security scanning, encrypted control, eigenanalysis for delayed NMs, active fault management, enhanced ordinary differential equations net (ODE-Net), and distributed neural Simplex controls.

At the core of AI-Grid are basic functions such as software-defined microgrid controls for voltage and frequency regulations and three-phase power flow and state estimation algorithms for tracking of operating states of microgrids. A series of model-based and data-driven methods are implemented, which allows us to accurately estimate NM states under hierarchical control effects, frequent network changes, and network outages. Overlaying on the fundamental modules,

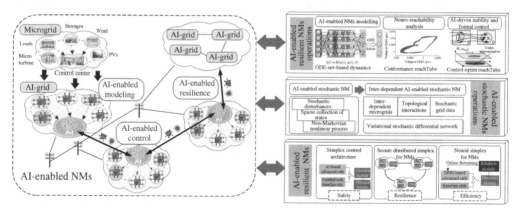

Figure 2.1 Schematic AI-enabled autonomous and resilient networked microgrids. Source: General Electric, diyanadimitrova / Adobe Stock, Duke Energy Corporation.

a series of AI-enabled functions are incorporated into our platform for enhanced resilience, robustness, and assurance guarantees. AI-Grid also offers a user-friendly, expandable 3D user interface. We will soon implement our team's mix-integer optimization engine, a zero-trust security framework, transactive energy functions, and more.

In the following, we will introduce a few AI-enabled functions, as summarized in Fig. 2.1, as well as the software platform of AI-Grid.

2.3 AI-Enabled, Provably Resilient NM Operations

NMs are highly susceptible to transient processes initiated by uncertain renewable generation, plug-and-play operations, and grid faults. Hence, verifying NMs' dynamic, stability, and control performance under heterogeneous uncertainties has become critically important. To enable provably resilient NM operations and address the data-rich, information-poor (DRIP) problem, we establish AI-enabled analytics for online model discovery, dynamic verification, situational awareness, fault management, and protection under various uncertainties and unidentified subsystems.

2.3.1 Neuro-Reachability: AI-Enabled Dynamic Verification of NMs Dynamics

A key innovation of AI-Grid is **Neuro-Reachability**, i.e. neural ODE-Net-enabled reachability methods. Neuro-reachability allows for online dynamic model discovery and data-driven formal verification of NM dynamics under a wide range of uncertainties.

First, we devised ODE-Net-enabled dynamic model discovery for NMs, which can best preserve the continuous-time dynamic behaviors of NMs without assuming any specific dynamic modes. Experiments in four-microgrid NMs demonstrate the proposed method's accuracy and generalizability. Second, we developed ODE-Net-based neuro-reachability (see Fig. 2.2) to formally verify

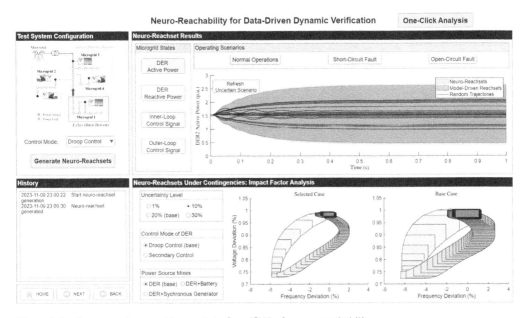

Figure 2.2 Demo-version graphic user interface (GUI) of neuro-reachability.

NMs dynamics under inaccessible physics models, uncertain renewables, and unforeseen faults. We empower reachability analysis with the conformance theory and use an optimization approach to estimate the neural model error set. Through experiments, we verified the accuracy and conservativeness of the method.

In addition, we are in the process of deploying reachable power flow (ReachFlow) which can efficiently provide fast state monitoring for NMs even when the operating points have "random walks" driven by renewables and disturbances. Reachable eigenanalysis (ReachEigen) is another reachability tool that can provide small-signal stability for NMs under "random walks" of equilibriums.

2.3.2 Neuro-DSE: AI-Enabled Dynamic State Estimation

AI-Grid also deploys a novel AI-enabled situational awareness technique (see Fig. 2.3), i.e. the ODE-Net-based neural dynamic state estimation (neuro-DSE). We incorporate Kalman filters into ODE-Net and design a self-refining process to perform ODE-Net training and dynamic state tracking jointly. Simulation validates that neuro-DSE provides satisfactory accuracy even for tracking unobservable states under different sources of noise. Neuro-DSE, therefore, provides a powerful tool for real-time, model-free situational awareness of NMs.

2.3.3 Neural-Adaptability: AI-Based Resilient Microgrid Control

Further, AI-Grid deploys a series of learning-based control synthesis and NM operations to jointly allow for ultra-resilience NMs.

AI-Grid incorporates AI-based active fault management (AI-AFM) to control NMs during faults. We innovatively replace optimization-based AFM with federated-learning-based AFM to meet the real-time needs for coordinating dozens of microgrids. Our hardware-in-the-loop (HIL) realtime

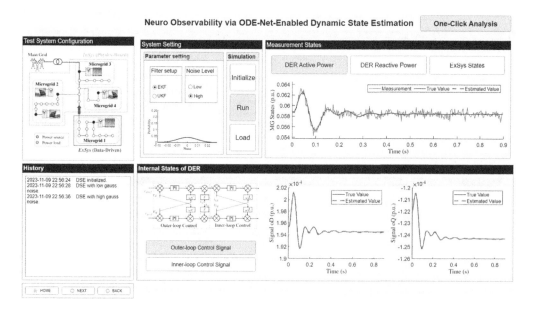

Figure 2.3 Demo-version GUI of neuro-DSE.

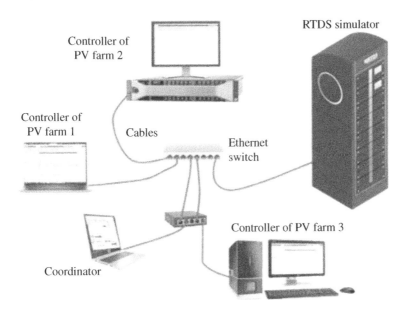

Figure 2.4 Hardware tesbed of AI-AFM.

simulation (see the testbed in Fig. 2.4) demonstrates that AI-AFM can output reference values within 10 ms irrespective of the number of microgrids.

Further, we established an AI-based traveling wave protection (AI-TWP) to provide ultraefficacious NM protection (see the testbed in Fig. 2.5). We leverage the Wavelet Kernel net convolutional neural network (WKN-CNN) to accurately and efficiently process the reflected traveling wave signals in the time-frequency domain. The method reports high validation accuracy of 95.83% in a typical 100%-renewable microgrid.

Figure 2.5 Hardware tesbed of AI-TWP.

2.4   Resilient Modeling and Prediction of NM States Under Uncertainty

The effective and reliable analysis and control of NMs rely on the accurate modeling and prediction of NM states. Practical NMs as a graph may experience complicated stochastic dynamics over time. The AI modeling of stochastic dynamics is prohibitively difficult due to the uncertain state transitions, sporadic data with possible irregular intervals as a result of loss or data corruption due to reasons such as unreliable communications or device malfunction.

In this section, we demonstrate our new tools for **modeling continuous dynamics under partial data and uncertainty**. More specifically, to tackle the challenges, we advance the ODE-net to more flexibly and generally model the nonlinear stochastic dependency among high-dimensional observations with possible irregular data samples. We propose a new continuous-time stochastic predictive model called adversarial graph-gated differential network (AGGDN) to forecast the microgrid states based on data samples observed at irregular intervals.

2.4.1   Hybrid Neural ODE-SDE Graph Modeling of NM Dynamics

AGGDN effectively learns the continuous graph dynamics from a sequence of spatially and temporally irregular observations of NMs. The model (Fig. 2.6) consists of two major modules: an ordinary differential equation (ODE) to model the topological relationship of microgrid nodes and properties of the microgrid and a stochastic differential equation (SDE) to capture the process uncertainty. The two modules form the hybrid ODE–SDE structure to evolve simultaneously. Besides, to better adapt to the partial input, we propose a soft-masking function in our ODE module, which also provides a soft-mask trajectory to modulate extracted features from irregular observations. As optimizing the model with stochastic terms is difficult, we further introduce a Wasserstein adversarial training objective to efficiently train our model so that it can more scalably and accurately learn the process uncertainty.

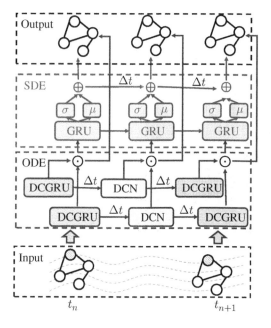

Figure 2.6   AGGDN-enhanced ODE-Net.

We denote a NM graph with time-variant signals by $\mathbb{G} = \{\mathcal{V}, \mathcal{E}, \{X_n, \mathcal{M}_n, t_n\}_{n=0}^N\}$, where $\mathcal{V} = \{v_i\}_{i=1}^{|\mathcal{V}|}$ is the set of microgrid nodes which can be loads or generators and $\mathcal{E} = \{(v_i, v_j)\}$ is the set of edges (e.g. buses and cables) between nodes. The cardinalities $|\mathcal{V}|$ and $|\mathcal{E}|$ denote the number of elements in $\mathcal{V}$ and $\mathcal{E}$. We further denote the adjacent matrix of a graph as A. In a given NM with the graph topology $\{\mathcal{V}, \mathcal{E}\}$, the temporal dynamics are observed as a sequence of N data frames with the d-dimensional time-variant multivariate signals $X_n \in \mathbb{R}^{|\mathcal{V}| \times d}$ at time $t_n \in \mathbb{R}_+$. Signals can be voltages, currents, or powers. Since there are often missing samples in real-world systems caused by sensing or transmission problems, in each frame, a mask $\mathcal{M}_{t_n} = \{0,1\}^{|\mathcal{V}| \times d}$ is used to indicate the existence of values in the corresponding dimensions. Therefore, the actual observation sequence $\mathcal{O} = \{X_n \odot \mathcal{M}_n\}_{n=0}^N$ fed into the model is a sporadic time series, possibly with irregular data in both temporal and spatial domains, and $\odot$ is an element-wise multiplication.

Given a collection of NM data $D = \{\mathbb{G}^{(m)}\}_{m=1}^{|D|}$, i.e. the measurements of NM signals, our objective is to learn a continuous-time recurrent predictive model $\mathcal{G}$ that maximizes the masked log-likelihood:

$$\mathcal{L}_{ll}(\mathcal{G}) = \mathbb{E}_{\mathbb{G} \in D} \sum_{n=1}^N \mathcal{M}_n \otimes \log P_{\mathcal{G}}(X_n | \mathcal{O}_{0:n-1}, t_{0:n}, A) \tag{2.1}$$

where $\otimes$ is the sum of element-wise products of two matrices. $P_{\mathcal{G}}$ is the model-defined probability density of each element in the feature matrices. To avoid introducing more errors and uncertainty, we only evaluate the log-likelihood in the observed training data indicated by the binary masks.

2.4.2   NeuralODE with Soft-Masking to Adapt to Spatially Partial Observations

The ODE module is applied to extract topological relation and the signal property from the measurements of NMs at discrete points and embed them into the continuous hidden feature $H_t \in \mathbb{R}^{|\mathcal{V}| \times d_h}, t \in \mathbb{R}_+$, where d_h is the dimension of the feature space. Our ODE module includes three functions: (1) the nonlinear mapping to directly integrate the data information into the hidden feature at the observation time, (2) the differential equation to update the values of hidden feature at the interval of observations, and (3) a soft-masking function for our network to adapt to the positions of missing values. As observations $\mathcal{O}_n = X_n \odot \mathcal{M}_n$ in each frame n are irregular in the spatial domain with missing values from some nodes; accordingly, the hidden feature H_t includes two factors, the feature $H_{t,f}$ that extracts the topological relation and data property, and the masking $H_{t,m} \in (0,1)^{\mathcal{V} \times d_h}$ that modulates the values of the feature: $H_t = \text{Sigmoid}(H_{t,m}W_m + b_m) \odot H_{t,f}$, where $\text{Sigmoid}(\cdot)$ denotes the activation function, and $\{W_m, b_m\}$ are the weight and bias parameters of the auxiliary feed-forward networks.

To parameterize the dynamics of $H_{t,f}$ and $H_{t,m}$ during the interval of the observation time (t_{n-1}, t_n), we apply ODE with

$$\frac{dH_{t,m}}{dt} = F_m(H_{t,m}, A), \quad \frac{dH_{t,f}}{dt} = F_f(H_{t,f}, A) \tag{2.2}$$

$$H_{t,m} = H_{t_{n-1},m} + \int_{t_{n-1}}^t F_m(H_{t,m}, A)dt$$

$$H_{t,f} = H_{t_{n-1},f} + \int_{t_{n-1}}^t F_f(H_{t,f}, A)dt \tag{2.3}$$

where $F_m(\cdot)$ and $F_f(\cdot)$ are the first-order derivatives for the mask and features, which are computed by the graph neural network.

2.4.3 NeuralSDE with Wasserstein Adversarial Training for Efficient Learning of Process Uncertainty

In practical NMs, the observations are influenced by the process uncertainty inside the systems. For instance, the uncertainty in the controlling signal of DER will cause large oscillations in the microgrid current flows. We introduce the latent state of SDE, a random matrix represented as $Z_t \in \mathbb{R}^{|\mathcal{V}| \times d_z}$, $t \in \mathbb{R}_+$, to embed the process uncertainty of the underlying dynamics of each node. In order to capture the complicated stochastic process of NMs, we parameterize the latent state by a nonlinear SDE:

$$dZ_t = \mu(Z_t, H_{\leq t})dt + \sigma(H_{\leq t})dB_t, \quad \text{where} \quad Z_t = Z_{t_0} + \int_{t_0}^{t} \mu(Z_t, H_{\leq t})dt + \int_{t_0}^{t} \sigma(H_{\leq t})dB_t$$

$$(2.4)$$

where μ and σ are the drift and diffusion functions of an SDE. B_t denotes the standard Brownian motion. μ is the function of both the current latent state and historical ODE features, but σ will only take the historical ODE features as the input, because including latent state into σ may bring additional noise into the gradient computation to degrade the training process.

Signals of NMs can be predicted with the combination of two components: a trajectory to capture the data evolution trend and a residual term: $\hat{X}_t = \hat{X}_t^0(H_t) + \hat{X}_t^{(res)}(H_t, Z_t)$, where $\hat{X}_t^0$ is the function of the hidden feature H_t to predict the smooth trend of the signal and $\hat{X}_t^{(res)}$ further incorporates the latent state Z_t to estimate the residual term that captures the large variation of signals. ODE features embed the spatial relation of the graph structure, and the SDE component is used to synthesize a latent state trajectory from the ODE features to capture the process uncertainty. A popularly used gated recurrent unit (GRU) can be applied to integrate the historical part of the ODE feature trajectory, with feed-forward networks to take the GRU features to realize the drift and diffusion functions.

With the incorporation of the latent state to capture system uncertainly, both $P_{\mathcal{G}}(Z_{t_n}|\mathcal{O}_{0:n-1}, A)$ and the log-likelihood function $\mathcal{L}_{ll}$ in Eq. (2.1) will not have closed-form solutions. Rather than simplifying the log-likelihood function and applying a variational inference to obtain an evidence lower bound and the Monte Carlo method to estimate $P_{\mathcal{G}}(Z_{t_n}|\mathcal{O}_{0:n-1}, A)$, given a graph structure with a large number of NM nodes and signals, we introduce Wasserstein adversarial training to increase the efficiency and the accuracy of learning our proposed model.

2.4.4 Experiments

We demonstrate the effectiveness of each component in our proposed model and analyze the robust performance of our model under different conditions through experiments on several microgrid cases. During experiments, we mainly evaluate the following metrics for the comparison between our model and the selected baselines:

- Mean absolute error (MAE)
- Rooted mean squared error (RMSE)
- Mean absolute percentage error (MAPE)

All these metrics are evaluated between the predicted trajectories and the ground-truth, no matter whether they are observable or not.

We use IEEE-33 bus system (see Fig. 2.7) for performance evaluation. As a typical microgrid instance, it contains 5 electricity sources and 28 load nodes. The 2D DQ current signals going

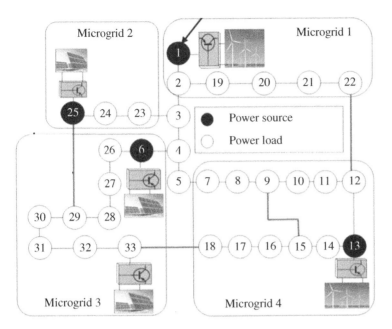

Figure 2.7 IEEE 33-bus system.

through each node are generated using a hardware-based tool real time digital simulator (RTDS) produced by [1], the world standard to create real-time data of a power system for HIL testing of equipment protection and control. To capture the impact of topology diversity and better learn the relationship among nodes, we randomly cut some connections among nodes during generation to form 21 different network structures. For each structure, we collect 50 trajectories with signals sampled at the interval of 3 ms. For each trajectory, we split the sequence of samples into segments of 100 frames for training and testing. Before being fed into the model, all data are normalized with the mean and the standard derivation in the temporal domain of the corresponding node.

2.4.4.1 Overall Performance
To synthesize the scenario of the sporadic observations, we randomly select a ratio p_t of the data frames in the temporal dimension as observed data. For each selected frame, we also assume only signals from p_s of the nodes are observed. Therefore, only $p_t \times p_s$ fraction of the data are fed into our model, and the data set can be regarded as sparse and also irregular in both temporal and spatial domains due to the random selection. Table 2.1 shows the performance comparison between our method and some other representative literature works in the case of $(p_t = 0.5, p_s = 0.8)$. From the table, we can see that our proposed method has the best performance throughout.

- **Continuous-time modeling**: From Table 2.1, we can see that no matter our proposed model or other continuous-time models are superior to traditional discrete-time models. Different from discrete models, which only update features at the observation time instants, continuous ones make the updates $\Delta T/\delta t$ times between two neighboring observations, where the propagation step δt is much smaller than the sampling interval ΔT (we set $\delta t = 0.1\Delta T$ in the experiment). When used with graph convolutions, the integrated information from adjacent nodes will continuously help correct and update the states of the current node. Therefore, even though we use the simplest Euler-Method to approximately calculate the integration, the predicted trajectory fits the original one better than the trajectory provided by discrete models. Moreover,

Table 2.1 Testing performance of different models on IEEE 33-bus system ($p_t = 0.5, p_s = 0.8$).

	MAE ($\downarrow$)	RMSE ($\downarrow$)	MAPE ($\downarrow$)
Discrete			
STGCN [2]	0.0812	0.1273	18.18%
GCGRU [3, 4]	0.0349	0.0643	9.65%
Continuous			
Graph-ODE-RNN [5]	0.0306	0.0578	8.36%
Graph-GRU-ODE [6]	0.0313	0.0607	8.49%
Ours			
AGGDN	0.0243	0.0457	7.03%

Table 2.2 Performance comparison of our ODE module with and without the soft-masking function on IEEE 33-Bus System ($p_t = 0.5, p_s = 0.8$).

	MAE ($\downarrow$)	RMSE ($\downarrow$)	MAPE ($\downarrow$)
AGGDN (w/o soft-masking)	0.0264	0.0504	7.49%
AGGDN (w/ soft-masking)	0.0250	0.0471	7.21%

since the propagation step can be arbitrarily small, the continuous-time model can provide predictions at any time to enable timely control, rather than only discretely on observation time points.

- **Soft-masking function**: We introduce a soft-masking function to modulate the hidden features in our ODE module for better adapting to the missing data cases. The comparison results in Table 2.2 have proved the role of such a design.
- **SDE module**: To better capture uncertainties existing in all the real-world systems, we propose to incorporate stochastic modules, i.e. SDE, in our model. Different from ODE models, which make strong Gaussian assumptions about the data distribution, the real distribution is learned by Monte Carlo sampling process within the SDE propagation. In order to demonstrate the effect of our SDE module, we also implemented a simplified version, AGGDN(ODE), which does not have the SDE part. The experiment results based on both schemes running on the same data are shown in Table 2.3, and the improvement brought by the SDE module is obvious.

Table 2.3 Performance comparison of our full model and the simplified model without SDE on IEEE 33-bus system ($p_t = 0.5, p_s = 0.8$).

	MAE ($\downarrow$)	RMSE ($\downarrow$)	MAPE ($\downarrow$)
AGGDN (ODE)	0.0250	0.0471	7.21%
AGGDN (full)	0.0243	0.0457	7.03%

Table 2.4 Performance comparison of our model on IEEE 33-bus system ($p_t = 0.5, p_s = 0.8$) before and after using adversarial training.

	MAE ($\downarrow$)	RMSE ($\downarrow$)	MAPE ($\downarrow$)
AGGDN (w/o adversarial training)	0.0250	0.0473	7.20%
AGGDN (w/ adversarial training)	0.0243	0.0457	7.03%

Table 2.5 Performance of our model on IEEE 33-bus system with and without noise on DERs.

	MAE ($\downarrow$)	RMSE ($\downarrow$)	MAPE ($\downarrow$)
W/O missing, W/O noise	0.0012	0.0062	1.68%
W/ missing, W/O noise	0.0040	0.0168	5.31%
W/O missing, W/ noise	0.0101	0.0189	8.69%
W/ missing, W/ noise	0.0203	0.0405	18.81%

- **Adversarial training**: Since we incorporate stochastic terms in our model, the training process becomes more difficult. To better train our model, we utilize an adversarial training strategy to avoid the possible gradient explosion problem. For reference, we also attach the comparison results before and after using the adversarial training in Table 2.4.

2.4.4.2 Performance in the Presence of Noise

DERs in the microgrid generate power and transmit them to the whole network. The power system states are affected by the input signals to DERs, which may contain noise in real-world scenarios. The dynamics of sources can be considered as a kind of uncertainty discussed previously. To demonstrate the robustness of our model in dealing with noise and representing the system dynamics, we conduct experiments in the no-missing case and the missing case ($p_t = 1.0, p_s = 0.7$) separately, with the numerical results given in Table 2.5.

To better illustrate the performance of our model in different cases, we show some temporal curves in Fig. 2.8. From the table and plots, we can see that the noise on DERs only reduces the performance slightly. Since the noise is transmitted to the whole microgrid as well, its effects on different nodes are also correlated.

2.4.4.3 Prototyping

We further prototype the proposed AGGDN model with two interfaces shown in Figs. 2.9 and 2.10, one for the training and the other for the testing. Users are allowed to flexibly configure testing scenarios. The right panel of the training interface shows the progress of the system over time, including the error between predictions and ground-truth, as well as the randomly chosen example curves of both. The loss between the predicted system states and the true data reduces with the training going, and the curves become more and more consistent. On the testing interface, the ground truth data samples are shown on the right panel, with observed samples denoted with dots while missing ones are represented as "x"s, and the predicted system states are shown as a continuous curve below.

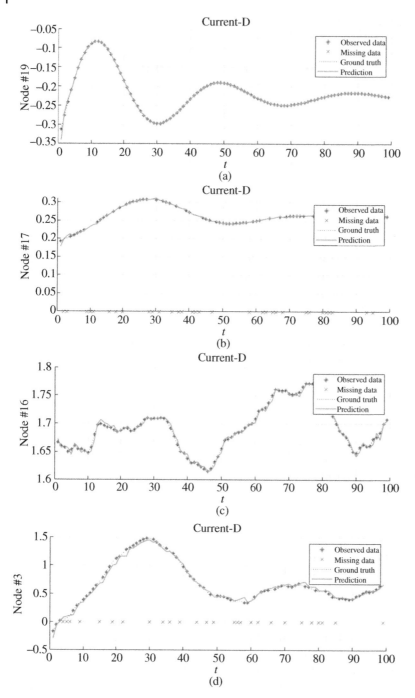

Figure 2.8 An illustration of the predicted DQ current curves in the D-axis of randomly chosen nodes in different cases on IEEE 33-bus system. (a) Without missing, without noise. (b) With missing, without noise. (c) Without missing, with noise. (d) With missing, with noise.

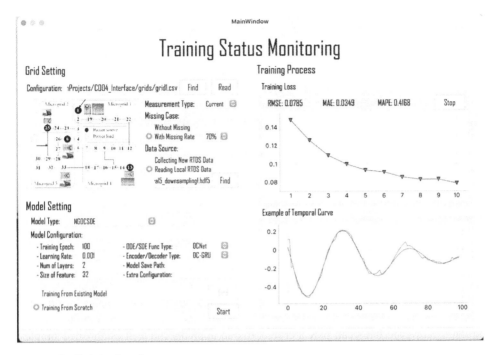

Figure 2.9 Training interface.

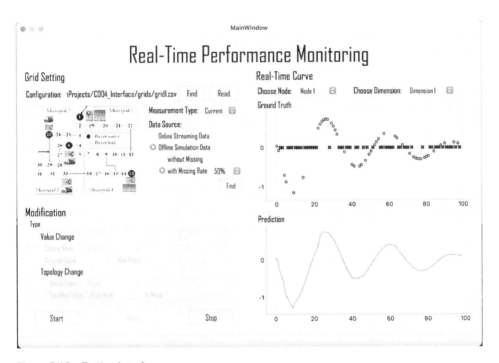

Figure 2.10 Testing interface.

2.5 Runtime Safety and Security Assurance for AI-Grid

In AI-Grid, we deliver a deployable Three Lines of Defense model that integrates Stony Brook's unique techniques – Neural Simplex Architecture (NSA), programmable active security scanning, and encrypted control – to enable self-protecting, cyber-physical-resilient, and intelligent NMs. In this section, we focus on **Neural Simplex** techniques for runtime safety/security assurance for AI-enabled NMs. The NSA is a runtime assurance framework for neural controllers (NCs). We have developed a number of NCs for microgrids, each of which is a deep neural network (DNN) trained using the deep deterministic policy gradient (DDPG) algorithm with the safe-learning strategy of penalizing unrecoverable actions. We demonstrate through simulations that the neural controller outperforms traditional droop controllers and generalizes well, and that the decision module (DM) and baseline controller (BC) ensure safety in the face of errors by the advanced controller (AC).

2.5.1 Introduction

Barrier certificates (BaCs) are a powerful method for verifying the safety of continuous dynamical systems without explicitly computing the set of reachable states. Proving safety of plants with complex controllers, such as AI-based controllers, is difficult with any formal verification technique, including BaCs. However, BaCs can play a crucial role in applying the well-established simplex control architecture [7, 8] to provide provably correct runtime safety assurance for systems with complex controllers. It is imperative to have a runtime safety assurance framework for AI-based systems such as our AI-Grid Platform, since AI-based controllers, e.g. deep neural networks (DNNs), are difficult to verify and may be vulnerable to adversarial attacks.

This section presents *Barrier-based Simplex* (Bb-Simplex), a provably correct design for runtime assurance of continuous dynamical systems. Bb-Simplex is part of the AI-Grid Platform. Bb-Simplex is centered around the simplex control architecture, which consists of a high-performance *AC* that is not guaranteed to maintain safety of the plant, a verified-safe *BC*, and a *DM* that switches control of the plant between the two controllers to ensure safety without sacrificing performance. In Bb-Simplex, *BaCs* are used to prove that the BC ensures safety. Furthermore, Bb-Simplex features a new scalable (relative to existing methods that require reachability analysis, e.g. [9–11]) and automated method for deriving, from the BaC, the conditions for switching between the controllers. Our method is based on the Taylor expansion of the BaC and yields computationally inexpensive switching conditions.

We demonstrate Bb-Simplex by applying it to a microgrid modeled in RTDS, an industry-standard high-fidelity, real-time power systems simulator. The microgrid features an advanced controller for PV DER voltage control, in the form of a DNN trained using reinforcement learning (RL). Accordingly, we use Bb-Simplex in conjunction with the *NSA* [12], where the AC is an AI-based *neural controller* (NC). NSA also includes an *adaptation module* (AM) for online retraining of the NC while the BC is in control. Our results demonstrate that Bb-Simplex can automatically derive switching conditions for the AI-Grid system, the switching conditions are not overly conservative, and Bb-Simplex ensures safety even in the presence of adversarial attacks on the neural controller.

2.5.1.1 Architectural Overview of Bb-Simplex

Figure 2.11 shows the overall architecture of the combined barrier-based NSA. The left part of the figure depicts our design methodology; the right part illustrates NSA. Given the BC, the required safety properties, and a dynamic model of the plant, our methodology generates a BaC and then

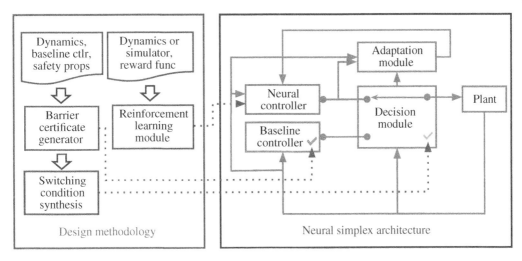

Figure 2.11 Overview of the barrier certificate-based neural simplex architecture.

derives the switching condition from it. The RL module learns a high-performance NC based on the performance objectives encoded in the reward function.

The structure of this section is as follows. Section 2.5.2 provides background material on BaCs. Section 2.5.3 features our novel approach for deriving switching conditions from BaCs. Section 2.5.4 introduces our microgrid case study and the associated controllers used for microgrid control. Section 2.5.5 presents the results of our microgrid case study. Section 2.5.6 extends our Bb-Simplex framework to handle approximate knowledge of the system dynamics. Section 2.5.7 extends it to hybrid systems, i.e. systems with multiple modes having different dynamics. Section 2.5.8 discusses related work.

2.5.2 Preliminaries

We use BaCs to prove that the BC ensures safety. A BaC is a function of the state satisfying a set of inequalities on the value of the function and value of its time derivative along the dynamic flows of the system. Intuitively, the zero-level-set of a BaC forms a "barrier" between the reachable states and unsafe states. The existence of a BaC assures that starting from a state where the BaC is positive, safety is forever maintained [13–15]. Moreover, there are automated methods to synthesize BaCs, e.g. [16–19]. We implemented two automated methods for BaC synthesis from the literature. As discussed next, one of the methods is based on sum-of-squares optimization (SOS) and the other uses deep learning. Our design methodology for computing switching conditions (see Section 2.5.3) requires a BaC, but is independent of how the BaC is obtained.

2.5.2.1 BaC Synthesis Using SOS Optimization
This method first derives a Lyapunov function V for the system using the expanding interior-point algorithm in [20]. It then uses the SOS-based algorithm in [17] to obtain a BaC from V. Note that the largest super-level set of a Lyapunov function within a safety region is a BaC. The algorithm in [16, 17] computes a larger BaC by starting with that sub-level set and then expanding it, by allowing it to take shapes other than that of a sub-level set of the Lyapunov function. This method involves a search of Lyapunov functions and BaCs of various degrees by choosing different candidate polynomials and parameters of the SOS problem. It is limited to systems with polynomial

dynamics. In some cases, non-polynomial dynamics can be recast as polynomial using, e.g. the techniques in [20].

2.5.2.2 BaC Synthesis Using Deep Learning

We also implemented *SyntheBC* from [21], which uses deep learning to synthesize a BaC. First, training samples obtained by sampling different areas of the state space are used to train a feedforward ReLU neural network with two hidden layers as a candidate BaC. Second, the validity of this candidate's BaC must be verified. The NN's structure allows the problem of checking whether the neural network (NN) satisfies the defining conditions of a BaC to be transformed into mixed-integer linear programming (MILP) and mixed-integer quadratically-constrained programming (MIQCP) problems, which we solve using the Gurobi optimizer. If the verification fails, the Gurobi optimizer provides counter-examples which can be used to guide retraining of the NN. In this way, the training and verification steps can be iterated as needed.

2.5.3 Deriving the Switching Condition

We employ our novel methodology to derive the switching logic from the BaC. The DM implements this switching logic for both forward and reverse switching. When the forward-switching condition (FSC) is true, control is switched from the NC to the BC; likewise, when the reverse-switching condition (RSC) is true, control is switched from the BC to the NC. The success of our approach rests on solving the complex problems discussed in this section to derive an FSC. Consider a continuous dynamical system of the form:

$$\dot{x} = f(x, u) \tag{2.5}$$

where $x \in \mathbb{R}^k$ is the state of the plant at time t and $u \in \Omega$ is the control input provided to the plant at time t. The set of all valid control actions is denoted by Ω. The set of *unsafe states* is denoted by $\mathcal{U}$. Let $x_{lb}, x_{ub} \in \mathbb{R}^k$ be *operational bounds* on the ranges of state variables, reflecting physical limits and simple safety requirements.

The set $\mathcal{A}$ of *admissible states* is given by: $\mathcal{A} = \{x : x_{lb} \leq x \leq x_{ub}\}$. A state of the plant is *recoverable* if the BC can take over in that state and keep the plant invariably safe. For a given BC, we denote the *recoverable region* by $\mathcal{R}$. Note that $\mathcal{U}$ and $\mathcal{R}$ are disjoint. The safety of such a system can be verified using a BaC $h(x) : \mathbb{R}^k \to \mathbb{R}$ of the following form [14–17]:

$$
\begin{aligned}
h(x) &\geq 0, \quad &&\forall x \in \mathbb{R}^k \backslash \mathcal{U} \\
h(x) &< 0, \quad &&\forall x \in \mathcal{U} \\
(\nabla_x h)^T f(x, u) + \sigma(h(x)) &\geq 0, \quad &&\forall x \in \mathbb{R}^k
\end{aligned}
\tag{2.6}
$$

where $\sigma(.)$ is an extended class-$\mathcal{K}$ function. The BaC is negative over the unsafe region and non-negative otherwise. $\nabla_x h$ is the gradient of h w.r.t x and the expression $(\nabla_x h)^T f(x, u)$ is the time derivative of h. The zero-super-level set of a BaC h is $\mathcal{Z}(h) = \{x : h(x) > 0\}$. In [17], the invariance of this set is used to show $\mathcal{Z}(h) \subseteq \mathcal{R}$.

Let η denote the control period a.k.a. time step. Let $\hat{h}(x, u, \delta)$ denote the nth-degree Taylor approximation of BaC h's value after time δ, if control action u is taken in state x. The approximation is computed at the current time to predict h's value δ time units later and is given by:

$$\hat{h}(x, u, \delta) = h(x) + \sum_{i=1}^{n} \frac{h^i(x, u)}{i!} \delta^i \tag{2.7}$$

where $h^i(x, u)$ denotes the ith time derivative of h evaluated in state x if control action u is taken. The control action is needed to calculate the time derivatives of h from the definition of h and

Eq. (2.5) by applying the chain rule. Since we are usually interested in predicting the value one time step in the future, we use $\hat{h}(x, u)$ as shorthand for $\hat{h}(x, u, \eta)$. By Taylor's theorem with the Lagrange form of the remainder, the remainder error of the approximation $\hat{h}(x, u)$ is:

$$\frac{h^{n+1}(x, u, \delta)}{(n+1)!} \eta^{n+1} \text{ for some } \delta \in (0, \eta) \tag{2.8}$$

An upper bound on the remainder error, if the state remains in the admissible region during the time interval, is:

$$\lambda(u) = \sup \left\{ \frac{|h^{n+1}(x, u)|}{(n+1)!} \eta^{n+1} : x \in \mathcal{A} \right\} \tag{2.9}$$

The FSC is based on checking recoverability during the next time step. For this purpose, the set $\mathcal{A}$ of admissible states is shrunk by margins of μ_{dec} and μ_{inc}, a vector of upper bounds on the amount by which each state variable can decrease and increase, respectively, in one time step, maximized over all admissible states. Formally,

$$\mu_{\text{dec}}(u) = |\min(0, \eta \dot{x}_{\text{min}}(u))| \\ \mu_{\text{inc}}(u) = |\max(0, \eta \dot{x}_{\text{max}}(u))| \tag{2.10}$$

where $\dot{x}_{\text{min}}$ and $\dot{x}_{\text{max}}$ are vectors of solutions to the optimization problems:

$$\dot{x}_i^{\text{min}}(u) = \inf\{\dot{x}_i(x, u) : x \in \mathcal{A}\} \\ \dot{x}_i^{\text{max}}(u) = \sup\{\dot{x}_i(x, u) : x \in \mathcal{A}\} \tag{2.11}$$

The difficulty of finding these extremal values depends on the complexity of the functions $\dot{x}_i(x, u)$. For example, it is relatively easy if they are convex. In our case study of a realistic microgrid model, they are multivariate polynomials with degree 1, and hence convex. The set $\mathcal{A}_r$ of *restricted admissible states* is given by:

$$\mathcal{A}_r(u) = \{x : x_{\text{lb}} + \mu_{\text{dec}}(u) < x < x_{\text{ub}} - \mu_{\text{inc}}(u)\} \tag{2.12}$$

Let $\text{Reach}_{=\eta}(x, u)$ denote the set of states reachable from state x after exactly time η if control action u is taken in state x. Let $\text{Reach}_{\leq \eta}(x, u)$ denote the set of states reachable from x within time η if control action u is taken in state x.

Lemma 2.1 *For all $x \in \mathcal{A}_r(u)$ and all control actions u, $\text{Reach}_{\leq \eta}(x, u) \subseteq \mathcal{A}$.*

Proof: The derivative of x is bounded by $\dot{x}_{\text{min}}(u)$ and $\dot{x}_{\text{max}}(u)$ for all states in $\mathcal{A}$. This implies that μ_{dec} and μ_{inc} are bounds on the amounts by which the state x can decrease and increase, respectively, during time η, as long as x remains within $\mathcal{A}$ during the time step. Since $\mathcal{A}_r(u)$ is obtained by shrinking $\mathcal{A}$ by μ_{dec} and μ_{inc} (i.e. by moving the lower and upper bounds, respectively, of each variable inwards by those amounts), the state cannot move outside of $\mathcal{A}$ during time η. ∎

2.5.3.1 Forward Switching Condition

To ensure safety, a FSC should switch control from the NC to the BC if using control action u proposed by NC causes any unsafe states to be reachable from the current state x during the next control period, or causes any unrecoverable states to be reachable at the end of the next control period. These two conditions are captured in the following definition:

Definition 2.1 *Forward Switching Condition:* A condition $\text{FSC}(x, u)$ is a FSC if for every recoverable state x, every control action u, and control period η, $\text{Reach}_{\leq \eta}(x, u) \cap \mathcal{U} \neq \emptyset \vee \text{Reach}_{=\eta}(x, u) \not\subset \mathcal{R}$ implies $\text{FSC}(x, u)$ is true.

Theorem 2.1 *A Simplex architecture whose FSC satisfies Definition 2.1 keeps the system invariably safe provided the system starts in a recoverable state.*

Proof: Our definition of an FSC is based directly on the switching logic in Algorithm 1 of [22]. The proof of Theorem 1 in [22] shows that an FSC that is exactly the disjunction of the two conditions in our definition invariantly ensures system safety. It is easy to see that any weaker FSC also ensures safety. ∎

We now propose a new and general procedure for constructing a switching condition from a BaC and prove its correctness.

Theorem 2.2 *Given a barrier certificate h, the following condition is a FSC:* $FSC(x, u) = \alpha \vee \beta$ *where* $\alpha \equiv \hat{h}(x, u) - \lambda(u) \leq 0$ *and* $\beta \equiv x \notin \mathcal{A}_r(u)$.

Proof: Intuitively, $\alpha \vee \beta$ is an FSC because (1) if condition α is false, then control action u does not lead to an unsafe or unrecoverable state during the next control period, provided the state remains admissible during that period; and (2) if condition β is false, then the state will remain admissible during that period. Thus, if α and β are both false, then nothing bad can happen during the control period, and there is no need to switch to the BC.

Formally, suppose x is a recoverable state, u is a control action, and that $\text{Reach}_{\leq\eta}(x, u) \cap \mathcal{U} \neq \emptyset \vee \text{Reach}_{=\eta}(x, u) \not\subseteq \mathcal{R}$, i.e. there is an unsafe state in $\text{Reach}_{\leq\eta}(x, u)$ or an unrecoverable state in $\text{Reach}_{=\eta}(x, u)$. Let x' denote that unsafe or unrecoverable state. Recall that $\mathcal{Z}(h) \subseteq \mathcal{R}$, and $\mathcal{R} \cap \mathcal{U} = \emptyset$. Therefore, $h(x', u) \leq 0$. We need to show that $\alpha \vee \beta$ holds. We do a case analysis based on whether x is in $\mathcal{A}_r(u)$.

Case 1: $x \in \mathcal{A}_r(u)$. In this case, we use a lower bound on the value of the BaC h to show that states reachable in the next control period are safe and recoverable. Using Lemma 2.1, we have $\text{Reach}_{\leq\eta}(x, u) \subseteq \mathcal{A}$. This implies that $\lambda(u)$, whose definition maximizes over $x \in \mathcal{A}$, is an upper bound on $\hat{h}(x, u, \delta)$ for $\delta \leq \eta$. This implies that $\hat{h}(x, u) - \lambda(u)$ is a lower bound on value of BaC for all states in $\text{Reach}_{\leq\eta}(x, u)$. As shown above, there is a state x' in $\text{Reach}_{\leq\eta}(x, u)$ with $h(x', u) \leq 0$. $\hat{h}(x, u) - \lambda(u)$ is lower bound on $h(x', u)$ and hence must also be less than or equal to 0. Thus, α holds.

Case 2: $x \notin \mathcal{A}_r(u)$. In this case, β holds. Note that in this case, the truth value of α is not significant (and not relevant, since $FSC(x, u)$ holds regardless), because the state might not remain admissible during the next control period. Hence, the error bound obtained using Eq. (2.9) is not applicable. ∎

2.5.3.2 Reverse Switching Condition

The RSC is designed with a heuristic approach, since it does not affect safety of the system. To prevent frequent switching between the NC and BC, we design the RSC to hold if the FSC is likely to remain false for at least m time steps, with $m > 1$. The RSC, like the FSC, is the disjunction of two conditions. The first condition is $h(x) \geq m\eta|\dot{h}(x)|$, since h is likely to remain non-negative for at least m time steps if its current value is at least that duration times its rate of change. The second condition ensures that the state will remain admissible for m time steps. In particular, we take:

$$RSC(x) = h(x) \geq m\eta|\dot{h}(x)| \wedge x \in \mathcal{A}_{r,m} \tag{2.13}$$

where the m-times-restricted admissible region is:

$$\mathcal{A}_{r,m} = \{x : x_{\text{lb}} + m\mu_{\text{dec}} < x < x_{\text{ub}} - m\mu_{\text{inc}}\} \tag{2.14}$$

where vectors μ_{dec} and μ_{inc} are defined in the same way as $\mu_{\text{dec}}(u)$ and $\mu_{\text{inc}}(u)$ in Eqs. (2.10) and (2.11) except with optimization over all control actions u.

2.5.3.3 Decision Logic

The DM's switching logic has three inputs: the current state x, the control action u proposed by the NC, and the name c of the controller currently in control (as a special case, we take $c = NC$ in the first time step). The switching logic is defined by cases as follows: $DM(x, u, c)$ returns BC if $c = NC$ $\wedge$ FSC(x, u), returns NC if $c = BC \wedge$ RSC(x), and returns c otherwise.

2.5.4 Application to Microgrids

A *microgrid* (MG) is an integrated energy system comprising DERs and multiple energy loads. DERs tend to be *renewable* energy resources and include solar panels, wind turbines, batteries, and emergency diesel generators. By satisfying energy needs from local renewable energy resources, MGs can reduce energy costs and improve energy supply reliability for energy consumers. Some of the major control requirements for an MG are power control, load sharing, and frequency and voltage regulation.

An MG can operate in two modes: grid-connected and islanded. When operated in grid-connected mode, DERs act as constant source of power which can be injected into the network on demand. In contrast, in islanded or autonomous mode, the DERs form a grid of their own, meaning not only do they supply power to the local loads, but they also maintain the MG's voltage and frequency within the specified limits [23]. For our case study, we focus on voltage regulation in both grid-connected and islanded modes. Specifically, we apply Bb-Simplex to the controller for the inverter for a PV DER.

Applying Bb-Simplex to other DERs, which have inverter interfaces such as battery is straightforward. Of the three controllers necessary for diesel generator DER, our methodology can be applied to voltage and frequency controllers straightforwardly. The exciter system controls the magnetic flux flowing through the rotor generator, and its dynamics are coupled with that of the diesel engine. We plan to explore using the approach presented in [24] to handle these coupled dynamics and apply Bb-Simplex to the exciter system.

2.5.4.1 Baseline Controller

For our experiments, we used the SOS-based methodology described in Section 2.5.2 to derive a Barrier Certificate (as a proof of safety) for the BC. We use a droop controller as the BC. A droop controller is a type of proportional controller, traditionally used in power systems for control objectives such as voltage regulation, power regulation, and current sharing [25–28]. The droop controller tries to balance the electrical power with voltage and frequency. Variations in the active and reactive powers result in frequency and voltage magnitude deviations, respectively. The dynamic model for a voltage droop controller for an inverter has the form $\dot{v} = v^* - v + \lambda_q(Q^* - Q)$, where v^*, v, Q^*, Q are voltage reference, voltage, reactive power reference and reactive power of inverter, respectively, and λ_q is the controller's droop coefficient. Detailed dynamic models for an MG with multiple inverters connected by transmission lines and with droop controllers for frequency and voltage are given in [16, 20].

Consider the following model of an MG's droop-controlled inverters:

$$\dot{\theta}_i = \omega_i \tag{2.15}$$

$$\dot{\omega}_i = \omega_i^0 - \omega_i + \lambda_i^p(\mathbf{P_i} - P_i) \tag{2.16}$$

$$\dot{v}_i = v_i^0 - v_i + \lambda_i^q(\mathbf{Q_i} - Q_i) \tag{2.17}$$

where θ_i, ω_i, and v_i are the phase angle, frequency, and voltage of the ith inverter, respectively. $\mathbf{P_i}$ and $\mathbf{Q_i}$ are the inverter's active and reactive power set-points, and λ^p and λ^q are the droop controller's coefficients. The values of set-points $\mathbf{P_i}$ and $\mathbf{Q_i}$ of an inverter depend upon local loads

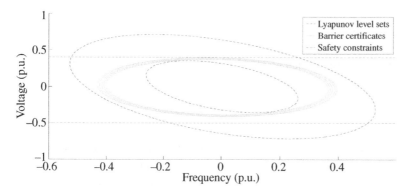

Figure 2.12 Lyapunov-function level sets (dashed ellipses). Innermost ellipse also indicates initial BaC, which is optimized iteratively (solid-line ellipses). Dashed lines are voltage safety limits.

and power needed by the rest of the MG. The loads are not explicitly modeled here. In our case studies, we vary these power set-points to simulate changing loads. Let $\mathcal{M}$ be the set of all inverter indices. The active power P_i and reactive power Q_i are given by:

$$
\begin{aligned}
P_i &= v_i \sum_{j \in \mathcal{N}_i} v_k (G_{i,j} \cos \theta_{i,j} + B_{i,j} \sin \theta_{i,j}) \\
Q_i &= v_i \sum_{j \in \mathcal{N}_i} v_k (G_{i,j} \sin \theta_{i,j} - B_{i,j} \cos \theta_{i,j})
\end{aligned}
\tag{2.16}
$$

where $\theta_{i,j} = \theta_i - \theta_j$, and $\mathcal{N}_i \subseteq \mathcal{M}$ is the set of neighbors of inverter i. $G_{i,j}$ and $B_{i,j}$ are respectively the conductance and susceptance values of the transmission line connecting inverters i and j. As shown in [20], the stability of such a system can be verified using Lyapunov theory. Detailed dynamic models for an MG with multiple inverters connected by transmission lines and with droop controllers for frequency and voltage are given in [16, 20].

Figure 2.12 shows this process of incrementally expanding the Lyapunov function to obtain the BaC. SOS-based algorithms apply only to polynomial dynamics so we first recast our droop controller dynamics to be polynomial using a DQ0 transformation to AC waveforms as shown in [29]. This transformation is exact; i.e. it does not introduce any approximation error. In our experimental evaluation (Section 2.5.5), we obtain the BaCs for BCs in the form of droop controllers for voltage regulation, in the context of MGs containing up to three DERs of different types. Note that battery DERs operate in two distinct modes, charging and discharging, resulting in a hybrid system model with different dynamics in different modes. For now, we consider only runs in which the battery remains in the same mode for the duration of the run. Extending our framework to hybrid systems is future work.

2.5.4.2 Neural Controller
To help address the control challenges related to microgrids, the application of *neural networks for microgrid control* is on the rise as documented in [30]. Increasingly, RL is being used to train powerful DNNs to produce high-performance MG controllers.

We present our approach for learning NCs in the form of DNNs representing deterministic control policies. Such a DNN maps system states (or raw sensor readings) to control inputs. We use RL in form of DDPG algorithm, with the safe learning strategy of penalizing unrecoverable actions from [12]. DDPG was chosen because it works with deterministic policies and is compatible with continuous action spaces. An advantage of off-policy algorithms such as DDPG is that the problem of exploration can be treated independently from the learning algorithm. Off-policy learning is advantageous in our setting because it enables the NC to be (re-)trained using actions taken by the BC rather than the NC or the learning algorithm. The benefits of off-policy retraining are further considered in Section 2.5.4.3.

We consider a standard RL setup consisting of an agent interacting with an environment in discrete time. At each time step t, the agent receives a (microgrid) state x_t as input, takes an action a_t, and receives a scalar reward r_t. The DDPG algorithm employs an *actor–critic framework*. The actor generates a control action and the critic evaluates its quality. In order to learn from prior knowledge, DDPG uses a replay buffer to store training samples of the form (x_t, a_t, r_t, x_{t+1}). At every training iteration, a set of samples is randomly chosen from the replay buffer. For further details regarding the implementation of the DDPG algorithm, please refer to Algorithm 1 in [31].

To learn an NC for DER voltage control, we designed the following reward function, which guides the actor network to learn the desired control objective.

$$r(x_t, a_t) = \begin{cases} -1000 & \text{if } \text{FSC}(x_t, a_t) \\ 100 & \text{if } v_{\text{od}} \in [v_{\text{ref}} - \epsilon, v_{\text{ref}} + \epsilon] \\ -w \cdot \left(v_{\text{od}} - v_{\text{ref}}\right)^2 & \text{otherwise} \end{cases} \tag{2.17}$$

where w is a weight ($w = 100$ in our experiments), v_{od} is the d-component of the output voltage of the DER whose controller is being learned, v_{ref} is the reference or nominal voltage, and ϵ is the tolerance threshold. We assign a high negative reward for triggering the FSC, and a high positive reward for reaching the tolerance region, i.e. $v_{\text{ref}} \pm \epsilon$. The third clause rewards actions that lead to a state in which the DER voltage is close to its reference value.

Adversarial Inputs Controllers obtained via deep RL algorithms are vulnerable to *adversarial inputs* (AIs): those that lead to a state in which the NC produces an unrecoverable action, even though the NC behaves safely on very similar inputs. NSA provides a defense against these kinds of attacks. If the NC proposes a potentially unsafe action, the BC takes over in a timely manner, thereby guaranteeing the safety of the system. To demonstrate NSA's resilience to AIs, we use a gradient-based attack (Algorithm 4) from [32] to construct such inputs, and show that the DM switches control to the BC in time to ensure safety.

The gradient-based algorithm takes as input the critic network, actor network, adversarial attack constant c, parameters a, b of beta distribution $\beta(a, b)$, and the number of times n noise is sampled. For a given (microgrid) state x, the critic network is used to ascertain its Q-value and the actor network determines its optimal action. Once the gradient of the critic network's loss function is computed using the Q-value and the action, the l_2-constrained norm of the gradient (*grad_dir*) is obtained. An initial (microgrid) state x_0, to be provided as input to the actor network, is then perturbed to obtain a potential adversarial state x_{adv}, determined by the sampled noise in the direction of the gradient: $x_{\text{adv}} = x_0 - c \cdot \beta(a, b) \cdot \textit{grad\_dir}$.

We can now compute the Q-value of x_{adv} and its (potentially adversarial) action a_{adv}. If this value is less than $Q(x_0, a_0)$, then x_{adv} leads to a sub-optimal action. The gradient-based attack algorithm does not guarantee the successful generation of AIs every time it is executed. The success rate is inversely related to the quality of the training of the NC. In our experiments (see Section 2.5.5.5), the highest success rate for AI generation that we observed is 0.008%.

2.5.4.3 Adaptation Module

The AM retrains the NC in an online manner when the NC produces an unrecoverable action that causes the DM to failover to the BC. With retraining, the NC is less likely to repeat the same or similar mistakes in the future, allowing it to remain in control of the system more often, thereby improving performance. We use RL with the reward function defined in Eq. (2.17) for online retraining.

As in initial training, we use the DDPG algorithm (with the same settings) for online retraining. When the NC outputs an unrecoverable action, the DM switches control to the BC, and the AM computes the (negative) reward for this action and adds it to a pool of training samples. As in [12], we found that reusing the pool of training samples (DDPG's experience replay buffer) from initial training of the NC evolves the policy in a more stable fashion, as retraining samples gradually replace initial training samples in the pool. Another benefit of reusing the initial training pool is that retraining of the NC can start almost immediately, without having to wait for enough samples to be collected online.

There are two methods to retrain the NC:

1. **Off-policy retraining**: At every time step while the BC is active, the BC's action is used in the training sample. The reward for the BC's action is based on the observed next state of the system.
2. **Shadow-mode retraining**: At every time step while the BC is active, the AM takes a sample by running the NC in shadow mode to compute its proposed action, and then simulates the behavior of the system for one time step to compute a reward for it.

In our experiments, both methods produce comparable benefits. Off-policy retraining is therefore preferable because it does not require simulation (or a dynamic model of the system) and hence is less costly.

2.5.5 Implementation and Experiments

We apply our Bb-Simplex methodology to a model of a microgrid [33] with three DERs: a battery, PV (a.k.a. solar panels), and diesel generator. The three DERs are connected to the main grid via bus lines. As depicted in Fig. 2.13, the three DERs are connected to the main grid via bus lines. We are primarily interested in PV control, since we apply Bb-Simplex to PV voltage regulation. The PV control includes multiple components, such as "three-phase to DQ0 voltage and current" transformer, average voltage and current control, power and voltage measurements, inner-loop dq current control, and outer-loop maximum power point tracking (MPPT) control. Our experimental evaluation of Bb-Simplex was carried out on RTDS, a high-fidelity power systems simulator.

We ran experiments for three configurations of the microgrid: Configuration 1: grid-connected mode with only the PV DER connected within the MG; Configuration 2: islanded mode with PV and diesel generator DERs connected within the MG; Configuration 3: islanded mode with PV, diesel generator, and battery (in discharging mode) DERs connected within the MG. All configurations also include a load. These configurations demonstrate Bb-Simplex's ability to handle a wide variety of MG configurations involving various types of DERs. We did not perform experiments with the battery in charging mode, because in this mode, the battery is simply another load, and the configuration is equivalent to Configuration 1 or 2 with a larger load.

The state of the MG plant is given by $[i_d\, i_q\, i_{od}\, i_{oq}\, v_{od}\, v_{oq}\, i_{ld}\, i_{lq}\, m_d\, m_q]$, where i_d and i_q are the d- and q-components of the dq current measured at the local load of the inverter, i_{od} and i_{oq} are the d- and q-components of the output current of the inverter measured at point of coupling to the main grid, v_{od} and v_{oq} are the d- and q-components of the output voltage of the inverter measured at point of coupling to the main grid, i_{ld} and i_{lq} are the d- and q-components of the input current to the current controller, m_d and m_q are the d- and q-components of the output voltage from the current controller used to generate the next state.

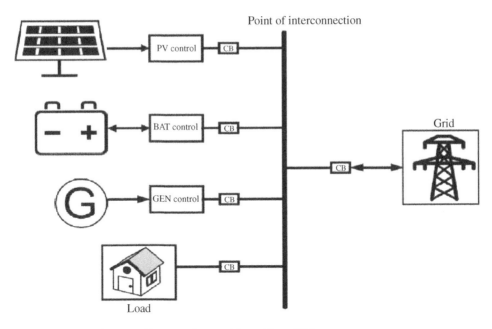

Figure 2.13 RTDS microgrid model. Source: Adapted from [33].

We use Bb-Simplex to ensure the safety property that the d-component of the output voltage (v_{od}) of the inverter for the PV DER is within $\pm 3\%$ of the reference voltage $v_{ref} = 0.48$ kV. We adopted a 3% tolerance for the voltage, based on the discussion in [33]. Bb-Simplex could similarly be used to ensure additional desired safety properties. The BC is the droop controller described in [33]. All experiments use runs of length 10 s, with the control period, RTDS time step, and simulation time step in MATLAB all equal to 3.2 milliseconds (ms), the largest time step allowed by RTDS.

2.5.5.1 Integration of Bb-Simplex in RTDS

The BC is implemented in RTDS using components in the RTDS standard libraries. The DM is implemented as an RTDS *custom component* written in C. For an MG configuration, expressions for the BaC, λ and μ (see Section 2.5.3) are derived in MATLAB, converted to C data structures, and then included in a header file of the custom component. The BaCs are polynomials comprising 41, 67, and 92 monomials, respectively, for configurations 1, 2, and 3.

The NC is trained and implemented using Keras, a high-level neural network Python application programming interface (API) developed by [34], running on top of TensorFlow from [35]. For training, we customized an existing skeleton implementation of DDPG in Keras, which we then used with the Adam optimizer [36] Hyperparameters used during training involved a learning rate lr = 0.0001, discounting factor $\gamma = 0.99$, and target network update weight $\tau = 0.001$.

RTDS imposes limitations on custom components that make it difficult to implement complex NNs within RTDS. Existing NN libraries for RTDS, such as those developed by [37, 38], severely limit the NN's size and the types of activation functions. Therefore, we implemented the NC external to RTDS, following the *software-defined microgrid control* approach in [39]. Figure 2.14 shows our setup. We used RTDS's GTNET-SKT communication protocol to establish a transmission control protocol (TCP) connection between the NC running on a PC and an "NC-to-DM" relay component in the RTDS MG model. This relay component repeatedly sends the plant state

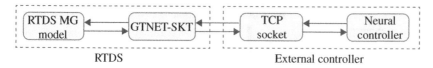

Figure 2.14 Integration of external NC with RTDS.

to the NC, which computes its control action and sends it to the relay component, which in turn sends it to the DM.

Running the NC outside RTDS introduces control latency. We measured the round-trip time between RTDS and NC (including the running time of NC on the given state) to be 4.34 ms. Since the control period is 3.2 ms, each control action is delayed by one control period. The latency is mostly from network communication, since the PC running the NC was off-campus. We plan to reduce the latency by moving the NC to a PC connected to the same LAN as RTDS.

2.5.5.2 User Interface

The AI-Grid user interface includes a panel that displays information related to Bb-Simplex. The screenshot in Fig. 2.15 shows the upper half of that panel for the PV DER, which is represented by the large black marker with the PV icon in the map-view panel; clicking on that marker opens the Bb-Simplex panel. The state of the system is shown partway through a simulated execution of the system starting from an adversarial input, described below in the paragraph on adversarial input attacks. The same execution is illustrated in Fig. 2.16(b).

The first line of data in the panel shows the percentage of time that the NC has been in control (NC Time), the average deviation of the voltage from the reference voltage (Avg Deviation), and the number of samples that have been used for online retraining of the NC (Retraining Samples). The next line shows which controller is currently active (i.e. in control), by highlighting its name.

Figure 2.15 User interface showing information related to Bb-Simplex.

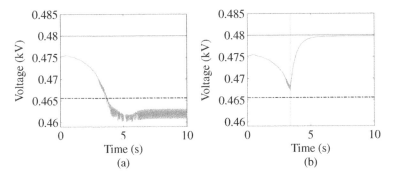

Figure 2.16 NC with adversarial inputs. (a) Without NSA. (b) With NSA.

The graph below that shows the voltage as a function of time, with the central horizontal line indicating the reference voltage, and the dashed/dotted lines delimiting the safety region.

Scrolling to the lower half of the panel (not shown in the figure) reveals three more graphs, which show the following quantities as functions of time: the deviation of the actual voltage from the reference voltage, the active controller, and the number of retraining samples.

2.5.5.3 Consistency of RTDS and MATLAB Models

Our methodology requires an analytical model of the microgrid dynamics to derive a BaC for the BC and a switching condition for the DM. We therefore developed an analytical model in MATLAB based on the RTDS model and the description given in [33]. To verify consistency of MATLAB and RTDS models, we compared trajectories obtained from them under various operating conditions.

Tables 2.6–2.8 report deviations in output voltage and current trajectories of the PV DER between the two models under the control of the BC. The results are based on 100 trajectories starting from random initial states.

As expected, the two models are in close agreement. The small deviations are due to a few factors: (1) the RTDS model uses realistic dynamic models of transmission lines including their noise,

Table 2.6 Performance comparison between output of PV DER in RTDS and MATLAB models for Configuration 1.

	VD (kV)	VD (%)	CD (Amp)	CD (%)
Avg	0.000214	0.04	0.000129	0.028
Min	0.000187	0.03	0.000124	0.015
Max	0.000378	0.08	0.000181	0.036

Table 2.7 Performance comparison between output of PV DER in RTDS and MATLAB models for Configuration 2.

	VD (kV)	VD (%)	CD (Amp)	CD (%)
Avg	0.000348	0.07	0.000126	0.032
Min	0.000103	0.02	0.000104	0.019
Max	0.000493	0.10	0.000187	0.052

Table 2.8 Performance comparison between output of PV DER in RTDS and MATLAB models for Configuration 3.

	VD (kV)	VD (%)	CD (Amp)	CD (%)
Avg	0.001041	0.12	0.000238	0.047
Min	0.000119	0.02	0.000133	0.019
Max	0.001403	0.21	0.000187	0.102

whereas the MATLAB model ignores transmission line dynamics; and (2) the RTDS model uses average-value modeling to more efficiently simulate the dynamics in real-time [33], whereas in MATLAB, trajectories are calculated by solving ordinary differential equations of the dynamics at each simulation time-step.

2.5.5.4 Evaluation of Forward Switching Condition

We derive a BaC using the SOS-based methodology presented in Section 2.5.2, and then derive a switching condition from the BaC, as described in Section 2.5.3.1. To find values of λ and μ, we use MATLAB's fmincon function to solve the constrained optimization problems given in Eqs. (2.10) and (2.11).

An ideal FSC triggers a switch to BC only if an unrecoverable state is reachable in one time step. For systems with complex dynamics, switching conditions derived in practice are conservative, i.e. may switch sooner. To show that our FSC is not overly conservative, we performed experiments using an AC that continuously increases the voltage and hence soon violates safety. The PV voltage controller has two outputs, m_d and m_q, for the d and q components of the voltage, respectively. The dummy AC simply uses constant values for its outputs, with $m_d = 0.5$ and $m_q = 1e-6$.

These experiments were performed with PV DER in grid connected mode, with reference voltage and voltage safety threshold of 0.48 and 0.4944 kV, respectively, and a FSC derived using a fourth-order Taylor approximation of the BaC. We averaged over 100 runs from initial states with initial voltage selected uniformly at random from the range 0.48 kV $\pm 1\%$. The mean voltage at switching is 0.4921 kV (with standard deviation of 0.0002314 kV), which is only 0.46% below the safety threshold. The mean numbers of time steps before switching, and before a safety violation if Bb-Simplex is not used, are 127.4 and 130.2, respectively. Thus, our FSC triggered a switch about three time steps, on average, before a safety violation would have occurred.

We also derived a neural network-based BaC using deep learning and verified it using the Gurobi optimizer as discussed in Section 2.5.2. We then derived the switching conditions from the verified neural BaC, again using a 4th-order Taylor approximation. We performed the same experiments as above to determine the conservativeness of this FSC. The mean voltage at switching is 0.4923 kV (with standard deviation 0.0002132 kV). The mean numbers of time steps before switching, and before a safety violation if Bb-Simplex is not used, are 128.1 and 130.2, respectively. Thus, our neural FSC triggered a switch about two time steps, on average, before a safety violation would have occurred.

2.5.5.5 Evaluation of Neural Controller

The NC for a microgrid configuration is a DNN with four fully connected hidden layers of 128 neurons each and one output layer. The hidden layers and output layer use the ReLU and tanh activation function, respectively. The input state to the NC (DNN) is the same as the inputs to the BC (droop controller) i.e. $[i_{ld}\ i_{lq}]$, where i_{ld} and i_{lq} are the d- and q-components of the input current to the droop controller. Thus, the NC has same inputs and outputs as the BC. The NC is trained on

Table 2.9 Performance comparison for Configuration 1.

Controller	CR	CT	σ (CT)	δ	$\sigma(\delta)$
NC	100	67.5	5.8	$1.1e-4$	$1.0e-5$
BC	100	102.3	8.2	$4.2e-4$	$3.7e-5$

Table 2.10 Performance comparison for Configuration 2.

Controller	CR	CT	σ (CT)	δ	$\sigma(\delta)$
NC	100	76.8	6.1	$1.3e-4$	$1.2e-5$
BC	100	108.8	8.3	$5.1e-4$	$3.8e-5$

Table 2.11 Performance comparison for Configuration 3.

Controller	CR	CT	σ (CT)	δ	$\sigma(\delta)$
NC	100	81.1	7.7	$1.5e-4$	$1.3e-5$
BC	100	115.7	9.8	$5.8e-4$	$3.8e-5$

1 million samples (one-step transitions) from MATLAB simulations, processed in batches of 200. Transitions start from random states, with initial values uniformly sampled from $[0.646, 0.714]$ for i_{ld} and $[-0.001, 0.001]$ for i_{lq} [33]. Training takes approximately two hours. The number of trainable parameters in the actor and critic networks are 198,672 and 149,111, respectively.

Performance We evaluate a controller's performance based on three metrics: convergence rate (CR), the percentage of trajectories in which the DER voltage converges to the tolerance region $v_{ref} \pm \epsilon$; average convergence time (CT), the average time required for convergence of the DER voltage to the tolerance region; and mean deviation (δ), the average deviation of the DER voltage from v_{ref} after the voltage enters the tolerance region. We always report CR as a percentage, CT in milliseconds, and δ in kV.

We show that the NC outperforms the BC. For this experiment, we used RTDS to run the BC and NC starting from the same 100 initial states. The CR is 100% for the NC and BC. Tables 2.9–2.11 compare their performance, averaged over 100 runs, with $\epsilon = 0.001$. We observe that for all three configurations, the NC outperforms the BC both in terms of average convergence time and mean deviation. We also report the standard deviations (σ) for these metrics and note that they are small compared to the average values. The FSC was not triggered even once during these runs, showing that the NC is well-trained.

Generalization Generalization refers to the NC's ability to perform well in contexts beyond the ones in which it was trained. First, we consider two kinds of generalization with respect to the microgrid state:

- **Gen 1**: The initial states of the DERs are randomly chosen from a range outside of the range used during training.
- **Gen 2**: The power set-point $P^\star$ is randomly chosen from the range $[0.2, 1]$, whereas all training was done with $P^\star = 1$.

Table 2.12 Generalization performance of NC for Configuration 1.

	CR	CT	σ (CT)	δ	$\sigma (\delta)$
Gen 1	100	108.7	9.8	1.7e − 4	1.5e − 5
Gen 2	100	77.1	6.9	1.3e − 4	1.1e − 5

Table 2.13 Generalization performance of NC for Configuration 2.

	CR	CT	σ (CT)	δ	$\sigma (\delta)$
Gen 1	100	118.2	10.1	2.1e − 4	1.9e − 5
Gen 2	100	81.2	6.2	1.5e − 4	1.4e − 5

Table 2.14 Generalization performance of NC for Configuration 3.

	CR	CT	σ (CT)	δ	$\sigma (\delta)$
Gen 1	100	120.4	10.8	2.2e − 4	1.9e − 5
Gen 2	100	88.5	7.3	1.6e − 4	1.4e − 5

Tables 2.12–2.14 present the NC's performance in these two cases, based on 100 runs for each case. We see that the NC performs well in both cases.

Second, we consider generalization with respect to the microgrid configuration. Here, we evaluate how the NC handles dynamic changes to the microgrid configuration during runtime. For the first experiment, we start with all the three DERs connected, but the diesel generator DER is disconnected after the voltage has converged. For the second experiment, we again start with all the three DERs connected, but both the diesel generator and battery DER are disconnected after the voltage has converged. For both instances, the NC succeeded in continuously keeping the voltage in the tolerance region ($v_{ref} \pm \epsilon$) after the disconnection. The disconnection caused a slight drop in the subsequent steady-state voltage, a drop of 0.114% and 0.132%, averaged over 100 runs for each case.

Finally, we consider generalization with respect to the microgrid configuration. We perform two sets of experiment for this. Let NC-i denote the NC trained for Configuration i. In the first set of experiments, we test the performance of NC-1 for Configuration 2 and NC-2 for Configuration 1 on 100 runs from random initial states. In both cases, the CR was 100%. However, the mean deviation for NC-1 was 4.7 times larger than when it was used with Configuration 1. The mean deviation for NC-2 was 2.4 times larger than when it was used with Configuration 2. We conclude that an NC trained on a more complex microgrid generalizes better than one trained on a simpler microgrid.

In the second set of experiments, we evaluate how NC-1 and NC-2 handle dynamic changes to the microgrid configuration, even though no changes occurred during training. Each run starts with the PV and diesel generator DERs both connected, and the diesel generator DER disconnected after the voltage has converged. Both NCs succeed in continuously keeping the voltage in the tolerance region ($v_{ref} \pm \epsilon$) after the disconnection. The disconnection causes a slight drop in the subsequent steady-state voltage, a drop of 0.195% for NC-1 and 0.182% for NC-2.

Adversarial Input Attacks We demonstrate that RL-based neural controllers are vulnerable to adversarial input attacks. We use the gradient-based attack algorithm described in Section 2.5.4.2 to generate adversarial inputs for our NCs. We use an adversarial attack constant $c = 0.05$ and the parameters for the beta distributions are $a = 2$ and $b = 4$. From 100,000 unique initial states, we obtain 8, 6, and 5 adversarial states for Configurations 1, 2, and 3, respectively. In these experiments, we perturb all state variables simultaneously. In a real-life attack scenario, an attacker might have the capability to modify only a subset of them. Nevertheless, our experiments illustrate the fragility of RL-based neural controllers and the benefits of protecting them with NSA.

We confirmed with simulations that all generated adversarial states lead to safety violations when the NC alone is used, and that safety is maintained when Bb-Simplex is used. Figure 2.16(a) shows one such case, where the NC commits a voltage safety violation. The solid horizontal line shows the reference voltage $v_{ref} = 0.48$ kV. The dashed horizontal line shows the lower boundary of the safety region, 3% below v_{ref}. Figure 2.16(b) shows how Bb-Simplex prevents the safety violation. The dotted vertical line marks the switch from NC to BC.

We also confirmed that for all generated adversarial states, the forward switch is followed by a reverse switch. The time between forward switch and reverse switch depends on the choice of m (see Section 2.5.3.2). In the run shown in Fig. 2.16(b), they are five time steps (0.016 seconds) apart; the time of the reverse switch is not depicted explicitly, because the line for it would mostly overlap the line marking the forward switch. For $m = 2,3,4$ with Configuration 1, the average number of time steps between them are 7 (0.0244 seconds), 11 (0.0352 seconds), and 16 (0.0512 seconds), respectively. For $m = 2,3,4$ with Configuration 2, the average time steps between them are 7 (0.0244 seconds), 13 (0.0416 seconds), and 17 (0.0544 seconds), respectively. For $m = 2,3,4$ with Configuration 3, the average time steps between them are 8 (0.0256 seconds), 14 (0.0448 seconds), and 19 (0.0608 seconds), respectively.

2.5.5.6 Evaluation of Adaptation Module

To measure the benefits of online retraining, we used the adversarial inputs described above to trigger switches to BC. We used the switching conditions derived using the SOS-based methodology. For each microgrid configurations, we ran the original NC from the first adversarial state for that configuration, performed online retraining while the BC is in control, and repeated this procedure for the remaining adversarial states for that configuration except starting with the updated NC from the previous step. As such, the retraining is cumulative for each configuration. We performed this entire procedure separately for different RSCs corresponding to different values of m. After the cumulative retraining, we ran the retrained controller from all of the adversarial states, to check whether the retrained NC was still vulnerable (i.e. whether those states caused violations).

For Configuration 1, the BC was in control for a total of 56, 88, and 128 time steps for $m = 2,3,4$, respectively. For Configuration 2, the BC was in control for a total of 42, 78, and 102 time steps for $m = 2,3,4$, respectively. For Configuration 3, the BC was in control for a total of 40, 70, and 95 time steps for $m = 2,3,4$, respectively. For $m = 2$, the retrained controllers were still vulnerable to some adversarial states for each configuration. For $m = 3,4$, the retrained controllers were not vulnerable to any of the adversarial states, and voltage always converged to the tolerance region. Performance comparison of the original and retrained NCs, averaged over 100 runs starting from random (non-adversarial) states shows a slight improvement in the performance of the retrained NC (13.4% for CT and 6% for δ). Thus, retraining improves both safety and performance.

Tables 2.15–2.17 compare the performance of the original and retrained NCs for each configuration, averaged over 100 runs starting from random (non-adversarial) states. The retraining shows

Table 2.15 Performance comparison of original NC and NC retrained by AM for Configuration 1.

NC	CR	CT	σ (CT)	δ	$\sigma(\delta)$
Retrained	100	60.4	5.6	$1.0e-4$	$1.0e-5$
Original	100	67.5	5.8	$1.1e-4$	$1.0e-5$

Table 2.16 Performance comparison of original NC and NC retrained by AM for Configuration 2.

NC	CR	CT	σ (CT)	δ	$\sigma(\delta)$
Retrained	100	69.4	5.3	$1.1e-4$	$1.0e-5$
Original	100	76.8	6.1	$1.3e-4$	$1.2e-5$

Table 2.17 Performance comparison of original NC and NC retrained by AM for Configuration 3.

NC	CR	CT	σ (CT)	δ	$\sigma(\delta)$
Retrained	100	70.2	5.7	$1.4e-4$	$1.3e-5$
Original	100	81.1	7.7	$1.5e-4$	$1.3e-5$

a slight improvement in the performance of the NC; thus, retraining improves both safety and performance.

A potential concern is whether with online retraining can be done in real-time; i.e. whether a new retraining sample can be processed within one control period, so the retrained NC is available as soon as the RSC holds. In the above experiments, run on a laptop with an Intel i5-6287U CPU, retraining is done nearly in real-time: on average, the retraining finishes 0.285 ms (less than one-tenth of a control period) after the RSC holds.

2.5.6 Extension to Approximate Dynamics

In this section, we extend our Bb-Simplex framework to handle approximate knowledge of the system dynamics, due to approximate knowledge of the dynamic equations or of the values of parameters in the dynamic equations. Note that we continue to assume the dynamics is deterministic; our framework can be further extended to handle uncertainty due to nondeterministic dynamics. Approximate dynamics may be obtained, for example, by learning it from execution traces. We extend the concepts of BaC and switching condition to take into account the inaccuracy in the dynamics, using a given bound on the approximation error. The bound may be learned together with the approximate dynamics or determined later while checking conformance between the approximate dynamics and the observed behavior.

We continue to let f denote the (unknown) actual dynamics, as in Eq. (2.5). Let f^* denote the approximate dynamics.

Definition 2.2 *Approximation Error ϵ:* The approximation error ϵ is a bound on the difference between the actual dynamics and the approximate dynamics over the domains of the system state and the control action.

$$\epsilon \geq \sup\{|f(x, u) - f^*(x, u)|, x \in \mathcal{A}, u \in \mathcal{U}\} \tag{2.18}$$

We first consider the impact of the approximation error on the definition of the BaC and the *SyntheBC* algorithm [21] for deriving a neural BaC. We then consider its impact on our methodology for deriving switching conditions.

2.5.6.1 Impact of Approximate Dynamics on BaC

Suppose a BaC h^* is derived using the approximate dynamics f^*. Thus, the conditions in Eq. (2.6) hold with h replaced with h^*, and f replaced with f^*. However, h^* is not necessarily a BaC for the actual dynamics f, since it may not satisfy the derivative condition, namely, $f(x)\frac{\partial h^*(x)}{\partial x} \geq 0, \forall x : h^*(x) = 0$. To address this issue, we use a modified (stronger) version of the derivative condition when learning a BaC from the approximate dynamics. The modified condition ensures that the resulting BaC h^* is also a BaC for the actual dynamics. The modified derivative condition is:

$$y\frac{\partial h^*(x)}{\partial x} \geq 0, \quad \forall x : h^*(x) = 0, \quad \forall y \in [f^*(x) - \epsilon, f^*(x) + \epsilon] \tag{2.19}$$

Theorem 2.3 *If a function $h^* : \mathbb{R}^d \longrightarrow \mathbb{R}$ satisfies the definition of a BaC given in Eq. (2.6), except with the derivative condition (the last condition) replaced with Eq. (2.19), for an approximate dynamics f^* with error bound ϵ, then h^* is a BaC for the actual dynamics f.*

Proof: The dynamics f does not occur in the first two conditions of Eq. (2.6), so h^* satisfies them for both dynamics. It remains to show that h^* also satisfies the final condition, the derivative condition, for the actual dynamics. This holds because the condition in Eq. (2.19) ensures that the derivative of h^* is nonincreasing for all possible values that $f(x)$ can assume, specifically, for all values within ϵ of $f^*(x)$. ∎

Learning a Neural BaC from Approximate Dynamics We modify the *SyntheBC* algorithm [21] for learning a candidate BaC in the form of an NN, and verifying the candidate BaC, so that it uses our modified version of the derivative condition.

The training that seeks to ensure that the NN satisfies the first two conditions in the definition of a BaC, and the verification that the candidate BaC satisfies these two conditions, remains unchanged. The training that aims to make the NN satisfy the derivative condition needs to be modified, to take into account the changes to that condition. In particular, we modify the loss function l_3 used in that training so that it takes into account all of the values that the actual dynamics f could have, consistent with the approximate dynamics f^*. The original definition of l_3 ensures that there is a loss (i.e. l_3 is positive) if the Lie derivative of N, given by the dot product $\frac{\partial N}{\partial x}f(x)$, is positive for any point x in a dataset D_3 created earlier during training, where the NN N is the current candidate BaC.

We modify l_3 so that there is a loss if the Lie derivative of N is positive anywhere in the box of size ϵ around $f^*(x)$. Thus, l_3 is now based on the maximum value of that Lie derivative in that box. Since, for a given value of x, the dot product $\frac{\partial N}{\partial x}f^*(x)$ is a multi-linear function of the components of the vector $f^*(x)$, that maximum can be computed efficiently by exploiting the fact that it must occur at one of the corners of the box.

Verification of Neural BaC To verify that a candidate BaC satisfies the modified derivative condition, we modify the MIQCP optimization problem in [27] by considering an extra constraint on the values of the dynamics. The original MIQCP optimization problem checks that the maximum value of the dot product $vf(x)$ is nonpositive, subject to a set of constraints on v and x. We modify the problem to check whether the maximum value of the dot product vy is non-positive, where y is a fresh variable and the set of constraints is extended with the constraints $y \geq f^*(x) - \epsilon$ and $y \leq f^*(x) + \epsilon$.

2.5.6.2 Impact of Approximate Dynamics on FSC

Recall from Theorem 2.2 that the FSC derived by our methodology is of the form $\text{FSC}(x, u) = \alpha \vee \beta$, where $\alpha \equiv \hat{h}(x, u) - \lambda(u) \leq 0$ and $\beta \equiv x \notin \mathcal{A}_r(u)$. We consider the impact on each disjunct.

Impact on α Recall that $\hat{h}(x, u)$, defined in Eq. (2.7) is a Taylor approximation of BaC h's value after time η, if control action u is taken in state x, and $\hat{h}(x, u) - \lambda(u)$ is a lower bound on h's value after time η. We have already discussed the impact of approximate dynamics on the BaC. Now, we analyze its impact on the Taylor approximation of the BaC. For concreteness, we consider the Taylor approximation of degree 2, and we expand the derivatives in Eq. (2.7) using the chain rule:

$$\hat{h}(x, u) = h(x, u) + \frac{\partial h}{\partial x} f(x, u)\eta + \frac{1}{2} \left(f^T(x, u)\frac{\partial^2 h}{\partial x^2} f(x, u) + \frac{\partial h}{\partial x}\frac{\partial f}{\partial x} f^*(x, u) \right) \eta^2 \tag{2.20}$$

The approximation error ϵ by itself does not imply a bound on the difference between the derivatives of f and the derivatives of f^*. For example, if f^* oscillates within a small range, and f maintains a steady value within that range, then their derivatives may differ significantly, making $\hat{h}^*(x, u)$ a poor estimate of $\hat{h}(x, u)$. To deal with this problem, we introduce bounds on the approximation error in the derivatives of f; these can be obtained in a similar way as the approximation error in f.

Definition 2.3 *Derivative Approximation Error ϵ_i:* The derivative approximation error ϵ_i is a bound on the maximum difference between the ith derivatives of the actual dynamics and the approximate dynamics with respect to state x over the domains of the system state and the control action.

$$\epsilon_i = \sup \left\{ \left| \frac{\partial^i f(x, u)}{\partial x^i} - \frac{\partial^i f^*(x, u)}{\partial x^i} \right|, x \in \mathcal{A}, u \in \mathbb{U} \right\} \tag{2.21}$$

Using these bounds, we define $\hat{h}^*(x, u, \epsilon, \epsilon_1)$ in a way that ensures it is a lower bound on $\hat{h}(x, u)$.

$$\hat{h}^*(x, u, \epsilon, \epsilon_1) = \inf \left\{ h^*(x) + \frac{\partial h^*}{\partial x}(f^*(x) + \delta)\eta \right.$$
$$+ \frac{1}{2}((f^{*T}(x) + \delta^T)\frac{\partial^2 h^*}{\partial x^2}(f^*(x) + \delta)$$
$$\left. + \frac{\partial h^*}{\partial x}\left(\frac{\partial f^*}{\partial x} + \delta_1\right)(f^*(x) + \delta))\eta^2 : \delta \in [-\epsilon, \epsilon], \delta_1 \in [-\epsilon_1, \epsilon_1] \right\} \tag{2.22}$$

We define a remainder error $\lambda^*(u, \epsilon, \epsilon_2)$ that is an upper bound on the remainder error $\lambda(u)$ by modifying the definition in Eq. (2.9) in a similar way.

This analysis can be generalized to higher-degree Taylor approximations. When using approximate dynamics with a degree-n Taylor approximation, α is defined by $\hat{h}^*(x, u, \epsilon, \epsilon_1, \ldots, \epsilon_{n-1}) - \lambda^*(u, \epsilon, \epsilon_n) \leq 0$.

Impact on β Recall that the β disjunct of the FSC derived using the actual dynamics checks whether the state is in a shrunken admissible region $\mathcal{A}_r$. When using approximate dynamics, we modify the definition of the shrunken admissible region to further shrink it by an amount proportional to the error bound ϵ. This ensures that the analog of Lemma 2.1 holds; i.e. if β is true in the current state, then the system remains within the actual admissible region during the next control period.

The modified restricted admissible region is denoted $\mathcal{A}_r^*(u, \epsilon)$ and is given by:

$$\mathcal{A}_r^*(u, \epsilon) = \{x : x_{\text{lb}} + \mu_{\text{dec}}^*(u, \epsilon) < x < x_{\text{ub}} - \mu_{\text{inc}}^*(u, \epsilon)\} \tag{2.23}$$

where vectors $\mu_{\text{dec}}^*(u, \epsilon), \mu_{\text{inc}}^*(u, \epsilon)$ are defined by:

$$\begin{aligned}
\mu_{\text{dec}}^*(u, \epsilon) &= \eta(|\min(0, \dot{x}_{\text{min}}^*(u))| + \epsilon) \\
\mu_{\text{inc}}^*(u, \epsilon) &= \eta(|\max(0, \dot{x}_{\text{max}}^*(u))| + \epsilon)
\end{aligned} \tag{2.24}$$

where $\dot{x}_{\text{min}}^*$ and $\dot{x}_{\text{max}}^*$ are vectors of solutions to the optimization problems:

$$\begin{aligned}
\dot{x}_{\text{min}}^*(u) &= \inf\{f^*(x, u) : x \in \mathcal{A}\} \\
\dot{x}_{\text{max}}^*(u) &= \sup\{f^*(x, u) : x \in \mathcal{A}\}
\end{aligned} \tag{2.25}$$

Lemma 2.2 *For all $x \in \mathcal{A}_r^*(u, \epsilon)$ and all control actions u, $\text{Reach}_{\leq \eta}(x, u) \subseteq \mathcal{A}$.*

The proof is similar to the proof of Lemma 2.1.

When using approximate dynamics, β is defined by $\beta \equiv x \notin \mathcal{A}_r^*(u, \epsilon)$.

(i.e. by moving the lower and upper bounds, respectively, of each variable inwards by those amounts), the state cannot move outside of $\mathcal{A}_r$ during time η.

2.5.7 Extension to Hybrid Systems

In this section, we briefly discuss how we extend our Bb-Simplex framework to hybrid systems. A *hybrid system* [15] is a system with both discrete and continuous variables and with multiple modes, each with a different dynamics (for the continuous variables). Given a continuous state space $\mathcal{X}$ and finite set of modes M, the overall state space of the hybrid system is denoted by $(m, x) \in M \times \mathcal{X}$. Discrete mode-transitions occur instantaneously in time. A transition $((m, x), (m', x'))$ indicates that the system can undergo a discrete (instantaneous) transition from the state (m, x) to the state (m', x'). We define *Guards* (G) and *Reset maps* (R) for mode-transitions as follows: $G(m, m') = \{x \in \mathcal{X} : ((m, x), (m', x')) \in \mathcal{T} \text{ for some } x' \in \mathcal{X}\}$ and $R(m, m') : x \mapsto \{x' \in \mathcal{X} : ((m, x), (m', x')) \in \mathcal{T}\}$, whose domain is $G(m, m')$.

2.5.7.1 Switching Logic for Hybrid Systems

Consider the plant dynamics:

$$\dot{x} = f(m, x, u) \tag{2.26}$$

The set of *unsafe states* in mode m is denoted by $\mathcal{U}_m$. For a given BC, we denote the *recoverable region* for mode m by $\mathcal{R}_m$. BaCs are functions that capture the following safety requirements of a hybrid system: (1) the state must remain safe during the continuous time evolution of the system in the current mode, and (2) a discrete transition $((m, x), (m', x'))$ from mode m to m' must reset a safe state $(m, x) \notin \mathcal{U}_m$ to another safe state $(m', x') \notin \mathcal{U}_{m'}$. Given a hybrid system Π, the safety of Π

can be verified using a collection of BaCs $h_m : \mathbb{R}^d \to \mathbb{R}, \forall m \in M(x)$, where $M(x)$ is the set of modes consistent with current state x; i.e. the state x is invariant in the set of modes $M(x)$.

$$
\begin{aligned}
h_m(x) &< 0, &\forall x \in \mathcal{U}_m \\
h_m(x) &\geq 0, &\forall x \in \mathbb{R}^d \backslash \mathcal{U}_m \\
(\nabla_x h_m)^T f(m, x, u) + \sigma(h_m(x)) &\geq 0, &\forall x \in \mathbb{R}^d \\
h_{m'}(x') &\geq 0, &\forall ((m, x), (m', x')) \in \mathcal{T}, \ h_m(x) \geq 0, \\
& & x \in G(m, m'), \ x' \in R(m, m')(x)
\end{aligned}
\tag{2.27}
$$

The first three clauses are similar to the BaC defined in Eq. (2.6). The last clause guarantees h_m cannot become negative after Π performs a discrete transition. If all the clauses in Eq. (2.27) are satisfied for all modes $m \in M(x)$, h_m is a barrier certificate of Π, where the safety of Π is certified. For a further exploration of Eq. (2.27), refer to [22].

In hybrid systems, mode switches can occur at any point in continuous time, not limited to multiples of a fixed time period. Thus, we have to ensure that the computed set of reachable states is always safe and recoverable even if a mode transition happens within time η. The reachset computations should take into account all the states reachable from the current state within time η, for all of the modes consistent with the current state. We compute $\text{Reach}_{\leq \eta}(x, u)$ using the methodology described in Section 3.5.2 of [40]. Their approach is directly applicable to ours, where their time interval $[0, r]$, state $x(0)$, mode $y(0)$, and control action $u(0)$ are analogous to our control period η, state x, mode $m \in M(x)$ and control action u, respectively.

2.5.7.2 FSC for Hybrid Systems

To define the *FSC* for hybrid systems, we extend Definition 2.1. Let $M(x, u, \eta)$ be the set of all modes reachable from state x in time interval η using control action u. Note that $M(x, u, \eta) \subseteq M$.

Definition 2.4 *Forward Switching Condition:* A condition FSC(x, u) is a FSC for hybrid systems if for every recoverable state x, control action u, control period η, and modes $m \in M(x, u, \eta)$, $\text{Reach}_{\leq \eta}(x, u) \bigcap \mathcal{U} \neq \emptyset \vee \text{Reach}_{= \eta}(x, u) \not\subseteq \bigcap_{m \in M(x, u, \eta)} R_m$ implies FSC(x, u) is true.

Theorem 2.4 *A Simplex architecture whose FSC satisfies Definition 2.4 keeps the hybrid system invariably safe provided the system starts in a recoverable state.*

Extending Theorem 2.2, we now propose a new procedure for constructing a switching condition for hybrid systems from a BaC and prove its correctness. We extend our existing definitions from Section 2.5.3 to hybrid systems by replacing h with h_m to obtain $\hat{h}_m(x, u)$ and $\lambda_m(u)$. Similarly, the sets of admissible states and restricted admissible states for mode m are denoted by $\mathcal{A}_m$ and $\mathcal{A}_m^r$, respectively.

Theorem 2.5 *Given a barrier certificate h_m, the following condition is a FSC:* FSC$(x, u) = \alpha \vee \beta$ *where,*

$$
\alpha \equiv \wedge_{m \in M(x)} \left\{ \hat{h}_m(x, u) - \lambda_m(u) \leq 0 \right\} \quad \text{and} \quad \beta \equiv x \notin \bigcap_{m \in M(x)} \mathcal{A}_m^r(u).
$$

2.5.8 Related Work

The use of BaCs in the Simplex architecture originated in [22]. There are, however, significant differences between their method for obtaining the switching condition and ours. Their switching logic involves computing, at each decision period, the set of states reachable from the current state within one control period, and then checking whether that set of states is a subset of the zero-level set of the BaC. Our approach avoids the need for reachability calculations by using a Taylor approximation of the BaC, and bounds on the BaC's derivatives, to bound the possible values of the BaC during the next control period and thereby determine recoverability of states reachable during that time. Our approach is computationally much cheaper: a reachability computation is expensive compared to evaluating a polynomial. Their framework can handle hybrid systems. Extending our method to hybrid systems is a direction for future work.

Mehmood et al. [41] propose a distributed Simplex architecture with BCs synthesized using control barrier functions (CBFs) and with switching conditions derived from the CBFs, which are BaCs satisfying additional constraints. A derivation of switching conditions based on Taylor approximation of CBFs is briefly described but does not consider the remainder error, admissible states, or restricted admissible states, and does not include a proof of correctness (which requires an analysis of the remainder error).

Kundu et al. [16] and Wang et al. [17] use BaCs for safety of microgrids, and Prajna et al. [42] propose an approach for stochastic safety verification of continuous and hybrid systems using BaCs. These approaches are based on the use of verified-safe controllers; they do not allow the use of unverified high-performance controllers, do not consider switching conditions, etc.

The application of neural networks for microgrid control is gaining in popularity [30]. Amoateng et al. [43] use adaptive neural networks and cooperative control theory to develop microgrid controllers for inverter-based DERs. Using Lyapunov analysis, they prove that their error-function values and weight-estimation errors are uniformly ultimately bounded. Tan et al. [44] use recurrent probabilistic wavelet fuzzy neural networks (RPWFNNs) for microgrid control, since they work well under uncertainty and generalize well. We used more traditional DNNs, since they are already high performing, and our focus is on safety assurance. Our Bb-Simplex framework, however, allows any kind of neural network to be used as the AC and can provide the safety guarantees lacking in their work. Unlike our approach, none of these works provide safety guarantees.

2.6 Software Platform for AI-Grid

The AI-Grid project has multiple functionalities, including software defined control (SDC), power flow, reachability analysis, and neuron network-based data prediction. The Center of Excellence in Wireless and Information Technology (CEWIT) team is responsible for putting all these different components together and making a software platform for the AI-Grid project. When we were in the design and development stage, we realized that all the functionality works on the Bronzeville community microgrid (BCM), but each team focuses on different aspects, and has different data requirements. In order to put things together while keeping each component apart from affecting each other, the software team in CEWIT created a generic architecture, which includes both back end data flow and a front end user interface (UI). Then, this generic architecture is customized for each team for different input, algorithms, and output. Different dnp3 outstation, structured query language (SQL) server database, dnp3 master, and a panel in the front end are developed for different components.

In this chapter, we will first discuss the infrastructure hardware of the system. Second, we will talk about the overall generic architecture, then discuss the role for each component. And finally discuss the customization for each project. Since this platform is still being developed, we will only talk about the SDC, power flow, and simplex project, which is more mature right now.

2.6.1 Infrastructure Overview

2.6.1.1 Real Time Digital Simulator (RTDS)

The RTDS, shown in Fig. 2.17 is a special purpose multiprocessor computer system that is optimized for power system simulations. It is designed for real-time simulation, which means that the computation of the simulated system advances one moment in each moment of wall-clock time.

In our system, we used the RTDS to run the BCM model. We can get the signal from the RTDS system, compute our own control signal, and then send it back to RTDS.

We currently have two systems of RTDS in the power lab, one in Suffolk Hall in Stony Brook campus and one in CEWIT. However, because the BCM model requires four cores in two racks to run, we currently only use the Suffolk Hall RTDS to run the simulation.

2.6.1.2 Windows

In order to run our SDC control system, we need one machine to host the programs to communicate with RTDS, another program to save the data, and another to run the SDC control program. We think MS Windows 11 is best to work with.

2.6.1.3 Network

The windows machine resides in Suffolk Hall, sits right next to the RTDS. This machine is also directly connected to RTDS using network cable for a fast communication speed. This machine has an IP of 172.20.28.169, which is also in the architecture (Fig. 2.17). All machines in the Suffolk Hall site are in a local powerlab network, which is accessible from outside by virtual private network (VPN) only.

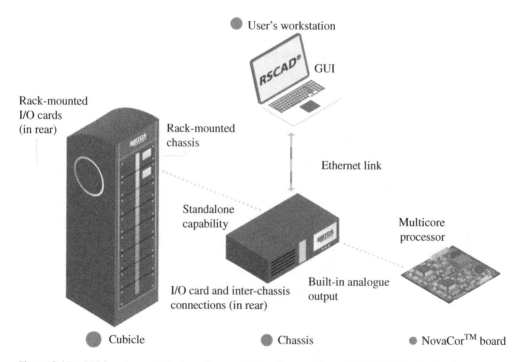

Figure 2.17 RTDS system architecture. Source: (https://www.rtds.com/) RTDS Technologies Inc.

2.6.1.4 SQL Server

Microsoft SQL Server is a relational database management system developed by Microsoft. As a database server, it is a software product with the primary function of storing and retrieving data as requested by other software applications – which may run either on the same computer or on another computer across a network (https://www.microsoft.com/en-us/sql-server/sql-server-2022).

In our project, we used MS SQL Server 2022 (16.0) Community version as the data storage because of its free and easy-to-use feature.

We also installed SQL Server Management Studio 18.0 to access and manipulate the database.

2.6.1.5 Python

We use python 3.11 in all the AI-Grid algorithms functions and data manipulation. Multiple python libraries would be used in the project, like numpy, pytorch, and pyodbc (https://pypi.org/project/pyodbc/).

2.6.1.6 DNP3

DNP3 (aka IEEE 1815) is the most common supervisory control and data acquisition (SCADA) protocol used in electric power in North America. It is event-based and efficiently transfers measurement data between outstation and master components (https://stepfunc.io/products/libraries/dnp3/).

Since we are required to use DNP3 to communicate with RTDS and real SCADA device in the future, we used C# version open source DNP3 library (https://github.com/stepfunc/dnp3) from Stepfunc in this project.

2.6.1.7 Asp.net Web Server

In this project, we use the asp.net website to visualize the result. Note this component may not be necessary in some projects. For example, in the SDC project, since the SDC is mainly a backend control process, we do not need to visualize the control signal.

2.6.2 Software Architecture Overview

Our overall AI-Grid software architecture is shown in Fig. 2.18. The overall workflow is as follows, which action is described in steps, note each bullet step number corresponds to the same number in the figure.

(1) Send data out using DNP3 outstation in RTDS; (2) Receive data using DNP3 master; (3) Save data into Database; (4) Run AI-Grid code to get updated value; (5) Save computed value back to database (6) C program read value from database; (7) Send control value back to RTDS using DNP3 master; (8) Receive value in RTDS using outstation; (9) Update front end with database value; (10) Save updated values from user input.

2.6.3 Software Architecture Component

2.6.3.1 RTDS

We received the BCM model from ComEd, the runtime view of the microgrid as shown in Fig. 2.19, we can see the connection topology, turn switch on/off, display system status, like voltage, active power and reactive power in this dashboard.

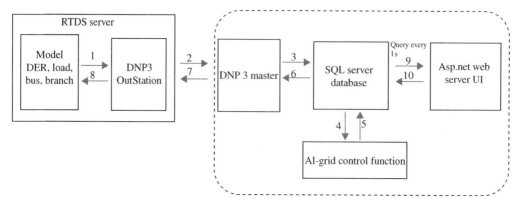

Figure 2.18 Software architecture overview.

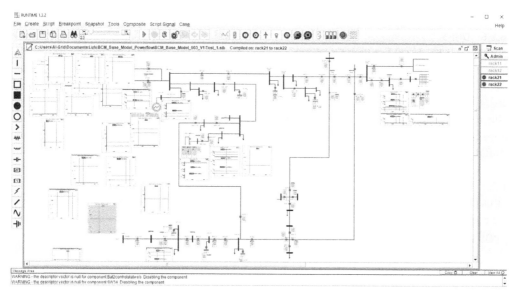

Figure 2.19 RTDS run time.

And in order to compute different AI-Grid functions, we need different multiple inputs from the BCM model, e.g. active power, reactive power, voltage, etc., from the DER, Bus, Load, Branch. Note in this microgrid, there are three DERs: one Solar PV, one diesel generator, one battery. The original microgrid has 90+ loads, but in our project, we will just select 15 most important loads for easier demonstration.

2.6.3.2 DNP3

We first configured a local version of C# master and outstation to test the speed of the communication. It would take 15 ms to send one float value in a round trip, averaged from 1000 test times. We also tested C++ version, which would take only 3 ms. However, since C++ is hard to integrate with our other components, we still stick to the C# for easier usage.

GTNet-DNP model (shown in Fig. 2.20) from RTDS is used to send the data out. In this model, we can configure the outstation to specify what data to be sent. From outside, we need to configure a master file to receive the data. A speed test is also run to test the communication delay, our result

Figure 2.20 Giga-transceiver network communication (GTNET) module in RTDS.

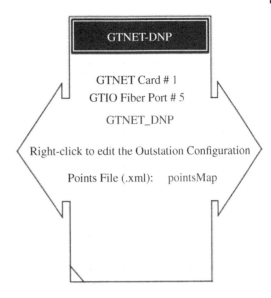

is around 25 ms including the database write and read for a single float value, averaged from 1000 test times.

2.6.3.3 Special Encoding/Decoding Process

During the testing, we realized that the analog values can be sent and received roundtrip only at a 100 ms interval, which is a limit set by RTDS. Figure 2.21 shows this, the *y*-axis is the value, *x*-axis is the time, unit in second. The straight line is the signal generated from RTDS, the staircase line is the data received by RTDS after it is sent out and then sent back and received.. We configure the DNP3 to sample every 1 ms, so DNP3 itself is not the bottleneck. In this figure, we can see 10 steps in 1 s, which means the delay is 0.1 s or 100 ms. This speed is ok for some projects, like power flow, but not fast enough for SDC controllers, which is preferred to be at 10 ms level. In RTDS, the binary value can be sent every 50 μs, so in order to accomplish this 10 ms control speed this, we implemented a binary encoder/decoder.

The data encoding/decoding from RTDS to database is illustrated in Fig. 2.22, with following steps: (1) Get values from RTDS; (2) Encode data into binary in RTDS outstation; (3) Send Data

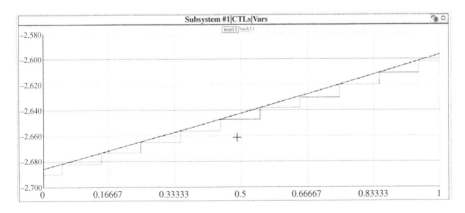

Figure 2.21 Analog encoding/decoding speed performance.

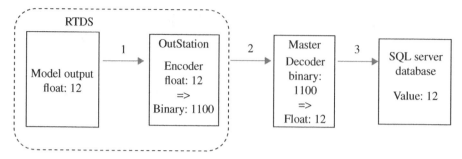

Figure 2.22 Customized encoding process.

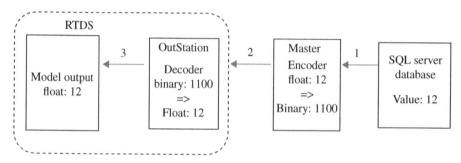

Figure 2.23 Customized decoding process.

using DNP3; (4) Decode data in Master file; (5) Save to database. Here, an example is if we want to send a value 12, it will be encoded into 1100 ($12 = 2^3 + 2^2 + 2^0 + 2^0$), and sent to the master file, decoded as 12 and saved into the database. Note in our actual SDC implementation, we actually used 16 bits to represent the value and 1 extra bit for the positive/negative sign.

The data encoding/decoding from database or controller to RTDS is illustrated in Fig. 2.23, with following steps: (1) Read values from database/Controller; (2) Encode data into binary in master file; (3) Send data using DNP3; (4) Decode data in RTDS Outstation; (5) Receive by RTDS model.

Figures 2.24 and 2.25 show the result of the encode/decode performance using sine wave and ladder wave, the black line is the data generated from RTDS, the sinusoidal line is the data generated from RTDS, the staircase line represents the data received by the RTDS after the encoding and decoding process. y-axis is the data value, the x value is the time, in milliseconds. The horizontal distance between the two lines at the same y-axis height is the delay. Based on both charts, the delay is around 10 ms, which is good for SDC control.

2.6.3.4 SQL Server

All the data are saved into the SQL database. C# and python code is used to access the data. For C#, we use EF Core to encapsulate the data access. Like the code snippet shown in Fig. 2.26:

For python, we use pyodbc to access the data, Fig. 2.27 shows a code example to execute a SQL query.

The data access speed is fast in our local network. All queries can be done in less than 1 ms. The speed is shown in Fig. 2.28. Note in our database, we only work with 15 loads, this data is constantly refreshed in real time, and since we only need to compute/display the same set of data, so there won't be any data query performance issue.

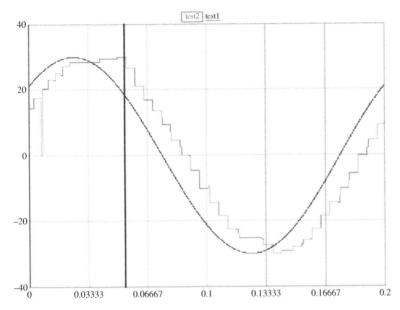

Figure 2.24 Speed test using sine wave for the encoding and decoding process.

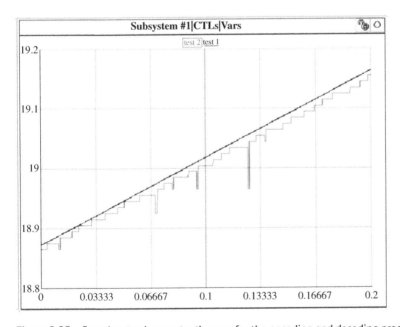

Figure 2.25 Speed test using sawtooth wave for the encoding and decoding process.

2.6.3.5 AI-Grid Control Function

All teams would require different algorithms to work upon the microgrid. All the programs would follow the similar steps like shown in the following: (1) Read input data from Database; (2) Compute AI-Grid functions; (3) Save result data into the database.

```csharp
public class SDCContext : DbContext
{
    2 references
    public DbSet<DER> DER { get; set; }
    1 reference
    public DbSet<DER_History> DER_History { get; set; }

    0 references
    public DbSet<Switch> Switch { get; set; }
    1 reference
    public DbSet<ControlSignal> ControlSignal { get; set; }

    0 references
    protected override void OnConfiguring(DbContextOptionsBuilder optionsBuilder)
    {
```

Figure 2.26 C# data access code.

```python
def db_execute(logger, sql):
    try:
        con = pyodbc.connect(DB_Connection)
        cursor = con.cursor()
        cursor.execute(sql)
        con.commit()
        cursor.close()
        con.close()
        return True
    except Exception as e:
        con.close()
```

Figure 2.27 Python data access code.

	Python insert ×1	Python insert ×1 (TLS)	Python select ×1	Python select ×1 (TLS)	C# select ×1
Local	0.608090 ms	0.623411 ms	0.614586 ms	0.629767 ms	2.58627 ms

Figure 2.28 Database query speed performance.

2.6.4 Customization for Each Team

Each team would use different input, algorithms, refresh speed, output, we would follow the generic design and create a separate workflow for each team.

2.6.4.1 SDC

Nowadays, microgrid controllers are often embedded in specialized hardware such as programmable logic controller (PLC) and DSP. The hardware-dependency and fit-and-forget design make it difficult and costly for microgrid controllers to evolve and upgrade under frequent changes such as plug-and-play of microgrid components. Furthermore, different DERs in a microgrid require customized controllers, leading to long development cycles and high operational costs for deploying microgrid services. To tackle the challenges, a software-defined control (SDC) architecture for microgrid is devised, which virtualizes traditionally hardware-dependent microgrid control functions as software services decoupled from the underlying hardware infrastructure, fully resolving hardware dependence issues and enabling unprecedentedly low costs.

Architecture We currently implemented SDC using the following architecture as shown in Fig. 2.29, with the following steps (see the illustrations in Figs. 2.30 and 2.31): (1) Send PV data out using DNP3 outstation in RTDS; (2) Receive data using DNP3 master; (3) Save data into Database; (4) Run SDC code to get updated PV Control Value; (5) Save SDC control value computed value back

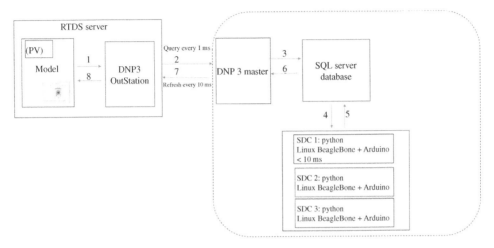

Figure 2.29   SDC architecture.

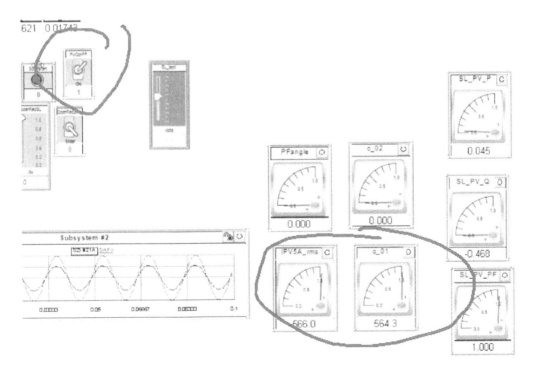

Figure 2.30   SDC test switch on.

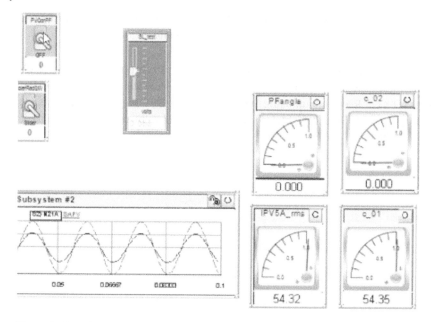

Figure 2.31 SDC switch turn off PV switch.

to Database; (6) C# program read value from database; (7) Send control value back to RTDS using DNP3 master; (8) Receive value in RTDS using Outstation.

Note since SDC is mainly fast backend control, there is no need to display data onto the UI.

On the controller side, the shown above is a PV controller, but the same idea can be easily expanded to other types of DER like diesel generator or battery. Note the SDC1, SDC2, and SDC3 on the chart, we can have multiple controllers on our server for different devices or different algorithms. We can also migrate this control to a Syndem/DSP device in the real environment if the control signal is harder to reach in the long distance.

A separate python project is created, the workflow for this python is shown:

(1) Read active power, reactive power, etc. from Database
(2) Compute control signal values
(3) Save data into the database

Result: As we can see from Fig. 2.29, and Fig. 2.30, when the PV is turned on and off, the current root mean square (RMS) value from our result and RTDS value stay very close to each other (IPV 5Arms is the value from RTDS and c01 is the value from our program).

2.6.4.2 Power Flow

We use power flow to compute the steady state of the power system, including voltage magnitudes and angles, the directions of branch flows. It is efficient and convenient to visualize the operational status on the platform.

The architecture of the power flow is shown in Figs. 2.32 and 2.33, steps are shown: (1) Save updated load on/off values, island mode on/off from user input into database; (2) Send values back to RTDS using DNP3 master; (3) Receive data using DNP3 outstation; (4) Compute updated

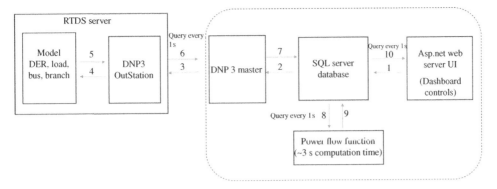

Figure 2.32 Powerflow architecture.

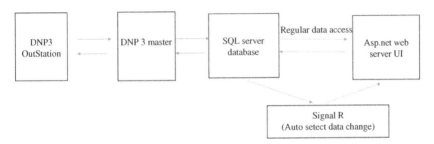

Figure 2.33 Dashboard SignalR architecture.

DER data in RTDS; (5) Send data using DNP3 Outstation; (6) Receive data using DNP3 Master; (7) Save data into Database; (8) Run Powerflow code to get updated power flow direction, bus data and branch data; (9) Save computed value back to Database; (10) Update front end with database value.

We can successfully run the power flow code to get the update flow direction and bus voltage information.

2.6.4.3 Digital Twin User Interface

Background: Developing the user-interface for a digital twin of a microgrid system has several requirements. First, the data pipeline requires the collection of data from perhaps hundreds of sensing objects. Additionally, this collected sensor data must be presented to the user in near real-time. The system must also be able to show a representation of physical assets and allow the user to monitor them.

Data ingestion and integration: The data flow of the digital twin is expected to provide real-time data from multiple sensing devices. To achieve this, a publisher–subscriber data pipeline was created. Publisher–subscriber is a scalable messaging paradigm that allows a publisher to deliver messages to interested subscribers by sending messages as soon as they become available. As a result, subscribers do not have to periodically check if new data has become available, allowing for near real-time communication. The open-source SignalR from the .NET ecosystem was chosen as the publisher–subscriber implementation. Data from various sensors is made available through DNP3 and as new data points are collected SignalR publishes these changes as

Figure 2.34 An overview of the AI-Grid platform.

messages to interested subscribers. An overview of the AI-Grid platform is shown in Fig. 2.34. A full diagram of the setup is provided below:

User-interface: The user-interface overlays the assets of a microgrid on a map allowing an operator to see the geospatial layout of DER's, loads and buses (see illustrations in Figs. 2.35 and 2.36). The map tiles are provided through the Google Maps Javascript API (https://developers.google .com/maps/documentation/javascript).

Monitoring: A key function of a digital twin is to allow remote monitoring of an existing physical system. Here each asset is represented on the map by a marker. When clicked, the marker expands to reveal the critical information regarding that asset. An example of an asset representing a high school is shown below:

As each asset has different critical data points to monitor, the supporting dashboard shows different visualizations depending on what is clicked. For example, when a bus is expanded, power flow readings are presented to the user:

WebGL visualizations: WebGL is a JavaScript API for rendering both 2D and 3D visualizations. The Google map display grants access to the WebGL context. This allows for control of tilting the camera in three dimensions and properly occluding rendered objects.

To improve interpretability of visualizations, 3D WebGL graphics are used where appropriate. The deck.gl framework (https://deck.gl/) was used for creating the 3D visualizations. For example, the following column visualization shows the voltage at each of the buses in a microgrid:

The magnitude of the voltage is represented by the color and height of the columns (see Fig. 2.37). The geospatial location and scale of the largest bus voltage is more obvious with the 3D visualization than with a table of latitude/longitude locations and voltages. An important aspect in monitoring a microgrid is knowing the magnitude and flow direction of power throughout the network. Again, useful 3D exploratory visualizations can quickly reveal this information to the user.

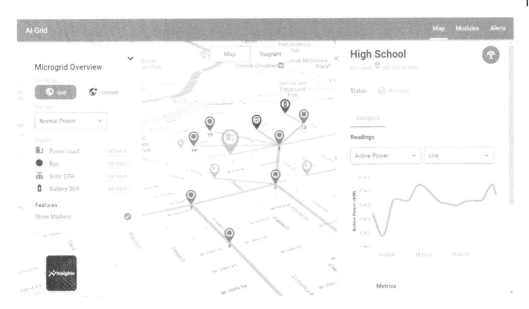

Figure 2.35 Monitoring view for a load microgrid asset.

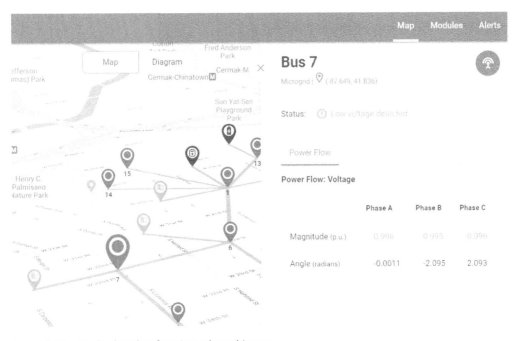

Figure 2.36 Monitoring view for a bus microgrid asset.

Here, the direction of the flow of power is indicated by animating particles traveling between different buses. The magnitude of active power of the flow is indicated by the color or grayscale level, with bright being a high value and dark being a low one (see an illustration in Fig. 2.38).

One-line diagram: To provide operators of the application with a more traditional view of the system, an equivalent one-line diagram representation of the system was created as well.

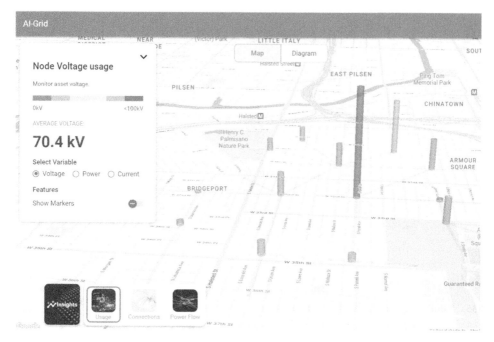

Figure 2.37 Deck.gl powered voltage visualization.

Figure 2.38 Deck.gl powered power-flow visualization.

2.7 AI-Grid for Grid Modernization

Our team is further developing AI-Grid into a next-generation distribution/microgrid/building management system. It will further include advanced fault management control and optimization to enable enhanced DERs, electric vehicles and microgrid penetration without compromising reliability. It will feature software defined networking (SDN)-based architectures and techniques to enable secure, reliable and fault-tolerant algorithms for resilient networked systems. It will provide reachability techniques to facilitate achieving provable resilience on the fly with high penetration of renewables. We will deploy AI-Grid and demonstrate how the data-intensive, self-configurable management system enables resilient distribution grids, smart communities, and smart infrastructures for electric vehicles and DER aggregations. We will verify how AI-Grid will enable extreme renewable energy hosting and improve resilience against attacks, faults, and disasters in an affordable, lightweight, and secure way.

As a whole, our team aims to provide an interoperable and high-precision visualization software platform, which incorporates various AI functions and a real-time data stream processing engine with multiple sensing modalities and operation attributes. We are working with communications infrastructure solution providers which will assist infrastructure design and system deployment. We are integrating an edge computing platform to implement real-time monitoring and control solutions. Leveraging the expertise and resources of our partners allows us to minimize and mitigate risk and achieve deployment at scale.

2.8 Exercises

Consider a thermostat gradually increasing the room temperature to reach its target of $20\,°C$. The current temperature is denoted by x. The control input u measures how much heat the radiator is emitting. The system dynamics is $\dot{x} = f(x, u) = \max(0, \min(u, 2))$. The BC is defined by $g(x) = \max((20 - x)/10, 0)$. The initial temperature is in the range $16–18\,°C$.

1. Derive a barrier certificate for this system showing that it maintains a safe temperature between 16 and $22\,°C$.
2. Using the approach in Section 2.5.3, derive from the BaC a forward switching condition that ensures this property holds when an AC is used.
3. Design a reward function for using the DDPG algorithm to train a neural controller to maintain the target room temperature of $20\,°C$.
4. The proofs of Theorems 2.4 and 2.5 in Section 2.5.7.2 are left as exercises. Try to prove these two theorems. Hint: Look at the proofs of Theorems 2.1 and 2.2.

References

1 RTDS-Technologies-Inc. (2022). Power hardware-in-the-loop (PHIL). https://www.rtds.com/applications/power-hardware-in-the-loop/ (accessed 1 November 2023).

2 Yu, B., Yin, H., and Zhu, Z. (2018). Spatio-temporal graph convolutional networks: a deep learning framework for traffic forecasting. *Proceedings of the 27th International Joint Conference on Artificial Intelligence (IJCAI)*.

3 Seo, Y., Defferrard, M., Vandergheynst, P., and Bresson, X. (2018). Structured sequence modeling with graph convolutional recurrent networks. *The 25th International Conference on Neural Information Processing*, 362–373.

4 Yu, B., Yin, H., and Zhu, Z. (2019). ST-UNet: a spatio-temporal u-network for graph-structured time series modeling. *arXiv*, abs/1903.05631.

5 Poli, M., Massaroli, S., Rabideau, C.M. et al. (2021). Continuous-depth neural models for dynamic graph prediction. https://arxiv.org/abs/2106.11581.

6 De Brouwer, E., Simm, J., Arany, A., and Moreau, Y. (2019). GRU-ODE-Bayes: continuous modeling of sporadically-observed time series. In: *Advances in Neural Information Processing Systems 32 (NeurIPS 2019)*, 7379–7390.

7 Seto, D., Krogh, B., Sha, L., and Chutinan, A. (1998). The Simplex architecture for safe online control system upgrades. *Proceedings of the 1998 American Control Conference. ACC (IEEE Cat. No. 98CH36207)*, volume 6, 3504–3508.

8 Sha, L. (2001). Using simplicity to control complexity. *IEEE Software* 18 (4): 20–28.

9 Bak, S., Greer, A., and Mitra, S. (2010). Hybrid cyberphysical system verification with Simplex using discrete abstractions. In *16th IEEE Real-Time and Embedded Technology and Applications Symposium*, 143–152.

10 Bak, S., Manamcheri, K., Mitra, S., and Caccamo, M. (2011). Sandboxing controllers for cyber-physical systems. In *Proceedings of the IEEE/ACM International Conference on Cyber-Physical Systems (ICCPS 2011)*, 3–12, Apr 2011.

11 Johnson, T.T., Bak, S., Caccamo, M., and Sha, L. (2016). Real-time reachability for verified Simplex design. *ACM Transactions on Embedded Computing Systems* 15 (2): 26:1–26:27.

12 Phan, D., Grosu, R., Jansen, N. et al. (2020). Neural Simplex architecture. In: *NASA Formal Methods Symposium*, 97–114. Springer International Publishing.

13 Borrmann, U., Wang, L., Ames, A.D., and Egerstedt, M. (2015). Control barrier certificates for safe swarm behavior. In: *Analysis and Design of Hybrid Systems, IFAC-PapersOnLine*, vol. 48 (ed. M. Egerstedt and Y. Wardi), 68–73. Elsevier.

14 Prajna, S. (2006). Barrier certificates for nonlinear model validation. *Automatica* 42 (1): 117–126.

15 Prajna, S. and Jadbabaie, A. (2004). Safety verification of hybrid systems using barrier certificates. In: *Proceedings of the 7th International Workshop on Hybrid Systems: Computation and Control (HSCC 2004), Lecture Notes in Computer Science*, vol. 2993 (ed. R. Alur and G.J. Pappas), 477–492. Springer.

16 Kundu, S., Geng, S., Nandanoori, S.P. et al. (2019). Distributed barrier certificates for safe operation of inverter-based microgrids. *2019 American Control Conference*, 1042–1047.

17 Wang, L., Han, D., and Egerstedt, M. (2018). Permissive barrier certificates for safe stabilization using sum-of-squares. *2018 Annual American Control Conference*, 585–590.

18 Zhao, H., Zeng, X., Chen, T., and Liu, Z. (2020). Synthesizing barrier certificates using neural networks. In: *Proceedings of the 23rd International Conference on Hybrid Systems: Computation and Control (HSCC 2020)*, 1–11. Association for Computing Machinery.

19 Sha, M., Chen, X., Ji, Y. et al. (2021). Synthesizing barrier certificates of neural network controlled continuous systems via approximations. *2021 58th ACM/IEEE Design Automation Conference*, 631–636.

20 Anghel, M., Milano, F., and Papachristodoulou, A. (2013). Algorithmic construction of Lyapunov functions for power system stability analysis. *IEEE Transactions on Circuits and Systems I: Regular Papers* 60 (9): 2533–2546.

21 Zhao, Q., Chen, X., Zhang, Y. et al. (2021). Synthesizing ReLU neural networks with two hidden layers as barrier certificates for hybrid systems. In: *Proceedings of the 24th International*

Conference on Hybrid Systems: Computation and Control (HSCC 2021), 1–11. Association for Computing Machinery.

22 Yang, J., Islam, M.A., Murthy, A. et al. (2017). A Simplex architecture for hybrid systems using barrier certificates. In: *Proceedings of the 36th International Conference on Computer Safety, Reliability, and Security (SAFECOMP 2017), Lecture Notes in Computer Science*, vol. 10488, 117–131. Springer.

23 Pogaku, N., Prodanovic, M., and Green, T.C. (2007). Modeling, analysis and testing of autonomous operation of an inverter-based microgrid. *IEEE Transactions on Power Electronics* 22 (2): 613–625.

24 Krishnamurthy, S., Jahns, T.M., and Lasseter, R.H. (2008). The operation of diesel gensets in a CERTS microgrid. *2008 IEEE Power and Energy Society General Meeting - Conversion and Delivery of Electrical Energy in the 21st Century*, 1–8.

25 Guerrero, J.M., Vasquez, J.C., Matas, J. et al. (2011). Hierarchical control of droop-controlled AC and DC microgrids \emdash a general approach toward standardization. *IEEE Transactions on Industrial Electronics* 58 (1): 158–172.

26 Lasseter, R.H. and Paigi, P. (2004). Microgrid: a conceptual solution. *2004 IEEE 35th Annual Power Electronics Specialists Conference (IEEE Cat. No.04CH37551)*, volume 6, 4285–4290.

27 Zhou, Y. and Ho, C.N.-M. (2016). A review on microgrid architectures and control methods. *2016 IEEE 8th International Power Electronics and Motion Control Conference (IPEMC-ECCE Asia 2016)*, 3149–3156.

28 Mehrizi-Sani, A. (2017). Distributed control techniques in microgrids. In: *Microgrid: Advanced Control Methods and Renewable Energy System Integration* (ed. M.S. Mahmoud), 43–62. Butterworth-Heinemann. ISBN: 978-0-08-101753-1.

29 O'Rourke, C.J., Qasim, M.M., Overlin, M.R., and Kirtley, J.L. (2019). A geometric interpretation of reference frames and transformations: dq0, Clarke, and Park. *IEEE Transactions on Energy Conversion* 34 (4): 2070–2083.

30 Lopez-Garcia, T.B., Coronado-Mendoza, A., and Dom\'{\i}nguez-Navarro, J.A. (2020). Artificial neural networks in microgrids: a review. *Engineering Applications of Artificial Intelligence* 95 (103894): 1–14.

31 Lillicrap, T.P., Hunt, J.J., Pritzel, A. et al. (2016). Continuous control with deep reinforcement learning. *4th International Conference on Learning Representations*, 1–14.

32 Pattanaik, A., Tang, Z., Liu, S. et al. (2017). Robust deep reinforcement learning with adversarial attacks.

33 Nzimako, O. and Rajapakse, A. (2016). Real time simulation of a microgrid with multiple distributed energy resources. *International Conference on Cogeneration, Small Power Plants and District Energy (ICUE 2016)*, 1–6.

34 Chollet, F. (2015). Keras. https://github.com/keras-team/keras.git (accessed 1 November 2023).

35 Abadi, M., Agarwal, A., Barham, P. et al. (2015). TensorFlow: large-scale machine learning on heterogeneous systems. https://www.tensorflow.org/ (accessed 28 October 2023).

36 Kingma, D.P. and Ba, J. (2015). Adam: a method for stochastic optimization. *3rd International Conference on Learning Representations*, 1–15.

37 Luitel, B. and Venayagamoorthy, G.K. (2013). Neural networks in RSCAD for intelligent real-time power system applications. *2013 IEEE Power Energy Society General Meeting*, 1–5.

38 Luitel, B., Venayagamoorthy, G.K., and Oliveira, G. (2013). Developing neural networks library in RSCAD for real-time power system simulation. *2013 IEEE Computational Intelligence Applications in Smart Grid (CIASG 2013)*, 130–137.

39 Wang, L., Qin, Y., Tang, Z., and Zhang, P. (2020). Software-defined microgrid control: the genesis of decoupled cyber-physical microgrids. *IEEE Open Access Journal of Power and Energy* 7: 173–182.

40 Althoff, M. (2010). Reachability analysis and its application to the safety assessment of autonomous cars. PhD thesis. Technische Universität M\"{u}nchen.

41 Mehmood, U., Stoller, S.D., Grosu, R. et al. (2021). A distributed Simplex architecture for multi-agent systems. In: *Proceedings of the Symposium on Dependable Software Engineering: Theories, Tools and Applications (SETTA 2021), Lecture Notes in Computer Science*, vol. 13071, 239–257. Springer.

42 Prajna, S., Jadbabaie, A., and Pappas, G.J. (2007). A framework for worst-case and stochastic safety verification using barrier certificates. *IEEE Transactions on Automatic Control* 52 (8): 1415–1428.

43 Amoateng, D.O., Al Hosani, M., Elmoursi, M.S. et al. (2018). Adaptive voltage and frequency control of islanded multi-microgrids. *IEEE Transactions on Power Systems* 33 (4): 4454–4465.

44 Tan, K.-H., Lin, F.-J., Shih, C.-M., and Kuo, C.-N. (2020). Intelligent control of microgrid with virtual inertia using recurrent probabilistic wavelet fuzzy neural network. *IEEE Transactions on Power Electronics* 35 (7): 7451–7464.

3

Distributed Power Flow and Continuation Power Flow for Steady-State Analysis of Microgrids

Fei Feng, Peng Zhang, and Yifan Zhou

3.1 Background

Microgrid has proved to be effective in ensuring electricity resiliency for customers. A most important and indispensable foundation for microgrid operation and management is the power flow analysis. Power flows in microgrids largely depend on the operating modes (e.g. grid-connected and islanded mode) and control modes of distributed energy resources (DERs) (e.g. droop control and secondary control) [1, 2]. In the grid-connected mode, power flow analysis of microgrids follows the same approach as that of distribution networks because those microgrids' voltages and frequency are supported by the main grid [3]. However, power flow analysis of the islanded microgrid is different from the grid-connected power flow because: (1) a swing bus no longer exists, (2) DERs are operated by hierarchical controls, and (3) microgrid is subject to frequent changes in structure and operating modes.

For an individual islanded microgrid, different control modes have been proposed to support stable operations [4]. Correspondingly, derivative-based power flow methods are developed to consider droop-based power sharing (PS) in islanded microgrids [5, 6]. Newton-trust-region-based algorithm is introduced to account for the microgrid frequency [7]. Similar droop-based algorithms have been extended to DC and hybrid microgrids [8–12]. Recently, a microgrid power flow considering secondary control modes (e.g. reactive power sharing [RPS], voltage regulation [VR], and smart tuning [ST]) has been developed by the authors [13]. Meanwhile, nonderivative power flow approaches (e.g. direct back/forward sweep [DBFS], implicit Z_{bus}) are developed for islanded microgrids by introducing an adaptive swing bus to update voltage and frequency [7, 14]. The modified backward/forward sweep (FS) algorithms, however, apply only to radial or weakly meshed microgrids and cannot analyze meshed microgrids commonly adopted in densely populated areas [14, 15]. The revisited implicit Z_{bus} Gauss (GRev), although capable of handling meshed microgrids, suffers from large number of iterations under specific conditions, which may affect its scalability [16].

However, an individual microgrid, because of limited capacity, could hardly serve as a dependable resiliency resource for densely populated communities in urban areas such as New York City. Networked microgrids, a cluster of microgrids integrated with interactive support and coordinated energy management, offer a promise to relieve power deficiencies in individual microgrids, support more local customers, and even black start neighboring energy infrastructures upon occurrence of a blackout. The distributed networked microgrids power flow [17] and compositional power flow [18] allow networked microgrids to adjust boundary bus voltages and regulate power interchanges through tie lines.

This chapter first introduces different power flow methods of individual islanded microgrid. Then, hierarchical-based networked microgrids power flow is shown.

3.2 Individual Microgrid Power Flow

3.2.1 Enhanced Newton-Type Power Flow

The most important and indispensable foundation for microgrid operation and management is the power flow analysis [15]. However, power flow of islanded microgrid has yet to be addressed because: (1) a swing bus no longer exists, 2) DERs are operated by hierarchical controls, and (3) microgrid is subject to frequent changes in structure and operating modes [18].

Thus, an enhanced Newton-type microgrid power flow (EMPF) fully adapts to both meshed and radial structures. The main contributions of EMPF lie in: (1) an augmented Newton type formulation of microgrid power flow which supports plug-and-play and allows future extensions into networked microgrids power flow as well as (2) a new Jacobian matrix formulation which is able to incorporate hierarchical control effects and thus precisely considers PS and VR in a modular fashion.

3.2.1.1 EMPF Formulation

In EMPF, in addition to the traditional PV and PQ buses, we introduce a bus type called *DER buses* to which those DERs equipped with droop and/or secondary control are connected. Generally, a slack bus no longer exists because none of the DERs in the droop-based microgrids is able to provide constant voltage and frequency. We can pick an arbitrary DER bus and use its voltage angle as the reference for the rest of the buses.

For an N-bus microgrid with ζ DER buses, the power injections from DERs are determined by a two-layer hierarchical control system [4]. Considering PV, PQ, and DER buses, we can derive the EMPF power flow equations as follows:

$$\mathbf{F}(\theta, \mathbf{V}, f) = \begin{bmatrix} \mathbf{S}(\mathbf{V}, f)^{\mathbf{G}} - \mathbf{S}^{\mathbf{L}} - \overline{\mathbf{Y}}(\theta) \cdot \mathbf{V} \circ \mathbf{V} \\ \mathbf{P}(f)^{\mathbf{Gs}} - \mathbf{P}^{\mathbf{sum}} \end{bmatrix} \tag{3.1}$$

where $\mathbf{S}(\mathbf{V}, f)^{\mathbf{G}} = [\mathbf{P}(f)^{\mathbf{G}}, \mathbf{Q}(\mathbf{V})^{\mathbf{G}}]^T \in \mathbb{R}^{(2N-1)\times 1}$ and $\mathbf{S}^{\mathbf{L}} = [\mathbf{P}^{\mathbf{L}}, \mathbf{Q}^{\mathbf{L}}]^T \in \mathbb{R}^{(2N-1)\times 1}$ are the generation and load matrices, respectively, $\mathbf{P}(f)^{\mathbf{Gs}}$, is the total real power from generators, ∘ means Hadamard product, and $\mathbf{P}^{\mathbf{sum}}$ is the sum of real power consumption including load and losses. Different from traditional power flow, frequency f is a variable in the EMPF formulation. $\overline{\mathbf{Y}}(\boldsymbol{\theta}) \in \mathbb{R}^{(2N-1)\times N}$ is the extended admittance matrix defined as:

$$\overline{\mathbf{Y}}(\theta) = \begin{bmatrix} \left|\mathbf{Y}_{ij}\right| \cos(\theta_i - \theta_j - \alpha_{ij}) \\ \left|\mathbf{Y}_{ij}\right| \sin(\theta_i - \theta_j - \alpha_{ij}) \end{bmatrix} i, j \in N \tag{3.2}$$

where $\theta \in \mathbb{R}^{(N-1)\times 1}$ is a voltage angle matrix and α_{ij} is the admittance angle of branch $i - j$.

3.2.1.2 Modified Jacobian Matrix

The modified Jacobian matrix $\mathbf{J} \in \mathbb{R}^{2N \times 2N}$ that incorporates DER behaviors under hierarchical control can be derived from Eq. (3.1), as follows:

$$\mathbf{J} = \begin{bmatrix} \dfrac{\partial \mathbf{F}(\theta, \mathbf{V}, f)}{\partial \theta}, & \dfrac{\partial \mathbf{F}(\theta, \mathbf{V}, f)}{\partial \mathbf{V}}, & \dfrac{\partial \mathbf{F}(\theta, \mathbf{V}, f)}{\partial f} \end{bmatrix} \tag{3.3}$$

where

$$\frac{\partial \mathbf{F}(\theta, \mathbf{V}, f)}{\partial \theta} = \left[-\frac{\partial \overline{\mathbf{Y}}(\theta) \cdot \mathbf{V} \circ \mathbf{V}}{\partial \theta}, \mathbf{0} \right]^T \tag{3.4}$$

$$\frac{\partial \mathbf{F}(\theta, \mathbf{V}, f)}{\partial \mathbf{V}} = \left[\frac{\partial \mathbf{S}(\mathbf{V}, f)^G}{\partial \mathbf{V}} - \frac{\overline{\mathbf{Y}}(\theta) \cdot \partial \mathbf{V} \circ \mathbf{V}}{\partial \mathbf{V}} - \frac{\overline{\mathbf{Y}}(\theta) \cdot \mathbf{V} \circ \partial \mathbf{V}}{\partial \mathbf{V}}, \mathbf{0} \right]^T \tag{3.5}$$

$$\frac{\partial \mathbf{F}(\theta, \mathbf{V}, f)}{\partial f} = \left[\frac{\partial \mathbf{S}(\mathbf{V}, f)^G}{\partial f}, \frac{\partial \mathbf{P}(f)^{Gs}}{\partial f} \right]^T \tag{3.6}$$

Here, the elements in $\mathbf{J}$ matrix are functions of different control modes. For the droop control mode, the P/f and Q/V droop coefficients are defined as $\mathbf{m} \in \mathbb{R}^{\zeta \times 1}, \mathbf{n} \in \mathbb{R}^{\zeta \times 1}$, respectively. Real PS among DERs is achieved through the P/f droop control, as shown in Eqs. (3.7) and (3.8).

$$\frac{\partial \mathbf{S}(\mathbf{V}, f)^G}{\partial f} = \begin{cases} -\frac{1}{m_i}, & \text{for DER bus} \\ 0, & \text{otherwise} \end{cases} \tag{3.7}$$

$$\frac{\partial \mathbf{P}(f)^{Gs}}{\partial f} = \sum_{i=1}^{\zeta} -\frac{1}{m_i} \tag{3.8}$$

The DER var behaviors and corresponding $\mathbf{J}$ elements under three typical secondary control modes [4] are expressed below:

Reactive power sharing (RPS) mode: RPS aims to realize proportional RPS, where the var injection from a leader bus Q_1 is updated through Q/V droop control and the rest of DER buses follow. Mathematically, the var outputs of DER buses and the corresponding $\mathbf{J}$ elements are

$$\mathbf{Q}_{\mathbf{DER}} = \left[Q_1(V_1), \rho \cdot \mathbf{Q_F}^* \right]^T \tag{3.9}$$

$$\frac{\partial \mathbf{S}(\mathbf{V}, f)^G}{\partial \mathbf{V}} = \begin{cases} -\frac{1}{n_1}, & \text{for leader DER bus} \\ 0, & \text{otherwise} \end{cases} \tag{3.10}$$

where ρ is the reactive power ratio defined by Q_1/Q_1^*, and $\mathbf{Q_F}^*$ denotes the rated var outputs of follower buses.

Voltage regulation (VR) mode: VR mode aims to recover the DER bus voltages to their rated values by adjusting the DER reactive power injections. Thus, the var outputs of DER buses and the corresponding $\mathbf{J}$ elements are updated by

$$\mathbf{Q}_{\mathbf{DER}} = \text{diag}(\mathbf{V}) \cdot \text{diag}(\mathbf{Z}_d^{-1}) \cdot (\mathbf{V}_d + \mathbf{V}^* - 2\mathbf{V}) + \mathbf{Q}_0 \tag{3.11}$$

$$\frac{\partial \mathbf{S}(\mathbf{V}, f)^G}{\partial \mathbf{V}} = \begin{cases} (Z_d^{-1})(V_d + V^* - 4V), & \text{for DER bus} \\ 0, & \text{otherwise} \end{cases} \tag{3.12}$$

Similar to [15], a dummy bus vector with voltages $\mathbf{V}_d$ is created for DER buses associated with a sensitivity vector $\mathbf{Z}_d$ representing the reactive power differences with respect to the voltage differences between dummy buses and the corresponding DER buses. Here, $\mathbf{V}^*$ denotes rated voltages, and the detailed procedure to update $\mathbf{V}_d$ can be found in [15], $\xi_{\Delta V_d}$ is voltage magnitude error between DER buses and its rated value.

Smart tuning (ST) mode: The leader DER bus follows the VR mode to recover back to its rated value, while other DER buses are adjusted for proportional RPS. Therefore, in this mode, the leader DER bus var output and corresponding **J** elements follow Eqs. (3.11) and (3.12), whereas the rest of DER buses follow Eqs. (3.9) and (3.10).

Once **J** and Δ**F** are evaluated at the end of each iteration, the microgrid variables θ, **V**, and f can be updated for the next iteration by solving the following equation:

$$\Delta\mathbf{F}(\theta, \mathbf{V}, f) = \mathbf{J} \cdot \left[\Delta\theta, \Delta\mathbf{V}, \Delta f\right]^T \tag{3.13}$$

The EMPF iterations continue until the errors in those variables reach the tolerance ξ.

The Newton-type power flow is sensitive to the starting point and relies on high-quality initial values for a fast convergence. To ensure the robustness of EMPF incorporating the hierarchical control, it is initialized by the values obtained by running a power flow with droop controls only. Once the convergence criterion is satisfied, all the voltages and branch power flows can be obtained. Because no assumption of microgrid architectures is utilized in EMPF, it can be used to solve power flows for arbitrary types of microgrids such as radial, meshed, or honeycomb configurations.

3.2.2   Revisited Implicit Z_{bus} Power Flow

Implicit Z_{bus} Gauss (Z_{bus}) is a fixed-point algorithm that exploits the sparse Y_{bus} matrix and equivalent current superposition to solve the network equations [19]. However, the traditional method is based on a slack bus which no longer exists. Thus, a revisited implicit Z_{bus} Gauss algorithm (GRev) is established which is able to handle arbitrary structures and allows incorporating DER hierarchical controls as well as load droops. Besides, GRev is more robust than Newton's method because of its insensitivity to initial values. Thus, GRev is particularly useful for meshed microgrids for urban and populated communities or mission-critical microgrids requiring higher reliability and resilience.

Similar to the aforementioned Section 3.2.1, *DER buses* are used to absorb the mismatch between generation and demand in an islanded microgrid by regulating the system frequency and bus voltages. One of the DER buses is selected as a leader bus for updating the frequency.

3.2.2.1   Basic GRev Formulation

The DER power injections are determined by a two-layer hierarchical control system [4]. Therefore, the GRev power flow for an N-bus microgrid with DER hierarchical control and load droops are formulated as:

$$\begin{bmatrix} \mathbf{I}_E(\mathbf{V}_E, f) \\ \mathbf{I}_S(\mathbf{V}_S, f) \end{bmatrix} = \begin{bmatrix} \mathbf{Y}_{EE} & \mathbf{Y}_{ES} \\ \mathbf{Y}_{SE} & \mathbf{Y}_{SS} \end{bmatrix} \begin{bmatrix} \mathbf{V}_E \\ \mathbf{V}_S \end{bmatrix} \tag{3.14}$$

$$\mathbf{I}_E(\mathbf{V}_E, f) = \left[\text{conj}(\mathbf{S}_E(\mathbf{V}_E, f)/\mathbf{V}_E)\right] \tag{3.15}$$

where $\mathbf{I}_S$ and $\mathbf{V}_S$ are the current and voltage of the DER bus designated for updating the frequency, respectively; $\mathbf{I}_E$ and $\mathbf{V}_E$ are those of the remaining buses; $\mathbf{Y}_{EE}$, $\mathbf{Y}_{SS}$, $\mathbf{Y}_{SE}$, and $\mathbf{Y}_{ES}$ are the admittance submatrices; $\mathbf{S}_E(\mathbf{V}_E, f) = (\mathbf{P}_G(f) - \mathbf{P}_L(f)) + j(\mathbf{Q}_G(\mathbf{V}_E) - \mathbf{Q}_L(\mathbf{V}_E)) \in \mathbb{C}^{(N-1)\times1}$ is the power injection (generation minus load) vector. Different from the Z_{bus} method, in the GRev algorithm $\mathbf{I}_{E,S}$ vary with $\mathbf{V}_{E,S}$ and the system frequency f, and $\mathbf{V}_S$ is no longer constant.

The iterative formula for updating $\mathbf{V}_E$ can then be derived from (3.14) and (3.15), as follows:

$$\mathbf{V}_E = \left[\mathbf{Y}_{EE}^{-1} \cdot (\mathbf{I}_E(\mathbf{V}_E, f)) - \mathbf{Y}_{ES} \cdot \mathbf{V}_S\right] \tag{3.16}$$

3.2.2.2 GRev with Hierarchical Control

The GRev algorithm consists of a double-loop iteration process. Whenever the nodal power injections update, the bus voltages will subsequently be refreshed. The power updating loop depends upon the hierarchical control modes. Assuming that the P/f and Q/V droop coefficients of DERs and loads are $\mathbf{m_G}$, $\mathbf{n_G}$, $\mathbf{m_L}$, $\mathbf{n_L}$, respectively. Under the droop control, nodal injections can be updated as follows:

$$\Delta \mathbf{S_E} = \left[\left(\mathbf{A} \circ \frac{1}{\mathbf{m_G}} - \mathbf{B} \circ \frac{1}{\mathbf{m_L}}\right)\Delta f + j\left(\mathbf{A} \circ \frac{1}{\mathbf{n_G}} - \mathbf{B} \circ \frac{1}{\mathbf{n_L}}\right)\Delta \mathbf{V_V}\right] \tag{3.17}$$

where $\mathbf{A}$ and $\mathbf{B}$ are the 0/1-status matrices of DERs and loads, respectively; the 0/1-status matrices indicate the connections between DERs/loads and microgrid. When the DER is connected with microgrid, the corresponding matrix element is set as 1; otherwise, 0; and $\circ$ denotes the Hadamard product.

On top of the droop scheme, the secondary control accounts for PS and voltage adjustment. Real power increments can be redistributed among DERs when loads recover to the nominal status, following $\Delta \mathbf{P_E} = \Delta f / \mathbf{m_G}$. Without loss of generality, the updating of the DER var outputs under three typical secondary control modes is expressed below:

Reactive power sharing (RPS) mode: In the RPS mode, all the other DERs adopt the reactive power ratio of the leader DER. $\mathbf{V_S}$ is refreshed according to the Q/V droop. The incremental var outputs $\Delta \mathbf{Q_E}$ can then be expressed by:

$$\Delta \mathbf{Q_E} = \left[\rho \mathbf{A} \circ \mathbf{Q_E}^* - \mathbf{B} \circ \frac{1}{\mathbf{n_L}}\Delta \mathbf{V_V} - \mathbf{Q_{E0}}\right] \tag{3.18}$$

where the reactive power ratio $\rho = Q_S / Q_S^*$, and $\mathbf{Q_E}^*$ denotes the rated var outputs of the follower buses.

Voltage regulation (VR) mode: VR aims to recover the DER bus voltages back to their rated values. Mathematically, the var updates of the DER buses are below:

$$\Delta \mathbf{Q_{E,S}} = \left[\mathbf{A} \circ \mathrm{diag}(\mathbf{V_{E,S}})\mathrm{diag}(\mathbf{Z}_d^{-1})(\mathbf{V}_d + \mathbf{V_{E,S}}^* - 2\mathbf{V_{E,S}}) - \mathbf{B} \circ \frac{1}{\mathbf{n_L}}\Delta \mathbf{V_{E,S}}\right] \tag{3.19}$$

In this mode, a dummy bus vector with voltages $\mathbf{V}_d$ is used to adjust var injections for the DER buses. Dummy bus voltage considers the difference between the DER voltage and its rated value. If the DER voltage deviates from the rated value, the deviation will be added up to the dummy bus voltage to adjust the DER var output based on Eq. (3.19). A sensitivity vector $\mathbf{Z}_d$ represents the var differences with respect to the voltage differences between dummy buses and the corresponding DER buses. Here, $\mathbf{V_{E,S}}^*$ denotes the rated voltages, and the detailed procedure to update $\mathbf{V}_d$ can be found in [15].

Smart tuning (ST) mode: One DER bus adopts the VR mode, while others follow the RPS mode. As for the bus voltage updating, once $\mathbf{S_E}(\mathbf{V_E}, f)$ is updated, $\mathbf{V_E}$ can be calculated by (3.15) and (3.16). Then the leader DER bus will update its voltage and the system frequency, as follows:

$$\Delta f = \left[\mathrm{real}(\mathrm{conj}(\mathbf{V_S}(\mathbf{Y_{SS}}\mathbf{V_S} + \mathbf{Y_{ES}}\mathbf{V_E})) - \mathbf{P_{loss}})/\mathbf{m_G}\right] \tag{3.20}$$

$$\Delta \mathbf{V_S} = \left[\mathrm{imag}(\mathrm{conj}(\mathbf{V_S}(\mathbf{Y_{SS}}\mathbf{V_S} + \mathbf{Y_{ES}}\mathbf{V_E})) - \mathbf{Q_{loss}})/\mathbf{n_G}\right] \tag{3.21}$$

The advantage of GRev is its insensitivity to the choice of initial conditions. All initial values for bus voltage and frequency can be set as unit vectors. This is largely because GRev is a fix-point iteration. Besides, GRev does not rely on any assumption of the microgrid structure, making it suitable for microgrids with arbitrary architectures no matter whether they are radial, meshed, or honeycomb.

3.2.3 Generalized Back/Forward Sweep Power Flow

3.2.3.1 Direct Back/Forward Sweep Method

For a conventional distribution grid, DBFS method is a matrix-based BFS which requires only one matrix operation for backward sweep (BS) and another one for FS. Using the concept of bus injection to branch current (BIBC) matrix and the branch current to bus voltage (BCBV) matrix, the basic equations are:

$$I_{bus} = (S/U_{bus})^* \tag{3.22}$$

$$I_{branch} = (BIBC \cdot I_{bus}) \tag{3.23}$$

$$\Delta U = (BCBV \cdot I_{branch}) \tag{3.24}$$

$$U_{bus} = (U^0 - \Delta U) \tag{3.25}$$

The BS and FS can be represented as (3.22) and (3.23) and (3.24) and (3.25), respectively. The power injection of the swing bus ($S_1 = P_1 + jQ_1$) is calculated after the convergence of DBFS.

3.2.3.2 Generalized Microgrid Power Flow Algorithm

The generalized microgrid power flow (GMPF) follows a double-loop process. Here, the outer loop is to update the reactive power until the secondary control objective (e.g. RPS, VR, and ST modes) is reached, whereas the inner loop is to update the active power. The GMPF iterations are specified below.

GMPF is first initialized using the power flow results for droop-controlled microgrid, specifically active power can always achieve accurate sharing by updating $\Delta P_i = -m_i \Delta f$. For reactive power regulation, under RPS mode, ΔQ_i is updated by (3.18). Under VR mode, ΔQ_i is updated by (3.19). Under ST mode, one DER bus adopts the VR mode while others follow the RPS mode.

3.3 Networked Microgrids Power Flow

An augmented Newton-type networked microgrids power flow (APF) solution that is able to not only incorporate various hierarchical control modes but also process arbitrary grid topology including radial and meshed structures, which consists of two steps: (1) A distributed, augmented Newton-type power flow formulation is derived to incorporate hierarchical control in networked microgrids. APF establishes a thorough power flow formulation of droop/secondary-controlled networked microgrids. Modified Jacobian matrices are derived in a distributed manner to allow for automatic modeling under various control schemes, such as power sharing, voltage regulation, and interface power exchange. APF eliminates the reliance on an adaptive swing bus to update voltage and frequency, fully supporting voltage adjustments of multiple tie lines. RTDS experiments are designed to verify the fidelity of APF. (2) A programmable distributed platform is constructed for networked microgrids power flow analysis. The platform provides a software-defined architecture to achieve interface power coordination for APF. The distributed manner of this platform fully supports plug-and-play among microgrids and protects local customers' privacy [17].

3.3.1 Networked Microgrids Architecture

Generally, hierarchical control for networked microgrids consists of two layers to realize power management and VR (see Fig. 3.1).

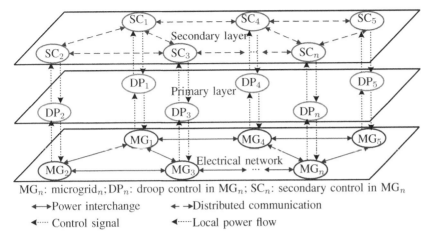

Figure 3.1 Architecture of networked microgrids.

Primary layer: When operating in islanded situations, DERs respond to the real and reactive power demand following the droop controls in the primary layer.

Secondary layer: The main functionality of secondary control is to mitigate the steady-state errors in droop control, while it sometimes can also be used to achieve global PS among microgrids. When one microgrid suffers from power deficiency, neighboring microgrids could provide power support by communicating with each other [4].

3.3.2 Distributed NMPF Formulation

In APF, *DER buses* follow the hierarchical control logic to jointly achieve expected PS and VR [13].

3.3.2.1 Power Sharing (PS) Mode

PS aims to realize a set of scheduled power interchanges between neighboring microgrids when a specific microgrid suffers power deficiency. The voltages at boundary buses connected with tie lines need to achieve certain target values.

To establish a fully distributed APF, leader DER buses are selected whose power outputs are regulated to eventually control the power interchanges through the tie lines.[1] The scheduled power interchange is realized by adjusting the boundary bus voltages of each microgrid. Without loss of generality, we assume that DER buses in microgrid i are to be regulated to achieve the expected PS between microgrid i and microgrid j through η tie lines. The iterative rule for updating the bus voltages of leader DERs in microgrid i is as follows:

$$\mathcal{V}_{(i)}^{G,l} = \mathcal{Z}_{(i,j)} \cdot \mathrm{conj}\left(\frac{\mathcal{S}_{(i,j)}^{\mathrm{In}}}{\mathcal{V}_{(j)}^{B}} \right) + \mathcal{V}_{(j)}^{B} + \tilde{\mathcal{V}}_{(i)}^{G,l} - \tilde{\mathcal{V}}_{(i)}^{B} \tag{3.26}$$

Here, $\mathcal{V}_{(i)}^{G,l} \in \mathbb{C}^{\eta \times 1}$ denotes the complex bus voltages of η leader DERs corresponding to η tie lines (calligraphic fonts will be used to denote complex numbers); $\mathcal{Z}_{(i,j)} \in \mathbb{C}^{\eta \times \eta}$ denotes the impedance matrix between microgrid i and j with diagonal elements being the impedance of the tie lines; $\mathcal{S}_{(i,j)}^{\mathrm{In}} \in \mathbb{C}^{\eta \times 1}$ is the vector of scheduled power injections from microgrid j to i; $\mathcal{V}_{(j)}^{B} \in \mathbb{C}^{\eta \times 1}$ denotes

---

1 The rule of thumb is to choose the leader DER buses to be the ones with the shortest distance from target boundary buses, as this normally leads to faster convergence.

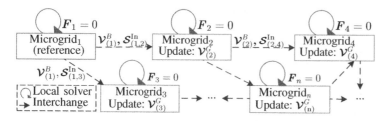

Figure 3.2 Outline of PS-APF.

the complex voltages of boundary buses in microgrid j which are connected with microgrid i; $\tilde{\mathbf{V}}_{(i)}^{G,l}$ and $\tilde{\mathbf{V}}_{(i)}^{B}$ are, respectively, the voltages of leader DER buses and boundary buses in microgrid i at the previous iteration. The rationale behind (3.26) is that, when a boundary bus's voltage is lower than its target value, a leader DER's bus voltage will be increased following (3.26), meaning the DER will output more power to increase the voltage profile of the network. The opposite will be performed when the boundary bus voltage is higher than the target value. Such a negative feedback process, if convergent, will achieve the target power interchanges.

Consequently, Fig. 3.2 presents the outline of the PS-mode APF. An arbitrary DER bus can be chosen as the angle reference in the networked microgrids. The APF procedure starts from microgrid 1, where the reference DER is located. First, the power flow status of microgrid 1 is updated locally with the scheduled power interchanges $\mathbf{S}_{(1,2)}^{\text{In}}$ and $\mathbf{S}_{(1,3)}^{\text{In}}$. Then, microgrid 1 releases the boundary bus voltages and power interchanges to its neighbors, i.e. microgrids 2 and 3. Microgrids 2 and 3 therefore update their own power flows independently with their leader DERs' voltages adjusted by (3.26) to achieve the required boundary conditions. Subsequently, microgrids 4 and 5 calculate their power flow status separately with the refreshed boundary conditions released by microgrids 2 and 3. The aforementioned process repeats until all the microgrids are traversed. In the whole procedure, each microgrid interacts with its neighbors using only the boundary information (i.e. boundary bus voltages and tie line powers). Therefore, the local power flow calculation of each individual microgrid is performed distributively and no private information will be disclosed.

3.3.2.2 Voltage Regulation (VR) Mode
The target of VR control mode is to maintain the leader DERs' bus voltages to specific values by adjusting the power injections of the DERs and the power interchanges through tie lines. We observe that, if we solely rely on the power adjustments of the leader DER buses, the power flow calculation may suffer from poor convergence as the scales of microgrids increase. To address this issue, we employ an outer loop for updating the power interchanges on top of the PS mode to achieve voltage restoration.

In our approach, when a leader DER's bus voltage is lower than its rated value, reactive power from the power interchange is increased to support the voltage magnitude. Meanwhile, the active power interchange is reduced to adjust the voltage angle. Similarly, when a leader DER's bus voltage is higher than the rated value, a reversed control logic of power interchange is executed to help the leader DER to restore voltage. The convergence of such a negative feedback loop will realize the target voltage profile. As such, the power interchanges through tie lines are updated as follows:

$$\mathbf{S}_{(i,j)}^{\text{In}} = \left(\tilde{\mathbf{P}}_{(i,j)}^{\text{In}} - \epsilon_p \left(\mathbf{V}_{(i)}^{B} - \mathbf{V}_{(i)}^{G,l} \cos\theta_{(i)}^{G,l} \right) \right) + j \left(\tilde{\mathbf{Q}}_{(i,j)}^{\text{In}} + \epsilon_q \left(\mathbf{V}_{i}^{B} - \mathbf{V}_{(i)}^{G,l} \sin\theta_{(i)}^{G,l} \right) \right) \tag{3.27}$$

Here, $\mathbf{P}_{(i,j)}^{\text{In}} = \text{Re}(\mathbf{S}_{(i,j)}^{\text{In}})$ and $\mathbf{Q}_{(i,j)}^{\text{In}} = \text{Im}(\mathbf{S}_{(i,j)}^{\text{In}})$, respectively, denote the active and reactive power injections through tie lines from microgrid i to j; $\tilde{\mathbf{P}}_{(i,j)}^{\text{In}}$ and $\tilde{\mathbf{Q}}_{(i,j)}^{\text{In}}$, respectively, denote the values of $\mathbf{P}_{(i,j)}^{\text{In}}$

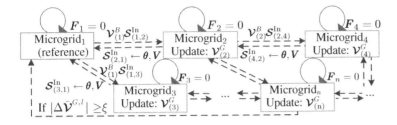

Figure 3.3 Outline of VR-APF.

and $Q_{(i,j)}^{\text{In}}$ inherit from the previous iteration; $V_{(i)}^{G,l} = |\mathcal{V}_{(i)}^{G,l}|$ and $\theta_{(i)}^{G,l} = \arg(\mathcal{V}_{(i)}^{G,l})$, respectively, denote the voltage amplitudes and angles of the leader DER buses; ϵ_p and ϵ_q are step sizes for updating real and reactive power. Proper values of ϵ_p and ϵ_q can be set as the real and imaginary parts of $S^{\text{LS}}/V_{\text{base}}$ in a local microgrid, where S^{LS} is the total load consumption, V_{base} is the base voltage. In this circumstance, a faster convergence is achieved for the VR outer loop.

The outline of the VR-mode APF is established in Fig. 3.3. First, power flow status of networked microgrids are initialized by PS-APF with power interchanges $S_{(i,j)}^{\text{In}}$ scheduled as zero. Then, the power interchanges are adjusted according to (3.27) to achieve the target voltages of leader DERs within each microgrid. Next, each microgrid updates the local power flow by PS-APF with the refreshed power interchanges. Aforementioned VR procedure continues until the leader DERs' voltages recover to nominal values $\mathcal{V}^{G,l^*}$, which indicates $|\Delta\overline{\mathcal{V}}^{G,l}| = |\mathcal{V}^{G,l^*} - \mathcal{V}^{G,l}| \leq \xi$.

The following establishes the distributed power flow model of each microgrid with a thorough formulation of hierarchical control. Without loss of generality, we study the distributed power flow formulation of an arbitrary microgrid i. Denote the number of buses, DERs, and leader DERs as N, n, and n_l. Denote the power interchanges between microgrid i and all its neighboring microgrids as S^{In}. Here, S^{In} is modeled as a vector so that APF is capable of handling PS among multiple tie lines.

The active PS among DERs in microgrid i follows the hierarchical control logic as:

$$f - f_k^{\text{ref}} = -\alpha_k^G (P_k^G - P_k^{\text{ref}} + \Omega_k^G) \tag{3.28}$$

Here, f denotes the frequency of the networked microgrid, which will be adjusted to achieve the nominal frequency f^* (i.e. 60 Hz) via hierarchical control of all microgrids; f_k^{ref} denotes the reference frequency signal of DER k; α_k^G denotes the P/f droop coefficient of DER k; P_k^G and P_k^{ref}, respectively, denote the active power generation and the reference active power of DER k; Ω_k^G denotes the secondary control signal for frequency regulation. Specifically, the follower DERs jointly share a frequency unbalance amount $\hat{f}$ according to their droop coefficients to assist frequency recovery:

$$\Omega_k^G = \hat{f}/\alpha_k^G \,, \quad k \in \mathbb{S}_f^G \tag{3.29}$$

where $\mathbb{S}_f^G$ denotes the set of follower DERs.

Integrating (3.28) and (3.29) yields the active power characteristics of *DER buses*:

$$\boldsymbol{P}^G = \boldsymbol{P}^{\text{ref}} - \frac{1}{\boldsymbol{\alpha}^G}\left(f^* - \boldsymbol{f}^{\text{ref}}\right) - \boldsymbol{\Omega}^G = \begin{cases} P_k^G(\Omega_k^G), & k \in \mathbb{S}_l^G \\ P_k^G(\hat{f}), & k \in \mathbb{S}_f^G \end{cases} \tag{3.30}$$

where $\boldsymbol{P}^G$, $\boldsymbol{P}^{\text{ref}}$, $\boldsymbol{\alpha}^G$, $\boldsymbol{f}^{\text{ref}}$, and $\boldsymbol{\Omega}^G \in \mathbb{R}^{n \times 1}$ are, respectively, the vector forms of the corresponding variables; $\mathbb{S}_l^G$ denotes the set of leader DERs.

For RPS, leader DERs adopt hierarchical control (see (3.31)) to achieve the scheduled $\boldsymbol{V}^G$, while follower DERs adopt droop control (see (3.32)) to assist leader DERs:

$$V_k^{G,l} - V_k^{\text{ref}} = -\beta_k^G(Q_k^G - Q_k^{\text{ref}}) - e_k^G, \quad k \in \mathbb{S}_l^G \tag{3.31}$$

$$V_k^G - V_k^{\text{ref}} = -\beta_k^G(Q_k^G - Q_k^{\text{ref}}), \quad\quad k \in \mathbb{S}_f^G \tag{3.32}$$

where $V_k^{G,l}$ is the target bus voltage of leader DER k pre-determined by the PS/VR mode as detailed in Sections 3.3.2.1 and 3.3.2.2; V_k^G and V_k^{ref}, respectively, denote the voltage amplitude of DER k and its reference value; β_k^G denotes the Q/V droop coefficient; Q_k^G and Q_k^{ref}, respectively, denote the reactive power generation and its reference value; e_k^G denotes the secondary control signal for VR.

Integrating (3.31) and (3.32) leads to a unified formulation of the reactive power characteristics of *DER buses*:

$$\boldsymbol{Q}^G\left(\hat{\boldsymbol{V}}^G\right) = \boldsymbol{Q}^{\text{ref}} - \frac{1}{\boldsymbol{\beta}^G}\left(\boldsymbol{V}^* - \boldsymbol{V}^{\text{ref}} + \hat{\boldsymbol{V}}^G\right) \tag{3.33}$$

Here, $\boldsymbol{Q}^G$, $\boldsymbol{Q}^{\text{ref}}$, $\boldsymbol{\beta}^G$, and $\boldsymbol{V}^{\text{ref}} \in \mathbb{R}^{n\times 1}$ are, respectively, the vector forms of the corresponding variables; $\hat{\boldsymbol{V}}^G \in \mathbb{R}^{n\times 1}$ and $\boldsymbol{V}^* \in \mathbb{R}^{n\times 1}$ are defined as follows:

$$\hat{\boldsymbol{V}}^G = \begin{cases} e_k^G, & k \in \mathbb{S}_l^G \\ V_k^G, & k \in \mathbb{S}_f^G \end{cases}, \quad \boldsymbol{V}^* = \begin{cases} V_k^{G,l}, & k \in \mathbb{S}_l^G \\ 0, & k \in \mathbb{S}_f^G \end{cases}$$

Subsequently, the summation of DER power generations is derived based on (3.30) and (3.33):

$$P^{\text{Gs}} = \sum_k\left(P_k^{\text{ref}} + \frac{f_k^{\text{ref}}}{\alpha_k^G}\right) - \left(\sum_k \frac{1}{\alpha_k^G}\right)(f^* + \hat{f}) \tag{3.34}$$

$$Q^{\text{Gs}} = \sum_k Q_k^{\text{ref}} - \sum_k \frac{1}{\beta_k^G}\left(V_k^* - V_k^{\text{ref}} + \hat{V}_k^G\right) \tag{3.35}$$

With the formulation of DER buses, the following power flow states are to be solved for microgrid i:

- $\hat{\boldsymbol{\theta}} \in \mathbb{R}^{(N-n_l)\times 1}$ assembling the voltage angles of non-leader-DER buses[2];
- $\hat{\boldsymbol{V}} \in \mathbb{R}^{N\times 1}$ assembling the voltage amplitudes of non-DER buses and $\hat{\boldsymbol{V}}^G$ of DER buses (see (3.33));
- $\hat{f}$ denoting the frequency unbalance variable defined in (3.29).

To achieve the voltage angles $\arg(\boldsymbol{V}_{(i)}^{G,l})$ determined by the PS/VR regulation (see (3.26)), we let the leader DERs adjust their active power outputs freely with (3.34) still holds. Therefore, Ω_k^G for $k \in \mathbb{S}_l^G$ need not to be involved in power flow calculation.

Consequently, we establish the APF power flow formulation of microgrid i as follows:

$$\boldsymbol{F}_i(\hat{\boldsymbol{\theta}}, \hat{\boldsymbol{V}}, \hat{f}) = \begin{bmatrix} \boldsymbol{\phi}^{nl} \cdot \boldsymbol{P}^G - \boldsymbol{P}^L - \boldsymbol{Y}_p \cdot \boldsymbol{V}_b \circ \boldsymbol{V}_b - \boldsymbol{P}^{\text{In}} \\ \boldsymbol{\phi}^{nr} \cdot \boldsymbol{Q}^G - \boldsymbol{Q}^L - \boldsymbol{Y}_q \cdot \boldsymbol{V}_b \circ \boldsymbol{V}_b - \boldsymbol{Q}^{\text{In}} \\ P^{\text{Gs}} - P^{\text{LS}} - P^{\text{Ins}} - P^{\text{loss}} \\ Q^{\text{Gs}} - Q^{\text{LS}} - Q^{\text{Ins}} - Q^{\text{loss}} \end{bmatrix} = \boldsymbol{0} \tag{3.36}$$

In (3.36), the first and second row blocks, respectively, formulate the nodal active power balance at non-leader-DER buses and nodal reactive power balance at non-reference buses, where $\boldsymbol{\phi}^{nl} \in \mathbb{R}^{(N-n)\times n}$ is the incidence matrix between non-leader-DER buses and DERs, whose (k, j)-element is

---

2 Note that voltage angles of leader-DER buses are pre-determined by the PS/VR regulation (see Sections 3.3.2.1 and 3.3.2.2).

1 when DER j is connected to a non-leader-DER bus j, and 0 otherwise; $\boldsymbol{\phi}^{nr} \in \mathbb{R}^{(N-1) \times n}$ is the incidence matrix between non-reference buses and DERs; $\boldsymbol{P}^L$ and $\boldsymbol{P}^{In}$, respectively, denote the active power loads and power interchanges at non-leader-DER buses; $\boldsymbol{Q}^L$ and $\boldsymbol{Q}^{In}$, respectively, denote the reactive power loads and power interchanges at non-reference buses; $\boldsymbol{Y}_p$ and $\boldsymbol{Y}_q$ denote the modified admittance matrices[3]; $\boldsymbol{V}_b$ denotes the bus voltage amplitudes. The last two row blocks in (3.36), respectively, formulate the active and reactive power balance in microgrid i, where P^{Gs}, P^{Ls}, P^{Ins}, and P^{loss}, respectively, denote the total active power generations, loads, power interchanges and losses; similarly, Q^{Gs}, Q^{Ls}, Q^{Ins}, and Q^{loss}, respectively, denote the reactive components.

With the distributed APF model of each microgrid, we summarize the PS/VR-based procedure at a high level:

- In the PS mode, the power interchange $\boldsymbol{S}^{In}_{(i,j)}$ is scheduled. APF first refreshes $\mathcal{V}^{G,l}$ based on (3.26). Then, power flows of each individual microgrid, i.e. $\boldsymbol{F}_i = \boldsymbol{0}(\forall i)$ in (3.36), are solved. The Newton–Raphson iteration is applied here until the errors in $\hat{\theta}, \hat{V}$, and $\hat{f}$ reach the tolerance ξ:

$$\Delta \boldsymbol{F}_i(\hat{\theta}, \hat{V}, \hat{f}) = \boldsymbol{J}_i \cdot \left[\Delta \hat{\theta}, \Delta \hat{V}, \Delta \hat{f}\right]^T \tag{3.38}$$

where $\boldsymbol{J}_i$ denotes the Jacobian matrices of the APF model. Due to page limitation, details of $\boldsymbol{J}_i$ are omitted.

- In the VR mode, the power interchange $\boldsymbol{S}^{In}_{i,j}$ is updated based on (3.27) at first. Then, $\boldsymbol{V}^G_i$ is refreshed based on (3.26), followed by the power flow calculation of each individual microgrid based on (3.36). The APF calculation continues until the bus voltages reach the target values.

3.3.3 Distributed NMPF Algorithm

A distributed framework with a triple loop process is developed for APF algorithm, mainly for the purposes of scalability and privacy-preserving.

(1) *Tertiary loop* is to achieve VR control objectives. The VR mode modifies power injections based on (3.27) to facilitate DER bus voltages recovery to nominal values.

(2) *Secondary loop* updates bus voltages of leader DERs and thus regulating power interchanges through multiple tie lines to realize PS mode. Once power interchanges $\boldsymbol{S}^{In}$ are scheduled, boundary bus voltages will be tuned to the target values by adjusting leader DER buses based on (3.26).

(3) *Primary loop* is to update voltage profiles of networked microgrids, which performs distributed APF computations for the individual microgrids. The power flow solver of each microgrid calculates the power flow states locally. Microgrids exchange only boundary bus information with the neighboring microgrids. The computing sequence is as follows. The power flow of microgrid i that contains angle reference bus is required to be calculated first. (1) If microgrid i is connected with neighboring AC microgrid j, the boundary bus voltages $\boldsymbol{V}^B_{(i)}$ will be passed on to the neighboring AC microgrid. Then, target voltages for boundary buses in microgrid j can be scheduled. Subsequently, once power interchange $\boldsymbol{S}^{In}_{(i,j)}$ and leader bus voltages $\mathcal{V}^{G,l}_{(j)}$ are updated, power flows of the neighboring microgrid j can be solved based on its own power flow solver.

---

3 $\boldsymbol{Y}_p$ and $\boldsymbol{Y}_q$ are the modified admittance matrices of microgrid i:

$$\begin{cases} Y_{p,kl} = |Y_{kl}| \cos(\theta_k - \theta_l - \delta_{kl}), & \forall k, l \\ Y_{q,kl} = |Y_{kl}| \sin(\theta_k - \theta_l - \delta_{kl}), & \forall k, l \end{cases} \tag{3.37}$$

where Y_{kl} and δ_{kl} are the admittance magnitude and angle of branch k–l.

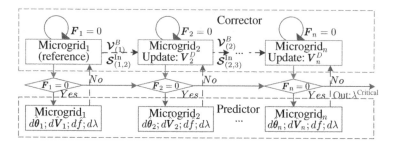

Figure 3.4 Outline of CPF⁺.

The aforementioned distributed APF platform is both programmable and secure in that

- Jacobian matrices of all microgrids are updated separately and flexibly, fully supporting plug-and-play of microgrids and microgrid components.
- Various hierarchical control modes [20] of networked microgrids can be programmed in APF's outer loop/inner loop, and the switching of control modes can be readily simulated in APF.
- Only boundary buses data (e.g. voltages and interface flows through tie lines) need to be shared with neighboring microgrids, which can protect the local customers' privacy.

3.3.4 APF-Based Continuation Power Flow

In this section, an APF-based continuation power flow algorithm (CPF⁺) is further developed to facilitate steady-state voltage stability analysis of networked microgrids under various hierarchical control modes. The CPF⁺ algorithm is again established in a fully distributed manner to support the plug-and-play and privacy preserving in networked microgrids.

Without loss of generality, we consider droop coefficients and power interchanges as the influencing factors λ. The power changes can be simulated by:

$$
\begin{cases}
\boldsymbol{P}^G(\lambda_1) = \boldsymbol{P}^{\text{ref}} - \dfrac{1}{(1+\lambda_1)\boldsymbol{\alpha}^G}\left(\boldsymbol{f}^* - \boldsymbol{f}^{\text{ref}}\right) - \boldsymbol{\Omega}^G \\[2mm]
\boldsymbol{Q}^G(\lambda_1) = \boldsymbol{Q}^{\text{ref}} - \dfrac{1}{(1+\lambda_1)\boldsymbol{\beta}^G}\left(\boldsymbol{V}^* - \boldsymbol{V}^{\text{ref}} + \hat{\boldsymbol{V}}^G\right)
\end{cases}
\tag{3.39}
$$

$$
\begin{cases}
\boldsymbol{P}^{\text{In}}(\lambda_2) = \boldsymbol{P}_0^{\text{In}} + \lambda_2(S_{\text{base}}\cos\varphi_{Li}) \\[2mm]
\boldsymbol{Q}^{\text{In}}(\lambda_2) = \boldsymbol{Q}_0^{\text{In}} + \lambda_2(S_{\text{base}}\sin\varphi_{Li})
\end{cases}
\tag{3.40}
$$

where $\boldsymbol{P}_0^{\text{In}}$ and $\boldsymbol{Q}_0^{\text{In}}$ are the original real powers and reactive powers flowing through tie lines; S_{base} is the power base; φ_{Li} is the power factor angle of tie line flow.

As presented in Fig. 3.4, CPF⁺ consists of a predictor and a corrector, both implemented distributedly. The predictor and the corrector jointly solve the power flow of networked microgrids with successively increased impact factors (i.e. droop coefficients or microgrid power interchanges) until the critical point is reached, which indicates a voltage collapse of the networked microgrids.

The predictor step is used to provide an approximate point when the specified factors change to the next step. The prediction can be calculated by choosing a non-zero value for one of the components of the tangent vector, as follows:

$$
\begin{bmatrix}
\dfrac{\partial \boldsymbol{F}_i}{\partial \hat{\boldsymbol{\theta}}} & \dfrac{\partial \boldsymbol{F}_i}{\partial \hat{\boldsymbol{V}}} & \dfrac{\partial \boldsymbol{F}_i}{\partial \hat{\boldsymbol{f}}} & \dfrac{\partial \boldsymbol{F}_i}{\partial \lambda} \\[2mm]
\boldsymbol{0} & \boldsymbol{0} & \boldsymbol{0} & 1
\end{bmatrix}
\begin{bmatrix}
d\hat{\boldsymbol{\theta}} \\
d\hat{\boldsymbol{V}} \\
d\hat{\boldsymbol{f}} \\
d\lambda
\end{bmatrix}
=
\begin{bmatrix}
\boldsymbol{0} \\
\pm 1
\end{bmatrix}
\tag{3.41}
$$

where $\lambda \in \{\lambda_1, \lambda_2\}$. Once the tangent vector is obtained by solving (3.41), the prediction can be made as follows:

$$[\hat{\theta}; \hat{V}; \hat{f}; \lambda] + \epsilon[d\hat{\theta}; d\hat{V}; d\hat{f}; d\lambda] \longrightarrow [\hat{\theta}; \hat{V}; \hat{f}; \lambda] \tag{3.42}$$

where ϵ is the predictor step size.

The corrector step aims to calculate APF with the predicted initial values. It first updates the droop coefficient or power interchange and then calculates voltages. For instance, if the goal is to identify the safe operating region of droop coefficients, the ith microgrid's variables can be calculated by solving the following equation

$$\Delta F_i(\hat{\theta}; \hat{V}; \hat{f}; \lambda_1) = J_i \cdot \left[\Delta\hat{\theta}; \Delta\hat{V}; \Delta\hat{f}; \Delta\lambda_1\right]^T \tag{3.43}$$

If the goal is to pinpoint feasible power interchanges, the aforementioned distributed APF framework can be exploited to perform a traversal of power interchanges among microgrids.

3.4 Numerical Tests of Microgrid Power Flow

3.4.1 Validity of Individual Microgrid Power Flow

The effectiveness of EMPF is verified on a 33-bus microgrid with 5 DERs in Fig. 3.5. For comparison purposes, all system parameters are adopted from [15] except that $\mathbf{Z}_d = 0.001$. By flipping the five normally open switches, the microgrid configuration can be toggled from radial to meshed one. EMPF calculations are then performed on the radial microgrid (Test I) and the meshed microgrid (Test II). EMPF is implemented in Matlab on a 64-bit, 2.50 GHz PC.

3.4.1.1 Power Flow Results for Different Microgrid Configurations

Voltages obtained from Tests I and II are shown in Figs. 3.6 and 3.7, respectively. It can be observed that

- Results in Test I (radial microgrid) are identical to those in [15], which validates the correctness of EMPF.
- Generally, voltages in the meshed microgrid are smoother than those in the radial system. For instance, in the droop mode (EMPF_DP), the voltage at bus 30 in the meshed system is 0.41% higher than that in the radial system. This is because DER 25, once the switch 25–29 is closed, will help boost the voltages at neighboring buses including buses 26–33.

Figure 3.5 The 33-bus islanded microgrid with 5 DERs (leader bus: 1).

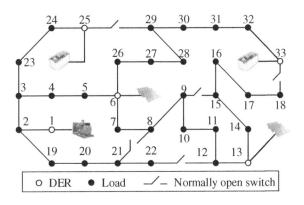

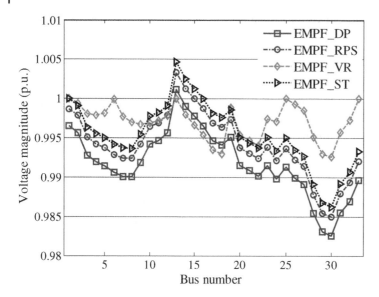

Figure 3.6 Test I: voltage magnitudes of radial microgrid.

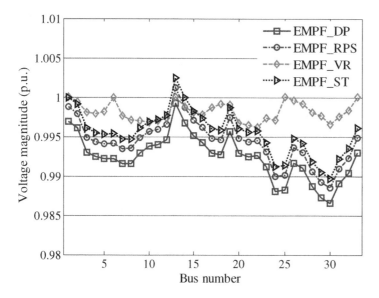

Figure 3.7 Test II: voltage magnitudes of meshed microgrid.

- Under EMPF_DP, however, the voltage at DER 13 in the meshed microgrid is lower than that in its radial counterpart because DER 13 has to supply heavy loads at buses 7 and 8 after the switches between 22–12 and 9–15 are closed.

3.4.1.2   Power Flow Results Under Various Control Modes

Table 3.1 summarizes DER power injections for both the radial and meshed microgrids under the four control modes. Table 3.2 presents the iteration performance for the radial and meshed microgrids

Table 3.1 Power injections from DERs (p.u.).

Test	DER#	DP	RPS	VR	ST
I	1	$2.50 + 0.97i$	$2.50 + 0.93i$	$2.50 - 0.90i$	$2.50 + 0.93i$
	6	$0.98 + 0.91i$	$0.98 + 0.93i$	$0.98 + 2.99i$	$0.98 + 0.93i$
	13	$1.70 + 0.89i$	$1.70 + 0.93i$	$1.70 + 0.01i$	$1.70 + 0.93i$
	25	$0.98 + 0.91i$	$0.98 + 0.93i$	$0.98 + 1.55i$	$0.98 + 0.93i$
	33	$1.30 + 0.95i$	$1.30 + 0.93i$	$1.30 + 0.99i$	$1.30 + 0.93i$
II	1	$2.50 + 0.96i$	$2.50 + 0.92i$	$2.50 - 1.18i$	$2.50 + 0.92i$
	6	$0.98 + 0.91i$	$0.98 + 0.92i$	$0.98 + 2.13i$	$0.98 + 0.92i$
	13	$1.70 + 0.91i$	$1.70 + 0.92i$	$1.70 - 0.22i$	$1.70 + 0.92i$
	25	$0.98 + 0.91i$	$0.98 + 0.92i$	$0.98 + 3.08i$	$0.98 + 0.92i$
	33	$1.30 + 0.94i$	$1.30 + 0.92i$	$1.30 + 0.93i$	$1.30 + 0.92i$

Table 3.2 CPU time and iteration numbers.

Parameter	DP(I)/(II)	RPS(I)/(II)	VR(I)/(II)	ST(I)/(II)
CPU time(s)	0.50/0.48	0.55/0.54	0.82/0.87	0.80/0.83
Iteration	5/4	10/10	16/16	15/15

- Generally, microgrid voltage profiles are improved by applying the secondary control, compared with those with droop control only. For instance, bus 27 voltage under the VR control is 0.9981 which is close to its rated value and is 1.44% better than that under DP mode only.
- In the RPS mode, the var injections from all DERs are equal because the follower buses share the same reactive power ratio with the leader bus. For instance, in Test I, the var injections of follower DERs 6, 13, 25, and 33 are 0.93 p.u. (base power: 500 kVA) which are equal to the var contribution from the leader bus 1. Therefore, EMPF can realize the proportional RPS.
- In the ST mode, the leader bus is controlled to fully restore its voltage, as shown in Figures 3.6 and 3.7. Meanwhile, the var contribution of each DERs is 0.93 and 0.92 p.u. for the radial and meshed microgrids, respectively, because in this mode the follower buses still follow the RPS mode.
- In the VR mode, the voltages at DER buses can be recovered to the nominal values. However, compared with the RPS and ST modes, it often leads to irregular PS among DERs. Therefore, it indicates that the VR mode is only feasible when DERs have adequate reactive power capacity.

3.4.2 Validity of Networked Microgrids Power Flow

In this section, we study the performance of APF using an eight-microgrid test system. The efficacy and versatility of the devised methods and the convergence performance will be thoroughly evaluated. As illustrated in Fig. 3.8, the eight-microgrid test system is established by interlinking and modifying the low voltage network of the IEEE 342-bus distribution system [21] into networked microgrids which consist of 4 meshed AC microgrids (Microgrids 1, 3, 4, and 6), 2 radial

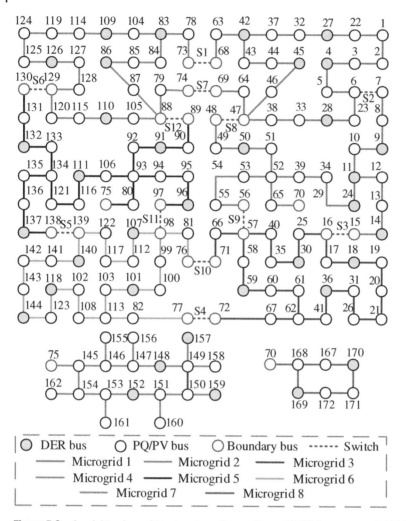

Figure 3.8 An eight-microgrid test system. Base voltage: 120 V. Base power: 5.15 MVA.

AC microgrids (Microgrids 2 and 5), and 2 DC microgrids (Microgrids 7 and 8). Table 3.3 shows the droop coefficients of DERs in the test system. The parameters of DC microgrids can be found in [7, 11].

3.4.2.1 APF Results Under Droop Control

We first validate the droop-based networked microgrid operations by opening all the switches. Table 3.4 presents the microgrid frequency and power outputs of partial DERs in Microgrids 1, 2, 3 and Microgrid 8.

The simulation results show that the generation and demand in each microgrid are balanced at a cost of deviated frequency and voltages. For example, the system frequency drops down to 0.9971 p.u. in Microgrid 3 when DERs boost real power injections to support loads and line losses (see Table 3.4). The voltage of bus 59 is decreased to 0.9966 p.u. for the increase of reactive power injections. The voltage of bus 170 declines to 0.998 p.u. to balance the active power consumption.

Table 3.3 Droop coefficients of DERs in the eight-microgrid system (p.u.).

Bus	4	45	28	42	27	11	9	14
m/n	1	1	1	1	0.8	0.8	1	1
Bus	24	50	54	18	59	36	30	109
m/n	1	1	1	0.8	1	1	1	1
Bus	83	86	110	126	132	137	91	96
m/n	1	1.5	0.9	1	1	1.2	1.5	0.8
Bus	111	107	118	101	140	144		
m/n	1	1	1	1	1	1		
Bus	157	159	169	170	148	152		
k	0.5	0.4	0.6489	0.6489	0.5	0.4		

Table 3.4 DER adjustments under droop control.

MG	Bus	Power of DERs	Initial power of DERs	Bus voltage	Frequency
1	4	$0.0281 + j0.0217$	$0.0276 + j0.0169$	0.9952	0.9995
	28	$0.0281 + j0.0195$	$0.0276 + j0.0169$	0.9974	
	45	$0.0281 + j0.0173$	$0.0276 + j0.0169$	0.9996	
2	11	$0.0399 + j0.0162$	$0.0388 + j0.0194$	1.0026	0.9991
	14	$0.0203 + j0.0114$	$0.0194 + j0.0117$	1.0002	
	24	$0.0203 + j0.0054$	$0.0194 + j0.0058$	1.0004	
3	59	$0.0363 + j0.0208$	$0.0334 + j0.0175$	0.9966	0.9971
	18	$0.0350 + j0.0264$	$0.0276 + j0.0233$	0.9976	
	36	$0.0330 + j0.0248$	$0.0276 + j0.0194$	0.9946	
8	169	0.0017	0	0.9989	—
	170	0.0031	0	0.9980	

The P/f and Q/V droop characteristics enable local microgrids to balance the power mismatches when each microgrid operates autonomously.

3.4.2.2 APF Results Under PS Control

Under PS control, we interconnect those 8 microgrids by closing all switches. To validate the capability of PS-based APF to maintain boundary buses' voltages, 100 stochastic scenarios are generated with randomized power loads in Microgrids 1 and 3.

Figure 3.9 compares the voltage amplitudes under PS control and droop control. It can be observed from Fig. 3.9 that when neighboring microgrids are interconnected, PS-based APF can adjust local boundary buses' voltages to achieve scheduled power interchanges. For example, the boundary bus 6 can always remain at 0.9932 p.u. under PS mode, regardless of how the local power consumption changes. In contrast, under droop control, the voltage of boundary buses can not be

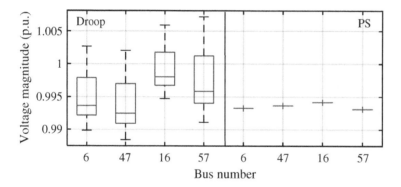

Figure 3.9 Comparison of boundary bus voltage amplitudes: droop control versus PS control.

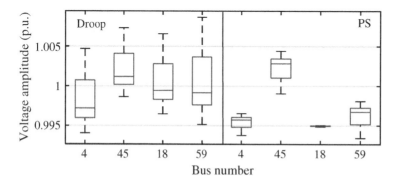

Figure 3.10 Comparison of DER bus voltage amplitudes: droop control versus PS control.

maintained. The voltage amplitudes of DER buses in Fig. 3.10 also show that the PS mode has a lesser effect on voltage variations of DER buses. For example, the voltage variation of bus 4 under the PS mode is smaller than that under droop control. Meanwhile, the voltage variation of bus 18 is smaller than other DER buses because of smaller droop coefficients.

Furthermore, Fig. 3.11 presents the voltage profiles under different power interchanges. Those results indicate that PS control can effectively relieve power deficiencies in individual microgrids during a utility side outage. For example, when Microgrid 2 requests for extra power ($S^{\text{In}} = 0.0086 + j0.0056$ p.u.) from neighboring microgrids, Microgrids 1 and 3 can boost boundary buses' voltages to transfer the scheduled power interchanges in Fig. 3.11.

3.4.2.3 APF Results Under VR Control

Figures 3.12 and 3.13 illustrate the efficacy of APF under VR control. Simulation results prove that under the VR mode, the leader DER voltages can be recovered to the nominal values through the coordination of power interchange between microgrids. For example, bus 4's voltage in Fig. 3.12 is adjusted to the nominal value by regulating the tie line S2 power to $-0.0035 + 0.0056i$ p.u. DER buses in DC microgrids also reach the nominal values, such as voltages at buses 169 and 170 in Microgrid 8 (see Fig. 3.13).

3.4.2.4 Convergence Performance of APF

Table 3.5 summarizes the convergence performance in different control modes.

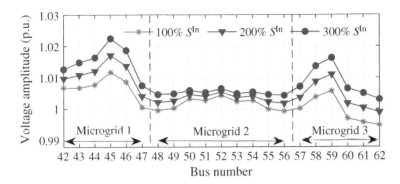

Figure 3.11 PS-APF voltage profiles under different power interchanges.

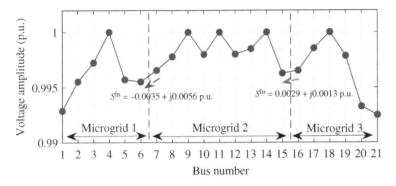

Figure 3.12 VR-APF-based bus voltages in AC microgrids.

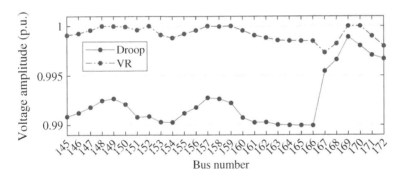

Figure 3.13 VR-APF-based bus voltages in DC microgrids.

Table 3.5 CPU time and iteration numbers.

	Grid	Droop	VR	PS0	PS1
Iter	AC1–6	4	44/51	4/4	5/5
	DC7	3	33/33	3/3	3/3
	DC8	3	3/3	3/3	3/3
Time(s)		—	0.34	62.1/71.24	3.02/2.24	2.93/2.54

3.5 Exercises

1. What are the advantages and disadvantages of Newton method and Gauss method for solving power flow?
2. Please explain the reasons why modified power flow methods are required for islanded microgrids with hierarchical control modes.
3. For the Newton-Raphson method to work, one should make sure the reciprocal of Jacobian matrix exists. (True/False)
4. How does the droop control affect the power flow results?
5. For the iterative solutions to droop-based microgrid power flow with the Newton-Raphson Method, the Jacobian matrix consists of different elements from the traditional one. Write down the elements of the different elements.
6. The number of iterations required for convergence is dependent/independent of the dimension of microgrids. Choose one.

References

1 Elrayyah, A., Sozer, Y., and Elbuluk, M.E. (2014). A novel load-flow analysis for stable and optimized microgrid operation. *IEEE Transactions on Power Delivery* 29 (4): 1709–1717. https://doi.org/10.1109/TPWRD.2014.2307279.

2 Lee, J., Kim, Y., and Moon, S. (2019). Novel supervisory control method for islanded droop-based AC/DC microgrids. *IEEE Transactions on Power Systems* 34 (3): 2140–2151.

3 Guerrero, J.M., Chandorkar, M., Lee, T.-L., and Loh, P.C. (2013). Advanced control architectures for intelligent microgrids–Part I: Decentralized and hierarchical control. *IEEE Transactions on Industrial Electronics* 60 (4): 1254–1262. https://doi.org/10.1109/TIE.2012.2194969.

4 Simpson-Porco, J.W., Shafiee, Q., Dörfler, F. et al. (2015). Secondary frequency and voltage control of islanded microgrids via distributed averaging. *IEEE Transactions on Industrial Electronics* 62 (11): 7025–7038. https://doi.org/10.1109/TIE.2015.2436879.

5 Mumtaz, F., Syed, M.H., Hosani, M.A., and Zeineldin, H.H. (2016). A novel approach to solve power flow for islanded microgrids using modified Newton Raphson with droop control of DG. *IEEE Transactions on Sustainable Energy* 7 (2): 493–503. https://doi.org/10.1109/TSTE.2015.2502482.

6 Kryonidis, G.C., Kontis, E.O., Chrysochos, A.I. et al. (2018). Power flow of islanded AC microgrids: revisited. *IEEE Transactions on Smart Grid* 9 (4): 3903–3905.

7 Allam, M.A., Hamad, A.A., and Kazerani, M. (2019). A sequence-component-based power-flow analysis for unbalanced droop-controlled hybrid AC/DC microgrids. *IEEE Transactions on Sustainable Energy* 10 (3): 1248–1261. https://doi.org/10.1109/TSTE.2018.2864938.

8 Garcés, A. (2018). On the convergence of Newton's method in power flow studies for DC microgrids. *IEEE Transactions on Power Systems* 33 (5): 5770–5777. https://doi.org/10.1109/TPWRS.2018.2820430.

9 Hamad, A.A., Azzouz, M.A., and El Saadany, E.F. (2016). A sequential power flow algorithm for islanded hybrid AC/DC microgrids. *IEEE Transactions on Power Systems* 31 (5): 3961–3970. https://doi.org/10.1109/TPWRS.2015.2504461.

10 Gupta, A., Doolla, S., and Chatterjee, K. (2018). Hybrid AC–DC microgrid: systematic evaluation of control strategies. *IEEE Transactions on Smart Grid* 9 (4): 3830–3843.

11 Eajal, A.A., Abdelwahed, M.A., El-Saadany, E.F., and Ponnambalam, K. (2016). A unified approach to the power flow analysis of AC/DC hybrid microgrids. *IEEE Transactions on Sustainable Energy* 7 (3): 1145–1158. https://doi.org/10.1109/TSTE.2016.2530740.

12 Aprilia, E., Meng, K., Al Hosani, M. et al. (2019). Unified power flow algorithm for standalone AC/DC hybrid microgrids. *IEEE Transactions on Smart Grid* 10 (1): 639–649. https://doi.org/10.1109/TSG.2017.2749435.

13 Feng, F. and Zhang, P. (2020). Enhanced microgrid power flow incorporating hierarchical control. *IEEE Transactions on Power Systems* 35 (3): 2463–2466. https://doi.org/10.1109/TPWRS.2020.2972131.

14 Díaz, G., Gómez-Aleixandre, J., and Coto, J. (2016). Direct backward/forward sweep algorithm for solving load power flows in ac droop-regulated microgrids. *IEEE Transactions on Smart Grid* 7 (5): 2208–2217. https://doi.org/10.1109/TSG.2015.2478278.

15 Ren, L. and Zhang, P. (2018). Generalized microgrid power flow. *IEEE Transactions on Smart Grid* 9 (4): 3911–3913. https://doi.org/10.1109/TSG.2018.2813080.

16 Feng, F. and Zhang, P. (2020). Implicit Z_{bus} Gauss algorithm revisited. *IEEE Transactions on Power Systems* 35 (5): 4108–4111. https://doi.org/10.1109/TPWRS.2020.3000658.

17 Feng, F., Zhang, P., Zhou, Y., and Wang, L. (2022). Distributed networked microgrids power flow. *IEEE Transactions on Power Systems* 38 (2): 1405–1419. https://doi.org/10.1109/TPWRS.2022.3175933.

18 Zhang, P. (2021). *Networked Microgrids*. Cambridge: Cambridge University Press.

19 Zimmerman, R.D. (1995). Comprehensive distribution power flow: modeling, formulation, solution algorithms and analysis. PhD thesis. Cornell University New York.

20 Wang, L., Qin, Y., Tang, Z., and Zhang, P. (2020). Software-defined microgrid control: the genesis of decoupled cyber-physical microgrids. *IEEE Open Access Journal of Power and Energy* 7: 173–182. https://doi.org/10.1109/OAJPE.2020.2997665.

21 Schneider, K., Phanivong, P., and Lacroix, J.-S. (2014). IEEE 342-node low voltage networked test system. *2014 IEEE PES General Meeting — Conference Exposition*, 1–5. https://doi.org/10.1109/PESGM.2014.6939794.

4

State and Parameter Estimation for Microgrids

Yuzhang Lin, Yu Liu, Xiaonan Lu, and Heqing Huang

4.1 Introduction

In traditional power grids, energization from generation segment to end users is implemented in a centralized architecture. The major power generation resides in the "upstream" substations and provides power through transmission systems to the "downstream" loads across distribution systems. The operational resiliency of such a centralized structure has been challenged in recent decades. In 2019, Argentina was struck by a nationwide blackout due to the undesired operation of two 500 kV transmission lines [1]. In 2021, the historic winter storm in Texas left millions of customers without access to electricity for days due to the failure of major renewable generations [2]. In traditional power grids, only portions of the feeder could be affected by the outages and could be re-energized by the adjacent feeders through tie-lines. However, during extreme events where the upstream substations are unavailable, the entire distribution system may lose energization and be forced into blackout. The requirements for grid modernization and resiliency enhancement calls for localized and aggregated energy systems, such as microgrids, which can enhance the survivability of critical infrastructures.

In addition to grid operation resiliency enhancement, distribution system operation in the context of microgrids also represents a feasible approach to mitigating the power quality issues brought about by the increasing penetration of renewable energy sources. In traditional power grids, renewable energy generation is primarily installed within a power plant (e.g. wind/photovoltaic farm). Due to the intermittent nature of renewable energy resources, operational issues (e.g. voltage and frequency fluctuations, and harmonics) and planning challenges (e.g. limited renewable hosting capacity, and renewable energy resource siting) could be introduced across the system when the penetration level of renewable energy increases [3]. However, by adopting microgrids, the power balance between generation and consumption is achieved locally. Rather than being lumped at a single location and delivering power through long transmission lines, renewable generation can be distributed across the feeder and support local loads (e.g. residential houses). The reduced power loss and increased controllability make the utilization of MG a feasible option for future renewable integration.

It is worth mentioning that the need for situational awareness of microgrids poses significant interests in enhancing the monitoring of microgrids as well as the inverter-based resources (IBRs) therein. State estimation, as originally designed for bulk power grids and bulk generators, can be adjusted to address the need for increased visibility in microgrids. It is also noteworthy that the state

Microgrids: Theory and Practice, First Edition. Edited by Peng Zhang.
© 2024 The Institute of Electrical and Electronics Engineers, Inc. Published 2024 by John Wiley & Sons, Inc.

estimation should consider the interactions between the physical networks, controllers, and also communication networks, which provides future research directions that extend the applications of microgrids.

Also, similar to the case of bulk power grids, accurate models are one of the prerequisites for the security assessment and control of microgrids. However, as microgrid components are typically manufactured and owned by different entities, and system operating conditions are ever-varying, system operators do not often have complete and accurate information about the models under normal and fault conditions. This makes model parameter estimation an important task. With enough measurements, it is possible to estimate or calibrate the values of model parameters facilitating more effective operation and control of microgrids.

In this chapter, we will introduce state-of-the-art technologies for estimating the states and model parameters of microgrids. The most important components of microgrids are IBRs and network components such as transmission lines. The state and parameter estimation methods for IBRs and network components will be introduced in Sections 4.2 and 4.3, respectively.

4.2 State and Parameter Estimation for Inverter-Based Resources

4.2.1 Background and Motivation

IBRs, such as photovoltaics, wind power, and energy storage, serve as main power sources for microgrids, especially during their disconnection from a main grid due to a blackout. IBRs have significantly different dynamic characteristics than conventional synchronous generators (SGs). First, the dynamics of the IBRs are faster and have little inherent inertia [4]. Second, the dynamics of IBRs are heavily determined by their controllers, which are becoming increasingly diverse [5]. Third, IBRs create new mechanisms of instability, such as protection actions caused by DC-link voltage transients or phase-locked loop (PLL) [6]. Proper modeling and monitoring of IBRs' dynamic behaviors can help understand microgrid operating conditions and develop more secure and efficient operational solutions.

With the increasing deployment of phasor measurement units (PMUs), dynamic state estimation (DSE) has become a very popular technique for the monitoring of bulk power grids [7]. A variety of DSE methods have been proposed for tracking the dynamic states of SGs, such as extended Kalman filter (EKF), [8], unscented Kalman filter (UKF) [9], cubature Kalman filter (CKF) [10], ensemble Kalman filter (EnKF) [11], and particle filtering (PF) [12], to name a few.

Compared with the large volume of research on DSE for SGs, the research on DSE for IBRs remains relatively immature. References [13, 14] uses UKF and unscented particle filtering (UPF) for the DSE of doubly fed induction generator-based wind turbines (DFIG-WTs). Reference [15] proposes an EKF with unknown inputs (EKF-UI) to solve the DSE problem for DFIGs in the case of unknown mechanical input torque. References [16, 17] use EKF and EnKF to observe the dynamics of permanent magnet synchronous generator-based wind turbines (PMSG-WTs). Reference [18] proposes an adaptive cubature Kalman filter (ACKF) method to perform DSE on photovoltaic generation systems. A holistic WLS-form EKF-based estimation paradigm for grid state and battery state of charge (SoC) is proposed in [19]. Reference [20] proposes a DSE framework using the EKF method combined with consensus control for microgrid systems containing multiple battery energy storage systems. Although the DSE for IBRs remains an emerging topic requiring further investigation, it has already inspired many novel applications in the fields of power system modeling [21], monitoring [22], control [23], and protection [24].

Despite the increasing popularity of DSE for IBRs, existing methods largely follow the well-established paradigm of DSE for conventional SGs. However, a critical difference to be noted is that the dynamics of IBRs are heavily determined by controls. Unlike the state-space models of SGs which primarily reflect electrical or mechanical processes, a large part of IBR's state-space models represent the digital computation of controllers. In other words, the dynamics of IBRs are a heavy mix of physical dynamics and digital dynamics (i.e. cyber dynamics). Reference [25] notes this critical challenge but does not fully resolve it. Applying the classical DSE paradigm for SGs with a single state-space model bears several essential limitations.

(1) As the physical states and the digital states are blended in a single set of equations, the uncertainties of the two-way data flow between the inverter system and the controller (i.e. noise and bad data in measurement signals and control signals) cannot be explicitly modeled or addressed.
(2) With the blended state-space equations, the change of either the inverter physics or the control logic will affect all state variables and output signals, therefore cyber events (e.g. controller failures or attacks) and physical events (e.g. electric circuit faults) cannot be distinguished.
(3) The controls of IBRs are highly diverse. With the blended state-space equations, the DSE methods have to modify the entire algorithm when the control strategy changes. Note that even the same IBR may switch between different control modes, making it difficult to develop and implement DSE algorithms.

Then, we will present a cyber-physical dynamic state estimation (CPDSE) framework as a unified DSE paradigm for IBRs to address the challenges mentioned above. The framework can be readily extended to cyber-physical dynamic parameter estimation (CPDPE) for model calibration, which will also be briefly discussed.

4.2.2 Overview of CPDSE Framework

An IBR system can be generally divided into a physical subsystem and a control (cyber) subsystem.

(1) **Physical subsystem**: The physical subsystem of an IBR typically includes: a power source, such as a wind turbine, a solar PV panel, or a battery storage system; a power electronic interface, which can take the form of DC/AC, DC/DC/AC, or AC/DC/AC conversion; a DC-link capacitor/inductor; an AC-side filter.
(2) **Control (cyber) subsystem**: Digital controllers perform computation based on measurement signals and generate control signals to operate power electronic devices. They can be considered as the cyber subsystem of an IBR, as they involve information processing rather than physical processes. IBR control algorithms can be generally divided into grid following (GFL) controls and grid forming (GFM) controls [26, 27]. Both GFL and GFM controls can take a variety of forms based on the need for energy generation/conversion as well as microgrid operational needs.

Figure 4.1 presents the symmetrical structure of the cyber and physical subsystems of an IBR. Measurement and control signals are the data flows between the two subsystems – the inputs of one subsystem are the outputs of the other.

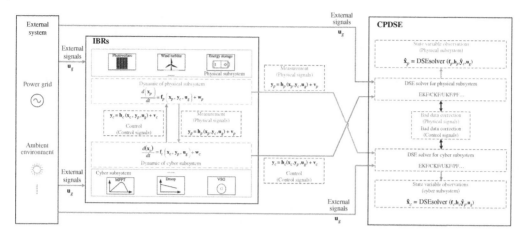

Figure 4.1 Generic cyber-physical state-space representation of IBRs and the CPDSE paradigm.

4.2.2.1 Cyber-Physical State-Space Representation of IBRs

With the cyber-physical representation shown in Fig. 4.1, an IBR's dynamic behavior can be represented by the following four sets of equations:

(1) State transition equations of the physical subsystem

The state transition equations of the physical subsystem are given by:

$$\frac{d\left(\mathbf{x}_p\right)}{dt} = \mathbf{f}_p\left(\mathbf{x}_p, \mathbf{y}_c, \mathbf{u}_g\right) + \mathbf{w}_p \tag{4.1}$$

where $\mathbf{x}_p \in \mathbb{R}^{n_p \times 1}$ is the state vector of the physical subsystem; $\mathbf{y}_c \in \mathbb{R}^{m_c \times 1}$ is the control vector from the cyber subsystem; $\mathbf{u}_g \in \mathbb{R}^{l \times 1}$ is the external input vector such as control references or ambient conditions; $\mathbf{w}_p \in \mathbb{R}^{n_p \times 1}$ is the process noise vector of the physical subsystem representing the uncertainty of state transition due to model parameter errors, and harmonics.

(2) Output equations of the physical subsystem

Measurement signals are the outputs of the physical subsystem, described by the following output equations:

$$\mathbf{y}_p = \mathbf{h}_p\left(\mathbf{x}_p, \mathbf{y}_c, \mathbf{u}_g\right) + \mathbf{v}_p \tag{4.2}$$

where $\mathbf{y}_p \in \mathbb{R}^{m_p \times 1}$ is the output vector of the physical subsystem, i.e. the measurement vector; $\mathbf{v}_p \in \mathbb{R}^{m_p \times 1}$ is the output noise vector of the physical subsystem representing the uncertainty of measurement signals due to sensing noise, drifts, or failures.

(3) State transition equations of the cyber subsystem

The state transition equations of the cyber subsystem are given by:

$$\frac{d\left(\mathbf{x}_c\right)}{dt} = \mathbf{f}_c\left(\mathbf{x}_c, \mathbf{y}_p, \mathbf{u}_g\right) + \mathbf{w}_c \tag{4.3}$$

where $\mathbf{x}_c \in \mathbb{R}^{n_c \times 1}$ is the state vector of the cyber subsystem; $\mathbf{w}_c \in \mathbb{R}^{m_c \times 1}$ is the process noise of the cyber subsystem representing the uncertainty of state transition due to discretization, computation error, and cyber attacks. Clearly, the dynamics of the cyber subsystem are influenced by the measurement signals from the physical subsystem, $\mathbf{y}_p$.

(4) Output equations of the cyber subsystem

Control signals are the outputs of the cyber subsystem, given by the following output equations:

$$\mathbf{y}_c = \mathbf{h}_c\left(\mathbf{x}_c, \mathbf{y}_p, \mathbf{u}_g\right) + \mathbf{v}_c \tag{4.4}$$

where $\mathbf{v}_c \in^{m_c \times 1}$ is the output noise vector of the cyber subsystem representing the uncertainty of control signals due to latency, actuation imperfection, or cyber attacks.

4.2.2.2 Comparison with Conventional Single-State-Space Representation

From (4.1) to (4.4), it is clear that the uncertainties of the output of each subsystem, represented by $\mathbf{v}_p$ and $\mathbf{v}_c$, respectively, will affect the state transition of the other subsystem. With a single state-space representation mixing the physical and digital state variables, such uncertainty cannot be represented.

$$\frac{d\left(\mathbf{x}\right)}{dt} = \mathbf{f}\left(\mathbf{x}, \mathbf{u}_g\right) + \mathbf{w} \tag{4.5}$$

$$\mathbf{y} = \mathbf{h}\left(\mathbf{x}, \mathbf{u}_g\right) + \mathbf{v} \tag{4.6}$$

where $x = \left[\left(\mathbf{x}_p\right)^T, \left(\mathbf{x}_c\right)^T\right]^T$, and $y = \left[\left(\mathbf{y}_p\right)^T, \left(\mathbf{y}_c\right)^T\right]^T$. For comparison purposes, Eqs. (4.5) and (4.6) can be rewritten to split the equations dictating the cyber and physical variables:

$$\frac{d\left(\mathbf{x}_c\right)}{dt} = \mathbf{f}'_c\left(\mathbf{x}_c, \mathbf{x}_p, \mathbf{u}_g\right) + \mathbf{w}'_c \tag{4.7}$$

$$\frac{d\left(\mathbf{x}_p\right)}{dt} = \mathbf{f}'_p\left(\mathbf{x}_c, \mathbf{x}_p, \mathbf{u}_g\right) + \mathbf{w}'_p \tag{4.8}$$

$$\mathbf{y}_c = \mathbf{h}'_c\left(\mathbf{x}_c, \mathbf{y}_p, \mathbf{u}_g\right) + \mathbf{v}'_c \tag{4.9}$$

$$\mathbf{y}_p = \mathbf{h}'_p\left(\mathbf{x}_c, \mathbf{y}_p, \mathbf{u}_g\right) + \mathbf{v}'_p \tag{4.10}$$

For the described model (4.7)–(4.10), substituting (4.10) into (4.7) and (4.9) into (4.8) will yield:

$$\frac{d\left(\mathbf{x}_p\right)}{dt} = \mathbf{f}_p\left(\mathbf{x}_p, \left(\mathbf{h}_c\left(\mathbf{x}_c, \mathbf{y}_p, \mathbf{u}_g\right) + \mathbf{v}_c\right), \mathbf{u}_g\right) + \mathbf{w}_p \tag{4.11}$$

$$\frac{d\left(\mathbf{x}_c\right)}{dt} = \mathbf{f}_c\left(\mathbf{x}_c, \left(\mathbf{h}_p\left(\mathbf{x}_p, \mathbf{y}_c, \mathbf{u}_g\right) + \mathbf{v}_p\right), \mathbf{u}_g\right) + \mathbf{w}_c \tag{4.12}$$

Comparing the conventional state-space representation (4.7)–(4.8) with the cyber-physical state-space representation (4.11)–(4.12), the most significant difference is that the conventional model (4.7)–(4.8) does not account for the uncertainties of the output signals, $\mathbf{v}_p$ and $\mathbf{v}_c$, and the state variables of the two subsystems affect each other directly. The cyber-physical representation (4.11)–(4.12) is much closer to reality where the two subsystems interact with each other through the measurement signals and the control signals $\mathbf{y}_p$ and $\mathbf{y}_c$ and the uncertainty $\mathbf{v}_p$ and $\mathbf{v}_c$ will enter the loop to affect state transition equations.

Besides the capability to address data flow uncertainties between the cyber and physical subsystems, the cyber-physical state-space model has other advantages. As each equation in (4.3)–(4.6) only involves the model of one subsystem, it is easier to check the inconsistency between the models and the outputs, such that cyber and physical events can be easily distinguished.

Furthermore, when the control algorithm changes, only $\mathbf{f}_c$ and $\mathbf{h}_c$ need to be changed, and $\mathbf{f}_p$ and $\mathbf{h}_p$ characterizing the physical subsystem dynamics remain valid. This would allow greater versatility and adaptability of the resulting DSE algorithm to various control algorithms, especially in the case of control mode changes in the same IBR.

4.2.2.3 CPDSE and CPDPE for IBRs

For microgrid operation, the states of the physical devices and the digital controllers should both be estimated as they are largely unattainable for the grid operators. Therefore, the objective of DSE for IBRs is to track the state variables of both the cyber subsystem and the physical subsystem in the presence of regular noise and even bad data in the outputs of the two subsystems, i.e. in the measurement signals and control signals. Noting the symmetry between the physical subsystem and the cyber subsystem, two dual estimators – a physical dynamic state estimator (PDSE) and a cyber dynamic state estimator (CDSE) – can be used to track the states of the two subsystems, respectively. At time step k, the generic form of CPDSE can be expressed as follows:

$$\text{PDSE: } \hat{\mathbf{x}}_{p(k)} = \text{DSESolver}\left(\mathbf{f}_p, \mathbf{h}_p, \hat{\mathbf{x}}_{p(k-1)}, \mathbf{y}_{c(k)}, \mathbf{u}_{g(k)}, \mathbf{y}_{p(k)}\right) \tag{4.13}$$

$$\text{CDSE: } \hat{\mathbf{x}}_{c(k)} = \text{DSESolver}\left(\mathbf{f}_c, \mathbf{h}_c, \hat{\mathbf{x}}_{c(k-1)}, \mathbf{y}_{p(k)}, \mathbf{u}_{g(k)}, \mathbf{y}_{c(k)}\right) \tag{4.14}$$

The two estimators form an elegant dual structure. PDSE aims to obtain the estimates of the physical subsystem state variables at time k, $\hat{\mathbf{x}}_{p(k)}$. The required information includes the state transition function and output function, $\mathbf{f}_p$ and $\mathbf{h}_p$, respectively; external input vector $\mathbf{u}_{g(k)}$; the state estimate vector of the previous step $\hat{\mathbf{x}}_{p(k-1)}$; the output vector of the physical subsystem $\mathbf{y}_{p(k)}$, and the output vector of cyber subsystem $\mathbf{y}_{c(k)}$. CDSE plays a dual role where functions/variables are replaced by their counterparts in the other subsystem, except for external input vector $\mathbf{u}_{g(k)}$ required by both subsystems. It should be noted that although the outputs of both subsystems, $\mathbf{y}_{p(k)}$ and $\mathbf{y}_{c(k)}$, appear in both PDSE and CDSE, their roles are opposite in the two estimators. In PDSE, measurement signals $\mathbf{y}_{p(k)}$ are the outputs of the subsystem of concern (physical), and control signals $\mathbf{y}_{c(k)}$ are the inputs. In CDSE, control signals $\mathbf{y}_{p(k)}$ are the outputs, and measurement signals $\mathbf{y}_{c(k)}$ are the inputs. Each estimator checks the consistency between the model and the outputs of the respective subsystem of concern; hence, PDSE/CDSE suppresses the noise and detects the bad data in the measurement/control signals, respectively.

The described CPDSE framework can be extended to perform CPDPE. The models of IBRs must accurately represent their dynamic behaviors such that security assessment can be conducted effectively. CPDPE can be used as a tool to calibrate the model of IBRs. The parameters to calibrate can be treated as augmented states to be estimated.

Denote the cyber and physical parameter vectors as $\boldsymbol{\tau}_c \in \mathbb{R}^{q_c \times 1}$ and $\boldsymbol{\tau}_p \in \mathbb{R}^{q_p \times 1}$, respectively. The physical parameter vector $\boldsymbol{\tau}_p$ can include machine inertia, inductance, and capacitance. The cyber parameter vector $\boldsymbol{\tau}_c$ can include PI gains, and droop control parameters. The augmented states of the two cyber and physical subsystems can be expressed as $\tilde{\mathbf{x}}_c = \left[\mathbf{x}_c, \boldsymbol{\tau}_c\right]$ and $\tilde{\mathbf{x}}_p = \left[\mathbf{x}_p, \boldsymbol{\tau}_p\right]$, respectively. The state transition equations can be augmented to include those for the parameters:

$$\frac{d\left(\boldsymbol{\tau}_p\right)}{dt} = \mathbf{0} + \mathbf{w}_{\tau p}, \quad \frac{d\left(\boldsymbol{\tau}_c\right)}{dt} = \mathbf{0} + \mathbf{w}_{\tau c} \tag{4.15}$$

where $\mathbf{w}_{\tau c} \in \mathbb{R}^{q_c \times 1}$ and $\mathbf{w}_{\tau p} \in \mathbb{R}^{q_p \times 1}$ are process noise vectors of the cyber and physical parameters, respectively. The estimation of physical parameters can be done by replacing $\mathbf{f}_p$ by $\tilde{\mathbf{f}}_p = \left[\mathbf{f}_p, \mathbf{0}\right]$, and replacing $\hat{\mathbf{x}}_{p(k-1)}$ by $\hat{\tilde{\mathbf{x}}}_{p(k-1)}$ in (4.13). Similarly, the estimation of cyber parameters can be done by replacing $\mathbf{f}_c$ by $\tilde{\mathbf{f}}_c = \left[\mathbf{f}_c, \mathbf{0}\right]$, and replacing $\hat{\mathbf{x}}_{c(k-1)}$ by $\hat{\tilde{\mathbf{x}}}_{c(k-1)}$ in (4.14).

The described CPDSE and CPDPE frameworks are generic and applicable to any type of IBR with various power sources, power electronic converters, and control algorithms as well as various choices of estimation algorithms. For the rest of the section, the discussion will focus on CPDSE but can be readily extended to CPDPE. In section 4.2.5, specific examples of both cyber and physical subsystems of IBRs will be given.

4.2.3 Examples of Cyber-Physical State-Space Models

This section will exemplify the construction of the cyber-physical state-space model with specific IBR systems. The physical subsystem is chosen as a three-phase photovoltaics (PV) generation system. For the cyber subsystem, we choose virtual synchronous generator (VSG), a common GFM control. Throughout the section, we use "z," "c," "x," "w," and "v" as superscripts denoting measurement signals, control signals, external signals, process noises, and output noises, respectively. State variables and model parameters are superscript-free.

4.2.3.1 Physical State-Space Model

A common structure of PV generation systems is shown in Fig. 4.2. It consists of a DC-link capacitor, a two-level three-phase insulated gate bipolar transistor (IGBT) inverter, and an AC-side LCL filter. The parameters of the physical model include the DC-link capacitor C_{dc}, AC-side filter resistances, and inductances R_i, L_i, R_g, L_g, R_c, C_f. The state variables of the physical subsystem are defined as $\mathbf{x}_p = \left[V_{dc}, I_{id}, I_{iq}, I_{gd}, I_{gq}, V_{cd}, V_{cq}\right]^T$, where V_{dc} is the voltage of DC-link capacitor; $I_{id}, I_{iq}, I_{gd}, I_{gq}, V_{cd}$, and V_{cq} are the dq-axis currents and voltages associated with the AC-side filter, respectively. The external input signals are defined as $\mathbf{u}_g = \left[V_{gd}^{ex}, V_{gq}^{ex}, I_{dc}^{ex}, \omega_g^{ex}\right]^T$, where V_{gd}^{ex} and V_{gq}^{ex} are the dq-axis voltages on the grid side; I_{dc}^{ex} is DC-link capacitor current; ω_g^{ex} is the system angular frequency. The input signals from the cyber subsystem are defined as $\mathbf{y}_c = \left[V_{id}^c, V_{iq}^c\right]^T$, where V_{id}^c and V_{iq}^c are dq-axis voltages to be generated by the IGBT inverter.

The state transition equations of the physical subsystem are derived as (4.16)–(4.22) [28] in Table 4.1. The outputs of the physical subsystem (i.e. measurement signals) can be defined as $\mathbf{y}_p = \left[I_{gd}^z, I_{gq}^z, V_{dc}^z\right]^T$, where V_{dc}^z is the DC-link capacitor voltage measurement; I_{gd}^z and I_{gq}^z are the dq-axis current measurements on the grid side are derived as (4.23)–(4.25).

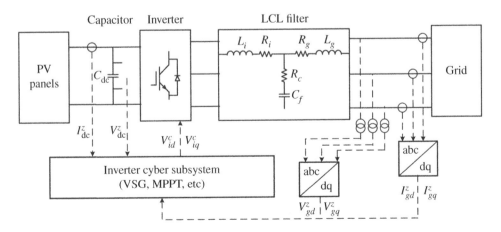

Figure 4.2 Structure of IBRs.

Table 4.1 Cyber-physical state-space of a PV generation system (VSG Mode).

Physical subsystem state transition equations

$$\frac{dV_{dc}}{dt} = \frac{1}{C_{dc}V_{dc}}\left(V_{dc}I_{dc}^z - V_{iq}^c I_{iq} - V_{id}^c I_{id}\right) + V_{dc}^w \tag{4.16}$$

$$\frac{dI_{id}}{dt} = \frac{1}{L_i}\left(V_{id}^c - R_i I_{id} - V_{cd} - \hat{\omega}L_i I_{iq}\right) + I_{id}^w \tag{4.17}$$

$$\frac{dI_{iq}}{dt} = \frac{1}{L_i}\left(V_{iq}^c - R_i I_{iq} - V_{cq} - \omega_g^{ex}L_i I_{id}\right) + I_{iq}^w \tag{4.18}$$

$$\frac{dI_{gd}}{dt} = \frac{1}{L_g}\left(V_{cd} - R_g I_{gd} - V_{gd}^z + \omega_g^{ex}L_g I_{gq}\right) + I_{gd}^w \tag{4.19}$$

$$\frac{dI_{gq}}{dt} = \frac{1}{L_g}\left(V_{cq} - R_g I_{gq} - V_{gq}^z - \omega_g^{ex}L_g I_{gd}\right) + I_{gq}^w \tag{4.20}$$

$$\frac{dV_{cd}}{dt} = -\left(\frac{R_c}{L_i} + \frac{R_c}{L_g}\right)V_{cd} + \left(\frac{1}{C_f} - \frac{R_c}{L_i}R_i\right)I_{id}$$
$$-\left(\frac{1}{C_f} - \frac{R_c}{L_g}R_g\right)I_{gd} + \frac{R_c}{L_i}V_{id}^c + \frac{R_c}{L_g}V_{gd}^z + \omega_g^{ex}V_{cq} + V_{cd}^w \tag{4.21}$$

$$\frac{dV_{cq}}{dt} = -\left(\frac{R_c}{L_i} + \frac{R_c}{L_g}\right)V_{cq} + \left(\frac{1}{C_f} - \frac{R_c}{L_i}R_i\right)I_{iq}$$
$$-\left(\frac{1}{C_f} - \frac{R_c}{L_g}R_g\right)I_{gq} + \frac{R_c}{L_i}V_{iq}^c + \frac{R_c}{L_g}V_{gq}^z - \omega_g^{ex}V_{cd} + V_{cq}^w \tag{4.22}$$

Physical subsystem state transition equations

$$I_{gd}^z = I_{gd} + I_{gd}^v \tag{4.23}$$

$$I_{gq}^z = I_{gq} + I_{gq}^v \tag{4.24}$$

$$V_{dc}^z = V_{dc} + V_{dc}^v \tag{4.25}$$

Cyber subsystem state transition equations

$$\frac{dP_{set}}{dt} = \frac{-k_{gp}\left(\omega_g^{ex} - \omega_0\right) - \left(P_{set} - P_0\right)}{T_d} + P_{set}^w \tag{4.26}$$

$$\frac{d\omega}{dt} = \frac{1}{J}\left(\frac{P_{set}}{\omega} + D_P\left(\omega_g^{ex} - \omega\right) - \frac{V_{gd}^{ex}I_{gd}^z + V_{gq}^{ex}I_{gq}^z}{\omega}\right) + \omega^w \tag{4.27}$$

$$\frac{d\Delta\theta}{dt} = \omega - \omega_g^{ex} + \Delta\theta^w \tag{4.28}$$

$$\frac{dV_Q}{dt} = k_{qi}\left(Q_c^{ex} - \left(V_{gq}^{ex}I_{gd}^z - V_{gd}^{ex}I_{gq}^z\right)\right) + V_Q^w \tag{4.29}$$

$$\frac{d\varphi_f}{dt} = \frac{1}{K}\left(k_{qp}\left(Q_c^{ex} - \left(V_{gq}^{ex}I_{gd}^z - V_{gd}^{ex}I_{gq}^z\right)\right)\right.$$
$$\left. + V_Q + D_q\left(\sqrt{\left(V_{gq}^{ex}\right)^2 + \left(V_{gd}^{ex}\right)^2} - \varphi_f\omega\right)\right) + \varphi_f^w \tag{4.30}$$

Cyber subsystem state transition equations

$$V_{id}^c = \omega\varphi_f\cos\left(\Delta\theta\right) + V_{id}^v \tag{4.31}$$

$$V_{iq}^c = \omega\varphi_f\sin\left(\Delta\theta\right) + V_{iq}^v \tag{4.32}$$

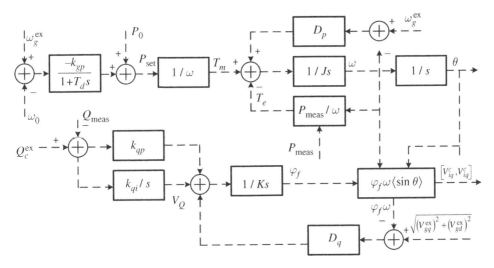

Figure 4.3 Grid-forming control example: VSG-based control.

4.2.3.2 Cyber State-Space Model

VSG adjusts the active and reactive power outputs according to the fluctuation of system voltage and frequency by mimicking the electromechanical characteristics of an SG, i.e. the swing equations. The control block diagram of VSG is shown in Fig. 4.3. ω_0 and P_0 are the control parameters for the $P - f$ droop; k_{gp} and $T - d$ characterize the delayed response of governor control; J and D_p simulate SG inertia and damping coefficient, respectively.

Define the state variables as $\mathbf{x}_c = \left[P_{set}, \omega, \Delta\theta, V_Q, \varphi_f \right]^T$, where the active power setpoint P_{set} is computed through the $P - f$ droop characteristic with a delay; ω is the virtual rotor speed; $\Delta\theta$ is the virtual power angle; V_Q is the integral output of reactive power control; φ_f mimics the flux linkage of a SG. Define the external signals as $\mathbf{u}_g = \left[\omega_g^{ex}, Q_c^{ex}, V_{gd}^{ex}, V_{gq}^{ex} \right]^T$. The state transition equations are derived as (4.26)–(4.30) in Table 4.1. Equation (4.26) is a delayed $P - f$ droop mimicking governor control. Eqs. (4.27)–(4.28) mimic the swing equations. Equations (4.29)–(4.30) mimics excitation system response. The output signals of the cyber subsystem are defined as $\mathbf{y}_c = \left[V_{id}^c, V_{iq}^c \right]^T$. The output equations are given by (4.31)–(4.32).

Although VSG is taken as the example, various types of IBR systems can be modeled in the generic cyber-physical state-space formulation, including conventional GFL controls such as maximum-power-point-tracking (MPPT) control.

4.2.4 CKF for Dynamic State Estimation and Bad Data Processing

This section will present a detailed algorithm for materializing the generic CPDSE framework described in Section 4.2.2.3. The algorithm can be used in any other IBR systems once their cyber-physical state-space models are derived. CKF is adopted as the DSE solver to propagate uncertainties with high accuracy in nonlinear systems. A bad data detection mechanism based on largest normalized residual (LNR) will also be described.

With the elegant duality of the cyber and physical subsystems, the CKF-based CDSE and PDSE algorithms can be presented in a uniform fashion. Denote a subscript index set $\Omega = \{p, c\}$, where "p" represents variables associated with the physical subsystem, and "c" represents those associated with the cyber subsystem.

Suppose $\phi \in \Omega$, and $\overline{\phi}$ is the complement of ϕ on Ω. The discretized state transition equations (derived from the continuous-time differential Eqs. (4.16)–(4.32)) and output equations of a subsystem can be expressed as follows:

$$
\begin{aligned}
\mathbf{x}_{\phi(k)} &= \mathbf{f}_\phi\left(\mathbf{x}_{\phi(k-1)}, \mathbf{y}_{\overline{\phi}(k-1)}, \mathbf{u}_{g(k-1)}\right) + \mathbf{w}_{\phi(k)}, \\
\mathbf{w}_{\phi(k)} &= N\left(\mathbf{0},\ \mathbf{Q}_{\phi(k)}\right)
\end{aligned}
\tag{4.33}
$$

$$
\begin{aligned}
\mathbf{y}_{\phi(k)} &= \mathbf{h}_\phi\left(\mathbf{x}_{\phi(k)}, \mathbf{y}_{\overline{\phi}(k)}, \mathbf{u}_{g(k)}\right) + \mathbf{v}_{\phi(k)}, \\
\mathbf{v}_{\phi(k)} &= N\left(\mathbf{0},\ \mathbf{R}_{\phi(k)}\right)
\end{aligned}
\tag{4.34}
$$

For a n-dimensional state space, the CKF algorithm uses the third-order spherical radial rule to generate a set of $2n$ cubature points and weighs them equally to approximate the probability distribution propagating through a nonlinear function, as described in [10]. The equations of the CKF-based CPDSE are as shown in Table 4.2.

Step 0: Initialization. Set $k = 1$. The estimated states $\hat{\mathbf{x}}_{\phi(0)}$ and the error variance $\mathbf{P}_{0|0}$ are initialized (4.35)–(4.36).

Step 1: State prediction. Using (4.33), the *a priori* state estimate can be obtained as (4.37), where the $\hat{\mathbf{x}}_{\phi(k-1)}$ is the *a posteriori* state estimate of the system at time k; $\mathbf{P}_{k|k}$ is the estimated error variance matrix of the corresponding state variables; $\chi_{\phi(i,k-1)}$ and $\chi^*_{\phi(i,k)}$ are the sampling point before and after propagating through the state equations. The *a priori* state estimate $\hat{\mathbf{x}}_{\phi(k|k-1)}$ and error covariance matrix $\mathbf{P}_{k|k-1}$ at time k are obtained as (4.38)–(4.39).

Step 2: Output prediction. Using (4.34), the predicted outputs and their autocovariance/covariance matrices are obtained as (4.40)–(4.42), where $\mathbf{y}_{\phi(i,k)}$ is a volume point corresponding to the predicted output at time k; $\hat{\mathbf{y}}_{\phi(k|k-1)}$ is the predicted output at time k; $\mathbf{P}_{yy,k}$ and $\mathbf{P}_{xy,k}$ are the autocovariance and covariance matrices of the predicted outputs, respectively.

Step 3: Kalman gain computation and state correction. The Kalman gain $\mathbf{K}_k$ at time k is evaluated as (4.43). Then, we make corrections to the *a priori* state estimates through the difference between measurements and their predicted values. The *a posteriori* state estimate and error covariance matrix are obtained as (4.44)–(4.45).

Step 4: Bad data detection. Bad data detection and correction can be achieved by the LNR approach [29]. Normalized residuals can be evaluated as (4.47). The LNR can be found (4.48). If the absolute value of LNR is greater than a set threshold, i.e. $\left|\mathbf{r}^u_k\right| > t$, the uth measurement is identified as a bad data. Assuming Gaussian distribution, t is typically set as 3.0, corresponding to 99.74% confidence level. Otherwise, go to Step 6.

Step 5: Bad data correction. The identified bad data is corrected as (4.49). After correction, go back to Step 3 and repeat the process.

Step 6: Set $k \leftarrow k + 1$ and return to Step 1.

For PDSE and CDSE, the CKF-based DSE solver described above takes different values of $\phi \in \Omega = \{c,\ p\}$ but follows the same six main steps described above. When both PDSE and CDSE are completed at time k, they move to the next time step $k + 1$.

The LNR-based bad data detection of each estimator checks the consistency between the corresponding subsystem model and output signals. PDSE detects and corrects bad data in measurement signals, and CDSE detects and corrects bad data in control signals.

4.2.5 Simulation Results

A modified IEEE 13-node test system is used to illustrate the CPDSE framework described in the Sections (4.2.2–4.2.4). This system can be viewed as a distribution-feeder-level microgrid or

Table 4.2 Formulas related to the CKF.

$$\hat{\mathbf{x}}_{\phi(0)} = E\left(\mathbf{x}_{\phi(0)}\right) \tag{4.35}$$

$$\mathbf{P}_{0|0} = E\left[\left(\mathbf{x}_{\phi(0)} - \hat{\mathbf{x}}_{\phi(0)}\right)\left(\mathbf{x}_{\phi(0)} - \hat{\mathbf{x}}_{\phi(0)}\right)^T\right] \tag{4.36}$$

$$\begin{cases} \chi_{\phi(i,k-1)} = \left(\mathbf{P}_{k-1|k-1}\right)^{1/2}\xi_i + \hat{\mathbf{x}}_{\phi(k-1)}, \\ \chi^*_{\phi(i,k)} = \mathbf{f}_\phi\left(\chi_{\phi(i,k-1)}, \mathbf{y}_{\overline{\phi}(k-1)}, \mathbf{u}_{g(k-1)}\right)+, \end{cases} \quad i = 1,2,\ldots 2n \tag{4.37}$$

$$\hat{\mathbf{x}}_{\phi(k|k-1)} = \frac{1}{2n}\sum_{i=1}^{2n}\chi^*_{\varepsilon(i,k)} \tag{4.38}$$

$$\mathbf{P}_{k|k-1} = \frac{1}{2n}\sum_{i=1}^{2n}\chi^*_{\phi(i,k)}\left(\chi^*_{\phi(i,k)}\right)^T - \hat{\mathbf{x}}_{\phi(k|k-1)}\left(\hat{\mathbf{x}}_{\phi(k|k-1)}\right)^T + \mathbf{Q}_{\phi(k)} \tag{4.39}$$

$$\begin{cases} \chi_{\phi(i,k)} = \left(\mathbf{P}_{k|k-1}\right)^{1/2}\xi_i + \hat{\mathbf{x}}_{\phi(k|k-1)}, \\ \mathbf{y}_{\phi(i,k)} = \mathbf{h}_\phi\left(\chi_{\phi(i,k)}, \mathbf{y}_{\overline{\phi}(k)}, \mathbf{u}_{g(k)}\right)+, \end{cases} \quad i = 1,2\ldots 2n \tag{4.40}$$

$$\hat{\mathbf{y}}_{\phi(k|k-1)} = \frac{1}{2n}\sum_{i=1}^{2n}\mathbf{y}_{\phi(i,k)} \tag{4.41}$$

$$\begin{cases} \mathbf{P}_{yy,k} = \frac{1}{2n}\sum_{i=1}^{2n}\mathbf{y}_{\phi(i,k)}\left(\mathbf{y}_{\phi(i,k)}\right)^T - \hat{\mathbf{y}}_{\phi(k|k-1)}\left(\hat{\mathbf{y}}_{\phi(k|k-1)}\right)^T + \mathbf{R}_{\phi(k)}, \\ \mathbf{P}_{xy,k} = \frac{1}{2n}\sum_{i=1}^{2n}\chi_{\phi(i,k)}\left(\mathbf{y}_{\phi(i,k)}\right)^T - \hat{\mathbf{x}}_{\phi(k|k-1)}\left(\hat{\mathbf{y}}_{\phi(k|k-1)}\right)^T \end{cases} \tag{4.42}$$

$$\mathbf{K}_k = \mathbf{P}_{xy,k}\left(\mathbf{P}_{yy,k}\right)^{-1} \tag{4.43}$$

$$\hat{\mathbf{x}}_{\phi(k)} = \hat{\mathbf{x}}_{\phi(k|k-1)} + \mathbf{K}_k\left(\mathbf{y}_{\phi(k)} - \hat{\mathbf{y}}_{\phi(k|k-1)}\right) \tag{4.44}$$

$$\mathbf{P}_{k|k} = \mathbf{P}_{k|k-1} - \mathbf{K}_k\mathbf{P}_{yy,k}\mathbf{K}_k^T \tag{4.45}$$

$$\begin{cases} \begin{cases} \chi_{\phi(i,k+)} = \left(\mathbf{P}_{k|k}\right)^{1/2}\xi_i + \hat{\mathbf{x}}_{\phi(k)}, \\ \mathbf{y}_{\phi(i,k+)} = \mathbf{h}_\phi\left(\chi_{\phi(i,k+)}, \mathbf{y}_{\overline{\phi}(k)}, \mathbf{u}_{g(k)}\right), \end{cases} \quad i = 1,2,\ldots 2n_\phi \\ \hat{\mathbf{y}}_{\phi(k+)} = \mathbf{h}_\phi\left(\hat{\mathbf{x}}_{\phi(k)}, \mathbf{y}_{\overline{\phi}(k)}, \mathbf{u}_{g(k)}\right)^T, \\ \mathbf{P}_{yy,k+} = \mathbf{R}_{\phi(k)} - \frac{1}{2n}\sum_{i=1}^{2n}\mathbf{y}_{\phi(i,k)}\left(\mathbf{y}_{\phi(i,k)}\right)^T - \hat{\mathbf{y}}_{\phi(k+)}\left(\hat{\mathbf{y}}_{\phi(k+)}\right)^T \end{cases} \tag{4.46}$$

$$\mathbf{r}_k = \left(\mathbf{P}_{yy,k+}\right)^{-1/2}\left(\mathbf{y}_{\phi(k)} - \mathbf{h}_\phi\left(\hat{\mathbf{x}}_{\phi(k)}, \mathbf{y}_{\overline{\phi}(k)}, \mathbf{u}_{g(k)}\right)\right) \tag{4.47}$$

$$u = \arg\max_j\left\{\mathbf{r}_k^j, j = 1,2\ldots n\right\} \tag{4.48}$$

$$\mathbf{y}_{\phi(k)}^u \leftarrow \mathbf{y}_{\phi(k)}^u - \frac{\mathbf{R}_{\phi(k)}^{uu}}{\mathbf{P}_{yy,k}^{uu}}\left(\mathbf{y}_{\phi(k)} - \mathbf{h}_\phi\left(\hat{\mathbf{x}}_{\phi(k)}, \mathbf{y}_{\overline{\phi}(k)}, \mathbf{u}_{g(k)}\right)\right) \tag{4.49}$$

networked microgrid. In the system, 5 PV generation systems are integrated into the grid – a three-phase one in VSG (GFM) control mode at node 633, two three-phase ones in MPPT (GFL) control mode at nodes 692 and 680, respectively, and two single-phase ones in MPPT (GFL) control mode at nodes 645 and node 684. The rated capacities of the three-phase IBRs and single-phase IBRs are 500 and 80 MVA, respectively. The true state of the system is simulated through Simulink (Version 10.5, R2022a), and the CPDSE algorithms are implemented in MATLAB (Version 9.12, R2022a). The measurement sampling frequency is set as 3840 Hz (64 points per cycle), which is readily satisfied by merging units (MUs), digital fault recorders (DFRs), or inverter sensors.

CDSE and PDSE are implemented for both GFM- and GFL-controlled IBRs. For illustration purposes, only the PDSE results for a GFL-controlled IBR (at node 692) and the CDSE results for a GFM-controlled IBR (at node 633) are presented.

For the GFL-controlled IBR at node 692, a step change of solar irradiance from 0.92 to 0.52 p.u. and a step change of reactive power reference from 0 to 0.25 p.u. are implemented at 2.5 and 1.5 seconds, respectively. Gaussian noise $N\left(0, 0.02^2\right)$ is added to the outputs of both the physical and cyber subsystems $(\mathbf{y}_p, \mathbf{y}_c)$, i.e. the measurement signals and the control signals, respectively. Process noise naturally exists due to the discretization and power electronic non-idealities.

Figure 4.4 shows the state tracking results of PDSE. In the physical subsystem, the active and reactive power is adjusted by the change of the dq-axis currents outputted by the inverter, I_{id}, I_{iq}. The PDSE tracks the state variables in the physical inverter system swiftly and accurately.

Table 4.3 presents the root mean square errors (RMSEs) of the raw and estimated measurement signals, respectively. The PDSE significantly filters out the errors in the measurement signals.

In order to verify the bad data processing function, data points in the measurement signals are randomly selected continuously (for 0.2 seconds) or individually and set to 0 to represent signal losses. Figure 4.5 shows the true, measured, and estimated values of the physical subsystem outputs. The raw measurements deviate from true values drastically. With the LNR algorithm

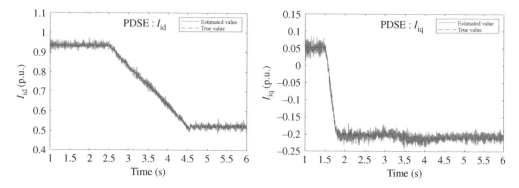

Figure 4.4 State estimate of PDSE (I_{id}, I_{iq}) for GFL-controlled IBR.

Table 4.3 RMSE of measurement signals (GFL).

	I^z_{gd}	I^z_{gd}	V^z_{dc}
Raw	2.00%	2.00%	2.00%
Estimated	0.68%	0.62%	0.31%

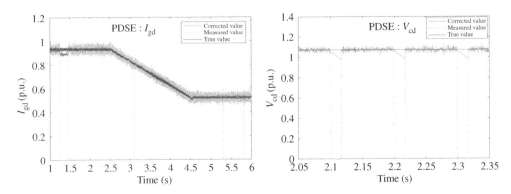

Figure 4.5 Bad Data correction results of PDSE (I_{id}, V_{cd}) for GFL-controlled IBR.

embedded in PSE, the bad data points are detected and corrected. The corrected values are very close to the true values.

For the GFM-controlled IBR at node 633, we introduce a step change to the grid frequency by -0.6 Hz at 3.5 seconds. Gaussian noise $N\left(0, 0.01^2\right)$ is added to the output of both the physical and cyber subsystems ($\mathbf{y}_p, \mathbf{y}_c$). The results of CDSE are shown in Fig. 4.6. It can be observed that the VSG behaves similar to an synchronous generator (SG) with power angle swings. The outer-loop $P - f$ droop control finally regulates the real power output to a higher steady-state level, which resembles SG's governor control. The CDSE closely tracks the virtual rotor speed, virtual power

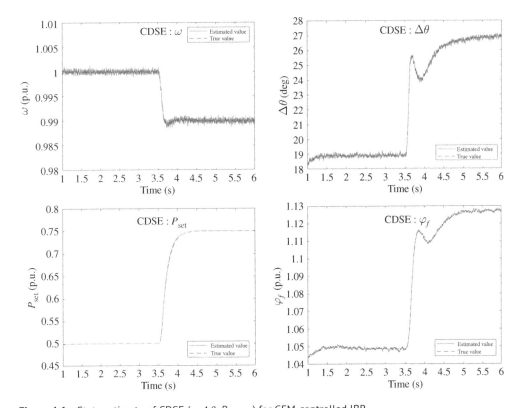

Figure 4.6 State estimate of CDSE ($\omega, \Delta\theta, P_{set}, \varphi_f$) for GFM-controlled IBR.

Table 4.4   RMSE of control signals (GFM).

	V_{id}^c	V_{iq}^c
Raw	1.00%	1.00%
Estimated	0.09%	0.18%

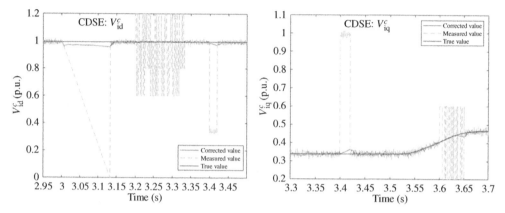

Figure 4.7   Bad data correction results of CDSE (V_{id}^c, V_{iq}^c) for GFM-controlled IBR.

angle, among other states of the VSG controller during the transient. Note that these variables are digital variables internal to the controller, which are typically unavailable to grid operators. The CDSE provides an effective means to observe the internal states of the digital controller.

Table 4.4 presents the RMSEs of the raw and estimated control signals, respectively. Clearly, the CDSE effectively reduces the uncertainty of the control signals.

To verify the robustness of CDSE, bad data are introduced into the control signals of the VSG. The results are shown in Fig. 4.7. Evidently, the CDSE effectively detects and corrects various types of bad data such as random errors, ramping errors, and errors due to channel swapping.

4.3   State and Parameter Estimation for Network Components

4.3.1   Background and Motivation

During the operation of microgrids, faults may occur within microgrid circuits. After the occurrence of the fault, there are two mandatory steps. The first step is protection – to isolate the faulted circuit after the fault occurs, to ensure minimum damage to electric equipment, and to the overall microgrid. The second step is fault location – after the faulted line is isolated from the rest of the system, to accurately locate the fault within the faulted line, to minimize the time spent searching for the fault, and therefore reduce power outage time and improve power supply reliability. With the increasing penetration of renewables, protection, and fault location for microgrid circuits will experience challenges. The fault currents could be limited by the controller, causing difficulty in detecting circuit faults. In addition, with less system inertia, the electromagnetic transients during faults could become more severe and unusual, resulting in challenges for legacy schemes. DSE is a promising tool to accurately track electromagnetic dynamics of the systems [30, 31]. In fact,

DSE has been applied to track electromagnetic transients in various components of power systems, including various transmission lines [32, 33], transformers [34], microgrid circuits [35], elements in power electronic circuits [22], to name a few. Specifically, DSE can be applied to tackle the protection and fault location challenges in microgrid circuits. DSE-based protection for microgrid circuits is to estimate the dynamic states of the microgrid circuits, and to formulate the trip logic based on the consistency between the available measurements and dynamic model of the microgrid circuit. DSE-based fault location for microgrid circuits introduces the fault location as an unknown parameter, and estimates the dynamic states as well as the parameter of the faulted microgrid circuits.

Next, the ideas of DSE-based protection and fault location for microgrid circuits will be presented in detail with demonstrative examples.

4.3.2 Dynamic State Estimation-Based Protection for Microgrid Circuits

DSE-based protection requires building a dynamic model of the healthy microgrid circuit under protection. The model is usually described via algebraic and differential equations (DAEs). A microgrid circuit can be represented by the multi-section π transmission line model, where n is the number of sections, as shown in Fig. 4.8. Here, a three-phase microgrid circuit is shown as an example.

The dynamic model of the healthy microgrid circuit in DAEs is shown in (4.50),

$$\begin{cases} \mathbf{i}(t) = \mathbf{A}_1 \cdot \mathbf{x}(t) + \mathbf{B}_1 \cdot d\mathbf{x}(t)/dt \\ \mathbf{0} = \mathbf{A}_2 \cdot \mathbf{x}(t) + \mathbf{B}_2 \cdot d\mathbf{x}(t)/dt \\ \mathbf{v}(t) = \mathbf{A}_3 \cdot \mathbf{x}(t) \end{cases} \tag{4.50}$$

where $\mathbf{x}(t) = [v_1(t), \dots, v_{n+1}(t), i_{L1}(t), \dots, i_{Ln}(t)]^T$ is the state vector; $\mathbf{i}(t) = [i_1(t), i_{n+1}(t)]^T$ and $v(t) = [v_1(t), v_{n+1}(t)]^T$ are current and voltage measurements; $\mathbf{A}_1, \mathbf{B}_1, \mathbf{A}_2, \mathbf{B}_2$ and are coefficient matrices,

$$\mathbf{A}_1 = \begin{bmatrix} \mathbf{G}/2 & \mathbf{0}_{3\times(3n)} & \mathbf{I}_3 & \mathbf{0}_{3\times(3n-3)} \\ \mathbf{0}_{3\times(3n)} & -\mathbf{G}/2 & \mathbf{0}_{3\times(3n-3)} & \mathbf{I}_3 \end{bmatrix},$$

$$\mathbf{B}_1 = \begin{bmatrix} \mathbf{C}/2 & \mathbf{0}_{3\times(3n)} & \mathbf{0}_{3\times(3n)} \\ \mathbf{0}_{3\times(3n)} & -\mathbf{C}/2 & \mathbf{0}_{3\times(3n)} \end{bmatrix} \tag{4.51}$$

$$\mathbf{A}_2 = \begin{bmatrix} \mathbf{Y}_1 & \mathbf{E}_1 \\ \mathbf{E}_2 & \mathbf{R}_n \end{bmatrix}, \mathbf{B}_2 = \begin{bmatrix} \mathbf{D}_1 & \mathbf{0}_{(3n-3)\times3n} \\ \mathbf{0}_{3n\times(3n+3)} & \mathbf{L}_n \end{bmatrix},$$

$$\mathbf{A}_3 = \begin{bmatrix} \mathbf{0}_{3\times0} & \mathbf{I}_3 & \mathbf{0}_{3\times(6n)} \\ \mathbf{0}_{3\times(3n)} & \mathbf{I}_3 & \mathbf{0}_{3\times(3n)} \end{bmatrix} \tag{4.52}$$

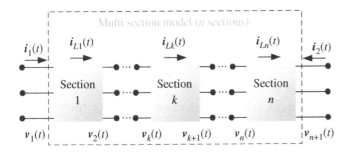

Figure 4.8 Healthy microgrid circuit model (multi-section π transmission line model).

where $\mathbf{I}_k \in \mathbb{R}^{k \times k}$ is an identity matrix, $\mathbf{0}_{k \times m} \in \mathbb{R}^{k \times m}$ is a zero matrix. $\mathbf{Y}_1 = [\mathbf{0}_{(3n-3) \times 3}, \mathbf{G}_{n-1}, \mathbf{0}_{(3n-3) \times 3}]$, $\mathbf{E}_1 = [\mathbf{0}_{(3n-3) \times 3}, \mathbf{I}_{3n-3}] - [\mathbf{I}_{3n-3}, \mathbf{0}_{(3n-3) \times 3}]$, $\mathbf{E}_2 = [\mathbf{0}_{3n \times 3}, \mathbf{I}_{3n}] - [\mathbf{I}_{3n}, \mathbf{0}_{3n \times 3}]$, and $\mathbf{D}_1 = [\mathbf{0}_{(3n-3) \times 3}, \mathbf{C}_{n-1}, \mathbf{0}_{(3n-3) \times 3}]$. $\mathbf{R}_n, \mathbf{L}_n, \mathbf{G}_{n-1}$, and $\mathbf{C}_{n-1}$ are block diagonal matrices with $n\mathbf{R}, n\mathbf{L}, (n-1)\mathbf{G}$ and $(n-1)\mathbf{C}$ at diagonals, respectively. Note that $\mathbf{R}, \mathbf{L}, \mathbf{G}$, and $\mathbf{C}$ matrices are the parameter matrices of each section and can be calculated as,

$$\begin{cases} \mathbf{R} = l/n \cdot \mathbf{R}_{pu}, \mathbf{L} = l/n \cdot \mathbf{L}_{pu} \\ \mathbf{G} = l/n \cdot \mathbf{G}_{pu}, \mathbf{C} = l/n \cdot \mathbf{C}_{pu} \end{cases} \tag{4.53}$$

where l is the entire length of the line of interest. $\mathbf{R}_{pu}, \mathbf{L}_{pu}, \mathbf{G}_{pu}$, and $\mathbf{C}_{pu}$ are the series resistance, series inductance, shunt conductance, and shunt capacitance parameter matrices per unit length.

Subsequently, the DSE algorithm can be applied to solve for the unknown states of the system and to formulate the protection logic. To solve (4.50), DAEs are converted to algebraic equations (AEs). Here the tool of trapezoidal integration (TI) is taken as an example. Equation (4.50) is integrated over the time interval $[t - h, t]$, where h is the time step of the DSE (sampling interval),

$$\mathbf{z}(t) = \mathbf{Y}_{eqx} \cdot \mathbf{x}(t) - \mathbf{B}_{eq} \tag{4.54}$$

where $\mathbf{Y}_{eqx} = \begin{bmatrix} \mathbf{A}_1 + 2\mathbf{B}_1/h \\ \mathbf{A}_2 + 2\mathbf{B}_1/h \\ \mathbf{A}_3 \end{bmatrix}, \mathbf{B}_{eq} = -\begin{bmatrix} \mathbf{A}_1 - 2\mathbf{B}_1/h \\ \mathbf{A}_2 - 2\mathbf{B}_1/h \\ \mathbf{0} \end{bmatrix} \mathbf{x}(t-h) + \begin{bmatrix} \mathbf{i}(t-h) \\ \mathbf{0} \\ \mathbf{0} \end{bmatrix}$, and $\mathbf{z}(t) = \begin{bmatrix} \mathbf{i}(t) \\ \mathbf{0} \\ \mathbf{v}(t) \end{bmatrix}$.

Next, DSE can be applied to solve (4.54). Here the weighted least squares (WLS) algorithm is taken as an example to solve (4.54). The optimization problem can be formulated as (4.55). It is worth noting that other DSE methods such as Kalman filter (KF) can also be applied to solve (4.55); however, it can be easily proved that KF and WLS type methods are equivalent under certain conditions [36].

$$\min_{\mathbf{x}(t)} J(t) = \mathbf{r}(t)^T \mathbf{W} \mathbf{r}(t) \tag{4.55}$$

where $\mathbf{r}(t) = \mathbf{Y}_{eqx} \mathbf{x}(t) - \mathbf{B}_{eq} - \mathbf{z}(t)$ is the residual vector. $\mathbf{W} = \text{diag}\left\{..., 1/\sigma_i^2, ...\right\}$ is the weight matrix, where σ_i is the noise standard deviation of the ith entry of the measurement vector $\mathbf{z}(t)$. $\text{diag}\{\cdot\}$ represents a diagonal matrix with diagonal elements $\{\cdot\}$.

The best estimates of the state vector $\hat{x}(t)$ is:

$$\hat{x}(t) = (\mathbf{Y}_{eqx}^T \mathbf{W} \mathbf{Y}_{eqx})^{-1} \mathbf{Y}_{eqx}^T \mathbf{W}(\mathbf{z}(t) + \mathbf{B}_{eq}) \tag{4.56}$$

Substitute the estimated $\hat{x}(t)$ back to (4.55) to obtain the chi-square value $\hat{J}(t)$. $\hat{J}(t)$ describes the consistency between the measurements and the dynamic model of the healthy microgrid circuit. During normal operation or external faults, the consistency is high and the chi-square value is small. During internal faults, the consistency is low and the chi-square value is relatively high. To quantify the consistency, a user-defined threshold J_{set} is provided to distinguish between internal and external faults,

$$\text{test}(t) = \begin{cases} 1, \hat{J}(t) \geq J_{set} \\ 0, \hat{J}(t) < J_{set} \end{cases} \tag{4.57}$$

To further ensure the security of protection during external faults, a user-defined delay T_{set} is introduced to the protection logic. As an example, here the trip decision is made, i.e. trip$(t) = 1$, if the value of test(t) satisfies (4.58).

$$\text{trip}(t) = \begin{cases} 1, \int_{t-T_{set}}^{t} \text{test}(\tau) \, d\tau > 0.5 T_{set} \\ 0, \int_{t-T_{set}}^{t} \text{test}(\tau) \, d\tau < 0.5 T_{set} \end{cases} \tag{4.58}$$

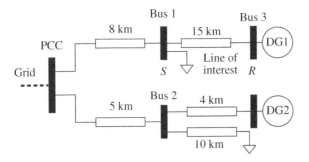

Figure 4.9 An example microgrid system.

Also, it is worth noting that there are other types of line models [37, 38], discretization methods [39], and DSE algorithms [22, 39] that can be adopted to formulate DSE-based protection.

The equivalent circuit of an example three-phase 35 kV microgrid system is shown in Fig. 4.9 and is built in PSCAD/EMTDC. The line of interest is 15 km. The nominal frequency of the system is 50 Hz. The dual terminal three-phase voltage and current instantaneous measurements are installed at both terminals (S and R) of the line of interest, with the sampling rate of 4 k samples/second according to IEC 61850-9-2 standard.

The dynamic line model is represented as the multi-PI section model with two sections. The trapezoidal integration is utilized, and the DSE time step h is the sampling interval 250 μs. The threshold of chi-square value is $J_{set} = 1000$ and the trip delay time window is $T_{set} = 5$ ms. Next, two fault events are studied as examples.

Event 1 is a high impedance internal A–G fault, which occurs at 7 km away from bus 1 and at time 0.3 seconds, with 100 ohm fault impedance. As examples, the actual, estimated, and residuals of the three-phase voltages at S end are depicted in Fig. 4.10. The residuals are small before 0.3 s and rise

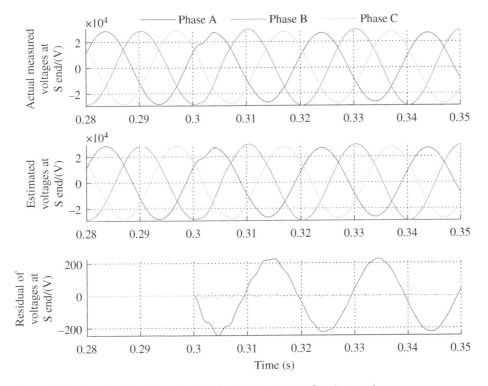

Figure 4.10 Actual, estimated, and residuals of voltage at the S end, event 1.

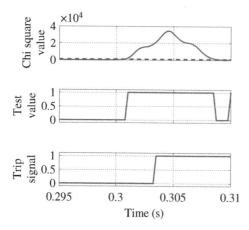

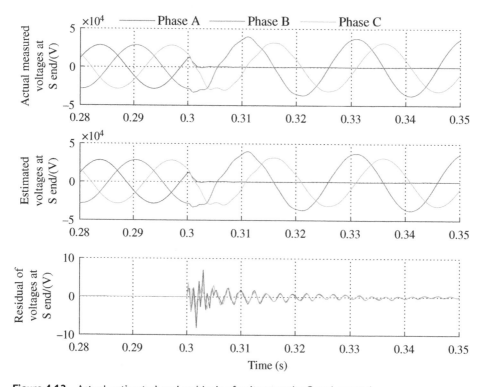

Figure 4.11 Results of trip decision, event 1.

dramatically during the fault. To quantify the residuals, the chi-square values are shown in Fig. 4.11. The light gray dotted line is the threshold J_{set}. The chi-square value exceeds the threshold 1.0 ms (4 samples) after the occurrence of the fault, at 0.301 s. As a result, the trip decision is made with 2.5 ms delay, at 0.3035 s. The protection correctly trips the line even during this high impedance internal fault.

Event 2 is a low impedance external A-G fault within the line at the left side of bus 1 (1 km from bus 1) and at time 0.3 s, with 0.01 ohm fault impedance. As examples, the actual, estimated,and residuals of the three-phase voltages at S end are depicted in Fig. 4.12. The chi-square values are shown in Fig. 4.13. The light gray dotted line is the threshold J_{set}. One can observe that, although

Figure 4.12 Actual, estimated, and residuals of voltage at the S end, event 1.

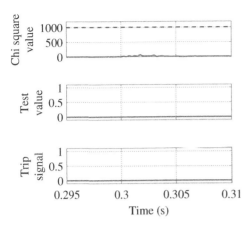

Figure 4.13 Results of trip decision, event 1.

there are slight transients during this external fault, the chi-square values are way below the threshold, and the relay correctly refuses to trip this external fault.

4.3.3 Dynamic State Estimation-Based Fault Location for Microgrid Circuits

Fault location can be viewed as a parameter estimation problem, and also requires to build a dynamic model. Different from the DSE-based protection which models a healthy microgrid circuit, DSE-based fault location models a microgrid circuit with fault. If the multi-section π line model is taken as an example, the faulted microgrid circuit can be represented by the combination of three parts, a healthy line at the left side of the fault (with m sections), a healthy line at the right side of the fault (with n sections), and the fault itself, as shown in Fig. 4.14.

The dynamic model of the faulted microgrid circuit in DAEs is shown in (4.59),

$$\begin{cases} \mathbf{z}(t) = \mathbf{Y}_{eqx1}\mathbf{x}(t) + \mathbf{D}_{eqx1}d\mathbf{x}(t)/dt \\ \mathbf{0} = \mathbf{Y}_{eqx2}\mathbf{x}(t) + \mathbf{D}_{eqx2}d\mathbf{x}(t)/dt \end{cases} \tag{4.59}$$

where $\mathbf{x}(t) = [\mathbf{v}_1^{(l)}(t), \ldots, \mathbf{v}_{m+1}^{(l)}(t), \mathbf{v}_2^{(r)}(t), \ldots, \mathbf{v}_{n+1}^{(r)}(t), \mathbf{i}_{L1}^{(l)}(t), \ldots, \mathbf{i}_{Lm}^{(l)}(t), \mathbf{i}_{L1}^{(r)}(t), \ldots,$ $\mathbf{i}_{Ln}^{(r)}(t)]^T$ is the state vector; $\mathbf{z}(t) = [\mathbf{v}_1^{(l)}(t), \mathbf{v}_{n+1}^{(r)}(t), \mathbf{i}_1^{(l)}(t), \mathbf{i}_2^{(r)}(t)]^T$ is the measurement vector; $\mathbf{Y}_{eqx1}, \mathbf{D}_{eqx1}, \mathbf{Y}_{eqx2}$, and $\mathbf{D}_{eqx2}$ are coefficient matrices,

$$\mathbf{Y}_{eqx1} = \begin{bmatrix} \mathbf{I}_3 & \mathbf{0}_{3\times(3m+3n)} & \mathbf{0}_{3\times(3m)} & \mathbf{0}_{3\times(3n)} \\ \mathbf{0}_{3\times(3m+3n)} & \mathbf{I}_3 & \mathbf{0}_{3\times(3m)} & \mathbf{0}_{3\times(3n)} \\ \mathbf{G}_l/2 & \mathbf{0}_{3\times(3m+3n)} & \mathbf{I}_3 & \mathbf{0}_{3\times(3m+3n-3)} \\ \mathbf{0}_{3\times(3m+3n)} & \mathbf{G}_r/2 & \mathbf{0}_{3\times(3m+3n-3)} & -\mathbf{I}_3 \end{bmatrix} \tag{4.60}$$

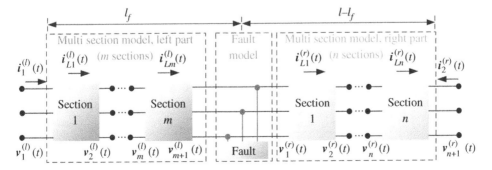

Figure 4.14 Microgrid circuit model with fault (multi-section π transmission line model).

$$
\mathbf{D}_{eqx1} = \begin{bmatrix} \mathbf{0}_{3\times(3m+3)} & \mathbf{0}_{3\times(3n)} & \mathbf{0}_{3\times(3m)} & \mathbf{0}_{3\times(3n)} \\ \mathbf{0}_{3\times(3m+3)} & \mathbf{0}_{3\times(3n)} & \mathbf{0}_{3\times(3m)} & \mathbf{0}_{3\times(3n)} \\ \mathbf{C}_l/2 & \mathbf{0}_{3\times(3m+3n)} & \mathbf{0}_{3\times(3m)} & \mathbf{0}_{3\times(3n)} \\ \mathbf{0}_{3\times(3m+3n)} & \mathbf{C}_r/2 & \mathbf{0}_{3\times(3m)} & \mathbf{0}_{3\times(3n)} \end{bmatrix} \tag{4.61}
$$

$$
\mathbf{Y}_{eqx2} = \begin{bmatrix} \mathbf{Y}_{11} & \mathbf{0}_{(3m-3)\times(3n+3)} & \mathbf{E}_{3m-3} & \mathbf{0}_{(3m-3)\times(3n)} \\ \mathbf{0}_{(3n-3)\times(3m+3)} & \mathbf{Y}_{22} & \mathbf{0}_{(3n-3)\times(3m+3)} & \mathbf{E}_{3n-3} \\ \mathbf{0}_{(3n)\times(3m)} & \mathbf{E}_{3n} & \mathbf{0}_{(3n)\times(3m)} & \mathbf{Y}_{44} \\ \mathbf{Y}_{51} & \mathbf{0}_{3\times(3m+3n-3)} & \mathbf{Y}_{53} & \mathbf{0}_{3\times(3n-3)} \end{bmatrix} \tag{4.62}
$$

$$
\mathbf{D}_{eqx2} = \begin{bmatrix} \mathbf{D}_{11} & \mathbf{0}_{(3m-3)\times(3n+3)} & \mathbf{0}_{(3m-3)\times(3m)} & \mathbf{0}_{(3m-3)\times(3n)} \\ \mathbf{0}_{(3n-3)\times(3m+3)} & \mathbf{D}_{22} & \mathbf{0}_{(3n-3)\times(3m+3)} & \mathbf{0}_{(3n-3)\times(3n)} \\ \mathbf{0}_{(3m)\times(3m+3)} & \mathbf{0}_{(3m)\times(3n)} & \mathbf{D}_{33} & \mathbf{0}_{(3m)\times(3n)} \\ \mathbf{0}_{(3n)\times(3m)} & \mathbf{0}_{(3n)\times(3n+3)} & \mathbf{0}_{(3n)\times(3m)} & \mathbf{D}_{44} \\ \mathbf{D}_{51} & \mathbf{0}_{3\times(3m+3n-3)} & \mathbf{0}_{3\times6} & \mathbf{0}_{3\times(3n-3)} \end{bmatrix} \tag{4.63}
$$

where $\mathbf{I}_j \in \mathbb{R}^{j\times j}$ is an identity matrix, and $\mathbf{0}_{j\times k} \in \mathbb{R}^{j\times k}$ is a zero matrix with the dimension of $j \times k$; $\mathbf{Y}_{11} = \begin{bmatrix} \mathbf{0}_{(3m-3)\times 3} & \mathbf{G} \end{bmatrix}$, where $\mathbf{G}$ is a block diagonal matrix with $m-1$ $\mathbf{G}_l$ at diagonals. $\mathbf{Y}_{51} = \begin{bmatrix} \mathbf{0}_{3\times(3m)} & \mathbf{M}_{fault} - (\mathbf{G}_l + \mathbf{G}_r)/2 \end{bmatrix}$, $\mathbf{Y}_{53} = \begin{bmatrix} -\mathbf{I}_2 & \mathbf{I}_2 \end{bmatrix}$. $\mathbf{Y}_{22}, \mathbf{Y}_{33}, \mathbf{Y}_{44}$ are block diagonal matrices with $n-1$ $\mathbf{G}_r$, m $\mathbf{R}_l$, and n $\mathbf{R}_r$ at diagonals, respectively. $\mathbf{E}_j = \begin{bmatrix} \mathbf{0}_{j\times 2} & \mathbf{I}_j \end{bmatrix} - \begin{bmatrix} \mathbf{I}_j & \mathbf{0}_{j\times 2} \end{bmatrix}$, $\mathbf{D}_{11} = \begin{bmatrix} \mathbf{0}_{(3m-3)\times 3} & \mathbf{C} \end{bmatrix}$. C is a block diagonal matrix with $m-1$ $\mathbf{C}_l$ at diagonals, respectively. $\mathbf{D}_{51} = \begin{bmatrix} \mathbf{0}_{3\times(3m)} & -(\mathbf{C}_l + \mathbf{C}_r)/2 \end{bmatrix}$. $\mathbf{D}_{22}, \mathbf{D}_{33}$, and $\mathbf{D}_{44}$ are block diagonal matrices with $n-1$ $\mathbf{C}_r$, m $\mathbf{L}_l$, and n $\mathbf{L}_r$ at diagonals, respectively. $\mathbf{R}_l = \mathbf{R}_{pu} \cdot l_f/m$, $\mathbf{L}_l = \mathbf{L}_{pu} \cdot l_f/m$, $\mathbf{G}_l = \mathbf{G}_{pu} \cdot l_f/m$, $\mathbf{C}_l = \mathbf{C}_{pu} \cdot l_f/m$, $\mathbf{R}_r = \mathbf{R}_{pu} \cdot (l - l_f)/n$, $\mathbf{L}_r = \mathbf{L}_{pu} \cdot (l - l_f)/n$, $\mathbf{G}_r = \mathbf{G}_{pu} \cdot (l - l_f)/n$, and $\mathbf{C}_r = \mathbf{C}_{pu} \cdot (l - l_f)/n$. $\mathbf{R}_{pu}, \mathbf{L}_{pu}, \mathbf{G}_{pu}$, and $\mathbf{C}_{pu}$ are the series resistance, series inductance, shunt conductance, and shunt capacitance parameter matrices per unit length, l is the total length of the line, and l_f is the fault location. M_{fault} is determined by the fault type. For three-phase power systems, the value of M_{fault} is shown in Fig. 4.15. R_f is the fault resistance.

Next, the dynamic model in DAEs should be discretized into AEs. Similarly, the trapezoidal integration can be adopted as an example, with the time step h,

$$
\mathbf{z}(t) = \mathbf{Y}_{eqx}\mathbf{x}(t) - \mathbf{B}_{eq} \tag{4.64}
$$

Type	M_{fault}	Type	M_{fault}	Type	M_{fault}	Type	M_{fault}
AG	$\dfrac{1}{R_f}\begin{bmatrix} 1 & 0 & 0 \\ 0 & 0 & 0 \\ 0 & 0 & 0 \end{bmatrix}$	AB	$\dfrac{1}{R_f}\begin{bmatrix} 1 & -1 & 0 \\ -1 & 1 & 0 \\ 0 & 0 & 0 \end{bmatrix}$	ABG	$\dfrac{1}{R_f}\begin{bmatrix} 1 & 0 & 0 \\ 0 & 1 & 0 \\ 0 & 0 & 0 \end{bmatrix}$	3-phase	$\dfrac{1}{R_f}\begin{bmatrix} 1 & 0 & 0 \\ 0 & 1 & 0 \\ 0 & 0 & 1 \end{bmatrix}$
BG	$\dfrac{1}{R_f}\begin{bmatrix} 0 & 0 & 0 \\ 0 & 1 & 0 \\ 0 & 0 & 0 \end{bmatrix}$	BC	$\dfrac{1}{R_f}\begin{bmatrix} 0 & 0 & 0 \\ 0 & 1 & -1 \\ 0 & -1 & 1 \end{bmatrix}$	BCG	$\dfrac{1}{R_f}\begin{bmatrix} 0 & 0 & 0 \\ 0 & 1 & 0 \\ 0 & 0 & 1 \end{bmatrix}$		Dimension of M_{fault}: 3×3
CG	$\dfrac{1}{R_f}\begin{bmatrix} 0 & 0 & 0 \\ 0 & 0 & 0 \\ 0 & 0 & 1 \end{bmatrix}$	CA	$\dfrac{1}{R_f}\begin{bmatrix} 1 & 0 & -1 \\ 0 & 0 & 0 \\ -1 & 0 & 1 \end{bmatrix}$	CAG	$\dfrac{1}{R_f}\begin{bmatrix} 1 & 0 & 0 \\ 0 & 0 & 0 \\ 0 & 0 & 1 \end{bmatrix}$		

Figure 4.15 Definitions of M_{fault} for different fault types for three-phase power systems.

where $\mathbf{Y}_{eqx} = \begin{bmatrix} \mathbf{Y}_{eqx1} + 2\mathbf{D}_{eqx1}/h \\ \mathbf{Y}_{eqx2} + 2\mathbf{D}_{eqx2}/h \end{bmatrix}$, $\mathbf{B}_{eq} = -\begin{bmatrix} \mathbf{Y}_{eqx1} - 2\mathbf{D}_{eqx1}/h \\ \mathbf{Y}_{eqx2} - 2\mathbf{D}_{eqx2}/h \end{bmatrix} \mathbf{x}(t-h) + \begin{bmatrix} \mathbf{z}(t-h) \\ \mathbf{0} \end{bmatrix}$, and $\mathbf{z}(t) = \begin{bmatrix} \mathbf{i}(t) \\ \mathbf{0} \end{bmatrix}$.

To solve the unknown states and parameters in (4.64), this section utilizes WLS formulation as an example, as shown in (4.65).

$$\min_{\mathbf{x}(t,t_m)} J(t) = \mathbf{r}(t)^T \mathbf{W} \mathbf{r}(t) \tag{4.65}$$

where $\mathbf{r}(t) = \mathbf{Y}_{eqx}\mathbf{x}(t) - \mathbf{B}_{eq} - \mathbf{z}(t)$ is the residual vector, and $\mathbf{W}$ is similarly defined as in (4.65).

In fact, there are several methods that can be applied to solve [40]. One option is to introduce unknown parameters as augmented state variables, with the additional information that those states are constants at different time instant. In this case, the problem is highly nonlinear [41]. However, solving a highly nonlinear DSE problem with a relatively short time window is challenging, and can potentially encounter issues such as non-convergence, large condition number, and high computational burden [42]. Therefore, the other option is as follows. Step 1: given l_f and R_f, the DSE problem becomes linear, and the chi-square value can be similarly generated to quantify the consistency between the measurement and the model. Step 2: the gradient descent method can be applied to find the actual l_f and R_f corresponding to the minimum chi-square value [42].

For step 1, since the problem becomes linear, the best estimates $\hat{x}(t)$ can be obtained without iteration,

$$\hat{x}(t) = \left(\mathbf{Y}_{eqx}^T \mathbf{W} \mathbf{Y}_{eqx}\right)^{-1} \mathbf{Y}_{eqx}^T \mathbf{W} \left(\mathbf{z}(t) + \mathbf{B}_{eq}\right) \tag{4.66}$$

Then, substitute $\mathbf{x}(t) = \hat{x}(t)$ into 4.65 to obtain the chi-square value $\hat{J}(t)$. One can observe that, throughout the calculation procedure at different time instants, the coefficient matrix $\mathbf{Y}_{eqx}$ is a constant with given l_f and R_f. Therefore, the inverse of matrix $\mathbf{Y}_{eqx}^T \mathbf{W} \mathbf{Y}_{eqx}$ is a constant and can be calculated prior to the DSE procedure.

For step 2, the chi-square value quantifies the consistency between the measurements and the dynamic model of the faulted microgrid circuit – the actual fault location and fault resistance should correspond to the minimum chi-square value. To describe the consistency over a time span, here average chi-square value during the fault (with the value y) can be selected. One can observe that y is a function of l_f and R_f. The fault location can be estimated by solving the following optimization problem,

$$\min_{l_f, R_f} y = \chi(l_f, R_f) \tag{4.67}$$

where $\chi(\cdot)$ describes the functional relationship among y, l_f, and R_f. Here the gradient descent method can be adopted for fault location,

$$[l_f^{(v+1)}, R_f^{(v+1)}] = [l_f^{(v)}, R_f^{(v)}] - \alpha^{(v)} \nabla \chi(l_f^{(v)}, R_f^{(v)}) \tag{4.68}$$

where $\alpha^{(v)}$ is the step size, and should satisfy the Armijo condition [40]. The value of $\nabla \chi(l_f^{(v)}, R_f^{(v)})$ can be calculated via numerical methods with small perturbations.

The example three-phase 35 kV microgrid system is the same as that in the protection application, as shown in Fig. 4.9. The line of interest is 15 km and the measurements are also installed at both terminals, with 4k samples/second sampling rate. The available time window during the fault is assumed to be 10 ms. An example A-G fault occurs at 6 km from the S end, with fault impedance of 10 ohm. The fault is still a A-G fault at 6 km from the S end. The chi-square value as a function of fault location and R_f are shown in Fig. 4.16. The minimum point is reached when the fault location

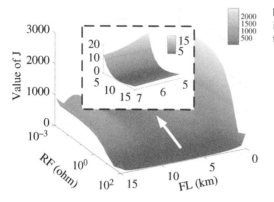

Figure 4.16 Chi-square value with different fault impedance and location: 10 ohm A-G fault, 6 km from the S end.

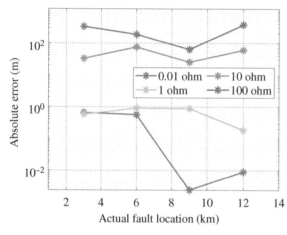

Figure 4.17 Fault location results: A to G fault with different fault impedances and locations.

is very close to 6 km and the fault resistance is very close to 10 ohm. Therefore, the fault location can be determined. The fault location error in this case is around 7.457×10^{-2} km.

To further validate the fault location accuracy of the described method, various faults with different fault resistance (0.01 ohm, 1 ohm, 10 ohm, and 100 ohm) and different fault locations (every 3 km) are studied. The results are shown in Fig. 4.17. One can observe that the described method can accurately locate faults, and the maximum error is around 1 m for low impedance (0.01, 1 ohm) faults and is around 0.35 km even for high impedance (100 ohm) faults.

4.4 Conclusion

This chapter addresses the motivation, concept, methodology, and validation of state-of-the-art microgrid state and parameter estimation technologies. IBRs and network components are the main components in microgrids for these applications.

For IBRs, the key difference noted between IBRs and conventional SGs is that IBRs introduce a heavy mix of physical and digital dynamics. In order to address this unique challenge, cyber-physical representation is developed which can explicitly distinguish between the physical dynamics in the inverter subsystems and the digital dynamics in the control subsystems. Based on this distinct representation, a general dual CPDSE framework is presented along with specific models and algorithms to materialize it. It is also shown how the CPDSE framework can be extended to CPDPE to address the model calibration issues.

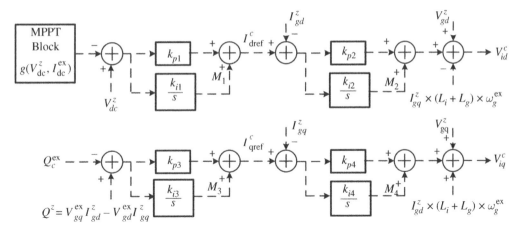

Figure 4.18 MPPT controller for PV generation system.

For network components such as transmission lines, the main application of state and parameter estimation is protection and fault location. DSE-based protection combines all available information from measurements and checks their consistency with the model. With the redundancy of measurements, it has the capability to suppress the impact of measurement errors and detect limited fault currents, which are difficult to achieve by conventional protection mechanisms. It is also shown that fault location can also be formulated as a joint state and parameter estimation problem, where the fault location is the parameter to be estimated.

4.5 Exercise

1. An GFL controller for PV generation systems is shown in Fig. 4.18. Treating the output of the MPPT block $g\left(V_{dc}^z, I_{dc}^{ex}\right)$ as an input of the cyber subsystem state space, write the state transition equations and output equations of this controller.
2. From the principle of DSE-based protection, will the DSE based protection wrongly trip the line under protection during system transients (the line under protection is still healthy)?
3. For fault location applications, should we select larger for smaller section numbers for relatively long microgrid circuits?

4.6 Acknowledgment

The authors appreciate the assistance of Mr. Binglin Wang, who is currently pursuing his Ph.D. degree at ShanghaiTech University, in preparing the manuscript of this chapter.

References

1 Nordrum, A. (2019). Transmission failure causes nationwide blackout in Argentina. https://spectrum.ieee.org/energywise/energy/the-smarter-grid/transmission-failure-causes-nationwide-blackout-in-argentina (accessed 14 November 2023).

2 U.S Department of Energy (2017). Hurricanes Maria, Irma and Harvey Situation Reports-Archived: August 26, 2017 - September 19, 2017. https://www.energy.gov/oe/downloads/hurricanes-maria-irma-and-harvey-situation-reports-archived-august-26-2017-september-19 (accessed 14 November 2023).

3 Liang, X. (2017). Emerging power quality challenges due to integration of renewable energy sources. *IEEE Transactions on Industry Applications* 53 (2): 855–866.

4 Peng, Q., Jiang, Q., Yang, Y. et al. (2019). On the stability of power electronics-dominated systems: challenges and potential solutions. *IEEE Transactions on Industry Applications* 55 (6): 7657–7670.

5 Li, Y., Gu, Y., and Green, T. (2022). Mapping of dynamics between mechanical and electrical ports in SG-IBR composite grids. *IEEE Transactions on Power Systems* 37 (5): 3423–3433.

6 Pico, H.N.V. and Johnson, B.B. (2019). Transient stability assessment of multi-machine multi-converter power systems. *IEEE Transactions on Power Systems* 34 (5): 3504–3514.

7 Wang, C., Qin, Z., Hou, Y., and Yan, J. (2018). Multi-area dynamic state estimation with PMU measurements by an equality constrained extended Kalman filter. *IEEE Transactions on Smart Grid* 9 (2): 900–910.

8 Kennedy, K., Lightbody, G., and Yacamini, R. (2003). Power system harmonic analysis using the Kalman filter. *2003 IEEE Power Engineering Society General Meeting (IEEE Cat. No.03CH37491)*, Volume 2, 752–757.

9 Qi, J., Sun, K., Wang, J., and Liu, H. (2018). Dynamic state estimation for multi-machine power system by unscented Kalman filter with enhanced numerical stability. *IEEE Transactions on Smart Grid* 9 (2): 1184–1196.

10 Li, S., Hu, Y., Zheng, L. et al. (2019). Stochastic event-triggered cubature Kalman filter for power system dynamic state estimation. *IEEE Transactions on Circuits and Systems II: Express Briefs* 66 (9): 1552–1556.

11 Lorentzen, R.J. and Naevdal, G. (2011). An iterative ensemble Kalman filter. *IEEE Transactions on Automatic Control* 56 (8): 1990–1995.

12 Emami, K., Fernando, T., Iu, H.H.-C. et al. (2015). Particle filter approach to dynamic state estimation of generators in power systems. *IEEE Transactions on Power Systems* 30 (5): 2665–2675.

13 Yu, S., Emami, K., Fernando, T. et al. (2016). State estimation of doubly fed induction generator wind turbine in complex power systems. *IEEE Transactions on Power Systems* 31 (6): 4935–4944.

14 Yu, S., Fernando, T., Emami, K., and Iu, H.H.-C. (2016). Dynamic state estimation based control strategy for DFIG wind turbine connected to complex power systems. *IEEE Transactions on Power Systems* 32 (2): 1272–1281.

15 Paul, A., Joos, G., and Kamwa, I. (2018). Decentralized dynamic state estimation of doubly fed induction generator using terminal measurements. In: *2018 IEEE Power & Energy Society Innovative Smart Grid Technologies Conference (ISGT)*, 1–5. IEEE.

16 Bakhtiari, F. and Nazarzadeh, J. (2019). Optimal estimation and tracking control for variable-speed wind turbine with PMSG. *Journal of Modern Power Systems and Clean Energy* 8 (1): 159–167.

17 Afrasiabi, S., Afrasiabi, M., Rastegar, M. et al. (2019). Ensemble Kalman filter based dynamic state estimation of PMSG-based wind turbine. In: *2019 IEEE Texas Power and Energy Conference (TPEC)*, 1–4. IEEE.

18 Huang, S., Wang, T., Ji, T., and Jin, M. (2021). Adaptive cubature Kalman filter based dynamic state estimation for grid-connected photovoltaic system. In: *2021 4th International Conference on Energy, Electrical and Power Engineering (CEEPE)*, 570–575. IEEE.

19 Song, S., Wei, H., Lin, Y. et al. (2022). A holistic state estimation framework for active distribution network with battery energy storage system. *Journal of Modern Power Systems and Clean Energy* 10 (3): 627–636.

20 Li, Y., Zhang, L., Lai, K., and Zhang, X. (2022). Dynamic state estimation method for multiple battery energy storage systems with droop-based consensus control. *International Journal of Electrical Power & Energy Systems* 134: 107328.

21 Wang, T., Huang, S., Gao, M., and Wang, Z. (2021). Adaptive extended Kalman filter based dynamic equivalent method of PMSG wind farm cluster. *IEEE Transactions on Industry Applications* 57 (3): 2908–2917.

22 Yue, K., Liu, Y., Zhao, P. et al. (2021). Dynamic state estimation enabled health indicator for parametric fault detection in switching power converters. *IEEE Access* 9: 33224–33234.

23 Liu, Y., Choi, S., Meliopoulos, A.P.S. et al. (2016). Dynamic state estimation enabled preditive inverter control. In: *2016 IEEE Power and Energy Society General Meeting (PESGM)*, 1–5. IEEE.

24 Choi, S. and Meliopoulos, A.P.S. (2016). Effective real-time operation and protection scheme of microgrids using distributed dynamic state estimation. *IEEE Transactions on Power Delivery* 32 (1): 504–514.

25 Song, S., Wu, P., Lin, Y., and Chen, Y. (2021). A general dynamic state estimation framework for monitoring and control of permanent magnetic synchronous generators-based wind turbines. *IEEE Access* 9: 72228–72238.

26 Fu, X., Sun, J., Huang, M. et al. (2020). Large-signal stability of grid-forming and grid-following controls in voltage source converter: a comparative study. *IEEE Transactions on Power Electronics* 36 (7): 7832–7840.

27 Zhong, Q.-C. and Weiss, G. (2010). Synchronverters: inverters that mimic synchronous generators. *IEEE Transactions on Industrial Electronics* 58 (4): 1259–1267.

28 Li, Y., Gu, Y., and Green, T. (2022). Revisiting grid-forming and grid-following inverters: a duality theory. *IEEE Transactions on Power Systems* 37 (6): 4541–4554.

29 Lin, Y. and Abur, A. (2018). A highly efficient bad data identification approach for very large scale power systems. *IEEE Transactions on Power Systems* 33 (6): 5979–5989.

30 Meliopoulos, A.P.S., Cokkinides, G.J., Myrda, P. et al. (2017). Dynamic state estimation-based protection: status and promise. *IEEE Transactions on Power Delivery* 32 (1): 320–330.

31 Liu, Y., Singh, A.K., Zhao, J. et al. (2021). Dynamic state estimation for power system control and protection. *IEEE Transactions on Power Systems* 36 (6): 5909–5921.

32 Liu, Y., Meliopoulos, A.P., Sun, L., and Fan, R. (2016). Dynamic state estimation based protection of mutually coupled transmission lines. *CSEE Journal of Power and Energy Systems* 2 (4): 6–14.

33 Liu, Y., Meliopoulos, A.P.S., Fan, R. et al. (2017). Dynamic state estimation based protection on series compensated transmission lines. *IEEE Transactions on Power Delivery* 32 (5): 2199–2209.

34 Fan, R., Liu, Y., Meliopoulos, S. et al. (2020). Comparison of transformer legacy protective functions and a dynamic state estimation-based approach. *Electric Power Systems Research* 184: 106301.

35 Liu, Y., Meliopoulos, A.P., Sun, L., and Choi, S. (2018). Protection and control of microgrids using dynamic state estimation. *Protection and Control of Modern Power Systems* 3 (1): 1–13.

36 Lin, Y., Liu, Y., and Yue, K. (2023). Modern power system state estimation methods. In: *Encyclopedia of Electrical and Electronic Power Engineering*. Elsevier.

37 Liu, Y., Wang, B., Zheng, X. et al. (2020). Fault location algorithm for non-homogeneous transmission lines considering line asymmetry. *IEEE Transactions on Power Delivery* 35 (5): 2425–2437.

38 Lu, D., Liu, Y., Xie, J. et al. (2022). Multi-layer model enabled fault location for underground cables in MMC-HVDC grids considering distributed and frequency dependent line parameters. *IEEE Transactions on Power Delivery* 37 (4): 3082–3096.

39 Wang, B., Liu, Y., Yue, K. et al. (2021). Improved dynamic state estimation based protection on transmission lines in MMC-HVDC grids. *IEEE Transactions on Power Delivery* 37 (5): 3567–3581.

40 Boyd, S. and Vandenberghe, L. (2004). *Convex Optimization*. Cambridge University Press.

41 Liu, Y., Meliopoulos, A.P.S., Tan, Z. et al. (2017). Dynamic state estimation-based fault locating on transmission lines. *IET Generation, Transmission & Distribution* 11 (17): 4184–4192.

42 Wang, B., Liu, Y., Lu, D. et al. (2021). Transmission line fault location in MMC-HVDC grids based on dynamic state estimation and gradient descent. *IEEE Transactions on Power Delivery* 36 (3): 1714–1725.

5

Eigenanalysis of Delayed Networked Microgrids

Lizhi Wang, Yifan Zhou, and Peng Zhang

5.1 Introduction

Networked microgrids (NMs) are emerging as a promising paradigm toward high electricity resiliency and reliability for communities. As NMs are prone to heterogeneous delays and frequent perturbations, reliably assessing small signal stability for the improvement of NMs' situational awareness, and controllability is key to determining whether NMs can be used as dependable resiliency resource [1].

Eigenanalysis for NMs is still an intractable challenge. First, for NMs equipped with distributed control, there exist ubiquitous and heterogeneous communication delays (i.e. local delays within MG and global delays among MGs) due to dynamical routing and disturbances in the communication throughput and measurement delays [2, 3]. Second, there is a lack of a unified NMs model indispensable for understanding the control interactions among microgrids and the evolution of system stability during the formation of a system of NMs. Finally, the best practice of validating NMs' eigenanalysis results using a real-time simulator is yet to be explored.

To tackle the challenges, this chapter devises a delayed NMs model considering heterogeneous delays and develops an ordinary-differential-equation-based solution operator discretization (ODE-SOD) method capable of computing the eigenspectra and delay margins (stability margins formed by critical delays). Furthermore, this chapter offers insights into the stability impacts of measurement delays, local communication delays among distributed energy resources (DERs), and global communication delays among microgrids and fully validates them on an RTDS-based cyber-physical testbed.

5.2 Formulation of Delayed NMs

Depending on costs, coverage requirements, and suitability for control, NMs adopt different networking technologies (e.g. Ethernet, WiFi, WiMax, cellular) resulting in various Quality of Service (QoS) and delays [1].

The dynamic formulation of an islanded microgrid is introduced first. A DER controller with droop is considered, which employs the power controller, current controller, and voltage controller:

$$\frac{d\delta_i}{dt} = \omega_i - \omega_{\text{ref}} \tag{5.1a}$$

$$\frac{dP_i}{dt} = \omega_{\text{c},i}(-P_i + v_{\text{od},i}i_{\text{od},i} + v_{\text{oq},i}i_{\text{oq},i}) \tag{5.1b}$$

Microgrids: Theory and Practice, First Edition. Edited by Peng Zhang.

$$\frac{dQ_i}{dt} = \omega_{c,i}(-Q_i + v_{oq,i}i_{od,i} - v_{od,i}i_{oq,i}) \tag{5.1c}$$

$$\frac{d\phi_{dq,i}}{dt} = v^*_{odq,i} - v_{odq,i} \tag{5.1d}$$

$$\frac{d\gamma_{dq,i}}{dt} = i^*_{Ldq,i} - i_{Ldq,i} \tag{5.1e}$$

$$\frac{di_{Ldq,i}}{dt} = \frac{-r_{f,i}}{L_{f,i}}i_{Ldq,i} \pm \omega_i i_{Lqd,i} + \frac{1}{L_{f,i}}(v^*_{idq,i} - v_{odq,i}) \tag{5.1f}$$

$$\frac{dv_{odq,i}}{dt} = \pm\omega_i v_{oqd,i} + \frac{1}{C_{f,i}}i_{Ldq,i} - \frac{1}{C_{f,i}}i_{odq,i} \tag{5.1g}$$

$$\frac{di_{odq,i}}{dt} = \frac{-r_{c,i}}{L_{c,i}}i_{odq,i} \pm \omega_i i_{oqd,i} + \frac{1}{L_{c,i}}(v_{odq,i} - v_{ndq,i}) \tag{5.1h}$$

where i denotes the DER index; t denotes the time. State variables of (5.1) include δ_i (DER angle), P_i (active power generation), Q_i (reactive power generation), $\phi_{dq,i} = [\phi_{d,i}; \phi_{q,i}]$ (output signal of the voltage controller in dq-axis), $\gamma_{dq,i} = [\gamma_{d,i}; \gamma_{q,i}]$ (output signal of the current controller in dq-axis), $i_{Ldq,i} = [i_{Ld,i}; i_{Lq,i}]$ (current after the output LC filter), $v_{odq,i} = [v_{od,i}; v_{oq,i}]$ (voltage output of DER in dq-axis), and $i_{odq,i} = [i_{od,i}; i_{oq,i}]$ (current output of DER in dq-axis).

The nonlinear ODE in (5.1) formulates the dynamics of the droop-controlled inverters, and this controller sketch is widely utilized to study the system transient responses under large disturbances [4–6]. Specifically, (5.1a)–(5.1c) formulate the power controller, where $\omega_i = \omega_{0,i} - m_{p,i}(P_i - P_{0,i})$ is the angular frequency; $m_{p,i}$ is the active power droop gain of DER i; $\omega_{0,i}$ is the nominal angular frequency; $P_{0,i}$ is the reference active power; ω_{ref} denotes the angular frequency of the reference DER [7]; $\omega_{c,i}$ is the filter parameter. Equation (5.1d) formulates the voltage controller, where $v^*_{od,i} = V_{n,i} - n_{q,i}Q_i$; $v^*_{oq,i} = 0$; $n_{q,i}$ is the reactive power droop gain of DER i; and $V_{n,i}$ is the nominal voltage parameter. Equation (5.1e) formulates the current controller, where $i^*_{Ldq,i} = F_i i_{odq,i} \mp \omega_{n,i}C_{f,i}v_{oqd,i} + K_{pv,i}(v^*_{odq,i} - v_{odq,i}) + K_{iv,i}\phi_{dq,i}$; and $F_i, K_{pv,i}, K_{iv,i}, C_{f,i}$ are controller parameters. Equations (5.1f)–(5.1h) models the LC filter for DER voltage and current output, where $v^*_{idq,i} = \mp\omega_{n,i}L_{f,i}i_{Lqd,i} + K_{pc,i}(i^*_{Ldq,i} - i_{Ldq,i}) + K_{ic,i}\gamma_{dq,i}$; and $F_i, K_{pc,i}, K_{ic,i}, L_{f,i}$ are controller parameters; $v_{nd,i}$ and $v_{nq,i}$ are dq-axis voltage at the bus connected to DER i.

Recently, it has been reported that neglecting the dynamics in power lines and loads leads to overly optimistic stability assessment. This motivated us to model the dynamics in the entire NMG. For each constant impedance load, the dynamics are described by the load resistance and inductance:

$$\frac{di_{lDQ,j}}{dt} = -\frac{r_{l,j}}{L_{l,j}}i_{lDQ,j} \pm \omega_{ref}i_{lQD,j} + \frac{1}{L_{l,j}}v_{lDQ,j} \tag{5.2}$$

where j denotes the load index; $i_{lDQ,j} = [i_{lD,j}; i_{lQ,j}]$ denotes the load current j in DQ-axis (i.e. the common reference frame [7]); $v_{lDQ,j} = [v_{lD,j}; v_{lQ,j}]$ denotes the DQ-axis voltage at the bus connected to load j; $r_{l,j}$ and $L_{l,j}$ are the load resistance and inductance.

For each branch in the microgrid, the dynamics are modeled by the branch resistance and inductance:

$$\frac{di_{bDQ,k}}{dt} = -\frac{r_{b,k}}{L_{b,k}}i_{bDQ,k} \pm \omega_{ref}i_{bQD,k} + \frac{1}{L_{b,k}}v_{bDQ,k} \tag{5.3}$$

where k denotes the branch index; $i_{\text{bDQ},k} = [i_{\text{bD},k}; i_{\text{bQ},k}]$ denotes the branch currents in DQ-axis; $v_{\text{bDQ},k} = [v_{\text{bD},k}; v_{\text{bQ},k}]$ is the DQ-axis voltage difference along the branch; and $r_{b,k}$ and $L_{b,k}$ are the branch resistance and inductance.

At each bus of the microgrid, the Kirchhoff's Current Law (KCL) should be satisfied:

$$0 = \sum_{i \in S_n^{\text{DER}}} i_{\text{oDQ},i} - \sum_{j \in S_n^l} i_{\text{lDQ},j} - \sum_{k \in S_n^b} i_{\text{bDQ},k} \tag{5.4}$$

where n is the bus index; S_n^{DER}, S_n^l and S_n^b denote the sets of DERs, loads, and branches connected to bus n; and $i_{\text{oDQ},i} = [i_{\text{oD},i}; i_{\text{oQ},i}]$ denotes the output current of DER i in the DQ-axis by the following transformation:

$$\begin{bmatrix} i_{\text{oD},i} \\ i_{\text{oQ},i} \end{bmatrix} = \begin{bmatrix} \cos \delta_i & -\sin \delta_i \\ \sin \delta_i & \cos \delta_i \end{bmatrix} \begin{bmatrix} i_{\text{od},i} \\ i_{\text{oq},i} \end{bmatrix} \tag{5.5}$$

By integrating (5.1)–(5.5), the microgrid model can be established, which is a nonlinear differential algebraic equation (DAE) model addressing the transient responses of each microgrid component. To solve it, a ODE model of microgrids can be converted from a DAE model [8].

Local controls at local devices or DERs introduce *measurement delays*. Within the territory of an individual microgrid, those controls relying on inter-DER communications introduce *local communication delays* [9]. Furthermore, global NM controls introduce *global communication delays*.

A general dynamic model of an arbitrary NM component considering heterogeneous delays can be established as:

$$\dot{x}^{(i)}(t) = f_i(x^{(i)}(t), y^{(i)}(t), x_m^{(i)}(t - \tau_i^m), y_m^{(i)}(t - \tau_i^m), \\ \{x_c^{(j)}(t - \tau_{ij}^c), y_c^{(j)}(t - \tau_{ij}^c) \mid j \in N_i\}) \tag{5.6}$$

Here, f_i denotes the component dynamics; $x^{(i)}$ and $y^{(i)}$, respectively, denote the non-delayed differential and algebraic states of component i; $x_m^{(i)}(t - \tau_i^m)$ and $y_m^{(i)}(t - \tau_i^m)$, respectively, denote the differential and algebraic states subject to measurement delays τ_i^m; $x_c^{(j)}(t - \tau_{ij}^c)$ and $y_c^{(j)}(t - \tau_{ij}^c)$, respectively, denote the differential and algebraic states subject to delays τ_{ij}^c of communications between i and j; N_i denotes the set of components communicating with component i.

The overall delayed NMs model is integrated as follows:

$$\begin{cases} \dot{x}^{(i)}(t) = f_i(x^{(i)}(t), y^{(i)}(t), x_m^{(i)}(t - \tau_i^m), y_m^{(i)}(t - \tau_i^m), \\ \qquad \{x_c^{(j)}(t - \tau_{ij}^c), y_c^{(j)}(t - \tau_{ij}^c) \mid j \in N_i\}), \ \forall i \\ 0 = g(x(t), y(t)) \\ 0 = g(x(t - \tau), y(t - \tau)), \ \forall \tau \in \{\tau_i^m | \forall i\} \cup \{\tau_{ij}^c | \forall i, j\} \end{cases} \tag{5.7}$$

Here, g represents the algebraic (power flow) equations; x and y indicate, respectively, all the differential and algebraic states.

Linearizing (5.7) at equilibrium point and eliminating y finally establish the small-signal stability model of delayed NMs:

$$\begin{cases} \Delta \dot{x}(t) = \tilde{A}_0 \Delta x(t) + \sum_i \tilde{A}_i^m \Delta x(t - \tau_i^m) \\ \qquad + \sum_i \sum_{j \in N_i} \tilde{A}_{ij}^c \Delta x(t - \tau_{ij}^c) \\ \Delta x(t) \triangleq \varphi, \ t \in [-\tau_{\max}, 0] \end{cases} \tag{5.8}$$

Here, $\tilde{A}_0$ is the non-delayed state matrix; $\tilde{A}_i^m$ and $\tilde{A}_{ij}^c$ are, respectively, the delayed state matrices caused by measurements and communications; $\tau_{\max}$ is the maximum delay in the NMs.

5.3 Delayed NMs Eigenanalysis

5.3.1 Solution Operator Basics

The solution operator $\mathcal{T}(h)$ associated with (5.9) maps the initial condition φ at t into the system state at a later time instant $t + h$, where h is the transfer step length satisfying $0 < h \leq \tau_{max}$:

$$(\mathcal{T}(h))\varphi(t) = \Delta x_h(t) = \Delta x(t + h), \quad t \in [\tau_{max}, 0] \tag{5.9}$$

$\mathcal{T}(h)$ is a piecewise function with the following two segments:

(1) **Time integration**: When $t \in [-h, 0]$ and $t + h \in [0, h]$, $\Delta x_h(t)$ is the solution of (3), whose existence is determined by the Picard's Existence and Uniqueness Theorem [10].
(2) **Shift**: For $t \in [-\tau_{max}, -h]$, $\Delta x_h(t)$ is always the initial condition part of (3) since $t + h \leq 0$. $\mathcal{T}(h)$ in this case is a shift. $\mathcal{T}(h)$ is explicitly formulated as:

$$
\Delta x_h(t)
$$
$$
= \begin{cases}
\varphi(0) + \int_0^t (\tilde{A}_0 \Delta x_h(t) + \sum_i \tilde{A}_i^m \Delta x_h(t - \tau_i^m) \\
\quad + \sum_i \sum_{j \in N_i} \tilde{A}_{ij}^c \Delta x_h(t - \tau_{ij}^c)) ds \ t \in [-h, 0] \\
\varphi(t + h) \qquad\qquad\qquad\qquad t \in [-\tau_{max}, -h].
\end{cases} \tag{5.10}
$$

5.3.2 ODE-SOD Eigensolver

Constructing the solution operator allows for mapping the eigen-spectrum λ of the original delayed NMs (see (5.8)) on the s plane into an eigen-spectrum μ of $\mathcal{T}(h)$ on the z-plane, which can be further discretized [11] to tractably obtain μ. The mapping is detailed as:

$$
\begin{cases}
\mu = e^{\lambda h}, \quad \mu \in \sigma(\mathcal{T}(h)) \backslash 0 \\
\lambda = \frac{1}{h} \ln \mu = \frac{1}{h}(\ln |\mu| + j \arg \mu)
\end{cases} \tag{5.11}
$$

where $|\mu|$ and $\arg \mu$ are modulus and principal argument of μ.

This procedure is thus called the ODE-SOD eigenanalysis method. The stability of delayed NMs can be directly determined by the largest modulus $|\mu_1|$. If $|\mu_1| > 1$, the system is unstable. If $|\mu_1| < 1$, the system is asymptotically stable. $|\mu_1| = 1$ denotes that the system is critical, and the corresponding delay τ is the delay margin of the NMs.

To further speed up the calculation, a coordinate rotation (i.e. $\tilde{A}^{0\prime} = \tilde{A}_0 e^{-j\theta}$, $\tilde{A}^{k\prime} = \tilde{A}^k e^{-j\theta}$, $\tau_k' = \tau_k e^{j\theta}$) is employed to calculate only those eigenvalues with a damping ratio less than $\zeta = sin(\theta)$. To discretize $\mathcal{T}(h)$, two meshes Ω_N and Ω_M are established using pseudospectral discretization [10, 12]. The mesh Ω_N consists of N scaled and shifted zeros of an Nth order Chebyshev polynomial of the first kind over $[0, h]$. The mesh Ω_M contains $QM + 1$ nodes, where the interval $[-\tau_{max}, 0]$ is first divided into $Q = [\tau_{max}/h]$ sub-intervals and each sub-interval is then discretized by $M + 1$ scaled and shifted zeros of an Mth order Chebyshev polynomial of the second kind. Correspondingly, $\mathcal{T}(h)$ is discretized as matrix $\mathbf{T}_{M,N}$ of $(QM + 1)n \times (QM + 1)n$ dimension:

$$\mathbf{T}_{M,N} = \Pi_M + \Pi_{M,N}(\mathbf{I}_{Nn} - \Sigma_N)^{-1}\Sigma_{M,N} \tag{5.12}$$

where $I_{Nn} \in \mathbb{R}^{Nn \times Nn}$ is an identity matrix; $\Sigma_N = \sum_{i=1}^{m} L_N^i \otimes \tilde{A}_i$ and $\Sigma_{M,N} = \sum_{i=1}^{m} L_{M,N}^i \otimes \tilde{A}_i$ are dense matrices; $\Pi_M = U_M \otimes I_N$ and $\Pi_{M,N} = U_{M,N} \otimes I_N$ are highly sparse matrices; $\otimes$ denotes the Kronecker product; $L_{M,N}^i$ and L_N^i are delay-dependent Lagrange coefficient matrices; and U_M and $U_{M,N}$ are both constant matrices. Algorithm 5.1 summarizes ODE-SOD.

Algorithm 5.1: ODE-SOD Algorithm for Delayed NMs

▷ **Initialization**: System state matrix $\tilde{A}^0$, delayed state matrices $\tilde{A}_{ij}^c$, $\tilde{A}_i^m$, delay τ_{ij}^c and τ_i^m, damping ratio ζ ;

▷ **Parameters tuning and matrices precondition**:

$h = \tau_{\max}/12$

if $\tau_i^m = \tau_j^m$ *or* $\tau_i^m = \tau_{ij}^c$ **then**

$\quad$ $\tilde{A}^k = \tilde{A}_i^m + \tilde{A}_{ij}^c$;

$\quad$ $\tau_k = \tau_i^m = \tau_j^m = \tau_{ij}^c, (i, j \in N)$;

else

$\quad$ $\tilde{A}^k = \tilde{A}_i^m, ..., \tilde{A}_{ij}^c$;

$\quad$ $\tau_k = \tau_i^m, ..., \tau_{ij}^c, (i, j \in N)$;

end

▷ **Eigenvalues computation**:

Coordinate rotation of $\tilde{A}^{0\prime}, \tilde{A}^{k\prime}, \tau_k^\prime$;

Discretization of solution operator $\mathcal{T}(h)$ using Nth order Chebyshev polynomial method;

Compute $\mu^\prime$ by implicitly restarted Arnoldi (IRA) iteration and $\hat{\lambda}$ by (5.11);

Refine $\hat{\lambda}$ to a accurate λ by Newton iterations;

▷ **Output**: Eigenvalues λ and μ with damping ratio less than ζ; largest modulus $|\mu_1|$ and the system's stability; delay margin $\{\tau_{ij}^c, \tau_i^m\}$ with $|\mu_1| = 1$;

5.4 Case Study

Extensive case studies are performed on a six-microgrid NM system, as shown in Fig. 5.1. A two-layer hierarchical control scheme is employed, where the inverter model includes the power-sharing controller dynamics, output filter dynamics, coupling inductor dynamics, voltage, and current controller dynamics, and global secondary controller (i.e. a distributed-averaging proportional-integral control) dynamics [1].

The ODE-SOD algorithm is developed in MATLAB R2020a on a 2.50 GHz PC. A high-fidelity RTDS model of the test system is built to verify the delayed eigenanalysis. The simulation time step in RTDS is set as 50 µs.

5.4.1 Methodology Validity

The efficacy of delayed eigenanalysis of NMs is verified with three typical cases.

Case I is a nondelayed scenario which allows for the use of the conventional QR method to verify the baseline performance. Figure 5.2 illustrates the correctness of the ODE-SOD-based solutions in that they are identical to those from QR. Meanwhile, a z-plane observation (see Fig. 5.2(b)) shows that all the mapped eigenvalues are enclosed within a unit circle, indicating a stable nondelayed

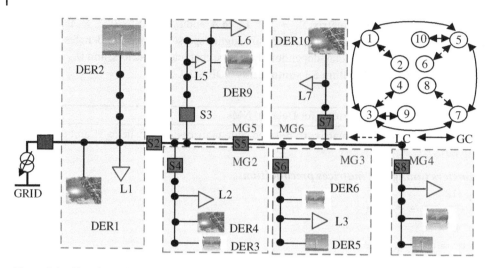

Figure 5.1 Six-microgrid test NMs system with communication topology. Source: Wang et al. [7]/IEEE.

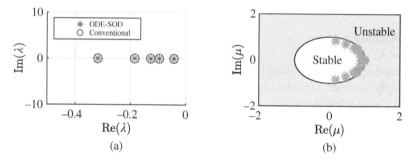

Figure 5.2 Case I: nondelayed NMs. (a) *s*-plane eigen-spectrum and (b) *z*-plane eigen-spectrum.

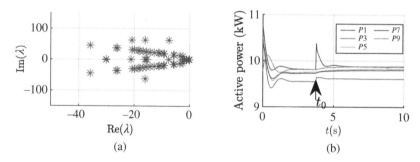

Figure 5.3 Case II: delayed yet stable NMs ($\tau^c = 100$ ms, $\tau^m = 0$). (a) Rightmost delayed eigenvalues and (b) RTDS simulation results.

NM system. Case II analyzes delayed yet stable NMs. The delayed eigenanalysis shows that all the eigenvalues are located on the left-half plane, which is coincident with the stabilized transients of the perturbed NMs (see the RTDS results in Fig. 5.3). In contrast, Case III demonstrates instabilities due to delays, where the delayed eigenanalysis reveals unstable eigenvalues entering the right-half plane. Those eigenresults, as Fig. 5.4 shows, are again verified by RTDS which reproduces the undamped transients triggered by a small disturbance.

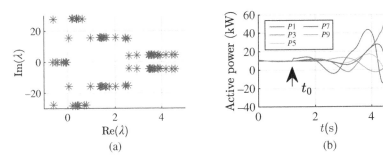

Figure 5.4 Case III: delayed and unstable NMs ($\tau^c = 100$ ms, $\tau^m = 100$ ms). (a) Rightmost delayed eigenvalues and (b) RTDS simulation results.

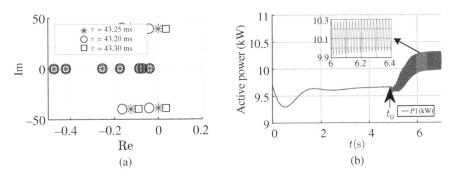

Figure 5.5 Small-signal stability of delayed NMs under a critical scenario ($\tau = \tau^c = 43.25$ ms $= \tau^m = 43.25$ ms). (a) Rightmost delayed eigenvalues and (b) RTDS simulation results.

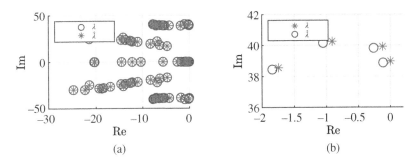

Figure 5.6 Small-signal stability of delayed NMs under critical scenarios. (a) Eigenvalue comparison and (b) three typical Eigenvalue comparison.

Figure 5.5 further demonstrates a critical case in which the rightmost multiple eigenvalues are located on the imaginary axis, with $\tau^c = 43.25$ ms and $\tau^m = 43.25$ ms. RTDS simulation results in Fig. 5.5(b) verify that a small disturbance causes undamped oscillations.

Additionally, Fig. 5.6(a) illustrates the necessity of using Newton's iterations in the ODE-SOD method. The estimated eigenvalues $\hat{\lambda}$ without Newton iterations are compared with the accurate eigenvalues λ (i.e. the final result of ODE-SOD). It can be seen that although λ and $\hat{\lambda}$ are relatively close, $\hat{\lambda}$ may not rigorously satisfy the transcendental characteristics equations of the delayed system. As shown in Fig. 5.6(b), for the estimated eigenvalue $\hat{\lambda}_i$, it takes several iterations to converge to λ_i.

5.4.2 Cyber Network's Impact on NMs Stability

5.4.2.1 Impact of Communication Delay

Two communication topologies are designed to investigate the impact of communication delays on NMs' stability: (a) a fully connected topology; and (b) a partially connected topology from which MG3's cyber network is separated.

Evolution of the rightmost mapped eigenvalue μ_1 with local communication delays τ_{local}^c as well as global communication delays τ_{global}^c indicates that:

- Decreases of communication delays generally enhance the NMs stability (i.e. decreased μ_1).
- A stronger communication connection tends to improve the NMs stability, i.e. NMs stability in Fig. 5.7(a) is superior to that in Fig. 5.7(b).
- Communication delay effect may "saturate." Even if communication delays become large locally and/or globally, they do not necessarily turn a stable NM system into an unstable one (i.e. μ_1 is always lower than 1 in Fig. 5.7).

5.4.2.2 Impact of Measurement Delay

Overlaying the impact of measurement delays with that of communication delays gives the following observations:

- Increases of measurement delays decrease the NMs stability as shown in Fig. 5.8.
- Measurement delays have significantly higher impacts on the NMs stability than communication delays do, as demonstrated in Figs. 5.7 and 5.8(a).
- An interesting finding is that the rightmost eigenvalue $\text{Re}(\lambda_1)$ decreases when τ^m changes from 2.5 to 5 seconds (see Fig. 5.8(b)). This means that, once measurement delays reach a threshold, further increase of delays no longer worsens the stability of NMs.
- The ODE-SOD-based eigenanalysis reveals that the measurement delay margin of the test NMs is about 50 ms, as shown in Fig. 5.8(b). Thus, this tool offers useful information for NM planning and operations by estimating the maximum delays that the system can withstand.

5.4.3 Electrical Network's Impact on NMs Stability

This section tries to answer two pressing questions: Does interconnecting more microgrids make an existing NM system more stable? How does the strength of the electric grid affect delayed

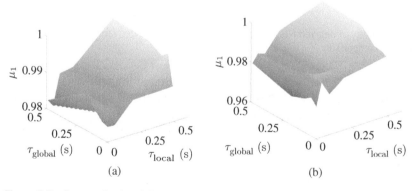

Figure 5.7 Communication delay and topology affecting NMs stability. (a) Trends of μ_1 under a fully connected topology and (b) trends of μ_1 under a partially connected topology.

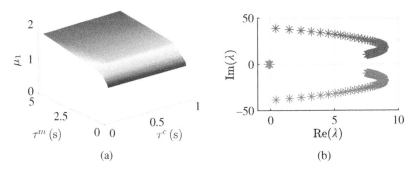

Figure 5.8 Impact of measurement delays on NMs stability. (a) Trends of μ_1 under heterogeneous delays and (b) trace of the rightmost eigenvalue with different τ^m (1 ms–5 s).

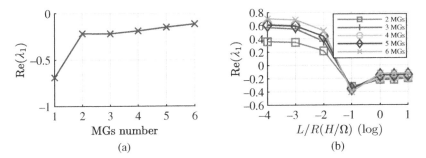

Figure 5.9 Impact of electrical connections on NMs stability. (a) Re(λ_1) with different number of MGs and (b) Re(λ_1) with different L/R.

NMs' stability? In this case, the time delay of NMs is the same as Case II, that is $\tau^c = 100$ ms. Figure 5.9(a) shows that with more microgrids interconnected, Re(λ_1) would further approach zero. The stability margin of a NMs system, therefore, likely decreases with more microgrids forming into NMs. Figure 5.9(b) illustrates the impact of the L/R ratio of the tie lines. The NMs stability deteriorates with decreased L/R, meaning a weakened connection likely induces instability of delayed NMs.

5.5 Conclusion

This chapter develops a delayed eigenanalysis tool capable of quantifying NMs' stability under various factors such as heterogeneous delays, communication topologies, microgrid plug-and-play, and electrical distances. The next step is to further extend the delayed eigenanalysis method to efficiently calculate eigenresults under time-varying and stochastic delays in NMs.

5.6 Exercises

1. What is a small-signal stability problem? Explain different categories of small-signal stability problems.
2. Explain the small signal stability of a single-machine infinite bus system.

Table 5.1 Inverter parameters (10 kVA rating).

Parameter	Value	Parameter	Value
f_s	8 kHz	m_p	9.4e−5
L_f	0.35 mH	n_q	0.3e−3
C_f	50 μF	k_{pv}	0.05
r_f	0.1 Ω	k_{iv}	390
L_c	0.35 mH	k_{pc}	10.5
r_{LC}	0.03 Ω	k_{ic}	16e3
ω_c	31.41	F	0.75
r_{load}	25		

3. We have a DER with droop control connecting to a constant RL load. The DER has a power controller, voltage controller, and current controller and is connected to the bus node through an output LC filter and coupling inductance. The parameter is listed in the Table 5.1.
   – Write the state equations of the power controller, voltage controller, current controller, LC and coupling inductance, line, and load.
   – Write the linearized state equations of the whole system.
   – Determine the eigenvalues, damped frequency of the oscillator in Hz, and damping ratio.

References

1 Zhang, P. (2021). *Networked Microgrids*. Cambridge University Press.
2 Wang, L., Qin, Y., Tang, Z., and Zhang, P. (2020). Software-defined microgrid control: the genesis of decoupled cyber-physical microgrids. *IEEE Open Access Journal of Power and Energy* 7: 173–182.
3 Luo, H., Hiskens, I.A., and Hu, Z. (2020). Stability analysis of load frequency control systems with sampling and transmission delay. *IEEE Transactions on Power Systems* 35 (5): 3603–3615.
4 Kabalan, M., Singh, P., and Niebur, D. (2017). Nonlinear Lyapunov stability analysis of seven models of a DC/AC droop controlled inverter connected to an infinite bus. *IEEE Transactions on Smart Grid* 10 (1): 772–781.
5 Dheer, D.K., Soni, N., and Doolla, S. (2016). Improvement of small signal stability margin and transient response in inverter-dominated microgrids. *Sustainable Energy, Grids and Networks* 5: 135–147.
6 Huang, L., Xin, H., Wang, Z. et al. (2017). Transient stability analysis and control design of droop-controlled voltage source converters considering current limitation. *IEEE Transactions on Smart Grid* 10 (1): 578–591.
7 Pogaku, N., Prodanovic, M., and Green, T.C. (2007). Modeling, analysis and testing of autonomous operation of an inverter-based microgrid. *IEEE Transactions on Power Electronics* 22 (2): 613–625.
8 Zhou, Y., Zhang, P., and Yue, M. (2020). An ODE-enabled distributed transient stability analysis for networked microgrids. *2020 IEEE Power Energy Society General Meeting (PESGM)*, 1–5. https://doi.org/10.1109/PESGM41954.2020.9282139.

9 Xu, L., Guo, Q., Wang, Z., and Sun, H. (2021). Modeling of time-delayed distributed cyber-physical power systems for small-signal stability analysis. *IEEE Transactions on Smart Grid* 12 (4): 3425–3437.

10 Breda, D., Maset, S., and Vermiglio, R. (2012). Approximation of eigenvalues of evolution operators for linear retarded functional differential equations. *SIAM Journal on Numerical Analysis* 50 (3): 1456–1483.

11 Ye, H., Mou, Q., and Liu, Y. (2017). Calculation of critical oscillation modes for large delayed cyber-physical power system using pseudo-spectral discretization of solution operator. *IEEE Transactions on Power Systems* 32 (6): 4464–4476. https://doi.org/10.1109/TPWRS.2017.2686008.

12 Ye, H., Mou, Q., and Liu, Y. (2017). Matlab code for the SOD-PS algorithm, February 2017.

6

AI-Enabled Dynamic Model Discovery of Networked Microgrids

Yifan Zhou and Peng Zhang

Networked microgrids (NMs) offer an effective platform to coordinate renewable energies, support communities, and empower power system resilience [1–3]. Because of the high penetration of inverter-interfaced distributed energy resources (DERs), NMs may constantly undergo frequent, rapid, and nonlinear transient processes induced by fluctuating, exogenous, and random uncertainties [4]. Conventional NMs analytics are physics based, which strongly rely on accurate dynamic models of the system. However, in real-world NMs, the thorough state-space model of each and every microgrid may not always be attainable [5, 6], especially the detailed models of DER controllers, due to the possible vacancy of microgrids parameters, the frequently changing control strategies and plug-and-play operations, the privacy of consumers, etc.

Constructing a reliable dynamic model for NMs from the Supervisory Control and Data Acquisition (SCADA) and μPMU (phasor measurement unit) data, therefore, becomes an indispensable basis for NMs dynamic analysis when accurate physics models are unattainable. This is a long-standing obstacle due to the "random walk" of the NM operating points disturbed by uncertainties as well as the strong nonlinearity of NMs models induced by the dynamics of massive DERs. Recently, the swift growth of artificial intelligence (AI) techniques has shed light on constructing learning-based dynamic models of the unidentified subsystems of NMs. The flexibility, expressibility, and generalization capability of machine learning (ML) techniques enable establishing strongly nonlinear microgrid dynamics without assuming any specific dynamic modes in prior.

This chapter focuses on the AI-enabled dynamic model discovery of NMs. A neural-ordinary-differential-equations (ODE-Net)-enabled method is presented to learn continuous-time dynamic models of the unidentified subsystems of NMs under heterogeneous uncertainties [7]. Specifically, Section 6.1 introduces the background knowledge on the physics-based ODE model for NMs; Section 6.2 establishes the physics-data-driven NMs dynamic model; Section 6.3 presents the ODE-Net-enabled dynamic model discovery method of microgrids; Section 6.4 further introduces a physics-informed learning for the ODE-Net-enabled dynamic models; and finally, Section 6.5 presents numerical experiments on typical test systems to validate the method.

6.1 Preliminaries on ODE-Based Dynamical Modeling of NMs

In general, a NM system can be functionality described by a differential algebraic equation (DAE) system, where the differential equations describe the component dynamics and the algebraic equations describe the network power flow. However, for the dynamics of NMs, two specific features stand out:

- **Inverter controllers**: A salient feature of NMs is the capability of coordinating DERs and microgrids via a hierarchical control scheme. Since DERs are connected to the grid via inverter-based controllers, different control modes (e.g. grid-forming control, grid-following control) can significantly impact the dynamics of NMs.
- **Network dynamics**: NMs are usually comprised small-scale local grids featured by small values of network impedance [8]. The dynamics of line branches have been found to play an unexpectedly important role in the NMs dynamics. Ignoring the network dynamics (e.g. simplifying them into algebraic power flow equations) can engender an overly optimistic assessment of the dynamic behavior of DER-dominated microgrids and may even lead to disastrous hazards in the operations of NMs.

Consequently, it is necessary to establish NMs dynamic models comprising the thorough dynamics of the inverter-interfaced DERs, the power loads, the transmission lines, etc. This section establishes the dynamic model of an NMs system based on its physics nature, which lays the foundation for constructing AI-enabled dynamic models of NMs from data. We will show that, analytically, the dynamics of DER-dominated NMs incorporating network transients are governed by a system of nonlinear ordinary differential equations (ODEs).

6.1.1 Formulation of DERs with Hierarchical Control

Specifically, we consider a two-layer hierarchical control in the NMs. As illustrated in Fig. 6.1, in the hierarchical control scheme, droop control performs frequency/voltage regulation based on local measurements of DERs and secondary control performs coordination among DERs to achieve power-sharing and frequency/voltage restoration.

The droop control of DERs can be functionally formulated as follows:

$$\boldsymbol{\omega} = \boldsymbol{\omega}^* - \boldsymbol{m}_p(\boldsymbol{P} - \boldsymbol{P}^*) + \boldsymbol{\Omega} \tag{6.1a}$$

$$\boldsymbol{E} = \boldsymbol{E}^* - \boldsymbol{n}_q(\boldsymbol{Q} - \boldsymbol{Q}^*) + \boldsymbol{e} \tag{6.1b}$$

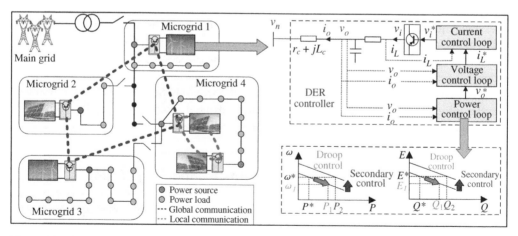

Figure 6.1 Schematic NMs with hierarchical control. Source: Sergiy Serdyuk/Adobe Stock Photos; European Bank for Reconstruction and Development (EBRD).

where vectors $\boldsymbol{\omega}$ and $\boldsymbol{E}$, respectively, denote DERs' angular speeds and voltage magnitudes; $\boldsymbol{P}$ and $\boldsymbol{Q}$, respectively, denote DERs' active/reactive power output; $\boldsymbol{m}_p$ and $\boldsymbol{n}_p$, respectively, denote the matrices of DERs' active/reactive droop gains; ω^*, $\boldsymbol{E}^*$, $\boldsymbol{P}^*$, and $\boldsymbol{Q}^*$, respectively, denote the nominal values of each signal; $\boldsymbol{\Omega}$ and $\boldsymbol{e}$ denote the secondary control signals.

As for secondary control, there are various methods to achieve control functionality. Without loss of generality, this section takes the distributed-averaging proportional-integral (DAPI) logic [9] as an example to introduce the NMs model with secondary control. In the DAPI scheme, each DER communicates with its neighbors to perform active/reactive power sharing and achieve the global control objectives. Correspondingly, the dynamics of DERs' secondary control signals are governed by:

$$\frac{d\boldsymbol{\Omega}}{dt} = -\boldsymbol{\alpha}(\boldsymbol{\omega} - \omega^*) - \boldsymbol{A}\boldsymbol{\Omega} \tag{6.2a}$$

$$\frac{d\boldsymbol{e}}{dt} = -\boldsymbol{\beta}(\boldsymbol{E} - \boldsymbol{E}^*) - \boldsymbol{B}\boldsymbol{Q} \tag{6.2b}$$

where $\boldsymbol{\alpha}$, $\boldsymbol{\beta}$, $\boldsymbol{A}$, and $\boldsymbol{B}$ are control parameters describing the control impact from each DER itself and neighbor DERs. The non-zero elements of $\boldsymbol{A}$ and $\boldsymbol{B}$ should be allied with the communication topology.

The aforementioned hierarchical control is implemented via the inner and outer control loops of each inverter. Based on the average model of the inverters, the DER controllers' dynamics are further formulated as follows [10]:

$$\frac{d\boldsymbol{\delta}}{dt} = \boldsymbol{\omega} - \omega_s \tag{6.3a}$$

$$\frac{d\boldsymbol{P}}{dt} = \omega_c(-\boldsymbol{P} + \boldsymbol{v}_{oD}\boldsymbol{i}_{oD} + \boldsymbol{v}_{oQ}\boldsymbol{i}_{oQ}) \tag{6.3b}$$

$$\frac{d\boldsymbol{Q}}{dt} = \omega_c(-\boldsymbol{Q} + \boldsymbol{v}_{oQ}\boldsymbol{i}_{oD} - \boldsymbol{v}_{oD}\boldsymbol{i}_{oQ}) \tag{6.3c}$$

$$\frac{d\boldsymbol{\phi}_D}{dt} = \boldsymbol{v}_{oD}^* - \boldsymbol{v}_{oD} - (\boldsymbol{\omega} - \omega_s)\boldsymbol{\phi}_Q \tag{6.3d}$$

$$\frac{d\boldsymbol{\phi}_Q}{dt} = \boldsymbol{v}_{oQ}^* - \boldsymbol{v}_{oQ} + (\boldsymbol{\omega} - \omega_s)\boldsymbol{\phi}_D \tag{6.3e}$$

$$\frac{d\boldsymbol{\gamma}_D}{dt} = \boldsymbol{i}_{LD}^* - \boldsymbol{i}_{LD} - (\boldsymbol{\omega} - \omega_s)\boldsymbol{\gamma}_Q \tag{6.3f}$$

$$\frac{d\boldsymbol{\gamma}_Q}{dt} = \boldsymbol{i}_{LQ}^* - \boldsymbol{i}_{LQ} + (\boldsymbol{\omega} - \omega_s)\boldsymbol{\gamma}_D \tag{6.3g}$$

$$\boldsymbol{L}_f\frac{d\boldsymbol{i}_{LD}}{dt} = -\boldsymbol{r}_f\boldsymbol{i}_{LD} + \omega_s\boldsymbol{L}_f\boldsymbol{i}_{LQ} + (\boldsymbol{v}_{iD}^* - \boldsymbol{v}_{oD}) \tag{6.3h}$$

$$\boldsymbol{L}_f\frac{d\boldsymbol{i}_{LQ}}{dt} = -\boldsymbol{r}_f\boldsymbol{i}_{LQ} - \omega_s\boldsymbol{L}_f\boldsymbol{i}_{LD} + (\boldsymbol{v}_{iQ}^* - \boldsymbol{v}_{oQ}) \tag{6.3i}$$

$$\boldsymbol{C}_f\frac{d\boldsymbol{v}_{oD}}{dt} = \omega_s\boldsymbol{C}_f\boldsymbol{v}_{oQ} + (\boldsymbol{i}_{LD} - \boldsymbol{i}_{oD}) \tag{6.3j}$$

$$\boldsymbol{C}_f\frac{d\boldsymbol{v}_{oQ}}{dt} = -\omega_s\boldsymbol{C}_f\boldsymbol{v}_{oQ} + (\boldsymbol{i}_{LQ} - \boldsymbol{i}_{oQ}) \tag{6.3k}$$

In (6.3a), $\boldsymbol{\delta}$ denotes the DERs' angles; $\boldsymbol{\omega}$ denotes the DERs' angle speeds as governed by the hierarchical control logic defined in (6.1a); ω_s denotes the NMs' frequency. In (6.3b) and (6.3c), $\boldsymbol{P}$ and $\boldsymbol{Q}$ denote DERs' active/reactive power as defined in (6.1); ω_c denotes the cut-off frequency of low-pass filters; $\boldsymbol{v}_{oD}$, $\boldsymbol{v}_{oQ}$, $\boldsymbol{i}_{oD}$, and $\boldsymbol{i}_{oQ}$, respectively, denote the current and voltage outputs of DERs in the DQ-axis (i.e. the common reference frame). In (6.3d) and (6.3e), $\boldsymbol{\phi}_D$ and $\boldsymbol{\phi}_Q$, respectively, denote the output signals of voltage controller in the DQ-axis; $\boldsymbol{v}_{oD}^* = \cos(\delta)\boldsymbol{E}$ and $\boldsymbol{v}_{oQ}^* = \sin(\delta)\boldsymbol{E}$. In (6.3f) and (6.3g), $\boldsymbol{\gamma}_D$ and $\boldsymbol{\gamma}_Q$, respectively, denote the output signals of the current controller in the DQ-axis; $\boldsymbol{i}_{LD}^* = \boldsymbol{F}\boldsymbol{i}_{oD} - \omega_n \boldsymbol{C}_f \boldsymbol{v}_{oQ} + \boldsymbol{K}_{pv}\left(\boldsymbol{v}_{oD}^* - \boldsymbol{v}_{oD}\right) + \boldsymbol{K}_{iv}\boldsymbol{\phi}_D$; $\boldsymbol{i}_{LQ}^* = \boldsymbol{F}\boldsymbol{i}_{oQ} + \omega_n \boldsymbol{C}_f \boldsymbol{v}_{oD} + \boldsymbol{K}_{pv}\left(\boldsymbol{v}_{oQ}^* - \boldsymbol{v}_{oQ}\right) + \boldsymbol{K}_{iv}\boldsymbol{\phi}_Q$. In (6.3h)–(6.3i), $\boldsymbol{i}_{LD}$, $\boldsymbol{i}_{LQ}$, $\boldsymbol{v}_{oD}$, and $\boldsymbol{v}_{oQ}$, respectively, denote the DQ-axis currents and voltages after the output LC filter; $v_{iD}^* = -\omega_n \boldsymbol{L}_f \boldsymbol{i}_{LQ} + \boldsymbol{K}_{pc}(\boldsymbol{i}_{LD}^* - \boldsymbol{i}_{LD}) + \boldsymbol{K}_{ic}\boldsymbol{\gamma}_D$; $v_{iQ}^* = \omega_n \boldsymbol{L}_f \boldsymbol{i}_{LD} + \boldsymbol{K}_{pc}(\boldsymbol{i}_{LQ}^* - \boldsymbol{i}_{LQ}) + \boldsymbol{K}_{ic}\boldsymbol{\gamma}_Q$. $\boldsymbol{r}_f$, $\boldsymbol{L}_f$, $\boldsymbol{C}_f$, $\boldsymbol{F}$, $\boldsymbol{K}_{pv}$, $\boldsymbol{K}_{iv}$, $\boldsymbol{K}_{pc}$, and $\boldsymbol{K}_{ic}$ are controller parameters.

6.1.2 Formulation of Network Dynamics

In the following, we further incorporate the dynamics of line branches and power loads into the dynamic model of NMs. Specifically, the dynamics in the NM branches can be modeled by the corresponding branch impedance:

$$L_b \frac{di_{bD}}{dt} = -r_b i_{bD} + \omega_s L_b i_{bQ} + M_b v_D \tag{6.4a}$$

$$L_b \frac{di_{bQ}}{dt} = -r_b i_{bQ} - \omega_s L_b i_{bD} + M_b v_Q \tag{6.4b}$$

where $\boldsymbol{i}_{bD}$ and $\boldsymbol{i}_{bQ}$ denote the DQ-axis branch current; $\boldsymbol{v}_D$ and $\boldsymbol{v}_Q$ denote the DQ-axis bus voltages; $\boldsymbol{r}_b$ and $\boldsymbol{L}_b$, respectively, denote the matrices of branch resistances and inductance; $\boldsymbol{M}_b$ denote the incidence matrix between branches and buses, whose (i,j)-element is 1 if branch i starts from bus j, -1 if branch i ends at bus j and 0 otherwise.

Similarly, DERs are connected to NMs through branches, whose dynamics are also described by their impedance:

$$L_o \frac{di_{oD}}{dt} = -r_o i_{oD} + \omega_s L_o i_{oQ} + v_{oD} - M_o v_D \tag{6.5a}$$

$$L_o \frac{di_{oQ}}{dt} = -r_o i_{oQ} - \omega_s L_o i_{oD} + v_{oQ} - M_o v_Q \tag{6.5b}$$

where $\boldsymbol{i}_{oD}$ and $\boldsymbol{i}_{oQ}$ denote the DQ-axis current outputs of DERs; $\boldsymbol{r}_o$ and $\boldsymbol{L}_o$ denote the impedance of DER branches; $\boldsymbol{M}_o$ denotes the incidence matrix between DERs and buses, whose (i,j)-element is 1 when DER i is at bus j, and 0 otherwise.

As for power loads, different types of loads (e.g. ZIP load and electric motor) lead to different formulation. Without loss of generality, the following takes the constant impedance load as an example to illustrate the load dynamic formulation in NMs:

$$L_l \frac{di_{lD}}{dt} = -r_l i_{lD} + \omega_s L_l i_{lQ} + M_l v_D \tag{6.6a}$$

$$L_l \frac{di_{lQ}}{dt} = -r_l i_{lQ} - \omega_s L_l i_{lD} + M_l v_Q \tag{6.6b}$$

where $\boldsymbol{i}_{lD}$ and $\boldsymbol{i}_{lQ}$ denote the DQ-axis load currents; $\boldsymbol{r}_l$ and $\boldsymbol{L}_l$, respectively, denote the matrices of load resistances and inductance; and $\boldsymbol{M}_l$ denotes the incidence matrix between power loads and buses.

6.1.3 ODE-Enabled NMs Dynamic Model

Differential Eqs. (6.1)–(6.6) jointly formulates the NMs incorporating both inverter dynamics and network dynamics. Furthermore, each component (e.g. DER, branch, and load) is integrated into the system via the current balance equation at each bus:

$$M_o^T i_{oD} - M_l^T i_{lD} - M_b^T i_{bD} = 0, \ \ M_o^T i_{oQ} - M_l^T i_{lQ} - M_b^T i_{bQ} = 0 \tag{6.7}$$

Consequently, the DAEs-enabled NMs model is established.

In the following, we will reformulate the aforementioned NMs dynamic model into an rigorously equivalent ODE system [11]. Denote $v = [v_D; v_Q]$ as the DQ-axis bus voltages and $i = [i_{oD}; i_{oQ}; i_{lD}; i_{lQ}; i_{bD}; i_{bQ}]$ as the DQ-axis component currents. Denote z as all the other state variables in the NMs model. Accordingly, (6.1)–(6.7) can be functionally expressed as:

$$\begin{cases} \dot{z} = g(i, z) & (6.8a) \\ \dot{i} = Av + h(i, z) & (6.8b) \\ 0 = Bi & (6.8c) \end{cases}$$

where (6.8a) is an abstraction of (6.1), (6.2), and (6.3); (6.8b) is an abstraction of (6.4), (6.5), and (6.6); (6.8c) is an abstraction of (6.7); and the expression of A, B, g, and h can be readily obtained from the NMs model detailed in (6.1)–(6.7).

As defined in (6.4), (6.5), and (6.6), B assembles the incidence matrices between each component and buses, which is supposed to be row full rank. We split matrix B into B_1 and B_0, where B_1 is constructed by the maximally linearly independent columns of B and B_0 is constructed by the other columns. Correspondingly, we split i into i_1 and i_0. Since B_1 is non-singular, (6.8c) can be rewritten as:

$$B_0 i_0 + B_1 i_1 = 0 \implies i_1 = -B_1^{-1} B_0 i_0 \tag{6.9}$$

Equation (6.9) indicates that component currents are not mutually independent as they are restricted by the Kirchhoff's current law (KCL). We split matrix A and function h corresponding to i_1 and i_0. Therefore, the dynamics of i_0, the vector of independent component currents, is given by:

$$\dot{i}_0 = A_0 v + h_0(i, z) \tag{6.10}$$

Substituting (6.9) into (6.8a) and (6.8b) yields the following:

$$\begin{cases} \dot{z} = g(i_0, i_1, z) = g(i_0, -B_1^{-1} B_0 i_0, z) =: \hat{g}(i_0, z) & (6.11a) \\ \dot{i} = Av + h(i_0, -B_1^{-1} B_0 i_0, z) =: Av + \hat{h}(i_0, z) & (6.11b) \end{cases}$$

Left-multiplying (6.11b) with B leads to:

$$BAv + B\hat{h}(i_0, z) = B\dot{i} \stackrel{(6.8c)}{=} 0 \implies v = -(BA)^{-1} B\hat{h} \tag{6.12}$$

As a conclusion, the bus voltages v of NMs can be expressed as a function of the state variables i_0 and z, when the network dynamics are incorporated into NMs model.

Finally, substituting v into (6.11) and (6.10) leads to the ODE model of the NMs:

$$\begin{cases} \dot{z} = \hat{g}(i_0, z) & (6.13a) \\ \dot{i}_0 = -A_0 (BA)^{-1} B\hat{h}(i_0, z) + \hat{h}_0(i_0, z) & (6.13b) \end{cases}$$

Equation (6.13) formulates the ODE-enabled NMs dynamic model. The DAE-ODE conversion processes are generic enough to incorporate different types of power mixes (such as DERs, synchronous generators (SGs), and batteries), control modes (such as grid-forming and grid-following control strategies), load types (such as ZIP loads, electric motors), etc. It should be highlighted that in the aforementioned derivation, no approximation or linearization is introduced, which is totally different from the widely known small-signal state-space model. Instead, the ODE-enabled NMs model in is rigorously equivalent to the original DAE model in (6.8), which ensures the accuracy and efficacy of the ODE model under arbitrary large disturbances.

The rationale behind the ODE-enabled NMs model is that with the network dynamics incorporated, the algebraic equations in the DAE-enabled NMs model degenerate from nonlinear power flow equations to linear KCL equations, which leads to an index-1 DAE system. Therefore, the algebraic variables can be totally eliminated and therefore the DAE system can be rigorously converted to an equivalent ODE system.

The ODE-enabled NMs model exhibits several advantages compared with the DAE-based NMs model. First, the network dynamics are explicitly incorporated into the NMs model, which addresses the over-optimistic assessment by algebraic equation-based power flow formulation. Second, the scale of the ODE-enabled NMs model is mainly impacted by the number of DERs and loads rather than the system scale because of the sparsity of power grids, which provides better scalability than the DAE types of models. Third, solving ODEs is numerically more stable than solving the DAE counterparts, which renders improved numerical stability in time-domain simulations.

6.2 Physics-Data-Integrated ODE Model of NMs

In this section, we further establish the NMs dynamic model with partially unidentified subsystems. A physics-data-integrated (PDI) ODE model will be developed to describe the NMs behaviors by combining both the physics-based subsystems and black-box subsystems.

6.2.1 Physics-Based *InSys* Formulation

As shown in Fig. 6.2, according to the attainability of physics models, we divide the NMs into an internal, physics-based subsystem (*InSys*), and an external, black-box subsystem (*ExSys*).

For *InSys*, it is assumed that its complete physics model is accurately known. Hence, *InSys* can be explicitly formulated by its physics natures. According to the physics-based NMs formulation discussed in Section 6.1, *InSys* can be functionally expressed as a DAE system as follows:

$$\dot{z}_{in} = g_{in}(z_{in}, i_{in}, x_{ex}) \tag{6.14a}$$

$$\dot{i}_{in} = h_{in}(z_{in}, i_{in}) + A_{in}v_{in} + A_{in,bd}v_{bd} \tag{6.14b}$$

$$B_{in,in}i_{in} + B_{in,ex}i_{ex} = 0 \tag{6.14c}$$

The formulation in (6.14) is in analogy with (6.8). Equations (6.14a) and (6.14b) jointly formulate the dynamics of each component in *InSys*, where z_{in} denotes the state variables of each component, i_{in} denotes the DQ-axis currents from each component; v_{in} denotes the DQ-axis bus voltages in *InSys*; v_{bd} denotes the DQ-axis bus voltages of boundary buses. Equation (6.14c) formulates the

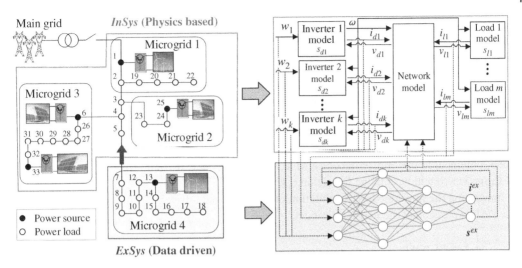

Figure 6.2 Illustration of NMs modeling with physics-based *InSys* and data-driven *ExSys*. Source: Sergiy Serdyuk/Adobe Stock Photos; elxeneize/Adobestock.

current balance constraint at each *InSys* bus, where i_{ex} denotes the current injections from *ExSys*. Because the hierarchical control and the power exchange in NMs possibly cause power- and cyber- interactions between *InSys* and *ExSys*, (6.14a) also incorporates the state variables of *ExSys*, i.e. x_{ex}, in the *InSys*'s dynamics (e.g. such as global control signals Ω and e in (6.2), boundary power injections i_{ex} from *ExSys* to *InSys*); A_{in}, $A_{in,bd}$, $B_{ex,in}$, and $B_{ex,ex}$ are corresponding incidence matrices.

6.2.2 Data-Driven *ExSys* Formulation

For *ExSys*, we assume its physics model is inaccessible. This happens frequently in NMs, such as the unknown parameters of loads and controllers in a microgrid, the unwillingness of microgrid owners to share the model with others, and the need for aggregating massive distributed/behind the meter devices. Since the dynamic modes of a power grid can not be known in prior, using machine learning techniques to capture such dynamics from measurements is a natural idea.

Theoretically, the *ExSys* model should also appear as a DAE system like arbitrary power grids (e.g. *InSys* model in (6.14)). However, based on the derivations in (6.1), we will show that *ExSys* can also be theoretically formulated as an ODE system.

Based on the differential equations of DERs, power loads, and branches, the *ExSys* model can also be functionally abstracted as:

$$
\begin{cases}
\dot{z}_{ex} = g_{ex}(i_{ex}, z_{ex}) & \text{(6.15a)} \\[2mm]
i_{ex} = f_{ex}(i_{ex}, z_{ex}) + A_{ex}v_{ex} + A_{ex,bd}v_{bd} & \text{(6.15b)} \\[2mm]
B_{ex,in}i_{in} + B_{ex,ex}i_{ex} = 0 & \text{(6.15c)}
\end{cases}
$$

The definition of (6.15) is in anomaly with (6.14). Since (6.15) is exactly in the same form with (6.8), therefore, it can be readily converted to an equivalent ODE system following the derivations in Section 6.1.3. As a conclusion, the final ODE-governed state-space model of *ExSys* is as follows:

$$
\begin{cases}
\dot{z}_{ex} = \hat{g}_{ex}(i_{ex0}, z_{ex}) & \text{(6.16a)} \\[2mm]
\dot{i}_{ex0} = \hat{h}_{ex0}(i_{e0}, z_{ex}) + A_{ex,bd0}v_{bd} - A_{ex0}(B_{ex,ex}A_{ex})^{-1} \\
\quad (B_{ex,ex}\hat{h}_{ex}(i_{ex0}, z_{ex}) + B_{ex,ex}A_{ex,bd}v_{bd} + B_{ex,in}\dot{i}_{in}) & \text{(6.16b)}
\end{cases}
$$

where i_{ex0}, A_{ex0}, $\hat{g}_{ex}$, $\hat{h}_{ex}$, and $\hat{h}_{ex0}$ are defined in analogy to those in (6.13).

Because the physics models of *ExSys* are unattainable, the explicit representation of (6.16) can not be obtained. To this end, we leverage AI techniques to establish a data-driven state-space model of *ExSys*, which will be detailed in Section 6.3.

Integrating the physics-based formulation of *InSys* and data-driven formulation of *ExSys* yields the DAE expression of the PDI-NMs model:

$$
\begin{cases}
\dot{z} = g(i, z, w) & \text{(6.17a)} \\[2mm]
\dot{i} = f(i, z, w) + Av & \text{(6.17b)} \\[2mm]
0 = Bi & \text{(6.17c)}
\end{cases}
$$

where z, i, and w, respectively, assemble the state variables, current injections, and inputs of *InSys* and *ExSys*. It is obvious that (6.17) is in exactly the same form as (6.8). Therefore, it can again be rigorously converted to an equivalent ODE form without any linearization assumption. Consequently, the PDI-ODE-enabled NMs dynamic model can be established.

6.3 ODE-Net-Enabled Dynamic Model Discovery for Microgrids

In this section, we introduce the data-driven state-space model of *ExSys* that can be established through AI techniques. As indicated by (6.16), *ExSys* can be formulated in a pure ODE form. Therefore, we specifically adopt the ODE-Net technique for discovering the continuous-time dynamic model of *ExSys* from measurement data.

6.3.1 ODE-Net-Based State-Space Model Formulation

According to the *ExSys* dynamic model depicted in (6.16), we functionally formulate *ExSys* as the following ODE system:

$$
\dot{x}_{ex} = \mathcal{N}_\theta(x_{ex}, s_{in}, \theta) \tag{6.18}
$$

Here, $\mathcal{N}$ represents the neural network (NN)-represented state-space model of *ExSys* depicted by its parameters θ, which will be learned from measurements; x_{ex} and s_{in}, respectively, denote the measurable/estimable state variables of *ExSys* (e.g. i_{ex} and z_{ex}) and input variables of *ExSys* (e.g. selected from *InSys* states i_{in} and z_{in}).

According to the conclusion in (6.2), the entire power grid integrating *ExSys* and *InSys* appears as a physics-neural hybrid system:

$$
\begin{cases}
\dot{x}_{in} = \mathcal{P}(x_{in}, x_{ex}) & \text{(6.19a)} \\[2mm]
\dot{x}_{ex} = \mathcal{N}_\theta(x_{ex}, s_{in}) & \text{(6.19b)}
\end{cases}
$$

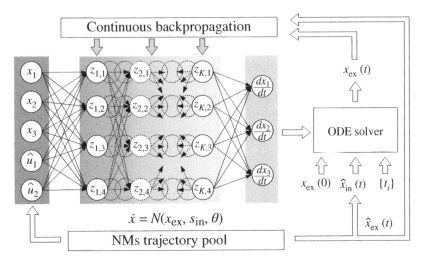

Figure 6.3 ODE-Net-enabled dynamic model discovery of *ExSys*.

where x_{in} and $\mathcal{P}$, respectively, denote the dynamic states and dynamic model of *InSys*, which are readily obtained from the *InSys* physics following (6.17).

Figure 6.3 illustrates the philosophy of ODE-Net-enabled dynamic model discovery. Taking $\boldsymbol{x}_{ex}$ and $\boldsymbol{s}_{in}$ at time t as the inputs, ODE-Net outputs the derivatives of $\boldsymbol{x}_{ex}$ with respect to time. Therefore, the forward propagation of this neural network explicitly describes the *ExSys* dynamic model in (6.18).

Denote the time-series measurements of NMs as:

$$\{(t_1, t_2, \dots, t_n), (\hat{\boldsymbol{x}}_{ex,1}, \hat{\boldsymbol{x}}_{ex,2}, \dots, \hat{\boldsymbol{x}}_{ex,n}), (\hat{\boldsymbol{s}}_{in,1}, \hat{\boldsymbol{s}}_{in,2}, \dots, \hat{\boldsymbol{s}}_{in,n})\} \tag{6.20}$$

where t_i denotes the ith measurement time point; $\boldsymbol{x}_{ex,i}$ and $\boldsymbol{s}_{in,i}$, respectively, denote the states and inputs of *ExSys* model at the time t_i. Here, we use $\wedge$ to represent measurements.

Obviously, measurements can only be made in discrete time as described in (6.20). However, power system dynamic models are always continuous time. Therefore, our target is to learn a continuous-time dynamic model in the form of (6.18) from the discrete-time measurements given in (6.20).

6.3.2 Continuous-Time Learning Model for ODE-Net

As illustrated in (6.18), a neural network can be regarded as a nonlinear function characterized by unknown variables θ. Machine learning methods optimize the neural network parameters by minimizing a loss function, which is usually constructed as the error between the ground truth and the neural network outputs. However, for the dynamic model discovery issue, the difficulty is that the output of ODE-Net is the time derivatives of $\boldsymbol{x}_{ex}$ (see (6.18)), while the measurements only provide $\boldsymbol{x}_{ex}$ (see (6.20)), indicating that the loss function can not be directly constructed.

Generally, two different paths exist regarding this issue: discrete-time learning and continuous-time learning.

6.3.2.1 Discrete-Time Learning
Conventional machine learning techniques for dynamic model discovery are mainly based on discrete-time learning [12, 13], which constructs the loss function by discretizing the

continuous-time differential equations into discrete-time difference equations. For example, based on the trapezoidal rule, (6.18) can be differentiated as:

$$\frac{x_{ex}(t) - x_{ex}(t-\tau)}{\tau} = \frac{1}{2}(\mathcal{N}(x_{ex}(t), s_{in}(t)) + \mathcal{N}(x_{ex}(t-\tau), s_{in}(t-\tau))) \tag{6.21}$$

where τ denotes the time step for numerical integration. Correspondingly, the loss function can be established and the neural network can be optimized by:

$$\min_{\theta} L_d = \sum_{i=1}^{n} \frac{1}{2}\eta_i \|y_i - \hat{y}_i\| \tag{6.22}$$

where $\hat{y} = \frac{1}{\tau}(\hat{x}_{ex}(t) - \hat{x}_{ex}(t-\tau))$ denotes the derivatives estimated from the measurements; and $y = \frac{1}{2}(\mathcal{N}(x_{ex}(t), \hat{s}_{in}(t)) + \mathcal{N}(x_{ex}(t-\tau), \hat{s}_{in}(t-\tau)))$ denotes the derivatives estimated from the neural network; η_i denotes the weighting factor at time point i.

However, the loss function in (6.22) indicates that discrete-time learning relies on the estimation of derivatives to perform neural network training, which can be sensitive to the measurements. This easily leads to biased training results because of non-ideal measurements or tiny residue errors during the training process. In other words, even if the discrete-time training produces a satisfactory fitting for the derivatives, it can not ensure the accuracy of the system states after performing numerical integration of (6.18).

6.3.2.2 Continuous-Time Learning

Motivated by the deficiency of discrete-time learning, our solution is an ODE-Net-enabled dynamic equivalence, which adopts a continuous-time learning philosophy.

The ODE-Net is trained by minimizing the error between the state measurements $\hat{x}$ and the numerical solution of (6.18):

$$\min_{\theta} L_c = \sum_{i=1}^{n} L_{ex,i} = \sum_{i=1}^{n} \frac{1}{2}\eta_i \|x_{ex,i} - \hat{x}_{ex,i}\|_2 \\ \text{s.t. } x_{ex,i} = \hat{x}_{ex,0} + \int_{t_0}^{t_i} \mathcal{N}_\theta(x_{ex}, \hat{s}_{in})dt \tag{6.23}$$

Comparing (6.22) with (6.23), an obvious distinction is that ODE-Net is capable of directly minimizing the difference between real dynamic states and trained dynamic states, which requires no discretization and fully respects the continuous-time characteristics of power system dynamics. Therefore, it is theoretically more invulnerable to non-ideal measurements and residue training errors.

6.3.3 Continuous Backpropagation

After establishing the continuous-time loss function, the next step is to perform the ODE-Net training, which is essentially an optimization problem as presented in (6.23).

The main difficulty in the optimization of (6.23) lies in the ODE integration operation in the constraints. In this subsection, the continuous propagation technique [14] is applied to handle the ODE integration in the ODE-Net training.

Lagrange multiplier λ is first introduced to (6.23) to remove the ODE constraints and build the following loss function:

$$\mathcal{L} = \sum_{i=1}^{n} L_{ex,i} - \int_{t_0}^{t_n} \lambda^T \left(\dot{x}_{ex} - \mathcal{N}_\theta(x_{ex}, s_{in})\right) dt \tag{6.24}$$

Backpropagation computes the gradient of the loss function with respect to the ODE-Net parameters to minimize the loss function [15]. With the loss function (6.24) involving the integration operator, the partial derivative of $\mathcal{L}$ with respect to θ is calculated as:

$$
\begin{aligned}
\frac{\partial \mathcal{L}}{\partial \theta} &= \sum_{i=1}^{n} \left(\frac{\partial L_{\mathrm{ex},i}}{\partial \boldsymbol{x}_{\mathrm{ex},i}} \frac{\partial \boldsymbol{x}_{\mathrm{ex},i}}{\partial \theta} - \int_{t_{i-1}}^{t_i} \lambda^T \left(\frac{\partial \dot{\boldsymbol{x}}_{\mathrm{ex}}}{\partial \theta} - \frac{\partial \mathcal{N}}{\partial \boldsymbol{x}_{\mathrm{ex}}} \frac{\partial \boldsymbol{x}_{\mathrm{ex}}}{\partial \theta} - \frac{\partial \mathcal{N}}{\partial \theta} \right) dt \right) \\
&= \sum_{i=1}^{n} \frac{\partial L_{\mathrm{ex},i}}{\partial \boldsymbol{x}_{\mathrm{ex},i}} \frac{\partial \boldsymbol{x}_{\mathrm{ex},i}}{\partial \theta} + \sum_{i=1}^{n} \left(\lambda^T(t_i^-) \frac{\partial \boldsymbol{x}_{\mathrm{ex},i}}{\partial \theta} - \lambda^T(t_{i-1}^+) \frac{\partial \boldsymbol{x}_{\mathrm{ex},i-1}}{\partial \theta} \right) \\
&\quad + \sum_{i=1}^{n} \int_{t_{i-1}}^{t_i} \left(\frac{d\lambda^T}{dt} \frac{\partial \boldsymbol{x}_{\mathrm{ex}}}{\partial \theta} + \lambda^T \frac{\partial \mathcal{N}}{\partial \boldsymbol{x}_{\mathrm{ex}}} \frac{\partial \boldsymbol{x}_{\mathrm{ex}}}{\partial \theta} + \lambda^T \frac{\partial \mathcal{N}}{\partial \theta} \right) dt
\end{aligned}
\tag{6.25}
$$

The Lagrange multiplier variables are given by [16]:

$$
\frac{d\lambda^T}{dt} = -\lambda^T \frac{\partial \mathcal{N}}{\partial \boldsymbol{x}_{\mathrm{ex}}}
\tag{6.26}
$$

where the boundary conditions are set as:

$$
\lambda^T(t_n^+) = 0, \quad \lambda^T(t_i^+) = \lambda^T(t_i^-) + \partial L_{\mathrm{ex},i}/\partial \boldsymbol{x}_{\mathrm{ex},i}
\tag{6.27}
$$

Then, (6.25) can be derived into:

$$
\frac{\partial \mathcal{L}}{\partial \theta} = \int_{t_0}^{t_n} \lambda^T \frac{\partial \mathcal{N}}{\partial \theta} dt
\tag{6.28}
$$

Collecting (6.26) and (6.28) leads to an ODE integration problem:

$$
\frac{d}{dt} \begin{bmatrix} \lambda^T \\ \partial \mathcal{L}/\partial \theta \end{bmatrix} = \begin{bmatrix} -\lambda^T \partial \mathcal{N}/\partial \boldsymbol{x}_{\mathrm{ex}} \\ \lambda^T \partial \mathcal{N}/\partial \theta \end{bmatrix}
\tag{6.29}
$$

Subsequently, $\partial \mathcal{L}/\partial \theta$ can be obtained from (6.29) by any ODE solver, e.g. Trapezoidal integration. Given the final value of $\lambda(t)$, i.e. $\lambda^T(t_n)$ in (6.27), rather than the initial value, (6.29) requires solving the ODEs backward in time, which leads to a reverse-mode integration [14]:

$$
\left. \frac{\partial \mathcal{L}}{\partial \theta} \right|_{t_1} = \left. \frac{\partial \mathcal{L}}{\partial \theta} \right|_{t_n} + \int_{t_n}^{t_1} \lambda^T \frac{\partial \mathcal{N}}{\partial \theta} dt = \sum_{i=2}^{n} \int_{t_i}^{t_{i-1}} \lambda^T \frac{\partial \mathcal{N}}{\partial \theta} dt
\tag{6.30}
$$

with $\lambda(t)$ also solved by the reverse-mode integration:

$$
\lambda^T(t_{i-1}^+) = \lambda^T(t_i^-) + \int_{t_i}^{t_{i-1}} \lambda^T \frac{\partial \mathcal{N}}{\partial \boldsymbol{x}_{\mathrm{ex}}} dt
\tag{6.31}
$$

Furthermore, consider *a set of time series* of NMs trajectories as $(\boldsymbol{t}^{(1)}, \hat{\boldsymbol{x}}_{\mathrm{ex}}^{(1)}, \hat{\boldsymbol{s}}_{\mathrm{in}}^{(1)}), \ldots, (\boldsymbol{t}^{(m)}, \hat{\boldsymbol{x}}_{\mathrm{ex}}^{(m)}, \hat{\boldsymbol{s}}_{\mathrm{in}}^{(m)})$. For the jth measurement, let $\partial \mathcal{L}^{(j)}/\partial \theta$ be the gradient of the loss function computed by (6.30). The overall gradient is obtained:

$$
\frac{\partial \mathcal{L}}{\partial \theta} = \sum_{j=1}^{m} \frac{\partial \mathcal{L}^{(j)}}{\partial \theta}
\tag{6.32}
$$

Consequently, the ODE-Net parameters are updated using gradient descent so that $\mathcal{L}$ can be decreased during training:

$$
\theta \longleftarrow \theta - r \frac{\partial \mathcal{L}}{\partial \theta}
\tag{6.33}
$$

where r denotes the learning rate. By iteratively performing the gradient descent shown in (6.33) until the loss function converges (e.g. smaller than a threshold), an optimized ODE-Net can be obtained.

The continuous backpropagation incorporates the "ODE solver" in the gradient descent for the ODE-Net parameter optimization, and hence effectively retains the intrinsic continuous differential structure of the dynamical NMs.

6.3.4 Further Discussion

Although (6.18) can be independently trained by the machine learning model presented in (6.23), a potential issue is that it only considers the impact from *InSys* on *ExSys* (i.e. (6.19b)) while the impact from *ExSys* on *InSys* (i.e. (6.19a)) is ignored.

As shown in (6.19), the numerical integration of the entire power grid requires assembling *InSys* and *ExSys* models. We name it as the *closed-loop simulation* of *InSys* and *ExSys* in the following discussion. Obviously, (6.23) only performs an *open-loop training* without involving the *closed-loop simulation*. Such an open-loop manner fails to explicitly monitor and control the closed-loop accuracy of (6.19) in its training process. Actually, since machine learning models are always trained from a finite set of data, they would not perfectly replicate the real dynamics under any circumstances, meaning even though (6.19b) is well trained, it contains certain training errors. On the other hand, even a tiny numerical error in the *ExSys* model can perturb the *InSys* dynamics (see (6.19a)). Without considering such errors accumulate during the interactions between *ExSys* and *InSys*, it is hard for the ODE-Net training to explicitly guarantee the accuracy of the dynamic simulation of the whole NMs.

6.4 Physics-Informed Learning for ODE-Net-Enabled Dynamic Models

Motivated by the aforementioned concerns of the purely data-driven ODE-Net learning, this section further establishes a physics-informed ODE-Net (PI-ODENet), which can explicitly monitor and actively control the closed-loop accuracy of *ExSys* and *InSys* through a physics-informed training process.

6.4.1 Physics-Informed Formulation for ODE-Net Training

While the *ExSys* dynamic model is inaccessible and will be learned by ODE-Net, the *InSys* dynamic model is well established by its physics nature. Therefore, the fundamental idea of PI-ODENet is to leverage the known physics model of *InSys* to assist the training of ODE-Net for *ExSys*.

The training model for PI-ODENet is formulated as:

$$\min_{\theta} \sum_{i=1}^{n} L_i = \sum_{i=1}^{n} \frac{\eta_i}{2} (\| \boldsymbol{x}_{\text{ex},i} - \hat{\boldsymbol{x}}_{\text{ex},i} \|_2 + \| \boldsymbol{x}_{\text{in},i} - \hat{\boldsymbol{x}}_{\text{in},i} \|_2) \tag{6.34a}$$

$$\text{s.t.} \quad \boldsymbol{x}_{\text{ex},i} = \hat{\boldsymbol{x}}_{\text{ex},0} + \int_{t_1}^{t_i} \mathcal{N}_{\theta}(\boldsymbol{x}_{\text{ex}}, s_{\text{in}}) dt \tag{6.34b}$$

$$\boldsymbol{x}_{\text{in},i} = \hat{\boldsymbol{x}}_{\text{in},0} + \int_{t_1}^{t_i} \mathcal{P}(\boldsymbol{x}_{\text{in}}, \boldsymbol{x}_{\text{ex}}) dt \tag{6.34c}$$

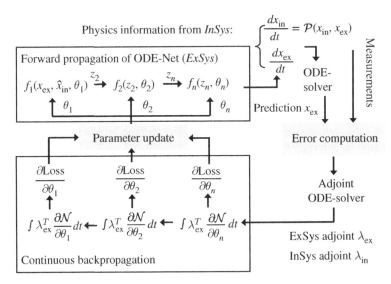

Figure 6.4 PI-ODENet-enabled dynamic model discovery of *ExSys*.

where $\hat{x}_{\text{ex},i}$ and $\hat{x}_{\text{in},i}$, respectively, denote the measured *ExSys* and *InSys* states at time point t_i.

Figure 6.4 illustrates the philosophy of PI-ODENet. Different from the ODE-Net training model in (6.23), training for PI-ODENet in (6.34) explicitly embeds the accuracy of both *ExSys* and *InSys* states into the training process in a closed-loop manner. Therefore, it theoretically ensures that, once the training converges, the learned dynamic model can faithfully generate dynamic responses in consistency with the NMs' real dynamics.

6.4.2 Physics-Informed Continuous Backpropagation

To perform the PI-ODENet training in (6.34), we derive physics-informed continuous backpropagation.

The Lagrange multipliers are again introduced to handle the ODE constraints in (6.34):

$$\mathcal{L} = \sum_{i=1}^{n} L_i - \int_{t_0}^{t_n} \left[\lambda_{\text{ex}}^T (\dot{x}_{\text{ex}} - \mathcal{N}_\theta) + \lambda_{\text{in}}^T (\dot{x}_{\text{in}} - \mathcal{P}) \right] dt \tag{6.35}$$

Note that compared with (6.24), (6.35) introduces two Lagrange multipliers, i.e. λ_{ex} and λ_{in}, to, respectively, serve as the adjoint states for *ExSys* and *InSys*.

Accordingly, the partial derivative of $\mathcal{L}$ with respect to θ is calculated as:

$$\begin{aligned}
\frac{\partial \mathcal{L}}{\partial \theta} = & \sum_{i=1}^{n} \left(\frac{\partial L_i}{\partial x_{\text{ex},i}} \frac{\partial x_{\text{ex},i}}{\partial \theta} + \frac{\partial L_i}{\partial x_{\text{in},i}} \frac{\partial x_{\text{in},i}}{\partial \theta} \right) \\
& - \sum_{i=1}^{n} \int_{t_{i-1}}^{t_i} \lambda_{\text{ex}}^T \left(\frac{\partial \dot{x}_{\text{ex}}}{\partial \theta} - \frac{\partial \mathcal{N}}{\partial x_{\text{ex}}} \frac{\partial x_{\text{ex}}}{\partial \theta} - \frac{\partial \mathcal{N}}{\partial x_{\text{in}}} \frac{\partial x_{\text{in}}}{\partial \theta} - \frac{\partial \mathcal{N}}{\partial \theta} \right) dt \\
& - \sum_{i=1}^{n} \int_{t_{i-1}}^{t_i} \lambda_{\text{in}}^T \left(\frac{\partial \dot{x}_{\text{in}}}{\partial \theta} - \frac{\partial \mathcal{P}}{\partial x_{\text{in}}} \frac{\partial x_{\text{in}}}{\partial \theta} - \frac{\partial \mathcal{P}}{\partial x_{\text{ex}}} \frac{\partial x_{\text{ex}}}{\partial \theta} \right) dt
\end{aligned} \tag{6.36}$$

In analogy with (6.26) and (6.27), by properly setting the boundary conditions, (6.36) leads to the following physics-informed gradient rules:

$$\frac{d}{dt}\begin{bmatrix} \lambda_{\text{ex}}^T \\ \lambda_{\text{in}}^T \\ \partial\mathcal{L}/\partial\theta \end{bmatrix} = \begin{bmatrix} -\lambda_{\text{ex}}^T \partial\mathcal{N}/\partial x_{\text{ex}} - \lambda_{\text{in}}^T \partial P/\partial x_{\text{ex}} \\ -\lambda_{\text{ex}}^T \partial\mathcal{N}/\partial x_{\text{in}} - \lambda_{\text{in}}^T \partial P/\partial x_{\text{in}} \\ \lambda_{\text{ex}}^T \partial\mathcal{N}/\partial\theta \end{bmatrix} \tag{6.37}$$

Equation (6.37) includes both the "adjoint dynamics" for λ_{ex} and λ_{in} and the "gradient dynamic" for $\partial\mathcal{L}/\partial\theta$, which can be solved by arbitrary ODE solvers. Consequently, the gradient descent for PI-ODENet can be performed readily.

Compared with the purely data-driven ODE-Net, PI-ODENet explicitly controls the closed-loop accuracy for dynamic model discovery of *ExSys* via the physics-informed gradient descent in (6.37).

6.5 Experiments

This section demonstrates the technical merit and efficacy of the devised methods in typical testing systems. The algorithm is implemented in MATLAB R2019b. The trapezoidal rule is employed for NMs dynamic integration. The Adaptive Moment Estimation (Adam) [17] algorithm is applied for ODE-Net training.

6.5.1 Case Design

6.5.1.1 Test System 1

The first test case is a 4-microgrid NMs [7] in Fig. 6.2, with the DER controller parameters modified from [10]. The DERs can be equipped with droop controllers and/or secondary controllers. The following cases are designed to verify the method. Each case shares the same settings of NMs topology and load condition, with different control strategies of DERs as well as power source mixes.

- **Case 1**: All the DERs are equipped with droop control; microgrid 4 is supposed to be model free and data driven, while other microgrids are physics based with formulations detailed in Section 6.1;
- **Case 2**: All the DERs, both in *InSys* and *ExSys*, are equipped with both droop and secondary controls) under an all-to-all communication among DERs; microgrid 4 is data driven similar as Case 1;
- **Case 3**: All settings are the same with Case 1, except that microgrid 3 (which comprises two DERs) is data driven (while microgrids 1, 2, and 4 are physics-based) and the DER at bus 6 in microgrid 3 is replaced by a SG;
- **Case 4**: All settings are the same with Case 4, except that the DER at bus 6 in microgrid 3 is replaced by an energy storage unit (ESU).

By default, the uncertainty of each DER is set as 20%, which means that the reference active power of DER controllers can randomly deviate from the prediction to address the impact of uncertain renewables. Consequently, the NMs undergo frequent transients due to the fluctuation of DERs. Thereby, measurements are acquired under uncertain perturbations to construct the training set for ODE-Net. Here, the training data contains time-series measurements lasting for 270 s. Figure 6.5

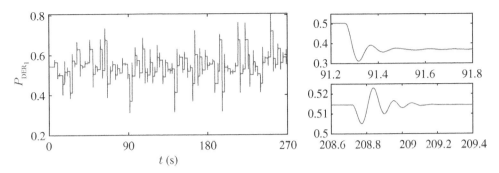

Figure 6.5 Illustration of training set: time-series measurements of NMs dynamics under perturbations.

shows the training data, taking a single dimension (i.e. power output from DER1) for illustration purpose.

6.5.1.2 Test System 2

The second test case is a large-scale test case integrating seven microgrids to validate the scalability of the method. As shown in Fig. 6.6, the backbone grid is modified from the Northeast Power Coordinating Council (NPCC) system, where the synchronous generators are simplified as constant

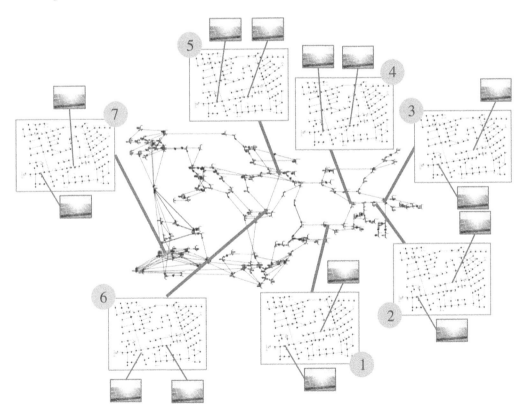

Figure 6.6 Topology of the 1001-bus, seven-microgrid test case.

voltage sources. Each microgrid has two DERs, and its topology is modified from the IEEE 123-bus feeder. All the DERs are equipped with droop control. Two cases are designed:

- **Case 1**: Microgrid 6 and microgrid 7 are model free, and their dynamic models will be discovered by ODE-Net;
- **Case 2**: Microgrids 4–7 are all assumed model-free and will be modeled by ODE-Net.

By default, the uncertainty of each DER is also set as 20% to induce transient processes in the system.

6.5.2 Method Validity

This subsection demonstrates the performance of ODE-Net in learning the state-space model of microgrids in **Test System 1**. A fully connected neural network architecture is used for ODE-Net, with two hidden layers and 40 neurons in each layer.

Figure 6.7 presents the ODE-Net training process for **Case 1**. At the starting stage, ODE-Net is randomly initialized and largely deviates from the real NMs trajectories, as illustrated in Fig. 6.7(b). Then, after the neural network training via continuous backpropagation, ODE-Net converges to a perfect match of the NMs trajectories on the training set, as illustrated in Fig. 6.7(c). Additionally, Fig. 6.7(d) presents the evolution of the loss function at the logarithmic scale. As a rule of thumb, 1500 iterations lead to convergence of the ODE-Net.

Furthermore, Figs. 6.8 and 6.9 illustrate the ODE-Net performance on the testing set for **Case 1** and **Case 2**, which verify its ability to generalize beyond the training set. Three types of scenarios are studied, i.e. no-fault, a short-circuit fault at bus 19, and an open-circuit fault at branches 2–3. Each fault occurs at 0.3 s and is cleared at 0.32 s. An interesting finding is that the ODE-Net-enabled NMs formulation accurately captures the uncertain NMs transients not only under the frequently fluctuating DER uncertainties but also under large disturbances. However, the latter scenarios never appear in the training set. This shows the robustness of the ODE-Net-enabled NMs formulation.

Finally, the impact of power source mixes is investigated through **Case 3** and **Case 4**. Different power sources influence the NMs dynamics by their diverse dynamic features. Figures. 6.10(a) and 6.11(a) study the quasi-static reachsets (i.e. with no fault but only DER uncertainties perturbing the NMs) for **Case 3** and **Case 4**. Comparing Fig. 6.10(a) with Fig. 6.9(a), it can be observed that the impact of uncertainties on the NMs dynamics reduces when the NMs are equipped with a SG, benefiting from the inertia and regulating ability of SG as well as its full dispachability compared with DERs. The uncertain impact is further restrained when the NMs are equipped with an ESU, as

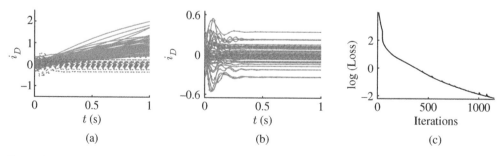

Figure 6.7 Training process of ODE-Net for NMs dynamic model discovery in test system 1. Dotted lines: real dynamics; Solid lines: learned dynamics. (a) ODE-Net performance at the starting stage, (b) ODE-Net performance at the final stage, and (c) Loss function evolution process.

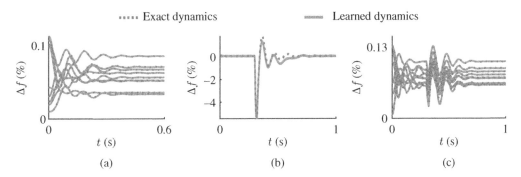

Figure 6.8 Testing performance of ODE-Net for test system 1 (4-microgrid NMs), case 1 (droop control). Δf: frequency deviation. (a) No fault occurs in the NMs, (b) under a short-circuit fault, and (c) under an open-circuit fault.

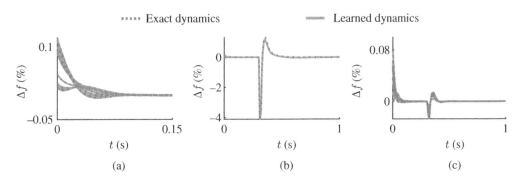

Figure 6.9 Testing performance of ODE-Net for test system 1 (4-microgrid NMs), case 2 (hierarchical control). (a) No fault occurs in the NMs, (b) under a short-circuit fault, and (c) under an open-circuit fault.

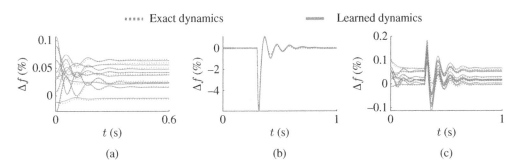

Figure 6.10 Testing performance of ODE-Net for test system 1 (4-microgrid NMs), case 3 (synchronous generator case). (a) No fault occurs in the NMs, (b) under a short-circuit fault, and (c) under an open-circuit fault.

presented by the results of **Case 4**, indicating enhanced robustness against the uncertainties with ESU. The results demonstrate that ODE-Net-enabled dynamic models can capture the dynamics features of different power sources. Furthermore, for fault cases shown in Figs. 6.10 and 6.11, ODE-Net-based models also exhibits satisfactory performances for reflecting the characteristics of the controllable power sources in restraining the uncertainties and damping the frequency/voltage dip/rise during the large disturbances.

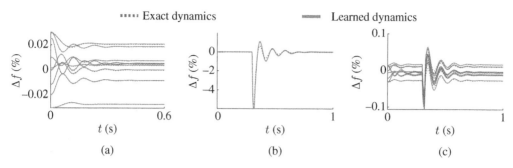

Figure 6.11 Testing performance of ODE-Net for test system 1 (4-microgrid NMs), case 4 (energy storage device case). (a) No fault occurs in the NMs, (b) under a short-circuit fault, and (c) under an open-circuit fault.

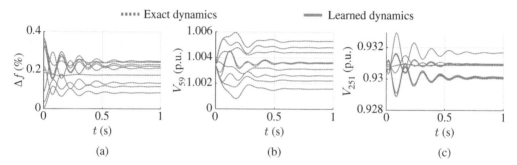

Figure 6.12 Testing performance of ODE-Net under uncertain DER generations for test system 2 (1001-bus NMs), case 1 (2 microgrids unknown). (a) Frequency of the system, (b) bus voltage (backbone system), and (c) bus voltage (microgrid 1).

6.5.3 Method Scalability

This subsection demonstrates the scalability of ODE-Net in **Test System 2**. A fully connected neural network architecture is used for ODE-Net, with two hidden layers and 32 neurons in each layer.

We first study case 1, where the dynamic models of microgrids 6–7 will be learned by ODE-Net. Figure 6.12 provides the ODE-Net learning results on the testing set, where new uncertain scenarios are generated to address the DER fluctuations (beyond the training set). It can be seen that ODE-Net provides a perfect match of the system dynamics both for the transmission grid and distribution grids, which again indicates our ODE-Net-based dynamic model learned from data can capture system dynamics in large-scale systems with inverter controllers. Furthermore, Fig. 6.12 studies the performance of the method under fault cases, which targets validating the generalization capability of the method in large-disturbance scenarios in large-scale systems. A short circuit fault is added at the point of common coupling (PCC) of microgrid 3. It can be seen that the ODE-Net-based dynamic models discovered from data still accurately capture the transient behaviors of the system after faults, indicating the model's effectiveness under contingency scenarios. This again illustrates the robustness and generalization capability of the ODE-Net-enabled dynamic formulation.

We then study **Case 2**, which is a tougher case because the dynamic models of microgrids 4–7 are all unknown, meaning over 50% of the microgrid dynamics will be modeled by ODE-Net.

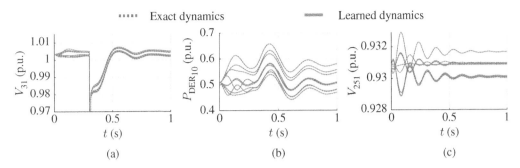

Figure 6.13 Testing performance of ODE-Net under both contingencies and uncertainties for test system 2 (1001-bus NMs), case 1 (2 microgrids unknown). (a) Voltage of PCC of microgrid 5, (b) active power of DER2 at microgrid 5, and (c) reactive power of DER2 at microgrid 5.

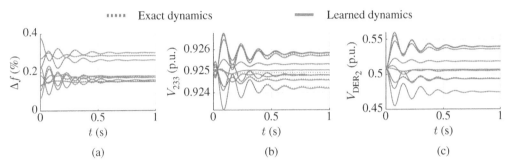

Figure 6.14 Testing performance of ODE-Net under uncertain DER generations for test system 2 (1001-bus NMs), case 2 (4 microgrids unknown). (a) Frequency of the system, (b) bus voltage (microgrid 1), and (c) bus voltage (microgrid 2).

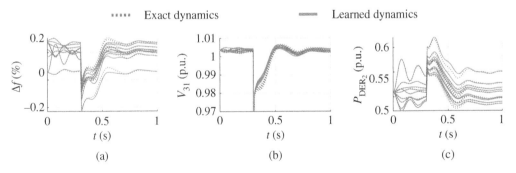

Figure 6.15 Testing performance of ODE-Net under both contingencies and uncertainties for test system 2 (1001-bus NMs), case 2 (4 microgrids unknown). (a) Frequency of the system, (b) active power of DER_2 at microgrid 1, and (c) reactive power of DER_2 at microgrid 1.

Figures 6.14 and 6.15 provide the ODE-Net learning results under both DER uncertainties and a short circuit fault at the PCC of microgrid 3 (again, testing cases are all beyond training cases). It can be seen that even though microgrids 4–7 are all modeled from data, ODE-Net still generates satisfactory results to capture system dynamics.

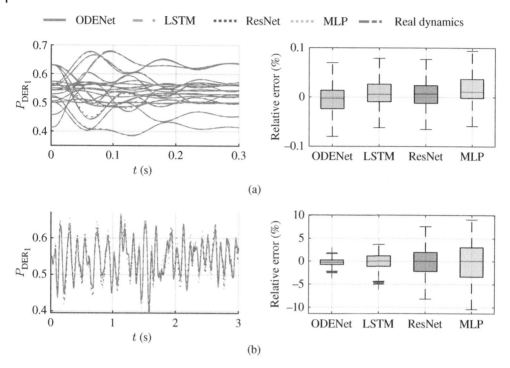

Figure 6.16 Comparison between ODE-Net and conventional NN methods for NMs dynamic analysis. (a) Comparison on training set (270 s measurement) and (b) comparison on test set (50 s predictions).

6.5.4 Method Superiority over Discrete-Time Learning

To further illustrate the superiority of ODE-Net in modeling continuous-time dynamics of NMs, ODE-Net is compared with three representative conventional neural networks (NN), i.e. multi-layer perceptron (MLP), residual neural network (ResNet), and long short-term memory (LSTM). The configuration of each NN is set as:

- An MLP comprised three hidden layers with 200 hidden units in each layer;
- A ResNet comprised 10 hidden layers with double-layer skips and 200 hidden units in each layer;
- An LSTM network with 200 hidden units.

Figure 6.16 presents the simulation results. Although both ODE-Net and NN methods exhibit satisfactory accuracy on the training set (see Fig. 6.16(a) for time-domain trajectories and boxplot statistics), Fig. 6.16(b) clearly illustrates that their performance differs largely on the test set containing frequently perturbed NMs dynamics. Among the selected conventional NN techniques, LSTM exhibits the best performance on the test set. Still, ODE-Net has a 50% higher precision than LSTM despite its simplest network architecture (i.e. a two-layer MLP). In addition, ODE-Net is able to take irregularly or sparsely sampled measurements since it directly handles the differential equations. In contrast, conventional NN usually requires the measurements to be well-aligned or synchronized to ensure proper discretization.

Another advantage of ODE-Net is that with the discovered state-space model, various power system analytics (e.g. transient simulation, dynamic verification, and stability analysis) can be conducted by regarding the ODE-Net-based *ExSys* as a special type of "dynamic component" and incorporating the data-driven model into the existing system model.

6.6 Summary

In this chapter, we introduce an AI-enabled dynamic model discovery method for NMs. The overarching goal of this section is to establish a data-driven method well suited to discovering the strongly nonlinear NMs dynamics under uncertainties. The key innovation is to integrate the neural ordinary differential equations network (ODE-Net) with a physics-informed machine learning technique to allow for establishing a continuous-time dynamic model from data. Extensive case studies demonstrate the efficacy of the ODE-Net-enabled method for NMs dynamic model discovery under multiple uncertainties and various operational scenarios.

6.7 Exercises

1. What is the purpose of developing learning-based dynamic models?
2. What is the difference between ODE-Net learning and conventional discrete-time learning? Explain which philosophy is more reasonable for microgrid dynamic model discovery.
3. In the ODE-Net-based dynamic model discovery process, please explain which part requires the ODE solver.

References

1 Zhang, P. (2020). *Networked Microgrids*. Cambridge University Press.

2 Wan, W., Bragin, M., Yan, B. et al. (2020). Distributed and asynchronous active fault management for networked microgrids. *IEEE Transactions on Power Systems* 35 (5): 3857–3868.

3 Zhou, Y. and Zhang, P. (2020). Reachable power flow: theory to practice. *IEEE Transactions on Power Systems* 36 (3): 2532–2541.

4 Zhou, Y., Zhang, P., and Yue, M. (2020). Reachable dynamics of networked microgrids with large disturbances. *IEEE Transactions on Power Systems* 36 (3): 2416–2427.

5 Wilding, T.A., Gill, A.B., Boon, A. et al. (2017). Turning off the DRIP ('Data-rich, information-poor')–rationalising monitoring with a focus on marine renewable energy developments and the benthos. *Renewable and Sustainable Energy Reviews* 74: 848–859.

6 Xu, Y., Dong, Z.Y., Guan, L. et al. (2012). Preventive dynamic security control of power systems based on pattern discovery technique. *IEEE Transactions on Power Systems* 27 (3): 1236–1244.

7 Zhou, Y. and Zhang, P. (2021). Neuro-reachability of networked microgrids. *IEEE Transactions on Power Systems* 37 (1): 142–152.

8 Vorobev, P., Huang, P.-H., Al Hosani, M. et al. (2017). A framework for development of universal rules for microgrids stability and control. *56th IEEE Conference on Decision and Control*, 5125–5130.

9 Simpson-Porco, J.W., Shafiee, Q., Dörfler, F. et al. (2015). Secondary frequency and voltage control of islanded microgrids via distributed averaging. *IEEE Transactions on Industrial Electronics* 62 (11): 7025–7038.

10 Pogaku, N., Prodanovic, M., and Green, T.C. (2007). Modeling, analysis and testing of autonomous operation of an inverter-based microgrid. *IEEE Transactions on Power Electronics* 22 (2): 613–625.

11 Zhou, Y., Zhang, P., and Yue, M. (2020). An ode-enabled distributed transient stability analysis for networked microgrids. In: *2020 IEEE Power & Energy Society General Meeting (PESGM)*, 1–5. IEEE.

12 Raissi, M., Perdikaris, P., and Karniadakis, G.E. (2018). Multistep neural networks for data-driven discovery of nonlinear dynamical systems. *arXiv preprint arXiv:1801.01236*.

13 Shakouri, H. and Radmanesh, H.R. (2009). Identification of a continuous time nonlinear state space model for the external power system dynamic equivalent by neural networks. *International Journal of Electrical Power & Energy Systems* 31 (7-8): 334–344.

14 Chen, T.Q., Rubanova, Y., Bettencourt, J., and Duvenaud, D.K. (2018). Neural ordinary differential equations. In: *Advances in Neural Information Processing Systems 31 (NeurIPS 2018)*, 6571–6583.

15 Goodfellow, I., Bengio, Y., and Courville, A. (2016). *Deep Learning*. MIT Press. http://www.deeplearningbook.org.

16 Sun, Y., Zhang, L., and Schaeffer, H. (2019). NeuPDE: neural network based ordinary and partial differential equations for modeling time-dependent data. *arXiv preprint arXiv:1908.03190*.

17 Kingma, D.P. and Ba, J. (2014). Adam: a method for stochastic optimization. *arXiv preprint arXiv:1412.6980*.

7

Transient Stability Analysis for Microgrids with Grid-Forming Converters

Xuheng Lin and Ziang Zhang

7.1 Background

Depending on the resources available to the microgrid, different energy conversion techniques are usually paired with various energy resources. The most popular two options are a generator or an inverter. Due to the size and application requirements, most energy conversion in a microgrid is done by inverters, where the energy source includes photovoltaic, inverter-based wind turbines, and most energy storage systems.

The stability of the bulk power system is dominated by the dynamics of synchronous generators (SGs). Extensive studies have been conducted to provide an understanding of the stability of such systems since the beginning of power system research [1–4]. However, microgrids usually have a high penetration of inverter-based resources (IBRs). In addition, the collective inverter dynamics will dominate the dynamic of the microgrid.

The stability of IBR-dominated systems is an evolving research field. In this chapter, two aspects of microgrid stability are discussed:

- Steady-state stability for microgrids refers to the microgrid dynamics against small loads or generation change. The microgrid has been linearized around the operating state, and Eigen analysis has been used to study the stability of the system.
- Transient stability for microgrids considers microgrid dynamics after a fault. Due to the system's nonlinearity, the time-domain simulation approach is used for stability analysis.

Unlike synchronous generators, inverters do not have a standard set of equations that describes their dynamic characteristics. However, the transient stability of an inverter-dominated system (e.g. a microgrid) can still be considered as the ability of the inverters to maintain synchronism with synchronous generators against a large disturbance. Such disturbances include line faults, loss of large generation, or loss of large load.

Recent studies have provided different perspectives that further define the transient stability of a system with inverters. For example, if the inverter control is based on virtual synchronous machine (VSM) technology, the virtual angle stability can be directly used as a stability metric for such a system [5], the power angle differences between the inverter and a reference bus (e.g. an infinite bus) [6], or Lyapunov-based functions [7, 8].

Microgrids: Theory and Practice, First Edition. Edited by Peng Zhang.
© 2024 The Institute of Electrical and Electronics Engineers, Inc. Published 2024 by John Wiley & Sons, Inc.

7.2 System Modeling

Similar to the stability analysis process for a large power system, the transient stability of microgrids relies on the device and system model selected for the study. In this section, the two inverter models and the system model used for the rest of the chapter are introduced. Based on whether the inverter can maintain the voltage by itself, the inverter can be divided into two categories, grid-following and grid forming inverters. The grid-following inverter (GFL) is the most common inverter in current market. The grid-forming inverter (GFM) is still being tested in the lab although a few have been applied to the power system. Some types of GFMs control methods are droop control, VSM, and virtual oscillators.

- **The grid-following control**: This controller contains two main subsystems: a phase lock loop (PLL) that estimates the instantaneous angle of the measured converter terminal voltage and a current control loop that regulates the AC current injected into the grid [9]. This is often referred to as current control because the current is the physical quantity that is regulated. In this setting, the PLL provides the angular reference of the current commands and carries out the "following" behavior. The grid-following AC terminals mimic a current source whose real and reactive output tracks the references. For fixed power commands, an inverter acts like a constant real-reactive power (PQ) source. This control strategy is called grid-following because its functionality depends on each inverter having a well-defined terminal voltage that its PLL can latch onto and follow. In this setting, the system voltage and frequency are regulated by resources external to each GFL. Today, nearly all grid-connected inverters are controlled with a grid-following controller such as this. Additional functionalities are typically layered on top of this baseline controller and vary between single-phase and three-phase implementations.
- **Droop control**: Its key feature is that it exhibits a linear trade-off between frequency and voltage versus real and reactive power, much like a typical synchronous machine does in steady state. These so-called "droop laws" are referred to as the P–ω (real power–frequency) and Q–V (reactive power–voltage) relationships [10], and they give rise to the following properties regardless of whether they are machines or inverters:
  - **System-wide synchronization**: All units reach the same frequency.
  - **Power sharing**: Each unit provides power in proportion to its capacity (or its programmable droop slope).
- **Virtual synchronous machines**: This approach is based on the emulation of a synchronous machine within the controls of an inverter. Specifically, inverter terminal measurements are fed as inputs into a digital synchronous machine model whose emulated dynamics are mapped to the inverter output in real time [11]. The complexity of the virtual machine can vary greatly, from detailed electromechanical models to simplified swing dynamics. Implementations that closely match machine characteristics, possibly even with virtual flux dynamics, have both Q–V and P–ω characteristics and are often called "synchronverters." On the other end of the spectrum, virtual inertia methods are simpler and capture only the dynamics of an emulated rotor and its steady-state P–ω droop.
- **Virtual oscillator controllers**: Another inverter control method based on the emulation of non-linear oscillators has emerged. Much like a VSM, real-time measurements are processed by the digitally implemented model whose output variables modulate the inverter power stage. The key difference is that the model takes the form of an oscillator circuit with a natural frequency that coincides with the nominal AC grid frequency, and its remaining parameters are tuned to adjust the nominal voltage and control bandwidth [12]. Although the virtual oscillator might appear radically different, it has been shown to exhibit the Q–V and P–ω droop laws in steady state.

7.2.1 Grid-Following Inverter

The block diagrams and differential equations of a GFL are shown in Figs. 7.1 and 7.2, and Eq. (7.1). These blocks are composed of various parts, such as a current control unit, a coordinate transformation unit, and a PLL.

$$e_d = k_{pc}(i_{gd}^* - i_{gd}) + k_{ic} \int_0^t (i_{gd}^* - i_{gd}(\tau))\, d\tau + v_{gd} - i_{gq}k * X_L \tag{7.1}$$

$$e_q - k_{pc}(i_{gq}^* \quad i_{gq}) + k_{ic} \int_0^t (i_{gq}^* - i_{gq}(\tau))\, d\tau + v_{gq} - i_{gd}k * X_L \tag{7.2}$$

where e_d and e_q are the voltages in dq coordinates. k_{pc} and k_{ic} are the parameters of the controller. $i_{gd}^*, i_{gq}^*, i_{gd}, i_{gq}$ are the current references and current in dq coordinates. v_{gd} and v_{gq} are the voltages in dq coordinates. X_L and k are the filter reactance and $2\pi * 60$. The current control loop rapidly controls the active and reactive currents pushed into the grid by quickly tuning the inverter's internal voltage $E\angle\theta$. The current output is maintained in the dq coordinate by the proportional–integral (PI) controller shown in Fig. 7.1. The grid voltage $V_g\angle\theta_g$ and current $I_g\angle\varphi_g$ are calculated from the xy coordinate to the dq coordinate based on the phase angle θ_{PLL} obtained by the PLL, as shown in Fig. 7.2. The current control loop outputs are the inverter's internal voltages in the dq coordinate, e_d and e_q. They are then calculated back to the xy coordinate, as shown in Fig. 7.1(c). Then the inverter model can be connected to the three-phase network solution.

The control objective of the PLL is to estimate and sync with the phase angle of the grid voltage. As shown in Fig. 7.2, the PLL uses a PI controller to regulate the element of the grid voltage on the

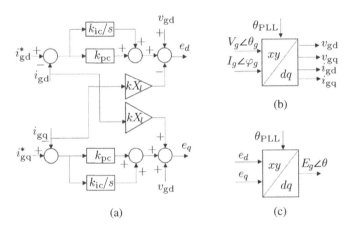

Figure 7.1 (a) Current loop and (b) and (c) coordinate transformation.

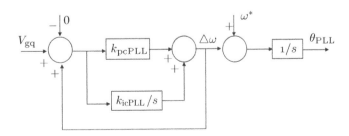

Figure 7.2 PLL control block.

Table 7.1 Case study model and control parameters.

IEEE 13-node test feeder					
S_b	10 MVA	v_b	4.16 kv	ω_b	$2\pi * 60$ rad/s
Synchronous generator (SG)					
S_r	6.25 MVA	v_r	4.16 kv	H	6.5
Grid-following inverter (GFL)					
S_r	10 MVA	k_{pc}	0.1 p.u.	k_{ic}	1.5 p.u.
k_{pPLL}	0.1 p.u.	k_{iPLL}	0.4 p.u.	k	0.15 p.u.
Grid-forming inverter (GFM)					
d_ω	0.1 p.u.	k_p	0.4 p.u.	k_i	0.15 p.u.

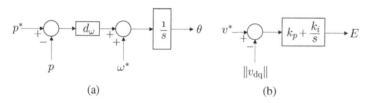

(a)   (b)

Figure 7.3 (a) Droop frequency control and (b) ac voltage control.

q-axis, v_{gq}, to be zero by increasing or decreasing the virtual angular frequency $\triangle\omega$. The integration of the sum of $\triangle\omega$ and frequency reference ω^* is the approximated phase angle θ_{PLL}, which is equal to θ_g in the steady state. K_{pcPLL} and k_{icPLL} are the proportional and integral gains of the controller. The parameters of the GFL are listed in Table 7.1.

7.2.2 Grid-Forming Inverter

There are various grid-forming control techniques proposed in recent years. However, this work chooses droop control, which was developed very early and tested extensively. The block diagram of a droop-controlled GFM is shown in Fig. 7.3. The droop control regulates the angular frequency ω of the inverter internal voltage $E\angle\theta$ according to the Pf droop control. p^* and ω^* are the reference values of active power and frequency. d_ω is the droop gain. p is the measured active power in per unit. The AC voltage control maintains the voltage magnitude E. v^* and $\|v_{dq}\|$ are the voltage reference values and measured voltage in per unit. The parameters of the GFM can be found in Table 7.1. The differential equations are

$$\dot{\theta} = \omega \tag{7.3}$$

$$E = k_p(v^* - \|v_{dq}\|) + k_i \int_0^t (v^* - \|v_{dq}(\tau)\|)d\tau \tag{7.4}$$

7.2.3 SG Model

The SG model includes a six-order electrical model and a two-order mechanical model. This SG is a balanced, symmetrical, three-phase model with field winding and damper windings on the rotor.

The differential equations are

$$V_d = -i_d R_s - \omega \psi_q + \frac{d\psi_d}{dt} \tag{7.5}$$

$$V_q = -i_q R_s - \omega \psi_d + \frac{d\psi_q}{dt} \tag{7.6}$$

$$V_0 = -i_0 R_0 + \frac{d\psi_0}{dt} \tag{7.7}$$

$$V_{fd} = \frac{d\psi_{fd}}{dt} + R_{fd} i_{fd} \tag{7.8}$$

$$0 = \frac{d\psi_{kd}}{dt} + R_{kd} i_{kd} \tag{7.9}$$

$$0 = \frac{d\psi_{kq1}}{dt} + R_{kq1} i_{kq1} \tag{7.10}$$

$$0 = \frac{d\psi_{kq2}}{dt} + R_{kq2} i_{kq2} \tag{7.11}$$

$$\begin{bmatrix} \psi_d \\ \psi_{kd} \\ \psi_{fd} \end{bmatrix} = \begin{bmatrix} L_{md} + L_l & L_{md} & L_{md} \\ L_{md} & L_{1kd} + L_{f1d} + L_{md} & L_{f1d} + L_{md} \\ L_{md} & L_{f1d} + L_{md} & L_{1fd} + L_{l1d} + L_{md} \end{bmatrix} \begin{bmatrix} -i_d \\ i_{kd} \\ i_{fd} \end{bmatrix} \tag{7.12}$$

$$\begin{bmatrix} \psi_q \\ \psi_{kq1} \\ \psi_{kq2} \end{bmatrix} = \begin{bmatrix} L_{mq} + L_l & L_{mq} & L_{mq} \\ L_{mq} & L_{mq} + L_{kq1} & L_{mq} \\ L_{mq} & L_{mq} & L_{mq} + L_{kq2} \end{bmatrix} \begin{bmatrix} -i_q \\ i_{kq1} \\ i_{kq2} \end{bmatrix} \tag{7.13}$$

$$\triangle\omega(t) = \frac{1}{2H} \int_0^t (T_m - T_e) - kd \triangle \omega dt \tag{7.14}$$

$$\omega(t) = \triangle\omega(t) + \omega_0 \tag{7.15}$$

where V_d, V_q, V_0 are the stator voltages in dq-coordinates. R_s is stator resistance. ω is frequency. ψ_d, ψ_q, ψ_0 are the rotor flux linkage in dq-coordinates. V_{fd} is the field winding voltage in dq-coordinates. ψ_{kd}, R_{fd}, i_{fd} are the field winding flux linkage, resistance, and current in d-axis. ψ_{kd}, R_{kd}, i_{kd} are the damper winding flux linkage, resistance, and current in d-axis. ψ_{kq1}, ψ_{kq2}, R_{kq1}, R_{kq2}, i_{kq1}, i_{kq2} are the flux linkage, resistance, and current in damper winding q-axis. $\triangle\omega$ is speed variation with respect to the speed of operation. H is a constant of inertia. T_m, T_e are the mechanical and electromagnetic torque. kd is the damping factor representing the effect of damper windings. $\omega(t)$, ω_0 are the mechanical speeds of the rotor and speed of operation (1 p.u.). The steam turbine and governor include a complete tandem-compound steam prime mover, a speed governing system, a four-stage steam turbine, and a shaft with up to four masses. The excitation is a DC type with an automatic voltage regulator. A PSS is also included.

7.2.4 Network Model

To research the transient dynamics of a low-inertia microgrid, we utilize Sim Power Systems to conduct a dynamic simulation of the modified IEEE 13-node test feeder shown in Fig. 7.4. The single-phase and two-phase loads and corresponding lines are removed. In addition, the rest of the loads are configured to be balanced three-phase loads so that we have a balanced microgrid. A synchronous generator (SG) replaces the original infinite bus. A GFL or GFM is placed at the

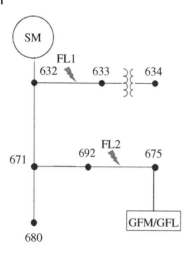

Figure 7.4 Modified 13-node test feeder with a SM, GFM, or GFL.

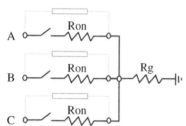

Figure 7.5 Fault model.

end of the network. This microgrid is assumed to be in islanded mode. The protection scheme is not implemented here as we focus on the system's post-fault transient behavior.

7.2.5 Fault Model

The three-phase-to-ground fault is modeled as a three-phase circuit breaker whose opening and closing times can be controlled. The block diagram is shown in Fig. 7.5.

For a three-phase-to-ground fault, the switches are on. Ron and Rg are the fault resistance and the ground resistance. Their values are 0.001 ohms and 0.01 ohms. The three-phase-to-ground fault locations are marked as FL1 and FL2 in Fig. 7.4. These locations are at the middle of the line.

7.3 Metric for Transient Stability

The critical clearing time (CCT) is the maximum time interval by which the fault must be cleared to preserve the system's stability. Therefore, the CCT in a microgrid is defined as the maximum fault-on time such that the system can remain stable after the fault is cleared. The concept of CCT is explained using Fig. 7.6. Suppose in a power system with SG, there is a fault that happens at 10 seconds. The fault-on times are 0.014 and 0.015 seconds in Fig. 7.6(a) and (b). In Fig. 7.6(a), the rotor speed oscillates after the fault and gradually converges back to the nominal value after clearing the fault. However, in Fig. 7.6(b), the rotor speed drops after 19 seconds. Then, the CCT is said to be 0.014 seconds.

This work focuses on frequency stability; therefore, voltage control and stability are ignored.

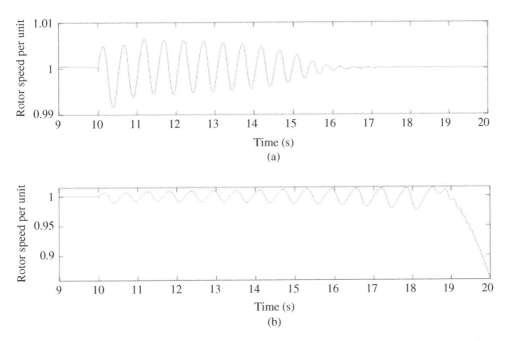

Figure 7.6 Rotor speed. (a) The stable case, fault-on time is 0.014 seconds. (b) The unstable case, fault-on time is 0.015 seconds.

7.4 Microgrid Transient Stability Analysis

In this section, we first demonstrate how different fault locations and different types of inverter controls can affect the stability of the microgrid. The stability metric introduced in the previous section has been used to quantify the stability of the microgrid. Then we provided a few examples on how to improve the stability of the microgrid through tuning the controller parameter.

7.4.1 Transient Stability of an Islanded Microgrid with Single SG

The transient stability is tested when no inverters are present in the microgrid. The three-phase-to-ground fault is applied at FL1 at 10 seconds with a fault duration of four seconds. Although this fault duration is too long, the purpose is to provide a benchmark to study the impact of adding a GFL or GFM on the transient stability of the microgrid. The rotor speed is plotted in Fig. 7.7. The result shows that the SG is stabilized after the fault, although the rotor speed is slightly higher than the nominal value.

7.4.2 Impact of GFL Inverter on Transient Stability of an Islanded Microgrid

In this section, we test the transient stability of the microgrid when the GFL is included in the network. The contingency event is the three-phase-to-ground fault that happens at 10 seconds. We also changed the power generation of the SG and the GFL to study the effect of different power sharing on the transient stability. The result is shown in Table 7.2. In Case 1, the power generations of the SG and the GFL are 5 and 3 MW, respectively. In Case 2, the power generations of the SG and the GFL are 6 and 2 MW, respectively.

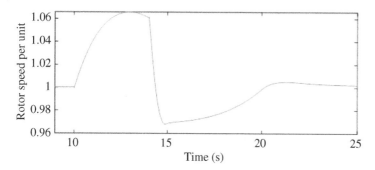

Figure 7.7 Rotor speed for the fault event.

Table 7.2 CCT of different cases.

	GFL case	
	Case 1	Case 2
	SG power generation: 5 MW	SG power generation: 6 MW
	GFL power generation: 3 MW	GFL power generation: 2 MW
FL1	0.014 s	0.077 s
FL2	0.011 s	0.067 s
	GFM case	
	Case 1	Case 2
	SG power generation: 5 MW	SG power generation: 6 MW
	GFL power generation: 3 MW	GFL power generation: 2 MW
FL1	3.09 s	3.218 s
FL2	2.958 s	3.046 s

Table 7.2 suggests that when the fault location is closer to the SG, and the SG takes more power sharing, the CCT is longer. This is because the generation device closer to the fault receives more disturbance; however, due to inertia, the SG can withhold the fault for a more extended period than the GFL. In Fig. 7.8, the rotor speed and angular frequency of the SG and the GFL are plotted for Case 1 and FL1. Figure 7.8(a) shows the dynamics when fault-on time is CCT, and Fig. 7.8(b) shows the dynamics when fault-on time is CCT plus 0.001 seconds.

7.4.3 Impact of GFM and Parameter Tuning on Transient Stability of an Islanded Microgrid

In this section, the GFM is placed in the network, as shown in Fig. 7.4. The simulations in the previous section are performed again, and the CCT results are given in Table 7.2. For Case I and FL1, we tuned the droop gain and recorded the CCT for each droop gain for the contingency. The result is shown in Fig. 7.9. Notice that when the GFM is in the network, the CCT extends significantly

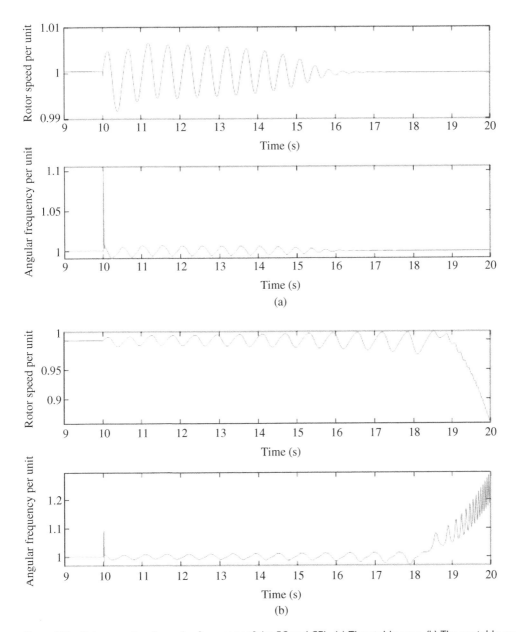

Figure 7.8 Rotor speed and angular frequency of the SG and GFL. (a) The stable case. (b) The unstable case.

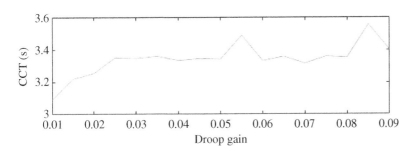

Figure 7.9 Droop gain versus CCT.

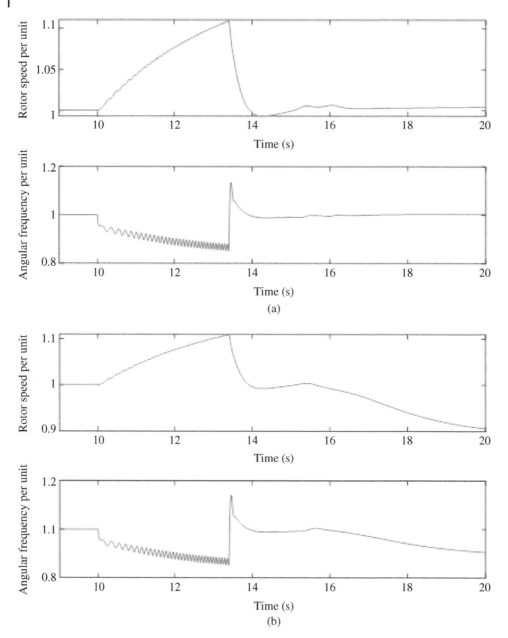

Figure 7.10 Rotor speed and angular frequency of the SG and GFM. (a) The stable case. (b) The unstable case.

compared to the GFL in the network. Because the GFM can regulate the frequency and voltage by itself, the GFM and the SG work together to pull their operation point back to the steady state. In addition, as the droop gain increases, the CCT also increases, although there are a few exceptions. The reason for the rising of CCT could be that a larger droop gain makes the rate of change (angular frequency) of the GFM's voltage angle response faster to the transient.

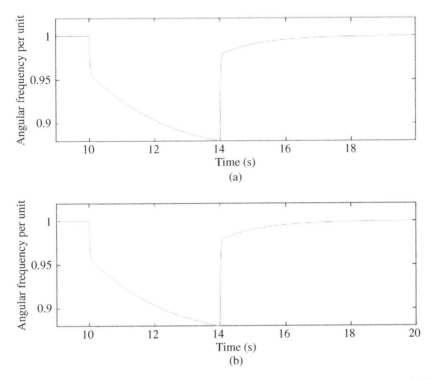

Figure 7.11 Angular frequency of the GFM. (a) Angular frequency of the GFM at node 632. (b) Angular frequency of the GFM at the end of the network.

Figure 7.10 shows the rotor speed and angular frequency of the SG and the GFM for the droop gain of 0.05. Figure 7.10(a) shows the dynamics when fault-on time is CCT, and Fig. 7.10(b) shows the dynamics when fault-on time is CCT plus 0.001 seconds.

7.4.4 The Transient Stability of an Islanded Microgrid with Only GFM

A three-phase-to-ground fault is tested when the SG is replaced with the GFM and another GFM at the end of the circuit. The fault location is F1, and the fault-on time is four seconds. These two GFMs share the power equally, and the parameter setting is listed in Table 7.1. The angular frequencies of these two GFMs are plotted in Fig. 7.11. The angular frequencies return to the nominal value even if the fault-on time is the largest compared to Table 7.2. The conclusion is that the transient stability of an islanded microgrid with only GFM is the best due to the low order of the system.

7.5 Conclusion and Future Directions

The addition of the GFL degrades the stability of the microgrid. It can be seen from Table 7.2 that the transient during the fault with inverter is larger than the transient without the addition of the inverter. Also, the CCT of the microgrid becomes shorter after adding the addition of the GFL. However, an improvement can be made by increasing the proportional gain of the current controller. As the primary energy interface in the future power systems and microgrids, inverters have shown a great potential to improve the stability of the system. However, because inverters could operate in few cycles, there are need for the coordination between inverter and protection systems. And because each inverter could have different controller design, there are also need for better understanding of the heterogeneous nonlinear system.

7.6 Exercises

1. List parameters that can affect the stability of a microgrid
   Note: The nature of the disturbance (magnitude, location, duration, etc.) has a direct impact on whether the microgrid can withstand this disturbance. However, the disturbances are not the design parameters of a microgrid.
2. Consider the microgrid model in Fig. 7.4
   (a) Create a three-phase fault at location node 671.
   (b) Find the critical clearing time of the three-phase fault at location node 671.
   (c) Find at least one way to improve the stability of this microgrid. Show the improvement through simulation.
   Solution:
   – Select the three-phase fault block from the Simulink library. Connect this block to the node 671.
   – Critical clearing time is 0.014 seconds.
   – Increase the generator power generation to improve the stability, or increase the PLL gain. The simulation result should look similar to Fig. 7.8.

References

1 Steinmetz, C.P. (1920). Power control and stability of electric generating stations. *Transactions of the American Institute of Electrical Engineers* 39 (2): 1215–1287.

2 Evans, R.D. and Bergvall, R.C. (1924). Experimental analysis of stability and power limitations. *Transactions of the American Institute of Electrical Engineers* 43: 39–58.

3 Sauer, P.W., Pai, M.A., and Chow, J.H. (2017). *Power System Dynamics and Stability: With Synchrophasor Measurement and Power System Toolbox*. Wiley.

4 Chiang, H.-D. (2011). *Direct Methods for Stability Analysis of Electric Power Systems: Theoretical Foundation, BCU Methodologies, and Applications*. Wiley.

5 Shuai, Z., Shen, C., Liu, X. et al. (2018). Transient angle stability of virtual synchronous generators using Lyapunov's direct method. *IEEE Transactions on Smart Grid* 10 (4): 4648–4661.

6 Yu, H., Awal, M.A., Tu, H. et al. (2020). Comparative transient stability assessment of droop and dispatchable virtual oscillator controlled grid-connected inverters. *IEEE Transactions on Power Electronics* 36 (2): 2119–2130.

7 Zhao, T., Wang, J., Lu, X., and Du, Y. (2021). Neural Lyapunov control for power system transient stability: a deep learning-based approach. *IEEE Transactions on Power Systems* 37 (2): 955–966.

8 Cui, W. and Zhang, B. (2021). Lyapunov-regularized reinforcement learning for power system transient stability. *IEEE Control Systems Letters* 6: 974–979.

9 Du, W., Tuffner, F.K., Schneider, K.P. et al. (2020). Modeling of grid-forming and grid-following inverters for dynamic simulation of large-scale distribution systems. *IEEE Transactions on Power Delivery* 36 (4): 2035–2045.

10 Tayyebi, A., Groß, D., Anta, A. et al. (2020). Frequency stability of synchronous machines and grid-forming power converters. *IEEE Journal of Emerging and Selected Topics in Power Electronics* 8 (2): 1004–1018.

11 Zhong, Q.-C. and Weiss, G. (2010). Synchronverters: inverters that mimic synchronous generators. *IEEE Transactions on Industrial Electronics* 58 (4): 1259–1267.

12 Johnson, B., Rodriguez, M., Sinha, M., and Dhople, S. (2017). Comparison of virtual oscillator and droop control. In: *2017 IEEE 18th Workshop on Control and Modeling for Power Electronics (COMPEL)*, 1–6. IEEE.

8

Learning-Based Transient Stability Assessment of Networked Microgrids

Tong Huang

Learning Objective:

- Understand the goal of transient stability assessment of networked microgrids and the conventional approaches to achieving it.
- Learn the basic stability concepts and why they are important in the context of microgrids.
- Explain the basic idea of the neural Lyapunov approach and identify its limitation.

8.1 Motivation

The electricity infrastructure is fragile during extreme weather-related events [1]. For example, the extreme heat wave in California caused rotating electricity outages in 2020. The winter storms in Texas resulted in local power crisis in 2021. Microgrids provide a promising way that enhance the resilience of the power grids to natural disasters. A microgrid, i.e. a small-scale electric energy system that contains both generation and load, can either connect to its host distribution system under normal conditions, or operate autonomously if the main grid is not able to supply electricity. In addition, a microgrid can network with its neighboring microgrids, and they can work collaboratively to supply load and improve the quality of electricity services. Reference [2] provides an example in which one microgrid can support its networked peers to restore the nominal frequency of electricity.

To design and operate networked microgrids, it is critical to ensure their security in both steady-state and transient time scales. The steady-state security analysis aims to ensure that loads can be balanced by generation resources without violating steady-state constraints, e.g. line thermal limits, and bus voltage limits, given a dispatch strategy that determines how much power each generation resource needs to produce. The objective of the transient stability analysis is to determine if the networked microgrids can be stabilized around a prescribed steady state under disturbances. This chapter focuses on the transient stability analysis of networked microgrids.

Based on the magnitudes of disturbances, the stability analysis can be categorized into the small-signal stability (SSS) analysis, and the transient stability analysis. The SSS analysis concerns the grid dynamics under small disturbances (e.g. small load or renewable fluctuations), and it is built upon the system dynamics that are linearized around certain operating conditions. By examining the eigenvalues of the system matrix of the linear model, the SSS can be concluded. Such an analysis can be carried out in a centralized or distributed manner [3]. However, the linear

Microgrids: Theory and Practice, First Edition. Edited by Peng Zhang.

model used in the SSS analysis only approximates microgrids' behaviors under small disturbances, and the model is not valid when a large disturbance occurs.

The transient stability analysis examines microgrids' responses under large disturbances (e.g. line tripping, and connection and disconnection of generation resources). There are two categories of methods for conducting the transient security analysis: the numerical methods and the analytical methods. The numerical methods determine the transient stability by numerically solving the differential and algebraic equations that describe microgrids' dynamics, given a finite number of scenarios [4]. While straightforward, the numerical methods are time-consuming when handling a large-scale system in the electromagnetic transient (EMT) time scale. Another drawback of the numerical methods is that they cannot certify stability rigorously. By definition, the stability certification requires one to check the system trajectories under an infinite amount of disturbances, which is impossible for the numerical methods. The analytical methods can certify the stability of a dynamical systems under large disturbances. A classic approach to certifying the transient stability in the field of power system research is the energy function approache [5–7]. These approaches certify the transient stability by finding a function that summarizes the behaviors of power grids. While the energy function-based methods can be applicable to bulk transmission systems where the resistance to inductance (R/X) ratios of transmission lines are small, the energy function for distribution networks that comprise distribution lines with large R/X ratios does not exist [5].

This chapter introduces a machine learning-based method to certify the transient stability of networked microgrids. The basic idea of the learning-based method is to construct a neural network structured Lyapunov function that summarizes the dynamical behaviors of the networked microgrids. The function learned leads to a RoA that can quantify the disturbances' magnitudes that the networked microgrids can tolerate.

The rest of this chapter is organized as follows. Section 8.2 presents the dynamics of networked microgrids. Section 8.3 illustrates how to find the neural network-structured Lyapunov function for transient stability assessment. Section 8.4 demonstrates the performance of the learning-based approach in networked microgrids. Section 8.5 summarizes this chapter and points out limitations of the learning-based method presented and future research directions.

8.2 Networked Microgrid Dynamics

The physical architecture of a future distribution system based on networked microgrids is shown in Fig. 8.1. The building block of the distribution system is a microgrid which contains both distributed energy resources (DERs) and loads. The loads in the microgrids can be supplied by the main distribution grid. If the main grid loses its function, the microgrid can supply the load autonomously by leveraging its internal generation resources. Each microgrid networks with others through distribution lines. The point where a microgrid connects to the main grid is called the point of common coupling (PCC). We also envision that there is a power-electronics (PE) interface between an individual microgrid and its host distribution system. For a DC microgrid, the PE interface is necessary as the interface converts the electricity from DC to AC. For an AC microgrid, having such a PE interface significantly reduces the management complexity of the distribution system operators (DSOs), since the DSO only needs to coordinate several PE interfaces, instead of tuning massive DERs' setpoints.

In the n networked microgrids shown in Fig. 8.1, the dynamics of each microgrid are determined by the control algorithms programmed in its PE interface. One way to design the control algorithms

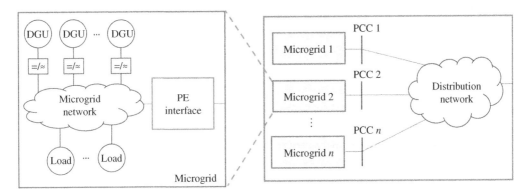

Figure 8.1 Networked microgrids. Source: [8]/IEEE.

is to enable the microgrid to behave like a conventional synchronous machine whose dynamics are governed by the swing equations:

$$\frac{d\delta_i(t)}{dt} = \omega_i \tag{8.1a}$$

$$\frac{d\omega_i(t)}{dt} = -\frac{D_i}{M_i}\omega_i + \frac{1}{M_i}(P_i^* - P_i) \tag{8.1b}$$

where δ_i is the deviation of voltage phase angle from its nominal value δ_i^*; ω_i is the deviation of frequency deviation from its nominal value ω_i^*; D_i and M_i are control parameters of the ith microgrid; and P_i and P_i^* are real power injection to the ith node and its nominal value, respectively. Another way to design the control algorithm in the PE interface is based on the angle droop law:

$$\frac{d\delta_i(t)}{dt} = -\frac{1}{M_i'}\delta_i + \frac{D_i'}{M_i'}(P_i^* - P_i) \tag{8.2}$$

where M_i' and D_i' are the control parameters of the ith PE interface equipped with the angle droop controller. While other control laws are discussed in literature [9], this chapter assumes that either the frequency droop law or the angle droop law is deployed at each PE interface in Fig. 8.1 for the sake of simplicity.

The nodal dynamics at node i described by either (8.1) or (8.2) interact with other microgrids via distribution lines that introduce algebraic coupling, i.e. the power flow equations:

$$P_i = \sum_{k \neq i} V_i V_k y_{ik} \cos(\delta_{ik} - \theta_{ik}) + g_{ii} V_i^2 \tag{8.3}$$

where V_i is the voltage magnitude at node i; g_{ii} is the ith diagonal entry of the admittance matrix Y of the distribution network; and $y_{ik} \angle \theta_{ik}$ is the polar form of the (i, k)th entry of Y.

Based on the frequency and/or angle droop equations as well as the power flow equations, with the configuration in Fig. 8.1, a distribution system with networked microgrids can be mathematically described by

$$\frac{d\mathbf{x}}{dt} = \mathbf{f}(\mathbf{x}) \tag{8.4}$$

where the state vector $\mathbf{x}$ collects δ_i or $\{\delta_i, \omega_i\}$ for $i = 1, 2, \ldots, n$; and function $\mathbf{f}(\cdot)$ defines how the state $\mathbf{x}$ evolves in term of time. Since all state variables in (8.4) are defined as the deviation from their nominal values, the origin $\mathbf{o}$ is one equilibrium point of the dynamic system (8.4).

The transient stability assessment aims to *quantify the magnitudes of disturbances that can be tolerated by the networked microgrids*. By the "disturbances that can be tolerated," we mean that after the disturbances occur, the system trajectory of (8.4) will converge to the equilibrium point **o**. The transient stability assessment boils down to how to certify the stability of the equilibrium point **o** and how to find a RoA. The concept of the stability and the RoA will be introduced in Section 8.3.

8.3 Learning a Lyapunov Function

This section first introduces the definition of stability for a nonlinear system and a Lyapunov function that leads to a RoA. Then we present the basis of neural networks in the field of machine learning, which is leveraged to find a Lyapunov function. Finally, the procedure of learning a Lyapunov function from state space is elaborated.

8.3.1 Stability of Equilibrium Points

Given a dynamical system (8.4) with an equilibrium point **o**, the definition of the asymptotic stability is introduced as follows:

Definition 8.1 (Asymptotic Stability) [10]: The equilibrium point **o** is asymptotically stable, if for all $\epsilon > 0$, there exists $\xi = \xi(\epsilon) > 0$ such that $\|\mathbf{x}(0)\| < \xi$ implies that $\|\mathbf{x}(t)\| < \epsilon$, and $\lim_{t \to \infty} \mathbf{x}(t) = \mathbf{0}$, for all $t \geq 0$.

The definition of the asymptotic stability essentially says that for an asymptotically stable equilibrium point, the state $\mathbf{x}(t)$ can stay arbitrarily close to the equilibrium point for time $t > 0$, and the system trajectory $\mathbf{x}(t)$ can converge to the equilibrium eventually, as long as the initial condition $\mathbf{x}(0)$ is sufficiently close to the equilibrium point. The stability is a desirable property for the networked microgrids. The equilibrium point corresponds to the steady-state operating condition dispatched by a DSO based on some economic studies. In addition, the DSO expects the microgrid system trajectory to be stabilized around the operating condition prescribed after a disturbance causes microgrid states **x** to deviate from the operating condition. For an asymptotically stable equilibrium point, if a large disturbance drives the initial condition very far away from the equilibrium point, it is still possible that the system trajectory will never converge to the equilibrium point, i.e. the prescribed operating condition.

Next, we introduce the region of attraction (RoA) that is used for quantifying the magnitude of disturbances that networked microgrids can tolerate.

Definition 8.2 (Region of Attraction) [10]: For an asymptotically stable equilibrium point **o**, region $\mathcal{R}$ is a RoA of the equilibrium point **o**, if $\mathbf{x}(0) \in \mathcal{R}$ implies $\mathbf{x}(t)$ converges to **o**, as time t tends to infinity.

Figure 8.2 visualizes an RoA. If the initial condition (the black point) is inside the RoA ($\mathbf{x}_0$), the system trajectory (dashed line) will converge to the equilibrium point (at the origin).

A natural question is how to certify the asymptotic stability and how to find the RoA for an operating condition of networked microgrids. These two questions can be answered by finding a Lyapunov function:

Figure 8.2 Illustration of an RoA.

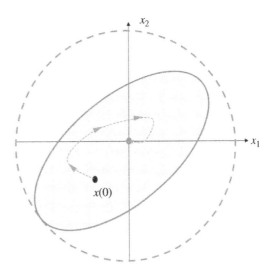

Definition 8.3 (Lyapunov Function) [10]: A differentiable, continuous function $V(\mathbf{x})$ is a Lyapunov function for the dynamics (8.4) with equilibrium $\mathbf{o}$, if the function satisfies the following two conditions in a ball, i.e. $\mathcal{B}_r := \{\mathbf{x} | \|\mathbf{x}\|_2^2 < r^2\}$:

1. $V(\mathbf{x})$ is positive definite[1] in $\mathcal{B}_r$;
2. $\frac{dV(\mathbf{x})}{dt}$ is negative definite in $\mathcal{B}_r$.

The existence of the Lyapunov function defined above certifies the asymptotic stability of the equilibrium $\mathbf{o}$. Furthermore, if a Lyapunov function V is found, an RoA can be estimated based on the Lyapunov function via

$$\mathcal{R}_d = \{\mathbf{x} | V(\mathbf{x}) < d\} \subseteq \mathcal{B}_r \qquad (8.5)$$

where $d > 0$. An RoA is visualized in Fig. 8.3. By increasing d, the RoA can be enlarged, as shown in Fig. 8.3. It is desirable for a DSO to obtain an RoA that is as large as possible, in order to quantify large disturbances. In a two-dimensional space, such a RoA can be found by gradually increasing d until the boundary of RoA (the solid circles in Fig. 8.3) touches the boundary of $\mathcal{B}_r$ (the the dashed circle in Fig. 8.3). However, it is hard to implement such an intuitive procedure, if the system contains more than three states. It turns out that the largest d that satisfies (8.5) can be found by calculating the "touching point" of the two boundaries. We refer readers to [8] for more details of a sub-procedure for obtaining the touching point.

It can be seen that the key to estimating an RoA is finding a legitimate Lyapunov function. However, it is challenging to find a Lyapunov function for a general dynamical system. Next, we introduce some preliminary knowledge of neural network in the machine learning field, in order to present the neural Lyapunov methods.

8.3.2 Neural Network Architecture

Figure 8.4 shows the architecture of a neural network with multiple inputs, one output, and one hidden layer. The computation unit of the neural network is a neuron. The neurons are interconnected with each other in an hierarchical manner. In the neural network shown in Fig. 8.4, the

---

1 A function $h(\mathbf{x})$ is positive definite in a region $\mathcal{B}$, if $h(\mathbf{0}) = 0$ and $h(\mathbf{x}) > 0$ in $\mathcal{B} - \{\mathbf{0}\}$. A function $h'(\mathbf{x})$ is negative definite in $\mathcal{B}$ if $-h'(\mathbf{x})$ is positive definite.

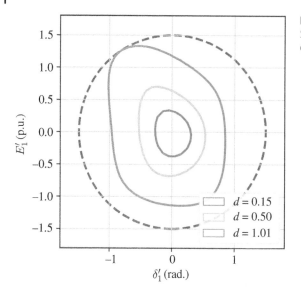

Figure 8.3 Enlarging an RoA by increasing d. Source: [8]/IEEE: δ_1' and E_1' are two states of a dynamical system.

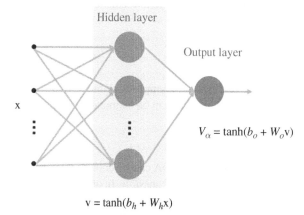

Figure 8.4 Neural network architecture. Source: [8]/IEEE.

neuron directly relating to the output constitutes an output layer, while the rest neurons constitute a hidden layer. In Fig. 8.4, denote by **x** and **v** the input vector and output vector of the hidden layer, respectively. The hidden layer relates its input to its output by

$$\mathbf{v} = \tanh(\mathbf{b}_h + W_h \mathbf{x}) \tag{8.6}$$

where $\mathbf{b}_h$ and W_h are a bias vectors, and a weighting matrix for the hidden layer; and $\tanh(\cdot)$ is an element-wised hyperbolic tangent function, i.e.

$$\tanh([a_1, \ldots, a_m]^\top) = [\tanh(a_1), \ldots, \tanh(a_m)]^\top.$$

The output layer in Fig. 8.4 relates **v** to V_α via

$$V_\alpha = \tanh(b_o + W_o \mathbf{v}) \tag{8.7}$$

where b_o and W_o are a bias and a weighting matrix for the output layer, respectively; and V_α is a scalar that depends on **v** as well as a set of parameters $\alpha = \{\mathbf{b}_h, W_h, b_o, W_o\}$. It is worth noting that the derivative $\partial V_\alpha / \partial \mathbf{x}$ can be derived analytically based on the chain rule. This observation will be used later to check the second condition of the Lyapunov function.

Suppose that the Lyapunov function possesses a neural network structure shown in Fig. 8.4. The input of the neural network corresponds to the state vector of dynamics (8.4), while the scalar output of the neural network can be interpreted as the value of the Lyapunov function evaluated at the input state vector of the neural network. The parameters α can be updated by a classic algorithm called backpropagation (BP). To achieve the transient stability analysis, one key question is how to tune the parameters in the set α such that the neural network satisfies the two conditions of Lyapunov functions. This will be answered in Section 8.3.3.

8.3.3 Neural Lyapunov Methods

We proceed to define an objective function called the empirical Lyapunov risk that guide the BP algorithm to tune α so that the neural network in Fig. 8.4 behaves like a legitimate Lyapunov function. Suppose that there are M state vectors $\mathbf{x}_1, \ldots, \mathbf{x}_M$ that are randomly drawn from state space. The empirical Lyapunov risk of the M state vectors is evaluated based on

$$L_M(\boldsymbol{\alpha}) = \frac{1}{M} \sum_{i=1}^{M} \left(\mathtt{ReLU}(-V_\alpha(\mathbf{x}_i)) + \mathtt{ReLU}(\dot{V}_\alpha(\mathbf{x}_i)) \right) \tag{8.8}$$

where $\mathtt{ReLU}(\cdot)$ is the rectified linear unit that will return its input if the input is non-negative, and zero if otherwise; and $\dot{V}_\alpha(\mathbf{x}_i)$ denotes the time derivative of V_α evaluated at $\mathbf{x}_i$. The first "$\mathtt{ReLU}$" term in (8.8) will lead to positive penalty if the first condition of the Lyapunov function is violated, while the second "$\mathtt{ReLU}$" term will increase the Lyapunov risk if the second condition of the Lyapunov function is violated. To embed the two conditions of the Lyapunov function into the neural network, parameter α is chosen such that the empirical Lyapunov risk is minimized, i.e.

$$\boldsymbol{\alpha}^* = \arg\min_\alpha L_M(\boldsymbol{\alpha}) \tag{8.9}$$

Note that the neural network with parameter α^*, i.e. V_{α^*}, may not be a legitimate Lyapunov function. This is because the two conditions are only evaluated based on the selected M state vectors. It is possible that the conditions are violated in some region in $\mathcal{B}_r$ that is not covered by the selected samples. Therefore, the samples used for tuning the neural network need to be augmented.

One way to augment the sample space that reinforces the two conditions of a Lyapunov function is by searching for counterexamples, i.e. the state vectors where the two conditions are violated, and by adding the counterexamples to the training set. The satisfiability modulo theories (SMT) can be used for finding the counterexamples. Formally, an $\mathbf{x}(\neq \mathbf{0})$ is a counterexample, if it satisfies

$$V_\alpha(\mathbf{x}) \leq 0 \quad \text{or} \quad \dot{V}_\alpha(\mathbf{x}) \geq 0 \tag{8.10}$$

Satisfying (8.10) suggests that at least one of condition for a legitimate Lyapunov function is violated. We can leverage the SMT tool such as $\mathtt{dReal}$ to find the counterexamples [11]. $\mathtt{dReal}$ has the delta-completeness property [12], i.e. if $\mathtt{dReal}$ concludes there is no solution that satisfies (8.10), this will guarantee that no solution exists.

The overall procedure to learn a Lyapunov function is summarized in Fig. 8.5. We start from the M state vectors randomly drawn from a region $\mathcal{B}_r$. Next, the parameter α is tuned by solving (8.9) based on the random state vectors. The neural network with the updated parameters serves as a Lyapunov function candidate. Then, the SMT tool is leveraged to examine the satisfiability of the two conditions of a Lyapunov function by finding counterexamples. If counterexamples are found, they will be added to the sample set and the neural network will be retuned. Otherwise, the Lyapunov function candidate is a legitimate Lyapunov function that can be used for estimating an RoA.

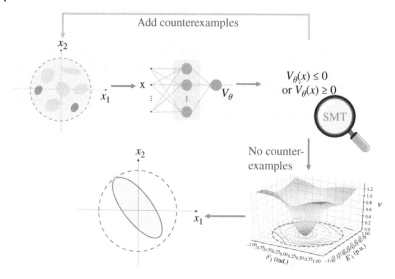

Figure 8.5 Overall procedure.

8.4 Case Study

In this section, the neural Lyapunov approach to transient stability assessment is demonstrated in networked microgrids. Figure 8.6 shows the topology of the networked microgrids. The Lyapunov function learned and its time derivative are visualized in Fig. 8.7 in which Fig. 8.7(a) shows the Lyapunov function learned is positive definite and Fig. 8.7(b) shows that the time derivative of the function is negative definite. Therefore, it can be observed that the function learned is indeed a Lyapunov function, since the two conditions of a Lyapunov function are satisfied.

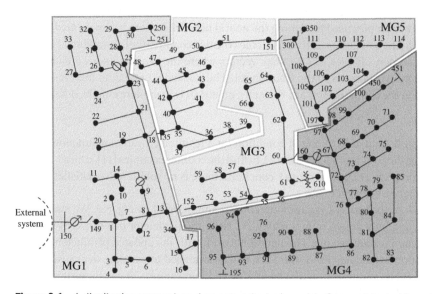

Figure 8.6 A distribution system based on networked microgrids. Source: Adapted from [13].

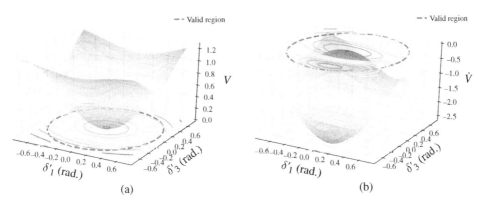

Figure 8.7 (a) Visualization of the Lyapunov function learned. (b) Visualization of the time derivative of the learned Lyapunov function.

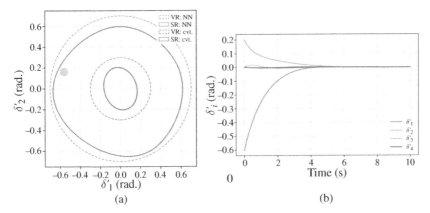

Figure 8.8 (a) Visualization of RoA in the $\delta'_1 - \delta'_2$ plane. (b) Time-domain response under a disturbance. Source: [8]/IEEE.

Based on the Lyapunov function in Fig. 8.7(a), an RoA can be estimated based on (8.5). The RoA estimated is shown in Fig. 8.8(a) where the solid circle and dashed circle are the boundaries of RoA (i.e. security region, SR) and $\mathcal{B}_r$ (the valid region, VR), respectively. Figure 8.8(a) also compares the RoAs from the neural-network Lyapunov method (NN) with that from a conventional (cvt.) approach. The conventional approach assumes the Lyapunov function used for the RoA estimation has a quadratic form. In Fig. 8.8(a), it can be seen that the transient stability assessment based on the neural Lyapunov approach is less conservative than that based on the conventional approach. For example, suppose that a disturbance drives the initial condition to the point in Fig. 8.8(a). The learning-based approach can conclude that the networked microgrids can tolerate such a disturbance without carrying out any simulations. However, the conventional approach can conclude nothing about the asymptotic behavior of the microgrids. Figure 8.8(b) shows the simulation result after the disturbance for the purpose of verification. It can be observed that the system trajectory indeed converges to the equilibrium point. The RoA estimated from the neural network-structured Lyapunov function draws the same conclusion almost instantaneously.

8.5 Summary

This chapter presents a learning-based approach to transient stability assessment of networked microgrids. The core idea is learning a neural network-structured Lyapunov function. The properties of a legitimate Lyapunov function are reinforced by defining an empirical Lyapunov risk that guides the backpropagation algorithm to tune a neural network. The training sample set is enriched by leveraging an SMT tool. The simulation study demonstrates how to use the RoA estimated by the neural network-structured function, and it also elaborates its less conservativeness compared with a conventional approach.

While this chapter shows the potential of using the neural Lyapunov method to assess stability of networked microgrids, the method is computationally expensive, which limits its application in a small-scale system. One research direction is how to speed up the computation. One potential way to accelerate the computation is to enable each microgrid to learn a part of the Lyapunov function, i.e. a storage function, and then to find the Lyapunov function by combining the storage functions learned by microgrids. We refer the interested readers to references [14, 15] for more details.

8.6 Exercises

1. Briefly explain why the analytical transient stability approaches are still needed, given that one can simulate the behaviors of networked microgrids.
2. Suppose that an initial condition falls out of the larger solid circle in Fig. 8.8(a). Can one conclude that the system trajectory $\mathbf{x}(t)$ will drift away from the equilibrium point (the origin in Fig. 8.8(a))?
3. To evaluate the empirical Lyapunov risk, one needs to know $\dot{V}_\alpha$. Based on control theory, we know that

$$\dot{V}_\alpha = \frac{\partial V_\alpha}{\partial \mathbf{x}} \mathbf{f}(\mathbf{x})$$

where $\mathbf{f}(\cdot)$ is the system dynamics defined in (8.4). Given the neural network defined by (8.6) and (8.7), please derive $\frac{\partial V_\alpha}{\partial \mathbf{x}}$ in terms of the weighting matrices W_h and W_o, based on the chain rule.

References

1 Huang, T., Sun, H., Kim, K.J. et al. (2020). A holistic framework for parameter coordination of interconnected microgrids against disasters. *2020 IEEE Power Energy Society General Meeting (PESGM)*, 1–5.

2 Huang, T., Wu, D., and Ilic, M. (2023). Cyber-resilient automatic generation control for systems of AC microgrids. *IEEE Transactions on Smart Grid*.

3 Zhang, Y. and Xie, L. (2016). A transient stability assessment framework in power electronic-interfaced distribution systems. *IEEE Transactions on Power Systems* 31 (6): 5106–5114.

4 Issicaba, D., Lopes, J.A.P., and da Rosa, M.A. (2012). Adequacy and security evaluation of distribution systems with distributed generation. *IEEE Transactions on Power Systems* 27 (3): 1681–1689.

5 Chiang, H. (1989). Study of the existence of energy functions for power systems with losses. *IEEE Transactions on Circuits and Systems* 36 (11): 1423–1429.

6 Chiang, H.-D., Chu, C.-C., and Cauley, G. (1995). Direct stability analysis of electric power systems using energy functions: theory, applications, and perspective. *Proceedings of the IEEE* 83 (11): 1497–1529.

7 Fouad, A.-A. and Vittal, V. (1988). The transient energy function method. *International Journal of Electrical Power & Energy Systems* 10 (4): 233–246.

8 Huang, T., Gao, S., and Xie, L. (2022). A neural lyapunov approach to transient stability assessment of power electronics-interfaced networked microgrids. *IEEE Transactions on Smart Grid* 13 (1): 106–118.

9 Xie, L., Huang, T., Kumar, P.R. et al. (2022). On a control architecture for future electric energy systems. *Proceedings of the IEEE* 110 (12): 1940–1962.

10 Slotine, J.-J.E. and Li, W. (1991). *Applied Nonlinear Control*, vol. 199. Englewood Cliffs, NJ: Prentice Hall.

11 Gao, S., Kong, S., and Clarke, E.M. (2013). *dReal*: an SMT solver for nonlinear theories over the reals. In: *International Conference on Automated Deduction*, 208–214. Springer.

12 Gao, S., Avigad, J., and Clarke, E.M. (2012). δ-complete decision procedures for satisfiability over the reals. In: *International Joint Conference on Automated Reasoning*, 286–300. Springer.

13 Huang, T., Gao, S., Long, X., and Xie, L. (2021). A neural lyapunov approach to transient stability assessment in interconnected microgrids. *Proceedings of the 54th Hawaii International Conference on System Sciences*, 3330.

14 Jena, A., Huang, T., Sivaranjani S. et al. (2021). Distributed learning-based stability assessment for large scale networks of dissipative systems. *2021 60th IEEE Conference on Decision and Control (CDC)*, 1509–1514.

15 Jena, A., Huang, T., Sivaranjani, S. et al. (2022). Distributed learning of neural Lyapunov functions for large-scale networked dissipative systems. *arXiv preprint arXiv:2207.07731*.

9

Microgrid Protection

Rômulo G. Bainy and Brian K. Johnson

9.1 Introduction

9.1.1 Motivation

Distributed energy resources (DERs) pose challenges for power systems protection and relaying schemes, especially at the distribution level. Inverter-based resources (IBR) are especially challenging due to their fast current limiting capabilities. Designing secure and reliable settings for commonly applied distribution system protection schemes with DERs is challenging by itself, and the challenge increases for microgrid systems capable of isolated operation. The emerging inverter control capabilities for an IBR to transition between grid-following control and grid-forming control when a local system becomes an islanded microgrid may simplify or complicate protection system settings, so this capability also needs consideration in protection systems settings and design.

This chapter begins by revisiting fundamental principles essential for understanding power systems protection schemes. It proceeds to present various protection elements commonly utilized in power systems, which when combined, formulate comprehensive protection schemes and systems. Following this, the chapter addresses the unique challenges introduced by DERs for traditional distribution protection schemes. It then explores the complexities arising from microgrid operations, focusing on requirements and obstacles encountered when protecting a microgrid.

Moreover, the chapter discusses how the solutions implemented to ensure reliable protection for microgrids are intricately tied to the costs associated with potential interruptions of loads within these systems. Higher costs due to load interruptions often justify the implementation of more sophisticated protection schemes. Lastly, the chapter provides insights into commonly employed solutions tailored for specific types of microgrids.

9.2 Protection Fundamentals

Most bulk electric power sources are in remote locations and electric power is distributed over long distances to reach consumers. The power transmission and distribution systems normally operate under sinusoidal steady-state conditions. But momentary disturbances such as short circuit faults and large load or generation variations impact the system's integrity. Power system protection schemes act to localize the impact of disturbances, to limit the scope of any interruption, and ensure a safe and reliable supply of energy.

Microgrids: Theory and Practice, First Edition. Edited by Peng Zhang.
© 2024 The Institute of Electrical and Electronics Engineers, Inc. Published 2024 by John Wiley & Sons, Inc.

A protection system consists of measurement devices, a protective relay, and a circuit interrupting device controlled by the protective relay. Measurement devices are typically current and voltage transducers that step current and voltage down to safe levels for input to the protective relay and provide electrical isolation. In most cases, iron core current transformers (CTs) and voltage transformers (VTs) provide measurements to protective relays and meters. Other technologies for current and voltage sensors include linear couplers (air core CTs), current shunts, optical CTs, and optical VTs.

A protective relay is a special-purpose device that processes measured voltages and currents, executes protection functions using those measurements, and initiates a trip command to the circuit breaker if it detects a faulty condition. These functions are protection elements. Many functions are available, each using characteristics of currents and voltages to identify faulty conditions.

Protection engineers determine setting thresholds for protective relays to allow the relay to differentiate between normal and abnormal operation by analyzing the power system over the full range of expected operation for each major component.

A defective protection element, or one with incorrect settings can lead to equipment damage, local outages, or a cascading event and may result in thousands of consumers losing service. Additionally, protection system malfunction can result in high economic loss. Therefore, protection engineers must design protection systems after thoroughly analysis of grid conditions and accounting for utility's requirements to optimize important aspects such as speed of operation, selectivity, and security [1].

Microgrids pose several challenges to protection due to modes of operation (i.e. islanded and grid-connected), the presence of DERs, connection of IBR, limited fault current contributions, and low inertia due to scarcity of synchronous machines. Conventional protection schemes heavily rely on high short-circuit currents, especially schemes commonly applied in distribution systems. However, microgrids experience drastic changes in fault current capacity when the microgrid interchanges between grid-connected and islanded operational modes [2]. The limited fault current levels in islanded-mode can cause the relays to be desensitized, and thus fail to detect faults. One solution is to use lower thresholds at the expense of risking that the relay misoperate when connected to bulk power system. Conventional protection schemes use symmetrical components to enhance protection performance (e.g. sensitivity and security); however, IBRs may produce balanced currents under asymmetrical faults, or produce negative sequence currents that differ from classic circuit laws due to their inherent dependency on inverter control logic. IBR fault contribution is determined by their internal control strategy, which varies between manufacturers and local grid codes. The negative sequence current injected by IBRs under unbalanced faults depends on firmware and enabled control mode and is challenge to protection due to its unpredictable behavior [3].

9.2.1 Big Picture

Power system protection relies on protective relays to execute vital protection functions. These relays have undergone significant evolution over the past four decades. Initially, protective relays were electromechanical devices that carried out protection functions by manipulating torques generated from interacting voltages and currents. Typically, a single relay handled one function, often on a single phase or within the neutral-to-ground path. Despite their effectiveness, these schemes had limitations and required frequent maintenance. Although some electromechanical relays are still in use, especially in distribution systems, most utilities have replaced them with modern ones.

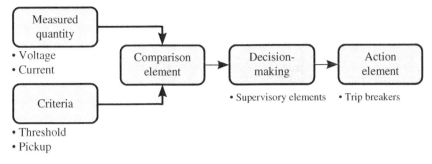

Figure 9.1 Basic functional scheme of a protective device.

A significant advancement beyond electromechanical relays was the introduction of solid-state relays. These relays utilized discrete electronic components complemented by passive circuit elements to execute protection functions. Each solid-state relay handled a specific function on a single phase.

Microprocessor relays have progressively become dominant in the protection market, especially for transmission-level protection and for safeguarding high-value equipment. They have also become ubiquitous in distribution systems. A single microprocessor relay can encompass all the functions of a panel containing a dozen or more electromechanical or solid-state relays. Additionally, these relays can perform supplementary tasks such as event logging, communication functions, and implementing automation schemes. These versatile devices are categorized as Intelligent Electronic Devices (IEDs). The essential components of an IED include analog-to-digital converters, input/output cards, a main board with a processor, firmware, and communication cards. Signal processing algorithms are employed to handle measured voltages and currents, and protection functions are programmed as computational algorithms embedded into the firmware [1]. Figure 9.1 shows the core functionality of a protective relay.

The flowchart in Fig. 9.1 displays two boxes on the left. The upper box represents electrical measurements from the system (such as current and voltage), which undergo analog-to-digital conversion and digital filtering. The lower box implements one or more thresholds, or pickup values, programmed into the function by a protection engineer. The next step involves comparing the measurement with the threshold. If the measurement exceeds the threshold (indicating a faulty condition), the subsequent box makes the decision, which may include starting a timer, logically combining this output with another function's output, or immediately sending a trip command to the designated circuit breaker [1].

Engineers design protection systems to detect and isolate faults within specific protection zones, which are areas defined by the location, typically a transmission line, feeder, transformer, bus, or generator. Measurement devices (such as current transformers and voltage transformers) and circuit breakers are positioned at the boundaries of these protection zones. Figure 9.2 illustrates part of a power system, where the dashed lines represent the boundaries of the protection zones.

The following list outlines key terms frequently used by protection engineers when designing and analyzing protection systems or configuring protective devices.

- **Coordination**: Appropriate settings ensuring selectivity while enabling multiple relays to collaborate, providing backup protection for a device or area. Proper calculation of fault-clearing times is essential for coordination.
- **Reliability**: Probability that the protection scheme will promptly trip in harmony with other schemes when required, such as during an evolving fault.

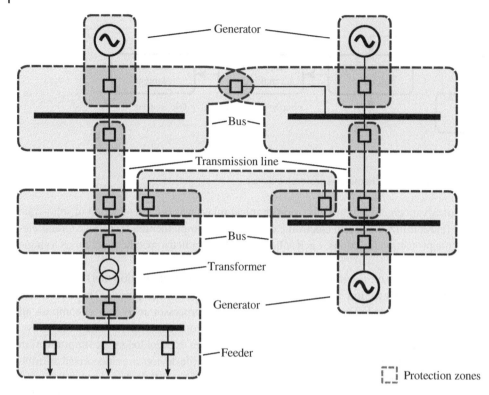

Figure 9.2 Diagram exemplifying protection zones.

- **Security**: Capability to avoid maloperations under both normal and unforeseen conditions, for example, CT saturation.
- **Selectivity**: Definition of zone boundaries and time delays enabling the scheme to disconnect the fewest consumers/generators to isolate the faulty device or area.
- **Sensitivity**: Operation exclusively for faults that the devices were configured for, with settings (e.g. pickup) optimized to detect any faults covered by the protection scheme.
- **Simplicity**: Clear logical schemes with a minimal number of relays, dependencies, delays, measuring points, and breakers.
- **Speed**: Ability to promptly detect faults and send a trip command. Speed often involves a trade-off with selectivity and security.
- **Cost**: More efficient protection systems often incur higher costs. Asset owners weigh the cost of slower or less selective protection against the cost of installing and maintaining a more capable scheme. Historically, distribution-level system operators have chosen lower-cost options.

A protection function cannot instantaneously detect and isolate internal faults. Delays are inherent in both the digital system (the relay) and the mechanical system (circuit breaker mechanism). The combined digital and mechanical response times yield the clearing time, T_c, of a protection function, defined by Eq. (9.1):

$$T_c = T_p + T_s + T_i \tag{9.1}$$

Here, T_p represents the processing time, encompassing the time required for input filtering, digitization, mathematical operations, comparison against predefined thresholds, and issuing the trip command. T_s involves time delays in device responses and allows for unblocking a function through the action of other supervisory protection functions, such as directional elements. This ensures safe operation and coordination. T_i, the isolation time, is typically the slowest, depending on the time required by the breaker to physically open its contacts. It denotes the interval from when the tripping commands are issued by the protective device until the protection zone is completely isolated [1].

9.2.2 Protection Systems and Actions

Designing a protection system requires combining a set of protection elements with supervising logic and performing protection studies to calculate proper settings [1]. Microgrids are usually meshed and their size varies from a neighborhood to an entire city; therefore, selective protection schemes rely heavily on communication systems, which represent extra cost when compared to classic distribution protection.

The primary objective of any power system is to deliver electric power from sources to loads. Historically, power distribution systems in many countries followed a radial pattern, where power flowed from the distribution substation through feeders and out to loads. This paradigm underpins distribution system protection, where relays and breakers assume unidirectional current flow. Relays and recloser controllers are time-coordinated to limit the number of loads interrupted when clearing a fault, with the breaker farthest down toward the fault opening first. Additionally, many systems employed fuses as the last resort protection element. Protective relay schemes coordinate with fuses, ensuring that the relay trips circuit breakers to interrupt current before a fuse blows. In systems with overhead lines, breakers or circuit reclosers briefly de-energize the line to clear the fault, and then reclose after a few hundred milliseconds. However, replacing a blown fuse requires a line crew to go on-site, increasing time to restore power the distribution system.

Modern distribution systems are designed to be reconfigurable, reducing the impact of permanent faults caused by damaged equipment or fallen trees. This reconfiguration can be done manually or through automated schemes after clearing a fault. The proliferation of DER has led to bidirectional current flows both in normal operation and in response to faults, posing challenges for protection schemes. The presence of additional fault current sources within the feeder complicates coordination and can lead to fuses blowing before protective relays can respond. Furthermore, the increased deployment of DER has raised concerns about the distribution system forming an unintentional island downstream of an open circuit breaker if the DER does not trip offline. IEEE standard 1547 [4] has defined protection responses to avoid this situation, as discussed in Section 9.3.3.

Microgrids function as effectively meshed systems rather than radial ones, leading to possible reverse current flows. Such reverse current flow might occur due to the high penetration of DER, dynamic grid reconfiguration, or weak power system conditions. Additionally, short-circuit capacities in parts of the microgrid can substantially change due to DER and IBR presence, especially during islanded operation. Figure 9.3 illustrates a microgrid containing photovoltaic (PV) sources, where a fault at node 4 results in the reversal of current direction in branches 4–3, 5–4, and 6–5.

Protection engineers need to analyze the system's possible configurations and power-flow conditions in line with available resources (e.g. DERs and IBRs) to define, configure, and establish an appropriate protection scheme. Configuring protection settings becomes challenging when DER outputs vary widely over time. Adaptive protection proves to be a powerful approach, enabling sensitivity adjustments at different generation levels. This adaptive scheme aims to identify the system's operational point and adjust protection settings accordingly for swift and reliable protection. Another approach involves central protection schemes with communication for data

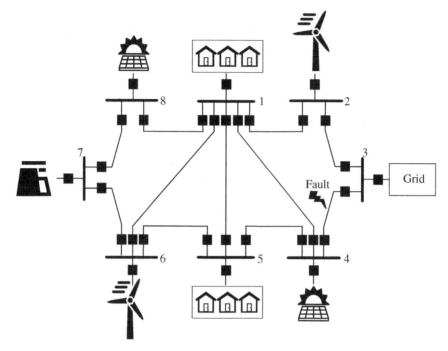

Figure 9.3 System with a fault event that can result in reverse current.

exchange between protection elements. Real-time data availability enhances fault detection while facilitating dynamic protection schemes that adapt settings according to the power system's current state. The chosen scheme sends trip commands to the relevant circuit breakers, isolating the fault while minimizing the number of interrupted customers [5].

A protection scheme comprises multiple standard protection elements with custom logic designed based on cost, technical requirements, and performance. Various protection relays from different vendors are equipped with firmware that supports standard protection functions and combines them using logical equations. Sections 9.2.2–9.2.6 offer a concise introduction to the most common protection elements employed in microgrid protection.

9.2.2.1 Overcurrent Element
The definite time and inverse time overcurrent elements (IEEE/ANSI device numbers 50/51, respectively) are widely used to protect distribution systems, especially radial systems. Inverse time overcurrent elements can easily coordinate with fuses or transformer thermal damage curves. Additionally, setting up these elements is a straightforward process by utilizing short-circuit analysis software tool.

In their simplest form, i.e. ANSI 50, definite time overcurrent elements only require the current to be measured. The protection element compares the measured current magnitude on each phase with a threshold value, also called a pickup current, to take action.

A definite time element can be set with multiple thresholds. Exceeding a lower threshold may start a timer. If the fault is still present when the timer expires a trip command will be sent to a circuit breaker. If the current exceeds a higher threshold, that implies a fault that is closer to the relay location, leading to an instantaneous trip with no intentional delay.

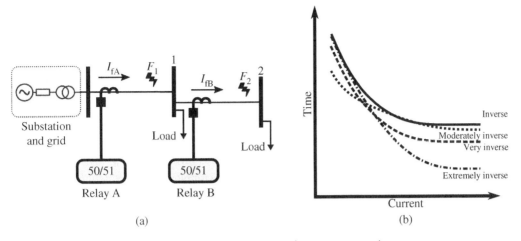

Figure 9.4 (a) Application of overcurrent relay. (b) Inverse-time-overcurrent element.

Figure 9.4(a) shows a radial distribution system protected by overcurrent relays A and B where A must trip instantaneously for faults out to node 1 and act as a backup in case relay B fails to trip or is not sensitized by the fault current caused by F_2. Similarly, relay B must trip instantaneously for faults out to node 2 in order to isolate the fault from the rest of the system. The pickups for each relay are calculated for relevant operational points of the system and after thoroughly short-circuit analysis. The scheme shown in Fig. 9.4(a) can be enhanced by adding a directional element (e.g. ANSI 32 or directional power element) able to successfully detect if the fault is forward or reverse.

The inverse time overcurrent element, ANSI device 51, has a curve following one of the standard characteristics shown in Fig. 9.4(b). A larger current result in a smaller response time, hence response time is inversely proportional to current [5]. This characteristic also lines up with the response of a fuse. The use of inverse time overcurrent functions allows easy coordination with multiple relays working down a radial distribution feeder. These relays are located across the system and can be pole or platform-mounted reclosers consisting of a CT, a relay, and a circuit breaker. In order to ensure coordination, each relay has the same curve type (e.g. moderately inverse), but the upstream relays' curves are spaced from the downstream ones. The spacing is based on coordination time and is set as a digital "time dial" setting in the relay [1]. As a result, the response time for the relay at the head of a radial feeder will have the slowest response. To improve protection response for faults in the first segment of the feeder, the inverse time overcurrent element is often combined with an instantaneous set definite time element. The pickup current and inverse-time characteristic(s) are calculated according to I_{fA} and I_{fB} for different fault conditions.

A drawback of overcurrent elements is that their response is highly dependent on the system configuration and current direction. If the system configuration changes, the effective impedance changes, reducing available fault current. This response is especially concerning for microgrids transitioning from grid-connected to islanded-mode.

There are several solutions available to allow overcurrent schemes to perform more effectively in distribution systems and microgrids where fault current levels change with distribution system reconfiguration. Common approaches are adaptive overcurrent protection schemes and voltage supervised overcurrent schemes.

In an adaptive protection scheme, the relay is programmed with multiple sets of overcurrent settings that can only be active in specific scenarios. The relay transitions between settings groups based on an external input communicated to the relay. One option for a case where opening a single circuit breaker transitions a microgrid to islanded operation is to track the status of the circuit breaker auxiliary open/close status contacts (ANSI 52a and 52b), and broadcast those over a communication network to all relays in the system. Another is for a central distribution management system or microgrid controller to track reconfigurations and send commands to the relays to switch to appropriate settings. A reliable communication system is essential.

One option to implement multiple settings configurations takes advantage of microprocessor relays with multiple settings groups. The transition between settings groups does leave the relay inactive for time periods of hundreds of milliseconds to seconds depending on the relay. Another option is to logically AND settings equation outputs with control bits in trip logic. The second option has the advantage of fast transitions but can lead to complex settings equations.

Another option for improved performance of overcurrent elements in microgrids with low fault currents is application of voltage-controlled overcurrent elements (ANSI 50V/51V). These elements have long been implemented in generator stations to provide backup coordination with overcurrent elements on lines leaving the station. The generator fault currents can approach steady state in the period of interest for coordination and may fall below the generator-rated current. However, when a fault is present, the voltage on the faulted phases will experience a voltage sag. The 50V/51V elements block the overcurrent response until the voltage falls below a set threshold. Some microgrid schemes apply voltage controlled-overcurrents elements to enable adaptive overcurrent protection without relying on communication.

Meshed and/or looped microgrids require combining overcurrent elements with directional elements to enable a selective protection scheme. On the other hand, overcurrent relays without directional supervision can be employed to protect line segments that are radial in relation to sources feeding the fault. Overcurrent elements cannot provide reliable protection when DERs present a large variation in fault current capacity. Another aspect negatively affected by insufficient or unpredictable fault response from DERs is coordination, which can result in tripping zones external to the fault within a microgrid [3].

9.2.2.2 Distance Element

Distance relays (i.e. ANSI 21) are widely used in transmission line protection, with one relay installed at each end of the line, looking into it. The addition of a communication path enables high-speed protection for the entire line, as illustrated in Fig. 9.5(a).

Voltage and current (V_R and I_R) are measured at the relay's location; an effective impedance ($Z_R = \frac{V_R}{I_R}$) is calculated with the ratio between measurements. The complex value ($Z_R = R_R + jX_R$) is plotted at a complex X–R plane. Instead of a threshold, the distance relay uses restraining characteristics, which are also referred to as zones. Figure 9.5(b) shows the retraining "Mho" characteristic for zones 1, 2, and 3. Each zone has a set distance that is commonly 80%, 120%, 150% of the line impedance, respectively. If the effective impedance falls inside one of the circles, the relay takes an action. Crossing zone 1 initiates an immediate trip. Zones 2 and 3 start timers, to allow downstream devices to trip first in case the relay is overreaching.

When zone 1 is set at 80% of the line impedance, faults occurring in the final 20% of the line are tripped by zone 2 for the relay located farthest from the fault, and by zone 1 for the one closest to the fault. The communication channel depicted in Fig. 9.5(a) is employed to protect the entire line. If the distance element at Relay A detects a fault in its zone 2, the relay sends an indication to Relay B via the communication channel. If Relay B detects a fault in its zone 2, it initiates an immediate trip. This method, called permissive overreaching transfer trip, enables high-speed protection

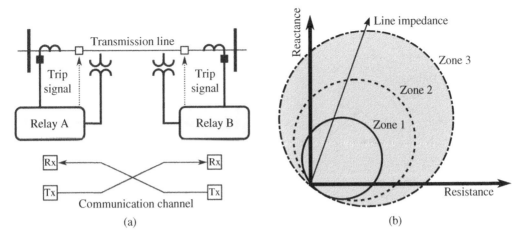

Figure 9.5 (a) Application of distance protection. (b) $R-X$ complex plane.

for the entire line. Other communication-aided distance approaches are also utilized in practical applications.

A major drawback of the distance element is its dependence on high values of line impedance which are common on transmission lines, but unusual in microgrids at the distribution level. On the other hand, this element can be employed in portions of the microgrid that have higher impedance such as transmission and subtransmission level microgrids [2].

9.2.2.3 Current Differential

Kirchhoff's current law (KCL) is the base concept for differential protection (e.g. ANSI 87) and can be applied anywhere from a single node, a device, or even an area provided that all currents are measured and communication latency is within reasonable limits. The algebraic sum of all currents is ideally zero in normal operation. The differential protection is commonly utilized to protect transformers and buses because the measurements for these devices are geographically located together; thus reliable communication is easily achievable. In case of transmission lines, the use of differential protection relies on exchanging measurements across long distances; therefore, communication outages have to be accounted by the protection engineers when designing the scheme [1].

Figure 9.6 shows a differential scheme with n terminals, each monitor current values in real time. The operating and restraining currents are calculated as Eqs. (9.2) and (9.3), respectively.

$$I_{op} = \sum_{i=1}^{n} I_i \tag{9.2}$$

$$I_{res} = \left| \sum_{i=1}^{n} I_i \right| \tag{9.3}$$

Under ideal conditions, I_{op} is equal to zero. When an internal fault is present, the operating current is equal to I_f. The differential scheme detects a fault if conditions (9.4) and (9.5) are fulfilled.

$$I_{op} > I_{pk} \tag{9.4}$$

$$I_{op} \geq K \cdot I_{res} \tag{9.5}$$

where I_{pk} is the minimum pickup current and K is the slope.

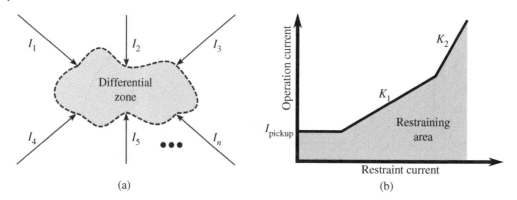

Figure 9.6 (a) Example of differential protection. (b) Operational plane.

The minimum pickup current compensates for CT ratio mismatch and also provides a minimum threshold value which avoids misoperation. The slope defines a restraining area in the chosen representation plane; a dual-slope approach is shown in Fig. 9.6(b). One can observe that the differential element becomes more sensitive (e.g. larger operation area) the smaller K is set. In Fig. 9.6(b) the set values for K_1 and K_2 are 0.4 and 0.8, respectively.

Differential elements are easy to apply within a substation to protect a transformer or a bus. Protecting a transmission line or distribution line requires a communication channel to exchange phasor data between each end. Measurements also need to be time-aligned to compensate for communication delay and lack of clock synchronization. Modern relays can also compensate for capacitive charging current in a transmission or distribution line.

Microgrid protection benefits from the sensitivity, customization, and zone creation enabled by the differential element. However, requirements such as reliable communication network, CT precision, and availability of all required measurements represent drawbacks that restrict the use of these elements in microgrids. Microgrids transitioning from radial distribution grids face high costs in implementing current differential protection, requiring significant modifications like CT and circuit breaker installations. Relying solely on current differential relays for every segment of a large microgrid is economically impractical. Additionally, unmeasured currents from shunt devices (e.g. DER) introduce current imbalances, compromising sensitivity, and increasing complexity when designing a differential scheme [5]. Differential protection finds more prevalence in microgrids within industrial settings where downtime carries substantial financial implications.

9.2.2.4 Directional Comparison

Directional comparison protection relies on the phase relationship between voltage and current measured at relay locations. For this method to safeguard a grid segment, a communication channel between direction elements is essential. Furthermore, the integration of positive, negative, and zero sequence detectors enhances selectivity. If a relay at one end identifies a forward fault, it communicates this indication to the other end of the line. If the receiving relay also detects a forward fault, it triggers a trip action.

Figure 9.7 illustrates a directional comparison scheme where an unblocking signal is exchanged between terminals to trip in the event of an internal fault.

One challenge in implementing this scheme in microgrids is the presence of IBRs across the grid. These devices lack a predictable phase relationship between voltage and current and do not adhere to the physics-based behavior of synchronous generators. For instance, an IBR injecting capacitive current to boost local voltage during a fault might appear as a reverse fault to a directional element

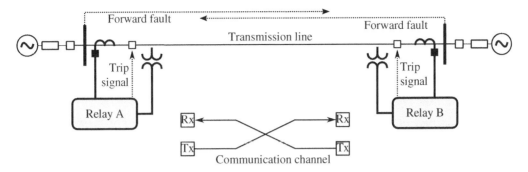

Figure 9.7 Example of directional comparison scheme.

set assuming fault current would lag the voltage based on the line impedance angle. Additionally, different IBR manufacturers produce devices that respond differently under fault conditions due to variations in their control loops and limiters [2].

9.2.3 Phasor-Based Protection

A sinusoidal signal can be represented as a phasor. The calculation process is called digital filtering or phasor estimation; a variety of algorithms are at the disposal of protection engineers. For example, full-cycle Fourier transform, modified-cosine filter, and half-cycle Fourier transform [6].

Non-recursive phasor estimation algorithms utilize a moving data window. The fundamental frequency (e.g. f_0) is the one utilized to size the window. However, harmonics of f_0 are also commonly employed for enhanced protection, CT saturation detection, and detecting power transformer magnetizing inrush current. For example full-cycle algorithms utilize an entire period of the fundamental frequency ($T_w = \frac{1}{f_0}$), while half-cycle ones employ a window half the size $T_w = \frac{1}{2f_0}$. The moving data window is a collection of samples which are collected at sampling frequency $N f_0$, where N is the number of samples per window. For a 60 Hz power system, N is normally set to 16; however, manufacturers tend to utilize proprietary phasor estimation algorithms; thus N may vary according to chosen relay.

Figure 9.8(a) shows a sinusoidal signal and a one-cycle moving window with the sampling frequency (i.e. each circle is a sample) set to $16f_0$; thus N is set to 16. The phasor-estimation

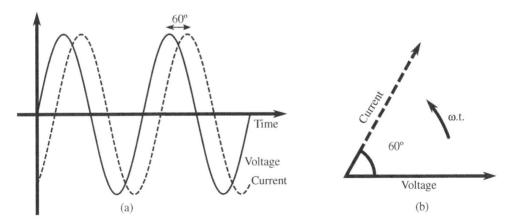

Figure 9.8 (a) Sinusoidal signal. (b) Estimated phasor.

algorithm calculates an amplitude A_p and an angle ϕ_p that is represented in polar form in Fig. 9.8(b). Under steady-state conditions, the phasor maintains a constant value for A_p and ϕ_p; however, the estimated magnitude and angle change in case of a fault (or other changes in the system). Phase, ground, and neutral voltage/current measurements are calculated, symmetrical components – i.e. positive, negative, and zero – are then obtained using the Fortescue linear symmetrical components transformation [6].

9.2.4 Full-Cycle Fourier Transformation

The fundamental frequency f_o phasor of a signal $x(t)$ is calculated by Eqs. (9.6) and (9.7).

$$Y_{\cos} = \int_t^{t+T_0} x(t) \cos \omega dt \tag{9.6}$$

$$Y_{\sin} = \int_t^{t+T_0} x(t) \sin \omega dt \tag{9.7}$$

The integrals are solved as Eqs. (9.8) and (9.9), with N samples and a moving window $X = \{x_0, x_1, \ldots, x_{N-1}\}$ for one period of fundamental frequency.

$$Y_{\cos} \approx \frac{2}{N\Delta t}(x_0 \cos \omega t_0 + x_1 \cos \omega t_1 + \ldots x_{N-1} \cos \omega t_0)\Delta t \tag{9.8}$$

$$Y_{\sin} \approx \frac{2}{N\Delta t}(x_0 \sin \omega t_0 + x_1 \sin \omega t_1 + \ldots x_{N-1} \sin \omega t_0)\Delta t \tag{9.9}$$

where Δt is the sampling period or the time difference between two consecutive samples.

The terms $Y_{\cos}$ and $Y_{\sin}$ are reorganized and lead to the Fourier Digital Filter Equations below:

$$Y_{\cos} = \frac{2}{N} \sum_{k=0}^{N-1} x_k \cos \left(\frac{2\pi k}{N} \right) \tag{9.10}$$

$$Y_{\sin} = \frac{2}{N} \sum_{k=0}^{N-1} x_k \sin \left(\frac{2\pi k}{N} \right) \tag{9.11}$$

Finally, the amplitude and angle are calculated as shown in Eqs. (9.12) and (9.13).

$$A_p = \sqrt{Y_{\cos}^2 + Y_{\sin}^2} \tag{9.12}$$

$$\phi_p = \tan^{-1} \left(\frac{Y_{\sin}}{Y_{\cos}} \right) \tag{9.13}$$

Phasor estimation does not cause a heavy processing burden on relays; as a result, they are widely employed in the algorithms developed by different manufacturers. Some conditions that compromise precision of phasor estimation algorithms are: presence of non-integer harmonics, CT saturation, and exponentially decaying DC components in fault currents. The size of the moving window affects speed and precision as well, shorter windows to get faster estimation also lead to less precise results. Therefore, the protection engineers have to balance this relation according to the application's requirements [6].

9.2.5 Superimposed Quantities

Superimposed quantities have been employed in transmission for many years. They have the advantage of not being affected by load conditions before or after a fault [7]. The principle of superposition enables the definition of pre-fault, pure-fault, and faulted networks, shown in Figs. 9.9(a)–(c), respectively.

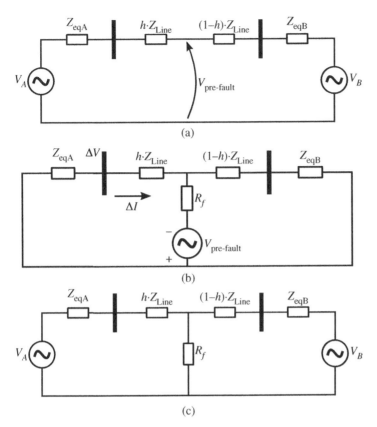

Figure 9.9 Superposition theorem. (a) Pre-fault equivalent network. (b) Pure fault equivalent network. (c) Faulted equivalent network.

Considering the fault f shown in Fig. 9.9(c), the superposition theorem states that faulted quantities (V_f and I_f) are expressed as the sum of pre-fault (V_{pre} and I_{pre}) and pure-fault circuits (ΔV and ΔI), as shown in Eqs. (9.14) and (9.15), respectively.

$$V_f = V_{pre} + \Delta V \tag{9.14}$$

$$I_f = I_{pre} + \Delta I \tag{9.15}$$

The pure-fault components (ΔV and ΔI) are superimposed quantities; thus a direct comparison between pre-fault and fault conditions. In the same way that phasors are utilized to calculate ΔV and ΔI, incremental sequence quantities and incremental impedances can also be utilized according to application and desired performance of protection. One main advantage is their speed when utilized as a time-domain directional elements not requiring phasor calculations and their inherent drawbacks. The utilization of this protection element in microgrids improved protection performance by positively affecting selectivity, sensitivity, and security; however, this scheme is affected by IBR fault response that does not sensitize directional elements properly [7].

9.2.6 Traveling Wave-Based Protection

Traveling wave-based protection operates at MHz speeds, offering rapid tripping times compared to kHz-based methods. A recently commercialized wave-based relay, SEL-T400L [8], is used for safeguarding transmission lines. This device utilizes wave reflections caused by faults to pinpoint their locations accurately, enabling quicker trip commands than methods like phasor-based protection. However, implementing this technology in densely meshed microgrids poses challenges due to unpredictable reflections and resonances, potentially leading to relay misoperations.

One implementation of this technology, known as directional waves, resembles directional elements. In this approach, the relay swiftly measures voltage and current, detecting fault-induced waves nearly at the speed of light in the conductor [9]. Traveling wave protection can be implemented at one or both ends of a transmission line. Single-ended methods require additional reflections to detect faults, making them slower than double-ended approaches. Double-ended applications precisely detect reflected wave arrival times at both ends, ensuring precise protection response if implemented with proper time synchronization. However, this method demands a fast and reliable communication channel.

Compared to traditional distance protection, traveling wave-based protection is immune to issues like over/underreaching or in/outfeed currents. However, parallel circuits, mutual impedance, and the response of current and voltage transformers can impact schemes relying on traveling waves [1]. Additionally, protection engineers must meticulously study slow-evolving faults and surge-arrester operations to maintain sensitivity and prevent misoperations. Further research is needed to apply the traveling wave method on a large scale in microgrids [9].

9.2.7 Centralized Protection

Another option for microgrid protection is implementation of a centralized protection scheme. In such a scheme, measurements from relays are communicated to a central controller using standard protocols over high-speed networks. The central controller performs protection calculations and generates trip comments. Such schemes are typically applied in tandem with protection elements that do not require communication to ensure reliability.

9.3 Typical Microgrid Protection Schemes

Microgrids may consist of lines at subtransmission voltage levels and distribution level feeders and loads. Much of the research literature concentrates on distribution-level applications. Sections 9.3.1 and 9.3.3 discuss protection schemes applied at these distinct levels. A protection system will incorporate different protection elements enabled in protective relays installed at locations within the subtransmission or distribution system. The relays will have primary and backup protection schemes to improve reliability and security. Protection engineers test the settings to verify proper coordination.

9.3.1 Subtransmission

Subtransmission systems are typically fed from both ends, which is drastically different for protection system design when compared to a radial system. In fact, current reversal is expected to occur and fault direction turns into a major concerning aspect that must be addressed by protection engineers. In many applications, subtransmission lines of a microgrid are protected by employing communication-aided distance elements or communication-aided directional overcurrent elements. Some applications apply current differential schemes, which require fast and reliable communication channels. Typical relay response times are between less than one cycle and two cycles.

9.3.2 Distribution

9.3.2.1 Radial

Distribution systems are typically built out as radial systems, resulting in protection schemes heavily dependent on definite time and inverse time overcurrent elements. Additionally, selectivity is often limited in distribution grids due to circuit breakers being located only at substations and limited number of pole or platform-mounted reclosers across the feeder. A typical protection scheme utilizes a combination of overcurrent elements supplemented by negative-sequence, and zero-sequence ground elements. Usually, negative sequence protection elements are only used for complex applications. The reasons for utilizing zero and negative sequence quantities is their immunity against changes in load level and ability to detect asymmetrical faults.

Many distribution system configurations include interconnections with neighboring feeders, allowing reconfiguration to improve reliability of supply when prolonged outages occur. These interconnections are made with normally open switches that can be closed to reconfigure the system. Some schemes require line crews to take the switching action, but an increasing number of systems can either be reconfigured using remote commands from a human operator or an automation scheme. Reconfiguring the system requires updating coordination of overcurrent elements. A common approach is to change relay settings groups when the system is reconfigured.

In radial systems where the fault current level does not vary sufficiently with fault location to coordinate inverse time overcurrent elements, definite time overcurrent elements can successfully perform protection. In these applications, the relay farthest from the substation has the fastest response time, and each upstream relay is set with an additional delay based on the predetermined response time of the downstream element.

9.3.2.2 Looped/Meshed

Some utilities apply looped or meshed distribution systems. These configurations are more commonly used in dense urban areas and in areas with critical loads. In many cases, circuit breakers are installed at each end of the lines. Some of these systems can transition to microgrid operation, a trend likely to increase. Protection schemes for looped/meshed distribution commonly rely on directional supervision of overcurrent elements. Directional comparison, line distance, and current differential schemes are also utilized in these if communication channels are available and enhanced protection performance is required. In addition, permissive overreaching transfer trip schemes based on definite time overcurrent elements can also be used to speed protection response. The scheme uses definite time elements with reach set beyond the far end of the line that would normally trip on a time delay. This approach is similar to the scheme for distance elements described in Section 9.2.2.2.

9.3.2.3 Typical Response Times

Response times of distribution protection schemes are often relatively slow, especially when they require coordinating inverse time overcurrent elements; as a result, tripping times can reach hundreds of milliseconds. Protection response times can be improved by adding negative and zero sequence protection elements. Another solution is to add instantaneous overcurrent elements set only to trip for close-in faults. In addition to the relay response time, circuit breakers and reclosers often present response times of four to five fundamental frequency cycles.

9.3.3 IEEE 1547 Guidelines for DERs

The installation of distributed generation on utility distribution systems gained significant interest in the late 1990s. However, concerns arose regarding the behavior of these resources during faults. Initial concerns centered around DER failing to disconnect when the substation breaker or appropriate recloser cleared. There were also concerns that fault currents from DERs could lead to misoperation of non-directional overcurrent protection on radial feeders. Fault current contributions from DERs necessitated updates to protection coordination with fuses. The initial DERs were predominantly synchronous generators, but standards related to their operation were limited. The initial version of IEEE 1547 addressed these concerns by specifying voltage or frequency conditions under which the DER should disconnect and remain offline until full restoration. Subsequent revisions of IEEE 1547 aimed to provide more flexibility for DERs to support system response [4]. Additionally, a guide focused on microgrid behavior was developed under IEEE 1547.

IEEE 1547 recommends the use of over/under frequency elements (ANSI/IEEE 81O/81U) where DERs are required to disconnect if the frequency deviation exceeds specified limits for set periods of time. Similar limits exist for over/undervoltage (ANSI/IEEE 59/27). Furthermore, provisions mandate the communication of a transfer trip command to larger DERs when an upstream breaker is tripped. It is crucial to design protection schemes in compliance with the most current active revision of IEEE 1547.

9.4 Challenges Posed by Microgrids

Designing a protection scheme for microgrids is challenging due to DER penetration, stochastic behavior of renewable resources, presence of IBR, and changes in operational modes of microgrids (e.g. grid-connected and islanded). The growing penetration of DERs represents a change in paradigm for traditional distribution systems designed for radial operation, and thus unidirectional flow of current. Consequently, microgrids derived from a radial distribution system will have their protection scheme compromised by DER penetration. The existence of DERs alongside loads results in bidirectional fault current flow, affecting fault current magnitude, and direction. Overcurrent elements assume fixed direction for the monitored current, and require directional supervision in cases where fault current can flow in either direction to properly protect the microgrid. The relay needs measured voltage for directional supervision, which may require adding voltage transformers at the relay location. Directional elements are set assuming the angle of the phase current relative to phase voltage depends on the angle of circuit impedances, often with current lagging voltage by close to 90°. Similar assumptions are made for negative and zero sequence elements.

DERs connected through IBR (e.g. renewable resources) provide low fault current contributions, with behavior dictated by converter controls rather than physical response of a rotating machine. IBR fault magnitude is often limited based on physical constraints of power electronic switches. The fault response is heavily dependent on control strategies and limiters implemented by the manufacturer of that IBR. As a result, most IBRs provide the same fault current regardless of electrical distance to the fault. IBRs behave as controlled current sources rather than voltage sources behind impedances. The phase angle of the fault current is a control variable, and IBR may provide fault currents in phase with voltage, or even leading the voltage, which can cause directional element misoperation. Moreover, unpredictable negative sequence behavior during a fault also represents a major challenge for protection schemes supervised by negative sequence directional elements. Some IBR vendors regulate negative sequence currents to zero, others regulate negative sequence current magnitude but not angle, and others regulate both. Another aspect of microgrids that critically impacts fault studies and characterization of fault currents are their looped/meshed topology, current capacity of DERs that varies according to renewable availability, and online change of operational mode, i.e. from grid-connected to islanded-mode.

The protection challenges posed by microgrid operation covered in this section are summarized as:

1. **Operational mode**: Short-circuit current characteristics change drastically with changes in microgrid operational mode. A microgrid operating in islanded-mode presents reduced short-circuit current; thus, causing common distribution protection to not detect internal faults if settings are not changed following the mode transition [5].
2. **DER operation**: Bidirectional fault current flow and intermittent nature of renewable resources may negatively affect the performance of protection (e.g. coordination, sensitivity, and selectivity) [5, 10].
3. **Presence of IBRs**: Conventional protection assumes that most generators are synchronous machines, where fault response behaves like a voltage behind an impedance. Therefore, fault contribution depends directly on the energy stored in a rotating mass, and the angle of the fault currents and the negative sequence current response follow electrical circuit concepts. IBRs act like a controlled current source with fault current angle and negative sequence response depending on IBR settings or vendor decisions. As a result, IBRs pose a severe challenge to conventional protection. Moreover, many IBR vendors do not disclose the details of the fault response behavior, especially for smaller IBRs [2, 5, 11, 12].

9.4.1 Challenges Posed by Changes in Operational Mode

Microgrids are operated in three different modes: grid-connected, islanded-mode, or mixed. Each mode is briefly described below:

1. **Grid-connected mode**: The microgrid is connected to a power system that has high fault current capacity (i.e. strong system). The energy flows mainly from the grid to local loads that are part of the microgrid. Export of power to the power system is allowed in case the microgrid is producing more energy than it consumes.
2. **Islanded-mode**: The microgrid is disconnected from the power system and feeds loads within the constraints of available power sources. Some loads remain online after a load-shedding strategy acts according to power availability and previously defined priority. This mode is often active if the utility is experiencing an outage and there is a desire to feed loads that are part of the microgrid without interruptions.
3. **Hybrid-mode**: Under this operational mode, the microgrid supports functions for both grid-connected and islanded-modes. A microgrid controller may manage generation and load within the microgrid to enhance resilience or to lower cost.

9.4.1.1 Short-Circuit Capacity

Faults are unpredictable events that may occur in any segment of the power system. Synchronous generators can supply large fault currents without need of self-protection during the time until the protection system isolates the faulted portion of the system. The amplitude of these currents can reach 5–10 times the normal operation current. In this way, microgrids operating under grid-connected mode can rely on the utility grid fault current capacity to feed the majority or even the entirety of fault current. Smaller generators provide less fault current than larger generators, and IBRs provide less fault current than small synchronous generators. If the microgrid is in islanded-mode, generators within the microgrid are the source of fault current. The generators are also responsible for maintaining reasonable ranges for voltage magnitude and frequency. This represents a challenge because these smaller generators have limited fault current capacity and may experience fast frequency changes (i.e. substantial changes in $\frac{Hz}{s}$) which directly affects traditional overcurrent sensing devices, some forms of phasor-based protection (e.g. distance), and frequency-based protection (e.g. rate of change of frequency [ROCOF]).

9.4.1.2 Current-Flow Direction

The operational mode of the microgrid affects the direction of fault current in segments of the system. Therefore, microgrid protection schemes that rely solely on overcurrent elements (e.g. 50/51 and fuses) provide inadequate performance and cannot properly protect microgrids. These schemes require high magnitude and unidirectional fault current which is not guaranteed in islanded-mode or if the microgrid contributes to fault current, i.e. generators remain connected under fault conditions. At a minimum, they need directional supervision, but coordination may be challenged by low fault current magnitudes.

9.4.2 Challenges Posed by DER Operation

When designing protection schemes for a microgrid, protection engineers should account for the DER type (e.g. primary source) and behavior during a fault. These devices may be configured to trip and stay offline following a fault. However, with increased penetration of DER, a microgrid cannot afford to have all DERs stay disconnected during a fault. Therefore, DER may have to ride through the fault or even contribute to fault current which drastically influences fault current magnitude, shape, and angle. Additionally, DER generation varies during the day according to availability of their renewable resource, thus affecting pre-fault conditions and direction of currents across the system.

9.4.2.1 Voltage Regulation and Stability

Voltage stability in microgrids, particularly in the context of integrating DER, is of critical importance. Technologies such as solar photovoltaic systems, wind turbines, and energy storage systems can introduce fluctuations in power supply due to their intermittent nature. Ensuring a stable voltage profile during these fluctuations is crucial for the stable operation of microgrids. Challenges arise from the intermittent output of renewable sources, complexities in inverter control, and the dynamic nature of loads, demanding innovative solutions to maintain voltage stability.

One emerging trend is the implementation of advanced inverter control systems. Smart inverters equipped with grid-supportive features play a significant role in regulating voltage and supporting grid stability, especially during transient events. Additionally, integrating energy storage systems into microgrids provides a buffer against voltage fluctuations because they can act as a virtual inertia with suitable control design. Batteries and other storage technologies store surplus energy during periods of high generation and release it during high demand, thereby stabilizing voltage levels. Predictive analysis and sophisticated algorithms are also utilized to forecast generation patterns and demand fluctuations. These predictive insights enable protection engineer to properly adjust relay operation, ensuring that the voltage remains stable by anticipating and mitigating potential imbalances between generation and consumption.

Furthermore, the design and planning of microgrid architectures are crucial in maintaining voltage stability. Properly sized conductors, transformers, capacitor banks, and protection mechanisms are essential components of a resilient microgrid system. Additionally, demand-side management strategies, such as demand response programs, are implemented to balance load and generation. By coordinating DER output and consumer demand in real time, microgrid operators can effectively stabilize voltage levels. Overall, an approach combining advanced technology, predictive analysis, and informed grid design is necessary to address the challenges and uphold voltage stability in microgrids powered by DERs [10].

9.4.2.2 Frequency Decay and Angular Stability

In microgrids, accounting for fast frequency decay following a disturbance and ensuring angular stability are vital aspects of protection design. The rapid fluctuations in generation and load characteristics, especially due to the integration of DERs, necessitate advanced protection such as adaptive protective relaying strategies.

Conventional power systems with synchronous machines present frequency decay rates considerably smaller than those of microgrids. In fact, a microgrid equipped with Diesel generators has a frequency decay of around $10\frac{\text{Hz}}{\text{s}}$ while the same grid equipped with high penetration of IBR interfaced renewable resources would have a frequency decay rate almost 10 times larger [5]. This behavior of microgrids is critical to angular stability of synchronous machines in the microgrid and to protection schemes that rely on frequency tracking. Such a protection scheme may not operate correctly for frequency decays faster than $20\frac{\text{Hz}}{\text{s}}$. Under-frequency relays are fundamental components of microgrid protection to counter frequency decay. These relays promptly detect a reduction in frequency, triggering actions such as load shedding by disconnecting non-essential loads. Load shedding rebalances generation and demand, thus stabilizing the system. Ensuring angular stability requires protection to check for synchronism by continuously monitoring phase angle differences across the grid. Upon detecting significant deviations indicative of potential loss of synchronism, relays initiate corrective actions, including sectionalizing the microgrid to prevent cascading failures.

Designing protective relaying systems in microgrids presents challenges, especially for those heavily reliant on IBR. These microgrids demand fast protection schemes capable of responding to frequency and angular stability issues. The integration of diverse advanced devices, such as renewable sources, energy storage systems, and grid-forming inverters, necessitates meticulous coordination among protection schemes due to varied response times and device characteristics.

Additionally, the rapid fluctuations of frequency and phase angles in microgrids significantly impact phasor estimation. Incorrect frequency values lead to phase angle errors and incorrect identification of oscillations, affecting phasor-based protection and even control actions. Bandpass filter desynchronization due to incorrect frequencies causes out-of-band errors, distorting phasor measurements. Accurate frequency estimation methods are pivotal, ensuring precise phasor estimation and the stability of microgrids. Addressing these challenges demands ongoing research, focusing on creating adaptive protection schemes utilizing advanced algorithms and field data. These innovations enhance the agility, reliability, and resilience of microgrid protection systems, enabling them to effectively handle frequency decay and angular stability challenges posed by the integration of DERs.

9.4.3 IBR Challenges

As noted above, IBRs present challenging short-circuit response when compared to conventional synchronous machines. Moreover, IBRs present unique transient and negative-sequence response [11] caused by limiters and control loops implemented by vendors. IBRs from different manufacturers will respond differently, thus further increasing complexity of protection design for microgrids. Newer IBRs are designed either to operate in grid-following (GFL) mode or grid-forming (GFM) mode, which are described below [12]. Some IBRs can transition from GFL to GFM when the controls detect a weak grid or islanded state.

1. **GFL**: IBR control relies on fast tracking of the grid phase angle to prioritize current control loops enabling them to operate as controlled sources of active and reactive power. GFL are often controlled to extract peak power from the wind or available sunlight. GFL are often not dispatchable.
2. **GFM**: IBR control allows the real and reactive power output to change to facilitate regulating voltage and frequency through different control schemes, such as droop, virtual synchronous machine (VSM), or dispatchable virtual oscillator control (dVOC). A system with multiple GFM IBRs requires implementation of frequency and voltage droop control loops across the microgrid in order to properly coordinate. GFM are dispatchable. Energy storage systems are good options for GFM.

9.4.3.1 Impacts of IBR Fault Current

As noted above, the IBR fault current signature is heavily dependent on limiters and control loops implemented into these devices. Some critical issues are limited fault current contribution, inconsistent angle between voltage and current for phase currents or negative sequence components, and fault current contribution that does not vary with fault location.

IBRs are equipped with semiconductor switches that have limited thermal capacity. Controls ensure that the devices stay within those limits to prevent irreparable damage to the power electronic switches. The IBR is designed to automatically limit the magnitude of currents through switches and voltage across the switches by physical and software-based limiters. For example, the DC link voltage is usually limited by the crowbar protection while the current injected by switches is limited by control loops designed by the IBR manufacturer. The fault current of IBRs is normally limited to around 1.1–1.5 times the inverter current rating. Some GFM inverters have this range extended to 2–3 p.u.; however, even with the increased capability, conventional protection struggles to differentiate normal events from actual faults [2].

A fault may cause voltage to sink in parts of the system, as a result, GFM and GFL inverter controls may supply reactive power to boost local voltage while meeting device limits. GFM inverters increase their reactive power injection in order to meet previously defined voltage set-points, thus, changing phase angle difference between current and voltage. GFL inverters, on the other hand, may supply reactive power to enhance the ability to ride-through faults (i.e. meet fault ride-through requirements) and provide limited voltage regulation according to the IBR reactive power limits [2].

Negative sequence current injection is normally suppressed by the control loops of GFL IBRs. As a result, the magnitude of negative sequence currents is typically under 10% of the positive sequence, thus, causing issues to sensitize protection schemes supervised by negative sequence elements (e.g. directional elements). Some GFLs are programmed to provide set percentages of negative sequence current to meet protective relay requirements. In some cases, the angle of the negative sequence current relative to the negative sequence voltage is uncontrolled, in others the angle is controlled, but takes at least one cycle at fundamental frequency to stabilize. GFM inverters are often capable of injecting higher magnitudes of negative sequence current. As is the case with GFL inverters, the control of the angle of the negative current relative to the negative sequence voltage varies by IBR vendor or even firmware revision. Moreover, proprietary control strategies dictate negative sequence behavior which is extremely challenging to predict or properly model for protection studies.

Conventional protection (e.g. phasor-based) relies on non-distorted sinusoidal voltage and current; however, IBRs' limiters result in nonlinear fault currents due to saturation in the controllers. Commonly applied phasor estimation schemes remove integer harmonics and switching frequency noise. But they are impacted by interharmonics and low frequency components below fundamental frequency. Harmonic currents pose more severe problems for electromechanical relays and some older microprocessor relays and recloser controllers. Protection that utilizes phasor estimation algorithms may malfunction because they are not designed to handle severe waveform changes posed by IBR response.

9.4.3.2 Impacts Posed by IBR to Protection Schemes

IBRs impact not only coordination of overall protection, but also affect the performance of conventional elements such as directional supervision. The main challenges posed by IBRs to conventional protection are listed below [2]:

- **Protection system coordination**: The existence of high penetration of IBRs across different parts of the microgrid impacts the coordination of relays, fuses, and auto-reclosers. Additionally, the IBR fault current depends on their limiters and do not vary significantly relative to the electrical distance from the fault location. This behavior forces protection engineers to utilize time-delays to ensure coordination which compromises overall performance of the protection performance and may lead to cascade failure.
- **Overcurrent protection**: Setting pickup and inverse-time characteristics suitable for grid-connected and islanded-mode is a challenge. The presence of IBRs escalates this issue due to the differences in their control loops when operating under a microgrid in islanded-mode. Consequently, selectivity is compromised and misoperation may occur under normal circumstances where the protection scheme fails to differentiate normal events from a fault, e.g. inrush current of a machine.
- **Directional protection**: The unpredictability of angle of the phase current and the negative sequence currents in the fault response critically impact directional elements. The issue arises because conventional protection schemes commonly utilize directional elements for supervision and/or detect asymmetrical faults, with many relays using negative sequence directional supervision. Therefore, protection based on sequence components may misoperate or not be sensitized by the fault current. Consequently, careful protection studies must be carried out before designing the protection system.

9.5 Examples of Solutions in Practice

9.5.1 Case 1: North Bay Hydro Microgrid

The first Canadian utility-scale microgrid was deployed in 2019 by North Bay Hydro Distribution (NBHD) in Ontario [13]. The system is rated at 600 V. The main goal of this microgrid is to provide continuous power to clients even under an outage of the utility grid. Additionally, two interconnections with the utility grid are available for redundancy purposes. The microgrid consists of three facilities that supply electricity and heat, with the following DERs: two combined heat and power (CHP) natural gas generators, rated at 265 kW each, one battery energy storage system (BESS), rated at 250 kW and with battery capacity of 274 kWh, and a total of 8 kW of PV units. The NBHD microgrid is designed for seamlessly transition between grid-connected and islanded-mode under planned and unplanned conditions.

9.5.1.1 Challenges

- **Bidirectional and variable-fault current**: This condition is caused by high penetration of DERs and the requirement for seamless transition between operational modes (i.e. grid-connected from/to islanded).

- **Power-export restrictions**: The utility company limits the total amount of power exported to the grid, thus increasing operation challenges.
- **Requirement for fast protection operation**: Slow microgrid protection and the resulting delay in restoration was not acceptable for the application.
- **Seamless transition between operational modes**: Protection and control must be capable of detecting changes in the microgrid grid, e.g. change in operational mode.
- **Protection against closing out of phase**: The controls at the interconnections to the grid have to monitor both sides of the breakers in order to avoid out-of-phase connection when restoring grid ties.

9.5.1.2 Protection Overview

Analyzing the specific characteristics of the NBHD microgrid topology is crucial for protection design. For example, transformers in this setup can experience magnetizing inrush currents up to 10 times their nominal values, posing challenges for directional overcurrent relays that rely on phase-angle differences between voltage and current. Moreover, faults within the system can cause frequency swings, disrupting phase-angle estimation using phasors.

Another essential consideration is minimizing power export back to the grid. Although DERs are dispatched to reduce power import, low-import levels might lead to unintended power export if there are decreases in load consumption. Additionally, the utility grid demands swift operation from the protection systems, ruling out protection design-based selectivity and coordination through time delays.

Furthermore, the microgrid protection system must account for transformers utilized for grid connection, particularly the $\Delta - Y_g$ configuration. These transformer connections might lead relays to detect line-to-ground faults at the subtransmission level and operate before the DERs are tripped, adding complexity to the protection strategy.

Lastly, the microgrid protection scheme should include synchronism check and anti-paralleling protection to prevent the closure of breakers or switches when voltage is present on both sides. This feature is crucial to maintaining the integrity of the microgrid system, ensuring safe operations under various conditions. Addressing these specific requirements is imperative for the efficient and reliable functioning of the NBHD microgrid.

9.5.1.3 Solution

The design of the protection solution for the NBHD microgrid is intricately programmed within the relays and switchgear control. Their communication is facilitated through one of the advanced protocols specified in IEC 61850, Generic Object-Oriented Substation Event (GOOSE) messages, ensuring seamless data exchange. In addition, the DERs come equipped with integral protection, provided by the manufacturer during their installation.

In the NBHD microgrid setup, these embedded protections in the DERs serve as a vital backup, complementing the primary protection systems integrated into the relays. Notably, an essential feature implemented is anti-islanding, mandated by the utility, enhancing the microgrid's safety and reliability. Moreover, NBHD has set a stringent requirement that mandates the disconnection of all DERs in the event of a fault or outage occurring within their grid.

In summary, the microgrid's protection scheme is a combination of advanced technologies and strategic backup plans. It leverages the inherent protections within the DERs while integrating sophisticated communication protocols (e.g. GOOSE), ensuring a robust and reliable protection system that guarantees the security and resilience of the NBHD microgrid. The main parts of the protection scheme are listed below:

- Directional and nondirectional overcurrent elements
- Undervoltage and overvoltage elements
- Communication network
- IEC 61850 GOOSE messaging
- Adaptive protection
- Fast directional overcurrent blocking scheme

9.5.2 Case 2: IIT Microgrid

The 4.16 kV microgrid at the Illinois Institute of Technology (IIT) is an integral part of the campus, with two connections to the utility grid and incorporating synchronized phasor measurements for enhanced reliability [14]. This microgrid comprises seven distribution loops, each equipped with DER, including gas turbines, PV, wind, and BESS, along with various loads.

To ensure robust protection, an adaptive scheme has been implemented. This approach employs differential protection within each of the seven loops, coupled with four coordination levels established through communication protocols, directional relays, and utility switches. This comprehensive strategy guarantees the microgrid's efficient and secure operation, safeguarding both the DERs and the connected loads.

9.5.2.1 Challenges
- **Fault current level**: The short-circuit capacity of this microgrid varies significantly based on its operational mode, necessitating adaptive protection.
- **DER penetration**: The location and fault contribution of DERs can compromise sensitivity and lead to sympathetic tripping for out-of-section faults, demanding precise protection strategies.
- **Microgrid topology**: The presence of loops and meshes in this microgrid impacts current magnitude and direction, requiring careful consideration within the protection system design.

9.5.2.2 Protection Overview
The microgrid system includes a synchronous generator that provides ample fault current during islanded-mode. Other DERs in the system supply much smaller fault currents. To achieve coordination, a localized differential protection scheme, implemented within each loop, is utilized with the aid of directional overcurrent relays. This approach reduces time delays significantly when compared to centralized protection solutions. Moreover, directional overcurrent relays are placed at each end of the loop, facilitating communication to assist the differential protection scheme effectively.

Adaptability is crucial in this system. The protection mechanisms must dynamically adjust their settings based on the operational mode. To achieve this, the IIT microgrid alters the settings upon detecting islanded-mode. Additionally, the relays incorporate instantaneous and time overcurrent elements, with appropriate settings to ensuring high sensitivity for islanded operations, even in the presence of lower fault currents. While the existence of loops enhances service reliability, meticulous coordination, and hierarchical structuring are required to enable proper selectivity and ensure seamless protection.

9.5.2.3 Solution

The IIT microgrid protection scheme is divided into four levels:

1. **Load-way protection**: This scheme includes four reclosing attempts for overhead feeders and implements load shedding along with other inherent functions. Additionally, it incorporates under and over-voltage elements, as well as under and over-frequency elements in the relays.
2. **Loop protection**: Adaptive protection is enabled at this level, integrating differential and directional functions with communication capability among the relays. It operates in case of faults within the loop using differential protection as primary and directional overcurrent as backup, employing a permissive overreach transfer trip (POTT) scheme. This level serves as backup protection for the load-way scheme.
3. **Loop-feeder protection**: This level is equipped with non-directional overcurrent relays, which operate slower than both load-way and loop relays but faster than utility protection. It also functions as a backup for the entire loop in cases of communication loss or relay failure.
4. **Microgrid-level protection**: This level employs non-directional overcurrent relays with adaptive settings, serving as the primary protection mechanism for the level. It acts as a backup for the other levels, ensuring comprehensive protection for the entire microgrid system.

9.5.3 Case 3: Duke Energy Microgrid

The Mount Holly 12.47 kV microgrid, situated in North Carolina, stands 4.5 miles away from the utility grid. Duke Energy, as per [15], has meticulously designed, constructed, and now operates this advanced microgrid. The system boasts a diverse array of DERs, including a 650 kVA BESS, 100 kVA of PV panels, a natural gas generator, and multiple microturbines.

Notably, the microgrid offers seamless transitions between two operational modes: grid-connected and islanded-mode. Additionally, it supports black start functionality, enhancing its resilience and reliability in various operating scenarios. This sophisticated design and operational flexibility make Mount Holly's microgrid a valuable example of protection design.

9.5.3.1 Microgrid Overview and Challenges

The microgrid presents several challenging characteristics for the protection scheme. First, the DERs' transformer configuration as $Y_g - \Delta$ poses potential issues for feeder protection and control schemes. More specifically, these transformers impact current measured by the feeder relay, which decreases as DER penetration increases and, as a result, may desensitize protection during asymmetrical faults. When operating in islanded-mode, overvoltage protection might fail, and problems related to ferroresonance and open-phase conditions may happen.

Additionally, the microgrid's grounding design depends on the utility's $\Delta - Y_g$ transformer. However, this ground is inaccessible in islanded-mode, requiring a grounding transformer. However, energizing the grounding transformer through a gang-operated 15 kV breaker results in high levels of inrush current and severe voltage sags (measured as 0.52–0.98 p.u.).

Islanded-mode activation is communicated to the rest of the microgrid via auxiliary contacts (52a and 52b) at the utility's breaker and through communication protocols. To enable seamless islanding, the microgrid must select either a DER equipped with a synchronous generator (e.g. microturbine) or a BESS to act as a voltage/frequency regulator. The synchronous machine transitions from baseload to isochronous mode upon receiving a command from the microgrid controller. Similarly, the BESS switches from PQ mode to ISO mode. However, these transitions often cause voltage and frequency oscillations more severe than in regular conditions, potentially leading to tripping parts of the microgrid. Addressing these challenges is essential for ensuring the stability and reliability of the microgrid in all operating scenarios.

9.5.3.2 Solution

The microgrid solution incorporates over/under voltage (e.g. ANSI 27 and 59) and over/under frequency (e.g. ANSI 81) protections, each with distinct settings for grid-connected, seamless transition, and islanded-mode operations. Additionally, an islanding-detection scheme, ANSI 85-RIO, equipped with a proprietary communication protocol called SEL Mirrored Bits, has been developed and deployed in both the utility's relay and the BESS.

Several critical modifications were made to ensure proper microgrid operation and protection:

1. **DERs' transformers**: The $\Delta - Y_g$ transformers were replaced by $Y_g - Y$ transformers to open the zero-sequence current network, thereby standardizing the fault current seen by the utility's relay at 1 p.u., regardless of DER size.
2. **Grounding under islanded-mode**: A grounding transformer, connected under islanded-mode, is in place. Duke Energy devised a control scheme that keeps the grounding transformer disconnected unless the microgrid switches to islanded-mode.
3. **Inrush currents**: The microgrid employs three different methods to reduce the impact of inrush currents. While these solutions are currently under analysis and require further improvements, they are integral to the microgrid's operational integrity.
4. **Islanded-mode protection**: Protection function settings dynamically adjust based on the microgrid's operational mode. Duke Energy has implemented three groups of settings that adapt according to signals exchanged through the communication network.
5. **Microgrid protection functions**: The BESS is equipped with frequency/voltage regulation control loops and has a fault current rating of 2.4 p.u., significantly higher than other IBR devices. Consequently, protection functions relying on negative and zero sequence components can remain enabled even during islanded-mode, ensuring robust protection.

These modifications enhance the microgrid's resilience and reliability, ensuring effective protection under various operational scenarios.

9.5.4 Protection of DC Microgrids

DC microgrids pose several challenges from a protection point of view. The first is the challenge in interrupting fault currents, since most circuit breakers rely on naturally occurring current zeros to clear faults. Low-voltage breakers capable of interrupting direct current have been on the market for decades, but they are relatively slow, and limited to low voltages. Much faster solid-state and hybrid mechanical/solid-state breakers are available, but are more costly.

DC microgrids based on power electronic converters often require fast fault clearing times to try to clear faults before the dc bus capacitors of the power converters are discharged to allow faster restoration of the system after the fault is cleared. Fault clearing times on the order of 3–5 ms are often desired. Longer-lasting faults will be fed from connections to ac systems through voltage source converters. Design choices need to be made in the converter to limit the dc fault current contribution from the ac network to meet the capability of the power semiconductor devices.

As is the case with ac microgrids, protection schemes can either be communication-aided or based only on local measurements. A combination of both is preferred to achieve the highest sensitivity, especially for resistive faults. Single-ended schemes may use dc current, rate of change of current, or rate of change of voltage. Some schemes may also use a product of voltage times current or a ratio of voltage divided by current.

Communication-aided schemes can use current differential protection schemes or directional comparison schemes. Traveling wave protection has seen application in high voltage direct current transmission protection and fault location for decades and has been explored for dc microgrids, especially ones with longer lines. But traveling wave protection can be challenging with short lines. Traveling wave schemes can be single-ended or implemented as communication-aided directional comparison schemes.

9.6 Summary

In this chapter, we explored essential protective relaying concepts and their adaptation in the era of DERs, with a focus on microgrid operations. Emphasizing practicality, we discussed common protective solutions and presented real-world case studies. This comprehensive overview equips readers with key insights into the dynamic realm of microgrid protection.

9.7 Exercises

1. Consider the table below of voltage and current values measured from a power system. The frequency is 60 Hz, and the signal window size is of $N = 16$ samples.

Sample	Voltage (V)	Current (A)
0	0.000	3.637
1	43.626	4.164
2	80.610	4.057
3	105.322	3.332
4	114.000	2.100
5	105.322	0.548
6	80.610	−1.087
7	43.626	−2.557
8	0.000	−3.637
9	−43.626	−4.164
10	−80.610	−4.057
11	−105.322	−3.332
12	−114.000	−2.100
13	−105.322	−0.548
14	−80.610	1.087
15	−43.626	2.557

(a) Calculate the phasor representation of voltage and current signals above.
(b) Determine the magnitude and phase angle of the voltage phasor.
(c) Determine the magnitude and phase angle of the current phasor.

Hints:

– To calculate the phasor representation of a signal, use the following formula:

$$\overline{X} = \frac{2}{N} \sum_{n=1}^{N} x[n] \cdot e^{-j\left(\frac{2\pi f}{N}\right)}$$

where $\overline{X}$ is the phasor signal, $x[n]$ represents the signal samples, f is the frequency, N is the number of samples in the signal window, and j is the imaginary unit.

– The magnitude of a phasor V can be calculated as $|V| = \sqrt{\text{Re}(V)^2 + \text{Im}(V)^2}$, where $\text{Re}(V)$ and $\text{Im}(V)$ represent the real and imaginary parts of the phasor, respectively.

– The phase angle of a phasor V can be calculated as $\angle V = \arctan\left(\frac{\text{Im}(V)}{\text{Re}(V)}\right)$, where $\text{Re}(V)$ and $\text{Im}(V)$ represent the real and imaginary parts of the phasor, respectively.

2. Consider a transmission line protected by a distance relay (i.e. function 21) with the following settings for "Mho" characteristics and line impedance:

- **Zone 1 settings**:
  – **Forward reach**: 80% of the line length
  – **Angle**: 60° with respect to the line impedance angle
- **Zone 2 settings**:
  – **Forward reach**: 120% of the line length
  – **Angle**: 60° with respect to the line impedance angle
- **Transmission line impedance**:
  – **Line impedance**: $0.002 + j0.1\ \Omega/\text{km}$
  – **Length**: 100 km

(a) Considering the following R–X diagram with "Mho" characteristics for zones 1 and 2. What are the differences between measured impedances Z_α, Z_β, and Z_γ in terms of protection decision-making? How does that affect protection speeds and reliance on additional functions?

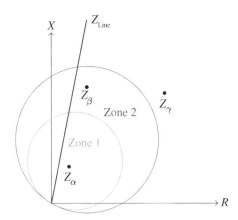

(b) Calculate the reach distances for both Zones 1 and 2 using the given settings and the line length of 100 km.

(c) Determine whether a fault at a distance of 85 km from the relay will cause the distance protection element to trip in Zone 1 or 2. Provide reasoning for your answer.

(d) For a fault occurring at a distance of 130 km from the relay, identify in which zone the fault lies, and whether the distance protection element will operate. Explain your reasoning.

Hints:
- The reach distance (R) for a "Mho" characteristic is calculated using the formula:

$$R = \frac{1}{\text{Impedance}} \times \text{Setting Distance}$$

where the impedance is given in ohms per kilometer and the setting distance is given as a fraction of the line length.
- To determine the zone of operation, calculate the "Mho" characteristic impedance using the given line impedance values. Then, compare the fault impedance (obtained by multiplying the fault distance by the line impedance) with the reach distances for Zones 1 and 2. If the fault impedance falls within the "Mho" characteristic, the element will trip.

3. An industrial process facility (with a load of 1 MVA) is currently served by a radial distribution feeder with a total feeder load of 4 MVA). The owner of the facility wishes to add a 500 kVA synchronous generator (the site is about 40% of the way down the feeder). Assume the feeder has coordinated overcurrent protection starting from relay and breaker at the main substation. There are reclosers and fused laterals both upstream and downstream from the distributed generator (DG) location.

   (a) Identify the minimum set of interconnect protection elements required (assume the site is located in a state that does not have specific interconnect requirements). What rules should these be based upon?

   (b) What impacts might the presence of the generator have on the coordination of feeder protection? Assume transformer at the point of interconnection is connected $Y_g - \Delta$ with the delta facing the feeder.

   (c) How do these impacts change if the generator transformer has the Y_g winding facing the distribution system?

   (d) What changes should be required for the protection if the rating of the generator is increased to 4 MVA?

4. Consider a distribution system capable of transitioning to microgrid operation.

   (a) How does a transition to microgrid operation fed by synchronous machines create problems for a distribution system protected using inverse time overcurrent elements?

   (b) How does that change if the generation is largely from inverter-based resources?

   (c) Discuss implementation of a scheme where relays change the time-overcurrent curve settings based on a logic input tied to opening of breaker to initiate microgrid operation. What additional resources or equipment must be purchased and installed?

   (d) Discuss implementation of a protection scheme using differential elements. What additional resources or equipment must be purchased and installed?

   (e) Discuss implementation of a protection scheme using 51V elements. What additional resources or equipment must be purchased and installed?

References

1 Anderson, P.M., Henville, C., Rifaat, R. et al. (2022). *Power System Protection*, IEEE Press Series on Power and Energy Systems, 2e. Wiley-IEEE Press. https://doi.org/10.1002/9781119513100.

2 Manson, S. and McCullough, E. (2021). Practical microgrid protection solutions: promises and challenges. *IEEE Power and Energy Magazine* 19(3): 58–69. https://doi.org/10.1109/MPE.2021.3057953.

3 Shi, S., Jiang, B., Dong, X., and Bo, Z. (2010). Protection of microgrid. In: *10th IET International Conference on Developments in Power System Protection (DPSP 2010)*, vol. 2010, 11. IET. ISBN: 978-1-84919-212-5. https://doi.org/10.1049/cp.2010.0209. https://digital-library.theiet.org/content/conferences/10.1049/cp.2010.0209.

4 PES-TR67.R1 (2020). *Impact of IEEE 1547 Standard on Smart Inverters and the Applications in Power Systems*. IEEE Power and Energy Society.

5 Venkata, S.S., Reno, M.J., Bower, W. et al. (2019). Microgrid Protection: Advancing the State of the Art. *Technical Report*. Sandia Report *SAND2019-3167*. Albuquerque, New Mexico: Sandia National Laboratories.

6 Phadke, A.G. and Thorp, J.S. (2009). *Computer Relaying for Power Systems*. Wiley.

7 Benmouyal, G. and Roberts, J. (1999). Superimposed quantities: their true nature and application in relays. *26th Annual Western Protective Relay Conference*, Spokane, October 1999.

8 SEL (2021). SEL-T400L-ultra-high-speed transmission line relay-traveling-wave fault locator-high-resolution event recorder. https://selinc.com/products (accessed 14 November 2023).

9 Li, X., Dysko, A., and Burt, G.M. (2014). Traveling wave-based protection scheme for inverter-dominated microgrid using mathematical morphology. *IEEE Transactions on Smart Grid* 5(5): 2211–2218. https://doi.org/10.1109/TSG.2014.2320365.

10 Lagos, D., Papaspiliotopoulos, V., Korres, G., and Hatziargyriou, N. (2021). Microgrid protection against internal faults: challenges in islanded and interconnected operation. *IEEE Power and Energy Magazine* 19(3): 20–35. https://doi.org/10.1109/MPE.2021.3057950.

11 Reno, M.J., Brahma, S., Bidram, A., and Ropp, M.E. (2021). Influence of inverter-based resources on microgrid protection: Part 1: Microgrids in radial distribution systems. *IEEE Power and Energy Magazine* 19(3): 36–46. https://doi.org/10.1109/MPE.2021.3057951.

12 Ropp, M.E. and Reno, M.J. (2021). Influence of inverter-based resources on microgrid protection: Part 2: Secondary networks and microgrid protection. *IEEE Power and Energy Magazine* 19(3): 47–57. https://doi.org/10.1109/MPE.2021.3057952.

13 Higginson, M., Payne, M., Moses, K. et al. (2021). North bay hydro microgrid: innovative protection of a complex system. *IEEE Power and Energy Magazine* 19(3): 70–82. https://doi.org/10.1109/MPE.2021.3057954.

14 Che, L., Khodayar, M.E., and Shahidehpour, M. (2014). Adaptive protection system for microgrids: protection practices of a functional microgrid system. *IEEE Electrification Magazine* 2(1): 66–80. https://doi.org/10.1109/MELE.2013.2297031.

15 Vukojevic, A., Lukic, S., and White, L.W. (2020). Implementing an electric utility microgrid: lessons learned. *IEEE Electrification Magazine* 8(1): 24–36. https://doi.org/10.1109/MELE.2019.2962887.

10

Microgrids Resilience: Definition, Measures, and Algorithms

Zhaohong Bie and Yiheng Bian

10.1 Background of Resilience and the Role of Microgrids

10.1.1 Essence of Resilience

The power system is the foundation for a modern society. In the last decades, the increasingly frequent natural disasters, such as hurricanes, ice storms, floods, etc. have imposed severe impacts on power systems and caused a series of blackouts. At the same time, risks from man-made attacks like terrorist attacks and cyber attack also pose a challenge to power supply security. Facing the increasing threats of these low-probability, high-impact extreme events, there is an urgent requirement to build more "resilient' power systems.

Resilience was first introduced as a concept in ecological system in 1972, which referred to "a measure of the persistence of systems and of their ability to absorb change and disturbances and still maintain the same relationships between populations or state variables" ([1]). Over the past few decades, resilience was widely adopted in research in environment science, economy, psychology, material science, disaster engineering, etc. For the energy infrastructure, the power system in particular, many definitions of similar essence have been put forward, which can be summarized as "the ability of an entity to anticipate, resist, absorb, respond to, adapt to, and recover from a disturbance" ([2]), as shown in Fig. 10.1(a). The process of a resilient power system in response to a disturbance is illustrated in Fig. 10.1(b).

Generally, the methods to boost resilience can be divided into two categories: infrastructure hardening and operational strategies. Hardening measures contain upgrading components, elevating substations, vegetation management and moving lines under ground, etc. While operational measures mainly focus on applying smart grid technologies, automation devices, and advanced restoration strategies to efficiently adapt to and recover from a disturbance.

Among the smart grid technologies, microgrids are able to maintain power supply to critical customers, or even support main grid restoration during a contingency. The utilization of microgrids in enhancing power system resilience has become a subject of great research value.

10.1.2 Microgrids in Resilience

According to the IEEE Standard 1547.4, distributed resource island systems, also referred to as microgrids, can improve the reliability of the electric power systems during an outage or disturbance ([3]). The resilience of microgrids has been proven during and after the Great East Japan Earthquake by the Sendai microgrid, the Roppongi Hills (Tokyo) microgrid, and a Smart Energy

Microgrids: Theory and Practice, First Edition. Edited by Peng Zhang.
© 2024 The Institute of Electrical and Electronics Engineers, Inc. Published 2024 by John Wiley & Sons, Inc.

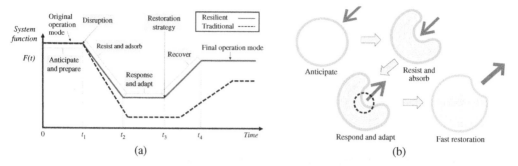

Figure 10.1 (a) Stages of power systems in response to a disturbance; (b) response process of a resilient power system.

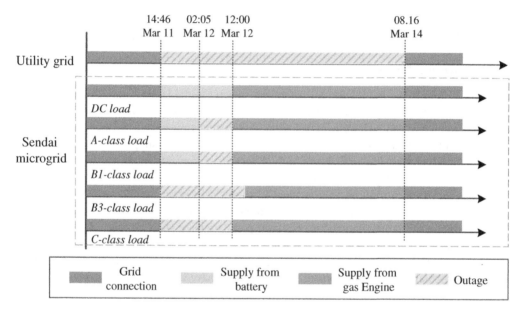

Figure 10.2 Microgrid operation during grid interruption.

home in Saitama, Japan ([4]). During the blackout, these microgrids succeeded in supplying the internal load or even providing power to the surrounding areas.

A specific case of Sendai microgrid is presented ([5, 6]). The Sendai microgrid was developed by Nippon telegraph and telephone (NTT) Facilities and was installed on the campus of Tohoku Fukushi University in Sendai City. It has several generation sources: two gas engines, a phosphoric acid fuel cell and a photovoltaic array. Both DC and AC power loads are connected. According to the power quality requirement, the AC load can be further divided into four classes: A, B1, B3, and C. The DC power load, A-class load, and B1-class load are supplied via an integrated power supply (IPS). The IPS is configured with four types of two-way converters: high-quality inverter, DC–DC converter, photovoltaics (PV)-connected converter, and a sealed lead-acid battery as an emergency backup, enabling the supply of high-quality power.

Influenced by the Great East Japan Earthquake, the electric power company in the Sendai area stopped supplying power to the area surrounding the Sendai microgrid, resulting in a three-day outage. The Sendai Microgrid switched to island mode. The microgrid operation during grid interruption is shown in Fig. 10.2. Immediately after the outage at 14:46 on 11 March, services of DC power load, A-class load and B1-class load continued to be supplied using energy from solar cells and storage batteries. At 02:05 on 12 March, as the remaining level of battery storage in the IPS fell and the voltage extended beyond the operating range, the microgrid operator stopped operation of the battery for safety reasons, resulting in the outage of A-class and B1-class load. At approximately 12:00, the gas engines were restarted manually and microgrid resumed power supply to load of A, B1, and C classes with the gas engines operated in island mode. Afterward, the B3-class load was served at about 14:00. Finally, on March 14, the external grid was restored and microgrid was reconnected to the distribution grid and returned to its normal operating mode.

The example from the Great East Japan Earthquake reveals microgrids are able to maintain supply even though the surrounding electric systems are inoperable for many days. Relying on a variety of distributed generation resources, microgrids provide backup support for critical loads and effectively mitigated the loss of life and property caused by interruption.

10.2 Enhance Power System Resilience with Microgrids

According to the stages of the power system response to extreme events, a series of measures associated with microgrids for resilience can be taken. A resilience-oriented framework involving some extensively studied measures in planning, preparation, and restoration stages with microgrids is shown in Fig. 10.3, and the measures are introduced in this section.

10.2.1 Investment Planning

10.2.1.1 Sitting and Sizing of Distributed Generators
The fast-developing microgrid technologies and increasing penetration of distributed generators (DGs) provide promising ways to make the grid more resilient. In addition to loss reduction, voltage stability maintenance, and reliability enhancement, DGs can also be managed by microgrids to restore the load after extreme events. Since DGs play a crucial role in resilience enhancement, their optimal locations and sizes with a certain investment budget are important for resilient power system planning.

1. Deterministic mathematical programming
In this part, a basic mathematical programming for dispatchable DGs allocation considering given component failure outcomes is introduced. Two assumptions are applied to simplify the modeling of restoration: (i) The loads are controllable and load shedding can be regarded as a continuous variable; (ii) Each distribution line is equipped with sectionalization switches at both line sides.

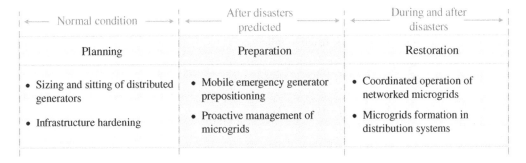

Figure 10.3 Framework of enhancing resilience with microgrids.

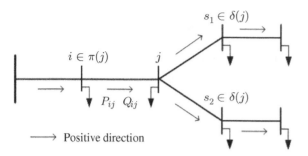

Figure 10.4 A typical radial distribution network.

The objective function is to minimize load curtailment through DG-based microgrids after major outage occurs in distribution systems.

$$\min \sum_{j \in \mathbf{V}} \rho_j P_{\text{shed},j} \tag{10.1}$$

where $\mathbf{V}$ is the set of distribution system nodes. ρ_j is the priority weight of load at node j. $P_{\text{shed},j}$ is load curtailment at node j.

(1) Investment budget constraints

The maximum DG number and aggregate sizes are limited by Eqs. (10.2) and (10.3).

$$\sum_{j=1}^{N^V} x_j^{\text{DG}} \le C_N \tag{10.2}$$

$$\sum_{j=1}^{N^V} \sum_{i=1}^{N^{\text{DG}}} x_j^{\text{DG}} k_{j,i}^{\text{DG}} s^{\text{DG}} \le C_S \tag{10.3}$$

where N^V is the number of distribution network nodes. x_j^{DG} is a binary variable which is equal to 1 if node j is chosen for DG placement. C_N, C_S are planning budgets for DG number and aggregate sizes. s^{DG} is one discrete increment of DG capacity. $k_{j,i}^{\text{DG}}$ is a binary variable denoting the ith increment is added to compose the DG size at node j, and N^{DG} is a parameter which ensures the maximum DG size should not exceed one increment multiplied by N^{DG}.

(2) Operational constraints

The linearized DistFlow equations are used to calculate the complex power flow and voltage profile. Figure 10.4 shows the typical radial distribution network. For each line, the positive direction is set and based on this pre-set direction the parent and child nodes are defined as the nodes located at the inflow lines and outflow lines respectively. Denote $\pi(j)$ and $\delta(j)$ as sets of parent and child nodes of node j. If the actual power flow is in the reverse direction, it is regarded as a negative value in the set direction.

The operational constraints can be listed as follows:

$$\begin{cases} \displaystyle\sum_{s \in \delta(j)} P_{js} - \sum_{i \in \pi(j)} P_{ij} = P_{\text{DG},j} - (P_{L,j} - P_{\text{shed},j}) \\[4mm] \displaystyle\sum_{s \in \delta(j)} Q_{js} - \sum_{i \in \pi(j)} Q_{ij} = Q_{\text{DG},j} - (Q_{L,j} - Q_{\text{shed},j}) \end{cases}, \forall j \in \mathbf{V} \tag{10.4}$$

$$\begin{cases} U_i - U_j - (r_{ij}P_{ij} + x_{ij}Q_{ij})/U_0 \le M(1 - c_{ij}) \\[2mm] U_i - U_j - (r_{ij}P_{ij} + x_{ij}Q_{ij})/U_0 \ge -M(1 - c_{ij}) \end{cases}, \forall(i, j) \in \mathbf{L} \tag{10.5}$$

$$\begin{cases} -S_{ij}^{\max}c_{ij} \le P_{ij} \le S_{ij}^{\max}c_{ij} \\ -S_{ij}^{\max}c_{ij} \le Q_{ij} \le S_{ij}^{\max}c_{ij} \end{cases}, \forall(i,j) \in \mathbf{L} \tag{10.6}$$

$$\begin{cases} 0 \le P_{\text{shed},j} \le P_{L,j} \\ 0 \le Q_{\text{shed},j} \le Q_{L,j} \end{cases}, \forall j \in \mathbf{V} \tag{10.7}$$

$$U_j^{\min} \le U_j \le U_j^{\max}, \forall j \in \mathbf{V} \tag{10.8}$$

$$0 \le P_{\text{DG},j} \le x_j^{\text{DG}} \sum_{n=1}^{N^{\text{DG}}} k_{j,n}^{\text{DG}} s^{\text{DG}}, \forall j \in \mathbf{V} \tag{10.9}$$

Constraints (10.4) represent the real and reactive power balance at each node, where $\mathbf{V}$ is the set of distribution nodes. P_{ij}, Q_{ij} are active and reactive power flow from node i to j. $P_{\text{DG},j}$, $Q_{\text{DG},j}$ are active and reactive DG output power. $P_{L,j}$, $Q_{L,j}$ are active and reactive loads. $P_{\text{shed},j}$, $Q_{\text{shed},j}$ are active and reactive load curtailment. Constraint (10.5) shows the relationship of voltage magnitudes between two adjacent nodes, where U_j is voltage magnitude of node j. r_{ij}, x_{ij} are resistance and reactance of line (i,j). U_0 is reference voltage. c_{ij} is a binary variable which is equal to 0 if the line is disconnected. $\mathbf{L}$ is the set of distribution lines. M is a large number. Constraint (10.6) is the line flow limit and forces the power flow to 0 for the disconnected lines, where $S_{ij}^{\max}$ is apparent power capacity of line (i,j). Constraint (10.7) ensures the load curtailment should not exceed load demands. Constraint (10.8) restricts the feasible range for voltage magnitude of each node, where $U_j^{\min}$ and $U_j^{\max}$ are lower and upper bounds of allowed voltage. Constraint (10.9) ensures active DG output should not exceed DG capacity.

(3) Topology constraints
During the restoration, multiple microgrids are formed based on dispatchable DGs. The following commodity flow-based constraints are applied to ensure the radial topology during reconfiguration.

$$\gamma_j \le \sum_{i \in \pi(j) \cup \delta(j)} (1 - c_{ij}), \forall j \in \mathbf{V} \tag{10.10}$$

$$-M\gamma_j - 1 \le \sum_{s \in \delta(j)} F_{js} - \sum_{i \in \pi(j)} F_{ij} \le M\gamma_j - 1, \forall j \in \mathbf{V} \backslash \text{Sub} \tag{10.11}$$

$$-Mc_{ij} \le F_{ij} \le Mc_{ij}, \forall(i,j) \in \mathbf{L} \tag{10.12}$$

$$\sum_{ij \in \mathbf{L}} c_{ij} = N^V - |\text{Sub}| - \sum_{j \in \mathbf{V}} \gamma_j \tag{10.13}$$

where F_{ij} is the fictitious power flow at the distribution line from node i to j. γ_j is a binary variable which is equal to 1 if node j is chosen as a root node. Sub is the set of substations and |Sub| is the number of islands powered by substations. The topology constraints are explained in detail in Section 10.2.3.2.

(4) Component failure constraints
Denote binary parameter u_{ij} which equals to 1 if line (i,j) is faulted. The faulted line should be disconnected, which is represented by (10.14).

$$c_{ij} \le 1 - u_{ij}, \forall(i,j) \in \mathbf{L} \tag{10.14}$$

The final form of deterministic mathematical formulations for DG sitting and sizing are:

> Objectives: (10.1)
> s.t. (10.2)–(10.14)

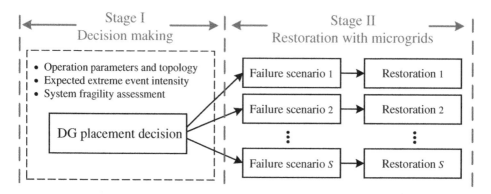

Figure 10.5 Two-stage stochastic programming for DG placement.

2. Stochastic and robust programming

A critical issue in resilience-oriented DGs allocation is to deal with the uncertain extreme events and system failure consequences. Related studies usually apply two types of modeling techniques: stochastic and robust modeling.

Stochastic programming deals with the uncertainties by constructing a probability-weighted scenario tree and makes decisions to improve the overall optimization effect on the expected mean. Figure 10.5 shows a two-stage stochastic programming framework, in which the first-stage DG placement decisions are evaluated by numbers of second-stage restoration problems corresponding to the generated scenarios of component failure outcomes. According to the form of two-stage stochastic programming, the DG placement problem is reformulated as follows:

Objectives:

$$\min \sum_{sn} \sum_{j \in \mathbf{V}} \pi_{sn} \rho_j P_{\text{shed},j}^{sn} \tag{10.15}$$

s.t.

$$\sum_{j=1}^{N^V} x_j^{\text{DG}} \le C_N \tag{10.16}$$

$$\sum_{j=1}^{N^V} \sum_{i=1}^{N^{\text{DG}}} x_j^{\text{DG}} k_{j,i}^{\text{DG}} s^{\text{DG}} \le C_S \tag{10.17}$$

$$\begin{cases} \sum_{s \in \delta(j)} P_{js}^{sn} - \sum_{i \in \pi(j)} P_{ij}^{sn} = P_{\text{DG},j}^{sn} - (P_{L,j} - P_{\text{shed},j}^{sn}) \\ \sum_{s \in \delta(j)} Q_{js}^{sn} - \sum_{i \in \pi(j)} Q_{ij}^{sn} = Q_{\text{DG},j}^{sn} - (Q_{L,j} - Q_{\text{shed},j}^{sn}) \end{cases}, \forall j \in \mathbf{V}, \forall sn \tag{10.18}$$

$$\begin{cases} U_i^{sn} - U_j^{sn} - (r_{ij} P_{ij}^{sn} + x_{ij} Q_{ij}^{sn})/U_0 \le M(1 - c_{ij}^{sn}) \\ U_i^{sn} - U_j^{sn} - (r_{ij} P_{ij}^{sn} + x_{ij} Q_{ij}^{sn})/U_0 \ge -M(1 - c_{ij}^{sn}) \end{cases}, \forall(i, j) \in \mathbf{L}, \forall sn \tag{10.19}$$

$$\begin{cases} -S_{ij}^{\max} c_{ij}^{sn} \le P_{ij}^{sn} \le S_{ij}^{\max} c_{ij}^{sn} \\ -S_{ij}^{\max} c_{ij}^{sn} \le Q_{ij}^{sn} \le S_{ij}^{\max} c_{ij}^{sn} \end{cases}, \forall(i, j) \in \mathbf{L}, \forall sn \tag{10.20}$$

$$\begin{cases} 0 \le P_{\text{shed},j}^{sn} \le P_{L,j} \\ 0 \le Q_{\text{shed},j}^{sn} \le Q_{L,j} \end{cases}, \forall j \in \mathbf{V}, \forall sn \tag{10.21}$$

$$0 \leq P_{DG,j}^{sn} \leq x_j^{DG} \sum_{n=1}^{N^{DG}} k_{j,n}^{DG} s^{DG}, \forall j \in \mathbf{V}, \forall sn \tag{10.22}$$

$$U_j^{min} \leq U_j^{sn} \leq U_j^{max}, \forall j \in \mathbf{V}, \forall sn \tag{10.23}$$

$$\gamma_j^{sn} \leq \sum_{i \in \pi(j) \bigcup \delta(j)} (1 - c_{ij}^{sn}), \forall j \in \mathbf{V}, \forall sn \tag{10.24}$$

$$-M\gamma_j^{sn} - 1 \leq \sum_{s \in \delta(j)} F_{js}^{sn} - \sum_{i \in \pi(j)} F_{ij}^{sn} < M\gamma_j^{sn} - 1, \forall j \in \mathbf{V}\backslash Sub, \forall sn \tag{10.25}$$

$$-Mc_{ij}^{sn} \leq F_{ij}^{sn} \leq Mc_{ij}^{sn}, \forall (i,j) \in \mathbf{L}, \forall sn \tag{10.26}$$

$$\sum_{ij \in \mathbf{L}} c_{ij}^{sn} = N^V - |Sub| - \sum_{j \in \mathbf{V}} \gamma_j^{sn}, \forall sn \tag{10.27}$$

$$c_{ij}^{sn} \leq 1 - u_{ij}^{sn}, \forall (i,j) \in \mathbf{L}, \forall sn \tag{10.28}$$

where sn denotes the scenario index. π_{sn} is the probability of scenario sn.

Unlike the stochastic approach which relies on a probability distribution of the uncertainties and sampled scenarios of the uncertainty realizations, robust optimization requires limited information of the uncertainty set and makes decisions against the worst cases. A tri-level robust defender-attacker-defender (DAD) model is introduced. Based on DAD framework, the planner determines optimal DG locations and sizes on the first level. According to the planning result, the attacker finds the worst-case attacks on the second level. On the third level, the operator minimizes the cost of load shedding. The three levels interacted with each other to improve the decision performance under the worst attacks, as shown in Fig. 10.6. The mathematic formulations based on the robust DAD model are listed as follows:

Objectives:

$$\min_{\mathbf{h}} \max_{\mathbf{u}} \min_{\mathbf{z,c}} \sum_{j \in V} P_{shed,j} \tag{10.29}$$

s.t.

$$\sum_{ij \in \mathbf{L}} u_{ij} \leq K \tag{10.30}$$

$$(10.2)-(10.13) \tag{10.31}$$

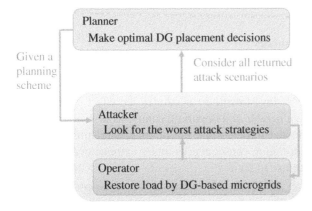

Figure 10.6 DAD model for DG placement.

Table 10.1 Resilience-oriented planning model and the applied algorithms.

References	Planning decisions	Model	Algorithm
Shi et al. [7]	DG siting and sizing	Two-stage stochastic	Progressive hedging (PH) algorithm
Ma et al. [8]	Line hardening, DG sitting, automatic switches deploying		Dual decomposition algorithm
Wang et al. [9]	Line hardening, DG sitting	Two-stage robust optimization	Greedy search
Yuan et al. [10]	Line hardening, DG sitting		Column and constraint generation (CCG) algorithm
Lin and Bie [11]	Line hardening		Nested CCG algorithm

where $\mathbf{h}$ denotes the DG placement decisions $\{x_j^{DG}, k_{j,i}^{DG}\}$, $\mathbf{u}$ denotes the attack strategies $\{u_{ij}\}$. $\mathbf{z}$ and $\mathbf{c}$ represent the continuous variables for power flow operation $\{P_{ij}, Q_{ij}, P_{\text{shed},j}, Q_{\text{shed},j}, P_{\text{DG},j}, Q_{\text{DG},j}, U_j\}$ and integer variables for topology reconfiguration $\{c_{ij}\}$. K is the maximal number of faulted lines.

Relevant algorithms have been developed to solve the stochastic or robust problem. Several literatures related to resilience-oriented planning models and the applied algorithms are listed in Table 10.1.

10.2.1.2 Infrastructure Hardening

The grid-connected microgrids can be divided into campus microgrids, military microgrids, community microgrids, and onboard microgrids which improve resilience for different types of customers. The extreme events could destroy components inside the microgrid or damage gas distribution pipelines so that microgrids cannot operate fully to sustain the power supply. Therefore, hardening infrastructures to enhance the robustness and resistance of microgrids to the impact of extreme events is important in the planning stage.

Some hardening practices are summarized as follows:

(1) upgrading components with stronger and more robust materials
(2) undergrounding overhead conductors
(3) elevating substations
(4) managing vegetation
(5) relocating facilities to areas less prone to shocks
(6) elevating substations

Although hardening measures can remarkably make microgrids less susceptible to extreme events, they usually require large amount of investment. To maximize the effect of resilience improvement with limited investment, the vulnerable and important components should be identified. Related researches have been conducted to determine the optimal hardening decisions considering the uncertainties in component failure consequences under extreme events, which are similar to the methods introduced in Section 10.2.1.1.

10.2.2 Pre-Event Preparation

10.2.2.1 Mobile Emergency Resource Prepositioning

Mobile energy resources (MERs) are critical flexibility resources for fast service restoration ([12]). Mobile emergency generators (MEGs), as one of the main types of MERs, are truck-mounted energy resources with the merits of mobility and large capacity ([13]). In case of a major outage, especially when customers are without complete power access to the main grid, MEGs can be routed to connect with parts of the network for supplying local load or forming microgrids to further supply the regional load in distribution systems.

During and after an extreme event, the available local resources for load restoration in distribution systems are usually limited and MEGs are difficult to deliver to site in time due to damaged transportation network. The resource limitation can be a bottleneck to carry out many restoration measures such as network sectionalization and microgrid formation. Despite the severe impacts on power and transportation systems, many extreme events such as wind-related weather disasters can be predicted in advance, which makes it possible for utility companies to make preparation. Therefore, the pre-disaster emergency MEGs prepositioning occupies a vital part in resilience improvement.

The objective of the prepositioning problem is to proactively allocate MEGs to candidate locations to reduce the post-disaster travel time and achieve rapid restoration. After disasters, MEGs are dispatched to some nodes and act as power sources to restore critical loads by forming multiple microgrids. The prepositioning model for MEGs can be formulated as a two-stage stochastic optimization problem. The objective function is to minimize the expected total load curtailment during the estimated restoration time:

$$\min \sum_{\text{sn}} \sum_{t \in T} \sum_{j \in \mathbf{V}} \pi_{\text{sn}} \rho_j P_{\text{shed},j,t}^{\text{sn}} \tag{10.32}$$

where sn is the scenario index. π_{sn} is the probability of scenario sn. T is the estimated restoration time duration and t is the index for time steps. $\mathbf{V}$ is the set of distribution system nodes. ρ_j is the priority weight of load at node j. $P_{\text{shed},j}$ is load curtailment at node j.

The MEG prepositioning and real-time allocation constraints can be formulated as follows ([13]):

$$\sum_{m \in \mathbf{M}} \chi_{m,\text{cl}} \leq X_{\text{cl}}, \forall \text{cl} \tag{10.33}$$

$$\sum_{\text{cl} \in \text{CL}} \chi_{\text{cl},m} = 1, \forall m \tag{10.34}$$

$$\sum_{k \in \mathbf{G}} \gamma_{\text{cl},m,k,\text{sn}} \leq \chi_{\text{cl},m}, \forall \text{cl}, \forall m, \forall \text{sn} \tag{10.35}$$

$$\sum_{\text{cl} \in \text{CL}} \sum_{m \in \mathbf{M}} \gamma_{\text{cl},m,k,\text{sn}} \leq 1, \forall k, \forall \text{sn} \tag{10.36}$$

where cl and CL are the index and set of candidate locations for MEG prepositioning. m and $\mathbf{M}$ are the index and set of MEGs. $\chi_{m,\text{cl}}$ is binary variable which is equal to 1 if MEG m is deployed at location cl. k and $\mathbf{G}$ are indexed and set for candidate nodes for MEG connection for restoration. $\gamma_{\text{cl},m,k,\text{sn}}$ is binary variable which is equal to 1 if MEG m is dispatched from location cl to node k in scenario sn. Constraint (10.33) limits the number of prepositioned MEGs at each candidate location. Constraint (10.34) ensures each MEG is prepositioned to exactly one of the candidate locations.

Constraint (10.35) ensures MEG m can be sent to node k from location cl if it is prepositioned at location cl. Constraint (10.36) ensures at most one MEG is dispatched to each candidate node.

If a MEG is sent to node k from candidate location cl after travel time $T_{cl,k}$, it can act as a DG with the following active and reactive power output constraints:

$$0 \leq P^{sn}_{DG,k,t} \leq \sum_{cl \in CL} \sum_{m \in M} \gamma_{cl,m,k,sn} P^{max}_m, \quad \forall k \in G, \forall t, \forall sn \tag{10.37}$$

$$0 \leq Q^{sn}_{DG,k,t} \leq \sum_{cl \in CL} \sum_{m \in M} \gamma_{cl,m,k,sn} Q^{max}_m, \quad \forall k \in G, \forall t, \forall sn \tag{10.38}$$

$$0 \leq P^{sn}_{DG,k,t} \leq M \left(1 - \gamma_{cl,m,k,sn}\right), \quad t \leq T_{cl,k} - 1, \quad \forall cl, \forall m, \forall k \in G, \forall sn \tag{10.39}$$

$$0 \leq Q^{sn}_{DG,k,t} \leq M \left(1 - \gamma_{cl,m,k,sn}\right), \quad t \leq T_{cl,k} - 1, \quad \forall cl, \forall m, \forall k \in G, \forall sn \tag{10.40}$$

where $P^{sn}_{DG,k,t}$ and $Q^{sn}_{DG,k,t}$ are active and reactive power output variables. P^{max}_m and Q^{max}_m are maximum active and reactive power output of MEG m. Constraint (10.37), (10.38) mean if MEG m is dispatched from location cl to node k, the active and reactive power outputs at node k are within the output ranges, otherwise the active and reactive power outputs at node k should be 0. Constraint (10.39), (10.40) ensure the MEG sent to node k from candidate location cl cannot output power until travel time $T_{cl,k}$ passes.

Multiple microgrids can be formed based on the MEGs allocated to distribution nodes. The operational constraints, topology constraints, and component failure constraints at each time step are the same as (10.18)–(10.22), (10.24)–(10.28) in Section 10.2.1.1. The final form of the MEG prepositioning problem is summarized as follows:

Objective:

$$\text{(10.32)} \tag{10.41}$$

s.t.

$$\text{(10.33)} - \text{(10.40)} \tag{10.42}$$

$$\text{(10.18)} - \text{(10.22)}, \text{(10.24)} - \text{(10.28)} \text{ for variables in each time step} \tag{10.43}$$

To improve the quality of generated component failure scenarios under wind-related weather disasters like hurricanes, after situational awareness, according to weather forecasting data, the geographic information system (GIS) can be integrated to evaluate the spatiotemporal impacts of disasters ([14]). The regional wind speed can be calculated according to the distance between the region cell and the hurricane eye ([15]), and the component failure probability is acquired according to the disaster intensity and component fragility curve.

10.2.2.2 Proactive Management of Microgrids

In order to mitigate the impact of severe disturbances on microgrids, the operators can take proactive operation measures to prepare various microgrid facilities. For instance, scheduling the commitment status of DGs, deploying demand side reserves and adjusting power transactions with the upstream grid to make microgrids more resourceful and flexible to disasters ([16]). Moreover, operators can maintain the state of charge of energy storage system to assure the survivability of critical loads for a period after the disasters ([17]).

In addition to proactively scheduling the energy resources, adjusting the network topology to maintain less vulnerable branches in service before disasters can also reduce the microgrid vulnerability at the disaster onset ([18]). To find the configuration with minimum vulnerable components while total load can be supplied, the operators start with the most conservative configuration, i.e. no vulnerable branch is kept in service, then gradually increase the number of vulnerable components in service. Every time increasing the vulnerable components, an optimal power flow (OPF) is run to find out the amount of load not supplied. At the same time, microgrid rescheduling, network reconfiguration, demand-side participation, etc. are considered to reduce unsupplied load. The first configuration with entire load supplied will be chosen as the proactive topology of the microgrid.

10.2.3 Efficient Restoration

10.2.3.1 Coordinated Operation of Networked Microgrids

In case of permanent faults and interruptions occur at the external grid, microgrids connected to distribution feeders through the points of common coupling (PCCs) can transformed into island modes and continually maintain the internal power supply. After islanding, the microgrids can be operated autonomously or cooperatively based on preferences of the stakeholders. Autonomous operation achieves a higher level of security since the power balance is met locally and independently in different microgrids, possible cascading faults can be avoided ([19]). However, it relies on sufficient generation and storage capacity in each microgrid to ensure stable power supply. On the contrary, operating microgrids in a coordinated manner and sharing the generation and storage capacity among microgrids is more economical and efficient ([20]). In this mode, the microgrids are physically interconnected and functionally interoperable to construct a system of networked microgrids ([21]).

To coordinate the operation of networked microgrids, a hierarchical optimization model based on model predictive control (MPC) method ([22]) is introduced.

1. Hierarchical optimization framework for networked microgrids

The two-layer hierarchical model is demonstrated in Fig. 10.7. The individual microgrid central controllers (MCCs) are on the lower layer, responsible for the operation of the corresponding microgrids. The cyber communication network and a common bus are on the upper layer, which are responsible for data communication and power exchange.

After a power interruption occurs, in the first stage, the individual microgrid schedules the internal DGs, energy storages, and controllable loads to minimize load curtailment and operation cost. In the second stage, microgrids with load curtailment will receive power support from those with excess power. A rolling-horizon method is employed for the two-stage scheduling. As illustrated in Fig. 10.8, the optimization horizon is set as multiple time slots in order to take into account the fluctuation of renewable energy, the ramp-up/down limits of DGs, and the charging/discharging of ESSs. After decisions are made, only the decision for the first time slot in the horizon window will be implemented. The above step will be repeated until the outage ends.

2. Consensus algorithm for power exchange plan determination

In the second stage, microgrids share excess power to further reduce load curtailment. Microgrids communicate mutually to acquire the global information for power exchange plan determination. A consensus algorithm is applied to exchange information in a decentralized way ([23]). Assume each microgrid has at least one cyber communication link to others, as shown in Fig. 10.9. According to the algorithm, each microgrid interchanges information with its adjacent microgrids and updates the information locally and iteratively. The global information is finally revealed to each microgrid.

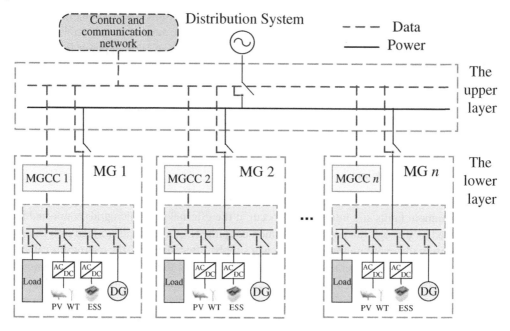

Figure 10.7 Hierarchical optimization model framework.

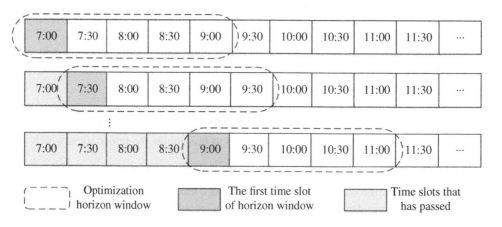

Figure 10.8 MPC-based scheduling method.

At the start of the process, a state variable of the wanted information, i.e. excess or deficient power, is initialized in each microgrid. The state variable can be updated as follows:

$$X_n(k+1) = X_n(k) + \tau \sum_{m \in NEI_n} \left(X_m(k) - X_n(k) \right) \tag{10.44}$$

where k represents the iteration times. τ is the step size and satisfies $0 < \tau < 1/\lambda$, where λ is the maximum node out-degree of the system digraph ([24]). $X_n(0)$ is the initial value of state variable of the nth microgrid and N represents the number of microgrids. After a finite number of iterations, the state variable of each microgrid will converge on a common value $\tilde{X}_n$:

$$\tilde{X}_n = \frac{1}{N} \sum_{n=1}^{n=N} X_n(0) \tag{10.45}$$

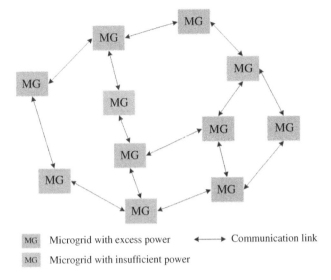

MG   Microgrid with excess power   ←——→ Communication link

MG   Microgrid with insufficient power

Figure 10.9 Communication network of microgrids.

To obtain the information of total excess power of the networked-microgrids system, denote P_n^{ex} as the excess power of microgrid n. For the microgrid without excess power or with load curtailment, the value of this variable is 0. Then each microgrid takes its P_n^{ex} as the initial state variable and implements the iteration using Eq. (10.44). The iteration variable of every microgrid will finally converge on $\frac{1}{N} \sum_{n=1}^{n=N} P_n^{ex}$. Through multiplying the converged result by N, total excess power of the whole system can be known to every microgrid. Similarly, to acquire the global information of deficient power, denote P_n^{dt} as the deficient power and P_n^{dt} takes the value of 0 if microgrid n does not lack power. Design a vector $\mathbf{P}_n^{dt}$ which has a length of N ([25]). The i th element of vector $\mathbf{P}_n^{dt}$ is initialized as:

$$\mathbf{P}_n^{dt}[i] = \begin{cases} N \times P_n^{dt}, i = n \\ 0, i \neq n \end{cases} \tag{10.46}$$

Through iterating the vector $\mathbf{P}_n^{dt}$ locally using Eq. (10.44), the deficient power of every microgrid which requires external power support can be known as follows:

$$\tilde{\mathbf{P}}_n^{dt} = \frac{1}{N} \sum_{n=1}^{n=N} \mathbf{P}_n^{dt}(0) = \left[P_1^{dt}, P_2^{dt}, \dots, P_N^{dt} \right]^T \tag{10.47}$$

After that, the microgrid with excess capacity can determine the power that it should export to the microgrids with load curtailment based on some pre-set agreements, e.g. assign the offered power of each microgrid according to the proportion of its excess power. The entire process can be implemented in a decentralized way, which requires lower complexity of communication systems and achieves a higher level of privacy protection.

3. Case studies

The decentralized information exchange is demonstrated with a test system comprised of five microgrids as shown in Fig. 10.10. Under normal conditions, the microgrids are connected to distribution network through PCCs. When extreme events cause failure in external power grid, microgrids will disconnect from external grid and connect to each other by closing the interconnection tie switches. Assume the communication network is of ring structure, i.e. microgrids can exchange information according to the route $1 \leftrightarrow 3 \leftrightarrow 5 \leftrightarrow 4 \leftrightarrow 2 \leftrightarrow 1$.

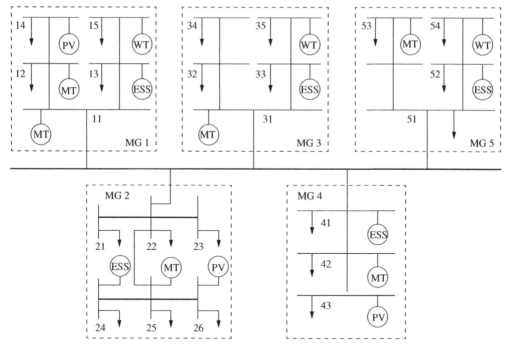

Figure 10.10 Test system of multi-microgrids.

Table 10.2 Schedule result of each microgrid.

Result (kW)	MG 1	MG 2	MG 3	MG 4	MG 5
P_n^{ex}	500.00	0	0	228.20	297.95
P_n^{dt}	0	96.63	24.60	0	0

Assume the excess or deficient power of each microgrid at a time step is shown in Table 10.2. It can be seen that microgrids 2 and microgrid 3 has load curtailment and need energy support from other microgrids. To obtain the value of power exchange in a decentralized way, the consensus method is implemented. The iteration processes of variable P_n^{ex} and the second element of vector $\mathbf{P}_n^{dt}$ are shown in Fig. 10.11. The converged values of two variable are 205.23 and 96.63, which indicates total excess power is 1026.15 and the deficient power of microgrid 2 is 96.63. Then with the agreements of assigning the offered power of each microgrid according to the proportion of its excess power, the allocated power (kW) that microgrids 1, 4, and 5 should exported to microgrid 2 is 47.08, 21.49, 28.06, respectively.

10.2.3.2 Microgrids Formation in Distribution Systems

The DG units provide an alternative approach to continue supplying loads after major faults of the main grids. With the emerging smart grid technology, many electric distribution utilities are planning to deploy the remote-controlled switches or update the existing manual switches at distribution feeders, which makes it possible to dynamically adjust the topology and microgrids boundaries by operating the associated switches which can serve as the temporary PCCs ([21]). In this context,

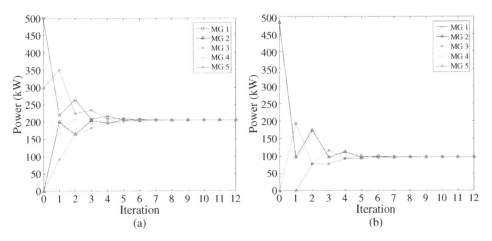

Figure 10.11 Examples of iteration process. (a) P_n^{sur}. (b) the second element of vector P_n^{dt}.

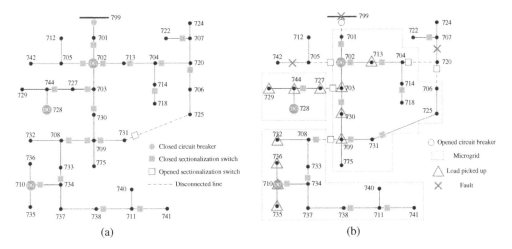

Figure 10.12 (a) Distribution system under normal condition. (b) Microgrid formation result after the disaster.

forming multiple microgrids energized by DGs with frequency and voltage regulation capabilities from the distribution system becomes an effective strategy to enhance resilience.

Despite devices such as bidirectional protective relays and soft open points (SOPs) support operating distribution systems in a ring or meshed manner ([26]), in most common cases distribution systems and microgrids are operated with a radial topology, i.e. no cycle exists in the graph, to facilitate the coordination and protection and to reduce the short-circuit current ([27]).

Figure 10.12 shows a case of microgrid formation based on the IEEE 37-node test system. The system operates with a radial topology under normal conditions as shown in Fig. 10.12(a). After disasters, the system is disconnected from the substation by opening the circuit breaker and faults are isolated by opening switches at lines 704–720, 706–720, and 702–705. After that, 3 microgrids can be formed by opening switches at line 703–727 and 708–709 and closing switches at line 725–731. The radiality is maintained, as shown in Fig. 10.12(b).

The methods expressing radiality constraints include loop-eliminating method, path-based method, commodity flow-based method, and parent-child relation-based method ([27]).

Among these methods, the commodity flow-based method is most commonly used due to its simplicity and efficiency. The basic form of this method ([28]) and method extensions ([29], [30]) are introduced in the following.

1. Basic form of commodity flow-based method for radiality guarantee
According to graph theory, a graph is radial if and only if the following two conditions are simultaneously satisfied: (1) the number of closed branches equals the number of nodes minus the number of islands; (2) the connectivity of each island is guaranteed.

Assume that one DG is selected to control the voltage and frequency for each microgrid. To achieve the first condition, we have

$$\sum_{ij \in L} c_{ij} = N^V - N^{DG} \tag{10.48}$$

where L is the set of distribution lines. c_{ij} is a binary variable which is equal to 1 if branch (i, j) is closed. N^V is the number of nodes. N^{DG} is the number of selected DGs and is also the number of microgrids to be formed.

To achieve the second condition, a fictitious network is designed with the same topology as the original distribution system. Define the selected DG nodes as "root nodes" which act as the fictitious sources in microgrids to be formed, while the other nodes have a fictitious unit load demand. Therefore, the satisfaction of power balance at each node in the fictitious network means that at least one path exists between the load node and the root node, so the islands must be connected. Thus, we have

$$\begin{cases} \sum_{s \in \delta(j)} F_{js} - \sum_{i \in \pi(j)} F_{ij} = W_j, \ j \in F \\ \\ \sum_{s \in \delta(j)} F_{js} - \sum_{i \in \pi(j)} F_{ij} = -1, \ j \in V \backslash F \end{cases} \tag{10.49}$$

$$-Mc_{ij} \le F_{ij} \le Mc_{ij} \forall (i, j) \in L \tag{10.50}$$

where $\pi(j)$ and $\delta(j)$ are sets of parent and child nodes of node j. F_{ij} is the fictitious power flow at the distribution line from node i to j. W_j is the fictitious power output at source node j. M is a sufficient large positive number. Constraint (10.49) ensures fictitious power balance at each node while constraint (10.50) denotes fictitious power can only flow at closed lines.

2. Extensions of the commodity flow-based method
The above basic formulations (10.48)–(10.50) ensure radiality based on pre-defined root nodes, therefore, the number of microgrids is fixed before optimization, which restricts the restoration effect. What's more, the de-energized island, i.e. the island does not possess energized DGs, is not permitted, which prevents the method from incorporating the process of fault isolation. In this part, an extension of the commodity flow-based method is introduced to tackle these drawbacks.

The first improvement is to determine the root nodes through optimization rather than pre-setting. In view of the fact that microgrids or de-energized islands are created by disconnecting lines, we regard the two nodes at the ends of a disconnected line as potential root nodes. Then constraints (10.48)–(10.50) can be reformulated as follows.

$$\gamma_j \le \sum_{i \in \pi(j) \bigcup \delta(j)} (1 - c_{ij}), \ \forall j \in V \tag{10.51}$$

$$-M\gamma_j - 1 \le \sum_{s \in \delta(j)} F_{js} - \sum_{i \in \pi(j)} F_{ij} \le M\gamma_j - 1, \forall j \in V \backslash \text{Sub} \tag{10.52}$$

$$-Mc_{ij} \leq F_{ij} \leq Mc_{ij}, \forall (i,j) \in \mathbf{L} \tag{10.53}$$

$$\sum_{ij \in \mathbf{L}} c_{ij} = N^V - |\text{Sub}| - \sum_{j \in \mathbf{V}} \gamma_j \tag{10.54}$$

where F_{ij} is the fictitious power flow at the distribution line from node i to j. γ_j is a binary variable which is equal to 1 if node j is chosen as a root node. Sub is the set of substations and $|\text{Sub}|$ is the number of island powered by substations. Constraint (10.51) indicates node j can be a root node if at least one of its linked line is disconnected. Constraint (10.52) is the nodal power balance constraint for the fictitious network. It means that, for each node except the substation, it should be either a root node with unlimited power output ($\gamma_{j,t} = 1$ and the power balance constraint for node j is relaxed) or a load node with a unit demand ($\gamma_{j,t} = 0$ and the injected power is 1). Constraint (10.53) prevents fictitious power flowing on a disconnected line. Constraint (10.54) states the number of lines is equal to the number of nodes minus the number of root nodes and islands powered by substations.

The second improvement is to incorporate fault isolation in microgrid formation process. Once a fault occurs on a line, power outage will propagate to nodes connected to the line until the nearest switches isolate it. The parts connected to the fault will remain interrupted until the fault is cleared. Take line (i,j) and node j as an example; as depicted in Fig. 10.13, node j will be influenced by interruption from line (i,j) under the following two scenarios. Scenario A: the fault occurs on line (i,j) and there is no switch to isolate the outage or the switch is closed at j side of the line. Scenario B: node i is influenced by interruption and there is no switch or the switch is closed at either side of line (i,j).

Denote x_j as a binary variable which is equal to 1 if node i is influenced by interruptions, and B_{ij} as a binary parameter which is equal to 1 if line (i,j) is faulted. Denote binary variable y_{ij1}/y_{ij2} equal to 1 if the switch at i/j side of line (i,j) is closed or the side is non-switchable. We have

$$\begin{cases} Mx_j \geq \displaystyle\sum_{i \in \pi(j)} y_{ij2} \cdot B_{ij} + \sum_{s \in \delta(j)} y_{js1} \cdot B_{js} \\ Mx_j \geq \displaystyle\sum_{i \in \pi(j)} y_{ij1} \cdot y_{ij2} \cdot x_i + \sum_{s \in \delta(j)} y_{js1} \cdot y_{js2} \cdot x_s \end{cases} \quad \forall j \in \mathbf{V} \tag{10.55}$$

$$\varepsilon \left(2 - y_{ij1} - y_{ij2}\right) \leq 1 - c_{ij} \leq M \left(2 - y_{ij1} - y_{ij2}\right) \forall (i,j) \in \mathbf{L} \tag{10.56}$$

where ε is a sufficient small positive number. Constraint (10.55) expresses the above scenarios A and B in which node j will be influenced by interruption. Constraint (10.56) represents the line is closed if switches at both sides are closed or non-switchable.

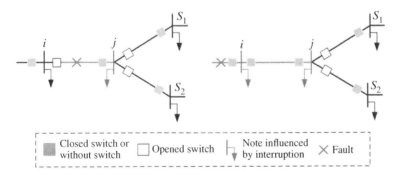

Figure 10.13 Scenarios where node j is influenced by outage propagation from (i,j).

3. Formulations for microgrids formation based on commodity flow method

Mathematic formulations for microgrids formation to maximize the restored load based on commodity flow method are listed as follows:

$$\max \sum_{j\in V} \rho_j En_j^L P_j^L \tag{10.57}$$

s.t.

$$(10.5), (10.6) - (10.8), (10.51) - (10.56) \tag{10.58}$$

$$\begin{cases} \sum_{s\in\delta(j)} P_{js} - \sum_{i\in\pi(j)} P_{ij} = P_j^{DG} - En_j^L P_j^L \\ \sum_{s\in\delta(j)} Q_{js} - \sum_{i\in\pi(j)} Q_{ij} = Q_j^{DG} - En_j^L Q_j^L \end{cases}, \forall j \in V \tag{10.59}$$

$$\begin{cases} En_j^G \le \left(1 - x_j\right) \\ 0 \le P_j^{DG} \le En_j^G P_j^{DG,\max} \quad \forall j \in V \\ 0 \le Q_j^{DG} \le En_j^G Q_j^{DG,\max} \end{cases} \tag{10.60}$$

where En_j^L, En_j^G are binary variables representing the pick-up status of load and energization status of DGs, respectively. Differing from Section 10.2.1.1, the loads are considered uncontrollable and can only be restored as a whole by closing the load switch at each node to achieve more practicality. Constraint (10.60) makes sure if node j is influenced by interruptions, the DG at it shouldn't be energized. Otherwise, the DG is operable and the output is limited by DG capacity.

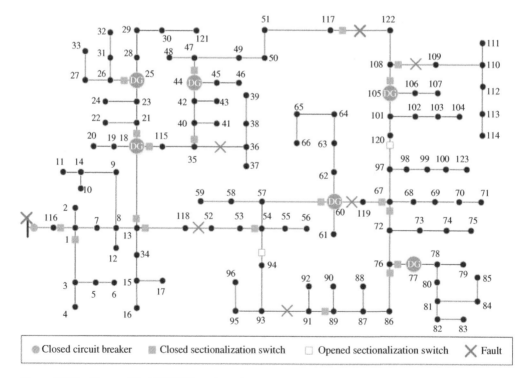

Figure 10.14 The modified IEEE 123-node system.

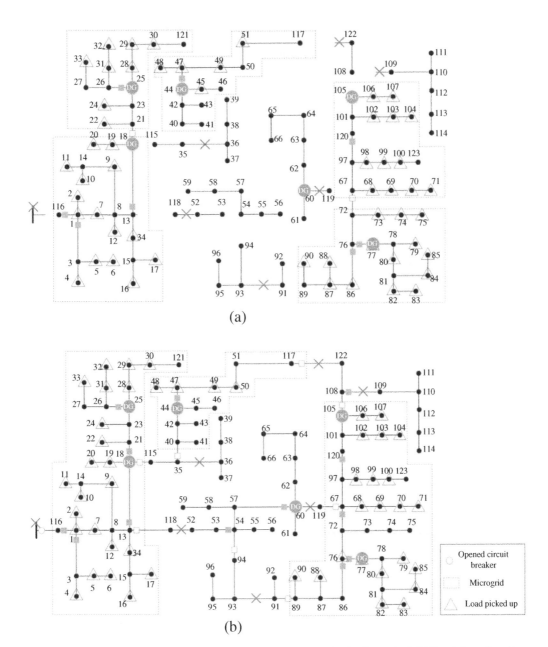

Figure 10.15 Microgrids formation results. (a) Result based on the basic method. (b) Result based on the extended method.

4. Case studies

The microgrids formation model is demonstrated on the IEEE 123-node system ([31]). As shown in Fig. 10.14, the modified IEEE 123-node system has 6 DGs rated at 500 kW and 250 kVAr. A circuit breaker and 21 remote-controlled switches are added. Switches at L97–120 and L54–94 are normally open. Total active load of the system is 3490 kW. Seven faults occur at the substation and line L35–36, L52–118, L60–119, L91–93, L108–109, L117–122.

First the microgrids formation based on the basic form of topology constraints (10.48)–(10.50) is tested. Before reconfiguring the network, faults are isolated artificially by opening the nearest switches, i.e. switches at L35–40, L18–115, L117–122, L108–109, L105–108, L13–118, L53–54, L57–60, L67–119 and the circuit breaker. Then the optimization problem for the remaining fault-free network is solved. Assigning each DG as a root node, 5 microgrids are formed and the result is shown in Fig. 10.15(a). The total restored load is 2175 kW. As a contrast, the extended topology constraints (10.51)–(10.56) are applied. Both fault isolation and reconfiguration are optimized at the same time. The result is shown in Fig. 10.15(b). The total restored load is 2240 kW. The restoration effect is improved because of a more flexible selection of root nodes for microgrids formation.

10.3 Future Challenges

Extreme events will remain a formidable challenge to the power systems in the long run. Future investment, policies, and new technologies related with microgrids are much needed to reinforce power system resilience. Besides the natural disasters, the response and prevention against cyber threats in microgrids is also of great importance for resilience improvement. Developments in distribution automation and control of microgrids have led to increasing number of digital devices in power systems. In spite of advantages in operation, communication, and control, these devices also introduce cyber vulnerabilities. To enhance cyber-physical microgrid resiliency, detailed modeling for both the physical system and the cyber components is required. The effective detection schemes and countermeasures for cyber-attacks is urgently needed. The application of deep reinforcement learning (DRL) in improving power system operation has been increasingly recognized. DRL provides another way to make decisions for the economic dispatch, energy management, and dynamic distribution network reconfiguration. Compared to mixed-integer programming, DRL avoids NP-hard and may mitigate computational burden in terms of large-scale systems. The DRL method can improve the adaptability of microgrids to changeable conditions, thus is helpful for realizing an online dynamic microgrid formation. From the view of industrial application, to employ the advanced resilience strategies such as microgrid formation, the grid codes and industry standards should be established to allow for the DG islanding function, and system hardening and system planning should be carried out to install DG, tie switches, and other necessary facilities. In addition, distribution automation efforts should speed up to increase the power system visibility to achieve higher installation rates of smart meters and automatic switches that allow for a more efficient response to disruptions.

10.4 Exercises

1. According to the stages of the power system response to extreme events, measures to improve the resilience can be divided as planning measures, preparation measures, and restoration

measures. Please list at least one measure associated with microgrids in each stage and give brief explanations about the way the measure influences the resilience curve in Fig. 10.1.

2. The extreme events could destroy components inside the microgrid or infrastructures on which the microgrid depends so that microgrids cannot operate fully to sustain the power supply. Please list some hardening measures to strengthen the components/infrastructures in the planning stage to enhance the robustness of microgrids against extreme events.

3. Analyze the different participation modes of DGs and MEGs in microgrid formation and distribution system restoration.

4. According to the consensus algorithm for power exchange plan determination, for the networked microgrids system in Fig. 10.10, supposing the excess power of 5 microgrids are 200 kW, 300 kW, 0, 0, 500 kW in a time slot, how many iterations will it take for the microgrids to obtain the total excess power of the system?

5. To apply the commodity flow-based method for radiality guarantee during microgrid formation, what's the necessary and sufficient condition to maintain the radial topology in the microgrids?

References

1 Holling, C.S. (1973). Resilience and stability of ecological systems. *Annual Review of Ecology and Systematics* 4 (1): 1–23.

2 Bie, Z., Lin, Y., Li, G., and Li, F. (2017). Battling the extreme: a study on the power system resilience. *Proceedings of the IEEE* 105 (7): 1253–1266.

3 IEEE Standard (2011). *IEEE Guide for Design, Operation, and Integration of Distributed Resource Island Systems with Electric Power Systems. IEEE Std 1547.4-2011*, 1–54.

4 IEC (2014). Microgrids for disaster preparedness and recovery-with electricity continuity plans and systems.

5 Irie, H., Hirose, K., Shimakage, T., and Reilly, J. (2013). The sendai microgrid operational experience in the aftermath of the tohoku earthquake: a case study.

6 Gholami, A., Aminifar, F., and Shahidehpour, M. (2016). Front lines against the darkness: enhancing the resilience of the electricity grid through microgrid facilities. *IEEE Electrification Magazine* 4 (1): 18–24.

7 Shi, Q., Li, F., Kuruganti, T. et al. (2021). Resilience-oriented DG siting and sizing considering stochastic scenario reduction. *IEEE Transactions on Power Systems* 36 (4): 3715–3727.

8 Ma, S., Su, L., Wang, Z. et al. (2018). Resilience enhancement of distribution grids against extreme weather events. *IEEE Transactions on Power Systems* 33 (5): 4842–4853.

9 Wang, X., Shahidehpour, M., Jiang, C., and Li, Z. (2019). Resilience enhancement strategies for power distribution network coupled with urban transportation system. *IEEE Transactions on Smart Grid* 10 (4): 4068–4079.

10 Yuan, W., Wang, J., Qiu, F. et al. (2016). Robust optimization-based resilient distribution network planning against natural disasters. *IEEE Transactions on Smart Grid* 7 (6): 2817–2826.

11 Lin, Y. and Bie, Z. (2017). Tri-level optimal hardening plan for a resilient distribution system considering reconfiguration and DG islanding. *Applied Energy* 210: 1266–1279.

12 Wang, W., Xiong, X., He, Y. et al. (2022). Scheduling of separable mobile energy storage systems with mobile generators and fuel tankers to boost distribution system resilience. *IEEE Transactions on Smart Grid* 13 (1): 443–457.

13 Lei, S., Wang, J., Chen, C., and Hou, Y. (2018). Mobile emergency generator pre-positioning and real-time allocation for resilient response to natural disasters. *IEEE Transactions on Smart Grid* 9 (3): 2030–2041.

14 Liu, X., Shahidehpour, M., Li, Z. et al. (2017). Microgrids for enhancing the power grid resilience in extreme conditions. *IEEE Transactions on Smart Grid* 8 (2): 589–597.

15 Muhs, J.W. and Parvania, M. (2019). Stochastic spatio-temporal hurricane impact analysis for power grid resilience studies. *2019 IEEE Power & Energy Society Innovative Smart Grid Technologies Conference (ISGT)*, 1–5.

16 Gholami, A., Shekari, T., Aminifar, F., and Shahidehpour, M. (2016). Microgrid scheduling with uncertainty: the quest for resilience. *IEEE Transactions on Smart Grid* 7 (6): 2849–2858.

17 Hussain, A., Bui, V.-H., and Kim, H.-M. (2018). A proactive and survivability-constrained operation strategy for enhancing resilience of microgrids using energy storage system. *IEEE Access* 6: 75495–75507.

18 Amirioun, M.H., Aminifar, F., and Lesani, H. (2018). Resilience-oriented proactive management of microgrids against windstorms. *IEEE Transactions on Power Systems* 33 (4): 4275–4284.

19 Wang, Z. and Wang, J. (2015). Self-healing resilient distribution systems based on sectionalization into microgrids. *IEEE Transactions on Power Systems* 30 (6): 3139–3149.

20 Wu, J. and Guan, X. (2013). Coordinated multi-microgrids optimal control algorithm for smart distribution management system. *IEEE Transactions on Smart Grid* 4 (4): 2174–2181.

21 Chen, B., Wang, J., Lu, X. et al. (2021). Networked microgrids for grid resilience, robustness, and efficiency: a review. *IEEE Transactions on Smart Grid* 12 (1): 18–32.

22 Farzin, H., Fotuhi-Firuzabad, M., and Moeini-Aghtaie, M. (2016). Enhancing power system resilience through hierarchical outage management in multi-microgrids. *IEEE Transactions on Smart Grid* 7 (6): 2869–2879.

23 Wang, Z., Chen, B., Wang, J., and Chen, C. (2016). Networked microgrids for self-healing power systems. *IEEE Transactions on Smart Grid* 7 (1): 310–319.

24 Olfati-Saber, R., Fax, J.A., and Murray, R.M. (2007). Consensus and cooperation in networked multi-agent systems. *Proceedings of the IEEE* 95 (1): 215–233.

25 Chen, C., Wang, J., Qiu, F., and Zhao, D. (2016). Resilient distribution system by microgrids formation after natural disasters. *IEEE Transactions on Smart Grid* 7 (2): 958–966.

26 Cao, W., Wu, J., Jenkins, N. et al. (2016). Operating principle of soft open points for electrical distribution network operation. *Applied Energy Barking Then Oxford*.

27 Lei, S., Chen, C., Song, Y., and Hou, Y. (2020). Radiality constraints for resilient reconfiguration of distribution systems: formulation and application to microgrid formation. *IEEE Transactions on Smart Grid* 11 (5): 3944–3956.

28 Ding, T., Lin, Y., Li, G., and Bie, Z. (2017). A new model for resilient distribution systems by microgrids formation. *IEEE Transactions on Power Systems* 32 (5): 4145–4147.

29 Lin, Y., Chen, B., Wang, J., and Bie, Z. (2019). A combined repair crew dispatch problem for resilient electric and natural gas system considering reconfiguration and DG islanding. *IEEE Transactions on Power Systems* 34 (4): 2755–2767.

30 Bian, Y., Chen, C., Huang, Y. et al. (2022). Service restoration for resilient distribution systems coordinated with damage assessment. *IEEE Transactions on Power Systems* 37 (5): 3792–3804.

31 IEEE PES AMPS DSAS Test Feeder Working Group (2014). 123-bus feeder. http://sites.ieee.org/pes-testfeeders/resources/ (accessed 1 November 2023).

11

In Situ Resilience Quantification for Microgrids

Priyanka Mishra, Peng Zhang, Scott A. Smolka, Scott D. Stoller, Yifan Zhou, Yacov A. Shamash, Douglas L. Van Bossuyt, and William W. Anderson Jr.

11.1 Introduction

11.1.1 Background and Motivation

Resilience quantification using probabilistic analysis [1] of high-impact low-probability (HILP) events [2] has been a mainstream approach for traditional interconnected power systems. These probabilistic approaches are not designed to be carried out in real time, which is necessary for a microgrid powered by uncertain distributed energy resources (DERs). In such a situation, a method that incorporates the time-varying operating conditions irrespective of the model is needed, and in situ resilience of the microgrid becomes critical [3].

"In situ resilience" refers to the ability of a system to reestablish functionality without relying on external resources [4]. To date, there do not exist any proactive operational strategies for quantifying the in situ resilience of microgrids. In [5], the concept of a "disturbance and impact resilience evaluation curve" is developed, which quantifies the resilience of a dynamical power system in terms of a *robustness degree*: the degree to which a microgrid can function correctly in the presence of stressed conditions post-disturbance [6]. However, the computation of the robustness degree of a microgrid is left as an open problem.

The literature contains several methods for evaluating the robustness of microgrids. One such approach [7] is to determine the size of the largest connected component post-disturbance. This is an offline approach that requires information on the number of generators or power plants connected to the microgrid. For the computation of robustness in real time, signal temporal logic (STL) is utilized in different ways. Standard STL robustness is used in [8] to quantify the extent to which a signal can be perturbed in space before affecting property satisfaction. In [9], STL time robustness is used to provide an equivalent notion of perturbation for Cyber-Physical Systems.

Inspired by these approaches, this chapter develops an STL-based technique for in situ resilience quantification of microgrids. Physically, resilience means robustness to disturbances (*invulnerability*) along with fast recovery (*recoverability*) by reducing non-robustness. Thus, a positive robustness degree corresponds to how long the property remains satisfied post-disturbance and can be used as a metric of invulnerability. Similarly, a negative robustness degree corresponds to how long the property remains unsatisfied post-violation and can be used as a recoverability metric.

Our main contribution is a novel STL-based method that captures system traces (e.g. time-series nodal voltages and active and reactive power), checks system robustness based on the STL requirements, and then determines in situ resilience of the microgrid system using the evaluation

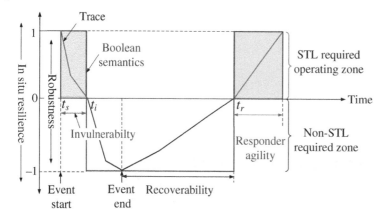

Figure 11.1 In situ resilience evaluation in terms of the positive degree of the robustness curve (Invulnerability) followed by the negative degree of robustness (Recoverability). Shaded areas indicate when the depicted system trace (measured signals) falls within the required operating zone. Upon an event occurrence (i.e. disturbance) at time t_s, the system seeks to behave resiliently by keeping the trace in the required operating zone. At t_i, the system can no longer withstand the disturbance, and thus the robustness degree becomes negative. After the event ends, the system succeeds at time t_r to recover to the desired operating zone.

procedure shown in Fig. 11.1. The novelty of this work lies in its: (1) incorporation of time-varying operating conditions; (2) independence from the microgrid model; and (3) its capability of quantifying in situ resilience in real time. The efficacy and effectiveness of our approach are demonstrated in the context of various operational scenarios in a representative microgrid.

11.2 STL-Enabled In Situ Resilience Evaluation

This section introduces in situ resilience quantification for microgrids, employing locally measured signals. First, for a given signal, the robustness degree is computed, which provides the extent to which the signal satisfies the STL requirement (Invulnerability), and subsequently violates the STL requirement (Recoverability). Then, a resilience quantification metric is devised based on the microgrid's robustness degrees of invulnerability and recoverability. The STL-based approach for resilience quantification is illustrated in Fig. 11.1 using a trace that is a collection of signals measured over time. The microgrid trace is passed through the STL monitoring process that evaluates it against a predefined formal requirement. This results in a positive robustness degree as long as the signal lies in the STL-required zone. Any violation of the STL requirement results in a negative robustness degree. The magnitude of the robustness degree indicates the invulnerability of the microgrid against a disturbance when positive, and recoverability when negative. These computed invulnerability and recoverability values can be combined to obtain an overall resilience quantification metric in one dimension.

11.2.1 Robustness Computation Using STL

The robustness degree for a given STL formula φ is evaluated over a trace σ for a given microgrid. STL formulas have formal syntax and semantics. The syntax describes the structure of STL formulas, while the semantics describes the meaning of the formulas and the rules to evaluate them.

The syntax of STL formulas is defined as [9]:

$$\varphi := p \mid \neg\varphi \mid \varphi_1 \wedge \varphi_2 \mid \varphi_1 \cup_\tau \varphi_2 \tag{11.1}$$

where p is an atomic proposition; φ_1 and φ_2 are STL formulae; $\neg$, $\wedge$, and $\cup$, respectively, denote Boolean negation, Boolean conjunction, and the until temporal operator. τ is an interval over $\mathbb{R}_{\geq 0}$, where $\mathbb{R}$ is the set of real numbers. Real-valued outputs measured over time are considered to be *signals*, whereas a collection of signals forms a *trace*. For example, if voltage $V(t)$, active power $P(t)$, and reactive power $Q(t)$ are three measured signals over time t, then $\sigma(t) = V(t), P(t), Q(t)$ is the corresponding trace. Intuitively, $\sigma(t)$ defines the behavior of the system over time t. φ_1, and φ_2 are STL formulas. STL formulas can be true or false as captured by STL's Boolean semantics, which provides the robustness degree for the satisfaction of STL formulas.

For a signal y in trace σ, let $y(\sigma(t))$ denote the value of y at time t. Given an STL formula φ and trace σ over time t, the *quantitative semantics* $\chi(\varphi, \sigma, t)$ is defined as:

$$\chi(y \geq 0, \sigma, t) = y(\sigma(t))$$

$$\chi(\neg\varphi, \sigma, t) = -\chi(\varphi, \sigma, t)$$

$$\chi(\varphi_1 \wedge \varphi_2, \sigma, t) = \min(\chi(\varphi_1, \sigma, t), \chi(\varphi_2, \sigma, t)) \tag{11.2}$$

$$\chi(\varphi_1 \cup_\tau \varphi_2, \sigma, t) = \max_{t' \in t+\tau} \min(\chi(\varphi_2, \sigma, t'), \min_{t \in t+t'} \chi(\varphi_1, \sigma, t''))$$

Such quantitative semantics provide a real value representing a quantitative measure of the satisfaction or violation of an STL formula φ. STL's Boolean semantics χ_B provide Boolean outcomes by capturing the satisfaction or violation of an STL formula φ.

Given an STL formula φ, trace σ, and time t, the *Boolean semantics* $\chi_B(\varphi, \sigma, t)$ is defined as [9]

$$\chi_B(\varphi, \sigma, t) = \begin{cases} 1, & (\sigma, t) \vDash \varphi \\ -1, & (\sigma, t) \nvDash \varphi \end{cases} \tag{11.3}$$

where $\vDash$ indicates that σ satisfies φ at time t; $\nvDash$ indicates σ does not satisfy φ at time t. See [9].

Given an STL formula φ and Boolean semantics χ_B for system trace σ over time t, *time robustness* θ^+ can be evaluated as [10]:

$$\theta^+(\varphi, \sigma, t) = \chi_B(\varphi, \sigma, t) \cdot \max\{\tau \geq 0 : \forall\, t' \in [t, t+\tau], \chi_B(\varphi, \sigma, t') = \chi_B(\varphi, \sigma, t)\} \tag{11.4}$$

where max is maximum. Time robustness, as defined by (11.4), will be positive/negative for as long as σ satisfies/violates φ starting from time t.

System requirements are formally defined in terms of STL formulas. Therefore, a negative robustness degree indicates the extent to which a system trace violates a given STL requirement. However, the robustness degree becomes positive as soon as the trace enters the STL-required zone. This is discussed further in Section 11.2.2.

11.2.2 STL Requirements for Microgrids

A requirement is a formal specification of the acceptable operation of a microgrid. Resiliency, which is related to stability, is the requirement that a microgrid always resumes stable operation post-disturbance. The notion of resiliency is temporal in nature and hence requires monitoring of the microgrid over time. STL, a temporal logic, is well-suited for reasoning about resiliency.

STL requirements of a microgrid are defined based on voltage stability criteria [11] during normal operation. Angle stability is not applicable to microgrids without rotating machines and long lines; it can be added if needed. Microgrid requirements can be formalized in STL based on their output signals: output voltage, active power, and reactive power. For the jth bus, STL requirement ϕ_j can be formulated as:

$$\phi_j = (P_j \leq P_{j_{max}}) \wedge (dP_j/dV_j < 0) \wedge (Q_{j_{min}} \leq Q_j)$$
$$\wedge (dQ_j/dV_j > 0) \wedge (V_{j_{min}} \leq V_j \leq V_{j_{max}}) \tag{11.5}$$

where $V_j(t)$, $P_j(t)$, and $Q_j(t)$ denote the jth bus voltage, output active power, and reactive power at time t, respectively; all of these values are tracked in trace σ_j. $P_{j_{max}}$, $Q_{j_{min}}$, $V_{j_{min}}$, and $V_{j_{max}}$ are the safe operating limits for P_j, Q_j, and V_j, respectively. Once the requirements are set, resilience can be quantified using robustness degree as discussed in Section 11.2.3.

11.2.3 In Situ Resilience Quantification Mechanism

In situ resilience is quantified in terms of invulnerability and recoverability (values I and R, respectively), both of which are based on the time robustness degree of STL requirements ϕ_j. In the process, resilience monitoring will be performed locally on a per-bus basis. For in situ resilience quantification, trace $\sigma(t) = V(t), P(t), Q(t)$ is passed through an STL monitoring process which upon the occurrence of an event/disturbance (at time t_s in Fig. 11.1) begins evaluating the robustness degree θ^+ of ϕ_j for the purpose of invulnerability and recoverability quantification. A positive θ^+ value indicates the microgrid's degree of invulnerability. As soon as θ^+ turns negative (at time t_i in Fig. 11.1), quantification of the microgrid's recoverability begins (ending at time t_r in Fig. 11.1). Collectively, a smaller (absolute value of) R and a larger I value represent greater resilience.

Invulnerability and recoverability values can be combined in a weighted sum to obtain an overall resilience metric ζ, with weight $\alpha \in (0, 1)$:

$$\zeta = \alpha \left(\frac{I}{I + |R|} \right) + (1 - \alpha) \left(\frac{I + |R|}{|R|} \right) \tag{11.6}$$

Modulus operator $| \cdot |$ is used to obtain a positive ζ, with $I + |R|$ serving as a normalization factor.

11.3 Case Study

11.3.1 Experimental Setup

We have implemented the proposed resilience monitoring algorithm in MATLAB R2021b using the Breach tool [12]. For a given trace, STL's Boolean semantics (11.3) of STL formula (11.5) is used, such that the measured outputs $V(t)$, $P(t)$, and $Q(t)$ are described as a finite sequence of time-stamped points. Such a sequence of points is considered to be piece-wise linear via interpolation in the Breach tool. The minima and maxima of these points are computed over a sliding time window through an optimal streaming algorithm. Breach can accommodate temporal aspects by employing the until operator in addition to the Boolean operators.

11.3.2 Experimental Results

The proposed in situ resilience evaluation method is applied to the CIGRE benchmark microgrid: a 12.47 kV microgrid with 8 DERs equipped with droop controllers [13], as shown in Fig. 11.2. The measured time-series data (i.e. trace) is collected at a time step of 1 ms. We set $\alpha = 0.5$, $V_{min} = 0.9$ p.u., and $V_{max} = 1.1$ p.u. Events such as DER and load disconnection represent considerable operational risks to the system. Quantification of the in situ resilience of the system to such disturbances is imperative so that the requisite action can be taken.

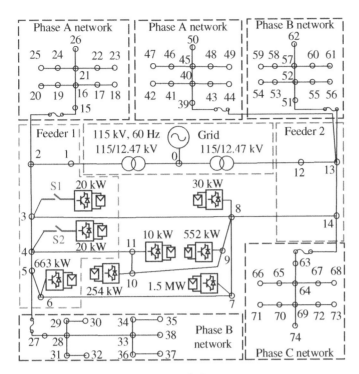

Figure 11.2 CIGRE microgrid with DERs.

11.3.2.1 DER Disconnection

Consider the event where two DERs, each with a 20-kW rating, are disconnected from buses 3 and 4, causing high-demand, low-generation conditions on the microgrid. The impact of this event on bus 3's V_3, dP_3/dV_3, and dQ_3/dV_3 in terms of their Boolean semantics is shown in Fig. 11.3(a)–(c), respectively. All of these signals are tracked in trace σ_3 and its Boolean semantics are shown in Fig. 11.3(e). The robustness of the system in terms of I is calculated while the Boolean semantics is 1; otherwise, R is calculated. For each I–$|R|$ pair, the resilience metric ζ is given in Fig. 11.3(d). ζ increases with an increase in I and attains a value $\zeta = 0.787$ for $I = 20$ and $|R| = 61$. This is reflective of the microgrid's resiliency and its ability to regain stability after the disconnection of the DERs.

11.3.2.2 Load Shedding

A step change of 27 kW in the pure resistive load (R_{Load}) and in the combined resistive and inductive load (RL_{Load}) of 16.8 kW and 12 kVAr, respectively, have been created at bus 3 to cause high-generation and low-demand conditions in the microgrid. Figure 11.4(a) shows the impact of several step load changes on bus 3's voltage levels and power flows. System robustness for these conditions is shown in Fig. 11.4(b). The microgrid requires low R to attain STL requirements (11.5) for RL_{Load}, indicating that the microgrid is more invulnerable to the sudden change in RL_{Load} than R_{Load}. However, the microgrid requires time to stabilize when a large load is suddenly disconnected. The results in Fig. 11.4(c)–(d) show that the microgrid exhibits a high I value (more resilient) and a high $|R|$ value (less resilient). This highlights the advantage of the STL-based method in quantifying resilience with changing loading conditions, rendering real-time resilience monitoring functionality for dynamical systems.

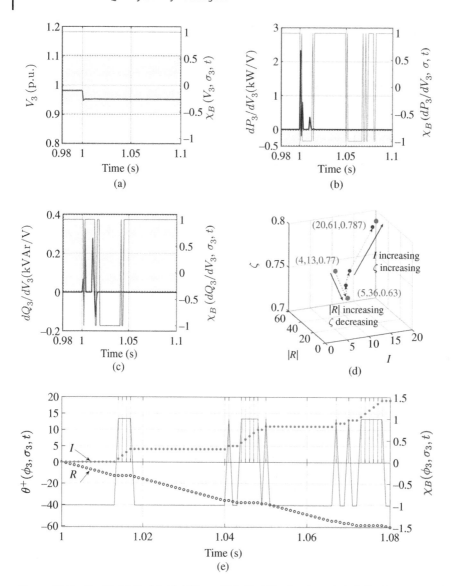

Figure 11.3 Bus 3's (a) V_3, (b) dP_3/dV_3, and (c) dQ_3/dV_3 signals and their Boolean semantics. (d) $(I, |R|, \zeta)$ triples. (e) Robustness degrees of I (filled circles) and $|R|$ (open circles) and Boolean semantics of trace σ_3. Shaded areas in (a)–(c) indicate STL-required operating zones. Gray vertical lines in (e) show +1 Boolean semantics.

11.3.3 Comparison with Existing Method

The performance of the proposed method is compared with the existing method [1] for DER disconnection in the given microgrid. For the quantification of the resilience metric (ξ) of the microgrid, the method in [1] computes invulnerability (I) and recoverability (R) using the total power delivered P_t and total power demand D_t at time t. Its definitions of I, R, and ξ are:

$$
\begin{aligned}
I &= \frac{P_{t_s}}{D_{t_s}} \\
R &= 1 - \frac{\sum_{t=t_i}^{t_r} D_t - P_t}{\sum_{t=t_i}^{t_r} D_t} \\
\xi &= \alpha I + (1 - \alpha)R
\end{aligned}
\tag{11.7}
$$

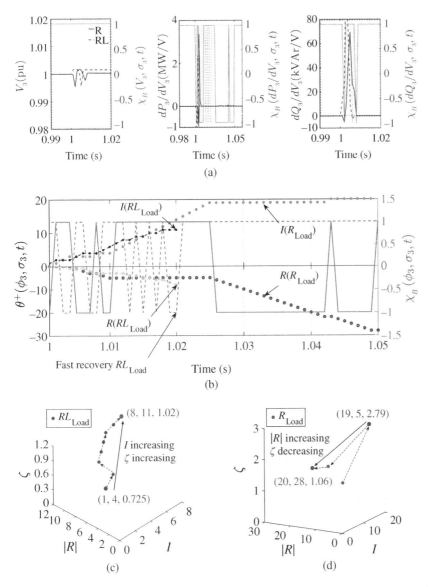

Figure 11.4 Bus 3's (a) V_3, dP_3/dV_3, and dQ_3/dV_3 signals and their Boolean semantics (dotted lines represent RL_{Load}); (b) Robustness degrees of I (filled octagonal stars R_{Load} and black, filled circles RL_{Load}) and $|R|$ (dark gray, open octagons R_{Load} and light gray, open octagons RL_{Load}), and Boolean semantics of trace σ_3; (c) $(I, |R|, \zeta)$ triples for RL_{Load} and (d) for R_{Load}. Shaded areas indicate STL-required operating zones.

where P_{t_s} and D_{t_s} are the total power delivered by the available sources and the demand, respectively, immediately after the disruption; t_s is the time of disruption. $t_r - t_i$ is the recovery period. The constant $\alpha \in [0, 1]$ is a weight. In situations where the information about power delivered or load demand for one or more nodes is not properly delivered to the microgrid operator, the computed I and R values will be erroneous. In contrast, the proposed method utilizes local data for resilience quantification. It thus performs local (per-bus) resiliency monitoring and monitored activities on one bus have no bearing on those of another bus. This highlights a significant difference between the two methods.

Performance comparison of the two methods is carried out using the microgrid of Fig. 11.2. A critical event is created, where two DERs, each with a 20 kW rating, are disconnected – from

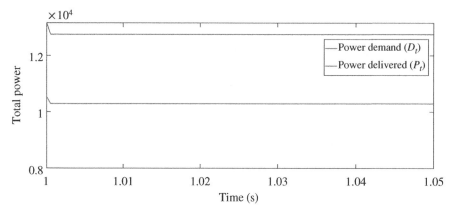

Figure 11.5 Total power and demand information acquired by microgrid operator in case-1.

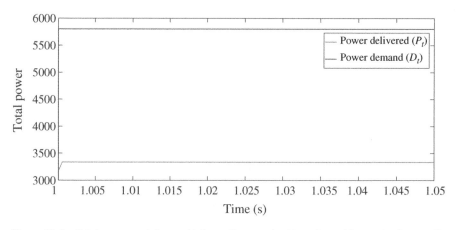

Figure 11.6 Total power and demand information acquired by microgrid operator in case-2.

buses 3 and 4, respectively – causing high-demand, low-generation conditions in the microgrid. Under such an event, two cases are considered. In case-1, the operator acquires all of the power-delivered and power-demanded information, as shown in Fig. 11.5. This results in $I = 0.806$, $R = 0.817$, and resilience metric value $\xi = 0.812$ for the microgrid (see Table 11.1). In case-2, the operator does not have access to the information on bus 3. Figure 11.6 is reflective of the smaller total-power-delivered and total-demand values acquired by the operator. Table 11.1 shows the computed $\xi = 0.59$, which is incorrect. These results confirm that the existing method incorrectly quantifies the resilience of the microgrid in the case of missing bus information.

The I, R, and resilience metric ζ values computed using the proposed method are given in Table 11.1. It is observed that for case-1, the values of $\zeta_3 = 0.787$ for bus 3 and $\zeta_4 = 0.81$ for bus 4. The average of the two ζ values is 0.7985, which is approximately equal to the resilience metric $\xi = 0.812$ obtained with [1]. This is because both methods are closely related to the duration of the power versus demand imbalance in the microgrid post-disturbance. The expression P_{t_s}/D_{t_s} for I increases as a function of the available power at the time of the disruption t_s. Presumably, the more available power at t_s, the longer the microgrid can remain invulnerable post-disturbance; i.e. the time robustness degree for I in the proposed formulation increases. Similarly, R depends on the expression $\sum_{t=t_i}^{t_r} D_t - P_t$. The faster P_t approaches D_t, the shorter the microgrid's recovery period; i.e.

Table 11.1 Comparative assessment of proposed and existing method [1].

Cases	Proposed method						Existing method		
	Bus 3 data			Bus 4 data			Global data		
	I_3	$|R_3|$	ζ_3	I_4	$|R_4|$	ζ_4	I	R	ξ
Case-1	20	61	0.787	21	58	0.81	0.806	0.817	0.812
Case-2	20	61	0.787	21	58	0.81	0.57	0.61	0.59

the time robustness degree for R in the proposed formulation decreases. The experimental results show that the two methods produce similar results when applied to the microgrid of [13]. This demonstrates the proposed approach's ability to accurately quantify the resilience of the microgrid.

Similarly, for case-2, the proposed method accurately calculates $\zeta_4 = 0.81$ using local bus 4 data and is not affected by the unknown state of bus 3. It is worth noting that [1] uses global information P_t and D_t, whereas the proposed approach employs local bus data for the ζ computation. The comparative performance analysis highlights the advantage of the proposed method over the existing technique in quantifying the resilience of the microgrid using only local data.

11.4 Conclusion

This chapter introduces an STL-based technique for estimating in real time in situ resilience of microgrids by computing the time robustness of an STL formula capturing the system's operational limits over a measured system trace. Case studies show the ability of the STL-based method not only to quantify in situ resilience but also to identify resilient microgrid trajectories in terms of invulnerability and recoverability. This makes it particularly useful for real-time resilience monitoring.

11.5 Exercises

1. Consider a variant of resilience in which a *recovery period* (from a violation of an STL property φ) is followed by a *durability period* (during which φ remains true). How would Fig. 11.1 change in this case?
2. Give quantitative semantics for atomic propositions of the form $y \geq y'$, where y and y' are signals in trace σ. Hint: The answer is similar to the quantitative semantics for $y \geq 0$ in Eq. (11.2).

11.6 Acknowledgment

This work is supported in part by U.S. Department of Navy award N00014-20-1-2858 and in part by U.S. Department of Navy award N00014-22-1-2001, both issued by the Office of Naval Research.

References

1 Giachetti, R.E., Bossuyt, D.L.V., Anderson, W.W., and Oriti, G. (2022). Resilience and cost trade space for microgrids on islands. *IEEE Systems Journal* 16 (3): 3939–3949.

2 Bie, Z., Lin, Y., Li, G., and Li, F. (2017). Battling the extreme: a study on the power system resilience. *Proceedings of the IEEE* 105 (7): 1253–1266.

3 Schneider, K.P., Tuffner, F.K., Elizondo, M.A. et al. (2017). Evaluating the feasibility to use microgrids as a resiliency resource. *IEEE Transactions on Smart Grid* 8 (2): 687–696.

4 Dell, B., Hopkins, A.J.M., and Lamont, B.B. (2012). *Resilience in Mediterranean-Type Ecosystems.* Dordrecht: Springer.

5 IEEE PES Task Force, Stanković, A.M., Tomsovic, K.L. et al. (2022). Methods for analysis and quantification of power system resilience. *IEEE Transactions on Power Systems* 38 (5): 4774–4787.

6 C/S2ESC (1990). *IEEE Standard Glossary of Software Engineering Terminology.*

7 Cuadra, L., Salcedo-Sanz, S., Del Ser, J. et al. (2015). A critical review of robustness in power grids using complex networks concepts. *Energies* 8 (9): 9211–9265.

8 Mehdipour, N. (2021). *Resilience for Satisfaction of Temporal Logic Specifications by Dynamical Systems.* Boston University ProQuest Dissertations Publishing.

9 Donzé, A. and Maler, O. (2010). Robust satisfaction of temporal logic over real-valued signals. In: *Proceedings of the Formal Modeling and Analysis of Timed Systems (FORMATS)*, Lecture Notes in Computer Science, vol. 6246 (ed. K. Chatterjee and T.A. Henzinger), 92–106. Berlin, Heidelberg: Springer-Verlag.

10 Maler, O. and Nickovic, D. (2004). Monitoring temporal properties of continuous signals. In: *Proceedings of the Formal Techniques, Modelling and Analysis of Timed and Fault-Tolerant Systems*, Lecture Notes in Computer Science, vol. 3253 (ed. Y. Lakhnech and S. Yovine), 152–166. Berlin, Heidelberg: Springer-Verlag.

11 Schiffer, J., Seel, T., Raisch, J., and Sezi, T. (2016). Voltage stability and reactive power sharing in inverter-based microgrids with consensus-based distributed voltage control. *IEEE Transactions on Control Systems Technology* 24 (1): 96–109.

12 Donzé, A., Ferrère, T., and Maler, O. (2013). Efficient robust monitoring for STL. In: *Proceedings of the Computer Aided Verification*, Lecture Notes in Computer Science, vol. 8044 (ed. N. Sharygina and H. Veith), 264–279. Berlin, Heidelberg: Springer-Verlag.

13 Strunz, K., Abbasi, E., Fletcher, R. et al. (2014). Benchmark systems for network integration of renewable and distributed energy resources. CIGRE, TF C6.04.02: TB 575, April 2014.

12

Distributed Voltage Regulation of Multiple Coupled Distributed Generation Units in DC Microgrids: An Output Regulation Approach

Tingyang Meng, Zongli Lin, Yan Wan, and Yacov A. Shamash

12.1 Introduction

Distributed generation (DG) units are usually small-capacity electric power generation units installed and operated in microgrids and connected directly to the loads rather than the power transmission network [1–4]. DG units can be used for peak shaving during the peak hours and as standby capacity during off-peak hours. They can also increase the reliability of power supply and the power quality, which are important for many industries. Compared to larger traditional power plants, construction of DG units is usually more flexible due to shorter construction time [5].

One of the fundamental power management problems for both DC and AC microgrids consisting of multiple DG units is to regulate the voltage while satisfying the desired load power (see, for example, [6–8]). In [8], an LMI-based decentralized voltage control scheme is proposed for DC islanded microgrids, and the design is demonstrated with constant resistance loads and a constant reference voltage. In [9], a voltage droop control strategy based on a consensus filter using communication among units is proposed. In [10], an optimal voltage regulator for DG units based on state feedback control and impedance shaping is presented. In [11], voltage stabilization controllers are proposed for DG units that are independent of the line parameters by solving an LMI problem. The aforementioned results mostly focus on constant loads. Compared to constant loads, time-varying loads, such as nonideal constant power loads (see, for example, [12, 13]) and loads from hourly power demand (see, for example, [14]), are more challenging for the control of DG units in microgrids. Time-varying loads may cause degradation of system performance and can even cause system instability [15].

In this chapter, we consider multiple DG units in an islanded DC microgrid. Each DG unit is assumed to have its own possibly time-varying load and is connected to and thus coupled with its neighboring units. Time-varying loads, such as pulsed loads, are considered due to the diverse DC microgrid application needs. Unlike most existing results, our design of control laws explicitly takes the time-varying loads into consideration and is shown to be able to maintain the output voltages of the units to desired reference voltages. More specifically, we formulate the voltage regulation problem of the multiple coupled DG units into a control problem and design the control law based on the output regulation theory in control theory (see, for example, [16, 17]). The ability to sustain time-varying loads, not just constant loads, also helps to increase the resilience of the microgrid.

This chapter is organized as follows. Section 12.2 establishes mathematical models for individual DG units and couplings among different units and presents the voltage regulation problem of

Microgrids: Theory and Practice, First Edition. Edited by Peng Zhang.
© 2024 The Institute of Electrical and Electronics Engineers, Inc. Published 2024 by John Wiley & Sons, Inc.

multiple coupled DG units to be solved in this chapter. Section 12.3 reviews the output regulation problem and its solution from control theory. Section 12.4 formulates the problem stated in Section 12.2 as an output regulation problem and presents a design of the control laws that solve the problem. Section 12.5 provides simulation examples to illustrate the implementation of the proposed control laws and verify their effectiveness. Section 12.7 contains exercise questions for the formulation of the control problem and the control design presented in the chapter. Section 12.6 concludes the chapter.

12.2 Problem Statement

In this section, we first establish the mathematical model for a single DG unit, which is coupled with other nearby DG units. We then formulate the distributed voltage regulation problem in a control theory framework.

Consider the electrical schematic of a single DG unit shown in Fig. 12.1. The unit consists of a DC power source (e.g. a solar panel, a wind turbine, and a battery), a buck converter, a possibly time-varying current load, and the transmission lines that couple the unit with its nearby DG units. The voltage of the DC power source is denoted by V_{DC}, the resistance and inductance of the line are denoted by R_i and L_i, respectively, the load current is denoted by $I_{\text{L}i}$, and the output capacitance is denoted by C_i.

By applying the Kirchoff's voltage and current laws, we have the following equations for DG unit i,

$$\begin{cases} C_i \dfrac{dV_i}{dt} = \displaystyle\sum_{i\in\mathcal{N}_i} I_{ij} + I_i - I_{\text{L}i}, \\[2mm] L_i \dfrac{dI_i}{dt} = U_i - I_i R_i - V_i \end{cases} \tag{12.1}$$

where V_i is the output voltage of the DG unit, I_i and U_i are respectively the output current and voltage of the buck converter, and I_{ij} is the current from the jth unit to the ith unit. The set $\mathcal{N}_i$ denotes the set of DG units that are directly connected to the ith unit.

Our objective in this chapter is to design a control law for each DG i that specifies the output voltage U_i of the buck converter so that the output voltage V_i of the DG unit is regulated to a reference output voltage, regardless of the time-varying current load $I_{\text{L}i}$ and the coupling current $\sum_{j\in\mathcal{N}_i} I_{ij}$ from neighboring units.

An electrical schematic of the coupling between DG unit i and DG unit j is shown in Fig. 12.2. The transmission line between the two units is assumed to have resistance R_{ij} and inductance L_{ij}. It is

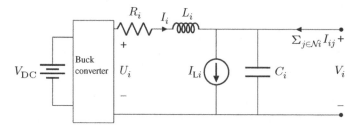

Figure 12.1 An electrical schematic of a DG unit.

Figure 12.2 The electrical scheme of the coupling between DG units i and j.

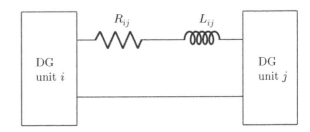

also assumed that $R_{ij} = R_{ji}$, and $L_{ij} = L_{ji}$. Given the direction of the current in the line connecting unit i and unit j, we have $I_{ji} = -I_{ij}$.

In order to reduce the complexity of the model, we apply the Quasi-Stationary Line approximation (see, for example, [8, 18]) to the transmission line coupling DG unit i and unit j, i.e. the inductance of the transmission line is assumed to be $L_{ij} = 0$. Then, we have

$$I_{ij}R_{ij} = V_j - V_i, j \in \mathcal{N}_i \tag{12.2}$$

where V_i and V_j are the output voltages of DG units i and j, respectively.

In view of Eqs. (12.1) and (12.2), the state space model for DG unit i is written as follows,

$$\begin{cases} \dot{x}_i = A_i x_i + \sum_{j \in \mathcal{N}_i} A_{ij} x_j + B_i u_i + M_i d_i, \\ y_i = H_i x_i \end{cases} \tag{12.3}$$

where $x_i = [V_i \, I_i]^{\mathrm{T}}$ is the state, $u_i = U_i$ is the control input, $y_i = V_i$ is the controlled voltage output, $d_i = I_{\mathrm{L}i}$ is the current load, and

$$A_i = \begin{bmatrix} \displaystyle\sum_{j \in \mathcal{N}_i} -\dfrac{1}{R_{ij}C_i} & \dfrac{1}{C_i} \\ -\dfrac{1}{L_i} & -\dfrac{R_i}{L_i} \end{bmatrix}, \quad A_{ij} = \begin{bmatrix} \dfrac{1}{R_{ij}C_i} & 0 \\ 0 & 0 \end{bmatrix},$$

$$B_i = \begin{bmatrix} 0 \\ \dfrac{1}{L_i} \end{bmatrix}, \quad M_i = \begin{bmatrix} -\dfrac{1}{C_i} \\ 0 \end{bmatrix}, \quad H_i = \begin{bmatrix} 1 & 0 \end{bmatrix}.$$

The time-varying current load and the reference voltage output of DG unit i are assumed to be generated by the following reference model, referred to as the exosystem,

$$\begin{cases} \dot{w}_i = S_i w_i, \\ d_i = D_i w_i, \\ r_i = -Q_i w_i \end{cases} \tag{12.4}$$

where $w_i \in \mathbb{R}^{s_i}$ is the state, $d_i \in \mathbb{R}$ represents the current load, and $r_i \in \mathbb{R}$ represents the reference output voltage and $S_i \in \mathbb{R}^{s_i \times s_i}$, $D_i \in \mathbb{R}^{1 \times s_i}$, and $Q_i \in \mathbb{R}^{1 \times s_i}$ are constant vectors.

Remark 12.1 The exosystem (12.4) is able to generate or approximate bound signals such as constant signals, sinusoidal signals, square waves, and pulses. In general, a larger size of S_i will provide a better approximation, with an increased computational cost in the construction and implementation of the control laws.

The distributed voltage regulation problem for multiple coupled DG units is formulated as follows.

Problem 12.1 Consider N coupled DG units in a DC microgrid, each subject to a time-varying current load generated by the exosystem (12.4). For each DG unit i, design a feedback control law u_i that utilizes its own state and the information from its neighboring units such that the output voltage $y_i = V_i$ of the closed-loop system is regulated to the reference voltage r_i generated by the exosystem.

12.3 Review of Output Regulation Theory

In this section, we recall the output regulation problem and its solution from control theory [16, 17, 19]. In Section 12.4, Problem 12.1 stated in Section 12.2 will be formulated and solved as an output regulation problem.

Consider a linear system

$$
\begin{cases}
\dot{x} = Ax + Bu + Pw, \\
\dot{w} = Sw, \\
e = Cx + Qw
\end{cases}
\tag{12.5}
$$

The first equation in (12.5) describes the plant to be controlled, with the state $x \in \mathbb{R}^n$, the control input $u \in \mathbb{R}^m$, the output $Cx \in \mathbb{R}^p$ and the disturbance $Pw \in \mathbb{R}^n$. The third equation defines the error $e \in \mathbb{R}^p$ between the output and the reference signal $-Qw$ that the plant output is required to track. The second equation represents an autonomous system, referred to as the exogenous system or the exosystem, with the state $w \in \mathbb{R}^s$. The exosystem models a class of disturbances and references taken into consideration. In particular, an unstable eigenvalue of S generates an exponentially or polynomially growing signal depending on whether its real part is strictly positive or not. A simple eigenvalue $\lambda = 0$ of S generates a step function with its magnitude determined by the initial condition $w(0)$. An eigenvalue $\lambda = 0$ with a corresponding Jordan block of size two generates a ramp function with its slope determined by the initial condition $w(0)$. A pair of simple eigenvalues of the form $\pm j\omega$ generates sinusoidal functions of frequency ω and various magnitudes and phases determined by different initial conditions $w(0)$. Finally, an eigenvalue of S with a negative real part generates signals that converge to zero with time. As a result, it is usually assumed, without loss of generality, that all eigenvalues of S have a non-negative real part.

The control problem is to achieve internal stability of the closed-loop system in the absence of the disturbance and the reference and, in the presence of the disturbance, to cause the plant output Cx to track the reference input $-Qw$ asymptotically, that is,

$$
\lim_{t \to \infty} e(t) = 0.
$$

Such a control problem is referred to as the output regulation problem. Output regulation rejects the disturbance and achieves output tracking. To solve the output regulation problem, we can adopt either a static state feedback law of the form

$$
u = Fx + Gw
\tag{12.6}
$$

or a dynamic error feedback law of the form

$$
\begin{cases}
\dot{\bar{x}} = \overline{A}\bar{x} + \overline{B}e, \ \bar{x} \in \mathbb{R}^l, \\
u = \overline{C}\bar{x} + \overline{D}e
\end{cases}
\tag{12.7}
$$

where $\overline{A}, \overline{B}, \overline{C}$, and $\overline{C}$ are constant matrices of appropriate dimensions.

The solution of the output regulation problem involves the following assumption.

Assumption 12.1 The system and the exosystem satisfy the following properties:

A1: All eigenvalues of S have a non-negative real part.

A2: The pair (A, B) is stabilizable.

A3: The pair

$$\left(\begin{bmatrix} C & Q \end{bmatrix}, \begin{bmatrix} A & P \\ 0 & S \end{bmatrix} \right)$$

is detectable.

A1 in Assumption 12.1 does not incur a loss of generality. As explained earlier, the presence of an eigenvalue with a negative real part contributes a component to the solution of the exosystem that converges to zero with time and thus does not affect the regulation of the output. A2 is necessary for stabilization of the plant by either state feedback or error feedback. A3 is needed for stabilization by error feedback.

The following result establishes necessary and sufficient conditions for the existence of a solution to the output regulation problem.

Proposition 12.1 Suppose that A1 and A2 in Assumption 12.1 hold. Then, the problem of output regulation is solvable by a state feedback law of the form (12.6) if and only if there exist matrices $\Pi \in \mathbb{R}^{n \times s}$ and $\Gamma \in \mathbb{R}^{m \times s}$ that solve the following linear matrix equations,

$$\begin{cases} \Pi S = A\Pi + B\Gamma + P, \\ \quad 0 = C\Pi + Q \end{cases} \tag{12.8}$$

Moreover, such a state feedback law is given by

$$u = Fx + (\Gamma - F\Pi)w \tag{12.9}$$

where F is any matrix such that $A + BF$ is Hurwitz.

If, in addition, A3 in Assumption 12.1 also holds, under the same condition, the output regulation problem is solvable by an error feedback of the form (12.7). Such an error feedback law is given by

$$\begin{cases} \begin{bmatrix} \dot{\hat{x}} \\ \dot{\hat{w}} \end{bmatrix} = \begin{bmatrix} A & P \\ 0 & S \end{bmatrix} \begin{bmatrix} \hat{x} \\ \hat{w} \end{bmatrix} + \begin{bmatrix} B \\ 0 \end{bmatrix} u \\ \qquad + \begin{bmatrix} L_A \\ L_S \end{bmatrix} \left(e - \begin{bmatrix} C & Q \end{bmatrix} \begin{bmatrix} \hat{x} \\ \hat{w} \end{bmatrix} \right) \\ u = F\hat{x} + (\Gamma - F\Pi)\hat{w}, \end{cases} \tag{12.10}$$

where F, L_A, and L_S are matrices such that $A + BF$ and

$$\begin{bmatrix} A + L_A C & P + L_A Q \\ L_S C & S + L_S Q \end{bmatrix}$$

are both Hurwitz.

Equation (12.8) are referred to as the regulator equations. The necessary and sufficient conditions for their solvability have been established. We recall the following result from [17].

Proposition 12.2 For any matrices P and Q, the regulator equations (12.8) are solvable for Π and Γ if and only if, for any eigenvalue λ of S,

$$\text{rank} \begin{bmatrix} A - \lambda I & B \\ C & 0 \end{bmatrix} = n + p \tag{12.11}$$

Finally, we will illustrate the solution of the output regulation problem with an example.

Example 12.1 Consider the longitudinal dynamics of an aircraft under certain flight condition

$$\begin{cases} \dot{x} = Ax + Bu + Mw, \\ y = Cx \end{cases} \tag{12.12}$$

with

$$A = \begin{bmatrix} 0 & 14.3877 & 0 & -31.5311 \\ -0.0012 & -0.4217 & 1.0000 & -0.0284 \\ 0.0002 & -0.3816 & -0.4658 & 0 \\ 0 & 0 & 1.0000 & 0 \end{bmatrix},$$

$$B = \begin{bmatrix} 0.7898 \\ -0.0059 \\ -0.2542 \\ 0 \end{bmatrix},$$

$$M = \begin{bmatrix} -0.6526 & -0.3350 & 0.4637 & 0.9185 & 0 \\ 0.0049 & 0.0025 & -0.0035 & -0.0068 & 0 \\ 0.2100 & 0.1078 & -0.1492 & -0.2956 & 0 \\ 0 & 0 & 0 & 0 & 0 \end{bmatrix},$$

$$C = \begin{bmatrix} 1 & 0 & 0 & 0 \end{bmatrix},$$

where the state consists of the velocity x_1 (feet/s, relative to the nominal flight condition), the angle of attack x_2 (degree), the pitch rate x_3 (degree/s), and the Euler angle rotation of aircraft about the inertial axis x_4 (degree), the control u (degree) is the elevator input, and d represents the external disturbance. The disturbance rejection problem for this system was studied in [20] as an output regulation problem with $Q = 0$. Suppose our design objective is now for the output of the system to track a constant input in the presence of external disturbances $w = \begin{bmatrix} w^{\mathrm{T}} & w_2^{\mathrm{T}} \end{bmatrix}^{\mathrm{T}}$, where $w_1 \in \mathbb{R}^2$ and $w_2 \in \mathbb{R}^2$ are sinusoidal with frequencies 0.1 and 0.3 rad/s, respectively. Clearly, this problem of output tracking in the presence of external disturbances can also be cast into an output regulation problem for system (12.5) with

$$S = \begin{bmatrix} 0 & -0.1 & 0 & 0 & 0 \\ 0.1 & 0 & 0 & 0 & 0 \\ 0 & 0 & 0 & -0.3 & 0 \\ 0 & 0 & 0.3 & 0 & 0 \\ 0 & 0 & 0 & 0 & 0 \end{bmatrix},$$

$$P = \begin{bmatrix} M & 0 \end{bmatrix},$$

$$Q = \begin{bmatrix} 0 & 0 & 0 & 0 & 1 \end{bmatrix}.$$

It is easy to verify that Assumptions A1, A2, and A3 are all satisfied. We next check the rank condition (12.11) as follows

$$\mathrm{rank} \begin{bmatrix} -\lambda & 14.3877 & 0 & -31.5311 & 0.7898 \\ -0.0012 & -0.4217 - \lambda & 1.0000 & -0.0284 & -0.0059 \\ 0.0002 & -0.3816 & -0.4658 - \lambda & 0 & -0.2542 \\ 0 & 0 & 1.0000 & -\lambda & 0 \\ 1 & 0 & 0 & 0 & 0 \end{bmatrix} = 5$$

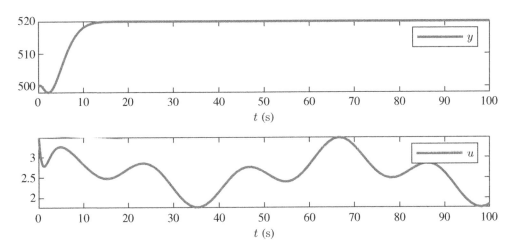

Figure 12.3 The control input and the output of the system under state feedback.

for all eigenvalues of S, that is, $\lambda = \pm j0.1, \pm j0.3, 0$. Thus, by Propositions 12.1 and 12.2, the output regulation problem is solvable.

Let us first construct a state feedback law that solves the output regulation problem. The solution to the output regulator equations (12.8) for this problem is given by

$$
\Pi = \begin{bmatrix} 0 & 0 & 0 & 0 & -1.0000 \\ 0.0001 & 0.0000 & -0.0001 & 0.0001 & 0.0028 \\ 0 & 0 & 0 & 0 & 0 \\ 0 & 0 & 0 & 0 & 0.0012 \end{bmatrix} \tag{12.13}
$$

$$
\Gamma = \begin{bmatrix} 0.8260 & 0.4241 & -0.5868 & -1.1630 & -0.0050 \end{bmatrix} \tag{12.14}
$$

The matrix F is designed as

$$
F = \begin{bmatrix} -0.0322 & -2.5525 & 5.1617 & 5.5793 \end{bmatrix} \tag{12.15}
$$

With the matrices Π, Γ, and F given, the state feedback law can be constructed as in (12.9). To illustrate the performance of this state feedback law, we simulate the closed-loop system with the initial conditions

$$
x(0) = \begin{bmatrix} 500 & 0 & 0 & 0 \end{bmatrix}^{\mathrm{T}}, \quad w(0) = \begin{bmatrix} 0.5 & 0 & -0.3 & 0 & -520 \end{bmatrix}^{\mathrm{T}}.
$$

Shown in Fig. 12.3 are the control input and the output of the system under the state feedback. Shown in Fig. 12.4 is the evolution of the states of the closed-loop system. Shown in Fig. 12.5 are the states of the exosystem.

Let us now construct an error feedback law that solves the same output regulation problem. The solution Π and Γ were given in (12.13) and (12.14). The matrix F was given in (12.15). The observer gains L_A and L_S are given as

$$
L_A = \begin{bmatrix} 2466.6337 \\ -6.6492 \\ 0.4745 \\ -2.4638 \end{bmatrix},
$$

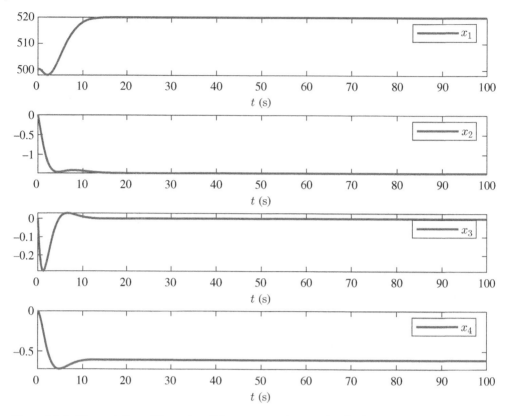

Figure 12.4 The evolution of the states of the system under state feedback.

$$
L_S = \begin{bmatrix} -0.6206 \\ -16.0569 \\ 1.0246 \\ 1.7754 \\ -2471.5062 \end{bmatrix}.
$$

With these matrices given, the error feedback law can be constructed as in (12.10). To illustrate the performance of this error feedback law, we simulate the closed-loop system with the initial conditions

$$
x(0) = \begin{bmatrix} 500 & 0 & 0 & 0 \end{bmatrix}^T, \quad w(0) = \begin{bmatrix} 0.5 & 0 & -0.2 & 0 & -520 \end{bmatrix}^T,
$$
$$
\hat{x}(0) = \begin{bmatrix} 500 & 0 & 0 & 0 \end{bmatrix}^T, \quad \hat{w}(0) = \begin{bmatrix} 0.45 & 0 & -0.15 & 0 & -520 \end{bmatrix}^T.
$$

Shown in Fig. 12.6 are the control input and the output of the system under the error feedback. Shown in Fig. 12.7 is the evolution of the states and the observed states of the system. Shown in Fig. 12.8 are the states and the observed states of the exosystem.

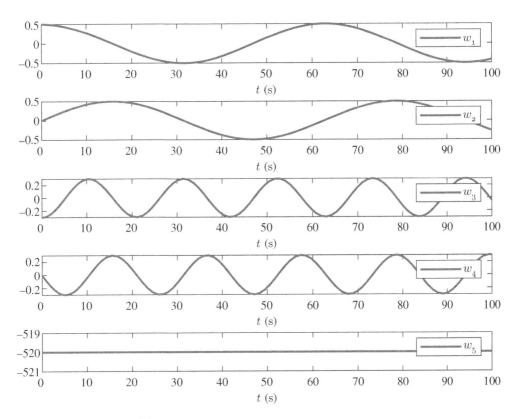

Figure 12.5 The states of the exosystem.

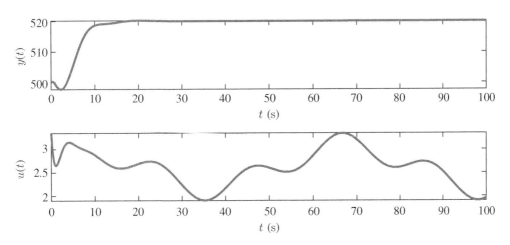

Figure 12.6 The control input and the output of the system under error feedback.

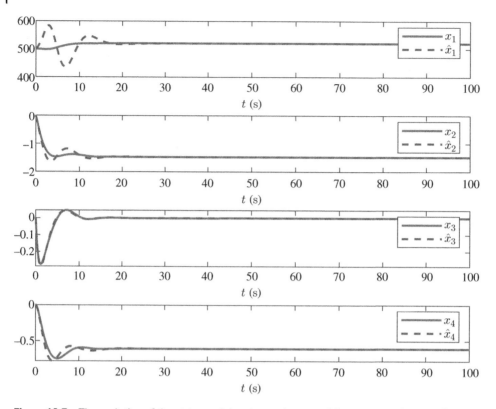

Figure 12.7 The evolution of the states and the observed states of the system under error feedback.

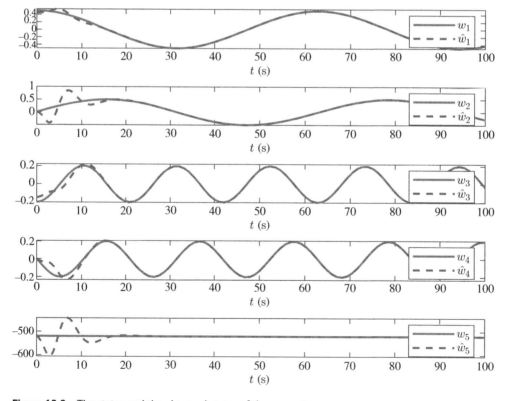

Figure 12.8 The states and the observed states of the exosystem.

12.4 Distributed Voltage Regulation in the Presence of Time-Varying Loads

Based on the state space model (12.3) of a single DG unit and its corresponding exosystem (12.4), the overall model of the microgrid consisting of N DG units is given as

$$\begin{cases} \dot{x} = Ax + Bu + Md, \\ y = Hx, \\ \dot{w} = Sw, \\ d = Dw, \\ r = -Qw \end{cases} \tag{12.16}$$

where

$$x = \begin{bmatrix} x_1 \\ x_2 \\ \vdots \\ x_N \end{bmatrix}, \quad u = \begin{bmatrix} u_1 \\ u_2 \\ \vdots \\ u_N \end{bmatrix}, \quad d = \begin{bmatrix} d_1 \\ d_2 \\ \vdots \\ d_N \end{bmatrix},$$

$$y = \begin{bmatrix} y_1 \\ y_2 \\ \vdots \\ y_N \end{bmatrix}, \quad w = \begin{bmatrix} w_1 \\ w_2 \\ \vdots \\ w_N \end{bmatrix}, \quad r = \begin{bmatrix} r_1 \\ r_2 \\ \vdots \\ r_N \end{bmatrix},$$

and

$$A = \begin{bmatrix} A_1 & A_{12} & \cdots & A_{1N} \\ A_{21} & A_2 & \cdots & A_{2N} \\ \vdots & \vdots & \ddots & \vdots \\ A_{N1} & A_{N2} & \cdots & A_N \end{bmatrix}, \quad B = \begin{bmatrix} B_1 & & & \\ & B_2 & & \\ & & \ddots & \\ & & & B_N \end{bmatrix},$$

$$M = \begin{bmatrix} M_1 & & & \\ & M_2 & & \\ & & \ddots & \\ & & & M_N \end{bmatrix}, \quad H = \begin{bmatrix} H_1 & & & \\ & H_2 & & \\ & & \ddots & \\ & & & H_N \end{bmatrix},$$

$$S = \begin{bmatrix} S_1 & & & \\ & S_2 & & \\ & & \ddots & \\ & & & S_N \end{bmatrix}, \quad D = \begin{bmatrix} D_1 & & & \\ & D_2 & & \\ & & \ddots & \\ & & & D_N \end{bmatrix},$$

$$Q = \begin{bmatrix} Q_1 & & & \\ & Q_2 & & \\ & & \ddots & \\ & & & Q_N \end{bmatrix}.$$

The distributed state feedback control laws for the DC units are designed in the following form,

$$u_i = F_i \left(x_i - \Pi_{ii} w_i - \sum_{j \in \mathcal{N}_i} \Pi_{ij} w_j \right) + \Gamma_{ii} w_i + \sum_{j \in \mathcal{N}_i} \Gamma_{ij} w_j,$$

$$i \in \{1, 2, \dots, N\} \tag{12.17}$$

where $F = \text{diag}\left\{F_1, F_2, \ldots, F_N\right\}$ with $F_i \in \mathbb{R}^{1 \times 2}, i \in \{1, 2, \ldots, N\}$, being the feedback gain matrices, is such that $A + BF$ is Hurwitz. Let $\Pi_{ij} \in \mathbb{R}^{2 \times s_i}$ and $\Gamma_{ij} \in \mathbb{R}^{1 \times s_i}, i, j \in \{1, 2, \ldots, N\}$, be such that

$$\Pi_{ij} = \begin{bmatrix} 0 & 0 & \cdots & 0 \\ 0 & 0 & \cdots & 0 \end{bmatrix}, \quad \Gamma_{ij} = \begin{bmatrix} 0 & 0 & \cdots & 0 \end{bmatrix},$$

if $j \notin \mathcal{N}_i \cup \{i\}$, for all $i \in \{1, 2, \ldots, N\}$. Let,

$$\Pi = \begin{bmatrix} \Pi_1 \\ \Pi_2 \\ \vdots \\ \Pi_N \end{bmatrix} = \begin{bmatrix} \Pi_{11} & \Pi_{12} & \cdots & \Pi_{1N} \\ \Pi_{21} & \Pi_{22} & \cdots & \Pi_{2N} \\ \vdots & \vdots & \ddots & \vdots \\ \Pi_{N1} & \Pi_{N2} & \cdots & \Pi_{NN} \end{bmatrix},$$

$$\Gamma = \begin{bmatrix} \Gamma_1 \\ \Gamma_2 \\ \vdots \\ \Gamma_N \end{bmatrix} = \begin{bmatrix} \Gamma_{11} & \Gamma_{12} & \cdots & \Gamma_{1N} \\ \Gamma_{21} & \Gamma_{22} & \cdots & \Gamma_{2N} \\ \vdots & \vdots & \ddots & \vdots \\ \Gamma_{N1} & \Gamma_{N2} & \cdots & \Gamma_{NN} \end{bmatrix}$$

be the solution to the regulator equations

$$\begin{cases} \Pi S = A\Pi + B\Gamma + MD, \\ 0 = H\Pi + Q \end{cases} \tag{12.18}$$

Based on the above discussion, we establish the following result on the solution of Problem 12.1.

Theorem 12.1 Consider the N-coupled DG units subject to time-varying current loads in a DC microgrid, as described by (12.16). The distributed state feedback control laws (12.17) solve Problem 12.1.

Proof: Under the control laws (12.17), the closed-loop system is given as

$$\dot{x} = Ax + BFx - BF\Pi w + B\Gamma w + MDw.$$

Consider the signal $\zeta = x - \Pi w$. Then,

$$\begin{aligned} \dot{\zeta} &= \dot{x} - \Pi\dot{w} \\ &= Ax + BF\zeta + B\Gamma w + MDw - \Pi Sw. \end{aligned}$$

Multiplying w to the right of both sides of the first equation of the regulator equations (12.18), we have

$$\Pi Sw = A\Pi w + B\Gamma w + MDw.$$

Thus,

$$\begin{aligned} \dot{\zeta} &= Ax + BF\zeta - A\Pi w \\ &= (A + BF)\zeta. \end{aligned}$$

Recalling that the choice of F is such that $A + BF$ is Hurwitz, we have

$$\lim_{t \to \infty} \zeta(t) = 0.$$

Since $Q = -H\Pi$ by the second equation of the regulator equations (12.18), the tracking error can be evaluated as

$$y - r = Hx + Qw$$
$$= Hx - H\Pi w$$
$$= H\zeta.$$

Therefore, we have

$$\lim_{t \to \infty} (y(t) - r(t)) = 0.$$

That is, in the presence of the loads, the output voltages of the DG units under the distributed state feedback control laws (12.17) will converge to the reference output voltages exponentially.

To implement the distributed state feedback control laws (12.17), we construct a state observer for each DG unit as follows,

$$\begin{cases} \dot{\hat{x}}_i = A_i\hat{x}_i + \sum_{j \in \mathcal{N}_i} A_{ij}\hat{x}_j + B_iu_i + M_id_i + L_{A,i}\left(\begin{bmatrix} C_i & 0 \\ 0 & Q_i \end{bmatrix} \begin{bmatrix} \hat{x}_i \\ \hat{w}_i \end{bmatrix} - \begin{bmatrix} y_i \\ r_i \end{bmatrix} \right), \\ \dot{\hat{w}}_i = S_i\hat{w}_i + L_{S,i}\left(\begin{bmatrix} C_i & 0 \\ 0 & Q_i \end{bmatrix} \begin{bmatrix} \hat{x}_i \\ \hat{w}_i \end{bmatrix} - \begin{bmatrix} y_i \\ r_i \end{bmatrix} \right), \\ \qquad\qquad i \in \{1, 2, \dots, N\}, \end{cases}$$

where $\hat{x}_i$ and $\hat{w}_i$ are the estimate of x_i and w_i, respectively, and $L_{A,i}$ and $L_{S,i}$ are such that $A_{ob} + L_{ob}C_{ob}$ is Hurwitz, with

$$A_{ob} = \begin{bmatrix} A_{ob,ij} \end{bmatrix}, \quad L_{ob} = \text{blkdiag}\{L_{ob,i}\}, \quad C_{ob} = \text{blkdiag}\{C_{ob,i}\},$$

$$A_{ob,ii} = \begin{bmatrix} A_i & M_iD_i \\ 0 & S_i \end{bmatrix}, \quad A_{ob,ij} = \begin{bmatrix} A_{ij} & 0 \\ 0 & 0 \end{bmatrix},$$

$$L_{ob,i} = \begin{bmatrix} L_{A,i} \\ L_{S,i} \end{bmatrix}, \quad C_{ob,i} = \begin{bmatrix} C_i & 0 \\ 0 & Q_i \end{bmatrix}.$$

We note that, differently from the error feedback design reviewed in Section 12.3, where only the error between the output and the reference is available for feedback, here, we have assumed that the output, the reference, and the load are all available for feedback. We refer to our design here as the output feedback design.

12.5 Simulation Results

In the simulation, we consider a DC microgrid consisting of three coupled DG units. Each unit has its own possibly time-varying current load and the three units are coupled through the transmission lines. The parameters of the components in this DC Microgrid are given as $C_1 = C_2 = C_3 = 2.2\,\text{mF}$, $L_1 = L_2 = L_3 = 1.8\,\text{mH}$, $R_1 = R_2 = R_3 = 0.2\,\Omega$, $L_{12} = L_{23} = 1.8\,\mu\text{H}$, $R_{12} = R_{23} = 0.05\,\Omega$. The output voltages of the DC voltage sources are assumed to be $V_{DC} = 100\,\text{V}$. The output voltages of the buck converters are specified by their duty cycles.

Given the parameters of the components, the mathematical models of the three DG units are calculated as

$$A_1 = A_3 = \begin{bmatrix} -9091 & 454.5 \\ -555.6 & -111.1 \end{bmatrix}, \quad A_2 = \begin{bmatrix} -18182 & 454.5 \\ -555.6 & -111.1 \end{bmatrix},$$

$$A_{12} = A_{21} = A_{23} = A_{32} = \begin{bmatrix} 9091 & 0 \\ 0 & 0 \end{bmatrix},$$

$$B_1 = B_2 = B_3 = \begin{bmatrix} 0 \\ 555.6 \end{bmatrix},$$

$$M_1 = M_2 = M_3 = \begin{bmatrix} -454.5 \\ 0 \end{bmatrix},$$

$$H_1 = H_2 = H_3 = \begin{bmatrix} 1 & 0 \end{bmatrix}.$$

The electrical schematic of the DC microgrid is constructed and simulated in Matlab Simulink by using the Simscape Electrical Block Libraries.

Simulation is performed for both state feedback and output feedback designs. For each design, three different scenarios are considered. In Scenario 1, each DG unit is subject to a constant current load. In Scenario 2, each DG unit is subject to a sinusoidal current load. In Scenario 3, each DC unit is subject to a pulsed current load. It is noted that the control laws are capable of regulating the output voltage to a time-varying reference voltage generated by the exosystem. However, since a constant output voltage is desired in most applications, we will only present simulation results for a constant reference voltage in this section.

12.5.1 State Feedback Design

12.5.1.1 Scenario 1: Constant Loads

In this scenario, the three DG units are subjected to constant current loads $I_{L1} = 5\,\text{A}$, $I_{L2} = 4\,\text{A}$, and $I_{L3} = 3\,\text{A}$. The reference output voltages are set to be $r_1 = r_2 = r_3 = 40\,\text{V}$. The exosystems are described by

$$S_1 = S_2 = S_3 = \begin{bmatrix} 0 & 0 \\ 0 & 0 \end{bmatrix},$$

$$Q_1 = Q_2 = Q_3 = \begin{bmatrix} 0 & -1 \end{bmatrix},$$

$$D_1 = D_2 = D_3 = \begin{bmatrix} 1 & 0 \end{bmatrix},$$

with initial conditions

$$w_1(0) = \begin{bmatrix} 5 \\ 40 \end{bmatrix}, \quad w_2(0) = \begin{bmatrix} 4 \\ 40 \end{bmatrix}, \quad w_3(0) = \begin{bmatrix} 3 \\ 40 \end{bmatrix}.$$

The solution to the regulator equations (12.18) is given as

$$\Pi_1 = \begin{bmatrix} 1.0000 & 0 & 0 & 0 & 0 & 0 \\ 20.0022 & 1.0000 & -20.0022 & 0 & 0 & 0 \end{bmatrix},$$

$$\Pi_2 = \begin{bmatrix} 0 & 0 & 1.0000 & 0 & 0 & 0 \\ -20.0022 & 0 & 20.0022 & 0.8000 & -20.0022 & 0 \end{bmatrix},$$

$$\Pi_3 = \begin{bmatrix} 0 & 0 & 0 & 1.0000 & 0 & 0 \\ 0 & 0 & -20.0022 & 0 & 20.0022 & 0.6000 \end{bmatrix},$$

$$\Gamma_1 = \begin{bmatrix} 4.9997 & 0.2000 & -3.9997 & 0 & 0 & 0 \end{bmatrix},$$
$$\Gamma_2 = \begin{bmatrix} -3.9997 & 0 & 4.9997 & 0.1600 & -3.9997 & 0 \end{bmatrix},$$
$$\Gamma_3 = \begin{bmatrix} 0 & 0 & -3.9997 & 0 & 4.9997 & 0.1200 \end{bmatrix}.$$

The feedback gain matrices are chosen as

$$F_1 = F_2 = F_3 = \begin{bmatrix} -470.3175 & -55.4336 \end{bmatrix}.$$

It is noted that the control input u_i of DC unit i specifies the output voltage of the buck converter. In order for the buck converter to reach the desired output voltage, we calculate the control input in the form of duty cycle $U_{\mathrm{DC}i}$ as

$$U_{\mathrm{DC}i} = \frac{U_i}{V_{\mathrm{DC}}} = \frac{1}{V_{\mathrm{DC}}} \left(F_i(x_i - \Pi_i w) + \Gamma_i w \right) \tag{12.19}$$

Simulation is performed with the initial conditions

$$x_1(0) = \begin{bmatrix} 40 \\ 5 \end{bmatrix}, \quad x_2(0) = \begin{bmatrix} 40 \\ 5 \end{bmatrix}, \quad x_3(0) = \begin{bmatrix} 40 \\ 5 \end{bmatrix}.$$

Shown in Figs. 12.9–12.11 are respectively the control inputs, the output voltages, and the currents in the DG units. The output voltages are regulated to 40V under the influence of the constant current loads.

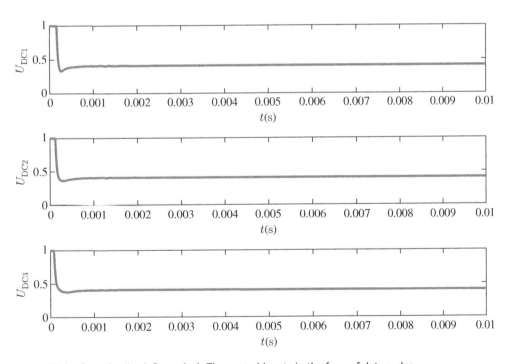

Figure 12.9 State feedback Scenario 1: The control inputs in the form of duty cycles.

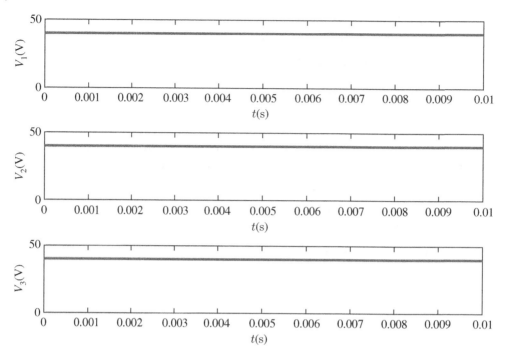

Figure 12.10 State feedback Scenario 1: The output voltages.

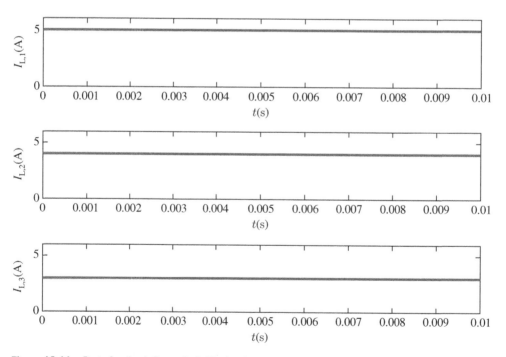

Figure 12.11 State feedback Scenario 1: The load currents.

12.5.1.2 Scenario 2: Sinusoidal Loads

In this scenario, the three DG units are subject to sinusoidal current loads, $I_{L1}(t) = 20 + 10\sin(5t)$, $I_{L2}(t) = 20 + 10\sin(4t)$, and $I_{L2}(t) = 20 + 10\sin(7t)$. The reference voltages are set to be $r_1 = r_2 = r_3 = 40$ V. The exosystems in this case are described by

$$
S_1 = \begin{bmatrix} 0 & 0 & 0 \\ 0 & 0 & 5 \\ 0 & -5 & 0 \end{bmatrix}, \quad S_2 = \begin{bmatrix} 0 & 0 & 0 \\ 0 & 0 & 4 \\ 0 & -4 & 0 \end{bmatrix}, \quad S_3 = \begin{bmatrix} 0 & 0 & 0 \\ 0 & 0 & 7 \\ 0 & -7 & 0 \end{bmatrix},
$$

$$
Q_1 = Q_2 = Q_3 = \begin{bmatrix} -1 & 0 & 0 \end{bmatrix},
$$

$$
D_1 = D_2 = D_3 = \begin{bmatrix} 0.5 & 2 & 0 \end{bmatrix},
$$

with the initial conditions

$$
w_1(0) = w_2(0) = w_3(0) = \begin{bmatrix} 40 \\ 0 \\ 5 \end{bmatrix}.
$$

We note that the zero in the $(1, 1)$ element in the exosystem matrices S_1, S_2, and S_3 generates the constant components in both the reference voltages and the current loads. A separate zero block can be amended to these matrices as blkdiag$\{0, S_1\}$, blkdiag$\{0, S_2\}$, and blkdiag$\{0, S_3\}$ to allow flexibility in generating different constant components in the reference voltages and the current loads.

The solution to the regulator equations (12.18) is given as

$$
\Pi_1 = \begin{bmatrix} 1.0000 & 0 & 0 & 0 & 0 & 0 & 0 & 0 \\ 20.1022 & 0.1000 & 0 & -20.0022 & 0 & 0 & 0 & 0 \end{bmatrix},
$$

$$
\Pi_2 = \begin{bmatrix} 0 & 0 & 0 & 1.0000 & 0 & 0 & 0 & 0 \\ -20.00220 & 0 & 20.1022 & 0.1000 & -20.0022 & 0 & 0 & 0 \end{bmatrix},
$$

$$
\Pi_3 = \begin{bmatrix} 0 & 0 & 0 & 0 & 0 & 0 & 1.0000 & 0 & 0 \\ 0 & 0 & 0 & -20.0022 & 0 & 0 & 20.1022 & 0.1000 & 0 \end{bmatrix},
$$

$$
\Gamma_1 = \begin{bmatrix} 5.0197 & 0.0200 & 0.0009 & -3.9997 & 0 & 0 & 0 & 0 \end{bmatrix},
$$

$$
\Gamma_2 = \begin{bmatrix} -3.9997 & 0 & 0 & 5.0197 & 0.0200 & 0.0007 & -3.9997 & 0 & 0 \end{bmatrix},
$$

$$
\Gamma_3 = \begin{bmatrix} 0 & 0 & 0 & -3.9997 & 0 & 0.0013 & 5.0197 & 0.0200 & 0 \end{bmatrix},
$$

and the feedback gain matrices F_1, F_2, and F_3 are chosen the same as in Scenario 1 above.

Simulation is performed with the initial conditions

$$
x_1(0) = \begin{bmatrix} 40 \\ 5 \end{bmatrix}, \quad x_2(0) = \begin{bmatrix} 40 \\ 5 \end{bmatrix}, \quad x_3(0) = \begin{bmatrix} 40 \\ 5 \end{bmatrix}.
$$

Shown in Figs. 12.12–12.14 are respectively the control inputs in the form of duty cycles, the output voltages, and the currents in the DG units. It can be observed that the voltages are regulated to 40V under the influence of the sinusoidal current loads.

12.5.1.3 Scenario 3: Pulsed Loads

In this scenario, the three DG units are subject to pulsed load currents, as shown in Fig. 12.17, and a constant reference voltage $r_1 = r_2 = r_3 = 40$ V. The pulsed current loads are approximated by the

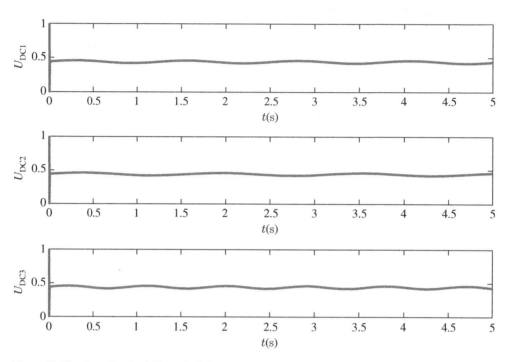

Figure 12.12 State feedback Scenario 2: The control inputs in the form of duty cycles.

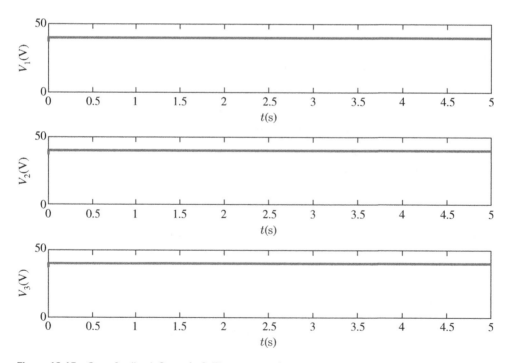

Figure 12.13 State feedback Scenario 2: The output voltages.

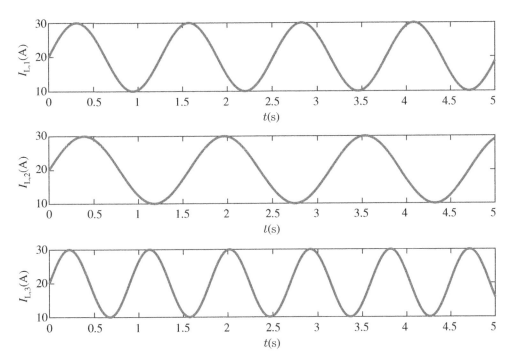

Figure 12.14 State feedback Scenario 2: The load currents.

first five frequency components in its Fourier series, i.e.

$$I_{L1}(t) = 8 + 6\frac{4}{\pi}\sum_{k=1}^{5}\frac{1}{2k-1}\sin((2k-1)\pi t),$$

$$I_{L2}(t) = 6 + 3\frac{4}{\pi}\sum_{k=1}^{5}\frac{1}{2k-1}\sin((2k-1)\pi t),$$

$$I_{L3}(t) = 10 + \frac{4}{\pi}\sum_{k=1}^{5}\frac{1}{2k-1}\sin((2k-1)\pi t),$$

each of which is generated by an exosystem. The matrices S_1, S_2, and S_3 that describe the exosystems, in this case, are constructed as block diagonal matrices, with their first block being a scalar zero, and their kth block, $k - 2, 3, \ldots, 6$, being

$$\begin{bmatrix} 0 & (2k-3)\pi \\ -(2k-3)\pi & 0 \end{bmatrix}.$$

The other matrices that describe the exosystems are

$$Q_1 = Q_2 = Q_3 = \begin{bmatrix} -40\ 0\ 0\ 0\ 0\ 0\ 0\ 0\ 0\ 0\ 0 \end{bmatrix},$$

$$D_1 = \begin{bmatrix} 8\ 6\ 0\ 6\ 0\ 6\ 0\ 6\ 0\ 6\ 0 \end{bmatrix},$$

$$D_2 = \begin{bmatrix} 6\ 3\ 0\ 3\ 0\ 3\ 0\ 3\ 0\ 3\ 0 \end{bmatrix},$$

$$D_3 = \begin{bmatrix} 10\ 4\ 0\ 4\ 0\ 4\ 0\ 4\ 0\ 4\ 0 \end{bmatrix},$$

and the initial conditions of the exosystems are given as

$$w_1(0) = w_2(0) = w_3(0) = \begin{bmatrix} 1\ 0\ \dfrac{4}{\pi}\ 0\ \dfrac{4}{3\pi}\ 0\ \dfrac{4}{5\pi}\ 0\ \dfrac{4}{7\pi}\ 0\ \dfrac{4}{9\pi} \end{bmatrix}^{\mathrm{T}}.$$

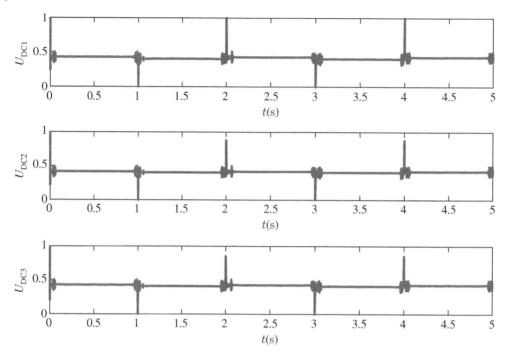

Figure 12.15 State feedback Scenario 3: The control inputs in the form of duty cycles.

With the solution to the regulator equations (12.18) and the same feedback gain matrices as in Scenario 1, the duty cycles $U_{\mathrm{DC}i}$, the control inputs, are calculated in the same way as in (12.19).

Simulation is performed with the initial conditions

$$x_1(0) = \begin{bmatrix} 40 \\ 5 \end{bmatrix}, \quad x_2(0) = \begin{bmatrix} 40 \\ 5 \end{bmatrix}, \quad x_3(0) = \begin{bmatrix} 40 \\ 5 \end{bmatrix}.$$

Shown in Figs. 12.15–12.17 are respectively the control inputs in the form of duty cycles, the output voltages, and the pulsed load currents and their approximations. It can be observed that the voltages are regulated to 40V even under the influence of the time-varying current loads.

12.5.2 Output Feedback Design

In this subsection, we present simulation results for output feedback design. As in the state feedback design, we will consider the same three different scenarios.

12.5.2.1 Scenario 1: Constant Loads

In this scenario, we consider the case where the three DG units are subject to constant current loads $I_{L1} = 1.5\,\mathrm{A}, I_{L2} = 1\,\mathrm{A}$, and $I_{L3} = 2\,\mathrm{A}$. The reference output voltages are set to be $r_1 = r_2 = r_3 = 40\,\mathrm{V}$. The exosystems are the same as in Scenarios 1 in the state feedback design with same initial conditions

$$w_1(0) = \begin{bmatrix} 40 \\ 5 \end{bmatrix}, \quad w_2(0) = \begin{bmatrix} 40 \\ 5 \end{bmatrix}, \quad w_3 = \begin{bmatrix} 40 \\ 5 \end{bmatrix}.$$

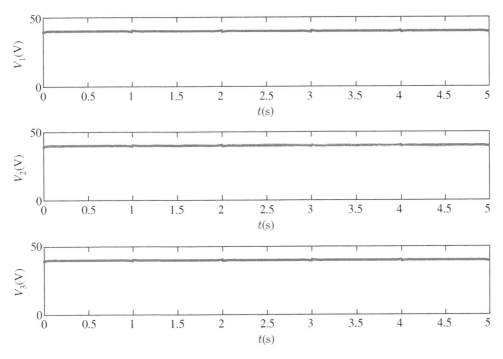

Figure 12.16 State feedback Scenario 3: The output voltages.

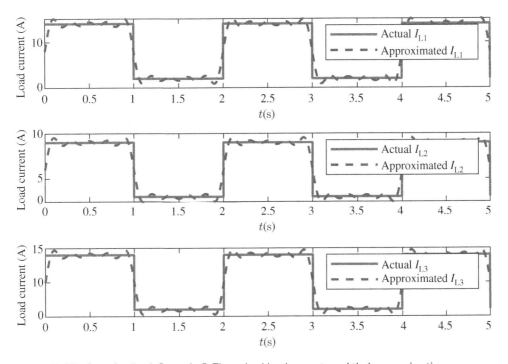

Figure 12.17 State feedback Scenario 3: The pulsed load currents and their approximations.

The feedback gains F_1, F_2, and F_3 are chosen the same as in Scenario 1 of the state feedback design. The observer gains are chosen as

$$L_1 = \begin{bmatrix} L_{A,1} \\ L_{S,1} \end{bmatrix} = 10^6 \times \begin{bmatrix} -0.0046 & 0 \\ 0.2812 & 0 \\ 0 & 0.0020 \\ 1.1618 & 0 \end{bmatrix},$$

$$L_2 = \begin{bmatrix} L_{A,2} \\ L_{S,2} \end{bmatrix} = 10^6 \times \begin{bmatrix} -0.0046 & 0 \\ 0.2812 & 0 \\ 0 & 0.0020 \\ 1.7427 & 0 \end{bmatrix},$$

$$L_3 = \begin{bmatrix} L_{A,3} \\ L_{S,3} \end{bmatrix} = 10^5 \times \begin{bmatrix} -0.0460 & 0 \\ 2.81116 & 0 \\ 0 & 0.0200 \\ 8.7137 & 0 \end{bmatrix}.$$

Simulation is performed with the initial conditions

$$x_1(0) = \begin{bmatrix} 40 \\ 5 \end{bmatrix}, \quad x_2(0) = \begin{bmatrix} 40 \\ 5 \end{bmatrix}, \quad x_3(0) = \begin{bmatrix} 40 \\ 5 \end{bmatrix},$$

$$\hat{x}_1(0) = \begin{bmatrix} 50 \\ 0 \end{bmatrix}, \quad \hat{x}_2(0) = \begin{bmatrix} 50 \\ 0 \end{bmatrix}, \quad \hat{x}_3(0) = \begin{bmatrix} 50 \\ 0 \end{bmatrix},$$

$$\hat{w}_1(0) = \begin{bmatrix} 40 \\ 5 \end{bmatrix}, \quad \hat{w}_2(0) = \begin{bmatrix} 40 \\ 5 \end{bmatrix}, \quad \hat{w}_3(0) = \begin{bmatrix} 40 \\ 5 \end{bmatrix}.$$

Shown in Fig. 12.18–12.23 are respectively the control inputs in the form of duty cycles, the output voltages, the states of the DG units, the states of the exosystems, the observed states of the DG unites, and the observed states of the exosystems.

12.5.2.2 Scenario 2: Sinusoidal Loads

In this scenario, the three DG units are subject to sinusoidal current load, $I_{L1}(t) = 8 + 5\sin(5t)$, $I_{L2}(t) = 8 + 5\sin(4t)$, and $I_{L2}(t) = 8 + 5\sin(7t)$. The reference voltages are set to be $r_1 = r_2 = r_3 = 40$ V. The exosystems in this case can be described by

$$S_1 = \begin{bmatrix} 0 & 0 & 0 \\ 0 & 0 & 5 \\ 0 & -5 & 0 \end{bmatrix}, \quad S_2 = \begin{bmatrix} 0 & 0 & 0 \\ 0 & 0 & 4 \\ 0 & -4 & 0 \end{bmatrix}, \quad S_3 = \begin{bmatrix} 0 & 0 & 0 \\ 0 & 0 & 7 \\ 0 & -7 & 0 \end{bmatrix},$$

$$Q_1 = Q_2 = Q_3 = \begin{bmatrix} -1 & 0 & 0 \end{bmatrix},$$

$$D_1 = D_2 = D_3 = \begin{bmatrix} 0.2 & 1 & 0 \end{bmatrix},$$

with the initial conditions

$$w_1(0) = w_2(0) = w_3(0) = \begin{bmatrix} 40 \\ 0 \\ 5 \end{bmatrix}.$$

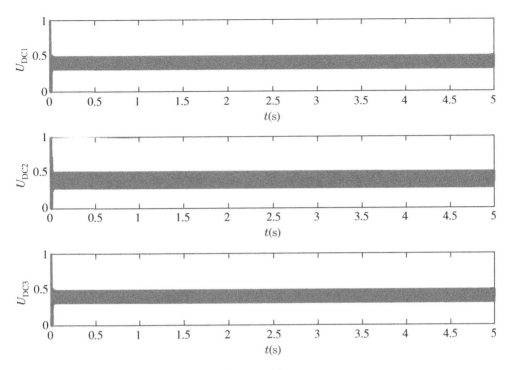

Figure 12.18 Output feedback Scenario 1: The control inputs.

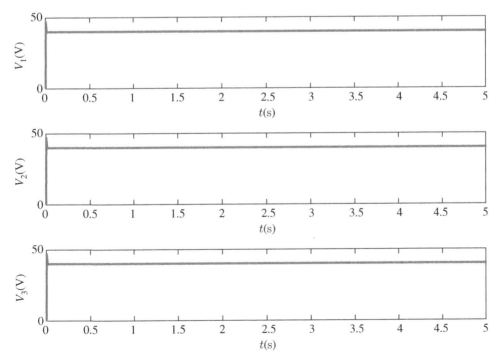

Figure 12.19 Output feedback Scenario 1: The output voltages.

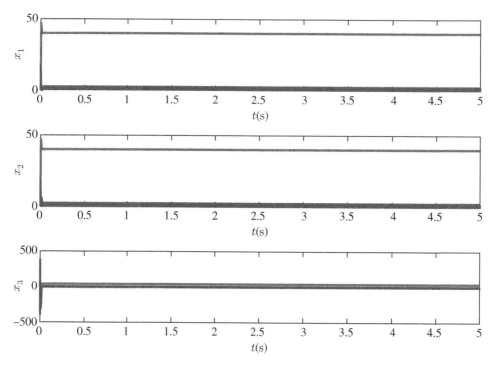

Figure 12.20 Output feedback Scenario 1: The states of the DG units.

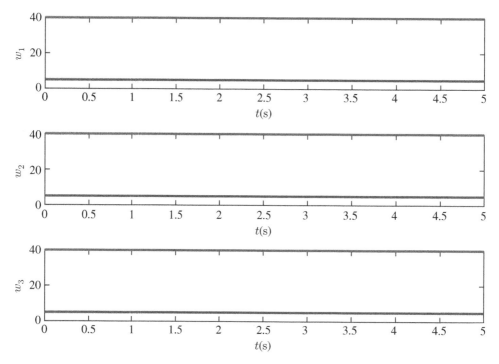

Figure 12.21 Output feedback Scenario 1: The states of the exosystems.

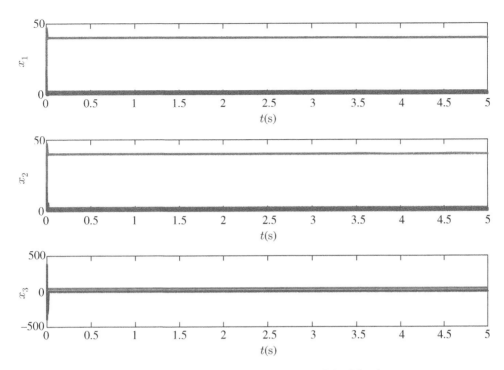

Figure 12.22 Output feedback Scenario 1: The observed states of the DG units.

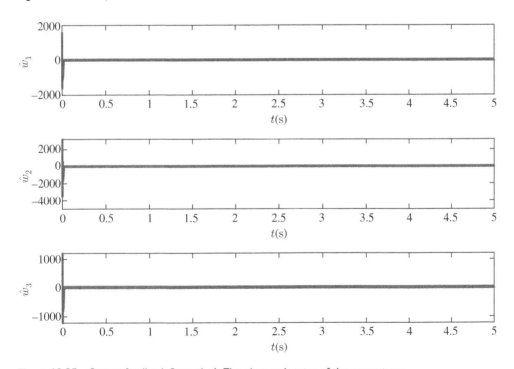

Figure 12.23 Output feedback Scenario 1: The observed states of the exosystems.

The solution to the regulator equations (12.18) is given as

$$\Pi_1 = \begin{bmatrix} 1.0000 & 0 & 0 & 0 & 0\ 0\ 0\ 0\ 0 \\ 20.2022 & 1.0000 & 0 & -20.0022 & 0\ 0\ 0\ 0\ 0 \end{bmatrix},$$

$$\Pi_2 = \begin{bmatrix} 0 & 0 & 0 & 1.0000 & 0 & 0 & 0\ 0\ 0 \\ -20.00220 & 0 & 40.2044 & 1.0000 & 0 & -20.0022 & 0\ 0\ 0 \end{bmatrix},$$

$$\Pi_3 = \begin{bmatrix} 0\ 0\ 0 & 0 & 0\ 0 & 1.0000 & 0 & 0 \\ 0\ 0\ 0 & -20.0022 & 0\ 0 & 20.2022 & 1.0000 & 0 \end{bmatrix},$$

$$\Gamma_1 = \begin{bmatrix} 5.0397 & 0.2000 & 0.0090 & -3.9997 & 0\ 0\ 0\ 0\ 0 \end{bmatrix},$$

$$\Gamma_2 = \begin{bmatrix} -3.9997 & 0\ 0 & 9.0394 & 0.2000 & 0.0072 & -3.9997 & 0\ 0 \end{bmatrix},$$

$$\Gamma_3 = \begin{bmatrix} 0\ 0\ 0 & -3.9997 & 0\ 0 & 5.0397 & 0.2000 & 0.0126 \end{bmatrix}.$$

The feedback gains F_1, F_2, and F_3 are chosen the same as in Scenario 1 of the state feedback design. The observer gains are chosen as

$$L_1 = \begin{bmatrix} L_{A,1} \\ L_{S,1} \end{bmatrix} = 10^8 \times \begin{bmatrix} -0.0001 & 0.0000 \\ -0.0174 & -0.0665 \\ 0.0000 & 0.0000 \\ -0.0165 & -0.0657 \\ 0.5587 & 1.8354 \end{bmatrix},$$

$$L_2 = \begin{bmatrix} L_{A,2} \\ L_{S,2} \end{bmatrix} = 10^8 \times \begin{bmatrix} 0.0000 & 0.0000 \\ -0.0174 & -0.0717 \\ 0.0000 & 0.0000 \\ -0.0165 & -0.0708 \\ 0.6989 & 2.4736 \end{bmatrix},$$

$$L_3 = \begin{bmatrix} L_{A,3} \\ L_{S,3} \end{bmatrix} = 10^8 \times \begin{bmatrix} -0.0001 & 0.0000 \\ -0.0174 & -0.0561 \\ 0.0000 & 0.0000 \\ -0.0165 & -0.0554 \\ 0.3984 & 1.1059 \end{bmatrix},$$

Simulation is performed with the initial conditions

$$x_1(0) = x_2(0) = x_3(0) = \begin{bmatrix} 40 \\ 5 \end{bmatrix},$$

$$\hat{x}_1(0) = \hat{x}_2(0) = \hat{x}_3(0) = \begin{bmatrix} 50 \\ 0 \end{bmatrix},$$

$$\hat{w}_1(0) = \hat{w}_2(0) = \hat{w}_3(0) = \begin{bmatrix} 40 \\ 0 \\ 5 \end{bmatrix}.$$

Figure 12.24 shows the control inputs of the DG units. Figure 12.25 shows the output voltages of the DG units. Figure 12.26 shows the load currents of the DG units. Figures 12.27 and 12.28 show the observed states of the DG units and the observed states of the exosystems, respectively.

12.5.2.3 Scenario 3: Pulsed Loads

In this scenario, the three DG units are subject to the same pulsed loads and reference voltages as in Scenario 3 of the state feedback case. Thus, the exosystems and their initial conditions are the

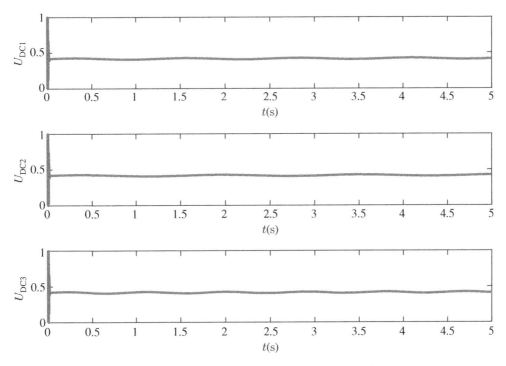

Figure 12.24 Output feedback Scenario 2: The control inputs in the form of duty cycles.

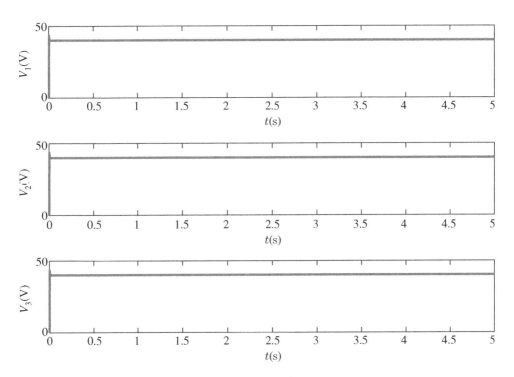

Figure 12.25 Output feedback Scenario 2: The output voltages.

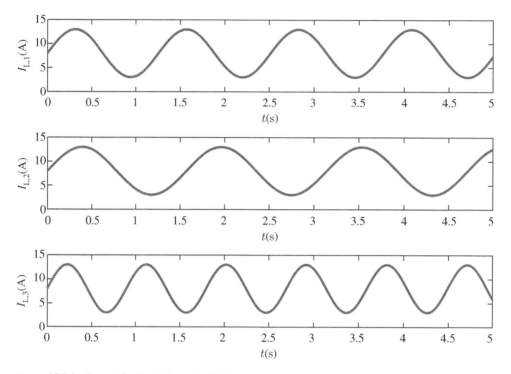

Figure 12.26 Output feedback Scenario 2: The load currents.

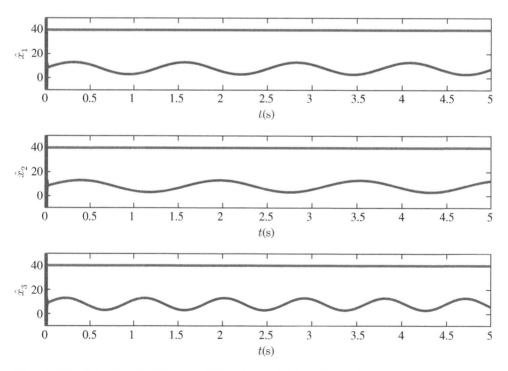

Figure 12.27 Output feedback Scenario 2: The observed states of the system.

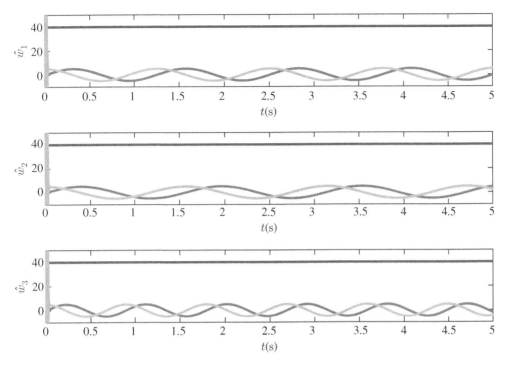

Figure 12.28 Output feedback Scenario 2: The observed states of the exosystem.

same as given there. Accordingly, we will adopt the same feedback gains F_1, F_2, and F_3 constructed there and only need to construct the state observers with the observer gains

$$L_1 = \begin{bmatrix} L_{A,1} \\ L_{S,1} \end{bmatrix} = 10^{12} \times \begin{bmatrix} -0.0000 & -0.0000 \\ 0.0000 & 0.0000 \\ -0.0000 & 0.0000 \\ 3.0046 & 0.7627 \\ 5.4508 & 1.3789 \\ -4.9620 & -1.2571 \\ -0.4862 & -0.1176 \\ 2.1149 & 0.5320 \\ -2.0714 & -0.5287 \\ -0.0477 & -0.0092 \\ 1.1179 & 0.2835 \\ -0.1098 & -0.0284 \\ -0.1622 & -0.0408 \end{bmatrix},$$

$$
L_2 = \begin{bmatrix} L_{A,2} \\ L_{S,2} \end{bmatrix} = 10^{12} \times \begin{bmatrix} 0.0000 & -0.0000 \\ 0.0000 & -0.0000 \\ 0.0000 & 0.0000 \\ 5.0636 & -0.7203 \\ 9.1530 & -1.4055 \\ -8.3446 & 1.2428 \\ -0.7790 & 0.2366 \\ 3.5299 & -0.6094 \\ -3.5103 & 0.4391 \\ -0.0603 & 0.0711 \\ 1.8817 & -0.2748 \\ -0.1886 & 0.0159 \\ -0.2707 & 0.0471 \end{bmatrix},
$$

$$
L_3 = \begin{bmatrix} L_{A,3} \\ L_{S,3} \end{bmatrix} = 10^{12} \times \begin{bmatrix} -0.0000 & -0.0000 \\ 0.0000 & 0.0000 \\ -0.0000 & 0.0000 \\ 3.0046 & 0.7627 \\ 5.4508 & 1.3789 \\ -4.9620 & -1.2571 \\ -0.4862 & -0.1176 \\ 2.1149 & 0.5320 \\ -2.0714 & -0.5287 \\ -0.0477 & -0.0092 \\ 1.1179 & 0.2835 \\ -0.1098 & -0.0284 \\ -0.1622 & -0.0408 \end{bmatrix}.
$$

Simulation is performed with the initial conditions

$$
x_1(0) = x_2(0) = x_3(0) = \begin{bmatrix} 40 \\ 5 \end{bmatrix},
$$

$$
\hat{x}_1(0) = \hat{x}_2(0) = \hat{x}_3(0) = \begin{bmatrix} 0 \\ 0 \end{bmatrix},
$$

$$
\hat{w}_1(0) = \hat{w}_2(0) = \hat{w}_3(0) = \begin{bmatrix} 1.0000 \\ 0 \\ 1.2732 \\ 0 \\ 0.4244 \\ 0 \\ 0.2546 \\ 0 \\ 0.1819 \\ 0 \\ 0.1415 \end{bmatrix}.
$$

Figures 12.29–12.31 show the control inputs in the form of duty cycles, the output voltages, and the load currents. Figures 12.32 and 12.33 show the observed states of the DG units and the observed states of the exosystems, respectively.

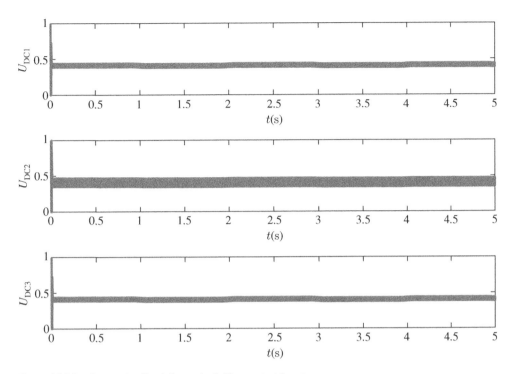

Figure 12.29 Output feedback Scenario 3: The control inputs.

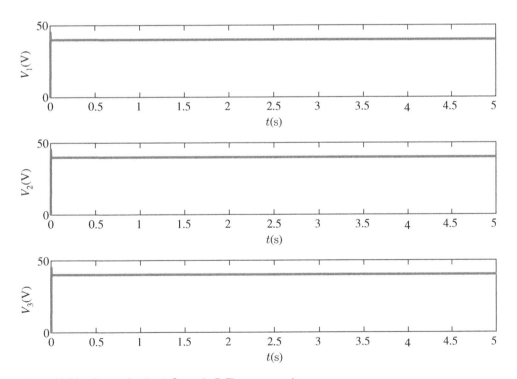

Figure 12.30 Output feedback Scenario 3: The output voltages.

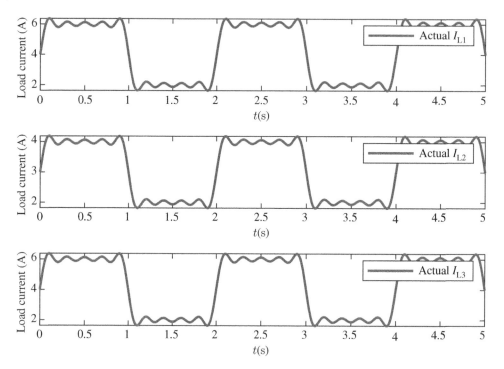

Figure 12.31 Output feedback Scenario 3: The load currents.

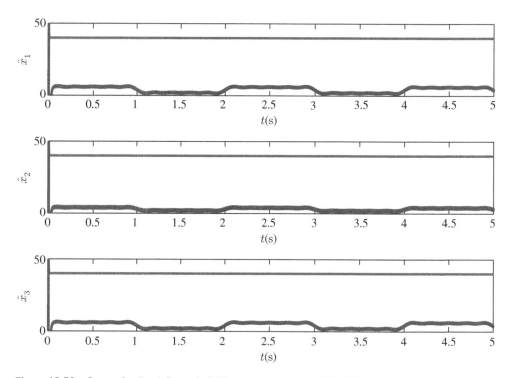

Figure 12.32 Output feedback Scenario 3: The observed states of the DG units.

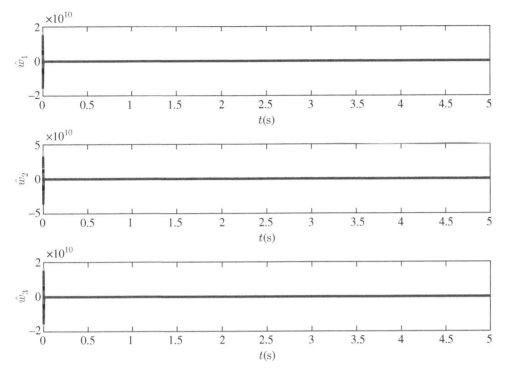

Figure 12.33 Output feedback Scenario 3: The observed states of the exosystems.

12.6 Conclusions

In this chapter, we developed distributed voltage regulation control laws for multiple coupled DG units in a DC microgrid. The designs are based on the classical output regulation theory. The proposed control laws were shown to be able to regulate the output voltages of the units to possibly time-varying reference voltages, regardless of the influence of local time-varying current loads and the coupling among the units. Extensive simulation results were presented to illustrate the design procedure and verify the effectiveness of the resulting control laws.

12.7 Exercises

1. Consider the exosystems in Scenario 2 of the state feedback design in Section 12.5.1. Suppose that the three DG units are subject to sinusoidal current loads,

$$I_{L1}(t) = 24 + 12 \sin\left(5t + \frac{\pi}{4}\right),$$

$$I_{L2}(t) = 24 + 12 \sin\left(4t + \frac{\pi}{3}\right),$$

$$I_{L2}(t) = 24 + 12 \sin\left(7t + \frac{\pi}{2}\right),$$

and the reference voltages are set to be $r_1 = r_2 = r_3 = 48\,\text{V}$. Determine the initial conditions $w_1(0)$, $w_2(0)$, and $w_3(0)$ so that the exosystems generate the current loads and the reference voltages.

2. Consider the scenario described in Exercise 12.1 and the state feedback laws designed in Scenario 2 of the state feedback design in Section 12.5.1. Simulate the closed-loop system with the initial conditions

$$x_1(0) = x_2(0) = x_3(0) = \begin{bmatrix} 46 \\ 5 \end{bmatrix}.$$

Plot the control inputs, the output voltages, and the load currents in the DG units.

3. Consider Scenario 2 in the output feedback design in Section 12.5.2 and the output feedback laws designed there. Simulate the closed-loop system with the following initial conditions,

$$x_1(0) = x_2(0) = x_3(0) = \begin{bmatrix} 39 \\ 5 \end{bmatrix},$$

$$\hat{x}_1(0) = \hat{x}_2(0) = \hat{x}_3(0) = \begin{bmatrix} 45 \\ 0 \end{bmatrix},$$

$$\hat{w}_1(0) = \hat{w}_2(0) = \hat{w}_3(0) = \begin{bmatrix} 42 \\ 1 \\ 4 \end{bmatrix}.$$

Plot the control inputs of the DG units, the output voltages of the DG units, the load currents of the DG units, the observed states of the DG units, and the observed states of the exosystems.

12.8 Acknowledgment

This work relates to Department of Navy awards N00014-20-1-2858, N00014-22-1-2001, and N00014-23-1-2124 issued by the Office of Naval Research. The United States Government has a royalty-free license throughout the world in all copyrightable material contained herein.

References

1 Ackermann, T., Andersson, G., and Söder, L. (2001). Distributed generation: a definition. *Electric Power Systems Research* 57 (3): 195–204.

2 El-Khattam, W. and Salama, M.M.A. (2004). Distributed generation technologies, definitions and benefits. *Electric Power Systems Research* 71 (2): 119–128.

3 Lasseter, R.H. (2007). Microgrids and distributed generation. *Journal of Energy Engineering* 133 (3): 144–149.

4 Lopes, J.A.P., Hatziargyriou, N., Mutale, J. et al. (2007). Integrating distributed generation into electric power systems: a review of drivers, challenges and opportunities. *Electric Power Systems Research* 77 (9): 1189–1203.

5 Pepermans, G., Driesen, J., Haeseldonckx, D. et al. (2005). Distributed generation: definition, benefits and issues. *Energy Policy* 33 (6): 787–798.

6 Bollen, M.H.J. and Sannino, A. (2005). Voltage control with inverter-based distributed generation. *IEEE Transactions on Power Delivery* 20 (1): 519–520.

7 Katiraei, F. and Iravani, M.R. (2006). Power management strategies for a microgrid with multiple distributed generation units. *IEEE Transactions on Power Systems* 21 (4): 1821–1831.

8 Tucci, M., Riverso, S., Vasquez, J.C. et al. (2016). A decentralized scalable approach to voltage control of DC islanded microgrids. *IEEE Transactions on Control Systems Technology* 24 (6): 1965–1979.

9 Zhao, J. and Dörfler, F. (2015). Distributed control, load sharing, and dispatch in DC microgrids. In: *2015 American Control Conference*, 3304–3309. IEEE.

10 Eskandari, M., Li, L., Moradi, M.H. et al. (2020). Optimal voltage regulator for inverter interfaced distributed generation units Part I: Control system. *IEEE Transactions on Sustainable Energy* 11 (4): 2813–2824.

11 Tucci, M., Riverso, S., and Ferrari-Trecate, G. (2017). Line-independent plug-and-play controllers for voltage stabilization in DC microgrids. *IEEE Transactions on Control Systems Technology* 26 (3): 1115–1123.

12 Liu, J., Zhang, W., and Rizzoni, G. (2017). Robust stability analysis of DC microgrids with constant power loads. *IEEE Transactions on Power Systems* 33 (1): 851–860.

13 Sumsurooah, S., Odavic, M., and Bozhko, S. (2017). μ approach to robust stability domains in the space of parametric uncertainties for a power system with ideal CPL. *IEEE Transactions on Power Electronics* 33 (1): 833–844.

14 Marzband, M., Alavi, H., Ghazimirsaeid, S.S. et al. (2017). Optimal energy management system based on stochastic approach for a home microgrid with integrated responsive load demand and energy storage. *Sustainable Cities and Society* 28: 256–264.

15 Yousefizadeh, S., Bendtsen, J.D., Vafamand, N. et al. (2018). EKF-based predictive stabilization of shipboard DC microgrids with uncertain time-varying load. *IEEE Journal of Emerging and Selected Topics in Power Electronics* 7 (2): 901–909.

16 Francis, B.A. (1977). The linear multivariable regulator problem. *SIAM Journal on Control and Optimization* 15 (3): 486–505.

17 Huang, J. (2004). *Nonlinear Output Regulation: Theory and Applications*. SIAM.

18 Venkatasubramanian, V., Schattler, H., and Zaborszky, J. (1995). Fast time-varying phasor analysis in the balanced three-phase large electric power system. *IEEE Transactions on Automatic Control* 40 (11): 1975–1982.

19 Isidori, A. and Byrnes, C.I. (1990). Output regulation of nonlinear systems. *IEEE Transactions on Automatic Control* 35 (2): 131–140.

20 Hu, T. and Lin, Z. (2004). Output regulation of linear systems with bounded continuous feedback. *IEEE Transactions on Automatic Control* 49 (11): 1941–1953.

13

Droop-Free Distributed Control for AC Microgrids

Sheik M. Mohiuddin and Junjian Qi

With the increased penetration of renewable energy-based distributed power-generating sources into the grid, the concept of microgrids is emerging. A microgrid can be defined as a cluster of distributed generators (DGs), energy storage systems (ESSs), and loads that are grouped as a single controllable entity to operate independently or in conjunction with the grid [1]. Microgrids can provide strong support to the grid by alleviating stresses, reducing feeder losses, and improving reliability, efficiency, and expandability [2–5].

For the integration of renewable energy resource-based DGs and ESSs into the grid, power electronic (PE) converters are required due to their DC nature. During the grid-connected operation of the microgrid, the PE converters usually operate in the grid following (GFL) mode. In this mode, a phase-locked loop is employed to track the grid voltage and frequency [6]. In the GFL mode, the PE converters work as a current source and regulate the current injection into the grid. One limitation of GFL converters is that they need to be connected to a strong grid or in parallel with the sources that can regulate the grid's voltage and frequency.

Alternatively, grid forming (GFM) converters can operate in weak and islanded grid conditions. A GFM inverter acts as a voltage source behind an impedance, and can independently regulate the microgrid's voltage and frequency [7]. Traditionally, for regulating voltage and frequency through GFM converters, hierarchical control by appropriately coordinating the primary, secondary, and tertiary control layers is used [5, 8].

This chapter discusses voltage and frequency regulation of islanded microgrids with GFM inverters. For control implementation, the microgrid needs to be modeled as a cyber-physical system which is discussed next.

13.1 Cyber-Physical Microgrid Modeling

Figure 13.1 shows the cyber-physical representation of a microgrid system which includes the power network and the communication network. The power network includes DGs, transmission lines, loads, and other physical devices. Utilizing the communication network at the cyber layer, the DGs exchange information with the neighboring DGs in a centralized or distributed fashion. More details about the physical network and the communication network in a microgrid are given below.

Microgrids: Theory and Practice, First Edition. Edited by Peng Zhang.
© 2024 The Institute of Electrical and Electronics Engineers, Inc. Published 2024 by John Wiley & Sons, Inc.

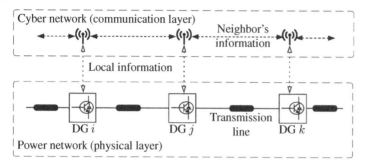

Figure 13.1 Cyber-physical representation of a microgrid.

13.1.1 Power Network

In the physical modeling of the microgrid power network, all zero injection and load buses modeled as constant impedances are eliminated by Kron reduction [9], and only the buses at the outputs of the LC filters of the sources are kept. The corresponding reduced bus admittance matrix is denoted by $\mathbf{Y}$. Then the active and reactive power injections at bus i, P_i, and Q_i, can be written as:

$$P_i = v_i \sum_{j \in \mathcal{W}_i} v_j \left(G_{ij} \cos \theta_{ij} + B_{ij} \sin \theta_{ij} \right) \tag{13.1}$$

$$Q_i = v_i \sum_{j \in \mathcal{W}_i} v_j \left(G_{ij} \sin \theta_{ij} - B_{ij} \cos \theta_{ij} \right) \tag{13.2}$$

where $\mathcal{W}_i$ is the set of buses that connect with bus i (including bus i), v_i is the voltage magnitude of bus i per unit, θ_i is the phase angle of bus i, $\theta_{ij} = \theta_i - \theta_j$, and G_{ij} and B_{ij} are the real and imaginary parts of the element in $\mathbf{Y} = \mathbf{G} + j\mathbf{B}$. Note that in the reduced network each bus is connected to all the other buses. The normalized real and reactive powers for inverter i are:

$$\lambda_{P_i} \triangleq \frac{P_i}{\overline{P}_i} \tag{13.3}$$

$$\lambda_{Q_i} \triangleq \frac{Q_i}{\overline{Q}_i} \tag{13.4}$$

where $\overline{P}_i$ and $\overline{Q}_i$ are, respectively, the rated upper limits of the real and reactive power for inverter i. Note that all values here are in per unit.

13.1.2 Communication Network

For the modeling of the cyber layer, a sparse communication network is considered by which data is exchanged among the agents of the sources. The communication network can be modeled as a directed graph (digraph) $\mathcal{G}$ in which the nodes are agents and the edges are the communication links connecting nodes. As in [4, 10], the communication links may exchange data with different gains and the edge weight between nodes i and j is a_{ij}.

The communication graph can be represented by an adjacency matrix $\mathbf{A} = [a_{ij}] \in \mathbb{R}^{N \times N}$ where N is the number of dispatchable sources. As in [4], it can be assumed that the adjacency matrix is a time-invariant and scalar matrix. The Laplacian matrix is defined as $\mathbf{L} = \mathbf{D}^{in} - \mathbf{A}$ where $\mathbf{D}^{in} = \text{diag}\{d_i^{in}\}$ is the in-degree matrix, $d_i^{in} = \sum_{j \in \mathcal{N}_i} a_{ij}$, and $\mathcal{N}_i$ is the set of the neighbors of node i in the communication network. Similar to the in-degree matrix, an out-degree matrix $\mathbf{D}^{out} = \text{diag}\{d_i^{out}\}$ can also be defined with $d_i^{out} = \sum_{j \in \mathcal{N}_i} a_{ji}$.

The Laplacian matrix is balanced if the in-degree and out-degree matrices are equal. A graph is said to have a spanning tree if it contains a root node from which there exists at least one direct path to every other node. A direct path from node i to node j is a sequence of edges that connect i and j. It is assumed that the communication graph with an adjacency matrix $\mathbf{A}$ has at least a spanning tree and a balanced Laplacian matrix.

13.2 Hierarchical Control of Islanded Microgrid

The secondary control in a microgrid is usually implemented in coordination with the zero-level and primary controls. In some literature, the secondary control implementation without relying on the primary level control is also proposed [4, 11]. In the following, the zero-level control, primary control, and secondary control are discussed.

13.2.1 Zero-Level Control

The zero-level control is at the lowest level of the control hierarchy. The upper level control (primary/secondary control) decides the voltage magnitude reference v_i^* and the frequency reference ω_i^* which are used to decide the output voltage reference as:

$$v_{o,i}^*(t) = v_i^*(t) \sin\left(\omega^r t + \theta_i^*(t)\right) \tag{13.5}$$

where ω^r is the nominal frequency and $\theta_i^*(t) = \int_0^t \delta\omega_i(\tau)d\tau$ is the phase angle reference. Then, the three-phase voltage reference is calculated as:

$$\begin{bmatrix} v_{oa,i}^*(t) \\ v_{ob,i}^*(t) \\ v_{oc,i}^*(t) \end{bmatrix} = \begin{bmatrix} v_i^*(t) \sin\left(\omega^r t + \theta_i^*(t)\right) \\ v_i^*(t) \sin\left(\omega^r t + \theta_i^*(t) - 2\pi/3\right) \\ v_i^*(t) \sin\left(\omega^r t + \theta_i^*(t) + 2\pi/3\right) \end{bmatrix} \tag{13.6}$$

which is sent to the zero-level controller to generate the desired pulse width modulation (PWM) signals at the output of the inverter [12, 13].

The zero-level controller is designed in stationary reference frame and includes inner voltage and current control loops as shown in Fig. 13.2. These control loops include proportional plus resonant (PR) terms which are tuned at fundamental frequency.

13.2.2 Primary Control

In an islanded microgrid, traditionally droop-based primary control [14] is employed to ensure frequency and voltage performance without relying on communication [15–17]. Droop control can help suppress the impact of circulating reactive currents due to line impedance mismatch in distribution feeders [18], prevent overloading of DGs by proportional power sharing [19], and retain inertia of the microgrid by mimicking the behavior of synchronous generators [20–24].

In droop control, the voltage and frequency references v_i and ω_i are generated based on a droop law as follows [25]:

$$v_i^* = v^r - m_{Q_i} Q_i \tag{13.7}$$

$$\omega_i^* = \omega^r - m_{P_i} P_i \tag{13.8}$$

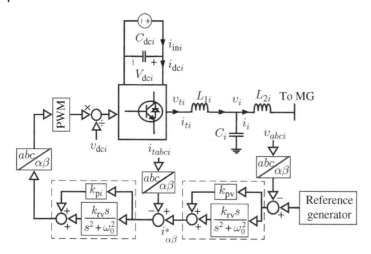

Figure 13.2 Schematic diagram of proportional resonant inner voltage and current controller for DG *i*.

where v^r and ω^r are the rated voltage and frequency of the microgrid. P_i and Q_i are the active and reactive power injected by DG *i*. m_{Q_i} and m_{P_i} are the voltage and frequency droop-gains of DG *i*, which is usually selected as [14]:

$$m_{Q_i} = \frac{v^{\max} - v^{\min}}{Q_i^{\max}} \tag{13.9}$$

$$m_{P_i} = \frac{\omega^{\max} - \omega^{\min}}{P_i^{\max}} \tag{13.10}$$

Typically, single-loop or multi-loop droop control structures are used for GFM inverters [26]. Figures 13.3 and 13.4 respectively show the control block diagram of the single-loop and multi-loop droop controllers.

In the case of a single-loop structure, there are no cascaded inner voltage and current control loops as in the multi-loop control. For the single-loop droop control, the generated voltage and

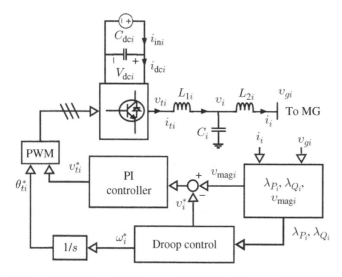

Figure 13.3 Schematic diagram of single-loop inner control for DG *i*.

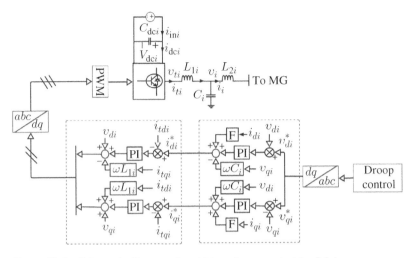

Figure 13.4 Schematic diagram of multi-loop inner control for DG *i*.

phase angle references are directly sent to the PWM; therefore, this control regulates inverter terminal voltage v_{ti}. On the other hand, for multi-loop droop control, the voltage and phase angle references are sent to the inner voltage loop. Therefore, the multi-loop droop control regulates the capacitor voltage v_i.

As in the proportional resonant controller presented in Fig. 13.2, the single- and multi-loop droops can also be used with the secondary controller. When the secondary control is used, the voltage and frequency references will be generated by the secondary controller instead of the droop controller.

Albeit the operational simplicity of the droop-based control, it has several limitations [27], such as poor dynamic performance [2], high sensitivity to measurement noises [2], inappropriate reactive power-sharing due to line impedance mismatch, non-identical bus voltages [28–30], performance degradation in the presence of nonlinear loads [4, 31], and voltage/frequency deviation due to load variation [4]. Improved droop controls such as adaptive droop control [16] and optimization-based virtual impedance method [20] are thus proposed. However, their performance can be affected by the mismatch in feeder impedance and nonlinear/unbalanced loading conditions [3].

13.2.3 Distributed Secondary Control with Droop

To address the limitations of the droop control, secondary control has been proposed to ensure a stable frequency and voltage profile in the microgrid [2, 15, 18, 24]. The secondary control can be structured centrally or distributively [27]. In the centralized control architecture [32], a high-bandwidth, point-to-point communication is required between the central controller and local DG control units, increasing the communication and computational cost [17]. The central controller also suffers from the risk of single point of failure [8, 17, 33] and could have reliability and scalability issues [23, 24, 34].

By contrast, the distributed secondary control that utilizes a sparse communication network provides a promising solution [4, 35–37], in which each DG only has access to the information of its neighboring DGs, reducing the computational complexity and improving the reliability and resiliency to faults or unknown system parameters [4, 5, 24, 38]. Distributed sparse networked master–slave mechanism is proposed in [39, 40]. In [15, 17, 23] fully distributed secondary controls

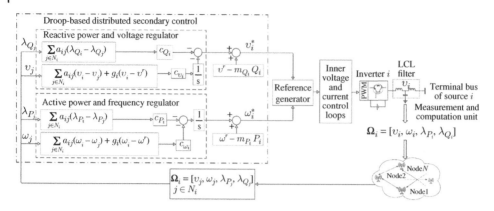

Figure 13.5 Droop-based distributed secondary control for DG *i*.

are proposed for frequency and voltage regulation in islanded microgrids. A distributed coordinated control with containment and consensus-based approach is proposed in [33]. In [24], a robust finite-time secondary control is presented. However, these control approaches still utilize droop technique and have the inherent aforementioned drawbacks.

Figure 13.5 shows a generalized droop-based distributed secondary voltage and frequency control as presented in [15, 23, 41, 42].

First, to improve the performance of decentralized droop control, a distributed secondary control with voltage and reactive power corrections can be designed as:

$$v_i = v^r - m_{Q_i} Q_i + \lambda_i + h_i \tag{13.11}$$

where λ_i and h_i are the voltage and reactive power correction terms which can be decided based on the following distributed control laws:

$$\frac{d\lambda_i}{dt} = -c_{v_i} \left(\sum_{j \in \mathcal{N}_i} a_{ij}(v_i - v_j) + g_i(v_i - v^r) \right) \tag{13.12}$$

$$\frac{dh_i}{dt} = -c_{Q_i} \sum_{j \in \mathcal{N}_i} a_{ij}(\lambda_{Q_i} - \lambda_{Q_j}) \tag{13.13}$$

where c_{v_i} and c_{Q_i} are positive gains, $g_i \geq 0$ is the pinning gain by which the controller of node *i* is connected to the reference, v^r is the voltage reference signal, λ_{Q_i} is the normalized reactive power of DG *i*, and $\mathcal{N}_i$ is the set of the neighbors of node *i* in the communication network.

Also, the distributed secondary frequency control with active power-sharing correction can be designed as follows:

$$\omega_i = \omega^r - m_{P_i} P_i + \Omega_i \tag{13.14}$$

where the frequency correction term Ω_i can be decided as:

$$\frac{d\Omega_i}{dt} = -c_{\omega_i} \left(\sum_{j \in \mathcal{N}_i} a_{ij}(\omega_i - \omega_j) + g_i(\omega_i - \omega^r) \right) - c_{P_i} \sum_{j \in \mathcal{N}_i} a_{ij}(\lambda_{P_i} - \lambda_{P_j}) \tag{13.15}$$

where c_{ω_i} and c_{P_i} are positive gains, ω^r is the nominal frequency, and λ_{P_i} is the normalized active power of DG *i*.

13.3 Droop-Free Distributed Control with Proportional Power Sharing

Although the distributed secondary voltage and frequency control gives improved voltage and frequency response than the traditional droop-based control, they still depend on the droop gains and their performance suffers from the limitation of the traditional droop controller [4].

To solve this issue, droop-free distributed control has been proposed. In [10] and [4], distributed cooperative controls that do not rely on the droop mechanism are proposed for DC and AC microgrids. In voltage regulation, the average voltage of the system is controlled to be the rated voltage. In [43], a consensus-based harmonic power-sharing approach is proposed to eliminate the impact of line impedance mismatch and regulate the average value of the DG voltages. An optimal distributed secondary control is proposed in [44] to achieve accurate reactive power sharing and regulate the average voltages. In [37], a distributed adaptive virtual impedance (DAVI) control method is presented, which can also achieve accurate reactive power sharing and regulate the DG average voltage.

As shown in Fig. 13.6, the droop-free control in [4] generates voltage and frequency correction terms based on secondary control law instead of using fixed droop gains. In this approach, the voltage and frequency reference for DG i can be selected as:

$$v_i^* = v^r + \Delta v_i \tag{13.16}$$

$$\omega_i^* = \omega^r + \Delta \omega_i \tag{13.17}$$

where Δv_i and $\Delta \omega_i$ are voltage and frequency references for DG i which are selected by the secondary controllers. More details will be provided next.

13.3.1 Distributed Average Voltage Estimation

Figure 13.7 demonstrates the average voltage estimation policy at each node. The distributed observer of the average value of the voltage magnitude is designed in [10] as:

$$\bar{v}_i(t) = v_i(t) + \int_0^t \sum_{j \in \mathcal{N}_i} a_{ij}(\bar{v}_j - \bar{v}_i) d\tau \tag{13.18}$$

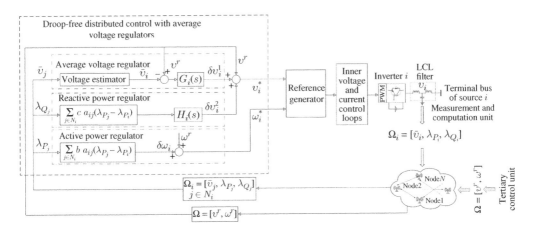

Figure 13.6 Droop-free distributed secondary control for DG i.

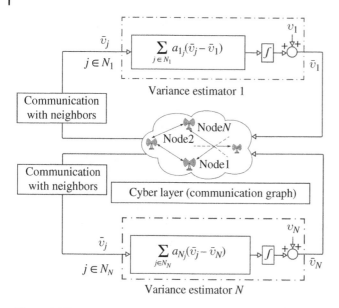

Figure 13.7 Average voltage estimation policies. Source: Reproduced with permission from [11]/IEEE.

where v_i and $\bar{v}_i$ are, respectively, the local measurement of the output voltage on the capacitor in the LCL filter and the estimated average voltage at node i. Let $\mathbf{v} = [v_1, v_2, \dots, v_N]^\top$ and $\bar{\mathbf{v}} = [\bar{v}_1, \bar{v}_2, \dots, \bar{v}_N]^\top$ be the voltage measurements and estimated average voltage vectors.

It is proved in [10] that all elements of $\bar{\mathbf{v}}$ converge to a consensus value, which is the true average voltage. Mathematically this can be written as:

$$\bar{\mathbf{v}}^{ss} = \mathbf{M}\mathbf{v}^{ss} = \langle \mathbf{v}^{ss} \rangle \mathbf{1} \tag{13.19}$$

where $\mathbf{M} \in \mathbb{R}^{N \times N}$ is the averaging matrix whose elements are all $1/N$. Denote a N-dimensional vector by $\mathbf{x} \in \mathbb{R}^N$. $\mathbf{x}^{ss}$ is the vector of the steady-state value of $\mathbf{x}$. $\langle \mathbf{x} \rangle$ is the average value of the elements in $\mathbf{x}$ and is a scalar. $\mathbf{1} \in \mathbb{R}^{N \times 1}$ is a column vector whose elements are all one.

13.3.2 Voltage Regulation

Voltage regulation and perfect reactive power sharing is achieved through an average voltage regulator and reactive power regulator. The average voltage regulator and reactive power regulator generate two voltage correction terms which are added to the microgrid-rated voltage to generate reference voltage for the lower level controllers.

The average voltage regulator at DG i utilizes the average voltage estimator in Section 13.3.1 to compute the global average voltage of all DGs in the microgrid in a fully distributed way. The estimated average voltage $\bar{v}_i$ is then compared with the microgrid-rated voltage v^r and the error term is fed into a PI controller G_i to generate the first voltage correction term δv_i^1. This is used to regulate the average of the voltage magnitude of all sources to the rated voltage of the microgrid.

To achieve proportional reactive power sharing, the reactive power regulator adjusts the voltage reference with an additional voltage correction term δv_i^2. The regulator at DG i computes reactive power mismatch as:

$$mQ_i = \sum_{j \in \mathcal{N}_i} c\, a_{ij}(\lambda_{Q_j} - \lambda_{Q_i}) \tag{13.20}$$

where the coupling gain c is a design parameter, and λ_{Q_i} and λ_{Q_j} are the normalized reactive power of the ith source and jth source, respectively.

The computed reactive power mismatch mQ_i is then fed into a PI controller to adjust the second voltage correction term δv_i^2.

13.3.3 Frequency Regulation

The proportional sharing of active power in microgrids is usually achieved through frequency regulation. In this approach, the secondary control sets the frequency references to achieve proportional active power sharing utilizing an active power regulator [4]. The active power regulator at source i calculates the neighborhood active power mismatch to assign the frequency correction term $\delta\omega_i$:

$$\delta\omega_i = \sum_{j \in \mathcal{N}_i} b\, a_{ij} \left(\lambda_{P_j} - \lambda_{P_i} \right) \tag{13.21}$$

where the coupling gain b is a design parameter, and λ_{P_i} and λ_{P_j} are, respectively, the normalized real power of the ith source and jth source.

Then the frequency correction term $\delta\omega_i$, is added to at he rated frequency to generate the frequency set point as

$$\omega_i^* = \omega^r + \delta\omega_i. \tag{13.22}$$

13.4 Droop-Free Distributed Control with Voltage Profile Guarantees

The droop-free control with proportional reactive power sharing discussed in Section 13.3 cannot guarantee that the voltage profile is always acceptable. For example, as in distributed energy resource (DER) interconnection standards such as IEEE 1547, the voltage should be maintained within $\pm 5\%$ of the rated voltage [45]. Regulating only the average voltage cannot guarantee that these requirements are always satisfied.

A droop-free distributed control is developed in [11] to guarantee an acceptable voltage profile and avoid any violation of voltage deviations. This is made possible by a distributed voltage variance observer, an additional voltage variance regulator, and relaxed reactive power sharing.

Figure 13.8 shows the schematic of the control policy for Node i. The frequency regulation is the same as that in Section 13.3.3. In voltage regulation, the voltage regulator, voltage variance regulator, and reactive power regulator modules adjust the set-point of the voltage magnitude for node i, v_i^*, respectively by adding the voltage correction terms δv_i^1, δv_i^2, and δv_i^3 to the rated voltage of the microgrid, v^r, as:

$$v_i^*(t) = v^r + \delta v_i^1(t) + \delta v_i^2(t) + \delta v_i^3(t) \tag{13.23}$$

The average voltage regulator is the same as that in Section 13.3.2. Next, the distributed voltage variance estimation policy, the voltage variance regulator, and the reactive power regulator will be discussed.

13.4.1 Distributed Voltage Variance Estimation

A distributed observer for the variance of the voltage magnitude is developed in [11]. Figure 13.9 illustrates the voltage variance estimation policy at each node. The variance observer at node i

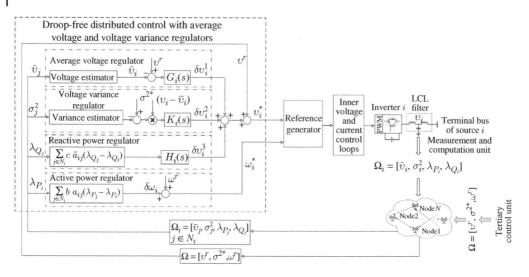

Figure 13.8 Droop-free distributed control with voltage profile guarantees for source *i* of the AC microgrid. Source: Reproduced with permission from [11]/IEEE.

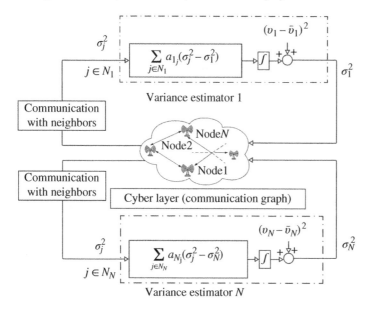

Figure 13.9 Voltage variance estimation policies. Source: Reproduced with permission from [11]/IEEE.

receives its neighbors' estimate σ_j^2's ($j \in \mathcal{N}_i$). The observer updates its estimate σ_i^2 by processing its neighbors' estimates of variance, local voltage measurement v_i, and its average voltage estimate $\bar{v}_i$ as:

$$\sigma_i^2(t) = \left(v_i(t) - \bar{v}_i(t)\right)^2 + \int_0^t \sum_{j \in \mathcal{N}_i} a_{ij} \left(\sigma_j^2(\tau) - \sigma_i^2(\tau)\right) d\tau \tag{13.24}$$

Let $\boldsymbol{\sigma}^2 = [\sigma_1^2, \sigma_2^2, \ldots, \sigma_N^2]^\top$ denote the voltage variance estimation vector from the observer in (13.24). In order to prove the convergence of the voltage variance observer, we need the following lemma from [4].

Lemma 13.1 *Assume that the digraph* **G** *has a spanning tree and the Laplacian matrix* **L** *is balanced. Then,*

$$\lim_{s \to 0} s(s\mathbf{I}_N + \mathbf{L})^{-1} = \mathbf{M} \tag{13.25}$$

where $\mathbf{I}_N \in \mathbb{R}^{N \times N}$ *is an identity matrix.*

Theorem 13.1 *Assume that the communication graph used in a cooperative control has a spanning tree and the associated Laplacian matrix* **L** *is balanced. Then, using the observer in (13.24), all estimates of the variance of the voltage magnitude converge to the true global variance of the voltage magnitude.*

Proof: [Proof of Theorem 13.1] By differentiating (13.24) we have:

$$\dot{\sigma}_i^2 = 2(v_i - \overline{v}_i)(\dot{v}_i - \dot{\overline{v}}_i) + \sum_{j \in \mathcal{N}_i} a_{ij}(\sigma_j^2 - \sigma_i^2)$$

$$= 2(v_i - \overline{v}_i)(\dot{v}_i - \dot{\overline{v}}_i) + \sum_{j \in \mathcal{N}_i} a_{ij}\sigma_j^2 - d_i^{\text{in}}\sigma_i^2 \tag{13.26}$$

Let $\mathbf{e} = \mathbf{v} - \overline{\mathbf{v}}$, the global dynamics of the observer in (13.24) can be written as:

$$\dot{\boldsymbol{\sigma}}^2 = 2 \operatorname{diag}(\mathbf{e}) \dot{\mathbf{e}} - \mathbf{L} \boldsymbol{\sigma}^2 \tag{13.27}$$

where $\overline{\mathbf{v}}$ is the average voltage estimation vector from (13.18) and $\operatorname{diag}(\mathbf{x})$ creates a diagonal matrix whose diagonal elements are from vector $\mathbf{x}$. In frequency domain, we have:

$$s\boldsymbol{\Sigma}^2 = 2\mathcal{L}\left(\operatorname{diag}(\mathbf{e})\dot{\mathbf{e}}\right) - \mathbf{L}\boldsymbol{\Sigma}^2 \tag{13.28}$$

where $\mathcal{L}$ takes the Laplace transform and there is:

$$\mathcal{L}\left(\operatorname{diag}(\mathbf{e})\dot{\mathbf{e}}\right) = s \cdot \frac{1}{2\pi i} \lim_{T \to \infty} \int_{c-iT}^{c+iT} \operatorname{diag}\left(\mathbf{E}(\tau)\right) \mathbf{E}(s-\tau) d\tau \tag{13.29}$$

where $i = \sqrt{-1}$. Then we have:

$$\boldsymbol{\Sigma}^2 = 2\left(s\mathbf{I}_N + \mathbf{L}\right)^{-1} \mathcal{L}\left(\operatorname{diag}(\mathbf{e})\dot{\mathbf{e}}\right) \tag{13.30}$$

According to Lemma 13.1 of [10], all poles of the term $s(s\mathbf{I}_N + \mathbf{L})^{-1}$ lie in the open left-hand plane (OLHP). Thus all poles of $(s\mathbf{I}_N + \mathbf{L})^{-1}$ also lie in the OLHP. As in [10] both **V** and $\overline{\mathbf{V}}$ are type-1 vectors (i.e. it has a single pole at the origin and all other poles lie in the OLHP), thus $d\left(v_i(t) - \overline{v}_i(t)\right)^2/dt$ converges to zero and all poles of $\mathcal{L}\left(\operatorname{diag}(\mathbf{e})\dot{\mathbf{e}}\right)$ lie in the OLHP. Therefore, we can apply the final value theorem to obtain:

$$\lim_{t \to \infty} \boldsymbol{\sigma}^2(t) = \lim_{s \to 0} s\boldsymbol{\Sigma}^2$$

$$= \lim_{s \to 0} 2s(s\mathbf{I}_N + \mathbf{L})^{-1} \mathcal{L}\left(\operatorname{diag}(\mathbf{e})\dot{\mathbf{e}}\right)$$

$$= \lim_{s \to 0} s(s\mathbf{I}_N + \mathbf{L})^{-1} \times 2 \lim_{s \to 0} \mathcal{L}\left(\operatorname{diag}(\mathbf{e})\dot{\mathbf{e}}\right) \tag{13.31}$$

According to Lemma 13.1 [10], we have:

$$\lim_{t \to \infty} \boldsymbol{\sigma}^2(t) = \lim_{s \to 0} s\boldsymbol{\Sigma}^2$$

$$= \mathbf{M} \times 2 \lim_{t \to \infty} \int_0^t \operatorname{diag}(\mathbf{e})\dot{\mathbf{e}} \, d\tau = \mathbf{M} \times \left. \left(\operatorname{diag}(\mathbf{e})\mathbf{e}\right) \right|_0^\infty$$

$$= \mathbf{M} \times \left(\operatorname{diag}(\mathbf{e}_\infty)\mathbf{e}_\infty - \operatorname{diag}(\mathbf{e}_0)\mathbf{e}_0\right) \tag{13.32}$$

In (13.18) it is implied that $\overline{\mathbf{v}}_0 = \mathbf{v}_0$ and thus $\mathrm{diag}(\mathbf{e}_0)\,\mathbf{e}_0 = \mathbf{0}$. Similar to [4, 10], it is assumed that the system parameters are designed to stabilize the microgrid. We denote the steady-state voltage as $\mathbf{v}_\infty = \mathbf{v}^{ss}$. As has been proved in [10], the estimated average voltage converges to the true global average voltage, and thus $\overline{\mathbf{v}}_\infty = \langle \mathbf{v}^{ss} \rangle \mathbf{1}$. Therefore, we have:

$$\lim_{t \to \infty} \sigma^2(t) = \mathbf{M} \times \left(\mathrm{diag}(\mathbf{e}_\infty)\,\mathbf{e}_\infty \right)$$
$$= \mathbf{M} \, \mathrm{diag}\left(\mathbf{v}^{ss} - \langle \mathbf{v}^{ss} \rangle \mathbf{1} \right) \left(\mathbf{v}^{ss} - \langle \mathbf{v}^{ss} \rangle \mathbf{1} \right) \tag{13.33}$$

Equation (13.33) implies that all estimates of the variance converge to the true global variance of the voltage magnitude. Equivalently, for $1 \le i \le N$ we have:

$$\lim_{t \to \infty} \sigma_i^2(t) = \frac{1}{N} \sum_{i=1}^{N} \left(v_i(t) - \overline{v}_i(t) \right)^2 \tag{13.34}$$

∎

13.4.2 Voltage Variance Regulator

In order to generate the second voltage correction term, δv_i^2, $(\sigma^{2*} - \sigma_i^2)(v_i - \overline{v}_i)$ is fed into the second PI controller, K_i. This voltage variance regulator is designed to regulate the global voltage variance σ_i^2 to the reference value σ^{2*}. In this control design, $v_i - \overline{v}_i$ is multiplied in order to decide the control direction:

- When σ_i^2 is greater than σ^{2*}, if v_i is greater than $\overline{v}_i$ then v_i needs to be reduced; otherwise v_i needs to be increased.
- When σ_i^2 is less than σ^{2*}, if v_i is greater than $\overline{v}_i$ then v_i needs to be increased; otherwise v_i needs to be decreased.

Without multiplying $(v_i - \overline{v}_i)$, the control cannot decide the control direction of v_i only with the error between the estimated voltage variance and the variance reference.

The reference of the variance σ^{2*} can be determined by the tertiary control. Voltage regulation requirements are decided in DER interconnection standards such as IEEE 1547, and usually the voltage should be maintained within $\pm 5\%$ of the rated voltage [45]. In order to meet the voltage regulation requirement, we can set up an upper bound for the variance of the voltage that corresponds to the case in which the voltage of exactly one DER has a 5% deviation from the rated voltage while the voltages of all the other DER are maintained at the rated voltage:

$$\sigma^{2,\max} \le \frac{(0.05\,v^r)^2}{N} \tag{13.35}$$

where v^r is the rated voltage of the microgrid. Note that (13.35) is a sufficient but not necessary condition for meeting the voltage regulation requirement. Therefore, as long as we choose $\sigma_i^{2*} \le \sigma^{2,\max}$ we can always guarantee that the voltage regulation requirement is satisfied. If some DG units are disconnected, the number of DG units will become $N' < N$, in which case we have:

$$\sigma^{2,\max} \le \frac{(0.05\,v^r)^2}{N} < \frac{(0.05\,v^r)^2}{N'} \tag{13.36}$$

which implies that the upper bound determined by using the number of DG units under normal operating can always guarantee to satisfy the voltage regulation requirement even when some DG units are disconnected.

13.4.3 Relaxed Reactive Power Regulator

To generate the third voltage correction term δv_i^3, mQ_i is calculated and fed into another PI controller, H_i. Choose $k \in \{1, \dots, N\}$, and the kth source will not be implemented with reactive power sharing. In reactive power regulator, we modify $\mathbf{A}$ to be $\tilde{\mathbf{A}} = [\tilde{a}_{ij}]$ by setting the kth row and kth column to be zero. The corresponding Laplacian matrix becomes $\tilde{\mathbf{L}} = \tilde{\mathbf{D}}^{in} - \tilde{\mathbf{A}}$ where $\tilde{\mathbf{D}}^{in} = \text{diag}\{\tilde{d}_i^{in}\}$ is the in-degree matrix with $\tilde{d}_i^{in} = \sum_{j \in \mathcal{N}_i} \tilde{a}_{ij}$. Then mQ_i as the neighborhood reactive power mismatch can be calculated as:

$$mQ_i = \sum_{j \in \mathcal{N}_i} c\,\tilde{a}_{ij} \left(\lambda_{Q_j} - \lambda_{Q_i} \right) \tag{13.37}$$

In this approach, the kth source does not proportionally share reactive power with the other sources. Whether its normalized reactive power is greater or less than those of the other sources depends on the loading conditions and the voltage variance set-point. In a community microgrid, multiple classes of assets can be owned by single or multiple parties, including the community, a utility, or other public or private companies [46]. The special DG could be a community-owned DG that is controlled to inject or absorb reactive power in order to guarantee the performance of the overall system.

13.5 Steady-State Analysis for the Control in Section 13.4

In steady state, the average voltage estimators converge to the true average voltage of the microgrid. Equivalently, $\bar{\mathbf{v}}^{ss} = \mathbf{M}\mathbf{v}^{ss} = \langle \mathbf{v}^{ss} \rangle \mathbf{1}$. Besides, the voltage variance estimators also converge to the true voltage variances of the microgrid, i.e. $\sigma^{2,ss} = \sum_{i=1}^{N} \left(v_i^{ss} - \langle \mathbf{v}^{ss} \rangle \right)^2 / N$. Then based on the control methodology in Fig. 13.8, we have:

$$\delta \mathbf{v}^1 = \delta \mathbf{v}_0^1 + \left(\mathbf{G}_P + \mathbf{G}_I(t - t_0) \right) \left(v^r - \langle \mathbf{v}^{ss} \rangle \right) \mathbf{1} \tag{13.38}$$

$$\delta \mathbf{v}^2 = \delta \mathbf{v}_0^2 + \left(\mathbf{K}_P + \mathbf{K}_I(t - t_0) \right) (\sigma^{2*} - \sigma^{2,ss}) \left(\mathbf{v}^{ss} - \langle \mathbf{v}^{ss} \rangle \mathbf{1} \right) \tag{13.39}$$

$$\delta \mathbf{v}^3 = \delta \mathbf{v}_0^3 + \left(\mathbf{H}_P + \mathbf{H}_I(t - t_0) \right) \left(-c \tilde{\mathbf{L}} \lambda_Q^{ss} \right) \tag{13.40}$$

where $\delta \mathbf{v}_0^1$, $\delta \mathbf{v}_0^2$, and $\delta \mathbf{v}_0^3$ are, respectively, column vectors that carry the integrator outputs in G_i's, K_i's, and H_i's at $t = t_0$, and λ_Q^{ss} is the vector of λ_{Q_i}'s. Then we have:

$$\begin{aligned}
\mathbf{v}^{*ss} &= \mathbf{v}^r + \delta \mathbf{v}^1 + \delta \mathbf{v}^2 + \delta \mathbf{v}^3 \\
&= v^r \mathbf{1} + \delta \mathbf{v}_0^1 + \delta \mathbf{v}_0^2 + \delta \mathbf{v}_0^3 + \mathbf{G}_P \left(v^r - \langle \mathbf{v}^{ss} \rangle \right) \mathbf{1} \\
&\quad + \mathbf{K}_P(\sigma^{2*} - \sigma^{2,ss}) \left(\mathbf{v}^{ss} - \langle \mathbf{v}^{ss} \rangle \mathbf{1} \right) - c\,\mathbf{H}_P \tilde{\mathbf{L}} \lambda_Q^{ss} \\
&\quad + \left(\mathbf{G}_I \left(v^r - \langle \mathbf{v}^{ss} \rangle \right) \mathbf{1} - c\,\mathbf{H}_I \tilde{\mathbf{L}} \lambda_Q^{ss} \right. \\
&\quad + \left. \mathbf{K}_I(\sigma^{2*} - \sigma^{2,ss}) \left(\mathbf{v}^{ss} - \langle \mathbf{v}^{ss} \rangle \mathbf{1} \right) \right)(t - t_0)
\end{aligned} \tag{13.41}$$

where $\mathbf{G}_I$ ($\mathbf{G}_P$), $\mathbf{K}_I$ ($\mathbf{K}_P$), and $\mathbf{H}_I$ ($\mathbf{H}_P$) are the diagonal matrices carrying the integral (proportional) gains of the controller matrices $\mathbf{G}$, $\mathbf{K}$, and $\mathbf{H}$ such that $\mathbf{G}_P + \mathbf{G}_I/s = \mathbf{G}$, $\mathbf{K}_P + \mathbf{K}_I/s = \mathbf{K}$, and $\mathbf{H}_P +$

$\mathbf{H}_I/s = \mathbf{H}$. Since (13.41) holds for all $t \geq t_0$ and provides a constant voltage set-point vector v^{*ss}, the time-varying part of (13.41) is zero:

$$\left(v^r - \langle \mathbf{v}^{ss} \rangle\right)\mathbf{U}_1 \mathbf{1} - \tilde{\mathbf{L}} \lambda_Q^{ss} + (\sigma^{2*} - \sigma^{2,ss})\mathbf{U}_2 \left(\mathbf{v}^{ss} - \langle \mathbf{v}^{ss} \rangle \mathbf{1}\right) = \mathbf{0} \tag{13.42}$$

where $\mathbf{U}_1 = c^{-1}\mathbf{G}_I \mathbf{H}_I^{-1} = \mathrm{diag}(u_{1i})$ and $\mathbf{U}_2 = c^{-1}\mathbf{K}_I \mathbf{H}_I^{-1} = \mathrm{diag}(u_{2i})$ are diagonal matrices. Note that $\mathbf{M}\left(\mathbf{v}^{ss} - \langle \mathbf{v}^{ss} \rangle \mathbf{1}\right) = \mathbf{0}$. Premultiplying both sides of (13.42) by $\mathbf{M}\mathbf{U}_2^{-1}$, we have:

$$\left(v^r - \langle \mathbf{v}^{ss} \rangle\right)\mathbf{M}\mathbf{U}_2^{-1}\mathbf{U}_1 \mathbf{1} = \mathbf{M}\mathbf{U}_2^{-1}\tilde{\mathbf{L}}\lambda_Q^{ss} \tag{13.43}$$

which is equivalent to:

$$\left(v^r - \langle \mathbf{v}^{ss} \rangle\right)\mathbf{U}_2^{-1}\mathbf{U}_1 \mathbf{1} = \mathbf{U}_2^{-1}\tilde{\mathbf{L}}\lambda_Q^{ss} \tag{13.44}$$

Premultiplying both sides of (13.44) by $\mathbf{U}_2$, we have:

$$\left(v^r - \langle \mathbf{v}^{ss} \rangle\right)\mathbf{U}_1 \mathbf{1} = \tilde{\mathbf{L}}\lambda_Q^{ss} \tag{13.45}$$

Assume that after modifying $\mathbf{A}$ to be $\tilde{\mathbf{A}} = [\tilde{a}_{ij}]$ by setting the kth row and kth column to be zero, the corresponding Laplacian matrix $\tilde{\mathbf{L}}$ is still balanced. Note that the elements of the kth row and kth column of $\tilde{\mathbf{L}}$ are also all zeros. Given the balanced Laplacian matrix, $\mathbf{1}^T\tilde{\mathbf{L}} = \mathbf{0}^T$. Premultiplying (13.45) by $\mathbf{1}^T$, we have:

$$\left(v^r - \langle \mathbf{v}^{ss} \rangle\right) \sum_{i=1}^{N} u_{1i} = 0 \tag{13.46}$$

Because all entries of the matrix $\mathbf{U}_1$ are positive, from (13.46) we have $\langle \mathbf{v}^{ss} \rangle = v^r$, implying that the controllers can successfully regulate the averaged voltage magnitude of the microgrid at the rated voltage. As in [4], substituting $v^r - \langle \mathbf{v}^{ss} \rangle = 0$ into (13.45) we have:

$$\tilde{\mathbf{L}}\lambda_Q^{ss} = \mathbf{0} \tag{13.47}$$

Because all of the elements of the kth row and kth column of $\tilde{\mathbf{L}}$ are zero, (13.47) is reduced to:

$$\check{\mathbf{L}}\check{\lambda}_Q^{ss} = \mathbf{0} \tag{13.48}$$

where $\check{\mathbf{L}}$ is obtained by eliminating the kth row and kth column of $\tilde{\mathbf{L}}$ and $\check{\lambda}_Q^{ss}$ is obtained by removing the kth element. As has been shown in [4, 10], (13.48) leads to $\check{\lambda}_Q^{ss} = m\mathbf{1}$ where m is any real number, indicating that the controller can share the total reactive load among the sources except source k in proportion to their ratings. Substituting $v^r - \langle \mathbf{v}^{ss} \rangle = 0$ and (13.47) into (13.42), we can get:

$$(\sigma^{2*} - \sigma^{2,ss})\mathbf{U}_2 \left(\mathbf{v}^{ss} - v^r\mathbf{1}\right) = \mathbf{0} \tag{13.49}$$

Since all entries of the matrix $\mathbf{U}_2$ are positive, (13.49) requires:

$$(\sigma^{2*} - \sigma^{2,ss})(v_i^{ss} - v^r) = 0, \quad \forall i = 1, \ldots, N \tag{13.50}$$

Since the controllers can only guarantee $\langle \mathbf{v}^{ss} \rangle = v^r$ while $v_i^{ss} = v^r$ for $\forall i = 1, \ldots, N$ is a much stronger condition that corresponds to a zero variance of the voltage magnitude, which, for most cases, cannot hold. Therefore, in order to satisfy (13.49), there is $\sigma^{2,ss} = \sigma^{2*}$, indicating that the controllers can successfully regulate the variance of the voltage to the reference value σ^{2*}.

Therefore, the mismatch inputs to the three controllers for voltage control in Section 13.4 all decay to zero in the steady-state, which results in successful tracking of average voltage set-point, tracking of voltage variance set-point, and sharing of reactive power among all sources except the selected one.

13.6 Microgrid Test System and Control Performance

The performance of the distributed control is evaluated through real-time simulations in OPAL-RT OP5600 platform. All simulations are performed at a time-step of 50 μs using the fixed-step discrete ODE-5 (Dormand–Prince) solver. We consider a microgrid with radially connected four DGs as shown in Fig. 13.10. Similar test systems have been used in [4, 5, 41].

The parameters of the test system and the control are given in Table 13.1. To eliminate the effect of inverter switching harmonics, LCL filter is considered at the output of each inverter. The inverters are connected by RL-lines.

A ring communication topology shown in Fig. 13.10 is considered to provide a sparse communication network with a spanning tree and a balanced Laplacian matrix [4]. The adjacency matrix $\mathbf{A}$ and the coupling gains b and c for the cyber network are selected as:

$$\mathbf{A} = \begin{bmatrix} 0 & 2.8 & 0 & 2.8 \\ 2.8 & 0 & 2.8 & 0 \\ 0 & 2.8 & 0 & 2.8 \\ 2.8 & 0 & 2.8 & 0 \end{bmatrix}, \quad b = 2, \quad c = 0.025$$

The variance reference σ^{2*} is set as 16, unless otherwise noted. For reactive power regulation, the DG4 is considered as the special DG that does not participate in reactive power sharing. In order to track the references from the secondary control, PR voltage and proportional current control

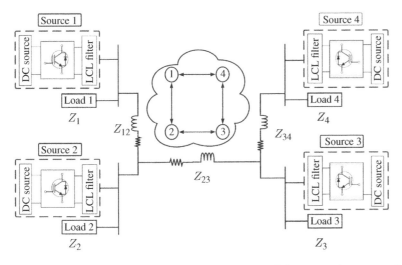

Figure 13.10 Schematic diagram of the considered radial connected test network with ring communication topology. Source: Reproduced with permission from [11]/IEEE.

Table 13.1 Parameters of the test system and control.

	Parameters		Value
	Symbol	Quantity	
DGs	v^r	Rated voltage	$\left(208 \times \frac{\sqrt{2}}{\sqrt{3}}\right)$ V
	R_f	LCL filter resistance	0.1 Ω
	L_f	LCL filter inductance	1.8 mH
	C_f	LCL filter capacitance	25 μF
	PDG1–PDG3	DG1–DG3 rated active power	1300 W
	PDG4	DG4 rated active power	2600 W
	QDG1–QDG3	DG1–DG3 rated reactive power	1300 VAr
	QDG4	DG4 rated reactive power	2600 VAr
Loads	Z_1	Load 1	$(25 + 18.85i)\,\Omega$
	Z_2	Load 2	$(50 + 37.7i)\,\Omega$
	Z_3, Z_4	Load 3, Load 4	$(100 + 74.4i)\,\Omega$
Lines	Z_{12}	Line impedance 1, 2	$(0.8 + 1.357i)\,\Omega$
	Z_{23}	Line impedance 2, 3	$(0.4 + 0.678i)\,\Omega$
	Z_{34}	Line impedance 3, 4	$(0.7 + 0.565i)\,\Omega$
Control in Section 13.4	G_P, G_I	Voltage control P, I term	0.0008, 3
	K_P, K_I	Variance control P, I term	0.0008, 3
	H_P, H_I	Reactive power control P, I term	0.02, 15
Control in Section 13.3	G_P, G_I	Voltage control P, I term	0.0008, 3
	H_P, H_I	Reactive power control P, I term	0.02, 8

loops are used as the zero-level control to generate the desired signals at the output of the PWM inverter [12].

Load Z_1 is increased at 10 and 60 seconds. Figure 13.11 shows the performance of the distributed control in Sections 13.4 and 13.3. It is seen that active power sharing is achieved by both methods. The control methods in Section 13.3 achieve reactive power sharing among all four DGs but fail to bind the voltage deviations within 5% of the rated voltage. In contrast, the distributed control in Section 13.4 limits the voltage deviation within 5% of the rated voltage by relaxing reactive power sharing of the special DG and regulating the voltage variance to its reference value.

The three voltage correction terms and the frequency response are shown in Fig. 13.12. Small transients are generated after load change but the distributed control with voltage variance rapidly adjusts the frequency to the nominal value. Tracking of the global voltage and global voltage variance are also depicted which validate the controller performance on retaining the voltage and voltage variance irrespective of load change.

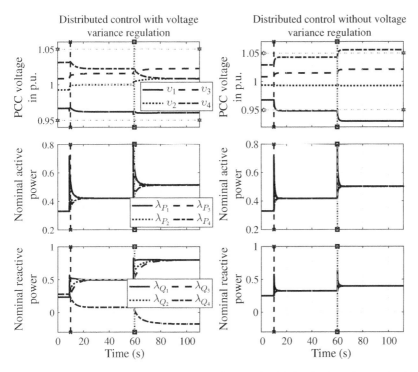

Figure 13.11 Comparison of the control in Section 13.4 (left, distributed control with voltage variance regulation) and that in Section 13.3 (right, distributed control without voltage variance regulation). Source: Reproduced with permission from [11]/IEEE.

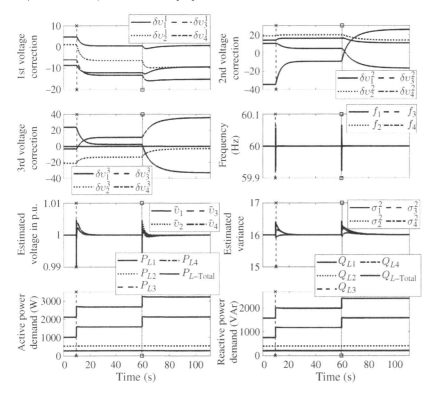

Figure 13.12 Detailed performance of the control in Section 13.4. Source: Reproduced with permission from [11]/IEEE.

13.7 Steady-State Performance Under Different Loading Conditions and Controller Settings

Choose any $l \in \{1, \ldots, N\}$. The control in Section 13.3 is equivalent to solving the following non-linear equations:

$$\begin{cases} \sum_{j=1}^{N} v_j/N = v^r & \text{(13.51a)} \\[2mm] Q_l/\overline{Q}_l = Q_i/\overline{Q}_i, \quad \forall i \neq l & \text{(13.51b)} \\[2mm] P_l/\overline{P}_l = P_i/\overline{P}_i, \quad \forall i \neq l & \text{(13.51c)} \end{cases}$$

Here, P_i and Q_i are functions of the voltage magnitudes and phase angles as defined in (13.1)–(13.2). Note that in (13.51) there are a total of $2N - 1$ equations with $2N - 1$ variables $v_1, v_2, \ldots, v_N, \delta_2, \ldots, \delta_N$ to be solved for (assume δ_1 is the reference), and thus the equations in (13.51) will have a unique practical solution if there is any.

As for the control in Section 13.4, we first choose $k \in \{1, \ldots, N\}$ and the kth source will not be implemented in reactive power regulator. Usually, the source with the largest capacity is selected because it has the largest room to regulate its reactive power. Then choose any $l \in \{1, \ldots, N\} \backslash k$. The control in Section 13.4 is equivalent to solving the following nonlinear equations:

$$\begin{cases} \sum_{j=1}^{N} v_j/N = v^r & \text{(13.52a)} \\[2mm] \sum_{u=1}^{N} \left(v_u - \sum_{j=1}^{N} v_j/N \right)^2 / N = \sigma^{2*} & \text{(13.52b)} \\[2mm] Q_l/\overline{Q}_l = Q_i/\overline{Q}_i, \quad \forall i \neq l \text{ or } k & \text{(13.52c)} \\[2mm] P_l/\overline{P}_l = P_i/\overline{P}_i, \quad \forall i \neq l & \text{(13.52d)} \end{cases}$$

Note that in (13.52), there are also a total of $2N - 1$ equations with $2N - 1$ variables to be solved for, and thus these equations will have a unique practical solution if there is any.

Simulating microgrid control in Simulink with detailed inverter models and very small time steps can be very time consuming. Therefore, evaluating the performance of a microgrid control for a large system with a large number of DG units by using detailed Simulink simulation is very challenging. By contrast, the equivalent problems provide an alternative and more computationally tractable way to evaluate the performance of the distributed controls.

For the same case in Section 13.6, we have calculated the steady-state solutions and compared them with those from the Matlab/Simulink simulation, and found that they are consistent. Then only by solving these nonlinear equations, the performance of the distributed control with voltage variance can be evaluated for a wide range of loading conditions and controller settings.

In Fig. 13.13, we show the point of common coupling (PCC) voltage and nominal reactive power of the distributed control with voltage variance regulation and the droop-free control in [4] under different loading conditions with fixed power factor of 0.8. The 2.656 kVA load corresponds to the same loading condition before the load change in Section 13.6. For the distributed control with voltage variance regulation, the variance reference is set as $\sigma^{2*} = 16$. From Fig. 13.13, it is clear that with the load changes, the voltage deviations in the droop-free control in [4] increases while the distributed control with voltage variance can regulate the voltage variance to its reference value.

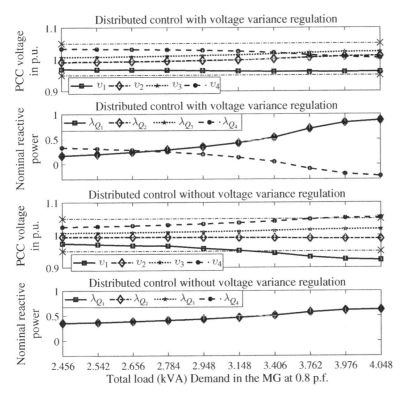

Figure 13.13 Performance comparison of the controls in Section 13.4 (distributed control with voltage variance regulation) and Section 13.3 (distributed control without voltage variance regulation) under different loading conditions. For the control in Section 13.4 fixed variance reference $\sigma^{2*} = 16$ is used. Source: Reproduced with permission from [11]/IEEE.

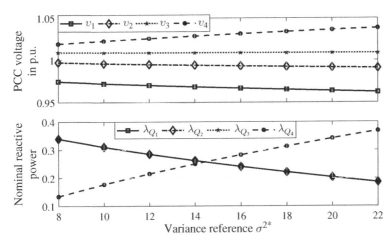

Figure 13.14 Performance of the control in Section 13.4 under different variance reference settings with a total load of 2.12×10^3 W and 1.6×10^3 Var. Source: Reproduced with permission from [11]/IEEE.

Under 4.048 kVA load, the maximum PCC voltage deviation in the droop-free control is 7.67%, while for the control in Section 13.4 it is only 3.89%.

In Fig. 13.14, the controller outputs are shown for a 2.656 kVA load with power factor of 0.8 under different settings of the variance reference. It is seen that the controller can regulate the voltage variance to a wide range of variance settings by utilizing the flexibility of the special DG while achieving equal reactive power sharing among the other DGs.

13.8 Exercises

1. What is the typical control structure of an AC microgrid?
2. What are the similarities and differences between the control in Section 13.4 and that in Section 13.3?
3. Given the 4-DG test network in Fig. 13.10, compare the voltage magnitudes and reactive power outputs of the DGs with the controls in Sections 13.2.3, 13.3, and 13.4 under changing load in Z_1, Z_2, Z_3, and/or Z_4.

References

1 Olivares, D.E., Mehrizi-Sani, A., Etemadi, A.H. et al. (2014). Trends in microgrid control. *IEEE Transactions on Smart Grid* 5 (4): 1905–1919.

2 Wang, Z., Wu, W., and Zhang, B. (2018). A distributed quasi-newton method for droop-free primary frequency control in autonomous microgrids. *IEEE Transactions on Smart Grid* 9 (3): 2214–2223.

3 Han, Y., Li, H., Shen, P. et al. (2017). Review of active and reactive power sharing strategies in hierarchical controlled microgrids. *IEEE Transactions on Power Electronics* 32 (3): 2427–2451.

4 Nasirian, V., Shafiee, Q., Guerrero, J.M. et al. (2016). Droop-free distributed control for AC microgrids. *IEEE Transactions on Power Electronics* 31 (2): 1600–1617.

5 Bidram, A., Nasirian, V., Davoudi, A., and Lewis, F.L. (2017). *Cooperative Synchronization in Distributed Microgrid Control*. Switzerland: Springer.

6 Lin, Y., Eto, J.H., Johnson, B.B. et al. (2020). Research Roadmap on Grid-Forming Inverters. *Technical Report*. Golden, CO (United States): National Renewable Energy Lab.(NREL).

7 Du, W., Tuffner, F.K., Schneider, K.P. et al. (2021). Modeling of grid-forming and grid-following inverters for dynamic simulation of large-scale distribution systems. *IEEE Transactions on Power Delivery* 36 (4): 2035–2045.

8 Bidram, A. and Davoudi, A. (2012). Hierarchical structure of microgrids control system. *IEEE Transactions on Smart Grid* 3 (4): 1963–1976.

9 Kersting, W.H. (2006). *Distribution System Modeling and Analysis*. CRC Press.

10 Nasirian, V., Moayedi, S., Davoudi, A., and Lewis, F.L. (2015). Distributed cooperative control of DC microgrids. *IEEE Transactions on Power Electronics* 30 (4): 2288–2303.

11 Mohiuddin, S.M. and Qi, J. (2020). Droop-free distributed control for AC microgrids with precisely regulated voltage variance and admissible voltage profile guarantees. *IEEE Transactions on Smart Grid* 11 (3): 1956–1967.

12 Macana, C.A. and Pota, H.R. (2017). Adaptive synchronous reference frame virtual impedance controller for accurate power sharing in islanded AC-microgrids: a faster alternative to the conventional droop control. *2017 IEEE ECCE*, 3728–3735, October 2017.

13 Vasquez, J.C., Guerrero, J.M., Savaghebi, M. et al. (2013). Modeling, analysis, and design of stationary-reference-frame droop-controlled parallel three-phase voltage source inverters. *IEEE Transactions on Industrial Electronics* 60 (4): 1271–1280.

14 Pogaku, N., Prodanovic, M., and Green, T.C. (2007). Modeling, analysis and testing of autonomous operation of an inverter-based microgrid. *IEEE Transactions on Power Electronics* 22 (2): 613–625.

15 Dehkordi, N.M., Sadati, N., and Hamzeh, M. (2017). Fully distributed cooperative secondary frequency and voltage control of islanded microgrids. *IEEE Transactions on Energy Conversion* 32 (2): 675–685.

16 Mahmood, H., Michaelson, D., and Jiang, J. (2015). Reactive power sharing in islanded microgrids using adaptive voltage droop control. *IEEE Transactions on Smart Grid* 6 (6): 3052–3060.

17 Shafiee, Q., Nasirian, V., Vasquez, J.C. et al. (2018). A multi-functional fully distributed control framework for AC microgrids. *IEEE Transactions on Smart Grid* 9 (4): 3247–3258.

18 Simpson-Porco, J.W., Shafiee, Q., Dörfler, F. et al. (2015). Secondary frequency and voltage control of islanded microgrids via distributed averaging. *IEEE Transactions on Industrial Electronics* 62 (11): 7025–7038.

19 Fan, Y., Hu, G., and Egerstedt, M. (2017). Distributed reactive power sharing control for microgrids with event-triggered communication. *IEEE Transactions on Control Systems Technology* 25 (1): 118–128.

20 Zhu, Y., Fan, Q., Liu, B., and Wang, T. (2018). An enhanced virtual impedance optimization method for reactive power sharing in microgrids. *IEEE Transactions on Power Electronics* 33 (12): 10390–10402.

21 Nutkani, I.U., Loh, P.C., and Blaabjerg, F. (2014). Droop scheme with consideration of operating costs. *IEEE Transactions on Power Electronics* 29 (3): 1047–1052.

22 Rowe, C.N., Summers, T.J., Betz, R.E. et al. (2013). Arctan power–frequency droop for improved microgrid stability. *IEEE Transactions on Power Electronics* 28 (8): 3747–3759.

23 Wu, X., Shen, C., and Iravani, R. (2018). A distributed, cooperative frequency and voltage control for microgrids. *IEEE Transactions on Smart Grid* 9 (4): 2764–2776.

24 Dehkordi, N.M., Sadati, N., and Hamzeh, M. (2017). Distributed robust finite-time secondary voltage and frequency control of islanded microgrids. *IEEE Transactions on Power Apparatus and Systems* 32 (5): 3648–3659.

25 Rocabert, J., Luna, A., Blaabjerg, F., and Rodríguez, P. (2012). Control of power converters in AC microgrids. *IEEE Transactions on Power Electronics* 27 (11): 4734–4749.

26 Du, W., Chen, Z., Schneider, K.P. et al. (2020). A comparative study of two widely used grid-forming droop controls on microgrid small-signal stability. *IEEE Journal of Emerging and Selected Topics in Power Electronics* 8 (2): 963–975.

27 Sun, H., Guo, Q., Qi, J. et al. (2019). Review of challenges and research opportunities for voltage control in smart grids. *IEEE Transactions on Power Apparatus and Systems* 34 (4): 2790–2801.

28 Maknouninejad, A. and Qu, Z. (2014). Realizing unified microgrid voltage profile and loss minimization: a cooperative distributed optimization and control approach. *IEEE Transactions on Smart Grid* 5 (4): 1621–1630.

29 Shafiee, Q., Nasirian, V., Guerrero, J.M. et al. (2014). Team-oriented adaptive droop control for autonomous AC microgrids. *IECON 2014 - 40th Annual Conference of the IEEE Industrial Electronics Society*, 1861–1867, October 2014.

30 Xu, Y. and Sun, H. (2018). Distributed finite-time convergence control of an islanded low-voltage AC microgrid. *IEEE Transactions on Power Apparatus and Systems* 33 (3): 2339–2348.

31 Etemadi, A.H., Davison, E.J., and Iravani, R. (2012). A decentralized robust control strategy for multi-DER microgrids–Part I: Fundamental concepts. *IEEE Transactions on Power Delivery* 27 (4): 1843–1853.

32 Micallef, A., Apap, M., Spiteri-Staines, C. et al. (2014). Reactive power sharing and voltage harmonic distortion compensation of droop controlled single phase islanded microgrids. *IEEE Transactions on Smart Grid* 5 (3): 1149–1158.

33 Han, R., Meng, L., Ferrari-Trecate, G. et al. (2017). Containment and consensus-based distributed coordination control to achieve bounded voltage and precise reactive power sharing in islanded AC microgrids. *IEEE Transactions on Industry Applications* 53 (6): 5187–5199.

34 Bidram, A., Davoudi, A., and Lewis, F.L. (2014). A multiobjective distributed control framework for islanded AC microgrids. *IEEE Transactions on Industrial Informatics* 10 (3): 1785–1798.

35 Qu, Z. (2009). *Cooperative Control of Dynamical Systems: Applications to Autonomous Vehicles*. London: Springer.

36 Yazdanian, M. and Mehrizi-Sani, A. (2014). Distributed control techniques in microgrids. *IEEE Transactions on Smart Grid* 5 (6): 2901–2909.

37 Zhang, H., Kim, S., Sun, Q., and Zhou, J. (2017). Distributed adaptive virtual impedance control for accurate reactive power sharing based on consensus control in microgrids. *IEEE Transactions on Smart Grid* 8 (4): 1749–1761.

38 Cheng, Z., Duan, J., and Chow, M. (2018). To centralize or to distribute: that is the question: a comparison of advanced microgrid management systems. *IEEE Industrial Electronics Magazine* 12 (1): 6–24.

39 Zhang, Y. and Ma, H. (2012). Theoretical and experimental investigation of networked control for parallel operation of inverters. *IEEE Transactions on Industrial Electronics* 59 (4): 1961–1970.

40 Bidram, A., Davoudi, A., Lewis, F.L., and Qu, Z. (2013). Secondary control of microgrids based on distributed cooperative control of multi-agent systems. *IET Generation, Transmission and Distribution* 7 (8): 822–831.

41 Bidram, A., Davoudi, A., Lewis, F.L., and Guerrero, J.M. (2013). Distributed cooperative secondary control of microgrids using feedback linearization. *IEEE Transactions on Power Apparatus and Systems* 28 (3): 3462–3470.

42 Wu, X., Xu, Y., Wu, X. et al. (2020). A two-layer distributed cooperative control method for islanded networked microgrid systems. *IEEE Transactions on Smart Grid* 11 (2): 942–957.

43 Zhou, J., Kim, S., Zhang, H. et al. (2018). Consensus-based distributed control for accurate reactive, harmonic, and imbalance power sharing in microgrids. *IEEE Transactions on Smart Grid* 9 (4): 2453–2467.

44 Xu, Y., Sun, H., Gu, W. et al. (2019). Optimal distributed control for secondary frequency and voltage regulation in an islanded microgrid. *IEEE Transactions on Industrial Informatics* 15 (1): 225–235.

45 IEEE P1547.4/D11 (2011). *IEEE Draft Guide for Design, Operation, and Integration of Distributed Resource Island Systems with Electric Power Systems*, March 2011, 1–55.

46 Gui, E.M., Diesendorf, M., and MacGill, I. (2017). Distributed energy infrastructure paradigm: community microgrids in a new institutional economics context. *Renewable and Sustainable Energy Reviews* 72: 1355–1365.

14

Optimal Distributed Control of AC Microgrids

Sheik M. Mohiuddin and Junjian Qi

In Chapter 13, a distributed control scheme with an average voltage regulator, a voltage variance regulator, and a relaxed reactive power-sharing regulator is introduced. However, to achieve the control objectives, a special distributed generator (DG) is required which may have a large variation in reactive power-sharing from the other DGs. Furthermore, the technical constraints on voltage magnitude and reactive power output capacity of the DGs have not been systematically considered in the control of islanded microgrids. Also, the distributed microgrid controls extensively use proportional integral (PI) controllers which may not always guarantee theoretical convergence.

Consequently, a generalized control framework is required for optimally coordinating voltage regulation and reactive power-sharing objectives while obeying all necessary technical constraints. The authors in [1, 2] propose optimal distributed voltage control with limited communication among the neighboring agents where voltage and reactive power constraints are considered and an objective function to minimize the power loss and reactive power operation cost is defined. A hybrid voltage control strategy with limited communication is also proposed in [3]. However, the controls in [1–3] are developed using the linearized DistFlow method of the radial distribution network and work with the grid-following inverters under grid-connected mode where reactive power from the DGs is considered as the control variable. In this chapter, a generalized optimization problem is formulated for the grid-forming (GFM) inverters in islanded AC microgrids.

Notations: Define $\mathbb{R}^N$ as the N dimensional real vector space and $\mathbb{R}_+^N$ as the nonnegative orthant in $\mathbb{R}^N$. Denote $\mathbf{0} \in \mathbb{R}^N$ as a column vector with all zeros and $\mathbf{1} \in \mathbb{R}^N$ as a column vector with all ones. For a vector $\mathbf{x} \in \mathbb{R}^N$, $\mathbf{x} \leq \mathbf{0}$ means each component of $\mathbf{x}$ is less than or equal to zero. Denote $\text{col}(\mathbf{x}_1, \ldots, \mathbf{x}_N) = (\mathbf{x}_1^\top, \ldots, \mathbf{x}_N^\top)^\top$ as the column vector stacked with column vectors $\mathbf{x}_1, \ldots, \mathbf{x}_N$. Given a matrix $\mathbf{X} \in \mathbb{R}^{M \times N}$, we let $\mathbf{x}_{*,i}$ denote the ith column-vector of $\mathbf{X}$. For a set $\Omega \subset \mathbb{R}^N$, its relative interior is $\text{rint}(\Omega)$. A projection operator is defined as $\mathcal{P}_C(\mathbf{z}) \triangleq \text{argmin}_{\mathbf{x} \in C} \|\mathbf{x} - \mathbf{z}\|$.

14.1 Optimization Problem for Secondary Control

For frequency regulation, the same control as in Section 13.3 [4] can be used. For voltage and reactive power regulation in the secondary control, the following design principles need to be considered:

1. **Distributed structure**: The algorithm should be fully distributed and each agent should update its local variables only based on its local data and the information from its neighbors.

Microgrids: Theory and Practice, First Edition. Edited by Peng Zhang.
© 2024 The Institute of Electrical and Electronics Engineers, Inc. Published 2024 by John Wiley & Sons, Inc.

2. **Optimality**: The distributed algorithm to be developed should be guaranteed to converge to the same optimal solution as the original optimization problem for which a proper trade-off is made between voltage regulation and reactive power-sharing among the DGs.
3. **Constraints**: The technical constraints on voltage magnitude and reactive power capacity should always be satisfied.

14.1.1 Formulated Optimization Problem

The voltage and reactive power regulation in the secondary control is formulated as the following optimization problem:

$$\min_{\mathbf{v}} \ f(\mathbf{v}) = \sum_{i=1}^{N} f_i(\mathbf{v}) \tag{14.1a}$$

$$\text{s.t.} \quad \mathbf{1}^\mathsf{T}\mathbf{v}/N - v^r = 0 \tag{14.1b}$$

$$-\lambda_{\mathbf{Q}}(\mathbf{v}, \theta) + \underline{\lambda}_{\mathbf{Q}} \leq \mathbf{0} \tag{14.1c}$$

$$\lambda_{\mathbf{Q}}(\mathbf{v}, \theta) - \overline{\lambda}_{\mathbf{Q}} \leq \mathbf{0} \tag{14.1d}$$

$$-\mathbf{v} + \underline{\mathbf{v}} \leq \mathbf{0} \tag{14.1e}$$

$$\mathbf{v} - \overline{\mathbf{v}} \leq \mathbf{0}, \tag{14.1f}$$

where v^r is the rated voltage in per unit, $\mathbf{v} = [v_1, v_2, \dots, v_N]^\mathsf{T}$ with lower and upper bounds as $\underline{\mathbf{v}} = [\underline{v}_1, \underline{v}_2, \dots, \underline{v}_N]^\mathsf{T}$ and $\overline{\mathbf{v}} = [\overline{v}_1, \overline{v}_2, \dots, \overline{v}_N]^\mathsf{T}$ (usually $\underline{v}_i$ and $\overline{v}_i$ are chosen as 0.95 and 1.05, respectively), $\theta = [\theta_2, \dots, \theta_N]^\mathsf{T}$ ($\theta_1 = 0$ is selected as the reference), $\lambda_{\mathbf{Q}} = [\lambda_{Q_1}, \lambda_{Q_2}, \dots, \lambda_{Q_N}]^\mathsf{T}$, and $\underline{\lambda}_{\mathbf{Q}} = [\underline{\lambda}_{Q_1}, \underline{\lambda}_{Q_2}, \dots, \underline{\lambda}_{Q_N}]^\mathsf{T}$ and $\overline{\lambda}_{\mathbf{Q}} = [\overline{\lambda}_{Q_1}, \overline{\lambda}_{Q_2}, \dots, \overline{\lambda}_{Q_N}]^\mathsf{T}$ are the lower and upper limit vectors for normalized reactive power. The objective function for DG i, $f_i(\mathbf{v})$, is defined as:

$$f_i(\mathbf{v}) = \frac{\alpha|\mathcal{N}_i|^2}{2}\sigma_i^2 + \frac{\beta}{2}\sum_{j\in\mathcal{N}_i} a_{ij}(\lambda_{Q_i} - \lambda_{Q_j})^2$$

$$\triangleq f_i^1(\mathbf{v}) + f_i^2(\mathbf{v}), \tag{14.2}$$

where

$$\sigma_i^2 = \sum_{j\in\mathcal{N}_i} \frac{\left(v_j - \sum_{k\in\mathcal{N}_i} v_k/|\mathcal{N}_i|\right)^2}{|\mathcal{N}_i|} \tag{14.3}$$

$\alpha \geq 0$ and $\beta \geq 0$ ($\alpha\beta \neq 0$) as design parameters, $\mathcal{N}_i$ is the set of the in-neighbors of DG i in the communication network (including node i), and $|\mathcal{N}_i|$ is the cardinality of $\mathcal{N}_i$. Note that the first term in (14.2) and the equality constraint in (14.1b) together will guarantee admissible voltage profiles while only one of them cannot. If α and β are chosen properly, the voltage profile should be admissible for usual operating conditions. However, considering many different possible operating conditions, it is still possible that the voltage magnitude of some DGs may violate their upper or lower bounds. For that reason, the constraints (14.1e)–(14.1f) are still needed.

14.1.2 Convex Optimization Problem

With the nonlinear reactive power injection functions in both objective function and inequality constraints, (14.1) is a nonlinear, nonconvex optimization problem, which is very challenging to solve. In order to address this problem, the nonlinear reactive power injection functions can be approximated by linear functions to convert problem (14.1) to a convex optimization problem.

Specifically, it has been shown that the reactive power injection can be approximated as [5]:

$$\tilde{Q}_i = -\sum_{j=1}^{N} B_{ij} v_j - \sum_{j=1}^{N} G_{ij} \theta_j \tag{14.4}$$

It is clear that λ_{Q_i} is approximately a linear function of v_j for $j = 1, \ldots, N$. As shown in both [5] and our own numerical experiments on AC microgrids, such an approximation has acceptable accuracy compared with the nonlinear reactive power injection. Using $\tilde{\lambda}_{Q_i} = \tilde{Q}_i / \overline{Q}_i$ to approximate λ_{Q_i}, the objective function becomes:

$$\tilde{f}_i(\mathbf{v}) = \frac{\alpha |\mathcal{N}_i|^2}{2} \sigma_i^2 + \frac{\beta}{2} \sum_{j \in \mathcal{N}_i} a_{ij} (\tilde{\lambda}_{Q_i} - \tilde{\lambda}_{Q_j})^2$$

$$\triangleq f_i^1(\mathbf{v}) + \tilde{f}_i^2(\mathbf{v}). \tag{14.5}$$

Then the optimization problem (14.1) is modified to be:

$$\min_{\mathbf{v}} \tilde{f}(\mathbf{v}) = \sum_{i=1}^{N} \tilde{f}_i(\mathbf{v}) \tag{14.6a}$$

$$\text{s.t.} \quad \mathbf{1}^{\mathsf{T}} \mathbf{v} / N - v^{\mathsf{r}} = 0 \tag{14.6b}$$

$$-\tilde{\mathbf{B}} \mathbf{v} - \tilde{\mathbf{G}} \theta + \underline{\lambda}_{\mathbf{Q}} \leq \mathbf{0} \tag{14.6c}$$

$$\tilde{\mathbf{B}} \mathbf{v} + \tilde{\mathbf{G}} \theta - \overline{\lambda}_{\mathbf{Q}} \leq \mathbf{0} \tag{14.6d}$$

$$-\mathbf{v} + \underline{\mathbf{v}} \leq \mathbf{0} \tag{14.6e}$$

$$\mathbf{v} - \overline{\mathbf{v}} \leq \mathbf{0}, \tag{14.6f}$$

where the elements in $\tilde{\mathbf{B}}$ and $\tilde{\mathbf{G}}$ are obtained by $\tilde{B}_{ij} = -B_{ij} / \overline{Q}_i$ and $\tilde{G}_{ij} = -G_{ij} / \overline{Q}_i$.

Let $h(\mathbf{v}) = \mathbf{1}^{\mathsf{T}} \mathbf{v} / N - v^{\mathsf{r}}$. (14.6e)–(14.6f) are defined as local constraint set $\Omega_i = \{v_i \in \mathbb{R} | \underline{v}_i \leq v_i \leq \overline{v}_i\}$ for $i = 1, \ldots, N$. The reactive power constraints (14.6c)–(14.6d) are denoted by $\mathbf{g}(\mathbf{v}) \leq \mathbf{0}$ where,

$$\mathbf{g}(\mathbf{v}) \triangleq \begin{bmatrix} \mathbf{g}^{\mathrm{L}}(\mathbf{v}) \\ \mathbf{g}^{\mathrm{U}}(\mathbf{v}) \end{bmatrix} = \begin{bmatrix} -\tilde{\mathbf{B}} \mathbf{v} - \tilde{\mathbf{G}} \theta + \underline{\lambda}_{\mathbf{Q}} \\ \tilde{\mathbf{B}} \mathbf{v} + \tilde{\mathbf{G}} \theta - \overline{\lambda}_{\mathbf{Q}} \end{bmatrix} \tag{14.7}$$

Then the optimization problem (14.6) is equivalently written as:

$$\min_{\mathbf{v} \in \Omega} \tilde{f}(\mathbf{v}) = \sum_{i=1}^{N} \tilde{f}_i(\mathbf{v}) \tag{14.8a}$$

$$\text{s.t.} \quad h(\mathbf{v}) = 0 \tag{14.8b}$$

$$\mathbf{g}(\mathbf{v}) \leq \mathbf{0}, \tag{14.8c}$$

where $\Omega \triangleq \prod_{i \in \mathcal{V}} \Omega_i \subset \mathbb{R}^N$ denotes the local constraints of the N agents and its relative interior is $\mathrm{rint}(\Omega)$.

14.1.3 Convexity of Problem (14.8)

Note that

$$\frac{\partial f_i^1(\mathbf{v})}{\partial v_i} = \alpha \sum_{j \in \mathcal{N}_i} (v_i - v_j). \tag{14.9}$$

Similarly, the partial derivatives of f_i^1 with respect to v_j $(j \in \mathcal{N}_i)$ and v_k $(k \notin \mathcal{N}_i)$ can be obtained as:

$$\frac{\partial f_i^1(\mathbf{v})}{\partial v_j} = \alpha \sum_{k \in \mathcal{N}_i} (v_j - v_k) \text{ and } \frac{\partial f_i^1(\mathbf{v})}{\partial v_k} = 0.$$

The partial derivative of $\tilde{f}_i$ with respect to v_i is:

$$\frac{\partial \tilde{f}_i(\mathbf{v})}{\partial v_i} = \sum_{j \in \mathcal{N}_i} \left(\alpha(v_i - v_j) + \beta a_{ij}(\tilde{\lambda}_{Q_i} - \tilde{\lambda}_{Q_j}) \frac{\partial(\tilde{\lambda}_{Q_i} - \tilde{\lambda}_{Q_j})}{\partial v_i} \right) \tag{14.10}$$

where

$$\frac{\partial(\tilde{\lambda}_{Q_i} - \tilde{\lambda}_{Q_j})}{\partial v_i} = \frac{-B_{ii}}{\overline{Q}_i} + \frac{B_{ij}}{\overline{Q}_j} \triangleq d_i \tag{14.11}$$

It is clear that f_i^1 is a convex function of $\mathbf{v}$. To check the convexity of $\tilde{f}_i$ with respect to $\mathbf{v}$, we only need to check the convexity of $F_{ij} = (\tilde{\lambda}_{Q_i} - \tilde{\lambda}_{Q_j})^2/2$ for $j \in \mathcal{N}_i$.

For $j \in \mathcal{N}_i$ we have

$$\frac{\partial(\tilde{\lambda}_{Q_i} - \tilde{\lambda}_{Q_j})}{\partial v_j} = \frac{-B_{ij}}{\overline{Q}_i} + \frac{B_{jj}}{\overline{Q}_j} \triangleq d_j$$

For $k \neq i, j \in \mathcal{N}_i$, we have

$$\frac{\partial(\tilde{\lambda}_{Q_i} - \tilde{\lambda}_{Q_j})}{\partial v_k} = \frac{-B_{ik}}{\overline{Q}_i} + \frac{B_{jk}}{\overline{Q}_j} \triangleq d_k.$$

Then there is

$$\frac{\partial^2 F_{ij}}{\partial v_l v_m} = d_l d_m, \ l, m \in \{i, j, k\}. \tag{14.12}$$

Let $\mathbf{d} = [d_1, d_2, \ldots, d_N]^\top$, the Hessian matrix for F_{ij} with respect to $\mathbf{v}$ can be written as:

$$\nabla^2 F_{ij} = \mathbf{d}\mathbf{d}^\top \tag{14.13}$$

For all $\mathbf{z}$, since

$$\mathbf{z}^\top \nabla^2 F_{ij} \mathbf{z} = \mathbf{z}^\top \mathbf{d}\mathbf{d}^\top \mathbf{z} = (\mathbf{z}^\top \mathbf{d})^2 \geq 0, \tag{14.14}$$

$\nabla^2 F_{ij}$ is positive semidefinite, indicating that F_{ij} is a convex function of $\mathbf{v}$ and further $\tilde{f}_i$ is a convex function of $\mathbf{v}$ since nonnegative weighted sums preserve convexity. Therefore, problem (14.8) is a convex optimization problem.

14.1.4 Distinct Features of the Formulation

1. Different from most distributed optimization-based voltage control methods developed for grid-following inverters, such as in [2], the formulation is for GFM inverters for which the control variables are voltage instead of reactive power injection.

2. The formulation is more general than the existing methods and makes possible a proper trade-off between voltage regulation and reactive power-sharing, and also obeying the technical constraints on both voltage magnitude and reactive power capacity of all sources. Typical existing methods can be considered as special cases of the formulation. For example, the droop-free control in [4] can be obtained by setting α to be zero and removing all inequality constraints.

3. Voltage regulation objective is achieved by the first term in the objective function for voltage variance, the equality constraint for average voltage, and the inequality constraints (14.6e)–(14.6f). If there is only the average voltage equality constraint, it is not possible to guarantee that the voltage profile is good and the voltage deviation will be within ±5% of the rated voltage, as shown in [6].

4. The voltage control problem is explicitly formulated as an optimization problem for which distributed algorithms can be developed by adapting the standard methods such as primal–dual gradient algorithm to theoretically guarantee convergence and optimality. This is distinct from most existing methods such as in [4, 6] in which PI controllers are extensively used.

5. By utilizing the linearized reactive power injection function, the optimization problem has been mathematically proven to be a convex optimization problem which makes possible an efficient and reliable algorithm for solving such a problem.

14.2 Primal–Dual Gradient Based Distributed Solving Algorithm

The following assumptions are needed to ensure the well-posedness of problem (14.8) [7].

1. (Convexity and continuity) For $i \in \mathcal{V}$, Ω_i is compact and convex. On an open set containing Ω_i, $\tilde{f}_i$ is strictly convex, h and g are convex, and $\tilde{f}_i$, h, and g are locally Lipschitz continuous.
2. (Slater's constraint qualification) There exists $\check{v} \in \text{rint}(\Omega)$ such that $h(\check{v}) = 0$ and $g(\check{v}) < 0$.
3. (Communication topology) The communication network $\mathcal{G}$ has a spanning tree and a balanced Laplacian matrix.

These assumptions are quite mild and similar ones are widely used in the literature, such as [4, 6–9].

Remark 14.1 Although condition 3 in the assumption only requires a directed communication network that has a spanning tree and a balanced Laplacian matrix to make the control work, in this chapter, in the practical implementation we require the communication network to be undirected in addition to satisfying condition 3. Then when there are communication link (bidirectional) losses, as long as the communication network still has a spanning tree, it will still have a balanced Laplacian matrix so that condition 3 is still satisfied, which will significantly enhance the resiliency of the control algorithm against potential communication link losses compared with the case with a directed communication network.

14.2.1 Augmented Lagrangian

We introduce Lagrangian multipliers μ for the equality constraint (14.8b), and $\xi = \text{col}(\underline{\xi}, \overline{\xi})$ with $\underline{\xi} = [\underline{\xi}_1, \dots, \underline{\xi}_N]^\top$ and $\overline{\xi} = [\overline{\xi}_1, \dots, \overline{\xi}_N]^\top$ for the reactive power inequality constraints $g^L \leq 0$ and $g^U \leq 0$. For the optimization problem in (14.8), the augmented Lagrangian is defined as [10, 11]:

$$L(\hat{v}, \mu, \xi) \triangleq \tilde{f}(\hat{v}) + \mu|h(\hat{v})| + \underline{\xi}^\top g^L(\hat{v}) + \overline{\xi}^\top g^U(\hat{v}). \tag{14.15}$$

Augmented Lagrangian instead of the standard Lagrangian is used because the primal–dual gradient algorithm associated with the augmented Lagrangian has better convergence properties [2]. The equality constraints are treated in a similar way as in [11] in which a penalty function is defined.

14.2.2 Standard Primal–Dual Gradient Algorithm

The standard primal–dual gradient algorithm [2, 7, 12] for solving (14.8) can be written as:

$$
\hat{v}_i[n+1] = \mathcal{P}_{\Omega_i}\left(\hat{v}_i[n] - \tau \frac{\partial L(\hat{\mathbf{v}}[n], \mu[n], \xi[n])}{\partial v_i}\right)
$$

$$
= \mathcal{P}_{\Omega_i}\left(\hat{v}_i[n] - \tau\left(\sum_{j=1}^{N} \frac{\partial \tilde{f}_j(\hat{\mathbf{v}}[n])}{\partial v_i}\right.\right.
$$

$$
\left.\left. + \mu[n]\mathcal{D}_{v_i}|h(\hat{\mathbf{v}}[n])| + \sum_{j=1}^{N} \tilde{B}_{ji}\left(\overline{\xi}_j[n] - \underline{\xi}_j[n]\right)\right)\right) \tag{14.16}
$$

$$
\mu[n+1] = \mu[n] + \gamma \frac{\partial L(\hat{\mathbf{v}}[n], \mu[n], \xi[n])}{\partial \mu}
$$

$$
= \mu[n] + \gamma|h(\hat{\mathbf{v}}[n])| \tag{14.17}
$$

$$
\underline{\xi}_i[n+1] = \mathcal{P}_{\mathbb{R}_+}\left(\underline{\xi}_i[n] + \varphi \frac{\partial L(\hat{\mathbf{v}}[n], \mu[n], \xi[n])}{\partial \underline{\xi}_i}\right)
$$

$$
= \mathcal{P}_{\mathbb{R}_+}\left(\underline{\xi}_i[n] + \varphi\, g_i^{L}(\hat{\mathbf{v}}[n])\right) \tag{14.18}
$$

$$
\overline{\xi}_i[n+1] = \mathcal{P}_{\mathbb{R}_+}\left(\overline{\xi}_i[n] + \varphi \frac{\partial L(\hat{\mathbf{v}}[n], \mu[n], \xi[n])}{\partial \overline{\xi}_i}\right)
$$

$$
= \mathcal{P}_{\mathbb{R}_+}\left(\overline{\xi}_i[n] + \varphi\, g_i^{U}(\hat{\mathbf{v}}[n])\right) \tag{14.19}
$$

where $\mathcal{D}_{v_i}$ is the operator for subgradient with respect to v_i, g_i^{L} (g_i^{U}) is the ith element of $\mathbf{g}^{L}$ ($\mathbf{g}^{U}$), and τ, γ, and φ are positive scalar design parameters. Here, $\hat{v}_i[n]$ is the voltage set point decided by the primal–dual gradient algorithm at time step n. Since the zero-level control of inverter i will track $\hat{v}_i[n]$ and control the output voltage $v_i[n]$ to be $\hat{v}_i[n]$, the actual measured values $v_i[n]$ instead of $\hat{v}_i[n]$ will be used for the update.

However, there are the following major challenges to implementing the primal–dual gradient algorithm in an efficient, distributed manner.

1. **Non-separable objective function**: The objective function in (14.8) is $\tilde{f} = \sum_{j=1}^{N} \tilde{f}_j(\mathbf{v})$ in which $\tilde{f}_j(\mathbf{v})$ is a function of $\mathbf{v}$ instead of only a function of v_j. Therefore, when calculating $\partial \tilde{f}/\partial v_i$ in (14.16), the terms $\partial \tilde{f}_j/\partial v_i$ for $j \notin \mathcal{N}_i$ is not available for agent i, and thus (14.16) is not fully distributed.
2. **Unavailable global average voltage**: For the μ update in (14.17), the global average voltage is needed but is not available for each agent.
3. **Globally coupled reactive power constraints**: The reactive power injection in the reduced network is a function of all v_i's. Consequently, (14.16) needs information of the whole ξ and (14.18)–(14.19) needs information of the whole $\mathbf{v}$, which, however, are not available for each agent.

14.2.3 Non-Separable Objective Function

For agent i, the objective function $\tilde{f}(\mathbf{v})$ can be approximated around $v_i[n]$ as [13]:

$$\tilde{f}(\mathbf{v}) = \tilde{f}_i(\mathbf{v}) + \pi_i[n](v_i - v_i[n]), \tag{14.20}$$

where $\pi_i[n]$ is the partial derivative of $\sum_{j \neq i} \tilde{f}_j(\mathbf{v})$ with respect to v_i at $v_i[n]$:

$$\pi_i[n] \triangleq \sum_{j \neq i} \frac{\partial \tilde{f}_j(\mathbf{v}[n])}{\partial v_i} \tag{14.21}$$

Note that the evaluation of $\pi_i[n]$ requires all $\partial \tilde{f}_j(\mathbf{v}[n])/\partial v_i$ which may not be available locally at node i. To solve this problem, consider [13]:

$$\pi_i[n] = N \underbrace{\left(\frac{1}{N} \sum_{j=1}^{N} \frac{\partial \tilde{f}_j(\mathbf{v}[n])}{\partial v_i} \right)}_{D_i[n]} - \frac{\partial \tilde{f}_i(\mathbf{v}[n])}{\partial v_i} \tag{14.22}$$

where $D_i[n]$ can be estimated by a distributed observer based on dynamic consensus [4, 9, 14] as:

$$\hat{D}_i^j[n] = \frac{\partial \tilde{f}_j(\mathbf{v}[n])}{\partial v_i} + \sum_{t=0}^{n} \sum_{k \in \mathcal{N}_j} a_{jk} \left(\hat{D}_i^k[t] - \hat{D}_i^j[t] \right) \Delta t$$

$$j = 1, \dots, N \tag{14.23}$$

where $\hat{D}_i^j$ is the estimate of D_i by DG j and Δt is the step size. The distributed estimator in (14.23) can be equivalently written in continuous time domain as [4, 6, 9]:

$$\hat{D}_i^j(t) = \frac{\partial \tilde{f}_j(\mathbf{v}(t))}{\partial v_i} + \int_0^t \sum_{k \in \mathcal{N}_j} a_{jk} \left(\hat{D}_i^k(\tau) - \hat{D}_i^j(\tau) \right) d\tau$$

$$j = 1, \dots, N \tag{14.24}$$

By differentiating (14.24), we have the following equation for DG j:

$$\dot{\hat{D}}_i^j = \dot{\tilde{D}}_i^j + \sum_{k \in \mathcal{N}_j} a_{jk} \left(\hat{D}_i^k - \hat{D}_i^j \right)$$

$$= \dot{\tilde{D}}_i^j + \sum_{k \in \mathcal{N}_j} a_{jk} \hat{D}_i^k - d_j^{\text{in}} \hat{D}_i^j \tag{14.25}$$

where $\tilde{D}_i^j = \partial \tilde{f}_i(\mathbf{v}(t))/\partial v_i$ and $d_j^{\text{in}} = \sum_{k \in \mathcal{N}_j} a_{jk}$ is the in-degree of DG j. Accordingly, the global dynamics for DG i estimator $\hat{D}_i^i$ can be written as:

$$\dot{\hat{\mathbf{D}}}_i = \dot{\tilde{\mathbf{D}}}_i + \mathbf{A}\hat{\mathbf{D}}_i - \mathbf{D}^{\text{in}}\hat{\mathbf{D}}_i = \dot{\tilde{\mathbf{D}}}_i - \mathbf{L}\hat{\mathbf{D}}_i, \tag{14.26}$$

where $\mathbf{D}^{\text{in}}$ is the in-degree matrix, $\hat{\mathbf{D}}_i = [\hat{D}_i^1, \hat{D}_i^2, \dots, \hat{D}_i^N]$, $\tilde{\mathbf{D}}_i = [\partial \tilde{f}_1(\mathbf{v}(t))/\partial v_i, \partial \tilde{f}_2(\mathbf{v}(t))/\partial v_i, \dots, \partial \tilde{f}_N(\mathbf{v}(t))/\partial v_i]$, and $\mathbf{L}$ is the Laplacian matrix of the communication network. Then in frequency domain, (14.26) can be represented as:

$$s\hat{\mathbb{D}}_i - \hat{\mathbf{D}}_i(0) = s\tilde{\mathbb{D}}_i - \tilde{\mathbf{D}}_i(0) - \mathbf{L}\hat{\mathbb{D}}_i \tag{14.27}$$

where $\hat{\mathbb{D}}_i$ and $\tilde{\mathbb{D}}_i$ represent the Laplace transform of $\hat{\mathbf{D}}_i$ and $\tilde{\mathbf{D}}_i$. Also, from (14.24) we have $\hat{\mathbf{D}}_i(0) = \tilde{\mathbf{D}}_i(0)$. Then (14.27) can be rewritten as:

$$\hat{\mathbb{D}}_i = s(s\mathbf{I}_N + \mathbf{L})^{-1}\tilde{\mathbb{D}}_i \tag{14.28}$$

where $\mathbf{I}_N \in \mathbb{R}^{N \times N}$ is an identity matrix. Applying the final value theorem to (14.28), we have

$$\lim_{t \to \infty} \hat{\mathbf{D}}_i(t) = \lim_{s \to 0} s \hat{\mathbb{D}}_i = \lim_{s \to 0} s^2(s\mathbf{I}_N + \mathbf{L})^{-1} \tilde{\mathbb{D}}_i \tag{14.29}$$

According to the Lemma A.2 of [9], if the communication network has a spanning tree and a balanced Laplacian matrix, then we have:

$$\lim_{s \to 0} s(s\mathbf{I}_N + \mathbf{L})^{-1} = \mathbf{M} \tag{14.30}$$

where $\mathbf{M} \in \mathbb{R}^{N \times N}$ is an averaging matrix with all the elements as $(1/N)$. Then Eq. (14.29) becomes:

$$\lim_{t \to \infty} \hat{\mathbf{D}}_i(t) = \mathbf{M} \lim_{s \to 0} s \tilde{\mathbb{D}}_i = \mathbf{M} \lim_{t \to \infty} \tilde{\mathbf{D}}_i(t) = \lim_{t \to \infty} \frac{1}{N} \sum_{j=1}^{N} \frac{\partial \tilde{f}_j(\mathbf{v}(t))}{\partial v_i} \mathbf{1} \tag{14.31}$$

which implies that the estimated $\hat{D}_i^j$ for $j = 1, \dots, N$ in (14.24) will converge to the true average value. Then $\pi_i[n]$ is replaced by $\tilde{\pi}_i[n]$ as:

$$\tilde{\pi}_i[n] = N\hat{D}_i^i[n] - \frac{\partial \tilde{f}_i(\mathbf{v}[n])}{\partial v_i} \tag{14.32}$$

14.2.4 Distributed Average Voltage Estimation

In $h(\mathbf{v}) = \mathbf{1}^\mathsf{T}\mathbf{v}/N - v^r$, the average voltage of all inverter output buses, $(\mathbf{1}^\mathsf{T}\mathbf{v}[n]/N)$, can be estimated by agent $i = 1, \dots, N$ as $v_i^{\mathrm{av}}[n]$ using the following distributed observer based on dynamic consensus [4, 9]:

$$v_i^{\mathrm{av}}[n] = v_i[n] + \sum_{t=0}^{n} \sum_{j \in \mathcal{N}_i} a_{ij} \left(v_j^{\mathrm{av}}[t] - v_i^{\mathrm{av}}[t] \right) \Delta t \tag{14.33}$$

where Δt is the step size. It has been proven in [9] that for $\forall i = 1, 2, \dots, N$, v_i^{av} converges to a consensus value which is the true global average voltage when the communication network has a spanning tree and a balanced Laplacian matrix.

Since each agent will have its own estimate of $h(\mathbf{v})$, it will have its own μ. We denote the μ for agent i by μ_i. With v_i^{av} converging to the true global average voltage, from (14.17) it is clear that the μ_i's will achieve a consensus.

14.2.5 Coupled Reactive Power Inequality Constraints

In (14.16), the term related to the reactive power inequality constraints (14.8c) is $\sum_{j=1}^{N} \tilde{B}_{ji} \left(\overline{\xi}_j - \underline{\xi}_j \right)$. It is clear that it depends on $\overline{\xi}_j$ and $\underline{\xi}_j$ for all agents, mainly because the reactive power inequality constraints are coupled. This will make the distributed implementation very challenging.

In order to address this problem, we write the coupled inequality constraints as $\mathbf{g}(\mathbf{v}) \triangleq \sum_{i \in \mathcal{V}} \mathbf{g}_i(v_i) \leq \mathbf{0}$, where $\mathbf{g}_i : \Omega_i \to \mathbb{R}^{2N}$ can be written as:

$$\mathbf{g}_i(v_i) = \begin{bmatrix} \mathbf{g}_i^L(v_i) \\ \mathbf{g}_i^U(v_i) \end{bmatrix} = \begin{bmatrix} -\tilde{B}_{1i}v_i - \tilde{G}_{1i}\theta_i + a'_{1i}\underline{\lambda}'_{Q_1} \\ \vdots \\ -\tilde{B}_{Ni}v_i - \tilde{G}_{Ni}\theta_i + a'_{Ni}\underline{\lambda}'_{Q_N} \\ \tilde{B}_{1i}v_i + \tilde{G}_{1i}\theta_i - a'_{1i}\overline{\lambda}'_{Q_1} \\ \vdots \\ \tilde{B}_{Ni}v_i + \tilde{G}_{Ni}\theta_i - a'_{Ni}\overline{\lambda}'_{Q_N} \end{bmatrix} \tag{14.34}$$

where θ_i can be obtained locally by the installed phasor measurement unit (PMU) and for $j = 1, \dots, N$ there are:

$$\underline{\lambda}'_{Q_j} = \begin{cases} \dfrac{\underline{\lambda}_{Q_j}}{|\mathcal{N}_j|}, & \text{if } a'_{ji} \neq 0 \\ 1, & \text{otherwise} \end{cases} \quad \text{and} \quad \overline{\lambda}'_{Q_j} = \begin{cases} \dfrac{\overline{\lambda}_{Q_j}}{|\mathcal{N}_j|}, & \text{if } a'_{ji} \neq 0 \\ 1, & \text{otherwise} \end{cases}$$

Note that for every $i \in \mathcal{V}$ agent i only has access to $\mathbf{g}_i(v_i)$ in (14.34) rather than $\mathbf{g}(\mathbf{v})$, and $\mathbf{g}_i(v_i)$ only depends on the data of agent i and the information from its neighbors.

To achieve the actual constraint of (14.7) based on the local $\mathbf{g}_i(v_i)$ in (14.34), the dynamic consensus method can be used. Specifically, the average $\sum_{i=1}^{N} \mathbf{g}_i(v_i)/N$ can be estimated as:

$$\mathbf{g}_i^{\mathrm{av}}[n] = \mathbf{g}_i(v_i[n]) + \sum_{t=0}^{n} \sum_{j \in \mathcal{N}_i} a_{ij} \left(\mathbf{g}_j^{\mathrm{av}}[t] - \mathbf{g}_i^{\mathrm{av}}[t] \right) \Delta t \tag{14.35}$$

where $\mathbf{g}_i^{\mathrm{av}}[n]$ is the average value estimated by DG i. Thus, $\mathbf{g}(\mathbf{v}[n])$ can be obtained as $N\mathbf{g}_i^{\mathrm{av}}[n]$. Then for agent i we replace $\mathbf{g}(\mathbf{v}[n])$ by $\tilde{\mathbf{g}}_i[n]$ as:

$$\tilde{\mathbf{g}}_i[n] \triangleq \begin{bmatrix} \tilde{\mathbf{g}}_i^{\mathrm{L}}[n] \\ \tilde{\mathbf{g}}_i^{\mathrm{U}}[n] \end{bmatrix} = N\mathbf{g}_i^{\mathrm{av}}[n] \tag{14.36}$$

14.2.6 Solving Algorithm

Based on the above discussion, in order to solve the problem in (14.8) fully distributedly, we construct a modified Lagrangian function for the problem in (14.8) as:

$$\tilde{L}(\mathbf{v}, \mu, \xi) \triangleq \tilde{f}(\mathbf{v}) + \mu |h(\mathbf{v})| + \underline{\xi}^{\top} \sum_{i \in \mathcal{V}} \mathbf{g}_i^{\mathrm{L}}(v_i) + \overline{\xi}^{\top} \sum_{i \in \mathcal{V}} \mathbf{g}_i^{\mathrm{U}}(v_i).$$

Different from (14.18)–(14.19), in which agent i only updates the Lagrangian multiplier ξ_i for its own reactive power inequality constraint, here we let agent i maintain a collection of local multipliers $\xi_i \triangleq \mathrm{col}(\underline{\xi}_i, \overline{\xi}_i) \in \mathbb{R}^{2N}$ for all reactive power inequality constraints. We develop the following primal–dual gradient-based algorithm utilizing the techniques in Sections 14.2.3–14.2.5. For agent i, it can be written as:

$$\begin{aligned} \hat{v}_i[n+1] &= \mathcal{P}_{\Omega_i} \left(v_i[n] - \tau \frac{\partial \tilde{L}(\mathbf{v}[n], \mu_i[n], \xi_i[n])}{\partial v_i} \right) \\ &= \mathcal{P}_{\Omega_i} \left(v_i[n] - \tau \left(N\hat{D}_i^i[n] + \mu_i[n]D_{v_i}|v_i^{\mathrm{av}}[n] - v^{\mathrm{r}}| + \tilde{\mathbf{b}}_{*,i}^{\top} \left(\overline{\xi}_i[n] - \underline{\xi}_i[n] \right) \right) \right) \end{aligned} \tag{14.37}$$

$$\begin{aligned} \mu_i[n+1] &= \mu_i[n] + \gamma \frac{\partial \tilde{L}(\mathbf{v}[n], \mu_i[n], \xi_i[n])}{\partial \mu} \\ &= \mu_i[n] + \gamma |v_i^{\mathrm{av}}[n] - v^{\mathrm{r}}| \end{aligned} \tag{14.38}$$

$$\begin{aligned} \underline{\xi}_i[n+1] &= \mathcal{P}_{\mathbb{R}_+^N} \left(\underline{\xi}_i[n] + \varphi \nabla_{\underline{\xi}} \tilde{L}(\mathbf{v}[n], \mu_i[n], \xi_i[n]) \right) \\ &= \mathcal{P}_{\mathbb{R}_+^N} \left(\underline{\xi}_i[n] + \varphi \tilde{\mathbf{g}}_i^{\mathrm{L}}[n] \right) \end{aligned} \tag{14.39}$$

$$\begin{aligned} \overline{\xi}_i[n+1] &= \mathcal{P}_{\mathbb{R}_+^N} \left(\overline{\xi}_i[n] + \varphi \nabla_{\overline{\xi}} \tilde{L}(\mathbf{v}[n], \mu_i[n], \xi_i[n]) \right) \\ &= \mathcal{P}_{\mathbb{R}_+^N} \left(\overline{\xi}_i[n] + \varphi \tilde{\mathbf{g}}_i^{\mathrm{U}}[n] \right) \end{aligned} \tag{14.40}$$

where $\tilde{\mathbf{b}}_{*,i}$ is the ith column vector of the $\tilde{\mathbf{B}}$ matrix. Then, each agent i commits its own optimization variable by setting the voltage magnitude reference for inverter i, $v_i^{\mathrm{ref}}[n+1]$, to be $\hat{v}_i[n+1]$.

The zero-level control of inverter i will track this reference and control the output voltage $v_i[n + 1]$ to be $\hat{v}_i[n + 1]$, and $v_i[n + 1]$ will be used for the next time step. Note that in the algorithm, the number of DGs N is needed, which may not be known to each DG. One potential approach is to distributedly estimate it such as by a distributed L2-norm estimation method based on dynamic average consensus [15].

The parameters τ, γ, and φ impact the convergence of the algorithm. Too large values may affect the stability while too small values may make the convergence very slow. In [1], a theoretical upper bound is presented for similar parameters, which, however, are conservative. In [1], these parameters actually still need to be selected using some trial and error methods. Similar to [1], we have also selected these parameters based on some numerical simulations and testing to guarantee acceptable convergence performance.

The algorithm is presented as Algorithm 14.1. The information flow of the control is shown in Fig. 14.1.

Algorithm 14.1: Distributed Solving Algorithm

Initialization: Set $v_i[0] = v_i^{\text{meas}}$, $\mu_i[0] = 0$, $\xi_i[0] = \mathbf{0}$, $v_i^{\text{av}}[0] = v_i[0]$, $\hat{D}_i^i[0] = \partial \tilde{f}_i(\mathbf{v}[0])/\partial v_i$, $\mathbf{g}_i^{\text{av}}[0] = \mathbf{g}_i(v_i[0])$, $\tilde{\mathbf{g}}_i[0] = N\mathbf{g}_i^{\text{av}}[0]$ for $\forall i = 1, \dots, N$. Set $n = 0$.

(S.1) Each agent i updates its variables $\hat{v}_i[n + 1]$, $\mu_i[n + 1]$, $\underline{\xi}_i[n + 1]$, and $\overline{\xi}_i[n + 1]$ based on (14.37)–(14.40);

(S.2) Each agent i sets voltage magnitude reference for inverter i as $v_i^{\text{ref}} = \hat{v}_i[n + 1]$ and sends it to zero-level control which controls the voltage $v_i[n + 1]$ to v_i^{ref};

(S.3) Each agent i updates $\hat{D}_i^i[n]$ and $\tilde{\pi}_i[n]$ based on (14.23)–(14.32);

(S.4) Each agent i updates $\mathbf{g}_i^{\text{av}}[n]$ and $\tilde{\mathbf{g}}_i[n]$ based on (14.35)–(14.36);

(S.5) Increase n by 1 and go to (S.1).

Different from [7], for which there are only inequality constraints, in (14.8) there is a coupled equality constraint which requires global information about the average voltage of the system.

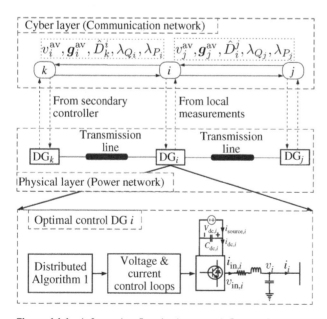

Figure 14.1 Information flow in the control. Source: Reproduced with permission from [16]/IEEE.

First, the method in Section 14.2.4 is needed to obtain the global information in a distributed manner. Second, in the algorithm both coupled inequality constraints and coupled equality constraints are considered. Third, the algorithm uses a dynamic consensus-based approach to deal with coupled inequality constraints, which is different from the nonsmooth penalty function-based approach in [7].

The parameters in $\tilde{\mathbf{G}}$ and $\tilde{\mathbf{B}}$ ($\tilde{G}_{ki}$ and $\tilde{B}_{ki}$ for $k = 1, \ldots, N$ and $\tilde{B}_{jk}$ for $j \in \mathcal{N}_i$, $k = 1, \ldots, N$) are assumed to be known to DG i. If they are not known, they can be estimated distributedly by adapting the alternating direction method of multipliers (ADMM) based distributed robust estimation method in [17] or the data driven parameter estimation method presented in [18].

14.3 Microgrid Test Systems

The performance of the optimal distributed control is evaluated through simulations on the 4-DG microgrid test system in [6] (Fig. 14.2), and the modified IEEE 34-bus distribution test system (Fig. 14.3).

Real-time simulations for the 4-DG system are performed on OPAL-RT OP4510 at 50 μs fixed time step using the ODE-5 (Dormand–Prince) solver. The updates of the distributed algorithm are also performed at 50 μs time step and we consider only a single consensus iteration at each time step. We use the 2-level power electronic switching model for the inverters [19]. The switching frequencies of the pulse width modulation (PWM) generators are considered as 16,200 Hz. To damp out the undesired low-frequency harmonics, the cutoff frequencies of the measurement filters are selected as low as 3 Hz.

For the IEEE 34-bus test system, we have implemented the controls in Matlab without detailed zero-level control loops. The Matlab simulations are performed as phasor simulations without consideration of detailed electromagnetic transients of the system [20].

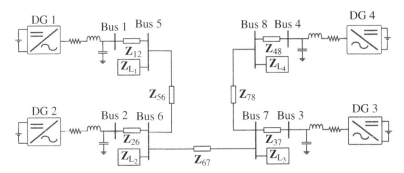

Figure 14.2 Schematic diagram of the 4-DG test system. Source: Reproduced with permission from [16]/IEEE.

Figure 14.3 Modified IEEE 34-bus distribution test system. The stars indicate DGs and the triangles indicate loads. Source: Reproduced with permission from [16]/IEEE.

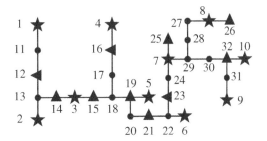

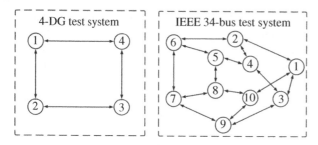

Figure 14.4 Communication networks of the 4-DG test system and the IEEE 34-bus test system. Source: Reproduced with permission from [16]/IEEE.

The parameters τ, γ, and φ are selected as 0.0005, 0.005, and 0.01. Different values of α and β are used for different cases to provide flexibility. The reactive power limits are selected as $\underline{\lambda}_{Q_i} = 0.2$ and $\overline{\lambda}_{Q_i} = 0.8$ for all DGs. $\underline{v}_i$ and $\overline{v}_i$ for all DGs are chosen as 0.95 and 1.05.

The communication networks used for the 4-DG test system and the IEEE 34-bus test systems are shown in Fig. 14.4 which both satisfy the condition 3 of the assumptions in Section 14.2. For the nonzero elements in **A**, we choose $a_{ij} = 0.25$ for the 4-DG system and $a_{ij} = 0.0625$ for the IEEE 34-bus system based on our numerical tests.

14.4 Control Performance on 4-DG System

Real-time simulations on OPAL-RT OP4510 are performed for the 4-DG islanded microgrid test system. The parameters of this system are adopted from [6]. For the implementation of the algorithm, the DG output voltage and phase angle measurements from Simulink measurement blocks are used. The $(\tilde{\lambda}_{Q_i} - \tilde{\lambda}_{Q_j})$ term in (14.10) is replaced by $(\lambda_{Q_i} - \lambda_{Q_j})$ which can be obtained from the Simulink reactive power measurement blocks.

14.4.1 Performance Under Different α and β

The controller performance under normal loading conditions with different α and β are demonstrated in Fig. 14.5. The uniqueness of the control is that it allows flexibility in coordinating voltage regulation and reactive power-sharing by adjusting these two parameters. From the left-hand side of Fig. 14.5, it can be seen that the controller can almost achieve proportional reactive power-sharing with small α. When α is increased, the voltage bounds become tighter at the cost of increased deviations in reactive power-sharing which is clear from the right-hand side subfigures of Fig. 14.5.

14.4.2 Performance Under Heavy Load Scenario

The distributed control in [6] guarantees an admissible voltage profile at the cost of relaxed reactive power-sharing for one special DG and the control in [4] achieves proportional reactive power-sharing without imposing bounds on voltages. By contrast, the optimal control achieves a trade-off between voltage regulation and reactive power-sharing objectives.

In Fig. 14.6, we have considered an extreme scenario where, at 5 seconds, a large load is connected to the bus close to DG1. α and β are chosen as 1 and 0.065, respectively. For the

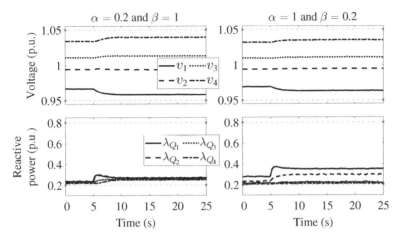

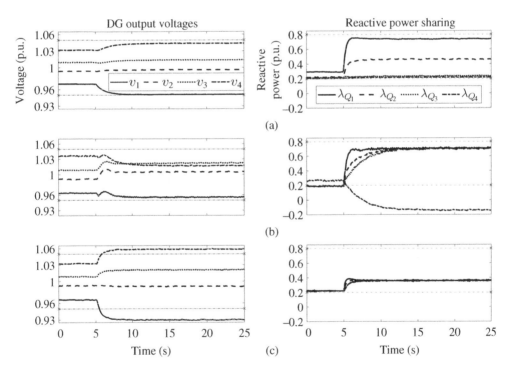

Figure 14.5 Performance evaluation of the control under different α and β. Load change is applied at 5 seconds. Source: Reproduced with permission from [16]/IEEE.

Figure 14.6 Comparison between the optimal distributed control in this chapter (a) and the controls in Section 13.4 (b) and Section 13.3 (c). Load change applied at 5 seconds. (a) Optimal distributed control, (b) distributed control with voltage variance regulation, and (c) distributed control without voltage variance regulation. Source: Reproduced with permission from [16]/IEEE.

control in this chapter, the voltages range between 0.95 and 1.042, while for the control in Section 13.4 [6], the voltages range between 0.955 and 1.023, and for the control in Section 13.3 [4] between 0.927 and 1.06 which violates the IEEE standard acceptable limits (0.95–1.05). In the case of reactive power-sharing, the reactive power under the control in this chapter ranges between 0.218 and 0.738 whereas for the control in Section 13.4 [6], it ranges between −0.165 and 0.747.

14.5 Control Performance on IEEE 34-Bus System

The performance of the optimal distributed control is validated by Matlab simulations on the modified IEEE 34-bus distribution test system as shown in Fig. 14.3. The line parameters are adopted from [21] and the locations of DGs and loads are, respectively, indicated by stars and triangles in Fig. 14.3. For the droop-free control in [4], the α in (14.5) is set as 0 and the inequality constraints are removed, and then the standard gradient decent algorithm [22] is used to solve the optimization problem.

14.5.1 Performance Under Load Change

Figure 14.7 shows the DG output voltages and reactive power under the optimal distributed control with load change at 10 seconds. Figure 14.8 shows the performance of the control in Section 13.3 [4] for the same case. For the control in Section 13.3 [4], the voltages range between 0.909 and 1.076 whereas the optimal distributed control can bind the voltages between 0.95 and 1.05 and the normalized reactive power between 0.2 and 0.8.

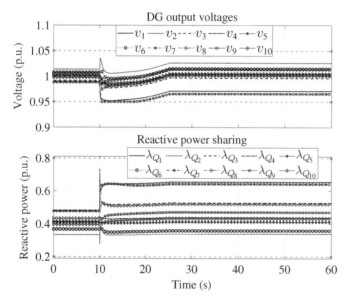

Figure 14.7 Performance evaluation on the IEEE 34-bus test system for the optimal distributed control. Load change is applied at 10 seconds. Source: Reproduced with permission from [16]/IEEE.

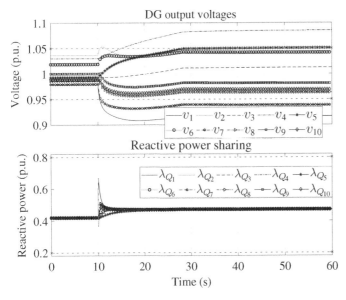

Figure 14.8 Performance evaluation on the IEEE 34-bus test system for the control in Section 13.3. Load change is applied at 10 seconds. Source: Reproduced with permission from [16]/IEEE.

14.5.2 Performance Under A Wide Range of Load Scenarios

The performance of the optimal distributed control and the distributed control in Section 13.3 [4] is compared under a wide range of load scenarios. A total of 100 test cases are generated in which in each load is added a random change that follows a normal distribution with zero mean and a standard deviation as 10% of the initial load.

The mean absolute deviation $\sum_{i=1}^{N} |v_i - 1|/N$ and the maximum absolute deviation $\max_{i=1,...,N} |v_i - 1|$ of the DG output voltages are calculated for each case, which are shown in Fig. 14.9. For all test scenarios, the mean and maximum absolute deviations with the optimal

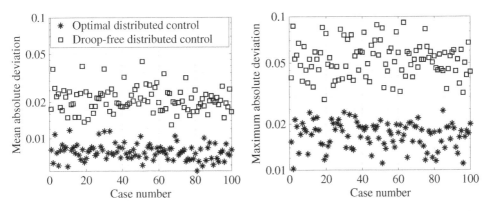

Figure 14.9 Mean absolute deviations and maximum absolute deviations of DG output voltages for the optimal distributed control in this chapter and the droop-free distributed control in Section 13.3. Source: Reproduced with permission from [16]/IEEE.

distributed control are much less than those for the control in Section 13.3 [4]. With the control in Section 13.3 [4], the maximum voltage deviations for many cases are greater than 0.05, while the optimal distributed control can always guarantee that the maximum voltage deviation is less than 0.05.

14.5.3 Optimality Comparison

The final steady state obtained from the optimal distributed control is compared with a centralized optimization approach for different load scenarios. For the centralized approach, we solve the optimization problem (14.8) using CPLEX [23] in Matlab YALMIP toolbox [24].

The root mean square error (RMSE) of the final steady states for voltage and reactive power obtained from the distributed and the centralized approach under 100 load scenarios is shown in Fig. 14.10.

In Fig. 14.11, the voltage and reactive power obtained from the optimal distributed control and the centralized optimization approach are given respectively for case 43 and case 58 in Fig. 14.10 for which the voltage and reactive power RMSEs have the highest values. The results obtained from the distributed control are almost identical to those from the centralized approach.

14.5.4 Performance Under Line Parameter Uncertainty

In real power systems, the line parameters may not be obtained accurately. Thus, the performance of the optimal distributed control needs to be investigated under uncertainty in the line parameters. To validate robustness against line parameter uncertainty, we consider 20 different load scenarios. For each of the test cases, the line parameters are increased/decreased by 10% or 20% from the true values, and the corresponding solutions are compared with the actual solution without line parameter uncertainty.

In Figs. 14.12 and 14.13, the RMSE of the steady state voltage and reactive power respectively under 10% and 20% changes in the line parameters is shown. It can be seen that with higher line

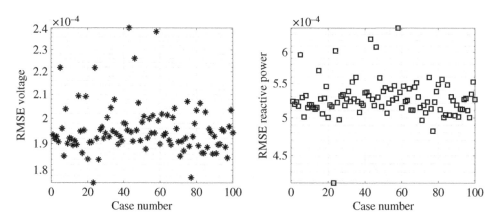

Figure 14.10 RMSE of voltage and reactive power between centralized and distributed solution for 100 test cases under different load scenarios. Source: Reproduced with permission from [16]/IEEE.

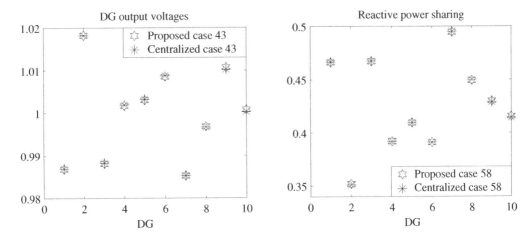

Figure 14.11 DG output voltage and reactive power comparison between the centralized optimization and the optimal distributed control respectively for case 43 and case 58 in Fig. 14.10. Source: Reproduced with permission from [16]/IEEE.

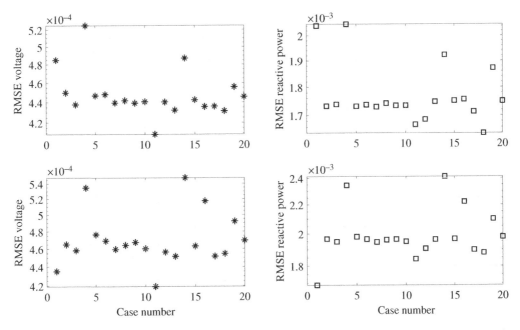

Figure 14.12 RMSE of voltage and reactive power with 10% decrease or increase in line parameters under different loading conditions. (a) Voltage RMSE with 10% parameter decrease; (b) Reactive power RMSE with 10% parameter decrease; (c) Voltage RMSE with 10% parameter increase; and (d) Reactive power RMSE with 10% parameter decrease. Source: Reproduced with permission from [16]/IEEE.

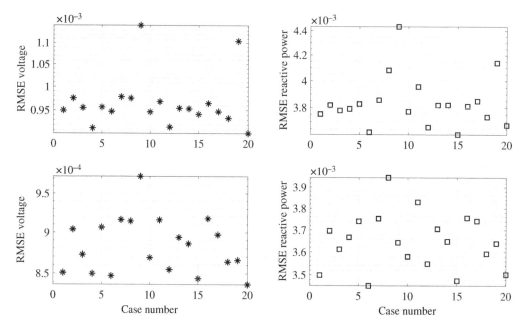

Figure 14.13 RMSE of voltage and reactive power with 20% decrease or increase in line parameters under different loading conditions. (a) Voltage RMSE with 20% parameter decrease; (b) Reactive power RMSE with 20% parameter decrease; (c) Voltage RMSE with 20% parameter increase; and (d) Reactive power RMSE with 20% parameter decrease. Source: Reproduced with permission from [16]/IEEE.

parameter uncertainty, the RMSE will increase but is still small, indicating that the optimal distributed control has robustness against line parameter uncertainties.

14.6 Exercises

1. What are the differences between the control in Section 13.4 and the optimal distributed control in this chapter?
2. How does the optimization problem in (14.6) achieve voltage profile guarantees?
3. Given the 4-DG system in Fig. 14.2, use Algorithm 14.1 to generate voltage references and use the droop-free control in Section 13.3 to generate frequency references. Compare voltage magnitudes and reactive power outputs of the DGs under changing load in Z_{L_1}, Z_{L_2}, Z_{L_3}, and/or Z_{L_4}.

References

1 Magnússon, S., Qu, G., Fischione, C., and Li, N. (2019). Voltage control using limited communication. *IEEE Transactions on Control of Network Systems* 6 (3): 993–1003.

2 Qu, G. and Li, N. (2020). Optimal distributed feedback voltage control under limited reactive power. *IEEE Transactions on Power Apparatus and Systems* 35 (1): 315–331.

3 Liu, H.J., Shi, W., and Zhu, H. (2019). Hybrid voltage control in distribution networks under limited communication rates. *IEEE Transactions on Smart Grid* 10 (3): 2416–2427.

4 Nasirian, V., Shafiee, Q., Guerrero, J.M. et al. (2016). Droop-free distributed control for AC microgrids. *IEEE Transactions on Power Electronics* 31 (2): 1600–1617.

5 Yang, J., Zhang, N., Kang, C., and Xia, Q. (2017). A state-independent linear power flow model with accurate estimation of voltage magnitude. *IEEE Transactions on Power Apparatus and Systems* 32 (5): 3607–3617.

6 Mohiuddin, S.M. and Qi, J. (2020). Droop-free distributed control for AC microgrids with precisely regulated voltage variance and admissible voltage profile guarantees. *IEEE Transactions on Smart Grid* 11 (3): 1956–1967.

7 Liang, S., Zeng, X., and Hong, Y. (2018). Distributed nonsmooth optimization with coupled inequality constraints via modified Lagrangian function. *IEEE Transactions on Automatic Control* 63 (6): 1753–1759.

8 Yi, P., Hong, Y., and Liu, F. (2016). Initialization-free distributed algorithms for optimal resource allocation with feasibility constraints and application to economic dispatch of power systems. *Automatica* 74: 259–269.

9 Nasirian, V., Moayedi, S., Davoudi, A., and Lewis, F.L. (2015). Distributed cooperative control of DC microgrids. *IEEE Transactions on Power Electronics* 30 (4): 2288–2303.

10 Nedić, A. and Ozdaglar, A. (2009). Subgradient methods for saddle-point problems. *Journal of Optimization Theory and Applications* 142 (1): 205–228.

11 Zhu, M. and Martinez, S. (2012). On distributed convex optimization under inequality and equality constraints. *IEEE Transactions on Automatic Control* 57 (1): 151–164.

12 Nesterov, Y. (2013). *Introductory Lectures on Convex Optimization: A Basic Course*, vol. 87. Springer Science & Business Media.

13 Di Lorenzo, P. and Scutari, G. (2016). NEXT: in-network nonconvex optimization. *IEEE Transactions on Signal and Information Processing over Networks* 2 (2): 120–136.

14 Spanos, D.P., Olfati-Saber, R., and Murray, R.M. (2005). Dynamic consensus on mobile networks. *Proceedings of the 16th International Federation of Automatic Control*, 1–6.

15 Zhang, S., Tepedelenlioglu, C., Spanias, A., and Banavar, M.K. (2015). Node counting in wireless sensor networks. *2015 49th Asilomar Conference on Signals, Systems and Computers*, 360–364.

16 Mohiuddin, S.M. and Qi, J. (2022). Optimal distributed control of AC microgrids with coordinated voltage regulation and reactive power sharing. *IEEE Transactions on Smart Grid* 13 (3): 1789–1800.

17 Kekatos, V. and Giannakis, G.B. (2013). Distributed robust power system state estimation. *IEEE Transactions on Power Apparatus and Systems* 28 (2): 1617–1626.

18 Xu, H., Domínguez-García, A.D., Veeravalli, V.V., and Sauer, P.W. (2020). Data-driven voltage regulation in radial power distribution systems. *IEEE Transactions on Power Apparatus and Systems* 35 (3): 2133–2143.

19 Yazdani, A. and Iravani, R. (2010). *Voltage-Sourced Converters in Power Systems: Modeling, Control, and Applications*. Wiley.

20 Huang, Q. and Vittal, V. (2018). Advanced EMT and phasor-domain hybrid simulation with simulation mode switching capability for transmission and distribution systems. *IEEE Transactions on Power Apparatus and Systems* 33 (6): 6298–6308.

21 Mwakabuta, N. and Sekar, A. (2007). Comparative study of the IEEE 34 node test feeder under practical simplifications. *Proceedings of the 39th North American Power Symposium*, 484–491.

22 Boyd, S., Parikh, N., Chu, E. et al. (2010). Distributed optimization and statistical learning via the alternating direction method of multipliers. *Foundations and Trends in Machine Learning* 3: 1–122.

23 IBM ILOG CPLEX (2009). V12. 1: User's manual for CPLEX. *International Business Machines Corporation* 46 (53): 157.

24 Löfberg, J. (2004). YALMIP: a toolbox for modeling and optimization in MATLAB. *Proceedings of the CACSD Conference*, Taipei, Taiwan.

15

Cyber-Resilient Distributed Microgrid Control
Pouya Babahajiani and Peng Zhang

15.1 Push-Sum Enabled Resilient Microgrid Control

Microgrids, equipped with distributed control schemes based on averaged consensus [1], are prone to random link failures resulting in unbalanced communication networks and therefore, Laplacian matrices that are no longer doubly stochastic, which jeopardize the stability of microgrids. The reason is, traditional average consensus protocols cannot achieve the exact average under directed and time-varying communication network [2]. Consequently, the stability of microgrids is impacted under unbalanced scenarios [3]. Existing attempts in the literature to address the unbalanced issues [4], unfortunately, apply only to discrete-time systems. Their system dynamics are commonly assumed first order such that the agents need to only carry one state and they adopt time-varying stepsizes, making those algorithms unsuited for microgrid control.

To tackle the above challenges, a continuous-time push-sum algorithm is devised to enable resilient distributed control for microgrids. The new push-sum-enabled microgrid control can provably achieve average consensus at every node even under a time-varying, directed communication network. It enables flexible switching of communication topologies and thus leads to unprecedented cyber-physical resilience in microgrids.

15.1.1 Push-Sum-Based Resilient Distributed Control

Recently, push-sum has been emerging to enhance the resilience of distributed averaging approaches for discrete-time systems [5], [6]. We extend the push-sum method [7] to continuous-time domain to devise a resilient distributed control scheme which is resilient against abrupt changes in communication topology and cyberattack on communication links. The overarching goal is to enable unprecedented resilient control for real-life microgrids abstracted as time-varying graphs, which cannot be attained by conventional distributed control relying on the existence of spanning trees [8]. Here time-varying graphs mean unreliable communications because of the unavoidable random failures in distributed energy resource (DER) communications.

In AC microgrids, a predominantly inductive network naturally decouples the load-sharing process; the reactive power regulator must handle the reactive load sharing by adjusting voltage magnitude while the active power regulator would handle the active load sharing through adjusting the frequency. A common approach for inverter-interfaced DER is to connect the power electronic inverter with an LC filter. Therefore, the predominantly inductive line is either from the natural line/cable characteristics or implemented with virtual impedance. The locally deployed LC filter in each DER makes the output impedance inductive dominant [9]. Hence, we consider an

Microgrids: Theory and Practice, First Edition. Edited by Peng Zhang.
© 2024 The Institute of Electrical and Electronics Engineers, Inc. Published 2024 by John Wiley & Sons, Inc.

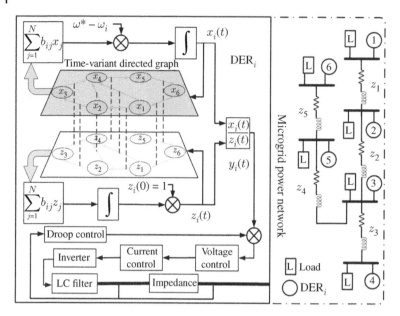

Figure 15.1 Schematic diagram of the push-sum-enables microgrid control.

islanded AC microgrid with inductive lines and inverter-interfaced DERs, as shown in Fig. 15.1. Without loss of generality, a push-sum-enabled frequency controller is formulated as follows:

$$\omega_i = \omega^* - n_i P_{e,i} + y_i,$$

$$\dot{x}_i = \omega^* - \omega_i + \sum_{j=1}^{n} b_{ij} x_j,$$

$$\dot{z}_i = \sum_{j=1}^{n} b_{ij} z_j, \quad z(0) = 1,$$

$$y_i = \frac{x_i(t)}{z_i(t)} \tag{15.1}$$

where ω_i is the actual grid frequency, ω^* is the rated frequency, $P_{e,i}$ is the active power injection from DER inverter i, n_i is the droop coefficient, and y_i is the secondary control signal.

In push-sum, the secondary control signal for microgrid DER i is determined by the cumulative estimate of the sum $x_i(t)$ and a weight $z_i(t)$. In other words, the push-sum algorithm solves the distributed averaging problem on networks with one additional variable per node such that nodes not only record a linear combination of other nodes, but also keep track of their relative importance in the system through the scaling factor "z" such that its magnitude is directly affected by the number of incoming links and inversely by the number of outgoing links. At the beginning, the weights are initialized as $z_i(0) = 1$. The algorithm also works for time-varying communication graphs. If a new node is added to the network or an existing node is removed, all z_i are again reset and initialized to 1.

The Laplacian matrix L for an underlying time-variant directed graph $\mathcal{G}(t)$ is defined as $L = D_{in} - A$, where A and D_{in} are adjacency and in-degree matrices, respectively. b_{ij} is the ijth entry of $-L(t)$.

It can be shown that after communication network changes, $z_i(t) \to n v_i$ as $t \to \infty$ such that n is the number of DERs and the vector v is the normalized eigenvector of matrix e^{-Lt} associated with the simple eigenvalue 1 or the zero eigenvalue of $-L$. Figure 15.1 illustrates the new control scheme where the communication network is modeled by a directed graph.

To formally prove the convergence of the push-sum-based method, the microgrid is assumed stable in the steady state. Since the DERs' frequency must be equal in steady state, we have $\omega_1 = \omega_2 = \cdots = \omega_n$ and thus $n_1 P_{e,1} - y_1 = n_2 P_{e,2} - y_2 = \cdots = n_n P_{e,n} - y_n$. This leads to

$$
\begin{aligned}
\dot{x}_i &= \omega^* - \omega_i + \sum_{j=1}^{n} b_{ij} x_j \\
&= n_i P_{e,i} - y_i + \sum_{j=1}^{n} b_{ij} x_j = 0
\end{aligned}
\tag{15.2}
$$

As each column-sum of the Laplacian matrix equals zero, i.e. $\sum_{i=1}^{n} \sum_{j=1}^{n} b_{ij} x_j = 0$, it can be found that

$$
\sum_{i=1}^{n} \dot{x}_i = \sum_{i=1}^{n} (n_i P_{e,i} - y_i) = n(n_i P_{e,i} - y_i) = 0
\tag{15.3}
$$

Consequently, $n_i P_{e,i} = y_i$. To study the convergence, we define

$$
\bar{x}(t) := \frac{1}{n} \sum_{i=1}^{n} x_i(t)
\tag{15.4}
$$

Substituting (15.3) to (15.4) yields

$$
\dot{\bar{x}}(t) = \frac{1}{n} \sum_{i=1}^{n} (n_i P_{e,i} - y_i)
\tag{15.5}
$$

Let us construct a Lyapunov function $V(x) = \frac{1}{2}(x - x^*)^2$. Then we have

$$
\begin{aligned}
\dot{V}(\bar{x}(t)) &= \frac{1}{n} \sum_{i=1}^{n} (n_i P_{e,i} - y_i)(\bar{x}(t) - x^*) \\
&= \frac{1}{n} \sum_{i=1}^{n} (\omega^* - \omega_i)(\bar{x}(t) - x^*)
\end{aligned}
\tag{15.6}
$$

To analyze the behavior of (15.6), suppose that system is in steady state and then experiences a power deficiency, e.g. a sudden step load or a loss of generation. At this instance, system's frequency drops as demand exceeds the generation. Hence, in order to compensate the power imbalance, DERs need to increase their generation, and considering the power-sharing, this results to a higher $n_i P_{e,i}$ which means the new equilibrium point will be a higher value. Therefore,

$$
\omega^* - \omega_i > 0, \quad \bar{x}(t) - x^* < 0
\tag{15.7}
$$

If the total active power produced outweighs the total demand, e.g. following a load reduction, system's frequency increases and the new equilibrium point will be a lower value, then

$$
\omega^* - \omega_i < 0, \quad \bar{x}(t) - x^* > 0
\tag{15.8}
$$

Considering (15.7) and (15.8), $\dot{V}(\bar{x}(t)) < 0$ and thus $\bar{x}(t) \to x^*$ as $t \to \infty$. It can be shown that $y_i(t)$ is an observer for $\bar{x}(t)$ and $\lim_{t \to \infty} \|y_i(t) - \bar{x}(t)\| = 0$ [7]. Therefore, eventually, $n_i P_{e,i}, y_i$ and $\bar{x}$ all converge to x^*, and ω_i converges to ω^*.

15.1.1.1 Case Studies

The performance of the push-sum-enabled distributed frequency control is tested on a microgrid with six inverter-interfaced DERs simulated in MATLAB/SIMPOWER environment (see Fig. 15.1). The nominal voltage and frequency are 380 V and 60 Hz, respectively. Rated active powers for

DERs 1/2, 3/4, and 5/6 are 100, 80, and 116 kW, respectively. Droop coefficients for DERs 1/2, 3/4, and 5/6 are 8×10^{-4}, 10^{-3}, and 7×10^{-4}, respectively. Line impedances are $Z_1(0.65\Omega, 1.3\,\text{mH})$, $Z_2(0.5\Omega, 1\,\text{mH})$, $Z_3(0.58\Omega, 1.2\,\text{mH})$, $Z_4(0.6\Omega, 1\,\text{mH})$, and $Z_5(0.55\Omega, 1.3\,\text{mH})$.

Time-Varying Unbalanced Communication This case verifies the performance of the push-sum scheme under unbalanced communication and switching topology (random graph). Initially, the communication network operates as an undirected connected graph $\mathcal{G}_1$ (see Fig. 15.2(a)). Then the cyber network is changed at $t = 10\,\text{s}$, $t = 14\,\text{s}$, and $t = 20\,\text{s}$ to directed unbalanced graphs $\mathcal{G}_2$, $\mathcal{G}_3$, and $\mathcal{G}_4$, respectively. Eventually a step load of 13 kW is applied at $t = 30\,\text{s}$.

As illustrated in Fig. 15.2(b), when the communication switches, the scaling factor z_i tracks the imbalances and prevents the secondary controller y_i from being disturbed and therefore, frequency remains intact.

Empowering Plug-and-Play This case verifies the push-sum-based controller's feature of plug-and-play capability. This merit is investigated by detaching DER6 at $t = 15\,\text{s}$ and plugging it in again at $t = 22\,\text{s}$. The exploited communication graph is shown in Fig. 15.3. As can be seen, if DERi is disconnected, the remaining graph is still strongly connected.

As depicted, after disconnection of DER6, the power deficiency reallocated among the remaining DERs and they manage to share the loads. As Fig. 15.3 shows, accurate active power sharing and frequency restoration are maintained during plug-and-play operation. The reason of small transient

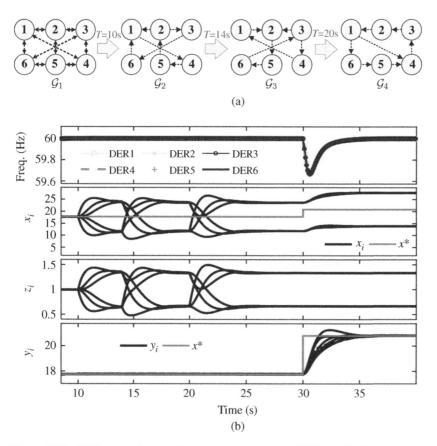

Figure 15.2 (a) Time-varying unbalanced communications. (b) Microgrid's frequency and controller parameters under the condition of time varying unbalanced communications.

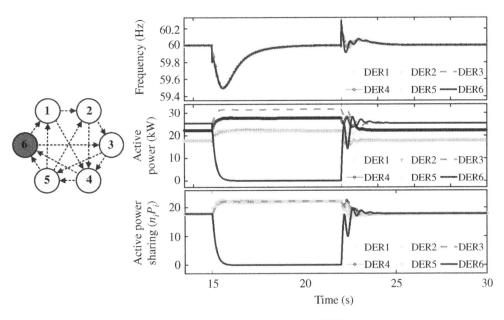

Figure 15.3 Frequency and active power sharing after removal of DER6.

oscillations after reconnection of DER6 is that, no presynchronization is implemented ahead of reconnection.

Cyberattacks Immunity Microgrid communication networks are exposed to potential cyber-attacks such as False-Data-Injection that would jeopardize the overall microgrid performance in terms of efficiency and stability. In this work, the attack model assumed to compromise the communication link between DERs. The attack model involves the following assumptions:

1. The attacker has knowledge of the microgrid and the topology $\mathcal{G}$.
2. The attacker is capable of relaying and altering the communications between two DERs.
3. The attacker is not able to predict the communication direction between DERs

In case cyberattack on communication link happens, when topology switches, all DERs know where to send data and which adjacent DERs are supposed to send them data. Hence, if a received signal does not match with the switching pattern, that link is identified as a corrupted communication link. The attack is modeled as

$$\dot{x}_i = n_i P_{e,i} - y_i + b_{ii}x_i + \sum_{j=1,\, j\neq i}^{n} b_{ij}(x_j + \mu_j) \tag{15.9}$$

Suppose that, the malicious signal μ_j is injected into the information transferred from DER_j to DER_i. Summation over all $\dot{x}_i$s yields

$$\sum_{i=1}^{n} \dot{x}_i = \sum_{i=1}^{n} (n_i P_{e,i} - y_i)$$

$$+ \left[\sum_{i=1}^{n} b_{i,1}, \dots, \sum_{i=1}^{n} b_{i,n} \right] [x_1, \dots, x_n]^T + b_{ij}\mu_j \tag{15.10}$$

$$= \sum_{i=1}^{n} (n_i P_{e,i} - y_i) + \sum_{i=1}^{n}\sum_{j=1}^{n} b_{ij}x_j + b_{ij}\mu_j$$

Now, if the communication topology switches such that DER$_j$ does not send information to DER$_i$, since DER$_i$ knows all the incoming and outgoing links at each instant, the term $b_{ij}\mu_j$ in (15.10) will be disregarded, then

$$\sum_{i=1}^{n}\dot{x}_i = \sum_{i=1}^{n}(n_iP_{e,i} - y_i) + \sum_{i=1}^{n}\sum_{k=1,\ k\neq j}^{n}b_{ik}x_k. \tag{15.11}$$

Since the communication graph is still strongly connected and each column-sum of the Laplacian matrix equals zero,

$$\sum_{i=1}^{n}\sum_{k=1,\ k\neq j}^{n}b_{ik}x_k = \mathbf{1}Bx = 0, \tag{15.12}$$

where $\mathbf{1}$ is the vector whose entries all equal one, and

$$\sum_{i=1}^{n}\dot{x}_i = \sum_{i=1}^{n}(n_iP_{e,i} - y_i). \tag{15.13}$$

Comparing (15.13) and (15.3), it can be readily obtained that (15.2)–(15.8) can be followed to complete the proof of the system's stability as after attack elimination, the analyses are essentially the same.

Furthermore, since DER$_i$ knows there is no information from DER$_j$ but still receives $b_{ij}\mu_j$, the link from DER$_j$ to DER$_i$ is identified and isolated as a corrupted link. Therefore, after the communication topology switches, corrupted links (and consequently malicious signals) cannot jeopardize the convergence and stability as they are not parts of the communication graph anymore. This concept is also shown in Fig. 15.4 (a).

In the third scenario, the communication networking starts from $\mathcal{G}_1$ in Fig. 15.4 (a) and then links from DER$_2$ to DER$_1$ and DER$_4$ to DER$_3$ are attacked at $t = 10$ s and $t = 15$ s, respectively. A step load of 20 kW is also applied at $t = 35$ s.

Figure 15.4 (b) shows the impacts of time-varying cyberattacks on frequency and active power sharing, where the malicious signal μ_j is considered as Sine waves with an offset. Performance of the push-sum strategy against such cyberattack is also demonstrated. After the attack, the communication topology switches at $t = 20$ s and $t = 30$ s.

It is worth noting that, speed of attack detection depends on the speed of topology switching and, in this scenario, the reason for choosing the switching and attack times is to better illustrate the impacts of attacks and performance of the push-sum enabled scheme.

According to the communication topology at $t = 20$ s, there is no link from DER$_4$ to DER$_3$ and the data flow through this link is supposed to be zero which is not due to the attack. Hence, this link is identified as a corrupted link and it is isolated from the network. The same happens for the link from DER$_2$ to DER$_1$ at $t = 30$ s. Consequently, the attacked links are identified and isolated, active power sharing is again synchronized among DERs and frequency is restored at the rated 60 Hz. The push-sum scheme thus guarantees ultra-resilience of microgrid operations and its speed of attack detection depends on the switching speed.

Regarding identification of cyberattack on communication links, there are some existing methods in the literature [10, 11]. To detect and isolate false data injections, [10] proposes a distributed cyberattack control strategy for islanded MGs with distributed control systems, through turning on and off the communication links aperiodically. However, this action tangibly slows down the convergence.

To mitigate attacks on communication links, a trust/confidence-based control protocol is proposed in [11] such that each inverter monitors the information it receives from its neighbors, updates its local confidence factor, and sends to its neighbors. Data received at each inverter from

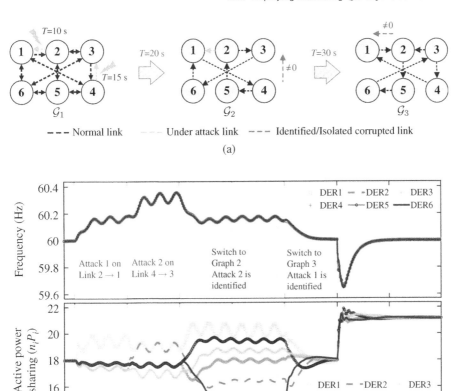

Figure 15.4 (a) Attacked link identification/isolation procedure. (b) Impact of time-varying cyberattack on frequency and active power sharing and the performance of the push-sum strategy.

neighbors is weighted by its trust factor. One shortcoming of this method is, the selection of the trust factor is done empirically, as it depends on several factors like network connectivity, speed of convergence of the consensus algorithm and other gains in the consensus algorithm. So if the communication topology changes, the trust factor needs to be reset accordingly and if anything happens to the system (e.g. inertia drop, communication noise, changes of network connectivity, etc.), there is no guarantee that the current trust factor does not jeopardize the system stability.

Compared to the above references, to identify attack on communication links, our developed method relaxes assumptions like constraints on the type of attack signals and empirical setting of trust factors that requires having knowledge about the network. With push-sum enabled scheme, corrupted link is identified if a received signal does not match with the topology switching pattern and rate of convergence depends on the network connectivity.

15.2 Employing Interacting Qubits for Distributed Microgrid Control

Distributed control of multi-inverter microgrids can achieve the combined goals of flexible plug-and-play architecture, guaranteeing frequency, and voltage regulation while preserving precise power sharing among nonidentical DERs [12]. However, security of communication among distant parties is an indispensable criterion for evaluating the performance of any communication network [13] and distributed control of microgrids is not an exception. While distributed control strategies can enhance microgrids resilience, the openness brought by the corresponding communication networks may cause cybersecurity challenges since they can be susceptible to cyberattacks on communication links. Adversarial attacks from third-party agents can drive the microgrid toward inconsistent performance and impair the operation and control functions of participating DERs and stability of the whole system [14].

Finding solutions to encounter cyber manipulation in microgrids with distributed control strategies is an ongoing research [15–20]. However, the existing solutions may become insecure due to the rapid development of supercomputers and the emergence of quantum computers [21–23] and so they can make traditional/classical methods obsolete. On the other hand, utilizing principles of quantum mechanics, quantum communication offers provable security of communication and is a promising solution to counter such threats [23].

Quantum physics principles give rise to novel capabilities unattainable with classical transmission media [13, 24–31]. Such a quantum communication enables secure communication between any two points and will connect quantum processors in order to achieve capabilities that are provably impossible by using only classical information [32]. Several major applications have already been reported, including secure communication, quantum distributed computation, simulation of quantum many-body systems, and exponential savings in communication [13, 28]. However, central to all these applications is the ability to transmit quantum bits (qubits) which cannot be copied, and any attempt to do so can be detected. This feature makes qubits well-suited for security applications [13]. Promising findings on quantum internet have even led some researchers to believe that all secure communications will eventually be done through quantum channels [30].

Regarding this revolutionary step in secure communication, many models including quantum key distribution (QKD) [33], quantum teleportation [34], and quantum secure direct communication [35–37] have been developed. Based on QKD technology, many different types of quantum communication networks have been proposed [38–40]. However, these communication networks based on QKD technology only transmit the key, but do not directly transmit information. On the other hand, quantum secure direct communication is a kind of information carrier with quantum state in communication. In this method, secret information is directly transmitted over a secure quantum channel and, in contrast to QKD schemes, they do not require key distribution and key storage [35].

In this effort, we answer the questions that how is it possible to establish synchronization through exchanging qubits. How can we exploit a quantum communication infrastructure for distributed control of microgrids while control objectives like frequency/voltage regulation and power/current sharing in AC and DC microgrids are guaranteed? Therefore, inspired by the aforementioned developments, specifically those with quantum states in communication, we aim to devise a scalable quantum distributed controller (QDC) for AC and DC microgrids within which, the information carrier is quantum states and the transmission media is a quantum channel, i.e. information is encoded into quantum states which are directly sent over quantum channels among participating DERs. Quantum states are then processed and measured at each DER and the measurement outcomes are exploited as control signals [41–43].

One of the primary objectives in distributed control and coordination is to drive a network to reach a consensus, where all agents hold the same value for some key parameter(s), by local interactions [44, 45]. Several efforts have been made to investigate consensus problems in the quantum domain [46–48]. To describe quantum state evolution of quantum systems with external inputs (open quantum systems), a so-called Lindblad equation can be used [49, 50]. Authors in [51] show that quantum consensus can be obtained through a Lindblad master equation with the Lindblad terms generated by swapping operators among the qubits, giving rise to the dynamical evolution of the quantum network. The swapping operations also introduce an underlying interaction graph for the quantum network, which leads to a distributed structure for the master equation.

One existing approach to achieve quantum consensus is to model the quantum network's state evolution through the quantum synchronization master equation [51]. Another approach is to appeal to the gossip-type interaction between neighboring quantum computing devices [47]. Existing literature only considers two special cases: (i) Under a non-zero Hamiltonian with the swapping as jump operators, each qubit tends to the same trajectory corresponding to a network Hamiltonian and initial states; (ii) Under a zero Hamiltonian, the network's final state is the average of the initial states. In both cases, however, it is difficult to derive the explicit trajectory of each qubit as a function of the Hamiltonian, and the synchronization orbit is certainly no longer the one determined by the Hamiltonian for most choices of the Hamiltonian [51]. Furthermore, in the existing approaches, measurement is not considered, i.e. the existing frameworks are valid as long as the corresponding quantum system is not measured, which makes them impractical for realistic distributed control of microgrids.

Toward the goal of devising the QDC, considering the above challenges and potential to design a quantum synchronization scheme, we first formulate the quantum synchronization problem using a quantum master equation and identify and characterize suitable jump operators to drive the quantum network to synchronization. The protocol we construct gives rise to a differential equation that allows us to analyze the convergence. We utilize proper observables and show that all the corresponding expectation values (averaging measurement outcomes) will eventually converge to a possibly time-varying target value, and finally exploit these expectation values to construct the control signals to drive a network of DERs to synchronization.

Hence, our devised QDC gives rise to a novel quantum communication scheme for distributed control of microgrids and enables microgrids to utilize the existing quantum communication frameworks as communication infrastructure, and also paves the way for more advanced quantum-secure communication frameworks for microgrids, unattainable with classical transmission media.

The rest of the chapter is organized as follows. Section 3.2 provides some preliminaries including relevant concepts in graph theory and quantum systems along with notations and conventions. The developed QDC together with a numerical example and proof of convergence are presented in Section 3.3. Section 3.4 is devoted to explain the devised quantum distributed frequency controller and voltage controller for AC and DC microgrids, respectively. Simulation results are also provided. Section 3.5 provides a discussion on realization of QDC, summarizes the main results, and gives an outlook on possible further developments and applications.

15.2.1 Preliminaries

In this section, we introduce some fundamental concepts from graph theory [52] and quantum systems [53].

15.2.1.1 Graph Theory

Some basic concepts from graph theory [52] are provided here. A simple graph $G = (V, E)$ consists of a set of n nodes (or agents), $V = \{v_1, v_2, \ldots, v_n\}$, and a set of edges, $E \subset V \times V$. An edge $(v_i, v_j) \subset E$ represents that agents v_i and v_j can exchange information with each other. A sequence of non-repeated edges $(v_i, v_{p_1}), (v_{p_1}, v_{p_2}), \ldots, (v_{p_{m-1}}, v_{p_m}), (v_{p_m}, v_j)$ is called a path between nodes v_i and v_j. If there exists a path between any two different nodes $v_i, v_j \in V$, G is said to be connected. An agent v_j is called a neighbor of agent v_i if $(v_j, v_i) \subset E$. The set of neighbors of agent v_i is denoted as $N_i = \{v_j \in V \mid (v_j, v_i) \subset E\}$. The adjacency matrix of graph G, denoted as A, is an $n \times n$ matrix whose entries $a_{i,j} = 1$ if $v_j \in N_i$ and $a_{i,j} = 0$ otherwise. The degree matrix D of graph G, denoted as D, is defined as an $n \times n$ diagonal matrix whose ith diagonal entry equals the degree of node v_i, i.e. $\sum_{v_j \in N_i} a_{i,j}$. The Laplacian matrix of graph G, denoted as L, is defined as $D - A$. Note that A, D, L are all symmetric. The node-edge incidence matrix $B \in R^{V \times E}$ is defined component-wise as $B_{i,j} = 1$ if edge j enters node i, $B_{i,j} = -1$ if edge j leaves node i, and $B_{i,j} = 0$ otherwise. For $x \in R^V$, $B^T x \in R^E$ is the vector with components $x_i - x_j$, with $\{i, j\} \in E$. If $\text{diag}(\{a_{i,j}\}_{\{i,j\} \in E})$ is the diagonal matrix of edge weights, then the Laplacian matrix is given by $L = B \text{diag}(\{a_{i,j}\}_{\{i,j\} \in E})B^T$.

15.2.1.2 Quantum Systems and Notations

Throughout this report, the (adjoint) † symbol indicates the transpose-conjugate in matrix representation, and the tensor product symbol ⊗ is the Kronecker product.

The mathematical description of a single quantum system starts by considering a complex Hilbert space $\mathcal{H}$. We utilize Dirac's notation, where $|\psi\rangle$ denotes an element of $\mathcal{H}$, called a ket which is represented by a column vector, while $\langle \psi | = |\psi\rangle^\dagger$ is used for its dual, a bra, represented by a row vector, and $\langle \psi | \varphi \rangle$ for the associated inner product. We denote the set of linear operators on $\mathcal{H}$ by $\mathfrak{B}(\mathcal{H})$. The adjoint operator $X^\dagger \in \mathfrak{B}(\mathcal{H})$ of an operator $X \in \mathfrak{B}(\mathcal{H})$ is the unique operator that satisfies $(X |\psi\rangle)^\dagger |\chi\rangle = \langle \psi | (X^\dagger | \chi \rangle)$ for all $|\psi\rangle, |\chi\rangle \in \mathcal{H}$. The natural inner product in $\mathfrak{B}(\mathcal{H})$ is the Hilbert–Schmidt product $\langle X, Y \rangle = \text{tr}(X^\dagger Y)$, where tr is the usual trace functional which is canonically defined in a finite-dimensional setting. We denote by I the identity operator. $[A, B] = AB - BA$ is the commutator and $\{A, B\} = AB + BA$ is the anticommutator of A and B.

A quantum bit (qubit), defined as the quantum state of a two-state quantum system, is the smallest unit of information, and it is analogous to a classical bit. The state of a qubit, represented by $|\psi\rangle = \alpha |0\rangle + \beta |1\rangle$, is a superposition of the two orthogonal basis states $|0\rangle$ and $|1\rangle$, where α and β are complex numbers in general, where $|\alpha|^2 + |\beta|^2 = 1$. We will simplify the notation of a n-qubit state $|q_1\rangle \otimes \cdots \otimes |q_n\rangle \in \mathcal{H}^{\otimes n}$ as $|q_1 \ldots q_n\rangle$.

In the case of mixed state, the state of a quantum system is represented by a *density operator* ρ, which is a self-adjoint positive semi-definite operator with trace one. Moreover, the state $|\psi\rangle \in \mathcal{H}$ with $\langle \psi | \psi \rangle = 1$ in the above is called a pure state, which can also be written in the form of a density matrix $\rho = |\psi\rangle \langle \psi|$. For further information on qubits see [53, 54].

15.2.2 Quantum Distributed Control

Distributed control problems of microgrids are typically modeled as networked differential equations over a simple, connected graph $G = (V, E)$ whose node set $V = \{v_1, v_2, \ldots, v_n\}$ represents microgrids and edge set E depicts allowable communication among the microgrids. As an illustrative example, the problem of distributed frequency control and power sharing in AC microgrids can be formulated as

$$\begin{aligned}
\omega_i &= \omega^* - n_i P_i + \Phi_i, \\
\dot{\Phi}_i &= f(\Phi_i, P_i, \Phi_j, j \in N_i)
\end{aligned} \tag{15.14}$$

where ω_i represents the derivative of the voltage phase angle of DER$_i$ (i.e. the frequency at DER$_i$) with respect to time, ω^* is a nominal network frequency, P_i is the measured active power injection at

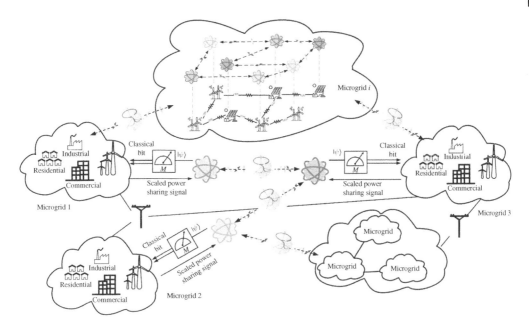

Figure 15.5 QDC framework for microgrids – Quantum communication is established among the DERs – Power sharing signals, $n_i P_i$, are scaled to be encoded into quantum information.

DER$_i$, n_i is the gain of the droop coefficient, N_i denotes the set of neighbors of DER$_i$ (i.e. $N_i = \{v_j \in V \mid (v_j, v_i) \subset E\}$), and the dynamics of Φ_i represents the secondary control, which is a function (f) of its current value, P_i, and Φ_j's of its neighbors. The goal of the DER problem is to ensure that the network frequency will be regulated to the rated value ω^* and that active power sharing is guaranteed (i.e. $\omega_i = \omega^*$ and $n_i P_i = n_j P_j$ for all i, j). It is worth emphasizing that Eq. (15.14) has a universal form which can be used, with simple modification, to describe the problem of distributed voltage control and power/current sharing in DC microgrids, as we will elaborate in Section 15.4.

Recent development in quantum algorithms for solving linear/nonlinear/partial differential equations suggests potential efficiency and capability of quantum devices in solving these class of equations [55–61]. Therefore, we aim to construct a quantum distributed framework to control a network of DERs, as shown in Fig. 15.5. In this framework, each DER is equipped with or connected to a quantum computing (QC) device, which prepares a quantum state to be manipulated and measured and then seeks a consensus among all the QCs in a distributed manner.

The state of each quantum device can be described by a positive Hermitian density matrix ρ. Since synchronization requires interaction among all quantum devices, let us assume that each device can be considered as a quantum system and has access to the (quantum) information of its neighbors. The following Lindblad master equation is a suitable way to describe the dynamics of a system with dissipation:

$$\dot{\rho}(t) = -\frac{\imath}{\hbar}[H, \rho] + \sum_{i=1}^{n} \left(C_i \rho C_i^\dagger - \frac{1}{2}\{C_i^\dagger C_i, \rho\} \right) \tag{15.15}$$

where H is the effective Hamiltonian as a Hermitian operator over the underlying Hilbert space, $\hbar$ is the reduced Planck constant, $\imath$ denotes the imaginary unit (i.e. $\imath^2 = -1$), and C_i's are jump operators. For more information on Markovian master equations in Lindblad form, see [62]. To anticipate later discussions, it should be emphasized that our goal is to leverage the Lindblad master equation in order to construct the network of differential equations, such as those in Eq. (15.14), in which, in contrast to the classical synchronization, quantum bits are what is exchanged among the nodes. We next demonstrate that utilizing suitable jump operators and observers for each quantum node would lead the average expectation values of all the observers in the corresponding quantum setting to converge to a possibly time-varying target value and the synchronization rule follows the Kuramoto model modified by the presence of a sinusoidal driving [63].

15.2.2.1 Algorithm

Let us update the state of each quantum node at each time step as follows:

$$
|q_i(t)\rangle = \begin{pmatrix} \cos\frac{\pi}{4} \\ e^{i\phi_i(t)} \sin\frac{\pi}{4} \end{pmatrix}, \quad t \in \{0, 1, 2, \ldots\} \tag{15.16}
$$

which is the general state in polar coordinates set on the xy-plane in the Bloch sphere, where $\phi_i(0) \in (0, \pi/2)$ and each $\phi_i(t)$, $t \geq 1$, is the averaged measurement outcome which can be obtained by simply averaging measurement outcomes of many realizations of a single experiment for node i and will be discussed in detail shortly. Let $|\psi\rangle = |q_1 q_2 \cdots q_n\rangle$ be the state of the whole quantum network and $\rho = |\psi\rangle\langle\psi|$. We introduce the following master equation:

$$
\dot{\rho}(t) = \sum_{i=1}^{n} \left(C_i \rho C_i^\dagger - \frac{1}{2}\{C_i^\dagger C_i, \rho\} \right)
$$
$$
+ \sum_{\{i,j\} \in E} \left(C_{i,j} \rho C_{i,j}^\dagger - \frac{1}{2}\{C_{i,j}^\dagger C_{i,j}, \rho\} \right) \tag{15.17}
$$

where $C_{i,j}$ is the swapping operator that specifies the external interaction between quantum computing devices i and j such that

$$
C_{i,j}(|q_1\rangle \otimes \cdots \otimes |q_i\rangle \otimes \cdots \otimes |q_j\rangle \otimes \cdots \otimes |q_n\rangle)
$$
$$
= |q_1\rangle \otimes \cdots \otimes |q_j\rangle \otimes \cdots \otimes |q_i\rangle \otimes \cdots \otimes |q_n\rangle. \tag{15.18}
$$

Let us define the jump operator, C_i, by

$$
C_i = I^{\otimes(i-1)} \otimes R_z(\phi) \otimes I^{\otimes(n-i)} \tag{15.19}
$$

with $R_z(\phi)$ being the rotation-Z operator which is a single-qubit rotation through angle ϕ radians around the Z-axis (Fig. 15.6):

$$
R_z(\phi) = \begin{pmatrix} e^{-i\phi/2} & 0 \\ 0 & e^{i\phi/2} \end{pmatrix}. \tag{15.20}
$$

By definition, the operator C_i acts only on $|q_i\rangle$ without changing the states of other qubits, i.e.

$$
C_i \rho C_i^\dagger = C_i \left(\left| \overbrace{q_1 \cdots}^{i-1} q_i \overbrace{\cdots q_n}^{n-i} \right\rangle \left\langle \overbrace{q_1 \cdots}^{i-1} q_i \overbrace{\cdots q_n}^{n-i} \right| \right) C_i^\dagger
$$
$$
= \left| \overbrace{q_1 \cdots}^{i-1} q_i' \overbrace{\cdots q_n}^{n-i} \right\rangle \left\langle \overbrace{q_1 \cdots}^{i-1} q_i' \overbrace{\cdots q_n}^{n-i} \right| \tag{15.21}
$$

Figure 15.6 Qubit state representation on the Bloch sphere. Rotations around the positive X, Y, and Z axes are represented by the dashed black arrows.

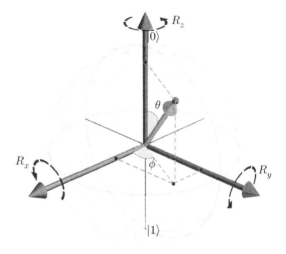

To see the impact of the introduced jump operator on $|q_i\rangle$, by selecting $\phi = \phi_{t,i} - \phi_i$ we have

$$
R_z(\phi) |q_i\rangle \langle q_i| R_z^\dagger(\phi) =
$$
$$
\begin{pmatrix}
\cos^2 \frac{\pi}{4} & e^{-\iota(\phi_i+\phi)} \cos \frac{\pi}{4} \sin \frac{\pi}{4} \\
e^{\iota(\phi_i+\phi)} \cos \frac{\pi}{4} \sin \frac{\pi}{4} & \sin^2 \frac{\pi}{4}
\end{pmatrix}
\tag{15.22}
$$
$$
= \begin{pmatrix}
\cos \frac{\pi}{4} \\
e^{\iota(\phi_{t,i})} \sin \frac{\pi}{4}
\end{pmatrix}
\begin{pmatrix}
\cos \frac{\pi}{4} & e^{-\iota(\phi_{t,i})} \sin \frac{\pi}{4}
\end{pmatrix}
$$

which is considered as the target state for quantum node i. As can be seen, the jump operators C_i's are state-dependent and updated based on the target values $\phi_{i,t}$ and the measured $\phi_i(t)$. Furthermore, as mentioned, at the beginning of each time step, all the qubits are initialized as (15.16) based on the measurement outcome of the previous step. Therefore, at each time step, the master equation components are updated based on the target values and the obtained measurement signals. Thus, the density matrix at time $t + dt$ can be decomposed into $\rho(t + dt) = \rho(t) + d\rho_t$, where $d\rho(t)$ is defined in (15.17).

In order to obtain the angles ϕ_i, we introduce the following observables:

$$
A_{1,i} = I^{\otimes(i-1)} \otimes \sigma_x \otimes I^{\otimes(n-i)}
\tag{15.23}
$$

$$
A_{2,i} = I^{\otimes(i-1)} \otimes \sigma_y \otimes I^{\otimes(n-i)}
\tag{15.24}
$$

The operator $I^{\otimes(i-1)} \otimes \sigma_{x/y} \otimes I^{\otimes(n-i)}$ acts only on $|q_i\rangle$ where node-wise means, having σ_x and σ_y which are Pauli matrices as observables at each node,

$$
\sigma_x = \begin{pmatrix} 0 & 1 \\ 1 & 0 \end{pmatrix}, \qquad \sigma_y = \begin{pmatrix} 0 & -\iota \\ \iota & 0 \end{pmatrix}
\tag{15.25}
$$

The expectation value of an observable A via measurement on a system described by a density matrix ρ is given by $\langle A \rangle = \text{tr}(\rho A)$ [54]. For a general one qubit state ρ, $\text{tr}(\rho\sigma_x) = r \sin\theta \cos\phi$, $\text{tr}(\rho\sigma_y) = r \sin\theta \sin\phi$ and $\text{tr}(\rho\sigma_z) = r \sin\theta$, where r, θ, and ϕ are the parameters that describe ρ in the Bloch sphere, which is essentially the spherical coordinate but with $r \leq 1$. Generally, the Lindblad equation results in states becoming more mixed; however, we only let the system evolve in a short time and reinitialize the system in a product of pure qubit states. Therefore, we can consider $r = 1$ and $\theta = \pi/2$ and hence

$$\mathrm{tr}(\rho\sigma_x) = \cos\phi_i, \qquad \mathrm{tr}(\rho\sigma_y) = \sin\phi_i \tag{15.26}$$

which are equivalent to $\mathrm{tr}(\rho A_{1,i}) = \cos\phi_i$ and $\mathrm{tr}(\rho A_{2,i}) = \sin\phi_i$, respectively. Since both C_i and $C_{i,j}$ are unitary, we have

$$d\rho(t) = \sum_{i=1}^{n} (C_i \rho C_i^\dagger - \rho)dt + \sum_{\{i,j\} \in E} (C_{i,j} \rho C_{i,j}^\dagger - \rho)dt \tag{15.27}$$

If we repeat the procedure of the Lindblad evolution in a short duration, measurement, and reinitialization, we can obtain approximated equations for ϕ_i's in the limit $dt \to 0$. The goal is to obtain the dynamic of the phase angles ϕ_i. Note that $\frac{d}{dt}\langle A \rangle = \frac{d}{dt}\mathrm{tr}(\rho A) = \mathrm{tr}(\dot{\rho}A)$. From (15.27),

$$\mathrm{tr}(\dot{\rho}A_{1,i}) = \cos\phi_{t,i} - \cos\phi_i + \sum_{j=1}^{n} a_{i,j}(\cos\phi_j - \cos\phi_i)$$
$$\mathrm{tr}(\dot{\rho}A_{2,i}) = \sin\phi_{t,i} - \sin\phi_i + \sum_{j=1}^{n} a_{i,j}(\sin\phi_j - \sin\phi_i) \tag{15.28}$$

where $a_{i,j} = 1$ if $C_{ij} \neq 0$ and $a_{i,j} = 0$ otherwise. Utilizing $\mathrm{tr}(\rho A_{1,i})$ and $\mathrm{tr}(\rho A_{2,i})$, we have the dynamic of ϕ_i as follows:

$$\begin{aligned}
\dot{\phi}_i &= \frac{d}{dt}\arctan\left(\frac{\mathrm{tr}(\rho A_{2,i})}{\mathrm{tr}(\rho A_{1,i})}\right) \\
&= \left\{\frac{\mathrm{tr}(\dot{\rho}A_{2,i})\mathrm{tr}(\rho A_{1,i}) - \mathrm{tr}(\dot{\rho}A_{1,i})\mathrm{tr}(\rho A_{2,i})}{\cos^2\phi_i}\right\}\cos^2\phi_i \\
&= \left(\sin\phi_{t,i} - \sin\phi_i + \sum_{j=1}^{n} a_{i,j}(\sin\phi_j - \sin\phi_i)\right)\cos\phi_i \\
&\quad - \left(\cos\phi_{t,i} - \cos\phi_i + \sum_{j=1}^{n} a_{i,j}(\cos\phi_j - \cos\phi_i)\right)\sin\phi_i \\
&= \sin(\phi_{t,i} - \phi_i) + \sum_{j=1}^{n} a_{i,j}\sin(\phi_j - \phi_i)
\end{aligned} \tag{15.29}$$

It is worth mentioning that both $\mathrm{tr}(\rho A_{1,i})$ and $\mathrm{tr}(\rho A_{2,i})$ are used in (15.29) to find the trajectory that ϕ_i traverses along the time; however, either arccos $(\mathrm{tr}(\rho\sigma_x))$ or arcsin $(\mathrm{tr}(\rho\sigma_y))$ gives ϕ_i. In the 'Analysis' subsection, we will show how the pinning term $\sin(\phi_{t,i} - \phi_i)$ forces the phase ϕ_i to stick at the value $\phi_{t,i}$ and the coupling mechanism $\sum_{j=1}^{n} a_{i,j}\sin(\phi_j - \phi_i)$ helps to synchronize the entire system.

The basic outline of the algorithm is drawn schematically in Fig. 15.7 and is summarized as follows:

1. Initialize qubits as a point on the first quarter of the equator of the Bloch Sphere, i.e. $0 < \phi_i(0) < \pi/2$, using Eq. (15.16).
2. Teleport information throughout the network such that each quantum node receives the quantum information from its adjacent nodes.
3. At each node, update the rotation-Z (R_z) operator's argument based on the pinner ($\phi_{t,i}$) and the current value of the phase angle ϕ_i.

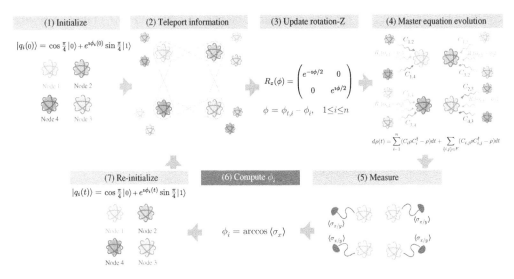

Figure 15.7 Schematic depiction of the QDC - At step (1), $\phi_i(0) \in (0, \pi/2)$ – If at step (5) σ_y operator is used as observer to measure the qubit at each node, at step (6), ϕ_i is calculated by $\phi_i = \arcsin \langle \sigma_y \rangle$ - At step (7), the averaged measurement outcome of step (6) is utilized to reinitialize the qubits.

4. Evolve the master Eq. (15.27) for one-time step δt by means of the swapping and rotation-Z operators.

5. Measure the expectation value of the σ_x or σ_y operator as the observer at each node. Repeating this multiple times and averaging gives the $\cos \phi_i$ or $\sin \phi_i$, depending on the exploited observable.

6. On classical hardware at each node, compute $\arccos \langle \sigma_x \rangle$ or $\arcsin \langle \sigma_y \rangle$ to obtain the phase angle ϕ_i.

7. Reinitialize the state of each quantum node according to Eq. (15.16).

8. Go back to step 2.

15.2.2.2 Analysis

For synchronization of the system, it is critical that the pinner for all of the oscillators be the same, i.e. $\phi_{t,i} = \phi^*$, otherwise, synchronization cannot be achieved in general. To study whether quantum node i is synchronized to the pinner, it is convenient to study the phase deviation of quantum node i from the pinner. We introduce the following change of variables,

$$\phi_i = \phi^* + \zeta_i \tag{15.30}$$

where ζ_i denotes the phase deviation of the ith oscillator from the pinner ϕ^*. Substituting (15.30) into (15.29), we have

$$\dot{\zeta}_i = \sum_{j=1}^{n} a_{i,j} \sin(\zeta_j - \zeta_i) - \sin(\zeta_i) \tag{15.31}$$

By studying the properties of (15.31), we can obtain the condition for synchronization. If all ζ_i's converge to 0, then we have $\phi_i = \phi^*$ as $t \to \infty$, indicating that all nodes are synchronized to the pinner. Let $B = [B_{i,j}]_{n \times m}$ be the incidence matrix [52] of the communication graph G with m being the number of edges. Then, (15.31) can be recast in a state form:

$$\dot{\zeta} = -\sin \zeta - BW \sin(B^T \zeta) \tag{15.32}$$

where $W = \text{diag}(\{a_{i,j}\}_{\{i,j\}\in E})$ is the diagonal matrix of edge weights and $\sin(\cdot)$ takes entrywise operation for a vector.

To proceed, set $\varepsilon = \max_{1\leq i\leq n}|\zeta_i|$. When $\varepsilon < (\pi/2)$, if $\zeta_i = \varepsilon$, we have $-\pi < -2\varepsilon \leq \zeta_j - \zeta_i \leq 0$ for $1 \leq j \leq n$. Hence, in (15.31), $\sin(\zeta_j - \zeta_i) \leq 0$ and $\sin\zeta_i > 0$ hold, and hence $\dot{\zeta}_i < 0$ hold. Therefore, the vector field is pointing inward in the set, and no trajectory can escape to values larger than ε. Similarly, it can be obtained that, when $\zeta_i = -\varepsilon$, $\dot{\zeta}_i > 0$ holds. Thus no trajectory can escape to values smaller than $-\varepsilon$. Therefore, $\zeta \in [-\varepsilon, \varepsilon] \times \cdots \times [-\varepsilon, \varepsilon] = [-\varepsilon, \varepsilon]^n$ is positively invariant when $\varepsilon < \pi/2$, where $\times$ denotes Cartesian product. Define a Lyapunov function $V = (1/2)\zeta^T\zeta$, which equals zero only if all ζ_i are zero, meaning the synchronization of all nodes to the pinner. Differentiating V along the trajectories of (15.32) yields

$$\dot{V} = \zeta^T\dot{\zeta} = -\zeta^T\left(\sin\zeta + BW\sin B^T\zeta\right)$$
$$= -\zeta^T S_1 \zeta - \zeta^T BW S_2 B^T \zeta \qquad (15.33)$$

where $S_1 \in R^{n\times n}$ and $S_2 \in R^{m\times m}$ are given by

$$S_1 = \text{diag}\left\{\text{sinc}(\zeta_1), \dots, \text{sinc}(\zeta_n)\right\},$$
$$S_2 = \text{diag}\left\{\text{sinc}(B^T\zeta)_1, \dots, \text{sinc}(B^T\zeta)_m\right\} \qquad (15.34)$$

where $\text{sinc}(x) \equiv \sin(x)/x$ and $(B^T\zeta)_i$ denotes the ith element of $m \times 1$ dimensional vector $B^T\zeta$. When all ζ_i are within $[-\varepsilon, \varepsilon]$ with $0 \leq \varepsilon < (\pi/2)$, $(B^T\zeta)_i$ is in the form of $\zeta_k - \zeta_l$ $(1 \leq k, l \leq n)$, and hence is restricted to $(-\pi, \pi)$. Given that in $(-\pi, \pi)$, $\text{sinc}(x) > 0$ holds, it follows that S_1 and S_2 satisfy the following inequalities:

$$S_1 \geq \sigma_1 I, \qquad \sigma_1 = \text{sinc}(\varepsilon),$$
$$S_2 \geq \sigma_2 I, \qquad \sigma_2 = \text{sinc}(2\varepsilon) \qquad (15.35)$$

So, we have $S_1 + BW S_2 B^T \geq \sigma_1 I + \sigma_2 BW B^T$, which in combination with (15.33) yields

$$\dot{V} \leq -\zeta^T(\sigma_1 I + \sigma_2 BW B^T)\zeta \qquad (15.36)$$

Note that $BW B^T$ is the Laplacian matrix of the underlying graph G, which is always positive semidefinite. Since σ_1 and σ_2 are positive, $\sigma_1 I + \sigma_2 BW B^T$ must be positive definite. It follows that, when $0 \leq \varepsilon < (\pi/2)$, we have $\dot{V} \leq -2\mu V$, where

$$\mu = \lambda_{\min}(\sigma_1 I + \sigma_2 BW B^T) > 0 \qquad (15.37)$$

which implies that all the nodes will synchronize to the pinner exponentially fast at a rate no less than μ, which is dependent on the network connectivity.

15.2.2.3 Numerical Example

We consider a network composed of two quantum nodes with the following initial states:

$$|q_1\rangle = \frac{1}{\sqrt{2}}\begin{pmatrix} 1 \\ 1 \end{pmatrix}, \qquad |q_2\rangle = \begin{pmatrix} 1/\sqrt{2} \\ 0.5 + 0.5\imath \end{pmatrix}.$$

In this example, the state $\frac{1}{\sqrt{2}}[1, \ \imath]^T$ is the first target, then at $t = 5.5$ the target changes to the state $\frac{1}{\sqrt{2}}[1, \ e^{\pi/10\imath}]^T$ and hence, the pinner is $\phi^* = \pi/2$ and then $\phi^* = \pi/10$, respectively. The two qubits $|q_1\rangle$ and $|q_2\rangle$ interact through a swapping operator, forming a connected interaction graph. The trajectories of ϕ_1 and ϕ_2, i.e. phase angles of $|q_1\rangle$ and $|q_2\rangle$ respectively, are sketched in Fig. 15.8 utilizing the Python-based open-source software QuTiP [64]. As illustrated, both phase angles converge to ϕ^*. Therefore, the final state of the quantum network is $|qq\rangle$, where we denote by $|q\rangle$ the state $\frac{1}{\sqrt{2}}[1, \ e^{\pi/10\imath}]^T$.

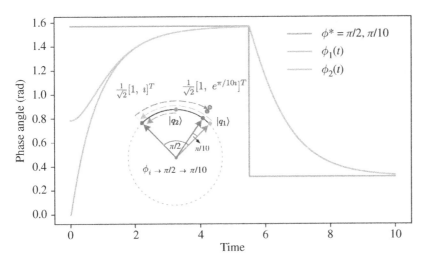

Figure 15.8 Quantum state tracking – Exponential synchronization of phase angles – The selected time step for the numerical example is 10^{-4} – The final state is $\frac{1}{\sqrt{2}}[1, e^{\pi/10 i}]^T$.

15.2.3 Quantum Distributed Controller for AC and DC Microgrids

15.2.3.1 Quantum Distributed Frequency Control

In AC microgrids, a predominantly inductive network naturally decouples the load-sharing process; the reactive power regulator must handle the reactive load sharing by adjusting voltage magnitude while the active power regulator would handle the active load sharing through adjusting the frequency. A common approach for inverter-interfaced DER is to connect the power electronic inverter with an LC filter. Therefore, the predominantly inductive line is either from the natural line/cable characteristics or implemented with virtual impedance. The locally deployed LC filter in each DER makes the output impedance inductive dominant [9], then the power-sharing control laws that allow the active power to be shared based on DER units' rated capacities according to the droop setting, can be written as [65]

$$\omega_i = \omega^* - n_i P_i \tag{15.38}$$

As discussed, the problem of distributed frequency control and power sharing would take a form such as Eq. (15.14) where Φ_i as the secondary controller is a synchronization rule consisting of pinning terms and coupling mechanism and is a function of its current value, P_i, and its neighbors' values Φ_j's. Looking at Eq. (15.29), it can be seen that there are pinning terms, produced by the rotation-Z operators and coupling mechanism, produced by swapping operators. Therefore, in order to apply the QDC, we need to define the target for (15.29) which is done through scaling $n_i P_i$. Here, we call $n_i P_i$ the power-sharing signal. Specifically, $n_i P_i$ is scaled to be restricted to in the range $(0, \pi/2)$; thus, we select k such that $k < \frac{\pi/2}{\max(n_i P_i)}$ so that, $k n_i P_i$ is ready to be incorporated into the argument of the rotation-Z operator at node i and then the process follows the steps explained in Fig. 15.7. Hence, our developed QDC for AC microgrids is formulated as follows

$$
\begin{aligned}
\omega_i &= \omega^* - n_i P_i + \frac{\phi_i}{k}, \\
\dot{\phi}_i &= \sin\left(k n_i P_i - \phi_i\right) + \sum_{j=1}^{n} a_{i,j} \sin\left(\phi_j - \phi_i\right)
\end{aligned}
\tag{15.39}
$$

where ϕ_i/k is the secondary control variable.

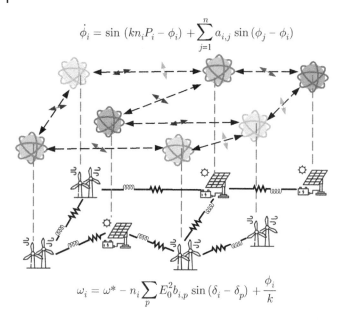

$$\dot{\phi}_i = \sin\left(kn_iP_i - \phi_i\right) + \sum_{j=1}^{n} a_{i,j} \sin\left(\phi_j - \phi_i\right)$$

$$\omega_i = \omega^* - n_i \sum_p E_0^2 b_{i,p} \sin\left(\delta_i - \delta_p\right) + \frac{\phi_i}{k}$$

Figure 15.9 Coupling of the physical microgrid to the network of quantum controllers can be considered as coupling of Kuramoto models.

In a typical AC microgrid with distributed line impedances, since the susceptance of line impedance is usually much larger than its conductance, and also due to the small angle difference between each bus voltage, the active power and reactive power are decoupled and the output active power of each DER can be expressed as [66]

$$P_i = \sum_{p=1}^{n} E_iE_p|Y_{i,p}| \sin\left(\delta_i - \delta_p\right) = \sum_{p=1}^{n} g_{i,p} \sin\left(\delta_i - \delta_p\right) \tag{15.40}$$

where E_i is the nodal voltage magnitudes $E_i > 0$, $-Y_{i,p}$ is the admittance of the line between DER_i and DER_p and δ_i is the voltage phase angle and its dynamic characteristic is $\dot{\delta}_i = \omega_i - \omega^* = \phi_i/k - n_iP_i$.

From (15.40), the physical power network can be treated as a connected network whose entries of its adjacency matrix are $g_{i,p} = E_iE_p|Y_{i,p}|$ and hence, considering (15.39), it can be readily obtained that, the coupling of the network of QDCs and the physical microgrid is the coupling of a forced Kuramoto model with a Kuramoto model (Fig. 15.9). At the steady state, the microgrid is assumed stable. Since the DERs' frequency must be equal, we have $\omega_i = \omega_j$ and thus $n_iP_i - \phi_i/k = n_jP_j - \phi_j/k$ $\forall i, j$. As shown before, ϕ_i converges to the pinner as $t \to \infty$. Thus, $n_iP_i = \phi_i/k$ and $n_iP_i = n_jP_j$ $\forall i, j$ and ω_i converges to ω^*.

15.2.3.2 Verification on an AC Networked-Microgrid Case Study

The performance of the developed QDC is tested on a networked microgrids with five AC microgrids each one has 3 DERs (Fig. 15.10). The nominal voltage and frequency are 380 V and 60 Hz, respectively. All other parameters can be seen in Fig. 15.10. For the sake of simulation, two scenarios are examined. In the first scenario, the system is examined in the face of a step load change. To verify the QDC's feature of plug-and-play capability, as the second scenario, plug-and-play of DERs and microgrids are tested.

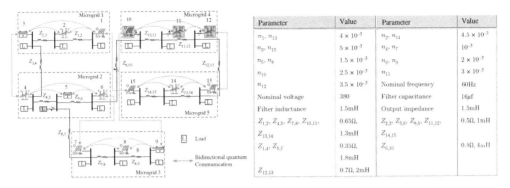

Parameter	Value	Parameter	Value
n_1, n_{13}	4×10^{-3}	n_2, n_{14}	4.5×10^{-3}
n_3, n_{15}	5×10^{-3}	n_4, n_7	10^{-3}
n_5, n_8	1.5×10^{-3}	n_6, n_9	2×10^{-3}
n_{10}	2.5×10^{-3}	n_{11}	3×10^{-3}
n_{12}	3.5×10^{-3}	Nominal frequency	60Hz
Nominal voltage	380	Filter capacitance	16μf
Filter inductance	1.5mH	Output impedance	1.5mH
$Z_{1,2}, Z_{4,5}, Z_{7,8}, Z_{10,11}$,	0.65Ω,	$Z_{2,3}, Z_{5,6}, Z_{8,9}, Z_{11,12}$,	0.5Ω, 1mH
$Z_{13,14}$	1.3mH	$Z_{14,15}$	
$Z_{1,4}, Z_{5,7}$	0.35Ω,	$Z_{6,10}$	0.8Ω, 4μH
	1.8mH		
$Z_{12,13}$	0.7Ω, 2mH		

Figure 15.10 Networked AC microgrids diagram and parameters – Bidirectional arrows represent the undirected quantum communications – To show the controller performance, the highlighted load at Microgrid 2 is attached and detached – At the case of plug-and-play of DERs, the highlighted DER_{10}, DER_{11} and DER_{12} in Microgrid 4 are disconnected and reconnected again.

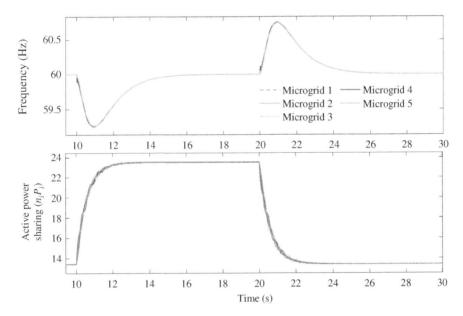

Figure 15.11 DERs' frequencies throughout the network after attaching and detaching the step load.

Controller Performance Studies in this section illustrate the performance of the QDC under a step load change applied to microgrid 2 at $t = 10$ s and results are depicted in Fig. 15.11. The exploited communication graph is shown in Fig. 15.10. As can be seen, frequency regulation is maintained throughout the step load change and active power is accurately shared among the heterogeneous DGs throughout the entire runtime.

Plug-and-Play Functionality – Plug-and-Play of DERs Due to the availability of renewable generators, microgrid's physical and communication topologies can be time-varying. In this case, we demonstrate that to support plug-and-play functionality, our developed QDC provides a robust secondary control framework that works effectively in spite of time-varying communication networks. Thus, this case verifies the QDC's feature of plug-and-play capability. This merit is investigated, by detaching DER_{10}, DER_{11}, and DER_{12} at $t = 10$ s and plugging them in again at $t = 20$ s.

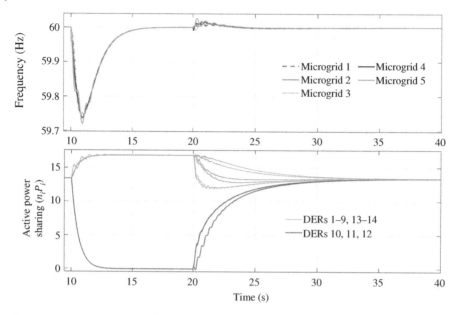

Figure 15.12 Frequency regulation and active power sharing after plug-and-play of DERs 10, 11, and 12.

As depicted in Fig. 15.12, after disconnection of the DERs, the power deficiency reallocated among the remaining DERs and they manage to share the loads. As shown, accurate active power sharing and frequency restoration are maintained during plug-and-play operation.

Plug-and-Play of Microgrids As the second plug-and-play scenario, microgrids 4 and 5 are disconnected from microgrids 1, 2, and 3 at $t = 10$ s and reconnected again at $t = 20$ s. Afterward, at $t = 24$ s, the communication between microgrids 1–5 and 3–4 is reestablished. Fig. 15.13 shows how after disconnection of microgrids frequency is regulated in both microgrids 1–3 and 4–5 to the rated 60 Hz. Furthermore, after disconnection, the active power is shared among DERs 1–9 and DERs 10–15, and then among all the DERs after reconnection of microgrids 4–5, starting from $t = 24$ s. The reason of transient oscillations after reconnection is that, no presynchronization is implemented ahead of reconnection.

Comparison with the Classical Benchmark In order to benchmark the QDC better, its performance is compared with the distributed-averaging PI (DAPI) controller (with the positive constant $k_i = 1$ [67]). For both DAPI and QDC, the communication graph is the same as Fig. 15.10. In this case, DER$_{11}$ is unplugged at t = 10s followed by the step load of 40 kW at $t = 20$ s, and then reconnected at $t = 30$ s. Results depicted in Fig. 15.14 demonstrate that, regarding restoring time at the events of load disturbance and plug-and-play, both controllers have close performances. However, our devised QDC enables encoding information into quantum states, directly sent over quantum channels among participating DERs, and thus allows microgrids to be profited from quantum communication advantages.

15.2.3.3 Quantum Distributed Voltage Control for DC Microgrids

In DC microgrids, droop control function is mainly utilized to provide decentralized power sharing. It generates the voltage reference V_i^{ref} as [68]

$$V_i^{\text{ref}} = V^* - d_i I_i \tag{15.41}$$

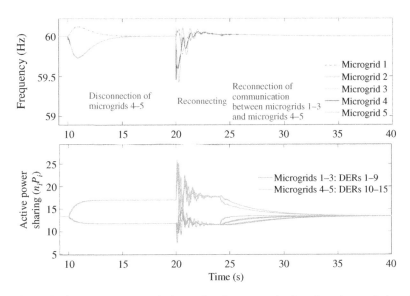

Figure 15.13 Frequency regulation and active power sharing after plug-and-play of microgrids 4–5.

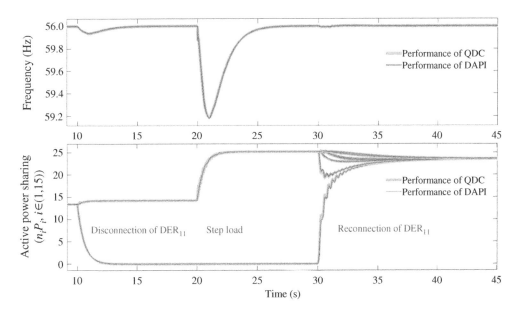

Figure 15.14 Comparison between QDC and DAPI at the face of step load and plug-and-play events.

where V^* is the nominal dc voltage, d_i is the current droop gain, and I_i is the output current of DER_i. Consider the DC microgrid depicted in Fig. (15.15), ignoring the inductance effect of lines, the DC bus voltage V_b can be determined as

$$V_b = V_i^{\text{ref}} - R_i I_i \tag{15.42}$$

It can be easily shown that, if the current droop gain d_i is set much larger than the line resistance R_i, $\frac{I_i}{I_j} \approx \frac{d_i}{d_j}$ and $V_b \approx V_i^{\text{ref}} \, \forall i, j$. The larger d_i is chosen, the more accurate power sharing can be obtained, however, larger d_i may cause the dc bus voltage V_b to deviate more from the nominal value V^*.

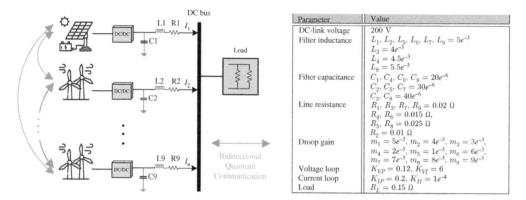

Parameter	Value
DC-link voltage	200 V
Filter inductance	$L_1, L_2, L_5, L_6, L_7, L_9 = 5e^{-3}$
	$L_3 = 4e^{-3}$
	$L_4 = 4.5e^{-3}$
	$L_8 = 5.5e^{-3}$
Filter capacitance	$C_1, C_4, C_6, C_9 = 20e^{-6}$
	$C_2, C_5, C_7 = 30e^{-6}$
	$C_3, C_8 = 40e^{-6}$
Line resistance	$R_1, R_3, R_7, R_9 = 0.02\ \Omega$
	$R_4, R_6 = 0.015\ \Omega$,
	$R_5, R_8 = 0.025\ \Omega$
	$R_2 = 0.01\ \Omega$
Droop gain	$m_1 = 5e^{-3}, m_2 = 4e^{-3}, m_3 = 3e^{-3},$
	$m_4 = 2e^{-3}, m_5 = 1e^{-3}, m_6 = 6e^{-3},$
	$m_7 = 7e^{-3}, m_8 = 8e^{-3}, m_9 = 9e^{-3}$
Voltage loop	$K_{VP} = 0.12, K_{VI} = 6$
Current loop	$K_{IP} = 0.2, K_{II} = 1e^{-4}$
Load	$R_L = 0.15\ \Omega$

Figure 15.15 DC microgrid model quipped with the QDC and parameters.

Therefore, we aim to attain both power-sharing and precise voltage restoration, simultaneously, by adding the QDC. To equip the DC microgrid with the QDC, we follow the same implementation procedure explained for distributed frequency control and hence, the droop function (15.41) is modified as

$$
\begin{aligned}
V_i^{\text{ref}} &= V^* - d_i I_i + \frac{\phi_i}{c}, \\
\dot{\phi}_i &= \sin\left(cd_i I_i - \phi_i\right) + \sum_{j=1}^{n} a_{i,j} \sin\left(\phi_j - \phi_i\right)
\end{aligned}
\tag{15.43}
$$

which can be rewritten as

$$
\begin{aligned}
V_i^{\text{ref}} &= V^* - m_i P_{dc,i} + \frac{\phi_i}{c}, \\
\dot{\phi}_i &= \sin\left(cm_i P_{dc,i} - \phi_i\right) + \sum_{j=1}^{n} a_{i,j} \sin\left(\phi_j - \phi_i\right)
\end{aligned}
\tag{15.44}
$$

where $P_{dc,i} = V_b I_i$ and m_i is the power droop gain. Again, we select c such that $c < \frac{\pi/2}{\max\left(m_i P_{dc,i}\right)}$, so that, it is ready to be incorporated into the argument of the rotation-Z operator at node i and then the process follows the steps explained in Fig. 15.7. As can be seen (15.44) has a similar form as (15.39). Obviously, the first part in (15.44) is to drive the dc bus voltage V_b to the nominal value V^* while the second part is to guarantee that $\phi_i = \phi_j$ is satisfied, i.e. the current/power sharing is achieved which demonstrates that the QDC is also applicable to distributed voltage control in DC microgrids.

15.2.3.4 Verification on a DC Microgrid Case Study

This case verifies the universality of the QDC. This merit is investigated by equipping a 9 DER DC microgrid case study with the QDC (see Fig. 15.15) and applying a step load of 267 kW at $t = 10$ s. Results are depicted in Fig. 15.16. The exploited communication graph is shown in Fig. 15.16.

As can be seen, voltage regulation is guaranteed throughout the step load disturbance and power/current is accurately shared among the heterogeneous DGs throughout the entire runtime.

15.2.4 Discussion and Outlook

15.2.4.1 Realization

In an abstract sense, a quantum network is a network of quantum processors as nodes on specific locations, that are connected via links [69]. Like a quantum network/internet, realization of the

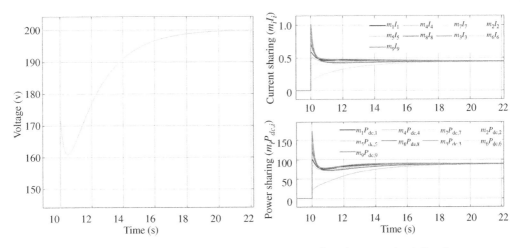

Figure 15.16 Voltage regulation, power sharing and current sharing after a step load disturbance at $t = 10\,\text{s}$.

QDC requires essential quantum hardware/software elements. First, a physical link (quantum channel) that is able to transmit qubits is needed. Standard telecom fibers are of suitable choices since they are currently used to communicate classical light and so far, photons are known as the ideal physical carrier of information to implement intrinsically secure quantum communications, specifically, for long-distance communications [13]. Various required building blocks for the links such as photonic quantum channels between ground stations or between ground stations and satellites, quantum repeaters, quantum memory, etc., have recently been experimentally demonstrated [70, 71].

Second, a quantum algorithm is required to simulate Eq. (15.27) at each node. Several methods have been recently proposed for the problem of simulating open quantum systems represented by either the operator sum representation or the Lindblad master equation [72–74]. The overall approach in these algorithms is first, transforming the open dynamic into Kraus formalism in the operator sum form (if it is in Lindblad representation), which is the most general form of the time evolution for a density matrix, second, converting the Kraus operators into unitary matrices and third, decomposing the unitary matrices into unitary quantum gates. This procedure allows the evolution of the initial state through unitary quantum gates.

The third element is measurement. In a typical quantum algorithm, we need to estimate expectation values of a set of operators/observables in a quantum state ρ that can be prepared repeatedly using a programmable quantum system [75]. As discussed, each $\phi_i(t)$ is the averaged measurement outcome which is obtained by averaging measurement outcomes of many realizations of a single experiment for node i. The reason is, an informative quantum measurement is demolishing (i.e. causing the wave function to collapse) and gives probabilistic outcomes [76]. Hence, to obtain precise estimates, each operator must be measured many times.

Several solutions have been introduced to overcome this fundamental scaling problem including matrix product state tomography, neural network tomography, shadow tomography, and classical shadow [76–81]. Among these methods, the derandomized Pauli measurement approach in [75] is of our interest, where authors describe a deterministic protocol for estimating Pauli-operator expectation values using very few copies of ρ. Developments like these are of particular importance since they are proposing promising solutions to the issues of large shot counts and high number of times required for transmitting a particular quantum state among two nodes for measurement purposes.

15.2.4.2 Outlook

While we are on the verge of quantum internet, planning for future smart power grids, as the largest man-made systems, based on classical communications seems obsolete and may fail to address the new requirements and security challenges. Therefore, keeping up with the quantum technology seems essential. The potential to design a quantum synchronization scheme motivated us to study the feasibility of realizing a QDC.

In this work, we introduce a new synchronization mechanism by leveraging the quantum properties of qubits. Since the distributed control problems of microgrids can be modeled as networked differential equations, we leverage a proposed master equation to construct the network of differential equations. Then, we demonstrate that by characterizing proper observables, expectation values of all the observers at all nodes will eventually get synchronized to a possibly time-varying target value and the synchronization rule follows the forced Kuramoto model. We show how our proposed quantum synchronization scheme can be exploited to regulate AC microgrids' frequency and DC microgrids' voltage and guarantee precise power sharing. Then, our theoretical derivations are complemented by a series of numerical and simulation results, which have fully confirmed the validity and generality of the QDC scheme. Due to the superposition feature of qubits, the QDC provides a foundation for introducing more enhanced quantum-secure distributed control for microgrids through randomizing the θ angle of qubits in the initialization step, which could finally resultin unprecedented security for distributed control of AC and DC microgrids.

References

1 Zhang, P. (2021). *Networked Microgrids*. Cambridge: Cambridge University Press.

2 Li, T. and Zhang, J. (2010). Consensus conditions of multi-agent systems with time-varying topologies and stochastic communication noises. *IEEE Transactions on Automatic Control* 55 (9): 2043–2057.

3 Rezaienia, P., Gharesifard, B., Linder, T., and Touri, B. (2020). Push-sum on random graphs: almost sure convergence and convergence rate. *IEEE Transactions on Automatic Control* 65 (3): 1295–1302.

4 Cao, Q., Song, Y., Guerrero, J.M., and Tian, S. (2016). Coordinated control for flywheel energy storage matrix systems for wind farm based on charging/discharging ratio consensus algorithms. *IEEE Transactions on Smart Grid* 7 (3): 1259–1267.

5 Kempe, D., Dobra, A., and Gehrke, J. (2003). Gossip-based computation of aggregate information. *44th Annual IEEE Symposium on Foundations of Computer Science*, 482–491, October 2003.

6 Gerencsér, B. and Hendrickx, J.M. (2019). Push-sum with transmission failures. *IEEE Transactions on Automatic Control* 64 (3): 1019–1033.

7 Touri, B. and Gharesifard, B. (2015). Continuous-time distributed convex optimization on time-varying directed networks. *54th IEEE Conference on Decision and Control*, 724–729, December 2015.

8 Su, S. and Lin, Z. (2016). Distributed consensus control of multi-agent systems with higher order agent dynamics and dynamically changing directed interaction topologies. *IEEE Transactions on Automatic Control* 61 (2): 515–519.

9 Lu, X., Yu, X., Lai, J. et al. (2018). A novel distributed secondary coordination control approach for islanded microgrids. *IEEE Transactions on Smart Grid* 9 (4): 2726–2740.

10 Zhou, Q., Shahidehpour, M., Alabdulwahab, A., and Abusorrah, A. (2020). A cyber-attack resilient distributed control strategy in islanded microgrids. *IEEE Transactions on Smart Grid* 11 (5): 3690–3701.

11 Abhinav, S., Modares, H., Lewis, F.L. et al. (2018). Synchrony in networked microgrids under attacks. *IEEE Transactions on Smart Grid* 9 (6): 6731–6741.

12 Babahajiani, P., Wang, L., Liu, J., and Zhang, P. (2021). Push-sum-enabled resilient microgrid control. *IEEE Transactions on Smart Grid* 12 (4): 3661–3664.

13 Wehner, S., Elkouss, D., and Hanson, R. (2018). Quantum internet: a vision for the road ahead. *Science* 362 (6412): https://doi.org/10.1126/science.aam9288.

14 Sahoo, S., Dragičević, T., and Blaabjerg, F. (2021). Multilayer resilience paradigm against cyber attacks in DC microgrids. *IEEE Transactions on Power Electronics* 36 (3): 2522–2532.

15 Sadabadi, M.S., Sahoo, S., and Blaabjerg, F. (2021). A fully resilient cyber-secure synchronization strategy for AC microgrids. *IEEE Transactions on Power Electronics* 36 (12): 13372–13378.

16 Chen, Y., Qi, D., Dong, H. et al. (2021). A FDI attack-resilient distributed secondary control strategy for islanded microgrids. *IEEE Transactions on Smart Grid* 12 (3): 1929–1938.

17 Sahoo, S., Yang, Y., and Blaabjerg, F. (2021). Resilient synchronization strategy for AC microgrids under cyber attacks. *IEEE Transactions on Power Electronics* 36 (1): 73–77.

18 Zuo, S., Beg, O.A., Lewis, F.L., and Davoudi, A. (2020). Resilient networked AC microgrids under unbounded cyber attacks. *IEEE Transactions on Smart Grid* 11 (5): 3785–3794.

19 Bidram, A., Poudel, B., Damodaran, L. et al. (2020). Resilient and cybersecure distributed control of inverter-based islanded microgrids. *IEEE Transactions on Industrial Informatics* 16 (6): 3881–3894.

20 Zuo, S., Altun, T., Lewis, F.L., and Davoudi, A. (2020). Distributed resilient secondary control of DC microgrids against unbounded attacks. *IEEE Transactions on Smart Grid* 11 (5): 3850–3859.

21 Fano, G. and Blinder, S.M. (2019). Chapter 11 - Quantum chemistry on a quantum computer. In: *Mathematical Physics in Theoretical Chemistry*, Developments in Physical & Theoretical Chemistry (ed. S.M. Blinder and J.E. House), 377–400. Elsevier.

22 Wright, K., Beck, K.M., Debnath, S. et al. (2019). Benchmarking an 11-qubit quantum computer. *Nature Communications* 10 (1): 1–6.

23 Qi, R., Sun, Z., Lin, Z. et al. (2019). Implementation and security analysis of practical quantum secure direct communication. *Light: Science & Applications* 8 (1): 1–8.

24 Popkin, G. (2021). The internet goes quantum. *Science* 372 (6546): 1026–1029.

25 Kimble, H.J. (2008). The quantum internet. *Nature* 453 (7198): 1023–1030.

26 Yu, Y., Ma, F., Luo, X.-Y. et al. (2020). Entanglement of two quantum memories via fibres over dozens of kilometres. *Nature* 578 (7794): 240–245.

27 Castelvecchi, D. (2021). Quantum network is step towards ultrasecure internet. *Nature* 590 (7847): 540–541.

28 Azuma, K., Mizutani, A., and Lo, H.-K. (2016). Fundamental rate-loss trade-off for the quantum internet. *Nature Communications* 7 (1): 1–8.

29 Pirandola, S. and Braunstein, S.L. (2016). Physics: unite to build a quantum internet. *Nature News* 532 (7598): 169.

30 Castelvecchi, D. (2018). The quantum internet has arrived (and it hasn't). *Nature* 554 (7690): 289–293.

31 Dowling, J.P. (2020). *Schrödinger's Web: Race to Build the Quantum Internet*. CRC Press.

32 Jiang, Z., Tang, Z., Qin, Y. et al. (2021). Quantum internet for resilient electric grids. *International Transactions on Electrical Energy Systems* 31 (6): e12911.

33 Park, B.K., Woo, M.K., Kim, Y.-S. et al. (2020). User-independent optical path length compensation scheme with sub-nanosecond timing resolution for a 1× N quantum key distribution network system. *Photonics Research* 8 (3): 296–302.

34 Luo, Y.-H., Zhong, H.-S., Erhard, M. et al. (2019). Quantum teleportation in high dimensions. *Physical Review Letters* 123 (7): 070505.

35 Qi, Z., Li, Y., Huang, Y. et al. (2021). A 15-user quantum secure direct communication network. *Light: Science & Applications* 10 (1): 1–8.

36 Hu, J.-Y., Yu, B., Jing, M.-Y. et al. (2016). Experimental quantum secure direct communication with single photons. *Light: Science & Applications* 5 (9): e16144–e16144.

37 Pirandola, S., Mancini, S., Lloyd, S., and Braunstein, S.L. (2008). Continuous-variable quantum cryptography using two-way quantum communication. *Nature Physics* 4 (9): 726–730.

38 Joshi, S.K., Aktas, D., Wengerowsky, S. et al. (2020). A trusted node–free eight-user metropolitan quantum communication network. *Science Advances* 6 (36): eaba0959.

39 Wengerowsky, S., Joshi, S.K., Steinlechner, F. et al. (2018). An entanglement-based wavelength-multiplexed quantum communication network. *Nature* 564 (7735): 225–228.

40 Chen, Y.-A., Zhang, Q., Chen, T.-Y. et al. (2021). An integrated space-to-ground quantum communication network over 4,600 kilometres. *Nature* 589 (7841): 214–219.

41 Babahajiani, P., Zhang, P., Wei, T.-C. et al. (2022). Employing interacting qubits for distributed microgrid control. *IEEE Transactions on Power Systems* 38 (4): 3123–3135.

42 Babahajiani, P. and Zhang, P. (2022). Quantum distributed microgrid control. In: *2022 IEEE Power & Energy Society General Meeting (PESGM)*, 1–5. IEEE.

43 Babahajiani, P. and Zhang, P. (2022). Quantum-secure distributed frequency control. In: *2022 IEEE Power & Energy Society General Meeting (PESGM)*, 1–5. IEEE.

44 Jadbabaie, A., Lin, J., and Morse, A.S. (2003). Coordination of groups of mobile autonomous agents using nearest neighbor rules. *IEEE Transactions on Automatic Control* 48 (6): 988–1001.

45 Olfati-Saber, R. and Murray, R.M. (2004). Consensus problems in networks of agents with switching topology and time-delays. *IEEE Transactions on Automatic Control* 49 (9): 1520–1533.

46 Sepulchre, R., Sarlette, A., and Rouchon, P. (2010). Consensus in non-commutative spaces. *49th IEEE Conference on Decision and Control (CDC)*, 6596–6601.

47 Mazzarella, L., Sarlette, A., and Ticozzi, F. (2015). Consensus for quantum networks: symmetry from gossip interactions. *IEEE Transactions on Automatic Control* 60 (1): 158–172.

48 Mazzarella, L., Ticozzi, F., and Sarlette, A. (2015). Extending robustness and randomization from consensus to symmetrization algorithms. *SIAM Journal on Control and Optimization* 53 (4): 2076–2099.

49 Breuer, H.-P. and Petruccione, F. (2002). *The Theory of Open Quantum Systems*, 1e. Oxford University Press.

50 Lindblad, G. (1976). On the generators of quantum dynamical semigroups. *Communications in Mathematical Physics* 48 (2): 119–130.

51 Shi, G., Dong, D., Petersen, I.R., and Johansson, K.H. (2016). Reaching a quantum consensus: master equations that generate symmetrization and synchronization. *IEEE Transactions on Automatic Control* 61 (2): 374–387.

52 Godsil, C. and Royle, G. (2001). *Algebraic Graph Theory*. Springer-Verlag.

53 Nielsen, M.A. and Chuang, I.L. (2010). *Quantum Computation and Quantum Information: 10th Anniversary Edition*. Cambridge University Press.

54 Preskill, J. (1998). *Lecture Notes for Physics: Quantum Information and Computation*. California Institute of Technology.

55 Lloyd, S., De Palma, G., Gokler, C. et al. (2020). Quantum algorithm for nonlinear differential equations. *arXiv preprint arXiv:2011.06571*.

56 Childs, A.M., Liu, J.-P., and Ostrander, A. (2020). High-precision quantum algorithms for partial differential equations. *arXiv preprint arXiv:2002.07868*.

57 Joseph, I. (2020). Koopman–von Neumann approach to quantum simulation of nonlinear classical dynamics. *Physical Review Research* 2 (4): 043102.

58 Berry, D.W., Childs, A.M., Ostrander, A., and Wang, G. (2017). Quantum algorithm for linear differential equations with exponentially improved dependence on precision. *Communications in Mathematical Physics* 356 (3): 1057–1081.

59 Engel, A., Smith, G., and Parker, S.E. (2019). Quantum algorithm for the Vlasov equation. *Physical Review A* 100 (6): 062315.

60 Costa, P.C.S., Jordan, S., and Ostrander, A. (2019). Quantum algorithm for simulating the wave equation. *Physical Review A* 99 (1): 012323.

61 Harrow, A.W., Hassidim, A., and Lloyd, S. (2009). Quantum algorithm for linear systems of equations. *Physical Review Letters* 103 (15): 150502.

62 Wiseman, H.M. and Milburn, G.J. (2009). *Quantum Measurement and Control*. Cambridge University Press.

63 Ott, E. and Antonsen, T.M. (2008). Low dimensional behavior of large systems of globally coupled oscillators. *Chaos: An Interdisciplinary Journal of Nonlinear Science* 18 (3): 037113.

64 Johansson, J.R., Nation, P.D., and Nori, F. (2013). QuTiP 2: a python framework for the dynamics of open quantum systems. *Computer Physics Communications* 184 (4): 1234–1240.

65 Pogaku, N., Prodanovic, M., and Green, T.C. (2007). Modeling, analysis and testing of autonomous operation of an inverter-based microgrid. *IEEE Transactions on Power Electronics* 22 (2): 613–625.

66 Shi, M., Chen, X., Shahidehpour, M. et al. (2021). Observer-based resilient integrated distributed control against cyberattacks on sensors and actuators in islanded AC microgrids. *IEEE Transactions on Smart Grid* 12 (3): 1953–1963.

67 Simpson-Porco, J.W., Shafiee, Q., Dörfler, F. et al. (2015). Secondary frequency and voltage control of islanded microgrids via distributed averaging. *IEEE Transactions on Industrial Electronics* 62 (11): 7025–7038.

68 Nasirian, V., Davoudi, A., Lewis, F.L., and Guerrero, J.M. (2014). Distributed adaptive droop control for DC distribution systems. *IEEE Transactions on Energy Conversion* 29 (4): 944–956.

69 Hansenne, K., Xu, Z.-P., Kraft, T., and Gühne, O. (2022). Symmetries in quantum networks lead to no-go theorems for entanglement distribution and to verification techniques. *Nature Communications* 13 (1): 1–6.

70 Yin, J., Li, Y.-H., Liao, S.-K. et al. (2020). Entanglement-based secure quantum cryptography over 1,120 kilometres. *Nature* 582 (7813): 501–505.

71 Liao, S.-K., Cai, W.-Q., Handsteiner, J. et al. (2018). Satellite-relayed intercontinental quantum network. *Physical Review Letters* 120 (3): 030501.

72 Hu, Z., Xia, R., and Kais, S. (2020). A quantum algorithm for evolving open quantum dynamics on quantum computing devices. *Scientific Reports* 10 (1): 1–9.

73 Head-Marsden, K., Krastanov, S., Mazziotti, D.A., and Narang, P. (2021). Capturing non-Markovian dynamics on near-term quantum computers. *Physical Review Research* 3 (1): 013182.

74 Schlimgen, A.W., Head-Marsden, K., Sager, L.A.M. et al. (2021). Quantum simulation of open quantum systems using a unitary decomposition of operators. *Physical Review Letters* 127 (27): 270503.

75 Huang, H.-Y., Kueng, R., and Preskill, J. (2021). Efficient estimation of Pauli observables by derandomization. *Physical Review Letters* 127 (3): 030503.

76 Huang, H.-Y., Kueng, R., and Preskill, J. (2020). Predicting many properties of a quantum system from very few measurements. *Nature Physics* 16 (10): 1050–1057.

77 Huang, H.-Y. (2022). Learning quantum states from their classical shadows. *Nature Reviews Physics* 4: 81. https://doi.org/10.1038/s42254-021-00411-5.

78 Cramer, M., Plenio, M.B., Flammia, S.T. et al. (2010). Efficient quantum state tomography. *Nature Communications* 1 (1): 1–7.

79 Carrasquilla, J., Torlai, G., Melko, R.G., and Aolita, L. (2019). Reconstructing quantum states with generative models. *Nature Machine Intelligence* 1 (3): 155–161.

80 Torlai, G., Mazzola, G., Carrasquilla, J. et al. (2018). Neural-network quantum state tomography. *Nature Physics* 14 (5): 447–450.

81 Aaronson, S. and Rothblum, G.N. (2019). Gentle measurement of quantum states and differential privacy. *Proceedings of the 51st Annual ACM SIGACT Symposium on Theory of Computing*, 322–333.

16

Programmable Crypto-Control for Networked Microgrids

Lizhi Wang, Peng Zhang, and Zefan Tang

16.1 Introduction

Networking a group of microgrids to form networked microgrids (NMs) greatly enhances the grid resiliency [1, 2]. With the increasing integration of distributed energy resources (DERs) in NMs, distributed-consensus-based secondary controls are currently attracting more and more attention due to their enhanced flexibility and resiliency over centralized controls [3]. With a distributed consensus-based secondary control, some information has to be transmitted to and is later processed at neighboring agents. For instance, to achieve certain control functions, some DERs need to share their individual information with their neighbors. However, this inevitably poses a privacy threat, i.e. the information (which can be sensitive) sent from one DER to its neighbors is revealed at the neighbors' sides. In other words, the DER who sends information to its neighbors may not want the information to be revealed; meanwhile, the information should be processed at the neighbors' sides [4, 5].

A traditional method to preserve the privacy in NMs (i.e. the so-called differential-privacy method) adds a certain noise to each sharing message [6]. By introducing a random perturbation to each sharing message, neighbors cannot obtain true messages [7]. Some examples that have adopted this method include the optimal power flow privacy protection [8] and transmission lines and transformers parameters privacy protection [9]. However, with noises added, the iterated control results are inevitably impacted, and the more the noises are added (which improves the privacy), the more seriously the system operation is affected [10].

An alternative method to address the privacy issue is reducing the number of sensitive sharing parameters through redesigning the control strategy [11, 12]. For instance, in [11], a privacy-preserving distributed control strategy is proposed for active power sharing in islanded microgrids, where DERs only need to exchange their frequencies while other sensitive parameters are kept privately. Wang et al. [12] re-designs the energy management strategy for microgrids, where each customer's schedulable demand remains private (other non-sensitive parameters are shared instead). However, while this method can preserve the privacy of certain parameters, it is based on specific problems, i.e. for a different problem, a different control strategy needs to be carried out. Further, this method still requires certain parameters to be transmitted. As different parameters in NMs have certain correlations, by checking the pattern of the received information, each neighbor can still obtain the pattern of the sensitive information, meaning that privacy is not fully preserved.

Recently, the partial homomorphic encryption (PHE) method has been widely used to preserve the privacy [13–16]. With PHE, multiple communicating parties send their encrypted messages

Microgrids: Theory and Practice, First Edition. Edited by Peng Zhang.
© 2024 The Institute of Electrical and Electronics Engineers, Inc. Published 2024 by John Wiley & Sons, Inc.

(namely, ciphertexts) to their neighbors, respectively, and when each party receives all the neighbors' ciphertexts, the product of all the ciphertexts is decrypted. The unique feature of PHE is that the product of multiple ciphertexts decrypts to the sum of their corresponding plaintexts. However, although PHE provides a unique solution to protect each plaintext from being observed directly, the privacy of each communicating party still relies on the honesty of the neighbors, as each party has the capability of decrypting each neighbor's ciphertexts [13]. In addition, PHE relies on long keys to ensure a high-security level; this however inevitably leads to heavy computation and communication burdens [16], causing crypto-controllers to work slowly. Further, when any communicating party fails to work, the PHE-based system operates abnormally. A provably private, fast, and flexible cryptography-based control scheme for NMs is therefore needed but does not yet exist.

To bridge the gaps, a cryptography-based, programmable control (crypto-control) scheme is designed to provably preserve the privacy of DERs while ensuring fast, flexible distributed control in NMs. Specifically, we devise a programmable crypto-control-based NMs (PCNMs) architecture, where crypto-controllers are fully virtualized in IoT devices and software-defined networking (SDN) is utilized to manage the network. A novel dynamic encrypted weight addition (DEWA) approach, which integrates a new switching-keys-enabled PHE and a secret sharing scheme, is developed to preserve each DER's privacy while the real-time computation is ensured. Different from PHE, DEWA uses switching, lightweight keys to provide fast, distributed control in NMs (while the security level remains high), and adopts the zero secret sharing technique to preserve each DER's privacy. We establish a real-time DEWA-based PCNMs testbed incorporating DEWA, real SDN switches, and IoT devices, validate the hardware testbed in a real-time digital simulator (RTDS) environment, and mathematically and experimentally analyze the DEWA privacy-preserving property. Test results validate the effectiveness, benefits, and superiority of DEWA-based PCNMs, and the small signal stability result is also provided.

16.2 PCNMs and Privacy Requirements

In this section, we first present PCNMs including their architecture and the control strategy and then describe the privacy requirements in PCNMs.

16.2.1 Architecture of PCNMs

The PCNMs architecture is illustrated in Fig. 16.1. It consists of three layers: (1) a physical layer, (2) a network layer, and (3) a programmable crypto-control layer. The components in the physical layer contain various DERs (such as solar panels, wind turbines, diesels, and storages), and others like smart meters, loads, and transformers. The measurements of each DER (e.g. power generation, frequency, etc.) are sent to its crypto-controller with a regular frequency via a communication network. The function of the crypto-controller is to realize consensus-based secondary control while preserving DERs' privacy. Each DER communicates with its programmable crypto-controller through an IP-address-assigned interface that can adopt different communication protocols.

The network layer utilizes intelligent SDN switches to achieve a fast and flexible communication environment. The decoupling of control and data planes and the centralization of the control logic in the SDN controller make SDN switches simple forwarding devices, where the SDN controller obtains a global knowledge of network states, enabling the fast development of sophisticated applications. In this study, the communication data paths for crypto-controllers are regulated by the SDN controller, for handling the communication congestion to achieve a reliable communication network [17, 18].

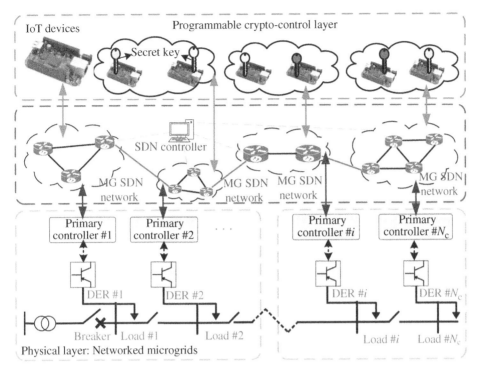

Figure 16.1 The PCNMs architecture.

In PCNMs, each crypto-controller is installed in an IoT device. Crypto-controllers communicate with each other through IP-address assigned interfaces. The SDN- and IoT-based crypto-controller structure is illustrated in Fig. 16.2, where each crypto-controller is a program developed in Python containing multiple threads. Secure keys are sent from the SDN controller to each crypto-controller every a certain time period. Specifically, each crypto-controller has a private key (i.e. k_{si} for crypto-controller i) and some public keys (i.e. k_{pj}, where $j \in \mathcal{N}_i$ and $\mathcal{N}_i$ denotes the set of crypto-controller i's neighbors). k_{si} is used as a decryption key to decrypt each ciphertext sent from a neighbor. k_{pj} is used to encrypt a plaintext. Each crypto-controller encrypts DER state variables and sends them to neighboring crypto-controllers using the public keys of neighbors. When each crypto-controller receives each message sent from a neighboring crypto-controller, it decrypts the message using its own private key and processes the consensus-based secondary control.

The uniqueness of the PCNMs is that the secondary controllers are pushed to the edge of the system and are virtualized in IoT devices with great scalability and flexibility. Moreover, SDN is leveraged to make the communication network suitable for the deployment of crypto-controllers and capable of handling the heavy communication burden caused by encryption. Meanwhile, instead of introducing a centralized controller, the SDN controller is in charge of the keys management and distribution without affecting crypto-controller operations. Note that the security of SDN itself can be guaranteed by various comprehensive solutions (see [19] for a detailed description).

16.2.2 Formulations of Programmable Control

Without loss of generality, a two-layer hierarchical control strategy is adopted in this work. Specifically, the two-layer hierarchical control strategy consists of a droop control and a

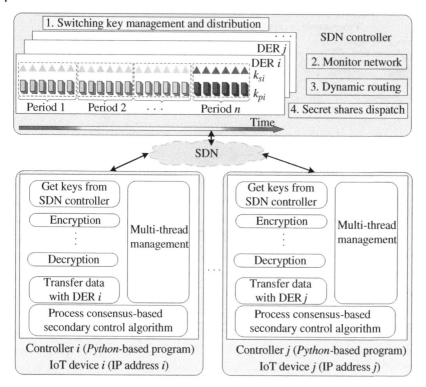

Figure 16.2 The SDN- and IoT-based crypto-controller structure.

consensus-based secondary control. The droop control is utilized to regulate the frequency and voltage. For the secondary control, each DER communicates with its neighboring DERs through transmitting required control signals to achieve power sharing and voltage and frequency restorations in PCNMs. This section presents the formulations of this control strategy and demonstrates the privacy leakage issue.

For the droop control, voltage, and frequency droop characteristics are given as follows:

$$\begin{cases} \boldsymbol{\omega} = \boldsymbol{\omega}^* - \boldsymbol{m}_p(\boldsymbol{P} - \boldsymbol{P}^*) \\ \boldsymbol{E} = \boldsymbol{E}^* - \boldsymbol{n}_q(\boldsymbol{Q} - \boldsymbol{Q}^*) \end{cases} \tag{16.1}$$

where $\boldsymbol{\omega}$ denotes a vector containing all the DERs' frequencies, and $\boldsymbol{E}$ denotes a vector containing all the DERs' output voltage magnitudes. $\boldsymbol{P}$ and $\boldsymbol{Q}$ denote two vectors containing the active and reactive power outputs of all the DERs, respectively. $\boldsymbol{\omega}^*, \boldsymbol{E}^*, \boldsymbol{P}^*$, and $\boldsymbol{Q}^*$ respectively denote vectors containing nominal values of each signal. $\boldsymbol{m}_p$ and $\boldsymbol{n}_q$ are two vectors containing the active and reactive power droop gains of all the DERs, respectively.

When only the droop control is employed, $\boldsymbol{\omega}^*$ and $\boldsymbol{E}^*$ are fixed, resulting in that any load change causes voltage and frequency to deviate from their set points. A secondary control that eliminates the frequency and voltage deviations caused by the primary droop control can address this issue. In this study, we introduce a distributed-consensus-based secondary control that allows the voltage, frequency, and shared power to converge to their reference values. Specifically, this controller introduces a secondary frequency control variable Ω_i for frequency regulation and a secondary voltage control variable e_i for voltage regulation. Mathematical formulations of this controller are as follows:

With droop and secondary controls, DER i's frequency ω_i can be regulated as follows:

$$\begin{cases} \omega_i = \Omega_i - m_{p,i}(P_i - P_i^*) \\ \Omega_i = \int \left(k_i^\omega \left[\sum_{j \in \mathcal{N}_i} a_{ij}(\omega_j - \omega_i) + g_i(\omega^{\text{ref}} - \omega_i) \right] \right. \\ \qquad\qquad \left. + k_i^P \left[\sum_{j \in \mathcal{N}_i} a_{ij}(m_j P_j - m_i P_i) \right] \right) dt \end{cases} \tag{16.2}$$

where $j \in \mathcal{N}_i$ refers to the jth neighboring DER of DER i. ω^{ref} is the reference value of ω_i. a_{ij} is the fixed communication weight for DERs i and j. The finite-time consensus can be achieved through setting the value of a_{ij} [20]. g_i is assigned to 1 if DER i knows the value of ω^{ref}, and otherwise it is set to zero. k_i^ω and k_i^P are control gains.

DER i's voltage E_i can be regulated as follows:

$$\begin{cases} E_i = e_i - n_{q,i}(Q_i - Q_i^*) \\ e_i = \int k_i^E \left[\sum_{j \in \mathcal{N}_i} a_{ij}(E_j - E_i) + g_i(E^{\text{ref}} - E_i) \right] dt \end{cases} \tag{16.3}$$

where E^{ref} refers to the voltage reference, and k_i^E is the control gain.

Specifically, DER i's neighboring DERs can be divided into two groups, i.e. DERs in the same microgrid (with DER i) and those in other microgrids. Let $\mathcal{N}_i = \mathcal{N}_i^1 \cup \mathcal{N}_i^2$, where $\mathcal{N}_i^1$ denotes the set of DER i's neighboring DERs in the same microgrid (i.e. Intro-MG), and $\mathcal{N}_i^2$ denotes the set of DER i's neighboring DERs in other microgrids (i.e. Inter-MGs). In this study, we only consider the privacy protection of parameters from Inter-MGs. To clearly demonstrate the privacy leakage issue, we reformulate (16.2) and (16.3). For simplicity, three control variables (i.e. u_i^ω, u_i^E, and u_i^P) are used to represent the three complex components in (16.2) and (16.3), respectively, as follows:

$$\begin{cases} u_i^\omega = k_i^\omega \left[\sum_{j \in \mathcal{N}_i} a_{ij}(\omega_j - \omega_i) + g_i(\omega^{\text{ref}} - \omega_i) \right] \\ u_i^E = k_i^E \left[\sum_{j \in \mathcal{N}_i} a_{ij}(E_j - E_i) + g_i(E^{\text{ref}} - E_i) \right] \\ u_i^P = k_i^P \left[\sum_{j \in \mathcal{N}_i} a_{ij}(m_j P_j - m_i P_i) \right] \end{cases} \tag{16.4}$$

As each DER sends out one data packet (which contains all the states, i.e. ω_i, E_i, and P_i) each time, we use an array X_i to represent all the states being sent out, i.e. $X_i = [\omega_i, E_i, P_i]$. We first rewrite (16.3) in a compact form as follows:

$$\begin{cases} u_i^\omega = k_i^\omega \left(-L_i \boldsymbol{\omega} + g_i(\omega^{\text{ref}} - \omega_i) \right) \\ u_i^E = k_i^E(-L_i \boldsymbol{E} + g_i(E^{\text{ref}} - E_i)) \\ u_i^P = -k_i^P L_i \boldsymbol{P} \end{cases} \tag{16.5}$$

where L_i is the ith row of the Laplacian matrix L ($L = [l_{ij}] \subseteq \mathbb{R}^{N_c \times N_c}$ with each element $l_{ij} = \sum_{i=1}^{N_c} a_{ij} - a_{ij}$) and $\boldsymbol{\omega}$, $\boldsymbol{E}$, and $\boldsymbol{P}$ are arrays, i.e. $\boldsymbol{\omega} = [\omega_1, \omega_2, \ldots, \omega_{N_c}]$, $\boldsymbol{E} = [E_1, E_2, \ldots, E_{N_c}]$, and $\boldsymbol{P} = [P_1, P_2, \ldots, P_{N_c}]$, where N_c is the total number of communicating DERs. For simplicity (16.5), can be expressed as

$$\boldsymbol{u}_i = K_i[-\hat{K}X + G(X^{\text{ref}} - X_i)] \tag{16.6}$$

where $\boldsymbol{u}_i = [u_i^\omega, u_i^E, u_i^P]$, $\boldsymbol{K}_i = \mathrm{diag}(k_i^\omega, k_i^E, k_i^P)$, $\boldsymbol{X} = [\boldsymbol{\omega}, \boldsymbol{E}, \boldsymbol{P}]$, $\boldsymbol{X}^{\mathrm{ref}} = [\omega^{\mathrm{ref}}, E^{\mathrm{ref}}, 0]$, $\hat{\boldsymbol{K}} = \mathrm{diag}(\boldsymbol{L}_i, \boldsymbol{L}_i, \boldsymbol{L}_i)$, and $\boldsymbol{G} = \mathrm{diag}(g_i, g_i, 0)$. To clearly demonstrate the privacy leakage issue, $\boldsymbol{u}_i$ for DER i can be decoupled into $\boldsymbol{u}_i = \boldsymbol{u}_i^1 + \boldsymbol{u}_i^2$, where $\boldsymbol{u}_i^1$ and $\boldsymbol{u}_i^2$, respectively, represent control variables using the information from Intro-MG and Inter-MGs. Let $\hat{k}_{ii}$ and $\hat{k}_{ij}$ ($i, j \in \{1, 2, \ldots, N_c\}$) denote the ith and jth elements of $\boldsymbol{L}_i$, respectively. $\boldsymbol{u}_i^1$ and $\boldsymbol{u}_i^2$ can be represented as follows:

$$
\begin{cases}
\boldsymbol{u}_i^1 = \underbrace{-\boldsymbol{K}_i \boldsymbol{G} \boldsymbol{X}_i - \boldsymbol{K}_i \hat{k}_{ii} \boldsymbol{X}_i - \boldsymbol{K}_i \sum_{j \in \mathcal{N}_i^1} \hat{k}_{ij} \boldsymbol{X}_j}_{\text{Intro–MG}} \\
\qquad + \underbrace{\boldsymbol{K}_i \boldsymbol{G} \boldsymbol{X}^{\mathrm{ref}}}_{\text{Reference}} \\
\boldsymbol{u}_i^2 = \underbrace{-\boldsymbol{K}_i \sum_{j \in \mathcal{N}_i^2} \hat{k}_{ij} \boldsymbol{X}_j}_{\text{Inter–MGs}}
\end{cases}
\tag{16.7}
$$

In (16.7), the variables sent from Intro-MG and Inter-MGs are clearly separated, i.e. the privacy of each neighboring DER in Inter-MGs is contained in $\boldsymbol{X}_j$ (where $j \in \mathcal{N}_i^2$). This setup enables that DEWA (as will be discussed in Section 16.3) can be applied to $\boldsymbol{X}_j$ instead of ω_j, E_j, and P_j in different equations.

16.2.3 Privacy Requirement in PCNMs

From (16.7), each crypto-controller, $i \in \{1, 2, \ldots, N_c\}$, computes $\boldsymbol{u}_i$ using the states from Intro-MG and Inter-MGs, i.e. it locally aggregates the contributions of its neighbors $j \in \mathcal{N}_i^2$, the privacy of which has to be preserved. Meanwhile, the weights $\hat{k}_{ij}$ and $\hat{k}_{ji}$ are commonly the same in the practical applications of microgrids, and this inevitably poses challenges for the privacy-preserving. For PCNMs, we consider that weights are unknown to all the crypto-controllers, and the following privacy requirements need to be satisfied:

- Crypto-controller i can obtain $\boldsymbol{u}_i^0$ and $\sum_{j \in \mathcal{N}_i^2} \hat{k}_{ij} \boldsymbol{X}_j$ within each time period, while $\boldsymbol{X}_j$, $\hat{k}_{ij}$, and $\hat{k}_{ij} \boldsymbol{X}_j$ remain unrevealed.
- Any other participating crypto-controllers cannot obtain $\boldsymbol{X}_j$ of other participants in the computation.
- Even if crypto-controller i collaborates with some other crypto-controllers, all the crypto-controllers' $\boldsymbol{X}_j$s cannot be revealed.
- The public-key encryption scheme should be lightweight to achieve a real-time computation with tolerant computation delays, ensuring high stability of PCNMs.

16.3 Dynamic Encrypted Weighted Addition

DEWA combines PHE with a secret sharing scheme to enable privacy-preserving of distributed algorithms. Here PHE is redesigned with switching keys to preserve each DER's privacy while ensuring real-time computations. In PHE scheme, all the communicating DERs send their encrypted messages (i.e. encrypted $\hat{k}_{ij} \boldsymbol{X}_j$ from DER j to DER i) to their neighbors, respectively, and when each party receives all the neighbors' ciphertexts, the product of all the ciphertexts is decrypted. A unique feature of PHE is that the product of multiple ciphertexts decrypts to the sum of their corresponding plaintexts (i.e. $\sum_{j\in\mathcal{N}_i^2}\hat{k}_{ij}X_j$, to be used in (16.7)). However, in PHE, each communicating DER needs to know the communication weights between itself and neighbors, i.e. $\hat{k}_{ij}$ for DER j to encrypt $\hat{k}_{ij}X_j$. This makes that although PHE provides a unique solution to protect each plaintext from being observed directly (i.e. only $\sum_{j\in\mathcal{N}_i^2}\hat{k}_{ij}X_j$ is observed), each party still has the capability of decrypting each neighbor's ciphertexts (i.e. DER i can decrypt the encrypted $\hat{k}_{ij}X_j$ individually to obtain X_j).

To address this challenge, in DEWA, each communication weight between two DERs is not revealed to DERs. Instead of sending the encrypted $\hat{k}_{ij}X_j$ from DER j to DER i, DER j sends $\mathcal{E}(\hat{k}_{ij})^{X_j}\mathcal{E}(s_{ji})$ to DER i, where $\mathcal{E}(\hat{k}_{ij})$ is the encrypted $\hat{k}_{ij}$ using DER i's public key, and is sent from the SDN controller to DER j. $\mathcal{E}(s_{ji})$ is the encrypted s_{ji} (i.e. the so-called zero secret share, a random number designed within the SDN controller) using DER i's public key, and s_{ji} is sent from the SDN controller to DER j. Since DER j does not have DER i's private key, it cannot decrypt $\mathcal{E}(\hat{k}_{ij})$ to obtain $\hat{k}_{ij}$. When DER i receives $\mathcal{E}(\hat{k}_{ij})^{X_j}\mathcal{E}(s_{ji})$ from DER j, it cannot obtain $\hat{k}_{ij}$ and X_j, as the random number s_{ji} is not known to DER i. The private information of DER j (i.e. X_j) is thus protected from being observed by DER i.

Meanwhile, the Paillier cryptosystem (i.e. a type of PHE) [21] is adopted in DEWA. In addition to having the common feature of a traditional PHE (i.e. the product of multiple ciphertexts decrypts to the sum of their corresponding plaintexts), Paillier has another silent feature, i.e. $\mathcal{E}(\hat{k}_{ij})^{X_j}=\mathcal{E}(\hat{k}_{ij}X_j)$. With these two features, after decrypting the product of all the neighbors' ciphertexts, DER i obtains $\sum_{j\in\mathcal{N}_i^2}\hat{k}_{ij}X_j + \sum_{j\in\mathcal{N}_i^2}s_{ji}$. Further, in DEWA, we utilize the additive zero secret sharing technique [22] such that s_{ji} is designed in a way that $\sum_{j\in\mathcal{N}_i^2}s_{ji}=0$. Therefore, with DEWA, DER i can successfully obtain $\sum_{j\in\mathcal{N}_i^2}\hat{k}_{ij}X_j$, while both $\hat{k}_{ij}$ and X_j remain unrevealed.

In this section, we first present some backgrounds of the Paillier cryptosystem and the additive zero secret sharing, and then describe our DEWA algorithm with more details.

16.3.1 The Paillier Cryptosystem

16.3.1.1 Overview of The Paillier Cryptosystem

The Paillier cryptosystem contains three components, namely, a key generation part (denoted as Keygen()), an encryption part $c = \mathcal{E}(m)$ (where m and c denote the plaintext and the ciphertext, respectively), and a decryption part $m = \mathcal{D}(c)$. These three components are described in detail as follows:

- **Keygen()**: The objective of Keygen() is to output a public key k_p (used to encrypt the plaintext) and a private key k_s (used to decrypt the ciphertext). The procedures of Keygen() are given below:
  (1) Two large prime numbers (i.e. p and q) with the same bit length are selected in a way that the greatest common divisor of pq and $(p-1)(q-1)$ is one.
  (2) Let $n = pq$ and λ be the least common multiple of $p-1$ and $q-1$.
  (3) Select a random integer $g \in \{1, 2, \ldots, n^2 - 1\}$.
  (4) Let $\mu = \frac{n}{(g^\lambda \bmod n^2)-1}$.
  (5) Then, the public key k_p can be represented by a group of two integers, n and g, i.e. k_p: (n, g), and the private key k_s is the set of λ and μ, i.e. k_s: (λ, μ).
- $\mathcal{E}(\mathbf{m})$: With the public key k_p, $\mathcal{E}(m)$ is used to encrypt the plaintext m and generate the ciphertext c. Specifically, a random integer $r \in \{1, 2, \ldots, n-1\}$ is first selected, in a way that the greatest common divisor of r and n is one. Then, the ciphertext c is obtained as $c = g^m r^n \bmod \text{n}^2$, where the plaintext m has to satisfy $0 \le m < n$.

- $D(\mathbf{c})$: $D(c)$ is used to decrypt the ciphertext c and obtain the plaintext m using the private key k_s. Specifically, m can be obtained as $m = \mu \frac{(c^\lambda \bmod n^2)-1}{n} \bmod n$.

16.3.1.2 Homomorphism

A great benefit of the Paillier cryptosystem lies in its homomorphic properties, i.e. the product of multiple ciphertexts decrypts to the sum of their corresponding plaintexts and $\mathcal{E}(\hat{k}_{ij})^{X_j} = \mathcal{E}(\hat{k}_{ij}X_j)$, as described before. Mathematically, with two plaintexts m_1 and m_2, the two salient features can be represented as follows:

$$\begin{cases} D(\mathcal{E}(m_1)\mathcal{E}(m_2)) = m_1 + m_2 \\ D(\mathcal{E}(m_1)^{m_2}) \quad = m_1 m_2 \end{cases} \tag{16.8}$$

16.3.2 Additive Zero Secret Sharing

Directly applying the Paillier cryptosystem in PCNMs is not well suited, as the ciphertext $\mathcal{E}(\hat{k}_{ij})^{X_j}$ sent from controller j can be inferred with controller i's private key. To address this challenge, the additive zero secret sharing strategy is adopted. With this strategy, a secret message (i.e. zero) is splitted into multiple random shares, which are then sent to a number of parties, respectively. Let s_{ji} denote each zero secret share sent to the jth neighbor of DER i. Instead of sending the encrypted $\hat{k}_{ij}X_j$ from DER j to DER i, DER j sends $\mathcal{E}(\hat{k}_{ij})^{X_j}\mathcal{E}(s_{ji})$ to DER i. Mathematically, all the zero secret shares distributed to DER i's neighbors have to satisfy $\sum_{j\in\mathcal{N}_i} s_{ji} = 0$.

16.3.3 The DEWA Algorithm

With the Paillier cryptosystem and additive zero secret sharing, DEWA is developed in this chapter. The algorithm is given in Algorithm 16.1, and the procedures are as follows:

1. **Key pairs generation**: Each key pair (i.e. k_p and k_s) for each crypto-controller is generated within the SDN controller. Key pairs are distributed to corresponding crypto-controllers, and are updated regularly.
2. **Zero secret shares generation**: Zero secret shares are generated using $\sum_{j\in\mathcal{N}_i} s_{ji} = 0$ within the SDN controller. They are then distributed to corresponding crypto-controllers, and are updated regularly.
3. **Weight encryption**: Each communication weight (i.e. $\hat{k}_{ij}$) is encrypted within the SDN controller using controller i's public key k_{pi}. The encrypted weight $\mathcal{E}_{k_{pi}}(\hat{k}_{ij})$ is then sent to controller j.
4. When controller j receives k_{pi}, k_{sj}, s_{ji}, and $\mathcal{E}_{k_{pi}}(\hat{k}_{ij})$ from the SDN controller, it computes the ciphertext c_j as follows and sends it to controller i:

$$\begin{aligned} c_j &= \mathcal{E}_{k_{pi}}(\hat{k}_{ij})^{X_j}\mathcal{E}_{k_{pi}}(s_{ji}) \\ &= \mathcal{E}_{k_{pi}}(\hat{k}_{ij}X_j + s_{ji}) \end{aligned} \tag{16.9}$$

5. When controller i receives all the ciphertexts from its neighbors, it computes their product α_i as follows:

$$\alpha_i = \prod_{j\in\mathcal{N}_i^2} c_j = \prod_{j\in\mathcal{N}_i^2} \mathcal{E}_{k_{pi}}(\hat{k}_{ij})^{X_j}\mathcal{E}_{k_{pi}}(s_{ji}) \tag{16.10}$$

Algorithm 16.1: The DEWA Algorithm

▷ **Input**: The number of crypto-controllers N_c, the key pairs update period T.

▷ **Offline key pairs generation**:

Output: Public key k_{pi} and private key k_{si}.

for $i \leftarrow 1$ *to* N_c **do**
| Use Keygen() to generate several key pairs;
end

▷ **Online** u_i **calculation**:

for *each period* T **do**
| SDN controller generates zero secret shares s_{ji};
| SDN controller encrypts $\mathcal{E}_{k_{pi}}(\widehat{k}_{ij})$;
| Send k_{pi}, $\mathcal{E}_{k_{pi}}(\widehat{k}_{ij})$ and s_{ji} to controller j;
| Send k_{si} to controller i;
| **while** *True* **do**
| | Controller j listens to packets of X_j sent from RTDS;
| | Controller i calculates u_i^1;
| | Controller j calculates c_j using (16.9), and sends to i;
| | Controller i calculates product α_i using (16.10);
| | Controller i decrypts α_i and computes u_i^2 using (16.11);
| | Controller i computes $u_i = u_i^1 + u_i^2$ using (16.17);
| | Controller i sends u_i to RTDS;
| | **if** *current period ends* **then**
| | | **Break**;
| | **end**
| **end**
end

▷ **Output**: The control signal u_i, $(i \in \{1, 2, ..., Nc\})$.

---

6. Controller i then decrypts α_i and computes the control signal u_i^2 (see (16.7)) as follows:

$$
\begin{aligned}
u_i^2 &= -K_i D_{k_{si}}(\alpha_i) \\
&= -K_i \left(\sum_{j \in \mathcal{N}_i^2} \hat{k}_{ij} X_j + \sum_{j \in \mathcal{N}_i^2} s_{ji} \right) = -K_i \sum_{j \in \mathcal{N}_i^2} \hat{k}_{ij} X_j
\end{aligned}
\tag{16.11}
$$

7. **Error correction**: During the key pairs dynamic updating process, it is likely to have a mismatch issue, i.e. k_{pi} and k_{si} are unmatched. To address this issue, we further implement an error correction scheme in DEWA. Specifically, when a mismatch issue is detected (e.g. an abnormal decrypted value is observed), the private key from the previous updating cycle is utilized.

16.4 DEWA Privacy Analysis

The DEWA privacy-preserving property is mathematically analyzed in this section. Specifically, we classify the attacks into two categories, namely, internal attacks and external attacks. A major type of internal attack is the so-called honest-but-curious attack. We present how the privacy in DEWA-based PCNMs can be preserved against these attacks and also analyze the security level through a security level index.

16.4.1 Privacy Analysis

Following formal privacy analysis procedures [23], we present the DEWA privacy analysis in this subsection. We demonstrate that the private information, including crypto-controllers' states X_j and communication weights $\hat{k}_{ij}$ cannot be inferred under the honest-but-curious and external attacks.

According to the formal privacy analysis, privacy can be defined as a typical cryptographic game involving two players, namely, a challenger and an adversary. The challenger encrypts messages and the adversary tries to break the ciphertext. The game rules are described as follows:

- The adversary creates two messages with the same length and sends them to the challenger.
- The challenger randomly selects one of the messages, encrypts it, and sends the ciphertext to the adversary.
- When the ciphertext is received by the adversary, the adversary determines which message it is. If the probability that the adversary obtains the correct result is larger than 50%, the privacy is not preserved, and otherwise, the privacy is well protected.

16.4.1.1 Scenario 1

In this scenario, we consider the honest-but-curious attack (execute operations normally with no active attack launched) in DEWA-based PCNMs. We denote the set of corrupted (i.e. under the honest-but-curious attack) and uncorrupted crypto-controllers as C and $\mathcal{U}$. For a corrupted crypto-controller $i \in C$, some of its neighbors are also corrupted, while some are not. Crypto-controller i can collaborate with its corrupted neighbors, trying to obtain X_j and $\hat{k}_{ij}$ from an uncorrupted neighboring crypto-controller $j \in \mathcal{U}$.

For a corrupted crypto-controller $i \in C$, it knows the following information: k_{pj}, k_{pi}, k_{si}, $\{s_{ji}\}_{j \in C}$, and $\{\hat{k}_{ij}\}_{j \in C}$. Note that $\sum_{j \in \mathcal{U}} s_{ji} = -\sum_{j \in C} s_{ji}$ (where crypto-controller j is the jth neighbor of crypto-controller i). Following the formal privacy analysis, the adversary (i.e. the corrupted crypto-controller i) sends the challenger (i.e. the uncorrupted crypto-controller j) two messages (denoted as X_j^0 and X_j^1, where $j \in \mathcal{U}$). The challenger randomly selects a message, encrypts it (i.e. $c_j = \mathcal{E}(\hat{k}_{ij} X_j^b) \mathcal{E}(s_{ji})$, where $b \in \{0,1\}$ and $j \in \mathcal{U}$), and sends the ciphertext to the adversary.

When the adversary receives the ciphertext, its private key k_{si} is used to decrypt the ciphertext, and the adversary obtains $\hat{k}_{ij} X_j^b + s_{ji}$. Then, the adversary will determine which message (X_j^0 or X_j^1) is encrypted. Denote the adversary's output as $b' \in \{0,1\}$, which is the guess of b. The probability of successfully determining the message is denoted as $Pr[b' = b | i \in C]$. As the zero secret shares $\{s_{ij}\}_{j \in \mathcal{U}}$ are random and unknown and the probability of breaking the zero shares scheme is negligible, the probability that the adversary obtains the correct result is as follows:

$$Pr[b' = b | i \in C] \leqslant \frac{1}{2} \tag{16.12}$$

From the aforementioned analysis, we can observe that in DEWA-based PCNMs, the honest-but-curious attack is not able to break the privacy of an uncorrupted neighboring crypto-controller. Privacy is thus well protected.

16.4.1.2 Scenario 2

In this scenario, we consider a typical external attack, i.e. an external attacker eavesdrops on a communication channel and intercepts the information on the channel.

In this case, the adversary only knows public keys (e.g. k_{pi} for crypto-controller i). The adversary can use a public key k_{pi} to encrypt an arbitrary message. Following the formal privacy analysis, the adversary sends the challenger (i.e. an uncorrupted crypto-controller j) two messages (denoted as

X_j^0 and X_j^1). The challenger randomly selects a message, encrypts it (i.e. $c_j = \mathcal{E}(\hat{k}_{ij}X_j^b)\mathcal{E}(s_{ji})$, where $b \in \{0, 1\}$), and sends the ciphertext to the adversary. Denote the adversary's output as $b' \in \{0, 1\}$, which is the guess of b. The probability of successfully determining the message (denoted as $Pr[b' = b]$) is larger than 50% only when the adversary finds a polynomial time distinguisher for the nth residue modulo n^2 (where $n = pq$ as in Keygen()) [21], which is believed to be computationally hard, and therefore the probability of successfully determining the message is not larger than 50%.

In sum, it is computationally infeasible for the external attacker to infer a crypto-controller's states X_i and weights $\hat{k}_{ij}$ from ciphertexts, and the privacy is well protected.

16.4.2 Security Level Analysis

The setting of key length affects the security level in DEWA-based PCNMs. For the switching-key management system, the security level is presented as an index [24], which indicates the computational cost of decrypting the private keys. For the DEWA scheme, the security level index can be represented as:

$$\mathbb{I}_N = \frac{1}{\mathrm{NT}} \sum_{j \in \mathcal{I}_t} L_p^j(\theta, \zeta) + \eta \tag{16.13}$$

where $T \in \mathcal{Z}_N := \{1, 2, 3, \ldots, N\}$ is the update period of key pairs, $\mathcal{I}_t := \{1, 2, 3, \ldots, l_t\}$ (l_t is the number of key pairs), and $\eta \geq 0$ is the cost incurred to identify the keys update. $L_p(\theta, \zeta) = \exp\{(\theta + o(1))(\log p)^\zeta (\log p)^{1-\zeta}\}$, where p is the prime used in the cyclic group for the DEWA scheme, and θ and ζ are determined by an algorithm to solve a discrete logarithm problem.

The index (16.13) shows that $\exists T \leq N$, $\mathbb{I}_N \geq \frac{1}{N}L_p(\theta, \zeta)$, $\forall N \in \mathcal{Z}$, which means that the proposed DEWA is more secure than the static-key management in terms of the computational cost of decrypting the private keys as long as the selected update period T satisfies $\mathbb{I}_N \geq \frac{1}{N}L_p(\theta, \zeta)$.

16.5 Case Studies

16.5.1 The PCNMs Testbed

The real-time PCNMs testbed is illustrated in Fig. 16.3. It consists of six IoT devices (i.e. six BeagleBone devices, where programmable crypto-controllers are installed, respectively), four real SDN switches (i.e. Pica8), an SDN controller, GTNET×2 cards, and RTDS hardware. Specifically, the NMs system is developed and compiled in RSCAD, which is designed to interact with the RTDS hardware. The six BeagleBone devices are utilized to virtualize and install the crypto-controllers. These BeagleBone devices are respectively connected to different ports of SDN switches. SDN is used to manage the whole network. It allows for real-time packet monitoring and network configuration, enabling a programmable, reliable, and efficient network management environment. Four SDN switches, managed by an SDN controller through a traditional switch, are used in this study to form a realistic network environment. GTNET×2 cards are utilized for the RTDS hardware to communicate with crypto-controllers. The DEWA algorithm is running within each crypto-controller. The switching key pairs are managed and distributed to each crypto-controller by the SDN controller.

To validate the effectiveness, benefits, and superiority of DEWA-based PCNMs, experimental results produced with the PCNMs testbed are reported. A six-microgrid NMs system is considered in this study. The one-line diagram of the system is shown in Fig. 16.4. It contains six microgrids

Figure 16.3 The PCNMs testbed.

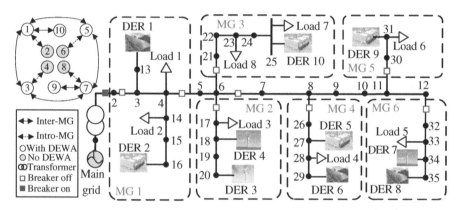

Figure 16.4 One-line diagram of the NMs model. Source: Gyula Gyukli/Adobe Stock; Sergiy Serdyuk/Adobe Stock.

including 10 DERs and eight power loads, and can operate in either grid-connected or islanded mode depending on whether the circuit breaker between buses 2 and 3 is closed or open, and the communication topology of the NMs is shown on the left side of Fig. 16.4. In this study, it operates in the islanded mode. The DEWA algorithm is implemented in each BeagleBone device. The real-time communication between each BeagleBone device and the RTDS hardware adopts the user datagram protocol (UDP). In this study, a ring-shaped communication topology in the cyber-layer is designed. The sampling rate in the RTDS is set at 35 Hz. The number of key pairs for each controller and the update period T of generated key pairs of k_s and k_p are set at 20 and 50, respectively.

The Sections 16.5.2-16.5.6 are organized into five studies. In the first study, we demonstrate that DEWA has little impact on the system's normal operations, while the traditional PHE method has a larger impact. In the second study, we illustrate that the delay produced by DEWA is only slightly higher than that without encryption. We also conduct the eigen analysis for both DEWA-based PCNMs and NMs without encryption, and results demonstrate that the DEWA-based PCNMs maintain high stability. In the third study, three types of ciphertexts (i.e. V_1, V_2, and V_3) are

Table 16.1 Power loads at each bus in Fig. 16.4.

Load 1	Load 2	Load 3	Load 4
$P1 = 10\,\mathrm{kW}$	$P2 = 5\,\mathrm{kW}$	$P3 = 10\,\mathrm{kW}$	$P4 = 5\,\mathrm{kW}$
$Q1 = 5\,\mathrm{kVAR}$	$Q2 = 2\,\mathrm{kVAR}$	$Q3 = 5\,\mathrm{kVAR}$	$Q4 = 2\,\mathrm{kVAR}$
Load 5	**Load 6**	**Load 7**	**Load 8**
$P5 = 5\,\mathrm{kW}$	$P6 = 15\,\mathrm{kW}$	$P7 = 10\,\mathrm{kW}$	$P8 = 15\,\mathrm{kW}$
$Q5 = 2\,\mathrm{kVAR}$	$Q6 = 7\,\mathrm{kVAR}$	$Q7 = 5\,\mathrm{kVAR}$	$Q8 = 7\,\mathrm{kVAR}$

Table 16.2 Droop control and DEWA parameters.

$m_{P\{1,3,5,7,9\}}$	$3e^{-5}$	$m_{P\{2,4,6,7,10\}}$	$1.5e^{-5}$
$n_{P\{1,3,5,7,9\}}$	0.04	$n_{P\{2,4,6,8,10\}}$	0.02
$k^{\omega}_{\{1,2,\dots,10\}}$	0.3	$k^{E}_{\{1,2,\dots,10\}}$	0.3
$k^{P}_{\{1,2,\dots,10\}}$	0.3	$\omega^{\mathrm{ref}} = 60\,\mathrm{Hz}$	$E^{\mathrm{ref}} = 253.9\,\mathrm{V}$

recorded respectively with and without DEWA to demonstrate DEWA's effectiveness. The fourth case study illustrates a benefit of using SDN in PCNMs, i.e. the dynamic routing function enabled by SDN greatly improves the resilience of PCNMs. In the fifth study, we present the superiority of DEWA-based PCNMs over the traditional differential-privacy method-based NMs.

16.5.2 The Impact of DEWA

In this case, we compare the impacts posed by DEWA and the traditional PHE method on the system's normal operations. The system configurations are the same as in Tables 16.1 and 16.2, and the communication topology is the same as in Fig. 16.4.

(1) The impact of DEWA: Figure 16.5 gives frequency and active power responses of different DERs (i.e. DERs 1, 3, 5, 7, 9, and 10) with DEWA, where at time $t = 3$ seconds, Load 1 (see Fig. 16.4) increases from $(10\,\mathrm{kW}, 5\,\mathrm{kVAR})$ to $(20\,\mathrm{kW}, 10\,\mathrm{kVAR})$. It can be observed that

- Before Load 1 changes, the system has a rated frequency, i.e. 60 Hz, meaning that the frequency is regulated well in the steady state.
- After experiencing a disturbance at time $t = 3$ seconds, the system is able to go back to the steady state, and the active powers from different DERs are regulated.

The above observations demonstrate that DEWA has little impact on the system's normal operation.

(2) Comparison of a controller without encryption, DEWA, and PHE: With an encryption algorithm, a certain amount of time typically needs to be consumed to encrypt and decrypt data. In this case, the impact of the key size is demonstrated. Specifically, private key sizes are selected as suggested by the NIST recommendation [25]. For the static encryption system, a larger key size commonly leads to improved security; in the meantime, it results in longer computation time, potentially affecting the real-time operation in PCNMs.

Figure 16.6 gives the comparison results of $\sum_{j \in \mathcal{N}_1^2} a_{1j}(m_j P_j - m_1 P_1)$ under the traditional control (i.e. without encryption), DEWA, and PHE (with a short key and a long one). DEWA has a key

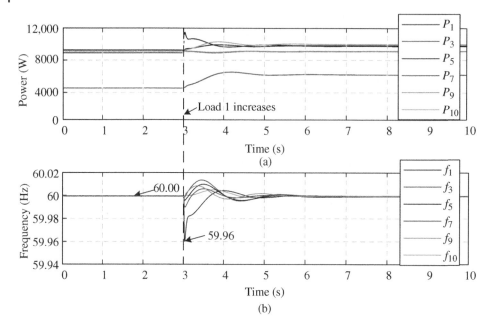

Figure 16.5 Frequency and active power responses of DERs with DEWA. (a) Active power generated by each DER and (b) frequency response of each DER.

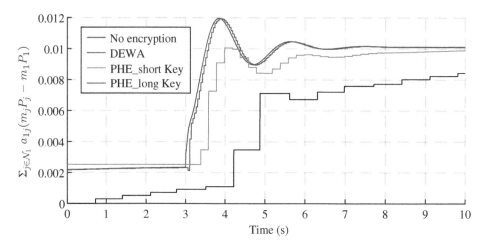

Figure 16.6 $\sum_{j\in\mathcal{N}_1} a_{1j}(m_j P_j - m_1 P_1)$ in different scenarios.

size of 160 bits and the update period T is set at 50 cycles. The short key for PHE has 512 bits and the long key has 1024 bits. The value of $\sum_{j\in\mathcal{N}_1^2} a_{1j}(m_j P_j - m_1 P_1)$ directly affects the power sharing performance, and the smaller the deviations between $\sum_{j\in\mathcal{N}_1^2} a_{1j}(m_j P_j - m_1 P_1)$ from a crypto-control (i.e. DEWA or PHE) and the traditional control (i.e. without encryption) is, the smaller the impact posed by the crypto-control will be. From Fig. 16.6, it can be seen that:

- The results obtained from the traditional control and DEWA are almost the same, indicating that DEWA has little impact on the system's normal operation.

Table 16.3 The number of packets between crypto-controller 3 and crypto-controller 1 for different time periods.

T (s)	n_s	n_r	Packet loss rate
5	178	168	5.62%
10	322	322	0
20	698	680	2.58%
30	1039	1037	0.19%
50	1720	1718	0.12%
60	2095	2078	0.14 %

- The result obtained from PHE with either a short key or a long one largely deviates from that obtained from the traditional control.
- For PHE, the increase of the key size largely affects the system's performance.

16.5.3 DEWA Delays and Eigen Analysis of DEWA-Based PCNMs

In this subsection, we first test the packet loss in the PCNM system, and compare the delays produced by DEWA-based PCNMs and those without encryption. We also conduct the eigenanalysis for both DEWA-based PCNMs and NMs without encryption.

(1) Packet loss test: In this case, we test the packet loss using UDP in the PCNMs system. As the microgrid measurements from DERs to crypto-controllers are not encrypted, we only consider the impact of packet loss between crypto-controllers. In this test case, the crypto-controller of DER 3 sends packets to the crypto-controller of DER 1. We record the number of packets sent (n_s) and received (n_r) for different time periods (T). Table 16.3 shows the number of packets between crypto-controller 3 and crypto-controller 1. We use Wireshark to capture packets for different time periods. It can be seen that the packet loss rate ranges from 0 to 5.62%, and PCNMs can work well even when the packet loss rate reaches 5.62%.

We also investigate the impact of packet loss on DEWA. We use a timestamp to generate a switching sequence of the keys. An advantage of using the timestamp is its simplicity, in that no extra information is communicated between different crypto-controllers to realize the dynamic key-encrypted control. In Step 7 of the DEWA algorithm, we consider the error correction process to handle the key mismatch problem in the case where packet loss and short delay occur at the time of key pairs change. Specifically, when a mismatch issue is detected (e.g. an abnormal decrypted value is observed), the private key from the previous updating cycle is utilized. Figure 16.7 shows the control signal generated in crypto-controller 1 with the key pairs changing every time period (T). It can be seen that error correction occurs with $T = 5$ and 10 in the test case, and the DEWA algorithm still works for PCNMs.

(2) Comparison of delays from DEWA and those without encryption: In this case, we compare the delays produced by DEWA-based PCNMs and those without encryption. Specifically, the delays investigated include measurement and computation delays. In DEWA-based PCNMs, each crypto-controller measures the required local physical state variables (i.e. output power, voltage, and frequency) through its measurement unit, which introduces measurement delays τ_m. Within each crypto-controller, control signals are encrypted and are sent to its neighbors. When each

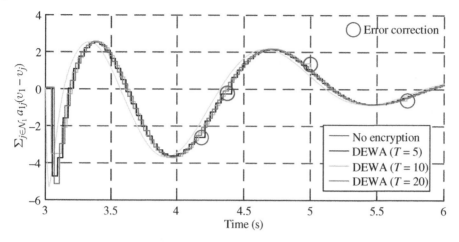

Figure 16.7 $\sum_{j\in\mathcal{N}_1} a_{1j}(v_1 - v_j)$ with different dynamic key update periods.

control signal is received by a neighbor, the signal is decrypted. Both the encryption and decryption processes inevitably introduce delays, i.e. the computation delays τ_c.

To clearly illustrate τ_m and τ_c, we add an incremental number, which starts from zero and increases by 1 every 0.02 seconds. This incremental number is sent from DER 3 with a frequency of 35 Hz, and is received by DER 1. At first, the incremental number is not encrypted. At around time $t = 1.55$ seconds, DEWA is enabled, meaning that the incremental number is first encrypted on the DER 3's side and is later decrypted on the DER 1's side. At around time $t = 1.93$ seconds, DEWA is removed. Figure 16.8 illustrates the incremental numbers sent from DER 3 (see the upper line) and those received by DER1 (see the bottom line) with and without DEWA. From Fig. 16.8, it can be seen that, the computational delay produced by DEWA is about $\tau_c = 38$ ms, and the measurement delay is about $\tau_m = 6$ ms. Moreover, we measure the measurement and computational delay between controllers 3 and 7 using Beaglebone devices, and Dell servers separately. The experimental results show that the measurement delay is 6.2 and 12 ms, respectively, and the computational delay is 39.9 and 25.1 ms, respectively.

(3) Eigen analysis for DEWA-based PCNMs and NMs without encryption: With measured τ_m and τ_c, we further conduct the eigen analysis for both DEWA-based PCNMs and NMs without encryption. Specifically, the dynamic model of the NMs considering heterogeneous delays can be found in our previous work [26]. As the equations of the small-signal stability model in delayed NMs become transcendental equations, an ODE-SOD [26] method is adopted to calculate the rightmost eigenvalues. Figure 16.9 illustrates the eigenvalues of DEWA-based PCNMs and NMs without encryption. It can be seen that the eigenvalues of DEWA-based NMs all have negative real parts (i.e. Re), indicating that, with $\tau_m = 6$ ms and $\tau_c = 38$ ms, DEWA-based NMs can maintain the stability. This eigen analysis result is consistent with the dynamic response in Fig. 16.5.

16.5.4 Privacy Evaluation for DEWA

The effectiveness of DEWA is validated in this subsection. In this case, the system configuration is the same as in Fig. 16.5. Three types of ciphertexts (i.e. V_1, V_2, and V_3) are recorded. At time $t = 3$ seconds, Load 1 (see Fig. 16.4) increases from (10 kW, 5 kVAR) to (20 kW, 10 kVAR). The recorded voltages without DEWA and with DEWA are illustrated in Figs. 16.10 and 16.11, respectively. It can be observed that,

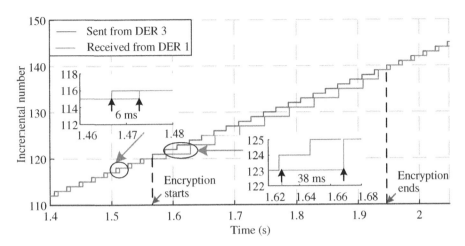

Figure 16.8 The incremental numbers sent from DER 2 and those received by DER1 with and without DEWA.

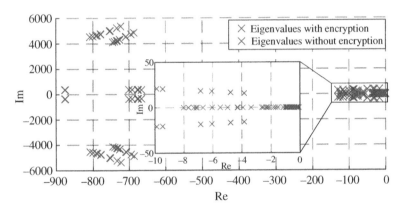

Figure 16.9 Eigenvalues of NMs with/without the encrypted control architecture.

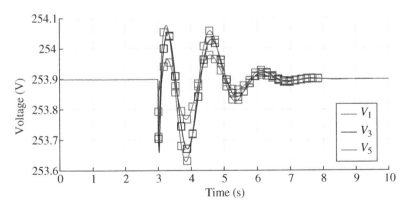

Figure 16.10 The recorded voltages without DEWA.

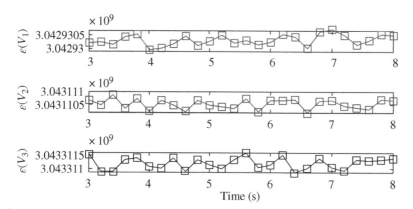

Figure 16.11 The encrypted voltages (i.e. ciphertexts) with DEWA.

- Without DEWA, each voltage eventually converges to a fixed reference value (see Fig. 16.10). This results in that the voltages can be readily identified.
- With DEWA, each encrypted voltage (i.e. ciphertext) is always varying (see Fig. 16.11). This makes identifying each voltage more difficult, and thus greatly enhances the privacy-preserving capability.

16.5.5 Benefit of SDN-Enabled Crypto-Control

In this subsection, we illustrate a benefit of using SDN in PCNMs, i.e. the dynamic routing function enabled by SDN greatly improves the resilience of PCNMs. Specifically, the system configuration is the same as in Fig. 16.5. Two subcases are developed. In the first subcase, at around time $t = 3.25$ seconds, communication congestion occurs in the channel between DER 1 and DER 3. This is achieved by adding a delay of 3 seconds in the DEWA algorithm. Figure 16.12 illustrates the responses of V_1, V_3, and V_5 before and after the congestion occurs when the dynamic routing function is disabled. It can be seen that the system soon collapses. This is due to the fact that the communication congestion causes the mismatch of public and private keys in the switching key management system.

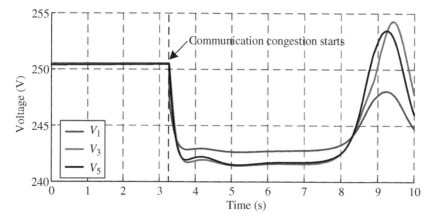

Figure 16.12 The responses of V_1, V_3, and V_5 before and after a communication congestion occurs when the dynamic routing function is disabled.

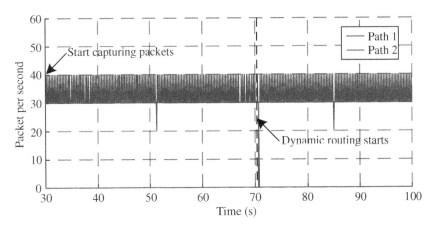

Figure 16.13 Data packets in the two paths before and after a communication congestion occurs at around time $t = 70.165$ seconds.

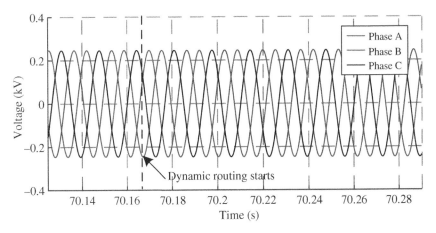

Figure 16.14 The responses of V_1, V_2, and V_3 before and after a communication congestion occurs when the dynamic routing function is enabled.

In the second subcase, we implement the SDN-enabled dynamic routing function in PCNMs. With dynamic routing, when a data path between DER 1 and DER 3 is subjected to congestion, another path is used in real-time. Wireshark is utilized to capture the traffic between DER 1 and DER 3. Figure 16.13 illustrates the data packets in the two paths before and after a communication congestion (i.e. the same with that in Fig. 16.12) occurs at around time $t = 70.165$ seconds. The responses of V_1, V_2, and V_3 with the dynamic routing function are given in Fig. 16.14. It can be seen from Figs. 16.12 and 16.14 that, with SDN, the dynamic routing function has been successfully accomplished and the system performance remains unaffected.

16.5.6 Comparison of DEWA with Existing Scheme

In this subsection, we present the superiority of DEWA-based PCNMs over the traditional differential-privacy method-based NMs and compare DEWA with existing privacy-preserving methods for IoT devices.

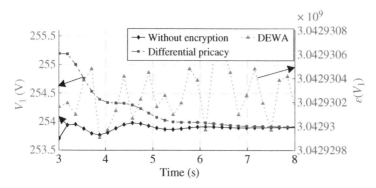

Figure 16.15 The response of V_1 in three different scenarios, i.e. without encryption, with the differential-privacy method, and with DEWA.

(1) **Comparison with differential-privacy method.**

The traditional differential-privacy algorithm is adopted from [27]. Its basic idea is that a random noise is added to each data message. This is widely used to preserve privacy in the control area. The system configuration is the same as in Fig. 16.5, where at around time $t = 3$ seconds, Load 1 increases from (10 kW, 5 kVAR) to (20 kW, 10 kVAR).

The response of V_1 in three different scenarios, i.e. without encryption, with the differential-privacy method, and with DEWA are shown in Fig. 16.15. It can be observed that,

- The differential-privacy method enhances the privacy during the transient process, as the variation of each parameter becomes larger when random noises are added. However, each varying parameter still eventually converges to a fixed reference value.
- With DEWA, each encrypted voltage is always varying, making identifying each voltage more difficult, and thus greatly enhancing the privacy-preserving capability.

(2) **Comparison with existing schemes for IoT devices.**

To show the effectiveness of our proposed scheme, we compare the proposed scheme with three existing encryption schemes for general IoT devices. We choose and compare the DEWA scheme with the Paillier's homomorphic encryption scheme and the ring signcryption scheme [28]. Table 16.4 presents our experimental results for selected privacy-preserving techniques. We compare the homomorphic encryption performance, which is suitable for cooperative control, and measure the time of main operations/phases such as the encryption time ($\mathcal{E}$), the decryption time (D). We give the maximum and minimum time values from 30 iterations. It can be seen that although the Paillier scheme is a PHE, it cannot be directly applied to a cooperative control system, and the operation time is longer than DEWA. For the widely used Ring Signcryption scheme, the long operation time makes it not suitable for microgrid control.

16.6 Conclusion

This chapter presents a PCNMs architecture, where crypto-controllers are fully virtualized and SDN is utilized to manage the network. DEWA is developed to preserve each DER's privacy while real-time computation is ensured. The DEWA privacy-preserving property is further mathematically analyzed, and a real-time DEWA-based PCNMs testbed incorporating DEWA,

Table 16.4 Minimal and maximal computation times (in ms) for evaluating DEWA and existing schemes on the PCNM platform.

Alg.	PHE	Cooperative control		C1 min	C1 max	C2 min	C2 max	C3 min	C3 max
DEWA	Y	Y	$\mathcal{E}$	17	22	19	32	20	23
			D	16	23	21	30	18	27
Paillier	Y	N	$\mathcal{E}$	43	49	39	52	44	53
(1024b)			D	75	88	80	91	78	90
Paillier	Y	N	$\mathcal{E}$	342	368	331	352	342	378
(2048b)			D	653	686	640	678	662	692
[28]	N	N	$\mathcal{E}$	12,698	13,362	—	—	—	—
			D	19,862	20,294	—	—	—	—

real SDN switches, and IoT devices is established in an RTDS environment. Test results validate the effectiveness, benefits, and superiority of DEWA-based PCNMs. Some future work includes adapting DEWA to other communication-based controls and developing methods to further enhance the privacy-preserving in PCNMs.

16.7 Exercises

1. Show that if c is an encryption of $m \in Z_n$ using Paillier encryption scheme with $g = n + 1$, then for any $u \in Z_n$, c^u mod $n2$ is an encryption of μ mod n. Use this to find a way to compute the encryption of $P(u)$ mod n, given $u \in Z_n$ and the ciphertexts $c_0, ..., c_d$ encrypting the coefficients of the polynomial P of degree at most d.
2. The Paillier cryptosystem, proposed by P. Paillier in 1999, relies on operations modulo n^2, where n is a RSA-type integer. Let p and q be two primes such that $p|q - 1$ and $q|p - 1$, and let $n = pq$.
   - Show that $x = x'$ mod n if and only if $x^n = x'^m$ mod n^2
   - Let $g = 1 + kn$ with $k^n = 1$. Show that g has order exactly n in $(Z/n^2Z)^2$ and that any element of order n is of this form
   - Explain how to solve the discrete logarithm problem in the subgroup generated by g.

References

1 Tang, Z., Zhang, P., and Krawec, W.O. (2021). A quantum leap in microgrids security: the prospects of quantum-secure microgrids. *IEEE Electrification Magazine* 9 (1): 66–73.

2 Zhang, P. (2021). *Networked Microgrids*. Cambridge University Press.

3 Tang, Z., Qin, Y., Jiang, Z. et al. (2020). Quantum-secure microgrid. *IEEE Transactions on Power Systems* 36 (2): 1250–1263.

4 Babahajiani, P., Wang, L., Liu, J., and Zhang, P. (2021). Push-sum-enabled resilient microgrid control. *IEEE Transactions on Smart Grid* 12 (4): 3661–3664.

5 Wang, L., Qin, Y., Tang, Z., and Zhang, P. (2020). Software-defined microgrid control: the genesis of decoupled cyber-physical microgrids. *IEEE Open Access Journal of Power and Energy* 7: 173–182.

6 Han, S., Topcu, U., and Pappas, G.J. (2016). Differentially private distributed constrained optimization. *IEEE Transactions on Automatic Control* 62 (1): 50–64.

7 Cortés, J., Dullerud, G.E., Han, S. et al. (2016). Differential privacy in control and network systems. In: *2016 IEEE 55th Conference on Decision and Control (CDC)*, 4252–4272. IEEE.

8 Dvorkin, V., Fioretto, F., Van Hentenryck, P. et al. (2020). Differentially private optimal power flow for distribution grids. *IEEE Transactions on Power Systems* 36 (3): 2186–2196.

9 Fioretto, F., Mak, T.W.K., and Van Hentenryck, P. (2020). Differential privacy for power grid obfuscation. *IEEE Transactions on Smart Grid* 11 (2): 1356–1366.

10 Pal, R., Hui, P., and Prasanna, V. (2018). Privacy engineering for the smart micro-grid. *IEEE Transactions on Knowledge and Data Engineering* 31 (5): 965–980.

11 Fan, B. and Wang, X. (2021). Distributed privacy-preserving active power sharing and frequency regulation in microgrids. *IEEE Transactions on Smart Grid* 12 (4): 3665–3668.

12 Wang, Z., Yang, K., and Wang, X. (2013). Privacy-preserving energy scheduling in microgrid systems. *IEEE Transactions on Smart Grid* 4 (4): 1810–1820.

13 Wu, T., Zhao, C., and Zhang, Y.-J.A. (2021). Privacy-preserving distributed optimal power flow with partially homomorphic encryption. *IEEE Transactions on Smart Grid* 12 (5): 4506–4521.

14 Darup, M.S., Redder, A., and Quevedo, D.E. (2018). Encrypted cooperative control based on structured feedback. *IEEE Control Systems Letters* 3 (1): 37–42.

15 Alexandru, A.B., Darup, M.S., and Pappas, G.J. (2019). Encrypted cooperative control revisited. In: *2019 IEEE 58th Conference on Decision and Control (CDC)*, 7196–7202. IEEE.

16 Hadjicostis, C.N. and Domínguez-García, A.D. (2020). Privacy-preserving distributed averaging via homomorphically encrypted ratio consensus. *IEEE Transactions on Automatic Control* 65 (9): 3887–3894.

17 Tang, Z., Zhang, P., Krawec, W.O., and Jiang, Z. (2020). Programmable quantum networked microgrids. *IEEE Transactions on Quantum Engineering* 1: 1–13.

18 Kreutz, D., Ramos, F.M.V., Verissimo, P.E. et al. (2014). Software-defined networking: a comprehensive survey. *Proceedings of the IEEE* 103 (1): 14–76.

19 Yang, S., Cui, L., Chen, Z., and Xiao, W. (2020). An efficient approach to robust SDN controller placement for security. *IEEE Transactions on Network and Service Management* 17 (3): 1669–1682.

20 Wang, L. and Xiao, F. (2010). Finite-time consensus problems for networks of dynamic agents. *IEEE Transactions on Automatic Control* 55 (4): 950–955.

21 Paillier, P. (1999). Public-key cryptosystems based on composite degree residuosity classes. In: *International Conference on the Theory and Applications of Cryptographic Techniques*, 223–238. Springer.

22 Cramer, R., Damgård, I.B., and Nielsen, J.B. (2015). *Secure Multiparty Computation and Secret Sharing*. Cambridge University Press.

23 Katz, J. and Lindell, Y. (2020). *Introduction to Modern Cryptography*. CRC Press.

24 Kogiso, K. (2018). Attack detection and prevention for encrypted control systems by application of switching-key management. *2018 IEEE Conference on Decision and Control (CDC)*, 5032–5037.

25 Farah, S., Javed, Y., Shamim, A., and Nawaz, T. (2012). An experimental study on performance evaluation of asymmetric encryption algorithms. *Recent Advances in Information Science, Proceeding of the 3rd European Conference of Computer Science (EECS-12)*, 121–124.

26 Wang, L., Zhou, Y., Wan, W. et al. (2021). Eigenanalysis of delayed networked microgrids. *IEEE Transactions on Power Systems* 36 (5): 4860–4863.

27 Huang, Z., Mitra, S., and Vaidya, N. (2015). Differentially private distributed optimization. *Proceedings of the 2015 International Conference on Distributed Computing and Networking*, 1–10.

28 Li, F., Zheng, Z., and Jin, C. (2016). Secure and efficient data transmission in the Internet of Things. *Telecommunication Systems* 62 (1): 111–122.

17

AI-Enabled, Cooperative Control, and Optimization in Microgrids

Ning Zhang, Lingxiao Yang, and Qiuye Sun

17.1 Introduction

Energy is an important material basis for the survival and development of modern society. With the development of industry from mechanization, electrification, informatization to networking, the energy system has entered the energy 5.0 era. The future energy reform is to deeply integrate advanced information and communication technology, intelligent control, and optimization technology with modern energy production, energy consumption, and user transactions. Specifically, it needs to fully develop and utilize renewable energy, improve the adjustment capacity of the energy system, enhance the ability to absorb renewable energy, develop high-efficiency and energy-saving energy technologies, realize the optimal allocation of resources, and ultimately realize the automation and intelligence of energy production and consumption. The microgrid is an independent energy supply, consumption, and distribution system. The pivotal link of the microgrid promotes the connection and integration of the energy system based on new energy technologies. Therefore, it is crucial to focus on the development of intelligent technologies in microgrids.

It is the core issue of microgrid to construct an advanced and reliable control and optimization system that matches the operation goals of microgrid. Its intelligent technology mainly includes two parts: (1) Coordinate control of microgrids by adjusting each device to realize the stable operation. (2) Optimized scheduling of microgrids to realize the safe and economical operation. The collaborative control of microgrids has attracted the attention of scholars. The power system has integrated distributed generation (DG) units. The droop control method which mimics the behavior of a synchronous generator was widely adopted in [1–3],. The droop control method that does not rely on external communication links enables "plug-and-play" of DG units. However, in practical situations, the droop control typically results in poor reactive power sharing due to the mismatched line impedances and causes the deviations of parameters like frequency and voltage. In this context, the consensus algorithm based on multi-agent control theory was utilized to resolve the problem [4–6]. However, the methods in all above references focus on the electricity only and applications to multi-energy complementary microgrids is ignored. The study on multi-energy complementary microgrids is important for including multiple kinds of energy and their couplings.

In addition, it is notable that the modeling and optimal operation of microgrids tend to be complicated nonlinearities and uncertainty. Moreover, it is difficult to obtain the accurate information and environment model due to the intrinsic randomness of microgrids such as DG units and loads. Therefore, reinforcement learning (RL) as one of machine learning algorithms is an adaptive learning and decision method which can guide behaviors for optimal decision-making

Microgrids: Theory and Practice, First Edition. Edited by Peng Zhang.

through rewards that interact with the environment [7]. RL has been investigated in various areas including distributed control, collaborative decision, transportation, and economics [8–10]. The model-free decision characteristic of RL makes it suitable for solving the above stochastic optimization problems in power system.

Therefore, this chapter focuses on the cooperative control and optimal operation of multi-energy complementary microgrids to achieve system intelligence, security, and economy.

17.2 Energy Hub Model in Microgirds

The microgrid is described as a controllable unit. On the one hand, it can meet the corresponding needs of the external transmission and distribution network in a very short time. On the other hand, the microgrid can also meet the specific needs of users and improve the power supply of local users, reliability, reduce line loss, and improve the utilization efficiency of waste heat.

As the basic energy unit in the microgrids, energy hub (EH) can receive multiple energy carriers and has the ability to convert and deliver energy. There are many elements such as connectors, converters, and storage facilities that can process multi-energy in energy hub. Owing to these elements, the input energy of hub can be converted to diversified forms or be stored in order to fulfill the energy demand at the output ports. Considering a common energy hub with multiple energy carriers, the energies transfer from input energies to output energies can be expressed as:

$$
\underbrace{\begin{pmatrix} L_\omega \\ \vdots \\ L_v \end{pmatrix}}_{L} = \underbrace{\begin{pmatrix} C_{\omega\omega} & \cdots & C_{\omega v} \\ \vdots & \ddots & \vdots \\ C_{\omega v} & \cdots & C_{vv} \end{pmatrix}}_{C} \underbrace{\begin{pmatrix} E_\omega \\ \vdots \\ E_v \end{pmatrix}}_{E} \tag{17.1}
$$

in which the various kinds of input energies and output energies are figured by $E = \left[E_\omega, \ldots, E_v\right]^T$ and $L = \left[L_\omega, \ldots, L_v\right]^T$, respectively. The matrix C is the forward coupling matrix which describes the conversion of energy from the input to the output. The elements of coupling matrix are coupling factors which are determined by the converters efficiencies and dispatch factors. The energy transfers from the ωth energy carrier to the vth energy carrier by a converter device with a coupling factor of $C_{\omega v}$ can be expressed as:

$$
L_v = C_{\omega v} E_\omega \tag{17.2}
$$

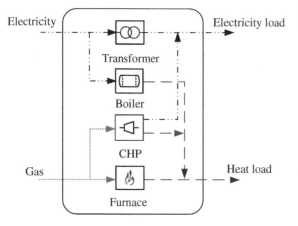

Figure 17.1 EH with electricity, natural gas, and heat systems.

where E_ω and L_v are energy input and output, respectively. For the EH with single converter, the coupling factor only corresponds to the efficiency of the element. While the coupling factors are determined by the converters efficiencies and dispatch factors for the EH with various converters.

Figure 17.1 depicts an EH model consisting of the transformer, the CHP, the boiler, and the furnace. The input energies of the hub are electricity and natural gas, EH can transform electricity and natural gas to two different energy formats i.e. heat and electricity by utilizing the four inner transformation devices like combined heat and power (CHP).

17.3 Distributed Adaptive Cooperative Control in Microgrids

17.3.1 Structure and Feature Analysis

To avoid overload problem and guarantee the security operation for the microgrids, the load power should be shared accurately and the system parameters of the microgrids should also be restored when the deviation occurs. To solve the issues, the distributed adaptive cooperative control (DACC) is presented. The controller structure utilized in the microgrids is designed on the basis of a multi-agent system. Figure 17.1 shows the architecture of the multi-agent system. The supply side and each EH are controlled by different agents which avail of the cooperative, intelligent, and adaptive features of the multi-agent system. Each agent has its own functions and goals. Based on the functions and goals, there are two agents designed for the proposed controller and another two normal agents in the multi-agent system. The agents designed for the controller include the energy supply agent (ESA) and the EH agent (EHA). Another two necessary agents are the directory facilitator agent (DFA) and the agent management service agent (AMSA). Moreover, the microgrids also installs the fault isolation device (FID) and the intelligent fault management (IFM) which are not argued. The agents are described as follows.

(1) Energy Supply Agent

The ESA is mainly utilized to dispose the information from EHAs and calculate the power flow between EH and energy supply side. In the multi-agent system of the microgrids, the ESA is an agent to ensure the energy supply for the EHs.

(2) Energy Hub Agent

The EHA is used to achieve the distributed coordinated control among different EHs. Meanwhile, the control objectives of each EH are realized by the agent. Since the EH is the basic unit of the microgrids, the EH controlled by the agent plays a decisive role in the microgrids operation.

(3) Agent Management Service Agent

The function of the AMSA is like a white page, which can maintain the directory of the agents in the microgrids.

(4) Directory Facilitator Agent

The function of the DFA is like a yellow page, which can maintain the directory of the agents and the services they can offer to other agents.

17.3.2 The Control of Outputs Based on Consensus Algorithm

Since the proportional power sharing and the restoration of the microgrid parameters cannot be reached spontaneously, it is necessary to propose a control strategy for the outputs of EHs. The model of the microgrids is a classical model. All system parameters like pressure and frequency are important and common parameters in theory and practice. Based on the characteristics of the district heating sub-network, a control strategy for heat output is proposed and can be expressed as follows:

$$L_h^2 = L_{hN}^2 + k_p \left(p_N - p\right) \tag{17.3}$$

where L_h and L_{hN} are heat output power and nominal heat output power of EH, respectively. The adjustment coefficient for heat output is denoted by k_p. The outlet pressure and nominal outlet pressure of EH are represented by p and p_N, respectively. Moreover, the adjustment coefficient k_p for each EH has the relationship to share the power proportionally:

$$\frac{L_{hN1}^2}{k_{p1}} = \frac{L_{hN2}^2}{k_{p2}} = \cdots = \frac{L_{hNn}^2}{k_{pn}} \tag{17.4}$$

As shown in (17.3), the control method is a flow control which can improve the respond speed of the system. But the presented strategy can not reach the proportional power sharing since the resistance of transmission pipeline for each EH is different. Furthermore, the pressure of the network is deviated which will badly impact the performance of the system by using the strategy.

An adaptive power control based on consensus is designed to achieve accurate power sharing. The heat output power L_h^2 is utilized to construct the first-order and linear multi-agent system dynamic, and to design corresponding consensus control protocol. Let $L_h^2 = H$. The first-order and linear multi-agent system dynamic can be expressed as:

$$\frac{\dot{H}_i}{k_{pi}} = u_{Hi} \tag{17.5}$$

In view of the consensus control, the u_{Hi} as the mismatch of the heat power is chosen, according to the information of neighbors for each EH and its own information. The auxiliary control u_{Hi} can be written as:

$$u_{Hi} = -R_H q_{Hi} \tag{17.6}$$

in which the coupling gain of heat power is denoted by R_H and the local neighbor heat power-sharing error is represented by q_{Hi}. To proportionally allocate the output of energy, the consensus protocol based on (17.4) should be designed that $\frac{H_i}{k_{pi}}$ of each EH should be equal. Namely, the heat power sharing error q_{Hi} can be formulated as:

$$q_{Hi} = \sum_{j=1, j \neq i}^{n} a_{ij} \left(\frac{H_i}{k_{pi}} - \frac{H_j}{k_{pj}}\right) \tag{17.7}$$

in which the elements in the adjacency matrix are described by a_{ij}, and the connected situation variation of EH agents can be reflected by the elements. Then the entire system can be expressed as:

$$\frac{\dot{H}}{k_p} = u_H \tag{17.8}$$

$$u_H = -R_H q_H \tag{17.9}$$

$$q_H = D \frac{H}{k_p} \tag{17.10}$$

in which the global variables are described as $\frac{H}{k_p} = \left[\frac{H_1}{k_{p1}}, \ldots, \frac{H_n}{k_{pn}}\right]^T$, $\frac{H}{k_p} = \left[\frac{H_1}{k_{p1}}, \ldots, \frac{H_n}{k_{pn}}\right]^T$, $q_H = \left[q_{H1}, \ldots, q_{Hn}\right]^T$, and $u_H = \left[u_{H1}, \ldots, u_{Hn}\right]^T$.

Then the mismatch need to feed to a proportional–integral (PI) controller $D_i(s)$ and a correction term δH_i is generated by the PI controller. The correction term δH_i which is utilized to update the constant term in (17.3) can be expressed as:

$$\delta H_i = D_i(s) u_{Hi} \tag{17.11}$$

To restore the pressure deviation introduced by the traditional control method, the dynamic consensus-based pressure control is utilized. The control method can compensate the estimated pressure which is the evaluated value during the consensus control. Since the pipeline resistance of each EH is different, the hub outlet pressure cannot be controlled at an identical value. On the other hand, the outlet pressures for all EHs should be controlled within an acceptable range of the rated pressure. The consensus protocol of the proposed control is presented as:

$$\overline{p}_i(t) = p_i(t) + R_p \int \sum_{j=1, j\neq i}^{n} a_{ij} \left(\overline{p}_j(t) - \overline{p}_i(t)\right) dt \tag{17.12}$$

where $\overline{p}_i$ and p_i are the estimated pressure and outlet pressure of ith EH, respectively. R_p is the coupling gain.

Then the estimated pressure $\overline{p}_i$ is in comparison with the reference pressure of district heat network. The pressure mismatch is expressed as:

$$u_{p_i} = p_{\text{ref}} - \overline{p}_i \tag{17.13}$$

where u_{p_i} is the pressure mismatch between estimated pressure and reference pressure. The reference pressure is denoted by p_{ref}. To generate the pressure correction term δp_i, the pressure mismatch u_{p_i} is fed to a PI controller G_i. The correction term δp_i is shown as:

$$\delta p_i = G_i(s) u_{pi} \tag{17.14}$$

In order to restore the estimated pressure to the reference pressure of the heating system, the local pressure set point needs to be updated by utilizing the pressure correction term δp_i obtained from the pressure controller.

Similar to the control for heat output which is shown in (17.3), a control method based on the droop characteristics in the modern power systems and the features of hub is investigated to decide the electricity output of hub. The control method can be denoted as:

$$L_e = L_{eN} + k_q \left(f_N - f\right) \tag{17.15}$$

where the electricity output power and nominal electricity output power of EH are represented by L_e and L_{eN}, respectively. k_q is the adjustment coefficient for electricity output. The measured frequency and nominal frequency of electricity network are denoted by f and f_N, respectively. With the existing low-pass filter, the control method for ith EH can be deduced for:

$$\tau_i \dot{L}_{ei} = -L_{ei} + L_{eiN} - k_q \left(f_N - f\right) \tag{17.16}$$

where the time constant of the low-pass filter is denoted by τ_i. Just like (17.4), the adjustment coefficient k_q for each EH has the relationship to share the power proportionally:

$$\frac{L_{eN1}}{k_{q1}} = \frac{L_{eN2}}{k_{q2}} = \cdots = \frac{L_{eNn}}{k_{qn}} \tag{17.17}$$

The electricity loads power can be shared by EHs that utilize the control method shown in (17.3) in view of the characteristic of the frequency in electricity network. However, the frequency of the network will be deviated during the controller operation. Since the grid is sensitive to changes in frequency, the frequency deviation may cause great influence on the security operation of electricity network.

For purpose of compensating the frequency deviation, the dynamic consensus-based frequency control is proposed. Combing with (17.16) and the frequency control, the control of electricity output for EH can be expressed as:

$$\tau_i \dot{L}_{ei} = -L_{ei} + L_{eN} - k_q \left(f_N - f \right) + u_{L_{ei}} \tag{17.18}$$

where $u_{L_{ei}}$ is the input for the electricity output control. To realize the purpose of frequency control, the dynamic consensus-based controller is shown as:

$$\lambda_{f_i} \dot{u}_{f_i} = -\alpha_i e_{f_i} - \beta_i \sum_{j=1, j \neq i}^{n} a_{ij} \left(\frac{L_{ei}}{L_{eiN}} - \frac{L_{ej}}{L_{ejN}} \right) \tag{17.19}$$

$$e_{f_i} = \sum_{j=1, j \neq i}^{n} a_{ij} \left(f_i - f_j \right) + g_{f_i} \left(f_i - f_{\text{ref}} \right) \tag{17.20}$$

in which λ_{f_i}, α_i, and β_i are all the proportional gain in the proposed controller. The mismatch of local frequency tracking for adjacent nodes is represented by e_{f_i}. The gain for virtual leader frequency is denoted by g_{f_i}.

According to (17.19) and (17.20), the controller can be divided into two parts: The first part is to track the frequency mismatches between ith EH and adjacent nodes meanwhile to follow the frequency of the virtual leader. By tracking the frequency mismatches, EHs can spontaneously adjust the outputs until reach the same frequency to make the system stability. Moreover, a virtual leader is built and the frequency of the leader always keep at the reference value. The leader is followed by every hub in order to make the frequency back to the reference frequency f_{ref}. The first part of the controller can ensure the security operation and restore the deviation of the frequency. However, the output of EH cannot be proportionally allocated by only using the first part of the controller, since the u_f for hubs converges to different value. Therefore, the second part of the controller is proposed to achieve the proportional allocation of the EH output as shown in (17.19). The electricity loads will be proportionally shared if $L_{ei}/L_{eiN} = L_{ej}/L_{ejN}$. For each EH, the derivative of u_{f_i} trending to 0 means that the u_{f_i} converges to value that makes the frequency back to reference value and the loads proportionally shared.

The stability of the proposed method can be proven by utilizing the Lyapunov stability theory and the Laplace transformation [11].

In this context, the control of EH outputs based on consensus algorithm can proportionally allocate the outputs of EH without requiring the information of the network parameters. Meanwhile, the system parameters which will be deviated from the nominal values due to the droop action can also be restored by utilizing the control method.

17.3.3 The Control of Devices Based on Improved Equal Increment Principle

The electricity and heat outputs can be determined by utilizing the proposed control method. Nevertheless, only the outputs unable to decide the operating of EH because of the complex coupling of internal equipment. Actually, most objectives of the EH are achieved by adjusting the constituent devices of hub. Since a crucial role of hub is to reduce the energy cost, it is necessary to investigate a control method for devices to decrease the energy consumption.

The equal incremental principle is a classic method which is widely applied in the engineering practices. Based on the characteristics of the devices in the EH, the principle is improved to solve the optimal operation of hub if the output is decided. The improved equal increment principle (IEIP) is shown as:

$$
\begin{cases}
\sigma_{z,r} = \sigma_z^*, \ Q_r^{\min} < Q_r < Q_r^{\max} \\
\sigma_{z,r} \leq \sigma_z^*, \ Q_r = Q_r^{\max} \\
\sigma_{z,r} \geq \sigma_z^*, \ Q_r = Q_r^{\min}
\end{cases}
\tag{17.21}
$$

where Q is the input of inner device. z represents various energies and $\forall z \in e, h$. σ^* is the standard incremental ratio of cost, and σ_r is the incremental ratio of cost for device r.

Generally, there are multiple converters in EH model. As shown in Fig. 17.1, the converters can be divided into two categories: the converter only outputs one kind of energy (COE) and the converter outputs multiple kinds of energies (CME). Different COEs have different consumption. Nevertheless, the fuel cost usually can be expressed by a quadratic form. The consumption function of COE is denoted as:

$$
Q_b \left(L_b \right) = \varepsilon_b L_b^2 + \mu_b L_b + \gamma
\tag{17.22}
$$

where ε, μ, and γ are the cost parameters of the COE. Q_b and L_b state the corresponding input and output of COE b. The incremental ratio of COE can be shown as:

$$
\sigma_b = \frac{dQ_b \left(L_b \right)}{dL_b}
\tag{17.23}
$$

The cost function of CMEs for electricity and heat generation like CHP can be denoted as:

$$
Q_v \left(L_{e,v}, L_{h,v} \right) = \overline{\alpha}_v L_{e,v}^2 + \overline{\beta} L_{e,v} + \overline{\tau} L_{h,v}^2 + \overline{\varepsilon} L_{h,v} + \overline{\lambda} L_{e,v} L_{h,v} + \overline{\gamma}
\tag{17.24}
$$

in which $\left[\overline{\alpha}, \overline{\beta}, \dots, \overline{\gamma} \right]$ stands for the cost parameters for CME. The index for CME is represented by v.

The incremental ratio of fuel consumption function for CME is described as:

$$
\sigma_{z,v} = \frac{\partial Q_v \left(L_{e,v}, L_{h,v} \right)}{\partial L_{z,v}}
\tag{17.25}
$$

The entire incremental ratios of component elements in EH are decided by (17.23) and (17.25). Hence, the elements outputs can be controlled based on the IEIP. Furthermore, each device input can be obtained according to (17.22) and (17.24). As shown in Fig. 17.1, the inputs of EH are equal to the total inputs of the inner devices. Likewise, the outputs of energy share the same feature. As denoted in (17.21)–(17.25), the control approach in terms of the IEIP of microgrids can determine the operation of the devices in EH.

The DACC can be realized by combining the consensus-based output control and the control of devices based on IEIP. In this regard, the complete running process of the microgrids utilized the DACC is described by Fig. 17.2. The EHs obtain the energies from supply side and satisfy the demand side. When loads vary, system parameters of the microgrids like pressure and frequency will be changed in view of the features of the microgrids. The contents of communication between each EH agent also contain the necessary information for the EH control. As shown in Fig. 17.2, the energy supplied by EH needs to go through the corresponding network to satisfy the load. And the parameters of the network lines will influence the results of system operation. The EH controlled by DACC only needs to detect the variations of the parameters and the information of neighbors without the information of the line resistances, which is hard to obtain.

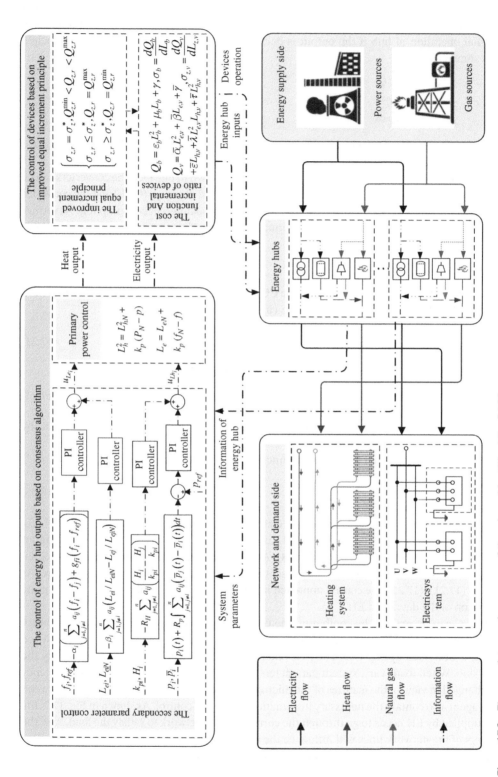

Figure 17.2 The complete running process of the microgrids utilized the DACC.

In the microgrids, different energy networks in the microgrids are coupled by the conversion devices and the energy transmission of each energy network is independent because of the particularity of different network lines. Each energy network has its own characteristics. The power system needs rapid response to ensure the stability. Since some equipment in EH responds slowly, EH will immediately obtain the energy from energy sources like the DG and battery to satisfy the load changes by the control of outputs, so that EH can complete the response in milliseconds and ensure the security operation of the system. At this point, other equipment start to be adjusted based on the control of devices. The inputs of EH will also be regulated. The operation of EH can meet the requirements of internal control and the demand of electricity and heat. In this way, the changes of power load and thermal load will not affect each other and avoid the instability of the system. The energy coupling is reflected in the internal and input terminals of EH.

Security operation is to achieve power sharing and restore the parameters. The DACC can realize the proportional sharing of load supply of the microgrids and reduce the energy consumption based on the EH and the different characters of different kinds of energies. The proposed DACC can proportionally allocate the outputs of EH without requiring the information of the network parameters and adjust the system parameters of microgrids. All EHs will not overload or work in extreme cases so that some problems such as the overheating and excessive wear of EHs can be avoided. Meanwhile, the parameters can be restored to avoid the system instability caused by the deviations of the system parameters. Based on the IEIP which can be proved by Lagrange multiplier method, the issue of reducing energy consumption is transferred to achieve the equal incremental of each device. The constituent equipment of EH can also be appropriately regulated and the energy cost can be reduced by utilizing the IEIP-based inner devices control of DACC.

17.3.4 Example

In order to demonstrate the application effect of cooperative control in microgrid, some examples are given. The microgrid with communication links among EHs is shown in Fig. 17.3. Each EH

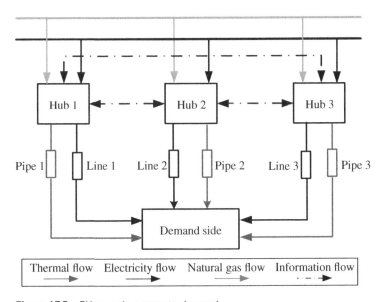

Figure 17.3 EHs supply energy to demand.

Table 17.1 Efficiency and capacity of devices.

Hubs	Devices	ε	μ	γ	Capacity (kW)
1	The gas furnace	0.00097	0.98615	23.38	640
1	The electric boiler	0.00096	0.98825	25.62	640
1	The transformer	0.00061	0.99662	30.13	2800
2	The gas furnace	0.00031	0.87824	22.63	320
2	The electric boiler	0.00025	0.90625	24.33	320
2	The transformer	0.00013	0.93125	28.35	1400
3	The gas furnace	0.00026	0.96922	30.51	160
3	The electric boiler	0.00028	0.95351	22.23	160
3	The transformer	0.00011	0.93652	21.63	700

Table 17.2 Efficiency and capacity of CHP.

Hub	$\bar{\alpha}$	$\bar{\beta}$	$\bar{\tau}$	$\bar{\varepsilon}$	$\bar{\lambda}$	$\bar{\gamma}$	Capacity (e/h) (kW)
1	0.00331	0.98022	0.00311	0.94215	0.00201	21.67	320/520
2	0.00213	0.89365	0.00196	0.79631	0.00184	21.13	160/260
3	0.00221	0.92151	0.00180	0.76286	0.00188	19.82	80/130

Table 17.3 Operation of hubs before the 300th second.

Hub	Output(e) (kW)	Output(h) (kW)	Input(e) (kW)	Input(g) (kW)	Loss(e) (kW)	Loss(g) (kW)
1	1861	897	3753.34	1247.17	1745.73	496.74
2	929	445	1168.11	320.12	84.11	30.14
3	463	224	590.65	174.11	36.19	41.55

shares the same construction as Fig. 17.1. In order to satisfy the requirement of component elements, all EHs are connected to gas and electricity networks.

The capacities and the coefficients of COEs like furnaces and boilers are denoted in Table 17.1. Table 17.2 describes the coefficients and the capacities of CHPs. The reference frequency is 50 Hz and the reference pressure is 1.1 MPa.

All examples share the same loads before the 300 s. There are 3300 kW electricity loads and 1600 kW thermal loads integrated into the energy system. The consequence data before the 300 s is represented in Table 17.3. The consequence data after the 300 s is presented in Table 17.4.

17.3.4.1 Example 1
In this example, the performance of the proposed DACC is shown when the loads decrease. The electricity loads decrease by 630 kW and the thermal loads decrease by 280 kW at 300 s.

Table 17.4 Operation of hubs after the 300th second.

Case/Hub	Output(e) (kW)	Output(h) (kW)	Input(e) (kW)	Input(g) (kW)	Loss(e) (kW)	Loss(g) (kW)
1/1	1541	764	2891.52	993.57	1219.97	360.18
1/2	769	386	963.49	283.33	62.94	28.84
1/3	383	189	495.35	152.01	33.89	41.46
2/1	2179	1096	4771.21	1590.24	2398.32	687.69
2/2	1092	548	1407.41	378.46	113.25	32.65
2/3	545	273	696.16	203.98	40.17	41.97

Tables 17.3 and 17.4 show the energy cost of each EH. More energy like gas and electricity are used by hub 1 to satisfy the loads since the inner devices of hub 1 need more fuel to meet loads as described in Tables 17.1 and 17.2. For other objectives such as carbon emissions, different energies have different effects on the objective. After the corresponding normalization, the DACC based on the IEIP can still be used to resolve the issue. As depicted in Fig. 17.4(a), the thermal and electricity loads are proportionally shared by utilizing the proposed DACC. The stabilization time for electricity outputs is just about 0.2 s. However, the heat outputs need more than 150 s to be stabilized. This difference indicates that the heat supply network spends more time to achieve stabilized state than grid.

The variations in electricity and thermal loads induce the variations in the frequency and the outlet pressure. Based on the DACC, the frequency is restored to the reference value after electricity loads change as shown in Fig. 17.4(b). Similarly, the EHs can also restore the estimated outlet pressure to the reference pressure by using the proposed method as described in Fig. 17.4(c). It validates that the DACC can reach the control objectives successfully without the information of lines.

17.3.4.2 Example 2

The effectiveness of DACC is tested under the situation in which the two kinds of loads both increase. In this case, electricity loads increase by 560 kW and thermal loads increase by 350 kW.

As denoted in Fig. 17.5(a), the thermal and electricity loads are shared upon the corresponding capacity of EH. Other diagrams in Fig. 17.5 expresses that the changing of parameters are negatively correlated with the variations of EH outputs. Meanwhile, all the parameters like frequency and pressure are restored to the reference by utilizing the proposed method. Table 17.3 describes that both the loss of electricity and gas are increased with loads growth. Since the cost of the component element can be expressed by quadratic functions, it means more fuel is used to produce each unit of energy as the load increases. Namely, the loss will also increase.

17.4 Optimal Energy Operation in Microgrids Based on Hybrid Reinforcement Learning

Collaborative control is used to ensure the safe operation of the microgrids at the first time when the load changes, and optimization is to further realize the economical operation of the microgrid.

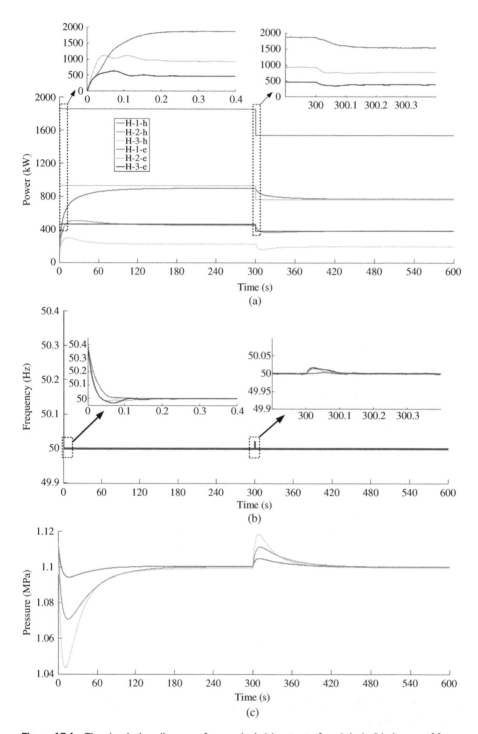

Figure 17.4 The simulation diagram of example 1. (a) output of each hub, (b) change of frequency, (c) change of pressure

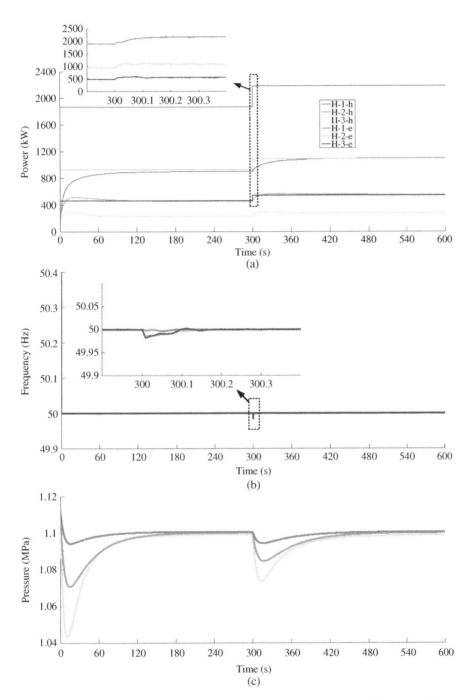

Figure 17.5 The simulation diagram of example 2. (a) output of each hub, (b) change of frequency, (c) change of pressure

17.4.1 Multi-objective Optimization Model Formulation

At present, the structure of microgrids containing multiple EHs tends to be decentralized and flat. Under the development of intelligent function for each EH, EH as an independent individual can intelligently control and make decisions. Each EH combines the energy conversion rules of internal equipment to perform economic operation. However, each EH pursues the maximum benefit in the process of EH operation, which decreases the reliability and security of EH. The objective of maximizing economic benefit will make many devices in EH run in the limit state. In this case, the security margin of the system is greatly reduced, and the failure risk increases.

17.4.1.1 Objective

A multi-objective optimization model is constructed to highlight the importance of the economy and security of EH.

The first objective is considered to minimize the energy cost of each EH at different energy carriers so as to make the best use of renewable energy. In the market mechanism, EH can purchase energy from each energy network. Meanwhile, due to the ability of EH to produce and convert energy, energy networks can also buy energy from energy sources. The total energy operation cost J of the EH in a scheduling cycle T_c is the difference between cost of energy consumption $\overline{C}_t$ and revenue of energy sales $\overline{R}_t$ which is expressed as:

$$J = \sum_{t=1}^{T_c} \left(\overline{C}_t - \overline{R}_t \right) \tag{17.26}$$

The energy consumption cost C_t is the sum of the multi-fuel consumed by EH multiplied by the fuel cost which can be calculated as:

$$C_t = P_{e,t}^{\text{in}} \times \text{Pr}_{e,t}^{\text{in}} + P_{g,t}^{\text{in}} \times \text{Pr}_{g,t}^{\text{in}} + P_{h,t}^{\text{in}} \times \text{Pr}_{h,t}^{\text{in}} \tag{17.27}$$

where variables $\text{Pr}_{e,t}^{\text{in}}$, $\text{Pr}_{g,t}^{\text{in}}$, and $\text{Pr}_{h,t}^{\text{in}}$ stand for the tariff prices of electricity, natural gas and heat, variables $P_{e,t}^{\text{in}}$, $P_{g,t}^{\text{in}}$, and $P_{h,t}^{\text{in}}$ are the supply of energy networks including electricity, natural gas, and heat in an hour.

The income function R_t in an hour is expressed as:

$$R_t = P_{e,t}^{\text{out}} \times \text{Pr}_{e,t}^{\text{out}} + P_{g,t}^{\text{out}} \times \text{Pr}_{g,t}^{\text{out}} + P_{h,t}^{\text{out}} \times \text{Pr}_{h,t}^{\text{out}} \tag{17.28}$$

where variables $\text{Pr}_{e,t}^{\text{out}}$, $\text{Pr}_{g,t}^{\text{out}}$, and $\text{Pr}_{h,t}^{\text{out}}$ are the sale prices for electricity, heat, and natural gas. Variables $P_{e,t}^{\text{out}}$, $P_{g,t}^{\text{out}}$, and $P_{h,t}^{\text{out}}$ are the energy outputs on the network side including electricity, natural gas, and heat.

The tariff prices and the sale prices of energy carriers depend on the real-time prices.

The aim of second objective is to maintain security of each EH based on constraints. On the basis of economic benefits, the operating status of each device in EH will affect the overall security of EH. The security is focused on avoiding the EH working in extreme cases. Therefore, the security metric which considers the operation risk is calculated by the threshold power of each device in EH. It can be presented as follows:

$$G = \sum_i \left| \frac{P_{i,t} - P_{i,\text{max}}}{P_{i,\text{max}}} \right|^{\xi} \tag{17.29}$$

where $P_{i,\text{max}}$ is the maximum power of the ith device in EH. $P_{i,t}$ is the power value of the ith device at time t, ξ is the parameter which stands for the influence level for internal devices on system security.

Above all, the dual-objective optimization model can be formulated as:

$$\min \{ \overline{H} = F_1(S)J + F_2(S)\overline{\omega}G \} \tag{17.30}$$

where $\overline{H}$ is hybrid objective of EH, $F_1(S)$ and $F_2(S)$ indicate the weight function of economic benefits and security operation in optimizing operational strategies. S stands for the system and environmental states. $\overline{\omega}$ is the security corresponding parameter to eliminate the dimensional difference between objectives. The selection of weight has a great influence on the operation effect. With the influence of the environment and the operating conditions of the system, the pursuit of economy and security will change accordingly. In order to adapt to the system requirement, the weights should be adaptively adjusted based on environment and operating status. But existing methods usually set $F_1(S)$, $F_2(S)$ to constants which cannot achieve optimal operation of the system. Therefore, a decision algorithm based on HITL is proposed which is described in detail later and the weight $F_1(S) = \lambda_1$, $F_2(S) = \lambda_2$ can be obtained. In normal circumstances, the human will evaluate the optimization strategy from MCRL and form a knowledge base. Then, Q-learning algorithm is used to optimize the $F_1(S)$, $F_2(S)$ with expert evaluation results. The results of HITL and machine learning are coordinated to determine the final optimization strategy. In abnormal circumstances, HITL will start up and operator needs to participate in adjusting the output of the devices in WE or the target weight to control the state of the system based on the information obtained. We use $\vec{\lambda} = [\lambda_1, \lambda_2]$ to represent the weight vector and the sum of λ_1 and λ_2 is 1.

17.4.1.2 Constraints

The constraints of a EH model should contain the energy balance equality constraints, the technical capacity constraints, and the technical storage constrains.

The energy balance constraints of a EH can be expressed as follows:

$$P_E^{\text{in}} = \sum_m P_{m,E}^{\text{in}} \tag{17.31}$$

where $P_{m,E}^{\text{in}}$ denotes energy equipment such as transformer and electric boiler. E stands for energy carriers including electricity, natural gas, and heat. m expresses the number of energy equipment.

$$P_E^{\text{out}} = L_E^{\text{out}} - P_E^d - P_E^{\text{st}} \tag{17.32}$$

where variables L_E^{out} denotes energy outputs of each EH. P_E^d is the load demands. P_E^{st} is power storage of storage devices. And P_E^{out} is the amount of energy sold to the corresponding energy networks.

The technical capacity constraints corresponding to the operation ranges of each energy device.

$$P_{n_1}^{\text{Con min}} \leq P_{n_1}^{\text{Con}} \leq P_{n_1}^{\text{Con,max}}, \forall n = 1, 2, \ldots, N_{\text{Dev}} \tag{17.33}$$

where n_1 describes the different energy conversion devices in WE. $P_{n_1}^{\text{Con min}}$ and $P_{n_1}^{\text{Con,max}}$ are the lower and upper bounds of the capacity for each energy conversion device.

$$-P_{n_2}^{\text{Charge}} \leq P_{n_2}^{\text{st}} \leq P_{n_2}^{\text{Discharge}} \tag{17.34}$$

where n_2 expresses energy storage devices, $P_{n_2}^{\text{Charge}}$ is the maximum boundary for device charging. $P_{n_2}^{\text{Discharge}}$ is the maximum boundary for device discharging.

17.4.2 Multi-Policy Convex Hull Reinforcement Learning with Human-in-the-Loop

With regard to the microgirds with EHs, each EH aims to achieve the real-time optimal operation based on objective function. Considering the randomness of renewable energy generation and loads in EH, accurate model of environment is difficult to obtain. The problem described above is transformed to stochastic optimization problem which is tremendously practical to solve this problem by leveraging historic trial-and error interactions with the dynamic environment.

According to the model of EH for economic and security requirements, RL is a powerful paradigm to find the optimal decision sequence for EH. In the basic structure of RL, each state is the perception of agent for the environment. The agent can only affect the environment through actions. When the agent performs an action, the environment will transition to another state with a probability. And the environment will feedback a reward to the agent based on the reward function. Through the trial-and-error learning experience to maximize the accumulated sum of rewards, the agent can learn optimal strategy.

In the current multi-objective decision-making model, the limitations of artificial intelligence technology may lead to system decision-making risk and system out of control. A hybrid strategy incorporating human-in-the-loop (HITL) has been proposed to achieve two-way collaboration between humans and machines.

17.4.2.1 Formulation of Multi-Policy Convex Hull Reinforcement Learning

MCRL algorithm is utilized by considering two objectives at the same time and to find a set of policies that approximate the Pareto front. As described above, the hybrid optimal energy operation process can be achieved by Markov decision process (MDP) modeling which is the basic model for MCRL. Therefore, four fundamental elements $(S, A, P, \overrightarrow{r})$ are defined to describe the approach, where S expresses a set of limited environment states, A stands for a set of limited actions, $P(\bar{s}, \bar{a}, \overrightarrow{s}')$ describes the state transition probability sequence from state $\bar{s}$ to state $\overrightarrow{s}'$ and $\overrightarrow{r}$ is a vector of immediate reward signals obtained from environment over time which is different from standard formulation.

In this book, each EH is regarded as an autonomous agent, and the state of system is evaluated after an action without an environmental model. The operating states of internal equipment in EH are considered in optimal operation process. The actions can be defined based on the decision variables of EH. It involves the inputs of each equipment and power storage of energy storage equipment in EH which are is related to the constraints (17.31) and (17.33).

$$\bar{a} = \{P_{m,E}^{in}, P_E^{st}\} \tag{17.35}$$

The reward $\overrightarrow{r} = [\bar{r}_1, \bar{r}_2]$ is designed to express the dual-objective. Economic objective is set as the negative of objective function which means the smaller objection function, the more reward feedback under the condition of satisfying the constraints.

$$\bar{r}_1 = \begin{cases} 0, & \text{if constraints are violated} \\ -J, & \text{otherwise} \end{cases} \tag{17.36}$$

where $\bar{r}_1$ stands for the economic reward obtained through interaction between the WE and the environment.

$$\bar{r}_2 = \begin{cases} 0, & \text{if constraints are violated} \\ \overline{\omega}G, & \text{otherwise} \end{cases} \tag{17.37}$$

where $\bar{r}_2$ means the security operation reward.

Therefore, the reward for an action is set as $\bar{r}(\bar{s}, \bar{a}) = \overrightarrow{\lambda} \cdot \overrightarrow{r}(\bar{s}, \bar{a})$ with some $\overrightarrow{\lambda}$. For a fixed $\overrightarrow{\lambda}$, the traditional Q function iteration method is used to get to the optimal value function. But in this paper, in order to dynamically adjust weights, the MCRL should care about any $\overrightarrow{Q}s$ that are maximal for some $\overrightarrow{\lambda}$. The convex hull operation of the Q values is extended to present the optimal policy over the average expected reward so that all preferences can be solved at once.

A simple maximum is performed to extract the best Q-value for a given $\vec{\lambda}$

$$\overline{Q}_{\vec{\lambda}}(\bar{s}, \bar{a}) \equiv \max_{\vec{q} \in Q(\bar{s}, \bar{a})} \vec{\lambda} \cdot \vec{q} \qquad (17.38)$$

Thus the convex hull value operation iteration rule for EH is defined as:

$$\overset{\circ}{Q}(s, a) = E[\vec{r}(s, a) + \gamma_1 \overset{\circ}{V}(s)] \qquad (17.39)$$

$$\overset{\circ}{V}(s) = \text{hull} \bigcup_{a'} \overset{\circ}{Q}(s', a') \,|\, s, a \qquad (17.40)$$

where s is the current state and s' is the state of the next moment. a is the action at state s, and γ_1 describes the discount factor. $\overset{\circ}{Q}(s, a)$ is defined to express the vertices of the convex hull of possible Q-value vectors for taking action a at state s. The convex hull describes the smallest convex set that contains all of a set of points and mixture policies which lie along the boundaries of the hull are formed. In this way, the set of expected rewards that are maximal for some $\vec{\lambda}$ can be backed up. All optimal policies can be viewed and the preferences can be changed at run time without relearning.

As the state-action space of the energy system is larger and complex, the Q value function stored in the form of look-up table will be large and increases computational cost. One of the promising solutions is to use the fitting to approximate the Q value function which can establish a mapping from the parameter space to the Q-valued function. We use radial basis function neural network (RBFNN) for function estimation. By using generalization performance of RBFNN, the continuous state space problem can be effectively solved. The first layer is the input layer which receives the state s of the system environment. The output layer of the network corresponds to Q value of the state-action pair. Thus, the approximation function of the Q value is calculated as follows:

$$Q(s, a) = \sum_{j=1}^{M} w_j h_j \qquad (17.41)$$

where M stands for the number of radial base neurons in the hidden layer. w_j is the weight between the hidden layer and output layer. h_j means the radial basis activation functions which depends on the distance of the point from the center of that unit. Gaussian function is used here.

$$h(\|s - c_j\|) = \exp\left(-\frac{\|s - c_j\|^2}{2\sigma^2}\right) \qquad (17.42)$$

where c_j is the Gaussian function center point of the ith node of the hidden layer. σ is the width parameter.

In this way, RBFNN parameters can be adjusted and updated using the direct gradient descent method.

$$\Delta w_j(t) = \theta_1 (\vec{\lambda}_t \cdot \vec{r}_t + \gamma_1 Q(s_{t+1}, a_{t+1}) - Q(s_t, a_t)) \frac{\partial Q(s_t, a_t)}{\partial w_j(t)} \qquad (17.43)$$

where θ_1 is learning rate of network weight.

17.4.2.2 Two-Channel Human-in-the-Loop Mechanism

HITL means that the role of human is introduced into the intelligent system to form a hybrid intelligent system. When in a complex dynamic environment, the confidence of the strategy only relying on the above artificial intelligence technology decreases for the current multi-target task, which may lead to decision risk. Human actively intervenes to adjust the parameters to give reasonable decisions, thus forming a feedback loop to improve the level of intelligence. Combining human and machine intelligence systems to form a two-way information exchange and control can effectively utilize human knowledge and optimally balance human intelligence and calculate ability. In order to be able to make correct understanding and decision-making on current multi-objective complex tasks, an improvement strategy combining HITL mechanism with machine learning is proposed to improve the efficiency of the optimization strategy.

The multi-policy decision-making process with HITL is described in Fig. 17.6. In the multi-policy convex hull reinforcement learning mechanism (MCRLM), each EH generates multiple sets of optimal operation strategies according to their different preferences of economic and security objectives through MCRL. A two-channel HITL mechanism (HITLM) is added into optimal decision model to realize real-time regulation of human under abnormal conditions and online assessment under normal circumstances separately. In normal circumstances, the HITLM will conduct a confidence evaluation and determine the confidence of the given decision. Then HITLM will extract new rules from the effect of decision execution and feedback to MCRLM. Thus, MCRLM and HITLM form a circulation loop. In unnormal circumstances, HITLM will directly adjust the system strategy from MCRLM. The MCRLM and HITLM coordinate with each other to improve the accuracy of the system decision.

The one channel of HITL is designed to evaluate the selected optimization strategy of the system from MCRL each time under normal circumstances. According to the objective function, multiple sets of policy information will be transmitted to the monitoring platform and will continue to appear on the display at the operator side. A knowledge base will be built based on expert evaluation of decision-making effectiveness. The human will evaluate the optimization strategy selected by system every time and score each strategy to form a knowledge base. Then, the score of each strategy $(Sc_{\pi_1}, Sc_{\pi_2}, Sc_{\pi_3}, \ldots, Sc_{\pi_n})$ from the human is applied to guide the next decision. In the decision mechanism, Q-learning algorithm is applied to further use human evaluation to correct machine learning algorithm. Using Q-learning algorithm to extract new rules to update the machine learning system. The action is set as the multi-target correspondence preference $A(t) = \{\lambda_1(t), \lambda_2(t)\}$ and the reward is the human score $R(t) = Sc(t)$. For each state-action pair, a Q function is used to update Q matrix elements to get the optimal strategy.

$$Q_{k+1}(S, A) = Q_{k+1}(S, A) + \theta_2[R_k(S, A) \\ + \gamma_2 \max_{A'} Q_k(S', A') - Q_k(S, A)] \tag{17.44}$$

where (S, A) stands for the state-action pair in time slot k. (S', A') is the possible state-action pair in the next time slot. $R_k(S, A)$ is the immediate reward with action A at state S on time slot k. θ_2 is the learning rate and γ_2 is the discount factor. Through the decision mechanism, the optimal decision from the human is obtained. HITL and MCRL coordinate to enhance the decision ability of MCRL and realize the safe and economic operation of WE. If the strategy is consistent with machine decision, the system will continue to execute the strategy by machine. If not, the system will change the strategy according to the decision mechanism.

The other channel of HITL is to perform real-time regulation under abnormal conditions by the monitoring and adjustment platform. The operating conditions of the system have changed and MCRL cannot make decisions with high confidence, system operators or experts will rely on their knowledge and experience to make HITL decisions. In this case, if the operator does not supervise the decision, the decision system will select the strategy which is set by machine controller in advance. However, once an emergency occurs e.g. a device fails, extreme weather or the system has specific requirements such as equipment overhaul, the direct selection strategy may result

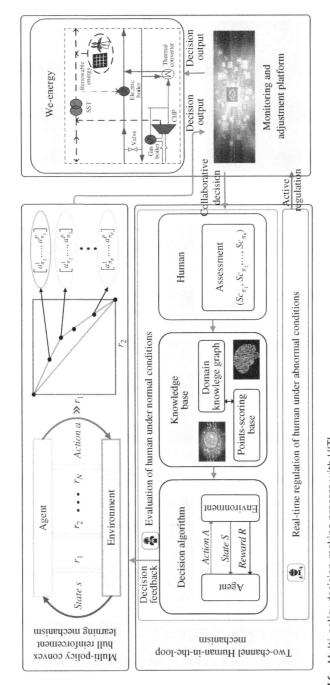

Figure 17.6 Multi-policy decision-making process with HITL.

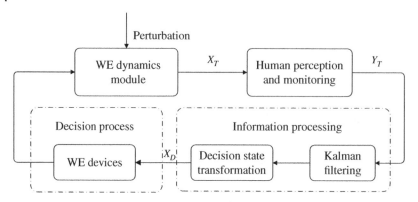

Figure 17.7 The structure of human decision module.

in lower security goals of the system and is not suitable for the operational requirements of the current system. Therefore, in emergency mode, HITL mechanism will start up and operator need to participate in adjusting the output of the devices in WE or the target weight to control the state of the system based on the information obtained. Then, according to the instructions of the control center, the information is fed back to the information link. At this time, the control center of microgrids unifies the economic benefits of EH.

Moreover, in order to improve the decision-making efficiency of human in the proposed algorithm, the decision time T_A of an operator is determined by the security risk of the current strategy. As security risk decreases, the decision time available for people becomes longer. Thus the system will evaluate the current system status and give the decision time based on 17.45.

$$T_A = \beta G^{\frac{1}{2\alpha}} + T_K \tag{17.45}$$

where α, β are the parameters determined by the actual situation. T_K is the fundamental decision time constant.

Above all, the structure of HITL module is expressed in Fig. 17.7. X_T stands for the policy status output by EH dynamics module. State variable X_T is converted to Y_T by the human perception and monitoring module. Then the decision variable X_D is obtained through the Kalman filtering module and decision state transformation module. The decision variable X_D is described as:

$$X_D = [T_R, T_A]^T \tag{17.46}$$

where T_R is the policy state variable which describes the time required to complete the task. And T_A is the processing time allowed by the task.

The MCRL with HITL will dynamically select multi-target tasks which further affects the overall system performance optimization. The algorithm flow of the MCRL with HITL is shown in Table 17.5.

The assessment channel of HITL mainly integrates the analysis of the current system by human with the machine intelligence system to improve the confidence of the overall system. The regulation channel of HITL allows human actively to participate in the adjustment to give a reasonable strategy for emergencies in which human has high decision-making authority.

17.4.3 Example

17.4.3.1 System Initialization

In order to evaluate the performance that the proposed MCRL with HITL method can be used for the optimal energy operation for EH in microgrids. The internal energy conversion equipment of

Table 17.5 Algorithm flow.

Algorithm MCRL Algorithm With HITL

Input: learning rate θ_1, θ_2, discount rate γ_1, γ_2, exploration factor ε_1, ε_2, weight of objective function λ_1, λ_2, Gaussian density parameter σ, RFBNN weight w

(1)   **for** $t = 0, 1, 2, \ldots$ **do**

(2)   Initialize state s

(3)   Compute the π for WE

(4)   Take action a_t and observe the new state s'

(5)   Calculate $\overset{\circ}{Q}(s, a)$ by 17.39–17.42, $\forall a \in A$

(6)   Updates network weight by 17.43

(7)   **until** s is a terminal state

(8)   **end for**

(9)   Output π by controller

Normal conditions

(10)   Initialize Q matrix, state $\forall S$, $\forall A$

(11)   **for** $T = 0, 1, 2, \ldots$ **do**

(12)   Execute action A, observe next continuous state S' and reward R

(13)   Update Q-value by 17.44

(14)   **until** S is a terminal state

(15)   **end for**

(16)   Selection the λ_1, λ_2 by HITL and Output π'

Abnormal conditions

(17)   **if** π' is not confident

(18)   Adjust the outputs of the devices or λ_1, λ_2

(19)   **end**

EH includes a transformer, a boiler, a gas furnace, and a CHP. There is a photovoltaic unit accessed in EH. Thus the inputs of EH in the test are electricity as well as natural gas and the outputs of EH include electricity and heat. We consider the operation horizon from 0:00 AM to the next 0:00 AM. Figure 17.8 depicted the load demands of electricity and heat over 24 hours. Moreover, Fig. 17.9 shows the photovoltaic unit corresponding to the EH for a working day. Natural gas price is 0.17 RMB/m³. Heat price is 0.6 RMB/kWh. The electricity price is a curve that fluctuates with time as shown in Fig. 17.10.

For the proposed MCRL above, the exploration factor ε was set to 0.1, the learning rate θ_1 and θ_2 were set to 0.05 and the discount rate γ_1 was set to 0.99 and γ_2 was set to 0.95. The learning factor θ_1 and θ_2 stand for the learning level in each iteration. Therefore, in order to achieve better overall rewards, it can be set as a smaller value to gain long-term benefits of the algorithm.

Considering the weight $\vec{\lambda}$ in the dual-objective function of EH, we divided the value of λ_1 and λ_2 into 10 gears according to actual system preferences which is from 0.1 to 1 and the interval is 0.1 in MCRL with HITL.

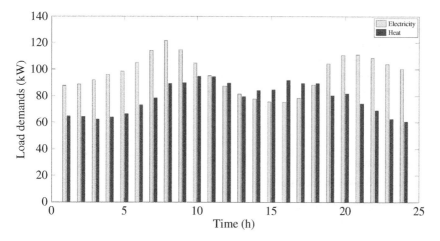

Figure 17.8 Demand power of electricity and heat over 24 h.

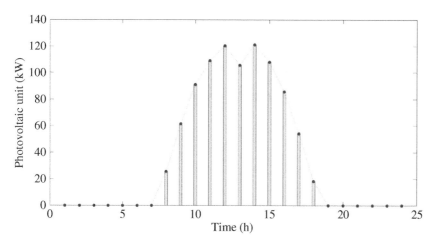

Figure 17.9 The photovoltaic unit corresponding to the WE for a working day.

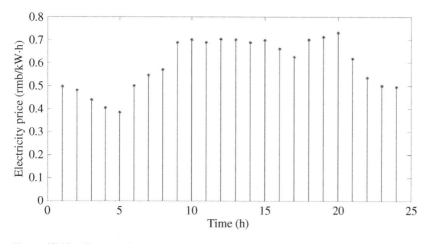

Figure 17.10 The electricity price for a working day.

17.4.3.2 Example 1

In example 1, we apply the MCRL with HITL approaches to complete optimization task of EH under normal conditions. After executing the simulation, EH can converge to the maximum reward. The converge result of the HITL algorithm under normal conditions is shown in Fig. 17.11. It can be seen that the EH chose poor actions yielding lower rewards at the outset, however, the reward increased as the EH discovered the actions yielding higher rewards by learning them through trials and errors with each successive iteration and achieving the maximum reward after 1800 steps finally.

The operating status of EH with HITL has been expressed by Fig. 17.12. Figure 17.12(a)–(c) show the electricity, natural gas, and heat energy interaction with corresponding network during 24 h. Figure 17.12(d)–(f) describe the 3D map of operation states with paces over 24 h. Different colors represent different values. It can be seen that the 24-hour results reach convergence after 1800 steps. Figure 17.12(g)–(i) show energy interaction with corresponding network with paces in the 5th hour. The convergence results of the section in the y0z axis of Fig. 17.12(d)–(f) correspond to Fig. 17.12(a)–(c). Figure 17.12(g)–(i) correspond to the quantities in the form of paces that select the x0z section of Fig. 17.12(d)–(f) at the 5th hour. From Fig. 17.12(a)–(c), it is apparent that the full-duplex transaction mode of WE has been expressed in the operation process. Noted that from the 9th hour to the 16th hour, the amount of electricity purchase has been reduced and the amount of electricity sale has increased which is due to the power outputs of solar energy increasing. The modes that purchasing the natural gas as well as heat and selling electricity are operated for EH.

In order to show the performance of the HITL algorithm under normal conditions, the MCRL with HITL approach are compared with the traditional RL optimization. The traditional method is computed with the same structure of EH to solve the optimization problem described in section III. The λ_2 in the dual-objective function was selected to 0.5 in the traditional RL without HITL.

Figure 17.13 shows the security weight λ_2 over 24 h with HITL and the security of EH with HITL and without HITL. Once the maximum Q-value is obtained, the optimal energy consumption of each device can be determined. The security weight is obtained by comprehensive evaluation of human based on environmental factors such as electricity price, load demands, and solar power generation.

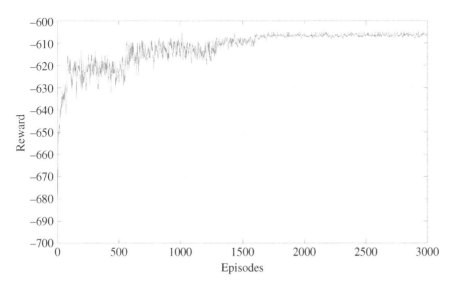

Figure 17.11 Convergence of the reward for a working day.

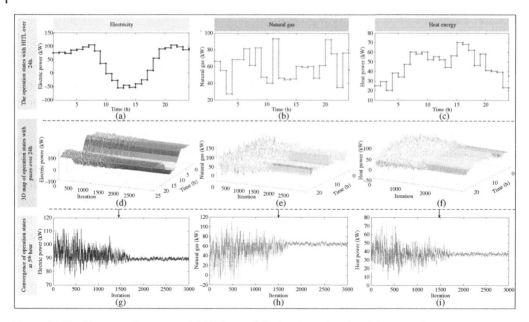

Figure 17.12 The operation states with HITL over 24 h. (a–c) are the electricity, natural gas and heat energy interaction with corresponding network during 24h respectively. (d), (e) and (f) are the 3D map of operation states of electricity, natural gas and heat with paces over 24h respectively. (g), (h) and (i) are the electricity, natural gas and heat energy interaction with corresponding network with paces in the 5th hour respectively.

Based on the above results, the single most striking observation to emerge from the security comparison is that when the security weight in HITL is greater than the security weight in traditional method, the security of the traditional method is generally lower than the security of HITL approach. Otherwise, in the period that the security weight in HITL is lower than the security weight in traditional method, it can be seen the security of the traditional method is higher than the security of HITL approach. At the 12th hour, the security weight in HITL approach is 0.5 which is same as the weight set by traditional method, the security of the traditional method and HITL are very similar.

Meanwhile, in order to compare the effect of the proposed method, we select the most used equipment (MUE) of all EH equipment for comparison. Figure 17.14 shows the usage percentage of MUE in EH with HITL and without HITL over 24 h. The equipment utilization rate to reach 80% is chosen as a warning line. It can be clearly seen that in the case of no HITL, the utilization rate of MUE was higher than 80% for many times, such as the 14th, 17th, 19th, 21st hours and even reached full operation at the 14th hour. The method with HITL effectively improved the security of equipment operation. In addition, the cumulative security for a day of HITL approach and traditional method are 25.6 and 24.6, respectively. The overall security indicators of the two methods are relatively close, but HITL approach can control the operation of the equipment at more critical times.

Combined with the weight result of Fig. 17.13, in Fig. 17.14, at some moment such as 4th hour to 8th hour, the security weight of method with HITL is lower than the method without HITL and the usage percentage of MUE in the method without HITL is less than the method with HITL. It can be seen that at this time, the output of the equipment is low, and the method with HITL can ensure the security operation of the system while more in pursuit of economic.

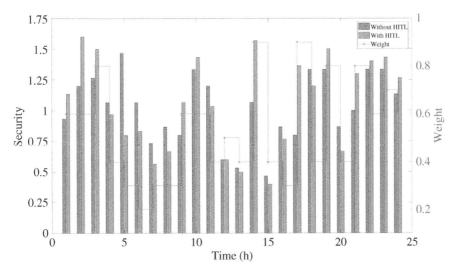

Figure 17.13 λ_2 with HITL and security of EH during a day with HITL and without HITL.

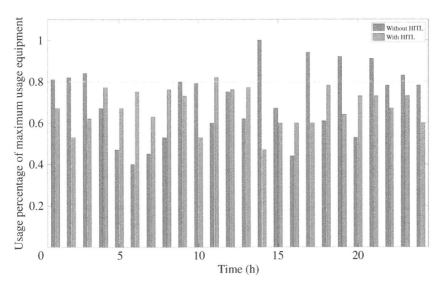

Figure 17.14 The usage percentage of MUE in EH with HITL and without HITL over 24 h.

Therefore, online assessment strategies of human under normal conditions can effectively reduce system risk.

17.4.3.3 Example 2

We deal with the abnormal conditions of microgirds in example 2. We choose a day with abnormal conditions that the boiler in EH malfunctioned at 13th, 14th, and 15th hour. Operating status of CHP during a day under abnormal conditions with HITL is shown in Fig. 17.15.

Based on the above results, it can be seen obviously that at 13th, 14th, and 15th hour, the electrical power of CHP is 28 kW which is set by human. In the example that the boiler cannot run, human first consider adjusting the internal equipment of EH to satisfy the load demand. So operating

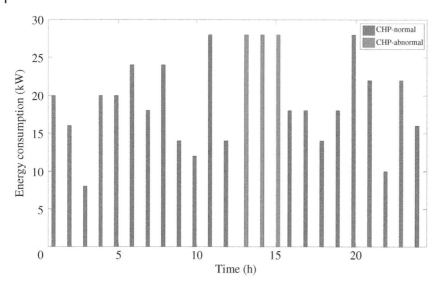

Figure 17.15 Operating status of CHP during a day under abnormal conditions with HITL.

power of CHP is increased to a larger value and exchange power with the heating network as well as equipment such as energy storage will change accordingly.

Above all, simulation results revealed that the improved RL optimization algorithm with HITL has a better performance for solving the optimization operation of EH while avoiding decision risk.

17.5 Conclusion

This chapter focuses on the AI-enabled cooperative control and optimization of microgrids with multi-energy complement. For cooperative control, an DACC that employs both consensus algorithm and IEIP is proposed. The deducing process of consensus-based control of EH outputs has been investigated. The control of outputs enables EHs to accurately share the loads power without knowing the information of the line resistances which is hard to know. Furthermore, the proposed control strategy realizes the restoration of frequency and average pressure to recover the changing of system parameters induced by the droop action. For intelligent optimization, a dual-objective optimal model that can drive each energy unit to have a proper trade-off between operation costs and operation risks is proposed. The presented MCRL with HITL mechanism algorithm makes the artificial intelligence technology and manual intervention adapt to each other and work together to form two-way information exchange and control, such that each EH could undertake the task independently to reach the coordinated system, especially considering the randomness of renewable energy sources and effectively avoid decision-making risks caused by the limitations of artificial intelligence technology.

17.6 Exercises

1. Please try to prove the stability of the proposed method according to the reference.
2. Communication is the core of collaborative control. Please try to think how to reduce communication frequency.

3. Please describe the advantages and disadvantages of model-free reinforcement learning and model-based reinforcement learning.
4. Please describe the advantages of reinforcement learning in energy system optimization decision-making.
5. For goal-driven reinforcement learning, the goal is to achieve a certain state. When the energy system optimization decision task has multiple objectives, please try to set the reward function for the task, and discuss the function of different reward function.

References

1 Bidram, A. and Davoudi, A. (2012). Hierarchical structure of microgrids control system. *IEEE Transactions on Smart Grid* 3 (4): 1963–1976.

2 De Brabandere, K., Bolsens, B., Van den Keybus, J. et al. (2007). A voltage and frequency droop control method for parallel inverters. *IEEE Transactions on Power Electronics* 22 (4): 1107–1115.

3 Simpson-Porco, J.W., Dörfler, F., and Bullo, F. (2013). Synchronization and power sharing for droop-controlled inverters in islanded microgrids. *Automatica* 49 (9): 2603–2611.

4 Guo, F., Wen, C., Mao, J., and Song, Y.-D. (2014). Distributed secondary voltage and frequency restoration control of droop-controlled inverter-based microgrids. *IEEE Transactions on Industrial Electronics* 62 (7): 4355–4364.

5 Simpson-Porco, J.W., Shafiee, Q., Dörfler, F. et al. (2015). Secondary frequency and voltage control of islanded microgrids via distributed averaging. *IEEE Transactions on Industrial Electronics* 62 (11): 7025–7038.

6 Zhang, H., Kim, S., Sun, Q., and Zhou, J. (2016). Distributed adaptive virtual impedance control for accurate reactive power sharing based on consensus control in microgrids. *IEEE Transactions on Smart Grid* 8 (4): 1749–1761.

7 Wei, Q., Lewis, F.L., Sun, Q. et al. (2016). Discrete-time deterministic Q-learning: a novel convergence analysis. *IEEE Transactions on Cybernetics* 47 (5): 1224–1237.

8 Huang, Z., Xu, X., He, H. et al. (2017). Parameterized batch reinforcement learning for longitudinal control of autonomous land vehicles. *IEEE Transactions on Systems, Man, and Cybernetics: Systems* 49 (4): 730–741.

9 Wu, H., Song, S., You, K., and Wu, C. (2018). Depth control of model-free AUVs via reinforcement learning. *IEEE Transactions on Systems, Man, and Cybernetics: Systems* 49 (12): 2499–2510.

10 Ding, L., Li, S., Gao, H. et al. (2018). Adaptive partial reinforcement learning neural network-based tracking control for wheeled mobile robotic systems. *IEEE Transactions on Systems, Man, and Cybernetics: Systems* 50 (7): 2512–2523.

11 Qu, Z. (2009). *Cooperative Control of Dynamical Systems: Applications to Autonomous Vehicles*. Springer Science & Business Media.

18

DNN-Based EV Scheduling Learning for Transactive Control Framework

Aysegul Kahraman and Guangya Yang

18.1 Introduction

Interest in distributed energy sources (DES), including electric vehicles (EVs) and alternative technologies such as renewable generation, fuel cells, and combined heat and power, has increased in recent decades due to a growing desire to reduce greenhouse gas emissions. The growth of EVs and interest in these technologies have been strong due to the lower cost of electricity, which varies throughout the day and provides opportunities to reduce charging costs and adjust grid usage to avoid peak loads. To mitigate the effects of climate change, the European Union (EU) has taken actions to decrease its emissions through implementing regulations. In 2021, the number of EVs doubled from the previous year [1]. The EU is working to decrease emissions by implementing regulations, with the goal of reducing CO_2 emissions from new cars by 55% by 2030 and 100% by 2035, compared to 2021 levels. This shift highlights the growing role of EVs as part of the main grid. With all the changes, traditional grid operation necessitates a change, as do smart grid implementations.

Transactive energy management (TEM) is a technology that optimizes energy generation, distribution, and consumption through market-based mechanisms and advanced systems. TEM aims to create a more efficient and resilient energy system by enabling real-time management of energy resources. TEM systems can help utilities, grid operators, and other energy market participants make informed decisions by integrating DES, as the main focus of this study, EVs specifically. Transactive control optimizes the use of DERs and supports the reliability of the electric power system by reducing greenhouse gas emissions. It has several advantages for the electric power system and has been proposed for demand response, renewable energy integration, and distribution network management. Transactive control allows for DES integration into the main grid and economic and systematic optimization by deriving dynamic market price bids, which makes it more advantageous than other grid operation solutions [2]. The contribution of transactive control is investigated for high DES penetration into power systems using one-time and iterative information exchange [3]. Another study focuses on the sharing of energy between EVs and solar panels in a parking lot, using TEM to optimize energy use and reduce costs, making it possible to integrate more renewable energy sources into the power grid [4]. Transactive control in commercial building heating, ventilation, and air conditioning (HVAC) systems allows for real-time monitoring and control, achieving similar goals [5]. Additionally, transactive control extends equipment lifetime and reduces active power losses [6].

The first part of this study was completed previously [7], which was proposing transactive control for EV charging schedules with the introduction of a non-beneficiary price coordinator to converge on a reasonable point with the iterations between distribution system operators (DSOs) and aggregators. Since this research area has been pursued meaningfully in recent years and found promising results [4–8], we continue to use the same method and improve the weaker sides with the addition of deep neural networks (DNNs). Despite its potential and benefits, transactive control also faces limitations, including uncertainties in renewables, load, electricity price, and user behaviors, which negatively impact the reliable, secure, and efficient operation and management of the power system, including EV charging schedules. Another challenge is finding a solution in a limited amount of time. The overall approach to solving this problem is to predict the uncertain inputs. However, deterministic scheduling solutions that are based on point forecasts are not sufficient against stochastic solutions. The solution to point forecasts is to use multiple scenarios to account for the possibility of the uncertain inputs being realized [8]. Because even solving one deterministic problem has a relatively long solution time with the possibility of not converging and producing a good scenario set is not an easy and straightforward job to do, the nature of transactive control does not allow us to employ multiple scenarios with a time constraint. To address these uncertainties, machine learning techniques, primarily artificial neural networks, are proposed as a solution since they allow for learning.

In various fields, including the electric power system, deep learning methods are gaining attention for tasks such as renewable energy generation forecasting, load forecasting, disturbance detection and classification, fault detection, energy management, and optimization [9]. Due to the increasing global interest in deep learning and the availability of more data, despite its uncertain nature, the technology employing deep learning (DL) is being increasingly investigated. Although traditional approaches exist to solve these problems, DNNs have proven their effectiveness in capturing simple and complex nonlinear relationships. One mature area of using DL is predicting stochastic inputs to address problem uncertainties, while other cutting-edge applications include demand-side management, failure analysis, cyber security, and addressing economic dispatch and planning under uncertainty [10–12]. In the smart grid, the three-state energy approach is implemented with expandable deep learning for multiple output real-time economic dispatch [13]. Another study solves the energy dispatch problem by using a stochastic optimization algorithm to schedule energy production and consumption and a machine learning algorithm to predict the power output of renewable energy sources and charging patterns of EVs in industrial zones with renewable energy sources and EVs [14]. Another study balances energy supply and demand in the electricity market by using network-constrained transactive energy, operating with an aggregator in the energy market, and minimizing costs by trading flexible energy without causing network issues [15].

In this chapter, we propose a DNN-based scheduling of EV charging to overcome the limitations of the transactive control approach in the day-ahead energy market. Our main contribution is a DNN-based learning framework for EV charging schedules in the day-ahead market which takes into account the uncertainty of the problem by using transactive control solution data. We evaluate the performance of the proposed DNN with respect to all power system and EV constraints in a case study, and observe that there were no violations in the entire test set, proving that the proposed solution can efficiently solve energy management and scheduling problems within the limited time frame.

18.2 Transactive Control Formulation

This section covers the proposed methodology for a transactive control model. This methodology is largely derived from [7] to produce the data set that will be used for the learning frame that is given in the Section 18.3. The method utilized in this study can be subdivided into three components:

(1) the price coordinator, (2) the aggregators, and (3) the DSO are the main components of the transactive control for finding the non-violating charging scenario of EVs. These three components interact with each other and produce a data set. The transactive control system with network constraints for EV integration is given in Fig. 18.1.

The proposed system coordinates aggregators, which have several EVs, a DSO, and the price coordinator, to eliminate grid congestion and prevent voltage violations while finding a charging schedule for EVs. The reason for checking the grid congestion is that the aggregator aims to minimize the EVs' charging costs with respect to the day-ahead market prices. The priority of using the grid when the market prices are lower can create congestion or violations on the external grid. Thus, the price coordinator, which is a nonprofit entity, helps the aggregator adjust the initial charging schedule with respect to the DSO response based on grid congestion and voltage limits on each bus. The price coordinator produces a shadow price, which organizes each component of the current system. To be able to explain further, mathematical formulas are given based on the previous work [7].

The whole procedure starts with aggregators generating the initial charging schedule of EVs by using linear programming solver. For the sake of simplicity, reactive power is not considered here and assumed the increment in reactive power is zero since it is relatively smaller in low-voltage networks.

The objective of the aggregator is minimizing charging cost by having earlier assumptions such as knowing the arriving details and daily path of EVs, having a proper battery knowledge for state of charge (SoC) and charging behavior, using the day ahead market prices as they are without focusing

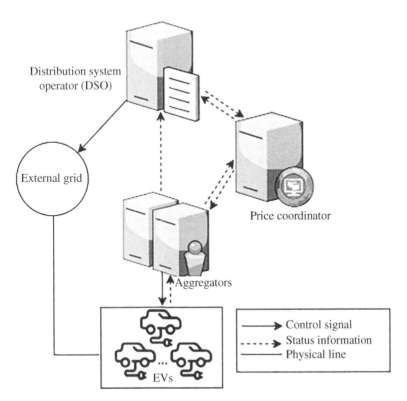

Figure 18.1 A network-constrained transactive control operation for EVs with price coordinator and distribution system operator.

on degradation cost. Aggregator minimization at step ω is given in Eq. (18.1).

$$\text{min} \quad \left(\sum_{k=1}^{N_F} \sum_{i=1}^{N_T} \sum_{l=1}^{N_B} C_{k,i,l} \left(\tilde{P}_{k,i,l} - P_{k,i,l}^E \right)^2 + \sum_{i=1}^{N_T} \sum_{l=1}^{N_B} \lambda_{\omega}^*(i,l) + \sum_{k=1}^{N_A} \tilde{P}_{k,i,l} \right) \tag{18.1a}$$

$$\text{s.t.} \quad \sum_{i=1}^{N_T} \tilde{P}_{k,i,l} t_i = \sum_{j \in I} \left(\text{SOC}_{\text{max},j} - \text{SOC}_{0,j} \right) E_{\text{cap},j} \tag{18.1b}$$

where,

N_F	Number of aggregators
N_T	Number of time slot in the scheduling period
N_B	Number of buses
k	Index for number of aggregators, $k = 1, \ldots, N_F$
i	Index for time slot in the scheduling period, $i = 1, \ldots, N_T$
l	Index for bus of the distribution network, $l = 1, \ldots, N_B$
j	Index for number of EVs under the aggregator, $i = 1, \ldots, N_k^E$
t	Length of each time slot
N_k^E	Number of EVs under aggregator k
$P_{k,i,l}^E$	Summation of $P_{j,i}$ of aggregator k, in time slot i, at bus l
$C_{k,i,l}$	Weighting factor referring the power difference
$\tilde{P}_{k,i,l}$	Optimization variable of aggregator k, in time slot i, at bus l
$\text{SOC}_{\text{max},j}$	Maximum SOC of EV j at the end of the charging period
$\text{SOC}_{0,j}$	Initial SOC of EV j
$E_{\text{cap},j}$	Capacity of EV j
λ_{ω}^*	Converged lagrangian multiplier

The prior relation between the variables are $P_{k,i,l}^E = \sum_{j \to l} P_{j,i,l}, k = 1, \ldots, N_F, i = 1, \ldots, N_T, l = 1, \ldots, N_B$ and $\sum_{k=1}^{N_A} \tilde{P}_{k,i,l} = P_{\text{trans}}(i,l), i = 1, \ldots, N_T$ to make sure will be checking with respect to buses.

The first part of the aggregator objective function tries to decrease the difference between optimized schedule and optimization variable which corresponds to desired power schedule by multiplying with the $C_{(k,i,l)}$ weighting factor. While the second part of the equation is referring the lagrangian multiplier, the last part is corresponding to the aggregator. The constraint of the aggregator shown in Eq. (18.1b) ensures that the requested energy will be available to each individual EV at the end of the charging period.

DSO minimization at step ω is written by Eq. (18.2).

$$\text{min} \quad a \sum_{i=1}^{N_T} \sum_{l=1}^{N_B} \left(P_{\text{trans}}(i,l) - \sum_{k=1}^{N_A} P_{k,i,l}^E \right)^2 + b P_{\text{loss}} - \sum_{i=1}^{N_T} \sum_{l=1}^{N_B} \lambda_{\omega}^*(i,l) P_{\text{trans}}(i,l) \tag{18.2a}$$

$$\text{s.t.} \quad \sum_{l=1}^{N_B} P_{\text{trans}}(i,l) \leq \overline{P}_{\text{trans}}(i) \tag{18.2b}$$

$$U_0(i,l) + \Delta U(i,l) \geq U_{\text{min}}(i,l) \tag{18.2c}$$

where,

a, b	Weighting factors
$P_{\text{trans}}(i, l)$	Optimization variable, desirable power of DSO
$\overline{P}_{\text{trans}}$	Maximum power transformer for the aggregators
$U_0(i, l)$	The initial voltage of the buses
$U_{\min}(i, l)$	The minimum allowable voltage of the buses

The part which has multiplication by the weighting factor as specified by Eq. (18.2a) represents that the operational restrictions such transformer thermal capacity and voltage limitations, the DSO's goal is to monitor and manage the aggregators' power schedule while minimizing network losses. $P_{\text{trans}}(i, l)$ can be found by using CVX and MATPOWER together. Equation (18.2b) is the constraints for DSO to be able to keep the network constraints of power transfer capacity and voltage.

Price coordinator function is given by Eq. (18.3) for updating the lagrangian multiplier for the next step through the iterations.

$$\lambda_{\omega+1}(i, l) = \lambda_{\omega}^*(i, l) + \alpha_{\omega} \left(\sum_{k \in l} \tilde{P}_{k,i,l}^* - P_{\text{trans}}(i, l)^* \right) \tag{18.3}$$

where ω is the index for every iteration, $\tilde{P}_{k,i,l}^*$ is the convergence of optimization variable, $P_{\text{trans}}(i, l)^*$ is the one found in DSO minimization, and α_{ω} is the positive constant which represents the step size for leading the next iteration for finding the converged λ for every for in each time step.

The prior steps until writing the final objective functions and constraint of each main component of this problem can be found in the earlier work in [7] which has been adopted here, the formulation is kept as summarized since this part is already studied in detail. The main issue with transactive control is the time for finding a converged solution in a reasonable time. To handle this issue, we apply to DNN by aiming to decrease the solution time and prepare the solution framework for larger systems.

18.3 Proposed Deep Neural Networks in Transactive Control

Artificial intelligence (AI) has proven its value in many fields. With this, DNNs have emerged as a valuable solution for dealing with nonlinear complex systems by learning the relationship between given input and output via their layers [4]. Even though this learning might be seen as a black box, there is still a need for good judgment in model selection, adjustment, training, data processing, and more. In contrast to traditional models, DNNs can be personalized based on the problem itself. The DNN structure used in this study includes a multi-input, multi-output (MIMO) design (multi-step ahead forecasting) with more layers (one input, one output, one concatenate layer, and many hidden layers since the size of the input and output data is relatively larger). Employing MIMO strategy is valuable especially for multistep forecasting in supervised learning [16–18]. Figure 18.2 shows each layer of the created structure as well as the dimensions of input and output sizes per sample.

In general, the estimation of the weights is performed by minimizing a loss function, which is taken as the mean squared error here. For the weight estimation, a back-propagation algorithm that uses a steepest-descent technique based on the computation of the gradient of the loss function is employed. Finally, the sigmoid function is used as the activation function by keeping all data in the

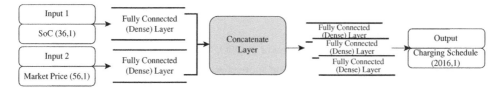

Figure 18.2 Structure of the proposed DNN for learning charging schedule.

interval [0, 1]. Adam optimizer is selected. For the stochastic variables here, explanatory variables and historical values based on both the available data and trial and error are needed to achieve better learning performance. In detail, the artificial neural networks (ANN) inputs are vectors of electricity price, represented as a matrix with dimensions 1 row and 56 columns (denoted as 1×56), and initial State of Charge (SoC), represented as a matrix with dimensions 1 row and 36 columns (denoted as 1×36). The notation 'N $\times$ M' is used to represent a matrix with 'N' rows and 'M' columns. Even though this input is the same for all scenarios, we still feed it as an input; however, it did not show any progress with this addition.

As can be seen from Fig. 18.2, since the dimensions of the data sets vary and some are vectors while others are matrices, we flatten the matrices before inputting them into the model. For instance, a matrix with dimensions 96×18 is flattened to become a vector of 1728. We then create two different individual inputs, combine them after adding a fully connected layer for each input, and continue to add more fully connected layers until reaching the output layer. The output matrix has dimensions 56×36, which corresponds to charging for 14 hours at 15-minute intervals for each car (each aggregator manages 18 cars). Thus, the output is flattened into a vector of 2016×1, making the model ready for learning. DNNs take longer time, depending on the size of the training set, due to the complexity of the network and adjusting the weights through the layers.

We created the neural networks (NN) model to take inputs and learn the relation between inputs and outputs, which is not a direct and easy task to do, and it requires some assumptions to be able to solve the problem. The assumption is that all EVs have the same initial SoC. Even though this is not mandatory, it is easily implementable with the different initials, and it increases the potential number of scenarios. Here, we initially want to see that our approach contributes to the solution getting implemented. It is later possible to increase the number of scenarios with respect to different initial conditions and loads or to include more input. One reason for not seeing any progress by including the driving path is that it might have the same path or demand every day, so it did not bring any improvement throughout the training. We run the TEM code to find the EV charging schedule by minimizing the cost as well as preventing grid congestion and voltage violations, which also refers to the data that will be used with DNN. During the training, the created DNN model is run, and the weights are adjusted by minimizing the mean squared error (MSE). After completion of the training, we tested our model with different inputs and saved the test results to see if the schedule obtained through DNN would be suitable for network constraints in terms of any violations. Also, a general comparison between the NN and TEM schedule performances is made in terms of solution time, cost, and meeting demand; further details are under the case study and simulation results. The total sample size is now up to 900. About 80% is used for training, and 10% is used for validation. The rest of the 10%, which is equal to 180 samples, is used for testing.

18.4 Case Study

In this section, we apply the transactive control approach for finding the EV charging schedule with two aggregators, each of which has 18 EVs assigned. After running the problem many times with respect to different initial SOCs, which are in the range of 10–100%, and day-ahead market prices, the solution, which has no congestion or voltage violations caused by the charging, has been collected to train the DNN, which is constituted to learn how to make a schedule for charging EVs without having any congestion or violations.

The performance of the proposed DNN-based policy is evaluated using transmission networks of varying sizes. The simulation scripts are written in Python and run on a laptop computer with a 2:4 GHz 8-Core Intel Core i7 processor and 64 GB of random access memory (RAM). The training data is produced by CVX, a package for specifying and solving convex programs within the parallel usage of MATPOWER, a MATLAB power system simulation package. The distribution network with the connected EVs is taken from the previous study and given in Fig. 18.3. The scheduling period is between 16:00 and 06:00 with a 15-minute interval, 56 steps in total.

The predicted day-ahead prices are taken from the market data and are not separately forecasted in this study. Aggregators know the forecasted market prices and use them to make the initial schedule in the transactive control frame. The market data is kept separately and used as one of the main inputs to train the network. After training is complete, the network uses the preserved energy prices to set the algorithm, and Fig. 18.4 shows some daily samples from the test set.

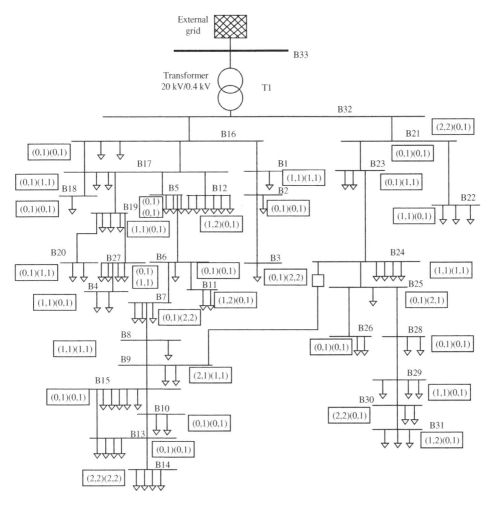

Figure 18.3 The representative distribution network [7].

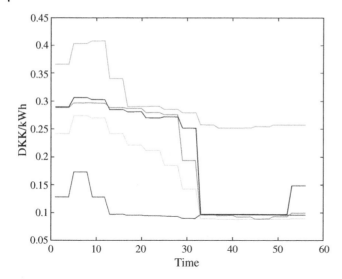

Figure 18.4 Day-ahead market prices for different days from the test set to visualize the variety between days.

Running and saving the transactive control part is the most time-consuming part if we do not already have a saved data set that we keep when this system is already running. However, the system can be imagined as running and working with no problem, and improving this study is post-processing, which means we had the transactive model and previous cases ready. On the other hand, if there is no previously saved transacting solution for individual EV charging schedules that could be used with a data set for training, there will be a need to run the model as many days as we can, which is computationally expensive. That is another reason for us to focus on learning. With the collected data set, training the network is the next step. Following training, testing or real-time running occurs quickly, bringing a viable solution.

18.5 Simulation Results and Discussion

To investigate the performance of the DNN, we compare its results with the network-constrained transactive control method for various cases. According to the EV charging schedules obtained through DNN as an output of the test set, there could not be detected any violations. About 10% of the test set refers to 180 different cases or days. In this case, we test and compare our approach 180 times. Figure 18.5 visualizes only some sample EV costs from different aggregators across all days in the test set. The light gray line represents a TEM solution, and the dark gray line represents a DNN solution.

Similarities between charging cost of EVs are expected since the assumptions for the initial SoC and their paths are assumed to be the same in this study. It is not difficult to remove this assumption, but it must be granted because it will necessitate more scenarios. If we do not have an available collected schedule that is found with transactive control with respect to initial conditions within many days, we must produce the solution first, then learn from it later. Divergence in EV behaviors necessitates more scenarios to provide a chance for the network to learn. After it learns more divergent behaviors, the model can work with different paths and initial conditions for individual EVs as well. Since both the number of EVs on each aggregator and individual EVs' behavior are

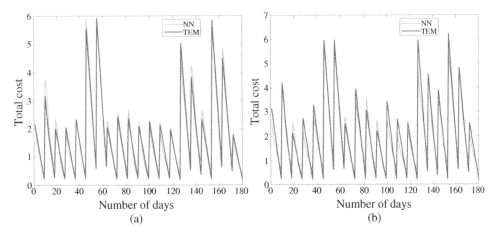

Figure 18.5 The daily charging cost for different EVs through the entire test set. (a) Daily charging cost for EV3 during 180 days. (b) Daily charging cost for EV24 during 180 days.

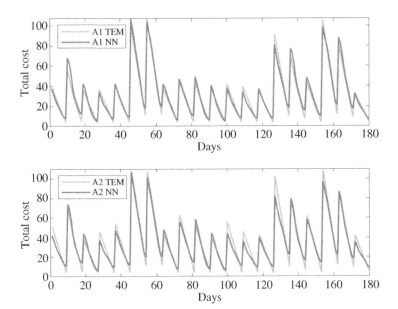

Figure 18.6 The daily charging cost for all EVs on aggregators 1 and 2 through the entire test set. (a) Total daily charging cost for the EVs on aggregator 1 through the test set. (b) Total daily charging cost for the EVs on aggregator 2 through the test set.

the same, similar cost lines with both DNN and TEM are inevitable. The visualization of this fact is shown below in Fig. 18.6.

Besides, the comparison of charging cost with respect to all days, it can be zoom in and focus on a day. Figure 18.7 shows the results of the charging cost of each EV for a specific day, here is given for 85th, 110th, 139th, and 168th. It can be seen that is TEM and DNN solution is not exactly the same, even if the trend seems to be different; however, this difference does not cause any violation according to the evaluation. It can be better to investigate power for the same day which helps to elaborate more on the cost plots. Figure 18.8 shows the power decisions for the 15th and 30th EVs in a day 85 and 110. First thing to realize both methods never go beyond 3.7 kW which is the

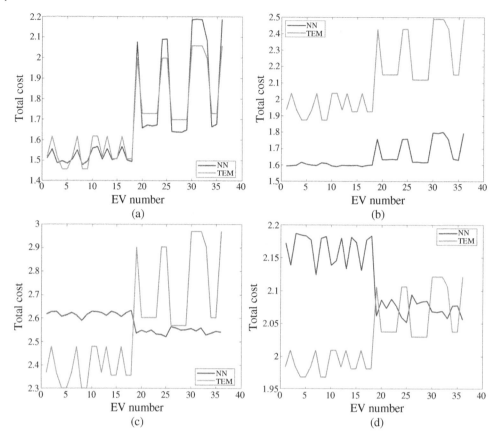

Figure 18.7 The daily charging cost for each EV on different days. (a) Daily charging comparison for all EVs for day 85. (b) Daily charging comparison for all EVs for day 110. (c) Daily charging comparison for all EVs for day 139. (d) Daily charging comparison for all EVs for day 168.

maximum charging power. Transactive control solutions show a peak when the electricity prices are getting cheaper to charge; on the other hand, network model keeps power lower by charging the EV after the prices are getting cheaper. It seems that both solutions give a similar total cost which is acceptable.

The charging power schedules for different days of the same EV and different EV for the same day, which can be seen in Fig. 18.8, produce solutions with a similar trend by realizing variances. Here, the point that needs to be understood is that the results have less variety because of assumptions that we made and the nature of the problem.

18.6 Conclusion

This chapter proposes using deep learning to create a control framework and reduce the real-time solution for the larger systems in smart grids. The data set is created first by running a transactive control approach for the problem, and the output of this part becomes the data that DNN needs to learn. The solution of the proposed method is checked to be sure there are no network violations. The issue is that the initial schedule did not consider demand when solving the schedule with

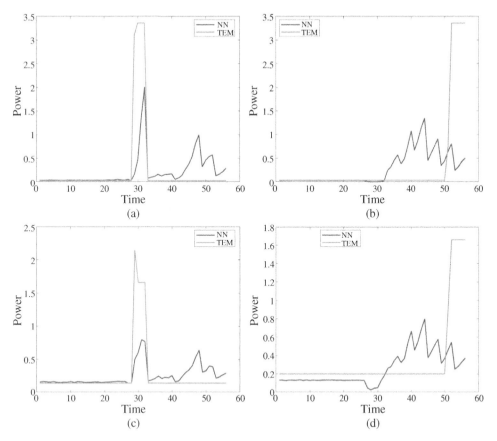

Figure 18.8 The daily charging cost for a specific EV on a specific day (same two days from Fig. 18.7) by comparing change of the day and EV. (a) Daily charging schedule for the EV15 for day 85. (b) Daily charging schedule for the EV15 for day 110. (c) Daily charging schedule for the EV30 for day 85. (d) Daily charging schedule for the EV30 for day 110.

respect to battery limits and market prices. Here, the problem might involve less energy than will be required. Normally, the same problem is present when it is solved with the TEM approach, but the problem is solved with respect to the energy capacity and power limit of the battery. Thus, the solution has not been affected by the load or the driver's schedule.

The employing of DNNs for scheduling EV charging and discharging processes can significantly improve performance in terms of decreasing the solution time of transactive control frameworks. The results of this study demonstrate the potential of this method to reduce costs in the energy market and make computationally less expensive real-time solutions. This approach offers a promising solution to the challenge of finding an acceptable solution in a limited time for a EV charging schedule problem.

The use of DNNs allows for the modeling of complex relationships between demand, the initial SOC, and electricity prices. By incorporating this information into the scheduling process, the proposed approach can provide more accurate and reliable predictions of energy needs and optimize the scheduling of EV charging and discharging processes based on those predictions. This, in turn, can lead to reduced energy costs and improved performance in the energy market. However, the effectiveness of this approach in real-world applications has yet to be fully validated.

Further research is necessary to assess the scalability and robustness of DNN-based EV scheduling in different energy market scenarios, especially with less assumptions. Additionally, it may be necessary to develop more advanced algorithms to address issues such as data privacy and security, which are critical in the energy sector. In conclusion, the study provides valuable insights into the potential of deep learning methods for solving energy management problems in the electricity market. The results of this study offer a promising direction for future research and development in this field.

18.7 Exercises

- What input data is required to train a DL model for prosumers?
- Define the three main components of the transactive control framework for finding non-violating EV charging scenarios.
- Explain the purpose Eq. (1.1a) in the aggregator's objective function. How does it contribute to the optimization process?
- If we replace DL with Reinforcement Learning (RL) to learn the behavior of EV users, what are the main differences, and what advantages and disadvantages would this bring?
- If the number of prosumers or grid topology changes, how should the model be updated to reflect these changes?
- How can we determine if the amount of training data is sufficient for the DL model?
- What are alternative solutions to this problem that do not involve learning?
- What metrics can be used to evaluate the quality and effectiveness of a DL model, and how can we determine if the amount of training data is sufficient?
- How does the choice between DL and statistical methods affect the accuracy and performance of a model for EV users, and what factors should be considered when making this choice?
- What kind of potential communication problem might happen when using DL to predict the EV schedules and updating the learning model?

References

1 IEA (2022). Global EV Outlook. https://www.iea.org/data-and-statistics/data-product/global-ev-outlook-2022 (accessed 13 November 2023).

2 Council of the EU (2022). First "Fit for 5" proposal agreed: the EU strengthens targets for CO_2 emissions for new cars and vans. https://www.consilium.europa.eu/en/press/press-releases/2022/10/27/first-fit-for-55-proposal-agreed-the-eu-strengthens-targets-for-co2-emissions-for-new-cars-and-vans/ (accessed 13 November 2023).

3 Hu, J., Yang, G., Kok, K. et al. (2017). Transactive control: a framework for operating power systems characterized by high penetration of distributed energy resources. *Journal of Modern Power Systems and Clean Energy* 5 (3): 451–464.

4 Mohammad, A., Zamora, R., and Lie, T.T. (2020). Transactive energy management of PV-based EV integrated parking lots. *IEEE Systems Journal* 15 (4): 5674–5682.

5 Corbin, C.D., Makhmalbaf, A., Huang, S. et al. (2016). Transactive Control of Commercial Building HVAC Systems. Technical report. Richland, WA (United States): Pacific Northwest National Lab. (PNNL).

6 Gray, M.K. and Morsi, W.G. (2018). A novel transactive energy framework for prosumers with battery storage and electric vehicles. In: *2018 IEEE Electrical Power and Energy Conference (EPEC)*, 1–6. IEEE.

7 Hu, J., Yang, G., Bindner, H.W., and Xue, Y. (2016). Application of network-constrained transactive control to electric vehicle charging for secure grid operation. *IEEE Transactions on Sustainable Energy* 8 (2): 505–515.

8 Liu, Z., Wu, Q., Ma, K. et al. (2018). Two-stage optimal scheduling of electric vehicle charging based on transactive control. *IEEE Transactions on Smart Grid* 10 (3): 2948–2958.

9 Ozcanli, A.K., Yaprakdal, F., and Baysal, M. (2020). Deep learning methods and applications for electrical power systems: a comprehensive review. *International Journal of Energy Research* 44 (9): 7136–7157.

10 Kong, W., Dong, Z.Y., Jia, Y. et al. (2017). Short-term residential load forecasting based on LSTM recurrent neural network. *IEEE Transactions on Smart Grid* 10 (1): 841–851.

11 Ibrahim, M.S., Dong, W., and Yang, Q. (2020). Machine learning driven smart electric power systems: current trends and new perspectives. *Applied Energy* 272: 115237.

12 Dabbaghjamanesh, M., Kavousi-Fard, A., and Zhang, J. (2020). Stochastic modeling and integration of plug-in hybrid electric vehicles in reconfigurable microgrids with deep learning-based forecasting. *IEEE Transactions on Intelligent Transportation Systems* 22 (7): 4394–4403.

13 Yin, L., Gao, Q., Zhao, L., and Wang, T. (2020). Expandable deep learning for real-time economic generation dispatch and control of three-state energies based future smart grids. *Energy* 191: 116561.

14 Zhang, K., Li, J., He, Z., and Yan, W. (2018). Microgrid energy dispatching for industrial zones with renewable generations and electric vehicles via stochastic optimization and learning. *Physica A: Statistical Mechanics and its Applications* 501: 356–369.

15 Hu, J., Yang, G., Ziras, C., and Kok, K. (2018). Aggregator operation in the balancing market through network-constrained transactive energy. *IEEE Transactions on Power Systems* 34 (5): 4071–4080.

16 Ouyang, Z., Ravier, P., and Jabloun, M. (2022). Are deep learning models practically good as promised? A strategic comparison of deep learning models for time series forecasting. In: *2022 30th European Signal Processing Conference (EUSIPCO)*, 1477–1481. IEEE.

17 Sanguinetti, L., Zappone, A., and Debbah, M. (2018). Deep learning power allocation in massive MIMO. In: *2018 52nd Asilomar Conference on Signals, Systems, and Computers*, 1257–1261. IEEE.

18 Yu, X., Guo, J., Li, X., and Jin, S. (2021). Deep learning based user scheduling for massive MIMO downlink system. *Science China Information Sciences* 64 (8): 182304.

19

Resilient Sensing and Communication Architecture for Microgrid Management*

Yuzhang Lin, Vinod M. Vokkarane, Md. Zahidul Islam, and Shamsun Nahar Edib

19.1 Introduction

19.1.1 Background and Motivation

Microgrids contain distributed energy resources (DERs), controllable loads, and energy storage elements requiring fast, efficient, and secure control strategies for their coordination. The three levels of available control mechanisms for the DERs in the microgrid are primary, secondary, and tertiary control [1]. As the primary control e.g. droop control works with the local information of the DERs to regulate operating conditions, for e.g. voltage, current, and power flow, they allow deviations from the nominal values requiring a secondary control for resetting the reference points. The secondary control requires the exchange of information among the DERs or between DERs and energy management system (EMS). In addition to secondary control, tertiary control aims to optimize the operation of grid in terms of cost, power loss, and voltage deviation requiring information about the microgrid and its interaction with the distribution grid and/or other microgrids. Therefore, a fast and resilient sensing and communication network (CN) is required to provide system operating information for secondary and tertiary control of the microgrid.

Other than control purposes, many applications such as fault detection/location, equipment health monitoring, generation/load forecasting, service restoration, and protection require measurements across a microgrid or networked microgrids [2, 3]. For example, the protection mechanism for a traditional distribution network becomes ineffective for microgrids because of their bidirectional power flow and varying fault currents due to grid-connected mode and islanded mode of operation. As a result, adaptive protection is proposed to adjust protection settings dynamically based on the observed operating state of the microgrid [4].

In summary, all the aforementioned applications require the state of the grid by installing sensors and configuring real-time communication. As sensor installation for all variables of concern is uneconomical, state estimation can be run to infer unmeasured variables based on available measurements from optimally installed sensors [5, 6]. Therefore, the sensing and communication architecture must be properly designed to satisfy the need for microgrid observability and operation. As cyber-physical microgrids are often exposed to internal failures and external disturbances such as cyberattacks and natural disasters, the performance of the sensing and

* This work is supported in part by U.S. Department of Navy award N00014-20-1-2858 and in part by U.S. Department of Navy award N00014-22-1-2001, both issued by the Office of Naval Research.

Microgrids: Theory and Practice, First Edition. Edited by Peng Zhang.

communication architecture must be tolerant to various types of failures, such as power grid branch failures, sensor failures, communication link failures, and central data processor failures. This is referred to as the resiliency requirements of system design and operation.

This chapter will address planning and operational issues in sensing and communication networks for fast, economical, and resilient monitoring of microgrids and networked microgrids. Section 19.1 first provides an overview of various sensing and communication technologies for microgrids and networked microgrids. Section 19.2 addresses the planning stage problems, where sensors and communication links/routers are to be placed to build the monitoring system. For a planning problem, the objective is to minimize the capital cost of the system while satisfying resiliency and latency requirements. Subsequently, Section 19.3 addresses the operation stage problems, where the sensor locations and communication topology are given, and data routing paths are to be found. In the operation stage, the objective is to minimize data transfer latency while satisfying resiliency requirements. Section 19.4 will conclude the chapter.

19.1.2 Overview of Microgrid Sensing Technology

19.1.2.1 Remote Terminal Units (RTUs) and Intelligent Electronic Devices (IEDs)

In a (SCADA) system for microgrids, RTUs act as a gateway toward the EMS. It performs two main tasks: (1) acquisition of raw data from field sensors and their conversion according to specified communication protocol (e.g. DNP3, IEC 61850), (2) conversion of digital command received from remote control center into analog outputs to control field devices (e.g. relay). RTUs can also generate local automation in the event of certain field conditions. Unlike the RTUs, modern IEDs integrate the sensing/control device with built-in support for standard communication protocols. The communication protocols, for e.g. Modbus, DNP3, IEC 61870-5, IEC 61850, ICCP, and Downstream-DNP3 can be used for communication among RTUs/IEDs and between RTU/IED and control center [7]. However, the sensor measurements are gathered every 2–4 seconds and they are not time-synchronized. Therefore, SCADA measurements offer a fair approximation of the system's state, assuming that the system is in quasi-steady state [8].

19.1.2.2 Phasor Measurement Units (PMUs)

Unlike SCADA measurements, PMUs record grid variables, for e.g. current phasor, voltage phasor, frequency, and rate of change of frequency (ROCOF) in high resolution and synchronously over the grid with the help of global positioning system (GPS). The real-time system state can be observed via the granular (e.g. 120 samples per second) and synchronous PMU measurements. They enhance the controllability and observability of microgrids and networked microgrids. Today, many devices, for e.g. relay, and digital fault recorder, also have built-in PMU functionality. There are standards, for e.g. IEEE Std C37.118.2.2011 for PMU data transfer [8]. The typical PMU features are adjusted to meet the needs of monitoring and control in microgrids and networked microgrids, and are referred to as μPMU or distribution-level PMU (D-PMU) [9, 10].

19.1.2.3 Smart Meters

Advanced Metering Infrastructure (AMI) is an integrated system of smart meters, communications networks, and data management systems that allow utilities and consumers to communicate. Smart meters collect real and reactive power consumption, voltage magnitude, and current magnitude profiles at the consumer end at certain intervals and communicate the data in the range of 1 minute to 60 minutes. The large volume of smart meter data is processed to enable demand-response program, behind-the-meter renewable generation prediction, fault and outage detection, load control, theft detection, and so on [11].

19.1.2.4 Smart Inverter

Unlike conventional inverters serving as a pure energy conversion device, smart inverters perform a variety of extra activities over the cyber domain. Smart inverters support grid functionalities related to voltage, frequency, and control enabling reliable and stable operation of microgrids and networked microgrids. According to IEEE Std 1547-2018, DERs need to provide specific grid supportive functionalities that can be achieved via smart inverters [12]. Mirafzal and Adib [13] defines some specific features of smart inverters such as self-governing, self-healing, and self-security features that enhance the capability of DERs in microgrids. Smart inverters also monitor the state of the DERs and support a wide range of communication protocols to exchange local information among DERs or DERs and EMS to enable different modes of control mechanisms.

19.1.3 Overview of Microgrid Communication Technology

Communication technology can be broadly classified into two categories: wired and wireless communication. Wireless communication is less expensive, but it has limitations such as poor data rates and significant interference. Wired communication, on the other hand, is more resistant to interference and can accommodate higher data rates, but it comes with a higher infrastructure cost.

19.1.3.1 Wired Communication

Optical fiber, power line communication (PLC), digital subscriber lines (DSL), coaxial cable, and Ethernet are wired communication technologies that can be utilized for microgrids [14]. The appropriate technology for a certain section of microgrid communication can be chosen based on available resources, required data rate, installation cost, and coverage. For example, optical fiber can be used for the backbone network where a large amount of data needs to be exchanged in real-time; whereas PLC can be utilized for data transfer from end devices to the backbone networks with less infrastructure cost and a moderate data rate.

19.1.3.2 Wireless Communication

Wireless technology provides a viable alternative for microgrid communication. The available protocols for enabling wireless communication are Zigbee, LoRa, WLAN (IEEE 802.11x standard), cellular network communication (e.g. LTE, WiMAX), and a few others [15]. For instance, ZigBee technology is based on the IEEE 802.15.4 standard and enables short-range, low-rate wireless communication, which is ideally suited for periodic/intermittent data transfer or a single signal transmission from a sensor. WLAN can be used for distribution substation automation and protection, and monitoring and control of DERs. Cellular networks can be utilized for sensor's data gathering in EMS along with other applications due to their fast data rate, wide coverage, and low latency [14].

To give a concrete example of the sensing and communication architecture planning and operation, in the following two sections we will describe the concepts and methodologies in terms of PMUs connected via an optical network. The described concepts and methodologies are generally applicable to most other types of sensors and communication networks reviewed above.

19.2 Resilient Sensing and Communication Network Planning Against Multidomain Failures

This section will introduce the concepts and methodologies for the planning of resilient sensing and communication networks. As planning problems are long-term and time-insensitive, rigorous mathematical programming problems can be formulated and solved even though they have relatively higher computational costs. Mixed integer linear programming (MILP) is the main optimization model adopted to solve planning-stage problems. Before formulating the problems, the basic concepts regarding resilient observability and essential design principles will be introduced first.

19.2.1 Observability Analysis and PMUs

Maintaining full observability is a prerequisite for tracking the real-time operating state of the system. Microgrid observability can be defined as the ability to recover the voltage phasors of all buses uniquely based on the available measurements. In the case of PMU measurements, the observability concept can be explained via Fig. 19.1. There is a PMU installed at bus B_1, which can measure the voltage phasor at bus B_1 ($V_{B_1} \angle \theta_{B_1}$) and the current phasors along all the branches incident to this bus (for e.g. the current from bus B_1 to bus B_2, $I_{B_1 B_2} \angle \delta_{B_1 B_2}$). If the line impedances are known, the voltage phasors at the neighboring buses (for example, the voltage phasor at bus B_2) can be calculated based on Ohm's law and Kirchhoff's voltage law.

In order to make a bus observable, the measurement data obtained by the PMU must be sent to a phasor data concentrator (PDC) via a communication path. For this reason, the following three conditions need to be fulfilled to make a given bus observable.

1. **PMU measurability**: A PMU is installed at the bus or at least one of the neighboring buses.
2. **Communication network (CN) connectivity**: A communication path is established between the PMU installed at the bus or the neighboring bus and a PDC.
3. **PDC processability**: A PDC is installed at the control center where the measurement data sent by the PMU can be processed.

In a microgrid or networked microgrid, there may be some junction buses not connected to any generator or load, which are referred to as zero-injection buses (ZIBs). These buses can be helpful for achieving grid observability as they reduce the number of required PMUs to make the grid fully observable. In Fig. 19.1, assume that bus B_2 is a ZIB and bus B_3 is a normal load bus. As bus B_2 is a

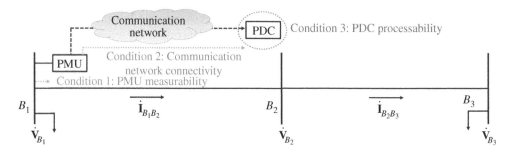

Figure 19.1 Conditions for observability of microgrid measured by PMUs.

ZIB, the current phasor $I_{B_1B_2} \angle \delta_{B_1B_2}$ should be equal to the current phasor $I_{B_2B_3} \angle \delta_{B_2B_3}$ by Kirchhoff's current law. As the voltage phasor at bus B_2 and the current phasor along branch B_2–B_3 are known, the voltage phasor at bus B_3 can be calculated using Ohm's law and Kirchhoff's voltage law. Thus, if bus B_2 is a ZIB, then buses B_1, B_2, and B_3 can be observed by sending the measurement data obtained by the PMU located at bus B_1 to a PDC through a communication path.

19.2.2 Concept Definition

1. **Essential PMU and redundant PMU**: For attaining full observability of a given microgrid, a minimum number of PMUs are required. They are defined as essential PMUs (EPMUs). Besides, additional PMUs are required to maintain grid observability if one of the EPMUs fails. They are defined as redundant PMUs (RPMUs). The complete set of EPMUs and that of RPMUs are defined as EPMU set and RPMU set, respectively.
2. **Candidate PMU configuration**: A candidate PMU configuration is a combination of EPMUs and RPMUs which ensure the full observability under normal operating conditions as well as under any single contingency (any single PMU failure or single power branch failure).
3. **Primary path and backup path**: In case of any single link failure along the data transfer path, at least two link-disjoint paths are required from a PMU to the PDC(s). The shorter of these two link-disjoint paths is defined as the primary path, while the longer is defined as the backup path for this particular PMU.
4. **Primary PDC and backup PDC**: To protect against any single PDC failure, the primary path and the backup path of a PMU are connected to two different PDCs. The PDC connected to the primary path is defined as the primary PDC, while the one connected to the backup path is defined as the backup PDC.

19.2.3 Fundamental Design Principles

As the configurations of sensors and CN have mutual impact, the sensing and communication planning problem should take a cross-domain approach for achieving multidomain resiliency as well as global minimum cost.

Taking PMU as the sensor example, the PMU configuration design should achieve resiliency against a single PMU failure or a single power grid (PG) branch failure. Therefore, an optimal PMU placement problem should be formulated to ensure that the observability of the microgrid or networked microgrid is maintained under these two types of failures. In addition, it should be noted that a PMU configuration with the minimum cost does not ensure that the total cost is minimal, as it may not correspond to the communication design with the minimum cost. Therefore, multiple candidate PMU configurations should be generated to allow the evaluation of communication design costs.

Taking the optical network as the CN example, the network should achieve resiliency against a single communication link failure or a single PDC failure. The awareness of PMUs' roles in grid observability is helpful for customizing the network design toward lower cost. Specifically, the communication link design should follow the policies below:

1. **Policy 1**: There should be two link-disjoint data transfer paths from each EPMU to two different PDCs.
2. **Policy 2**: There should be one data transfer path from each RPMU to one PDC.

A system can maintain full observability under any single component (branch/PMU/link/PDC) failure if these design policies are fulfilled. The following example will show the system's resilience under three possible PMU relationships.

1. **Relationships between two (or more) EPMUs**: Suppose by sending measurement data to either of the two PDCs, D_1 and D_2, either of the two EPMUs, E_1 and E_2, can make bus B_1 observable, as illustrated in Fig. 19.2(a). According to Policy 1, there should be two link-disjoint paths from each of these EPMUs to two PDCs. As two EPMUs can share their data transfer paths to the same PDC, part of the data transfer path between PMU E_1 and PDC D_1 can overlap with the path between PMU E_2 and PDC D_1. In the case of the failure of one of these two EPMUs (or the corresponding power branch incident to one of them), bus B_1 can still be observed by the remaining PMU. In the case of a link failure in the overlapped data transferred path to a PDC (e.g. to PDC D_1), the observability of bus B_1 can still be maintained by sending the measurement data of both PMUs to the other PDC (e.g. to PDC D_2) using the link-disjoint paths. In the case of a PDC (e.g. PDC D_1) failure, bus B_1 is still observable by sending the measurement data of both PMUs to the remaining PDC (e.g. PDC D_2). When taking into account more than two EPMUs, similar findings still hold.

2. **Relationships between one (or more) EPMU and one (or more) RPMU**: Suppose by sending measurement data to either of the two PDCs, D_1 and D_2, either an EPMU E_1 or an RPMU R_1 can make bus B_1 observable, as illustrated in Fig. 19.2(b). According to Policy 1, there should be two link-disjoint paths from an EPMU to two PDCs. According to Policy 2, there should be only one path from an RPMU to one PDC (e.g. to PDC D_1). As they can share their data transfer paths to the same PDC, part of the data transfer path between PMU E_1 and PDC D_1 can overlap with the path between PMU R_1 and PDC D_1. In the case of EPMU E_1 failure (or the corresponding power branch failure), bus B_1 can still be observed by the RPMU R_1, and vice versa. In the case of a link failure in the overlapped data transfer paths to PDC D_1, the observability of bus B_1 can still be maintained by sending the measurement data of the EPMU E_1 to PDC D_2 through the link-disjoint backup path. In the case of PDC D_1 failure, the connection to PMU R_1 is lost; however, the bus observability can still be maintained as the measurement data of PMU E_1 can be transferred to PDC D_2. When taking into account more than one EPMU and more than one RPMU, similar findings persist.

3. **Relationships between two (or more) RPMUs**: Suppose by sending measurement data to PDC D_1, either of the two RPMUs, R_1 and R_2, can make bus B_1 observable, as illustrated in Fig. 19.2(c). According to Policy 2, there should be only one path from each of these RPMUs to one PDC (e.g. to PDC D_1). As two RPMUs can share their data transfer paths to the same PDC, part of the data transfer path between PMU R_1 and PDC D_1 can overlap with the path between PMU R_2 and PDC D_1. In the case of one or both RPMU failures, the observability of bus B_1 is

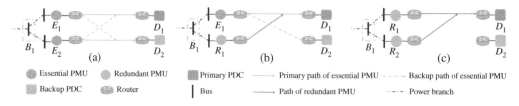

Figure 19.2 Covering observability of a bus during each type of component failure: (a) two or more EPMUs; (b) one (or more) EPMU and one (or more) RPMU; and (c) two (or more) RPMUs.

unaffected, as the system can afford lose any number of RPMUs. As a result, even in the event that one of the RPMUs fails, one of the links in their overlapped data transfer paths fails, or the PDC to which they are sending their measurement data fails, observability is still maintained.

As shown in the analysis above, the design policies guarantee power grid observability in the event of any single PMU, power branch, communication link, or PDC failure. Note that the same objective would have required two link-disjoint paths from every PMU to two PDCs if the communication design was unaware of the grid observability theory, which would have resulted in a substantially higher CN installation cost.

19.2.4 Resilient PMU Placement and Essential Role Selection

In this section, resilient PMU placement problems considering any single PMU failure or any single power branch failure is formulated.

19.2.4.1 Resilient PMU Placement

Consider a PMU-measured power grid with n buses. The formulation of the PMU placement problem is given in Table 19.1. The objective of the resilient PMU placement problem is to minimize the PMU installation cost written as Eq. (19.1), where g_j is the installation cost PMU j. In order to maintain full grid observability during any single PMU failure, each bus should be observed by at least two PMUs. This can be ensured by constraint (19.2); where $C_{i,j}$ is the $i-j$th entry of the binary connectivity matrix $\mathbf{C}$. It is 1 if buses i and j are connected, and is 0 otherwise. In the event of a power grid branch failure, the current phasor measurement is lost which may result in unobservable neighboring bus. The constraints for maintaining grid observability considering resiliency against any single power branch failure can be written as Eq. (19.3), which ensures that each bus has a PMU or is observable by means of at least two adjacent PMUs. The effect of

Table 19.1 Problem formulation for PMU placement problem and essential role selection problem.

PMU placement problem	Essential role selection problem
Objective:	Objective:
$$\hat{\mathbf{x}} = \min \sum_{j=1}^{n} g_j x_j, \qquad (19.1)$$	$$\hat{\mathbf{y}} = \min \sum_{j=1}^{n} h_j y_j, \qquad (19.5)$$
Constraints:	Constraints:
$$\sum_{j=1}^{n} C_{i,j} x_j \geq 2, \forall i \qquad (19.2)$$	$$\sum_{j=1}^{n} C_{i,j} y_j \geq 1, \forall i. \qquad (19.6)$$
$$2x_j + \sum_{\substack{j=1 \\ j \neq i}}^{n} C_{i,j} x_j \geq 2, \forall i. \qquad (19.3)$$	$$\sum_{j=1}^{n} y_j^{T(\beta\gamma)} y_j^{(\beta\delta)} - \sum_{j=1}^{n} y_j^{T(\beta\gamma)} y_j^{(\beta\gamma)} \qquad (19.7)$$
$$\sum_{j=1}^{n} x_j^{T(\alpha)} x_j^{(\beta)} - \sum_{j=1}^{n} x_j^{T(\alpha)} x_j^{(\alpha)} \qquad (19.4)$$	$$\leq -b_y; \gamma = 1, \ldots, \delta - 1.$$
$$\leq -b_x; \alpha = 1, \ldots, \beta - 1.$$	

Table 19.2 Definitions of binary decision variables.

Variable	1	0
x_j	If a PMU is installed at bus j	Otherwise
$x_j^{(\beta)}$	If a PMU is installed at bus j for βth candidate	Otherwise
$x_j^{(\alpha)}$	If a PMU is installed at bus j for α candidate	Otherwise
y_j	If an EPMU is installed at bus j	Otherwise
$y_j^{(\beta\delta)}$	If an EPMU is installed at bus j for $\beta\delta$th candidate	Otherwise
$y_j^{(\beta\gamma)}$	If a PMU is installed at bus j for $\beta\gamma$th candidate	Otherwise
$a_{s,d,p}$	if a connection is established between PMU s and PDC d on path p	Otherwise
$r_{s,d,p,e,f}$	if path p from PMU s to PDC d is using link (e, f)	Otherwise

the ZIBs can be included in the problem formulation by modifying constraints (19.2) and (19.3) (see [16] for detailed modifications). The definitions of the binary decision variables are explained in Table 19.2.

The solution vector $\hat{\mathbf{x}}$ provides the optimal locations of PMU configuration with the minimum PMU installation cost. However, the solution to the PMU placement problem is often not unique, and different PMU configurations result in different CN topologies. As a result, the PMU configuration associated with the minimum PMU installation cost may not be the global optimum when considering the communication link placement cost. To obtain a globally optimal solution, multiple candidate PMU configurations with minimum or near-minimum installation costs should be generated. This can be done by solving the optimal PMU placement problem repeatedly after adding a constraint to ensure that the newly obtained solution is different from all previously obtained solutions. The multiple candidate configuration generation constraint for the βth run of the problem can be written as Eq. (19.4), where the first term specifies the number of common PMU locations between these two candidate configurations and the second term provides the number of PMUs placed in the αth candidate configuration. The scalar b_x defines the minimum number of different PMUs required between any two different configurations.

19.2.4.2 Essential Role Selection

The candidate PMU configurations obtained by solving the resilient PMU placement problem consist of both EPMUs and RPMUs. As discussed in Section 19.2.3, to minimize the CN installation cost, the network design problem needs to have knowledge about different roles of PMUs. For this purpose, an essential role selection problem is formulated, which distinguishes between EPMUs and RPMUs for each candidate PMU configuration.

The essential role selection problem formulation is provided in Table 19.1. The objective of the essential role selection problem for each candidate PMU configuration can be written as Eq. (19.5), where h_j is the installation cost of PMU j. The constraint for identifying an EPMU set can be developed as Eq. (19.6). The solution vector $\hat{\mathbf{y}}$ provides the EPMU set, and the difference between the solution vectors $\hat{\mathbf{x}}$ and $\hat{\mathbf{y}}$ provides the RPMU set. The definitions of the binary decision variables are explained in Table 19.2.

Similar to the PMU placement problem, the solution to the essential role selection problem is not unique, and there could be multiple ways to select EPMU set and RPMU set for each candidate PMU configuration. For this reason, multiple EPMU sets are generated by solving the essential role selection problem repeatedly while ensuring that the new EPMU set is different from all previously obtained solutions (Eq. (19.7)).

19.2.5 Observability-Aware Resilient Communication Link Placement

In this section, a communication link placement problem will be formulated with the objective of minimizing the communication link installation cost while maintaining resiliency against any single link failures and PDC failures. As discussed before, the observability of a microgrid or networked microgrid can be maintained against the failure of any single PMU, power branch, communication link, or PDC by establishing two link-disjoint paths from every EPMU to two different PDCs and one path from every RPMU to a PDC. Assume that each microgrid bus is a potential location for a router, and a link can be installed between two routers as long as their geographical distance from one another is less than a threshold value. Any possible location for installing a link is defined as a *potential link*. The goal of the link placement problem is to minimize the installation cost by selecting a set of links to be installed from the set of potential links.

Let us consider a CN with F nodes. The set of sources (PMUs) and destinations (PDCs) are defined as S and D, respectively; SD is the set of all (s, d) pairs, where $s \in S$, $d \in D$, and $s \neq d$. The index of communication path between (s, d) is represented as p. According to their roles in grid observability, the source nodes are divided into two subsets, the EPMU set X_E and the RPMU set X_R. If any $s \in X_E$, then it has two paths, a primary one and a backup one, to two different destinations. If any $s \in X_R$, then it has only one path to one destination.

The formulation of the link placement problem of an EPMU set is provided in Table 19.3. The objective is to minimize the installation cost of the CN and can be written as Eq. (19.8), where $O_{e,f}$ is the installation cost of link (e, f). The constraints (19.9) and (19.10) ensure that there are two link-disjoint paths from each EPMU to two PDCs. Meanwhile, only one path is needed between each RPMU and a PDC, which can be ensured by the constraint (19.11). By following general CN routing problem, the constraints corresponding to flow directions are covered in Eqs. (19.12)–(19.19) (detailed meaning of these constraints can be found in [17]). A CN is considered congested when any link of the network carries more data than its capacity. The CN congestion can be avoided by enforcing constraint (19.20), where $cap_{e,f}$ is the maximum capacity of link (e, f). The definitions of the binary decision variables are given in Table 19.2.

The optimal locations of the links for the EPMU set can be obtained from the solution vector $\hat{\mathbf{r}}$. The optimal link placement problem (19.8) is solved repeatedly until the optimal CNs for all the EPMU sets in all candidate PMU configurations are achieved.

19.2.6 Optimal PMU-Communication Link Selection

The solutions to the PMU placement problem, essential role selection problem, and link placement problem provide multiple combinations of candidate PMU configurations and CN configurations. The final step of the optimization framework is to select the combined PMU-communication link configuration with the minimum overall cost as the final solution.

Although PMUs and optical networks are taken as examples, the methodologies for resilient sensing and CN planning described above can be readily modified for applications to other types of sensors and communication networks for microgrid monitoring. Based on observability analysis, sensors can be classified into essential ones and redundant ones, communication path requirements can be set differently based on the sensors' different roles in microgrid observability. For example, the role selection of SCADA or AMI measurements can follow the methodology proposed in [18].

Table 19.3 Problem formulation for communication link placement problem.

Communication link placement problem

Objective:

$$\hat{\mathbf{r}} = \min \sum_{(s,d)}^{SD} \sum_{p}^{P} \sum_{e,f}^{F} r_{s,d,p,e,f} * O_{e,f}, \quad (19.8)$$

Constraints:

$$\sum_{d}^{D} a_{s,d,p} = 2, \forall (s,d) \in SD, \quad (19.9)$$
$$s \in X_E$$

$$\sum_{d}^{D} r_{s,d,p,e,f} = 1, \forall s \in X_E, \quad (19.10)$$
$$\forall d \in D, \forall p \in P, \forall e, f \in F$$

$$\sum_{d}^{D} a_{s,d,p} = 1, \quad (19.11)$$
$$\forall (s,d) \in SD, s \in X_R$$

$$r_{s,d,p,e,f} \leq C'_{e,f} a_{s,d,p}, \quad (19.12)$$
$$\forall (s,d) \in SD, p \in P, e, f \in F$$

$$\sum_{e}^{F} r_{s,d,p,e,f} - \sum_{w}^{F} r_{s,d,p,s,w} \quad (19.13)$$
$$= -a_{s,d,p}, \forall (s,d) \in SD, p \in P$$

$$\sum_{e}^{F} r_{s,d,p,e,f} - \sum_{w}^{F} r_{s,d,p,d,w} \quad (19.14)$$
$$= a_{s,d,p}, \forall (s,d) \in SD, p \in P$$

$$\sum_{e}^{F} r_{s,d,p,e,f} - \sum_{w}^{F} r_{s,d,pf,w} = 0, \quad (19.15)$$
$$\forall (s,d) \in SD, p \in P, f \in F, f \notin SD$$

$$r_{s,d,p,e,s} = 0, \forall (s,d) \in SD, \quad (19.16)$$
$$p \in P, e \in F$$

$$\sum_{f}^{F} r_{s,d,p,e,f} \leq 1, \quad (19.17)$$
$$\forall (s,d) \in SD, p \in P, e \in F$$

$$\sum_{e}^{F} r_{s,d,p,e,f} \leq 1, \quad (19.18)$$
$$\forall (s,d) \in SD, p \in P, e \in F$$

$$r_{s,d,p,e,f} + r_{s,d,pf,e} \leq 1 \quad (19.19)$$
$$\forall (s,d) \in SD, p \in P, e \in F$$

$$\sum_{p}^{P} \sum_{d}^{D} \sum_{s}^{S} r_{s,d,p,e,f} \leq \text{cap}_{e,f}, \quad (19.20)$$
$$\forall e, f \in F$$

19.2.7 Case Study

19.2.7.1 Simulation Setup

The performance of the optimization framework is evaluated on the IEEE 57-bus test system. For CN, it is assumed that two PDCs are located at buses 7 and 40. A total of 228 potential links are placed based on assumed geographical proximity of buses.

The result of the proposed framework is compared with a baseline model, which is developed by mimicking the uncoordinated PMU and communication link placement processes. As the baseline model does not coordinate between the PMU placement problem and the communication link placement problem, the PMU placement problem for the baseline model only focuses on the minimization of PMU installation cost. Without the knowledge of microgrid observability and the different roles of PMUs, the communication link placement problem for the baseline model needs to establish two link-disjoint paths from every PMU to two different PDCs in order to achieve resiliency against any single communication link failure or any single PDC failure.

19.2.7.2 Simulation Results

By solving the candidate PMU configuration generation problem, a total of 59 different candidate PMU configurations are generated. From these candidate configurations, a total of 385 EPMU sets and their corresponding RPMU sets are generated by solving the essential role selection problem. For each combination of candidate PMU configuration and EPMU set, communication link placement problem is solved to find the communication configuration sets. The 188th candidate is selected as the final solution as it has the minimum overall cost (combined PMU and communication link cost).

The optimal PMU and CN configuration is illustrated in Fig. 19.3(a) and (b). A total of 23 PMUs and 58 communication links are placed. In order to illustrate how the sensing and communication design will provide resilient observability under various types of failures, a specific example will be discussed as follows. As can be seen from Fig. 19.3(a), bus 10 is observed by the EPMU located at bus 10 and the RPMUs located at buses 9, 12, and 51. For the EPMU at bus 10, the PDC at bus 40 is considered the primary PDC, and the PDC at bus 7 is considered a backup PDC, as illustrated in Fig. 19.3(b). In the case of any single PMU failure (e.g. EPMU at bus 10 fails), the observability of bus 10 is maintained by sending the measurement data of either of the remaining PMUs (e.g. the RPMUs located at buses 9, 12, and 51) to the PDC at bus 40. In the case of any single communication link failure on the primary path of the EPMU (e.g. data transfer path between the EPMU at bus 10 to the primary PDC at bus 40), observability is maintained by sending the measurement data of the EPMU to the backup PDC (e.g. the PDC at bus 7) using the link-disjoint backup path. In the case of any single (or multiple) communication link failure on the paths of the RPMUs, the observability of bus 10 is not affected as the loss of any number of RPMUs does not break down observability. In the case of any single PDC failure (e.g. the primary PDC at bus 40 for the EPMU at bus 10), the observability of bus 10 is maintained by sending the measurement data of the EPMU at bus 10 to the backup PDC at bus 7.

In terms of cost-effectiveness, the described method is compared with a baseline model where PMU placement and communication link placement are performed separately. In this case, the

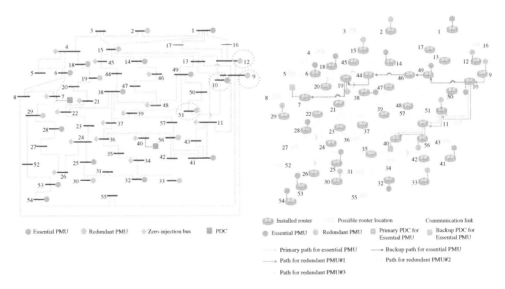

Figure 19.3 (a) Microgrid design and (b) communication path for EPMU located at bus 10 and the RPMUs located at buses 9, 12, and 51.

link placement is unaware of the roles of the PMUs in grid observability, hence to make the design resilient against any single link failure or PDC failure, it is necessary to deploy two disjoint paths from each PMU to two PDCs based on general routing theory. It is found that the described method reduces the total installation cost by 15.7% without affecting the resiliency of grid observability. It is also found that the baseline method takes a PMU configuration with fewest PMUs (22), yet it does not correspond to the design with the lowest cost overall. These observations confirm the importance of taking a cross-domain optimization approach in the sensing and communication architecture design.

19.3 Observability-Aware Network Routing for Fast and Resilient Microgrid Monitoring

This section introduces the concepts and methodologies for the operation of the sensing and communication network for microgrids or networked microgrids. In the operation stage, the sensor configuration and communication network topology are fixed, and the optimal routing paths should be found between sensors and central data processors (or between sensors and controllers). The end-to-end data transfer and resiliency against failures are of concern. This is generally referred to as the routing problem in networking domain. As the operation-stage problems are time-sensitive, there is often no time to solve a rigorous mathematical programming model to find the exact optimal solution. Instead, much more computationally efficient heuristic algorithms are developed to find near-optimal solutions. This section will start with a formal statement of the problem, followed by introduction to key concepts and routing principles based on grid observability analysis. Then, the heuristic routing algorithm will be introduced.

19.3.1 Problem Statement

Define a graph $G_p(\mathcal{V}_p, \mathcal{E}_p)$ representing the power grid (PG), where $\mathcal{V}_p$ is the set of buses and $\mathcal{E}_p$ is the set of branches. Some of the buses $v \in \mathcal{V}_p$ are equipped with PMUs $P = \{u_1, u_2, \ldots, u_p\}$ to make the grid observable, and some of them $v \in \mathcal{V}_p$ are equipped with PDCs $Q = \{c_1, c_2, \ldots, c_q\}$. Define another digraph $G_c(\mathcal{V}_c, \mathcal{E}_c)$ representing the communication network (CN), where $\mathcal{V}_c$ is the set of nodes/routers and $\mathcal{E}_c$ is the set of edges/communication links. Each PMU $u_p \in \mathcal{V}_p$ is collocated with one node $v \in \mathcal{V}_c$ creating the tuple $(u_p \in \mathcal{V}_p, v \in \mathcal{V}_c)$ forwarding the PMU measurements toward the router collocated with PDC $(c_q \in \mathcal{V}_p, v \in \mathcal{V}_c)$. For maintaining grid observability, each of the PMUs $u_p \in \mathcal{V}_p$ needs to be connected to one of the PDCs $c_q \in \mathcal{V}_c$ using the routers and links of the digraph G_c. The objective of PMU network routing is to find paths from $u_p \in \mathcal{V}_p$ to $c_q \in \mathcal{V}_c$ in digraph G_c such that the end-to-end data transfer delay is minimal while ensuring real-time observability. In addition, as the components in both digraphs may fail, another objective is to maintain grid observability in real-time even if a given number of components fail.

19.3.2 Shared Observability PMU Group (SOPG)

Based on the observability concept introduced in Section 19.2.1, if the observability of bus i should be kept under r_i PMU failures, the bus should be observed by at least $r_i + 1$ PMUs, i.e. there should be $(r_i + 1)$ PMUs placed among this bus itself and the neighboring buses. Suppose this requirement is satisfied in the planning stage, then each bus would be observable by a group of PMUs. The PMU group monitoring a bus i is referred to as shared observability PMU group (SOPG)

ω_i since they share the same goal – making bus i observable. Selecting neighboring buses of a bus i equipped with PMUs yields the SOPG ω_i for bus i.

Note that similar concept can be developed for other types of sensors or even multiple types of sensors in microgrids or networked microgrids, such that a more generic shared observability sensor group (SOSG) can be identified for each bus.

The PMUs in an SOPG have some properties to be noted: (1) as long as a SOPG ω_i has more than one member, there is redundancy in the measurement devices observing bus i, (2) a PMU can be a member of multiple SOPGs, as it can observe multiple buses. With the key concept of SOPG, an observability-aware PMU routing algorithm could be developed which fulfills the power grid's resilience requirements while minimizing data transfer delay, as will be described next.

19.3.3 Fundamental Routing Principles

The resilience against communication path failure can be achieved by either protection or dynamic restoration mechanisms. In the protection mechanism, the backup communication path is preset for PMU routing in the event of a primary path failure, but in dynamic restoration, the backup path is established during operation in the event of a failure [19]. The protection mechanism is adopted in the suggested framework to provide resilience against communication path failure due to its faster reaction during failure, which maintains the continuity of real-time PMU-based applications for monitoring microgrid dynamics. It should be noted that due to limited resources, providing protection against the failure of numerous components is infeasible. One realistic option is to give varying protection levels for different buses based on their importance to the microgrid. The proposed framework takes into account the resilience requirement of each bus independently and offers protection to meet those requirements. For instance, when the resilience requirement r_i of a bus is 2, the proposed framework maintains observability of that bus despite the failure of any combination of two components (i.e. PMU and communication link), as shown in Table 19.4.

The SOPGs are used to configure fast and resilient communication for routing PMU data to the PDCs. Figure 19.4 depicts a conceptual diagram of the routing principles, where the PMUs are organized into SOPGs in the microgrid, and the data transfer path in CN is organized into several link-disjoint communication trees (CTs) depicted in different colors. In Fig. 19.4, the PMUs belonging to each SOPG are connected to the corresponding bus with dotted lines, and the links in the CN are denoted by solid lines. We first initialize the CTs from the PDCs as roots and then grow the trees by connecting the PMUs as leaves. PMU u, a member of SOPG ω_i, finding its communication path to one of the PDCs via CT j, is denoted by u_i^j and the CTs connect all the PMUs to PDC c. To make the model resilient against the failure of a total of k components, including PMUs and communication links, then $(k+1)$ PMUs of SOPG ω_i must be connected to $(k+1)$ CTs in order to transfer data to the PDCs, as illustrated in Fig. 19.4. Because each bus i is made observable by leveraging a total of

Table 19.4 Resilience requirements against failure set $\mathcal{F}$.

No. of components failure $\equiv$ resilience requirements, r_i	The system needs to be resilient against failure set, $\mathcal{F}$
1	{1 PMU} OR {1 Link}
2	{2 PMUs} OR {2 Links} OR {1 PMU, 1 Link}

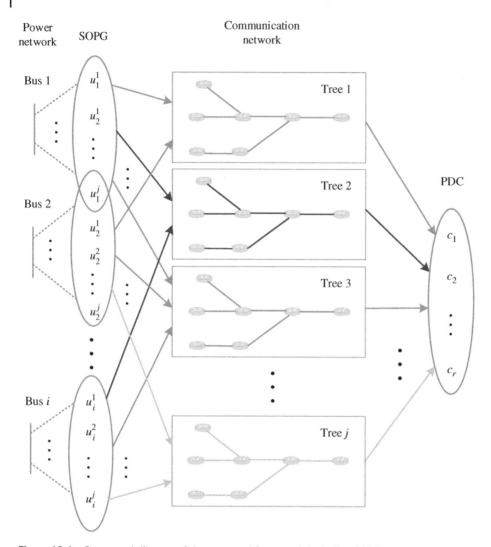

Figure 19.4 Conceptual diagram of the proposed framework including SOPG and communication trees.

$(k + 1)$ connections between SOPG ω_i and the PDCs, the observability of the entire microgrid will be preserved even if any combination of k PMUs and links fail. The PMU routing heuristic builds a feasible number of link-disjoint trees in the CN to transfer real-time PMU data to PDCs while minimizing end-to-end CN latency and meeting individual bus observability resilience criteria.

The algorithm described above can be readily modified for routing problems of other types of sensors or communication media in microgrids or networked microgrids. The key is to identify the set of sensors delivering observability to each bus of the grid and assign them to disjoint paths/trees for data transfer.

Next, the strategies for creating a communication forest (CF) comprising link-disjoint trees will be described. Each tree links multiple PMUs from various SOPGs to the lowest-cost PDC while adhering to link capacities. A cost metric will be defined below to help minimize PMU-PDC data transfer delay.

19.3.4 Cost Metric Definition

Because of the strict delay requirements of PMU-based applications such as state estimation, the PMU measurements must be transferred to the PDC with minimum delay; otherwise, the data will be obsolete to the state estimation function and observability would be lost. As a result, the cost metric is designed to reduce the end-to-end data transfer delay from the PMUs to the PDCs. Delays in a CN are classified into four types: propagation delay, transmission delay, queuing delay, and processing delay. Propagation delay is the time required to travel through a link between two routers, transmission delay is the time required to push a packet from the router to the link, queuing delay is the time required for the packet to wait in the router buffer, and processing delay is the routers' packet processing time. As technology advances, processing and propagation delays become insignificant. Queuing delay and transmission delay, on the other hand, are the most crucial aspects to consider toward the end-to-end delay estimate, which is primarily related to two main factors. The first factor is the amount of data to be transmitted via a link. When the volume of data-load traveling via a link approaches the link capacity, packets approaching the link will experience a longer queuing delay in the router's buffer before being able to access the link. As a result, it is critical to balance data-load among links in order to reduce queuing delays. The second factor is the hop count between the data source and the destination. When the number of hops between the source and the destination grows, transmission delay increases. However, the requirements for reducing queuing delays and transmission delays could be conflicting. A lower queuing delay can be obtained by balancing load across the network, which often requires higher hop count among sources-destinations increasing transmission delay [20]. Therefore, a trade-off between data-load balancing and hop count should be reflected in the cost metric for guiding the formation of the CF, as defined below:

$$\lambda_u^j = \alpha \cdot h_u^j + (1 - \alpha) \cdot v_u \tag{19.21}$$

$$v_u = \frac{1}{q} \sum_{\forall j} (dl^j - \bar{dl})^2 \tag{19.22}$$

where λ_u^j is the cost and h_u^j is the shortest hop count when PMU u reports its measurement to a PDC using jth tree, respectively; v_u is the variance of data-load of all trees after connecting PMU u to the jth tree; $\bar{dl}$ is the mean data-load of all trees, and α is a factor determining the trade-off between data-load balancing and hop count.

19.3.5 Routing Algorithm

The proposed PMU routing algorithm is summarized below:

- **Step 1**: Initialize a tree by linking the nearest PMU to each adjoint edge of a PDC. This helps balance the data-load among different trees in the CF.
- **Step 2**: Filter a PMU set from the unconnected PMUs for each initialized tree using the resilience requirement rule. According to this rule, a PMU from the SOPG can be linked to a tree as long as the bus's resilience criteria are met by the SOPG. For instance, if bus i having a resilience requirement ($r_i = 2$) is observed by SOPG ω_i, the resilience-requirement rule ensures that the PMUs in ω_i get connected to a total of $r_i + 1 = 3$ different trees.
- **Step 3**: Among the filtered PMUs to be attached to a tree, pick the one with the lowest cost based on Eq. (19.21). Here, the bandwidth constraint of the links is considered, and previously used links are reused to minimize the usage of resources.

- **Step 4**: Repeat Steps 2 and 3 for each of the initialized trees and extend the tree with the lowest possible cost.
- **Step 5**: Repeat Steps 2–4 until all PMUs are linked to a tree.

The principle of the described routing concept is illustrated by a small 11-bus microgrid system example in Fig. 19.5. In the example, the resilience-requirements of all buses are set to 1 ($r_i = 1$, $\forall i$). The grid connectivity among the buses is shown by dashed lines. PMUs are installed at some buses with $r_i + 1 = 2$ meaning that any bus i is observed by at least two PMUs; the PDC location is at bus 1. The given CN topology is considered the same as the PG topology, and the communication links used for routing are shown in solid lines. The detailed process is explained below. From the given PG topology, the SOPG can be found for bus i. For instance, a PMU set $\{u_6, u_8, u_9\}$ observing bus 7 and thus belonging to SOPG ω_7 is marked in Fig. 19.5. As the PDC c_1 has three adjacent links, three link-disjoint CTs are initialized from PDC c_1. Here, the three closest PMUs $\{u_2, u_6, u_{10}\}$ are connected to the three CTs, as seen in the figure. Next, the CTs are expanded based on the cost metric of Eq. (19.21). For the CT expansion, a set of potential PMUs are found from the unattached PMUs attachable to each CT based on the resilience requirement rule. Here, as $r_i = 1$, the PMUs belonging to SOPG ω_7 should join at least $r_i + 1 = 2$ CTs. Therefore, they should not all join the same CT, CT-3. This is ensured by the resilience requirement rule. Finally, from the potential PMU sets attachable to individual CTs, one PMU is picked to join the least-cost CT based on Eq. (19.21) and bandwidth constraints of the links. In the final design of this example, as shown in Fig. 19.5, all the buses are protected against any single device failure ($r_i = 1$). For instance, bus 7 is resilient against any PMU or link failure, because the three PMUs of SOPG ω_7 are connected to two different link-disjoint CTs (CT-2 and CT-3) and therefore, bus 7 will still be observable if any single PMU or link fails. Though the example is given for ($r_i = 1$), the described concept and algorithm can provide resilience against any predefined number of component failure, r_i.

19.3.6 Case Study

The described routing principles and algorithm are further illustrated on IEEE 30-bus test system. The power grid and CN have different topologies, but their mapping is one-to-one,

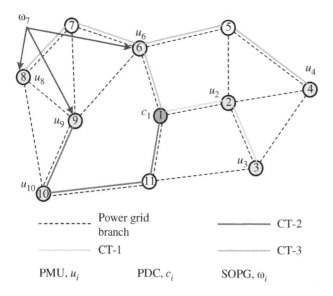

Figure 19.5 An illustration of PMU-PDC communication forest for an 11-bus power system.

meaning that each bus in the grid corresponds to a node in the CN. Based on the number of PMU measurement channels (W_i), the data rate of the PMU installed at bus i can be determined as follows [21]:

$$dl_i \propto W_i \qquad (19.23)$$

For simplicity, it is assumed that the data rate of each measurement channel is one unit, and W_i is equal to the number of buses observed by the PMU. All communication channels are assumed to have the same bandwidth, which is set to a multiple of the average data-load of PMUs so that one link can carry data from several PMUs.

19.3.6.1 Test Case for IEEE 30-Bus System

The proposed framework, as detailed in Section 19.3.3, establishes PMU-PDC connectivity based on the resilience requirements of individual buses in the power grid, as exemplified in Table 19.4. To demonstrate this, the grid's critical buses are given higher resilience than others. In the test system, both buses 2 and 27, as illustrated in Fig. 19.6, contain a generator and four adjacent transmission lines, whereas bus 6 has seven adjacent transmission lines. As a result, buses {2,6,27} are designated as critical buses with r_i of 2, whereas other common buses have r_i of 1.

A PMU configuration is determined as shown in Fig. 19.6, where each bus is observable by adequate PMUs in the corresponding SOPG to ensure resilience. For example, SOPG ω_{27} observes bus 27, which has three PMUs {25,29,30}. As a result, if any two PMUs fail, bus 27 is still observed by another PMU. Now, the PMU routing paths to the PDC are obtained by the algorithm as described in Section 19.3.3, ensuring that resilience is maintained in the event of any failure in set $\mathcal{F}$. Figure 19.7 depicts the synthesized CN for the IEEE 30-bus system, along with the potential links to be used for routing. The proposed framework is used to build a CF with seven link-disjoint trees, as shown in Fig. 19.7(a). With the PDC installed at node 17, the number of trees is set equal to the adjacent links of node 17 for generating balance trees according to Eq. (19.23). Otherwise, some of the trees would

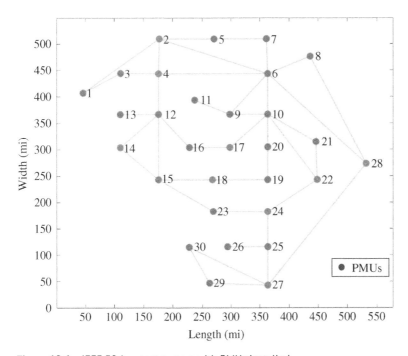

Figure 19.6 IEEE 30-bus test system with PMUs installed.

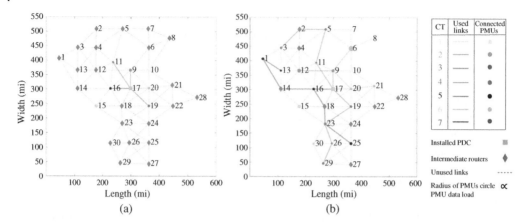

Figure 19.7 Communication forest for the CN of IEEE 30-bus system. Seven different trees are constructed from the PDC 17 (root) connecting PMUs as leaves, represented in seven different gray-scaled colors. The sizes of circles representing PMUs are proportional to their data load i.e., smaller/larger circle means relatively lower/higher data load, respectively. (a) Tree initialization from PDC 17; (b) Complete communication forest after all PMUs get connected to the trees.

spread over the graph, lengthening the delay. The individual trees on the graph are represented by seven distinct gray-scaled colors (varying from lighter to darker gray). At the same time, the PMUs accessing PDC via different trees are indicated in the respective tree's color. The diameter of the PMU nodes is set proportional to their data-load, i.e. the larger the diameter, the higher the load. Figure 19.7(b) depicts the whole constructed forest, which is generated using the steps described in Section 19.3.5. It can be verified that the resilience requirement r_i of bus i is met by comparing the grid topology in Fig. 19.6 and its PMU routing paths in Fig. 19.7(b). For example, as bus 2 requires resilience $r_2 = 2$, the SOPG of bus 2 contains three PMUs i.e. PMUs $\{1,5,6\}$, as seen in Fig. 19.6 and the PMUs are connected to three different trees (CT-5, CT-3, and CT-2, respectively), as seen in Fig. 19.7(b). As a result, even if any two components fail, as defined in set $\mathcal{F}$, bus 2 will still be observable. This is one of the key features of the described framework, in which resilience is acquired for grid bus observability rather than for individual sensing devices. Table 19.5 shows a few more examples of various buses maintaining their resiliency. As can be observed from the PMU-tree connection of each bus in Table 19.5, the resilience of bus observability is acquired in accordance with the resilience requirement rule. It can be observed from Table 19.5 that, to fulfill the resilience of r_i, bus i is observed by SOPG ω_i with at least $r_i + 1$ PMUs, where the PMUs are connected to at least $r_i + 1$ link-disjoint trees.

Table 19.5 Examples for explaining how an individual bus i is observable against the failure set $\mathcal{F}$.

Bus, i	Required resiliency, r_i	SOPG ω_i, observing bus i	Connection among PMUs of ω_i and tree in communication forest
2	2	PMUs $\{1,5,6\}$	$\{1\} \to$ CT-5, $\{5\} \to$ CT-3, $\{6\} \to$ CT-2
6	2	PMUs $\{6,7,8,9,10\}$	$\{6\} \to$ CT-2, $\{7,8,10\} \to$ CT-1, $\{9\} \to$ CT-4
22	1	PMUs $\{10,21\}$	$\{10\} \to$ CT-1, $\{21\} \to$ CT-2

To fulfill the resilience of r_i, bus i is observed by SOPG ω_i with at least $r_i + 1$ PMUs, where the PMUs are connected to at least $r_i + 1$ trees. Therefore, when r_i components in set $\mathcal{F}$ fails, the bus i be observable by the additional PMU connected to PDC.

19.3.6.2 Performance Comparison with Baseline

For the performance comparison, a baseline method is employed providing similar resilience as the proposed framework. In the baseline method, PMUs are treated as individual and unrelated data sources, comparable to the literature on PMU routing, without taking into account their interdependent roles in grid observability determined by grid topology. Treating PMUs as unrelated data sources, the baseline's aim is to transfer data from the PMUs to the PDCs with independent protection for each PMU against the failure set $\mathcal{F}$. Now, to ensure the similar resilience as the observability-aware algorithm described in this chapter, PMUs observing bus i need $r_i + 1$ link-disjoint paths to PDC, which is achieved by Bhandari's algorithm [22]. The performances of the described and baseline methods are compared using the end-to-end delay as obtained in [20]. Figure 19.8 compares the methods in terms of end-to-end delay for achieving full observability (100%) using minimum PMUs. The described method has around 1-unit end-to-end delay, whereas the baseline has approximately 3-unit end-to-end delay to achieve full observability, indicating that the proposed framework significantly outperforms the baseline. The delay for reporting measurements from all PMU channels is also depicted in Fig. 19.8(b), where it can also

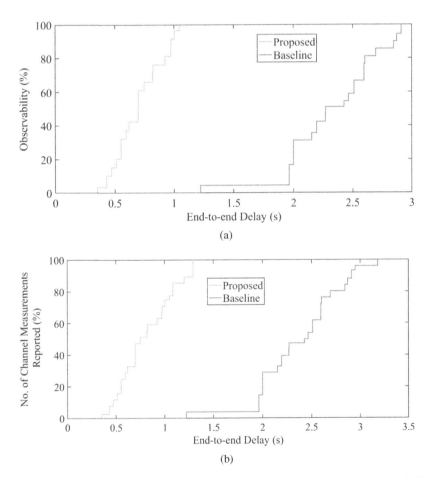

(a)

(b)

Figure 19.8 Performance comparison between the proposed and baseline methods in terms of communication delay: (a) Percentage of observability versus delay. (b) Percentage of PMU channels versus delay.

be seen that the described method has much lower delay than the baseline. The described method performs much better for two main reasons: (1) incorporating grid bus observability into the routing algorithm; (2) defining a cost metric to minimize both queuing and transmission delay. The described routing algorithm requires reduced volume of redundant data transfer from the sensors to provide resilience due to its awareness of grid observability.

19.4 Conclusion

This chapter addresses the sensing and communication problems in microgrid or networked microgrid monitoring. Having a fast, economical, and resilient sensing and communication network is the prerequisite for many advanced applications in microgrid management. The chapter starts by reviewing the motivations and existing sensing and communication technologies for microgrids and networked microgrids. Then, the planning-stage problem and operation-stage problem are separately addressed with methods involving different levels of computational efficiency. The resiliency of grid observability under various types of component failures is given special attention in the formulation of and solution to the problems. It is shown how the mutual awareness between the power grid domain and the communication network can significantly increase the performance of the architecture. Although PMUs and optical networks are taken as examples, the presented concepts and methodologies can be generically applied to most sensing and communication networks for microgrids and networked microgrids.

19.5 Exercises

1. Describe a few important applications of fast and resilient sensing and communication networks in microgrids and networked microgrids.
2. A PMU-measured microgrid with 30-bus is illustrated in Fig. 19.9. Assume that each grid bus is a potential location for a PMU, and the installation cost of a PMU is $45,000 per unit. For CN, each grid bus is a potential location for a router and each power grid branch is a possible location for a communication link. The installation cost of the communication links is shown in Fig. 19.9. Your objective is to find a PMU-communication link configuration which is resilient against any single multidomain failure.
   (a) Find two candidate PMU configurations with minimum PMU installation cost according to the resilient PMU placement problem described in Section 19.2.4.1. The two configurations must have at least 3 different PMUs, i.e. in Eq. (19.4), set the parameter b_x to 3.
   (b) For each of the candidate PMU configurations, find one EPMU set and corresponding RPMU set based on the essential role selection problem discussed in Section 19.2.4.2.
   (c) Assume that two PDCs are located at buses 6 and 15. Find the optimal locations of the links for each of the candidate PMU configurations according to the link placement problem described in Section 19.2.5. In Eq. (19.20), set the parameter $cap_{e,f}$ to 5.
   (d) Calculate the total PMU-communication network cost for the two candidate PMU configurations and find the better configuration between them.
3. Assume a random 21-bus microgrid system with installed PMUs and PDC location, as shown in Fig. 19.10. Answer the following questions:
   (a) Find the SOPG of the buses {5,7,16,21}.
   (b) Find the communication path from the SOPG obtained in Problem-3(a) to the PDC such that the resilience requirement, $r_i = 1$ is ensured for the buses {5, 21}, whereas $r_i = 2$ is ensured for buses {7, 16}.

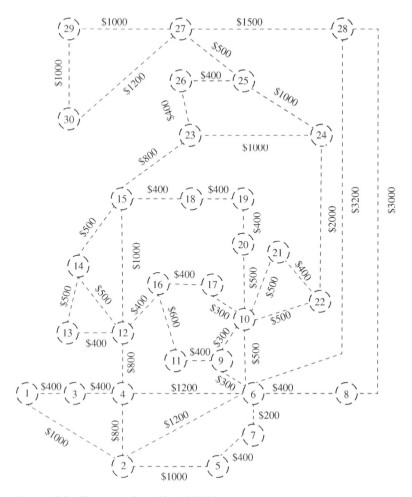

Figure 19.9 Topology of modified IEEE 30-bus system.

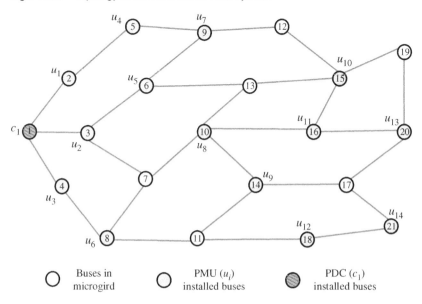

Figure 19.10 Topology of a random 21-bus microgrid system with installed PMUs and PDC at some buses.

(c) Assume the generated traffic from the PMUs are equal to their measurement channels i.e. [2, 3, 2, 2, 3, 3, 3, 4, 3, 4, 3, 2, 3, 2] units per second, respectively, and a communication link can be established between any two microgrid buses connected by a branch. Find the routing paths from all the PMUs to the PDC according to the described routing algorithm in Section 19.3.5 such that $r_i = 1$ is satisfied for all buses. In Eq. (19.21), set the parameter α to 0.5.

References

1 Olivares, D.E., Mehrizi-Sani, A., Etemadi, A.H. et al. (2014). Trends in microgrid control. *IEEE Transactions on Smart Grid* 5 (4): 1905–1919.

2 Wang, Z., Shen, C., Xu, Y. et al. (2019). Risk-limiting load restoration for resilience enhancement with intermittent energy resources. *IEEE Transactions on Smart Grid* 10 (3): 2507–2522.

3 Wen, L., Zhou, K., Yang, S., and Lu, X. (2019). Optimal load dispatch of community microgrid with deep learning based solar power and load forecasting. *Energy* 171: 1053–1065.

4 Oudalov, A. and Fidigatti, A. (2009). Adaptive network protection in microgrids. *International Journal of Distributed Energy Resources* 5 (3): 201–226.

5 Fang, Z., Lin, Y., Song, S. et al. (2021). State estimation for situational awareness of active distribution system with photovoltaic power plants. *IEEE Transactions on Smart Grid* 12 (1): 239–250.

6 Song, S., Wei, H., Lin, Y. et al. (2022). A holistic state estimation framework for active distribution network with battery energy storage system. *Journal of Modern Power Systems and Clean Energy* 10 (3): 627–636.

7 Mohagheghi, S., Stoupis, J., and Wang, Z. (2009). Communication protocols and networks for power systems-current status and future trends. *2009 IEEE/PES Power Systems Conference and Exposition*, 1–9.

8 Bobba, R.B., Dagle, J., Heine, E. et al. (2012). Enhancing grid measurements: wide area measurement systems, NASPInet, and security. *IEEE Power and Energy Magazine* 10 (1): 67–73.

9 von Meier, A., Culler, D., McEachern, A., and Arghandeh, R. (2014). Micro-synchrophasors for distribution systems. *ISGT 2014*, 1–5.

10 Liu, Y., Wu, L., and Li, J. (2020). D-PMU based applications for emerging active distribution systems: a review. *Electric Power Systems Research* 179: 106063.

11 Mohassel, R.R., Fung, A., Mohammadi, F., and Raahemifar, K. (2014). A survey on advanced metering infrastructure. *International Journal of Electrical Power & Energy Systems* 63: 473–484.

12 IEEE Std 1547-2018 (2018). *IEEE Standard for Interconnection and Interoperability of Distributed Energy Resources with Associated Electric Power Systems Interfaces (Revision of IEEE Std 1547-2003)*, 1–138.

13 Mirafzal, B. and Adib, A. (2020). On grid-interactive smart inverters: features and advancements. *IEEE Access* 8: 160526–160536.

14 Kumar, S., Islam, S., and Jolfaei, A. (2019). Microgrid communications - protocols and standards. In: *Variability, Scalability and Stability of Microgrids*, IET Energy Engineering, vol. 139 (ed. S.M. Muyeen, S.M. Islam, and F. Blaabjerg), 291–326. Institution of Engineering and Technology. https://doi.org/10.1049/PBPO139E_ch9.

15 Safdar, S., Hamdaoui, B., Cotilla-Sanchez, E., and Guizani, M. (2013). A survey on communication infrastructure for micro-grids. *2013 9th International Wireless Communications and Mobile Computing Conference (IWCMC)*, 545–550.

16 Edib, S.N., Lin, Y., Vokkarane, V.M., and Fan, X. (2022). A cross-domain optimization framework of PMU and communication placement for multi-domain resiliency and cost reduction. *IEEE Internet of Things Journal* 10 (9): 7490–7504.

17 Davis, D.A.P. and Vokkarane, V.M. (2020). Failure-aware protection for many-to-many routing in content centric networks. *IEEE Transactions on Network Science and Engineering* 7 (1): 603–618.

18 Abur, A. and Magnago, F.H. (1999). Optimal meter placement for maintaining observability during single branch outages. *IEEE Transactions on Power Systems* 14 (4): 1273–1278.

19 Liu, B.J., Yu, P., Xue-song, Q., and Shi, L. (2020). Survivability-aware routing restoration mechanism for smart grid communication network in large-scale failures. *EURASIP Journal on Wireless Communications and Networking* 2020 (1): 1–21.

20 Kateb, R., Akaber, P., Tushar, M.H.K. et al. (2019). Enhancing WAMS communication network against delay attacks. *IEEE Transactions on Smart Grid* 10 (3): 2738–2751.

21 Zhu, X., Wen, M.H.F., Li, V.O.K., and Leung, K.-C. (2019). Optimal PMU-communication link placement for smart grid wide-area measurement systems. *IEEE Transactions on Smart Grid* 10 (4): 4446–4456.

22 Bhandari, R. (1999). *Survivable Networks: Algorithms for Diverse Routing*. Springer Science & Business Media.

20

Resilient Networked Microgrids Against Unbounded Attacks

Shan Zuo, Tuncay Altun, Frank L. Lewis, and Ali Davoudi

20.1 Introduction

Section 20.1 first provides an overview of the background and motivation for studying attack-resilient control for microgrids. Sections 20.2 and 20.3 address the fully distributed attack-resilient secondary control problems for AC and DC microgrids, respectively. Section 20.4 will conclude the chapter.

20.1.1 Background and Motivation of Attack-Resilient Secondary Control of AC Microgrids

Distributed cooperative control of AC microgrids relies on consensus and containment approaches to accomplish frequency regulation [1] and voltage containment [2], respectively. The distributed communication network among inverters poses security concerns as individual inverters lack the global perspective with limited information exchanged among neighboring inverters [3–6]. Some existing methods detect, identify, and then isolate or recover the compromised inverters [7–9] but would require a number of inverters to be healthy. Moreover, stealthy attacks launched by intelligent attackers are generally undetectable. Specifically, attackers could exploit the intrinsic characteristics or internal dynamics of the system modeling and/or configuration to launch deliberately designed attacks without being detected by existing attack-detection algorithms [10]. The vulnerability assessment and consequences of power system state estimation with respect to such unobservable or undetectable false data injection (FDI) attacks are presented in [10–13]. Protection and prevention against stealthy and intelligent attackers are not always possible using attack-detection methods, and a paradigm shift to enhance the self-resilience of the large-scale networked microgrids by developing attack-resilient control protocols is the overarching objective for safeguarding the nation's critical infrastructures.

Distributed resilient control protocols are investigated recently in [14–21] to provide self-resilience against external attacks without detecting and identifying the compromised agents. The above-mentioned resilient control protocols for microgrids mainly deal with disturbances, noises, and/or faults that are unintentionally caused and are assumed to be bounded. However, in practice, malicious attackers could launch unknown and unbounded FDI attacks to maximize their damage, distorting cooperative performance and even leading to system instability [22]. To address unbounded FDI attacks for AC microgrids, an attack-resilient control framework, using observer-based techniques, is studied in [2] to maintain the bounded frequency regulation and voltage containment at the cost of additional communication channels among observers.

Microgrids: Theory and Practice, First Edition. Edited by Peng Zhang.
© 2024 The Institute of Electrical and Electronics Engineers, Inc. Published 2024 by John Wiley & Sons, Inc.

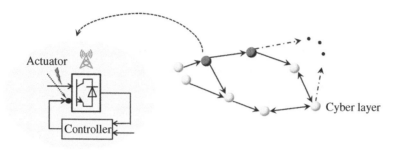

Figure 20.1 A networked multi-inverter system under actuator attacks.

This chapter will explore adaptive techniques to address unknown and unbounded attacks on input signals of the control loops, which are referred to as actuator attacks. Section 20.2 will consider the unbounded actuator attacks on both frequency and voltage control loops of an inverter, as illustrated in Fig. 20.1, which could severely destabilize the synchronization mechanism among microgrid inverters.

20.1.2 Background and Motivation of Attack-Resilient Secondary Control of DC Microgrids

Direct current (DC) microgrids are gaining popularity given the DC nature of emerging distributed energy resources, storage units, and modern loads [23–25]. In their control hierarchy [26, 27], the secondary control adjusts the voltage setpoints for the primary control (usually, implemented through droop mechanisms) and eliminates the steady-state voltage drift and/or loading mismatch. The control objectives are global voltage regulation, where the average voltage over the distribution line is regulated at a reference value, and proportional load sharing, where power electronics converters share the total load among themselves based on their power ratings. The distributed implementation of the secondary control level has become popular due to its improved efficiency, scalability, and robustness [27–33]. For example, the cooperative control framework in [29] supplements the voltage reference for each converter by two voltage correction terms. The average voltage across the distribution line is estimated by a voltage observer and then compared with the global reference voltage. The difference is then given to a proportional–integral (PI) controller to produce the first voltage correction term. The mismatch between the current of a converter and its neighbors, fed into a PI controller, computes the second voltage correction term. Enhanced dynamic performance [30], finite settling time [31], or reduced communication traffic [32, 33] have been studied. All such cooperative control methods rely on message passing among power electronics converters on a communication network, local sensing of voltage/current, and decentralized droop mechanisms.

Cyber manipulation has already been a subject of research in power systems. DC microgrids have become a cyber-physical system by integrating the physical layer of power electronics and the distribution network with the cyber layer of distributed local controllers and a sparse communication network. The distributed cooperative control framework makes these microgrids potentially vulnerable to malicious attacks and infiltration, as this control framework depends on local sensing and networked control on a sparse communication network [3, 10, 34]. Cyber-risk assessment for industrial control systems, including supervisory or distributed control systems, has been detailed in [35] to quantify the attack probability and its impact. The vulnerability assessments of the FDIs on the state estimation process are studied in [11, 12] for power systems.

To address attacks on cyber-physical systems, misbehaving agents are detected, identified, and then removed/overcome in [22, 36–42]. Recently, [7, 43, 44] have investigated attack detection strategies for DC microgrids. This line of approach usually has strict requirements on the number of misbehaving agents. For example, [22, 38] show the impracticality involved in reconstructing the system state if less than half of the sensors are healthy. Moreover, simply removing/isolating the compromised agents could potentially damage the connectivity of the communication network and hence compromise the consensus performance. Sufficient and necessary conditions, for undetectable attacks to exist, are discussed in [10, 34, 45]. Malicious attacks can bypass existing detection methods by exploring the configuration of the power system. The fundamental limitations of existing attack-detection methods are thoroughly analyzed in [36].

In general, it will be virtually impossible to eliminate every potential attack threat for microgrids, and a path toward resilience is deemed indispensable. Recently, distributed resilient control protocols deal with external disturbances/noises without detecting, identifying, and removing/overcoming the compromised agents [29, 46–50]. A noise-resilient voltage observer is developed in [29] to estimate the average voltage across the DC microgrids. Recently, [51] mitigates attacks on communication links and controller hijacking using a trust-based cooperative control paradigm. Note that the disturbances/noises/faults considered in the above literature have been treated as bounded signals. In practice, malicious attackers may intentionally launch unbounded injections at the cyber-physical system to maximize their damage [22]. Therefore, control strategies resilient to unknown unbounded attacks are of great importance to assure the reliability and security of DC microgrids. Section 20.3 will study the cooperative secondary control of DC microgrids in the presence of unknown unbounded attacks injected into the control input channels.

20.2 Adaptive Resilient Control of AC Microgrids Under Unbounded Actuator Attacks

This section will study unbounded attacks on the input channels of both frequency and voltage control loops of inverters that could deteriorate the cooperative performance and affect the microgrid stability. A fully distributed attack-resilient control framework using adaptive control techniques that, using stability analysis with Lyapunov techniques, are shown to preserve the uniformly ultimately bounded (UUB) consensus for frequency regulation and voltage containment. Moreover, the ultimate bound can be set by adjusting the tuning parameters. That is, the frequency and voltage terms can be tuned to converge to an arbitrarily small neighborhood around their respective reference values.

20.2.1 Preliminaries on Graph Theory and Notations

There are N inverters, with two leader nodes, mapped on a communication network, which is represented by a weighted digraph $\mathcal{G}$. The interactions among the inverters are represented by a subgraph $\mathcal{G}_f = (\mathcal{V}, \mathcal{E}, \mathcal{A})$ with a nonempty finite set of N nodes $\mathcal{V} = \{v_1, v_2, \ldots, v_N\}$, a set of edges or arcs $\mathcal{E} \subset \mathcal{V} \times \mathcal{V}$, and the associated adjacency matrix $\mathcal{A} = [a_{ij}] \in \mathbb{R}^{N \times N}$. Here, the digraph is assumed to be time-invariant, i.e. $\mathcal{A}$ is constant. An edge rooted at node j and ended at node i is denoted by (v_j, v_i), which means information can flow from node j to node i. a_{ij} is the weight of edge (v_j, v_i), and $a_{ij} > 0$ if $(v_j, v_i) \in \mathcal{E}$, otherwise $a_{ij} = 0$. Node j is called a neighbor of node i if $(v_j, v_i) \in \mathcal{E}$. The set of neighbors of node i is denoted as $\mathcal{N}_i = \{j | (v_j, v_i) \in \mathcal{E}\}$. Define the in-degree matrix as $D = \text{diag}(d_i) \in \mathbb{R}^{N \times N}$ with $d_i = \sum_{j=1}^{N} a_{ij}$ and the Laplacian matrix as $\mathcal{L} = D - \mathcal{A}$. There are two leader nodes to issue the upper and lower reference values. g_{ik} is the pinning gain from the (upper/lower) kth leader to the ith inverter, brought together in the diagonal matrix $\mathcal{G}_k = \text{diag}\left(g_{ik}\right)$.

$\sigma_{\min}(\cdot)$ and $\sigma_{\max}(\cdot)$ are the minimum and maximum singular values of a given matrix, respectively. $\mathcal{F}$ and $\mathcal{L}$ denote the sets of $\{1, 2, \dots, N\}$ and $\{N+1, N+2\}$, respectively. $\mathbf{1}_N \in \mathbb{R}^N$ is a column vector where all entries are one. $\otimes$, $\text{diag}\{\cdot\}$, $\|\cdot\|$, and $|\cdot|$ denote the Kronecker product, a block diagonal matrix, the Euclidean norm of a given vector, and the absolute value of a given scalar, respectively.

20.2.2 Conventional Cooperative Secondary Control of AC Microgrids

Conventional secondary control acts as an actuator by providing the input control signals for tuning the setpoints of decentralized primary controls. These primary droop mechanisms are given by the following for the ith inverter

$$\omega_i = \omega_{n_i} - m_{P_i} P_i \tag{20.1}$$

$$v_{\text{odi}} = V_{n_i} - n_{Q_i} Q_i \tag{20.2}$$

where P_i and Q_i are the active and reactive powers, respectively. ω_i and v_{odi} are the operating angular frequency and terminal voltage, respectively. ω_{n_i} and V_{n_i} are the setpoints for the primary droop mechanisms fed from the secondary control layer. m_{P_i} and n_{Q_i} are $P - \omega$ and $Q - v$ droop coefficients selected for each inverter's power ratings.

Differentiate the droop relations in (20.1) and (20.2), with respect to time, to obtain

$$\dot{\omega}_{n_i} = \dot{\omega}_i + m_{P_i} \dot{P}_i = u_{f_i} \tag{20.3}$$

$$\dot{V}_{n_i} = \dot{v}_{\text{odi}} + n_{Q_i} \dot{Q}_i = u_{v_i} \tag{20.4}$$

where u_{f_i} and u_{v_i} are auxiliary control inputs. To synchronize the terminal frequency and voltage of each inverter to their respective references, the leader-follower containment-based secondary control is adopted [52]. The local cooperative frequency and voltage control protocols using the relative information with respect to the neighboring inverters and the leaders are given by

$$u_{f_i} = c_{f_i} \left(\sum_{j \in \mathcal{F}} a_{ij} \left(\omega_j - \omega_i\right) + \sum_{k \in \mathcal{L}} g_{ik} \left(\omega_k - \omega_i\right) + \sum_{j \in \mathcal{F}} a_{ij} \left(m_{P_j} P_j - m_{P_i} P_i\right) \right) \tag{20.5}$$

$$u_{v_i} = c_{v_i} \left(\sum_{j \in \mathcal{F}} a_{ij} \left(v_{\text{odj}} - v_{\text{odi}}\right) + \sum_{k \in \mathcal{L}} g_{ik} \left(v_k - v_{\text{odi}}\right) + \sum_{j \in \mathcal{F}} a_{ij} \left(n_{Q_j} Q_j - n_{Q_i} Q_i\right) \right) \tag{20.6}$$

where c_{f_i}, c_{v_i} are positive constant coupling gains. ω_k and v_k are the frequency and voltage reference values of the kth leader, respectively. The frequency reference for both leaders is set as ω_{ref}. The upper and lower leaders have their voltage reference values set as v_{ref}^u and v_{ref}^l, respectively. The setpoints for the primary-level droop control, ω_{n_i} and V_{n_i}, are then computed from u_{f_i} and u_{v_i} as

$$\omega_{n_i} = \int u_{f_i} dt \tag{20.7}$$

$$V_{n_i} = \int u_{v_i} dt \tag{20.8}$$

Using (20.5) and (20.6) to rewrite (20.3) and (20.4) yields

$$\dot{\omega}_{n_i} = c_{f_i} \left(\sum_{j \in \mathcal{F}} a_{ij} \left(\omega_{n_j} - \omega_{n_i}\right) + \sum_{k \in \mathcal{L}} g_{ik} \left(\omega_{n_k} - \omega_{n_i}\right) \right) \tag{20.9}$$

$$\dot{V}_{n_i} = c_{v_i} \left(\sum_{j \in \mathcal{F}} a_{ij} \left(V_{n_j} - V_{n_i} \right) + \sum_{k \in \mathcal{L}} g_{ik} \left(V_{n_k} - V_{n_i} \right) \right) \tag{20.10}$$

where $\omega_{n_k} = \omega_k + m_{P_i} P_i$ and $V_{n_k} = v_k + n_{Q_i} Q_i$. Define $\Phi_k = \frac{1}{2}\mathcal{L} + \mathcal{G}_k$. Then, the global forms of (20.9) and (20.10) are

$$\dot{\omega}_n = -\text{diag}\left(c_{f_i} \right) \sum_{k \in \mathcal{L}} \Phi_k \left(\omega_n - \mathbf{1}_N \otimes \omega_{n_k} \right) \tag{20.11}$$

$$\dot{V}_n = -\text{diag}\left(c_{v_i} \right) \sum_{k \in \mathcal{L}} \Phi_k \left(V_n - \mathbf{1}_N \otimes V_{n_k} \right) \tag{20.12}$$

where $\omega_n = [\omega_{n_1}^T, \ldots, \omega_{n_N}^T]^T$ and $V_n = [V_{n_1}^T, \ldots, V_{n_N}^T]^T$. Define the global frequency and voltage containment error vectors as

$$e_f = \omega_n - \left(\sum_{r \in \mathcal{L}} \Phi_r \right)^{-1} \sum_{k \in \mathcal{L}} \Phi_k \left(\mathbf{1}_N \otimes \omega_{n_k} \right) \tag{20.13}$$

$$e_v = V_n - \left(\sum_{r \in \mathcal{L}} \Phi_r \right)^{-1} \sum_{k \in \mathcal{L}} \Phi_k \left(\mathbf{1}_N \otimes V_{n_k} \right) \tag{20.14}$$

Definition 20.1 The secondary frequency control objective is to make the local frequency of each inverter converge to the range of the two frequency references issued by the upper and lower leaders. Since these two reference values are identical, frequency regulation is achieved.

Definition 20.2 The secondary voltage containment control objective is to make each inverter voltage converge to the range spanned by the two references of the upper and lower leaders.

The following assumption is needed for the communication graph topology to guarantee cooperative consensus.

Assumption 20.1 The communication graph $\mathcal{G}$ includes a directed path from, at least, one leader to each inverter.

Lemma 20.1 *[53] Suppose that Assumption* 20.1 *holds;* $\sum_{k \in \mathcal{L}} \Phi_k$ *is non-singular and positive-definite. Moreover, the frequency and voltage containment control objectives are achieved if* $\lim_{t \to \infty} e_f(t) = 0$ *and* $\lim_{t \to \infty} e_v(t) = 0$, *respectively.*

20.2.3 Problem Formulation

This section formulates the resilient secondary frequency and voltage control problems for a networked AC microgrid. In particular, the general unknown unbounded attack injections to the local input channels of both frequency and voltage control loops are considered, which modify (20.9) and (20.10) to

$$\dot{\omega}_{n_i} = c_{f_i} \left(\sum_{j \in \mathcal{F}} a_{ij} \left(\omega_{n_j} - \omega_{n_i} \right) + \sum_{k \in \mathcal{L}} g_{ik} \left(\omega_{n_k} - \omega_{n_i} \right) \right) + \varpi_{f_i} \tag{20.15}$$

$$\dot{V}_{n_i} = c_{v_i} \left(\sum_{j \in \mathcal{F}} a_{ij} \left(V_{n_j} - V_{n_i} \right) + \sum_{k \in \mathcal{L}} g_{ik} \left(V_{n_k} - V_{n_i} \right) \right) + \varpi_{v_i} \tag{20.16}$$

where ϖ_{f_i} and ϖ_{v_i} denote the unbounded attack signals injected to the input channels of frequency and voltage control loops at the ith inverter, respectively.

Assumption 20.2 $\dot{\varpi}_{f_i}$ and $\dot{\varpi}_{v_i}$ are bounded.

Remark 20.1 Assumption 20.2 is reasonable since attack signals, with an excessively large change in values, could be easily detected in practice. In the event that the attacker does launch an attack signal with an infinite magnitude of the rate of change, the microgrids can incorporate a defensive mechanism to detect and reject such an injection. Since the intentionally injected attacks could be unbounded, the bounded noises and/or disturbances that are unintentionally caused can also be addressed using the attack-resilient controller to be designed.

Since ϖ_{f_i} and ϖ_{v_i} are unbounded, conventional cooperative control protocols fail to regulate the frequency and voltage terms. One then needs attack-resilient control methods to preserve the frequency regulation and voltage containment performances and to ensure closed-loop stability. The following convergence definition is needed.

Definition 20.3 *[54]* Signal $x(t)$ is UUB with an ultimate bound b if there exist positive constants b and c, independent of $t_0 \geq 0$ and, for every $a \in (0, c)$, there exist $t_1 = t_1(a, b) \geq 0$, independent of t_0, such that $\|x(t_0)\| \leq a \Rightarrow \|x(t)\| \leq b, \forall t \geq t_0 + t_1$.

Now, the following distributed resilient secondary frequency and voltage control problems are defined.

Definition 20.4 *(Attack-Resilient Frequency Control Problem)* The goal is to design an input control signal u_{f_i} in (20.3) for each inverter such that e_f in (20.13) is UUB under unbounded attacks to the local frequency control loop. That is, the inverter frequency goes to a small neighborhood around the reference value.

Definition 20.5 *(Attack-Resilient Voltage Control Problem)* The goal is to design an input control signal u_{v_i} in (20.4) for each inverter such that e_v in (20.14) is UUB under unbounded attacks to the local voltage control loop. That is, each inverter voltage goes to a small neighborhood around the range spanned by the two upper and lower references.

20.2.4 Distributed Resilient Controller Design

In this section, a fully distributed control method to solve the attack-resilient frequency and voltage control problems is developed. For convenience, denote

$$\zeta_{f_i} = \sum_{j \in \mathcal{F}} a_{ij} \left(\omega_{n_j} - \omega_{n_i} \right) + \sum_{k \in \mathcal{L}} g_{ik} \left(\omega_{n_k} - \omega_{n_i} \right) \tag{20.17}$$

$$\zeta_{v_i} = \sum_{j \in \mathcal{F}} a_{ij} \left(V_{n_j} - V_{n_i} \right) + \sum_{k \in \mathcal{L}} g_{ik} \left(V_{n_k} - V_{n_i} \right) \tag{20.18}$$

Then, the following attack-resilient control framework for both frequency and voltage control loops is designed

$$\begin{cases} \dot{\omega}_{n_i} = \left(\rho_{f_i} + \dot{\rho}_{f_i} \right) \zeta_{f_i} + \varpi_{f_i}, \\ \dot{\rho}_{f_i} = \chi_{f_i} \left| \zeta_{f_i} \right| \end{cases} \tag{20.19}$$

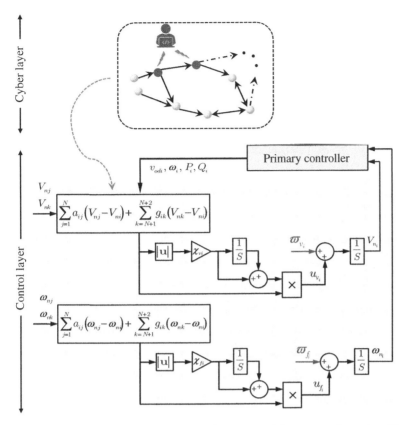

Figure 20.2 Communication layer among inverters and the proposed attack-resilient secondary control framework for an inverter.

$$\begin{cases} \dot{V}_{n_i} = \left(\rho_{v_i} + \dot{\rho}_{v_i} \right) \zeta_{v_i} + \varpi_{v_i}, \\ \dot{\rho}_{v_i} = \chi_{v_i} |\zeta_{v_i}| \end{cases} \tag{20.20}$$

where χ_{f_i} and χ_{v_i} are given positive constants, and ρ_{f_i} and ρ_{v_i} are time-varying coupling weights, with $\rho_{f_i}(0) \geq 0$ and $\rho_{v_i}(0) \geq 0$. Figure 20.2 shows the communication network among inverters and the proposed secondary control for an inverter.

Theorem 20.1 Under Assumptions 20.1 and 20.2 and using the cooperative resilient frequency control protocols consisting of (20.17) and (20.19), e_f in (20.13) is UUB. Furthermore, by increasing χ_{f_i} in (20.19), the ultimate bound of e_f can be adjusted to be an arbitrarily small value, i.e. inverter frequency converges to an arbitrarily small neighborhood around the reference value.

Proof: Consider the Lyapunov function candidate:

$$E = \frac{1}{2} \sum_{i=1}^{N} \int_0^{\zeta_{f_i}^2(t)} \left(\rho_{f_i}(s) + \dot{\rho}_{f_i}(s) \right) ds \tag{20.21}$$

Combine (20.19) and (20.21) to obtain

$$
\begin{aligned}
\dot{E} &= \frac{1}{2}\sum_{i=1}^{N}\left(\rho_{f_i}+\dot{\rho}_{f_i}\right)2\zeta_{f_i}\dot{\zeta}_{f_i} \\
&= \zeta_f^T \operatorname{diag}\left(\rho_{f_i}+\dot{\rho}_{f_i}\right)\dot{\zeta}_f \\
&= \zeta_f^T \operatorname{diag}\left(\rho_{f_i}+\dot{\rho}_{f_i}\right)\left(-\sum_{k\in\mathscr{L}}\Phi_k\dot{\omega}_n\right) \\
&= -\zeta_f^T \operatorname{diag}\left(\rho_{f_i}+\dot{\rho}_{f_i}\right)\left(\sum_{k\in\mathscr{L}}\Phi_k\right)\left(\operatorname{diag}\left(\rho_{f_i}+\dot{\rho}_{f_i}\right)\zeta_f+\varpi_f\right)
\end{aligned}
\tag{20.22}
$$

where $\zeta_f=[\zeta_{f_1}^T,\ldots,\zeta_{f_N}^T]^T$. Recalling the Sylvester's inequality and noting that $\sum_{k\in\mathscr{L}}\Phi_k$ is positive-definite, one then obtain

$$
\begin{aligned}
\dot{E} &\leq -\sigma_{\min}\left(\sum_{k\in\mathscr{L}}\Phi_k\right)\left\|\operatorname{diag}\left(\rho_{f_i}+\dot{\rho}_{f_i}\right)\zeta_f\right\|^2 \\
&\quad +\sigma_{\max}\left(\sum_{k\in\mathscr{L}}\Phi_k\right)\left\|\operatorname{diag}\left(\rho_{f_i}+\dot{\rho}_{f_i}\right)\zeta_f\right\|\left\|\varpi_f\right\| \\
&\leq -\sigma_{\min}\left(\sum_{k\in\mathscr{L}}\Phi_k\right)\left\|\operatorname{diag}\left(\rho_{f_i}+\dot{\rho}_{f_i}\right)\zeta_f\right\| \\
&\quad \times\left(\left\|\operatorname{diag}\left(\rho_{f_i}+\dot{\rho}_{f_i}\right)\zeta_f\right\|-\frac{\sigma_{\max}\left(\sum_{k\in\mathscr{L}}\Phi_k\right)}{\sigma_{\min}\left(\sum_{k\in\mathscr{L}}\Phi_k\right)}\left\|\varpi_f\right\|\right)
\end{aligned}
\tag{20.23}
$$

The next step is to prove that $\exists\tau>0$, such that

$$
\left\|\operatorname{diag}\left(\rho_{f_i}+\dot{\rho}_{f_i}\right)\zeta_f\right\|\geq\frac{\sigma_{\max}\left(\sum_{k\in\mathscr{L}}\Phi_k\right)}{\sigma_{\min}\left(\sum_{k\in\mathscr{L}}\Phi_k\right)}\left\|\varpi_f\right\|,\forall t\geq\tau
\tag{20.24}
$$

A sufficient condition to guarantee (20.24) is

$$
\left|\left(\rho_{f_i}+\dot{\rho}_{f_i}\right)\zeta_{f_i}\right|\geq\frac{\sigma_{\max}\left(\sum_{k\in\mathscr{L}}\Phi_k\right)}{\sigma_{\min}\left(\sum_{k\in\mathscr{L}}\Phi_k\right)}\left|\varpi_{f_i}\right|,\quad\forall t\geq\tau
\tag{20.25}
$$

Since both ρ_{f_i} and $\dot{\rho}_{f_i}$ are non-negative, the following sufficient condition is further obtained

$$
\rho_{f_i}\left|\zeta_{f_i}\right|\geq\frac{\sigma_{\max}\left(\sum_{k\in\mathscr{L}}\Phi_k\right)}{\sigma_{\min}\left(\sum_{k\in\mathscr{L}}\Phi_k\right)}\left|\varpi_{f_i}\right|,\quad\forall t\geq\tau
\tag{20.26}
$$

Note that (20.26) is guaranteed if both $\rho_{f_i}\geq\left|\varpi_{f_i}\right|$ and $\left|\zeta_{f_i}\right|\geq\frac{\sigma_{\max}\left(\sum_{k\in\mathscr{L}}\Phi_k\right)}{\sigma_{\min}\left(\sum_{k\in\mathscr{L}}\Phi_k\right)}$ hold. Since

$$
\frac{d\left|\varpi_{f_i}\right|}{dt}=\frac{\varpi_{f_i}\dot{\varpi}_{f_i}}{\left|\varpi_{f_i}\right|}\leq\left|\dot{\varpi}_{f_i}\right|
\tag{20.27}
$$

from Assumption 2, $\ddot{\varpi}_{f_i}$ is bounded. Hence, $\frac{d|\varpi_{f_i}|}{dt}$ is also bounded. Using (20.19) and choosing

$$
\left|\zeta_{f_i}\right| \geq \max \left\{ \frac{\sigma_{\max}\left(\sum_{k \in \mathscr{L}} \Phi_k\right)}{\sigma_{\min}\left(\sum_{k \in \mathscr{L}} \Phi_k\right)}, \ \frac{1}{\chi_{f_i}} \frac{d\left|\varpi_{f_i}\right|}{dt} \right\} \tag{20.28}
$$

one then obtains that $\exists \tau > 0$, such that (20.26) holds. Furthermore, (20.24) holds. Using (20.23), one now obtains that $\forall t \geq \tau$

$$
\dot{E} \leq 0, \quad \forall \left|\zeta_{f_i}\right| \geq \max \left\{ \frac{\sigma_{\max}\left(\sum_{k \in \mathscr{L}} \Phi_k\right)}{\sigma_{\min}\left(\sum_{k \in \mathscr{L}} \Phi_k\right)}, \ \frac{1}{\chi_{f_i}} \frac{d\left|\varpi_{f_i}\right|}{dt} \right\} \tag{20.29}
$$

Therefore, ζ_{f_i} is bounded. Note that

$$
\zeta_f = \sum_{k \in \mathscr{L}} \Phi_k e_f \tag{20.30}
$$

Hence, e_f is also bounded. Moreover, using LaSalle's invariance principle [55], it is seen from (20.29) that ζ_{f_i} is bounded by $\max \left\{ \frac{\sigma_{\max}(\sum_{k \in \mathscr{L}} \Phi_k)}{\sigma_{\min}(\sum_{k \in \mathscr{L}} \Phi_k)}, \frac{1}{\chi_{f_i}} \frac{d|\varpi_{f_i}|}{dt} \right\}$, where $\frac{\sigma_{\max}(\sum_{k \in \mathscr{L}} \Phi_k)}{\sigma_{\min}(\sum_{k \in \mathscr{L}} \Phi_k)}$ is a positive constant. Hence, the ultimate bound can be reduced by properly increasing the adaptive tuning parameter ϖ_{f_i} in (20.19). ∎

Theorem 20.2 Under Assumptions 20.1 and 20.2 and using the cooperative resilient voltage control protocols consisting of (20.18) and (20.20), e_v in (20.14) is UUB. Furthermore, by increasing χ_{v_i} in (20.20), the ultimate bound of e_v can be set arbitrarily small, i.e. the inverter voltage converges to an arbitrarily small neighborhood around the range covered by the two references.

Proof: The proof follows that of Theorem 20.1. ∎

Remark 20.2 To mitigate the propagated adverse effects caused by the unbounded actuator attacks, ϖ_{f_i} and ϖ_{v_i}, the time-varying coupling weights, ρ_{f_i} and ρ_{v_i}, are designed based on adaptive tuning laws. As seen from the proof of Theorem 20.1, such adaptively updated coupling weights can successfully compensate for the externally injected attack signals.

20.2.5 Case Studies

The proposed resilient control method is studied in the context of an IEEE 34-bus feeder system, islanded at bus 800, and augmented with four inverters and two leaders, as shown in Fig. 20.3. The specifications of inverters and their grid interconnections are adopted from [1, 56], respectively. Inverters 1 and 2 have twice the power ratings of inverters 3 and 4. The inverter droop gains are set as $m_{P_1} = m_{P_2} = 9.4 \times 10^{-5}$, $m_{P_3} = m_{P_4} = 18.8 \times 10^{-5}$, $n_{Q_1} = n_{Q_2} = 1.3 \times 10^{-3}$, and $n_{Q_3} = n_{Q_4} = 2.6 \times 10^{-3}$. Inverters communicate on a bidirectional communication network with the adjacency matrix of $\mathcal{A} = [0\ 1\ 0\ 1; 1\ 0\ 1\ 0; 0\ 1\ 0\ 1; 1\ 0\ 1\ 0]$. The pinning gains are $g_{15} = g_{36} = 1$.

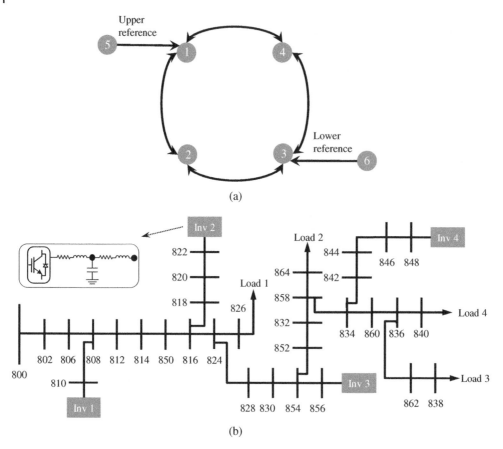

Figure 20.3 Cyber-physical microgrid system: (a) communication graph topology among four inverters and two leaders and (b) IEEE 34-bus system with four inverters.

The frequency reference, upper voltage reference, and lower voltage reference are 60 Hz, 340, and 330 V, respectively. The unbounded attack injections to the frequency and voltage control loops are set as $\varpi_{f_i} = 1\,\text{t}, i = 1, 2, 3, 4$ and $\varpi_{v_i} = 10\,\text{t}, i = 1, 2, 3, 4$, respectively. The performance of the resilient control protocols, (20.17)–(20.20), is compared with the conventional secondary control method in (20.5) and (20.6). The coupling gains for the conventional control protocols are set as $c_{f_i} = 10, c_{v_i} = 20, i = 1, 2, 3, 4$. The adaptive tuning parameters for the resilient control method are set as $\chi_{f_i} = 3, \chi_{v_i} = 3, i = 1, 2, 3, 4$.

Figure 20.4 compares the frequency response for the proposed and the conventional methods. Under ideal conditions (no attacks), inverters' frequencies synchronize to $f = 60\,\text{Hz}$ using both control methods. Once the unbounded attack to frequency control loops is initiated at $t = 4\,\text{s}$, the conventional method fails to preserve the system stability. By contrast, the proposed resilient method contains frequencies at a small neighborhood around 60 Hz. Figure 20.5 shows that, without attacks, both methods share active powers among inverters based on their droop gains. After initiating the unbounded attacks to frequency control loops at $t = 4\,\text{s}$, the active power performance from the conventional method becomes unstable. Meanwhile, the proposed method

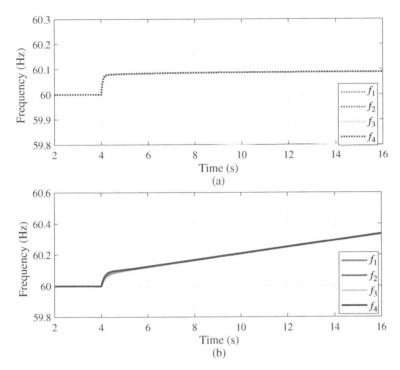

Figure 20.4 Frequency response under unbounded actuator attacks: (a) proposed resilient method and (b) conventional control method.

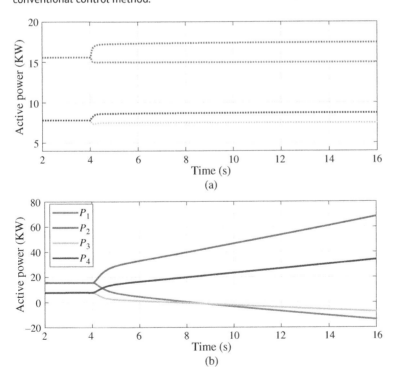

Figure 20.5 Active powers of inverters subjected to unbounded actuator attacks: (a) proposed resilient method and (b) conventional method.

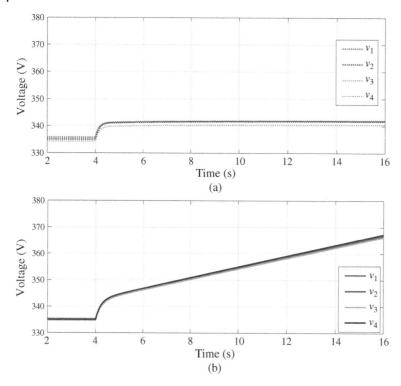

Figure 20.6 Voltage performance under unbounded actuator attacks: (a) proposed resilient method and (b) conventional control method.

contains active powers in a small neighborhood around the value of properly shared powers. Figure 20.6 compares inverters' voltages using both control methods. Without attacks, voltage values stay in the range of 330 to 340 V. After initiating the unbounded attacks to voltage control loops at $t = 4$ s, the voltage terms using the conventional method diverge, while those produced by the proposed method remain stable within $330 \sim 340$ V.

The ultimate bound of the UUB convergence can be adjusted to be an arbitrarily small value by increasing the adaptive tuning parameters. Figures 20.7 and 20.8 show the frequency and active

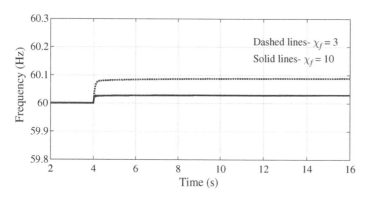

Figure 20.7 Comparative frequency performance under unbounded actuator attacks with different adaptive tuning parameters.

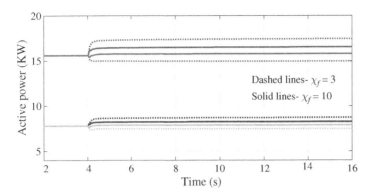

Figure 20.8 Comparative active power performance under unbounded actuator attacks with different adaptive tuning parameters.

power waveforms, where the performance with $\chi_{f_i} = 3$ and $\chi_{f_i} = 10$ are illustrated with solid and dashed lines, respectively. As seen, the ultimate bound can be reduced by increasing χ_{f_i}.

20.3 Distributed Resilient Secondary Control of DC Microgrids Against Unbounded Attacks

This section will study the cooperative secondary control of DC microgrids in the presence of unknown unbounded attacks injected to the control input channels. A fully distributed attack-resilient secondary control framework is established for DC microgrids to mitigate the adverse effect of unbounded attacks. Rigorous proofs, based on Lyapunov techniques, show that the proposed method guarantees the UUB convergence for both global voltage regulation and proportional load-sharing objectives under unbounded attacks. Moreover, the asymptotic stability of the overall closed-loop system is achieved in the presence of bounded attacks. Experimental results are illustrated in a hardware-in-the-loop environment to validate the effectiveness of the proposed approach.

20.3.1 Preliminaries

A physical islanded DC microgrid system is mapped to a communication digraph. This time-invariant communication digraph $\mathcal{G}$ is composed of N converters, one leader node, and some adversarial nodes (attackers). The leader node issues the reference value to the neighboring converters. Each converter receives direct relative information from its neighboring converters and possibly the leader on the sparse communication digraph. The connections among local converters are represented by $\mathcal{G}_f = (\mathcal{W}, \mathcal{E}, \mathcal{A})$ with a node set $\mathcal{W}$, an edges set $\mathcal{E} \subset \mathcal{W} \times \mathcal{W}$, and an adjacency matrix $\mathcal{A} = [a_{ij}]$. A graph edge, indicating the information flow from converter j to converter i, is shown by (w_j, w_i), with the weight of a_{ij}. If $(w_j, w_i) \in \mathcal{E}$, then $a_{ij} > 0$; Otherwise, $a_{ij} = 0$. Node j is considered as the neighbor of node i if $(w_j, w_i) \in \mathcal{E}$. The set of neighbors of node i is denoted as $\mathcal{N}_i = \{j | (w_j, w_i) \in \mathcal{E}\}$. $\mathcal{D} = \mathrm{diag}(d_i) \in \mathbb{R}^{N \times N}$, with $d_i = \sum_{j \in \mathcal{N}_i} a_{ij}$, is called the in-degree matrix. $\mathcal{L} = \mathcal{D} - \mathcal{A}$ represents the Laplacian matrix. $\mathcal{G}_f$ is assumed bidirectional, i.e. $a_{ij} = a_{ji}$. Hence $\mathcal{L}$ is symmetric. g_i is the pinning gain for the link from the leader to the ith converter. $g_i > 0$ if the leader links to the ith converter; otherwise, $g_i = 0$. $\mathcal{G} = \mathrm{diag}(g_i), \forall i = 1, \ldots, N$, represents a diagonal matrix composed of pinning gains. Digraph $\mathscr{G}$ is said to have a spanning tree, if there is a node i_r (called the root), such that there is a directed path from the root to every other node in the graph.

20.3.2 Standard Cooperative Secondary Control

Figure 20.9 illustrates the physical, control, and communication layers of the DC microgrids. As seen from the standard secondary control block, each converter transmits $X_i = \left[\overline{V}_i, R_i^{\mathrm{vir}} I_i\right]$ to its neighboring converters on a communication graph. $\overline{V}_i$ is the estimated average voltage across the microgrid, I_i is the output current of the ith converter, and R_i^{vir} is a virtual impedance tuned as $R_i^{\mathrm{vir}} = k / I_i^{\mathrm{rated}}$, where k is a design parameter and I_i^{rated} is the rated current of the ith converter. Note that $R_i^{\mathrm{vir}} I_i = k I_i / I_i^{\mathrm{rated}}$. Hence, achieving the proportional load sharing is equivalent to achieving the consensus on terms of $R_i^{\mathrm{vir}} I_i$.

A standard cooperative secondary control method for DC microgrids will be first presented, by transforming it into the consensus control for first-order linear MAS. The two main objectives of the secondary/primary control are to regulate the average voltage across the microgrids to a global reference value and proportionally share the load. As illustrated in Fig. 20.9, the standard secondary control uses the relative information of a converter with respect to the neighboring converters to adjust its local voltage setpoint V_i^*. The primary-level droop mechanism acts on local information and models the converter's output impedance with a virtual impedance R_i^{vir}. Cooperative secondary control among neighboring converters helps properly tune the local voltage setpoint V_i^* and attenuate the voltage and current residuals. The local voltage setpoint is given as

$$V_i^* = V_{n_i} + V_{\mathrm{ref}} - R_i^{\mathrm{vir}} I_i \tag{20.31}$$

where V_{n_i} is the reference for the primary control level and is selected at the secondary control level.

To achieve the global voltage regulation and proportional load sharing, the secondary control provides V_{n_i} locally for each converter by exchanging data with its neighboring converters. Using input–output feedback linearization techniques, differentiate the voltage droop mechanism in (20.31) to obtain

$$\dot{V}_{n_i} = \dot{V}_i^* + R_i^{\mathrm{vir}} \dot{I}_i = u_i \tag{20.32}$$

where u_i is the control input. Equation (20.32) computes the primary control reference V_{n_i} from u_i. The secondary control of DC microgrids is then transformed into a consensus problem for first-order linear MAS. In order to provide global voltage regulation and proportional load sharing, the cooperative secondary control law at each converter, based on the relative information with respect to the neighboring converters, is given by

$$u_i = c_i \left(g_i \left(V_{\mathrm{ref}} - \overline{V}_i \right) + \sum_{j \in \mathcal{N}_i} a_{ij} \left(R_j^{\mathrm{vir}} I_j - R_i^{\mathrm{vir}} I_i \right) \right) \tag{20.33}$$

where $c_i \in \mathbb{R} > 0$ is the coupling gain. $\overline{V}_i$ is the estimate of the global average voltage value at converter i and is given by

$$\dot{\overline{V}}_i = \dot{V}_i + c_i \sum_{j \in \mathcal{N}_i} a_{ij} \left(\overline{V}_j - \overline{V}_i \right) \tag{20.34}$$

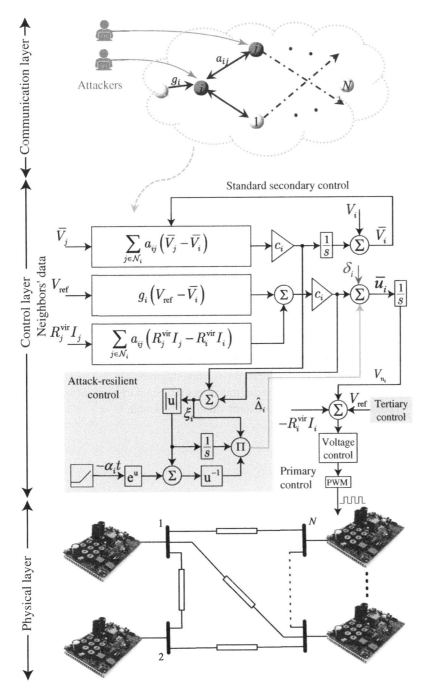

Figure 20.9 Cyber-physical DC microgrids: Communication layer for data exchange among converters, control layer including standard and resilient cooperative control modules, and the physical layer including converters, distribution line, and sources/loads.

where V_i is the local measured voltage. The secondary control setpoint for the primary control, V_{n_i}, is then computed from u_i as

$$V_{n_i} = \int u_i dt \tag{20.35}$$

Assume that the converter produces the demanded voltage, i.e. $V_i^* = V_i$. Combining (20.32), (20.43), and (20.34) yields

$$\dot{\overline{V}}_i + R_i^{\text{vir}} \dot{I}_i = c_i \left(\sum_{j \in \mathcal{N}_i} a_{ij} \left(\overline{V}_j - \overline{V}_i \right) + g_i \left(V_{\text{ref}} - \overline{V}_i \right) \right.$$
$$\left. + \sum_{j \in \mathcal{N}_i} a_{ij} \left(R_j^{\text{vir}} I_j - R_i^{\text{vir}} I_i \right) \right) \tag{20.36}$$

which are further formulated as

$$\dot{\overline{V}}_i + R_i^{\text{vir}} \dot{I}_i = c_i \left(\sum_{j \in \mathcal{N}_i} a_{ij} \left(\left(\overline{V}_j + R_j^{\text{vir}} I_j \right) - \left(\overline{V}_i + R_i^{\text{vir}} I_i \right) \right) \right.$$
$$\left. + g_i \left(\left(V_{\text{ref}} + R_i^{\text{vir}} I_i \right) - \left(\overline{V}_i + R_i^{\text{vir}} I_i \right) \right) \right) \tag{20.37}$$

[29] gives the detailed steady-state analysis to show that the cooperative secondary control achieves both voltage and current regulation objectives, due to the relationship between the supplied currents and the bus voltage through the microgrid admittance matrix. Similarly, one obtains that, in the steady state, $R_i^{\text{vir}} I_i$ converges to a certain constant value $k I_{ss}^{\text{pu}}$. Denote $\Theta_i = \overline{V}_i + R_i^{\text{vir}} I_i$ and $\Theta_{\text{ref}} = V_{\text{ref}} + k I_{ss}^{\text{pu}}$. Then,

$$\dot{\Theta}_i = c_i \left(\sum_{j \in \mathcal{N}_i} a_{ij} \left(\Theta_j - \Theta_i \right) + g_i \left(\Theta_{\text{ref}} - \Theta_i \right) \right)$$
$$= c_i \left(-\left(d_i + g_i \right) \Theta_i + \sum_{j \in \mathcal{N}_i} a_{ij} \Theta_j + g_i \Theta_{\text{ref}} \right) \tag{20.38}$$

The global form of (20.38) is

$$\dot{\Theta} = -\text{diag}\left(c_i \right) \left(\mathcal{L} + \mathcal{G} \right) \left(\Theta - \mathbf{1}_N \Theta_{\text{ref}} \right) \tag{20.39}$$

where $\Theta = [\Theta_1^T, \ldots, \Theta_N^T]^T$.
Define the following global cooperative regulation error

$$\varepsilon = \Theta - \mathbf{1}_N \Theta_{\text{ref}} \tag{20.40}$$

where $\varepsilon = [\varepsilon_1^T, \ldots, \varepsilon_N^T]^T$. The following assumption is needed for the communication network.

Assumption 20.3   The digraph $\mathcal{G}$ includes a spanning tree, where the leader node is the root.

Lemma 20.2   *[57]   Given Assumption 20.3, $(\mathcal{L} + \mathcal{G})$ is non-singular and positive-definite.*

Lemma 20.3   *Given Assumption 20.3, by designing the auxiliary control input as (20.43) and (20.34), the global voltage regulation and proportional load sharing are both achieved.*

Proof: Use (20.39) and (20.40) to obtain

$$\dot{\varepsilon} = -\text{diag}\left(c_i\right)\left(\mathcal{L} + \mathcal{G}\right)\varepsilon \tag{20.41}$$

From Lemma 20.2, $(\mathcal{L} + \mathcal{G})$ is positive-definite. Hence, $\varepsilon \to 0$ asymptotically, i.e. $\Theta_i \to \Theta_{\text{ref}}$. Due to the relationship between the supplied currents and the bus voltage, one obtains that $\overline{V}_i \to V_{\text{ref}}$ and $R_i^{\text{vir}} I_i \to k I_{ss}^{\text{pu}}$, simultaneously. Steady-state analysis in [29] shows that since $\mathcal{L}$ is symmetric, $\overline{V}_i$ converges to the global average voltage value by using observer (20.34). Hence, the average value of $V_l, \forall i = 1, \ldots, N$ converges to V_{ref}. That is, both the global voltage regulation and proportional load sharing are achieved. ∎

Remark 20.3 Using the input–output feedback linearization techniques, the secondary control problem of DC microgrids is transformed into the tracking synchronization problem of first-order linear MAS. Then, by using the standard cooperative control protocols for the first-order linear MAS, the standard secondary control protocols (20.43) and (20.34) are designed to achieve the global voltage regulation and proportional load sharing. Compared to the double PI controllers in [29], this formulation provides a faster dynamic response.

20.3.3 Resilient Secondary Control

In this section, the attack-resilient secondary control problems for DC microgrids under unbounded and bounded attacks are first formulated, respectively. Then, a fully-distributed adaptive control framework is developed to address these problems. Rigorous proofs based on Lyapunov techniques show that UUB and asymptotically stable (AS) convergences are achieved for voltage and current regulations against unbounded and bounded attacks, respectively.

20.3.3.1 Attack-Resilient Secondary Control Problem Formulation
As illustrated in Fig. 20.9, malicious attackers may inject unknown unbounded exogenous signals to perturb the local control input channel of each converter. Hence, instead of (20.32), one has

$$\dot{V}_{n_i} = \dot{V}_i^* + R_i^{\text{vir}} \dot{I}_i = \bar{u}_i = u_i + \delta_i \tag{20.42}$$

where $\bar{u}_i$ is the corrupted control input and δ_i denotes the potential unbounded injections into the local control input channel. The attacker aims at destabilizing the cooperative regulation system by inserting these unbounded attacks.

Assumption 20.4 For each converter, $\dot{\delta}_i$ is bounded.

Remark 20.4 The attackers' injections, δ_i, can be any unbounded signals satisfying Assumption 20.4 or any bounded signals. Compared with the noise-resilient control protocols for DC microgrids in [29, 51], which deal with bounded noises/disturbances, the more practical and challenging case of unbounded attack injections is considered.

Consider the attack injections in (20.42) and use the standard secondary control protocols (20.43) and (20.34). Then, instead of the closed-loop error dynamics in (20.41), one obtains

$$\dot{\varepsilon} = -\text{diag}\left(c_i\right)\left(\mathcal{L} + \mathcal{G}\right)\varepsilon + \delta \tag{20.43}$$

where $\delta = [\delta_1^T, \ldots, \delta_N^T]^T$ is the attack vector. Since δ is unbounded, $\varepsilon \to \infty$. That is, the standard secondary control fails to preserve the stability of the DC microgrids in the presence of unbounded attacks. It is hence important to develop an advanced attack-resilient control approach to address such unbounded attacks for microgrids.

Now, the attack-resilient secondary control problems for DC microgrids against unbounded and bounded attacks are formulated, respectively.

Problem 20.1 Under the unknown unbounded cyberattacks on local control input channels, design local control protocols u_i in (20.42) for each converter using only the local measurement such that, for all initial conditions, ε in (20.40) is UUB. That is, the bounded global voltage regulation and proportional load sharing are both achieved.

Problem 20.2 Under the unknown bounded attacks on local control input channels, design local control protocols u_i in (20.42) for each converter using only the local measurement such that, for all initial conditions, ε in (20.40) is AS. That is, the exact global voltage regulation and proportional load sharing are both achieved.

20.3.3.2 Distributed Attack-Resilient Secondary Controller Design
For convenience, denote

$$
\xi_i = c_i \left(\sum_{j \in \mathcal{N}_i} a_{ij} \left(\overline{V}_j - \overline{V}_i \right) + g_i \left(V_{\text{ref}} - \overline{V}_i \right) \right. \\
\left. + \sum_{j \in \mathcal{N}_i} a_{ij} \left(R_j^{\text{vir}} I_j - R_i^{\text{vir}} I_i \right) \right)
\tag{20.44}
$$

To ensure bounded global voltage regulation and proportional load sharing under unknown unbounded attacks, the following attack-resilient secondary control protocols are developed

$$
u_i = c_i \left[g_i \left(V_{\text{ref}} - \overline{V}_i \right) + \sum_{j \in \mathcal{N}_i} a_{ij} \left(R_j^{\text{vir}} I_j - R_i^{\text{vir}} I_i \right) \right] + \hat{\Delta}_i
\tag{20.45}
$$

$$
\hat{\Delta}_i = \frac{\xi_i \vartheta_i}{|\xi_i| + \exp\left(-\alpha_i t \right)}
\tag{20.46}
$$

$$
\dot{\vartheta}_i = \gamma_i |\xi_i|
\tag{20.47}
$$

where $\hat{\Delta}_i$ is an adaptive compensational term, ϑ_i is an adaptive updating parameter, α_i and γ_i are positive constants. For convenience, set $\gamma_i \geq 1$. The uniform continuous function $\exp\left(-\alpha_i t \right)$ constructs a smooth control approach for practical implementation purposes.

Next, the main result of solving the attack-resilient secondary control problems for DC microgrids is given.

Theorem 20.3 Given Assumptions 1 and 2, under the unknown unbounded attacks in (20.42), let the resilient control protocols consist of (20.34), (20.45), (20.46), and (20.48), then the cooperative regulation error ε in (20.40) is UUB. That is, Problem 20.1 is solved.

Proof: Use (20.34), (20.42), and (20.45) to obtain the time derivative of (20.44)

$$
\dot{\xi}_i = c_i \left(- \left(d_i + g_i \right) \dot{\Theta}_i + \sum_{j \in \mathcal{N}_i} a_{ij} \dot{\Theta}_j \right) \\
= -c_i \left(d_i + g_i \right) \left(\xi_i + \delta_i + \hat{\Delta}_i \right) + c_i \sum_{j \in \mathcal{N}_i} a_{ij} \left(\xi_j + \delta_j + \hat{\Delta}_j \right)
\tag{20.48}
$$

Denote $\Delta_i = \delta_i - \frac{1}{(d_i + g_i)} \sum_{j \in \mathcal{N}_i} a_{ij} \left(\xi_j + \delta_j + \hat{\Delta}_j \right)$. Given Assumption 20.4, $\dot{\Delta}_i$ is bounded. (20.48) is then written as

$$\dot{\xi}_i = -c_i \left(d_i + g_i \right) \left(\xi_i + \Delta_i + \hat{\Delta}_i \right) \tag{20.49}$$

Consider the following Lyapunov function candidate

$$V_i = \frac{1}{2} \left(|\xi_i| - \frac{d |\Delta_i|}{dt} \right)^2 \tag{20.50}$$

and its time derivative is given as

$$\dot{V}_i = \left(|\xi_i| - \frac{d |\Delta_i|}{dt} \right) \left(\frac{d |\xi_i|}{dt} - \frac{d^2 |\Delta_i|}{dt^2} \right) \tag{20.51}$$

Since $\dot{\Delta}_i$ is bounded, and note that $\frac{d |\Delta_i|}{dt} = \frac{\Delta_i \dot{\Delta}_i}{|\Delta_i|} \le |\dot{\Delta}_i|$, both $\frac{d |\Delta_i|}{dt}$ and $\frac{d^2 |\Delta_i|}{dt^2}$ are bounded. Note that

$$\frac{d |\xi_i|}{dt} = \frac{\xi_i \dot{\xi}_i}{|\xi_i|} = \frac{-c_i \left(d_i + g_i \right) \xi_i \left(\xi_i + \Delta_i + \hat{\Delta}_i \right)}{|\xi_i|}$$

$$= -c_i \left(d_i + g_i \right) \left(|\xi_i| + \frac{\xi_i \Delta_i}{|\xi_i|} + \frac{\xi_i \hat{\Delta}_i}{|\xi_i|} \right) \tag{20.52}$$

Substituting (20.52) into (20.51) yields

$$\dot{V}_i = \left(|\xi_i| - \frac{d |\Delta_i|}{dt} \right) \left(-c_i \left(d_i + g_i \right) |\xi_i| - \frac{d^2 |\Delta_i|}{dt^2} - c_i \left(d_i + g_i \right) \left(\frac{\xi_i \Delta_i}{|\xi_i|} + \frac{\xi_i \hat{\Delta}_i}{|\xi_i|} \right) \right) \tag{20.53}$$

Use (20.46) to obtain

$$-c_i \left(d_i + g_i \right) \left(\frac{\xi_i \Delta_i}{|\xi_i|} + \frac{\xi_i \hat{\Delta}_i}{|\xi_i|} \right)$$

$$= -c_i \left(d_i + g_i \right) \frac{\xi_i \Delta_i}{|\xi_i|} - c_i \left(d_i + g_i \right) \frac{|\xi_i| \vartheta_i}{|\xi_i| + \exp \left(-\alpha_i t \right)}$$

$$\le c_i \left(d_i + g_i \right) |\Delta_i| - c_i \left(d_i + g_i \right) \frac{|\xi_i| \vartheta_i}{|\xi_i| + \exp \left(-\alpha_i t \right)} \tag{20.54}$$

$$\le c_i \left(d_i + g_i \right) \frac{|\xi_i| |\Delta_i| + \exp \left(-\alpha_i t \right) |\Delta_i| - |\xi_i| \vartheta_i}{|\xi_i| + \exp \left(-\alpha_i t \right)}$$

Choosing $|\xi_i| > \frac{d |\Delta_i|}{dt}$ yields $\vartheta_i > \frac{d |\Delta_i|}{dt}$. Since $\frac{d |\Delta_i|}{dt}$ is bounded, $\exp \left(-\alpha_i t \right) |\Delta_i| \to 0$. Hence, $\exists \tau_1 > 0$, such that for all $t \ge \tau_1$,

$$|\xi_i| |\Delta_i| + \exp \left(-\alpha_i t \right) |\Delta_i| - |\xi_i| \vartheta_i \le 0 \tag{20.55}$$

Then, use (20.54) and (20.55) to obtain

$$-c_i \left(d_i + g_i \right) \left(\frac{\xi_i \Delta_i}{|\xi_i|} + \frac{\xi_i \hat{\Delta}_i}{|\xi_i|} \right) \le 0, \quad t \ge \tau_1 \tag{20.56}$$

Combining (20.53) and (20.56) yields

$$\dot{V}_i \le \left(|\xi_i| - \frac{d |\Delta_i|}{dt} \right) \left(-c_i \left(d_i + g_i \right) |\xi_i| - \frac{d^2 |\Delta_i|}{dt^2} \right), \quad \forall t \ge \tau_1 \tag{20.57}$$

Choosing $|\xi_i| \geq -\frac{1}{c_i(d_i+g_i)} \frac{d^2|\Delta_i|}{dt^2}$ yields

$$\dot{V}_i \leq 0, \quad \forall t \geq \tau_1 \tag{20.58}$$

That is, ξ_i is UUB. Moreover, using the LaSalle's invariance principle, ξ_i is bounded by

$$\max \left\{ \frac{d|\Delta_i|}{dt}, -\frac{1}{c_i(d_i + g_i)} \frac{d^2|\Delta_i|}{dt^2} \right\} \tag{20.59}$$

Note that

$$\xi = -\mathrm{diag}(c_i)(\mathcal{L} + \mathcal{G})\varepsilon \tag{20.60}$$

where $\xi = [\xi^T, \dots, \xi_N^T]^T$. Since ξ_i is UUB and $(\mathcal{L} + \mathcal{G})$ is non-singular, ε is UUB. This completes the proof. ∎

Finally, it is shown that the global voltage regulation and proportional load sharing are maintained under bounded attacks. That is, the adverse effects of the bounded attacks are completely compensated by the proposed protocols.

Theorem 20.4 Consider the unknown bounded attacks in (20.42). Let the resilient-control protocols consist of (20.34), (20.45), (20.46), and (20.48). Then, the cooperative regulation error ε in (20.40) is AS. That is, Problem 20.2 is solved.

Proof: Consider the error dynamics (20.49). Since δ_i is bounded, Δ_i is also bounded. Denote the supremum value of $|\Delta_i|$ as $\chi_i = \sup_{t \geq 0} |\Delta_i(t)|$. Note that χ_i is constant. Hence, $\dot{\chi}_i = 0$. Let $\tilde{\vartheta}_i = \chi_i - \vartheta_i$, and consider the following Lyapunov function candidate

$$V_i' = \frac{1}{2}\xi_i^2 + \frac{1}{2}\frac{c_i(d_i + g_i)}{\gamma_i}\tilde{\vartheta}_i^2 \tag{20.61}$$

Then, its time derivative is given as

$$\begin{aligned}
\dot{V}_i' &= \xi_i \dot{\xi}_i - \frac{c_i(d_i+g_i)}{\gamma_i}\tilde{\vartheta}_i \dot{\vartheta}_i \\
&= -c_i(d_i + g_i)(\xi_i^2 + \Delta_i \xi_i + \hat{\Delta}_i \xi_i) - c_i(d_i + g_i)(\chi_i - \vartheta_i)|\xi_i| \\
&\leq -c_i(d_i + g_i)|\xi_i|^2 - c_i(d_i + g_i)(\chi_i |\xi_i| - |\Delta_i||\xi_i| + \hat{\Delta}_i \xi_i - \vartheta_i |\xi_i|)
\end{aligned} \tag{20.62}$$

Let $\varpi_i = \chi_i - |\Delta_i|$. Then,

$$\begin{aligned}
\dot{V}_i' &\leq -c_i(d_i + g_i)|\xi_i|^2 - c_i(d_i + g_i)\left(\varpi_i|\xi_i| + \frac{|\xi_i|^2 \vartheta_i}{|\xi_i| + \exp(-\alpha_i t)} - \vartheta_i |\xi_i|\right) \\
&\leq -c_i(d_i + g_i)|\xi_i|^2 - c_i(d_i + g_i)\left(\varpi_i|\xi_i| - \frac{\exp(-\alpha_i t)\vartheta_i |\xi_i|}{|\xi_i| + \exp(-\alpha_i t)}\right)
\end{aligned} \tag{20.63}$$

Note that $\varpi_i \geq 0$ and $\exp(-\alpha_i t)\vartheta_i \rightarrow 0$. Hence, $\exists \tau_2 > 0$, such that for all $t \geq \tau_2$, $\dot{V}_i' \leq 0$. $\dot{V}_i' = 0$ if and only if $|\xi_i| = 0$. Hence, ξ_i is AS. Furthermore, $\lim\limits_{t \rightarrow \infty} \varepsilon_i(t) = 0$. This completes the proof. ∎

Remark 20.5 The proposed controller design is fully distributed, i.e. it does not need any information on the communication graph topology or the number of converters. Hence, the proposed approach is scalable and can be applied in a plug-and-play manner.

20.3.4 Hardware-in-the-Loop Validation

20.3.4.1 DC Microgrid HIL Testbed

In this section, the proposed attack-resilient cooperative control protocols are implemented on a DC microgrid testbed as illustrated in Fig. 20.10. This testbed consists of eight DC–DC converters emulated on a Typhoon HIL 604 system [58] and passing messages on a bidirectional communication graph, the distributed local controllers implemented on a dSPACE DS 1202 MicroLabBoxes [59], and a desktop computer that has an Intel Xeon 3.6 GHz processor and 64GB RAM. DC–DC buck

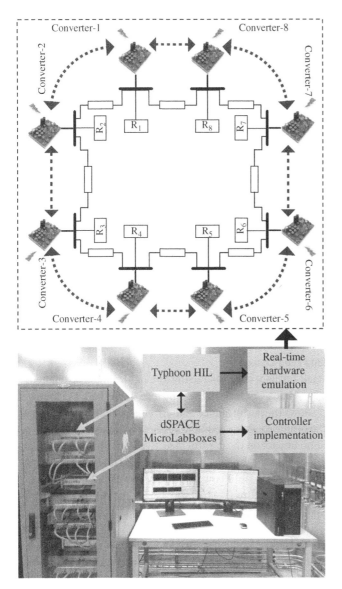

Figure 20.10 The microgrid has eight DC–DC converters on a bidirectional ring communication network and distributed local controllers, implemented on a Typhoon HIL604 system and a dSPACE MicroLabBoxes, respectively.

converter parameters are $C = 2.2$ mF, $L = 2.64$ mH, $R_L = 10\ \Omega$, $f_s = 60$ kHz, and $V_{ref} = 48$ V. The adjacency matrix and the diagonal matrix of pinning gains are chosen as

$$
\mathcal{A} = \begin{bmatrix}
0 & 1 & 0 & 0 & 0 & 0 & 0 & 1 \\
1 & 0 & 1 & 0 & 0 & 0 & 0 & 0 \\
0 & 1 & 0 & 1 & 0 & 0 & 0 & 0 \\
0 & 0 & 1 & 0 & 1 & 0 & 0 & 0 \\
0 & 0 & 0 & 1 & 0 & 1 & 0 & 0 \\
0 & 0 & 0 & 0 & 1 & 0 & 1 & 0 \\
0 & 0 & 0 & 0 & 0 & 1 & 0 & 1 \\
1 & 0 & 0 & 0 & 0 & 0 & 1 & 0
\end{bmatrix},
$$

$$
\mathcal{G} = \begin{bmatrix}
0 & 0 & 0 & 0 & 0 & 0 & 0 & 0 \\
0 & 1 & 0 & 0 & 0 & 0 & 0 & 0 \\
0 & 0 & 0 & 0 & 0 & 0 & 0 & 0 \\
0 & 0 & 0 & 0 & 0 & 0 & 0 & 0 \\
0 & 0 & 0 & 0 & 0 & 0 & 0 & 0 \\
0 & 0 & 0 & 0 & 0 & 0 & 0 & 0 \\
0 & 0 & 0 & 0 & 0 & 0 & 0 & 0 \\
0 & 0 & 0 & 0 & 0 & 0 & 0 & 0
\end{bmatrix},
$$

respectively. The control system parameters are chosen as

$$
I_1^{rated} = I_4^{rated} = I_5^{rated} = I_8^{rated} = 1,
$$
$$
I_2^{rated} = I_3^{rated} = I_6^{rated} = I_7^{rated} = 2,
$$
$$
R_1^{vir} = R_4^{vir} = R_5^{vir} = R_8^{vir} = 2,
$$
$$
R_2^{vir} = R_3^{vir} = R_6^{vir} = R_7^{vir} = 1,
$$
$$
c_i = 10, \quad \alpha_i = 0.1, \quad \gamma_i = 5, \quad \forall i = 1, 2, \dots, 8.
$$

20.3.4.2 Response to the Link Failure and Load Change

Herein, the response of the proposed attack-resilient controller to the link failure and load change is studied. First, the communication link $3 \leftrightarrow 4$ fails at $t = 1$ s. Then, the load at bus two, R_2, changes in step between 10 and 5Ω. Figure 20.11 shows the performance of the average voltage, the terminal voltages, and supplied currents using the proposed attack-resilient secondary controller. As shown in Fig. 20.11, the link failure has almost no impact on the control objectives since the communication digraph stays connected after link $3 \leftrightarrow 4$ fails. Moreover, the controller readjusts the voltages and properly updates the load sharing in case of load change to satisfy the global voltage regulation and proportional load sharing objectives. Figure 20.11(a) shows that the average voltage is regulated at the set point, i.e. $V_{ref} = 48$V.

20.3.4.3 Performance Assessment Under Unknown Unbounded Attacks

In this section, the attack model described in (20.42) is considered, where the attacks are injected at the local control input of each converter by selecting $\delta_i = 3t$, $\forall i = 1, 2, \dots, 8$. As seen, these attacks

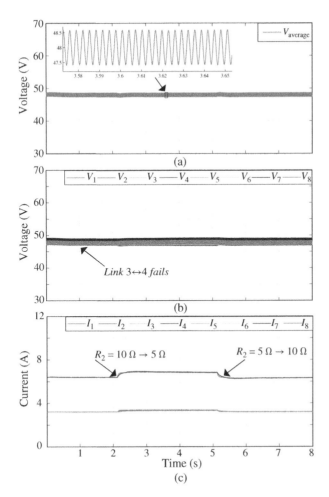

Figure 20.11 Response of the proposed attack-resilient secondary controller to the link failure and load change: (a) average voltage, (b) terminal voltages, and (c) supplied currents.

are unbounded. The proposed control protocols for each converter using (20.34), (20.45), (20.46), and (20.48) are constructed. For comparison, the experiment using the standard secondary control protocols consisting of (20.43) and (20.34) is also conducted.

Note that the uniform continuous function $\exp\left(-\alpha_i t\right)$ in (20.46) is used only to construct a smooth control method when $\left|\xi_i\right| = 0$. Hence, the influence of different values of α_i on the convergence rate can be neglected. Whereas, in addition to tuning the coupling gain c_i, the adaptive tuning parameter in (20.48), γ_i, can tune the convergence rate of the system performance. This behavior is then verified by running the experiment for different values of γ_i.

The experimental results, using the standard and the proposed attack-resilient methods, are comparatively illustrated in Figs. 20.12 and 20.13, respectively. Figure 20.12 shows that the terminal voltages and the supplied currents, using the standard control protocols, diverge when unbounded attacks are applied. Whereas, Figs. 20.13(a) and 20.13(b) show that, by using the proposed resilient control method, the average voltage stays bounded around 48 V and the tight

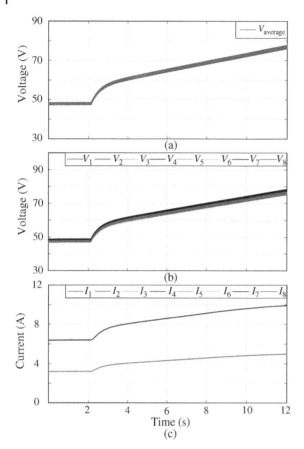

Figure 20.12 Performance of the standard secondary control in the case of unbounded attack signals $\delta_i = 3\,t$: (a) average voltage, (b) terminal voltages, and (c) supplied currents.

voltage regulation is maintained for each converter. Figure 20.13(c) shows that the supplied currents are properly shared despite the unbounded attacks. These studies verify the effectiveness of the proposed resilient approach in solving Problem 20.1, i.e. maintaining bounded global voltage regulation and proportional load sharing under unbounded attacks. Moreover, as seen in Fig. 20.13, the convergence rate is increased by properly increasing γ_i.

To verify the capabilities of the proposed method in handling large attack signals, the comparative experiments are conducted by selecting $\delta_i = 30t$, $\forall i = 1, 2, \ldots, 8$. The experimental results using the standard and the proposed resilient methods are comparatively illustrated in Figs. 20.14 and 20.15, respectively. Figure 20.14 shows that the terminal voltages and the supplied currents, using the standard secondary control, diverge after initiating the large unbounded attack signals and reach the saturation value fairly quickly. Whereas, Fig. 20.15 shows that the terminal voltages stay bounded around 48 V, and the proportional load sharing is still carried out, using the proposed method, even under fairly large unbounded injections.

20.3.4.4 Performance Assessment Under Unknown Bounded Attacks

In this section, the bounded attack signals are considered by selecting $\delta_i = 15$, $\forall i = 1, 2, \ldots, 8$. The experimental results using the standard method and the proposed resilient method under bounded attack injections are comparatively illustrated in Figs. 20.16 and 20.17, respectively. Figure 20.16 shows that the terminal voltages and the supplied currents, using the standard secondary control, stay bounded. The upper bounds for the local voltage and current are determined by the values

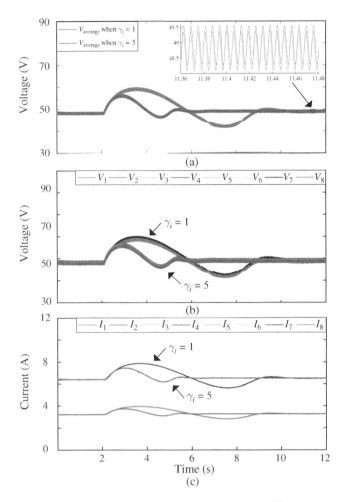

Figure 20.13 Performance of the proposed attack-resilient control in the case of unbounded attack signals $\delta_i = 3\,t$: (a) average voltage, (b) terminal voltages, and (c) supplied currents.

of attack signals. For example, the upper bound for the average voltage is almost 65 V under the current attack injections. Without implementing the resilient-control protocol, the voltage and current performance will further deteriorate under larger attack values. As shown in Fig. 20.17(a), the average voltage is properly regulated at 48 V, i.e. the adverse effects of bounded attacks are completely compensated. This is also validated by observing the current waveforms in Fig. 20.17(c).

20.4 Conclusion

This chapter addresses the distributed attack-resilient secondary control problems for AC and DC microgrids against unbounded attacks. The chapter starts by reviewing the motivations and existing resilient-control methods for the secondary control of microgrids. Then, the distributed attack-resilient secondary control problems for AC and DC microgrids are formulated and addressed, in the presence of unbounded actuator attacks. Fully distributed attack-resilient control

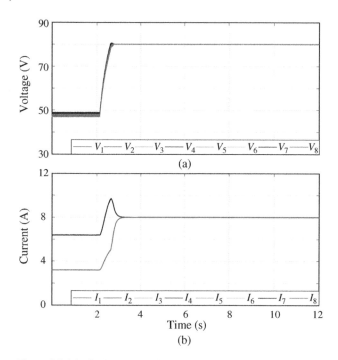

Figure 20.14 Performance of the standard secondary control in the case of unbounded attack signals $\delta_i = 30\,t$: (a) terminal voltages and (b) supplied currents.

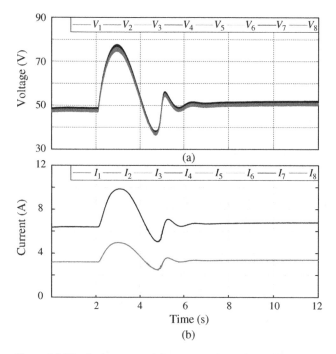

Figure 20.15 Performance of the proposed attack-resilient control in the case of unbounded attack signals $\delta_i = 30\,t$: (a) terminal voltages and (b) supplied currents.

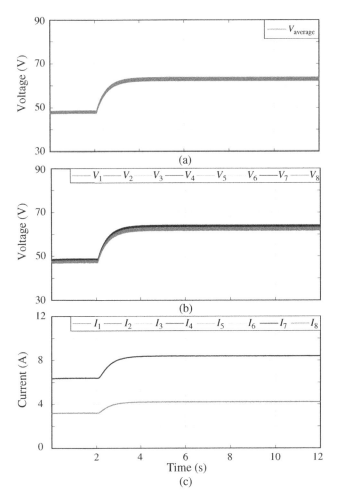

Figure 20.16 Performance of the standard secondary control in the case of bounded attack signals $\delta_i = 15$: (a) average voltage, (b) terminal voltages, and (c) supplied currents.

frameworks are designed using adaptive control techniques to ensure the UUB stability of the closed-loop microgrid system. Experimental results and comparisons between the attack-resilient and the standard secondary controllers under unbounded attacks demonstrate the efficacy of the enhanced resilient performance.

20.5 Acknowledgment

This work was supported in part by the ONR grant N00014-17-1-2239.

20.6 Exercises

1. A communication digraph is illustrated in Fig. 20.18, which consists of 1 leader node and 8 follower nodes. Assume the edge weight is 1. Answer the following questions:

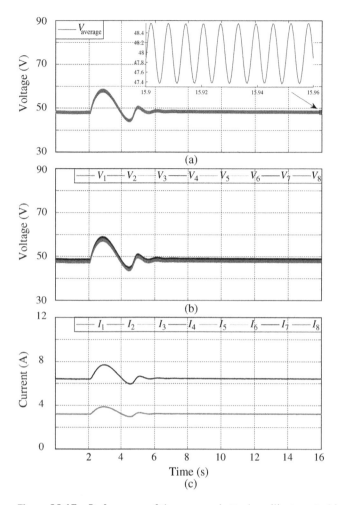

Figure 20.17 Performance of the proposed attack-resilient control in the case of bounded attack signals $\delta_i = 15$: (a) average voltage, (b) terminal voltages, and (c) supplied currents.

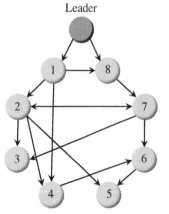

Figure 20.18 A communication digraph of multi-agent systems.

(a) Find the adjacency matrix $\mathcal{A}$.

(b) Find the in-degree matrix $\mathcal{D}$.

(c) Find the Laplacian matrix $\mathcal{L}$.

(d) Check whether matrix $\mathcal{L} + \mathcal{G}$ is positive-definite.

2. Describe the primary droop mechanisms of AC and DC microgrids, respectively.

References

1 Bidram, A., Lewis, F.L., and Davoudi, A. (2014). Distributed control systems for small-scale power networks: using multiagent cooperative control theory. *IEEE Control Systems Magazine* 34 (6): 56–77.

2 Zuo, S., Beg, O.A., Lewis, F.L., and Davoudi, A. (2020). Resilient networked AC microgrids under unbounded cyber attacks. *IEEE Transactions on Smart Grid* 11 (5): 3785–3794.

3 Kosut, O., Jia, L., Thomas, R.J., and Tong, L. (2011). Malicious data attacks on the smart grid. *IEEE Transactions on Smart Grid* 2 (4): 645–658.

4 Sridhar, S., Hahn, A., and Govindarasu, M. (2011). Cyber–physical system security for the electric power grid. *Proceedings of the IEEE* 100 (1): 210–224.

5 He, H. and Yan, J. (2016). Cyber-physical attacks and defences in the smart grid: a survey. *IET Cyber-Physical Systems: Theory & Applications* 1 (1): 13–27.

6 Deng, R., Xiao, G., Lu, R. et al. (2016). False data injection on state estimation in power systems – attacks, impacts, and defense: a survey. *IEEE Transactions on Industrial Informatics* 13 (2): 411–423.

7 Beg, O.A., Nguyen, L.V., Johnson, T.T., and Davoudi, A. (2018). Signal temporal logic-based attack detection in DC microgrids. *IEEE Transactions on Smart Grid* 10 (4): 3585–3595.

8 Zhang, J., Sahoo, S., Peng, J.C.-H., and Blaabjerg, F. (2021). Mitigating concurrent false data injection attacks in cooperative DC microgrids. *IEEE Transactions on Power Electronics* 36 (8): 9637–9647.

9 Tan, S., Guerrero, J.M., Xie, P. et al. (2020). Brief survey on attack detection methods for cyber-physical systems. *IEEE Systems Journal* 14 (4): 5329–5339.

10 Liu, Y., Ning, P., and Reiter, M.K. (2011). False data injection attacks against state estimation in electric power grids. *ACM Transactions on Information and System Security (TISSEC)* 14 (1): 1–33.

11 Hug, G. and Giampapa, J.A. (2012). Vulnerability assessment of AC state estimation with respect to false data injection cyber-attacks. *IEEE Transactions on Smart Grid* 3 (3): 1362–1370.

12 Liang, J., Sankar, L., and Kosut, O. (2015). Vulnerability analysis and consequences of false data injection attack on power system state estimation. *IEEE Transactions on Power Systems* 31 (5): 3864–3872.

13 Pan, K., Teixeira, A., Cvetkovic, M., and Palensky, P. (2018). Cyber risk analysis of combined data attacks against power system state estimation. *IEEE Transactions on Smart Grid* 10 (3): 3044–3056.

14 Chen, Y., Qi, D., Dong, H. et al. (2020). A FDI attack-resilient distributed secondary control strategy for islanded microgrids. *IEEE Transactions on Smart Grid* 12 (3): 1929–1938.

15 Liu, X.-K., Wen, C., Xu, Q., and Wang, Y.-W. (2021). Resilient control and analysis for DC microgrid system under DoS and impulsive FDI attacks. *IEEE Transactions on Smart Grid* 12 (5): 3742–3754.

16 Lu, J., Zhang, X., Hou, X., and Wang, P. (2022). Generalized extended state observer-based distributed attack-resilient control for DC microgrids. *IEEE Transactions on Sustainable Energy* 13 (3): 469–1480.

17 Deng, C., Wang, Y., Wen, C. et al. (2020). Distributed resilient control for energy storage systems in cyber–physical microgrids. *IEEE Transactions on Industrial Informatics* 17 (2): 1331–1341.

18 Abhinav, S., Modares, H., Lewis, F.L. et al. (2017). Synchrony in networked microgrids under attacks. *IEEE Transactions on Smart Grid* 9 (6): 6731–6741.

19 Dehkordi, N.M., Baghaee, H.R., Sadati, N., and Guerrero, J.M. (2018). Distributed noise-resilient secondary voltage and frequency control for islanded microgrids. *IEEE Transactions on Smart Grid* 10 (4): 3780–3790.

20 Shahab, M.A., Mozafari, B., Soleymani, S. et al. (2019). Distributed consensus-based fault tolerant control of islanded microgrids. *IEEE Transactions on Smart Grid* 11 (1): 37–47.

21 Afshari, A., Karrari, M., Baghaee, H.R. et al. (2019). Cooperative fault-tolerant control of microgrids under switching communication topology. *IEEE Transactions on Smart Grid* 11 (3): 1866–1879.

22 Fawzi, H., Tabuada, P., and Diggavi, S. (2014). Secure estimation and control for cyber-physical systems under adversarial attacks. *IEEE Transactions on Automatic Control* 59 (6): 1454–1467.

23 Kwasinski, A. (2011). Quantitative evaluation of DC microgrids availability: effects of system architecture and converter topology design choices. *IEEE Transactions on Power Electronics* 26 (3): 835–851.

24 Gu, Y., Xiang, X., Li, W., and He, X. (2014). Mode-adaptive decentralized control for renewable DC microgrid with enhanced reliability and flexibility. *IEEE Transactions on Power Electronics* 29 (9): 5072–5080.

25 Kwasinski, A. and Onwuchekwa, C.N. (2011). Dynamic behavior and stabilization of DC microgrids with instantaneous constant-power loads. *IEEE Transactions on Power Electronics* 26 (3): 822–834.

26 Guerrero, J.M., Vasquez, J.C., Matas, J. et al. (2011). Hierarchical control of droop-controlled AC and DC microgrids – a general approach toward standardization. *IEEE Transactions on Industrial Electronics* 58 (1): 158–172.

27 Dragičević, T., Lu, X., Vasquez, J.C., and Guerrero, J.M. (2016). DC microgrids – Part I: A review of control strategies and stabilization techniques. *IEEE Transactions on Power Electronics* 31 (7): 4876–4891.

28 Anand, S., Fernandes, B.G., and Guerrero, J. (2013). Distributed control to ensure proportional load sharing and improve voltage regulation in low-voltage DC microgrids. *IEEE Transactions on Power Electronics* 28 (4): 1900–1913.

29 Nasirian, V., Moayedi, S., Davoudi, A., and Lewis, F.L. (2015). Distributed cooperative control of DC microgrids. *IEEE Transactions on Power Electronics* 30 (4): 2288–2303.

30 Wang, P., Lu, X., Yang, X. et al. (2016). An improved distributed secondary control method for DC microgrids with enhanced dynamic current sharing performance. *IEEE Transactions on Power Electronics* 31 (9): 6658–6673.

31 Sahoo, S. and Mishra, S. (2019). A distributed finite-time secondary average voltage regulation and current sharing controller for DC microgrids. *IEEE Transactions on Smart Grid* 10 (1): 282–292.

32 Pullaguram, D., Mishra, S., and Senroy, N. (2018). Event-triggered communication based distributed control scheme for DC microgrid. *IEEE Transactions on Power Systems* 33 (5): 5583–5593.

33 Han, R., Meng, L., Guerrero, J.M., and Vasquez, J.C. (2018). Distributed nonlinear control with event-triggered communication to achieve current-sharing and voltage regulation in DC microgrids. *IEEE Transactions on Power Electronics* 33 (7): 6416–6433.

34 Liang, G., Zhao, J., Luo, F. et al. (2017). A review of false data injection attacks against modern power systems. *IEEE Transactions on Smart Grid* 8 (4): 1630–1638.

35 Ralston, P.A.S., Graham, J.H., and Hieb, J.L. (2007). Cyber security risk assessment for SCADA and DCS networks. *ISA Transactions* 46 (4): 583–594.

36 Pasqualetti, F., Dörfler, F., and Bullo, F. (2013). Attack detection and identification in cyber-physical systems. *IEEE Transactions on Automatic Control* 58 (11): 2715–2729.

37 Teixeira, A., Shames, I., Sandberg, H., and Johansson, K.H. (2014). Distributed fault detection and isolation resilient to network model uncertainties. *IEEE Transactions on Cybernetics* 44 (11): 2024–2037.

38 Tang, Z., Kuijper, M., Chong, M.S. et al. (2019). Linear system security – detection and correction of adversarial sensor attacks in the noise-free case. *Automatica* 101: 53–59.

39 Manandhar, K., Cao, X., Hu, F., and Liu, Y. (2014). Detection of faults and attacks including false data injection attack in smart grid using Kalman filter. *IEEE Transactions on Control of Network Systems* 1 (4): 370–379.

40 Zhao, J., Zhang, G., La Scala, M. et al. (2017). Short-term state forecasting-aided method for detection of smart grid general false data injection attacks. *IEEE Transactions on Smart Grid* 8 (4): 1580–1590.

41 He, Y., Mendis, G.J., and Wei, J. (2017). Real-time detection of false data injection attacks in smart grid: a deep learning-based intelligent mechanism. *IEEE Transactions on Smart Grid* 8 (5): 2505–2516.

42 Ashok, A., Govindarasu, M., and Ajjarapu, V. (2018). Online detection of stealthy false data injection attacks in power system state estimation. *IEEE Transactions on Smart Grid* 9 (3): 1636–1646.

43 Beg, O.A., Johnson, T.T., and Davoudi, A. (2017). Detection of false-data injection attacks in cyber-physical DC microgrids. *IEEE Transactions on Industrial Informatics* 13 (5): 2693–2703.

44 Sahoo, S., Mishra, S., Peng, J.C., and Dragičević, T. (2019). A stealth cyber-attack detection strategy for DC microgrids. *IEEE Transactions on Power Electronics* 34 (8): 8162–8174.

45 Kim, J. and Tong, L. (2013). On topology attack of a smart grid: undetectable attacks and countermeasures. *IEEE Journal on Selected Areas in Communications* 31 (7): 1294–1305.

46 Arabi, E., Yucelen, T., and Haddad, W.M. (2016). Mitigating the effects of sensor uncertainties in networked multiagent systems. *2016 American Control Conference (ACC)*, 5545–5550, July 2016.

47 De La Torre, G., Yucelen, T., and Peterson, J.D. (2014). Resilient networked multiagent systems: a distributed adaptive control approach. *53rd IEEE Conference on Decision and Control*, 5367–5372, December 2014.

48 Xie, C.-H. and Yang, G.-H. (2017). Decentralized adaptive fault-tolerant control for large-scale systems with external disturbances and actuator faults. *Automatica* 85: 83–90.

49 Jin, X., Haddad, W.M., and Yucelen, T. (2017). An adaptive control architecture for mitigating sensor and actuator attacks in cyber-physical systems. *IEEE Transactions on Automatic Control* 62 (11): 6058–6064.

50 Feng, Z., Wen, G., and Hu, G. (2017). Distributed secure coordinated control for multiagent systems under strategic attacks. *IEEE Transactions on Cybernetics* 47 (5): 1273–1284.

51 Abhinav, S., Modares, H., Lewis, F.L., and Davoudi, A. (2019). Resilient cooperative control of DC microgrids. *IEEE Transactions on Smart Grid* 10 (1): 1083–1085.

52 Han, R., Meng, L., Ferrari-Trecate, G. et al. (2017). Containment and consensus-based distributed coordination control to achieve bounded voltage and precise reactive power sharing in islanded AC microgrids. *IEEE Transactions on Industry Applications* 53 (6): 5187–5199.

53 Zuo, S., Song, Y., Lewis, F.L., and Davoudi, A. (2017). Output containment control of linear heterogeneous multi-agent systems using internal model principle. *IEEE Transactions on Cybernetics* 47 (8): 2099–2109.

54 Khalil, H.K. (2002). *Nonlinear Systems*, Pearson Education. Prentice Hall.

55 LaSalle, J. (1960). Some extensions of Liapunov's second method. *IRE Transactions on Circuit Theory* 7 (4): 520–527.

56 Mwakabuta, N. and Sekar, A. (2007). Comparative study of the IEEE 34 node test feeder under practical simplifications. In: *2007 39th North American Power Symposium*, 484–491. IEEE.

57 Fax, J.A. and Murray, R.M. (2004). Information flow and cooperative control of vehicle formations. *IEEE Transactions on Automatic Control* 49 (9): 1465–1476.

58 Typhoon HIL, Inc. (2017). *Typhoon HIL 603 Technical Manual*. Somerville, MA: Typhoon HIL, Inc.

59 MicroLabBox (2017). *Product Information*. Paderborn, Germany: dSPACE GmbH.

21

Quantum Security for Microgrids

Zefan Tang and Peng Zhang

Communication has always played a vital role in microgrids in maintaining reliable operations and achieving great benefits and will be even more critical with the increasing deployment of renewable energies, information technologies, and real-time automation and control systems. However, existing classical cryptographic methods for securing microgrid communication are based on mathematical assumptions, which are vulnerable to attacks from quantum computers. This chapter reviews the current status of developing quantum-secure microgrids, namely, microgrids that are secure against attacks from quantum computers. Specifically, it introduces why implementing quantum security is important, how quantum security can be integrated into a single microgrid and networked microgrids (NMs), respectively, and how quantum-secure microgrid and NMs testbeds are established. It also discusses some potential issues associated with applying existing quantum cryptography methods in the context of microgrids and provides future perspectives to make quantum security more practical in microgrids.

21.1 Background

21.1.1 Securing Microgrid Data Transmission

Communication systems are required in microgrids for various purposes, e.g. obtaining a consistent view of a wide-scale system state, coordinating restoration efforts according to real-time situational awareness, providing timely decision updates to distributed resource controllers and neighboring microgrids, and enabling advanced functions with the participation of distributed resources and/or neighboring microgrids [1–6].

The implementation of existing microgrid communication systems is mostly achieved through a standard Internet communication protocol suit, e.g. the Transmission Control Protocol/Internet Protocol (TCP/IP), which contains four layers, namely, application, transport, network, and link layers, from the highest layer to the lowest. The data are passed from the application layer through intermediate layers to the link layer, where each layer adds information. At the link layer, the accumulated data is sent out into the physical network, and when arrived, the data is passed up through the layers to its destination [7–10].

The security in this process is achieved through the use of security controls at one or more layers. A security control at the application layer offers high flexibility but involves extensive application

Microgrids: Theory and Practice, First Edition. Edited by Peng Zhang.
© 2024 The Institute of Electrical and Electronics Engineers, Inc. Published 2024 by John Wiley & Sons, Inc.

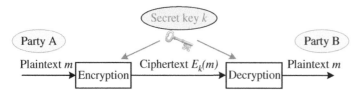

Figure 21.1 Illustration of a symmetric-key cryptographic system, where a single key is used for encryption and decryption.

customization and cannot protect the information at lower layers. Security controls at the transport layer, e.g. the Transport Layer Security (TLS), and at the network layer, e.g. the Internet Protocol Security (IPsec), are more convenient to add controls than controls at the application layer, because they do not require understanding the functions or characteristics of the application. A security control at the link layer is applied on a physical link, where the implementation is simple; but as the control is specific to a particular physical link, it is not feasible in most cases when there are multiple links [11].

For TLS, IPsec, and other similar protocols, it is important to preshare a secret key between two parties for the use of a cryptographic algorithm, e.g. the Advanced Encryption Standard (AES) [12]. For a symmetric-key cryptographic system like AES, a key is used to encrypt the data, and only the person who has the same key can decrypt the received ciphertext (see Fig. 21.1). Therefore, it requires that the keys distributed to two parties have to be generated securely, i.e. unknown to attackers.

This secure key-exchange process is achieved by a key generation system, e.g. mostly public-key cryptographic methods such as the Diffie–Hellman (DH) key exchange and Rivest–Shamir–Adleman (RSA) [13, 14]. The security of DH is guaranteed on the basis of the computational difficulty that the discrete logarithm problem cannot be efficiently solved, and the security of RSA requires that factoring a product of two large prime numbers is difficult to solve. In fact, all the classical public-key systems rely on mathematical assumptions, i.e. a certain mathematical problem cannot be efficiently solved even by the most powerful modern supercomputer using any existing algorithm. Once there is a way to efficiently address these mathematical problems, the existing microgrid communication systems will no longer be secure.

21.1.2 The Quantum Era is Coming

Those mathematical problems, while difficult to solve with modern computers, are at risk of being addressed by quantum computers. Quantum computing exploits quantum-mechanical properties such as superposition and entanglement to provide the probability of improving the computational capability. Since 1982 when a quantum machine was envisioned by Richard Feynman to simulate quantum physics, tremendous progress has been made on developing quantum computing hardware and algorithms. In the early 1990s, Dr. Peter Shor presented an algorithm, now well known as "Shor's Algorithm," which was able to effectively factor a product of two large prime numbers with the help of a quantum computer equipped with enough stable qubits, thus defeating the RSA encryption algorithm [15].

Today, quantum computing has been a hot and strategic research area among academia, government, and industry, including big companies like Google, IBM, Microsoft, Intel, Alibaba, and many others, who are actively pursuing the goal of developing the first large-scale universal quantum computer. Google reported in 2019 that a quantum computer with 53 qubits was able to achieve quantum supremacy over certain problems; and in August 2020, Google announced another big breakthrough by performing the largest chemical simulation on a quantum computer. IBM, the long-established leader in quantum computing, released a roadmap in September 2020 that the IBM team is developing a quantum computer with 1000-plus qubits targeted for the end of 2023. As the U.S. Department of Energy announced in August 2020, it has decided to allocate US$625 million funds to support multidisciplinary quantum information science research, especially quantum computing, over the next five years. Note that although the occurrence of quantum computers powerful enough to break current cryptographic systems is perhaps still far away, the high risk is forthcoming [16–18].

21.2 Quantum Communication for Microgrids

Unlike the classical communication that relies on mathematical assumptions, quantum communication is based on fundamental laws of quantum physics, which provide a more solid foundation in the quantum era as they have been fairly heavily tested. This section first presents an overview of quantum cryptography techniques that offer some promising candidates for securing microgrids, and then introduces some basics of the most mature technique in quantum cryptography, namely the quantum key distribution (QKD).

21.2.1 Overview of Quantum Cryptography

A potent solution to tackle the threat posed by quantum computers is to use quantum cryptography, which takes advantage of the principles of quantum mechanics to perform cryptographic tasks such as encryption and decryption, authentication, and key distribution. A number of quantum cryptography techniques have been proposed in the literature. The most well-known and mature one to date is the QKD, whose performance has been demonstrated in laboratories and relevant real-world applications such as computer networks, online banking, private clouds, and critical infrastructures for government and industry. Beyond the realm of QKD, there are some other quantum cryptography techniques, including quantum random number generation, quantum coin flipping, quantum money, quantum bit commitment, and position-based quantum cryptography. They utilize quantum information in different cryptography contexts, e.g. unclonable money. However, although there are a variety of quantum cryptography techniques, their applications in microgrids have not yet been extensively investigated. An important reason for this stems from the fact that most microgrid stakeholders have not been aware of the importance of developing quantum-secure microgrids. Only a few efforts have been made recently to implement QKD systems in microgrids [7, 19–23]. However, other quantum cryptography techniques also have considerable potential to contribute to the securing of microgrids in certain scenarios. An overview of quantum cryptography techniques is provided in Table 21.1, and the details are described below.

Quantum key distribution: QKD, as the most mature and well-known example of quantum cryptography, exploits the fundamental laws of quantum mechanics to establish information-theoretically secure keys for two communicating parties without relying on mathematical assumptions. Due to the no-cloning theorem in quantum mechanics, i.e. making a perfect copy of a quantum state is impossible, any eavesdropper attempting to obtain the information of the keys is detectable. It therefore guarantees that the keys obtained by the two parties are theoretically secure, at least on a protocol level.

Table 21.1 Characteristics of different quantum cryptography techniques.

Type of quantum cryptography	Characteristics	Applied in microgrids?
Quantum key distribution	Generate and distribute keys for two parties; eavesdropping on the keys is detectable	Yes
Quantum random number generation	Generate a random number; unbiased and unpredictable	Not yet
Quantum coin flipping	Decide which party wins the coin toss; the two parties do not trust each other	Not yet
Quantum money	Cannot be forged	Not yet
Quantum bit commitment	Solve the bit commitment problem	Not yet
Position-based quantum cryptography	Position specified	Not yet
Others	Quantum oblivious transfer, device-independent quantum cryptography, quantum digital signatures, post-quantum cryptography, etc.	Not yet

Quantum random number generation: Generating random numbers is critical in cryptography. The generated numbers function as cryptographic keys that are used to encrypt and later decrypt messages. The classical random number generation methods are based on deterministic algorithms, thus predictable, and therefore the generated numbers are pseudorandom, i.e. not truly random. A quantum random number generator takes advantage of quantum physics to generate random numbers. The generated numbers are unbiased and unpredictable, i.e. truly random, as quantum physics is fundamentally random. Quantum random number generation differs from QKD in that the generated random numbers are not distributed to two parties.

Quantum coin flipping: Quantum coin flipping uses quantum-mechanical properties to decide which party wins the coin toss. It plays a vital role when two remote parties need to perform a binary selection but do not trust each other. Unlike QKD, it is used for two distrustful parties. However, the protocols of quantum coin flipping are more difficult to accomplish in experiments.

Quantum money: Quantum money leverages quantum mechanics to create money that is impossible to forge. The security of traditional digital money like bitcoin relies on the computational difficulty of a certain function, such as the public–private key cryptography function, which is vulnerable to quantum computers with a particular algorithm. Quantum money, due to the no-cloning theorem, cannot be forged.

Quantum bit commitment: Quantum bit commitment exploits quantum physics to solve the bit commitment problem, i.e. one party, Alice, sends another party, Bob, a bit (0 or 1) but does not want to reveal the bit until she decides to do it; meanwhile, Alice should not be able to change the bit after it has been sent to Bob. Unlike QKD, where perfect security is, in theory, possible (e.g. security without mathematical assumptions on the adversary), this is not the case for quantum bit commitment.

Position-based quantum cryptography: It is used to achieve the goal that one party sends a message to another party at a specific location, and only when the receiving party is located in that particular position can the message be read.

Other techniques: Other techniques include quantum oblivious transfer, device-independent quantum cryptography, quantum digital signatures, and post-quantum cryptography.

21.2.2 Basics of Quantum Key Distribution

QKD is a key-growing method used to generate and distribute secret keys for two distant parties. It utilizes fundamental properties of quantum mechanics, which are resilient against increasing computational power, improved computational algorithms, and quantum computers. A QKD system, in general, consists of a quantum channel, i.e. a fiber or free space, and a classical one, i.e. a public but authenticated communication link. Quantum states are first transferred through the quantum channel between the sender and the receiver for generating raw keys. After the transmission and measurement of quantum states, corresponding post-processing steps are performed over the classical channel to distill the final secret keys. With the generated keys, data messages are encrypted and transmitted over the classical channel.

Over the years since QKD first exhibited its promise, it has attracted considerable interest, and a variety of QKD protocols have emerged, many of which have been experimentally demonstrated in real-world scenarios. In general, there are two types of QKD. One is the discrete-variable QKD (DV-QKD), where quantum information is encoded in discrete variables and single-photon detectors are used to measure the received quantum states, and thus, the measured results are discrete. Examples of DV-QKD are the BB84 protocol and the E91 protocol. Alternatively, the other type, continuous-variable QKD (CV-QKD), encodes quantum information in the phase and amplitude quadratures of a coherent laser, and uses homodyne detectors to measure the received quantum states where the measured results are continuous. Examples include the Gaussian-modulated CV-QKD.

21.2.2.1 Quantum States

Instead of using binary bits to encode information as in classical communication systems, quantum communication utilizes quantum states, or "qubits." A qubit is a two-state quantum-mechanical system whose state is commonly represented by the spin of an electron or the polarization of a photon. Unlike a binary bit, which has to be in one state or the other, a qubit can be in a coherent superposition of both states. For QKD systems, photons are the primary practical implementation of qubits. For the QKD system considered in this chapter, the polarization of the photon will be used to encode a quantum state. Two bases, namely horizontal polarization (denoted as the Z basis later) and diagonal polarization (denoted as the X basis later), will be utilized. If a source and its receiver operate on the same basis, the information can be transmitted deterministically; however, if different bases are used, the information received will be uncorrelated with the transmitted information. The security of a QKD protocol, in a way, takes advantage of this: by encoding a classical bit string using different, randomly-chosen bases, an adversary who is unaware of the basis choice can never be truly certain of the information being transmitted. Further, any attempt to learn this information causes noise in the quantum channel which may be detected by users later.

21.2.2.2 General Setting of a QKD System

The general setting of a QKD-based communication system consists of a quantum channel and a classical one (see Fig. 21.2). The quantum channel allows two parties to share quantum signals to create a secure and secret key. With the created key, the information to be transmitted is encrypted and later decrypted over the classical channel. The *key generation rate* of a QKD protocol is an important statistic and is affected by numerous parameters, most importantly the noise in the

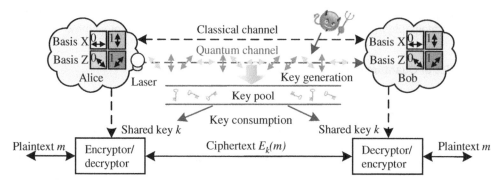

Figure 21.2 The general setting of a QKD system.

quantum channel (caused, perhaps, by an adversary or natural noise) and the distance between the two parties.

An important and unique property of QKD is that the two parties can detect when an eavesdropper is trying to gain knowledge of the keys. This is due to the quantum-mechanical property that measuring an unknown quantum state will, in general, change that state. It ensures that a non-secret key will never be used, making QKD-based encryption and authentication theoretically secure. It is worth noting that QKD is only used to generate keys through the quantum channel; data messages are still transmitted using classical encryption methods over the classical channel. In reality, QKD can be associated with symmetric key algorithms such as the one-time pad (OTP) or AES.

21.2.2.3 The Practical Decoy-State QKD Protocol

Different protocols have been proposed to implement QKD such as the well-known BB84, decoy-state, six-state, Ekert91, and BBM92. As an example, in this chapter, we consider a practical decoy-state QKD protocol. This protocol has been one of the most widely used schemes in the QKD community because of its ability to tolerate high channel loss and to operate robustly even with today's hardware. Its security and feasibility have been well demonstrated by several experimental groups, and theoretical security analyses including the evaluation of concise and tight finite-key security bounds have been provided.

The idea of this protocol is as follows. The information is encoded into qubits and then sent out by one party, commonly named Alice, using weak coherent laser pulses. With today's technology, the production of a single qubit is not practical; instead, weak coherent laser pulses are used. However, these pulses contain, with non-zero probability, multiple-qubit signals that would cause a break in security. To tackle this challenge, the decoy-state protocol randomly varies the intensity of each laser pulse using one of three intensities k_1, k_2, and k_3, which are the intensities of the signal state, the decoy state, and the vacuum state, respectively. Two bases, X and Z, are selected with probabilities p_x and $1 - p_x$, respectively. Recall that these bases refer to the polarization setting of the qubit. The other party, named Bob, measures the qubits by randomly selecting bases from X and Z. If Alice and Bob choose the same basis, they share information since sending and receiving qubits on the same basis, as mentioned, leads to a deterministic outcome; otherwise, the iteration is discarded. By repeating this numerous times, the two parties share the so-called *raw-key*, which is partially correlated and partially secret. Then, error correction is performed (leaking additional information to the adversary that must be taken into account), followed by privacy amplification, which produces a secret key of size ℓ.

21.2.2.4 Benefits of Using QKD for Microgrids

QKD has been envisioned as one of the most secure and practical instances of quantum cryptography. Specifically, using QKD provides the following benefits for microgrids [19, 20]:

- Keys generated by QKD are almost impossible to steal even in the face of an adversary with infinite supplies of time and processing power, because by encoding a classical bit using a randomly-chosen basis, an adversary unaware of the basis choice can never be truly certain of the information being transmitted.
- QKD is particularly well-suited to produce a long random key, which makes the OTP more realistic in practice. When QKD is combined with OTPs, both key generation and encryption are unconditionally secure.
- A QKD-enabled microgrid is capable of detecting the presence of an eavesdropper trying to gain knowledge of keys, whereas existing communication systems without this ability will inevitably require additional detection mechanisms. This is because any attempt to learn keys causes noise in the quantum channel that can be detected by users.
- QKD systems have the advantage of automatically generating provably secure keys over those manually distributing keys. This is needed in microgrids to meet various continuous data transmission requirements.

Note that there are also post-quantum ways to distribute keys. However, the security of post-quantum systems is always based on the assumptions that solving certain mathematical problems (not the discrete logarithm problem or factoring problem, but other problems for quantum computers) is hard. QKD, in contrast, does not require these assumptions.

21.3 The QKD Simulator

Before constructing real QKD systems in microgrids in practice, it is critical to establish a real-time QKD-enabled microgrid testbed to evaluate the performance of QKD-enabled microgrids under different conditions. Tang et al. [19] and [20] have developed real-time QKD-enabled microgrid and NMs testbeds, respectively, where the QKD performance was modeled in a simulator capable of simulating QKD protocols. The flow chart of the simulator to simulate the decoy-state protocol is shown in Fig. 21.3. Note that this simulator is easily extensible for different QKD protocols.

In this simulator, we use time as an indicator to determine whether the laser has sent a sufficient number of key signals to generate the secret key of size ℓ. Let the current time be t_c, and the last calling time be t_p. Then, within the interval $(t_c - t_p)$, the number of signals that have been sent by the laser, N_t, can be obtained as

$$N_t = v_s(t_c - t_p) \tag{21.1}$$

where v_s is the speed of the laser sending signals, a constant value assumed in this study.

The post-measured signals received by Bob are temporarily stored in a classical "buffer." When a sufficient block size of signals has been received, the post-processing will start. Let the block size for post-processing be B, which is set by the users, and the number of signals needed to be sent before post-processing can start be N_r. Then,

$$N_r = \frac{B - N_b}{R_c} \tag{21.2}$$

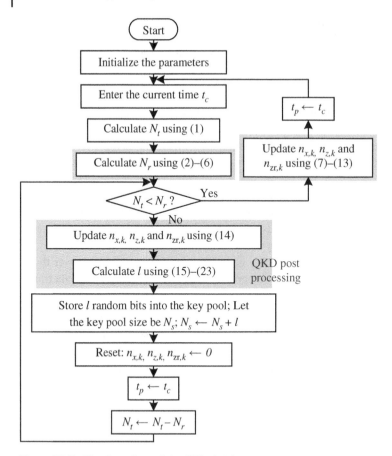

Figure 21.3 The flow chart of the QKD simulator.

where N_b is the number of raw-key signals in the "buffer," and R_c is the rate of correctly-received raw-key signals, i.e. the ratio of the number of correctly-received signals (leading to a useful raw key) and the number of signals actually sent. Specifically, R_c can be calculated as follows:

$$R_c = \sum_{k \in \{k_1, k_2, k_3\}} p_k p_x^2 p_{d_k} \tag{21.3}$$

where p_{d_k} is the probability that a signal with intensity k is received by Bob. It can be expressed as

$$p_{d_k} = (1 + p_{ap}) r_k, \quad \forall k \in \{k_1, k_2, k_3\} \tag{21.4}$$

where p_{ap} is the after-pulse probability. r_k is the expected detection rate (excluding after-pulse contributions) for intensity k, and can be calculated as follows:

$$r_k = 1 - (1 - 2p_{dc}) e^{-\eta_{tr} \eta_{Bob} k}, \quad \forall k \in \{k_1, k_2, k_3\} \tag{21.5}$$

where p_{dc} is the dark count probability and η_{Bob} is Bob's detection efficiency. η_{tr} is the transmittance that is related to the fiber length L as follows:

$$\eta_{tr} = 10^{-0.2L/10} \tag{21.6}$$

where the fibers are assumed to have an attenuation coefficient of 0.2 dB/km.

When the simulator is called, N_t and N_r are calculated and compared. Based on the comparison result of N_t and N_r, two cases exist as described below.

If N_t is smaller than N_r, the post-processing will not start, and the value of t_c will be assigned to t_p. Note that t_c is continuously increasing. Meanwhile, a certain number of signals within the time interval $(t_c - t_p)$ will be added into the "buffer." Let $n_{X,k}$ be the number of X signals received using intensity k. Then, of course, n_X, the size of the raw key in the "buffer" with the X basis, is simply the sum of all $n_{X,k}$ over all the intensities used. Specifically, $n_{X,k}$ can be updated as follows:

$$n_{X,k} \leftarrow n_{X,k} + N_t R_{X,k}, \quad \forall k \in \{k_1, k_2, k_3\} \tag{21.7}$$

where $R_{X,k}$ is the expected transmission rate of X signals for intensity k. It can be expressed as

$$R_{X,k} = p_k p_x^2 p_{d_k}, \quad \forall k \in \{k_1, k_2, k_3\} \tag{21.8}$$

Similarly, the number of Z signals received using intensity k, $n_{Z,k}$, can be updated as follows:

$$n_{Z,k} \leftarrow n_{Z,k} + N_t R_{Z,k}, \quad \forall k \in \{k_1, k_2, k_3\} \tag{21.9}$$

where $R_{Z,k}$ is the expected transmission rate of Z signals for intensity k, and can be expressed as

$$R_{Z,k} = p_k (1 - p_x)^2 p_{d_k}, \quad \forall k \in \{k_1, k_2, k_3\} \tag{21.10}$$

The size of the raw key in the "buffer" with the Z basis, n_Z, is the sum of all $n_{X,k}$ over all intensities used.

For this simulation, we assume a standard fiber channel and practical settings for the devices. In this case, the probability of having a bit error for intensity k, b_k, is as follows:

$$b_k = p_{dc} + e_{mis}(1 - e^{-\eta_{tr}k}) + \frac{p_{ap}r_k}{2}, \quad \forall k \in \{k_1, k_2, k_3\} \tag{21.11}$$

where e_{mis} is the error rate due to optical errors. Then, the number of erroneous bits in the Z basis for intensity k, $n_{Zr,k}$, can be updated as follows:

$$n_{Zr,k} \leftarrow n_{Zr,k} + N_t R_{Zr,k}, \quad \forall k \in \{k_1, k_2, k_3\} \tag{21.12}$$

where $R_{Zr,k}$ is the expected transmission error rate in the Z basis for intensity k and can be expressed as

$$R_{Zr,k} = p_k (1 - p_x)^2 b_k, \quad \forall k \in \{k_1, k_2, k_3\} \tag{21.13}$$

When all X, Z, and erroneous signals with all intensities have been added, the simulator goes back to the "listening" mode. As mentioned above, t_p becomes t_c, and t_c continuously grows.

If N_t is greater than or equal to N_r, post-processing will start. The simulator will then add all the X, Z, and erroneous signals with all the intensities to $n_{X,k}$, $n_{Z,k}$, and $n_{Zr,k}$, respectively. Specifically, $n_{X,k}$, $n_{Z,k}$, and $n_{Zr,k}$ can be updated in the following way:

$$\begin{cases} n_{X,k} \leftarrow n_{X,k} + N_r R_{X,k} \\ n_{Z,k} \leftarrow n_{Z,k} + N_r R_{Z,k} \quad \forall k \in \{k_1, k_2, k_3\} \\ n_{Zr,k} \leftarrow n_{Zr,k} + N_r R_{Zr,k} \end{cases} \tag{21.14}$$

After the post-processing is completed, the key is established and can be used by Alice and Bob. The simulator simulates the process by calculating the length ℓ of the secret key extracted that would be generated under the same condition in practice. The length ℓ of the extracted secret key can be obtained as follows:

$$\ell = \left\lfloor \xi_{X,0} + \xi_{X,1} - \xi_{X,1} h(\phi_X) - \lambda_{ec} - 6\log_2 \frac{21}{\varepsilon_s} - \log_2 \frac{2}{\varepsilon_c} \right\rfloor \tag{21.15}$$

where $h(x) = -x\log_2 x - (1 - x)\log_2(1 - x)$ is the binary entropy function. $\xi_{X,0}, \xi_{X,1}$, and ϕ_X are the number of vacuum events, the number of single-photon events, and the phase error rate of the single-photon events in the raw key with the X basis, respectively. ε_c is the probability that the keys extracted by the two parties are not identical, and ε_s is the user-specified maximum failure probability. λ_{ec} specifies how much information is leaked during error correction. It is set to $n_X \eta_{\text{ec}} h(\phi_X)$, where η_{ec} is the error-correction efficiency.

These parameters cannot be directly observed; however, by using the decoy-state protocol, they can be bounded. Basically, $\xi_{X,0}$ satisfies

$$\xi_{X,0} \geq \chi_0 \frac{k_2 n_{X,k_3}^- - k_3 n_{X,k_2}^+}{k_2 - k_3} \tag{21.16}$$

where χ_n is the probability that Alice sends a n-photon state. This value, using a weak-coherent laser, follows a Poisson distribution and is found to be:

$$\chi_n = \sum_{k \in \{k_1, k_2, k_3\}} e^{-k} k^n p_k / n! \tag{21.17}$$

and

$$n_{X,k}^\pm = \frac{e^k}{p_k} \left(n_{X,k} \pm \sqrt{\frac{n_X}{2} \ln \frac{21}{\varepsilon_s}} \right), \quad \forall k \in \{k_1, k_2, k_3\} \tag{21.18}$$

The number of single-photon events in the raw key with the X basis, $\xi_{X,1}$, satisfies

$$\xi_{X,1} \geq \frac{\chi_1 k_1 \left[n_{X,k_2}^- - n_{X,k_3}^+ - \frac{k_2^2 - k_3^2}{k_1^2} \left(n_{X,k_1}^+ - \frac{\xi_{X,0}}{\chi_0} \right) \right]}{k_1(k_2 - k_3) - k_2^2 + k_3^2} \tag{21.19}$$

Similarly, by using (21.16)–(21.19) with statistics from the basis Z, the number of vacuum events in Z_A, $\xi_{Z,0}$, and the number of single-photon events in the raw key with the Z basis, $\xi_{Z,1}$, can also be obtained.

The phase error rate of the single-photon events in the raw key with the X basis, ϕ_X, satisfies

$$\phi_X \leq \frac{\delta_{Z,1}}{\xi_{Z,1}} + f\left(\varepsilon_s, \frac{\delta_{Z,1}}{\xi_{Z,1}}, \xi_{Z,1}, \xi_{X,1} \right) \tag{21.20}$$

where

$$f(a, b, c, d) = \sqrt{\frac{(c + d)(1 - b)b}{cd \log 2} \log_2 \left(\frac{c + d}{cd(1 - b)b} \frac{441}{a^2} \right)} \tag{21.21}$$

and $\delta_{Z,1}$ is the number of bit errors of the single-photon events in the raw key with the Z basis. It is given by

$$\delta_{Z,1} \leq \chi_1 \frac{m_{Z,k_2}^+ - m_{Z,k_3}^-}{k_2 - k_3} \tag{21.22}$$

where

$$m_{Z,k}^\pm = \frac{e^k}{p_k} \left(m_{Z,k} \pm \sqrt{\frac{m_Z}{2} \ln \frac{21}{\varepsilon_s}} \right), \quad \forall k \in \{k_1, k_2, k_3\} \tag{21.23}$$

and $m_Z = \sum_{k \in \{k_1, k_2, k_3\}} m_{Z,k}$. Here, $m_{Z,k}$ is the number of error events in the Z basis. The initial values of the parameters from (21.1) to (21.23) are given in Table 21.2.

In summary, this simulator simulates the probabilities of various events occurring, such as multiple-photon emission, photon loss in the channel, phase errors, and detector imperfections.

Table 21.2 Initial values of the parameters in the QKD simulator.

k_1	k_2	k_3	p_{k_1}	p_{k_2}	p_{k_3}	$n_{X,k}$
0.4	0.1	0.0007	1/3	1/3	1/3	0
B	p_x	p_{dc}	ε_c	η_{Bob}	e_{mis}	$n_{Z,k}$
10^7	0.8	6×10^{-7}	10^{-11}	0.1	5×10^{-4}	0
t_p (s)	η_{ec}	p_{ap}	ε_s	L (km)	v_s (bit/s)	$n_{Zr,k}$
0	1.16	4×10^{-2}	10^{-11}	5	4×10^7	0

The simulator assumes quantum signals are continually being sent from end-nodes building a raw-key pool. When the simulator is called, it determines how many signals could have been sent from the last call (based on the speed of the simulated laser source and detector dead times), what the user's choices were for those signals (e.g. basis and intensity choices), and whether the receiver got a measurement outcome. If a sufficient number of signals have been sent, the error correction and privacy amplification results are simulated, leading to the generation of a simulated secret key of the actual size that would be generated under the same condition in practice. These secret key bits are added to the corresponding key pool.

Note that this QKD simulator is able to flexibly alter QKD parameters to simulate different scenarios, e.g. with different fiber lengths and noise levels. The simulator is also easily extensible for different QKD protocols and quantum channels. With a different QKD protocol, only the steps within the shaded areas in Fig. 21.3 need to be changed accordingly.

21.4 Quantum-Secure Microgrid

This section introduces how QKD can be integrated into a single microgrid. First, it presents a general quantum communication architecture designed for a single microgrid. Then, a real-time quantum-secure microgrid testbed is discussed, which includes a high-level introduction, communication network design, and microgrid modeling and operation. For a more detailed description, readers are referred to Tang et al. [19].

21.4.1 Quantum Communication Architecture for Microgrid

A general QKD-based communication architecture for a single microgrid is illustrated in Fig. 21.4. In this architecture, the microgrid control center (MGCC) collects data from different loads (denoted as the first type of communication) and sends control signals to local controllers (denoted as the second type of communication). As building a quantum channel is costly, it is practical and reasonable to implement QKD for only critical communications in a microgrid. Compared to the first type of communication, the second type is arguably more critical because a malicious control signal can directly lead to fateful consequences. The first type of communication is less critical because when the data are received from different loads by the MGCC, they will typically be dealt with by some anomaly detection methods.

As an example, a QKD system is implemented between the MGCC and the local controller for battery storage. This battery uses a P–Q control to adjust its power output based on the real power

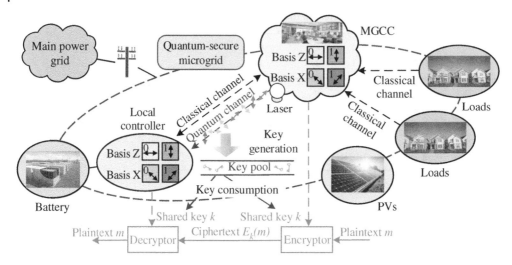

Figure 21.4 A general QKD-enabled quantum-secure microgrid communication architecture. Source: Tang et al. [7, 19–23]/IEEE.

reference received from the MGCC. It should be noted that QKD is only used to generate keys for two parties in an unconditionally secure way; the data encryption process is still achieved using classical cryptographic methods such as AES or OTP. Using AES to encrypt data is considered quantum secure as long as the key used for this process is kept secret. OTP is even more secure (or more accurately, unconditionally secure) because it uses a random key only once and then discards the key. But this requires that the key be as long as the plaintext. The keys generated by a QKD link are stored in a key pool, and when there is a need to transfer data, a certain number of key bits are extracted for encryption.

21.4.2 Quantum-Secure Microgrid Testbed

21.4.2.1 High-Level Design

The testing environment for a quantum-secure microgrid is illustrated in Fig. 21.5. The microgrid model is developed and compiled in real-time structured computer aided design (RSCAD), a power system simulation software designed to interact with the real-time digital simulator (RTDS). The RTDS in the testbed consists of three racks, which can be used separately for small-scale power systems or combined to provide more cores for a large-scale system. In this simulation, rack 2 is utilized to simulate the microgrid model in real time.

The measurements from the RTDS simulator are transmitted through a GTNETx2 card and sent to the MGCC through a communication network. A GTNETx2 card can either receive data from the RTDS and send them to external equipment, or it can receive data from the network and send them back to the RTDS, depending on whether the GTNETx2 card was designed to be in the sending or receiving mode. The MGCC runs on a remote server, which can receive load measurements and send signals back to RTDS with an Ethernet connection.

The high-level design of the testbed is illustrated in Fig. 21.6. Two GENETx2 cards are utilized for network communication. Note that although only one quantum channel is established in this case, the principle can be easily extended to cases with multiple quantum channels. The GTNETx2 card #2 is used to transmit data from the RTDS to the MGCC, which models the classical communication (represented in ① in Fig. 21.6) in real time, i.e. collecting load measurements for the MGCC. When

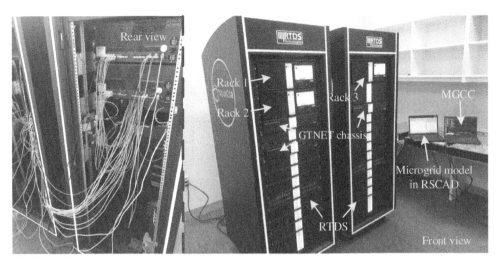

Figure 21.5 Testbed setup for a quantum-secure microgrid in an RTDS environment. Source: Tang et al. [7, 19–23]/IEEE.

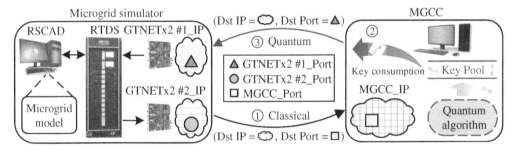

Figure 21.6 High-level design of the quantum-secure microgrid testbed. Source: Tang et al. [7, 19–23]/IEEE.

the MGCC receives the data, an analysis of the data is performed and appropriate control signals are sent to the local controller. Before a control signal is sent out, a key with the same length is extracted from the key pool. This process (represented in ② in Fig. 21.6) succeeds only when there are enough key bits.

The GTNETx2 card #1 is utilized to receive signals from the MGCC (represented in ③ in Fig. 21.6) and transfer them to the RTDS. The simulation results with the updated control signals are demonstrated in RSCAD. Note that the QKD system is modeled using the QKD simulator in Fig. 21.3. The keys are generated continuously by the QKD algorithm and are stored in a key pool. This real-time communication between the RTDS microgrid simulator and the MGCC using the QKD algorithm is the salient feature of this testbed.

21.4.2.2 QKD-Based Microgrid Communication Network
The network connection of key components in the RTDS simulator and a flow chart of the algorithm running in the MGCC are illustrated in Fig. 21.7. As shown on the left side of Fig. 21.7, each RTDS rack is connected to one or more GTNETx2 cards using fiber optic cables. All GTNETx2 cards are connected with an edge switch through Ethernet cables to transmit and receive data over the network. The User Datagram Protocol (UDP) is used in the simulation.

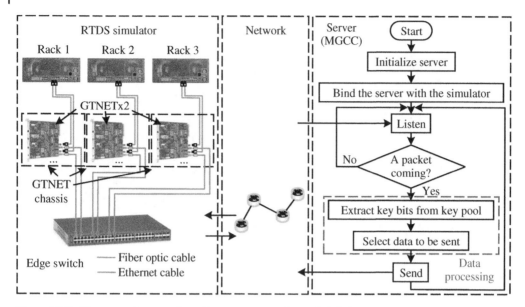

Figure 21.7 The network connection of key components in the RTDS simulator and a flow chart of the algorithm running in the MGCC. Source: Tang et al. [7, 19–23]/IEEE.

On the MGCC side, as shown on the right side of Fig. 21.7, the server enters the *listening* mode after being connected to the simulator. At this stage, the server is receiving any UDP packet whose destination IP and port match those of the server, respectively. Once a packet arrives, a quantum key with the same length of the received data is extracted from the key pool, and the corresponding control signals are generated. The server then enters the *sending* mode and starts to send out control signals whose destination IP and port are the IP and port of the GTNETx2 card #1 in the RTDS simulator (see Fig. 21.6), respectively. After sending the control signals, the server returns to the *listening* mode.

21.4.2.3 Microgrid Modeling and Operation

A typical microgrid system shown in Fig. 21.8 is used to evaluate the performance of the QKD-enabled quantum-secure microgrid. The buses within the microgrid are rated at 13.2 kV, and the microgrid is connected to the 138 kV main grid through a 138/13.2 kV transformer and a circuit breaker. The microgrid can operate in islanded mode or in grid-connected mode, depending on the state of the circuit breaker. The transformer is $\Delta - Y$ connected and is rated at 25 MVA with a 8% impedance.

DERs in the microgrid include a 5.5 MVA diesel generator, a 1.74 MW photovoltaic (PV) system, and a 2 MW doubly-fed induction generator wind turbine system. The diesel generator uses the droop control to regulate the microgrid frequency in islanded operation and to provide real and reactive powers in both grid-connected and islanded modes. Both the PV system and the wind turbine use maximum power point tracking (MPPT) control to maximize their power output. Three switched capacitors are connected on bus 1 to facilitate voltage synchronization in the microgrid.

A lithium-ion battery storage is further connected on bus 2 to provide a backup power supply and store extra energy when the microgrid is in islanded operation. The battery model consists of 250 stacks connected in parallel and each having 250 cells in series. A single cell has a capacity of 0.85 AH (ampere hour), and the initial state of charge in a single cell is set at 85%. A P–Q control is designed to regulate the output power of the battery, the value of which is determined by the

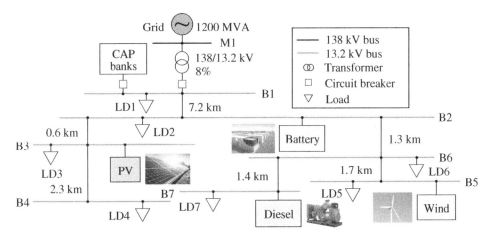

Figure 21.8 One-line diagram of the microgrid model. Source: Tang et al. [7, 19–23]/IEEE.

Table 21.3 Power loads at different buses in Fig. 21.8.

Load	Phase A KVA	Phase B KVA	Phase C KVA	Power factor
LD 1	506	506	506	0.9
LD 2	367	367	367	0.95
LD 3	344	344	344	0.9
LD 4	356	356	356	0.9
LD 5	325	625	100	0.95
LD 6	125	725	300	0.95
LD 7	275	625	150	0.95

real and reactive power references transferred from the MGCC via a communication channel. The initial values of the real and reactive power references are both set at zero.

The resistance and inductance of a unit length of the lines in the microgrid are $0.2322\,\Omega$/km and 2.355×10^{-3} H/km, respectively, and the lengths of the lines are given in Fig. 21.8. Power loads at different buses are given in Table 21.3.

21.5 Quantum-Secure NMs

21.5.1 Quantum Communication Architecture for NMs

A QKD-based quantum-secure NMs architecture is shown in Fig. 21.9 [20]. The NMs system consists of multiple interconnected microgrids. Within each microgrid, the MGCC collects information from customers and sends appropriate control signals to local controllers. In this architecture, QKD is utilized to generate keys for communications between each MGCC and local controllers in the same microgrid (MG), while communications between each MGCC and customers in the same MG are established over classical channels. The keys used for communication between two

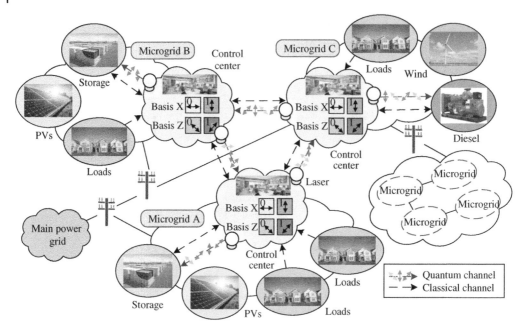

Figure 21.9 An overview of the QKD-enabled quantum-secure NMs architecture. Source: Tang et al. [20]/IEEE.

MGCCs in different microgrids are also generated using QKD. Note that this design is practical and reasonable, because, from an economic perspective, building a quantum link is costly; therefore, quantum channels are only allocated for important communications. The keys generated by different quantum channels are stored in separate key pools.

21.5.2 Quantum-Secure NMs Testbed

21.5.2.1 QKD-Enabled Testing Environment

The design of the testbed for the QKD-enabled quantum-secure NMs is illustrated in Fig. 21.10. Similarly, the NMs model is developed and compiled in RSCAD and runs in real time in the RTDS hardware.

The measurements from the RTDS simulator are sent to a remote server using GTNETx2 cards. The MGCCs in the NMs run on the same remote server (they can also run on different servers). The server receives load measurements from the RTDS and sends signals back to RTDS, with a 1 Gbps Ethernet connection.

The QKD simulator capable of simulating the decoy-state BB84 protocol is developed using Python on the remote server. As presented in Section 21.3, the simulator simulates the probabilities of various events occurring such as multiple-photon emission, photons being lost in the channel, phase errors, and detector imperfections. It assumes that quantum signals are continually being sent from end nodes building a raw-key pool. When the simulator is called, it determines how many signals could have been sent from the last call (based on the speed of the simulated laser source and detector dead times), what the user's choices were for those signals (e.g. basis and intensity choices), and whether the receiver got a measurement outcome. If a sufficient number of

Figure 21.10 Testbed for QKD-enabled quantum-secure NMs in RTDS. Source: Tang et al. [20]/IEEE.

signals have been sent, the error correction and privacy amplification results are simulated leading to the generation of a simulated secret key of the actual size that would be generated under the same condition in practice. The secret key bits are added to the respective key pool.

21.5.2.2 Quantum-Secure NMs Communication Network

The network topology of the QKD-enabled NMs is illustrated in Fig. 21.11. In this testbed, the keys used for communications between two microgrids and between an MGCC and a local controller are generated using separate QKD algorithms and are stored in separate key pools. When there is a need to use keys, a certain number of bits are consumed from the corresponding key pool. Two

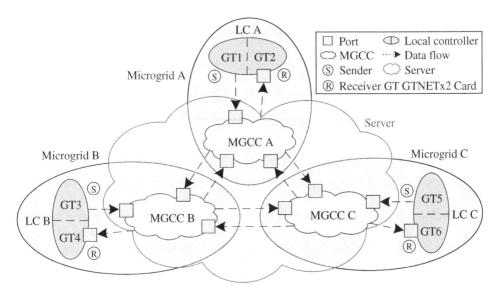

Figure 21.11 Network topology of the QKD-enabled quantum-secure NMs.

GTNETx2 cards are utilized for communication between a local controller in the RTDS simulator and each MGCC on the remote server. UDP is used to transmit and receive data.

Measurements in each microgrid are transmitted to its MGCC through a GTNETx2 card with a fixed speed set in RSCAD. The destination IP address is set as the server IP address, and the destination port is a specific number for the MGCC. Another GTNETx2 card is used for the local controller to receive MGCC data. Any UDP packet whose destination IP and port match those of this GTNETx2 card will be received.

From the MGCC side, each MGCC is receiving any UDP packet whose destination IP is the server's IP, and destination port matches the MGCC's port. When each MGCC receives a data packet from the RTDS, it sends messages to the other two MGCCs and a control signal to its local controller. Two other ports are set for each MGCC to receive UDP packets from the other two MGCCs. When each MGCC receives a data packet from another MGCC, a certain number of bits in the key pool between the two MGCCs are deduced.

21.6 Experimental Results

This section provides some experimental results generated by the QKD-enabled microgrid testbed in Section 21.4. Specifically, it demonstrates the impact of data transmission speed, the effectiveness of QKD-enabled communication, the performance of the QKD-enabled microgrid when quantum keys are exhausted, the impact of QKD on real-time microgrid operations, the performance of the quantum key generation speed under different conditions, and a comparison of two different QKD protocols.

To model the dynamic characteristics of the loads, a time-varying load with magnitude of 2 MW and frequency of 0.05 Hz is added to Load 2 (see Fig. 21.8). In this study, the value of the varying load is continuously sent from the RTDS to the remote server with a user-specified frequency. When the remote server receives the data packet, it calculates the value of the real power reference and sends it to the local P–Q controller for the battery at bus 2, such that the total power generation matches the sum of loads. The value of the reactive power reference is fixed at zero.

21.6.1 Impact of Data Transmission Speed

Data transmission speed is a critical statistic in a QKD-based microgrid. A speed greater than the key generation speed can result in the exhaustion of key bits in a key pool, eventually causing the failure of data communication.

To monitor traffic in the system, Wireshark, an open-source packet analyzer, can be used. Specifically, two types of packets are captured: packets sent from RTDS (GTNETx2 #2) to the MGCC and from the MGCC to RTDS (GTNETx2 #1). The transmission speeds of the two types of packets are set as the same. That is, once a packet is received by the MGCC, a packet is sent out from the MGCC.

The impact of the data transmission speed is illustrated in Fig. 21.12, where the fiber length L (between the MGCC and the local controller) is set at 50 km. The other parameters are the same as those in Table 21.2. Each packet sent from the MGCC to the RTDS consists of 64 binary bits, meaning that 64 key bits are consumed from the key pool when a packet is sent out.

From Fig. 21.12, it can be observed that:

- The data transmission speed has a large impact on the QKD-based microgrid. With the setting in Fig. 21.12, a speed larger than 20 packets/s will lead to the exhaustion of key bits in the key pool.

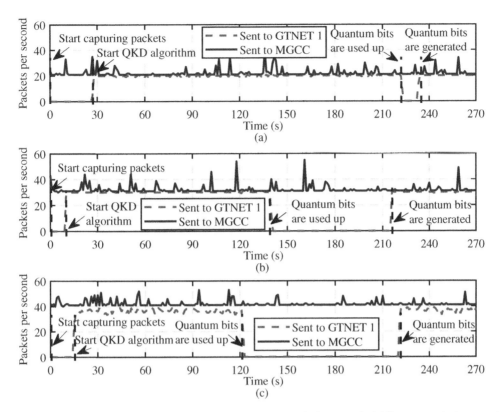

Figure 21.12 Traffic monitoring under different data transmission speeds. (a) Data transmission speed is 20 packets/s, (b) data transmission speed is 30 packets/s, and (c) data transmission speed is 40 packets/s.

- The larger the data transmission speed, the sooner the quantum-generated keys will be consumed. With the setting in Fig. 21.12, for a speed of 40 packets/s, the exhaustion lasts around 100 seconds within the key generation period. This long shortage can cause serious damage to microgrid operations, as there is no key in the key pool for encryption and authentication of data messages.

21.6.2 Effectiveness of QKD-Enabled Communication

Figure 21.13 illustrates the performance of the microgrid before and after communication starts working during the grid-connected mode. Before time $t = 3$ seconds, communication is disabled. The balance of total power generation and the sum of loads is achieved mainly by the main grid. When communication is enabled at time $t = 3$ seconds, storage starts to respond to changing loads and balance can be well maintained. Similar results can be observed in Fig. 21.14 where the microgrid switches from the grid-connected mode to the islanding mode.

21.6.3 Performance of QKD-Enabled Microgrid When Quantum Keys Are Exhausted

A QKD system mainly poses two impacts on microgrid operations: (1) keys generated by the QKD system are exhausted, and (2) the delay introduced by a QKD system affects real-time data transmission.

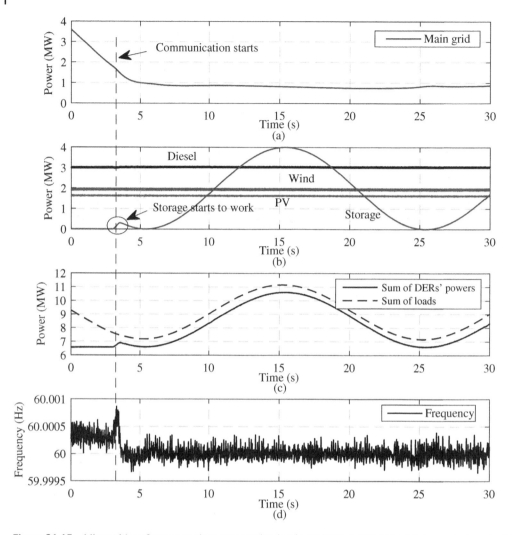

Figure 21.13 Microgrid performance when communication is enabled during the grid-connected mode. (a) Power from the main grid. (b) Output power of each DER. (c) Sums of DERs' powers and loads. (d) System frequency.

In this case, the first impact is evaluated. Figure 21.15 demonstrates the microgrid performance when keys are exhausted at time $t = 3$ seconds during the islanding mode. It can be seen that the system eventually collapses at time $t = 21$ seconds. The console's interface on the remote server is shown in Fig. 21.15(c).

21.6.4 Impact of QKD on Real-Time Microgrid Operations

A QKD system consists of a quantum channel and a classical one. The classical channel is shared by quantum key generation and normal data transmission. Adding a QKD system into microgrid therefore inevitably introduces more traffic into the classical channel. In this case, the impact of the delay caused by a QKD system is evaluated.

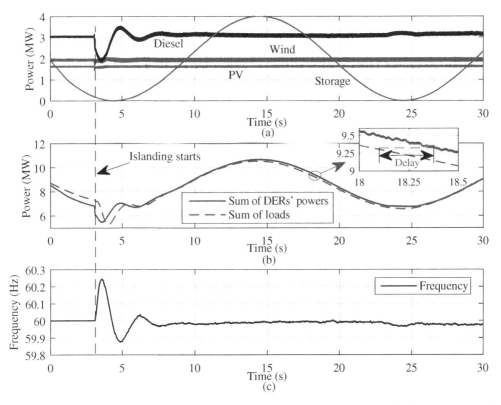

Figure 21.14 Microgrid performance during islanding mode. (a) Power from each DER. (b) The sums of DERs' powers and loads. (c) System frequency.

Specifically, we manually add a delay in the QKD algorithm on the remote server, meaning that when a data packet arrives, the control signal will be sent out with a certain time delay. Figure 21.16 gives the output power of each distributed energy resource (DER) when the delay is one second. It can be seen that, the storage updates every one second, and due to the delay, the output of the diesel varies significantly. The comparison of the impacts caused by different delays is given in Fig. 21.17, where the delay is set to be 0, 1, and 2 seconds, respectively. It can be seen that the larger the delay is, the more unstable the system is. An even larger delay, i.e. three seconds, directly leads to collapse of the microgrid during the islanding mode as shown in Fig. 21.18.

21.6.5 Evaluation of Quantum Key Generation Speed Under Different Fiber Lengths and Noise Levels

The speed of quantum key generation determines the maximum data transmission speed in a QKD-based microgrid. The larger the key generation speed, the higher the maximum data transmission speed. However, it was unclear which levels of key generation speed the QKD system could provide for the microgrid under different conditions. In this case, an evaluation of key generation speed under different fiber lengths Ls and noise levels e_{mis}s, is provided. The noise can be either natural or caused by an adversary. A strong attack on the quantum optic equipment is simulated by setting a large e_{mis}.

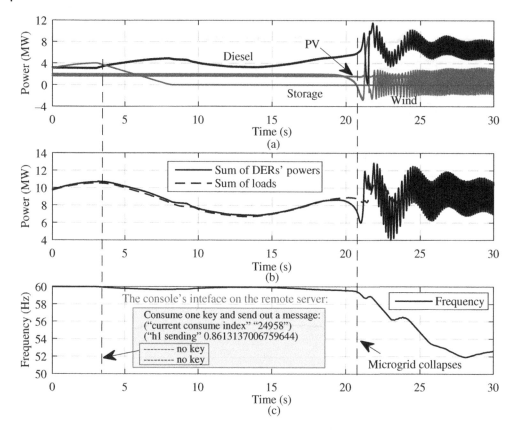

Figure 21.15 Microgrid performance when quantum keys are exhausted. (a) Power from each DER. (b) The sums of DERs' powers and loads. (c) Frequency.

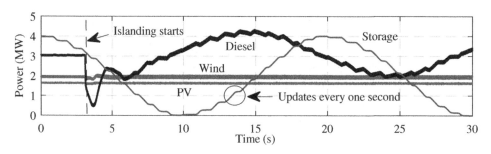

Figure 21.16 The output power from each DER when the delay is one second before and after microgrid islands.

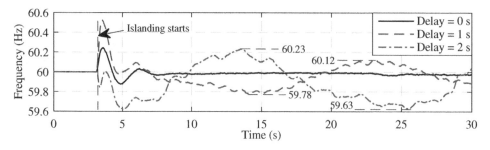

Figure 21.17 Comparison results of system frequencies with different delays.

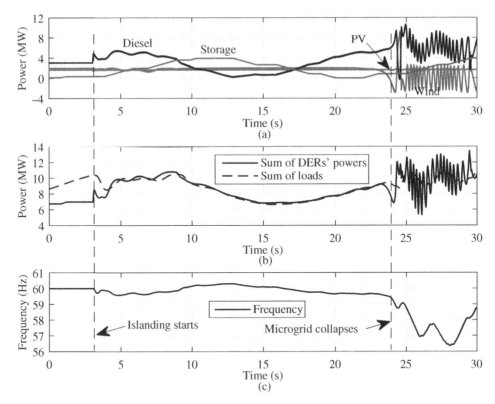

Figure 21.18 Microgrid performance when the delay is three seconds before and after microgrid islands. (a) Output power of each DER. (b) Sums of DERs' powers and loads. (c) System frequency.

The real-time simulation results are given in Fig. 21.19, where L is set from 1 to 80 km, e_{mis} is set from 5×10^{-4} to 9×10^{-4} with a step of 1×10^{-4}, and each packet consists of 64 binary bits. The other parameters are the same as those in Table 21.2. The key generation speed is calculated as the fraction of the generated key's size ℓ (see (21.15)) and the time required.

It can be observed that:

- A small L exhibits great superiority over a large L under the same e_{mis}, which gives valuable insight into the idea that the MGCC and the local controller should be close to each other in a QKD-based microgrid.
- The key generation speed is sufficient with a small L and a small e_{mis}. But it decreases dramatically when e_{mis} increases. A proper strategy therefore has to be carried out to improve the system's cyberattack resilience.
- Importantly, Fig. 21.19 gives valuable resources on which levels the data transmission speed should be set at under different Ls and e_{mis}s. With the setting in Fig. 21.19, any data transmission speed that is below the corresponding curve (with respect to a certain e_{mis}) in Fig. 21.19, will have sufficient key bits in the key pool under that e_{mis}.

21.6.6 Evaluation of Quantum Key Generation Speed Under Different Receiver's Detection Efficiencies

The detection efficiency of the receiver, η_{Bob}, is critical in a QKD system. Detection efficiency refers to the probability that the receiver can successfully detect photons, which is largely determined by the quality of the detection devices.

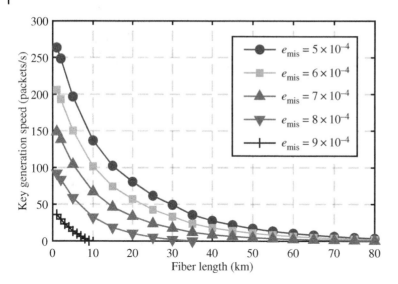

Figure 21.19 Quantum key generation speeds under different Ls and e_{mis}s.

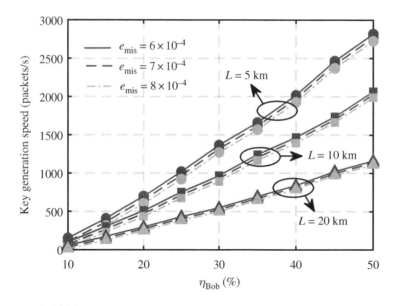

Figure 21.20 Quantum key generation speeds at different η_{Bob}s.

The impact of η_{Bob} is evaluated in our real-time testbed. The results are illustrated in Fig. 21.20, where L is set at 5, 10, and 20 km, respectively; e_{mis} is set at 6×10^{-4}, 7×10^{-4}, and 8×10^{-4}, respectively; and η_{Bob} is from 10% to 50% with a step of 5%. The other parameters are the same as in Table 21.2.

It can be seen that η_{Bob} has a significant impact on key generation speed. With a given L and a given e_{mis}, a small increase of η_{Bob} results in a great improvement in speed. This indicates that it is worth improving the quality of detection devices in a QKD-based microgrid.

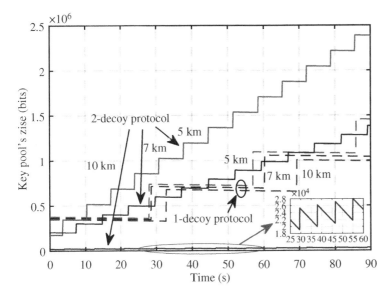

Figure 21.21 Comparison results of the 2-decoy-state protocol and the 1-decoy-state protocol with different fiber lengths using the testbed.

21.6.7 Comparison of Different QKD Protocols

In this test case, we use the testbed to compare the performances of two different QKD protocols, namely the 2-decoy-state protocol (as described above in this chapter) and the 1-decoy-state protocol. Specifically, key pool #1 stores key bits generated by the 2-decoy-state protocol and key pool #2 stores key bits generated by the 1-decoy-state protocol. The fiber length is set at 5, 7, and 10 km, respectively. Other parameters are the same for the two protocols. The comparison results are given in Fig. 21.21.

It can be observed that the 1-decoy-state protocol is more sensitive to the fiber length than the 2-decoy-state protocol, and outperforms the 2-decoy-state protocol when the fiber length is small. This is reasonable, as the 1-decoy-state protocol is more efficient in that there are no "vacuum" decoys (which are useless for key-rates); however, due to the lack of vacuum decoys, it is more sensitive to noise and loss.

21.7 Future Perspectives

The work of developing quantum-secure grids has just started. In the path toward a wider acceptance of quantum cryptography in power grids, further crucial issues need to be addressed and research projects should be carried out from a variety of aspects regarding both quantum cryptography and power grids. The research directions listed below seek to fulfill the promises of QKD to realize quantum-secure grids.

Resilience enhancement against side-channel attacks: The security properties of QKD protocols are captured by theoretical security proofs based on certain models. However, as practical devices tend to deviate from their specifications, the implementation of QKD may not perfectly comply with the model used in the security proof. An adversary can exploit the imperfections of critical devices and algorithms in QKD to launch side-channel attacks (e.g. the time-shift attack, the phase-remapping attack, and the detector-control attack), and learn a part of the secret key, causing security vulnerabilities. To improve the resilience of QKD-based microgrids against attacks in practice, both theoretical security and practical security of the QKD protocol should be considered in its implementation. A practical alternative to improve the resilience of QKD is the measurement-device-independent QKD (mdiQKD), which addresses the side-channel attacks by removing the measurement device, i.e. the most vulnerable part in a QKD system. The research work on developing mdiQKD-based microgrids is essential but has not yet been carried out.

Resilience enhancement against denial-of-service attacks: Another critical challenge faced by QKD is its vulnerability to denial-of-service (DoS) attacks. In QKD, when an adversary eavesdrops on the quantum channel, the two communicating parties can detect its occurrence by monitoring the bit error rate (BER). Once BER exceeds a predetermined threshold, the key establishment sessions are aborted. If the attack exists for a while, the key generation will continuously be disrupted. Proper strategies (e.g. reserving redundant rerouting configurations) are required to enhance the DoS-resilience of QKD systems in power grids.

Experimental demonstration of QKD in microgrids: The existing QKD-based microgrid testbeds exploit theoretical models to simulate QKD protocols generating ideal simulation results of certain QKD protocols. One of the future work is to implement real QKD systems in a real-time power system simulator, e.g. RTDS, thus providing a more realistic testing environment for QKD-enabled microgrids. More realistic testing results can be generated to experimentally demonstrate the performance of QKD under different microgrid scenarios.

Semi-quantum grid: QKD requires two communicating parties both be "quantum," meaning one has the capability of preparing and sending qubits and the other receiving and measuring qubits. In other words, any party who wants to participate in quantum communication has to gain the knowledge of operating qubits, which is however complex. This inevitably increases the operational complexity from the microgrid customer's side, and hence largely discourages the adoption of quantum communication in microgrids. An alternative to overcome this challenge is to use semi-QKD. Semi-QKD only requires one party be "quantum" with the capability of preparing, sending, receiving, and measuring qubits. The other party, on the other hand, can be "classical," meaning it only needs to interact with the quantum channel in a limited, classical manner. This largely reduces the operational complexity of quantum communication from the customer's side and can greatly promote the adoption of quantum communication in microgrids.

Point-to-multipoint QKD for microgrids: Most existing realizations of QKD are point-to-point systems. Microgrid however has a variety of communications, such as the ones between the control center and customers and between the control center and DERs. Neighboring microgrids also need to communicate with each other. Proper strategies are desired to make QKD more practical in microgrids. One solution is to use point-to-multipoint QKD systems. It is important to design and integrate the point-to-multipoint QKD in a proper way in accordance with different microgrid scenarios considering different configurations such as the data transmission frequency, the distance between two communicating parties, the number of communication channels, and the noise level on each quantum channel.

Quantum networks for microgrids: In addition to the development of point-to-multipoint QKD, another promising alternative to inspire the adoption of QKD in microgrids with multiple communication channels is to establish a quantum network. A quantum network distributes quantum keys for any two nodes in microgrids by connecting a number of facilities, fully utilizing quantum resources. However, the current status of quantum networks is far from practical applications, and critical issues exist. A proper design of quantum-network-based microgrids is required and some practical issues have to be resolved.

Novel QKD protocols and methods for microgrids: To handle a large amount of data in microgrids, it is required to have a high key generation rate. The existing QKD protocols and techniques can be unsuited in microgrids under more strict conditions. Novel QKD protocols and methods for large quantum channel multiplexing and fast key distillation are desired.

21.8   Summary

Driven by the flourishing progress in quantum computing software and hardware, the topic of developing quantum-secure microgrids has become increasingly important. Replacing existing cryptography systems with quantum ones in microgrids however faces substantial challenges from different aspects. This field is currently still in its infancy, and has considerable potential for growth. In this chapter, we reviewed the current status of recent research efforts on developing quantum-secure microgrids. Specifically, we introduced in detail an existing design of a QKD-based microgrid, including the basics of quantum communication, a practical QKD simulator, and the QKD-based microgrid architecture. By integrating QKD features into a real-time power system simulator, this testbed offers a flexible and programmable testing environment to evaluate the performance of QKD-enabled microgrids in a variety of scenarios. This is an important step toward constructing a real QKD system in microgrids in practice. In addition, we also extended the QKD-enabled microgrid to QKD-enabled NMs. With the testbeds, more research work could be done in the future. Some examples include exploiting the feasibility of more advanced and practical QKD protocols for microgrids, evaluating the QKD-enabled microgrid's performance under more scenarios, and developing methods to further enhance the cyber resilience of the QKD-enabled microgrid. We hope that this chapter will stimulate escalating awareness, provoke vigorous discussions, and inspire further research ideas. More research is needed in this emerging field, and we believe all those efforts will play a vital role and eventually benefit power system communities in the coming quantum era.

21.9   Exercises

1. Why is developing quantum-secure microgrids important?
2. Describe how a decoy-state BB84 QKD system works.
3. What are the benefits of using QKD for microgrids?
4. Briefly describe how a decoy-state BB84 QKD system is simulated.
5. QKD is only used to generate keys for two communicating parties in an unconditionally secure way. However, the data encryption process is still achieved using classical cryptographic methods such as AES or OTP; explain why this can be considered quantum secure.
6. Summarize how a simulated QKD system is integrated into an RTDS environment.
7. Describe how a quantum-secure NMs testbed can be established.
8. What are the impacts of QKD on microgrids?
9. What parameters can affect the performance of QKD? And how can they affect it? Give some examples.

References

1 Jiang, Z., Tang, Z., Zhang, P., and Qin, Y. (2021). Programmable adaptive security scanning for networked microgrids. *Engineering* 7 (8): 1087–1100.

2 Wang, L., Qin, Y., Tang, Z., and Zhang, P. (2020). Software-defined microgrid control: the genesis of decoupled cyber-physical microgrids. *IEEE Open Access Journal of Power and Energy* 7: 173–182.

3 Etingov, D.A., Zhang, P., Tang, Z., and Zhou, Y. (2022). AI-enabled traveling wave protection for microgrids. *Electric Power Systems Research* 210: 108078.

4 Zhang, P., Tang, Z., Yang, J. et al. (2018). PV extreme capacity factor analysis. In: *2018 IEEE Power & Energy Society General Meeting (PESGM)*, 1–5. IEEE.

5 Tang, Z., Zhang, P., Muto, K. et al. (2018). Extreme photovoltaic power analytics for electric utilities. *IEEE Transactions on Sustainable Energy* 11 (1): 93–106.

6 Wang, J., Qin, Y., Tang, Z., and Zhang, P. (2020). Software-defined cyber-energy secure underwater wireless power transfer. *IEEE Journal of Emerging and Selected Topics in Industrial Electronics* 2 (1): 21–31.

7 Tang, Z., Zhang, P., and Krawec, W.O. (2021). A quantum leap in microgrids security: the prospects of quantum-secure microgrids. *IEEE Electrification Magazine* 9 (1): 66–73.

8 Tang, Z., Jiao, J., Zhang, P. et al. (2019). Enabling cyberattack-resilient load forecasting through adversarial machine learning. In: *2019 IEEE Power & Energy Society General Meeting (PESGM)*, 1–5. IEEE.

9 Jiao, J., Tang, Z., Zhang, P. et al. (2019). Ensuring cyberattack-resilient load forecasting with a robust statistical method. In: *2019 IEEE Power & Energy Society General Meeting (PESGM)*, 1–5. IEEE.

10 Jiao, J., Tang, Z., Zhang, P. et al. (2021). Cyberattack-resilient load forecasting with adaptive robust regression. *International Journal of Forecasting* 38 (3): 910–919.

11 Wang, L., Zhang, P., Tang, Z., and Qin, Y. (2022). Programmable crypto-control for IoT networks: an application in networked microgrids. *IEEE Internet of Things Journal* 10 (9): 7601–7612.

12 Heron, S. (2009). Advanced encryption standard (AES). *Network Security* 2009 (12): 8–12.

13 Shor, P.W. (1994). Algorithms for quantum computation: discrete logarithms and factoring. In: *Proceedings 35th Annual Symposium on Foundations of Computer Science*, 124–134. IEEE.

14 Mollin, R.A. (2002). *RSA and Public-Key Cryptography*. CRC Press.

15 Ekert, A. and Jozsa, R. (1996). Quantum computation and Shor's factoring algorithm. *Reviews of Modern Physics* 68 (3): 733.

16 Zhou, Y., Tang, Z., Nikmehr, N. et al. (2022). Quantum computing in power systems. *iEnergy* 1 (2): 170–187.

17 Feng, F., Zhang, P., Zhou, Y., and Tang, Z. (2022). Quantum microgrid state estimation. *Electric Power Systems Research* 212: 108386.

18 Tang, Z., Zhang, P., and Zhou, Y. (2022). Quantum renewable scenario generation. In: *2022 IEEE Power & Energy Society General Meeting (PESGM)*, 1–5. IEEE.

19 Tang, Z., Qin, Y., Jiang, Z. et al. (2021). Quantum-secure microgrid. *IEEE Transactions on Power Systems* 36: 1250–1263.

20 Tang, Z., Qin, Y., Jiang, Z. et al. (2020). Quantum-secure networked microgrids. In: *2020 IEEE Power & Energy Society General Meeting (PESGM)*, 1–5. IEEE.

21 Tang, Z., Zhang, P., Krawec, W.O., and Jiang, Z. (2020). Programmable quantum networked microgrids. *IEEE Transactions on Quantum Engineering* 1: 1–13.

22 Tang, Z., Zhang, P., Krawec, W., and Wang, L. (2022). Quantum networks for resilient power grids: theory and simulated evaluation. *IEEE Transactions on Power Systems* 38 (2): 1189–1204.

23 Jiang, Z., Tang, Z., Qin, Y. et al. (2021). Quantum internet for resilient electric grids. *International Transactions on Electrical Energy Systems* 31 (6): e12911.

22

Community Microgrid Dynamic and Power Quality Design Issues

Phil Barker, Tom Ortmeyer, and Clayton Burns

22.1 Introduction

Community resilient microgrids are being considered as effective ways to provide electric power resiliency to communities in the aftermath of major events, such as hurricanes or ice storms. This resiliency benefit also provides added value to communities that are installing significant levels of local generation, including the intermittent renewable energy sources such as photovoltaics and wind. For planning purposes, a resilient microgrid is often designed to have the capability to serve critical loads for up to two weeks following a major event. Potential critical loads to be served can include:

- Hospitals
- Police, Fire, and EMS providers
- Shelters for displaced citizens
- Fresh and waste water plants
- Critical commercial businesses such as grocery, pharmacy, fuel, and banking providers.

In any given case, the final list of services to be supplied by the microgrid will be determined by local circumstances. It is expected that these microgrids would often be in the range of 5–30 MW in power rating.

By nature, these microgrids will be mostly connected by a primary distribution system, with voltages in the 4–34.5 kv range. Depending on the local circumstances, the microgrid may be designed to serve all of the customers in a given area or may serve only the critical loads in an area. In either case, the microgrid will include the local distribution system franchise holder, government entities such as the police department, non-profits, and for-profit businesses. The generation sources for these microgrids could be a combination of customer-owned behind-the-meter generation and generation directly connected to the primary system, owned by independent generation companies or perhaps by the distribution system franchise holder.

This chapter discusses the design process for several key dynamic and power quality aspects of these microgrids. These include voltage flicker, temporary overvoltages, harmonic distortion, frequency regulation, and black start capability. All power distribution systems are expected to meet the requisite standards and practices in these areas. By their nature, when microgrids are operating in the islanded mode (separated from the utility power grid), these dynamic and power quality limits can be challenging to meet. In islanded operation, the microgrid is fed only by its local generation. This local generation is inevitably less stiff than the utility power system, which leads to larger variations in voltage and frequency. Further, in microgrids with a large proportion of inverter-connected resources (IBR), the system is even less stiff during standalone operation.

The chapter includes case study details for the proposed Potsdam, NY, Microgrid. The Potsdam Microgrid design study focused on providing critical services to the Potsdam, NY community following an extreme event. For planning purposes, the extreme event was considered to be a major ice storm, such as the one that hit the area in 1998. While resilient microgrids often identify a specific extreme event in their design stage, they are expected to provide their service over the range of possible extreme events to the best of their capability. The Potsdam Community Resilient Microgrid design project was funded by the New York State Energy Research and Development Authority and National Grid (NYSERDA Report 18-13, 2018 [1]). It was conducted by Clarkson University, GE Energy Consulting and Nova Energy Specialists.

22.2 Potsdam Resilient Microgrid Overview

The resilient microgrid envisioned for Potsdam NY was based on the installation of a new primary distribution system that would serve only critical loads. Due to the planning scenario being based on major ice storms, the new distribution system would be strictly underground. The Potsdam microgrid would span a physical distance of about 2 miles from east to west and 1 mile north to south. Figure 22.1 shows the overall layout of the microgrid with the underground 13.2 kV cable

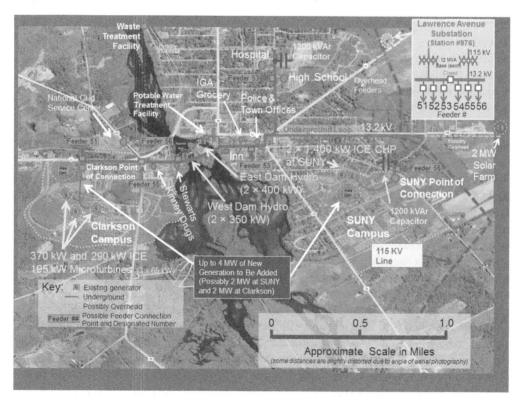

Figure 22.1 One-line layout diagram of the Potsdam Microgrid showing feed points, switches, customers and existing and proposed generation sites. Source: New York State Energy Research and Development Authority (NYSERDA).

paths superimposed on an aerial image of the Town of Potsdam. The microgrid was designed to serve the Clarkson University and SUNY-Potsdam campuses as well as the hospital, high school, police, water facilities, and several other key municipal loads and commercial loads in the town of Potsdam. The cables are buried in ducts and in an arrangement that allows a looped feeder configuration. With multiple switching vaults/cabinets present along the loop path, it is possible to create a redundant underground loop scheme that can be operated either as an open loop or closed-loop depending on various factors. A failed section can be sectionalized out of the system as needed to insure continuity of service to most or all loads.

National Grid's Lawrence Avenue Substation is the connection path to the bulk utility system source. The feeder labels (e.g. "Feeder 51" or "Feeder 53", etc.) that are shown in Fig. 22.1 represent possible alternative tie points to the conventional utility power system by any of several possible feeders that connect to the Lawrence Avenue substation. It is expected that Feeder 51, which is currently a dedicated express feeder that serves Clarkson University, is to be the typical key connection point.

The overall microgrid with the configuration of the looped scheme and positions of key switches and breakers is shown in a one-line diagram format in Fig. 22.2. As stated previously, Feeder 51 is the main feed point from the substation for this specific setup. During isolated microgrid operation, the utility power system would be separated by the opening of the breaker (or recloser) at switch position 1 that connects Feeder 51 to the microgrid. As stated previously, there are other possible

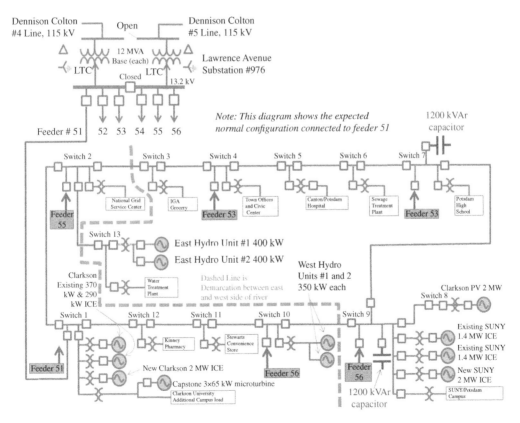

Figure 22.2 One-line layout diagram of the Potsdam Microgrid showing feed points, switches, customers and existing and proposed generation sites. (NYSERDA Report 18-13, 2018).

feed points and the microgrid can also be split in half if desired with a feed point to each side. A natural splitting demarcation line is that of the Raquette River, as shown in the diagram by the brown dashed line.

The Lawrence Avenue Substation is the key source of utility system power and will interact with the DG of the microgrid during normal system conditions when the microgrid is operated in parallel with the utility system. The substation characteristics play a large part in defining this interaction.

The station is a two-transformer design with six feeders. Each transformer is base rated at 12 MVA but has a full rating of 22.4 MVA with all stages of cooling activated. There are two 13.2 kV buses with three feeders on each. However, the bus tie switch is normally closed such that all feeders operate as if they are on one common bus, with both transformers operated in parallel.

The effective impedance of the power system at the 13.2 kV bus (with the tie switch closed) is 3.89% on a 12 MVA base. This makes the substation bus fault currents contributed by the utility system higher than average at about 13,507 A for a three-phase fault and 14,087 A for a line to ground fault. The substations' LTCs are set at 123 V and the line drop compensation is not activated. The lack of line drop compensation, along with the relatively low impedance of power system at this bus (due to both transformers being paralleled) makes the bus relatively insensitive to voltage changes caused by distributed generation current fluctuations. For example, fluctuations from the 2 MW of PV on this system when in parallel with the National Grid source would be inconsequential as discussed later in this report.

The minimum loading at the substation (with both transformers combined since they are operated in parallel) is about 3–4 MW. During minimum load periods if much of the planned generation is operating at or near rated capacity the substation might actually export energy into the 115 kV transmission system. On the other hand, the peak loading at both transformers combined is about 30–33 MW and the available DG is at maximum roughly 10 MW (see list of units later in report) which is about 1/3 of the substation peak load. While power export into the subtransmission system is unlikely much of the time, it is important to recognize that export is still a possibility if a large amount of DG were to be operated during minimum load periods. Provisions may be needed to handle this condition as discussed later in the report.

The peak load on the microgrid is approximately 10 MW, with the two campuses and the hospital being the largest loads and altogether representing over 80% of the total load.

22.3 Power Quality Parameters and Guidelines

The principal power quality parameters that relate to the operation of the microgrid that are:

1. Voltage regulation
2. Voltage flicker and short duration excursions above or below ANSI limits
3. Temporary overvoltage (short-duration severe overvoltage typically due to neutral shift, load rejection or ferroresonance usually lasting anywhere from $\frac{1}{2}$ cycle up to many seconds.
3. Transient overvoltage (lightning or switching voltage transients that are impulsive or ringing in nature with frequency much higher than 60 Hz.
4. Harmonic distortion
5. Frequency regulation (pertaining to deviations above or below the nominal system frequency of 60 Hz)

For electric utilities there are requirements for each of the parameters that utilities should achieve for their customers to assure satisfactory and safe operation of customer loads. Table 22.1 lists some

Table 22.1 Summary of power Quality criteria for microgrid operation.

Type of power quality concern	Applicable standard or method	Utility supplied grid with parallel DG (this is the normal mode)	Islanded microgrid mode (up to 2 weeks)
Voltage regulation (steady state)	ANSI Standard C84.1	Typical utility practice: Regulate voltage to *Range A service voltage* ($\pm$5%). Allow occasionally excursions into Range B (roughly +6% and −8%)	Range-A is still desirable but Range-B may be allowable given the infrequent and limited nature of islanded mode.
Voltage flicker, variations and brief duration excursions	**Flicker:**IEEE 1453 P_{st}, P_{lt} Flicker Criteria (Or IEEE 141 or 519) **Excursions/ Variations:** Tap changer cycling, load process sensitivity screen test	Limit voltage flicker to the *borderline of visibility* for LTC cycling and load process sensitivity, employ relatively strict screens for excess tap changer operations and load process dropouts or problems.	Borderline of irritation rather than visibility? Allow deeper dips on motor starts? Use "1% of time" IEEE-1453 violation criteria evaluated over annual basis? Relaxed sensitivity screen for tap changers and load process?
TOV: *Ground fault and load rejection overvoltage*	ITI curve, IEEE C62.92, Surge Arrester TOV Curves	Temporary overvoltage not recommended to exceed about 1.31 per unit for typical fault durations.	Consider a blend of the ITIC and the 1.31 per unit TOV requirements based on IEEE COG of 72%.
Harmonics	IEEE 519 and IEEE 1547 Guidelines	**Voltage harmonic**: Voltage THD up to 5% and individual voltage harmonic up to 3% **Current harmonics:** Use $I_{sc}/I_{load} < 20$ for current harmonics. Loads can have higher values at higher I_{sc}/I_{load} ratios (per IEEE 519-2014 and 1547-2018).	**Voltage harmonic**: Voltage THD up to 5% and individual voltage harmonic up to 3% **Current harmonics:** Both loads and DG should use the most strict $I_{sc}/I_{load} < 20$ allowances for current harmonics. The IEEE 519-2014 best applies to a microgrid but some relaxation of the limits may be allowed for short time periods (see discussion in Section 22.3.5).
Frequency regulation range (hertz)	Transformer and motor saturation limits, etc. (for motor loads see NEMA MG1)	Typically less than $\pm$0.5% deviation from 60 Hz nominal occurs for large scale utility grid systems.	Aim for $\pm$3% but up to $\pm$5% may be allowable for short periods depending on UPS responses, critical processes, transformer, and machine saturation curves.

key parameters and IEEE standards for electric operations under normal service conditions and suggests some possible relaxed allowances during islanded microgrid mode.

While a desirable operating objective is to aim to satisfy the normal IEEE and ANSI bulk system requirements for power quality, it must be understood that the practical limitations of the generation equipment when operating within an islanded microgrid are such that a relaxed level of voltage quality may at times be necessary. The microgrid has less generator inertia with respect to the size of load steps and a higher effective source impedance than the grid system; therefore, it is more susceptible to load step-induced frequency variations, load step-related voltage dips and harmonic distortion caused by load nonlinearity. In islanded microgrid mode, we can expect a certain amount of degradation of the voltage regulation, frequency regulation, and harmonic distortion due to the effects of loads on such weaker systems. The microgrid does not necessarily need to provide quite the same level of voltage quality as is provided by the utility under normal conditions since such conditions are not permanent.

In order to design a given microgrid, the operating voltage quality objectives for the islanded microgrid mode of operation must be set. The following points are worth considering:

1. Steady-state voltage should not be so low that excessive heating of motor devices or low-voltage-related cycling of UPS devices occurs.
2. Steady-state voltage must not be allowed to get too high that it causes overheating of resistive devices, failures of surge arresters due to too much leakage current, or saturation (overheating) of magnetic core devices such as motors/generators and transformers, or high-voltage-related cycling of UPS devices.
3. Steady-state voltage unbalance should not become too high as to cause overheating of three-phase motors/generators (due to negative sequence currents induced on rotors by unbalance).
4. Frequency variations (the absolute limits and rates of change of the microgrid frequency) must not be so great as to cause UPS systems to cycle, magnetic devices to saturate excessively, or generator frequency relays to operate.
5. Harmonic distortion must not be so high that it excites resonances and/or overheats devices or triggers the malfunction of devices.

There is precedent for relaxing the allowable power quality for short periods of time. For example, under today's existing ANSI standards, the voltage regulation limits on regular utility systems are often relaxed to allow occasional excursions outside of the ANSI C84.1 Range A and into the broader Range B limits. A higher degree of flicker is also allowable per IEEE 1453 if the conditions are not permanent and only last a few days or up to perhaps a few weeks in emergency conditions. Table 22.1 contrasts the conditions expected in grid parallel mode versus some possible relaxed conditions that could be allowable in an islanded microgrid mode for short periods.

22.3.1 Voltage Magnitude

When the microgrid is fed by the normal utility grid system, it is required that the steady-state voltage regulation should be within the ANSI C84.1 Range A voltage limits (ANSI C84.1-2020, 2020 [2]). These limits represent the voltage at the customer meter point. The utility also must provide proper voltage balance within 3% at all customer service connection points. Too much voltage unbalance can overheat motors and cause other problems. The Range-B voltage limits given in Table 22.2 are intended to be allowable only for infrequent conditions and with limited duration. When operating in the islanded microgrid mode, this type of operation also fits the "infrequent and

Table 22.2 ANSI C84.1 voltage limits (shown on a 120-volt base as well as in percent of nominal).

ANSI C84.1 Service Voltage Ranges		
Classification	Range A	Range B
Service voltage	114–126 volts (95–105%)	110–127 volts (91.7–105.8%)
Utilization voltage	110-125 volts (91.7–104.2%)	106–127 volts (88.3–105.8%)

The standard also specified that voltage between phases shall be balanced to with 3% at the service entrance under no load conditions.

Table 22.3 IEEE 1453 flicker guidelines (for grid parallel mode).

Type of flicker	Compatibility levels LV and MV applications	Planning levels (not to exceed >1% of time)	
		MV	HV and EHV
P_{st} **(10 minute interval)**	1.0	0.9	0.8
P_{lt} **(2 hour interval)**	0.8	0.7	0.6

limited duration" definition. So, a reasonable interpretation of the ANSI C84.1 standard, if used for an islanded microgrid, is that while Range-A is still the desired objective, Range-B conditions are allowable as long as the islanded mode is utilized infrequently.

22.3.2 Voltage Flicker

Flicker limits are developed based on the human perception of lamp flicker at noticeable and irritable levels. The standards IEC 61000-4-15-2010 (IEC 61000-4-15:2010 [3]) and IEEE 1453-2015 (IEEE 1453-2015 [4]) provide recommended practices for power grid flicker. This standard defines short-term P_{st} and long-term flicker P_{lt} metrics. P_{st} and P_{lt} are designed to have a value of 1.0 at flicker levels that produce irritation.

Table 22.3 shows recommended compatibility and planning levels for P_{st} and P_{lt}. As a general guideline for conventional power systems, IEEE 1453 recommends that P_{st} and P_{lt} should not exceed the planning levels more than 1% of the time (99% probability level). If this was taken over an ***annual basis***, it might be interpreted that up to several days, each year of flicker violations would be allowable as long as the rest of the year is well within limits. It makes sense to use this approach for the microgrid, but the question still remains as to how much of a violation above the limits in the table is allowable during such periods. As an educated suggestion, an increase in the allowable flicker P_{st} and P_{lt} thresholds by 25% to 50% above the regular values may be suitable for the microgrid during the violation period.

22.3.3 Temporary Overvoltage Guidelines

Temporary overvoltage (TOV) conditions are those that involve short-duration overvoltage lasting many tens of seconds in duration down to about $\frac{1}{2}$ cycle duration. These may be due to ground fault overvoltage or load rejection overvoltage due to DG or other causes. TOVs are overvoltages of the 60 Hz waveform. These are distinct from transient overvoltages, which refer to lightning or switch surges that typically last less than one cycle. A curve that represents the ability of information technology loads to survive short-term overvoltage conditions is the Information Technology Industry Council (ITIC) voltage tolerance curve (Information Technology Industry Council [5]). The ITI voltage tolerance curve shows a range of voltages that are suitable for such loads (see Fig. 22.3). It provides not only a high side boundary for withstanding damage but also a lower boundary for the ability of loads to handle voltage sags without dropout. The upper bound of the curve is felt to be quite conservative by many experts. In reality, many loads and devices can handle somewhat greater voltage without damage in most cases.

Another standard that can apply to temporary overvoltage limits is the IEEE effective grounding feeder design standard. This is aimed at limiting ground fault overvoltage due to neutral shift as specified in IEEE C62.92 (IEEE C62.92, 2016 [6]) parts 1 through 4. A key parameter is what is known as the coefficient of grounding (COG). The applicable standard for 4-wire multigrounded neutral distribution systems recommends that the COG be 72% or less on systems servicing line to neutral loads.

The COG value is the level of line-to-neutral voltage of the system during a ground fault on the unfaulted phases in relation to the line-to-line voltage. A COG of 72% means that the line-to-neutral voltage rises to 72% of the line-to-line voltage on the unfaulted phases during a ground fault on the faulted phase. That is the same thing as saying that the line to neutral voltage rises to about 125% of the pre-fault line-to-neutral voltage. If we factor in an extra 1.05 per unit to account for the ANSI C84.1 voltage regulation window, then the COG requirement of 72% suggests a TOV limit of about 131% for short periods (seconds or less) on typical systems.

The IEEE Standard for Interconnection of Distributed Energy Resources (IEEE-1547-2018) precisely defines the requirements for DER-related overvoltage on standard grid parallel systems in Section 7.4 of that document. It defines the overvoltage limit as being about 138% of nominal RMS for ground fault overvoltage and load rejection overvoltage. And also defines some allowable values of "instantaneous" overvoltage ranging from 2 per unit to 1.3 per unit depending on duration.

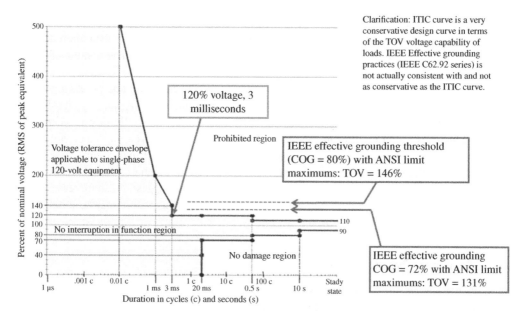

Figure 22.3 Information Technology Industry Council Voltage Tolerance Curve with IEEE Effective grounding Criteria Added. (Information Technology Industry Council).

Another type of overvoltage defining criteria is what is known as the "TOV capability curve" for surge arresters. This curve defines the voltage the arrester can withstand for short periods (from a few cycles duration up to several minutes) without being damaged. All surge arresters (whether utility company applied or consumer/customer applied) have known TOV capability curves that can be obtained from the manufacturer. There is not a standard curve for all arresters since there is a huge variety of surge arrester products, types, sizes and applications in use. However, what can be said is that these TOV curves, under essentially all situations, would be expected to be significantly higher than the ITIC curve or the voltages possible with COG of 72%. In fact, most, if not all utility arresters in use could easily withstand a COG of 80% as long as the duration was ten seconds or less. However, it is not clear that consumer load side devices have the capability to withstand a COG of 80% for both consumer arrester TOV limits and also electronic elements in the consumer appliance power supply devices.

While it is good if a microgrid can satisfy the ITIC curve, in practice, even regular utility systems are not designed to meet the stringent nature of ITIC. For the microgrid, an overvoltage limit that blends some parts of the ITIC curve with the COG = 72% IEEE guideline makes sense. This could consist of 125% for 2–10 seconds, 131% for 2 cycles to 2 seconds, 146% voltage for 2 cycles to 0.5 cycles, and merging with the ITIC boundary for anything shorter than 2 milliseconds. It is noteworthy that the IEEE 1547-2018 standard, which came after this microgrid study, now allows for a slightly different overvoltage limit curve than is defined by our suggested curve points and it would also make a very appropriate limit for microgrid overvoltage. However, the new IEEE curve only identifies temporary overvoltage limit conditions up to 10 cycles duration and so the longer duration TOV events need to be harmonized with limits we have already suggested for it to be applied as a power quality guideline for microgrids.

22.3.4 Disruptive Undervoltage Events

The ITIC curve of the earlier section can also be used to define the short-duration undervoltage boundary that may result in dropouts of sensitive devices. For example, 90% for 10 seconds or longer, 80% voltage for $\frac{1}{2}$ second to 10 seconds, and 70% voltage for about 1 cycle to up to 12 second. While short-duration undervoltage events due to motor starts, inrush, and load steps should be limited to an amount that does not cause unreasonable voltage flicker, it is also important that voltage dip variations should not cause disruption to operating devices on the microgrid as per the lower bound of the ITIC curve. A short-duration undervoltage can cause disruption by tripping out protective relays that control generators or load devices. For example, by dropping out motor, lighting or process control contactors, and by a deficiency of energy needed for stable operation of electronic power supplies of load devices. It is noteworthy that some devices may be more sensitive than the ITIC under voltage curve if they are equipped with protection relays. In particular, distributed generation devices, UPS systems and motors with undervoltage protection or contactors may be affected. While operating in microgrid mode these device relay settings will need to either be coordinated with the expected voltage dips or the severe dips (if any) limited such that they are not disruptive to microgrid operation.

22.3.5 Harmonics

The applicable standards for harmonics are IEEE 519 (IEEE Std. 519-2014) and IEEE 1547 (IEEE 1547-2018 [7]). The first standard applies broadly to power systems. The second standard, IEEE 1547, applies specifically to DG (see Table 22.4) applied to the system. The limits are for when the

Table 22.4 IEEE 519-2014 requirements for odd harmonics (IEEE Std. 519-2014).

Harmonic order	Allowed harmonic current level relative to fundamental current (Odd harmonics only)[a), b)]
$h < 11$th	4.0%
11th $\leq h < 17$th	2.0%
17th $\leq h < 23$rd	1.5%
23rd $\leq h < 35$th	0.6%
35th or greater	0.3%
Total harmonic current distortion	5%

a) The greater of the maximum load current integrated demand (15 or 30 minutes) at PCC without DG unit or the DG unit current capacity at PCC.
b) Even harmonics are limited to 25% of odd harmonic values per IEEE 519-2014. But IEEE 1547-2018 has slightly different limits than 519 for even harmonics.

DG is serving balanced linear loads. These standards also specify that voltage distortion on the distribution system should not exceed 5% THD.

For the purposes of the microgrid, while in grid parallel mode, the harmonic levels for DG (DER) and the power grid need to satisfy both the IEEE 1547 table and any respective IEEE 519 tables per the ratio of short circuit current to load current that exists in that mode. The DG harmonic limits of the earlier versions of IEEE 1547 were once exactly coordinated with the IEEE 519 limits for the case where the ratio of short circuit current to load current is less than 20 (meaning the Isc/Iload ratio). Today, the limits and methods of harmonic evaluation are slightly different per recent changes related to 1547-2018 standard. This leaves us considering which harmonic standard is best to apply for the case of a microgrid. It is our opinion that during the islanded microgrid mode, a suitable target for harmonic limits can be based on the existing IEEE 519-2014 limits table (which is based on a 20:1 ratio of short circuit to load current). Because, after all, we are dealing with a grid situation (albeit a microgrid) and not just a DER interconnection. The IEEE 519-2014 standard also has certain statistical criteria for the percentage of time and durations that harmonics are allowed to exceed steady-state values. These can be very useful to apply to microgrid situations where the operating hours per year are very limited.

In cases where the microgrid is operated for only short periods, there is room for some limited relaxation of the harmonic criteria owing to the infrequent nature of operation and limited duration of operation in an islanded microgrid mode. However, further study is needed to define how much relaxation could be allowed. Severe harmonics can cause disruption of customer load and generation devices in two main ways: misoperation of devices due to the waveform distortion and/or additional heating in wires, cables and rotating machinery. From a heating perspective, limited duration moderate violations of the harmonic criteria may be allowable since the effect of heating tends to be a gradual cumulative effect over time (meaning a gradual equipment life-shortening factor for each hour operated above-rated temperature) and may not be a severe issue for short periods. On the other hand, from a waveform disruption perspective (for example, misoperating relays caused by peak waveform and zero crossing distortion) the onset of effects could occur almost immediately and so more care would be needed in those cases.

22.3.6 Frequency Variations

The utility power system normally operates with tight frequency tolerances well within ± 0.5 Hz of the 60 Hz nominal frequency. Once the system has transitioned to microgrid mode, frequency variations will be larger. The question is how large can these variations be allowed to be before they pose a risk to loads and equipment and what should be the target range when operating in an islanded mode?

There are a number of factors that determine the frequency variations that are allowable on a microgrid from the perspective of customer loads and equipment. A key limiting factor on the low side is that lower than normal frequency may cause some magnetic core-containing devices (transformers and motors) to saturate. In general, a $\pm 5\%$ frequency deviation, while not ideal, is tolerable most of the time for magnetic core devices such as motors and transformers as long as the voltage is not higher than about $+5\%$ above nominal. However, in some cases the combination of higher voltage (near the top end of ANSI window) and low frequency (near -5%) can saturate and overheat devices with magnetic cores if they are heavily loaded so care must be exercised in evaluating the combination of higher than nominal voltage and lower than nominal frequency together. It is expected that the microgrid controller will operate to keep these large frequency variations from persisting for long periods of time.

Another factor of concern is that if the frequency is off nominal then line-connected motors may run at a speed that is either higher than normal (due to high frequency) or lower than normal (due to low frequency). This can change the loading on fans, pumps, etc. If any motor processes are highly sensitive to frequency in this manner, then a tighter tolerance than $\pm 5\%$ may be desirable. Some clocks and timing devices are also impacted by changes in the system frequency since they use the system frequency as their actual time reference. Error is accumulated in the time-keeping device as the integral of the "off-nominal" frequency condition. For example, a continuous high frequency of $+5\%$ over the course of 24 hours would advance some clocks by over 1 hour. To avoid time error on AC clocks that derive their reference from the grid frequency, the microgrid would need to regulate excursions of high and low frequency so as not to accumulate (integrate) unacceptable errors over time.

With regards to frequency, The National Electrical Equipment Manufacturers Association (NEMA) has several standards that may be applicable or helpful for the microgrid limits. The NEMA MG-1 standard deals with motors and specifies up to $\pm 5\%$ frequency for motors as long as voltage is within the normal ANSI limits, although its state performance may be degraded. NEMA also has a standard for uninterruptible power supplies (UPS) called NEMA PE 1-2012 that indicates a much tighter range ($\pm 0.5\%$ for systems above 2 kVA) for the output of UPS systems. This tight range would be expected since a UPS is generally a premium power device intended to offer a higher grade of power than needed for a regular microgrid condition. However, such a tight range suggests that some UPS architectures, such as line interactive type UPS or line preferred UPS systems, could have difficulty dealing with a broad frequency regulation range on the microgrid if they were still required to maintain their output frequency standard. This might result in them cycling back and forth between battery operating mode and grid supplied mode. It is noteworthy that the UPS standard also suggests a maximum rate of change of frequency (a slew rate) of no more than 1 Hz/s is allowable on the output of a UPS system.

Overall, based on the characteristic of loads, a targeted frequency range of $\pm 3\%$ when operating in islanded mode is a reasonable goal for the preferred operating frequency range for the microgrid that should not pose issues with the exceptions that certain types of UPS systems and DG equipment with tight protective relay settings might be upset by the broader frequency swings. Those devices may need to have tripping settings coordinated with the broader range of the microgrid.

22.3.7 IEEE 1547.4-2011

IEEE Standard 1547.4-2011, "IEEE Guide for Design, Operation, and Integration of Distributed Resource Island Systems with Electric Power Systems" is intended to offer guidance on various engineering factors apply to islanded microgrids. This would include power quality standards such as voltage, frequency, and TOV transients. However, the specific guidance offered in that document is limited in detail compared to the information presented in this chapter. Nonetheless, it offers background material for the various factors that must be considered in an islanded DG operating mode.

22.4 Microgrid Analytical Methods

22.4.1 Modeling Tools

A variety of analytical methods can be used to assess the performance of a microgrid design to meet the criteria presented in Section 22.3. A number of these power quality issues can be addressed with direct calculations using a spreadsheet-based analysis approach. This direct calculation approach is suitable for many elements of the analysis including calculating voltage changes with various generation fluctuations and general fault levels. For the more challenging aspects of the analysis, such as determining waveforms associated with faults, ground fault overvoltage plots in the time domain, ferroresonance and stability analysis, as well as checking the accuracy of some of the direct calculations, computer-based load flow, fault study, and electromagnetic and electromechanical transient simulations are necessary. There are commercial providers of these tools. In some cases, there are also open-source tools available as well. This section discusses analytical approaches for assessing microgrid power quality performance. The design studies for the proposed Potsdam Microgrid are used as case studies for these methods.

The one-line diagram for the Potsdam Microgrid is shown in Fig. 22.2. The existing and proposed generation for this microgrid are shown in Fig. 22.2. This design has eight natural gas or dual fuel synchronous generators, totaling 7.4, 1.5 MW of hydrogenation, a 2 MW photovoltaic installation, and three small microturbines. The load above 7.4 MW would be served by the hydro generation, PV generation and/or up to 2 MW of demand response. In the event of contingencies, the load would be prioritized, including the option to shut down buildings on the campuses if necessary. In a meeting with key stakeholders, it was decided not to include energy storage from the microgrid. Both planned separations and resynchronizations would be achieved seamlessly, but unplanned loss of the grid would require black start of the generation.

The models include the equivalent Lawrence Avenue Substation source impedance of the 115 kV transmission system and the impedance characteristics of the main connecting feeder (67,851) and the characteristics of the distribution system underground feeders. The system is modeled as a three phase, four-wire multigrounded neutral type of system.

The distributed generation plants located on the system were modeled using the. Each synchronous generator has an exciter model and a governor control representing the internal combustion engine (ICE) engine response. The machine data are based upon the typical time constants of response expected for ICE units operating on natural gas. There are also inverters present modeled as current sources. Induction machines are modeled as asynchronous generators with slip.

The large SUNY ICE generators are modeled as direct connected units (connected directly at 13.2 kV without step-up transformer) and other units are modeled as being interfaced by means of step-up transformers either from 480 V or 4.16 kV generator voltage up to 13.2 kV. The winding arrangements and generator neutral grounding are set to provide the neutral grounding conditions needed for each mode of operation. For example, there are two grounding transformers that are sized to provide effective grounding per IEEE standards with a COG well within effective grounding limits. Details on the modeling parameters are available in (NYSERDA Report 18-13, 2018).

For the loads in the model, these values are based on the data provided for the various microgrid customers that will be on the system and can be adjusted based on the scenario simulated. A real and reactive component of load is provided. This characteristic was particularly important in cases involving islanding analysis and ground fault overvoltage analysis where there is a need to properly characterize the load on the system to facilitate effectiveness of the islanding protection and the level of ground fault overvoltage during faults on certain types of generation islands that could impact the 115 kV system.

22.4.2 Simplified Spreadsheet Model Calculations

For many of the calculations, a simplified model was used for screening purposes, voltage sensitivity analysis, etc. The simplified one-line diagram of the Potsdam Microgrid is shown in Fig. 22.4. In this figure, all impedance values are given in percent to a 12 MVA, 13.2 kv base.

The analyses were conducted for both grid parallel and isolated operation of the microgrid. The microgrid generators will operate in two major modes – either as grid parallel generation with the utility power system acting as classical distributed generation resources (individually or as a group) or in an intentionally islanded mode (as a group of generators) to function as a standalone microgrid for "emergency" generation service to the loads on the microgrid in case of a utility grid power outage. For each of these two roles, the generator settings and operating characteristics are configured somewhat differently, causing the units to behave differently for each mode.

22.5 Analysis of Grid Parallel Microgrid Operation

Even though this is a microgrid project, the grid parallel role is still very important because more than 99% of the time, the microgrid generation will be operating in this manner. Proper operation in a grid parallel mode facilitates the ability to capture the classical DG economic benefits of localized power generation, T&D support and system ancillary services (such as bidding into ISO markets) that can help economically justify the presence of a microgrid for the rare times it is actually needed in an islanded mode. In aggregate, the microgrid will have up to about 10 MW of total generation capacity, with this generation all connected through Feeder 51. This is a large capacity for a distribution circuit at 13.2 kV rating, and so the grid parallel impacts must be considered in this mode of operation. Some key topics that relate to grid parallel operation include:

- Voltage regulation and flicker influence
- Fault levels
- Protection coordination
- Anti-Islanding protection

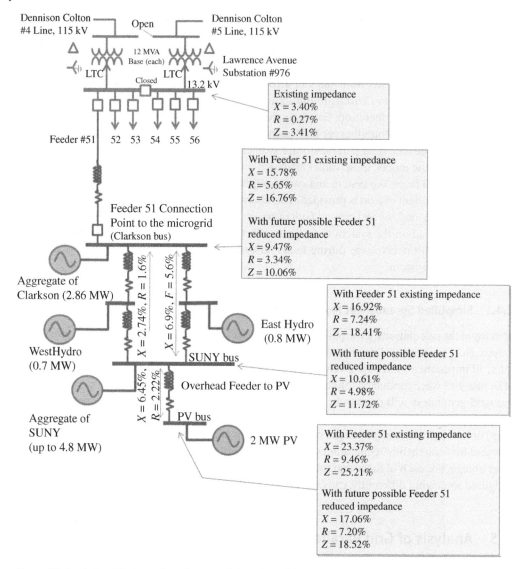

Figure 22.4 A simplified one-line diagram of the microgrid with connection to the bulk power system (impedances on 12 MVA base) (NYSERDA Report 18-13, 2018).

- Generator protection
- Ground fault overvoltage and grounding
- Dynamic behavior
- Power quality and harmonics.

22.5.1 Voltage Regulation (Grid Parallel)

22.5.1.1 Voltage Drop Due to Load

When the microgrid generation is operated in grid parallel mode as a distributed generation the generators will influence power flow on the system, which will cause changes in voltage on the system and directional changes in the power flow. It is important to screen the application to make sure that any voltage changes that occur during normal operation of the generators are within the proper ANSI C84.1 steady state operating guidelines discussed earlier in Section 22.3 and also will not subject customers to objectionable voltage flicker.

Figure 22.5 shows the one-line diagram of a simple radial distribution feeder. This diagram represents a three-phase feeder with a balanced three-phase load. The feeder is fed by the grid source at the substation and serves a single load through a per phase line impedance of the voltage drop equation for this feeder is

$$V_{load} = V_{source} - (R + jX)I_{load} \tag{22.1}$$

In this equation, the load and source voltages are the line-to-neutral voltages on the feeder, and the line current equals the load current in this simple case with a single load. This equation can be evaluated either in dimensional units (volts, amps, ohms), or in per unit.

The phasor diagram for this equation is also shown in Fig. 22.5, with the voltage drop parallel to and the reactive voltage drop shifted forward by the load voltage actually leads the source voltage in this case, with the line resistance greater than the line reactance.

An approximation of this equation is shown in Fig. 22.6. In this approximation, only the line drop components that are in phase with the substation voltage are considered. As shown in Fig. 22.6, when the angle between the load voltage and the source voltage is small, there is little difference between the load voltage estimated with this method and the load voltage calculated from the complex phasor equation. With this simplification, Eq. (22.1) becomes

$$V_{load} = V_{source} - I_{load}(X\sin(\theta) + R\cos(\theta)) \tag{22.2}$$

The angle θ is the angle at the line current lags the substation voltage, as shown in Fig. 22.6. Again, when the angle between load and source voltage is small, θ can be taken as the power factor angle at any point along the line.

The voltage <u>drop</u> on the line due to load is then

$$\Delta V = I_{load}(X\sin(\theta) + R\cos(\theta)) \tag{22.3}$$

Figure 22.5 Vectors of voltage drop due to a non-unity power factor load consuming watts and VARs.

Figure 22.6 Components of voltage drop projected to the horizontal axis to give the approximate voltage drop.

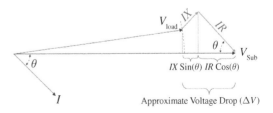

22.5.1.2 Voltage Rise Due to Generation

Figure 22.7 shows the simplified case where a single distributed generation (DG) unit is present on the feeder, with no load present. In this case, the generator current flows up the line from the DG unit to the substation. The voltage rise due to the DG unit is then

$$\Delta V = I_{DG}(X \sin(\theta) + R \cos(\theta)) \tag{22.4}$$

where now θ is the angle between the DG current and the line voltage. Note that the power factor of the generator plays a big role in voltage rise due to the DG. If a generator is producing only watts (unity PF) it will create a certain amount of voltage rise on the system due only to the product of I and R. If it is also producing VARs, then the reactive current associated with it causes an additional rise. On the other hand, if it is consuming VARs, the reactive current causes a drop. The generation situation that creates the greatest voltage change is one where the unit injects both real power and reactive power into the system. In many locations on the power system, the X/R ratio of line impedance is often greater than 1; therefore, the voltage sensitivity of the system to a given amount of VARs is greater than it is to the same amount of watts produced by a generator.

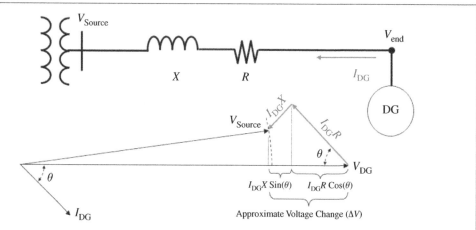

Note: The generator is producing reactive and real power in this example (Case 1), and using generator convention rules, the "$I_{DG}R$" and "$I_{DG}X$" voltage change vectors as shown are pointing in the opposite direction that they were for a load consuming reactive and real power in the earlier figures. The voltage is higher at the generator than at the utility source in this example. See inset table below for Case 1, as well as some other scenarios (Cases 2-4). The table is based on typical inductive and resistive feed point impedance situations and does not include effect of volage regulation equipment.

Type of Power Consumption/Production Situation	Real Power Component Effect on Voltage	Reactive Power Component Effect on Voltage	Total Combined Effect on Voltage
Case 1. Generator: *Produce Watts and VARs*	Increases the Voltage	Increases the Voltage	Increases the Voltage
Case 2. Generator: *Produce Watts and Consume VARs*	Increases the Voltage	Lowers Voltage	Net effect depends on ratio of real and reactive current and the X/R ratio of feed point impedance.
Case 3. Load: *Consume Watts and VARs*	Lowers Voltage	Lowers Voltage	Lowers the Voltage
Case 4. Load: *Consume Watts and Produce VARs*	Lowers Voltage	Increases the Voltage	Net effect depends on ratio of real and reactive current and the X/R ratio of feed point impedance.

Figure 22.7 Voltage change due to DG injected current into the system impedance (producing watts and VARs in this example).

22.5.1.3 Calculating Net Voltage Change on Practical Feeders

Practical distribution feeders have many loads and can have a number of DG units as well. The above methods can be used to predict the voltage drop or rise on a feeder section based on the load beyond that point. Separately, the voltage rise on a feeder segment due to DG units can be predicted from the net real and reactive powers injected by DG units downstream from the segment in question.

22.5.2 Voltage Sensitivity Test (Grid-Parallel Mode of Operation)

Using the aforementioned approach, we can calculate the voltage sensitivity of the Potsdam Microgrid at the intended normal point of common coupling (Clarkson bus in Fig. 22.4) due to the expected maximum current production of the generators feeding into the impedance of the system – the value we get is important to understand both the steady state and dynamic behavior of the system at the point of microgrid connection. A small value of a couple percent or less means there is not much to worry about voltage regulation impacts and a large value means that the system is sensitive to the amount of DG being connected and may need to be upgraded to reduce sensitivity. As part of the voltage sensitivity examination, it is worthwhile to evaluate the voltage change that occurs at the Lawrence Avenue Substation bus because that gives an indication of the broader area effects on multiple feeders emanating from that substation.

Example 22.1 *Voltage Sensitivity Test:* Figure 22.4 shows the Feeder 51 connecting the proposed Potsdam Microgrid to the utility power grid. Feeder 51 is a dedicated overhead feeder that connects the grid to the Potsdam Microgrid at the Clarkson bus, which is the point of common coupling (PCC) for the microgrid. The percent impedances are given in Fig. 22.4, with base voltage of 13.2 kv line to line and a 12 MVA base power. Figure 22.8 shows a simplified diagram, focusing on the impact of the microgrid on the PCC and distribution system feeding the microgrid. The impedance values in Fig. 22.4 are the total impedance from the designated point to the utility source. In this figure, all of the generation is represented as being at the PCC for the study of impacts on the overhead feeder and the electric grid.

Using the aforementioned approach, we can calculate the voltage sensitivity of the Potsdam Microgrid at the PCC due to the expected maximum current production of the generators feeding into the impedance of the system.

The maximum injected real power from all of the DG sources of the microgrid is roughly 10 MW, and the power factor will be 0.90 or higher.

a. Determine the impact of the sudden addition or loss of 10 MW of generation at unity power factor on the Feeder 51 voltage. Ans:

$$\Delta P_{DG} = \frac{10\,\text{MW}}{12\,\text{MWA}} = 0.833\,\text{p.u.}$$

Assume that the feeder voltage is $V_{PCC} = 1.0\,\text{p.u.}$
Then $\Delta I_{DG} = \frac{\Delta P_{DG}}{V_{PCC}} = 0.833\,\text{p.u.}$ At unity power factor,

$$\theta = 0°$$

Then, from Eq. (22.4),

$$\theta = 0°$$

Then, from Eq. (22.4),

$$\Delta V = 0.833(0.1578\sin(0°) + 0.0565\cos(0°)) = 0.0469\,\text{p.u.}$$

A complete loss of this generation would then cause a voltage drop of 0.0469 p.u. or 4.69% at the Clarkson Bus, the microgrid PCC. A sudden change of microgrid generation of this same magnitude would cause a similar voltage rise.

b. Determine the voltage impact if this same generation is operating at 0.9 power factor lagging. Ans With $\delta P_{DG} = 10\,\text{MW}$ or 0.833 p.u., the change in voltamps will be

$$\Delta S_{DG} = \frac{0.833}{0.9} = 0.926\,\text{p.u.}$$

At 0.9 lagging power factor, $\theta = \cos^{-1}(0.9) = 25.8°$ (a leading power factor has negative value of θ.) Then

$$\Delta V = 0.926(0.1578\sin(25.8°) + 0.0565\cos(25.8°)) = 0.1108\,\text{p.u.}$$

This result shows that with 10 MW power generation operating at 0.9 p.u. lagging, the increase in voltage from the generation will be $\Delta V_{gen} = 11.8\%$. This value is significantly greater than the 4.7% voltage rise when the generation is operating at unity power factor. It is also significantly higher than the 2% increase cited in Section 22.3 as indicating negligible impact on the system.

c. Determine the impact of these changes in voltage on the system feeding the Lawrence Avenue bus. Ans: From Fig. 22.8, the equivalent impedance of the two transformers and transmission lines is $Z_{eq} = 0.0027 + j0.034\,\text{p.u.}$ In the unity power factor case,

$$\Delta V = 0.833(0.034\sin(0°) + 0.0027\cos(0°)) = 0.0022\,\text{p.u.}$$

At 0.9 pf lagging,

$$\Delta V = 0.926(0.034\sin(28.5°) + 0.0027\cos(28.5°)) = 0.0144\,\text{p.u.}$$

It can be seen that the impact on the transmission voltage is relatively small compared to the distribution line drop in this case.

Notice that the specific load value of the microgrid does not matter in this example, since it involves the change in voltage due to generation for dynamics purposes. However, the load of the microgrid is necessary in order to calculate actual specific voltage level.

The Example 22.1 calculations show that the substation bus voltage change is insignificant (0.22%) even with a very large sudden $\Delta P = 10\,\text{MW}$ at unity PF. Even at 0.9 PF producing VARs, the voltage change is still limited to only 1.73% at that bus. This shows that the substation is sufficiently stiff (acts as a very low impedance feed point) such that even with relatively large changes in current and considerable reactive current injection, only a small voltage change occurs. Therefore, the microgrid will have little impact on the other feeders fed by that same bus. As there is potential for the microgrid generation to exceed the total load on the Lawrence Avenue bus, the LTC controllers must be set to handle reverse power flow into the 115 kV system by setting the mode of the regulation controller to always regulate in the forward direction no matter what the power flow direction. Note that some voltage regulation controllers can be set to either lock the tap changer or reverse the direction of regulation if reverse power flow occurs. This type of function would not be appropriate in this case. Also, if the line drop compensator is implemented on the LTC controls, that will need to be considered in this calculation.

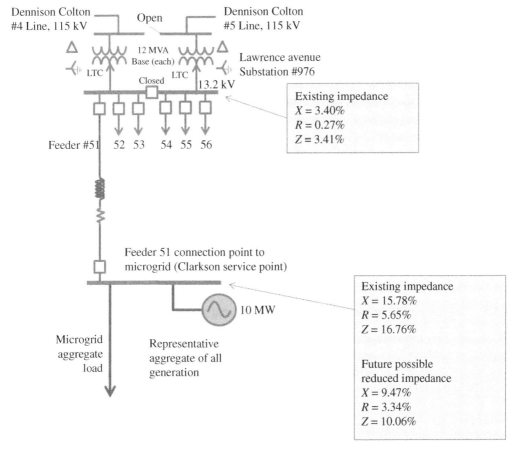

Figure 22.8 Voltage sensitivity test is based on feed point impedances at Lawrence Avenue Bus and Clarkson service point of Feeder 51 (impedance shown on a base of 12 MVA) (NYSERDA Report 18-13, 2018).

At the Clarkson PCC, the voltage change sensitivity is fairly large. For a 10 MW step, it is 4.69% at unity power factor and 11.08% at 0.9 power factor generating VARs. This is nearly half the ±5% ANSI C84.1 allowable range window at unity PF and more than the entire ANSI window if generation is at 0.9 PF (producing VARs). This tells us that a sudden step power change (or power swings) of all this generation at once has the potential to cause voltage excursions outside the ANSI limits. It also is an indication that keeping the feeder and microgrid within the ANSI C84.1 limits could be difficult without special provisions. It should be noted that the problem is not just with the generation. It is also that in order to build the microgrid as planned, putting the entire microgrid load onto the single Feeder 51 is really too much load for the impedance of that feeder.

In this example, there is no single 10 MW generator associated with the microgrid. However, it is useful to use this 10 MW value for an initial sensitivity test as it represents a possible power swing condition that can occur with a sudden switching on or off of all generations. The largest single-generation unit planned for this microgrid is 2 MW and that unit either coming on line or tripping off line could create a 2 MW power ramp instantly (for a trip at full load) to over several tens of seconds or a few minutes (for the unit coming on line and ramping to full load). The voltage sensitivity test calculation was therefore also done for a 2 MW generator at unity PF and 0.9 PF to observe the possible effect (see Table 22.5) of this size of generation step. The voltage

Table 22.5 Calculated voltage change due to 2 MW of generation (Rapid on/off step).

Location	Voltage change calculation results 2 MW step	
	Calculated voltage change (2 MW unity PF)	*Calculated voltage change (2 MW at 0.9 PF producing VARs)*
At the Lawrence Avenue Substation Bus	0.046%	0.32%
At the Clarkson PCC Service Point (existing impedance of feeder)	0.944%	2.21%
At the Clarkson PCC Service Point (upgraded feeder with lower impedance)	0.566%	1.33%

swings due to 2 MW steps at 0.9 PF producing VARs could be as great as 2.21% with the existing feeder impedance at the Clarkson point of common coupling and existing feeder impedance. With the reduced impedance feeder, the voltage change would be only 1.33%. At the substation bus the voltage change is insignificant at 0.32%.

22.5.3 Substation LTC Response and Feeder Regulator as a Solution

We can see from the voltage sensitivity results in the earlier section that if the DG plants are "stepped on suddenly" or have fast dynamic swings in power over the 10 MW range then this creates a large voltage change on the system at the PCC of the microgrid. This voltage change is solely due to the feed point impedance effect and is before the load tap changer (LTC) controller responds. After 30 seconds, the LTC begins to respond and will change its tap setting to compensate for any observed voltage change on the substation bus. With no line drop compensation, the effect on the tap changer controller on the voltage at the Clarkson PCC would be minimal since most of the voltage change occurs across the Feeder 51 impedance and is not at the Lawrence Avenue Substation bus. By using line drop compensation at Lawrence Avenue, the LTC could be made to compensate somewhat more for the voltage change due to the Feeder 51 impedance. However, this approach would not work well since the amount of compensation needed would adversely influence the adjacent feeders. Another option, since Feeder 51 is an express feeder to the microgrid tie point, would be to add a supplementary voltage regulator bank at the microgrid PCC (right near Clarkson) and use it to manage the voltage changes that occur during grid parallel DG operation. When used in that manner, a sudden step of 10 MW of generation (or the smaller 2 MW we discussed) creates a voltage rise initially, but after the time delay of the controller elapses it should correct that voltage rise to a lesser value. If a slowed ramp rate is utilized for bringing on generation, then that should allow the regulator time to correct the voltage change even for 10 MW of generation. While the regulator option can correct for slower "steady state" variations of many minutes or longer it would not be a suitable fix to correct for the voltage change due to fast dynamics (power angle swings), etc. A better approach is still to reduce the impedance of Feeder 51, although this could be done in combination with a regulator bank being added to produce a better overall solution.

22.5.4 Voltage Change Within the Microgrid

The voltage sensitivity test discussed so far was for the point of common coupling of the Feeder 51 with the microgrid. A lumped quantity of generation was used for that case to represent the voltage change for the worst case step of all generation (10 MW) and for a more typical step (2 MW). However, in reality there are many generators scattered around the microgrid that are electrically further away from the point of common coupling. They will see some additional impedance feeding in which will create additional voltage changes at their locations. The impedances on the primary (13.2 kV level) within the microgrid are as shown in Fig. 22.9.

As this microgrid uses underground cables for primary distribution, the impedance between locations within the microgrid is determined by the underground cable characteristics. Compared to the Feeder 51 Clarkson PCC impedance, the intra-microgrid impedance is relatively low owing to the use of large copper 500 MCM underground cables, which have both low reactance and resistance. This keeps the voltage drops along the cables relatively modest. Whereas, the existing overhead

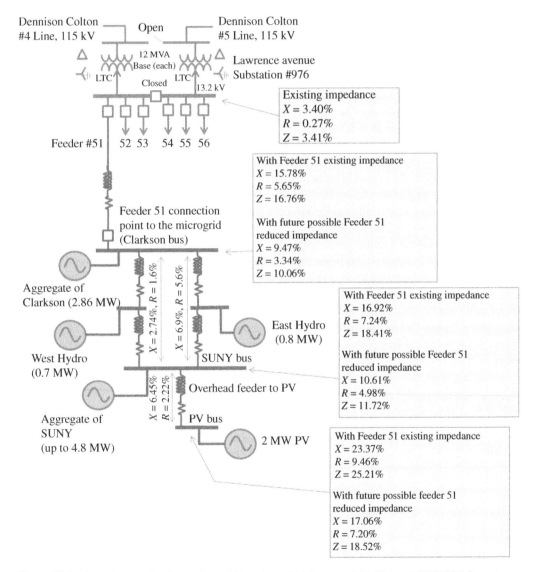

Figure 22.9 Impedance at key bus points within microgrid (shown on 12 MVA base) (NYSERDA Report 18-13, 2018).

Feeder 51 feed point connection to Clarkson has higher impedance between it and the Lawrence avenue substation owing to the fact that it is mostly an overhead line with higher X and R. While it is possible that the impedance of the Feeder 51 connection will be reduced in the future to facilitate operation of the microgrid during the buildout, even when upgraded it still will represent a large fraction of impedance that both the operating loads and DG must contend with while operating in grid parallel mode. The impedances marked "Future Possible Feeder 51 reduced Impedance" on the diagram are the ones that are most appropriate in our analysis. It is recommended to reduce the Feeder 51 impedance as part of the project and these estimates seem reasonable for what could likely be obtained.

The amount of additional voltage drop or rise within and across the microgrid during generation steps and dynamic swings while in grid parallel mode depends on which generating units are stepping/changing their power levels. The worst case location is the PV system feed point since it is electrically farthest from the Feeder 51 connection point and so has the highest feed point impedance looking all the way back to the Lawrence Avenue source. The PV system, if it becomes part of the microgrid, is to be connected by an overhead connection near the SUNY bus (as shown on the model) and will have a total $X = 17.06\%$ and $R = 7.2\%$ (on 12 MVA base) at its 13.2 kV feeder feed point (using the reduced feeder impedance design). This does not include the impedance of the PV step-up transformer since we are interested in primary side voltage changes affecting the feeder and other customers in that region.

A 2 MW variation of power level at the PV feed point would create a voltage change of 1.2% at unity PF and 2.57% producing VARs at 0.9 PF (see Table 22.6). This shows that operating the PV to produce enough VARs to lower the power factor to 0.9 lagging along with real watts could be problematic while in grid parallel mode by producing more voltage change than is desired on that overhead line section – especially during dynamic power output conditions caused by cloud shading effects. Cloud shading can result in up to nearly 100% swings in power relative to nominal PV rating in a short period of 2 minutes or even faster. The good news is that while the PF settings of the PV plant are not known at this time, it is considered extremely unlikely that it would actually be set to produce so many VARs as is currently configured. It is more than likely the PV is operating near unity PF or perhaps even absorbing some VARs if it is operating as a typical PV plant. Under those conditions, the calculations show the voltage change is well within reasonable limits and poses no issues for the system in grid parallel mode. Note also that the calculated variation in voltage at the Feeder 51 feed point (at Clarkson) is not high enough to be problematic regardless of whether the PV operates at unity PF or 0.9 PF producing reactive power as shown in Table 22.6.

Table 22.6 Voltage change due to 2 MW step at Clarkson Tie point and PV service feed point.

Location	Voltage change calculation results 2 MW step (based on PV feed point $X = 17.2\%$ and $R = 7.2\%$ at 12 MVA base)	
	Calculated voltage change (2 MW unity PF)	*Calculated voltage change (2 MW at 0.9 PF producing VARs)*
At the Clarkson PCC service point (upgraded feeder with lower impedance)	0.566%	1.33%
At the PV primary bus with PV fed by an overhead line	1.20%	2.57%

22.5.5 Low Generation Case: Low Voltage Issues

The previous Section 22.5.4 discussed voltage rise due to microgrid generation. The situation where there is little or no generation on the microgrid must also be considered. This can be analyzed using the simplified model of Eq. (22.3).

Example 22.2 Determine the voltage drop from the Lawrence Avenue Substation bus to the microgrid PCC for peak microgrid load.

The microgrid peak load is 10 MW at 0.9 power factor lagging. From Fig. 22.8, the Feeder 51 resistances and reactances are:

$$R_{51} = 5.65\% - 0.27\% = 0.0538\,\text{p.u.}$$
$$X_{51} = 15.78\% - 3.40\% = 0.1238\,\text{p.u.}$$

The impedance values for the line are used as the Lawrence Avenue bus is regulated, and this calculation is for long term voltages.

From Eq. (22.3), at peak load, the volt-amp magnitude at the point of common coupling is

$$S_{\text{PCC}} = \frac{10\,MW}{0.9} = 11.11\,\text{MVA}$$

The power factor angle is

$$\theta = \cos^{-1}0.9 = 25.8°$$

The complex power in per unit is then

$$\overline{S}_{\text{PCC}} = \frac{11.1\,\text{MVA}\underline{/25.8°}}{12\,\text{MVA}} = 0.926\underline{/25.8°}\,\text{p.u.}$$

Assuming that the PCC voltage is 1.0 p.u., the current at that point is then

$$\overline{I}_{51} = \frac{\overline{S}_{\text{PCC}}^{*}}{\overline{V}_{\text{PCC}}} = 0.926\underline{/-25.8°}\,\text{p.u.}$$

The asterisk on $\overline{S}_{\text{PCC}}^{*}$ indicates that this is the complex conjugate of the volt amps. From Eq. (22.3), the Feeder 51 voltage drop due to peak load current is

$$\Delta V = 0.926(0.1238\sin{(25.8°)} + 0.0538\cos{(25.8°)}\,) = 0.095\,\text{p.u.}$$

This calculation shows that the Lawrence Avenue bus would have to be at the ANSI limit of 1.05 p.u. to keep the voltage at the microgrid PCC above the lower limit of 0.95 p.u. This means that the other feeders on the bus would also have feeder head voltages of 1.05 p.u., which would impact them. This also does not account for the voltage drop within the microgrid.

One solution to this issue is to install power factor correction capacitors on the microgrid.

The voltage rise caused by capacitive current will be

$$\Delta V_{\text{CAP}} = I_{\text{CAP}}X_{51}$$

This equation reflects that the angle θ will be 90° when calculating voltage rise for the purely reactive current supplied by a capacitor bank.

To supply a voltage rise of 0.025 p.u., the capacitive current would be

$$I_{\text{CAP}} = \frac{0.025}{0.1238} = 0.202\,\text{p.u.}$$

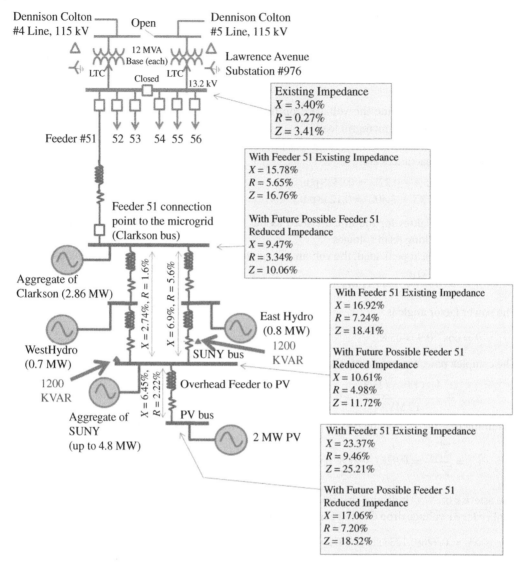

Figure 22.10 Approximate locations of the two proposed 1200 kVAR capacitor banks on the impedance model. (NYSERDA Report 18-13, 2018).

Again, assuming that the voltage at the PCC is 1.0 p.u., the needed size of the capacitor bank is

$$S_{CAP} = 0.202\,\text{p.u.} = 2.42\,\text{MVAR}$$

This is an approximate value of the necessary capacitor bank size, and both the size and location(s) on the microgrid would need to be established through load flow studies. This would be done to minimize reactive current flows on the underground cable of the microgrid, and to manage microgrid voltages at this peak load, zero generation operating point. It would be necessary that these are switched capacitor banks, as they would not be needed during high microgrid generation operation. They also may need to be switched in 2–4 steps, to maintain acceptable voltage change when they are switched. Figure 22.10 shows deployment of two capacitor banks with locations selected from the load flow studies. In addition to or as an alternative to capacitors, a voltage regulator installed at the microgrid of interconnection to the feeder could also be a means to help solve this issue. But compared to a regulator, capacitors at a fixed location on the microgrid would have the advantage that they could be useful with any of the alternative feeder tie points that the microgrid might use, whereas the regulator option would need to be located at each desired tie point.

22.5.6 Voltage Flicker Conditions (Grid Parallel Condition)

Rapid voltage changes on the distribution system due to variable generation output conditions (such as the PV system during cloud shading variations) should be limited so that they do not cause objectionable flicker that exceeds the allowable flicker curve limits. Microgrids should be designed and operated during grid parallel mode that can keep the voltage flicker below the borderline of visibility – and if not below that curve at least significantly below the borderline of irritation. For islanded operation of a few days or hours each year, the condition is so infrequent it is possible to relax the flicker standard somewhat. But there are no published guidelines on what may be considered "acceptable" for such short periods of a few days each year when done as an emergency condition to avoid an outage. However, the IEEE Std. 1453 flickermeter limits do have statistical methods that allow flicker limits to be exceeded 1% of the time, which may be a good approach in this case.

A convenient alternative for screening studies is to use the flicker curve in IEEE Std. 519-1992, shown in Fig. 22.11. Even though the IEEE 519-1992 curve is still widely used in the industry due

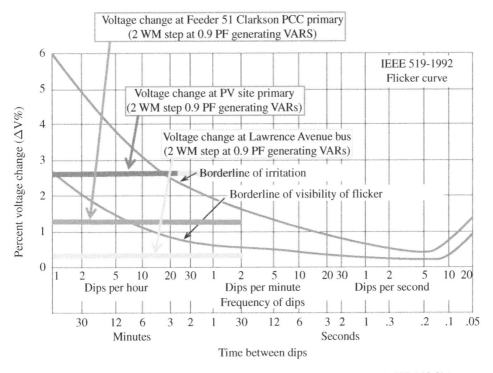

Figure 22.11 ΔV in percent for a 2 MW 0.9 PF power step as plotted on the old IEEE 519 flicker curve for rectangular voltage variations. On this curve the 2 MW PV site would not be expected to change at a frequency of more than roughly 2 times per minute.

to its legacy familiarity and ease of use, it is actually no longer the most up to date flicker standard. This curve reasonably approximates the IEEE 1453 limits in the case of voltage variations of repetitive magnitude and frequency. This curve is only suitable for **60 watt incandescent bulbs** and **relatively rectangular modulations** (step-changes) of the RMS voltage envelope. The IEEE 1453 methodology is more versatile and can be utilized to deal with a broader range of possible voltage change conditions and light source sensitivity factors. Sources such as PV, for example, do not have rectangular output modulations related to cloud shading. Rather, the changes are more gradual in nature. This makes PV that causes a given change in voltage to be far less likely to cause flicker than a conventional step load of the same frequency. In IEEE 1453, the short term (P_{st}) and long term (P_{lt}) flickermeter values of 1.0 are considered the borderline of irritation. These were developed from IEC 61000-4-15. A flicker screen used in many state interconnection requirements of DER as well as IEEE 1547-2018 suggests that PV sites should be screened for a P_{st} value of 0.35 or less. New York (New York State Standardized DER Interconnection Requirements, 2022) suggests the following screening formula for PV related flicker.

$$\Delta V = \left(\frac{d}{d_{P_{st=1}}} \right) \times F \leq 0.35 \text{ and } d = \left(\frac{R_L \times \Delta P + X_L \times \Delta Q}{V^2} \right)$$

$$\underline{\text{When}}: \frac{X_L}{R_L} < 5$$

$$\underline{\text{OR}}$$

$$\Delta V = \left(\frac{d}{d_{P_{st=1}}} \right) \times F \leq 0.35 \text{ and } d = \left(\frac{\Delta V}{V} \right) \approx \frac{\Delta S}{S}$$

$$\underline{\text{When}}: \frac{X_L}{R_L} \geq 5$$

$$\underline{\text{Whereby}}:$$

$$PV \text{ Plant Variability} = \left(\frac{F}{d_{P_{st=1}}} \right) = \frac{0.2}{2.56\%} = 7.8$$

Explanation of Variables & Acronyms

d = *the relative voltage change caused by the DER at the PCC*

dp_{st} = *1 (curve value) is the relative voltage change that yields a P_{st} value of unity when voltage fluctuations are rectangular*

P_{st} = *the short-term flicker emission limit for the customer installation (typically based on 10-minute time frame)*

X_L = *the line reactance in ohms*

R_L = *the line resistance in ohms*

Isc = *the maximum available 3-phase fault current at the PCC in amperes*

Ssc = *the maximum available fault apparent power at the PCC*

ΔS = *the change in apparent power in volt amperes*

ΔP = *the change in real power in watts of the DG*

ΔQ = *the change in reactive power in vars of the DG*

V = *the nominal line to line voltage*

ΔV = *the change in voltage at the PCC*

F = *the shape factor related to the shape of the expected voltage fluctuation*

Using the NYSIR recommended formula with a PV variability factor of 7.8, and the fact that the PV site point of connection will have an X/R ratio significantly less than 5, we can plug in the 2 MW at 0.9PF change in power (as per unit value relative to 12 MVA) and the impedance (R and

X also in per unit) at the PV point of connection. The estimated P_{st} value can then be calculated as follows:

$$P_{st} = \frac{R_L\left(\Delta P\right) + X_L\left(\Delta Q\right)}{\left(V_{\text{nominal per unit}}\right)^2}\,(7.8) = \frac{0.072\left(\frac{2}{12}\right) + 0.172\left(\frac{0.968}{12}\right)}{1^2}\,(7.8) = 0.20$$

With a P_{st} factor well under 0.35 we can see that the PV site is not expected to cause any flicker issues even at the point of connection (the most sensitive point). The calculated P_{st} would be even less farther up the system at the microgrid tie point and even less at the Lawrence Avenue Substation. In addition, the power factor is likely to be higher than 0.9, since most PV of that type operate near unity, leading to even less voltage change.

Using previous methods for calculating voltage change, the amount of ΔV at the substation bus, the Clarkson Feeder 51 PCC primary and the PV primary connection point for a 2 MW 0.9 PF rapid power change can also be plotted on the older out of date IEEE 519 flicker curve in Fig. 22.11. This just gives us a feel for what the levels look like relative to that older curve if we just assumed the changes were rectangular. The PV variations are not actually "rectangular" and are much more gradual in their rise and fall so this plot exaggerates the flicker severity. The conclusion is that at the PV primary feeder connection point, irritating flicker might be visible *if a very rapid 2 MW rectangular step change were to occur* and if the change includes a reactive component of PV output. But if we consider that in actuality the PV would be operated near unity PF (causing far less voltage change), and that it has a gradual rise and fall of the output, then there is no risk at all of visible flicker from the PV site during grid parallel operations. Thus this is very consistent with the P_{st} type analysis once those other considerations are factored in. This analysis shows how important it is to use the newer P_{st} methodology that takes into account the smoother variations of PV rather than attempt to use the old IEEE 519 curve.

In the case of conventional microgrid generation, such as gas fired ICEs, these units do not have uncontrolled running variations the way wind or PV sources do. Instead they will generally operate as *steady sources* over long periods of time or at the very most be ramped up/down at controlled slow rates per dispatch and/or load following requirements.

Two general guidelines for micrgrid ICEs (especially the largest unit blocks) are to be followed when generators are connected in the grid parallel mode. During the moment of interconnection between utility grid and the generator unit, in order to minimize the interconnection related power flow transients that can cause noticeable voltage flicker, the limits for voltage and frequency matching IEEE Std. 1547 should be followed. Also, when dispatching generators larger units should be ramped up slowly over a 2 minute or longer period and not suddenly slammed on the system at the fastest possible ramp rate. This makes the voltage transition more gradual (less noticeable) and also allows time for the upstream LTC and voltage regulator tap changers to adjust themselves to minimize the voltage change seen by the distribution system.

22.5.7 Voltage Flicker and Voltage Changes on 115 kV Transmission

There is also interest in knowing whether or not when operating in grid parallel mode if any problematic steady state or flicker related voltage fluctuations will occur on the 115 kV transmission system due to the DG running and/or connection currents. In the example of the Potsdam Microgrid, 10 MW of DG, even if suddenly fully stepped on at 0.9 PF is not able to cause more than a few tenths of a percent of variation in the 115 kV side voltage given the low relative impedance on that side of the system. Therefore, there is no impact on the 115kv system in this case.

22.5.8 Unusual Load and DG Interactions That Might Cause Flicker

Starting and stopping of the ICE DG as well as PV type cloud variations are not the only possible sources of DG induced voltage flicker. DG can cause flicker if there are load pulsations and/or ICE misfiring issues that excite certain types of sub-synchronous generator rotor angle oscillations. Such pulsations could be caused by misfiring engines (due to poor fuel or engine ignition control problems) or due to strange interactions between the machine and the system loads/equipment and system voltage sags. These pulsations and oscillations, if any occur, are not typically as severe as starting and stopping type voltage changes from a ΔV perspective but because they can occur on a more frequent basis they can be in a much more sensitive region of the flicker curve where a smaller ΔV, even as small as 0.5%, is visible to the naked eye as light flicker.

In the case of the Potsdam Microgrid, the loads at the customers have been characterized and there are no large pulsating loads such as industrial sized motors, rock crushers, cranes, and arc furnaces, that would cause issues. The largest line connected motors present on the system do not exceed 100 hp.

22.5.9 High Steady-State Voltage at Generator Terminals

It is important to avoid high terminal voltage at the generators because high voltage can cause saturation, which leads to additional heating that may damage the generators and require curtailment of output to avoid this problem. If high voltage occurs it may make it difficult to utilize the generator units to their full capability to provide real and reactive marketing services to the utility power system at all times. This could reduce the economic viability of the project.

The amount of voltage rise a generator can handle depends on many factors including loading, frequency conditions, presence of other heating effects (such as unbalance and harmonics) and ambient temperature conditions. NEMA/ANSI/IEEE standards rate most machines in a manner that allows up to 1.05 per unit steady-state voltage at the machines nominal rated frequency and power levels.

The impedances of the key feed point (Feeder 51 at Clarkson PCC) and the LTC settings (123 V) used at Lawrence Avenue are such that if the microgrid is lightly loaded and generation is attempting to operate (for system market purposes) at or near the full 10 MW level, the microgrid voltage may tend rise to much higher than 1.05 per unit on the primary feeder. In this condition the generator terminal voltages would be even higher due to the additional impedance of the step-up transformer and connecting cables for each generator where applicable. The ANSI Range A and B voltage limits would also be exceeded causing issues not just for the generators but for loads on the system too. Overall, this would mean that at certain times of light microgrid loading, the generator activity would need to be limited (curtailed) compared to its full capability. At heavy load the problem disappears due to the voltage drop caused by the load. To avoid this issue of the need for light load generator curtailment, it is recommended that the impedance of the Feeder 51 circuit as seen at the Clarkson PCC be made lower than its existing value. A voltage regulator may also be needed at that feedpoint to further enhance the voltage regulation capabilities. As an alternative to the voltage regulator, the generators may need to have reactive power generation limits imposed during periods of light microgrid load and some may even need to operate in a fixed power factor mode and absorb reactive power. These studies would need to be done using load flow software.

22.5.10 Operating Mode for Generator Controller (In Grid-Parallel State)

An important consideration for connected generators is the real/reactive power control operating mode. Choosing the correct mode will help insure proper voltage regulation on the power system and avoid possible problematic interactions with the utility's voltage regulation equipment.

When the units are operating in grid-parallel mode it is best that they should be operated in a manner where they do not attempt to directly regulate the voltage via a closed-loop voltage feedback method. For grid parallel DG an open-loop method of operation such as unity power factor, constant power factor or constant var are preferred because the substation transformer LTC and other utility system regulation equipment will not be fighting against the settings of the DG voltage control. This is in contrast to DG units operating in an *islanded mode* where the major units present need to be set in a closed-loop mode to help regulate the voltage if properly coordinated.

22.6 Fault Current Contributions and Grounding

The microgrid DGs will contribute fault current to the power system whenever there is a fault. The level of fault current contributed is a function of the impedance between the fault location and the generator, the ratings of the DG machines operating, the type of fault and the impedance characteristics of the machines. The initial current for a balanced fault at the machine terminals will be 4 to 11 times the rated current depending on the machine's subtransient reactance. On the other hand, inverters contribute fault current up to typically several tens of cycles of time that is typically 1 to 1.5 times the normal rated current of the inverter. The microgrid DG fault current calculation can have significant impact on distribution system and microgrid fault sensing, unintentional islanding, and temporary overvoltage during unbalanced faults. Fault current and TOV studies are done using commercial fault study software. A discussion of the fault current levels and protection impacts for the Potsdam Microgrid is include in (NYSERDA Report 18-13, 2018)

22.6.1 Unintended Islanding Protection (in Grid-Parallel Mode)

In this section we will discuss the anti-islanding protection issues associated with DG operation in the *grid parallel mode*. During the grid parallel mode of operation we want to avoid any unintentional islanding condition lasting for any significant period of time longer than what is considered safe from a reclosing dead time perspective and a voltage quality perspective.

For the distribution system of the Potsdam Microgrid, the reclosing sequence on Feeder 51 is a 1 shot (one attempted reclose) process, which occurs after 20 seconds of dead time. On the adjacent feeders the reclosing sequence is three shots at 15, 30, and 45 seconds dead time, respectively.

Regarding the distribution side of the system, since reclosing on any of the Lawrence Avenue distribution circuits is 15 seconds or longer, no matter which circuit is used to tie to the microgrid there is a significant amount of time to trip the DG offline before reclosing occurs. Per IEEE 1547, a 2-second limit is the required clearing time of the DG for basic default anti-islanding protection. However, a quicker islanding response can be achievable using the various direct transfer trips and relaying methods that are available. The key question is how quickly the DG island need to be disabled?

In fact, the answer to the question is more complex than just applying the standard 2 second IEEE limit. To understand how fast we need to trip the islands offline, we need to satisfy all of the following requirements:

1. Satisfy the tripping speed required for deep voltage sags and basic islanding conditions (per IEEE 1547 default values or utility-specified values)
2. De-energize the island well before utility reclosing occurs
3. De-energize the island before DG fault contributions cause unneeded fault current damage to conductors and other equipment (including the DG themselves).
4. Avoid damaging load rejection overvoltage or ground fault overvoltage.

The last point (item 4), can require an extremely fast tripping response that clears DG offline before the utility system actually even separates from the forming island. However, if *other means* are in place to limit the overvoltage conditions referred to in item 4 then this last point can be ignored leaving only the first three points to determine the tripping speed required. In the case of the Potsdam Microgrid the plan is to use effective grounding to limit the overvoltage and it is expected load rejection overvoltage will be kept in check by limiting export of energy outside the grid to a suitable level.

A superfast trip that clears the DG offline before overvoltage conditions arise due to generator load rejection and/or ground fault overvoltage can be accomplished with a *time coordinated direct transfer trip*. This goes beyond an ordinary DTT. This approach uses a fast DTT timed to trip the DG breaker just before the actual separation of the island from the utility system. The approach works by keeping the path to the substation transformer and utility's power system always available while the DG is present. In that case the "grounding source transformer effect" of the substation transformer is always present until DG is offline so ground fault overvoltage never has a chance to develop. And for load rejection, similarly, the path to the larger utility electric system to send excess generation is also available until the DG is tripped offline limiting load rejection overvoltage as well.

Fast trips coordinated to trip the DG offline before the feeder breaker opens can be difficult to achieve due to the clearing time considerations/objectives for the devices involved. The strategy on this project will be to try to avoid the need for the fast time coordinated DTT altogether by using *other means* to suppress transient and temporary overvoltage. As mentioned earlier, the use of effective grounding (by means of grounding transformers within the microgrid for the DG sources) will eliminate the risk of ground fault overvoltage. For load rejection overvoltage prevention, careful consideration of load to real time generation ratios in which we allow operation of the microgrid can reduce that risk and as long as these limits do not conflict with desired power market operating and load support objectives. These approaches should allow a slower form of DTT and/or local relay based islanding prevention approach for grid parallel operation. In the slower form many cycles of "islanded" operation are allowed following the formation of the island before the units are tripped (cleared) from the system. In the background, keep in mind that the project may still need to fall back on the fast time coordinated DTT approach pending on the final outcome of more detailed studies later in Phase II regarding desired operating modes for power export, etc., but it is hoped this can be avoided.

22.6.2 Unintentional Islands of Interest

Whether or not an island is at serious risk of developing depends greatly on the load to generation ratio on the island when it forms as well as the types of sources present and the degree of sophistication of the local islanding protection relaying scheme at each generator. If conditions are such that the relays cannot be counted on locally, then a direct transfer trip may be required. In particular, if a synchronous rotating machine generator is using only passive relay based islanding protection and if load can at times be balanced to generation on the island then a DTT is likely going to be needed to facilitate tripping. An effective screening approach is to examine the possible islands that can form upon breaker opening to see if minimum load on the island zone might be balanced to generation. If it is closely balanced then there is a danger. As a guideline, if the minimum load is at least twice the generation capacity then there is little danger and even relatively simple relay based protection should be sufficient.

Some unintentional islands that are of interest with regard to grid parallel operation mode are shown in Fig. 22.12. As an example, if the Feeder 51 tie breaker at the Clarkson point of common coupling (PCC) of the microgrid opens the load data show there could be an island of the microgrid where load is matched to generation. Another example would be if Feeder 51 breaker for the whole feeder at Lawrence Avenue Substation opens. The load in that case could also be matched to generation. An even larger island that might occur includes part of the 115 kV network to the nearest remote source breaker, but only if the load could match generation for that island.

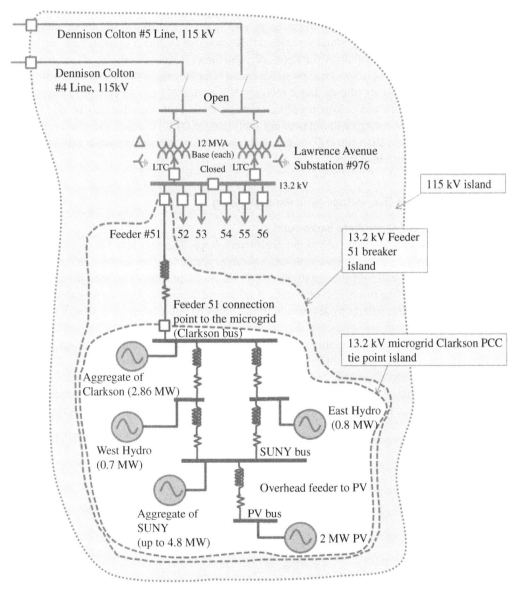

Figure 22.12 Some possible unintentional islands that can form at the Potsdam Microgrid project. (NYSERDA Report 18-13, 2018).

In addition to the islands of Fig. 22.12, there are other islands that can occur. However for the sake of clarity on that drawing, those are omitted from the diagram. Keep in mind that there are plans to have alternative feeds to the microgrid from feeders such as #53, #55, or #56. If those feeders are also used, then they too would each have their own set of islands that would be of concern.

22.6.3 Alternatives to DTT

DTT (even a slow one) is expensive to implement so the question arises as to whether or not *local DG-based relaying* alternatives to DTT are available that can be counted on to trip the larger rotating machine generators in a timely fashion. Since ideally we would like to trip them within no more than 5–10 cycles following island separation this can be very difficult to implement with just ordinary overvoltage, under-voltage, over-frequency, and under frequency relaying if load is nearly balanced to generation. There are some more advanced relay functions that could be considered such as special combinations of impedance relays (ANSI device 21), rate of change of frequency (ROCOF), and phase shift and/or synchro-phasor type protection schemes that may be able to trip the larger ICE units fast enough without needing a full-fledged DTT. However, these will probably still be somewhat slower than the DTT owing that one needs to average certain conditions over many cycles to avoid nuisance trips.

22.6.4 Ground Fault Overvoltage (Grid Parallel Mode)

22.6.4.1 Ground Fault Overvoltage Background
This section deals with the topic of ground fault overvoltage. The Potsdam Microgrid is dominated almost entirely by rotating synchronous generator type generation sources. Inverter based resources have only a small contribution to this microgrid. So the practices discussed in this section deal mostly with a rotating machine type system. A microgrid dominated by inverter based sources would behave somewhat differently and would utilize adjusted approaches for dealing with those types of resources.

Ground fault overvoltage is a condition caused by the neutral potential shifting its position away from the *center* of the three-phase voltage triangle toward one corner of it as shown in Fig. 22.13 during a line to ground fault. This condition will collapse the voltage on the faulted phase and increases the voltage on the unfaulted phases. The degree to which this neutral shift occurs depends on the ratio the zero sequence impedance to the positive sequence impedance of the source – the higher

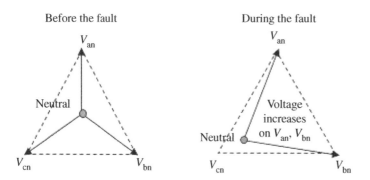

Figure 22.13 Neutral shift during ground fault overvoltage (this example shows a heavily but not fully shifted neutral. In worst case neutral can fully shift to one corner.)

the ratio the greater the shift. In the worst case for a source with no zero sequence return path (so called ungrounded connection) the neutral point shifts all the way to one corner of the triangle and the line-to-ground voltages on the unfaulted phases increase to as high as the line-to-line voltage. When a line-to-ground fault occurs on the distribution system and the utility source breaker opens, and if ungrounded DG is present feeding in, then the system source characteristic changes from a four-wire, effectively grounded system to a three-wire, ungrounded system source causing neutral shift and ground fault overvoltage.

Overvoltage on the unfaulted phases for an ungrounded power source can reach roughly 1.73 per unit of the normal maximum system voltage. That means the total maximum voltage the line might experience is 1.82 per unit as shown in Fig. 22.14 ($V_{max} = 1.73 \times 1.05 = 1.82$ per unit, 1.05 per unit is the top boundary of ANSI C84.1 Range A). If the power system neutral return path is "effectively grounded" as defined by IEEE C62.92, then this means is has a coefficient of grounding (COG) of 80% or less. COG is the ratio of line to neutral voltage divided by line to line voltage on the unfaulted phases. The higher the ratio, the greater the neutral shift toward one corner and the higher the line to ground overvoltage. Figure 22.14 shows the voltage levels of the Phase B line to neutral voltage for various shifts. The Phase A voltage also rises by a similar amount. Table 22.7 provides a summary of the voltage conditions with some different levels of neutral grounding.

22.6.5 IEEE Effective Grounding

A solution to avoid excessively high ground fault overvoltage in the line to neutral mode on power systems is to maintain an effective grounding of the system. In this case, the ratio of zero sequence and positive sequence impedances of the source transformer, wires, cables, etc., are selected in the system design such that a ground fault does not allow much neutral shift. A COG of 80% or less

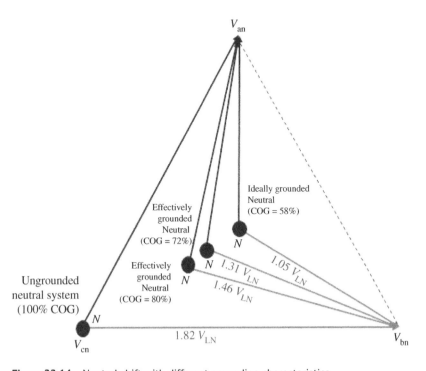

Figure 22.14 Neutral shift with different grounding characteristics.

Table 22.7 Neutral shift overvoltage (a.k.a. ground fault overvoltage) associated with various coefficients of grounding.

Type of neutral grounding	COG	Nominal line to ground overvoltage level (at nominal voltage level)	Maximum overvoltage level (includes ANSI 5% regulation factor)
Ungrounded	100%	1.73	1.82
Barely effectively grounded *(just barely meets the IEEE definition)*	80%	1.39	1.46
More deeply effectively grounded *(recommended good 4-wire practice per IEEE)*	≤72%	≤1.25	≤1.31
Ideally grounded	≤58%	≤1.0	≤1.05

occurs roughly when the ratio of the zero sequence reactance (X_0) to positive sequence reactance (X_1) as well as the ratio of zero sequence resistance (R_0) to positive sequence reactance (X_1) of the energy source(s) feeding-in meets the following criteria:

$$\frac{X_0}{X_1} \leq 3 \text{ and } \frac{R_0}{X_1} < 1 \tag{22.5}$$

An 80% COG is considered good for *transmission lines* that do not directly serve line to neutral loads and only have surge arresters connected line-to-neutral to worry about. Surge arrester ratings can easily be selected to handle the 80% COG voltage levels while still providing adequate surge protection. But at the distribution level because 80% COG still results in a fairly high 1.46 per unit voltage on a line-to-neutral basis (including ANSI 5% factor), any *distribution line* that serves line-to-neutral loads may need a lower COG for power quality purposes due to the overvoltage sensitivity of the loads. An 80% COG is considered good for *transmission lines* that do not directly serve line-to-neutral loads and only have surge arresters connected line-to-neutral to worry about. Surge arrester ratings can easily be selected to handle the 80% COG voltage levels while still providing adequate surge protection. But at the distribution level because 80% COG still results in a fairly high 1.46 per unit voltage on a line-to-neutral basis (including ANSI 5% factor), any *distribution line* that serves line-to-neutral loads may need a lower COG for power quality purposes due to the overvoltage sensitivity of the loads.

Even though 80% COG or less is the IEEE definition of the "borderline" of effective grounding, the fine print of IEEE C62.92 design guidelines for 4-wire multi-grounded neutral distribution systems recommend designing for 72% COG or less; this occurs roughly with $X_0/X_1 \leq 2$ and $R_0/X_1 \leq 0.7$. This lower COG puts the system deeper into effective grounding and limits the maximum overvoltage to a level of about 1.31 per unit line-to-ground voltage during a line-to-ground fault taking into account the ANSI 5% factor. A 1.31 per unit overvoltage is unlikely to cause any damage, even to the more sensitive load devices, for the short durations that the overvoltage conditions occur. Typically these last a few seconds or less until the fault is either cleared by system protection or the power tripped off altogether.

For power systems with distributed generation present the problem is that if a ground fault occurs, then the upstream utility source breaker can trip open leaving the DG feeding into the system as the only source(s) for a short period of time until islanding protection trips it offline. During this time there may be anywhere from a few cycles to a few seconds the danger if the "effective grounding path" to the utility source is lost and if there is no alternative zero sequence ground path on the local island – the COG can become 100% in this case. In many cases the DG neutral does not provide a suitable neutral current path (meaning the DG is not effectively grounded). Provisions need to be made to insure that the COG remains below 80% (and preferably at or below 72%) during conditions when DG islands occur, even for short periods of time.

22.6.6 Effective Grounding Status of Potsdam DG Sites

A review of the generators on the Potsdam Microgrid reveals that most or all of them are unlikely to be effectively grounded as operating currently with respect to the primary distribution system (Table 22.8). This is either because the generator interface transformer feeding into the system does not support effective grounding (has a delta high side winding) or that the neutral of the generation source itself, is not connected to the generation source with sufficiently low impedance to be considered as effectively grounded. For some units, such as the 2 MW PV site and smaller Clarkson generators the data were not immediately available to determine effective grounding status. Those will be assumed to be noneffectively grounded until data are available. Inverter-based generation can behave somewhat differently than rotating machines but is not significant enough in this case to drive the nature of the effective grounding solutions employed. Even though the characteristics of some of the existing sources are not known from an effective grounding perspective, the largest and most powerful sources feeding in from a fault current driving perspective are those that will dominate the effective grounding requirements for the project so the smaller and/or weaker sources do not actually matter that much in driving the requirements for this project. The main sources driving the requirements are the existing large SUNY ICE units and the proposed two new large generators (2500 kVA each for SUNY and Clarkson) and the hydro units. Since the new units are not installed yet, there is flexibility in the grounding design for them since the equipment has not been ordered yet.

Table 22.8 Effective grounding status of generators.

Generator name	Full nominal kVA rating of generator	Effective grounding status with respect to 13.2 kV distribution level
West Dam Hydro(s)	778	No
East Dam Hydro(s)	1000	No
Clarkson Existing A	463	Not known (assume no)
Clarkson Existing B	363	Not known (assume no)
Clarkson New	2500	Generator not installed yet[a]
SUNY Existing A	1750	No (high resistance grounding instead)
SUNY Existing B	1750	No (high resistance grounding instead)
SUNY New	2500	Generator not installed yet[a]
PV Array	2000 (unity PF inverter)	Not known (assume not effectively grounded)
Capstone Microturbine	195 (unity PF inverter)	Not known (but unlikely to be effectively grounded)

a) It will be recommended that these future units will not be effectively grounded directly themselves and instead a grounding transformer be installed near or adjacent to each unit to create effective grounding not only for them but for the whole microgrid system. Note 2: all generators are rotating machines except where noted.

The two existing SUNY units (1750 kVA each) are 13.2 kV rated generators directly interfaced to the primary feeder without a step-up transformer. They are not effectively grounded because they have their neutrals grounded through a 19 ohm resistor. Note that an inserted neutral impedance of this type appears as "3 times" its value mathematically from a zero sequence perspective. Therefore, the effective value of this resistor is 57 ohms. We can neglect machine resistance since it is normally very small in relation to the 57 ohm value. If the assumed X_1 impedance of the generator is between 16% and 25% [which would be 15.9 to 24.9 ohms on the generator base ohms] then the R_0/X_1 ratio is between 2.3 and 3.6 for this generator. This R_0/X_1 ratio value is too high for effective grounding as shown in Fig. 22.15. On the other hand, the X_0/X_1 ratio easily meets effective grounding requirements if X_0 is assumed to be in the usual range for generators. impedance. Overall, the grounding characteristics of the machines will fall roughly in the shaded box zone of Fig. 22.15 and they are not effectively grounded as currently installed.

The lack of effective grounding of the Potsdam Microgrid machines does not mean an overvoltage occurs at the instant a line to ground fault occurs. Since the Lawrence Avenue Substation is composed of two large typical grounding source type transformers (delta winding on 115 kV side and wye-ground on 13.2 kV side), the 13.2 kV distribution system remains effectively grounded up until the moment the feeder breaker (or any other breaker) opens that isolates those transformers from the microgrid. At that moment, this is when the ground fault overvoltage begins.

Figure 22.16 shows an EMTP simulation that is representative of a typical ICE generator in the size range of those to be employed in the microgrid project. The DG in this example is "just barely" effectively grounded with COG = 80%. The resulting overvoltage is roughly 1.3 to 1.4 per unit.

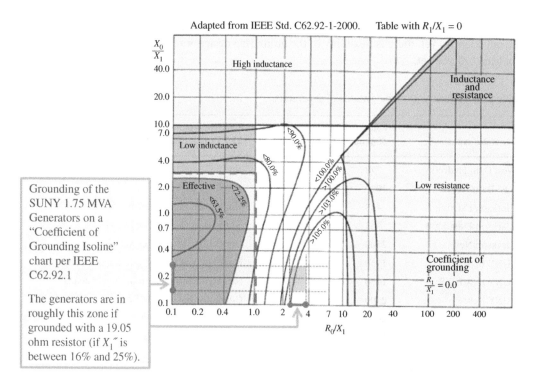

Figure 22.15 Coefficient of grounding of the two existing SUNY 1.75 MVA generators.

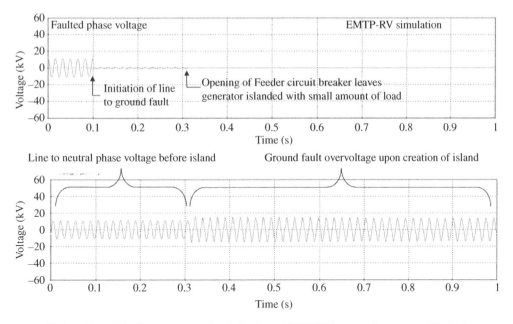

Figure 22.16 Ground fault overvoltage simulation for a 13.2 kV ICE generation source effectively grounded to a COG of about 80% (top trace is the line to neutral voltage on the faulted phase and bottom trace is the line to neutral voltage on an unfaulted phase).

Notice that the overvoltage does not occur until the feeder circuit breaker opens isolating the utility grounding source transformer from the island. In this case the overvoltage would not be as severe as totally ungrounded generator since this example we have a coefficient of grounding of about 80%. In the totally ungrounded case, the voltage would have risen to about 1.73 per unit.

22.6.7 Effective Grounding of DG with Respect to 115 kV System

Regardless of their grounding status at the distribution system level, none of the generators on the Potsdam Microgrid are effectively grounded with respect to the 115 kV system. This is because the effective grounding reference of the neutral/earth is lost in the pass through from the 13.2 kV side of Lawrence Avenue Substation to the 115 kV side. The high side transformer winding is a delta winding that does not provide a zero sequence path. In the event that one of the 115kv lines trips out while the microgrid generation is on line, the line will be energized but ungrounded. This means that if insufficient loading exists on the 115 kV island to suppress ground fault overvoltage at the 115 kV level, then mitigation may be required. The loading could be located anywhere on the islanded 115 kV zone – ideally on any of the feeders served by the Lawrence avenue substation that can be counted on to be present (still connected) during such an event.

22.6.8 Mitigation of Ground Fault Overvoltage

There are four main solutions to avoid ground fault overvoltage:

A. **High Load to Generation Ratio:** If the DG devices are not effectively grounded in the power system area of concern then the presence of load by itself may be enough to suppress the condition without any need for effective grounding. In general, a minimum load to aggregate generation ratio 5 or more within the islanded zone of the DG is enough to suppress ground fault overvoltage from all types of rotating machines (even those with the lowest impedance ratings of about 10%). Rotating machines with impedance at the higher end of the typical range (>25%) can be adequately suppressed with a minimum load to generation ratio of 3 or more. Inverter type generation is far weaker and a load to generation ratio of 1.5 or more is usually adequate for those devices. If there is confidence that the minimum loads are known and can be counted on to be present when the faulted condition occurs then the overvoltage will be suppressed even without effective grounding.

B. **Effective Grounding of the DG:** Effectively grounding the DG to 72% COG will suppress the overvoltage to a reasonable level by preventing the most severe neutral shift. Keep in mind that the effective grounding of the machine is not transferable upwards to higher system levels through any delta winding, so effective grounding in distribution may not solve a problem at a higher level in transmission. If the DG is effectively grounded at a system level of interest, then the presence of load is not critical to suppress ground fault overvoltage at that level. Effective grounding can be achieved at DG units by using the right type of interface transformer and/or the proper type of neutral connection to the DG depending on the specific type of generator.

C. **Separate Grounding Transformer:** A separate grounding transformer (not part of the DG itself) located within the islandable zone of concern can emulate the same effect as effectively grounding the DG. Even though not directly part of the DG plant, a separate grounding transformer can prevent neutral shift thereby preventing the overvoltage. The grounding transformer needs to be sized and rated according to the total DG in the zone it is attempting to suppress. The power rating of grounding transformer is roughly 25% of the rating machine capacity being suppressed (that is 250 kVA per 1 MVA of machine rating.) For PV inverters it is about 5% of the rating – that is 50 kVA per 1 MVA of PV inverter. The impedance of the transformer is sized to yield a COG of about 72% (or 80% if less conservative design). Effective grounding in a traditional rotating machine-like sense may not work for certain types of inverter-based resources due to the nature of operation of those devices – especially when they are functioning as current controlled balanced energy sources.

D. **Time Coordinated Direct Transfer Trip (DTT):** This method involves a time coordinated tripping with DTT such that the utility ground source is tripped after the DG is tripped offline. A time coordinated direct transfer trip can also suppress load rejection overvoltage within the broader island (although it imposes it on the DG site itself once that breaker trips). The EMTP simulation of Figure 22.16 is a good example showing how the overvoltage does not occur until the island forms (feeder breaker opens). So if the DG is tripped offline before the island forms there is no overvoltage.

So which of the four methods (solutions A, B, C, or D) will work best for the proposed Potsdam Microgrid? With regards to *Solution A*, the minimum load to generation ratios for the possible islands we have examined do not come close to meeting overvoltage suppression criteria discussed earlier. We do not have even a 3:1 ratio much of the time, and in fact some of the time the ratio could be approaching as low as 1:1 or even less if all generation is running during lighter load conditions and power is being exported out to the utility system. The loading is insufficient on the underground microgrid zone, the microgrid zone plus the connecting feeder (any of feeders 51, 55, 53, or 56 depending which one is doing the connecting) and also even the substation 13.2 kV bus zone itself at certain times. Overall, the loading solution at the distribution level is not viable for suppression of overvoltage. There may even be sufficient loading at the 115 kV transmission level for those possible islands pending further analysis at that level.

Regarding *Solution B*, that is effectively grounding a sufficient number of the DG. This could definitely be done but would require going back to several of the existing large ICE units and retrofitting them with effective grounding. The two new large 2 MW ICEs would also need to be effectively grounded. This approach is certainly workable, but it is more complex and difficult than it seems. This is because some generator installations already running may not want to be effectively grounded due to ground current flow and retrofit cost issues it imposes on their generators. And some may find it more difficulty to change out their interface transformer to provide effective grounding due to the configuration of the loads and busses at their facility. This is why a couple of strategically placed grounding transformers as discussed in Solution C may be the best overall approach.

Skipping ahead, *Solution D*, the time coordinated Direct Transfer Trip (DTT) is another possible solution for ground fault overvoltage. This will require a high speed communication link to trip the DG before or at the same instant the island forms. Further study of the speed of response of the switchgear and relaying scheme is needed to assess the viability of this option. However, this solution does not solve the need for suppressing ground fault overvoltage and providing neutral stability during islanded microgrid mode as well as grid parallel mode. Only Solution C solves an operating need for *both* modes of operation.

Solution C, the use of two separate grounding transformer banks is a workable solution and perhaps the easiest of all. Two grounding transformers (one on each side of the microgrid system in

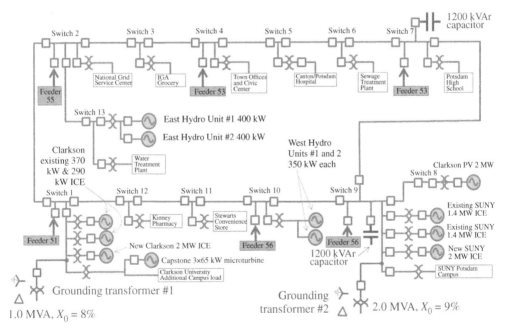

Figure 22.17 Grounding transformer locations needed to emulate effective grounding (these will be useful during both unintentional islands as well as intentional islanded microgrid mode for neutral stabilization and ground fault overvoltage mitigation). (NYSERDA Report 18-13, 2018).

case the system splits in half) would be capable of providing the needed effective grounding for all of the key generation scenarios and islanding modes (Fig. 22.17). Calculations show a needed rating of about 1 MVA near Clarkson and 2 MVA near SUNY each (3 MVA total) which together would be about 25% of the total generation capacity. An X_0 impedance of about 8% at Clarkson on the 1 MVA base rating and 9% at SUNY on the 2 MVA base rating would be enough to give an X_0/X_1 ratio that would achieve roughly 72% COG or better. Note that there is more grounding transformer capacity at SUNY than at Clarkson due to the larger concentration of generation at that side of the system.

One of the issues of adding grounding transformers of this size is to make sure that the current diversion effect is accounted for in the protection settings on the feeder. Otherwise, it may desensitize the ground fault relaying of any upstream protective devices (feeder breakers, microgrid intertie breakers, etc.)

A simple check on this is to compare the zero sequence impedance of the proposed grounding transformers to the zero sequence impedance of the utility source at the point of connection of the grounding transformer. Let us take the Clarkson grounding transformer site as an example. The expected utility source zero sequence impedance at this location is about 3.2% (on a 1 MVA base using the existing feeder impedance). On the other hand, the impedance of the proposed 1 MVA grounding transformer at Clarkson is 8% on a 1 MVA base. This transformer diverts roughly 30% of the current by itself. However, the SUNY-side grounding transformer will also divert current. In fact, it has even less impedance so its effect is more substantial. For the same L-G fault at the Clarkson grounding transformer site (and converting the SUNY transformer 9% impedance at 2 MVA to the same base as 1 MVA) we see that the SUNY transformer would be 4.5% + the impedances of the connecting underground cables which would add roughly another 1% on a 1 MVA base. Figure 22.18 shows the overall current splitting effect. Overall, the *combined* total current diversion effect of both transformers is about 66% (meaning 66% of the zero sequence current goes through the grounding transformers). So the effect is significant and will interfere with upstream protective relaying. The good news is that the effect is much smaller for faults closer to the substation. The analysis clearly shows some ground fault relaying adjustments will clearly be needed to account for this current.

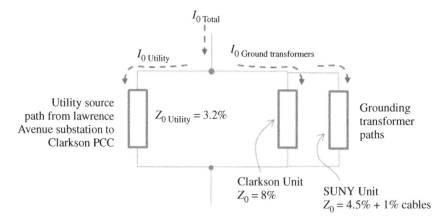

Figure 22.18 Zero sequence current divider effect of the grounding transformers in parallel with the utility source at the Clarkson Feeder 51 PCC (all impedances on a 1 MVA base, model excludes resistor grounded 2 × 1.4 MW SUNY generation).

The grounding transformer approach, while it will require some recalibration of ground fault relays controlling existing circuit breakers and some special provisions to deal with local ground potential rise caused by ground current flows, is the easiest approach to solving the ground fault overvoltage problem.

Note also that these grounding transformers are needed during islanded microgrid mode so using them to alleviate ground fault overvoltage in grid parallel mode is an attractive option since some sort of effective grounding or effective grounding emulation would be needed anyway during islanded microgrid mode operation.

22.6.9 Load Rejection Overvoltage

Load rejection overvoltage is another form of overvoltage. During grid parallel mode as long as the utility system remains connected to the Potsdam Microgrid and an export path is available there is not any danger of a load rejection over voltage event. However, if the microgrid is exporting into the utility's power system and suddenly that circuit breaker opens, then there will be more generation on the island at that instant than there is load. In the full generation scenario at time of light load it is possible there could be a 4–5 MW load rejection out of 10 MW of output. The voltage will suddenly rise on the system to balance conditions.

The synchronous generators can be thought of as voltage sources behind impedances and the inverter can be thought of as current controlled voltage sources (similar to current sources). Both of these will experience a voltage rise at their terminals when the loading is suddenly removed albeit the character and speed of response of the rise is somewhat different for each type of source.

For synchronous rotating machines the amount of voltage rise is a function of the machine impedance characteristics and time constants, the exciter response and the amount of load rejection that actually occurs (e.g. ΔP is 10% or 20% or 30% …etc.). The greater the power step the more severe the voltage change. The amount of surplus VARS on the system from actual capacitor banks as well as the reactive power operating state of the generator also play a role. So it is not just the real power (Watts) step change that matters. For small changes in real or reactive power (say 10–40%) the amount of voltage rise is not that great that the load generally will not have issues handling it. But for a full step load, the issues might become severe, especially if surplus VARs are available.

An example of an EMTP simulation of load rejection overvoltage on a bank of generators is shown in Fig. 22.19. In this case the generators are running near full power and suddenly the loading is removed. There are no capacitor banks present in this case. The overvoltage created is about 20% and will recover after a few seconds. Load rejection somewhat less severe than this is likely to be what will happen on the Potsdam Microgrid if power is being exported. In general, not an issue if lasts a couple seconds or less.

A more serious overvoltage is shown in Fig. 22.20. The overvoltage simulation was intentionally made extra severe by applying several contributing factors. This includes a ground fault overvoltage condition, a full load rejection and the presence of a large capacitor bank. The condition created is an extremely high magnitude distorted voltage wave on the primary distribution system. Under most situations we can envision, a condition this severe would not exist on the Potsdam Microgrid to create this bad of a situation. However, it is offered just as an illustration of the dangers that must be avoided.

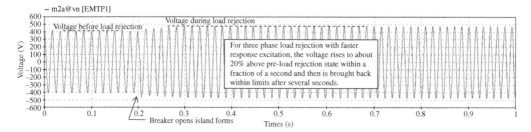

Figure 22.19 Mild load rejection overvoltage: Example EMTP simulation of a load rejection overvoltage on a bank of 480V rotating machines where the load rejection occurs at 0.2 seconds and ΔP=100% – this one is relatively benign due to a balanced VAR loading condition (so only real watts rejection) but voltage still increases by nearly 20%.

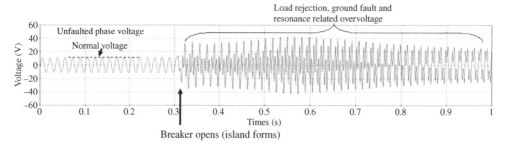

Figure 22.20 Simulated severe 'combination' overvoltage event: Example of EMTP simulation of a bank of rotating machines measured at primary level (ΔP=100%, ground fault overvoltage also present, plus extreme surplus VARs present, altogether resulting in severe high resonant over voltage.

As stated already, the danger of broader severe load rejection overvoltage across the entire microgrid is minor in this project given the way the grid is intended to be operated and the minimum loads expected. We probably will not see more that about a 30–40% ΔP load rejection event at most and usually it will be much less than this. But care does still need to be exercised to make sure that any sub-portions of the microgrid with a possible large surplus of real and reactive power together do not suddenly form due to the opening of circuit breakers. For example, care must be exercised in the region of SUNY (between switch 7 and 10) at times of peak ICE generation, peak hydro generation and peak PV plant generation. If that area should suddenly become isolated during periods of lightest load at peak generation, then there could be a large surplus of real power and reactive power available (see Fig. 22.21). A ground fault might simultaneously be present since faults often cause breakers to trip open. Note also the large capacitor banks on this island too. It is one of the greater island risks from the perspective of load rejection. By operating the generation and capacitors in a manner where VARs are always being imported into a zone with surplus real power production, this can help to a great extent to reduce load rejection if the zone tie breaker should open.

It is understood that the number of breakers and switches may be reduced in Phase II of the project for the final system layout, so some of the riskier possible system breakup scenarios will disappear in the final system design.

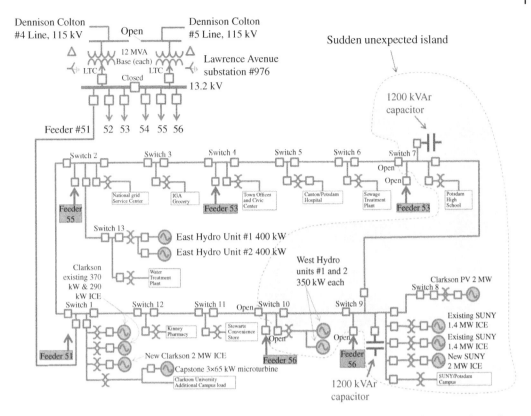

Figure 22.21 Example of a possible load rejection island that might form under the right conditions of opening breakers. (NYSERDA Report 18-13, 2018).

22.7 Microgrid Operation in Islanded Mode

This section focuses on technical analysis of the Potsdam Microgrid system when operating in the intentionally islanded mode. Operating in an islanded mode is a different environment compared to operating in "grid-parallel" mode. In islanded mode the DG units are the only source of energy on the island and have greater responsibility for system power quality than when operating in grid parallel mode. The key differences between grid-parallel and islanded modes of operation are outlined in Table 22.9. These differences include the fact that the generators as a group must provide voltage regulation, frequency regulation, load-following of real and reactive power, and sufficient fault current to allow proper operation of protective devices within the microgrid.

22.7.1 Initial Switching Transitions and Start-Up Procedures for Islanding Mode

The transition to an intentionally islanding mode for the microgrid occurs for any of the following primary reasons:

1. An unplanned outage has occurred on the upstream utility power system (due to lightning, wind, icing, etc.)

Table 22.9 Comparison of grid parallel and intentionally islanded mode of operation.

Parameter	Grid parallel mode(classical DG operation)	Intentional island mode(emergency power)
Voltage regulation	Generators "voltage follow." The utility LTC transformers/regulators control the voltage. Generators may contribute to regulation by fixed power factor mode or independent reactive dispatch control	Generators are totally responsible for regulating voltage on the island
Frequency regulation	Generators "frequency follow" leaving frequency regulation to the utility (unless provided as ancillary service to NYISO)	Generators have total responsibility to regulate frequency of the intentional island.
Real and reactive power	The generators operate at a level of real and reactive power per dispatch center needs that can be independent of the loading on the microgrid.	Generators will follow the load supplying the needed real and reactive power to exactly balance the load to generation.
FaultLevels	Fault current levels on the microgrid primary will be from 100% of the normal utility source levels (with no generation running) up to roughly double those levels when all generation is running.	Fault current from the generators on the primary will be anywhere from about 25% of the normal utility fault level if only partial generation is running, up to about 100% of the utility fault level in full generation scenario
Islanding protection and protective relaying	DTTs and anti-island protection is engaged at acceptable sensitivity levels. Overcurrent protection set to minimize interference with utility system fault protection. DG ride-through optimized for grid stability	DTTs disabled. Anti-island protection disabled or desensitized. Overcurrent protection compatible with islanded operation. DG ride-through settings optimized for microgrid stability

2. A planned outage occurs on the upstream power system.
3. The microgrid has separated proactively into an intentional islanding mode because a *possible* unplanned outage may occur in the near term (e.g. a thunderstorm, hurricane, or ice storm is approaching)
4. The microgrid has separated because it has been asked by utility system authorities to perform some sort of emergency demand response program and withdraw all microgrid loading from the main grid.
5. It is desired to perform a live test of the microgrid to make sure it will be ready for when a real outage occurs.

Depending on which of the above reasons the system transitions into microgrid mode the steps to perform the transition may be different. For any unplanned outage (Option 1) the transition may not be seamless – to be fully seamless, the microgrid must have sufficient battery or other energy storage to carry the microgrid load while the microgrid generation is brought on line (if necessary) and balance generation and load. The microgrid must also successfully ride through the disturbance that led to the separation. For example, a bolted fault on Feeder 51 in the Potsdam Microgrid case would have a significant transient on the microgrid. A design decision would also be required in this case—the microgrid could separate and scramble generation immediately on the feeder trip, and then wait to see if there is a successful reclose on Feeder 51. There is a discussion later in this section on that approach.

In the non-seamless transition, the power will go out and the island will then be established from a black start. However, for Options 2 through 5, these types of transitions could be done either as seamless transitions.

Table 22.10 provides a short list of the key steps needed to transition the microgrid from a grid parallel state into the islanded state where the DG is running as an intentionally islanded microgrid. The steps required for an unplanned intentional island and for a planned intentional island are shown.

22.7.2   Returning to Grid Parallel Mode from Intentional Island

Once the utility power is restored in a stable fashion on the utility source side of the **open** Potsdam Microgrid tie breaker (e.g. Feeder 51 Clarkson PPC circuit breaker), it is not necessary to incur an outage to the microgrid to reconnect to the utility system. The running and stable microgrid system can be connected back to the utility system simply by synchronizing the power on both sides of the PCC tie breaker to a suitable match of voltage and frequency, and then closing the breaker. After the connection is made, the generation instantly transitions to grid parallel mode meaning that it must voltage follow, frequency follow, have islanding protection and DTT trip modes activated and any other protection settings changed back to the appropriate grid parallel values. If it is not desired for the generation to continue to operate once back in grid parallel mode, the DG can be slowly ramped down and taken offline over 5–10 minutes. This will allow the utility voltage regulator tap changers to keep pace with the change in voltage due to the change in current flows.

22.7.3   Synchronization during Transitions and Aggregations

For synchronizing the microgrid with the utility power grid, or synchronizing the various parts of the microgrid, the requirements recommended by IEEE 1547 should be suitable for both the islanding modes and the grid parallel modes (see Table 22.11).

22.7.4   Coordinating Islanded Microgrid Operation with UPS Ride-Through Capability and UPS Frequency Limits

Any UPS systems serving a critical load on the microgrid that need to operate continuously during the expected *20–30 minute transition time* of a black start recovery. To do this, they must have sufficient battery ride-through time to bridge this time gap.

Another important factor about UPS compatibility with the microgrid when operating in the intentional islanding mode is the influence of the voltage and frequency settings of the various UPS systems on whether or not microgrid conditions will cause them to cycle back and forth between battery mode and grid supplied normal mode. This could also lead to customer dissatisfaction with the microgrid operating conditions.

The utility system is almost always well within ± 1 Hz frequency, but in microgrid mode we can expect the frequency variations to be on the order of ± 2 Hz or even occasionally ± 3 Hz. Many UPS are default setting programmed to hold the frequency tightly within a range of ± 1 Hz with a tight slew rate too and many would switch to battery mode if conditions go outside bounds. UPS may also switch back and forth to/from battery mode on if voltage deviations become excessive too. In multi-party microgrids, the microgrid designes will need to coordinate with UPS owners to ensure compatibility.

Table 22.10 Steps to transition to an intentionally islanded mode (full microgrid version) (NYSERDA Report 18-13, 2018).

	Type of intentional island	
Step number	Unplanned intentional island	Planned intentional island
Initial condition just before event	Utility power is stable and nominal (microgrid generation may or may not be running in classical DG mode). A power "outage event" is about to occur	Utility power is stable and nominal (microgrid generation may or may not be running in classical DG mode at this time). For whatever reason it is desired to separate the microgrid seamlessly to an intentional island mode
1	Utility power outage occurs due to ice storm, lightning, wind, etc. (upstream utility source breaker opens)	Available microgrid generation that is not already running that's needed is started, synchronized and put online in *grid parallel mode* (still classical DG mode at this step)
2	All running microgrid generation is tripped offline within about one second or less per normal DG methods (DTT, anti-islanding relays, etc.)	The total power output of DG units is adjusted until there is little to no power transfer (watts or VARs) at the microgrid tie breaker (such as Feeder 51 PCC In Potsam). Load shed might be needed to reach this state
3	Utility reclosing sequence completes its sequencing. If reclosing fails to successfully re-energize system, a decision is made after a suitable delay of minutes (manually or automatically) to go to intentional island mode	Microgrid controller sends signal to all generators to be *prepared* to transition to the intentional island mode (DTT disabled, anti-island disabled, frequency regulation, voltage regulation and load following, etc.)
4	A signal is sent to open microgrid tie breaker (Feeder 51 PCC) and begin microgrid start up procedure. Microgrid "subpart" splitting breakers open	The tie breaker (Feeder 51 PCC) is opened, generators switch instantly to voltage regulation, frequency regulation, and Watt/VAR following modes, DTT disabled, anti-islanding disabled, etc., and seamlessly carry load
5	Generation protection/control is set to intentional island mode (Anti-islanding disabled, DTT disabled, and voltage, frequency and watt and VAR modes set. Off-line generation is started or made ready to start (but not paralleled with microgrid loads just yet).	
7	Load shed signals are sent to trip non-essential loads depending on loading conditions and generation availability	
8	Signals sent to open switches at all major load distribution transformers (those greater than 500 kVA)	
9	One or more generators with black start capability are brought up and running on its bus and ready to pick-up local load).	
10	Black start units each pick-up some local load as allowed by generator ratings (some large transformers not online yet).	

Table 22.10 (Continued)

	Type of intentional island	
Step number	Unplanned intentional island	Planned intentional island
11	Tie breakers synchronize and connect microgrid subparts	
12	Additional generation and loads are brought on line generators brought online in a sequence	
13	Large transformers and large individual loads are brought on line in sequence as the generation capacity permits	
FinalState	Intentional Island is accomplished with loss of power up to about 20–30 minutes	Intentional island is achieved seamlessly without loss of power

Table 22.11 Synchronization parameters for tying together subparts of the Potsdam microgrid as well during transitions the partial or full microgrid to grid parallel mode. (NYSERDA Report 18-13, 2018).

Aggregate rating of DG units (kVA)	Frequency difference (Δf, Hz)	Voltage difference (ΔV, %)	Phase angle difference ($\Delta\theta$, degrees)
0–500	0.3	10	20
>500–1,500	0.2	5	15
>1,500–10,000	0.1	3	10

22.7.5 Power Mismatch Allowance for Seamless Intentional Island Formation

As was stated in the procedure of Table 22.10 for the seamless transition to a planned intentional island, it is important to have the real and reactive power transfer at a low or zero flow across the microgrid tie breaker at the moment of separation. The question is how close to "zero" is good enough?

A small mismatch on the order of a 5% or less of the generation output is often reasonable to achieve and should not lead to any serious voltage or frequency variations on the microgrid as long as the generation is set to load follow and has the appropriate disabling of anti-islanding, etc. A more conservative recommendation would be that the power transfer across the tie breaker be between 0% mismatch and a 5% over generation mismatch of the connected DG capacity at the moment of separation.

22.7.6 Inverter-Based Resource (IBR) Compatibility

Inverter-based resources such as photovoltaic generation and battery energy storage systems will have a different response characteristic than traditional synchronous generators. The 2 MW PV system is an inverter-based power source that uses active anti-islanding algorithms to detect islands and trip offline quickly if an island forms. The methods used vary from inverter to inverter and can be much faster than for synchronous generators.

The anti-islanding settings of the PV, which must be operative during grid parallel mode, need to be disabled or greatly relaxed to allow for microgrid operation. The microgrid has a greater rate of change of frequency, greater overall frequency variations, higher impedance, greater phase shifts, and larger harmonic distortion shifts than the utility sourced power system. If the settings are not relaxed it is possible the PV system may not be able to remain connected to the microgrid in a stable fashion. It might end up cycling on an off periodically in a disruptive fashion. Two recent reports on IBR performance on electric power grids (Sandia National Laboratories, 2020 [8]) (North American Reliability Corporation, 2018 [9]) provide useful information for microgrids as well. It is important to recognize that in the case of the Potsdam Microgrid, the inverter based resources are a fairly small component of the generation and do not necessarily even need to be connected when in islanded microgrid mode. Therefore, the Potsdam Microgrid is almost entirely rotating machine based. However, if the microgrid was dominated by IBR it would be necessary for these to operate in a grid forming mode with black start capability. A large amount of grid following IBR is not acceptable for use on a microgrid, albeit a small amount in proportion to total generation (such as 20% or less) can be successfully integrated depending on the source type specifics.

22.7.7 Avoiding Cable-Resonances During Startup Procedure

For microgrids the employ underground cables for their primary distribution, there can be many long 13.2 kV primary cables that might be energized from one end while the loading on the other cable end is not yet established (that is the loads have not yet been switched on at the destination). There will be significant cable capacitance. In addition, there can also be significant power factor correction capacitor banks on the primary system. While these configurations are necessary as part of the resilient underground design arrangement, there is concern that if the cables are fed from one end by the DGs with no load or utility connection present at the other end it could result in some significant resonances with the generators and transformers. To reduce the chance of any such harmonic resonance and/or ferroresonance occurring, it is recommended that the switching procedures for bringing certain loads, cables, and capacitors online follow switching policies that are antithetical to ferroresonance and that avoid excess capacitive VARs in any zone unless resistive damping is present. For each zone switched in, some load should be initially connected as part of any connecting cable segment, to suppress such resonances as the system is brought up and assembled from a black start.

22.7.8 Fault Levels (Islanded)

The fault levels on the microgrid, while islanded, will generally be considerably less than when operating in the grid-parallel mode with the utility source. This is the case even with rotating machine generators, but can be particularly the case when there are significant levels of IBR on the microgrid.

One of the more important factors regarding the operation of the microgrid is whether or not in an islanded state it will have sufficient fault current to operate protective devices *within customer premises at the secondary voltage level* (that is at 480V, 208V, or 120V). The degree to which reduced fault levels seen at the primary side of the distribution transformer influence the available fault level on the secondary side depends on the intervening impedance of the distribution transformer and the secondary cables. In cases where the transformer size is small, even though the available fault level on the primary side of the distribution transformer is reduced to a lower value, the dominating impedance of the situation is still the DT and its secondary cable, so the actual change in fault level at customer service entrances may not be that great. On the other hand, if the transformer is large, then the fault level may change significantly.

This must be considered when developing the protection system for the microgrid. In cases with large proportions of IBR, some inverters may need to be oversized in order to deliver sufficient fault current to the system.

22.7.9 Voltage Flicker Levels During Motor Starts (Islanded)

When the generators are operating in an islanded mode, the transient voltage regulation during motor starts, transformer inrush events, load steps, etc., will be somewhat more sensitive than when in grid parallel mode. Line connected motors can be one of the more problematic types of devices in this regard. In particular, larger motors of many hundreds of horsepower rating that are starting directly across the line (no soft starter) may result in noticeable voltage fluctuations (and will also cause frequency fluctuations).

Example 22.3 Motor starting study on the Potsdam Microgrid. Consider a 100hp 480v induction motor connected to the SUNY bus through a 13.2kv:480v step down transformer. The motor is a NEMA Code G machine, and does not have a soft starter. Determine the voltage dip on the 13.2kv bus and on the 480v bus at SUNY.

a. For a motor starting analysis, the NEMA Code G induction motor (without a soft starter), will have six times the full load running current at startup. There is an assumed 1 kVA of full load power consumption per horsepower of rating, so a 100hp motor uses about 100 kVA when running at full load and will have about 600 kVA of starting power with a locked rotor start.

b. The fault current study for the microgrid show that at minimum operating conditions for the microgrid operating in islanded mode, a bolted three phase fault will deliver 25 MVA to the SUNY 13.2kv bus.

c. Neglect the resistance of the generation and cables. Assume that the initial motor starting current is purely reactive. The per unit impedance of the primary system feeding the SUNY 13.2kv bus is

$$X_{source} = \frac{S_{base}}{S_{fault}} = \frac{12\,MVA}{25\,MVA} = 0.48\,p.u.$$

The per unit value of the motor starting current is

$$I_{start} = \frac{S_{start}}{S_{base}} = \frac{0.6\,MVA}{12\,MVA} = 0.05\,p.u.$$

The per unit voltage drop at the 13.2kv SUNY bus will then be

$$\Delta V = X_{source}I_{start} = 0.024\,p.u.$$

d. The three phase 13.2kv:480v step down transformer is rated 1 MVA, and has a 9% short circuit impedance to the transformer base. To the system base of 12 MVA, the transformer impedance in per unit is

$$X_{tr(sys)} = X_{tr(unit)}\frac{S_{base\ sys}}{S_{base\ tr}} = 0.09\frac{12\,MVA}{1\,MVA} = 1.08\,p.u.$$

The voltage drop on the 480v bus would then be

$$\Delta V = \left(X_{\text{source}} + X_{\text{tr}} \right) I_{\text{start}} = 0.078 \, \text{p.u.}$$

2. <u>Discussion</u> These results can be compared with the ITIC curve Fig. 22.3 and the flicker curve Fig. 22.11. From Fig. 22.3, the voltage should stay above 0.80 p.u. to avoid load disruption during the motor starting transient (assuming the motor will start in a time somewhere between 0.5 and 10 seconds). If the voltages are at 0.95 p.u. prior to motor starting, the 13.2 kv voltage will then dip to 0.926 and the 480 v bus voltage will dip to 0.872. Both of these values are above 0.80 per unit, so there should be no load disruption, and the motor should start successfully.

From Fig. 22.11, the voltage dip of 2.4% on the 13.2 kv bus will be just below the borderline of visibility with 1 start per hour. On the 480 v bus, the voltage dip of 7.8% will be above the borderline throughout the range of the curve. This motor starting would cause objectionable flicker on this bus. Unless there is little or no sensitive load on this bus when the motor starts, this motor will likely need to have a soft start unit installed.

22.7.10 Voltage Sags During Transformer Inrush (Islanded)

Voltage stability during start-up of the microgrid in intentional island mode is an important factor in the system's performance. When bringing the system up from a black start condition it will be required to "assemble" the island by energizing many distribution transformer units and picking up large load blocks served by these. The voltage dips, which occur during this process must be limited so as not to cause the microgrid to collapse. Distribution transformer "energization" is probably one of the most demanding aspects of islanded black start procedures. In the Potsdam Microgrid, the largest step down transformer has a 3000 kVA rating, and will require a large magnetizing inrush (with respect to the generation size).

It is difficult to gauge how much voltage dip will occur on a given time that a large transformer is energized since many parameters impact this. The transformer magnetizing inrush varies each time due to the trapped flux level of the core not always being the same and the angle of energization (breaker closing) is not always the same either. Furthermore, the angles and trapped flux state can be different on each phase resulting in different levels of inrush on each phase. If the transformer is energized with downstream load connected, that will further complicate the inrush.

To understand better the nature of how transformer inrush impacts the microgrid generation, a model for the worst case effect of energizing a transformer on the generator voltage is shown in Fig. 22.22. As a rule of thumb, the moment a transformer is energized, the inrush current associated with the first cycle is about half as large as what the current would be if we fully shorted the

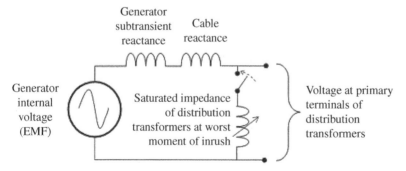

Figure 22.22 Simple model for worst case inrush related voltage dip at moment of peak transformer inrush.

transformer secondary. In other words, the effective value of the "saturated impedance" during the first cycle of inrush is about **twice the rated impedance** of the transformer on the transformer's base rating. Fortunately, this does not last long since the transformer quickly becomes unsaturated and the inrush current decays to a much lower value in subsequent cycles. During the inrush, a voltage divider effect is created as shown in the model whereby the voltage on the primary terminals of the transformer (the microgrid primary) can be estimated by the voltage divider equation using the ratio of the upstream source impedance and the saturated transformer impedance (see Fig. 22.22).

Example 22.4   Transformer Inrush. Consider the 3000 kVA 13.2kv:480v three phase transformer fed from the Clarkson bus.

The transformer series impedance is 5.5% on the transformer base. From the short study of the microgrid in islanded mode with all 10 MW of generation on line, the effective system impedance of the microgrid is $Z_{sys} = 0.20$ p.u. on the system 12 MVA base.

Using the rule of thumb cited in the previous paragraph, the saturated transformer impedance in per unit on the system base is estimated as

$$Z_{tr(sat)} = 0.055\,(2)\,\frac{12\,\text{MVA}}{3\,\text{MVA}} = 0.44\,\text{p.u.}$$

From Fig. 22.22, the voltage at the transformer 13.2kv terminals during the initial inrush is

$$V_{\text{Clarkson bus}} = V_{OC}\frac{Z_{\text{tran(sat)}}}{Z_{\text{tran(sat)}} + Z_{\text{sys}}} = 1.0\frac{0.44}{0.44 + 0.20} = 0.69\,\text{p.u.}$$

Here, the Clarkson bus voltage prior to transformer switching is take as $V_{oc} = 1.0$ p.u. and the impedances are assumed to be nearly purely reactive.

The model shows that the RMS voltage on the microgrid at the transformer will dip down to about 69% of nominal voltage during the first cycle of worst case inrush. While a dip to 69% seems severe, in practice the average voltage disturbance over many cycles and on all phases will not be nearly as bad because the inrush current decays rapidly with time and inrush conditions are not equally severe on all phases. Smaller transformers will cause proportionally less voltage dip. The load behind each transformer will cause an additional voltage dip, albeit the inrush is by far the largest component.

A big concern that we must be careful about is that the procedure for connecting some transformers will involve conditions where only *part of the microgrid generation* is online (not the full 10 MW). If there is only 5 MW or even 2 MW running in a zone and a 2 or 3 MVA transformer is energized, then the voltage dip effect is much larger. In those cases of limited generation the source impedance is effectively much higher relative to the inrush resulting in more voltage dip. The dips could easily become severe enough to disrupt sensitive loads during the microgrid assembly process.

22.7.11   Cold Load Pickup

On distribution feeders there is a phenomena called "cold load pickup." Unlike transformer magnetizing inrush or motor starts that last seconds at most, this is a longer duration current (lasting up to about 15 minutes) that can be up to twice (on residential feeders) the normal full load current when the feeder is energized after being dead for a long time (over 20 minutes). This temporary increase in "steady state" current is due to loss of diversity of thermostats within HVAC equipment and other devices that all want to be "on" when the feeder is first energized after a long outage. Cold load pickup on feeders servicing commercial/industrial loads has lower magnitude, perhaps up to 1.5 per unit of normal peak load. The cold load pickup for a given microgrid will depend on the type of microgrid load and other factors. A microgrid restoration plan will need to consider cold load in the load pickup procedure. Likely, the load restoration will need to be staged over time, delaying the energization of some sections of the microgrid by a few minutes as the cold load subsides in areas that are already up and running.

22.7.12   Microgrid Primary Voltage Target Level

During intentionally islanded operation of the microgrid, the voltage target of the 13.2 kV microgrid primary cables on a 120V basis should be somewhere in the 120–122V range. This is near the middle of the ANSI voltage regulation window but biased just a bit above the middle. This target zone will allow some head room for voltage rise across distribution transformers and secondary service cables at the customer sites that have DG present. Thus, DG injecting power at such sites under most/all expected operating states of real and reactive power will not create a voltage rise so large that exceeds ANSI Range-A limit (126V) at those sites. The 120–122V target range leaves 4–6 Volts of headroom which should be adequate for the expected power flows across the expected distribution transformer impedance and secondary cable impedance at generator sites.

For load sites on the microgrid *that have no generation* at the site, the voltage is always going to be lower on the secondary side of the transformer. The targeted 120–122V primary cable operating range will allow for up to 6–8 volts of drop across the distribution transformer impedance and secondary service conductors such that the voltage at those customers service entrances would be expected to always be higher than 114V (the lower ANSI range A limit) under all typical loading scenarios.

In the islanded mode, the target voltage range on the primary feeder can be maintained by having the microgrid control system continuously adjust the output voltage of each DER unit. As this action will control both the overall microgrid voltage level and the reactive power sharing among the DG units. For IBR, individual DERs can use the smart inverter settings of IEEE Std. 1547-2018. However, it is still likely that the microgrid controller will need to monitor voltage levels and var sharing, and readjust DER settings to account for the full range of operating conditions that will occur on a microgrid. Note that on geographically compact microgrids that connect assets at a primary distribution voltage, the voltage differential across the primary system can be relatively small, so reactive power sharing may be the primary focus.

22.7.13   Voltage Changes Due to PV Power Variations

Microgrid energy sources can range from strictly synchronous generators to having 100% inverter-based resources. It is therefore difficult to predict the voltage changes that would be imposed by one or more intermittent energy sources, such as photovoltaics or wind during islanded operation. It can be expected that individual PV installations will experience changes in output of up to 60–80% of nameplate rating, occurring over about 30 seconds. In PV systems, these occur on partly cloudy days. When there are several PV installations on a given system, there is a smoothing effect across these installations.

Unlike motor starting currents that are sudden instantaneous demand steps, mostly of a reactive current nature, the PV variations are slowly and smoothly changing power variations over several tens of seconds, minutes or even longer depending on the type of cloud shading conditions. Power may go up or down depending on whether the cloud shading is arriving or receding from the solar array field.

Another factor about PV is there is a tendency to run the PV inverters at/near unity power factor rather than in a state of high VAR production or absorption. Although VAR production or absorption cases are possible, we will assume for our PV system until further data are available that it is operating at a power factor near 1.0 ± 0.02.

These power output variations will be balanced either by synchronous generators or by energy storage systems.

The above factors imply the following in the analysis:

1. The slower 'smoother' power changes of PV will allow synchronous generator excitation controls to better keep up with these variations, as compared to sudden changes due to large motor starting. The governor response will be larger than is the case with motor starting, as the motor starting current is largely reactive.
2. Unity PF power variations from the PV installations work to minimize the voltage change effects on islanded network.
3. The PV power fluctuations also have an effect on the microgrid frequency, which indirectly influences voltage by the changing controllers of synchronous machines or grid forming inverter controllers, EMF as well as changing loads on the system that are frequency dependent.

It is not possible to model precisely the voltage variation effects until the equipment settings and more details of the project characteristics are finalized. When the equipment has been specified, theses fluctuations will need to be studies with dynamic simulation software.

22.7.14 Frequency Regulation (Islanded)

When the microgrid is connected to the utility power system, the impact of load steps is negligible on the frequency. However, once the system is islanded, the inertia and total rating of the generation on the island is much smaller. As a result, the frequency will be more significantly impacted by load steps. It is important to make sure that varying loads and load blocks connected to the grid do not seriously degrade the frequency regulation of the island. For satisfactory operation of the island, transient frequency dips during load steps should be no greater than $\pm 5\%$, with a design goal is to keep these to $\pm 3\%$ or less.

An electronic isochronous governor controller is recommended for the generators. This type of controller is the ideal because it allows the load sharing between generators to be managed while maintaining essentially constant frequency (in the steady state) without any droop. The alternative type of governing that is sometimes employed is droop control. However, droop control is based on the concept that the generator frequency slightly decreases as its load increases. That method allows a bank of generators (all with similar droop characteristics) to equally share load. The problem with droop governing is that as the load increases, the frequency of the entire generator bank decreases. With droop governing it may be more difficult to keep the frequency within a suitable band for sensitive devices or devices that have tight operating windows (such as some types of UPS equipment).

To insure that the frequency does not deviate too much from nominal and that the occurrences of such frequency variations are limited in their repetition rates, it is reasonable to establish guidelines for the largest load steps and repetition rates for those load steps. Table 22.12 is a first cut attempt to create a set of guidelines for the types of load steps that should be allowed on the islanded microgrid.

Table 22.12 Recommended load step rates versus size of load step.

Size of load step (L_{step}) as a fraction of operating internal combustion engine (ICE) generating capacity	Maximum recommended daily rate of recurrence
$L_{step} \leq 1\%$	No limit
$1\% < L_{step} \leq 2\%$	500
$2\% < L_{step} \leq 3\%$	200
$3\% < L_{step} \leq 5\%$	40
$5\% < L_{step} \leq 7\%$	25
$7\% < L_{step} \leq 10\%$	8
L_{step} exceeding 10%	Should be extremely rare or only during black start assembly of the microgrid

A 10% load step is considered the largest step that should be allowed frequently during steady operation of the microgrid. Base on the machine characteristics, this should limit the frequency change to within about a ±3% band around 60 Hz as loads cycle on/off (with isochronous governing). The 10% guideline is intended for cyclic loads that repeat up to 8 times per day (not hundreds of times per day). Small load steps can occur more frequently. Please also note the caveat that these are "real power steps" as related to frequency regulation, and any loads having significant magnetizing inrush, higher initial real currents, and/or high reactive current steps made need to be restricted per voltage flicker requirements discussed earlier. It is recognized that on rare occasions larger load steps than 10% will occur during islanding startup (to assemble the microgrid) and for other reasons.

As mentioned earlier in the report, UPS systems come with fixed or programmable operating windows for frequency (Hz tolerance in ±%) and rate of frequency change (Δ Hz/second). If frequency conditions go outside the window the UPS may transition to its local battery source thinking that the "grid" is outside acceptable limits. Having UPS equipment unnecessarily and excessively cycle back and forth between operating modes should be avoided. A little bit of occasional cycling is okay from time to time – but to be doing it constantly will wear out the equipment, the batteries and may cause possible disruptions of the critical loads the UPS units are serving. Table 22.12 step load guidelines were in part developed to help avoid this issue.

Prior to the creation of a microgrid, the frequency window of operation for a UPS could be set as tight as ±1% in some critical application cases or as wide as ±6 or even ±8% in others where equipment tolerances are not so critical. Another type of frequency tripping found on UPS system is the "slew rate." This is the rate of change of frequency. This function may trip at as low as 0.25 Hz/sec in some applications where high slew rates cause issues or as high as 2 Hz/sec or higher in others where slew rate is not so critical. There are hundreds of UPS units of all size scattered throughout the microgrid and they likely have a wide range of frequency trip windows and slew rate tripping conditions.

22.7.15 Harmonics (Islanded)

As discussed earlier in the report, DG devices must meet IEEE 1547 guidelines for harmonics (see Table 22.4 IEEE 1547 requirements for DG Harmonics (IEEE Std. 519-2014 [10])). It is not expected that any of the generation equipment, to be acquired or in use for the project already, will by itself necessarily exceed the limits. However, care should be exercised when a microgrid is created which will operated in islanded mode. The microgrid reactances (primarily due to lines, transformers and generators) will be much different than is the case when the same system is fed from a strong utility power grid. The microgrid may add power factor correction capacitors, perhaps with multiple switching states. The harmonic resonances of the system will therefore change significantly.

The lowest order harmonic resonance can be the most problematic. For screening purposes, the harmonic number of this resonance can be estimate from the per unit values of the capacitive reactance and the effective inductive reactance at that point. This is most accurate where there is a single dominant capacitor bank, but can be reasonably accurate with more than one bank electrically near to each other. In these cases, the harmonic order of the resonance is

$$N = \sqrt{\frac{X_{C(\text{p.u.})}}{X_{L(\text{p.u.})}}} \quad [22.5]$$

The values of capacitive reactance and effective inductive reactance depend on the mix of generation that is running and how many capacitors are switched on, etc. Also, the circuit loading will in general be enough to dampen the Q (quality factor) at the resonant frequency to a value to *well under 2* such that a problematic amplification of the harmonics is unlikely. A possible exception is during situations where the grid is being assembled from black start or during extremely light loading periods, such resonances might become more pronounced due to the lack of loading for damping during certain phases of the grid assembly.

There is little doubt that there will be non-linear loads on the microgrid. On many microgrids, there will also be inverter based resources that generator harmonics. If the all customers and all generation on a multi-owner microgrid meet the IEEE 1547 harmonic limits at their individual points of common coupling, it is unlikely that a harmonic voltage problem will occur unless there is a strong resonance right on one of the harmonic frequencies.

A factor that is helpful for harmonics on the microgrid is that the project will use separate grounding transformers (as opposed to grounded neutral generators). While the purpose of using separate grounding transformers was not related at all to harmonics, it was done for ground fault overvoltage mitigation purposes; this choice does have the ancillary benefit that it helps with harmonics by alleviating the line to neutral load serving duty from the DG equipment and shields those devices from seeing the "triplen" harmonics that appear in the line to neutral mode. Grounding transformers also provide a more linear source of ground current compared to many models of generators with grounded neutrals. Of course, harmonics in the line to line mode of the generators and the impacts of harmonic loads are still a factor to be watched, but at least from the line to neutral mode perspective the situation is helped by the use of grounding transformers.

Overall, at this point, there is not a particularly severe concern about harmonics other than in the lightly loaded situations where cable and power factor correction capacitances may lead to resonances. But it is clear that measurement data should be obtained to determine the background levels. If the current THD exceeds 10% with a broad spectrum then some derating of the generators may be needed. The amount of derating is small, just 10% for up to about 40% current THD (NYSERDA Report 18-13, 2018). It is unlikely that current THD will be that high based on the typical THD level seen at these types of commercial and campus loads. However, a high measured current THD would mean the voltage THD would be very likely outside IEEE limits of 5% owing to the relatively small I_{sc}/I_L ratio of the system and that a mitigation program might be needed.

22.7.16 Load Unbalance (Islanded)

Unbalance of the load is a larger issue for islanded synchronous generators and IBRs than for when they are operating in grid parallel mode. In grid parallel mode of operation, the voltage unbalance seen by the connected DER is limited by the strong utility power grid – thus the utility's system ends up supplying much of the unbalanced current flow. Even when the load *current* at a particular utility grid site is significantly unbalanced (say 20–25%), the *voltage* in a grid parallel situation remains fairly well balanced due to the relatively low negative and zero sequence impedances of the utility source. In general, both of these sequence components of voltage unbalance are rarely if ever beyond 2% at most substations. In fact, it is not unusual for these components to be under 1%. Overall, in grid parallel mode the utility source assumes the major role of supplying most of the unbalanced current to the load and the generators supply only a small residual amount based upon the ratio of the impedances of the machine and utility source.

Once the microgrid becomes islanded, the situation changes. Now the synchronous generator and IBR sources on the island must supply all of the unbalanced current to the load, and these sources are exposed more significantly to both the negative sequence and the zero sequence unbalance components. Delta–wye service transformers or grounding transformers can be used to isolate the flow of zero sequence currents in a microgrid. These do not shield the generators and IBRs from negative sequence phase unbalance. The negative sequence is the most important factor that causes heating on the face of the generator rotors that can lead to thermal damage.

Industry design practices generally allow that synchronous machines and induction generators of the size classes and types used in most DG scale applications are rated to handle up to about 10% continuous negative sequence current. If the negative sequence impedance of the machine is 0.20 per unit, then this translates into a negative sequence voltage of about 2%. For the islanded application the continuous negative sequence voltage at machine terminals should not be allowed to exceed 2% for any considerable length of time and the negative sequence current should not be allowed to exceed 10% of the rated phase current when at full load. Since the negative sequence voltage heats the rotor quickly, if the unbalance limit is exceeded then the machine can quickly overheat and be damaged if it is at or near rated load. If the machine is operating at reduced load, it can handle slightly more negative sequence current unbalance on a percent basis than it can at full load.

A curve from one particular manufacturer for generator steady state unbalance capability is shown in Figure 22.23 for illustration purposes. At full load the curve shows that the allowable unbalance is up to 10% (that is $100\% \times [1 - 90/100]$). At a loading of only 20% of rated load the curve shows that the difference between minimum and maximum phase current is up to 25% (that is $100\% \times [1 - 15/20]$). At even further reduced load, the current can fully unbalanced. Be advised that his curve is specific to one particular manufacturer's machine and serves only as a general illustration here. However, in the case of any specific generator utilized for a microgrid the manufacturer should always be consulted for the specific machine's capabilities. An excellent discussion on the unbalance capabilities of machines can be found in the IEEE Tutorial on Synchronous Machine Protection, 2nd Edition, published in 2011 [11]. The load-unbalance condition at the Potsdam microgrid load sites are not known at this stage. In the Phase II effort, data can be assessed to determine if the unbalance levels are a threat and require any sort of mitigation.

22.7.17 Microgrid Effective Grounding (Islanded)

The effective grounding scheme for the microgrid was discussed earlier in the grid parallel section of the report. However, we must also consider its operation in the intentionally islanded mode as

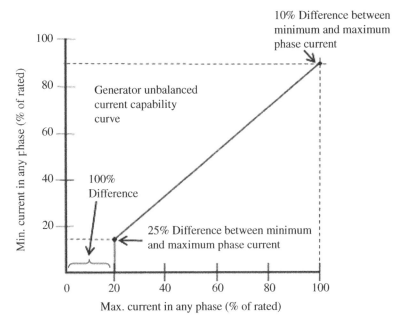

Figure 22.23 Negative sequence unbalance capability of a typical smaller salient pole synchronous generator.

well. As mentioned earlier in this chapter of the report, the two selected grounding transformers (Units #1, 1 MVA, $X_0 = 8$ and Unit #2, 2 MVA, $X_0 = 9\%$ will be sufficient to provide effective grounding on the system with all generation running with COG = 72% or less. During islanding mode (intentional or not) when ground faults occur, the maximum voltage on the island due to neutral shift on the unfaulted phases will not exceed 131% given this type of grounding. The 131% figure excludes the effects of any load rejection overvoltage which can add to it.

As shown in Fig. 22.24, the use of two grounding transformers sized for the generation on the SUNY side of the system and on the Clarkson side allows the microgrid to operate each side independently if desired and also serves the needs of initial black start and assembly of the grid from its subparts such that it can start with two separate effectively grounding islands and then be assembled together into a full microgrid. In the full grid mode the two grounding transformers and any other ground sources will share the zero sequence current in proportion to their relative impedances. The zero sequence impedance of the actual cables on the microgrid is roughly $1/10^{th}$ to $1/20^{th}$ the impedance of the two grounding transformers such that the cables impedances play little role in the division of current.

Also shown in Fig. 22.24 is the fact that the ground current flow due to loads and ground faults will mostly flow back through the grounding transformers and not through the generators (due to the winding configuration of their dedicated transformers)

As part of the control scheme of the microgrid, it is desirable to measure the current and voltage conditions at the grounding transformers and have some protection on them that is useful for certain functions. For one thing, it can indicate if they are online or not (such that microgrid operation can be disabled if they are not online). In addition, CTs in the grounding transformer neutral wires can sense the ground current flow and use that information to determine the total neutral current flow from a steady state loading and a transient fault current protection perspective. The flow levels will tell the controller how much zero sequence unbalance load current is present on the system,

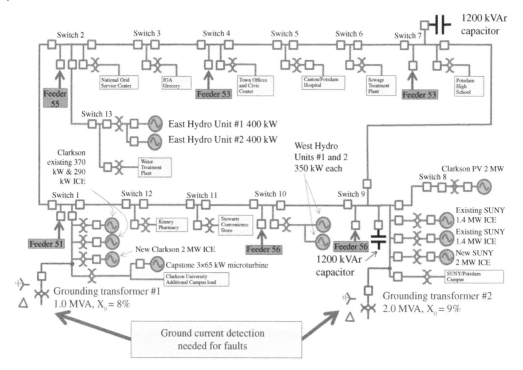

Figure 22.24 Microgrid effective grounding scheme provides two zones of neutral grounding from which to assemble the grid to the full grid or operate as two separate islands. (NYSERDA Report 18-13, 2018).

and we can detect ground faults and trip the major DG sites due to such faults if needed. We can also use voltage sensing combined with current sensing to detect certain types of malfunctions within the grounding transformer. The ground current waveforms measured from these points would also tell the controller something about nature "triplen" harmonics too; triplen harmonics being those that are the 3rd, 6th, 9th, etc. These are zero sequence harmonics that are carried in the neutral path.

An important factor that needs to be considered in the design and installation of the grounding transformers is that anytime there is a large grounding current source present there will be zero sequence currents coming through the earth/neutral wires to that source that concentrate in that location. During ground faults this can cause some potential rise locally and the transformer site will need to be treated somewhat like a small substation in that regard. There are several IEEE standards that apply to grounding of substations and energy sources. These include IEEE 80-2000 which is the "IEEE Guide for Safety in AC Substation Grounding." There is also IEEE 487-2007 which is the "IEEE Recommended Practice for the Protection of Wire-Line Communication Facilities Serving Electric Supply Locations," and also IEEE 1590-2009 the "IEEE Recommended Practice for the Electrical Protection of Communication Facilities Serving Electric Supply Locations Using Optical Fiber Systems."

The important factors that the IEEE standards address are to make sure the design of the installation avoids any step and/or touch potentials that could be dangerous and also that any telecommunication and data lines coming into the "zone of influence" are adequately protected if the potential rise exceeds a certain voltage threshold of concern. Usually that threshold of concern is a 300V rise or higher for zone modeled around the unit where some sort of isolation on telecommunication cables is needed if a metallic communication line should pass within the zone.

22.7.18 DG Plant Stability (Grid Parallel and Islanded)

An important parameter is how stable the DG plants of the microgrid will be when subjected to power system disturbances such as deep voltage sags. These disturbances could be the result of faults on distribution cables within the microgrid or faults originating on the distribution circuits external to the microgrid or out on the 115 kV transmission systems.

In grid parallel mode, it is beneficial if the microgrid generation can ride-through short duration disturbances, especially those associated with voltage sags due to faults on the 115 kV transmission system. This ride-through capability helps the plant continue to provide power system support during and immediately after a bulk power system fault has occurred and has been cleared by circuit breakers.

On the other hand, in islanded mode, stability is also important because if a cable section internally within the microgrid becomes faulted it is desirable for the nearest circuit breaker devices to sectionalize out the faulted section without the microgrid fully collapsing. For pickup of large loads (particular those with inrush or a large load block during grid assembly), it is essential to handle the largest permitted inrush/load steps without losing stability and collapsing.

Transient stability issues can occur following faults on the electrical system. During the fault, there is a power mismatch between the mechanical power delivered to a synchronous generator shaft and the electrical power that the system is able to accept from that generator. This power mismatch causes the shaft to accelerate, and as a result, the angle of the generator rotor can increase with respect to the mean angle of the power system. If that angle becomes too large, the generator will go out of step, and must be shut down. This in turn causes further issues, and can easily lead to a full shutdown of the microgrid. The rate of acceleration of a synchronous generator is governed by its inertia constant H. When there are a number of synchronous generators on a microgrid, the sum of the inertia constants of these generators governs acceleration of the group of generators, and the corresponding increase in frequency. On microgrids with significant levels of IBR, there is a lower percentage of synchronous generators relative to peak load, and the inertia of microgrid generation is reduced. This can have a negative impact on microgrid stability.

A screening criteria recommended by the Canadian Electrical Association (CEA) (Canadian Electrical Association, 1994 [12]) to help assess stability for DG installations. This criteria is based on the ratio of the available *utility-source short circuit MVA* at the DG connection point to the *rated MVA of the DG* (see Table 22.13).

For the Potsdam Microgrid project this ratio is calculated based on available fault level multiplied by nominal voltage at various points on the system. At the microgrid Feeder 51 tie point (near Clarkson) the utility short circuit MVA is about 70 MVA and the total aggregate rotational machine DG plant connected to that node is rated at about 10 MVA (excludes inverters and induction generators). The ratio between these two parameters is a factor of 7. But the "effective ratio" is closer to 10 when we consider that the various DG plants are operating at less than the sum of the full

Table 22.13 CEA stability screening criteria showing maximum recommended power angle swing angle for various ratios of utility system short circuit MVA to DG machine MVA.

Ratio of $MVA_{SCutility}/MVA_{rated-DG}$	Maximum recommended change in power angle ($\Delta\delta$ in degrees)	Stability condition
Ratio $\geq$ 20	50°	Less sensitive
5 $\leq$ Ratio < 20	35°	Moderately sensitive
3 $\leq$ Ratio < 5	25°	Most sensitive

nameplate power rating during grid-parallel operation. CEA guidelines suggest that for a ratio of between 5 and 20 rotor power angle swings up to $\Delta\delta = 35°$ can be tolerated (as measured starting from fault initiation to fault clearing) without losing stability.

A screening formula CEA recommends for calculating the increase in rotor power angle ($\Delta\delta$) for a sagged voltage condition is given by:

$$\Delta\delta \approx \frac{t^2 \cdot 21600 \cdot dP}{4H} \tag{22.6}$$

In the equation $\Delta\delta$ is the rotor power angle change (from fault initiation to fault clearing), t is the duration of the fault in seconds, dP is the change in generator power with respect to full rated apparent power of machine, and H is the inertia constant of machine. The factor 21,600 is a constant that takes into account various unit conversions and machine physics in order to provide an output ($\Delta\delta$) in "degrees."

Based on the machine parameters expected to be utilized for the microgrid generators, the CEA screening formula can be used to determine how long the machines will remain stable for a given fault condition on the system that causes a change in power (ΔP). Table 22.14 shows the critical clearing times of the faults needed to maintain stable operation at $\Delta\delta = 25$ and $35°$. As stated earlier, the maximum recommended change in power angle is $\Delta\delta = 35°$ according to the expected ratio of utility system short circuit MVA to the generator-rated MVA at the microgrid point of connection (dotted line in the table). However, to be more conservative for the Potsdam project, a $\Delta\delta = 25°$ or less limit is also shown (the fine dotted line in the table). If we utilize the $25°$ areas only, the generators can successfully ride through very deep voltage sags, causing a change in power (ΔP) of 90% and maintain stability if the voltage returns to near normal within about six cycles. It is also evident that the shallower voltage sags (such as those that cause ΔP of 50%, 20%, or only 10%) allow greater ride-through time before stability is lost. For example, with a shallow voltage sag that results in ΔP = 20%, there are about 13 cycles of ride-through before $\Delta\delta$ reaches $25°$.

Since this is a preliminary analysis and there are several unknowns that had to be assumed at this stage, these time durations should not be considered the final set of clearing times to use for the project. But they do give a guideline for the sort of clearing times that we will have to work with in the final more detailed design. Given the expected incoming transmission voltage sag durations and expected depth of sags, the allowable sag durations of the table show we can set the protection to ride through many of the shorter external disturbances that are most that are most critical for the bulk system ride-through. For an internal fault within the microgrid on the cables, the sag would be very deep and more difficult to ride through. But even in such a case, the numbers show that highspeed switchgear might be able to clear a faulted cable zone before the grid collapses. Whether this internal robustness is needed is a separate issue since internal faults would be rare.

In most cases, a complete transient stability study would be necessary as a part of the microgrid design process. As an example of a stability simulation, Fig. 22.25 shows an example simulation of a mild stability disturbance showing the power angle ($\Delta\delta$), the 60 Hz voltage waveform, and RMS voltage of that waveform. This results shows the impact of a 6-cycle duration mild voltage sag imposed on several megawatts of a single internal combustion engine (ICE) DG plant connected on a bus with impedance similar to the Clarkson Feeder 51 PCC. The initial machine power angle ($\Delta\delta$) is about $23°$ for the loaded machine and the change ($\Delta\delta$) in power angle remains well within the stability limit criteria.

Table 22.14 Stability analysis showing angle change of rotor for various fault durations (number of cycles needed for 25° and 35° is shown).

Cycles of Fault Duration	Δδ Rotor Angle in Degrees at Moment of Fault Clearing for the Indicated Voltage Sag Condition			
	Very Deep Voltage Sag to 10% of Nominal	Deep Voltage Sag to 50% of Nominal	Moderate Voltage Sag to 80% of Nominal	Mild Voltage Sags to 90% of Nominal
5	16.2	9.0	3.6	1.8
6	23.3	13.0	5.2	2.6
7	31.8	17.6	7.1	3.5
8	41.5	23.0	9.2	4.6
9	52.5	29.2	11.7	5.8
10	64.8	36.0	14.4	7.2
11	78.4	43.6	17.4	8.7
12	93.3	51.8	20.7	10.4
13	109.5	60.8	24.3	12.2
14	127.0	70.6	28.2	14.1
15	145.8	81.0	32.4	16.2
16	165.9	92.2	36.9	18.4
17	187.3	104.0	41.6	20.8
18	210.0	116.6	46.7	23.3
19	233.9	130.0	52.0	26.0
20	259.2	144.0	57.6	28.8
21	285.8	158.8	63.5	31.8
22	313.6	174.2	69.7	34.8
23	342.8	190.4	76.2	38.1
24	373.2	207.4	82.9	41.5
25	405.0	225.0	90.0	45.0
26	438.0	243.4	97.3	48.7
27	472.4	262.4	105.0	52.5

Extra conservative stability limit ($\Delta\delta < 25°$)

Screening guideline stability limit ($\Delta\delta < 35°$)

22.7.19 Energy Storage for Stability and Seamless Transition

There is interest in applying energy storage on this project for a number of different purposes. These include:

1. Enhancing frequency regulation
2. Improving stability
3. Mitigating voltage changes
4. Seamless transition to islanded status from the grid parallel mode.

Regarding frequency regulation, a relatively small amount of energy storage of no more than about 30 seconds up to a few minutes at 50–90% of the load step amount could be effective for the

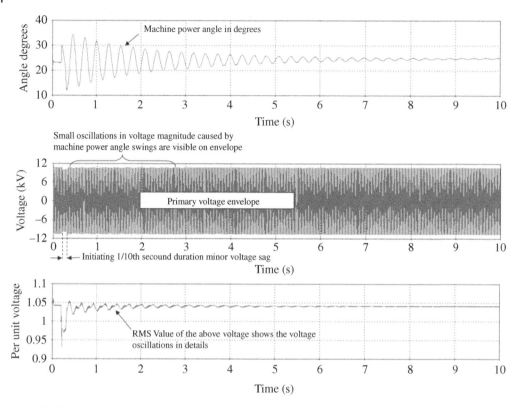

Figure 22.25 EMTP simulation of the stability response of an ICE generator to a mild voltage sag.

microgrid to mitigate changes in frequency due to loads steps and PV variations. This amount of energy storage would be particularly helpful during microgrid black start and assembly into a larger microgrid where significant load steps are occurring. For example, if we are looking to smooth the 2 MW load steps then the power rating of the device would need to between about 1–2 MW and it would need to have 30 seconds up to several minutes of energy storage deliverable at that power level.

Energy storage also could be used during load rejection events. A very fast acting storage device (with 1-2 cycle response time) could absorb excess energy and stabilize the system frequency so as to improve voltage and dynamic stability. When a synchronous generator suddenly loses its connection to all load, it can experience its upper frequency limit in as little as 0.4 seconds. A fast acting load absorption device with a rating near that of the generator could absorb most of the excess energy and slow down the frequency acceleration to such a degree that the governor would have time to reduce throttle and prevent the over frequency condition.

Regarding voltage changes and flicker due to typical loading steps, while energy storage can reduce these, it is usually more cost effective to use dynamic reactive power control rather than energy storage to mitigate voltage issues such as flicker. In microgrids with a significant level of IBRs that meet IEEE Std. 1547-2018 smart inverter requirements, this can be done with very little additional cost.

Perhaps the most demanding of the energy storage applications for microgrids is to do a seamless transition from grid parallel to islanded state. This requires a very large amount of energy storage (kWh) and storage power capacity (kW) such that it would not likely be practical on this project.

Such a device would likely cost millions of dollars at current market prices for a microgrid with about 10 MW of peak load. For a seamless transition in the event unexpedted loss of the power grid, this microgrid would have up to 10 MW of loading that needs to be picked up *instantly* and it may require 5–30 minutes of support at that level to make all the operating decisions and get all the generation available and ready to go.

Whether the storage needed is 5 minutes or 30 minutes would depend on how automated the microgrid generation start-up and switching process is and the various human response time and decision-making factors too. If it was highly automated and reliable perhaps only 5 minutes of storage at about 10 MW is needed for the process. A more manually oriented switching process that involves manual decision making could require up to 30 minutes of storage. Table 22.15 summarizes some of the storage applications we have discussed.

The seamless transition concept would involve a configuration like that shown in Fig. 22.26 for the Potsdam Microgrid. To do seamless transfer with the storage requires a large capacity storage device with fast switchgear (or static switch) that can deliver the expected generation to load mismatch at the time of the separation to microgrid mode. If we assume there is no generation running at the time and the system is at peak load then somewhere on the order of 10 MW or a bit more might be needed depending on speed of load shedding. As stated earlier, up to 30 minutes storage at that power level might be required if generation is expected to be slow to dispatch. On the other hand, as little as 5 minutes is possible with a highly automated dispatch and control system. ICE generators (if they are preheated and pre-lubricated) can be brought on line really fast (within 10 seconds or so), but 5 minutes is still a good margin of storage to have to give flexibility in dealing with reclosing and operating dispatch decisions, etc.

Regarding the term 'seamless transition' there are several degrees of seamless transition that could be considered depending on how much money is to be spent and how much complexity and power quality we want. These are illustrated in Table 22.16.

Table 22.15 Some energy storage applications for microgrids.

Type of energy storage application	Amount of power capacity recommended	Duration of storage needed at the recommended power level[a]
Enhance frequency regulation under mild disturbance conditions	50–90% of the load step of concern	30 seconds to a few minutes
Improve load rejection voltage and generator sag stability	25–75% of real kilowatts load on the island	5 seconds to 30 seconds
Mitigate voltage changes	Use reactive power instead *(although real power can help in low X/R ratio feed point applications)*	
Seamless transition	110% of the peak demand of load to be carried at the time seamless transfer is desired	5–30 minutes

a) The actual energy required is equal to the product of the time shown in this column multiplied by the power capacity needed in the adjacent column. However, that is for ***ideal*** storage devices with no limitations on discharge rate or charging rate and no internal losses. For nonideal devices, like batteries, losses can be high at high rates and the amount of energy (kWh) may be considerably higher to reach a stated power level due heating limits and other factors.

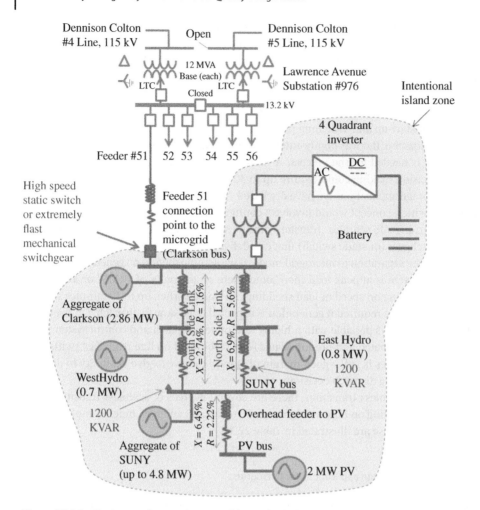

Figure 22.26 Equipment for seamless transition to intentionally islanded microgrid mode. (NYSERDA Report 18-13, 2018).

Table 22.16 Different grades and reliabilities of seamless transfer.

Type of seamless transfer	Waveform condition that gets through to the microgrid	Equipment required
Pure seamless transition	A minor switching perturbation 1/16th to 1/4th of a cycle duration.	13.2 kV rated high power static witch (SCRs or IGBT Energy Storage with 4-Quadrant inverter Step-up transformer Controls and ancillaries
Quasi-Seamless Transition	Up to three cycles of very deep voltage sag conditions or outage conditions (but DG if running can ride though) per stability analysis.	13.2 kV, three cycle total clearing time high speed mechanical switchgear Energy Storage with 4-Quadrant inverter Step-up transformer Controls and ancillaries.
Quasi-Seamless Transition Relying on High speed Load Shedding without Storage	Three cycles of deep voltage sag condition but risk of system collapse if generation to load mismatch not quickly corrected.	13.2 kV, three cycle total clearing time high speed mechanical switchgear High speed load shed controls and ancillaries.

The first approach in the table, the "pure seamless transition" would give the best waveform and employ a solid-state high-speed switch (often referred to as a static switch). It uses solid state switching elements such as SCR or IGBT devices, it would need to be rated at 13.2 kV and handle at least the full 10 MW power loading at the tie point. It also would need high speed controls. There would be a 10 MW four quadrant inverter with energy storage and related balance of system elements. Such a scheme could transition from grid parallel mode to islanded mode almost instantly with only a very tiny perturbation of the waveform (perhaps 1/16th to 1/4th of a cycle at most). The big drawbacks here are cost, complexity and equipment availability. The whole package would likely cost many millions of dollars – perhaps even $10 million considering development efforts. A big drawback is availability of components such as the 13.2 kV static switches. And there is a risk of nuisance transfer events that fail to successfully transition owing to the complexity of the technology.

Another approach (the second in the table) that would be less costly is to go with a conventional high speed switchgear device to do the isolation. This *Quasi-Seamless Transition* approach would be slower than the static switch but far less costly and more readily available. The drawback is that it would let through up to about three cycles of very deep voltage sag in the worst case. However, based on the stability discussion earlier in the report the generation if on line at the time could ride through that from a stability perspective. Some sensitive loads would drop out during the disturbance and the lights would visibly blink. The energy storage would pick-up any load mismatch or the full load if no generation was running at the time of the event. This option would still be very expensive – perhaps millions of dollars - but would likely cost perhaps 50% of the pure seamless transition.

The final option on the table is generally the lowest cost approach. It is to use high speed switchgear and a high-speed load-shedding scheme to balance load to generation. There is no energy storage used in this option. This scheme would always require some or much of the microgrid generation to be running and carrying the amount of the load that is desired to be sustained when the quasi-seamless transition occurs. As an example, for 4 MW of critical load, at least 4 MW of generation would always need to be running on the microgrid during grid parallel mode. For this approach, when an upstream outage occurs, the switchgear opens within 3 cycles and if load was greater than 4 MW, the high-speed load-shedding scheme begins tripping off loads in order of least critical to most critical to match the load to generation level. The load shed scheme must be fast enough to have the mismatched loading condition cleared up before the running generation goes unstable or trips off due to under frequency, etc. The tripping process of the loads is very fast and occurs within a few cycles of the event. This type of high-speed load-shedding scheme has been used successfully in many industrial facilities around the world. The cost is much less than the other approaches mentioned owing to the fact that there is not storage and nor is there a static switch. However, the big drawback is that the scheme forces a certain amount of generation to always be operating and so confines the project into a set of operating criteria that are not that may not be flexible or economic from a generation dispatch perspective.

22.8 Conclusions and Recommendations

This chapter discusses the range of power quality and dynamics factors associated with the design and development of multimegawatt primary voltage connected microgrids. The chapter includes case study examples related to the proposed Potsdam Microgrid, and multiparty resilient microgrid intended to serve critical service providers in Potsdam, NY in the event of a major disaster resulting in the long term unavailability of the bulk power grid. Both modes (grid parallel and islanded operation) of microgrid operation are covered.

Topics cover include:

1. Voltage regulation
2. Frequency regulation
3. Voltage flicker and short duration excursions above or below ANSI limits
4. Effective grounding and temporary overvoltage
5. Harmonic distortion
6. Transient stability
7. Phase current unbalance
8. Energy storage options
9. Seamless transition options
10. Black start procedures
11. Unintentional islanding

Several examples are included, that involve techniques suitable for initial scoping studies. Issues that must be resolved with more detailed computer based analysis and simulations are discussed.

22.9 Exercises

1. Example 22.1 predicts the voltage sensitivity of a microgrid during grid parallel operation. The example uses the existing impedance at the point of common coupling, as shown in Fig. 22.8.
   (a) Repeat this example using the reduced impedance value shown in Fig. 22.8.
   (b) Comment on any advantages/disadvantages of this reduced impedance option.
2. Example 22.2 calculates that at 2.4 MVA capacitor bank is needed to support the microgrid voltage during grid parallel operation.
   (a) Repeat this example using the reduced impedance value shown in Fig. 22.8.
   (b) Comment on any advantages/disadvantages of this reduced impedance option.
3. The design goal for the project includes a limit of 0.01 p.u. change in voltage due to capacitor switching. To achieve this goal, a capacitor installation could be installed with several subsystems that could be switched separately.
   (a) From the result of Example 2.2, determine if the 2.4 MVA capacitor size would need to be installed in subunits in order to meet the design goal. If this is the case, determined the number of subunits and the size of each subunit that would meet the requirement.
   (b) From the result of Exercise 2.2, determine if the capacitor required in this case would need to be installed in subunits in order to meet the design goal. If this is the case, determined the number of subunits and the size of each subunit that would meet the requirement.
4. In Example 22.3, Eq. (22.5) is used to predict a value of $P_{st} = 0.20$ for the PV array connected to the microgrid. This value assumes that the PV is operating at a power factor at 0.9.
   (a) Estimate the largest PV array that could be connected without exceeded a P_{st} of 0.35. Assume operation at 0.9 pf.
   (b) If the PV array were to always operated at unity power factor, could the PV array size be increased? If so, what would be the maximum size permitted?
   (c) As shown in Fig. 22.10, the PV bus is located some distance away from the SUNY bus. Comment on the issues that could arise if $P_{st} > 0.35$, for both the PV array and the customers at SUNY bus.

22.10 Acknowledgment

The authors gratefully acknowledge the support of the New York State Energy Research and Development Authority (NYSERDA) and National Grid for their support of the Potsdam microgrid project through NYSERDA project 41309.

References

1 NYSERDA Report 18-13 (2018). *Design of a Resilient Underground Microgrid in Potsdam, NY*. Albany, NY: New York State Energy Research and Development Authority.

2 ANSI C84.1-2020 (2020). *American National Standard for Electric Power Systems and Equipment - Voltage Ratings (60Hz)*.

3 IEC 61000-4-15:2010 (2010). *Electromagnetic Compatability (EMC) Part 4-15: Testing and measurement techniques-Flickermeter-Functional design specifications*. International Electrotechnical Commission.

4 IEEE 1453-2015 (2015). *IEEE Recommended Practice for the Analysis of Fluctuating Installations on Power Systems*. IEEE.

5 Information Technology Industry Council (1997). ITI (CBEMA) Curve Application Note.

6 IEEE C62.92 (2016). *IEEE Guide for the Application of Neutral Grounding in Electrical Utility Systems*. IEEE

7 IEEE 1547-2018 (2018). *IEEE Standard for Interconnection and Interoperability of Distributed Energy Resources with Associated Electric Power System Interfaces*. IEEE.

8 SAND2020-0266 (2020). *Momentary Cessation: Improving Dynamic Performance and Modeling of Utility-Scale INverter Based Resources During Grid Disturbances*. Sandia National Laboratories.

9 North American Electric Reliability Corporation (2018). *Reliability Guideline BPS-Connected Inverter-Based Resource Performance*.

10 IEEE Std. 519-2014 (2014). *IEEE Recommended Practice and Requirements for Harmonic Control in Electric Power Systems*. IEEE.

11 IEEE Power System Relaying Committee (2011). *IEEE Tutorial on Synchronous Machine Protection 2011 Edition*.

12 CEA128-D-767 (1994). *Connecting Small Generators to Utility Distribution Systems*. Canadian Electrical Association.

23

A Time of Energy Transition at Princeton University

Edward T. Borer, Jr.

23.1 Introduction

We are at a singular moment of change in the history of energy, one being shaped by the urgency of climate change combined with a national crisis of aging infrastructure.

Since before written history, humans have burned dry plant material to produce heat and light. At least since the mid-1700s, when James Watt refined the steam engine, the delivery of most forms of energy involved burning coal. We burned coal to heat water to make steam. Using steam, we delivered heat to buildings or used the expansion of steam to turn a shaft that could power machines, locomotives, ships, and eventually electric generators.

In the 1860s, Princeton University and many cities and peer institutions established central steam plants that used district heating systems to serve a collection of buildings. Steam was generated in a boiler house and distributed via a network of underground pipes to deliver heat to many buildings. Vestiges of the original systems remain in service today. In the 1890s, district heating systems began adding steam turbine-driven dynamos between the boiler and the buildings in order to produce electricity along with heat. These were some of the first combined heat and power (CHP) systems.

By the early 1900s, most transportation power relied on internal combustion engines. Highly-refined liquid fuels with their greater energy density, better controllability, and more consistent properties gained preference over natural solid fuels such as wood or coal. Nonetheless, moving people and stuff by air, by land, or by sea involved combustion.

During the twentieth century, utility companies grew and large regional power grids were established. Electric utilities were able to take advantage of the efficiency of scale, bulk fuel purchasing, diversity of location, and redundancy of equipment. They were able to produce power more reliably and at a lower levelized cost of energy (LCOE) than single grid-independent CHP systems. So, many of the earliest stand-alone CHP systems were eventually retired.

In 1902, Willis Carrier invented the modern air conditioner. Two world wars and the Great Depression delayed its wide adoption, and it was not until the early 1960s that Princeton University built its "refrigeration plant," establishing a campus district cooling system. Like steam, chilled water is piped underground from a central plant to each building where it is used to remove heat. At the time of construction, Princeton's boiler house pumps, fans, and all the chillers in the original refrigeration plant were powered by steam turbines, not electric motors. And behind that steam was coal combustion. During the 1960s, the boilers were converted to burn natural gas (Fig. 23.1).

Microgrids: Theory and Practice, First Edition. Edited by Peng Zhang.
© 2024 The Institute of Electrical and Electronics Engineers, Inc. Published 2024 by John Wiley & Sons, Inc.

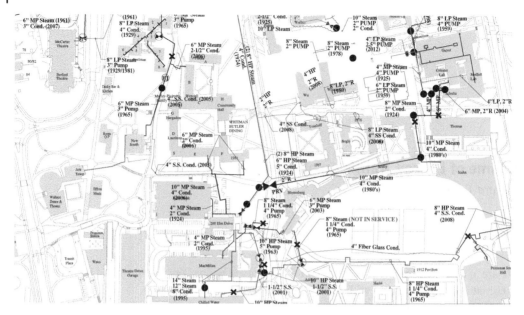

Figure 23.1 A schematic representation showing parts of Princeton University's district steam distribution system.

23.2 Cogeneration

By the late 1980s, the price of electricity was rising faster than the cost of natural gas. Once again it became financially attractive to produce electricity on site. Many district energy systems seeking high reliability and motivated by this increasing "spark spread," reestablished or added on-site power generation while remaining connected to the larger utility grid. These would be the first modern era microgrids. They normally operated synchronized to the regional utility grid. Many had the ability to sell excess energy into the grid, or provide "grid services" such as frequency regulation. Some included renewable energy sources, batteries, or thermal storage. These microgrids were also able to operate autonomously from the grid during regional emergencies. Most of these remain in operation today (Fig. 23.2).

Since district heating systems already had a demand for heat, the exhaust from gas turbines or reciprocating internal combustion engines could be used to make a valuable product. Like most CHP systems, Princeton's engine size was selected to match the minimum heat demand of the campus. The utility power grid could normally absorb any excess electric generation or make up for any electric deficiency. Even though their *engines* had only 30–45% efficiency, the combined production of heat and power (CHP) resulted in *system* efficiencies of 65–90%. This efficiency rating was better than that of electric-only utility companies that did not have thermal customers and therefore needed to dump low-grade heat back into the environment. Typically, central utility power plants reject about 1/3 of the fuel input energy through a chimney to the atmosphere and reject 1/3 to a lake, river, or ocean, or sometimes by evaporating water through cooling towers. Only about a third of their fuel energy is delivered to the power grid as useful electricity (Fig. 23.3).

Princeton's cogeneration system is comprised of a single gas turbine followed by a heat recovery steam generator (HRSG). Princeton uses the General Electric LM-1600 gas turbine to turn a

Benefits of micrigrids

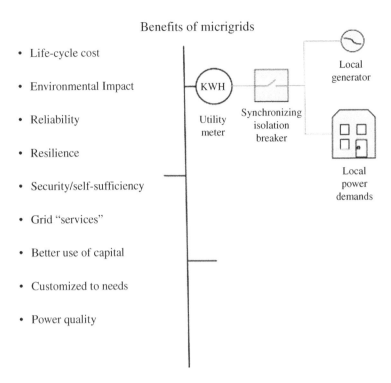

- Life-cycle cost

- Environmental Impact

- Reliability

- Resilience

- Security/self-sufficiency

- Grid "services"

- Better use of capital

- Customized to needs

- Power quality

Figure 23.2 Some key benefits of microgrids include a lower levelized cost of energy (LCOE) and self-sufficiency in emergencies.

Combined heat and power, "cogeneration"

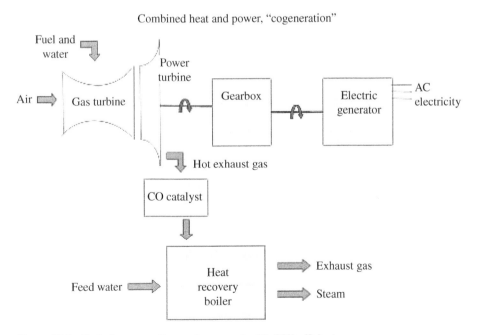

Figure 23.3 Typical cogeneration systems can be 75–85% efficient.

generator to produce electricity. Hot exhaust gas exits the engine at about 1385F. Exhaust gas passes through the HRSG giving up its heat to boil water. Water in the HRSG turns to steam. Exhaust gas leaving the HRSG is intentionally maintained above its dew point – to avoid acidic condensation which would be damaging. The exhaust temperature has dropped to about 200F when it leaves the chimney.

Because the local utility grid is highly reliable and not all electric users are "mission-critical," Princeton does not require 100% redundancy in on-site power generation. What's mission-critical? During emergency conditions, it is important to keep the energy plant, public safety, and health services running; to provide warm, dry places to eat and sleep; and to keep air exchanges, temperatures, and humidity stable in research laboratories. But, in a regional emergency, it is reasonable to shut off optional activities like clothing shops, performance halls, and athletic spaces.

Peak campus electrical demand is 27 MW. Princeton's gas turbine can produce up to 15 MW. During regional power outages such as Superstorm Sandy in 2012, some campus loads were curtailed so that the remaining electrical demands could be met within the gas turbine's capacity.

When the gas turbine or HRSG is out of service for maintenance, steam is produced by one or both Auxiliary Boilers located in the same facility as the cogeneration system. The auxiliary boilers can burn natural gas or #2 ultra low sulfur diesel (ULSD). They have a fuel-to-steam efficiency of approximately 78%.

For reliability, Princeton's campus microgrid runs synchronized with the local electric grid whenever the grid is available. Electric energy that is not supplied to the campus by cogeneration is provided by the on-campus solar arrays and by the local power utility. At times when the university has excess electricity, it is sold into the grid at a wholesale rate called the "locational marginal price" (LMP).

About the exercises:

Exercises in this chapter are to be worked sequentially as listed. They are intended to build on one another to help the reader understand the basis for converting Princeton's campus from combustion-based steam systems to heat-pump-based hot water systems with geoexchange. Understanding the method of calculating the answers is at least as important as the answers themselves.

There is a companion spreadsheet for this chapter. It includes solutions to all exercises and is available online through the book publisher.

"Efficiency" is energy output divided by energy input expressed as a percent.
To calculate efficiency, engineering units must match. Calculating the efficiency of an engine requires converting output energy and input energy to common units and then dividing the two. Engine efficiency can vary depending on load and environmental conditions. To understand real-world operating efficiency, it is helpful to total all energy output over a multi-year period and divide by all energy input over the same period. Two years of hourly operating data is available for the calculations in this chapter.

Similarly, to determine the efficiency of a combined heat and power (CHP, combined cycle, or cogeneration) system, it is necessary to convert both the heat output and the power output to common engineering units, add these together, and then divide by all energy input streams.

Exercise #1a

What is the average efficiency of Princeton's gas turbine?

A: The gas turbine ran at 30% average efficiency during the 2-year data period.

Exercise #1b

Gas turbine efficiency is sensitive to inlet air temperature and total power output. In the utility industry, it is common to refer to a generating system's "heat rate." Heat rate is the number of Btus required to produce 1 KWH of electricity. Heat rate is another way to describe a system's efficiency. Higher heat rates indicate a less efficient system. Plot the heat rate of Princeton's gas turbine against its power output in kilowatt. What is the typical kilowatt output range at which Princeton's turbine operates? At what power output is it most efficient?" What is its efficiency range?

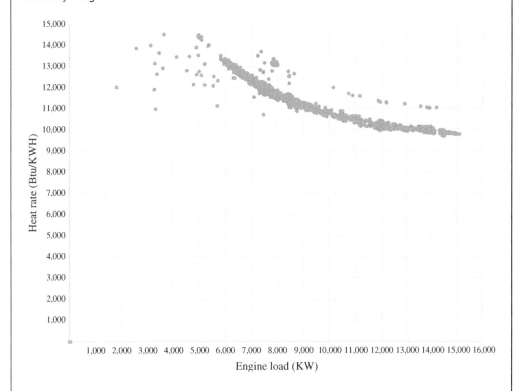

A: Princeton's gas turbine efficiency improves with increasing load. It is most efficient at full power and least efficient at low power. This is a second-order relationship, not linear. Princeton's engine is normally operated between 6000 KW and 15,000 KW. The engine's lowest heat rate is about 9800 Btu/KWH at 15,000 KW output, i.e. the engine's full-load efficiency is about 35%. Engine efficiency drops to about 25% at 6000 KW output.

Exercise #1c

How efficient is Princeton's Cogeneration system (electricity produced + HRSG steam) / (input energy)?

A: 69% average combined heat and power efficiency.

(Continued)

(Continued)

Exercise #1d

Much of the district steam system's insulation has become degraded in the century since it was first installed. If the district steam system loses 20% of the steam's energy before heat gets delivered to buildings, what is the combined steam *production and delivery* efficiency?

A: 55%.

Exercise #1e

What % of the Cogeneration System's input fuel energy is lost before heat or power leave the plant? Where did it go?

A: 31% of the input energy was lost to the environment; mostly up the exhaust stack as 200F exhaust gas.

Exercise #1f

Where do the distribution system losses go?

A: Heat lost from the distribution system is dissipated in the ground surrounding the steam distribution pipes.

Exercise #1g

What percentage of Princeton's annual steam needs were met by the cogeneration system versus the auxiliary boilers?

A: 65% of the campus's steam need was met by cogeneration. The balance of campus steam was produced by auxiliary boilers.

Exercise #1h

What percentage of Princeton's annual electric needs was met by the cogeneration system?

A: 37% of the campus electrical need was met by cogeneration.

The rest was met by on-site solar photovoltaics and the local utility grid.

Today we use jet engines with liquid or gaseous fuels and digital controls in district energy systems, but we are still burning stuff to make heat and power – just like we did in the 1890s. It has become painfully obvious that excessive combustion of carbon-based fuels around the world is causing disastrous changes to the environment. And that's unsustainable. To curb climate change, we need to stop burning stuff. Now.

23.3   The Magic of The Refrigeration Cycle

Let us hope for a moment, into the "way-back machine" and have a chat with Carrier, father of the air conditioner. There was something magical about the process he refined over a century ago, and as we travel back in time, he tells us that unlike boilers, gas turbines, and diesel engines – that *convert* energy from one form to another – his refrigeration machine *moves* energy from one place to another. We know that anytime we convert energy from one form to another, some of that energy is lost – usually dissipated as unproductive heat and/or sound. A boiler converts a fuel's chemical energy to heat and then transfers a portion of that heat to water to generate steam. Some of that heat is lost up the chimney. So, we talk about boiler efficiency, turbine efficiency, power-train efficiency, and generator efficiency – and all of these efficiencies are less than 100% because less energy is delivered in the form of useful heat and power than was input as fuel (Figs. 23.4 and 23.5).

But a refrigeration cycle, as it *moves* energy, also *concentrates* energy at a higher temperature. Refrigerators, air conditioners, industrial chillers, and heat pumps all use the same basic refrigeration cycle. One unit of energy put into a refrigerator compressor can move about 3 units of heat out of the cold box where the food is stored. Moving 3 units of heat energy out of a refrigerator with just 1 unit of energy input yields a "coefficient of performance" (COP) of 3. Refrigerators,

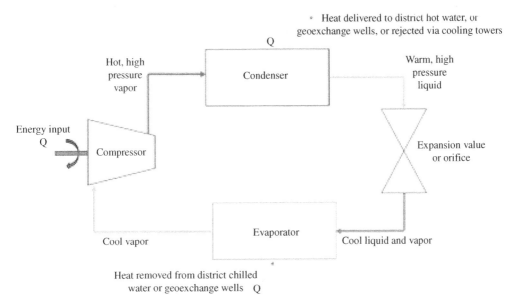

Figure 23.4   The basic vapor-compression refrigeration cycle.

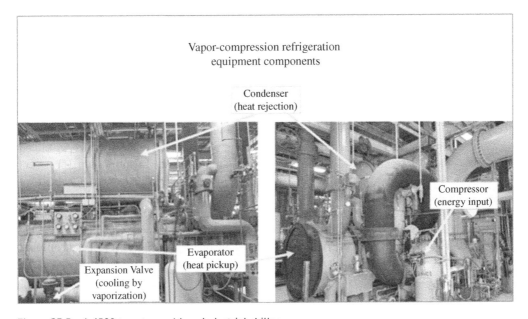

Figure 23.5   A 4500-ton steam-driven industrial chiller.

air conditioners, and industrial chillers all have a COP of greater than 1. Typically, their COPs are between 2 and 4. Some refrigeration cycles can exceed COPs of 7. They each remove heat from a cold place and reject it to a warmer place at a higher temperature. But where is that warmer place? A refrigerator pulls 3 units of heat from the food box and rejects that heat plus 1 unit of input energy to the kitchen. That is, 4 units of heat are dumped into the kitchen for every 3 units removed from the food. Carrier's air conditioner does the same thing by removing 3 units of heat from his family room and dumping that heat plus the input energy to a warmer outdoor environment. Princeton's modern industrial chillers pull heat from a campus chilled water loop and reject it into the atmosphere by evaporating water through cooling towers. In the future, Princeton will capture the energy that is currently being rejected, and use it to heat the campus.

Coefficients of Performance (COPs) are energy output divided by energy input expressed as a number

Exercise #2a
How efficient (what is the COP) of Princeton's Electric-Driven Chillers?
  A: Princeton's electric chiller's COP averaged 7.5.

Exercise #2b
Exercise #2a ignores other parasitic energy demands from the cooling system.
  In addition to the chiller energy input, energy is used by the cooling tower fans and cooling tower pumps. Assume that the cooling towers and pumps consume an additional 40% more electricity. What is the overall COP of the combined system of electric chillers, cooling towers, and pumps?
  A: The combined COP of the electric chillers, cooling towers, and cooling tower pumps was 5.4; i.e. approximately 5.4 times as much heat energy was moved out of campus buildings as was input to the central plant cooling system in the form of electricity. Additional pump and fan energy is consumed in the buildings to transfer a building's heat into the circulating chilled water.

23.4 Capturing Heat, Not Wasting It

When we install a refrigerator or air conditioner, we only want to keep the food or the family room cool. We do not intend to heat up the kitchen or add heat to the outdoors. Those are collateral effects. Wouldn not it make better sense to use that extracted heat in a productive way?

What if – after we have reboarded the way-back machine and returned to the present – we pulled the copper coils away from the back of our refrigerator and immersed them in a tank of water? The refrigerator would reject its heat to the water tank instead of the air in the kitchen. If we pumped domestic water through that tank the refrigerator could pre-warm the water and reduce the amount of input energy needed to make hot water for our house. I am not aware of anyone building such a system, but at Princeton, we are doing that on a campus-wide scale. The core "heat pump facilities" will be built and operational by the middle of 2023. It may take a decade or more to hook up all 200 buildings, but the bulk of our existing heating demands will be connected in the first 5 years (Fig. 23.6).

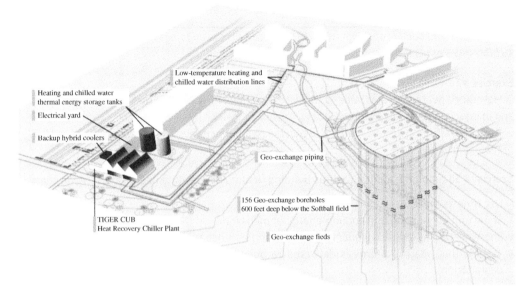

Figure 23.6 Princeton's district energy expansion project will include the use of bore holes situated beneath a softball field, new buildings, and other campus spaces. The illustration here captures the essence of the geoexchange component. Source: Courtesy of Introba.

Princeton needs heating and cooling 24/7/365 on our campus. In the winter, buildings are heated with steam. In the summer, they are cooled with chilled water. But, even on the coldest winter days, we still need to remove heat from the campus to cool lasers, electron microscopes, computed tomography (CT) scan machines, computer facilities, and some building cores. And, even on the hottest summer days, we need to deliver heat to the campus for domestic hot water, dishwashing, sterilizers, autoclaves, cage washers, and air-conditioning reheat for humidity control and comfort. There is a sizable need for simultaneous heating and cooling all year.

Annual energy data can inform us about how much energy is needed at peak moments, i.e. what equipment capacity is required. The data can also inform us about what seasonal storage capacity would be helpful.

Exercise #3a
How much energy was produced to heat the campus for a year? Express this in "pounds of steam" and convert to megawatt-hours. Assume 950 Btus can be delivered to the conditioned space for every pound of steam.

A: 588,240,373 pounds of steam were produced to heat the campus for an average year. This equates to 163,783 MWH of heat sent from the plant to the campus.

Exercise #3b
If the campus steam system insulation is severely degraded and loses 20% of the heat that's produced by the plant, how much energy did the campus buildings actually receive?

(Continued)

(Continued)

A: 470,592,299 pounds of steam were delivered to heat the buildings for a year. This equates to 131,027 MWH of heat supply.

Exercise #3c

How much heat removal is needed to cool the campus for a year? Express this in "ton-hours" and in equivalent megawatt-hours.

A: 34,835,489 ton-hours are required to cool the campus for a year. This equates to 122,516 MWH of cooling (heat removal).

Exercise #3d Graph the annual campus building heating and cooling needs on the same chart.

- Use time as the *x*-xis.
- Show units of heating and cooling energy in megawatt.
- Show heating as a positive quantity, and cooling as a negative quantity.
- Assume that heating energy is delivered to buildings without today's 20% losses due to degradation.

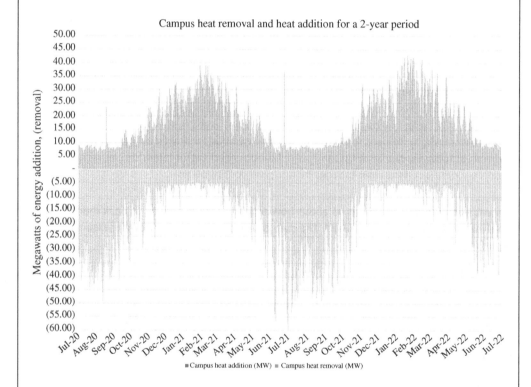

What is the peak heating need?
A: The campus buildings have a peak heating need of approximately 42 MW in mid-winter.
What is the peak cooling need?
A: The campus buildings have a peak cooling need (heat removal) of approximately 60 MW in mid-summer.

How much "simultaneous heating and cooling" is there? I.e., how much heating energy could be produced by the heat that is concurrently being removed in the process of cooling the campus?

A: Approximately 5 MW of simultaneous heating and cooling exists all year. This could be met by a heat pump (equivalent to about 1400 tons) running continuously without any energy storage.

Exercise #3e

Assume Princeton's annual cooling requirement, determined above, was produced with heat pumps that have a net cooling COP of 5.4, i.e. 1 unit of energy input can remove 5.4 units of heat. How much input energy is required (in megawatt-hours).

A: Using a heat pump system with a cooling COP of 5.4, it would take about 22,688 MWH to move 122,516 MWH of heat out of campus buildings.

How much heat energy is available to be stored for reuse (megawatt-hours)? I.e., how much heat will be removed from the campus in an entire summer? Add this to the input energy used by the heat pump.

A: Add the summer heat removed from campus to the energy input of the heat pump. This results in 145,205 MWH available for storage and later re-use. This is 11% more heat than the amount of heat energy required by the campus buildings in winter.

Princeton is actively converting each campus building from steam to hot water heating. We are installing a new district hot water system, at the heart of which will be heat pumps akin to our existing industrial chillers, except that instead of rejecting heat to the cooling tower at about 90°F (32°C) the heat pumps are designed to reject heat at up to 160°F, which is far hotter than we need to heat buildings or to heat domestic hot water. Heat pumps are a lot like industrial chillers, except that they operate at much higher condenser temperatures. Our existing chiller condenser temperatures are about 90 Fahrenheit. Our new heat pumps will be designed to operate at up to 160 Fahrenheit. This difference is part of what will enable us to use heat pumps for campus heating rather than having to waste the heat via a cooling tower. Put another way, a heat pump "rejects" heat. But it does so in a productive way by diverting it into a campus district hot water system.

The real magic of heat pumps is that while moving 3–4 units of heat energy out of a chilled water system, we can simultaneously deliver 4 or 5 units of heat energy to a hot water system – all with only 1 unit of energy input. Using electric heat pumps eliminates unnecessary combustion in boilers and eliminates the need to jettison waste heat via cooling towers. The heat pump removes 2–3 units on the cooling side and delivers 3–4 on the heating side. Including pumping energy, we anticipate a net annual COP of about 5.0. That compares very favorably to our current cogeneration system with its less impressive annual efficiency of 0.69! (Figs. 23.7 and 23.8).

By simultaneously producing heating and cooling energy and by converting buildings from steam to hot water we will be able to move heating and cooling to the places it is needed with about 20% the amount of total input energy we use today with separate heating and cooling systems.

23.5 Multiple Forms of Energy Storage

Of course, like most communities, Princeton requires more heat at night, more cooling during the day; more heat in the winter, and more cooling in the summer. This is where thermal storage and geoexchange come into play.

Separate heat removal (CHW) and addition (Steam)

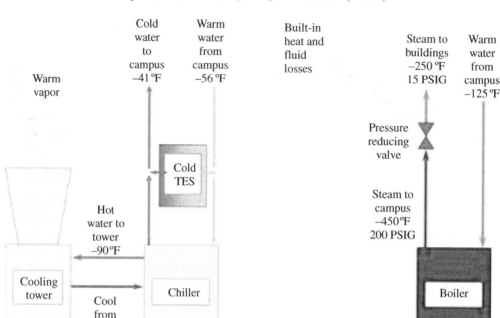

Figure 23.7 Currently, Princeton is producing steam to deliver heat to campus and concurrently producing chilled water to remove heat from campus and reject it to the environment by using cooling towers.

Any kind of energy storage offers at least two benefits. Energy storage separates the *time* of energy production from the time of energy use. Energy storage also decouples the *rate* of energy production from the rate of energy use. A cell phone battery may take 6 hours to charge overnight. But, it may be used for 18 or more hours the next day. An ice cube may take several hours to freeze on 1 day. But, it may melt in a drink the next day in a span of minutes. A hot water solar panel may take all day to heat its water storage tank. But the heated water may not be used to take a shower or wash clothes until a day or so later, and it may be used much faster than it was produced (Fig. 23.9).

Today we have one form of community-scale thermal storage on Princeton's campus. It is a 2.6 million-gallon tank of water. It separates the time of water cooling from the time of cool water delivery to campus. We can purchase electricity when it is very inexpensive at night. With that electricity, we can run electric chillers to cool 2.6 million gallons of water. Later in the day when electricity prices are higher, we can shut off the electric chillers and electric cooling tower pumps and fans. We can pump chilled water through the tank to cool the campus at a much lower cost than if we had run the chillers on a hot afternoon when electricity prices can be many times higher. In a volatile electric market, the addition of Chilled Water Thermal Storage can have payback periods as short as 3 or 4 years.

For commercial and industrial electricity customers who pay "real time" electric rates, there is always a high price and a low price every day. So, there is always an opportunity to purchase energy inexpensively and deliver the value from that energy when it would cost much more. The key is to pay attention to the electric market dynamics and *predictively* dispatch the chillers and

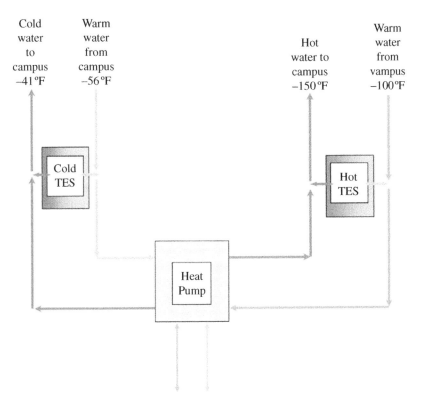

Figure 23.8 Princeton is moving toward a more efficient district energy system that will capture heat from the cooling process and reuse it on campus immediately or store heat below campus for use later in the season.

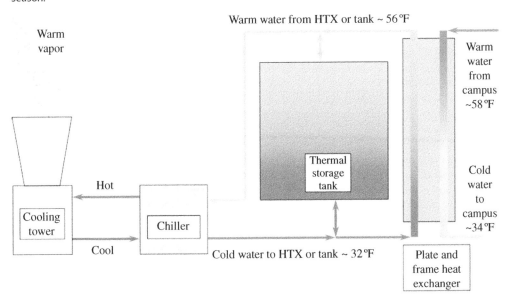

Figure 23.9 Any form of energy storage can be used to separate the time of production from the time of consumption and separate the rate of production from the rate of consumption. Thermal energy storage can be used to save energy cost, improve reliability of service, reduce power grid stress, and reduce stress for the plant operators.

thermal storage for greatest advantage. For these customers, thermal energy storage presents a savings opportunity every day of the year.

The main components of Princeton's "High Tension Service" electric bill are:

1. **Energy** (Locational Marginal Price US$/MWH), this primarily relates to the utility's fuel cost.
2. **Delivery** (about US$24/MWH added onto the LMP), this primarily relates to the utility's transportation (wires) maintenance cost.
3. **Capacity** (demand concurrent with the grid's 5 peak hours each year, US$/MW). This primarily relates to the utility's power generation and distribution capital cost (Fig. 23.10).

The mean energy price is about US$30/MWH. On normal days, the LMP may rise to US$100/MWH and drop to US$20/MWH. But, there are some unusual times when the grid will *pay* customers to take power, especially between 2:00 AM and 9:00 AM. These moments are rare. They usually precede extremely hot or cold weekdays. Many generators want to get started early on those days to capture the high prices predicted later in the day, so it forces the LMP down early in the morning before the grid demand rises.

The peak LMP usually remains below US$200/MWH. But, there are times when it can exceed US$1000/MWH – thirty times the mean price! These are quite rare. But, it is critical to predict them in advance and avoid buying power at those times. These are almost always between 4:00 PM and 7:00 PM on weekdays when the grid hits maximum demand.

Princeton's annual Capacity cost is set by the campus's demand (in Megawatts) during the 5 highest grid-demand hours of the year. Capacity currently costs about US$200,000 per megawatt per year. With the potential to use up to 27 MW, Princeton is careful to avoid this. Predicting and avoiding purchase of power during peak grid demand hours is critical. Princeton combines the following to avoid buying extremely high-cost energy and keep the capacity tag close to zero:

1. Run the gas turbine at full output.
2. Take advantage of whatever solar PV is available, though it drops off steeply that late in the day.
3. Shut off all electric-driven chillers and associated cooling towers.

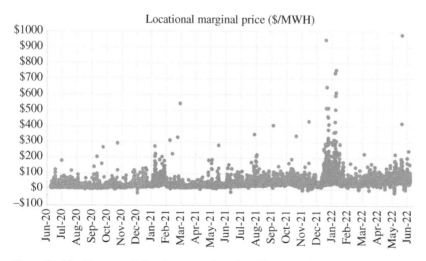

Figure 23.10 The cost of electric energy (Locational Marginal Price) varies throughout the year for commercial and industrial customers. Holidays, weekends, and days with moderate weather tend to be least expensive because less energy is needed to heat and cool buildings. Extremely hot and cold weekdays tend to be most expensive because that is when energy is most in demand. This incentivizes energy storage.

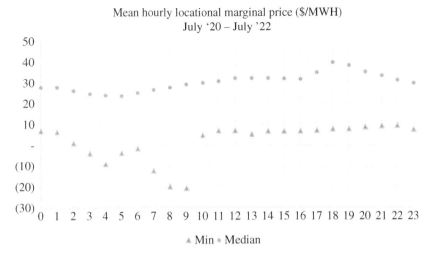

Mean hourly locational marginal price ($/MWH)
July '20 – July '22

▲ Min ● Median

Figure 23.11 The cost of electric energy (Locational Marginal Price) fluctuates continuously throughout the day for commercial and industrial customers. Hours from midnight to 9:00 AM tend to be least expensive. Hours from 3:00 PM until 11:00 PM tend to be most expensive. This incentivizes purchase of electricity at off-peak times and avoids buying electricity at peak times.

4. Run steam-driven chillers as needed.
5. Discharge the chilled water thermal storage over a multi-hour period spanning the hour predicted to be the peak.
6. Reduce campus energy demands by adjusting temperature setpoints or reducing outside air flowrates through buildings (Fig. 23.11).

23.6 Daily Thermal Storage – Chilled or Hot Water

To follow the most simplistic model for economic dispatch of thermal storage, you would charge storage systems between midnight and 8:00 AM. You would hold the storage through the day and then discharge the stored energy between 3:00 PM and 8:00 PM. This would provide modest financial benefits and would also help the grid by raising baseload demands and lowering peak demands. But, this model is based on *mean* prices. More sophisticated models predict each day and each hour of grid pricing. These can be far more lucrative.

Thermal storage can save money, but it also adds resilience to a district cooling system, benefits the customers, and makes the operator's job less stressful. With no thermal storage, the moment a chiller trips (shuts off automatically due to a problem), cooling customers get warmer water or they do not get enough chilled water. It may take 30 minutes or more to re-start a steam-driven chiller – if there was nothing damaged when it tripped. Insufficient cooling can cause damage to research equipment such as: lasers, CT-Scan machines, and electron microscopes. With thermal storage, when a chiller trips the energy plant operator can acknowledge that reliable service is more important than that day's economics; they can increase pump speed using a Variable Frequency Drive and immediately replace the lost chiller's capacity with stored chilled water from the tank. The customer's service continues uninterrupted. Then the plant operator has time to more thoughtfully troubleshoot, take corrective action, and restart the machine that tripped (Fig. 23.12).

(a)   (b)

Figure 23.12 Princeton's existing Chilled Water Thermal Energy Storage tank separates the process of producing cool water from the process of delivering cool water to the campus. Additional hot thermal storage and cold thermal storage tanks are being erected. (a) Is the existing chilled water thermal storage tank commissioned in 2005. (b) are the new hot water and chilled water thermal storage tanks being erected in 2023.

Thermal storage can have significant economic benefits, added resilience, and reduced Utility grid stress.

Exercise #4a

If Princeton's thermal storage tank is 2.6 million gallons, Chilled Water is stored at 33F, and chilled water returns from the campus to the plant at 56 °F, how much cooling energy can be stored? Express this in ton-hours and in kilowatt-hours.

A: With a 23° differential temperature, the 2.6-million-gallon thermal storage tank can store 41,511 ton-hours of cooling. This is equivalent to 146 MWh of heat removal potential.

Exercise #4b

Assume Princeton uses a 5000-ton chiller system to cool (charge) the 2.6 million gallons of water from 56 °F to 33 °F. How long does it take to cool the whole thermal storage volume? If the tank were discharged at a rate of 7500 tons, how long could that be sustained?

A: Using 5000 tons of chiller capacity to cool 2.6 million gallons by 23°F would take 8.3 hours to charge. Discharging the tank at a rate of 7500 tons, the tank could deliver cooling for 5.53 hours without requiring the use of any chillers or cooling towers.

Exercise #4c: Assume the 5000-ton chiller and 2.6-million-gallon thermal storage system has a net coefficient of performance of 3.7. Assume commercial electricity costs US$0.045 per kilowatt-hour at night, and US$0.088 per kilowatt-hour during the day. How much money could be saved each day by running the chillers to store cold water at night and avoid the purchase of electricity when cooling is needed during the day? If this is the average daily savings, how much could be saved in a year with this system?

A: Cooling 2.6 million gallons of water by 23°F with a coefficient of performance of 3.7 requires 39 MWh of input energy. With a difference between purchased power and avoided power costs of US$0.043/KWH, daily savings would be US$1697. If this represents typical daily savings, annual energy savings would be US$619,294.

Exercise #4d: Assume the Chilled Water Thermal Energy Storage tank is discharged at a rate of 6000 tons during all 5 grid-peak hours of the year. It is used to displace 6000 tons of electric-driven chillers with a net coefficient of performance of 3.7. Assume Capacity charges of US$200,000 per Megawatt per year. How much money could be saved by lowering peak demand and avoiding the annual Capacity charges?

A: By discharging the Chilled Water thermal energy storage (TES) tank at 6000 tons during all grid-peak hours to avoid running 6000 tons of electric-driven cooling with a net COP of 3.7, US$1.14 Million in annual Capacity savings could be realized.

Exercise #4e: What are the combined annual Chilled Water Thermal Energy Storage savings that can be achieved using the assumptions in 4c and 4d?

A: By dispatching the TES system to minimize daily energy costs and avoid peak-grid Capacity charges, annual savings of US$1.75 Million. There are corresponding benefits to the grid and to other ratepayers by raising demand on the Utility's baseload plants, reducing the need for the Utility to operate inefficient and expensive peaking generators and by reducing grid stress during peak demand hours.

Princeton will use big tanks of hot and cold water and big geoexchange wellfields for thermal energy storage. The thermal storage tanks hold a few million gallons each; large enough to deliver heating or cooling for several hours. The geoexchange wellfields are large enough to store an entire heating season of energy for the campus! Seasonal energy storage helps to address the mismatch between summer cooling demands and winter heating demands explored in Exercises 3a–3d.

Imagine it is a hot August day and we are running heat pumps to meet the campus cooling load. The heat pumps move heat out of the buildings and will have more than enough to satisfy all simultaneous heating needs. What will we do with the excess heat? First, we will use it to heat a few million gallons of water in a hot thermal storage tank that can be delivered to campus that night or several days later; whenever heating demand is higher. And what if we still need to remove heat from campus but do not need it for months to come? That's were seasonal storage plays a role.

23.7   Seasonal Thermal Storage – Geoexchange

Typical geoexchange wellfields include hundreds of wells. We are now drilling over a thousand geoexchange bores on campus, each 850 feet deep and each placed in a grid pattern spaced 20 feet apart. We are installing a 1.5-inch-diameter U-bend of HDPE plastic tube in each bore and grouting it in place for its entire depth. When we pump warm water from the heat pumps' condensers down through the HDPE tubes, it will give up its heat through the grout into the surrounding rock. The rock under our campus at that depth is naturally 50–55° Fahrenheit, so water entering at 90° will give up its heat to the rock and return to the surface at 60–70°. The earth becomes a heat sink all summer long. Over the course of the summer the rock gradually heats up, acting as a seasonal energy storage reservoir (Fig. 23.13).

We will harvest heat from campus buildings all summer, preserve it for several months underground, and reclaim much of it in winter. Some heat will dissipate through the surrounding rock, but billions of pounds of warm rock will store an ample supply of heat that can be moved back to campus when it is needed.

Photo Courtesy of Whiting Turner ↵   Photo Courtesy of Whiting Turner ↵

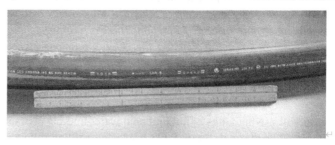

Figure 23.13 Princeton is installing over a thousand geoexchange bores on campus. The bores are between 600 and 850 feet deep. 1 ½" diameter HDPE tube is inserted in the wells in a U-bend to carry water down to the bottom and back to the surface again. As the water is pumped through the wells, it will give up heat to the rock (summer) or pick up heat from the rock (winter). Source: Courtesy of whiting-Turner.

The following exercises consider whether an entire season of heat removed from campus could be stored in geoexchange wells below campus. What volume is required, and what kind of temperature change is implied? Each of these calculations is simplified to demonstrate the general concept of campus-scale seasonal energy storage.

Exercise #5a: Using the campus's Total Annual Cooling energy requirement calculated in Exercise 3c, convert the amount of heat removed from campus in a year from Ton-hours to Btus.

A: 34.8 million ton-hours converts to 418 billion Btu heat rejected per year. This is how much energy is removed by the campus cooling system in a year.

Exercise #5b: Assume a rectangular geoexchange wellfield where the geoexchange wells are spaced in a grid pattern 20 feet from center-to-center and the geoexchange wells extend down 850 feet in the rock below campus. How many cubic feet of rock are associated with each geoexchange well?

A: The volume of rock associated with each geoexchange well is 340,000 cubic feet.

Exercise #5c: If the average density of rock is 165 pounds per cubic foot and there are 1000 wells in the geoexchange field, how many pounds of rock are associated with the wellfield?

A: The mass of rock associated with the wellfield is 56.1 billion pounds.

Exercise #5d: Based on your answers from 5a and 5c, how much heat energy would each pound of rock need to store to hold an entire year of heat removed from the campus?

   A: Each pound of rock in the geoexchange array would have to store 7.45 Btus in order to hold the entire amount of heat removed from campus in a year by the cooling system.

Exercise #5e: Assume the specific heat of rock is 0.47 Btu/pound/°F. How many degrees will the rock in the geoexchange array rise if an entire season of heat is stored? If the wellfield starts at 53°F in the spring, what temperature will it reach by fall?

   A: If all cooling energy removed from the campus in a year is stored in the geoexchange field with no heat removal and no losses, then the rock would rise 15.9°. If it began at 53°F in the early spring, it would rise to 68.9° by the fall. During the winter this heat would be removed and the geoexchange field would cool again.

Exercise #5f: What dimensions would a square geoexchange field take if it included 1024 geoexchange wells spaced 20 feet apart in a grid pattern? How many acres is this?

   A: A square geoexchange field with 1024 wells spaced 20 feet on center would require 409,600 square feet. This is equivalent to 9.4 acres or, a little more area than seven American football fields.

Comparing campus heating with heat pump-generated hot water instead of steam.

Exercise #6a: How much natural gas is required to heat Princeton's buildings for a year? Assume a fuel-to-delivered-heat efficiency of 55%. Exercises 1d and 3b may be helpful. Express your answer in dekatherms of natural gas (million Btus).

   A: With an annual building heat requirement of 470,592,299 pounds of steam and a fuel-to-delivered heat efficiency of 55%, 812,841 dekatherms of natural gas are required.

Exercise #6b: Assume natural gas costs US$6.00 per dekatherm and emissions associated with natural gas are 52.91 Kilograms of CO_2 per dekatherm of gas burned. What is the annual fuel cost and what are the CO_2 emissions associated with a year of heating the campus buildings? Express your answer in $ and Metric Tons.

   A: If 812 thousand dekatherms of natural gas are burned at US$6/DT, it would cost US$4.9 million to heat the campus for a year. The resulting emissions would be 43,007 Metric Tons of CO_2.

Exercise #6c: How much electricity would be required to heat Princeton's buildings using heat pumps for a year? Assume a heat pump COP of 5.0 and distribution losses of 5%.

   A: 27,515,581 Kilowatt-hours (27,516 MWH) of input energy would be required to produce a year of heat for campus buildings and overcome 5% distribution system losses using a heat pump with a COP of 5.0.

Exercise #6d: Assume commercial/industrial electricity cost averages US$0.055 per kilowatt-hour and emissions are 0.378metric tons of CO_2 per megawatt-hour. What is the annual energy cost and what are the CO_2 emissions associated with a year of heating the campus using electric heat pumps? Express your answer in $ and metric tons.

(Continued)

(Continued)

A: If 27,516 MWh of electricity are used at an average of US$0.055/KWH, it would cost US$1.5 million to heat the campus for a year with heat pumps having a heating COP of 5.0. The resulting emissions would be 10,401 metric tons of CO_2.

Exercise #6e: Comment on the cost and environmental impact of transitioning campus heating from natural-gas generated steam to electric heat pump-generated hot water.

A: Based on the calculations above, transitioning from natural-gas generated steam heating to electric heat pump-generated hot water heating, a savings of US$3,363,690 could be realized each year. This is a 69% reduction in heating cost.

Based on the calculations above, transitioning from natural-gas generated steam heating to electric heat pump-generated hot water heating, emissions reductions of 32,607 metric tons could be realized. This is a 76% reduction in heating emissions.

23.8 Moving to Renewable Electricity as the Main Energy Input

Electric heat pumps will make Princeton's campus heating system more energy-efficient, as discussed above. Heating costs can further be reduced by taking advantage of real-time electric pricing and hot water thermal storage. Similarly, by capturing the heat from campus and storing it, Princeton's cooling system will become more efficient. By avoiding the use of cooling towers, significant water and energy savings will be realized.

Transitioning away from burning natural gas to using electric-driven equipment will allow Princeton to use renewable electricity as the main source of energy. Because the new system will be much more efficient, we will need much less electric energy input. As part of our system upgrades, we have added 12 MW of solar photovoltaics to the original 4.5 MW that were installed in 2012. This should allow us to meet about 19% of our current electric needs with on-site renewable generation. Whatever electric demands we can not meet with on-campus renewable energy, we will procure from off-campus renewables.

Princeton has used two different contract models for solar PV procurement. The first 4.5 MW array was procured through an 8-year lease-buy contract. The more recent 12.5 MW of capacity have been procured using a power purchase agreement (PPA). The current PPA model allows an outside company to design, finance, build, own, operate, and maintain solar arrays on Princeton's property for a 15-year period. Princeton guarantees that the university will purchase all energy generated by the arrays. The outside company charges Princeton a fixed price of US$/KWH, with no escalation for 15 years. This offers the outside company a long-term guaranteed cashflow and allows Princeton long-term energy price certainty. To date, Princeton's solar procurements have delivered solar energy at a lower cost than the average price of energy purchased from the grid (Fig. 23.14).

Exercise #7a: Five of Princeton's solar arrays are mounted on building rooftops or above multi-level parking decks. But the large arrays are ground-mounted or erected as parking canopies above surface lots. The largest array has a nameplate rating of 5000 kW alternating current (AC). It is guaranteed to produce at least 7,849,620 kWh in its first year of operation. How many kilowatt-hours are generated per kilowatt of installed generation each year?

Figure 23.14   Two of Princeton's eight solar arrays can be seen here. Total installed capacity is over 16 MW.

A: 1570 KWH will be generated for each nameplate kilowatt of solar PV generation. This answer is location-specific. It would be different in different areas of the country.

Exercise #7b: If you assume the same ratio of annual energy production (kilowatt-hours) to nameplate capacity (kilowatt), how much solar generating capacity would be required to meet all of Princeton's heating and cooling needs with heat pumps? Use the annual heating and cooling energy needs you calculated in 3e and 6c.
   A: To generate 27,516 MWh for annual heating input energy and 22,618 MWh for annual cooling input energy would require an installed nameplate capacity of 32 MW of ground-mounted solar PV.

Exercise #7c: Princeton's 5000 KW ground-mounted solar array occupies a space of 15.5 acres. How many acres would be required to generate enough input energy to meet all of Princeton's heating and cooling needs using heat pumps?
   A: The existing ground-mounted array has a nameplate energy density of 323 kW/acre. An installation of 32 MW would require approximately 99 acres. This could generate enough input energy to meet the campus's heating and cooling needs with heat pumps. Additional power generation would be required to meet the university's other electric needs such as lighting, computers, and building air circulation.

23.9   Water Use Reduction

The changes Princeton is implementing do not just deliver energy savings. Few people realize how water-intense commercial-scale power generation and cooling are. To control NOx emissions, Princeton's gas turbine requires almost a 1 ton mass ratio of water and fuel. This equates to 483 gallons of water evaporated up the chimney for every megawatt-hour generated. Similarly, to

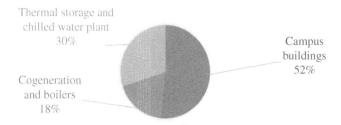

Total central campus water use
FY'17-FY'18

Thermal storage and
chilled water plant
30%

Cogeneration
and boilers
18%

Campus
buildings
52%

Figure 23.15 Nearly half the potable water used by Princeton's campus is evaporated by the energy plant.

produce chilled water about 2.1 gallons of water are evaporated through the cooling towers for every ton-hour of cooling. The result is that about 48% of all potable water supplied to Princeton's campus is evaporated at the energy plant. Only slightly more than half of the 228 million gallons purchased per year were used by people in the campus buildings. By transitioning to heat pumps that have no combustion and no cooling towers, millions of gallons of water can be saved annually (Fig. 23.15).

Exercise #8a: Princeton's recent addition of 12.5 MW of solar PV are guaranteed to produce at least 19,005,356 kWh of energy in their first year. If the cogeneration system uses 483 gallons of water for every megawatt-hour generated, how much can water use be avoided by displacing cogenerated electricity with energy from the new solar arrays? What percentage does this represent of the 110 million gallons of water used by the energy plant in a typical year?

A: By displacing 19,005 MWH of cogenerated energy with solar power generation, 9.1 million gallons of water use can be avoided each year. This results in an 8.3% reduction in total energy plant water use.

Exercise #8b: Princeton's cooling towers evaporate 2.1 gallons of water for every ton-hour of cooling delivered to the campus. Previously, we estimated that 15 million ton-hours are produced in a 2-year period. How much water use could be avoided by using heat pumps to deliver heating energy to a district hot water system rather than evaporating water to reject that heat to the environment? What percentage does this represent of the 110 million gallons of water used by the energy plant in a typical year?

A: By using heat pumps to capture heat and re-use it instead of rejecting it through cooling towers, 15.8 million gallons of water use can be avoided each year. This results in an additional 14.3% reduction in total energy plant water use.

And what about Princeton's existing cogeneration system and boilers? We plan to keep these valuable assets. But we will use them a lot less and we will need to fuel them with renewable natural gas or biodiesel, which are less environmentally damaging than more traditional fuels. Our cogeneration system runs more than 8000 hours a year now. Very soon the campus's solar PV array and its new heat pumps will become the least-cost sources of power and thermal energy. Cogeneration will remain available in the event of a grid emergency, as a reliable backup to the heat pumps, and for extreme hot and cold days when electricity prices are highest.

23.10   Closing Comments

By following the exercises in this chapter, you have seen that today's combustion-based steam heating systems can be vastly improved upon by converting them from steam generation and distribution to hot water generated by electric heat pumps combined with daily and seasonal energy storage. Far less input energy is required. You have also seen that it is feasible to store the summer's heat in geoexchange fields on a community scale and efficiently recover it for use during the winter. You understand that while solar PV requires a large land area per kilowatt, Princeton has been able to procure it at a lower average cost than local utility power and will enjoy long term price certainty as a result. The use of heat pumps instead of chillers with cooling towers, and the use of solar PV will help avoid the use of tens of millions of gallons of water each year.

Combining our heating and cooling systems into an electric heat pump-based system, and doing so with daily and seasonal thermal storage, enables Princeton to use renewable electricity as the main campus energy input. That is a big step toward a sustainable future.

Glossary:

CHP	Combined heat and power	The concurrent production of electricity or mechanical power and useful thermal energy (heating and/or cooling) from a single source of energy.
COP	Coefficient of performance	A ratio of useful heating or cooling provided to work (energy) required. Higher COPs equate to higher efficiency. Most refrigeration cycles have a COP greater than 1.0
Dekatherm	Dekatherm	Million Btus of Natural Gas.
HDPE	High density polyethelene	A thermoplastic polymer with outstanding tensile strength and large strength-to-density ratio, HDPE plastic has a high-impact resistance and melting point.
Heat rate	Heat rate	A measure of system efficiency in a steam power plant. It is defined as "the energy input to a system, divided by the electricity generated, in kW."
HRSG	Heat recovery steam generator	A boiler used to capture the exhaust heat from an engine
LCOE	Levellized cost of energy	A measure of a device's lifetime costs divided by energy production.
LMP	Location marginal price	The cost of electric energy at a specific location on the power grid at a specific moment.
Spark spread	Spark spread	The difference between the wholesale market price of electricity and its cost of production using natural gas.
TES	Thermal energy storage	
ULSD	Ultra low sulfur diesel fuel	Diesel fuel with substantially lowered sulfur content.

Edward T. Borer Jr. PE, CEM, LEED AP holds a BS and MS in Mechanical Engineering. He is Director, Energy Plant at Princeton University., He has 38 years of experience in electric utilities and district energy.

24

Considerations for Digital Real-Time Simulation, Control-HIL, and Power-HIL in Microgrids/DER Studies

Juan F. Patarroyo, Joel Pfannschmidt, K. S. Amitkumar, Jean-Nicolas Paquin, and Wei Li

24.1   Introduction

The increasing penetration of renewable energy and DERs has offered many advantages that help to reduce the carbon footprint and increase the robustness of conventional power networks. When a group of interconnected DERs and energy storage systems is integrated geographically close to one another, a microgrid system can be implemented with proper coordinated controls and protections. This evolution to distribute production sources and increase in inverter-based technologies calls for changes in analysis and calculation tools for engineers. For example, a microgrid may contain diverse types of DERs; some of them work in grid-forming mode, and some work in grid-following mode. In addition, inverter-based DERs have much faster dynamics than conventional generators, in the range of sub-cycle to sub-microseconds. To be able to maintain the balance across the microgrid, many different control layers need to be integrated into the simulation and the dynamics rely on the control parameters specified by the designer. In contrast to a typical distribution feeder, microgrids may operate independently, export power to the main grid, and have different topologies involving AC, DC, or a combination of both. Their control and protection systems can be complex (i.e. multi-level and distributed), involving equipment from multiple vendors making interoperability during key functions (e.g. islanding) a concern. With real-time simulation and HIL techniques, the problems associated with the integration of all these components can be overcome in the design phases.

   To be able to capture all the phenomena in simulation tests, it is necessary to rely on advanced models and hardware technologies that not only provide high accuracy, but also high computational speed. In addition, if the simulation is required to interact with real equipment or information systems, it is then required that it runs in real-time. Real-time simulation occurs when the simulation time runs at the same pace as the real world. Depending on the interaction with the physical world (electrical or signals), the real-time simulation can be divided into two main categories:

- **CHIL**: A controlling device is connected to a real-time simulator, which contains the electrical model. The controller sends the commands to the virtual actuators in the real-time simulation, and it receives the virtual signals of interest computed in the real-time simulation.
- **PHIL**: The virtual electrical model runs in a real-time simulator and, along with a power amplifier, interacts with external equipment using electrical signals that mimic the point of connection of a virtual node of the simulation model.

Microgrids: Theory and Practice, First Edition. Edited by Peng Zhang.
© 2024 The Institute of Electrical and Electronics Engineers, Inc. Published 2024 by John Wiley & Sons, Inc.

Depending on the type of real-time simulation, several factors and challenges must be considered. In this chapter, these challenges and factors are discussed for microgrid applications. In addition, some application scenarios are presented to help the reader understand the different situations where real-time simulation can be used to analyze complex microgrid systems.

24.2 Considerations and Applications for Real-Time Simulation

A common approach for testing the performance of a microgrid is the phasor-domain simulation. With this approach, the power balance of the network can be assessed, and the points of equilibrium can be found using load-flow algorithms. However, this method lacks precision when analyzing faster dynamics. To analyze fault or interconnection events, offline transient simulations are also used to assess instantaneous voltage/current values, harmonics, or any other relevant dynamics that occur in a few cycles of the fundamental frequency. As emerging technologies are being integrated into the conventional power grid, phasor, and transient domain simulations are required to study the integration of these technologies with the grid. It is common to find that manufacturers of these emerging technologies do not provide their internal models or calculations. Thus, they need to be integrated as a black box model to the network simulation using analog or digital interfaces. However, this integration needs to be in real time.

Real-time simulation is a method used when the timestep of the simulation needs to be synchronized with the real-world clock time. The simulation speed is determined by the computational load received by the CPU. If real-time performance is not achieved, untimely interactions between the device under test (DUT) and the real-time simulator will occur, making an HIL test erroneous. Depending on the computation performance, the simulation may run faster or slower than real time (see Fig. 24.1). To be able to run in real time, some considerations need to be kept in mind. First,

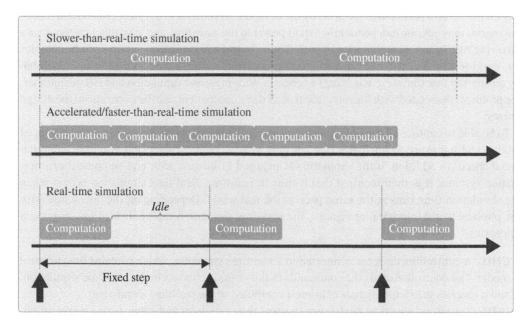

Figure 24.1 Comparison between regular PC offline simulation, accelerated/faster-than-real-time simulation, and real-time simulation.

the use of computational resources needs to be managed to avoid overrunning events, which occur when the computation time is greater than the timestep. Second, the solution of the mathematical equations needs to be optimized to reduce resource consumption. Finally, the real-time simulation can be improved by using dedicated hardware architectures that may solve the mathematical equations faster using FPGAs.

Different scenarios can be programmed in the real-time simulator, reducing costs along with prototyping time, and allowing the detection of potential design issues early in a project. For example, if the external device is a phasor-measurement unit (PMU), the real-time simulation can be used to mimic various environmental conditions such as the solar radiation or wind speed to analyze the characteristics of the voltage and frequency at any bus of a microgrid or conventional power network.

Real-time simulation can also be used to estimate the parameters or signals of a microgrid. If the model is accurate enough, the data from the real world can be fed to the real-time simulation to mimic the behavior virtually. This concept is known as a digital twin. The digital twin concept is commonly used to obtain a detailed estimation of the real-world signals and/or parameters to help grid or microgrid operators make better and quicker decisions compared to the conventional phasor-domain approach.

In this section, an analysis of the main challenges that may arise in real-time simulation is presented. These challenges include the overuse of resources and some strategies to overcome them. Also, some fundamentals regarding the modeling of power converters and their implementation in FPGA are presented. Finally, the use of smart inverters and the concept of digital twins for their use in microgrid real-time simulation are explained.

24.2.1 Challenges to Real-Time Simulation of Microgrids with Modern Converters

In a microgrid, multiple AC or DC DERs and loads are connected to AC or DC buses through power converters. Modern power converters have many advantages, while their control and protection systems (CPSs) are more sophisticated. Real-time simulation tools and HIL test benches are an effective and efficient way to study microgrids and validate their CPSs. However, there are several challenges that may arise when developing microgrid models for real-time simulation and HIL tests.

Firstly, the microgrid may contain a great number of electrical devices, such as various DERs, power converters, transformers, and loads. The simulation tools solve an electric circuit with the mathematical model matrices derived from the circuit using either the nodal voltage method or the state-space method. The more components the microgrid has the larger the scale of the circuit is, the greater the dimension of the matrices are, and thus it is more computationally intense to solve the equations. The real-time simulation tools use fixed timestep solvers. Whenever the time used to solve the equations is longer than the predetermined fixed timestep size, a so-called overrun occurs. Overruns introduce inaccuracies and may cause HIL and Power Hardware-in-the-Loop (PHIL) tests to fail. This is why it is very important to optimize models and use advanced mathematical techniques in order to achieve hard real-time performance.

One way to speed up simulation and achieve real-time performance is to partition the large matrices into multiple sets of smaller matrices. The dimension of the matrices is determined by the scale of the circuit. The computational load increases not in a linear way, but in a relationship between the square and cubical power to the matrices dimension. Therefore, solving multiple sets of partitioned smaller matrices is much easier and faster than solving the original one. Besides, the multiple sets of matrices can be solved using multiple calculation units (e.g. CPU cores) in parallel, which can further reduce the total computation time. Some power system device models, such as the distributed parameter line (DPL) model of long-distance power transmission lines, can be used as decoupling components to decouple the circuit and partition the matrices. The decoupling is done without any inaccuracies, thanks to the transmission line propagation delay being larger than the simulation timestep.

However, unlike a large transmission system, which contains long transmission lines to help partition the matrices, in microgrids the electric devices are physically located close to one another, which makes it more difficult to partition the matrices. In this regard, solving a microgrid can be more difficult than solving a transmission system if the microgrid model is considerably detailed or large.

The number of power converters in a microgrid increases as the number of DER and loads increase. Meanwhile, different power converters are used in microgrids, and some of the converter topologies may be complex and contain many switching devices. For example, a dual active bridge (DAB) based solid state transformer (SST) may contain hundreds of switches. In conventional simulation tools, a switch has a different resistance when its status changes, thus the topology and impedances of the overall circuit change, and the circuit matrices and their inverse need to be recalculated by the solver. In general, the more switches present in the circuit, the more computationally intense is the solving of the system equations.

One practical solution is to pre-calculate and store all the possible sets of matrices before the start of real-time simulation. During the real-time simulation, when the solver detects the change of a switch status, it picks the corresponding set of matrices to solve the equation. This method works for circuits with a small number of switches in each decoupled circuit. Nevertheless, the number of total possible sets of matrices increase exponentially. A circuit with N number of switches has 2^N combinations for the possible sets of matrices, which may require tremendous storage space and beyond the physical limits. Decoupling of the circuit and thus partition of the matrices can help to reduce the storage space. For example, assume a circuit with 20 switches needs to save 2^{20} sets of matrices. If this circuit is decoupled into two sub-circuits with 10 switches each, it needs to save 2^{10} sets of matrices for each sub-circuit, a total of 2×2^{10} sets of smaller matrices, which will take up a much smaller storage space. However, as explained, the microgrid does not contain decoupling components but has many switches, which makes it difficult for real-time simulation.

Another challenge is that compared to their counterparts in transmission systems, the new emerging converters in the microgrids have even higher switching frequencies, a few kHz to tens of kHz or even above. For the electromagnetic transient (EMT) simulation, the sampling frequency should be higher than the frequency of the fastest phenomena of interest. As a rule of thumb, 100 or more sample points in one switching cycle will be required to have accurate results of the switching phenomena. In other words, the simulation timestep needs to be at least 100 times smaller than the smallest possible switching period of the converter gate driver signals. Taking a converter with 10 kHz switching frequency as an example, the ideal simulation timestep is sub-microsecond, which is non-feasible for CPU models in real-time simulation and only feasible for FPGA-based models.

24.2.2 Modeling Power Converter for Real-Time Simulation and HIL Tests

Average converter models are generally used for studies focusing mainly on microgrid dynamics, interactions between DER controllers and high-level control of the microgrids. In the average models, the converter switching behavior is neglected, and the voltage source converter is represented by a controllable voltage source, which takes the value of the voltage reference from the VSC controller. Since no switching events are modeled, the average model can run fast to reach real-time simulation speed. However, the average converter model only gives the fundamental component of the voltage; but no PWM harmonics are simulated. This is considered an acceptable compromise for most dynamic studies, but for HIL tests, this kind of model cannot directly take gating signals of switches outputted from a real inverter controller and some low-level controls and protections, such as the voltage balance control in some modular converters or fast overvoltage protections, cannot be tested.

In recent years, many efforts have been dedicated to developing converter models for grid applications which will overcome the abovementioned challenges and enable real-time simulation and HIL test of microgrid components and their controllers. The goal is to have a model with high accuracy, fast real-time simulation speed, and an accurate representation of all components including every switch and capacitor inside the converter. This section will explain an equivalent-circuit-based switching function converter model, which meets the goal and has been used for microgrid studies and HIL tests. This modeling method can be applied to various topologies of VSC in microgrids, from simple 2-level VSC to DAB-based SST containing hundreds of switches [1, 2].

Although the VSC in microgrids covers various topologies, the basic unit is a bridge connected to the DC capacitor. The bridge can be in the form of a half-bridge (HB), full-bridge (FB), T-type bridge, or other varieties, while the DC capacitor can be either one capacitor connecting the positive and negative poles, or two capacitors with the mid-point grounded. The simplest example is a HB two-level converter as shown in Fig. 24.2. With two switches in the circuit, there are a total of four combinations of switch status, corresponding to four operation modes of the converter, namely insert mode, bypass mode, diode mode, and short-circuit fault mode. Figure 24.3 shows the paths of positive and negative currents flowing through the converter in different modes. By investigating the current paths in different modes, it can be determined if the capacitor is inserted or not, which is denoted as the capacitor insertion factor, K_p and K_n, for the positive or negative current path, respectively. Note for the case of two dc capacitors with midpoint being connected to external circuit, the insertion factor of the two capacitors may be different. Regardless of the current direction and operation mode, the circuit can be represented by the same equivalent circuit as in Fig. 24.3(d). The values of the controllable voltage sources, V_s, can be evaluated by Eq. (24.1) and the capacitor insertion factor values for HB converter with single DC capacitor are summarized in Table 24.1.

$$V_{s-i} = K_{i1} \times V_{c1} + K_{i2} \times V_{c2}, \quad i = n \text{ or } p \tag{24.1}$$

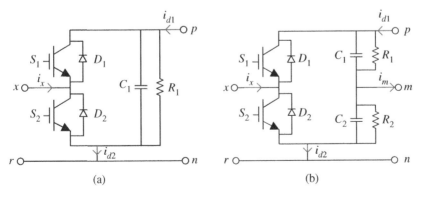

Figure 24.2 HB 2-level converter (a) without and (b) with midpoint connector.

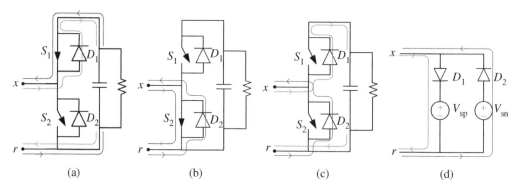

Figure 24.3 Current paths in HB-converter at different modes and equivalent circuit. (a) Insert mode, (b) bypass mode, (c) blocking mode, and (d) equivalent circuit.

Table 24.1 Equivalent circuit parameters for 2-level HB converter.

Mode	IGBT 1	IGBT 2	K_p	V_{s-p}	K_n	V_{s-n}
Insert	ON	OFF	1	V_c	1	V_c
Bypass	OFF	ON	0	0	0	0
Diode	OFF	OFF	1	V_c	0	0
fault	ON	ON	0	0	0	0

The equation to calculate the capacitor voltage is provided in Eq. (24.2), where i_{x-i}, with $i = p$ or n, are currents following through the positive or negative path. At any moment, only one of them can be a non-zero value depending on the actual current direction.

$$V_{cj} = \frac{1}{C_j} \int \left(\sum_{i=1,2} (K_{ij} \times i_{x-i}) + i_{dj} - \frac{V_{cj}}{R_j} \right) dt, \quad i = n, p \quad j = 1, 2 \tag{24.2}$$

Therefore, the HB converter circuit in Fig. 24.2 can be represented by its equivalent circuit as shown Fig. 24.4. In many cases in a microgrid, e.g. machine drive applications or non-grounded dc bus, no zero-sequence current flows from AC side to DC side. The ac and dc side equivalent circuit therefore can be decoupled as in Fig. 24.4(b), which is an example of a 3-phase converter.

Further investigation shows that the same equivalent circuit in Fig. 24.4 works not just for HB converters, but for all other converters. The only difference is that the tables for the capacitor insertion factor are different for each bridge. For example, the table for the 2-level FB or H-bridge converter with one DC capacitor is provided in Table 24.2. This table is more complex than the HB converter because the FB has four switches, meaning it has $2^4 = 16$ combinations of switch status and the capacitor insertion factor can be either positive or negative.

The same modeling method can be used to model more complex VSCs, since they are constructed from the same basic bridge units explained above. Taking MAB-based SST as an example, the one-phase scheme SST is shown in Fig. 24.5. The MAB includes m number of active bridges. For each active bridge, the AC side is connected to an m-winding transformer, and the DC side is connected to one DC capacitor each. Each DC capacitor is attached to another bridge where the AC side is connected to the external circuit, e.g. an AC grid. The bridges may be of different types (e.g. HB, T-type, etc.) and connected in a cascade structure.

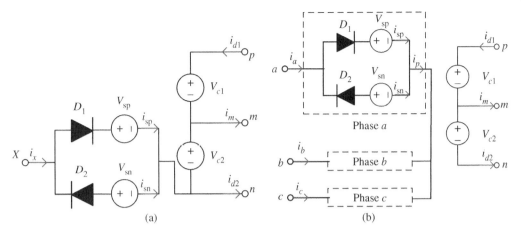

Figure 24.4 Converter equivalent circuits with ac and dc sides. (a) coupled and (b) decoupled.

Table 24.2 Equivalent circuit parameters for 2-level FB converter.

Mode	IGBT 1	IGBT 2	IGBT 3	IGBT 4	K_p	V_{s-p}	K_n	V_{s-n}
Positive insert	ON	OFF	OFF	ON	1	v_c	1	v_c
Negative insert	OFF	ON	ON	OFF	−1	$-v_c$	−1	$-v_c$
Diode	OFF	OFF	OFF	OFF	1	V_c	−1	$-v_c$
Bypass	ON	OFF	ON	OFF	0	0	0	0
	OFF	ON	OFF	ON				
Hybrid	OFF	OFF	ON	OFF	0	0	−1	$-v_c$
	OFF	ON	OFF	OFF				
	ON	OFF	OFF	OFF	1	v_c	0	0
	OFF	OFF	OFF	ON				

The SST can be represented by the equivalent circuit model as shown in Fig. 24.6. The bridges connecting the external circuit and the DC capacitors are represented by the equivalent node inside the left dash-lined box, and the MAB is represented by the equivalent circuit inside the right dash-lined box. The source voltages and the capacitor voltages are calculated using the Eqs. (24.1) and (24.2) explained above.

This converter model has several advantages to overcome the challenges mentioned in Section 24.2.1 and thus can be used in microgrid models for real-time simulation and HIL tests. First, the model can decouple the circuit into multiple sub-circuits by the converter, and thus allow parallel computation of the circuit to speed up the simulation speed. Second, this model reduces the number of switches in the circuit. For example, the equivalent circuit of the converter of any number of cascaded FB submodules contains only two diodes, as in Fig. 24.7. Third, this converter model allows an interpolation technique for intra-step converter switching events and thus it can achieve similar accuracy at a larger timestep (20–40 sampling points per switching cycle) compared to models without interpolation technique (100 sampling points per switching cycle) [1]. Lastly, this method can be implemented in FPGA for converters with even higher switching frequencies.

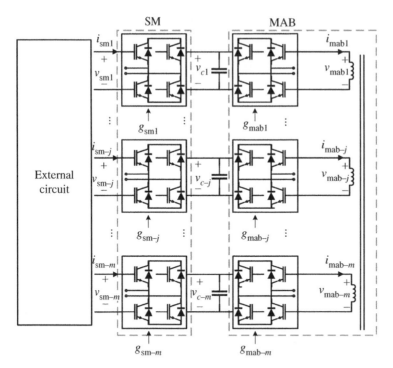

Figure 24.5 Schematic of a multi-active bridge based SST system.

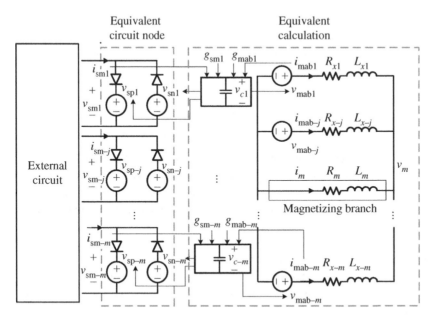

Figure 24.6 Multi-active bridge SST system model for real-time simulation and HIL tests.

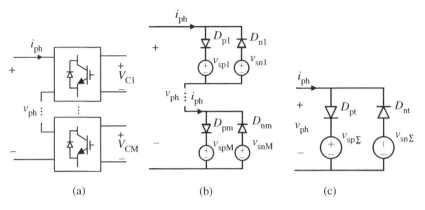

Figure 24.7 ac/dc converter of cascaded FB SMs and its equivalent model. (a) ac/dc converter of cascaded SMs, (b) cascaded equivalent circuits of SMs, and (c) equivalent circuit of ac/dc converter of cascaded SMs.

24.2.3 Solving Microgrid Models Through an FPGA

To solve a microgrid model with multiple high-speed switching power converters, and to handle many IOs, the CPU usually requires tens of microseconds to handle these tasks for each timestep. For real-time simulation and HIL test purposes, the timestep for such models cannot go much below 5 μs without over-runs when running on CPU. Moreover, modern power converters have a very high switching frequency, which requires much smaller timesteps, e.g. sub-microsecond timesteps. One practical idea is to solve the model using faster calculation units, such as an FPGA. The FPGA uses a fast clock which measures time in nanoseconds (e.g. 5 ns) and numerous calculation units for parallel calculation. Therefore, implementation of microgrid models in an FPGA-based simulator is a practical solution to obtain accurate results during real-time simulation for HIL tests.

The general process of FPGA implementation usually involves two steps. The first step is to use software to design the model to be implemented in FPGA. The model can run offline inside the software environment for result verification. The second step is that the model is used by software to generate firmware, i.e. the bitstream, which will be loaded into the FPGA for real-time execution.

Several difficulties of coding high level algorithms on FPGA prevent it from being more widely used. First, due to its extremely small timestep (e.g. a few microseconds or sub-microsecond), the logic model will have a prohibitively slow offline simulation speed running in the software, e.g. a few hours wall-clock time to simulate a 1 ms phenomenon. Second, the comparatively long time required to generate the FPGA firmware makes the test cycle long. The FPGA firmware generation process usually takes hours, which means that fixing any bug found in the firmware takes hours for firmware regeneration before testing. Third, due to resource limitations in FPGA, the signals are usually represented by either fixed-point arithmetic or single-precision floating-point arithmetic. Compared to double-precision floating-point arithmetic, the signals have narrower ranges of values and lower precision and thus are prone to error due to the data type. Last, advanced calculation functions on FPGA must be built from basic arithmetic and logic operators. It is more difficult to code complex algorithms on FPGA as compared to CPU. Because of the above-mentioned difficulties, it requires much FPGA knowledge and both great coding efforts and time to implement a microgrid model on the FPGA.

One way to overcome these difficulties is to develop a generic circuit solver in FPGA. The goal is that users need not implement their circuit models in FPGA by themselves, but that the generic solver will take advantage of the FPGA's features to solve the user's models.

The solver needs to meet the following requirements:

- The solver can solve microgrid models including power converters, motors, and other electrical devices with small timesteps, e.g. sub-microseconds.
- This FPGA solver can connect to a CPU model or to multiple FPGA solvers for multi-rate modeling of large-scale circuits, and it must further be capable of being connected to external devices through IO for HIL tests.
- The user can pre-define the circuit topology and parameters in a user interface and modify them without generating or regenerating FPGA bitstreams.
- Minimal FPGA coding knowledge is required from the user to use the solver.

The generic FPGA circuit solver, eHS, uses the Automated Nodal Electric Circuit Solver (ANECS) method to satisfy the above requirements and is widely used in many HIL tests by industrial users [3]. This method uses the Fixed Admittance Matrix Nodal (FAMN) technique [4], where a switch is represented by a small capacitor in the open state or a small inductor in the closed state. Using the Backward Euler discretizing method, either a capacitor or an inductor is represented as a resistor in parallel with a controllable current source containing historical information of the capacitor voltage or inductor current. The "discrete admittance" of the resistor, G, is given as in Eq. (24.3); where h is the simulation timestep, C is the capacitor value, L is the inductor value.

$$G = C/h = h/L \tag{24.3}$$

By choosing the capacitor and inductor value such that $C * L = h^2$, the discrete admittance G for the capacitor and for the inductor is the same. Therefore, regardless of its state, the switch can be represented by a fixed-value resistor with a parallel controllable circuit source using the backward Euler discretization method. Thus, the admittance matrix of the whole circuit containing multiple switches remains unchanged even if the switch status changes during the simulation. This method does not require recalculation of matrices or their inverse and thus is ideal for real-time simulation. The workflow of solving a circuit using eHS is illustrated in Fig. 24.8. The use of this FPGA solver for real-time simulation of microgrid is additionally reported in the literature [5, 6].

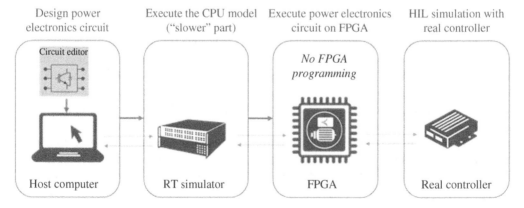

Figure 24.8 Workflow solving circuit on FPGA using eHS.

24.2.4 Smart Inverters for Microgrid Applications

Power quality is one of the most important factors to be maintained in a microgrid. If the power quality maintenance relies exclusively on the primary energy source, a big effort may be demanded from it, leading to disconnections, system degradation, and overpricing, among other issues. Thus, the DERs that inject power to the microgrid (solar, wind, etc.) are required to provide support to maintain the power quality in the microgrid buses and/or the POC. Standards such as the IEEE 1547-2018 [7] provide clear guidelines about how a DER should react to power quality deviations (frequency and/or voltage). These reactions usually imply changing the set point of the active and/or reactive power to compensate for frequency and/or voltage deviations. In addition, if the deviations are too large or take too much time, the DER may be disconnected depending on the ride-through configuration of the protection functions.

To test the integration of grid-support functions in a microgrid or a large network environment, it is necessary to use some of the standard functions of smart inverters. Some examples of standardized control libraries for smart inverter functions can be found in [8, 9]. A typical application for studying smart inverter integration is shown in Fig. 24.9 For instance, Rapid Control Prototyping (RCP) could be used, where the real-time simulator acts as the controller and is interfaced with a real power converter. Interfacing between smart inverter models and real devices can be supported via communication protocols, such as Modbus TCP, IEC 61850 GOOSE, etc. Also, smart inverter control and protection functions could be used to control a power amplifier in a PHIL experiment using the configuration presented in Section 24.4.

Smart inverters can be used in complex microgrid real-time simulations to maintain power quality and stability before deploying the microgrid controller in the field. To illustrate an example, the constant power factor (CPF) function is programmed in the secondary control block of a Photovoltaic Generation System (PVGS) connected close to the POC of a microgrid. The primary control block is programmed in DC-link control mode to modify power injection according to the bus RMS voltage. If the bus voltage decreases, the primary control block reduces the active current reference to increase the current injection in the DC-link capacitor. The Waveform Control block receives the current reference and generates a PWM signal or a voltage reference if the converter model

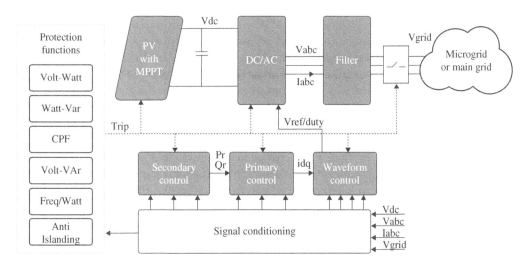

Figure 24.9 Typical implementation of a smart inverter control diagram.

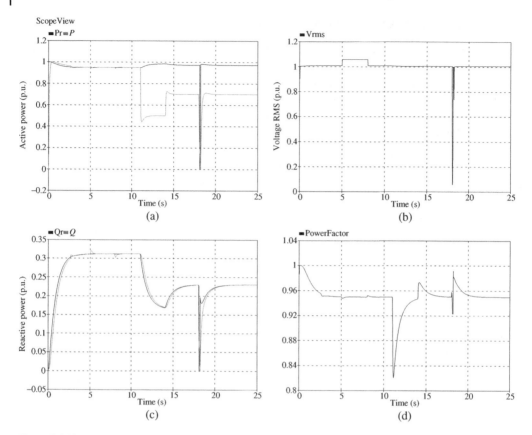

Figure 24.10 Test results for the CPF support function. (a) Active power "*P*" with reference "Pr" (b) RMS voltage at the POC (c) Reactive power "*Q*" with reference "Qr". (d) Power factor.

is a switching function model or an average model. As shown in Fig. 24.10, the RMS voltage at the POC is increased by 0.05 p.u. from $t = 5$ seconds to $t = 8$ seconds. Then, the solar irradiance decreased from 1000 to 500 W/m² at $t = 14$ seconds and increased to 700 W/m² at $t = 14$ seconds. Finally, a phase-ground fault is produced at the POC at $t = 18$ seconds and cleared 0.2 seconds later. It is remarked that the Secondary Control block adequately regulates the power factor setpoint of 0.95 despite changes in the solar irradiance, the voltage at the POC, and the presence of faults. In a microgrid, this configuration can be used to maintain the power factor at the POC of a microgrid according to the power authority regulations to avoid penalties or overcosts generated by the installation of oversized compensating capacitors.

Another example that shows the smart protection functions is presented in Fig. 24.11. For this scenario, the PV system injects the nominal power at $P = 1$ p.u. First, the frequency is increased right below the limit of 60.6 Hz. Then, it is increased to activate the frequency ride-through timer, which generates a trip command after the time limit of three seconds. Then, the voltage is decreased to 0.5 p.u, which is below the voltage limit of 0.7 p.u. At $t = 13$ seconds, the voltage right-through mode is activated to cease power injection, and the fault trip signal is activated at $t = 14$ seconds. Finally, an unintentional islanding event is generated at $t = 18$ seconds to activate the islanding protection scheme, which modifies current injection to activate the voltage protection function.

Smart inverters can also be used in a more complex environment where a mix of support functions and protections are distributed in a microgrid environment. There can also be multiple

ScopeView

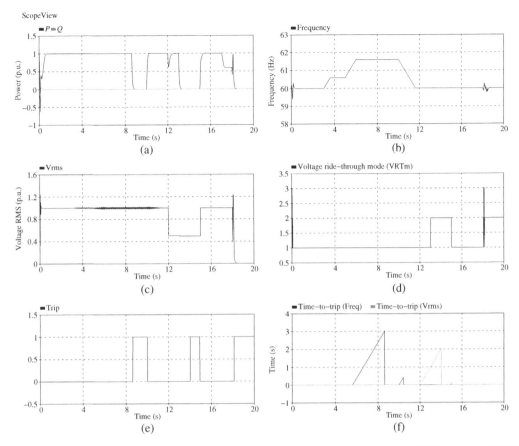

Figure 24.11 Test results for the protection functions. (a) Active and reactive powers (b) frequency (c) RMS voltage (d) Voltage ride-through mode (VRTm) (e) Trip signal (f) Time-to-trip for frequency and for voltage.

structures for implementing the functions in a HIL experiment. The controllers or protection blocks can be implemented in a real-time simulator to manage the power injection of a real DER. Also, the complete model with the power conversion circuit can be implemented using Power-HIL or in a pure digital real-time simulation to test the communication of the DER with upper control layers.

24.2.5 Digital Twins

A "digital twin" is a widely used concept with applications across many disciplines of the engineering realm. In general terms, a digital twin is referred to as the virtual representation of a plant or process that runs in the real world. However, this representation differs from a simulation model in the sense that it adapts its own states and/or parameters depending on the measured variables from the physical counterpart. In the context of power system studies, a digital twin might be a representation of a microgrid that uses the load demand, environmental variables (irradiance, wind speed, temperature, etc.), and/or breaker statuses to update the parameters or set points of the DERs. This improves the accuracy of the model and helps to estimate some other variables that cannot be read from the real microgrid due to the lack of sensors. In addition, the updated model can be used to predict future scenarios and/or suggest any future decisions to optimize the microgrid operation.

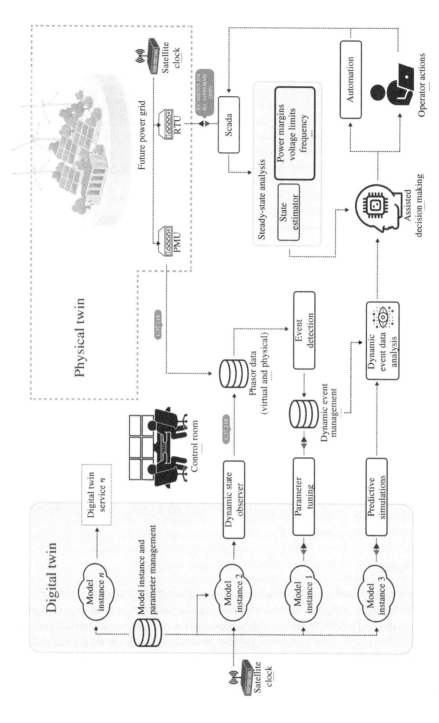

Figure 24.12 Typical functional diagram of a digital twin application (https://www.opal-rt.com/the-digital-twin-in-hardware-in-the-loop-hil-simulation-a-conceptual-primer/).

Given the above, a digital twin is characterized by, but is not subject to the following features:

- The digital twin may have a connection with the real world to read/write data to update the virtual model or make decisions in the physical counterpart.
- The digital twin has a series of parameters that may be estimated from the physical counterpart and be readjusted. This allows for improving the model accuracy under different situations.
- Depending on the level of detail of the digital twin, it can be used to estimate some of the interactions not measured in the physical counterpart. In other words, the digital twin can be used as an in depth observer.
- The digital twin is expected to react to the same stimuli produced in the physical counterpart. With this, it can be used to identify the source of a problem or identify the best decision to be made in a particular situation.
- The digital twin can be used to identify the response under extreme situations that may not be feasible to implement in the real counterpart. Thus, the digital twin can be used as a tool for predicting behavior under abnormal conditions.

A general overview of these features is shown in Fig. 24.12. In this example, the physical twin is a microgrid containing all types of DERs and loads whereas the digital twin may be composed of one or more models depending on the objective of the implementation. The real data measured from the physical twin is recorded using satellite clock measurements to avoid lags between twins. The measured data is stored in a database that compares it with the same data taken from the digital twin. This comparison is used to detect any event that may arise, and it can also be used to tune the parameters of the real model. For example, a sudden change in solar irradiance may cause a change in the injected power of a PV asset. If the measured time response of this injection differs from the digital twin, this information may be used to adjust the control parameters, the impedances, or the PV/wind panel characteristics of the virtual PV/wind asset.

24.3 Considerations and Applications of Control Hardware-in-the-Loop

Offline simulation is an important tool for preliminary design phases and controller development. During this stage, the controller is tested using an approximate simulation which may not consider all the components and phenomena involved such as the signal conditioning circuits, the processing power, the communication protocols, small signal noise, of the controller platform. This might make offline simulation fall short of covering all scenarios when the real controller is implemented.

CHIL simulation provides the advantages of a more realistic environment allowing integration of different technologies. For example, CHIL allows the consideration of real-world constraints such as firmware/software versions and bugs, hardware and computation capabilities, latencies and communication protocols, and IO configurations and signal conditioning, among others. In addition, the low-voltage interface between the HIL simulator and the controller provides a safe environment that may reduce costs and development time.

Furthermore, CHIL simulation allows the integration of the controller with other devices using advanced communication protocols such as IEC 61850 GOOSE, MODBUS, DNP3, etc. These communication protocols may be co-simulated as well to test the robustness of the controller when data is corrupted, modified, or delayed due to an unexpected event or cyberattack [10].

Compared to offline simulation or fully digital simulation, the CHIL implementation has some additional requirements to be considered. It is important to consider the timestep and select it

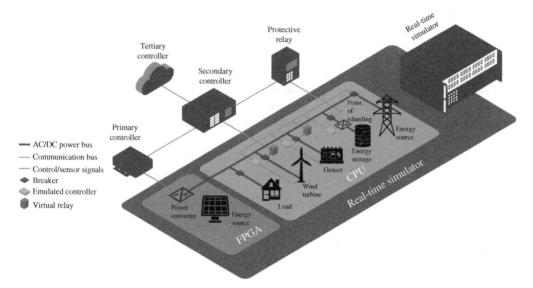

Figure 24.13 Typical interconnection scheme of a CHIL test.

according to the real-time simulation capabilities, the accuracy of the models, and the selected control strategy. Some control strategies contain integral functions that might be critically affected by variations in the timestep. If this timestep rate is too short and does not allow the simulator to finish the computations, some inaccuracies or instabilities may arise in the real-time simulation. Also, if the connection between the controller and the real-time simulator is made using a low-voltage interface, it is important to transform the signals to the acceptable ranges of the analog IO cards and consider voltage noise and signal latencies. Finally, the start-up sequence must be correctly designed. This implies that the CHIL experiment should be started in such a way that neither the real-time simulation nor the controller device becomes unstable during the start-up sequence, as in a real system.

Figure 24.13 shows a typical interconnection of a CHIL test of a microgrid. The primary controller device is directly connected to the power converter. Depending on the nature of the power converter, the control signals might be firing pulses for the simulated transistors or an average voltage level. From the virtual power converter, the controller might receive instantaneous voltages/currents from the AC side and DC-link voltage or any other relevant signal from the DC side. In addition, the primary controller might receive active and reactive power references from the secondary controller, which might be implemented in the real-time simulation, or it might be a real device. The same concept goes for the tertiary controller, which might be used to communicate signals on a larger scale using cloud-based services or coming from a central controller or the control center of the grid operator. Finally, the simulated circuit breakers might be controlled by real or simulated protective relays, which read the instantaneous measurements from the real-time simulation using a low-voltage interface connected to the DAC of the real-time simulator or interfaced through a power amplifier adapting the signal to levels of real measurement transformers secondaries, or through IEC 61850 communication protocols.

Once the CHIL components are integrated with the real-time simulator, a variety of scenarios can be executed to test the robustness of the controllers. The CHIL tests of the controllers with the virtual microgrid will be adequate for corrections and proper tuning of the control functions early in the engineering or design phase of the microgrid, giving a much higher level of confidence in the control scheme before the actual commissioning on-site. In Sections 24.3.1, 24.3.2, and 24.3.3, some application cases for CHIL tests are presented. The first application case is referred to the testing of a high-level microgrid controller. This application can also be used to test different communication protocols and their impact on the robustness of the microgrid controller. The second application case is about the implementation of a low-level inverter controller programmed in a microcontroller. The firing pulses that come out from the voltage controller are sent directly to the real-time simulator, which uses the FPGA-based simulation to obtain highly accurate results with small timesteps.

24.3.1 Microgrid Control System Testing

Microgrids are an excellent means to improve overall resilience of the grid and distribute more renewable energy generators with the goal of reducing the carbon footprint. They enclose geographically and electrically close generators and loads that need to be coordinated to ensure the highest continuity of service. Moreover, with potential faults and the variability of the renewable energy sources, backup solutions and adequate control and protection schemes need to be designed to ensure the highest resilience of the critical loads that could be served within the microgrid (e.g. hospitals, emergency services, etc.). Each microgrid is unique as a system and many different constraints make it very important to test microgrid control systems, including: the DER modes of operation (grid-following/forming), the stochastic nature and variability of the renewable energy resources, state of charge of the energy storage systems, and the various types of loads and equipment energized by the microgrid. Dispatch control functions for the energy management of assets within the microgrid often use multiple objectives for decision making such as the optimization of the energy cost, the minimization of the carbon emissions, the coordination of the energy storage charging and discharging to make the best use of renewable energy sources, etc. Some microgrids are fully islanded and independent, while others can be connected to the main grid, which requires intelligent transition processes with the microgrid control system to ensure seamless transition. The combination of these factors creates complex scenarios for which the control system needs to be validated and many of the possible scenarios may be difficult to test in the field.

CHIL represents a solution to validate microgrid control systems considering the aforementioned challenges. In the real-time simulator, the different scenarios in the microgrid can be swapped according to the objective of a particular test (See Fig. 24.14). However, since the objective of CHIL is using the same control system as the one used in the field, it is important to reinitialize the controller whenever a new scenario is generated to ensure the stability of the simulation. Sometimes, the scenarios can be organized in a sequence that can preserve the same control system's state without having to reinitialize it.

Standards such as the IEEE P2030.8 provide a methodology for testing microgrid controllers considering multiple scenarios [11]. The testing procedure shown in the IEEE P2030.8 standard can be used to analyze various performance metrics such as resiliency and reliability, power quality regulation, and economic operation, among others. As is shown in Fig. 24.15, the control functions are divided into dispatch functions and transition functions. The dispatch functions refer to states where the initial and final conditions of the grid-connection breaker remain the same whether in grid-connected mode or islanded mode, whereas the transition functions refer to the modes of operation where the microgrid changes from one set of dispatch function to another one, following a planned islanding, unplanned islanding, or reconnection. For each of these core functions, there might be multiple scenarios related to the DER and load states, the statuses of the breakers, weather conditions, SOC, power set points, etc. With these considerations, multiple complex

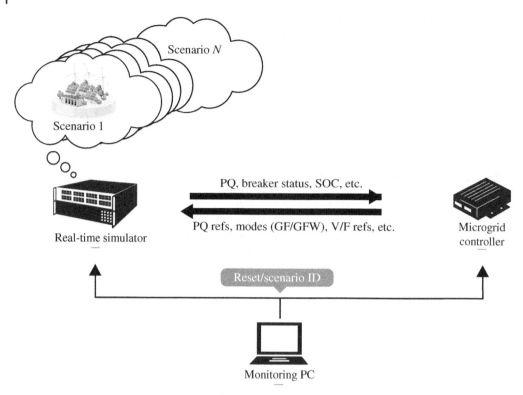

Figure 24.14 Microgrid control validation scheme using CHIL.

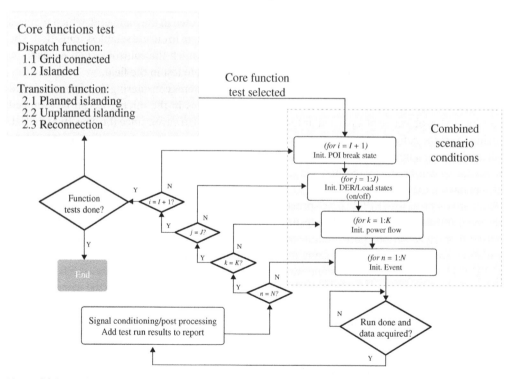

Figure 24.15 Microgrid control validation flowchart proposed in the IEEE P2030.8 std [11].

scenarios resulting from the combination of different states can be implemented. For example, some scenarios might require specific weather conditions or SOCs difficult to reproduce using field testing. Also, some fault conditions that might result in damaging the physical equipment can be implemented in CHIL to assess the reaction of the microgrid control system against abnormal and critical conditions in a safe environment.

24.3.2 GHOST Microgrid

The Grid Hardware Open-Source Testbed (GHOST), also formerly known as the BANSHEE microgrid model, was developed by the U.S. National Renewable Energy Laboratory (NREL) and the Massachusetts Institute of Technology (MIT) in 2017 [10, 12]. It comprises a series of DERs such as wind, solar, diesel generators, and battery energy systems. It also contains three POCs with the main grid and interconnection breakers between each of the three branches. This topology enables the performance of tests varying the points of connection with the main grid or analyzing the power flow between each of the three branches.

For this application scenario, the CHIL connection scheme is presented in Fig. 24.16. The GHOST microgrid is programmed in a real-time simulator. Due to the size of the network, the feeders are separated into three different networks, with each of them programmed in a dedicated CPU core. Solving the system in parallel avoids having the computation time being higher than the timestep and thus reaches the real-time simulation speed. The controllers that regulate waveforms and power injection/absorption are integrated within each DER in the real-time simulator. The microgrid controller that regulates operation and transition during/between the islanded and grid-connected modes is programmed externally in an industrial computer.

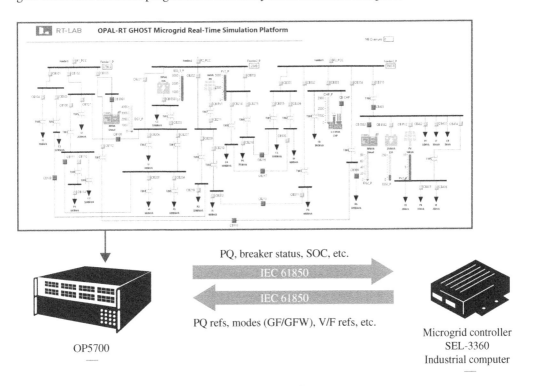

Figure 24.16 Connection scheme of the GHOST microgrid CHIL test.

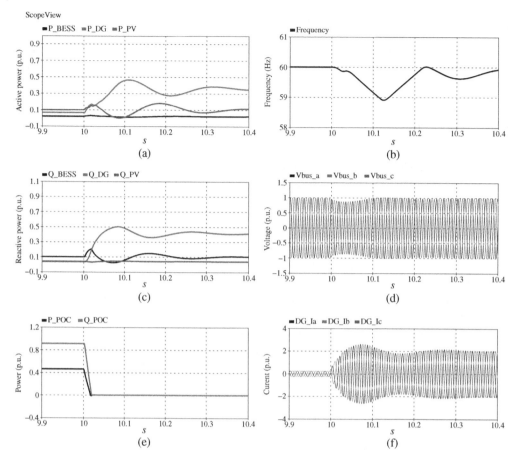

Figure 24.17 CHIL test results for unplanned islanding of the GHOST microgrid. (a) Active power (b) Frequency (c) Reactive power (d) RMS bus voltage (e) Active and reactive powers at the point of connection (f) Current of the diesel generator.

The controller communication with the virtual GHOST microgrid is implemented using the IEC61860 GOOSE protocol. The controller receives the active and reactive power measurements, the breakers statuses, the SOC of the batteries, and other relevant information to determine the optimal power dispatch in islanded or grid-connected mode. Also, the controller determines active and reactive power commands for each DER, the modes for islanding/grid-following operation, the voltage and frequency references for islanding operation, and the enable signals for each of the connection breakers.

One of the uses for this test application is that the controller can be initially connected to a real-time simulator to determine if the microgrid is stable under any event. For example, Fig. 24.17 shows the effect of opening the grid connection breaker (at $t = 10$ seconds) without having the net power close to zero at the POC. This can be interpreted as an unintentional islanding event. When the breaker opens, the diesel generator (DG) becomes the grid-forming generator, providing support for voltage and frequency. Since the net power is different from zero at the moment of disconnection, this net power becomes absorbed by the DG, which causes oscillations in voltage and frequency. In reality, this event may cause a collapse in the microgrid, which should be restarted through black start. However, when the disconnection occurs having zero net power at the POC

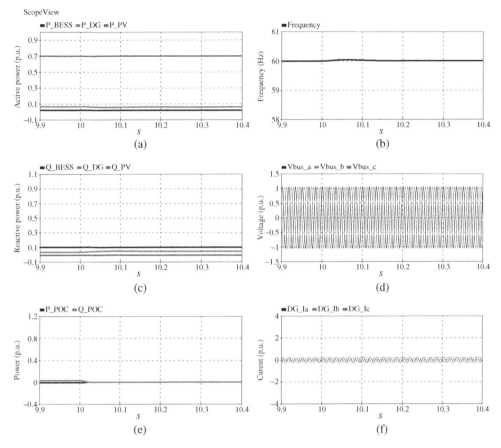

Figure 24.18 CHIL test results for planned islanding of the GHOST microgrid. (a) Active power (b) Frequency (c) Reactive power (d) RMS bus voltage (e) Active and reactive powers at the point of connection (f) Current of the diesel generator.

(intentional islanding), the transition between states is smooth and steady (Fig. 24.18). Also, the microgrid control system will dispatch the assets in a manner to ensure generator and load balance and close to zero power flow from/to distribution grid. It will then take over dispatch functions in the islanded mode.

Another important use of this application case is to test and configure the communication links, which may be affected by the quality of the network and/or the influence of data manipulation by a cyberattacker. This might help to improve the robustness of the controller under abnormal conditions. For example, if a data manipulation occurs, and an unintentional islanding event occurs, the controller needs to be programmed to react to the deviations in voltage and frequency despite not being informed of the sudden disconnection.

24.3.3 CHIL Using Microcontrollers

The CHIL testing scheme can also be used for low-level controller validation. With the use of FPGA-based simulation, the timestep can be reduced to the scale of sub-microseconds. This allows the implementation of highly detailed models such as power converters. For this application, the topology of a two-level DC–AC converter is implemented in an FPGA-based simulation along with

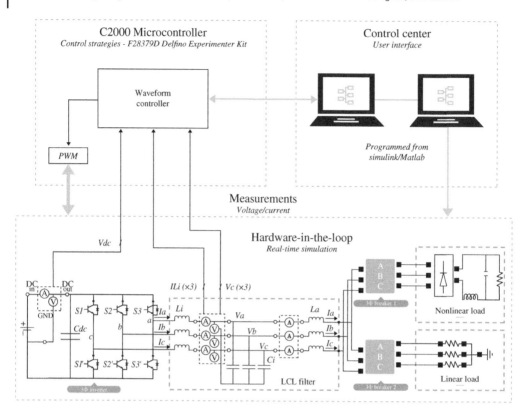

Figure 24.19 Connection scheme of the TI C2000 to the FPGA-based HIL simulation.

its output filter and loads as shown in Fig. 24.19 [13]. The controller is implemented externally in a Texas Instruments C2000 microcontroller. The DC–AC converter is driven by the PWM pulses generated by the controller inside the microcontroller. Additionally, the voltage and current measurements of the simulation are communicated to the microcontroller using the DAC of the real-time simulator and then physically read using the ADC of the microcontroller. The objective of this test is to analyze the current ripple and maximum values at the output of the two-level converter connected to a nominal load. Two different control strategies (proportional–resonant and proportional–integral) are implemented to see their impact on the transient response.

For the test, a three-phase voltage reference of 60 V RMS value and a frequency of 60 Hz is generated. Then, at $t = 2.28$ seconds, an increase of 60 V occurs in the voltage reference signal. The test is made using a linear load and the dynamics of the internal current (Iinv) are observed under changes in the voltage reference. This way, the protection circuits for the IGBT transistors of the DC/AC bridge can be estimated in a more realistic way depending on the selected control strategy. For this test, the values of the control constants are the same for both controllers.

Figures 24.20 and 24.21 show the results for the implementation of the Proportional–Resonant (PR) and the Proportional–Integral (PI) controller, respectively. The top subfigure shows the internal current, the middle subfigure shows the output current, and the bottom subfigure shows the capacitor voltage. It can be noted that the PI controller has a faster response, but it generates a higher overshoot. However, both controllers provide an adequate response considering that the switching noise does not cause the internal current to be higher than 3 A. The main difference between the PR and the PI controller is that the PI controller requires transformation of the

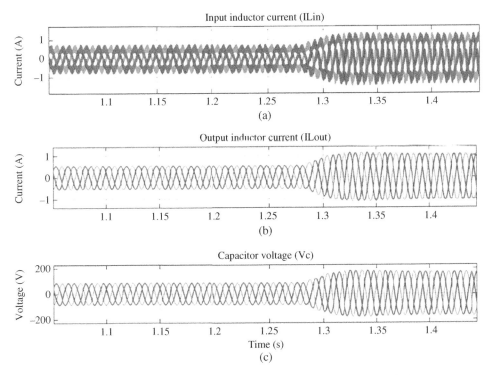

Figure 24.20 PR controller results in CHIL. (a) Input inductor current (b) Output inductor current (c) Capacitor voltage.

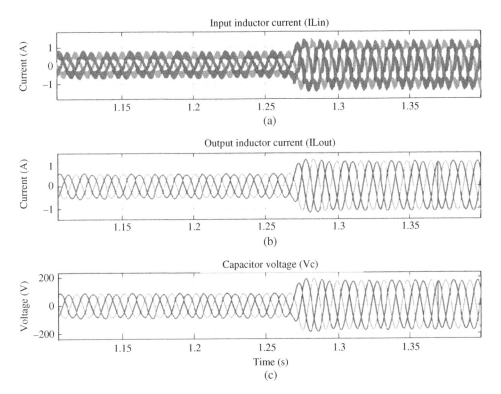

Figure 24.21 PI controller results in CHIL. (a) Input inductor current (b) Output inductor current (c) Capacitor voltage.

three-phase signals to the DQ frame. The PR controller operates in the alpha-beta or ABC frame, which does not require the use of a PLL. Also, the PR controller can be expanded to harmonics control. However, the PR controller is represented by a second-order transfer function and the PI controller only uses a first-order one.

These types of tests using FPGA-based simulation can be used for more complex experiments, such as analyzing the output current in the presence of nonlinear loads, analyzing harmonic rejection of each controller or testing different topologies and PWM modulation techniques. In the context of microgrids, this test can be useful for location of the converter close to the POC to analyze how the location and size of the DER can affect the quality of the power injected to the main grid or one of the microgrid buses.

24.4 Considerations and Applications of Power Hardware-in-the-Loop

Power Hardware-in-the-Loop (PHIL) based simulations and testing are increasingly being recognized as an effective approach for testing and system-level integration of DERs and other microgrid systems. The structure of a generic PHIL system is indicated in Fig. 24.22. The simulated system is implemented in a real-time simulator. With the help of D/A interfacing or digital fiber optic with small form-factor pluggable (SFP) transceiver interfacing, the simulated system is connected to a power amplifier. The power amplifier allows power interconnection and power exchange with a DUT. Thus, there is a hybrid virtual-analog connection of power between the simulated system and the DUT when a PHIL simulation is implemented.

PHIL enables researchers, academics, and engineers to test and study various areas related to microgrid and DERs. Some typical studies include DER integration studies, microgrid control and performance studies, fault studies pertaining to DERs and in microgrids, etc. The main benefit of PHIL simulations is a greater flexibility to apply tests that could not be done in practice on a real power system, in a non-destructive manner. For instance, if a DER is to be tested for grid sags/swells, PHIL simulations provide a means of emulating corresponding grid behavior, while also simulating the impact on the simulated grid for various DER transient responses. Thus, enabling a closed-loop interaction between the equipment being tested and the simulated network. Furthermore, with PHIL simulations, an accelerated setup of a microgrid system is possible since each DER can be individually integrated whilst the remainder of the microgrid is still being simulated. These are some of the reasons why PHIL simulations are gaining a lot of popularity in microgrids and DER studies.

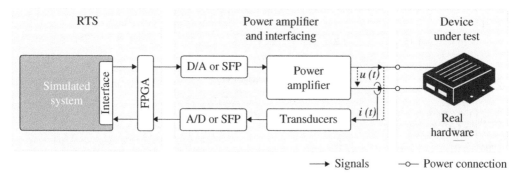

Figure 24.22 A generic PHIL system structure.

The stability and accuracy of a PHIL system are affected by several factors, such as the model detail (implemented on the real-time simulator), the power amplifier characteristics, and the interfacing between the real-time simulator and the power amplifier. Consider grid integration of a photovoltaic inverter (DUT) as a PHIL implementation example. The detail involved in the modeling of a grid in the real-time simulator defines the scope of PHIL studies to be performed. The choice of the amplifier primarily influences the power levels of the PV inverter to be integrated. Dynamic characteristics and interfacing of the power amplifier define the dynamic studies that can be performed on the PV inverter being integrated. Detailed considerations for performing PHIL simulations are provided in Section 24.4.1.

24.4.1 Selecting the Right Power Amplifier for a PHIL Application

To meet application requirements and test coverage when implementing a closed-loop PHIL system, the choice of a power amplifier and its corresponding specifications play a key role.

a. **Power, voltage, and current rating of the amplifier and the device under test**: The power, voltage, and current rating of the power amplifier chosen defines the test coverage concerning a PHIL application. Certain PHIL applications may require the power amplifier to have the capability to achieve a higher than nominal voltage, current, or power rating. Another factor that defines PHIL test coverage is the capability of the power amplifier to operate in all four quadrants, i.e. to sink and source active and reactive power, as shown in Fig. 24.23. This is a requirement in most PHIL application use cases.

   Moreover, some power amplifiers have the capability of regenerating the sinked power to the supplying grid, while others require additional dump loads to sink power, which needs consideration in terms of power density and energy consumption.

b. **Power amplifier open-loop bandwidth and closed-loop bandwidth**: The open-loop bandwidth of a power amplifier is the maximum frequency content a power amplifier can amplify to within $-3\,dB$ of magnitude attenuation. Power amplifier open-loop bandwidth greatly influences the closed-loop PHIL application bandwidth which in turn defines the frequency content of the dynamics to be effectively emulated. In certain cases, a power amplifier might have a different open-loop bandwidth in sinking mode versus sourcing mode, or a different bandwidth within versus above nominal voltage/current ratings. In such cases, the dynamics addressed by the PHIL application case might be affected based on the operating point of the power amplifier.

c. **Power amplifier latency**: Due to the non-zero response time of integrated power components, control logic, and several other reasons, a power amplifier has a defined latency, which essentially means that a power amplifier will amplify a received reference signal after a pre-defined delay. This delay results in a phase shift while implementing a closed-loop PHIL application use case. An amplifier with a large latency could significantly limit the achievable closed-loop PHIL bandwidth, which in turn would limit the PHIL application test coverage.

d. **Power amplifier slew-rate and linearity**: A power amplifier – due to limitations of its internal power components – could have a defined slew-rate, or a non-linearity. This could mean that the amplifier performance, particularly its open-loop bandwidth, could vary depending on the operating load or output voltage. Often, slew-rates can result in a power amplifier having two different sets of open-loop bandwidths: a large signal bandwidth and a small signal bandwidth. Care must be taken to account for this factor while developing a closed-loop PHIL application.

To illustrate the above, a power amplifier with its corresponding specifications is shown in Fig. 24.24. Along with the power amplifier subsystems such as busbars, meters are also shown in the test-bench, which are often required to perform PHIL studies related to microgrid applications.

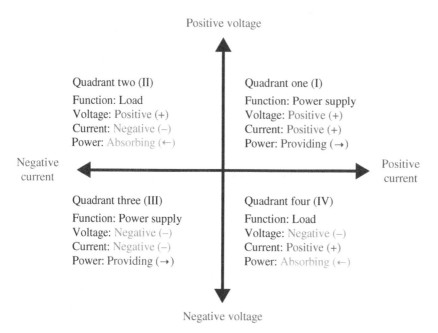

Positive voltage

Quadrant two (II)

Function: Load
Voltage: Positive (+)
Current: Negative (−)
Power: Absorbing (←)

Quadrant one (I)

Function: Power supply
Voltage: Positive (+)
Current: Positive (+)
Power: Providing (→)

Negative
current

Positive
current

Quadrant three (III)

Function: Power supply
Voltage: Negative (−)
Current: Negative (−)
Power: Providing (→)

Quadrant four (IV)

Function: Load
Voltage: Negative (−)
Current: Positive (+)
Power: Absorbing (←)

Negative voltage

Figure 24.23 Operating quadrants of the power amplifier.

Amplifier electrical specifications	
Rated voltage rating	120 V RMS per phase
Rated current rating	20 A RMS per phase
Sourcing/sinking capability	100% sinking and sourcing
PHIL performance	
Mode of operation	Voltage and current mode
Closed-loop-bandwidth	DC to 10 kHz
THD	≤0.75%
Latency	5.5–8.3 μs

(b)

(a)

Figure 24.24 A power amplifier testbench with its corresponding specifications. (a) A power amplifier testbench, (b) the corresponding specifications. Source: OPAL-RT TECHNOLOGIES, Inc.

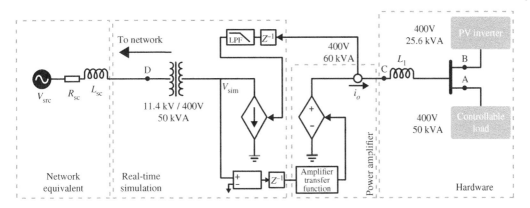

Figure 24.25 An illustration of PHIL interfacing.

24.4.2 PHIL Interfacing

The generic structure of a PHIL system is shown in Fig. 24.22. The simulated system sends out a reference to the power amplifier using an analog or a digital interface. The power amplifier in turn amplifies the provided reference to the voltage and current levels as required by the real hardware (in this case the DUT). Voltages and/or currents are sensed with the help of transducers and are transferred back to the simulated system using analog or digital interfaces. The resultant changes in the DUT operating conditions (for instance current drawn) are communicated back to the simulated system.

An example scenario to illustrate PHIL interfacing is shown in Fig. 24.25. The simulated system consists of a microgrid, whereas the real hardware, i.e. the DUT, is a PV inverter and a controllable load. The voltage reference generated from the simulated system is sent to the power amplifier (represented as an amplifier transfer function) resulting in a corresponding dynamic response during amplification. The sensed currents from the real hardware are fed back to the simulated system. A low-pass filter could represent the bandwidth of the feedback sensors, or in certain cases, digital filters implemented in the simulated system to stabilize the closed-loop response of the PHIL system.

The resulting block diagram of the closed-loop PHIL system is shown in Fig. 24.26. The diagram shows the transfer functions associated with the simulation and physical-hardware impedances, $Z_{sc}(s)$ and $Z_1(s)$, respectively. The delays associated with reference propagation (τ_1) and feedback propagation (τ_2) along with the amplifier transfer function, AF(s), are also shown in the block diagram. The stability of a closed-loop PHIL system can be investigated by analyzing either the frequency response or the time response of the block diagram shown in Fig. 24.26. As can be seen, the stability is influenced by the load-to-source impedance ratios, the amplifier transfer function, and the propagation delays. The filter shown in the block diagram can be used and tuned appropriately to stabilize a closed-loop PHIL system. The selection of the feedback filter would be a design exercise to compromise between PHIL closed-loop system stability and accuracy.

24.4.3 Network and DER Emulators

The targeted application has a significant impact on the power amplifier choice and the overall PHIL system design. One consideration is the operating mode of the amplifier. In network emulation, the simulated grid (or network) voltage is emulated by the power amplifier. The power amplifier in this application scenario operates in voltage control mode. In DER emulation, the simulated

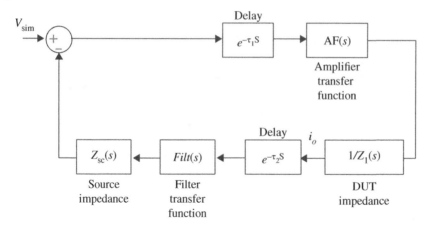

Figure 24.26 Block diagram of a closed-loop PHIL system.

DER power/current is provided as a reference to the power amplifier, which therefore, typically needs to operate in current control mode. In complex PHIL application scenarios, interconnection of multiple active components is also an important consideration. In the case of multiple active components directly connected to the same circuit bus, there must be exactly one voltage source with the rest being current sources. It is important to note, however, that such restrictions apply only to the high-level control of such sources; for example, two voltage-source amplifiers can be connected to the same circuit bus if the voltage reference for one amplifier is derived from a closed-loop current controller, and similarly two current-source amplifiers can be directly connected as long as the current reference for one is derived from a closed-loop voltage controller.

Another important application impact on the PHIL system design is regarding the implementation of measurement and feedback filtering. Network emulation typically requires DUT current measurement as a minimum. DER emulation typically requires DUT voltage measurement as a minimum. If the power amplifier does not provide the necessary measurement, additional sensors are required. Depending on the targeted application and performance requirements, feedback filtering is often required. This feedback filter can be implemented as a digital filter or an analog filter, each having its own set of advantages and disadvantages.

Another notable application consideration is an incompatibility arising from mismatches between the simulated voltage/current levels and that of the physical hardware voltage/current levels. Often the voltage and current levels in the simulated system are much higher than the practical voltages and currents emulated by a physical power amplifier. Appropriate gains for both the references and feedback in the PHIL interface can be designed to account for such voltage/current level mismatches between the simulated system and the physical hardware. For example, inverse scaling factors can be used for the references and feedback such that the interface emulates an ideal step-up or step-down transformer where equal power flow on both sides of the interface is maintained, yet the voltage or current is sufficiently reduced such that both are within hardware ratings.

The Sections 24.4.3.1. and 24.4.3.2 provide examples for network emulation and PVGS emulation using a PHIL system.

24.4.3.1 Network Emulator
A simple PHIL application case is that of an AC Network Emulator, with the basic concept shown in Fig. 24.27. The system consists of an AC source of voltage V_{abc} with source equivalent series

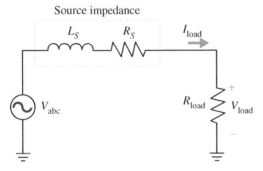

Figure 24.27 Network emulator base concept.

impedances L_s and R_s connected to a load of resistance R_{load}, with load current and voltage equal to I_{load} and V_{load}, respectively. The source impedances can be defined using both a Short-Circuit Ratio (SCR) and an X/R ratio using Eqs. (24.4)–(24.7).

$$\text{SCR} = \frac{\text{Short circuit VA}}{\text{Rated } W} = \frac{\left(\frac{v_{\text{nom}}^2}{z_s}\right)}{P_{\text{nom}}} \tag{24.4}$$

$$Z_s = \frac{v_{\text{nom}}^2}{\text{SCR} \cdot P_{\text{nom}}} \tag{24.5}$$

$$L_s = Z_s \frac{\sin\left(\tan^{-1}\left(\frac{X}{R}\right)\right)}{\omega} \tag{24.6}$$

$$R_s = Z_s \cos\left(\tan^{-1}\left(\frac{X}{R}\right)\right) \tag{24.7}$$

By simulating the AC source and source impedances, utilizing a physical resistive load, and implementing a PHIL interface to bridge them, the "Physical Load" configuration of the Network Emulator as shown in Fig. 24.28 is derived. In this case, the simulated network voltages are reproduced by the amplifier and applied across the physical load, with the physical currents measured and filtered before being used to control current sources in the simulation.

Alternately, the resistive load can be emulated as well if a second PHIL interface is implemented. This "Emulated Load" configuration is shown in Fig. 24.29, where the physical resistor is replaced by a second amplifier operating as a current source. To emulate this physical resistance, the necessary current references for this second amplifier are calculated through Ohms Law using the

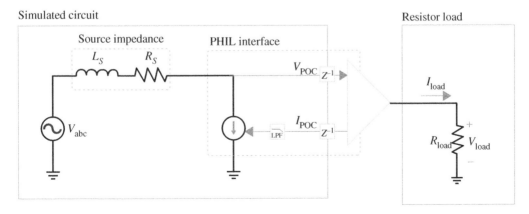

Figure 24.28 Network emulator configuration – physical load.

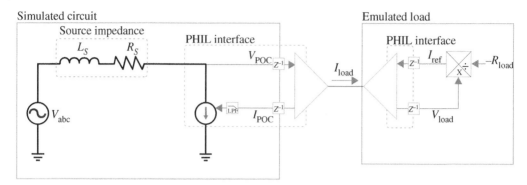

Figure 24.29 Network emulator configuration – emulated load.

measured physical voltages and a specified resistance. While this configuration will not be able to perfectly emulate a resistive load due to the unavoidable non-idealities that accompany the second PHIL interface, one advantage of it is the ability to freely define/manipulate the resistance value used. Note that when operating the system in this configuration, the voltage-source amplifier emulating the AC grid must be activated first; if the amplifier emulating the load (operated as a current source) is first activated without the voltage-source amplifier able to sink/source current, the system will temporarily be equivalent to one comprised entirely of current sources and will therefore be inherently unstable.

The scenario presented in this section can be easily extended. The network emulator can emulate a power network of a higher degree of complexity (e.g. multiple buses), and the load emulator can emulate a load of a higher degree of complexity (e.g. non-linear or switched loads).

To demonstrate the operation of the Network Emulator, tests in the Emulated Load configuration are conducted for both high and low SCR values under high loading. High SCR values correlate to strong AC grids, where high currents result in load voltages nearly equaling the source voltages due to minimal voltage drops across the source impedances. Conversely, low SCR values correlate to weak AC grids, where high currents result in load voltages being considerably lower than the source voltages due to greater voltage drops across the source impedances. Accordingly, for low or no current flow, there will be little to no difference in the load voltages for varying SCR values as the source impedance voltage drops will be significantly smaller.

Voltage and current waveforms of the emulated resistive load are provided in Figs. 24.30 and 24.31 for SCR values of 15 and 200, respectively, under high current flow. For a direct comparison, these tests were conducted with the same source voltage ($V_{abc} = 120 \ V_{RMS}$) and the same emulated load resistance ($R_{load} = 10 \ \Omega$). For an SCR of 15, the voltage drops across the source impedances are shown to be an average of 3.13 V, where for an SCR of 200 the voltage drops are reduced to an average of 2.33 V.

24.4.3.2 PVGS Emulator

The second PHIL application case discussed here is that of a PVGS Emulator, with its basic concept shown in Fig. 24.32. The main components of this system are the PV array, boost converter, inverter, and external AC grid. The PV array, along with a parallel capacitor, is connected to the input of the Boost Converter, which is used to optimize the power extraction of the array. The output of the Boost Converter is connected to a battery used to maintain a steady output voltage as well as an inverter used to perform the necessary DC to AC conversion such that the PVGS can control its power flow to the external grid/load.

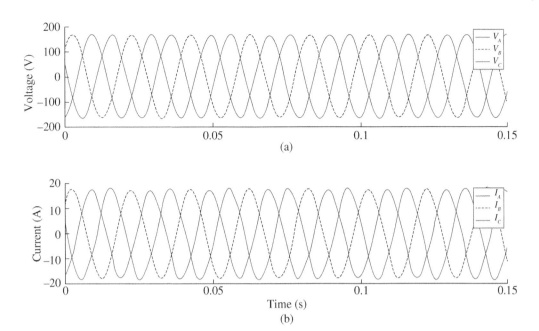

Figure 24.30 Network emulator – high current flow at an SCR of 15, (a) V_{RMS} = 116.87, (b) I_{RMS} = 12.34.

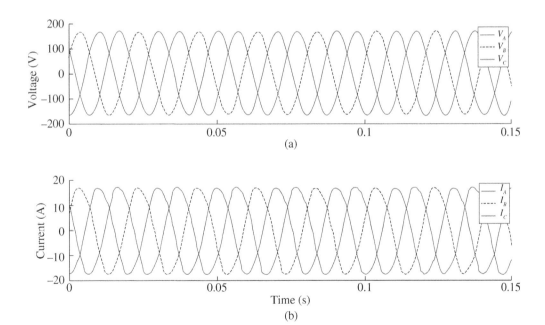

Figure 24.31 Network emulator – high current flow at an SCR of 200, (a) V_{RMS} = 117.67, (b) I_{RMS} = 12.34.

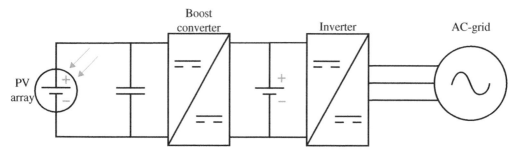

Figure 24.32 PVGS emulator basic concept.

The controller of the boost converter utilizes a Maximum Power Point Tracking (MPPT) algorithm designed to extract the maximum energy possible from the PV array. The controller achieves this by generating appropriate duty cycles for the Boost Converter such that the PV array is operated at the point on its Power–Voltage (*P–V*) characteristic curve where its power output is greatest, referred to as the maximum power point. A basic MPPT algorithm used for the PVGS Emulator, known as Perturb and Observe, functions by slightly adjusting the voltage output of the PV Array and measuring the resultant power output to determine the polarity of the next adjustment. If this power difference is positive, a further adjustment of the same polarity is made; if this difference is negative, the next adjustment made is of opposite polarity. While being simple to implement, notable drawbacks of this algorithm are the continuous output power oscillation of the PV array along with the potential for the algorithm to become stuck at a local maximum of the *P–V* curve rather than the global maxima.

In certain cases when the maximum power production of the PVGS is not desired – such as cases of power surpluses in the connected AC network – the boost converter can be controlled such that

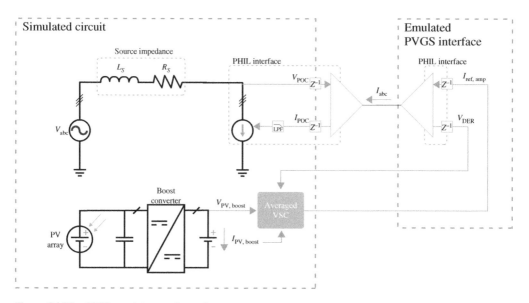

Figure 24.33 PVGS emulator configuration.

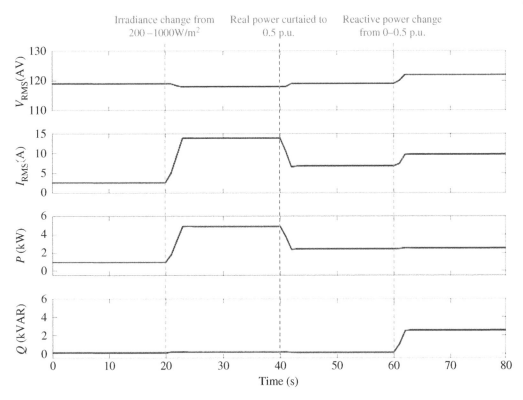

Figure 24.34 PVGS emulator – irradiance and real/reactive power setpoint changes (from top to bottom: RMS voltage, RMS current, 3-phase real power flow, 3-phase reactive power flow).

the PV array is intentionally operated away from its maximum power point in a mode referred to as curtailment. In this mode, the boost converter controller algorithm tracks a real power reference value along the P–V curve of the PV array instead of the maximum value.

PVGS emulation is illustrated using the setup described in Fig. 24.33. The setup consists of a power amplifier emulating a PVGS, while another power amplifier operates as a network emulator. The references sent to the corresponding power amplifiers and the feedback quantities are also described in the same figure.

To demonstrate the operation of the PVGS Emulator, this section contains results taken after various control input changes for a hardware configuration with a nominal power of 5 kVA. Figure 24.34 shows RMS voltage and current at the POC along with the total 3-phase real and reactive power flows during changes in the irradiance reference, curtail command, and reactive power references. At the time of 20 seconds, the irradiance is increased from 200 W/m^2 to the nominal value of 1000 W/m^2, resulting in a corresponding real power increase from approximately 1–5 kW (i.e. 20–100% of nominal power). At 40 seconds, the real power output of the PVGS is curtailed down to 0.5 p.u., resulting in a corresponding output of 2.5 kW. Finally, at 60 seconds, the reactive power reference of the PVGS is increased from 0 to 0.5 p.u., resulting in a corresponding output of 2.5 kVAR. Due to the source impedances of the emulated network along with the impedances of the physical wires in the hardware setup, the RMS voltage can be seen fluctuating as the various setpoints change.

24.5 Concluding Remarks

DERs and microgrid systems offer several benefits. These include increased robustness, improved operability, and stability of electric networks, and the possibility of integration of diverse types of renewable energy sources. DERs and microgrid systems are however complex; they consist of multiple power converters to facilitate renewable source integration. These systems also have multiple layers of control, locally for the individual DER and globally to manage the overall microgrid systems. Hence, to develop a reliable microgrid system, it is essential to have advanced models, simulation tools, and hardware technologies.

In Section 24.2, challenges, considerations, and applications of using digital real-time simulation in the context of microgrid systems were presented. For digital real-time simulation, it is important to consider the limitations regarding hardware and mathematical computations. The optimized power converter models and other circuit decoupling techniques are key factors to allow real-time simulation, CHIL, and PHIL tests for microgrid applications. For some DER converters with high switching frequencies (e.g. 10 kHz or more), the models need to be implemented and executed in FPGA with sub-microsecond timesteps to maintain the model accuracy on the converter switching behaviors. A general FPGA solver, which requires no FPGA knowledge of users but is capable of running models in FPGA with a few hundred nanosecond timesteps, is explained in this section. This section also presented the smart inverter concept, which can be used to analyze the impact of implementing grid-support functions and ride-through capabilities in a microgrid control system. Finally, real-time simulation was presented as a tool for the implementation of digital twins. This concept is useful not only to help designers and supervisors to make better decisions for the microgrid, but also can be used to estimate parameters or signals to improve the accuracy of the simulation model.

In Section 24.3, CHIL as tools to study microgrid systems were presented. CHIL is an approach used for testing control equipment connected to a virtual component or microgrid. It can be used to test the microgrid controllers according to standards such as the IEEE P2030.8, which provides guidelines for testing scenarios and features of a microgrid controller in an organized combinational way. Additionally, CHIL can be used to test low-level controllers programmed in microcontrollers. With this method, the peripheral configuration and control program can be validated for the microcontroller or digital system used to control a VSC.

Section 24.4 presented considerations and applications of PHIL simulations in the context of microgrid studies. While selecting a power amplifier, specifications that must be considered are discussed in detail. The amplifier voltage/current rating, mode of operation, sourcing/sinking requirement, bandwidth, slew rate, and latency, each have an impact on the achieved closed-loop PHIL performance. An overview of PHIL interfacing is discussed next, where the impact of the amplifier dynamic response, the feedback filter and DUT impedance, and their impact on closed-loop stability is discussed. Two applications are considered in the form of network emulation and PVGS emulation. The setup of PHIL systems to realize each of these applications is discussed in this section.

24.6 Exercises

1. What are the challenges when modeling microgrids for real-time simulation and HIL test applications?

2. Which type of real-time simulation is required in each of the following scenarios? (Digital real-time simulation, CHIL, PHIL):
   (a) A vendor wants to test a hardware device which contains a microgrid high-level controller that optimizes the degradation of batteries. The device is able to read low-level voltages and send control commands using TCP/UDP communication.
   (b) A research laboratory has developed a novel DC–AC converter and wants to determine how it reacts under abnormal conditions at the POC.
   (c) A student wants to analyze the behavior of a microgrid depending on the modification of the power set points of the grid-following DERs.
3. Assume a simple microgrid scheme that contains four DERs and two residential loads, all of them connected to a central node. A high-level microgrid controller needs to be implemented to ensure the correct power balance.
   (a) Draw a single-line diagram of the microgrid. Draw with a different color the information links that need to be connected to and from the microgrid controller (measurements, control commands, breaker commands, etc.)
   (b) Propose a scheme for CHIL experimentation by indicating which parts of the scheme drawn in part "a" need to be programmed in the real-time simulator and which of them should be programmed in an industrial computer.
4. Which of the following scenarios of voltage/current source amplifiers directly connected to a common bus are inherently unstable in a PHIL system? Assume no other components (e.g. a resistive load) are present.
   (a) Two voltage-source amplifiers and one current-source amplifier
   (b) One voltage-source amplifier and one current-source amplifier
   (c) Three current-source amplifiers
   (d) One voltage-source amplifier
   (e) One voltage-source amplifier and two current-source amplifiers.

References

1 Li, W. and Zhang, F. (2020). A general interpolated model of voltage source converters for real-time simulation and HIL test applications. In: *2020 IEEE Energy Conversion Congress and Exposition (ECCE)*, 6155–6161. IEEE.
2 Meng, X. and Li, W. (2022). Equivalent circuit modeling method for real-time simulation of multi-active bridge based solid-state transformer. In: *2022 IEEE Applied Power Electronics Conference and Exposition (APEC)*, 798–802. IEEE.
3 Dufour, C., Cense, S., Ould-Bachir, T. et al. (2012). General-purpose reconfigurable low-latency electric circuit and motor drive solver on FPGA. In: *IECON 2012-38th Annual Conference on IEEE Industrial Electronics Society*, 3073–3081. IEEE.
4 Pejovic, P. and Maksimovic, D. (1994). A method for fast time-domain simulation of networks with switches. *IEEE Transactions on Power Electronics* 9 (4): 449–456.
5 Bélanger, J., Yamane, A., Yen, A. et al. (2013). Validation of EHS FPGA reconfigurable low-latency electric and power electronic circuit solver. In: *IECON 2013-39th Annual Conference of the IEEE Industrial Electronics Society*, 5418–5423. IEEE.
6 Yamane, A., Rangineed, T.K., Gregoire, L.-A. et al. (2019). Multi-FPGA solution for large power systems and microgrids real time simulation. In: *2019 IEEE Conference on Power Electronics and Renewable Energy (CPERE)*, 367–370. IEEE.

7 IEEE Standard 1547:1547-2018 (2018). *IEEE Standard for Interconnection and Interoperability of Distributed Energy Resources with Associated Electric Power Systems Interfaces*. Distributed Generation Photovoltaics and Energy Storage.

8 Qoria, T., Cossart, Q., Li, C. et al. (2018). WP3-Control and Operation of a Grid with 100% Converter-Based Devices. Deliverable 3.2: Local Control and Simulation Tools for Large Transmission Systems. *MIGRATE Project*.

9 Song, X. (2023). Machine Learning assisted Digital Twin for event identification in electrical power system. PhD thesis. Universitätsverlag Ilmenau.

10 Koralewicz, P. (ed.) (2017). *Microgrid Controller Procurement Information Packet*, vol. 1.6. NREL.

11 IEEE Study, 2018:1-42 (2018). *IEEE Standard for the Testing of Microgrid Controllers*. IEEE.

12 Salcedo, R., Corbett, E., Smith, C. et al. (2019). Banshee distribution network benchmark and prototyping platform for hardware-in-the-loop integration of microgrid and device controllers. *The Journal of Engineering* 2019 (8): 5365–5373.

13 Plaza, J.D.V., Patarroyo-Montenegro, J.F., and Andrade, F. (2020). Development and implementation of a low-cost research platform for control applications for inverter-based generators. In: *2020 22nd European Conference on Power Electronics and Applications (EPE'20 ECCE Europe)*, 1–9. IEEE.

25

Real-Time Simulations of Microgrids: Industrial Case Studies

Hui Ding, Xianghua Shi, Yi Qi, Christian Jegues, and Yi Zhang

25.1 Universal Converter Model Representation

Universal converter models (UCMs) are represented differently in deblock mode and block mode. For the deblock mode, the descriptor state space (DSS) method is used; for the block mode, the predictive switching method is used. For a single-phase 2-level voltage-source converter (VSC) as shown in Fig. 25.1(a), the positive and negative DC bus voltages are v_P and v_N. The middle node voltage is v_M. The firing pulses of the top and bottom switches are S_T and S_B, which are always required to be complementary.

This single-phase converter can be represented in Fig. 25.1(b), where R_{on} is the switch conducting resistance, which is user-configurable to control the power loss of the converter. In this equivalent circuit, the three controlled sources can be calculated by Eq. (25.1).

$$\begin{cases} v_{cs} = S_T \cdot v_P + S_B \cdot v_N \\ i_{cs\_P} = S_T \cdot i_{ac} \\ i_{cs\_N} = S_B \cdot i_{ac} \end{cases} \tag{25.1}$$

The above equivalent circuit cannot be written in the format of Dommel's companion circuit. Therefore, it cannot be overlaid with the system conductance matrix. The easier way to solve the equivalent circuit is to decouple the AC and DC sides. However, this splitting method will introduce a time-step delay, which may introduce numerical instability issues and fictitious power imbalance on the AC and DC sides. To avoid these two issues, the equivalent converter model should be solved with the external circuit simultaneously. For a single-phase 2-level AC/DC converter, there is an existing AC inductor, AC resistor, and DC capacitors. Considering these three dynamic elements as shown in Fig. 25.2, the converter model can be represented by three differential equations (Eq. (25.2)).

$$\begin{cases} L_{ac}\frac{di_{ac}}{dt} + R_{eq}i_{ac} + v_{cs} = v_a \\ i_{c\_P} = C\frac{d(v_P - v_O)}{dt} \\ i_{c\_N} = C\frac{d(v_N - v_O)}{dt} \end{cases} \tag{25.2}$$

where R_{ac} and L_{ac} are the converter AC resistance and inductance; $R_{eq} = R_{ac} + R_{on}$. Besides these three dynamic equations, one additional algebraic equation to describe the summation of the currents flowing into the converter can be added as in Eq. (25.3).

$$i_{ac} + i_P + i_N + i_O = 0 \tag{25.3}$$

Microgrids: Theory and Practice, First Edition. Edited by Peng Zhang.

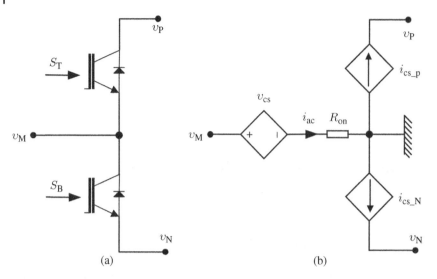

Figure 25.1 Single-phase 2-level converter representation. (a) Detailed switching model. (b) Equivalent circuit model.

With Eqs. (25.1)–(25.3), the single-phase 2-level converter model can be rewritten with the matrix format. This is the DSS representation of the single-phase 2-level converters. By solving the DSS equation, the algebraic and differential equations are solved simultaneously.

As the DSS representation is only valid for the deblock mode of converter operation, the block mode should be considered and solved separately. The predictive switching method developed in [2] can be highly reliable to predict the ON/OFF switch statuses for the next simulation time step in the fixed time-step RTS.

For the 2-level converters, the statuses of each leg have only four combinations, i.e. OFF/OFF, OFF/ON, ON/OFF, and ON/ON. Figure 25.3 shows Dommel's companion circuit of a single-phase 2-level converter in block mode representation.

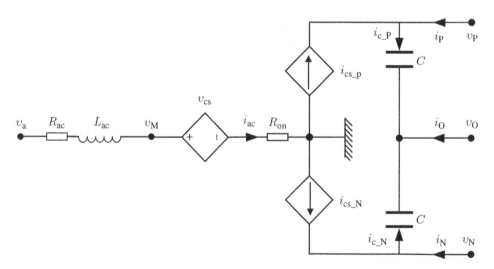

Figure 25.2 Single-phase 2-level converter representation considering the dynamic elements.

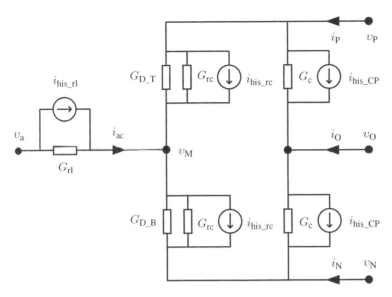

Figure 25.3 Single-phase 2-level converter companion circuit in block mode considering the dynamic elements.

In block mode, the diode in each leg is represented by a two-state resistor. The RC damper circuit is in parallel with each diode. The conductance of the RC damper is G_{rc}; the conductance of the diodes is G_{D_T}, G_{D_T}; the conductance of the AC inductor and DC capacitor is G_{rl}, G_c. With the predictive algorithm, a reasonable combination of the statuses for each leg can be chosen before solving the circuit. A maximum of four predictions is needed for each leg.

25.2 Practical Microgrid Case 1: Aircraft Microgrid System

The aircraft electrical system case is shown in Fig. 25.4. It contains three main buses, i.e. 115 V AC bus (phase voltage)/ 200 V (line-to-line voltage) with 400 Hz frequency, 270 V DC bus, and 28 V DC bus [3].

- To form the 115 V AC bus, a synchronous generator operating at variable-speed-constant-frequency (VSCF) mode is adopted via a diode rectifier and a two-level voltage source converter (2L-VSC).
- The conversion between the 115 V AC bus and the 270 V DC bus is a transformer rectifier unit (TRU), where 12-pulse TRU is used in the simulation.
- Dual-active-bridge (DAB) converter converts 270 V DC bus to 28 V DC bus. Both the 270 and 28 V DC buses have battery support.

All three buses have linear and nonlinear loads. For example, the 115 V AC bus has a linear resistive load, an induction machine, and a permanent-magnet synchronous machine (PMSM); the 28 V DC bus has a linear resistive load and a nonlinear DC machine.

25.2.1 Electrical System Modeling

For the aircraft electrical systems, only the converter-related electrical circuits are listed. Each detail of the converter and its controller design can be found in the manual [4]. The 2L-VSC UCM shown

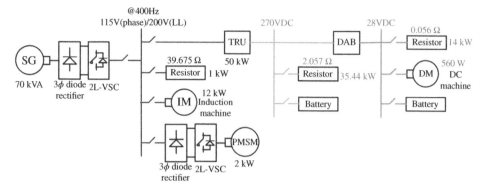

Figure 25.4 Aircraft electrical system architecture.

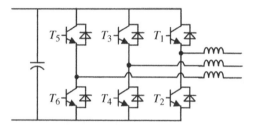

Figure 25.5 2L-VSC UCM converter topology.

in Fig. 25.5 is used in the simulation example. It can be set to a 2-phase or 3-phase converter. Note that:

- This converter model contains three inductors on the AC side and a capacitor on the DC side;
- All firing pulses are zeros to operate it as a diode rectifier;
- Improved firing pulses are used for the controlled converters.

The VSCF generator includes a synchronous generator, a three-phase diode rectifier, and a 2L-VSC inverter as shown in Fig. 25.6. The diode rectifier's DC voltage is controlled by the stator voltages, whereas the stator voltages are adjusted by regulating the field winding's voltage. In the topology shown in Fig. 25.6, the DC bus voltage is controlled by regulating the field winding's voltage, and the AC bus voltage is controlled by the 2L-VSC inverter.

TRU has the benefits of high reliability, low cost, and relatively high efficiency. To meet the aircraft power quality requirement, the multi-pulse uncontrolled rectifier is a preferred option. TRU is generally categorized into 12-pulse, 18-pulse, 24-pulse types, etc. Although a higher pulse number type could have better power quality, it increases the circuit complexity and the amount of power electronic devices. In this simulation example, a 12-pulse TRU in Fig. 25.7 is used to form

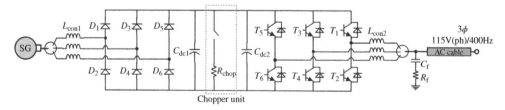

Figure 25.6 Synchronous generator with a three-phase diode rectifier and 2L-VSC.

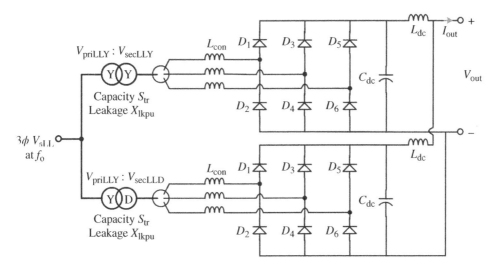

Figure 25.7 12-Pulse TRU circuit.

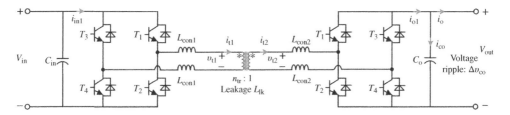

Figure 25.8 DAB circuit.

the 270 V DC bus. Each converter is in parallel to provide a high current output. It is important to select proper leakage reactance and turn ratio for the transformers to obtain the desired DC voltage.

DAB converter provides galvanic isolation, high-power density, fast power reversal, and buck–boost operation with the possibility of a high step ratio [5]. It is commonly used in aircraft electrical systems to convert a high-voltage DC bus to a low-voltage DC bus, e.g. the conversion between 270 and 28 V DC buses. As seen in Fig. 25.8, the DAB converter consists of two H-bridges at two sides of the transformer. The square-wave output of the two H-bridges with a duty ratio of 0.5 is commonly used to avoid the DC offset in the voltage across the transformer's windings.

A PMSM is considered to drive the ball screw actuator in aircraft. The PMSM is connected to the 115 V(phase) AC bus via a 2L-VSC and a diode rectifier. The topology for the PMSM load is shown in Fig. 25.9. To avoid overmodulation, the DC-link voltage should be properly selected, and then the inductance in the diode rectifier can be designed.

25.2.2 System Simulation

The main time step for this example is set to 50.0 μs and the sub-time step is set to 50 μs/6 = 8.333 μs. It includes 8 UCMs, 5 three-phase breakers, 7 single-phase breakers, one PMSM, one induction machine, one DC motor, and one synchronous generator. In the SubStep box, DAB is simulated at a 10 kHz switching frequency; The 2L-VSCs of the synchronous generator and PMSM is simulated at 15 and 4 kHz, respectively. The large resistors (i.e. 1.0 e5 Ω) are grounded at the TRU's converter

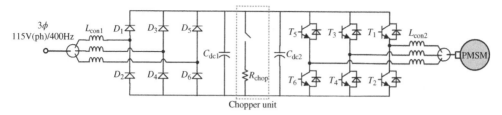

Figure 25.9 PMSM with three-phase diode rectifier and 2L-VSC.

terminals. These resistors are used to provide a reference point at the secondary side of the TRU circuit to avoid a time step overflow error. Similarly, the resistor ($1.0\,e4\,\Omega$) grounded at the DAB's output provides a reference point at the secondary side of the DAB circuit.

Figure 25.10 shows the Runtime layout for the example. Some push buttons, configuration switches, and sliders for reference settings are added. Some meters are also shown along with converters in the simple graph, including voltages, power, modulation index, and so on.

In this example, the normal speed range of the synchronous generator is set to 10,000–20,000 rpm, and the synchronous generator is in locked mode, i.e. the rotor's speed is locked to a certain value, to emulate the connection between the shaft of the synchronous generator and the aircraft engine.

If the synchronous generator's speed is out of the normal range, the machine will be offline and the 270 and 28 V batteries will provide the power to the loads. The batteries' initial state of charge (SOC) is set to 100%. Once the batteries' SOC decreases to 60%, they will be tripped and the system will be shut down. When the synchronous generator speed recovers to the normal range, it will charge the batteries again.

Figure 25.11 shows the simulation results under the synchronous generator at 15,000 rpm and all load breakers closed. It is seen that the system starts smoothly to establish the 200 V line-to-line (LL)/400 Hz AC bus, 270 V DC bus, and 28 V DC bus.

For the steady state of the synchronous generator at 15,000 rpm, the AC bus voltage is closely controlled to 200 V (LL), the TRU's output DC voltage stays close to 270 V, and the DAB output voltage is equal to 28 V. For the synchronous generator's diode rectifier, the DC bus voltage is controlled at 320 V as designed. The DC bus voltage for the PMSM's diode rectifier is 259V, which is close to the theoretical value (260.0 V). Therefore, all the operating points are matched well with the theoretical design.

Dynamic tests regarding the synchronous generator online or offline were conducted by changing the synchronous generator's speed. Figure 25.12 shows the results with speed changing from 15,000 to 9000 rpm. In this case, the synchronous generator should be offline and the batteries should be connected to power the loads. The simulation results show that the currents from the synchronous generator are zero. The currents of the batteries are negative, and their SOCs are decreased. Figure 25.13 shows the results under the synchronous generator's speed changing from 9000 rpm to 15,000 rpm. It is seen that the 115 V AC bus is reestablished by the synchronous generator and its converters, the currents of the batteries become positive, and their SOCs are increasing.

25.3 Practical Microgrid Case 2: Banshee Power System

The Banshee power system corresponds to a real-life, small industrial facility supplied via three utility radial feeders. The system resembles microgrids seen around the world and presents challenges found in a community microgrid, a small island, or industrial facilities, making it a solid benchmark

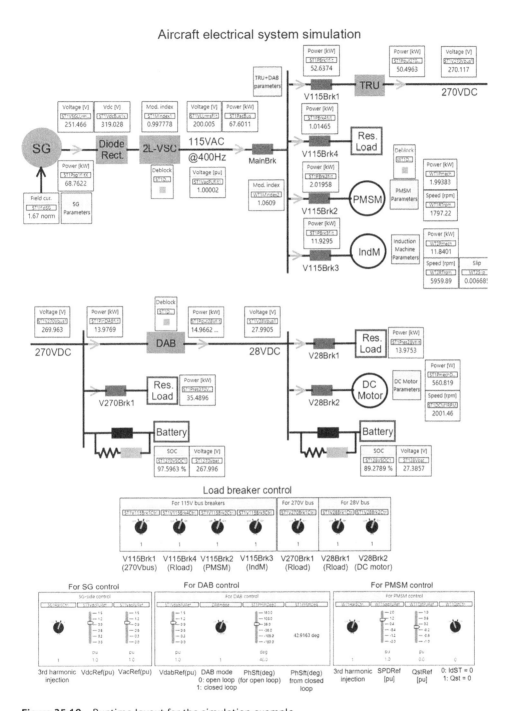

Figure 25.10 Runtime layout for the simulation example.

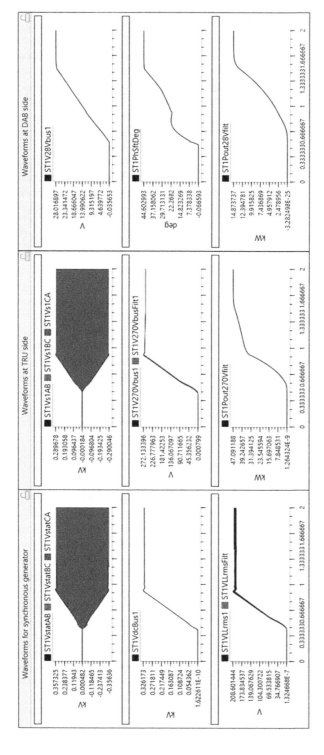

Figure 25.11 Waveforms for AC and DC buses generation with synchronous generator speed at 15,000 rpm.

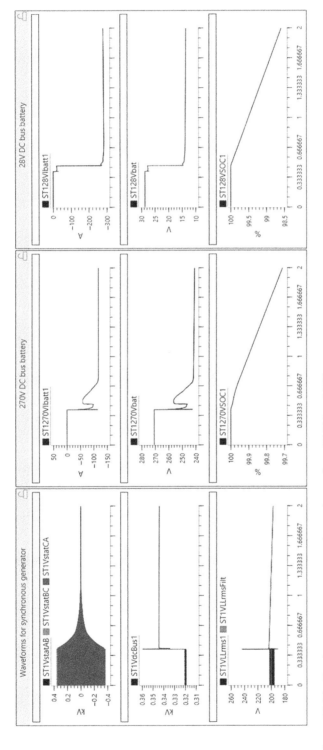

Figure 25.12 Synchronous generator speed step change from 15,000 to 9000 rpm.

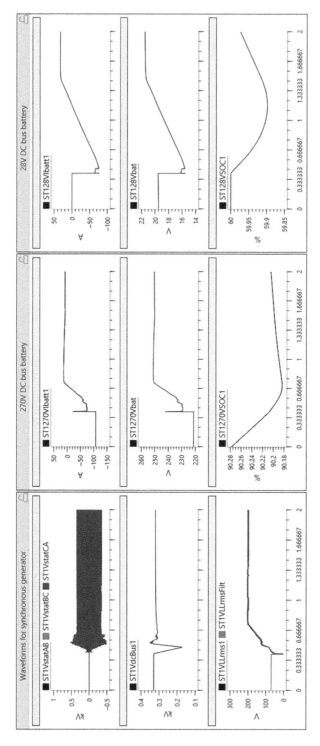

Figure 25.13 Synchronous generator speed step change from 9000 to 15,000 rpm.

for evaluating microgrid performance [6, 7]. Users can interface real industrial-grade microgrid controllers with a simulated microgrid, and observe the controllers' behavior under various operating conditions. The user can also test the various control functions of the microgrid controller such as (but not limited to) automatic generation control (AGC), voltage control, islanding controls, optimal power dispatch control, synchronization controls, black start, and load restoration controls etc. [8, 9].

25.3.1 System Modeling

The microgrid is composed of three adjacent feeders that may interconnect through normally open tie switches. System voltages include 13.8 kV at the distribution level and service voltages of 4.16 kV, 480 V, and 208 V. There are 18 aggregated loads which are classified as either critical, priority, or interruptible. Figure 25.14 shows the one-line diagram of the Banshee system. The major components of the Banshee system model include transformers, cables, breakers, aggregated Loads, motor loads, diesel generator, natural gas-fired combined heat and power generation, photovoltaic (PV) generation, battery energy storage system (BESS), and microgrid controls. The parameter calculation for each type of component can be found in the sample case manual [6].

The Banshee system includes a large number of circuit breakers used for connecting loads, generation, and various interconnection points between the three areas. Each breaker is controlled via a corresponding breaker control component, allowing for control via an external trip and close signals, or manually by use of push buttons within the Runtime environment.

There are two types of loads present in the Banshee system, aggregated loads and motor loads. About 18 aggregated loads are categorized as either critical, priority, or interruptible. The power factor and load select signals for each load are user-configurable. By default, all loads are selected and operated using a power factor of 0.9 lagging. Each of the aggregated loads has been modeled using a dynamic load component. The real and reactive power set points for each aggregated load are computed based on the load select and power factor signals.

The motor loads are modeled as induction motors driving a 200 hp chiller compressor. The torque–speed characteristic of the chiller compressor is modeled using a typical per-unitized torque–speed characteristic for a compressor load. There are four different types of generation assets available within the Banshee system:

- Diesel generator. The diesel generator model is based on a Caterpillar(CAT) C175-20. Simple governor (TGOV1) and exciter (IEEE Type 1) models are used alongside the synchronous generator model.
- Natural Gas Fired Combined Heat and Power (CHP) Generation. The natural gas-fired combined heat and power generation is based on a general electric (GE)\Jenbacher J620. Simple governor (TGOV1) and exciter (IEEE Type 1) models are used alongside the synchronous generator model.
- Photovoltaic Generation. The Banshee System also includes PV generation rated for 5 MVA. The PV generation by using the PV array component to supply the DC voltage for the converter. The converter is then modeled using a UCM with modulation waveform inputs. The controls for the PV generation utilize a maximum power point tracking (MPPT) algorithm which computes the DC voltage set point required for maximum power transfer based on the temperature and insolation levels of the PV array. This DC voltage set point then feeds into the outer loop DC-bus voltage control, which computes a corresponding real power set point. The real and reactive power is then fed into the inner current control loop operating in the DQ reference frame and a set of three-phase modulation waveforms is synthesized. These modulation waveforms are then used in the UCM converter component.

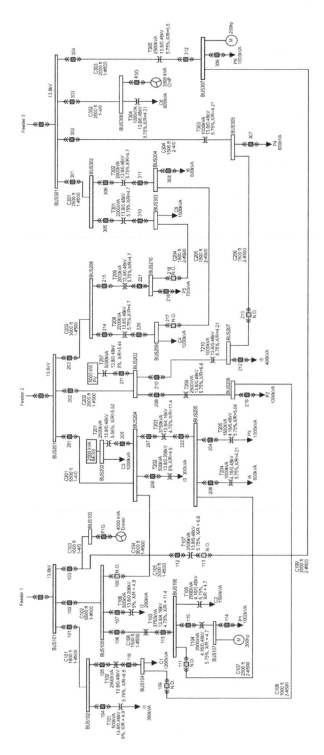

Figure 25.14 Banshee system single line diagram.

- Battery Energy Storage System. The Banshee system also includes a battery energy storage system (BESS) rated for 3 MVA. The batteries supply the DC voltage for the converter, which is modeled using a UCM with modulation waveform inputs. The BESS system has two control modes, one for grid-connected mode and another for islanded mode. In both modes of operation, the same inner current control loop is used with only the outer loop controls varying. For grid-following mode, an outer real/reactive power control loop is used and for islanded mode, an outer voltage control loop is used. Furthermore, in grid-following mode a PLL is used to generate the required phase information for the controls, while in grid-forming islanded mode a voltage-controlled oscillator (VCO) is used to generate the required phase information.

25.3.2 System Simulation

The Runtime environment for the Banshee system is contained within a group whose background image is the single-line diagram of the Banshee system as shown in Fig. 25.15. All of the major components within the Banshee system (i.e. Diesel generator, PV generation, BESS, Motor loads, etc.) each have their group containing all of the relevant controls for that component. One of the main points of interest when simulating a microgrid system is to study its behavior under various islanding conditions. In an islanded condition, the microgrid controls must maintain a balance between generation and load to preserve the system frequency. The following describes the behavior of the system when area 2 is islanded.

Note that as the frequency of area 2 begins to drop, interruptible loads I3 and I4 may be shed to maintain the balance between generation and load. Furthermore, recall that upon detecting an islanding condition for area 2, the BESS switches its control mode from grid-following or real/reactive power control to the grid-forming islanded or voltage control mode. Once in voltage control mode, the BESS can control the magnitude and frequency of the system voltage by injecting the required power flows into the network. This voltage control is critical because the controls for PV generation depend on the ability of its phase-locked loop (PLL) to accurately track the phase information of the voltage. Therefore, if the BESS is not able to accurately control the magnitude and frequency of the system voltage, then it could also cause the controls for PV generation to fail.

Run the simulation and take note of the initial states of circuit breakers 206 and 208 used to connect interruptible loads I3 and I4 to the feeder. The initial state of these breakers should be closed, meaning that both interruptible loads I3 and I4 are connected to the feeder. The default real and reactive power set points of the BESS system are 0.0 MW and 0.0 MVar. By tripping the POI breaker at the head of the feeder by pushing the trip push button, area 2 will be in island mode. Upon tripping the POI breaker, there are a few points to pause and ponder over (Fig. 25.16):

1. The conditional graphics for the POI breaker for the area has changed from close to open.
2. The decline in the system frequency as measured by the phasor measure unit (PMU) for area 2. This is visible by inspecting the frequency meter which has been displayed as a graph.
2. None of the loads have been shed as the BESS can maintain the voltage magnitude and frequency within acceptable limits.
4. The controls for the BESS have switched over from the grid-connected or real/reactive power control mode to the islanded or voltage control mode. Experiment with the islanded or voltage control mode of the BESS. Since our DQ reference frame is aligned along the d-axis, the d-axis component of the BESS voltage will directly correspond to the magnitude of the AC voltage. Recall that if the frequency of the islanded area drops below a certain point, our load-shedding controls will begin to shed off the interruptible loads I3 and I4.

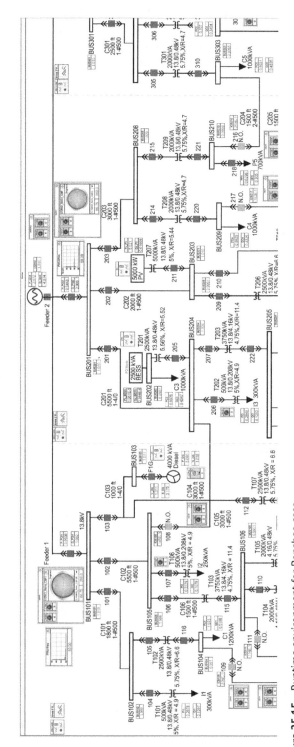

Figure 25.15 Runtime environment for Banshee case.

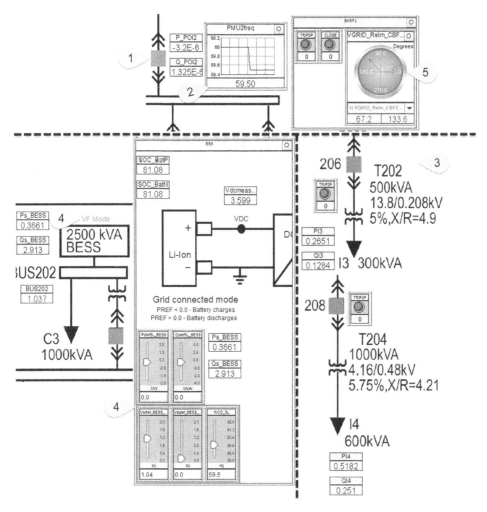

Figure 25.16 Pause and ponder points for area 2 islanding.

5. The vector display shows the voltage phasors on either side of the POI breaker. Since there is a difference in the system frequency of the island (∼59.5Hz) and that of the infinite bus (60Hz), the two phasors rotate with respect to one another. Note that by using the BESS to control the frequency of the islanded portion of the network, it is possible to change the relative of these speeds of the two phasors by virtue of the frequency difference between the two systems. Recall that when the BESS is in voltage control mode, the frequency of the islanded portion of the network can be controlled via the frequency slider.

Next, reclose the point of interconnection (POI) breaker for area 2 by pushing the close push button. Notice that upon pushing the close button, the POI breaker does not close immediately due to the synchro-check function within the breaker control component. The POI breaker does not reclose until all of the synchro-check requirements (i.e. maximum angle difference, slip frequency, voltage difference, etc.) have been met. Notice that once the POI breaker is closed, the BESS controller reverts to the grid-following or real/reactive power control mode.

25.4 Summary

For the microgrid system, controller hardware-in-the-loop (CHIL) testing on the real-time digital simulator ensures the correct functionality of the designed systems. With abundant power-electronic converters in the testing system, the RTS should provide precise converter models to reflect the steady state and dynamic response of the microgrid system. As the RTS cannot use the interpolation technique that has been widely used in non-real-time offline simulations, a new universal converter model with improved firing pulses or modulation waveform input directly is very important for system testing and studies. The proposed converter models in this chapter can precisely cover the interested frequency range without introducing non-characteristic harmonics. Also, the converter models have a smoothing transition between the blocked mode and the deblocked mode. This guarantees the correctness of the simulations in both the switching converter model and averaging converter model.

With the two typical microgrid studies for aircraft electrical systems and Banshee system simulation, either in a larger time step (tens of microseconds) or in a smaller time step (several microseconds), users could rely on the real-time simulators to conduct controller hardware-in-the-loop testing, and therefore significantly reduce risk and improve the performance before the deployment of microgrids.

25.5 Exercises

The typical DAB circuit includes two H-bridge converters. For the DAB circuit, the input voltage is 270.0 V; the output voltage is 28.0 V. The phase shift angle between the two H-bridges is set to 40° in the steady state. The switching frequency of the two converters is set to 10.0 kHz. The power rating of DAB is set to 14.56 kW. The high-frequency transformer rating is set to 15.0 kVA. If the AC inductance of the converter on the 270.0 V side is set to 1.0 μH and 0.1 μH on the 28.0 V side, what the leakage of the transformer should be?

To control the output voltage fluctuation to be less than 0.5 V, the DAB capacitance of the 28.0 V side should be properly chosen. If the total output current of DAB is 520.0 A, what is the minimum capacitance?

References

1 Lian, K.L. and Lehn, P.W. (2005). Real-time simulation of voltage source converters based on time average method. *IEEE Transactions on Power Systems* 20 (1): 110–118.

2 Maguire, T., Elimban, S., Tara, E., and Zhang, Y. (2018). Predicting switch ON/OFF statuses in real time electromagnetic transients simulations with voltage source converters. In: *2018 2nd IEEE Conference on Energy Internet and Energy System Integration (EI2)*, 1–7. IEEE.

3 RTDS Manual. *Simulation of Aircraft Electrical System*. In: Example Cases 17, Aircraft Electrical System.

4 Yazdani, A. and Iravani, R. (2010). *Voltage-Sourced Converters in Power Systems: Modeling, Control, and Applications*. Wiley.

5 Akagi, H., Kinouchi, S.-i., and Miyazaki, Y. (2016). Bidirectional isolated dual-active-bridge (DAB) DC-DC converters using 1.2-kV 400-A SIC-MOSFET dual modules. *CPSS Transactions on Power Electronics and Applications* 1 (1): 33–40.

6 RTDS Manual. *Banshee Microgrid Sample Case*. In: Example Cases 04, Microgrids.

7 MIT Lincoln Laboratory Electric Power HIL Controls Consortium Repository. https://github.com/PowerSystemsHIL/EPHCC/ (accessed 10 October 2023).

8 Manson, S., Nayak, B., and Allen, W. (2017). Robust microgrid control system for seamless transition between grid-tied and island operating modes. *44th Annual Western Protective Relay Conference*.

9 Salcedo, R., Corbett, E., Smith, C. et al. (2019). Banshee distribution network benchmark and prototyping platform for hardware-in-the-loop integration of microgrid and device controllers. *The Journal of Engineering* 2019 (8): 5365–5373.

26

Coordinated Control of DC Microgrids

Weidong Xiao and Jacky Xiangyu Han

Power distribution and transmission are currently dominated by alternative current (AC), thus, the recent R&D naturally focuses on AC microgrids and virtual power plants (VPP) to relieve the persisting difficulties of frequency stability, efficiency, power quality, synchronization, interconnection, and grid resilience. It becomes critical since the industry is facing a fast increase in renewable power generation to achieve the goal of zero carbon emissions. The economic benefit of AC has been slowly diminishing due to the fast development of power electronics, control engineering, distributed generations, and modular energy storage. Researchers often overlook the importance of DC microgrids in eliminating the long-term dilemma of AC.

DC is experiencing a fast increase in the digital age, widely being deployed to modern lighting devices, home appliances, digital devices, Internet-of-Things (IoT), solar photovoltaic (PV) power generation, and rechargeable batteries [1]. The growth results from the broad utilization and fast advancement in power electronics, control engineering, and IoT technologies. Since the 1970s, high-voltage direct current (HVDC) has been techno-economically preferred for long-distance and bulk power transmission supporting grid efficiency, stability, and resilience [2].

PV and batteries are DC-based and increasingly installed at the distributed level to minimize greenhouse gas emissions and support grid resilience around the world. The utilization of DC networks has been addressed with their advantages for industrial and commercial applications, such as the power supply systems for data centers and the HVDC for long-distance transmissions. Following the success, more applications can be expected to form future power grids based on DC in residential buildings at the low-voltage level and distribution at the median-voltage level to minimize energy loss and system cost. In general, the feasibility, reliability, energy savings, and economics of low-voltage DC power systems have been assessed and proven in many applications around the world.

Therefore, DC microgrids draw recent attention showing the advantages of high efficiency, flexibility, and expandability. The key to promoting DC microgrids lies in the advanced technology that enhances reliability during fault conditions, reduces overall costs and losses by removal of AC–DC conversion, as well as achieves user-friendly operations. Control and coordination play an important role in distributed generation (DG) and DC microgrids. The main research direction is to improve the grid stability, mitigate voltage deviation from the specified rating, and maintain optimal power sharing.

Microgrids: Theory and Practice, First Edition. Edited by Peng Zhang.

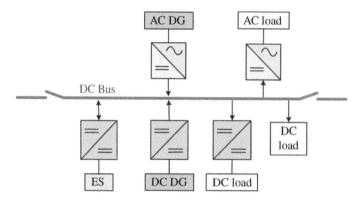

Figure 26.1 DC microgrid, including distributed generation and energy storage.

26.1 DC Droop

The concept of DC microgrids can be illustrated in Fig. 26.1, which includes DC distributed generators (DG), energy storage (ES), and AC DG. The DC bus couples the system interconnection to support AC or DC loads via power interfaces. The microgrid is also capable of being interconnected to other microgrids, AC or DC, via the DC bus and power converters. The system should be coordinated together to support a stable voltage of the DC bus considering generation intermittency, load disturbance, and nonideal factors. DC droop is the common scheme for microgrid coordination [3].

The performance index is defined in (26.1) to compare the effectiveness of different coordination strategies. It measures the deviation from the nominal DC voltage, V_{nom}.

$$eV = \frac{|V_{\text{nom}} - V_{\text{dc}}|}{V_{\text{nom}}} \tag{26.1}$$

where v_{dc} represents the measured voltage of the DC bus in Fig. 26.1.

26.1.1 Linear Droop Control Scheme

The droop method is broadly utilized for load sharing and voltage regulation among distributed generators and parallel converters. It shows the adaptive characteristics without a communication means among the distributed power generation units. A typical control structure is illustrated in Fig. 26.2 representing the DC droop function. All DC/DC converters interfacing power generation in a DC microgrid can be controlled by the same scheme. The DC droop function aims to coordinate the power output of all distributed generation and maintain bus voltage stability regardless of load variation and intermittency. The voltage and current of each DC/DC converter should be sensed

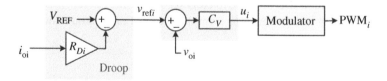

Figure 26.2 Primary control of individual converters coordinated by DC droop.

Figure 26.3 Equivalent circuit of parallel converters with DC droop coordination.

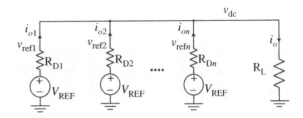

and used for its control scheme. The index i used in the diagram refers to distinguishing individual generation units.

The output voltage of the ith converter, v_{oi}, shall be regulated to follow the command signal, v_{refi}, which is determined by the droop function. The voltage controller is represented as C_V, as shown in Fig. 26.2. Its output is symbolized as u_i, which can be either the signal of duty ratio or phase-shift angle for pulse width modulation (PWM). The PWM signal is used to switch power semiconductors to achieve the required modulation and control. The droop function senses the current, i_{oi}, and determines the voltage command, v_{refi}, expressed by:

$$v = V_{REF} - R_{Di} \times i_{oi} \tag{26.2}$$

where V_{REF} is the specified level for the DC bus voltage. R_{Di} is treated as the droop gain or a virtue resistance to perform the sharing coordination. The voltage control loop determines the duty ratio (d_i) for the PWM signals by regulating the terminal output, v_{oi}, to follow the command signal, v_{refi}, in the distributed DC/DC converter.

The adaptive sharing feature of the droop function can be demonstrated by the equivalent circuit in a steady state by following the DC microgrid structure and the control scheme. When line loss is ignored, the power-sharing concept is illustrated in Fig. 26.3. The system specification defines the bus voltage to be V_{REF}. According to the droop scheme, the bus voltage shall be equal to V_{REF} when no load is applied, $i_o = 0$. Otherwise, the load current is contributed by the distributed generation, expressed by (26.3). The shared current for each unit is determined by the droop function, which is calculated by (26.4) When line loss is ignored. The value of R_{Di} determines the level of shared current for each DG.

$$i_o = \sum_{i=1}^{n} i_{oi} = i_{o1} + i_{o2} + \cdots + i_{on} \tag{26.3}$$

$$i_{oi} = \frac{V_{REF} - v_{dc}}{R_{Di}} \tag{26.4}$$

where v_{dc} is the common DC bus voltage, which can be sensed by each control unit. Thus, the shared current of distributed power generation is determined by the voltage difference or error between V_{REF} and v_{dc}. Figure 26.4 illustrates the voltage deviation (ΔV) between v_{dc} and V_{REF}, of which the highest level of ΔV appears at the maximum load condition. The deviation is caused by the DC droop, expressed in (26.4).

The rating of R_{Di} is determined by the power capacity of individual DG. If the maximum current of DG i is rated as I_{Maxi}, the virtual resistance can be computed by (26.5). Thus, each DG can automatically contribute based on its capacity and droop coordination.

$$R_{Di} = \frac{\Delta V}{I_{max\,i}} \tag{26.5}$$

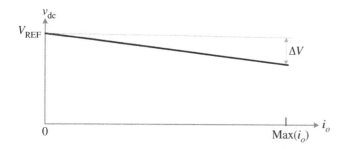

Figure 26.4 I–V characteristics of the DC bus caused by droop.

The major drawback of the droop method is the unavoidable voltage deviation between the light and full load conditions. The issue can be eased by improved algorithms or data communication among distributed generation (DG) units.

26.1.2 Piecewise Linear Formation of DC Droop

A piecewise linear formation of droop strategy (PLFDS) was proposed in 2019 to achieve high and balanced performance for both voltage regulation and current sharing in DC microgrids [4]. The PLFDS adopts the simple linear droop characteristics but divides into vertical segments uniformly, as shown in Fig. 26.5. In each segment, the equivalent nominal voltage and droop gain of the droop characteristics curve are updated locally and automatically with load conditions. The voltage reference of individual converters is computed by:

$$V_{\text{ref},i} = V_{\text{nom}} - R^j_{Di}\left(I_i - (j-1) \times I^j_{\text{seg}}\right) \tag{26.6}$$

where V_{nom} is the nominal voltage, $V_{\text{ref},i}$ and I_i are the ith converter output voltage reference and output current, and R^j_{Di} is the equivalent droop gain at the jth segment of the droop characteristics curve. The value of j increases discretely from 1 to n, where n is the number of segments. As the

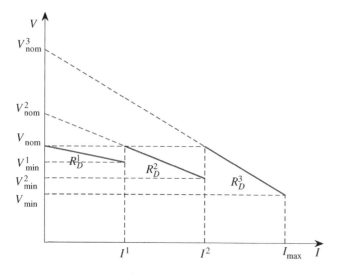

Figure 26.5 Characteristics of a piecewise droop controller ($n = 3$).

droop curve is divided uniformly, the width (I_{seg}^j) of all segments are identical, and the j can be derived by rounding up the result of converter output current divided by segment width:

$$I_{seg}^j = \frac{I_{max}}{n} \tag{26.7}$$

$$j = \left\lceil \frac{I_i}{I_{seg}^j} \right\rceil \tag{26.8}$$

The minimum voltage reference (V_{min}^j) of PLFDS should be the same as the voltage reference of linear droop controller when the current is located at the jth barrier, and the local voltage deviation (ΔV^j) of each segment can be obtained based on the fixed global voltage deviation (ΔV) and maximum current (I_{max}):

$$V_{min}^j = V_{nom} - j \times I_{seg}^j \frac{\Delta V}{I_{max}} \tag{26.9}$$

$$\Delta V^j = V_{nom} - V_{min}^j \tag{26.10}$$

The local equivalent droop gain (R_D^j) can be determined by the local voltage deviation (ΔV^j) and the width (I_{seg}^j) of a single segment:

$$R_D^j = \frac{\Delta V^j}{I_{seg}^j} \tag{26.11}$$

26.1.3 Case Study

A case study is carried out to compare the performance of the droop methods. Figure 26.6 shows two DGs supporting one common load. Two power converters are used to perform the bus voltage regulation and droop function. Table 26.1 gives the system parameters for simulation and performance comparison.

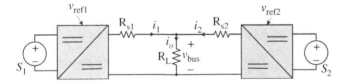

Figure 26.6 DC microgrid configuration for performance evaluation.

Table 26.1 Parameters for case study.

Parameters	Symbol	Value
Nominal bus voltage	V_{REF}	380 V
Bus voltage deviation	ΔV	5 V
Rated load current	$\max(i_o)$	18 A
Line resistance 1	R_{S1}	0.1 Ω
Line resistance 2	R_{S2}	0.3 Ω

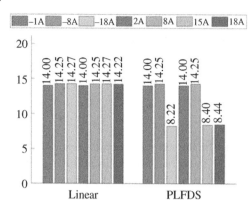

Figure 26.7 Current sharing accuracy comparison of droop schemes in percentage.

Line resistance cannot be completely neglected in a practical DC microgrid due to the distributed nature of generators and loads. The resistance depends on the length of the type of cables used in DG units and load distributions. The difference among the line resistances causes current-sharing inaccuracy since the local DG does not receive the accurate measure of the bus voltage. The DG connected via a low cable resistance contributes more current than the planned level. High resistance in the line lowers the current contribution for the connected DG. In the case study, the line resistances for DG1 and DG2 are represented by R_{s1} and R_{s2}, respectively, as shown in Fig. 26.6. The values are given in Table 26.1. Thus, one performance measure is the current share error expressed by:

$$ eI = \left| \frac{i_1 - i_2}{i_1 + i_2} \right| \tag{26.12} $$

The linear droop and piecewise linear formation are evaluated and compared according to the performance indices defined in (26.1) and (26.12). In this case, the linear droop gain is selected as 0.5 based on the design recommendation in Section 26.1.1. The droop gain for the PLFDS is divided into two sections with a boundary of 10 A. Figure 26.7 illustrates the accuracy of the current sharing performance using (26.12) expressed in percentage value. The current share error is reduced when $|i_o|$ is higher than 10 A since a new droop gain is applied.

The reduction of voltage deviation is demonstrated in Fig. 26.8 showing the advantage of the PLFDS. It is based on the voltage deviation index in (26.1). The improvement can be witnessed

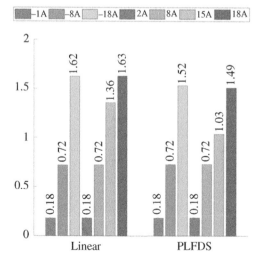

Figure 26.8 Relative voltage deviation comparison of droop schemes in percentage.

when $|i_o|$ is higher than 10 A since a new droop gain is applied. In summary, the PLFDS shows an improved balance in terms of both current-sharing accuracy and DC voltage steadiness. It can be flexibly implemented by combining with other technology to improve overall coordination performance further.

26.2 Hierarchical Control Scheme

The conventional droop method without centralized coordination makes it difficult to eliminate the voltage deviation or unbalanced current sharing [5]. The multitier control scheme is expected to solve the inaccuracy of DC link voltage and power sharing among DGs. Figure 26.9 shows the hierarchical control scheme used for DC microgrids.

A communication link is created to send a command from a centralized regulator, namely the secondary voltage controller (SVC). The high-level controller senses the bus voltage, v_{dc}, overviews the grid condition and determines the compensation value for individual DG set points, ΔV_i. Thus, the terminal voltage of all DGs can be individually adjusted by the primary controller to improve the accuracy of bus voltage and current sharing. A tertiary controller can also appear in the hierarchical structure, which forms a three-tier-level control structure, as shown in Fig. 26.9. Without it, the system is at the two-tier level including the secondary and primary controllers. Algorithms of optimization and artificial intelligence can be implemented in the centralized controller for improved performance.

26.3 Average Voltage Sharing

One drawback of the hierarchical control scheme lies in the reliance on high-bandwidth communication means. The research effort is to minimize the weight of the centralized controller and high-bandwidth communication link. The solution aims for a decentralized structure via low-bandwidth communication. The average voltage sharing (AVS) control scheme was presented to achieve the goal. Figure 26.10 demonstrates the control structure and operational principle.

Figure 26.9 Hierarchical control scheme for DC microgrids.

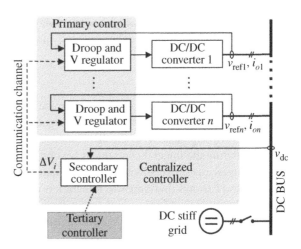

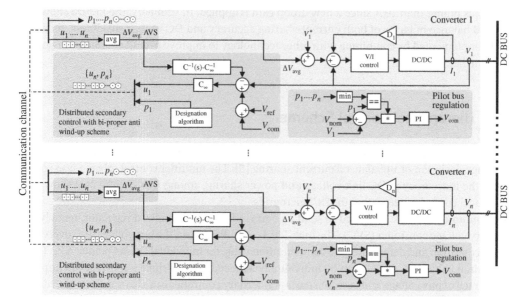

Figure 26.10 Diagram of average voltage sharing control scheme.

The distributed secondary control loop is implemented inside each converter unit, which is different from the conventional hierarchical control scheme. The controller output is then shared via the communication channel with other converters. By receiving all the control output signals from others, the average compensation value is calculated and then fed into the primary control loop. This idea is to emulate the same function of the centralized SVC that regulates the DC bus voltage by aggregating distributed secondary controllers, expressed by (26.13). The distributed secondary SVC is based on a standard proportional-integral (PI) format. All SVCs share the same parameters of the proportional gain of k_{p2} and integral gain of k_{i2}. Thus, the average compensation voltage is determined by (26.14).

$$\Delta V = \sum_{k=1}^{n} \frac{u_i}{n}, k \in \{1, 2 \cdots n\} \tag{26.13}$$

$$\Delta V_{\text{avg}} = k_{p2}(V_{\text{REF}} - V_{\text{avg}}) + k_{i2} \int (V_{\text{REF}} - v_{\text{avg}}) \tag{26.14}$$

where v_{avg} is the average voltage of the converter outputs. The integral term of the secondary controller aims to eliminate the error between V_{REF} and v_{avg} and improve the steady-state performance.

26.3.1 Anti Windup

The integral term in (26.14) aims to minimize the steady-state error between V_{REF} and v_{avg}. Individual input errors are not identical among the DG units. Hence, there is a high chance that the integral windup happens in the distributed secondary controllers. Integral windup is a process of accumulating the integral component beyond the physical limit of controller outputs [6]. The windup causes excess overshoots and accumulated errors leading to poor transient performance.

Figure 26.11 shows that opposite integration actions are induced when minimizing the error between the average reference and bus voltage. When the controller is implemented digitally, overflow occurs to the controller output register, failing the compensation scheme. To solve the

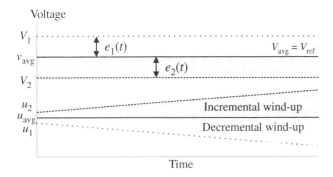

Figure 26.11 Illustration of windup effect in distributed PI controllers.

problem, an anti windup scheme should be applied. The controller in (26.14) can be modified by considering the average controller output effort as the actual output to the primary loop, shown in Fig. 26.10. The secondary controller is then reconstructed using the feedback form of the bi-proper controller, which can be derived as follows:

$$u_k = C_\infty e_k + \overline{C}(s)\hat{u}_k \tag{26.15}$$

where C_∞ is the DC gain of $C(s)$, $\hat{u}_k$ is the actual controller output, and u_k is the unconstrained controller output. $\overline{C}(s)$ is a strictly proper transfer function, expressed by (26.16).

$$\overline{C}(s) = C^{-1}(s) - C_\infty \tag{26.16}$$

The anti windup implementation can be derived the same as the regular average outputs of all the parallel controllers fed into their corresponding primary control loops. Therefore, the equivalent synthesized controller can be derived as (26.14) showing the correction function and elimination of steady-state errors. The feedback form of the bi-proper controller has been proven to be effective to prevent the integral windup effect.

26.3.2 Pilot Bus Regulation

The aforementioned control scheme displays only the regulation capability of the average terminal voltage. Adjusting the individual reference voltage set point reflects directly on the average reference magnitude, which shall be followed by the average terminal voltage. Therefore, an additional voltage regulator is included in the system configuration, as shown in Fig. 26.10. For selecting the pilot bus, additional bits, p_n, are assigned to be sent through the communication channel with the compensation signal to other converters. After receiving all the additional bits $p_1 \cdots p_n$, each controller compares the set with its designated index to determine whether its terminal voltage is selected to be regulated.

Figure 26.12 demonstrates the flowchart of the designation algorithm to select the pilot bus. The system can be initialized sequentially by the user based on the assigned priority. Also, when a failure occurs to the pilot converter, the role of performing pilot bus regulation can be transferred to other converters based on the given index to enhance the system tolerance against faults.

26.3.3 Dynamic Analysis

Figure 26.13 demonstrates the case study that two DGs support a common load to analyze the AVS scheme. A PI controller is used as C_V for the voltage regulation in the droop scheme, as

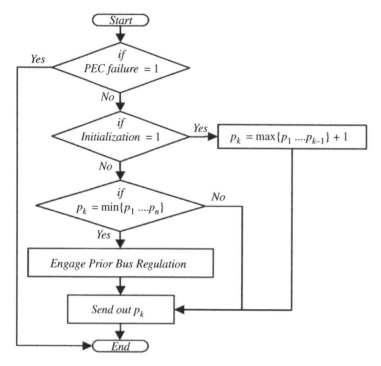

Figure 26.12 Flowchart of the designation algorithm.

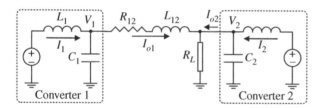

Figure 26.13 Equivalent circuit of two parallel converters sharing a load.

shown in Fig. 26.2. The controller transfer function is expressed by (26.17), where the voltage loop proportional gain is shown as K_p. The voltage loop integral gain is symbolized as k_i. The mathematical model of the system can then be described in the standard state-space representation in (26.18). The system parameters are shown in Table 26.2.

$$C_V(s) = K_p + \frac{K_i}{s} \tag{26.17}$$

$$\dot{X} = AX + BU \tag{26.18}$$

The development and effectiveness of the AVS scheme are demonstrated by a case study, of which The system transition matrix, **A**, can be found according to the system parameters. A pair of eigenvalues are derived as $-849 \pm 4574i$, which indicates resonant characteristics. To investigate the impact of the controller parameters and droop gains on the cable mode, the eigenvalue loci are plotted in Fig. 26.14.

The K_p and droop gains have a noticeable impact on the movement of the cable mode: increment of K_p moves the eigenvalue toward left-hand plain (LHP) and improves the damping performance;

Table 26.2 System parameters for case study.

Term	Value
Filter inductance	$L_1 = 2\,\text{mH}, L_2 = 2\,\text{mH};$
Filter capacitance	$C_1 = 670\,\mu\text{F}, C_2 = 670\,\mu\text{F};$
Cable inductance	$L_{12} = 600\,\mu\text{H};$
Cable resistance	$R_{12} = 280\,\text{m}\Omega;$
Load resistance	$R_L = 8\,\Omega.$

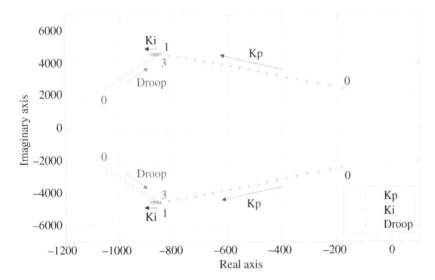

Figure 26.14 Eigenvalue loci of the cable mode with respect to parameter variations.

an increase of the droop gain shifts the mode toward right-hand plane (RHP) and lifts the oscillation frequency. The above observation can be further explained by ignoring the load resistance in the C-L-C circuit. Thus, the dynamic equation as seen from the left side can be derived in the s-domain in (26.19). Assuming both converters have the same proportional gain, K_p, the inductor current of converter 2 can be represented by (26.20).

$$V_1 = (I_1 - sC_1 V_1)(sL_{12} + R_{12}) + \upsilon_2 \tag{26.19}$$

$$I_2 = T_{12}(s)I_{\text{ref2}} = T_{12}(s)K_p(V_2^* - V_2) \tag{26.20}$$

where $T_{12}(s)$ is the closed-loop transfer function from the current reference to the inductor current. Considering the fast current tracking performance, the inductor current is assumed to be equal to the current reference. Thus, the current perturbation in the small-signal region can be derived as:

$$\Delta I_2 = -K_p \Delta V_2 \tag{26.21}$$

The proportional gain behaves like a virtual resistor with the resistance of $1/K_p$, as expressed in (26.21). Based on (26.19) and (26.21), the system dynamic representation is found, as expressed in (26.22).

$$\frac{\Delta V_1}{\Delta U_1} = \frac{b_2 s^2 + b_1 s + b_0}{a_3 s^3 + a_2 s^2 + a_1 s + a_0} \tag{26.22}$$

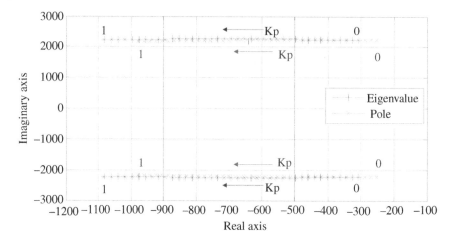

Figure 26.15 Comparison of the eigenvalue loci and pole movements of the cable mode by the variation of K_p while $K_i = 0$ and Droop $= 0$.

where $a_3 = L_{12}C_1C_2$, $a_2 = R_{12}C_1C_2 + K_pK_{12}(C_1 + C_2)$, $a_1 = K_pR_{12}(C_1 + C_2) + L_{12}K_p^2 + C_1 + C_2$, $a_0 = 2K_p + K_p^2R_{12}$, $b_2 = L_{12}C_2$, $b_1 = C_2R_{12} + L_{12}K_p$, and $b_0 = K_pR_{12} + 1$.

The pole movement of the characteristic equation is related to the value of the proportional gain, K_p, as illustrated in Fig. 26.15. The increase of the proportional gain results in a lower value of the virtual resistance, consequently, improving the system damping by shifting the cable resonance poles toward the LHP. Analyzing Fig. 26.13, the current is expressed by (26.23). A small perturbation leads to the s-domain transfer function expressed in (26.24).

$$I_{o1} = K_p(V_{ref} - D_1I_{o1} - V_{o1}) - C_1\frac{dv_{o1}}{dt} \tag{26.23}$$

$$\frac{\Delta I_{o1}}{\Delta V_{o1}} = -sC_{eff} - \frac{1}{R_{eff}} \tag{26.24}$$

where, $C_{eff} = \frac{C_1}{1+K_pD_1}$ and $R_{eff} = \frac{1+K_pD_1}{K_p}$.

The expression in (26.24) indicates the relations between the virtual shunt capacitance, C_{eff} and the droop gain, D_1. The virtual shunt resistance is symbolized as R_{eff}, which is related to the droop gain, D_1, and the proportional gain, K_p. It explains why increasing the droop gain obtains a higher resonance frequency and moves the real part of the cable mode toward the RHP.

To increase the system damping, a proportional–derivative (P–D) droop controller can be used adding a subtraction term, $-sD_{d1}I_{o1}$, into the voltage reference. Thus, the small perturbation effect is expressed by (26.25) derived from (26.23). Figure 26.16 illustrates the equivalent circuit. From (26.25), a high derive droop gain results in high values of L_{eff1} and R_{eff1}, indicating the increase of the damping and reduction of the resonance frequency.

$$\frac{\Delta I_{o1}}{\Delta V_{o1}} = -\frac{1}{X_{eff1}} - \frac{1}{X_{eff2}} \tag{26.25}$$

where, $X_{eff1} = R_{eff1} + sL_{eff1}$, $X_{eff2} = R_{eff2} + \frac{1}{sC_{eff2}}$, $R_{eff1} = \frac{1+K_pD_1}{K_p}$, $L_{eff1} = Dd1$, $R_{eff2} = \frac{K_pD_{d1}}{C_1}$, and $C_{eff2} = \frac{C_1}{1+D_1Kp}$

Figure 26.16 Equivalent circuit of two parallel converters with P–D droop control.

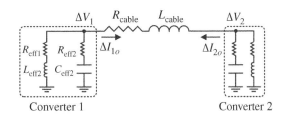

Converter 1   Converter 2

26.4 Bus Line Communication

DC microgrids show the advantage of accommodating battery-based energy storage via DC/DC conversions, which are highly efficient and low-cost [7]. This arises especially when coordinating between multiple energy storage units is required with minimal communication links. This section introduces a decentralized control technique, the bus line communication. It is named active DC bus signaling (A-DBS) to support the integration of distributed energy storage units. This technique aims at transmitting control signals among devices via the main DC bus without using dedicated communication links.

26.4.1 Dual Active Bridge

Bidirectional DC/DC conversion is required to support the charge and discharge operation of batteries. Thus, the topology of dual active bridge (DAB) is adopted as the main power interface. The converter provides a range of advantages including galvanic isolation, soft switching, and power flow control without current sensing. The circuit is demonstrated in Fig. 26.17 showing the two active bridges, the primary and secondary.

The power flow equation of the DAB is shown in (26.26), where P_{avg} is the average value of power flow in a steady state. The switching frequency is represented by ω. The direction and value of P_{avg} are determined by the phase shift value, Φ, between the active bridges. When $\Phi > 0$, the power flow is from the side of C_{DCP} to the right side. The reverse power flow is shown by $P_{avg} < 0$ and achieved by $\Phi < 0$. Other parameters in (26.26) refer to the labels in Fig. 26.17. The winding turns ratio is the transformer is expressed by $1 : n$. The power computation is based on the measurement of voltage only, which indicates another advantage of DABs.

$$P_{avg} = \frac{V_{DCP}V_{DCS}\phi(\pi - |\phi|)}{n\pi\omega L} \tag{26.26}$$

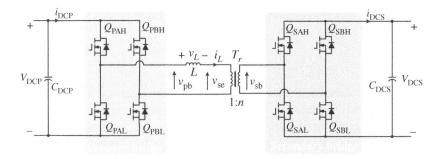

Figure 26.17 Circuit of dual active bridge (DAB).

26.4.2 Active DC Bus Signaling

The A-DBS technique uses the master-slave control framework for coordinating multiple devices. Its operation is described as follows:

- One power conversion unit is assigned as the master unit, which regulates the DC bus within a certain voltage range.
- Except for the master, other units are classified as slave units.
- A set of distinct voltage levels are defined to be "reference voltages," each level can be considered of as a carrier of the encoded message.
- The DC bus voltage is actively regulated at discrete levels, which in turn allows the control signals to be communicated among distributed units.
- The power-sharing coordination is based on the state of charge (SoC) of individual energy storage units.

Figure 26.18 demonstrates a DC microgrid formed by distributed battery energy storage units. It features a network that has three energy storage components, all interfaced by the DAB converters. A single DC source of i_{load} is used to model the combination of DC loads and generators, where the positive value indicates the electric current flowing from the current source into the DC bus. The DC link is equipped with a shunt-connected capacitor, C_{bus}, to mitigate voltage ripples. It is a case that all DABs are configured by the same parameters, L_0 and n, as shown in Fig. 26.18.

Each unit reports the local information including the SoC of batteries. The voltage and current are also locally measured and reported to others. The master unit is required to collect all necessary inputs from the slaves for an optimal decision. The command signal generated by the master unit can be sensed by the slaves via the voltage level of the bus.

26.4.3 Operation Principle and Control Logic

The DC bus voltage is denoted by V_{bus}, as shown in Fig. 26.18. The current equilibrium is expressed by (26.27), where i_{master}, i_{slave1}, and i_{slave2} are currents at the connection point of the master unit, slave unit 1 and slave unit 2, respectively.

$$i_{master} + i_{slave1} + i_{slave2} = i_{load} \tag{26.27}$$

The highest power is reached when the DAB is controlled by the phase shift is set to be $\frac{\pi}{2}$. Based on the power level expressed in (26.26), the maximum current of the DAB output can be calculated by (26.28), where V_{bat} is the battery voltage. The maximum current level of the power interfaces in Fig. 26.18 are the same since all DABs in this case show the same parameters, L_0, ω, and n. It should be noted that a DC microgrid can be configured by different sizes of power interfaces and battery capacities.

$$I_{max} = \frac{V_{bat}\pi}{4\omega L_0 n} \tag{26.28}$$

The amount of current to be dispatched by the slave units can follow the batteries' SoCs based on the maximum current of the power interface, I_{max}. Under the energy surplus condition, The slave unit should enter the charging mode. The set point of the charge current, $I_{charge,i}$ can be determined by (26.29). The index, i, distinguishes the individual slave unit shown as 1 or 2 in this case. When the battery energy is needed by the grid, the discharge current can be rated as $I_{discharge,i}$ according to (26.30). It shows that the lower SoC battery unit receives a higher charging current to make it

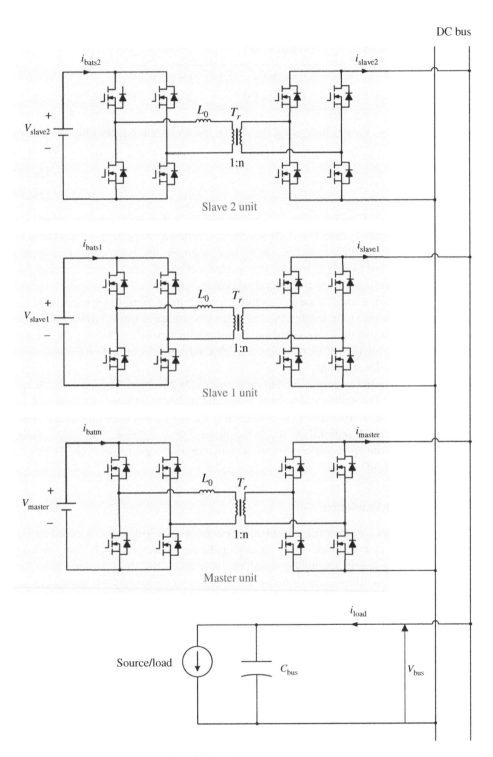

Figure 26.18 DC microgrid based on DABs and battery energy storage.

Table 26.3 Criteria for operation mode selection.

Unit	Discharge mode	Charge mode	Turned off
Slave 1	377.5–378.5 V	381.5–382.5 V	379.0–381.0 V
Slave 2	375.5–376.5 V	383.5–384.5 V	379.0–381.0 V

equal to others. Meanwhile, the discharge case shows that the lower SoC battery unit contributes less than others.

$$I_{\text{charge},i} = I_{\max} \times (\text{SoC}_i - 1) \tag{26.29}$$

$$I_{\text{discharge},i} = I_{\max} \times \text{SoC}_i \tag{26.30}$$

One important parameter shall be defined for V_{bus}, which is 380 V in this case study. The bus voltage can be sensed by all distributed units. The A-DBS technique is based on the levels of the bus voltage, V_{bus}, which follows predefined segments. The voltage variation signals the control command according to Table 26.3 in this case. It shows that the operation of on/off, discharging, and charging is explicitly activated by the concurrent DC bus voltage levels.

The master unit's task can be understood as "scanning," of which the control logic is illustrated in Fig. 26.19 to operate the master unit. It starts by setting the bus voltage at 380 V, leaving both slave units off. It is capable of sensing the load current and deciding which mode to activate, either charging or discharging. Adjusting the DC bus voltage, a specific slave unit is turned on for operation according to the predefined bus voltage.

The slave unit can determine its output current according to (26.29) or (26.30) for battery charge or discharge, respectively. The system can acquire information about any slave unit, recorded by the master unit. Figure 26.20 demonstrates the local control logic of the first slave unit based on the predefined bus voltage level. The control variable is the phase shift, ϕ, where ϕ_{cha} and ϕ_{dis} indicate the values determined by the operating conditions of charge or discharge, respectively. The value is determined by the local SoC.

26.4.4 Case Study and Simulation

A case study is performed to prove the effectiveness of the A-DBS technique in coordinating between multiple energy storage units. Three battery models are selected, which differ in type, capacity, and brand. The specifications are listed in Table 26.4. The three battery units are appropriately stacked up into three arrays with an accumulated capacity of 5 kWh. Following the system schematic, as shown in Fig. 26.18, the system specification is summarized in Table 26.5. The bidirectional power interface is based on DABs, of which the specifications are in Table 26.6.

26.4.4.1 Discharging Test

The first evaluation is performed to demonstrate the discharging operation of battery storage units. The test operation and parameters are summarized in Table 26.7. The simulation result of this test is shown in Fig. 26.21.

The master unit starts by regulating bus voltage at 380 V. The current is sensed to be positive showing $I_m > 0$. At $t = 0.25$ s, the discharging scanning mode is activated. The reference of bus voltage is reduced in steps. Following Fig. 26.19, Slave 1 is on when the bus voltage remains at 378 V, until $t = 0.5$ seconds. Then at the moment before the next adjustment at $t = 0.5$ seconds, the master current is registered and is denoted by $I_{m,1}$. The bus voltage is further lowered to 376 V at

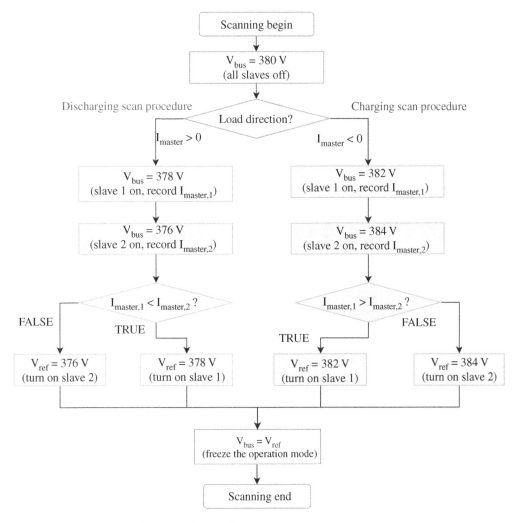

Figure 26.19 Flowchart of master unit operation.

$t = 0.75$, which signals the turn-on of Slave 2 according to the definition in Table 26.3. The master current is again registered and denoted by $I_{m,2}$. The changes in bus voltage over time can be seen in the "Bus Voltage" graph from Fig. 26.21.

The bus line communication virtually realizes that Slave 1 is less in SoC as opposed to Slave 2 since $I_{m,1} > I_{m,2}$. The master unit supplies more energy when it is coupled with slave 1, which implies it has a lesser capacity to give away. Hence, the operation mode is frozen at the one where Slave 2 is on for long-term discharging ($V_{bus} = 376\,V$), until the next round of scanning is called. By doing so, the SoC disparity between Slave 1 and Slave 2 can be gradually mitigated, and eventually, the energy storage balancing is achieved, as indicated in Fig. 26.21.

26.4.4.2 Charging Test

The second test is designed to examine the case of battery charging modes. The SoCs of both slave units are selected to be close to each other purposely. The initial condition is summarized in Table 26.8, and the simulation result of this test is shown in Fig. 26.22.

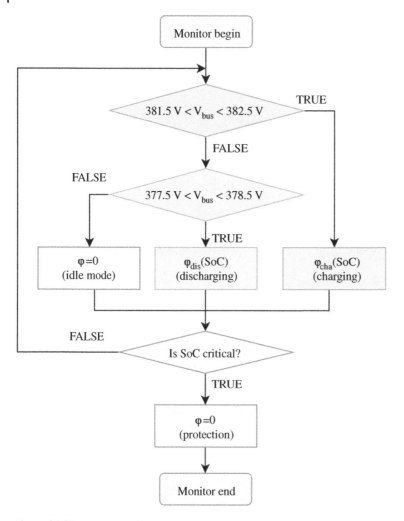

Figure 26.20 Flowchart for slave unit operation (slave 1).

Table 26.4 Battery units for case study.

Unit	Model	Type
Master	Panasonics UR18650RX	Lithium-Ion
Slave 1	SONY VTC5	Lithium-Ion
Slave 2	Panasonics LCX1228P	Lead acid

When the bus voltage is regulated to be 380 V, the current is sensed by the master unit as negative in value, supplying energy from the battery pack to the network. The current is measured and recorded as $I_{m,0}$. At $t = 0.25$ seconds, the bus voltage, V_{bus}, is controlled to be 382 V activating Slave 1 to contribute power. At $t = 0.5$ seconds, V_{bus}, is regulated to be 384 V turning on Slave 2. It is found that the magnitude of $I_{m,1}$ is less than $I_{m,2}$. It can be inferred that slave 1 is drawing more current

Table 26.5 System specification for case study.

Term	Value
Battery nominal voltage	36 V
Nominal DC bus voltage	380 V
DC bus capacitor	10 mF

Table 26.6 DAB specification for case study.

Term	Value
Winding turns ratio of transformer	1:11
Interleaving inductor	50 μH
Switching frequency	10 kHz
Parallel number of master interface:	5

Table 26.7 Initial status for discharging test.

Term	Value
SoC of the master unit	80%
SoC of slave 1 SoC	60%
SoC of slave 2 SoC	55%
Lumped current	$I_{\text{load}} = 2\,\text{A}$

the complement of $I_{m,1}$, which concludes that Slave 1 also has less SoC, as shown in Fig. 26.22. In this test scenario, even if Slave 1 is only 1% less than Slave 2 in SoC, a marginal percentage, the technique can correctly identify the difference. The SoC difference among the storage units should be gradually reduced by charging.

26.5 Summary

AC microgrids face the challenge of improving system efficiency, power quality, and grid resilience with the increasing number of distributed generation (DG) and energy storage units. The DC microgrid is booming since it can break the long-term dilemma of AC and represent the future grid network. Nowadays, DC has been widely used for renewable power generation, energy storage, and electronic loads. The chapter presents the important issues regarding the coordinated control of DC microgrids. The fundamental of DC droops is discussed at the beginning followed by its enhancement.

First, the control scheme adopts a piecewise linear formation for droop gains to balance the load-sharing accuracy with the deviation of bus voltage. Communication among the distributed units is unnecessary for the proposed operation. Second, The distributed AVS scheme is presented in this paper to maintain the terminal voltage at the nominal value and secure the uniform

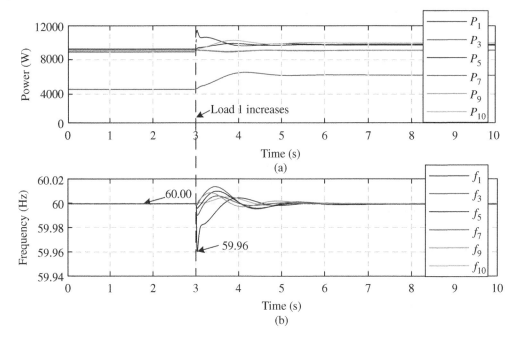

Figure 26.21 Simulation results for battery discharge. (a) Active power generated by each DER and (b) frequency response of each DER.

Table 26.8 Initial Status for charging test.

Term	Value
SoC of the master unit	80%
SoC of slave 1 SoC	20%
SoC of slave 2 SoC	21%
Lumped current	$I_{load} = -2.5\,A$

current sharing disregarding the variation of loading conditions. It is proven to be successful via low-bandwidth communication among power generator units without a centralized controller. Last, the chapter focuses on the energy storage usage for DC microgrids. It proposes a method of DBS to coordinate operations with the SoC balance for each battery unit. The method neglects the costly communication among the grid network but shares information along the system. The hierarchical control scheme shows the advantage of optimally coordinating all distributed power generation and energy storage units since a centralized can overview the system status, make effective commands, and communicate with local controllers. The drawback lies in the system complexity and high implementation cost.

In general, the coordinated control of DC microgrids sounds more straightforward than the strategy for AC systems. However, it opens more room to improve the performance regarding the system efficiency, reliability, cost-effectiveness, and integration of more renewable energy resources and energy storage thanks to the simple formation of DC. It also shows a bright picture to project future energy networks.

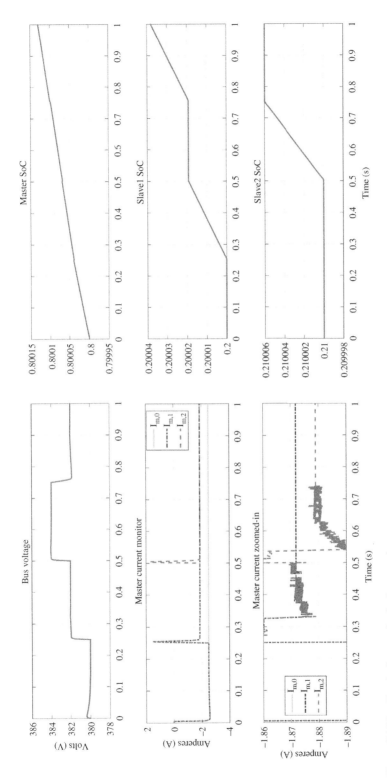

Figure 26.22 Simulation results for battery charge.

26.6 Exercises

1. Search online and find a real-world example of DC microgrid; discuss the ratings of voltage and capacity.
2. Compare the fundamentals of DC and AC droop and explain the difference.
3. Describe how the performance of DC droop is measured.
4. Follow the case study in Section 26.4.4 and simulate the system coordination based on the active DC bus signaing.

References

1 Xiao, W. (2017). *Photovoltaic Power System: Modeling, Design and Control*. Wiley.

2 Moawwad, A., El Moursi, M.S., Xiao, W., and Kirtley, J.L. (2014). Novel configuration and transient management control strategy for VSC-HVDC. *IEEE Transactions on Power Systems* 29 (5): 2478–2488.

3 Cingoz, F., Elrayyah, A., and Sozer, Y. (2017). Optimized settings of droop parameters using stochastic load modeling for effective DC microgrids operation. *IEEE Transactions on Industry Applications* 53 (2): 1358–1371.

4 Lin, Y. and Xiao, W. (2019). Novel piecewise linear formation of droop strategy for DC microgrid. *IEEE Transactions on Smart Grid* 10 (6): 6747–6755.

5 Huang, P.-H., Liu, P.-C., Xiao, W., and El Moursi, M.S. (2015). A novel droop-based average voltage sharing control strategy for DC microgrids. *IEEE Transactions on Smart Grid* 6 (3): 1096–1106.

6 Xiao, W. (2021). *Power Electronics Step-by-Step: Design, Modeling, Simulation, and Control*, 1e. McGraw-Hill Education.

7 Han, J.X. and Xiao, W. (2019). Advanced control scheme for DC microgrid via dual active bridge and bus signaling. *2019 IEEE 28th International Symposium on Industrial Electronics (ISIE)*, 2515–2520.

27

Foundations of Microgrid Resilience

William W. Anderson, Jr. and Douglas L. Van Bossuyt

27.1 Introduction

Resilient microgrids are of great importance to critical loads such as those found on military installations, in hospitals, at airports, and across many other critical infrastructure sectors. Many different natural and man-made events can challenge the ability of microgrids to continuously deliver energy to critical loads. Severe weather, earthquakes, flooding, and animal infestation are just a few potential natural causes of a failed microgrid. Man-made failures can range from accidental such as a crane boom extending into overhead power lines to wreckless such as an inebriated driver hitting a power pole to malicious and malevolent acts such as a sniper shooting the cooling fins on a transformer. In addition to physical attacks, humans can also cause cyber attacks with similar failure consequences to a microgrid.

Microgrids are specialized power systems that can be approached with a system engineering design perspective. Although there is a significant need to incorporate electrical engineering into the design, the selection of the most appropriate microgrid architecture lends itself to a systems engineering design methodology. This is especially true as renewable energy and energy storage is introduced into the microgrid.

While much work has been done to improve the reliability of microgrids, less work has been done to date to improve resilience. In this chapter, we adopt the P-602 definition of reliability: "the percentage of time energy delivery systems (utilities) can serve customers at acceptable regulatory standards" [1]. We suggest the following definition of resilience: "the microgrid's invulnerability and rapid and full recoverability from an improbable and severe disturbance." A key differentiator between reliability and resilience is that reliability deals with events that may be more frequent and more easily predictable while resilience deals with rare events that cause severe disturbances.

Regardless of how the resilience of a microgrid may be challenged, it is critically important to many critical infrastructure sectors that power be available to supply critical loads during times of crisis. In spite of a massive once-in-a-generation hurricane or a determined and aggressive adversary, power delivery must be assured in many cases. The loss of power can cause the loss of life, the loss of property and assets, and the loss of national defense capabilities.

This chapter also covers the related topic of climate resilience to help build an understanding of how increasing microgrid resilience and climate resilience of microgrids can be accomplished at a reasonable cost. Many microgrids currently rely upon carbon-intensive fuel sources such as diesel either for primary power generation or for backup power. We recommend including an analysis

Microgrids: Theory and Practice, First Edition. Edited by Peng Zhang.
© 2024 The Institute of Electrical and Electronics Engineers, Inc. Published 2024 by John Wiley & Sons, Inc.

of climate resilience to find microgrid configurations that can lower emissions while increasing resilience and keeping costs down.

The rest of this chapter provides a foundational understanding of microgrid resilience from the perspective of the United States Naval Facilities Command (NAVFAC) at the systems engineering level. While NAVFAC has a national defense focus, this chapter is intended to be broadly applicable to practitioners and researchers dealing with microgrids that must be resilient. A resilient microgrid means that power stays on during the worst crises to ensure lives are protected and saved.

27.2 Background/Problem Statement

Many Naval microgrids operate at islanded naval installations (INIs) where a connection to the regional grid is either not possible, or desirable. Installations such as San Nicolas Island, Diego Garcia, and others are isolated from grids by many kilometers of ocean. Other installations such as Guantanamo Bay are isolated from neighboring power infrastructure for a variety of reasons. Even in situations where a grid connection to an installation may exist, the grid connection may be severed for many reasons such as load shedding and demand curtailment, wildfire prevention, storms, etc. Oftentimes, NAVFAC and others assume that grid connections will be lost in situations that could challenge the resilience of a microgrid. Based on the recent history of installations being impacted by major weather events such as Camp LeJeune, this is a reasonable assumption. Thus, we suggest treating all microgrids as if they are islanded microgrids for the purposes of resilience analysis which is a conservative assumption.

INIs powered by microgrids generally must have power to perform their missions and the resilience of the power supply is important. Currently, INIs pay between US$0.21 and US$0.54/kWh for their mostly diesel-based energy supply which is approximately two to five times the Navy's average cost of power for other installations that have grid connections. In addition to having the highest cost of power generation across all Naval installations, INIs also have the greatest requirements for resilience. Until recently, no method of analyzing INI microgrid resilience was fit for purpose. We developed a tool to assess resilience with special focus on renewable energy (RE) to help reduce power costs to fill this gap [2, 3] and will present some of the details later in this chapter. Please confirm whether the "$" symbol presented in this chapter represents US dollar. If so, we will change the value as "US$ 0.21" (across the chapter both in texts and artworks).

The switch to RE for primary generation with diesel backup generators helps to reduce the risk of logistics supply being threatened by adversaries [4]. Many existing installations with grid connections and all INIs currently rely almost exclusively on diesel generators (DGs) as either the primary power source, the backup power source, or both. In a crisis where global fuel supplies are disrupted, relying on diesel may become an expensive proposition or impossible. Further, relying largely or entirely upon diesel fuel has negative impacts on the climate and reduces climate resilience of a microgrid [5].

It should be noted that in our experience while switching to RE generation sources can help to increase resilience in many scenarios and improve reliability from disturbances such as rapidly changing energy demands from attached loads, volatility and price increases for diesel fuel, and diesel fuel availability, the reliability and efficiency of a microgrid may decrease. This trade-off is important to understand and is discussed later in this chapter. Several methods from high-reliability industries such as the civilian nuclear power industry's defense in-depth philosophy may be useful to counteract this trade-off although they come with an increased price tag [6–8].

27.3   Defining Resilience

Many definitions of resilience exist in the literature and in practitioner lexicon. This section reviews several key definitions and supporting material to define resilience in the context of microgrids, especially for INI microgrids. First, we discuss energy security which, within the INI microgrid context, is the top-level requirement that is driven by the three pillars of energy security including reliability, resilience, and efficiency. Then we focus on resilience in particular to review existing definitions before advancing our own definition of resilience which we posit is most appropriate for INI microgrids. Finally, we INI microgrid resilience in the context of climate change, carbon dioxide emissions reduction, and increasing RE usage from the perspective of reducing dependency on diesel fuel.

27.3.1   Energy Security

The 10 USC 2924 US Code of Definitions defines energy security as "… having assured access to reliable supplies of energy and the ability to protect and deliver sufficient energy to meet mission essential requirements" [1]. This definition is refined and expanded into the three pillars of energy security (reliability, resilience, and efficiency) and is defined in NAVFAC P-602 as:

- **Reliability**: "the percentage of time energy delivery systems (utilities) can serve customers at acceptable regulatory standards. Reliability can be measured by the frequency and duration of service disruptions to customers."
- **Resiliency**: "the ability of a system to anticipate, resist, absorb, respond, adapt, and recover from a disturbance."
- **Efficiency**: "the use of the minimal energy required to achieve the desired level of service."

While reliability and resilience are considered the highest priorities for the Navy to improve energy security, resilience is the least understood. Specifically, disturbances and metrics for assessing resilience are not defined in Navy guidance documents. Further, we have found that some commercial microgrid analysis packages that contain resilience analysis components use a variety of resilience definitions, and sometimes these definitions can be confused or misunderstood in use between different installations.

Note that while NAVFAC P-602 and some other Navy and Department of Defense (DoD) issuances use the word "resiliency," we prefer the word "resilience" instead and unless in quoted text, we use "resilience" throughout this chapter and in our other scholarly work. This was a point of discussion at the Woodstalk Workshop (a resilience-focused workshop) held at the Naval Postgraduate School in March 2019 where the consensus of the workshop organizers was to promote the usage of "resilience" to the exclusion of "resiliency." However, we see a mix of the two words in the literature and in use with practitioners. Further, resilience is considered an "-ility" along with other "-ilities" such as reliability, supportability, maintainability, availability, and others which may add to the confusion. Therefore, we believe the most important aspect is to stay consistent in using either "resilience" or "resiliency" throughout a document.

27.3.2   Defining Resilience

There are numerous definitions for energy resilience in the literature and in use by practitioners. Of specific interest to INIs and similar microgrids are several complementary and competing definitions from the DoD, Department of Energy (DoE), Department of Navy (DoN), Federal Energy

Regulatory Commission (FERC), and others. This helps to frame the definition we suggest INIs and similar microgrids adopt.

The Resilience Engineering Institute states that "resilience is a word found in nearly a dozen academic disciplines, yet there is no consistent definition or approach shared among them ..." which we find to be true [9]. For instance, within the realm of INI microgrids, the following definitions are all in competition. The DoD's definition for energy resilience is "... the ability to avoid, prepare for, minimize, adapt to, and recover from anticipated and unanticipated energy disruptions in order to ensure energy availability and reliability sufficient to provide for mission assurance and readiness, including task critical assets and other mission essential operations related to readiness, and to execute or rapidly reestablish mission essential requirements" [10] The DoN's definition for resilience is "the ability of a system to anticipate, resist, absorb, respond, adapt and recover from a disturbance" [1]. The DoE energy resilience definition follows Presidential Policy Direction 21 [11] and is defined as "the ability to prepare for and adapt to changing conditions and withstand and recover rapidly from disruptions. Resilience includes the ability to withstand and recover from deliberate attacks, accidents, or naturally occurring threats or incidents." FERC's definition of resilience is the "ability to withstand and reduce the magnitude and/or duration of disruptive events, which includes the capability to anticipate, absorb, adapt to, and/or rapidly recover from such an event" [12].

We propose the following definition of resilience: "the microgrid's invulnerability and rapid and full recoverability from an improbable and severe disturbance" [3]. We believe this definition more accurately captures what should be measured for INIs and similar microgrids. Additionally, it should be emphasized that disturbances of interest are limited to High-Impact-Low-Probability (HILP) events upon the INI's microgrid because this is what resilience is intended to address. Historically, HILP disturbances for INIs have been mostly weather-related although human-caused HILPs such as sabotage, attack, or accident are also possible. Finally, it should be noted that the HILP disturbances found when assessing resilience are not to be confused with the Low-Impact-High-Probability (LIHP) disturbances commonly found when assessing reliability. LIHP disturbances include events that happen routinely such as routine afternoon thunderstorms at INIs located in the tropics.

A typical resilience curve (sometimes called a resilience trapezoid in the literature), shown in Fig. 27.1, contains several discrete regions defined by time and microgrid performance in relation to when a HILP disturbance occurs. Five distinct microgrid states exist including the pre-disturbance state, the degradation, state, the stabilization state, the recovery state, and the post-disturbance state. During the pre-disturbance state starting at t_p and ending at t_d, the microgrid is operating at expected performance (generally represented as 100% microgrid performance). The HILP disturbance occurs at t_d. After the HILP disturbance occurs, the microgrid degrades between t_d and t_s where microgrid performance stabilizes at a value lower than 100% performance. The degradation of the microgrid (the difference between pre-disturbance and stabilized microgrid performance) is the invulnerability measure (I). Next, the microgrid enters the stabilization state where microgrid performance has stabilized. This occurs between t_s and t_r. Then the microgrid begins to recover starting at t_r and progresses into the recovery state where it regains performance to t_{fr}. At t_{fr}, the microgrid now enters the post-disturbance state where performance has returned to nominal levels. The time between t_d and t_{fr} is the recovery measure (R). Later in this chapter, we introduce invulnerability and recovery metrics, and show a model that can calculate resilience.

It should be noted that in some situations there can be multiple plateaus during both the degradation state and the recovery state where it appears the situation has stabilized but it has not. There can also be false starts when entering the recovery state where performance increases only for it to

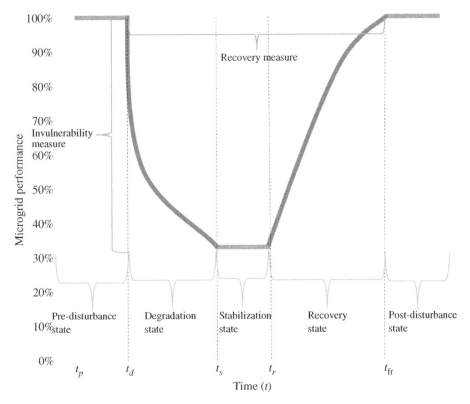

Figure 27.1 Microgrid resilience curve.

decrease once again. However, the resilience curve shown in Fig. 27.1 is generally what we observe happening in INI and similar microgrids.

27.3.3 Microgrid Cost

Evaluating the cost of a microgrid including capital construction, operations, and maintenance costs is important when working to improve microgrid resilience. While it is possible to develop an incredibly resilient microgrid, that design is likely to be prohibitively expensive. Instead, designs that balance the competing goals of increasing resilience and minimizing cost must be pursued.

In the literature, Levelized Cost of Energy (LCOE) is commonly used to understand the trade-off between reliability and energy costs [13]. We suggest a modification of LCOE: the Levelized Cost of Energy Demanded (*LCOED*) [2, 3]. *LCOED* considers energy demand versus power generation capacity. However, several other ways of accounting for cost in resilient microgrids have been proposed [14] and we observe this remains an open area of research especially in regards to balancing cost against resilience. Later in this chapter, we present how to calculate *LCOED* in the context of microgrid resilience.

27.3.4 Assessing Resilience and Cost

In order to assess INI and similar microgrid resilience and cost, we suggest using the five-step framework. Step 1 gathers data including microgrid power generation and demand data, and distributed energy resource (DER) data including power generation rating, type, and location. DERs include DGs and RE generation such as wind turbines (WT), photovoltaic (PV), and batteries (BESS) among others. Step 2 generates the HILP disturbance scenarios denoted as S_k, the probability of damage for each DER denoted as $P(d|S_k)$, and the disturbance start time. Step 3 establishes a baseline analysis of the existing microgrid to determine how the microgrid will respond to various HILP disturbance scenarios from both a resilience and a cost perspective. Step 4 then analyzes different potential microgrid configurations where the number and power rating of DER components to determine if a microgrid configuration that improves resilience while minimizing cost versus the baseline exists. Then finally in Step 5 a new preferred microgrid design is selected based on conducting a trade-off analysis between resilience and cost of the potential designs developed in Step 4.

In order to quantify resilience, we suggest measuring invulnerability (I) and recoverability (R). Our measurement of I and R is adapted from Giachetti et al. [2]. We measure invulnerability similarly to Francis and Bekera [15] and Vugrin et al. [16] as the drop in microgrid power generated after a HILP disturbance. Thus, invulnerability is the ratio of power generated at time t (P_t) to load demand (D_t) at time t.

$$I = \frac{P_t}{D_t} \tag{27.1}$$

It should be noted that the power generated is not the total rated power available. INI microgrids and many other microgrids used for mission-critical activities often have excess power generation capacity available. For instance, one INI we analyzed has more than five times the generation capacity it needs [2]. That INI microgrid could lose several of its DERs and still have excess generation capacity. Thus, instead of using the total generation capacity available which could cause $I > 1$, we instead advocate to use the actual power generated. This means that $P_t \leq D_t$ and $I \in [0, 1]$.

Next, we define recoverability as the ratio of the area bounded by the demand and the post-disturbance power generated between the time of the disturbance and full recovery [17]. This represents the portion and duration of the demand that is not met.

$$R = 1 - \frac{\sum_{t=t_d}^{t_r} D_t - P_t}{\sum_{t=t_d}^{t_r} D_t} \tag{27.2}$$

Recoverability is only analyzed when $P_t < D_t$. $R \in [0, 1]$ where 0 represents a microgrid that never recovers after the HILP disturbance and 1 represents a microgrid where the HILP disturbance never disrupts power delivery to the loads which can happen when the DERs have more generation capacity than the loads require.

Thus, we can develop a resilience measure ($\omega \in [0, 1]$) using invulnerability and recoverability.

$$\xi = \omega I + (1 - \omega)R \tag{27.3}$$

Further details of how to develop INI microgrid resilience models are available in [2, 3]. This includes aspects such as power balance, load shed strategies, and others.

Next, we examine *LCOED*. First, we determine the Net Present Value (NPV) of the cost of energy generation:

$$\text{NPV}_{\text{costs}} = \frac{\sum_{t=1}^{T} \sum_{i=1}^{N} (I_t + M_t + F_t - H_i^T)}{(1 + r)^t}$$

In the numerator, the cost of investment (I_t), maintenance (M_t), and fuel (F_t) minus the residual remaining value (H_i^T) of any equipment i with useful life past the planning horizon T is captured.

The planning horizon is set to be the expected life of the individual DER with the shortest expected life. The total costs are discounted by the discount rate r [2].

Fuel cost for DGs in time t is

$$F_t = \sum_{i \in I}(c_g f_i L_{it} \mu_{it}^{DG}) \quad \forall t \in T \tag{27.4}$$

where c_g is the price of diesel in \$/gal or \%/l depending upon the practitioner's desired units. Note that advanced DER such as hydrogen fuel cells or alternative fuels DER such as natural gas cogen facilities can also use the fuel cost for DG formula with slight modifications. f_i is fuel consumed by the DG and is measured in either gallons per hour or liters per hour. DG and many other DERs used on INI microgrids can have variable loading which is denoted by L_{it}. μ_{it}^{DG} indicates whether or not a specific DG is operational during the time period and can be used for any DER of relevance.

Now the NPV of the energy used by the loads attached to the microgrid can be calculated as:

$$\text{NPV}_{\text{energy}} = \frac{\sum_{t=1}^{T} D_t}{(1+r)^t}$$

and thus the *LCOED* is calculated as the costs per unit energy

$$\text{LCOED} = \frac{\text{NPV}_{\text{costs}}}{\text{NPV}_{\text{energy}}} \tag{27.5}$$

Further details of the *LCOED* calculation are available in [2, 3].

27.3.5 Climate Resilience

As we defined above, a resilient microgrid is one that is able to withstand and swiftly recover from HILP events. There is a group of HILP disturbances that are exacerbated by climate change and we suggest deserve their own consideration. Thus we define climate resilience as the ability to anticipate, prepare for, and respond to HILP disturbances due to climate change. As climate change accelerates, the concepts of microgrid resilience and climate resilience are intertwined at a fundamental level; in order for a microgrid to be truly resilient, it must be prepared for all types of HILP.

Climate change is beginning to demonstrate repeated, catastrophic events, such as heat waves and ice storms, that are greatly affecting the infrastructure of communities [18, 19]. Such events will undoubtedly test the resilience of INI and similar microgrids in addition to the rest of Naval base operations and civilian communities, and it is on track to become exponentially more detrimental as emissions continue to rise [20]. This outlook makes the future of mission assurance become less and less probable, as these HILP events increase in probability while the severity becomes less predictable. Improving the resilience of an INI microgrid while also investing in reducing its emissions allows for an investment in climate resilience. This effort to reduce emissions now will be a major contribution to keeping the climate suitable for human activities later.

A microgrid's ability to anticipate, prepare for, and swiftly recover from catastrophic events can vastly reduce the impact a community's response has on the environment via CO_2 emissions. In cases where the microgrid is islanded (an INI microgrid), the use of RE increases its resilience while also decreasing the emissions from the microgrid.

27.3.6 Assessing Climate Resilience

In order to assess climate resilience, we recommend understanding the amount of carbon dioxide (CO_2) being emitted by a microgrid during operations. This can then be used as a trade-off study variable when balancing against other major parameters such as resilience and cost. We prefer the method of evaluating CO_2 that is used in the commercial software package XENDEE [21, 22] which is as follows:

$$CO_{2\text{microgrid}} = CO_{2\text{utility}} + CO_{2\text{embedded}} + CO_{2\text{PPA}} + CO_{2\text{Fuel}} \qquad (27.6)$$

where $CO_{2\text{utility}}$ represents the CO_2 emissions from purchasing electricity from a utility, $CO_{2\text{embedded}}$ is the CO_2 emissions produced by manufacturing microgrid DERs, $CO_{2\text{PPA}}$ which represents the CO_2 emissions from any power purchase agreements that may be present where a 3rd party power plant produces power and the power is imported into the microgrid via the regional utility grid, and $CO_{2\text{fuel}}$ is the CO_2 emissions from burning fuel in the DERs for power.

The utility = CO_2 emissions from purchasing electricity, accounting for exports and reducing with clean energy. Embedded = CO_2 emissions from manufacture of DER, currently fixed at 15 kg/kW but needs more specificity for different DER similar to emissions factor of fuel, i.e. lithium-ion BESS CO_2 emissions are far greater than WT production C02 emissions, therefore they are not accurately represented by an unchanging 15 kg/kW value. PPA = CO_2 emissions from power purchase agreement (PPA), similar to utility but energy is sourced from 3rd party. And fuel = CO_2 emissions from burning fuel for power, related to emission factor of specific fuel [CO_2/kWh].

In the case of an INI microgrid, this formula can be simplified to:

$$CO_{2\text{microgrid}} = CO_{2\text{embedded}} + CO_{2\text{Fuel}} + CO_{2\text{transport}} \qquad (27.7)$$

where $CO_{2\text{transport}}$ represents the CO_2 emitted in transporting fuel, DERs, etc. to the microgrid location. For many INI microgrids, this is negligible. However, some INI microgrids can be many thousands of kilometers away from fuel supply sources thus requiring significant transport CO_2 emissions.

We recommend a five-step process to analyze the CO_2 the microgrid produces and examine how it can be reduced to improve climate resilience. The steps are: Step 1 gather load data for the facility, Step 2 model the microgrid in an appropriate software package, Step 3 calculate the *LCOED*, Step 4 generate trade-space graphics for ease of comparison, and Step 5 compare these results with the resilience versus *LCOED* tradespace, and identify correlations. Once the correlations are identified, a microgrid configuration can be chosen that helps to increase resilience, minimize *LCOED*, and increase climate resilience.

In addition to understanding the CO_2 emitted by an INI microgrid, it is also important to understand the potential new HILP disturbances or changes in the probability and severity of HILP disturbances as a result of climate change. For instance, more intense and frequent hurricanes may impact tropical regions. Some regions may experience severe drought and high temperatures more frequently. Some regions may have significantly higher and more frequent rain events which can lead to severe flooding. These sorts of climate change-related HILP disturbances must be cataloged and categorized for specific INI microgrid locations, and then assessed as described above for other HILP disturbances. Several sources of data are available in the literature to begin to identify climate change-related HILP disturbances [23].

27.4 Resilience Analysis Examples

In this section, we present some brief examples of implementing the resilience analysis methods we recommended and outlined above. While the microgrid INIs that we discuss are real microgrids supporting US Navy operations, the data are representative of reality and have been intentionally obfuscated to preclude the chance of any sensitive information being released. First, we explore an example of analyzing resilience and cost. Second, we turn our attention to an analysis of climate resilience.

27.4.1 Analysis of Resilience Versus Cost

San Nicolas Island is an island exclusively used by the US Navy 100 km off the coast of California. It has a variety of loads and DERs including DGs, WTs, PV, and BESS. The amount of generation capacity greatly exceeds the needs of the existing loads on the island. Diesel fuel is barged from the mainland as needed to refill storage tanks situated on the island.

We conducted experiments on excess power using the models described in [3] in order to better understand the relationship between the power/demand ratio and resilience and costs. We believe that this relationship may be more broadly applicable to more geographic locations and therefore be useful to future assessments. Figure 27.2 shows the relationship.

Of specific interest to the Navy is identifying the appropriate level of maintenance on microgrid performance. We investigated this question through a resilience and cost lens. Figure 27.3 shows the results of this analysis with further details being available in [3]. Surprisingly, we found that less maintenance can be beneficial to resilience in some scenarios. This may be attributable to maintenance errors that can cause DGs and other DERs to fail when called upon which is supported by Thompson et al. [24].

Through our analyses, we identified that there is a decreasing return on resilience improvement while cost increases linearly. We also identified that there is very little improvement in resilience beyond a power rating ratio of four. Further, invulnerability contributes more to resilience than recovery above a power rating ratio of 1.5.

Next, we investigated how DER redundancy can improve resilience and how increasing redundancy impacts cost. In particular, we investigated how additional DGs being added to San Nicolas Island's microgrid which consists of a mix of DGs, WTs, PVs, and BESS, could improve resilience. Figure 27.4 shows that increased redundant DGs do improve resilience to a point.

We investigated changing the mix of RE in the DER with increasing DG redundancy. Counter-intuitively, decreasing the mix of RE to only include PV or WT but not both improved

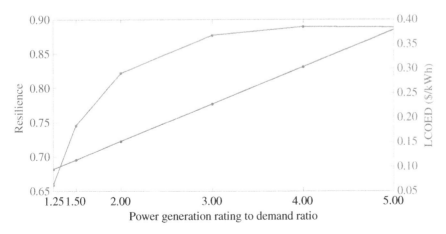

Figure 27.2 San Nicolas Island INI microgrid power generation rating to demand ratio.

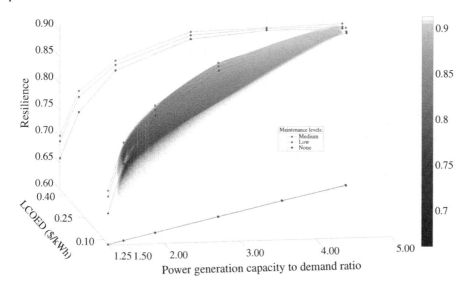

Figure 27.3 San Nicolas Island INI microgrid power generation rating to demand ratio with maintenance levels taken into account.

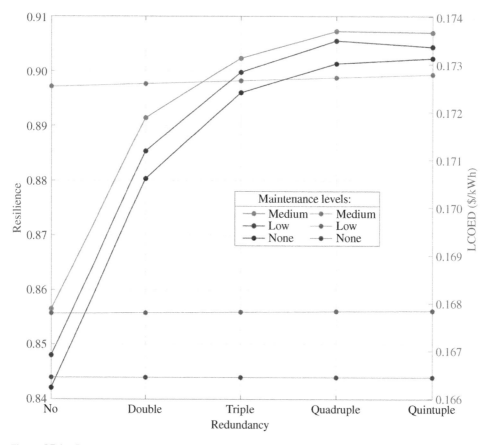

Figure 27.4 San Nicolas Island INI microgrid RE DER combined with DG redundancy.

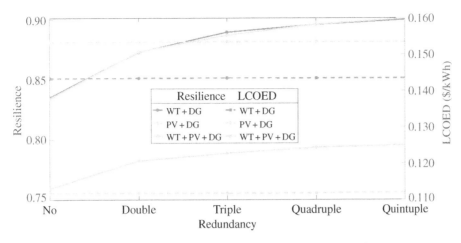

Figure 27.5 RE mix and DG redundancy for San Nicolas Island INI microgrid.

resilience. We investigated further and identified that having a high mix of RE can present more potential failures during a HILP disturbance. Further research and experimentation may prove this portion of our analysis to be in error and this should be further investigated. Figure 27.5 shows our findings.

We examine recovery for a variety of DER redundancy situations as seen in Fig. 27.6. Again, we see that resilience increases at a decreasing rate while cost linearly increases. We also note that WT and DG have the least costs while achieving satisfactory resilience likely because of the diurnal benefits of using WT versus PV.

Finally, we looked at the trade-off between *LCOED*, resilience, and the DER RE mix at San Nicolas Island. Figure 27.7 shows how there is more improvement at San Nicolas Island by adding WTs. However, adding more than four DGs and two WTs has little additional benefit.

Based on our analyses and research, we recommend that INI microgrid resilience be improved with the following in mind. We suggest implementing as much redundancy in DERs and other hardware as practical while recognizing the diminishing returns on resilience with increasing costs. The power rating to demand ratio should not exceed four. Microgrids that are moving to diversify from primarily DG DERs to include REs should give preference to WT over PV if the wind resources are sufficiently matched with loads. Maintenance investments to support resilience goals are sufficient at lower levels and maintenance above the manufacturer-recommended specification is unnecessary or even counter-productive. The exception to this guidance on maintenance is when a microgrid only has RE and no DG; in this case, maintenance should be increased beyond minimum levels. However, very few INI microgrids are or are expected to be 100% RE.

27.4.2 Analysis of Climate Resilience

We now turn our attention to analyzing San Nicolas Island for climate resilience. The INI microgrid on San Nicolas Island has a combination of RE and DG DER comprised of: five DG with power ratings of between 750 and 1250 kW per genset; seven WT rated at 100 kW each; and a 1.5 MW, 4 MWh lithium-ion battery storage system. The current annual CO_2 emissions of the microgrid are approximately 2254 metric tons and cost US\$2159.50 K. Figure 27.8 shows the energy dispatch for the microgrid on a peak day in June in a recent year.

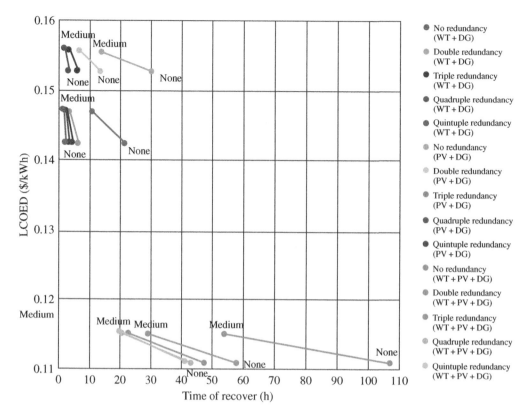

Figure 27.6 Recovery for a variety of DER redundancy situations for San Nicolas Island.

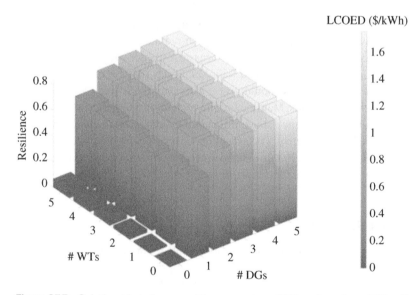

Figure 27.7 Relationship between resilience, number of WT, and number of DG at San Nicolas Island to *LCOED*. The color changes from dark gray to light gray as *LCOED* increases.

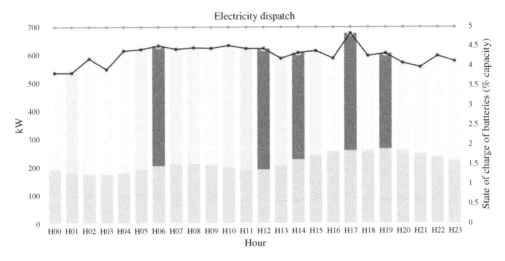

Figure 27.8 Energy dispatch for a peak load day on San Nicolas Island in a recent June.

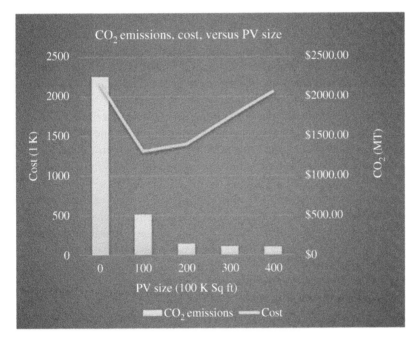

Figure 27.9 A steep decrease in emissions and cost at the investment of 9300 m² (100,000 ft²) of PV is observed in this model of the San Nicolas Island microgrid. The emissions equalized after a 18,600 m² (200,000 ft²) investment.

We conducted a study of CO_2 emissions reductions on San Nicolas Island by increasing PV system sizing up to 37,100 m² (400,000 ft²) and 4 MWh BESS. Figure 27.9 illustrates a steep decrease in emissions and cost at the investment of 9300 m² (100,000 ft²) of PV. The emissions equalized after a 18,600 m² (200,000 ft²) investment; therefore in this scenario, an ideal investment point is at the 9300 m² (100,000 ft²) of PV to balance CO_2 reduction and cost. Figure 27.10 shows the energy dispatch over a peak load day in June with the increased PV.

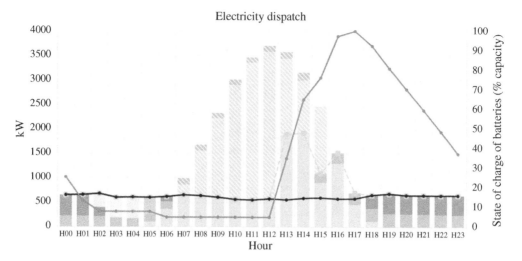

Figure 27.10 Energy dispatch for a peak load day on San Nicolas Island in a recent June after increasing PV to reduce CO_2 emissions.

If we now look at increasing the BESS size, there is a sharp decline in emissions with the investment of 9300 m^2 (100,000 ft^2) of PV. However, after this investment, the emissions are reduced to 0 with any extra 9300 m^2 (100,000 ft^2) investment in PV, alongside a 2.76 MWh increase in BESS capacity. If total emissions reduction is the goal, an investment at 18,600 m^2 (200,000 ft^2) of PV would be optimal. Figure 27.11 shows the energy dispatch for this scenario.

We suggest that a good way to increase resilience both in the microgrid and for the climate is to diversify DER sources and fortify the BESS. DER combinations of DG, WT, PV, and BESS result in the reduction of both long-term expenditures and CO_2 emissions while also improving resilience. Figure 27.12 shows one trade-off study between CO_2 emissions, the size of the PV, and *LCOED* for San Nicolas Island.

Comparing the tradespaces for resilience and CO_2 emissions at San Nicolas Island, there are several findings. First, we observe the sharpest decrease in emissions results between investing in 9300 m^2 (100,000 ft^2) of PV, and any number of DGs. The sharpest increase in resilience involves a similar investment, including one DG and any number of WTs. We also observe that the maximum resilience while keeping *LCOED* feasible can be found around the investment of two DGs and any WT, whereas minimal CO_2 can be found at the investment of 9300 m^2 (100,000 ft^2)–18,600 m^2 (200,000 ft^2) of PV, with one to two DGs.

We find the same can be found in a stronger fashion when combining the use of PV with an increase in the BESS size, as shown in Fig. 27.13. The *LCOED* increases proportionally with renewables and simultaneously decreasing emissions. Including an increase of the BESS by 2.74 MW, the relationship between PV size and DG number with emissions shows a dramatic decrease above 9300 m^2 (100,000 ft^2) of PV.

We observe a decrease in emissions between 0 and 9300 m^2 (100,000 ft^2 of PV, with any given amount of DG. *LCOED* is also shown to decrease with the use of more PV combined with a BESS. Overall, the diversity in DER sources combined with BESS is found to be a reliable method for reducing emissions while also increasing resilience.

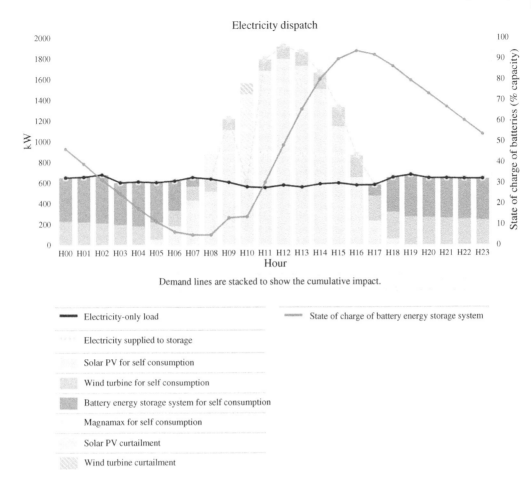

Figure 27.11 Energy dispatch for a peak load day on San Nicolas Island in a recent June after increasing PV and BESS to eliminate CO_2 emissions.

When correlating maximizing resilience with minimizing CO_2 emissions, the tradespaces convey a common ground of energy diversity, particularly favoring the availability of RE and DGs. Specifically, microgrid configurations where the resilience is the highest, along with emissions being the lowest, is a balance between three DGs, and $18{,}600\,\mathrm{m}^2$ ($200{,}000\,\mathrm{ft}^2$) of PV without a BESS increase or three DGs and $18{,}600\,\mathrm{m}^2$ ($200{,}000\,\mathrm{ft}^2$) of PV with a BESS increase.

It is expected that the increased availability of BESS capacity will vastly increase resilience, as the microgrid will no longer be solely reliant on RE availability that can fluctuate with weather and solar conditions, and instead allow the changing availability of RE resources to charge the BESS over time, making a constant flow of power readily available in times of need.

Indeed, we are observing these analyses performing as expected at a variety of INI microgrids that the US Navy operates. As more research is done to further refine these systems engineering analysis estimates and more technological innovations help to drive down CO_2 emissions, we expect to see increased resilience and increased climate resilience across many microgrids.

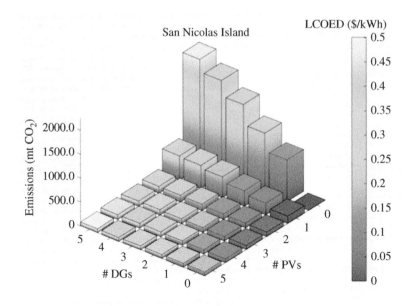

Figure 27.12 Relationship between CO_2 and size of PV in increments of $9300\,m^2$ ($100,000\,ft^2$), and number of DGs at San Nicolas Island. The color changes from dark gray to light gray as *LCOED* increases.

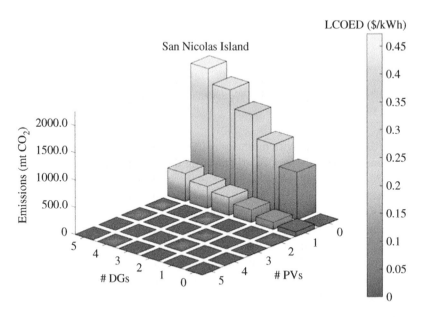

Figure 27.13 Relationship between CO_2 and size of PV in increments of $9300\,m^2$ ($100,000\,ft^2$), and number of DGs with an increased BESS sizing at San Nicolas Island. The color changes from dark gray to light gray as *LCOED* increases.

27.5 Discussion and Future Work

Assessing resilience is very important for the Navy and anyone that wants to ensure their investments are cost-effective whilst providing as much resilience as possible. More importantly, it is essential that an assessment is done of the microgrid architecture so as to not only compare the resilience of one option with another, but to better understand how these decisions to attain more resilience may impact reliability, emissions, etc.

We recommend that researchers wishing to conduct microgrid research such as for new power sources, new energy storage technologies, new cybersecurity strategies, new protection circuitry, etc. ensure that resilience is considered during the maturation of new technologies. Organizations such as the Office of Naval Research are now specifically looking for new technologies that will increase energy resilience. Similarly, climate resilience should also be considered with new microgrid-related technologies.

The methods we presented in this chapter to conduct resilience and climate resilience analysis while considering *LCOED* is one approach that we have found success with at the systems engineering level. These methods have helped us to improve microgrid resilience at a variety of INI microgrids and microgrids with grid connections. However, there are other types of resilience analysis that can be performed at the electrical engineering level and from a power engineering perspective among others that may be useful to make a case for resilience improvements that can be achieved with new technologies or with upgrades to an existing microgrid using existing commercial off-the-shelf hardware. Additionally, while we focused on invulnerability and recoverability, and believe these are the most important measures that contribute to a quantifiable measure of resilience, there are many other aspects of resilience that may need to be considered depending upon the specific loads attached to a microgrid.

Over the last several years, we have been working with a team of researchers and engineers at NAVFAC EXWC, the Naval Postgraduate School, and the University of Wisconsin-Milwaukee to develop microgrid resilience analysis tools at the systems engineering, electrical engineering, and power engineering levels. Some of these analysis tools are now available at https://microgrid.nsetti .nps.edu for public use and more are planned for inclusion. The tools available at the website and in our recent publications [2, 4, 5, 25–32] are specifically tailored to help Navy base energy managers and others make decisions about upgrading existing Navy microgrids to be more resilient. We continue to expand the tools available on the website to meet Navy and DoD needs and expect that these tools will also be useful for civilian microgrid operators.

There is still much future work to be done on better understanding and analyzing microgrid resilience and climate resilience. While we recommended that climate resilience include HILP disturbances specifically caused by climate change, we did not provide further guidance on the topic. This is because we are actively working on developing that guidance and an approach to incorporate climate resilience analysis directly into our existing resilience analysis methods. We are hopeful that others working on microgrid resilience will expand the community's ability to directly incorporate climate resilience analysis into tradespace studies, and expand the types of climate resilience analysis conducted. While we focused on CO_2 in specific, a variety of other climate-altering considerations likely need to be taken into account.

In future work, variability should be introduced in the intensity of the HILP disturbance both when considering the HILP disturbance as a point event and with further expansion of the resilience models to consider a HILP disturbance that can last hours or days such as flooding, a hurricane, or similar. Variability can also be further added to forecasting models for RE generation using predictive modeling. There may be utility in moving these models into a digital twin of the microgrid and using real-time and short-term predictive data to make updated resilience and reliability predictions for energy managers to use in preparing for HILP disturbances that can be seen coming several days in advance (heat waves, thunderstorms, potential tornadoes, etc.).

While we looked at recovery over a time period and ignored the specific shape of the resilience curve as portions of DER are recovered and loads regain power after a HILP disturbance, we believe a more nuanced approach to analyzing recovery is needed. For instance, a long, linear recovery may be preferable to a recovery that takes the same amount of time but goes from a very low total microgrid performance in a step-wise function to complete microgrid restoration. The same may be true of the invulnerability metric where there is a desire for the degradation state to be over many hours rather than to happen all at once. In such a scenario, microgrid energy managers may be able to conduct graceful shutdowns of critical loads or transfer the missions those loads do to other facilities that are not impacted by the HILP disturbance.

27.6 Conclusion

In this chapter, we suggested measuring invulnerability and recoverability to calculate resilience for INI and similar microgrids. Further, we suggested analyzing climate resilience primarily through the amount of CO_2 a microgrid produces. These measures of resilience and climate resilience were combined with *LCOED* to produce tradespace analyses to evaluate different microgrid configurations. The approach was demonstrated on a representative microgrid from San Nicolas Island where a US Navy installation is located offshore from southern California. All of these analyses were conducted at the systems engineering level and were sufficient to help inform decisions about microgrid upgrades for the installation.

We urge anyone who is developing new technologies for microgrids to include resilience and climate resilience in their analyses. This can help to justify the inclusion of new technologies in microgrids. Further, this will help to better guide technological development to meet the needs of microgrid operators who are supporting critical loads.

27.7 Acknowledgments

The research was partially funded by the Office of Naval Research and the Navy Shore Energy Technology Transition and Integration (NSETTI) program managed by Naval Facilities Engineering and Expeditionary Warfare Center in Port Hueneme, CA.

Any opinions or findings of this work are the responsibility of the authors and do not necessarily reflect the views of the Department of Defense or any other organizations. Approved for Public Release; distribution is unlimited.

We wish to thank Jennifer Chavez at the University of Texas at El Paso who assisted with some of the initial climate resilience analysis. Further, this work would not be possible without the numerous people involved in microgrid resilience research who we have worked with over the past five years.

We wish to thank Chris Babohgli for his work both helping prepare and assisting in conducting the first version of the labs that appear in this chapter at the US Naval Academy.

27.8 Exercises

1. Many remote communities maintain islanded microgrids to provide power to residents. Locations such as Deering, Alaska, Briceburg, California, the US Virgin Islands, and others rely upon their microgrids for electrical generation to supply critical loads.

   Research a microgrid near you and conduct a systems engineering-level resilience analysis. You will need to estimate the loads that the microgrid powers, the microgrid's generation and storage capabilities, and identify a HILP disturbance that the microgrid may endure.

   How resilient is the microgrid?

   Is the microgrid's resilience sufficient?

2. Many communities are starting to implement microgrids that can be islanded from the utility grid during emergencies.

   Develop a simple microgrid plan for your community or neighborhood including the major loads, potential generation and storage sources, and potential HILP disturbances.

   Calculate the *LCOED* of your proposed microgrid. This will require you to estimate the various costs associated with your microgrid.

   What can you do to improve the *LCOED* of your microgrid?

3. Many existing microgrids rely upon diesel fuel or other fossil fuel resources to operate. Bringing fuel to isolated, remote communities can be expensive and time-consuming, and during geopolitical events, it can become very difficult to acquire fuel. Further, the CO_2 emissions can exacerbate climate change.

   Identify a microgrid that primarily uses fossil fuel and analyze it for climate resilience from the perspective of CO_2 emitted.

   Redesign the microgrid to emit 50% less CO_2.

   Redesign the microgrid to emit no CO_2.

   Which of the three microgrid designs has the lowest *LCOED*?

   Can you find a design that minimizes CO_2 while also minimizing *LCOED*?

4. Identify HILP disturbances that could impact a microgrid in your community.

   Have any of the HILP disturbances happened in your community in the last 10 years? Describe what happened and how that impacted the utility grid or microgrid.

   Identify climate resilience-related HILP disturbances for a microgrid in your community. Have any of these disturbances happened in the last 10 years? What do the predictions indicate over the next 50 years? Will these climate resilience HILP disturbances become more common?

5. Examine Fig. 27.12. Which design do you think San Nicolas Island should use to reduce CO_2 while also keeping *LCOED* low and serving the loads on the island?

   Defend your answer with calculations.

6. Calculate the *LCOED* for a microgrid with:
   - Two 2 MW wind turbines
   - One 500 kW PV farm
   - One 50 kW DG
   - Two 2 MW, 4 MWh BESS

Use the following demand profile:

Time (h)	Load (kW)
0	2000
1	2400
2	2600
3	2800
4	3000
5	3400
6	3800
7	4000
8	4000
9	4000
10	3600
11	3000
12	2800
13	2800
14	2800
15	2600
16	2000
17	2000
18	2000
19	2000
20	2000
21	2000
22	2000
23	2000

Use the following parameters:
- WACC = 7.5%
- Fuel costs = US$2.60 gallon
- Fuel consumption = 0.06 gal/kWh
- Economic life of a:
  - WT = 20 years
  - PV = 25 years
  - DG = 30 years
  - BESS = 10 years
- Research and use reasonable capital expenditure (CAPEX) and operating expense (OPEX) for DER

7. In February 2021, an extreme ice storm and cold weather event hit the Texas grid. Research this HILP disturbance. What could the grid operator, Electric Reliability Council of Texas (ERCOT), have done differently to be more prepared?

How could microgrids help to prevent future widespread blackouts in Texas?

Labs

Lab 1: Resilience-Cost Assessment

Background
The intent of this Lab is to familiarize students with microgrid resilience and cost assessment methods presented in this chapter. The focus of this Lab is to study the trade-offs associated with redundancy levels and their associated cost in the form of *LCOED* versus resilience metrics.

Several Excel workbooks and MATLAB will be used to produce meaningful visualized depictions of the trade space in the form of 4D graphs of the data produced by the Excel tool, to guide designers to seek a microgrid whose architecture meets their standards of high resilience and low cost.

Required Documents
You will need the following documents to perform this lab:

- Resilience:Cost_Model_Lab.xlsm
- LOAD_DEMAND_PROFILE.xlsx
- RESILIENT_4D_CURVES_REDUNDANCY_LAB_NAVAL_ACADEMY.m
- 3D_REGRESSION_SCATTERPLOT_DATA – REDUNDANCY_LAB.xlsx

Step 1 Update the Resilience:Cost_Model_Lab.xlsm with the appropriate inputs.

Step 2 Once the model is established with all inputs for resilience and cost parameters, perform Monte Carlo simulations by varying redundancy levels (none, double, … , Quintuple) and maintenance levels (none, low, medium) as outlined in the procedure.
What findings can you make having observed the change in resilience in costs as you vary these parameters?

Step 3 Monte Carlo simulations will produce data sets for resilience, invulnerability, recovery, and time to recover, which will allow you to build 2D graphical representations of all the resilience metrics listed above versus *LCOED*. Please run these simulations now.

Step 4 Now do the following:
1. Open MATLAB and save the following files to Current Folder:
    (a) RESILIENT_4D_CURVES_REDUNDANCY_LAB_NAVAL_ACADEMY.m
    (b) 3D_REGRESSION_SCATTERPLOT_DATA – REDUNDANCY_LAB.xlsx
2. Using the data obtained from the Monte Carlo simulations, modify these files to generate a 4D graph in MATLAB for redundancy versus resilience versus *LCOED*.
3. Now you will perform another experiment with focus on proportionality of RE power to DG power. You will find the microgrid architectures to use on the tab labeled "PROPORTIONAL RE_DG" in "Resilience:Cost_Model_Lab.xlsm"
    (a) Generate another set of 2D graphs for recovery, invulnerability, resilience, and time to recover versus *LCOED*.

Step 5 What type of pattern do you see in each of the parameters of recovery, invulnerability, resilience, and time to recover when the proportionality of RE generation to DG generation (RE/DG) increases?

Step 6 Does the maintenance level play more of a factor when RE/DG is lower or when it is higher? Explain.

Step 7 Compile a professional technical report on the above.

Lab 2: Optimizing Renewable Energy Microgrids

Background

Now that you have a basic understanding of solar and wind energy, you are ready to explore how to optimize a RE microgrid architecture. You will have the opportunity to get the time series data needed to build and run optimization problems using a RE software modeling tool, XENDEE.

We will use the data collection tools to generate location-specific findings.

Step 1 Review the solar irradiance data collection on SODA's website: www.soda-pro.com.

Step 2 Use the Renewables Ninja website (www.renewables.ninja) to collect time series data for San Nicolas Island.

Step 3 Familiarize yourself with the PV_LIBMATLAB toolbox on how to produce a PV performance dataset.

Step 4 Conduct a step-by-step process to build the current (baseline) microgrid at San Nicolas Island and set up parameters using the XENDEE software package.
In your lab write-up, identify: assumptions, input variables, decision variables, and constraints.

Step 5 Conduct optimization using XENDEE with an objective function to minimize costs. What is the RE architecture? What are the total energy costs?

Step 6 Conduct optimization using XENDEE with an objective function to minimize emissions. What is the RE architecture? What are the total energy costs?

Step 7 How did the architectures change? What meaningful conclusions can you draw?

Step 8 Using the baseline you developed in Step 4, add PV and BESS. How did the architecture change? What are the total energy costs? How much savings do you get for total annual energy costs after adding PV and BESS?

Step 9 Explore variations on the RE microgrid architectures that you believe will minimize costs AND emissions. What happens when you triple the PV cost? Does the amount of recommended PV change?

Step 10 Keeping PV at triple the baseline cost, try reducing the diesel fuel cost to US$0.10/gal and run low-cost optimization. Demonstrate the change in results.

Step 11 Does the amount of PV recommended by XENDEE change significantly? Explain why.

Step 12 What do you now recommend for a RE architecture that provides the optimal RE architecture?

Step 13 What are your general observations about the behavior of a RE microgrid? What might you do differently for a location such as the military base at Diego Garcia?

Step 14 Compile a professional technical report on the above.

Lab 3: Optimizing Renewable Energy for McMurdo Station, Antarctica

Background

McMurdo Station is a National Science Foundation-run facility that supports research in Antarctica. The station is supported by a microgrid with a variety of DERs. There is interest in reducing the amount of fuel consumed by McMurdo Station to power the microgrid because there is a high cost to bringing fuel to Antarctica and also because of the consequences to the Antarctic environment from burning fossil fuel and the potential for fuel spillage.

Step 1 Research the existing McMurdo microgrid. Build the baseline (current existing) microgrid in XENDEE. Identify: assumptions, input variables, decision variables, and constraints.

Step 2 Run optimization using XENDEE with an objective function to minimize costs for the base-line model. Open up the results and write down the "Total Annual Energy Costs" and "Total Annual CO_2 Emissions" for "Investment scenario" (Both values will be used as a baseline reference to be compared when running optimization for RE).

Step 3 Now add PV, BESS, and WTs as "consider" into the baseline architecture using reasonable inputs. Run optimization using XENDEE with an objective function to minimize costs. What is the RE architecture? What are the total energy costs and savings?

Step 4 How did the architectures change? What meaningful conclusions can you draw?

Step 5 Now run optimization for reduced emissions. How did the architecture change? What are the total energy costs? How much savings do you get for total annual energy costs and emissions?

Step 6 What do you now recommend for a RE architecture that provides the optimal RE architecture?

Step 7 Using the Monte Carlo simulation tool for redundancy (see Labs 1 and 2 resources) optimize the RE architecture in the resilience & cost Excel tool (see Labs 1 and 2 resources) for McMurdo. Create the 4D tradespace in MATLAB with redundancy, resilience, and *LCOED* as axes.

Step 8 Which redundancy level provides the most resilience for the least cost? At which redundancy level do you begin to see diminishing returns?

Step 9 Compile a professional technical report on the above.

References

1 Department of Navy (2017). Three Pillars of Energy Security (Reliability, Resiliency, and Efficiency). Technical Report *P-602*. Washington Navy Yard: NAVFAC.

2 Giachetti, R.E., Van Bossuyt, D.L., Anderson, W.W., and Oriti, G. (2021). Resilience and cost trade space for microgrids on islands. *IEEE Systems Journal* 16 (3): 3939–3949.

3 Anderson, W.W. (2020). Resilience assessment of islanded renewable energy microgrids. PhD thesis. Defense Technical Information Center.

4 Anuat, E., Van Bossuyt, D.L., and Pollman, A. (2021). Energy resilience impact of supply chain network disruption to military microgrids. *Infrastructures* 7 (1): 4.

5 Kain, A., Van Bossuyt, D.L., and Pollman, A. (2021). Investigation of nanogrids for improved navy installation energy resilience. *Applied Sciences* 11 (9): 4298.

6 Papakonstantinou, N., Van Bossuyt, D.L., Linnosmaa, J. et al. (2021). A zero trust hybrid security and safety risk analysis method. *Journal of Computing and Information Science in Engineering* 21 (5): 050907.

7 Hale, B., Van Bossuyt, D.L., Papakonstantinou, N., and O'Halloran, B. (2021). A zero-trust methodology for security of complex systems with machine learning components. In: *International Design Engineering Technical Conferences and Computers and Information in Engineering Conference*, vol. 85376, V002T02A067. American Society of Mechanical Engineers.

8 Papakonstantinou, N., Linnosmaa, J., Bashir, A.Z. et al. (2020). Early combined safety-security defense in depth assessment of complex systems. In: *2020 Annual Reliability and Maintainability Symposium (RAMS)*, 1–7. IEEE.

9 Resilience Engineering Institute (2021). Resilience library. http://resilienceengineeringinstitute.org/resilience-library/ (accessed 26 November 2021).

10 Baxter, J. (2018). *Federal Utility Partnership Working Group Seminar*. OASD (Energy, Installations & Environment) Installation Energy. https://www.energy.gov/sites/prod/files/2018/04/f51/fupwg_spring_2018_20-baxter.pdf (Accessed 26 November 2021).

11 Sandia National Laboratories. Energy infrastructure resilience: framework and sector-specific metrics.

12 Emergency Preparedness Partnerships (2022). Defining resilience in the utility industry. https://emergencypreparednesspartnerships.com/defining-resilience-in-the-utility-industry/ (Accessed 25 October 2022).

13 Aldersey-Williams, J. and Rubert, T. (2019). Levelised cost of energy –a theoretical justification and critical assessment. *Energy Policy* 124: 169–179.

14 Hildebrand, J.P. (2020). Estimating the life cycle cost of microgrid resilience. Master's thesis. Defense Technical Information Center.

15 Francis, R. and Bekera, B. (2014). A metric and frameworks for resilience analysis of engineered and infrastructure systems. *Reliability Engineering & System Safety* 121: 90–103.

16 Vugrin, E., Castillo, A., and Silva-Monroy, C. (2017). Resilience Metrics for the Electric Power System: A Performance-Based Approach. *Report: SAND2017-1493*. USDOE National Nuclear Security Administration (NNSA).

17 USAID (2019). Fundamentals of Advanced Microgrid Design: Coursebook for Advancing Caribbean Energy Resilience Workshop. *Technical report*. Sandia National Labs.

18 Celik, S. (2020). The effects of climate change on human behaviors. In: *Environment, Climate, Plant and Vegetation Growth*, 577–589. Springer.

19 Goldstein, A., Turner, W.R., Gladstone, J., and Hole, D.G. (2019). The private sector's climate change risk and adaptation blind spots. *Nature Climate Change* 9 (1): 18–25.

20 Allen, M., Antwi-Agyei, P., Aragon-Durand, F. et al. (2019). Technical Summary: Global warming of 1.5 C. An IPCC Special Report on the impacts of global warming of 1.5 C above pre-industrial levels and related global greenhouse gas emission pathways, in the context of strengthening the global response to the threat of climate change, sustainable development, and efforts to eradicate poverty.

21 McJunkin, T.R., Stadler, M., and Mansoor, M. (2022). Microrred de la Montaña Feasibility Study–Executive Summary Extended. *Technical report*. Idaho National Lab.(INL), Idaho Falls, ID (United States).

22 Stadler, M. (2019). *XENDEE Design*. XENDEE.

23 Department of the Navy, Office of the Assistant Secretary of the Navy for Energy, Installations, and Environment (2022). *Department of the Navy Climate Action 2030*. Washington, DC: Government Report.

24 Thompson, C.C., Hale, P.S. Jr., and Gao, D.Y. (2018). An evaluation of HVAC failure and maintenance equipment data for facility resiliency and reliability. *ASHRAE Transactions* 124: 116–127.

25 Reich, D. and Oriti, G. (2021). Rightsizing the design of a hybrid microgrid. *Energies* 14 (14): 4273.

26 Varley, D.W., Van Bossuyt, D.L., and Pollman, A. (2022). Feasibility analysis of a mobile microgrid design to support DoD energy resilience goals. *Systems* 10 (3): 74.

27 Mallery, J., Van Bossuyt, D.L., and Pollman, A. (2022). Defense installation energy resilience for changing operational requirements. *Designs* 6 (2): 28.

28 Anglani, N., Oriti, G., Fish, R., and Van Bossuyt, D.L. (2021). Design and optimization strategy to size resilient stand-alone hybrid microgrids in various climatic conditions. In: *2021 IEEE Energy Conversion Congress and Exposition (ECCE)*, 210–217. IEEE.

29 Siritoglou, P., Oriti, G., and Van Bossuyt, D.L. (2021). Distributed energy-resource design method to improve energy security in critical facilities. *Energies* 14 (10): 2773.

30 Giachetti, R.E., Peterson, C.J., Van Bossuyt, D.L., and Parker, G.W. (2020). Systems engineering issues in microgrids for military installations. In: *INCOSE International Symposium*, vol. 30, 731–746. Wiley Online Library.

31 Peterson, C.J., Van Bossuyt, D.L., Giachetti, R.E., and Oriti, G. (2021). Analyzing mission impact of military installations microgrid for resilience. *Systems* 9 (3): 69.

32 Anglani, N., Oriti, G., Fish, R., and Van Bossuyt, D.L. (2022). Design and optimization strategy to size resilient stand-alone hybrid microgrids in various climatic conditions. *Open Journal of Industry Applications* 3: 237–246. https://doi.org/10.1109/OJIA.2022.3201161.

28

Reliability Evaluation and Voltage Control Strategy of AC–DC Microgrid

Qianyu Zhao, Shouxiang Wang, Qi Liu, Zhixin Li, Xuan Wang, and Xuan Zhang

28.1 Introduction

The microgrid can be classified into three categories based on the type of bus: Alternating current (AC) microgrid, direct current (DC) microgrid, and hybrid AC/DC microgrid [1]. AC–DC microgrid, as the dominant form of distributed generation and DC load access to the distribution system, has a complex internal structure and flexible operating mode. It plays a crucial role in utilizing renewable energy and fulfilling the specific reliability requirements of users, and is poised to be a key component of the future power distribution system [2–4].

The AC–DC microgrid architecture leverages the strengths of both AC and DC microgrids, yielding several key advantages [5–8]. (1) The system exhibits increased versatility in its connection to various forms of distributed generation, energy storage systems, and loads, thus reducing the need for AC/DC or DC/DC conversion links and mitigating commutation losses. (2) The AC and DC microgrids can be independently controlled and operated simultaneously, enhancing the reliability of the power supply. (3) The AC–DC microgrid architecture demonstrates increased flexibility and broadens the range of potential applications. (4) The architecture enables the provision of customized power supply to meet the diverse requirements of electricity users. (5) Energy storage systems can be directly connected to the DC or AC bus, simplifying control strategies. (6) In an AC microgrid, voltage can be regulated through transformer action, while in a DC microgrid, voltage level conversion can be performed through a DC/DC converter. Consequently, the AC–DC microgrid architecture is a crucial technology for the advancement of traditional power grids towards green smart grids, owing to its improved economy, safety, and reliability.

Reliability evaluation plays a vital role in the planning of AC–DC microgrids, as it serves to assess the capacity of the microgrid to meet the demands for power supply reliability as set forth by its customers [9–12]. In a performance-based pricing mechanism, the reliability level constitutes a critical factor for the cost-benefit analysis of the microgrid, influencing the investment decisions and planning scenarios of the electric utility. Despite this, the inherent intermittency of renewable energy distributed generation output, the complex power restoration mechanisms in the aftermath of failures, and the flexible operation of microgrids introduce novel difficulties and obstacles in the assessment of microgrid reliability.

In a master-slave controlled AC–DC microgrid configuration, the voltage-source converter (VSC) is equipped with a single control parameter, either reference voltage or reference power. On the other hand, in a droop-controlled AC–DC microgrid, each VSC operates with multiple control variables, including the droop factor, operating reference voltage, and operating reference power.

Microgrids: Theory and Practice, First Edition. Edited by Peng Zhang.
© 2024 The Institute of Electrical and Electronics Engineers, Inc. Published 2024 by John Wiley & Sons, Inc.

The manipulation of these control variables has a reciprocal effect on the VSC output voltage and system power flow. Furthermore, variations in the droop coefficients can impact the stability of the system beyond their influence on the droop control operating reference point [12–16].

The subsequent sections of this chapter are organized as follows. Section 28.2 presents a thorough evaluation of the typical topologies of AC–DC microgrids. Section 28.3 outlines a coordinated optimization methodology for the droop factor control in AC–DC microgrids. In Section 28.4, a case study is presented for illustration purposes. Finally, Section 28.5 concludes the chapter with a summary of the key findings.

28.2 Typical Topology Evaluation of AC–DC Microgrid

28.2.1 Typical Topology of AC–DC Microgrid

As with the traditional AC microgrid, the structural design of hybrid AC–DC microgrid affects the optimal configuration and coordinated operation. At present, there is no fixed form of AC–DC hybrid microgrid structure. The existing AC–DC microgrid topologies are mainly divided into four types, namely, AC bus grid-connected type, DC bus grid-connected type, AC–DC bus co-connected type and AC–DC bus co-connected interconnected type. The structure, characteristics and application scope of each of the above four topologies are introduced.

28.2.1.1 AC Grid-Connected Type
The AC grid-connected structure is rooted in the traditional AC microgrid architecture. The central component of this design is the AC bus, which serves as the connection and feedback mechanism for the entire system. The presence of energy storage systems in the DC microgrid serves to ameliorate the fluctuations in photovoltaic power output and maintain stability of the DC bus voltage, thereby ensuring power quality. The AC microgrid, in collaboration with the distribution network and the DC microgrid, upholds the stability of the AC bus voltage and frequency (Fig. 28.1).

28.2.1.2 DC Grid-Connected Type
The DC grid-connected type microgrid topology is depicted in Fig. 28.2. Its defining characteristics include: (1) The AC–DC microgrid is connected to the distribution network through a converter, which enables regulated power exchange between the two systems. However, it is unable to prevent the propagation of faults from the distribution network to the hybrid microgrid. (2) In cases where the AC load is relatively low, the interconnection converter with a smaller rated capacity may be selected, thereby reducing cost.

28.2.1.3 AC–DC Bus Co-Connected Type
The AC–DC grid-connected type microgrid topology is presented in Fig. 28.3. This topology features the separate connection of both the AC microgrid and DC microgrid to the distribution network, without interconnection between the two. Its key characteristics are: (1) The AC microgrid is electrically insulated from the distribution network, enabling regulated power exchange between the DC microgrid and the distribution network. (2) In the event of a distribution network failure resulting in microgrid islanding, power exchange between the AC microgrid and DC microgrid is not possible. This topology is suitable for areas with ample AC and DC loads, but with limited requirements for high reliability.

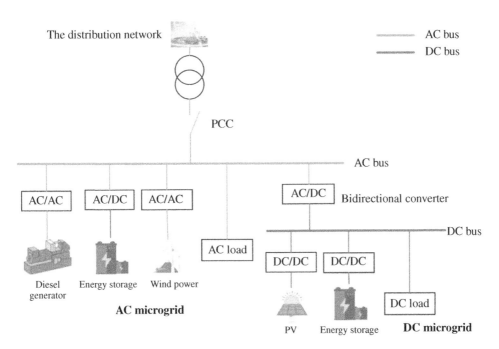

Figure 28.1 Topology of AC bus grid-connected type.

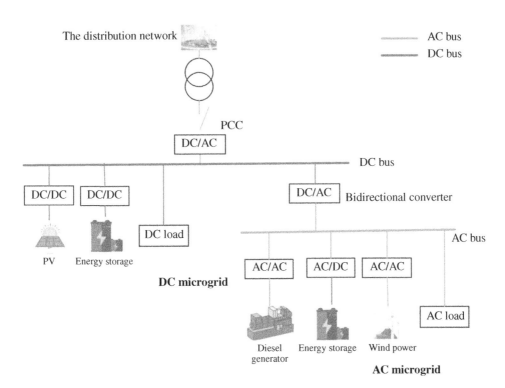

Figure 28.2 Topology of DC bus grid-connected type.

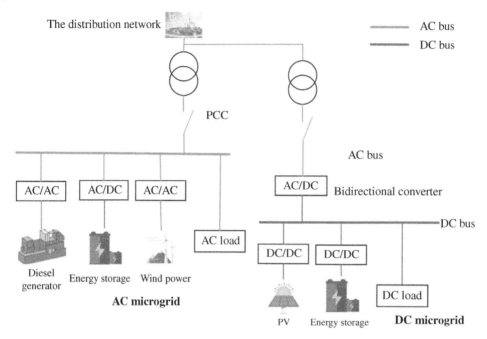

Figure 28.3 Topology of DC bus grid-connected type.

28.2.1.4 AC–DC Bus Co-Connected Interconnected Type

The AC–DC bus co-connected interconnected type microgrid topology is depicted in Fig. 28.4. This topology features the interconnection between the AC microgrid and DC microgrid, both of which are connected to the distribution network. Its defining characteristics include: (1) The interconnection of the AC microgrid, DC microgrid, and distribution network enhances the reliability of the entire AC–DC microgrid system. (2) The power flow within the AC–DC microgrid is diverse and the control of converters is complex, making this topology appropriate for areas with high AC–DC load and stringent reliability requirements.

28.2.2 Reliability Modeling of AC–DC Microgrid

28.2.2.1 Reliability Model of PV Power Generation System

A two-state Markov model is utilized to characterize the state transition process of the photovoltaic (PV) module's operational status, as depicted in Fig. 28.5. The parameters μ and λ represent the repair rate and failure rate of the PV module, respectively.

It is postulated that the operational and failure durations of each PV module are governed by exponential distributions, which can be estimated using the state duration sampling technique as outlined in Eq. (28.1).

$$\begin{cases} T_{\text{work}} = -\text{MTTF} \cdot \ln(\mu_1) \\ T_{\text{failure}} = -\text{MTTR} \cdot \ln(\mu_2) \end{cases} \tag{28.1}$$

Where mean time to failure (MTTF) represents the average trouble-free operational duration of the component, and mean time to repair (MTTR) represents the mean time to repair. The random variables μ_1 and μ_2 are uniformly distributed over the interval [0,1]. In accordance with the

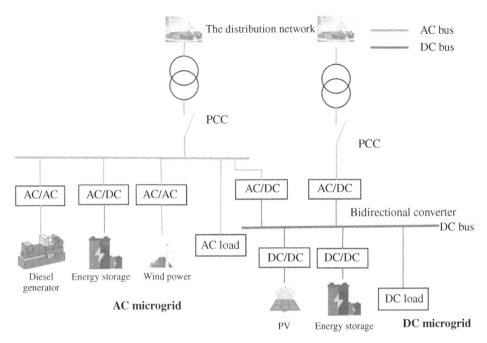

Figure 28.4   Topology of AC–DC bus co-connected interconnected type.

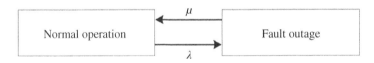

Figure 28.5   Markov two-state model.

exponential distribution assumption, the relationship between MTTF and λ, as well as between MTTR and μ, is depicted in Eq. (28.2).

$$\begin{cases} MTTF = \dfrac{1}{\lambda} \\[2mm] MTTR = \dfrac{1}{\mu} \end{cases} \tag{28.2}$$

28.2.2.2   Reliability Model of Wind Power Generation System

The state duration sampling technique is employed to simulate the operational state of the wind turbine. It is hypothesized that the operational and failure durations of each wind turbine follow exponential distributions, and their values can be estimated using Eq. (28.1).

The association between the actual power output and wind velocity can be depicted by the power output characteristic curve of the wind turbine. The wind velocity for each hour in the analysis can be simulated through a Weibull distribution.

Fluctuations in the wind turbine output can be mitigated through balancing with the power in the distribution system during grid-connected operation of the microgrid. Under these conditions, the entire output of the turbine can be utilized to meet the load demand. However, during the islanded operation mode of the AC–DC microgrid, it is often imperative to constrain the turbine output to maintain voltage and frequency stability in the system. In this study, we adopt the notion of turbine dispatch ratio and stipulate that the ratio of the turbine output to the dispatchable power on the island cannot exceed a predefined threshold value of DR, as outlined in Eq. (28.3).

$$
P_W = \begin{cases} P_{WA} \, P_{WA} \leq P_D \cdot DR \\ P_D \cdot DR \, P_{WA} \geq P_D \cdot DR \end{cases}
\tag{28.3}
$$

Where P_W denotes the permissible wind turbine output in the islanded state, P_{WA} represents the maximum attainable wind turbine output given the current turbine operating conditions and wind velocity. P_D symbolizes the total dispatchable power on the island, and DR represents the wind turbine dispatch ratio.

28.2.2.3 Reliability Model of Diesel Generator and Energy Storage

Diesel generators are based on conventional energy sources and feature controllable output capabilities. Similarly, the output of the energy storage system is amenable to regulation, with its available capacity contingent on its state of operation. Given the activation of conventional energy DGs and energy storage systems in the microgrid exclusively upon entry into islanded mode, and the ephemeral nature of microgrid islanded operation, it is reasonable to posit that the states of the diesel generators and energy storage systems are maintained during such operational phases. The initial states of the diesel generators and energy storage systems at the inception of islanded operation are ascertained through the state sampling methodology, as elucidated in Eq. (28.4).

$$
X_i = \begin{cases} 1 \, u \geq FOR_i \\ 0 \, u < FOR_i \end{cases}
\tag{28.4}
$$

Where X_i signifies the state of diesel generator i or energy storage i. When $X_i = 1$, it indicates that the element is in operation, and when $X_i = 0$, the element is inactive. FOR_i represents the probability of start-up failure for the conventional energy DG or the forced outage rate of the energy storage element.

It is assumed that at the initiation of islanded operation, the non-faulty energy storage devices are in a fully charged state. Throughout the islanded operation, when the aggregate output of renewable energy DGs surpasses the total load demand, the excess power is utilized to recharge the energy storage devices. Conversely, when the total output of all DGs is insufficient to meet the total load, the power deficit is compensated by discharging the energy storage devices. Unlike energy storage devices, diesel generators necessitate a start-up period after the microgrid enters islanded mode, during which only the energy storage devices and renewable energy DGs sustain power supply to the microgrid load. The charge and discharge states of the energy storage devices and the power are depicted in Eq. (28.5).

$$
\begin{aligned}
P_{charge}(t) &= \sum P_{RDG}(t) - P_L(t), \sum P_{RDG}(t) > P_L(t) \\
P_{discharge}(t) &= P_L(t) - \sum P_{DG}(t), \sum P_{DG}(t) < P_L(t)
\end{aligned}
\tag{28.5}
$$

Where $P_{charge}(t)$ and $P_{discharge}(t)$ denote the charging and discharging power of the energy storage device at time t, respectively. $P_{RDG}(t)$, $P_{DG}(t)$, and $P_L(t)$ symbolize the output power of the renewable energy DG, all types of DGs, and the load level at time t, respectively.

Moreover, the charging and discharging power of the battery pack must be within the upper bounds of its charging and discharging power capabilities. Additionally, when the residual battery power reaches the lower threshold of the state of charge (SOC), further discharge of the battery is prohibited, as outlined in Eq. (28.6).

$$
\begin{aligned}
&0 \leq P_{charge}, P_{charge} \leq P_{max} \\
&E_{reserve} \leq E_{remain} + E_{charge} - E_{discharge} \leq E_{max}
\end{aligned}
\tag{28.6}
$$

where P max signifies the maximum charge and discharge power of the energy storage device. E_{charge}, $E_{discharge}$, E_{remain}, and $E_{reserve}$ denote the charging power, discharging power, residual power, and minimum reserved power of the energy storage device, respectively. E_{max} represents the maximum power capacity of the energy storage device.

28.2.2.4 Reliability Model of Load

Considering the varying significance of loads within the AC–DC microgrid, the loads are classified into critical loads and interruptible loads, and the reduction of AC–DC microgrid loads is treated as a discontinuous process. In the event of inadequate islanded power, interruptible loads are prioritized for disconnection. During the microgrid reliability analysis, the load variation curve over time is presumed to follow the IEEE-RTS model. The model calculates the percentage of weekly peak load to the annual peak load over 52 weeks of the year, the percentage of daily peak load to the weekly peak load, and the percentage of hourly peak load to the daily peak load of the week, effectively encapsulating the variation of the hourly peak load throughout the year as an annual peak load scale, as depicted in Fig. 28.6.

28.2.3 Reliability Analysis for the AC–DC Microgrid

The AC–DC microgrid exhibits distinct differences from traditional distribution systems in terms of its composition, network topology, operational logic, and load restoration strategy. Monte Carlo simulations of microgrid operations and distribution system operations differ greatly in terms of their time scales, operational complexity, and system constraints. Simultaneous simulation of both systems would increase the difficulty of the simulation process, while separate and independent simulations would sever the intrinsic connection between the two systems, resulting in a distorted simulation outcome. To address this issue, this section introduces a PCC equivalent generator model for the AC–DC microgrid and divides the Monte Carlo simulation into two stages: distribution system simulation and microgrid simulation. This approach not only preserves the impact of the distribution system's operational characteristics on the microgrid, but also provides more nuanced information regarding microgrid operations through simulation with a smaller time step.

28.2.3.1 Equivalent Model of PCC

This chapter introduces the concept of equating the PCC of the microgrid as a conventional generator. The generator's operation is modeled using a repairable two-state model that transitions

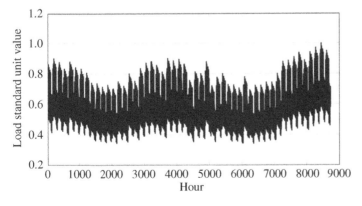

Figure 28.6 IEEE RBTS hourly load model.

between operating and faulty states only in response to random faults within the distribution system, and whose capacity is the maximum power the distribution system can provide to the microgrid under current operating conditions. It is assumed that the capacity of the generator in its operating state is sufficient to meet the demands of all loads within the AC–DC microgrid, without taking into consideration the capacity constraints of the distribution system.

When the generator is in operation, it signifies that the AC–DC microgrid is operating in parallel with the distribution system and that both the distribution system and the renewable energy DGs within the microgrid are jointly supplying power to the microgrid loads. Upon transition of the generator from its operational state to a faulty state, it indicates that the microgrid switches from grid-connected mode to islanded mode. At this point, the standby conventional energy DGs commence operation and provide power to the microgrid loads, in conjunction with the energy storage devices and renewable energy DGs. If the output of the renewable energy DGs exceeds the microgrid load, the surplus power is utilized to charge the energy storage devices.

28.2.3.2   Reliability Assessment for the AC–DC Microgrid Based on PCC Equivalent Generator Model

The Monte Carlo simulation of the microgrid can be executed in two stages by utilizing the equivalent generator model of the microgrid's PCC.

(1) Reliability analysis of external load point of PCC and AC–DC microgrid

It is assumed that the microgrid does not affect the reliability of its external load point. To determine the state transfer process of the PCC, the microgrid is equated to a special load point connected at its PCC, and the distribution system that encompasses the microgrid is considered as a conventional distribution system without DGs (Fig. 28.7).

The state duration sampling method is employed to simulate the random failures of components within the distribution system, thereby calculating the reliability index of the external load points of the microgrid and determining the state transition process of the PCC. The analysis presupposes that the failure rate and repair rate of the components within the distribution system, as well as the switch actuation time, are constant and that the switch actuation is entirely reliable.

(2) Reliability analysis of internal load points of AC–DC microgrid

Based on the state transition information of the PCC obtained in the first stage of the simulation, an equivalent generator model of the PCC is established. The reliability analysis of the AC–DC microgrid is then decoupled from the influence of the distribution system and transformed into the reliability analysis of a small isolated generation system composed of components such as the equivalent generator, diesel generator, and load within the AC–DC microgrid. The structure of the PCC equivalent AC–DC microgrid is depicted in Fig. 28.8, while the Monte Carlo simulation framework of the equivalent AC–DC microgrid is provided in Fig. 28.9.

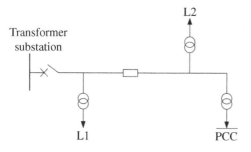

Figure 28.7   Schematic diagram of the local distribution system after AC–DC microgrid equivalence.

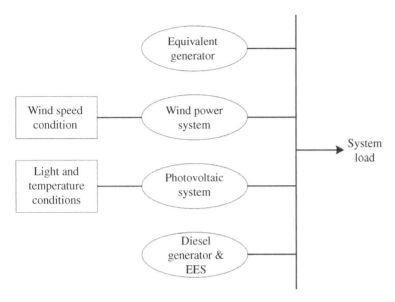

Figure 28.8 Schematic diagram of the microgrid after PCC equivalence.

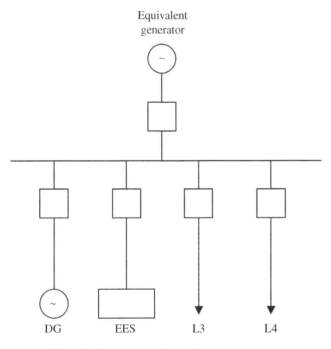

Figure 28.9 AC–DC microgrid Monte Carlo simulation framework.

The reliability analysis of the microgrid is conducted using a sequential Monte Carlo simulation method. This approach takes into account not only the time-varying characteristics of wind turbine and PV output, but also the effects of factors such as the success rate of AC–DC microgrid islanding conversion, the start-up time of conventional energy diesel generator, and the grid structure on the reliability of the AC–DC microgrid.

The Monte Carlo method is converged using the coefficient of variance as a criterion, as represented by Eq. (28.7).

$$\beta = \frac{\sigma[E(X)]}{E(X)} \leq \varepsilon \tag{28.7}$$

In this analysis, the reliability index X is calculated and its mean $E(X)$ and standard deviation of the mean $[E(X)]$ are obtained. The convergence tolerance, denoted as ε, is set to 0.01 in this study.

The Monte Carlo simulation for the AC–DC microgrid is executed in two stages, with the first stage requiring a sufficient number of simulated years to obtain information on the PCC's state transitions for the second stage, even if convergence has already been achieved within that timeframe.

28.3 Coordinated Optimization for the AC–DC Microgrid

In an AC–DC microgrid, an overly large droop factor can lead to system instability, while an overly small droop factor results in a low system damping ratio, causing prolonged transient transition times.

The optimization of the droop coefficient in the AC–DC microgrid system is aimed at striking a balance between stability and voltage regulation. By considering the impact of random changes in load and renewable energy output, the optimal droop coefficient is determined to ensure that the voltage and power flow distribution in the system remains stable and within acceptable limits, even under small disturbances. The goal is to effectively utilize the droop control mechanism for power distribution, while avoiding excessive voltage shifts at nodes caused by the droop control during random changes in load.

28.3.1 Small Signal Model of Droop Control AC–DC Microgrid

28.3.1.1 VSC Model

A comprehensive small-signal model for a droop-controlled AC–DC microgrid has been devised to assess its stability. In the model, a two-level three-phase voltage PWM converter is selected to represent the VSC, as depicted in Fig. 28.10. The control model of the VSC employs $d - q$ axis decoupling control for the current loop, under the assumption that there are no disturbances in the AC network voltage. Both AC and DC active power are conserved, with the reactive power on the AC side being set to zero.

The mathematical model for the AC side of the VSC can be represented as follows:

$$u_{sd} = L_{ac}\frac{di_d}{dt} + R_{ac}i_d + u_d - L_{ac}\omega i_q \tag{28.8}$$

Wherein, u_{sd} denotes the d-axis component of the AC-side supply voltage, R_{ac} and L_{ac} represent the AC-side resistance and reactance, respectively, i_d and i_q symbolize the d-axis and q-axis components of the AC-side current, respectively, ω stands for the angular frequency of the fundamental wave in the AC grid, and u_d represents the d-axis component of the input voltage to the power switch on the AC-side.

For the sake of simplicity and clarity, let's represent:

$$\lambda = \frac{i_d^* - i_d}{s} \tag{28.9}$$

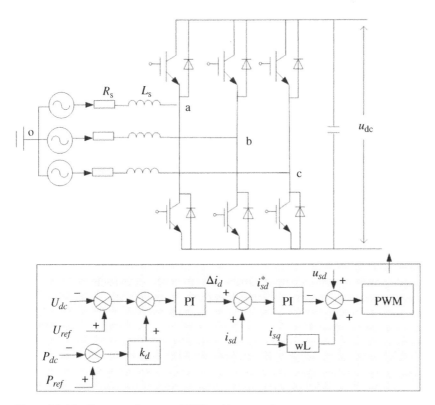

Figure 28.10 Structure diagram of VSC and its control system.

The state equation for the current internal loop control of the VSC is as follows:

$$u_d^* = u_{sd} - K_{p1}(i_d^* - i_d) - K_{i1}\lambda_d + l_{ac}\omega i_q \tag{28.10}$$

Wherein, u_d denotes the reference d-axis voltage for the AC side of the VSC, i_d represents the reference d-axis current for the AC side of the VSC, K_{p1} symbolizes the proportional parameter for the current internal loop control, and K_{i1} signifies the integral parameter for the current internal loop control.

Assuming that the AC-side reactive power command is set to zero and the q-axis current is small, the following expression can be derived:

$$u_{sd} = L_{ac}\frac{di_d}{dt} + R_{ac}i_d + u_d$$
$$u_d^* = u_{sd} - K_{p1}(i_d^* - i_d) - K_{i1}\lambda_d \tag{28.11}$$

When the effect of valve delay in the converter is neglected, it can be assumed that:

$$u_d^* = u_d \tag{28.12}$$

The active power on the DC side, as obtained through the filter, can be expressed as:

$$P_{dc} = \frac{\omega_s}{s + \omega_s}(u_{dc}i_{dc}) \tag{28.13}$$

In the active power external loop control, the deviation between the measured output power at the DC side of the VSC and the power reference value is utilized to calculate the deviation from the measured voltage at the DC side after the droop control connection.

$$u_{dc}^* = u_{dcref} - k(P_{dc} - P_{dcref}) \tag{28.14}$$

Where u_{dc}^* is the calculated DC reference voltage. u_{dcref} and P_{dcref} are the droop control operating reference voltage and operating reference power. u_{dc} and i_{dc} are the voltage and current on the DC side, respectively. k is the droop factor.

For representational convenience, make

$$\gamma_d = \frac{u_{dc}^* - u_{dc}}{s} \tag{28.15}$$

The VSC voltage outer loop control state equation as:

$$i_d^* = K_{p2}(u_{dc}^* - u_{dc}) + K_{i2}\gamma_d \tag{28.16}$$

Where K_{p2} is the voltage outer loop proportional parameter; K_{i2} is the voltage outer loop integral parameter.

From the DC-side circuit model, we can obtain

$$C\frac{du_{dc}}{dt} = i_x - i_{dc} \tag{28.17}$$

Where i_x is the DC-side power switching output current. C is the DC-side capacitance.

Assuming that the active power on the AC and DC sides of VSC is conserved, we can obtain

$$u_d i_d = u_{dc} i_x \tag{28.18}$$

Linearization of the above equation of state at the steady-state operating point yields.

$$\frac{d\Delta P}{dt} = -\omega_s \Delta P + \omega_s u_{dc,0} \Delta i_{dc} + \omega_s \Delta u_{dc} i_{dc,0} \tag{28.19}$$

$$\Delta\frac{d\lambda_d}{dt} = \Delta i_d^* - \Delta i_d \tag{28.20}$$

$$\Delta\frac{d\gamma_d}{dt} = \Delta u_{dc}^* - \Delta u_{dc} \tag{28.21}$$

$$\Delta u_d^* = -K_{p1}(\Delta i_d^* - \Delta i_d) + K_{i1}\Delta\lambda_d \tag{28.22}$$

$$\Delta i_d^* = K_{p2}(\Delta u_{dc}^* - \Delta u_d c) + K_{i2}\Delta\gamma_d \tag{28.23}$$

$$\Delta u_{dc}^* = -k\Delta P \tag{28.24}$$

$$\Delta\frac{di_d}{dt} = -\frac{R_{ac}}{L_{ac}}\Delta i_d - \frac{1}{L_{ac}}\Delta u_d \tag{28.25}$$

$$\Delta\frac{du_{dc}}{dt} = \frac{1}{C}\Delta i_x - \frac{1}{C}\Delta i_{dc} \tag{28.26}$$

$$\Delta i_x = \frac{u_{d,0}\Delta i_d}{u_{dc,0}} + \frac{u_d i_{d,0}}{u_{dc,0}} - \frac{\Delta u_{dc} i_{x,0}}{u_{dc,0}} \tag{28.27}$$

$$\Delta u_d^* = \Delta u_d \tag{28.28}$$

where $u_{dc,0}$ and $i_{dc,0}$ are the DC-side voltage and current of the system at the steady-state operating point, respectively. $i_{x,0}$ is the DC-side power switching output current of the system at the steady-state operating point and $i_{d,0}$ is the d-axis component of the ac-side current of the system at the steady-state operating point.

Figure 28.11 The line model.

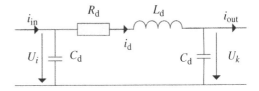

28.3.1.2 The Line Model

As shown in Fig. 28.11, a π-type equivalent circuit is used to describe the mathematical model of the DC distribution network line, and the circuit model is represented as:

$$
\begin{aligned}
i_{\text{in}} - C_d \frac{dU_i}{dt} &= i_d \\
U_i - U_k &= L_d \frac{di_d}{dt} + R_d i_d \\
i_d - C_d \frac{dU_k}{dt} &= i_{\text{out}}
\end{aligned}
\tag{28.29}
$$

where u_i and u_k are the voltages at node i and node k, respectively. R_d and L_d are the branch resistance and reactance, respectively. C_d is half of the equivalent capacitance of the branch. i_d is the current flowing through the inductor. i_{in} and i_{out} are the currents flowing into and out of the branch, respectively.

Linearization of the above equation of state at the steady-state operating point yields:

$$
\begin{aligned}
\Delta \frac{dU_i}{dt} &= \frac{1}{C_d} \Delta i_{\text{in}} - \frac{1}{C_d} \Delta i_d \\
\Delta \frac{di_d}{dt} &= \frac{1}{L_d} \Delta U_i - \frac{1}{L_d} \Delta U_k - \frac{R_d}{L_d} \Delta i_d \\
\Delta \frac{dU_k}{dt} &= \frac{1}{C_d} \Delta i_{\text{out}} - \frac{1}{C_d} \Delta i_d
\end{aligned}
\tag{28.30}
$$

28.3.1.3 The Load Model

To reduce the complexity of the model, the active load under normal conditions and the intermittent power supply using maximum power tracking can be considered as constant power sources. The distributed power output is considered as a negative load, while load-side capacitance is considered to further attenuate the effect of voltage fluctuations. As shown in Fig. 28.12, the DC/DC and its load are described using a ZIP model with equivalent capacitance, and the circuit model can be expressed as:

$$
i_L - C_L \frac{dU_L}{dt} = \frac{P_L}{U_L} + I_L + \frac{U_L}{R_L}
\tag{28.31}
$$

where i_L is the DC flowing into the load side. R_L denotes the power value of the constant power load. I_L denotes the current value of the constant current load. R_L denotes the resistance value of the constant resistance load. U_L is the voltage on the load side. C_L is the equivalent capacitance on the load side.

Figure 28.12 The load model.

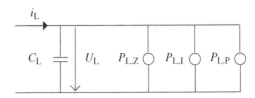

Linearization of the above equation of state at the steady-state operating point yields:

$$\Delta \frac{dU_L}{dt} = \frac{1}{C_L} \Delta i_L + \frac{P_{L,0}}{C_L U_{L,0}^2} \Delta U_L - \frac{1}{C_L R_L} \Delta U_L \tag{28.32}$$

where $P_{L,0}$ represents the power value of the constant power load at the steady-state operating point. $U_{L,0}$ represents the voltage on the load side of the system at the steady-state operating point.

28.3.2 Objective Function and Constraints for Droop Coefficient Optimization

28.3.2.1 Objective Function
For a certain moment, the optimal operating state of the system determined after the optimization of the operating reference power is taken as the stable operating point of the system. The state equation of the stable operation point is linearized to construct the differential state matrix, and the stability of the system is analyzed by finding the characteristic roots.

The maximum damping ratio and the maximum stability margin of the system are used as the objective function for droop factor optimization:

$$\min \ f_2 = -b_1 \xi + b_2 \alpha_d \tag{28.33}$$

where α_d is the real part of the dominant characteristic root of the system, ξ is the system damping ratio, and b_1 and b_2 are the weighting coefficients of the damping ratio and the stability margin index.

28.3.2.2 Constraints
The constraints for droop factor optimization include two types: one is the small disturbance stability constraint, which is used to ensure that the system is stable with small disturbances. The other is the node voltage magnitude constraint, VSC power constraint and other constraints after considering random changes in load, which are used to ensure that the system has a reasonable tide distribution after random changes in load.

(1) Small disturbance stability constraint
The real part of the dominant characteristic root of the system is less than zero indicating that the system is stable. Therefore, the real part of the dominant characteristic root at the stable operating point is less than zero as a constraint for the stability of small disturbances.

$$\alpha_d < 0 \tag{28.34}$$

(2) Node voltage and VSC power constraints
In various scenarios where the load and renewable energy output vary randomly, it is considered that as the load increases or the renewable energy output decreases, the node voltage decreases and the transmitted power of the VSC increases. The situation where the load is maximum and the renewable energy output is minimum is scenario 1, and the situation where the load is minimum and the renewable energy output is maximum is scenario 2, then scenario 1 is the situation to be considered as the node voltage minimum and the VSC power maximum constraint when all VSCs of the system are in rectification state. Similarly, taking scenario 2 as the scenario where the node voltage maximum constraint and the VSC power maximum constraint when all VSCs of the system are in the inverted state need to be considered. The node voltage constraint and VSC power constraint are expressed as:

$$\begin{cases} U_{i,\min} \leq U_{i,s1} \leq U_{i,\max} \\ U_{i,\min} \leq U_{i,s2} \leq U_{i,\max} \\ |P_{x,s1}| \leq P_{x,\mathrm{Cmax}} \\ |P_{x,s2}| \leq P_{x,\mathrm{Cmax}} \end{cases} \tag{28.35}$$

where $U_{i,s1}$ and $P_{x,s1}$ denote the voltage corresponding to node i and the power of the VSC located at node x in scenario 1. $U_{i,s2}$ and $P_{x,s2}$ are the voltage corresponding to node i and the power of the VSC located at node x in scenario 2.

(3) VSC power equalization constraint
In order to realize droop control power distribution, the VSC power equalization constraint is set. For node i, when the load or renewable energy output at that node changes by Δp_i, the difference between the ratio of each VSC power change and the capacity of that VSC and the ratio of the total power change to the sum of all VSC capacities (defined as the VSC power equalization index for the sake of description) is required to be less than a certain set value $\varepsilon_{\mathrm{rate}}$.

$$\left| \frac{\Delta P_{i,x}}{\Delta P_i} - \frac{P_{x,\mathrm{Cmax}}}{P_{\mathrm{sum,VSC}}} \right| \leq \varepsilon_{\mathrm{rate}} \; i = 1,\dots,h; x = h+1,\dots,h+m \tag{28.36}$$

where $\Delta P_{i,x}$ is the amount of change in the VSC power located at node x when the power of node i changes.

28.3.3 Optimal Solution Algorithm for Droop Coefficient

In this chapter, a genetic algorithm is used to optimize the droop coefficient. The power flow calculation is performed for each time period based on the results of the operational reference power optimization, so that power flow data such as node voltage and line power can be obtained. In calculating the system characteristic roots, for each individual generated in the genetic algorithm, a spatial state matrix is generated based on the power flow data corresponding to that individual, and the characteristic roots are derived by the QR decomposition method. When calculating the voltage and VSC power in the constraints, the power flow sensitivity model is used to calculate the node voltage and VSC power, considering that a large number of power flow calculations in the iterative process of the genetic algorithm will increase the computational effort. The power flow state obtained by running the reference power optimization is used as the reference state, and for each individual generated in the genetic algorithm, the power flow sensitivity matrix is calculated based on the sag coefficient corresponding to that individual. Based on this, the node voltage and VSC power are calculated using the power flow sensitivity based on the difference between the node power in the extreme scenario and the node power in the reference state.

According to the power flow sensitivity model, Eq. (28.35) is expressed as:

$$\begin{cases} U_{i,\min} \leq U_{i,0} + \sum_{n=1}^{h} S_{n,i}(P_{n,s1} - P_{n,0}) \leq U_{i,\max} \\ U_{i,\min} \leq U_{i,0} + \sum_{n=1}^{h} S_{n,i}(P_{n,s2} - P_{n,0}) \leq U_{i,\max} \\ |P_{x,0,\mathrm{VSC}} + \sum_{n=1}^{h} W_{n,x}(P_{n,s1} - P_{n,0})| \leq P_{x,\mathrm{Cmax}} \\ |P_{x,0,\mathrm{VSC}} + \sum_{n=1}^{h} W_{n,x}(P_{n,s2} - P_{n,0})| \leq P_{x,\mathrm{Cmax}} \end{cases} \tag{28.37}$$

where $P_{n,0}$ is the power of node n in the reference scenario. $P_{n,s1}$ and $P_{n,s2}$ are the power of node n in scenarios 1 and 2, respectively. $U_{i,0}$ is the voltage corresponding to node i after running the reference power optimization. $P_{x,0,VSC}$ is the transmitted power of the VSC located at node x after running the reference power optimization.

According to the power flow sensitivity model, Eq. (28.36) can be expressed as:

$$\left| W_{i,x} - \frac{P_{x,Cmax}}{P_{sum,VSC}} \right| \leq \varepsilon_{rate} i = 1, \ldots, h; x = h + 1, \ldots, h + m \tag{28.38}$$

In the genetic algorithm, a large penalty factor is applied to the objective function of individuals that do not satisfy the constraints, so that individuals that do not satisfy the constraints do not participate in the next evolution.

28.4 Case Study

28.4.1 Reliability Analysis of AC–DC Microgrid

The AC–DC microgrid system comprises of both an AC microgrid and a DC microgrid, as illustrated in Fig 28.1. The AC microgrid encompasses wind turbines, diesel generators, battery storage devices, and AC loads, while the DC microgrid encompasses PV systems, battery storage devices, and DC loads. The economic parameters for each DG and fault information are presented in Table 28.1, while the types of loads in the microgrid and the outage loss parameters are outlined in Table 28.2. To enable efficient analysis and comparison, three microgrid planning schemes have been proposed, with the corresponding parameters listed in Table 28.3. The environmental conditions at the microgrid site are based on the measured light and wind speed data, as well

Table 28.1 DG used in microgrid.

DG type	Reted power	MTTF(h)	MTTR(h)
Photovoltaic array[a]	92 kWp	900	20
Battery pack[b]	120 kWh	970	30
Wind turbine[c]	100 kW	1000	50
Diesel generator[d]	227 kW	600	50

a)   20 × 20 Photovoltaic module CNPV-230 Wp series and parallel connection.
b)   FOR = 0.03.
c) $V_{ci} = 3, V_r = 12, V_{co} = 40$.
d)   Startup failure probability = 0.2, Startup time is 0.1 h.

Table 28.2 Characteristics of microgrid users.

		Outage loss		
Load name	Load type	CPI (yuan/ (time kW))	IEAR (10^4 yuan/ kWh)	Number of users
DC load	Data center	3.53	0.05	1
AC load	Business load	0.22	0.002	10

Table 28.3 Microgrid planning scheme.

Scheme	Numbe of photovoltaic array	Number of wind turbines	Number of AC microgrid battery packs	Number of DC microgrid battery packs	Number of Ndiesel generators
1	4	8	1	1	2
2	8	4	1	1	2
3	8	4	1	2	3

as temperature data from Tianjin in 2020. For the purpose of the analysis, the following implicit assumptions have been made: (1) The success rate of the islanding conversion process is assumed to be 0.8. (2) The estimated start-up time for the diesel generator is 0.1 hours. (3) The turbine dispatch ratio is assumed to be 0.8.

Among the three proposed schemes, the characteristics of microgrid DGs are presented in Table 28.4. Furthermore, PV systems are less susceptible to module failures and system operating constraints, resulting in higher utilization rates. Scenarios 2 and 3 increase the proportion of PV modules based on Scenario 1, thereby doubling the amount of load supplied by PV systems in proportion to the total demand in the microgrid. In Scenario 3, the capacity of energy storage and conventional energy DG is increased from Scenario 2, leading to an increased emergency backup capacity for the microgrid's load demand, which can be quantitatively expressed through the storage duration (SD) index. Despite a slight difference in the output of renewable energy DG, the average generation load ratio of the microgrid is more influenced by its internal conventional energy DG.

The interconnectivity indicators between the microgrid and the distribution system, the reliability indicators of the microgrid system, and the reliability indicators related to the islanding operation of the microgrid are presented in Tables 28.5 and 28.6, respectively.

Table 28.4 Microgrid planning scheme.

Scheme	CC_WTG (%)	CC_PV (%)	REP_WTG (%)	REP_PV (%)	DGUR_WTG (%)	DGUR_PV (%)	SD (h)	AGLR (%)
1	14.04	19.94	10.45	7.01	0.9523	0.9782	0.1382	38.37
2	14.04	19.94	5.22	14.02	0.9523	0.9782	0.1382	39.41
3	14.04	19.94	5.22	14.02	0.9523	0.9782	0.2073	53.02

Table 28.5 Reliability index reflecting the relationship between microgrid and distribution system.

λ_{PCC} (time/year)	U_{PCC} (h)	r_{PCC} (h/time)	MIOP (%)	MP (%)
0.1873	0.92996	4.966	0.0084	8.299

Table 28.6 System reliability index of microgrid.

Scheme	SAIFI (time/ household a)	SAIDI (h/household a)	ENS (kWh/a)	CENI (%)	AREG (MWh/ household a)
1	0.243	3.567	3633	78.61	142.38
2	0.237	3.548	3630	79.02	156.99
3	0.203	3.549	3611	81.77	156.99

Since the relationship between the microgrid and the distribution system only depends on the location of the microgrid within the distribution system and the size of the microgrid's load, it is not affected by the composition of the power in the microgrid. Hence, the corresponding index values in Table 28.6 remain constant across all three scenarios. When a microgrid is established, as illustrated in Fig. 28.1, the reliability at the PCC of the microgrid is primarily determined by the fault condition of the 10 kV feeder. Due to the usually short rehabilitation time of the lines, the average duration of microgrid islanding operation and the total annual islanding operation time tend to be relatively short. Moreover, the microgrid penetration rate is high given the extensive scope of the 10 kV microgrid formation.

As can be seen from Tables 28.6 and 28.7, since PV generation is more stable and reliable than wind generation, the total installed capacity of renewable energy DG in the microgrid remains largely unchanged. Increasing the proportion of PV generation or increasing the capacity of backup units and energy storage can enhance the indicators related to system outage frequency. However, the impact on the outage time index and the outage power index is not substantial. With an increase in the capacity of conventional units and energy storage within the microgrid, all reliability indicators of islanding operation are significantly improved.

Both the island expected energy deficiency (IEED) and island energy deficiency interruption (IEDI) are indicators that reflect the energy shortfall of DG generation during the islanded operation of the microgrid. The distinction between the two lies in that the former is only determined by the relative size of the DG output to the load, while the latter also considers the adjustability of the load within the microgrid. When the microgrid can achieve continuous load adjustment, meaning that each load reduction in the microgrid is equal to the generation capacity shortfall, the IEED index is equivalent to the IEDI index. Given that the load shedding process is assumed to be discontinuous in the analysis, the total energy of load shedding (IEDI) resulting from the microgrid power deficit will be greater than the total energy deficit (IEED), and the extent of the difference often indicates the degree of adjustability of the microgrid load.

Table 28.7 Reliability index of isolated island operation of microgrid.

Scheme	ILOLP (%)	IENS (kWh/a)	IEED (kWh/time)	IEDI (kWh/time)	ILSE (kW/time)
1	19.99	212.48	38.36	65.10	700.57
2	19.62	208.65	37.29	64.28	710.87
3	14.45	205.27	17.72	36.49	537.58

The annual expected outage (island energy not supplied [IENS]) of the example microgrid during islanding represents only a small fraction of the annual expected outage (ENS) of the microgrid. This implies that the majority of outages experienced by the microgrid do not occur during its islanded operation. In reality, the most significant contributor to power outages at the microgrid load point in the example calculation is the failure of the 10kV distribution transformer. Such transformer failures often take a prolonged time to repair and are the primary reason for the unreliability of power supply to the load. Although a microgrid established on a MV feeder can protect its internal load from brief outages caused by external MV feeder failures, it does not provide backup for transformer failures within the microgrid, thus limiting its ability to enhance the reliability of power supply to the load.

The reliability of both AC load and DC load in an AC–DC microgrid is analyzed individually using Scheme 1 as an example (Table 28.8). It can be observed that the SAIFI, system average interruption frequency index (SAIDI), and ENS of DC load in the AC–DC microgrid are higher than that of AC load. This is because, in case of an AC/DC failure, AC load can still receive power supply from the grid, whereas the power supply of DC load is impacted. This indicates that in an AC grid-connected AC–DC microgrid, the failure of AC/DC equipment will affect the reliability of DC power supply, resulting in a higher number of DC load failures and longer outage time compared to AC load.

28.4.2 Voltage Control Strategy of AC–DC Microgrid

28.4.2.1 Analysis of Optimization Results

This section provides an analysis using an example of an AC–DC microgrid demonstration project in China. The topology of the DC microgrid section is presented in Fig. 28.13 and is configured in a pseudo-bipolar wiring form, operating with a handheld ring network. The voltage level on the medium voltage side is ±10 kV. The parameters for VSC droop control are presented in Table 28.9. The simulation was carried out on the Matlab 2016b simulation platform using a computer with the configuration of an Intel(R) Core(TM) i5-8350U CPU. The voltage results, obtained from the trend calculation voltage after five iterations, satisfy the convergence condition and are presented in Fig. 28.14.

28.4.2.2 Analysis of Droop Coefficient Optimization Results

In the optimization scenario for the droop coefficient boundary case, the upper and lower limits of load and PV output variation are set to 1.2 and 0.8 times the predicted values, respectively. The droop coefficients of VSC1 and VSC2 are selected from the range of [0.01 1] kV/MW. The system damping ratio weighting factor b1 and stability margin weighting factor b2 are set to 1 and 0.1, respectively. The genetic algorithm employs a population of 200 chromosomes and has a maximum number of evolutionary generations, with a crossover rate and mutation rate of 0.1

The optimization results of the droop coefficients for VSC1 and VSC2 are presented in Fig. 28.15. The updated reference power, calculated from the optimized droop coefficients, is shown in

Table 28.8 System reliability index of microgrid.

Load type	SAIFI (time/household*a)	SAIDI (h/household*a)	ENS (kWh/household*a)
DC load	0.252	3.699	334
AC load	0.242	3.554	330

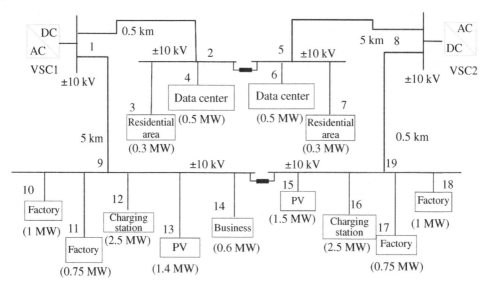

Figure 28.13 Topology diagram of an AC–DC microgrid demonstration project.

Table 28.9 Droop control parameters of the VSC.

VSC	VSC1	VSC2
Reference voltage (kV)	±10	±10
Reference power (kW)	1600	6600
Droop coefficient (kV/MW)	0.31	0.31

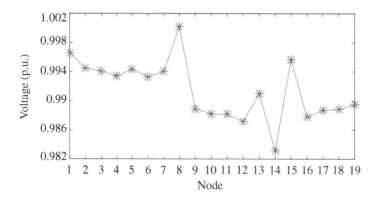

Figure 28.14 voltage results of the system.

Fig. 28.16. The load factor of the VSC under the scenario where the load is 1.2 times the predicted value and the PV output is 0.8 times the predicted value is presented in Fig. 28.17.

The optimization results show that the droop coefficient of VSC1 is larger than that of VSC2, which is in line with the fact that VSC1 has a smaller capacity than VSC2. This ensures an appropriate distribution of power between VSC1 and VSC2 as the load changes. Figure 28.17 demonstrates

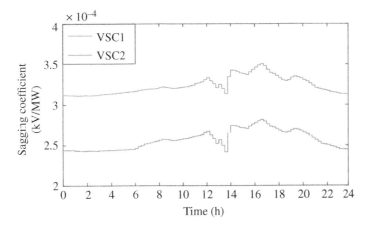

Figure 28.15 Optimization result of droop coefficient.

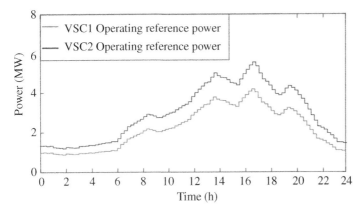

Figure 28.16 Updated operating reference power.

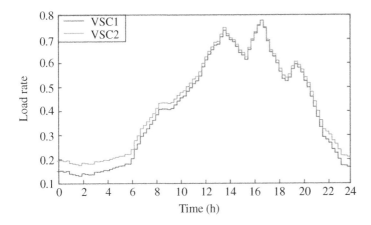

Figure 28.17 VSC load rate considering the change of load and PV output.

that the load of the VSC remains balanced when the load and PV output vary randomly. Figure 28.18 presents the maximum and minimum voltages for each moment in an extreme scenario, and it can be seen that the voltage constraints are generally satisfied throughout all time periods. The droop factor is temporarily reduced during the 12:00–14:00 time period to ensure that the node voltages meet the constraints. During this time period, the voltage slightly exceeds the limit of 0.0005 p.u. This is due to the use of a linear approximation in the optimization model to improve solution speed, which introduces a small error. However, these errors are minor and meet the accuracy requirements for optimal operation.

Figures 28.19 and 28.20 depict the real part of the dominant characteristic root and the system damping ratio for each time period, respectively, corresponding to the optimization results. It is evident that the real part of the dominant characteristic root is less than zero for all time periods, indicating that the system is stable. Figure 28.21 depicts the eigenvalues of the system with a real part greater than −200 at the 13:00 time period. All the eigenvalues have a real part less than zero, indicating that the system is stable. The points marked as circular dot correspond to the dominant characteristic roots of the system, with a damping ratio of 0.145 and a frequency of 38.07 Hz.

28.4.2.3 Model Comparison Analysis

The comparison model is designed to assess the effectiveness of the proposed coordinated optimization model for operating reference power and droop factor. The optimization variables in Model 1

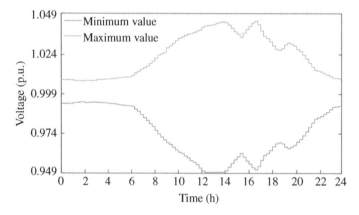

Figure 28.18 Maximum and minimum voltage considering changes in load and PV output.

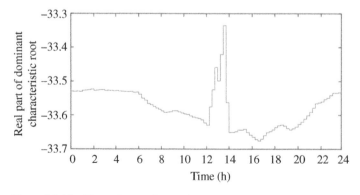

Figure 28.19 The real part of the dominant characteristic root of the system.

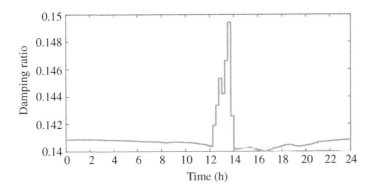

Figure 28.20 System damping ratio.

Figure 28.21 Zero-pole diagram of the system at 13:00.

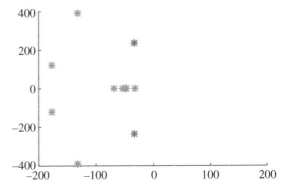

include operating reference power and droop factor, and the objectives are to enhance the voltage level and balance the VSC load. The constraints are the stability constraint in Eq. (28.34) and the steady-state power flow operation constraint in Eq. (28.35) and (28.36). In comparison to the proposed coordinated optimization model, Model 1 focuses solely on improving the voltage level and balancing the VSC load without taking into consideration the impact of the droop factor on system stability improvement.

The optimization variables in Model 2 are the operating reference power and droop factor, with the aim of enhancing system stability as described in Eq. (28.33). The constraints include stability constraints outlined in Eq. (28.34) and steady-state power flow operation constraints outlined in Eqs. (28.35) and (28.36). Unlike the proposed coordinated optimization model, Model 2 focuses solely on improving system stability through the operating reference power and droop factor, without taking into consideration the impact on voltage level and VSC load balancing.

Both Model 1 and Model 2 are optimized using a genetic algorithm, with a chromosome count of 400 and a maximum number of evolutionary generations. The crossover rate and mutation rate are set at 0.1. Power flow parameters such as voltage and VSC power are calculated through power flow analysis, and the QR decomposition method is employed to determine the system's characteristic roots.

(1) Optimization of Model 1

The optimization results of Model 1 are displayed in Figs. 28.22–28.27. It can be observed that the voltage level and VSC load balancing of Model 1 optimization results are largely comparable to those of the proposed coordinated optimization model. However, the small disturbance stability

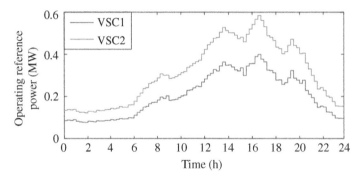

Figure 28.22 Optimization results of operating reference power.

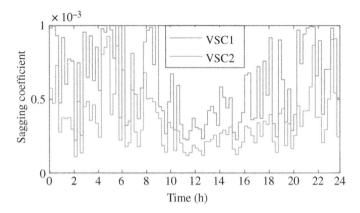

Figure 28.23 Optimization result of sag coefficient.

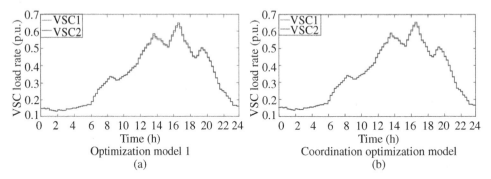

Figure 28.24 VSC load rate.

index of the system is not as favorable as that of the proposed coordinated optimization algorithm. This suggests that the goal of improving the voltage level and VSC load balancing can be achieved through reference power optimization, while the proposed coordinated optimization approach can enhance both the stability of the system and the voltage level and VSC load balancing.

(2) Optimization of Model 2
The results of the optimization process for Model 2 are presented in Figs. 28.28–28.33. It can be observed that the system's indices of small disturbance stability, as a result of the optimization of

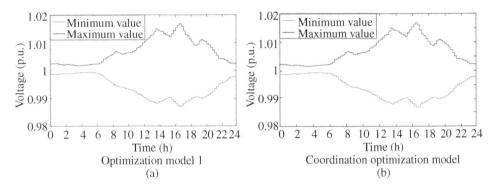

Figure 28.25 Maximum and minimum node voltage of the system.

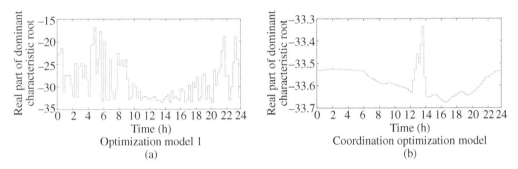

Figure 28.26 The real part of the dominant characteristic root of the system.

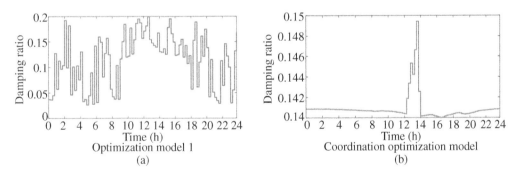

Figure 28.27 System damping ratio.

Model 2, are largely in agreement with those produced by the coordinated optimization algorithm that has been proposed. However, a comparison of the voltage level and balancing of the load for the VSC reveals that the results generated by Model 2 are not as favorable as those obtained through the application of the proposed coordinated optimization methodology.

The comparison between Model 1 and Model 2 reveals that the proposed coordinated droop control optimization algorithm in this chapter improves the voltage level and equalization of the VSC by optimizing the operating reference power, and enhances the stability of the system through the optimization of the droop coefficient. This leads to a substantial improvement in the overall performance of the system. In terms of algorithm efficiency, the increase in optimization variables results in an average optimization completion time of 13.74 seconds using the comparison algorithm. The

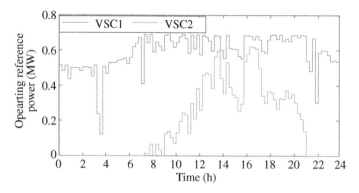

Figure 28.28 Optimization results of operating reference power.

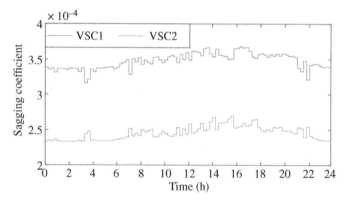

Figure 28.29 Optimization result of sag coefficient.

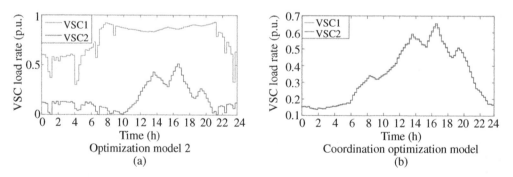

Figure 28.30 VSC load rate.

use of the least squares algorithm, with its faster computational speed, to optimize the operating reference power in this chapter results in a reduction of the number of optimization variables in the genetic algorithm, an improvement in the calculation speed of the algorithm, and the ability to meet the requirement for real-time optimization in DC distribution systems.

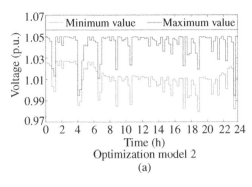
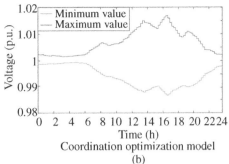

Figure 28.31 Maximum and minimum of voltage.

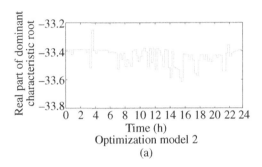
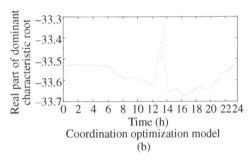

Figure 28.32 The real part of the dominant characteristic root of the system.

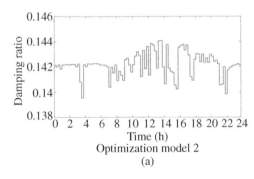
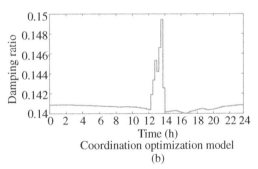

Figure 28.33 System damping ratio.

28.5 Actual Project Construction

The "Key Technology and Demonstration Application of High-Efficiency and High-Reliability Low-Voltage AC–DC Microgrid" project, funded by the Hebei Provincial Science and Technology Department, aimed to explore and demonstrate the crucial technology and practical application of AC–DC microgrid systems. The project sought to facilitate the flexible integration of DG resources, enhance the operational efficiency of power exchange equipment, support grid operations through peak-shaving and valley-filling strategies, and provide microgrid loads with a highly reliable power supply. In collaboration with the second phase of the Zhangjiakou National Scenery Storage and Transmission Demonstration Base's big data center project, a 500 kW PV+storage+IDC data center

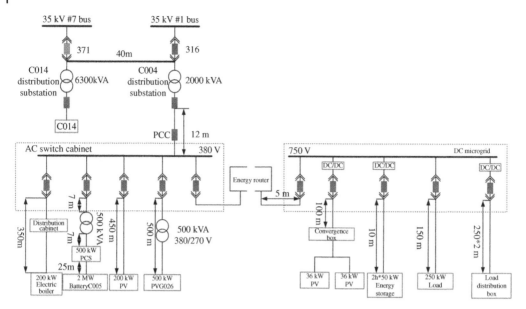

Figure 28.34 Construction wiring diagram.

demonstration project was established. The project was designed to optimize and demonstrate the implementation of a high-efficiency and high-reliability microgrid power supply system, which integrates rooftop-mounted PV systems and typical DC loads, as well as critical variable current equipment.

The detailed implementation of the wiring diagram is illustrated in Fig. 28.34. The voltage level of the DC microgrid is 750 V, while the voltage level of the AC microgrid is 380 V. The AC and DC microgrids are interlinked through an Energy Router. The AC microgrid is connected to a 2000 kVA distribution transformer through the PCC.

The AC–DC microgrid constructed in this project promotes the efficient utilization of local DC power sources and improves the conversion efficiency prior to integration into the distribution network. It also facilitates low-cost and high-efficiency integration of power storage, thereby improving the reliability and power quality of the microgrid. Advanced key converters, such as a bidirectional AC/DC converter, a photovoltaic unipolar DC/DC grid-connected converter, energy storage, and a high-efficiency resonant soft-switching bi-directional DC/DC power converter for V2G charging pile, have been developed. The combination of these converters with DC loads, such as an IDC, results in a high-efficiency power supply for the AC–DC microgrid and enhances its safety and stability. An energy management and monitoring system is utilized to enable the consumption of DGs and the flexible integration of energy storage, and the proposed methods and strategies have been validated through demonstration projects (Fig. 28.35).

28.6 Conclusion

In this chapter, a comparative analysis of the topologies of AC–DC microgrids is performed, and the equivalent generator model at the microgrid's Point of Common Coupling (PCC) is employed to evaluate the effect of random faults and islanding scenarios on the reliability of the AC–DC microgrid system. The utilization of renewable energy and the reliability conditions of three distinct

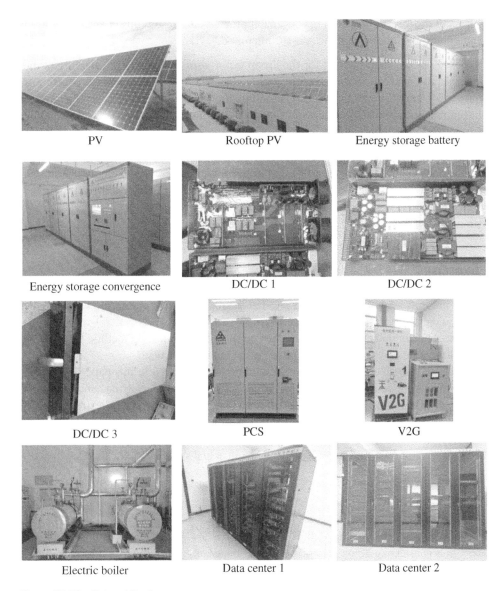

Figure 28.35 Related Devices.

DG planning strategies are compared using a microgrid evaluation index. The study also considers the impact of changes in the droop coefficient on bus voltage and converter power under random fluctuations of the system load, and an optimization method for the droop coefficient is developed with the goal of enhancing the stability of the system in the presence of small disturbances. The following conclusions have been derived.

(1) The equivalent generator model and approach proposed in this chapter provide a comprehensive evaluation of the impact of random faults in the distribution network on the AC–DC microgrid system. The methodology considers not only the time-varying attributes of the DGs within the AC–DC microgrid, but also the effects of random faults in transmission lines and transformers

and the impact of the microgrid islanding conversion process. As a result, this chapter presents a thorough analysis of the reliability of the AC–DC microgrid.

(2) The proposed optimization method for the droop factor aims to improve the stability of the system under small disturbances as the optimization objective. It considers voltage constraints and VSC power constraints when the load and output of the DG fluctuate randomly. The method enables the optimal operation of the droop-controlled AC–DC microgrid in the presence of random variations in load and renewable energy.

(3) The proposed coordinated optimization approach for droop control considers the unique characteristics of various control parameters. The optimization of the operating reference power and the droop coefficient in a coordinated manner enhances the stability of the system, improves the voltage level, and realizes the power distribution function of droop control.

28.7 Exercises

1. This article presents several structures for AC–DC microgrid. Which ones? And what are the advantages and disadvantages of each?
2. What are the steps for voltage control using the coordinated optimization method for the VSC droop coefficient and operating reference power proposed in this article?
3. The Zhangjiakou AC/DC microgrid demonstration project presented in this paper is ongoing and subject to ongoing construction. Certain challenges have arisen during the commissioning phase of the demonstration project, such as the selection of an appropriate grid structure that meets local resource characteristics and reliability requirements, and ensuring stable voltage levels after construction. The reader is encouraged to consider the various factors that need to be taken into account for effective voltage control in an actual system.

References

1 Li, C., Chaudhary, S.K., Savaghebi, M. et al. (2016). Power flow analysis for low-voltage AC and DC microgrids considering droop control and virtual impedance. *IEEE Transactions on Smart Grid* 8 (6): 2754–2764.
2 Khodabakhsh, J. and Moschopoulos, G. (2020). Simplified hybrid AC–DC microgrid with a novel interlinking converter. *IEEE Transactions on Industry Applications* 56 (5): 5023–5034.
3 Gupta, A., Doolla, S., and Chatterjee, K. (2017). Hybrid AC–DC microgrid: systematic evaluation of control strategies. *IEEE Transactions on Smart Grid* 9 (4): 3830–3843.
4 Zolfaghari, M., Abedi, M., and Gharehpetian, G.B. (2019). Power flow control of interconnected AC–DC microgrids in grid-connected hybrid microgrids using modified UIPC. *IEEE Transactions on Smart Grid* 10 (6): 6298–6307.
5 Zhou, J., Xu, Y., Sun, H. et al. (2019). Distributed power management for networked AC–DC microgrids with unbalanced microgrids. *IEEE Transactions on Industrial Informatics* 16 (3): 1655–1667.
6 Shen, X., Shuai, Z., Huang, W. et al. (2021). Power management for islanded hybrid AC/DC microgrid with low-bandwidth communication. *IEEE Transactions on Energy Conversion* 36 (4): 2646–2658.

7 Duan, J., Yi, Z., Di Shi, C.L. et al. (2019). Reinforcement-learning-based optimal control of hybrid energy storage systems in hybrid AC–DC microgrids. *IEEE Transactions on Industrial Informatics* 15 (9): 5355–5364.

8 Aprilia, E., Meng, K., Al Hosani, M. et al. (2017). Unified power flow algorithm for standalone AC/DC hybrid microgrids. *IEEE Transactions on Smart Grid* 10 (1): 639–649.

9 Youssef, K.H. (2014). A new method for online sensitivity-based distributed voltage control and short circuit analysis of unbalanced distribution feeders. *IEEE Transactions on Smart Grid* 6 (3): 1253–1260.

10 Lee, S.H. and Park, J.-W. (2016). Optimal operation of multiple DGs in DC distribution system to improve system efficiency. In: *2016 IEEE/IAS 52nd Industrial and Commercial Power Systems Technical Conference (I&CPS)*, 1–9. IEEE.

11 Haileselassie, T.M. and Uhlen, K. (2012). Impact of DC line voltage drops on power flow of MTDC using droop control. *IEEE Transactions on Power Systems* 27 (3): 1441–1449.

12 Pourbabak, H., Alsafasfeh, Q., and Su, W. (2020). A distributed consensus-based algorithm for optimal power flow in DC distribution grids. *IEEE Transactions on Power Systems* 35 (5): 3506–3515.

13 Ahmed, H.M.A. and Salama, M.M.A. (2019). Energy management of AC–DC hybrid distribution systems considering network reconfiguration. *IEEE Transactions on Power Systems* 34 (6): 4583–4594.

14 Ghahramani, M., Nazari-Heris, M., Zare, K., and Mohammadi-Ivatloo, B. (2019). Energy and reserve management of a smart distribution system by incorporating responsive-loads/battery/wind turbines considering uncertain parameters. *Energy* 183: 205–219.

15 Sun, F., Ma, J., Yu, M., and Wei, W. (2019). A robust optimal coordinated droop control method for multiple VSCs in AC–DC distribution network. *IEEE Transactions on Power Systems* 34 (6): 5002–5011.

16 Anand, S., Fernandes, B.G., and Guerrero, J. (2012). Distributed control to ensure proportional load sharing and improve voltage regulation in low-voltage DC microgrids. *IEEE Transactions on Power Electronics* 28 (4): 1900–1913.

29

Self-Organizing System of Sensors for Monitoring and Diagnostics of a Modern Microgrid

Michael Gouzman, Serge Luryi, Claran Martis, Yacov A. Shamash, and Alex Shevchenko

29.1 Introduction

It is well known that Modern Microgrids are dynamic in nature. The structure or the topology of the microgrid can change in response to an event, or it may grow in an unstructured manner based on the current needs. These changes must be monitored in real time to control the grid properly. Algorithms use data from various sensors to estimate the state or detect the grid's topology. The main three approaches for extracting information from the grid are as follows:

1. Sensing at the extremities of the grid. Ex: Smart Meters, PMU at substations.
2. Sensing from within the grid. Ex: Line monitors, Energy Flow Sensors.
3. Using data from both extremities and within the grid.

The best and most practical approach is using data from Smart Meters and Energy Flow Sensors to get the best situational awareness. Here we focus on identifying the topology of the grid from within the grid. Microgrids usually work on medium voltage (MV) or low voltage (LV), and the voltage drops tend to stay negligible in the small region. The direction of currents provides information about the grid's topology. Since the current from an ammeter gives us no information about the direction of current in an AC circuit, we need to use the data from an Energy Flow Sensor. The Energy flow sensor gives us information regarding the magnitude and the direction of the current in the powerline. The algorithm, outlined in Section 29.6, uses Kirchhoff's Current Law (KCL) to obtain the topology of the microgrid using the GPS coordinates, current magnitude, and phase. The Line Monitors available on the market also provide us with similar information as the Energy Flow Sensors described here. Therefore, making this algorithm less dependent on custom hardware and more universal.

29.2 Structures for Building Modern Microgrids

Microgrids are networks of distributed energy resources and loads. The similarities between modern microgrids and communication networks extend beyond their topologies and analysis. The power is also treated as packets similar to data packets in communication networks. When building a microgrid, various structures may be adopted based on the requirements and constraints.

Microgrids: Theory and Practice, First Edition. Edited by Peng Zhang.
© 2024 The Institute of Electrical and Electronics Engineers, Inc. Published 2024 by John Wiley & Sons, Inc.

Some basic structures listed below are fundamental blocks that repeat to make up the entire grid. These structures may be static like the traditional grid or dynamically change over a short period.

29.2.1 Fundamental Structures

The two fundamental structures that help us obtain any topologies are a split/branching and a loop. The radial structure of the traditional distribution network, which is also adopted in microgrids, uses branching to create a tree-like structure. Loops help create alternate or redundant paths for improved resiliency.

29.2.1.1 Branching
The radial structure is formed by creating a split or a branch from the power lines wherever necessary. Figure 29.1 shows an example of this structure. The power lines carry power from the generators to the load unidirectionally. They are widely used, as the network can easily be expanded while keeping the cost minimum. It is the most common and least expensive method in building a microgrid or a distribution network of the utility grid.

29.2.1.2 Loops
A ring structure, as shown in Fig. 29.2, is one of the simplest structures to help improve the resiliency of the network. This single loop provides two paths for the power to flow in case of faults or breakage in the power lines. The faults can easily be isolated while powering the other loads from both sides. Loops in a microgrid provide alternate paths for power to flow and can be implemented anywhere in the grid.

Figure 29.1   Radial structure with branching.

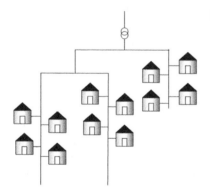

Figure 29.2   Ring topology as an example of loops.

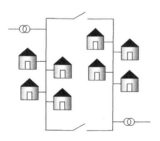

Figure 29.3 Hybrid structure with a combination of branching and loops.

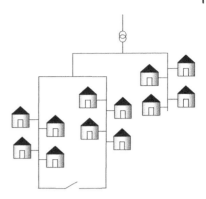

29.2.2 Dynamic Structure

The requirements of each microgrid are unique, and the structure reflects these requirements. The path for energy flow constantly changes due to the high penetration of renewables and control decisions taken for optimized operation. The power in a microgrid typically does not always flow in a single direction and is sometimes bidirectional. Modern microgrids are a network of prosumers and not purely producers and consumers. In the process of creation and development, microgrids can continuously or abruptly change their configuration. Thus, it is impossible to predict the network structure implemented in each microgrid at a given moment. Modern microgrids can be considered dynamic as they change their logical structure often. They are mainly made up of switches that change the path of current in branches and loops, forming a hybrid structure as shown in Fig. 29.3. Thus, to monitor the processes in a dynamic microgrid, the period between sensor readings in MGMS (Microgrid Monitoring System) should be no more than a period during which significant changes may occur. An alternative (or complementary) feature of the MGMS could be targeted scanning of part or all of the microgrid in on-demand mode or when sensors detect faults.

29.3 Requirements for the Monitoring and Diagnostics System of Modern Microgrids

Monitoring and diagnostics using various sensors help make the grid smarter. Monitoring the grid is vital for the following reasons:

1. Fault detection
2. Maintain power quality
3. Compliance
4. Topology identification

Fault detection is one of the essential features of a monitoring system. Various faults in an electrical network includes voltage dips, over-voltage, harmonic distortions, voltage fluctuations, phase unbalances, supply interruptions, transient over-voltage, etc. Each fault has one or more possible causes and consequences that the diagnostics system could predict and aid the operators in carrying out necessary repairs.

The **power quality** needs to be monitored to ensure the grid operates efficiently. For instance, large phase shifts in the network reduce the amount of real power and increase apparent power.

Identifying the phase shifts helps us appropriately compensate for them, reducing wastage of energy. Large commercial or industrial consumers generating significant phase shifts are required to compensate for the shift or face surge charges.

The grid operators must be in **compliance** with the rules and regulations in place while installing and maintaining the grid. Line Monitors help identify the sag in the overhead power lines to maintain it within safe conditions. The phase shifts and power quality must be maintained when networking microgrids with other microgrids or the utility grid.

Topology identification helps understand the structure of the grid when it is either unknown or has changed. The modern grids have redundant paths for power flow to increase resiliency. They also have cogenerators that automatically come online when required. These changes must be tracked in real time to control the grid effectively.

29.4 Communication Systems in Microgrids

Communication infrastructure plays a major role in the operation of the microgrid, both in the normal operation or during the transition from a grid-connected state to the islanded state or vice-versa. The communication can be carried out through wired or wireless technologies, depending on where the grid is deployed. A network of optical fibers is ideal for communication, but it is not extensively deployed in real grids due to its prohibitive costs. It is common to utilize existing communication infrastructure whenever it is available. The communication infrastructure provides multiple paths or alternate technologies as redundant paths in case of network link failures. These technologies may be managed through a software-defined network as explained by [1]. Some of the most popular technologies in the context of microgrids are as follows.

1. Power line communication
2. Dedicated wired communication link
3. Wireless communication
   (a) Radio technologies
   (b) Cell networks
4. Optical fiber communication

Power line communication reduces infrastructure costs since the data is sent to the power lines that carry the power in the grid. While this is a great way to send data within the grid for several reasons, achieving resilience in the grid would not be possible as the reliability of the data communication would be directly dependent on the power lines. If a power line is faulty or broken, we will not be able to control grid elements connected through that line either remotely or autonomously. Therefore, there is a need for a reliable communication channel that is not affected by the harsh and noisy transmission medium. This technology is most suited for controllable loads within a building, not long-distance communication.

A **dedicated wired communication** link would be more reliable than the power line transmission while trying to achieve resilience. A dedicated wire could be drawn along the power lines, or an existing communication network (like a telephone network) could be used. The existing communication network would be preferred when a neighborhood upgrades its existing infrastructure or creates a more resilient architecture. This strategy cannot be applied in temporary grids. The wired connection along the power lines introduces challenges such as electromagnetic interference, especially when switching generators/loads and when severe faults occur.

Wireless communication between sensors works the best in most cases due to minimal infrastructure requirements and the ability to have a mesh that ensures reliable transfer of information. This is the best method for communication between sensor nodes in most civil applications. Wireless technologies like LoRaWAN, ZigBee, WiFiMax, BLE, etc., help quickly create a network of sensors when there is no prior infrastructure in place. Cellular technology could be used when establishing a network with existing infrastructure. Since the sensors do not require high bandwidth, older generations of cellular technologies could be used for sensor communication rather than completely decommissioning them.

Optical fibers provide the best bandwidth and reliability while not being prone to interference. They are an ideal option for military grids where radio silence is required. They are generally used along with other technologies to reduce costs.

29.5 Sensors

The sensor technology for the electrical grids has evolved in the past few decades, giving rise to some of the most advanced sensors for optimally managing the electrical grid and bringing in automation. The basic parameters measured in any electrical system are:

1. Voltage
2. Current
3. Time

All other parameters are derived from these measurements. A potential transformer (PT) measures the voltage or potential. At lower voltages, the measurement units are connected directly to the wires. A current transformer (CT) is the most popular method to measure current. A resistive shunt and a Hall effect sensor are a couple of other methods. A local clock in the measurement device is used for time-based measurements. Due to its precision at a reasonable cost, a global positioning system (GPS)-based clock is the most popular way to obtain synchronized readings. The current and voltage readings taken over a certain time period are used to measure derived parameters. Some of them are as follows:

- True power and root mean square (RMS) values averaged over the cycle
- Apparent power, power factor, and volt-ampere reactive (VAR)
- Accumulated energy (watt-hours)
- Minimum and peak (e.g. voltage sag)
- Harmonics, subharmonics, and flicker
- Phase and frequency

Traditional transmission networks use supervisory control and data acquisition (SCADA) systems that record measurements once every 4-6s and time stamp them using the operator's local time. Phasor Measurement Units (PMUs) continuously report 30–60 samples every second. This data is time-stamped using the Coordinated Universal Time or UTC obtained from GPS units. The increasing demand for high-resolution time-synchronized data has led to the development and deployment of μPMU and point-on-wave (POW) measurement devices. POW measurement units report the sampled waveform at 256 samples/sec or higher. Figure 29.4 graphically summarizes the categories of sensors and their measurement sampling rate. The authors in [2] survey μPMU in distribution networks, and the authors in [3] provide the rationale for the use of POW devices and comparison with other measuring devices.

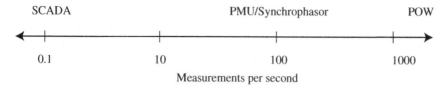

Figure 29.4 Sampling rate in measurements per second.

These sensors can be broadly classified as contact-based and noncontact-based, where contact refers to the galvanic connection with the power lines. The sensors like PMUs and power quality monitors that connect to the power lines directly or through potential transformers capture high-quality data with good time resolution. They are expensive and overkill for large-scale deployment in the distribution network or microgrids to achieve situational awareness. Cheaper, noncontact line monitors and fault circuit indicators are commonly used to identify the fault locations and achieve situational awareness. The rest of the chapter will consider a basic sensor with the ability to measure current magnitude and phase with GPS coordinates. All previously mentioned sensor technologies are able to provide us with the required information, and the choice of the sensor should be made based on additional information required at that location and the cost. Section 29.5.1 describes an energy flow sensor developed at Stony Brook University which was used to test and validate the monitoring system.

29.5.1 Energy Flow Sensor

An electrical energy/power flow sensor senses the magnitude and direction of power flow based on the principle of the Poynting theorem integrated over the entire period. Figure 29.5 shows a schematic of the Energy Flow Sensor and an early prototype is shown in Fig. 29.6. A capacitive cylindrical probe whose cross-section forms two concentric circles measures the electric field. An instrumental current transformer measures the current or the magnetic field in the line. The net direction of real and reactive power flow in a powerline is relative to the phase angle between voltage and current waveforms. The cosine of this angle is the power factor. A positive power factor indicates power flow in the direction of the sensor, and a negative power flow indicates that the power flow is in the opposite direction. This sensor forms the foundation of the system presented here. The power flow sensor integrates GPS and other sensors as desired by the grid operators to

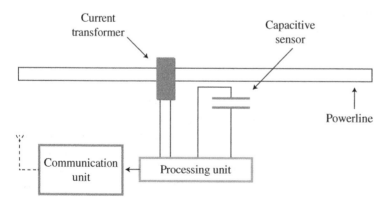

Figure 29.5 Schematic of energy flow sensor.

Figure 29.6 First version of the modified sensor with integrated current and voltage sensing probe.

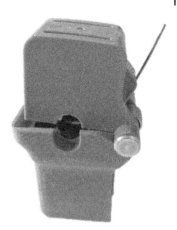

form a modular self-configured monitoring system. The Line Monitors currently on the market can measure the current magnitude and the phase shift along with the location. This is sufficient information required by the algorithm outlined in Section 29.6.

The voltage magnitude is not essential for identifying the direction of power flow. The phase of current with respect to the voltage is sufficient to determine this direction. The phase difference and the magnitude of current were proposed by [4] to identify topology by solving Kirchhoff's current law.

29.5.1.1 Construction of the Sensor
The use of current transformers is a well-known technique to noninvasively measure current in the power line. Resistive dividers and voltage transformers measure voltage in most commercially available devices [5]. Electro–optical sensors are also used in limited experimental work. The high cost of electro–optical devices discourages their use for achieving higher spatial awareness in the context of a dynamic microgrid. An electrostatic sensor is small and inexpensive, which makes it the best choice for most applications. An electrostatic sensor is a set of conductive plates placed in an electric field such that the difference in potential between the two plates is proportional to the strength of the electric field. The electric field from the power line induces a potential on an electrostatic sensor's sensing plate/film. The potential on this plate is read through different readout circuits. Passive circuits use an amplifier to amplify this charge to an appropriate level for further processing. Active circuits such as those based on the varactors provide better performance in broader conditions, including DC measurements. Varactor-based circuits as described by [5] are very stable and provide variable sensitivity of the sensor. Since the voltage levels in a power line do not vary much around the rated value, a simpler passive circuit is used. A reference is required to be able to read the potential on the cylindrical probe around the wire. Two concentric cylindrical probes, where one serves as a reference, are used. This provides the voltage waveform sufficient to measure the phase shift with the current waveform.

29.5.2 Why Energy Flow Sensors and not Ammeters?

An ammeter built for DC not only provides us with the magnitude of the current in the circuit but also tells us the direction. An AC ammeter only provides the RMS value of the current and does not provide any information about the direction of current flow. Therefore, an ammeter cannot be used to identify the direction of current flow and hence the grid's topology.

An Energy Flow Sensor, on the other hand, not only provides us with the current but also tells us about the direction of energy flow. A positive power factor implies that the power flow in the direction of the sensor is connected, and a negative power factor implies that the power is flowing in the opposite direction.

Figures 29.7 and 29.8 illustrate the limitation of AC ammeters in the presence of "Prosumers" (customers with cogeneration abilities, like solar or wind generators).

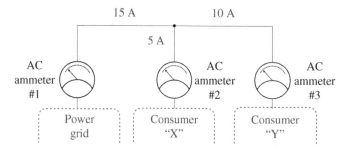

Figure 29.7 For Contemporary AC Grid Kirchhoff rule for AC circuit only partly supported by AC ammeters which show only magnitude (or RMS) of the current and carries no information about the direction of the energy flow.

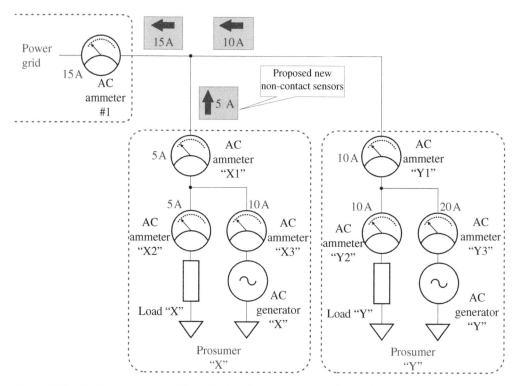

Figure 29.8 For Prosumers type of the grid, data from ammeters and even circuit diagrams are insufficient for understanding distribution of energy flows in the grid.

29.6 Network Topology Identification Algorithm

Topology identification, also called state estimation, is the first step to effectively monitoring and controlling the dynamically changing microgrid or static large-scale distribution grid. Also, automatically recognizing and configuring sensors is an important feature while monitoring such a network. When the grid structure is poorly documented or dynamically changed, additional challenges arise in deploying a reliable monitoring system. The proposed system uses previously described energy flow sensors to geographically locate and identify the electrical relationship between different sensor nodes. Kirchhoff's current law-based algorithm, as described in [4], is used to configure and monitor the sensors. Topology estimation provides a blueprint of the grid.

29.6.1 Algorithm

The minimum information required from the sensors to identify the grid's topology is as follows:

1. Current magnitude
2. The phase of current with respect to voltage
3. Geographic coordinates

The Energy Flow Sensors or similar sensors that measure the current magnitude and phase are directional in nature. Two sensors connected to the same powerline oriented in opposite directions have the same current magnitude, but the phase shifted by 180°. For example, $10\angle 6°$ and $10\angle -174°$ are sensor readings from the same line. All sensors at a splitting node are oriented toward the point of common connection.

To identify the grid's topology, starting from the root node, find a set of sensors where Kirchhoff's Current Law (KCL) is satisfied. KCL is satisfied when two sensors are on the same powerline or more than two are on a split.

- Check if two sensors are connected to the same line. [Consider that the phase may be shifted by 180°.]
  - If not connected on the same line, check if it is connected at a split node.
    - o If more than one set of sensors is found, choose the geographically closest nodes as a cluster
- If none of the above, the sensor is connected at a terminal node

Nodes identified as clusters or terminal nodes are marked connected and ignored during the search in the next iteration. If the network is known to have loops, the connected nodes need to be considered for further searches. In the next iteration, consider one of the already connected nodes as the root node. The above steps are repeated until all connections are identified. Since the entire grid would consist of hundreds or thousands of sensors, only a small set of sensors geographically close to each other is considered during each iteration of the algorithm. The same process is repeated with a different group of sensors with a bit of overlap to identify the structure of the entire grid. In the unlikely event of obtaining the same readings from geographically close nodes that are not connected, the previously known structure should be used to identify the anomaly. Therefore, data must be recorded and analyzed over time to draw conclusions for effective and reliable monitoring. Figure 29.9 graphically shows a single iteration of the algorithm.

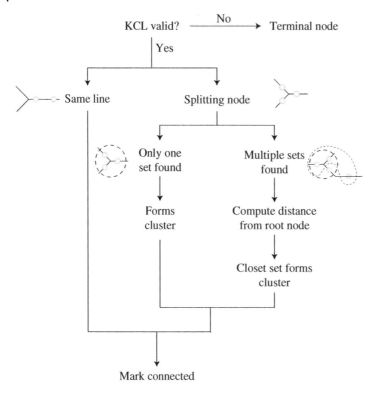

Figure 29.9 Single iteration of KCL-based topology identification algorithm.

29.6.2 Simplified Algorithm

The above algorithm can be greatly simplified if all the sensor nodes that form a cluster are closely placed. In practice, the sensors at a splitting node are so close that they return the same GPS coordinates. Therefore, it makes sense to have primitive sensors connected to a hub/data concentrator with a single GPS module that processes and transmits the data as illustrated in Fig. 29.10. Hence, the clusters can be easily identified by scanning the coordinates. In two connected clusters, one sensor from each cluster will be placed on the same line. Therefore, this pair of sensors should

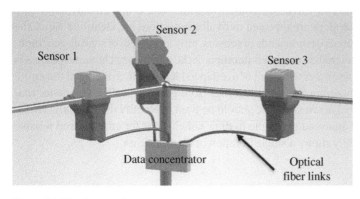

Figure 29.10 Sensors in a cluster connected to a hub.

have the same magnitude and may have the same phase or phase shifted by 180°. This reduces the algorithm's complexity by only having to check for sensors on the same line.

If a sensor in a cluster fails, data from that sensor is not available. The hub can easily identify the faulty sensor. Since KCL is still valid within a cluster, we could use the data from other available sensors in that cluster until it is replaced. The data from the sensor from another cluster that was previously identified as connected to the faulty sensor could be used to collect data and continue monitoring the grid until it is fixed. The same principle could be applied to cover for more than one faulty sensor at a cluster. These measures ensure that events in the grid are not missed due to sensor failure.

29.6.3   Illustration of the Algorithm

A sample grid, as shown in Fig. 29.11, is considered to illustrate the process of grid topology estimation. Table 29.1 shows the sample data from the illustrated network. The sensors at a split node are always oriented toward the node, while the sensors connected on the same line can be oriented arbitrarily. The steps of the algorithm are summarized below and graphically illustrated in Fig. 29.12.

Considering the example data in Table 29.1, we shall walk through the steps in identifying the network structure shown in Fig. 29.12.

1. Starting from root node #1, scan the table to check for the same current magnitude and phase. We notice that no other nodes have the same magnitude and phase (or phase shifted by 180°). However, we have two sets of nodes (1, 2, and 3, and 1, 3, and 5) that satisfy KCL.
2. Since we have two sets that satisfy KCL, we will choose the set closest to root node #1.

$$d_{ij} = \sqrt{(x_i - x_j)^2 + (y_i - y_j)^2}$$
$$D_1 = d_1 2 + d_1 3 = 4.47$$
$$D_2 = d_1 3 + d_1 5 = 6.71$$

Since $D_1 < D_2$ we record #1, #2, and #3 as a cluster and not use them in future searches.

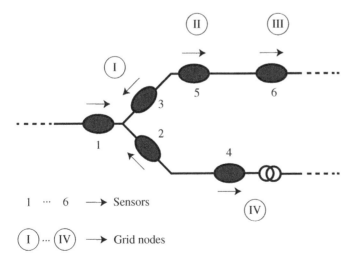

1 ⋯ 6 ⟶ Sensors

(I) ⋯ (IV) ⟶ Grid nodes

Figure 29.11   Sample network used to illustrate the topology identification algorithm.

Table 29.1 Sample data from sensors in the illustrated network.

Sensor	Current (Magnitude)	Current (Phase)	Geographical coordinated (x, y)
#1 (root)	29.9	8.33°	(1,10)
#2	20	−175°	(2,12)
#3	10	−165°	(2,8)
#4 (load)	10	15°	(3,6)
#5	20	−175°	(3,14)
#5	20	5°	(5,14)

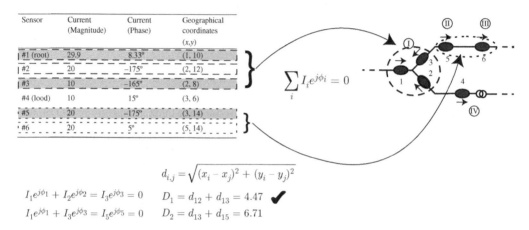

$$d_{i,j} = \sqrt{(x_i - x_j)^2 + (y_i - y_j)^2}$$

$$I_1 e^{j\phi_1} + I_2 e^{j\phi_2} = I_3 e^{j\phi_3} = 0 \qquad D_1 = d_{12} + d_{13} = 4.47 \; \checkmark$$

$$I_1 e^{j\phi_1} + I_3 e^{j\phi_3} = I_5 e^{j\phi_5} = 0 \qquad D_2 = d_{13} + d_{15} = 6.71$$

Figure 29.12 KCL-based topology identification: 1. Find the connections between sensors in the close vicinity, 2. Find clusters where KCL holds and choose the closely located group, 3. Identify the topology based on the relationship between clusters.

3. Considering #2 as the root node in this iteration, we notice that #5 reports the same magnitude and phase. Hence, they must be connected to the same line. We mark #5 as connected and move on to the next iteration.
4. Considering #5 as the root node, we see that #6 has the same magnitude, but the phase is shifted by 180°. Therefore, they are connected to the same line. #6 is marked as connected.
5. There is no other node with the same reading as #6, and there is also no group of nodes that satisfies KCL. #6 is marked as a terminal node.
6. #3 and #4 have the same magnitude, but the phase shifted by 180°. They are marked as connected.
7. #4 has no other nodes with the same reading or a group of nodes where KCL is satisfied. It is also the last node to be scanned in the table. #4 is marked as a terminal node.

The connection between all nodes has been identified, and this information can be used to graphically represent the connection between the nodes in the monitoring software. The GPS coordinates are used to overlay the nodes on a map for better interpretation.

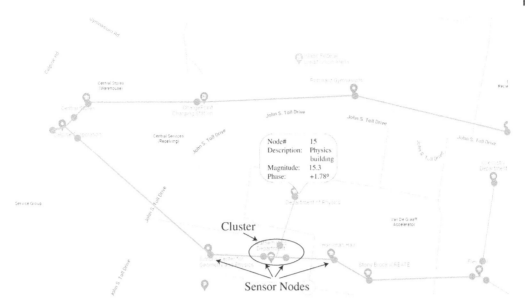

Node#   15
Description: Physics
                 building
Magnitude: 15.3
Phase: +1.78°

Cluster

Sensor Nodes

Figure 29.13   Dashboard screenshot showing the sensor nodes overlayed on a map with the interconnections shown graphically.

29.7   Implementation

An illustration of a monitoring system dashboard that uses this algorithm is shown in Fig. 29.13. The software is developed in collaboration with Dr. Rong Zhao and team at Stony Brook University for continuing research in the area of monitoring and control of microgrids. A dashboard showed the current status of the grid along with the topology overlayed on a geographical map. The data acquired by the monitoring system is recorded for analytics that could be used for predictive maintenance. The dashboard graphically shows the faults and warnings in areas that need the attention of the grid operator.

29.8   Exercise

1. What are the three approaches to extracting information from the microgrid?
2. List the communication technologies suitable for use in Microgrid monitoring.
3. What are the three parameters required to find the topology of the microgrid with KCL based algorithm?
4. Give examples of sensors that can measure the current magnitude and phase.
5. List some electrical parameters that can be calculated if the voltage and the current waveforms are available.
6. Why is it important to have a monitoring and diagnostics system in a microgrid?

Using the information provided in the table below, identify the interconnection of nodes graphically and show the interconnection.

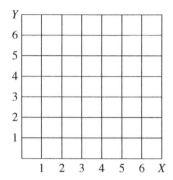

1.

ID	(X, Y)	Current (A)
1	(2,4)	230
2	(2,4)	−130
3	(2,4)	−100
4	(5,5)	130
5	(5,5)	−55
6	(5,5)	−75

2.

ID	(X, Y)	Current (A)
1	(2,4)	$187.1\angle 11.4°$
2	(2,4)	$85.12\angle -169.31°$
3	(2,4)	$102\angle -168°$
4	(5,5)	$85.12\angle 10.69°$
5	(5,5)	$32\angle 0°$
6	(5,5)	$54\angle -163°$

3.

ID	(X, Y)	Current (A)
1	(1,3)	$95.91\angle 2.3°$
2	(1,3)	$65.95\angle -176.6°$
3	(1,3)	$30\angle 0°$
4	(3,3)	$0^{a)}$
5	(3,3)	$30\angle 0°$
6	(5,4)	$69.95\angle 3.4°$
7	(5,4)	$21\angle 0°$
8	(5,4)	$45\angle -175°$
9	(3,3)	$30\angle 0°$

a) Note: if any nodes show reading as 0, ignore it from the search and only mark it on the map.

4.

ID	(X, Y)	Current (A)
1	(1,3)	0
2	(2,4)	$65.95\angle -176.6°$
3	(2,3)	$65.95\angle 3.4°$
4	(3,3)	$95.91\angle 2.3°$
5	(3,2)	$30\angle 0°$
6	(5,4)	$69.95\angle 3.4°$
7	(6,5)	$21\angle 0°$
8	(6,3)	$45\angle -175°$
9	(2,2)	$65.95\angle -176.6°$

References

1 Zhang, P. (2021). Networked microgrids. *Networked Microgrids.* https://doi.org/10 .1017/9781108596589. https://www.cambridge.org/core/books/networked-microgrids/ 271B5323E4E8588BF415A19EE576D130 (accessed 10 November 2023).

2 Dusabimana, E. and Yoon, S.G. (2020). A survey on the micro-phasor measurement unit in distribution networks. *Electronics (Switzerland)* 9 (2): https://doi.org/10.3390/electronics9020305.

3 Silverstein, A., Silverstein, A. Consulting, Follum, J. et al. (2020). High resolution, time synchro-nized, grid monitoring devices. *Naspi.*

4 Gavrilov, D., Gouzman, M., and Luryi, S. (2019). Monitoring large-scale power distribution grids. *Solid-State Electronics* 155: 57–64. https://doi.org/10.1016/j.sse.2019.03.012.

5 Noras, M.A. (2011). Solid state electric field sensor. *Proceedings of the ESA Annual Meeting on Electrostatics,* January 2011, 1–6. https://www.researchgate.net/profile/Maciej_Noras/publication/ 267986590_Solid_state_electric_field_sensor/links/54b948a60cf2d11571a34f0f.pdf (accessed 10 November 2023).

30

Event Detection, Classification, and Location Identification with Synchro-Waveforms

Milad Izadi and Hamed Mohsenian-Rad

30.1 Introduction

The power distribution system is the bulk of the power electric grid that delivers the electricity to the end-user customer. The electricity is mostly delivered in the form of *alternating current* (AC) power with the voltage and the current that vary *sinusoidally* in time. Power distribution system is the most vulnerable part of the grid since it is frequently exposed to a wide range of events [1]. An event is defined as any kind of change, including major or minor, in the voltage and/or the current. For example, when there is an electrical contact between an energized conductor, such as a power line, with a high grounding impedance surface, such as a tree branch, it causes a fault in the system. As another example, when a distribution network is disconnected from the main grid, it forms a microgrid, which significantly changes the state of the grid. Furthermore, the basic parameters of most distribution systems, such as line parameters, are often *unreliable* or even *not known* [2]. In addition, the integration of microgrids has added significant complexity to the structure of the distribution systems. Thus, it is *critical* to study and monitor power distribution systems. For the rest of this chapter, we will focus on analysis of events in power distribution systems.

The very first step to study and analyze events is to modernize the power distribution system to make it "smart" through the use of advanced technologies, such as smart grid sensors. Smart grid sensors provide information about the status of the grid which can help us gain a better *understanding* and *situational awareness* about the root cause of events happening across the grid. The three main classes of smart grid sensors that are available on power distribution systems are explained as follows.

- **Supervisory control and data acquisition (SCADA) systems**: They are the *old class* of smart grid sensors that have been integrated widely into distribution systems over the past 50 years ago [3]. SCADA systems report the root-mean-square (RMS) representation, a.k.a. magnitude representation, of the fundamental component of the AC voltage and AC current. They operate at low reporting rates, such as 1 sample per second. Further, their measurements are not synchronized, meaning that the measurements that are reporting simultaneously from two different SCADA systems from two different locations are not really aligned. SCADA measurements have been used to study events that create sustained changes in the RMS values of the fundamental component.

Microgrids: Theory and Practice, First Edition. Edited by Peng Zhang.

- **Phasor measurement units (PMUs)**: They are *another class* of smart grid sensors that have been deployed in distribution systems over the past two decades [4]. PMUs report the phasor representation of the fundamental component of the AC voltage and AC current. The phasor representation includes the magnitude representation and the phase angle representation. They are equipped with global positioning system (GPS) to provide time-synchronized voltage phasor and current phasor measurements, also known as *synchro-phasor measurements*. PMUs operate at high reporting rates, such as two samples per AC cycle, i.e. 120 samples per second. PMU measurements are used to study events that create sustained changes in the phasor values of the fundamental component.

- **Waveform measurement units (WMUs)**: They are the *new class* of smart grid sensors that have been emerging recently [5–13]. WMUs report the raw waveform of the AC voltage and AC current in time domain, also known as the point-on-wave and/or continuous-point-on-wave. They record the wave-shape of AC voltage and AC current, as opposed to only the RMS representation of the fundamental component as in SCADA systems; or only the phasor representation of the fundamental component as in PMUs. WMUs are equipped with GPS to provide time-synchronized voltage waveform and current waveform measurements, also known as *synchro-waveform measurements* [7, 8]. WMUs operate at very high reporting rates, such as 256 samples per AC cycle, i.e. 15,360 samples per second. This is much higher than the reporting rates of SCADA systems and PMUs [5, 12, 13]. WMU measurements are used to study any kind of event including but not limited to events that create temporary and/or sustained changes in the voltage and/or current.

SCADA systems and PMUs are both limited when AC voltage and/or AC current waveforms include *distortions* or take *non-sinusoidal* shape [7] or when the event has a very short duration [5], which are increasingly common situations in the modern power distribution systems and particularly microgrids. This is due to the fact that SCADA systems and PMUs are both designed to report representation measurements over a period of seconds or cycles, instead of the actual waveform measurements. Thus, they are both inherently incapable of properly capturing the events that cause distortions in AC voltage and/or AC current waveforms or events that last for a very short period of time, such as only for a fraction of a cycle, e.g. incipient faults [5]. Such sub-cycle events can be easily captured by WMUs due to their very high reporting rates and their access to synchronized waveform measurements. This is particularly necessary in microgrids, which typically suffer from dynamic issues. Such issues can cause serious outages if they not are addressed immediately.

The differences between the SCADA, PMU, and WMU can be better understood by examining their measurements during the same event.

Example 30.1 Consider the measurements in Fig. 30.1. They are captured by three different sensors at the upstream of a real-world distribution feeder. The sensors are a SCADA system, a PMU, and a WMU. They all observe *the exact same type of event* at *the exact same time* on the feeder. The event has occurred precisely at time 29.705 second. Figure 30.1(a)–(b) shows the fundamental voltage magnitude and current magnitude measurements collected by the SCADA system for the duration of one minute from 0 to 60 second. Figure 30.1(c)–(d) shows the fundamental voltage phasor and current phasor measurements collected by the PMU for the duration of three seconds from 28 to 31 second. Figure 30.1(e)–(f) shows the voltage waveform and current waveform measurements collected by the WMU for the duration of 100 μs (5 AC cycles) from 29.655 to 29.755 second. Only one phase is shown here. This example shows the waveform measurements recorded by the WMU provide *much more detail* about the event compared to the phasor measurements recorded

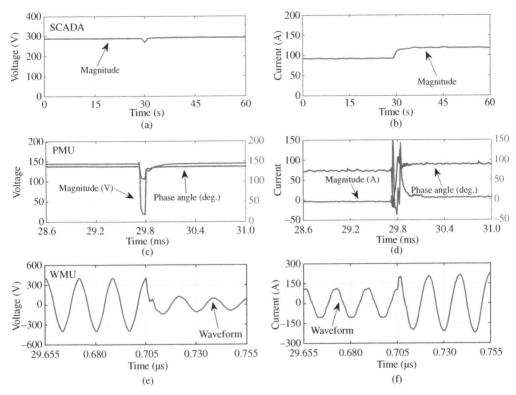

Figure 30.1 An example of WMU measurements compared to SCADA measurements and PMU measurements: (a)–(b) the voltage magnitude and current magnitude measurements by the SCADA; (c)–(d) the voltage magnitude and voltage phase angle measurements and the current magnitude and current magnitude measurements by the PMU; (e)–(f) the voltage waveform and current waveform measurements by the WMU.

by the PMU and the magnitude measurements recorded by the SCADA system. Thus, it is necessary to check the waveform measurements in studying events.

Example 30.2 Consider the measurements in Fig. 30.2. They are captured by two WMUs at two different locations of a real-world distribution feeder. WMU 1 is at the upstream of the feeder and WMU 2 is at the downstream of the feeder. They all observe *the exact same type of event* at *the exact same time* on the feeder that we saw in Fig. 30.1. Figure 30.2(a)–(b) shows the synchronized voltage waveform and current waveform measurements collected by WMU 1. Figure 30.2(c)–(d) shows the synchronized voltage waveform and current waveform measurements collected by WMU 2. Only one phase is shown here. The event immediately creates a sudden drop in the voltage waveform at both WMU 1 and WMU 2. It also causes a sudden rise in the current waveform. Furthermore, the event creates a momentary ringing distortion in the voltage waveform and current waveform due to the resonance formed between capacitor banks and inductive loads across the distribution system [1]. This example shows that synchro-waveform measurements carry *valuable and complementary information* about the state of the power system and the root cause of events.

Motivated by the discussion and examples in this section, our focus in this chapter is on the use of synchro-waveform data from WMUs for situational awareness in power distribution systems.

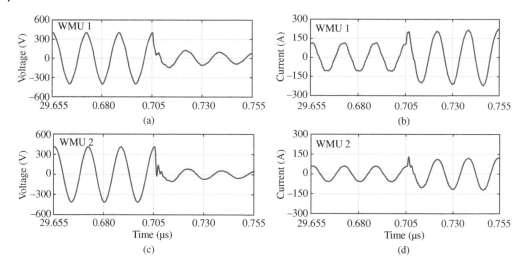

Figure 30.2 An example of synchro-waveform measurements at two different locations from two WMUs: (a)–(b) the voltage waveform and current waveform measurements by the WMU 1 at location 1; (c)–(d) the voltage waveform and current waveform measurements by the WMU 2 at location 2.

However, data availability in itself does not lead to enhanced situational awareness and grid intelligence. We need to translate the WMU data to insightful and actionable knowledge. This chapter proposes a *situational awareness framework* using synchro-waveform data from WMUs that includes *event detection*, *event classification*, and *event location identification*.

30.2 Event Detection

A very first step in any situational awareness is to detect events that occur in the grid. This section is about the use of synchro-waveform measurements for the purpose of event detection. Event detection is defined as when the event occurs. Thus, we propose a method to *detect* power quality events in power distribution systems by using synchro-waveform measurements from WMUs. The method is built upon a novel graphical tool, called *synchronized Lissajous curve* [7, 12, 13].

30.2.1 Synchronized Lissajous Curve

Consider a power distribution feeder, such as the one in Fig. 30.3. The distribution feeder can be either a passive network, an active network, or even a microgrid. Regardless of the type of the feeder, suppose two WMUs are installed on this feeder, where WMU 1 is installed at the beginning of the feeder and WMU 2 is installed at the end of the feeder. Let $v_1(t)$ denote the voltage waveform and $i_1(t)$ denote the current waveform that are measured by WMU 1. Also, let $v_2(t)$ denote the voltage waveform and $i_2(t)$ denote the current waveform that are measured by WMU 2. The waveform measurements are precisely time-synchronized. When an event occurs somewhere between WMU 1 and WMU 2 on the power distribution feeder, it creates signatures in the voltage waveform and/or current waveform that are captured by both WMUs 1 and 2. In this regard, we define the following two new waveforms [7, 12]:

$$v(t) = v_1(t) - v_2(t) \tag{30.1}$$

$$i(t) = i_1(t) - i_2(t) \tag{30.2}$$

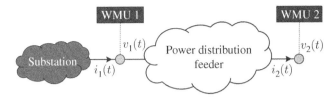

Figure 30.3 A distribution feeder that is equipped with two WMUs.

The waveform in (30.1) is the *difference* between the voltage waveforms at WMU 1 and WMU 2. The waveform in (30.2) is the *difference* between the current waveforms at WMU 1 and WMU 2. If WMU 1 and WMU 2 are not in the same nominal voltage levels, then we can define (30.1) and (30.2) in per unit.

The waveform measurements in (30.1) and (30.2) can be graphically represented as a Lissajous curve. A Lissajous curve is a graph that is constructed by plotting one waveform versus another waveform. It has various applications in signal and image processing; such as in electrocardiogram analysis and dielectric discharge analysis [14]. Furthermore, the Lissajous curves have had occasional applications also in power system engineering; such as to analyze non-linear single-phase circuits [15] or to identify fault location in transmission lines [16]. However, these existing applications have focused on the specific physical characteristics of the particular circuit or the particular equipment of interest.

In this chapter, we propose to plot the voltage waveform *difference* in (30.1) versus the current waveform *difference* in (30.2). We refer to such Lissajous curve as the *synchronized Lissajous curve* [7, 12] because it is constructed based on synchronized waveform measurements, which is indeed the case when WMUs are being used. We introduced the synchronized Lissajous curve for the first time in [12]. Next, we will discuss the synchronized Lissajous curve during the event condition.

During a *normal* operating condition, the synchronized waveforms in (30.1) and (30.2) are all purely sinusoidal. Thus, (30.1) and (30.2) include only the *fundamental* component and we have:

$$v(t) = V \cos(\omega t + \theta) \tag{30.3}$$

$$i(t) = I \cos(\omega t + \gamma) \tag{30.4}$$

where ω denotes the rotational frequency of the fundamental component; V denotes the magnitude of the fundamental component of the voltage; I denotes the magnitude of the fundamental component of the current; θ denotes the phase angle of the fundamental component of the voltage; and γ denotes the phase angle of the fundamental component of the current.

Next, we obtain the relationship between waveform $v(t)$ and waveform $i(t)$ by eliminating ωt from Eqs. (30.3) and (30.4). Accordingly, from (30.3), we have:

$$\omega t = \cos^{-1}\left\{\frac{v(t)}{V}\right\} - \theta \tag{30.5}$$

If we substitute (30.5) into (30.4), then we can rewrite (30.4) as follows:

$$
\begin{aligned}
i(t) &= I \cos\left(\cos^{-1}\left\{\frac{v(t)}{V}\right\} - \theta + \gamma\right) \\
&= I \cos\left(\cos^{-1}\left\{\frac{v(t)}{V}\right\}\right)\cos(\theta - \gamma) \\
&\quad + I \sin\left(\cos^{-1}\left\{\frac{v(t)}{V}\right\}\right)\sin(\theta - \gamma) \\
&= I \cos(\theta - \gamma)\frac{v(t)}{V} + I \sin(\theta - \gamma)\sqrt{1 - \frac{v(t)^2}{V^2}}
\end{aligned}
\tag{30.6}
$$

We square both sides and rearrange the terms to obtain:

$$Av(t)^2 + Bv(t)i(t) + Ci(t)^2 + D = 0 \tag{30.7}$$

where

$$\begin{aligned} A &= 1/V^2, & B &= -2\cos(\theta - \gamma)/VI, \\ C &= 1/I^2, & D &= -\sin^2(\theta - \gamma) \end{aligned} \tag{30.8}$$

Equation (30.7) always represents an *ellipse* because

$$B^2 - 4AC < 0 \tag{30.9}$$

Therefore, during normal operating conditions, the synchronized Lissajous curve is always an ellipse. Once an event occurs, the synchronized Lissajous curve deviates from its initial ellipse shape to a different shape. With that in mind, we propose an event detection method based on the synchronized Lissajous curve and the synchro-waveform measurements that come from WMUs. The proposed method relies on the changes in the area of the synchronized Lissajous curve during two successive cycles as we explain next.

30.2.2 Event Detection Methodology

Let us define the area of the synchronized Lissajous curve at time t over period T of the past AC cycle as follows:

$$\text{Area}(t) = \left| \int_{i(\tau=t-T)}^{i(\tau=t)} v(\tau)\, di(\tau) \right| \tag{30.10}$$

During normal operating conditions, there is little to no difference between two successive calculations of the areas in (30.10). However, once an event occurs, such difference suddenly becomes significant. This can help us detect the event. Suppose Area(t) and Area($t - \Delta t$) denote the areas of the synchronized Lissajous curves at times t and $t - \Delta t$, where Δt is the reporting interval of the WMUs, e.g. $\Delta t = 65\ \mu$s. We define the *similarity index* at time t as

$$S(t) = 1 - \left| \frac{\text{Area}(t) - \text{Area}(t - \Delta t)}{\max\{\text{Area}(t), \text{Area}(t - \Delta t)\}} \right| \tag{30.11}$$

If the areas of the two successive synchronized Lissajous curves are almost equal, then $S(t)$ is close to one. However, if the areas of the two successive synchronized Lissajous curves are considerably different, then $S(t)$ is close to zero, indicating that a sudden change has occurred in the synchronized Lissajous curve at time t. This means an event has occurred at time t.

Further, we propose an *adaptive* detection threshold by considering the past similarity indices to minimize the number of false alarms. In this regard, consider a window of time period W immediately before time t, i.e. from time $t - W$ to time $t - \Delta t$. The similarity indices of such window of duration W are

$$S(t - W),\ S(t - W + \Delta t), \dots,\ S(t - \Delta t) \tag{30.12}$$

Let us define $M(t)$ and $MAD(t)$ as the *median* and *median absolute deviation* of the similarity indices in (30.12), [17]. We define the adaptive threshold as follows:

$$\mathcal{T}(t) = \alpha\ (M(t) - \eta\, MAD(t)) \tag{30.13}$$

where α is a number between 0 and 1 to control the sensitivity of the event detection method. A common choice for η is 2.5 [17]. We use the median and median absolute deviation statistics because they are robust against outliers. We detect an event at time t if the following inequality holds:

$$S(t) < \mathcal{T}(t) \tag{30.14}$$

Importantly, the detection threshold must be revised after an event is detected. We *discard* the similarity index at event time from the next calculation of the adaptive threshold. That is, the similarity index at time t is used in the calculation of the next threshold *only if* time t is not an event time.

30.2.3 Event Detection Results

Here we assess the performance of the proposed event detection method by applying it to the IEEE 33-bus test system. The one-line diagram of the simulated test system is shown in Fig. 30.4. All simulations are done in PSCAD [18]. Two WMUs are assumed to be installed in the network, where WMU 1 is installed at bus 1 and WMU 2 is installed at bus 18, as marked on the figure. Each WMU captures the time-synchronized voltage waveform and current waveform at its location. To emulate real-world WMU measurements, white Gaussian noise is added to the simulated voltage waveform and current waveform measurements. Unless stated otherwise, we consider a signal-to-noise-ratio (SNR) of 80 dB in both voltage and current waveforms. The nominal system frequency is 60 Hz. Unless stated otherwise, the reporting rate of the WMUs is assumed to be 256 samples per cycle. The following events are simulated in various case studies: high-impedance faults, capacitor bank switching, and incipient faults.

We examine the performance of the proposed event detection method on three different classes of disturbances. Here, the sensitivity factor is set to 0.9; and the window duration is set to $W = 133$ ms. The results of event detection for the first class, i.e. the high-impedance fault, the second class, i.e. the capacitor bank switching, and the third class, i.e. the incipient fault, are shown in Fig. 30.5(a), (b), and (c), respectively.

In Fig. 30.5(a), the similarity index drops from almost 1 to 0.82 at time $t = 0.50$ second, indicating that an event occurs at this time, which is the correct event time. The similarity index fluctuates right after the event occurs, for about one cycle, from $t = 0.50$ second to $t = 0.517$ second, see the zoomed-in figure. The similarity index goes back to almost 1 after time $t = 0.517$ second. We can conclude that the event at time $t = 0.50$ second is the only event that occurs during this one-second period; and the event is a sustained event. Thus, the *profile* of the similarity index can help us identify the start time and the end time of each event. These are all useful parameters. For example, when it comes to event classification, the synchronized Lissajous curve from the start time of the event and for the duration of one cycle will be converted into an image. For the example in Fig. 30.5(a), we convert the synchronized Lissajous curve from time $t = 0.50$ second to time $t = 0.517$ second to an image. The image will be later used for event classification.

In Fig. 30.5(b), the similarity index drops from about 1 to nearly 0.6 at time $t = 0.40$ second, indicating that an event occurs at this time, which is correct. One cycle later, the similarity index again drops from almost 1 to about 0.9 at time $t = 0.417$ second, which is lower than the event

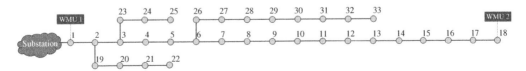

Figure 30.4 The IEEE 33-bus distribution system with two WMUs.

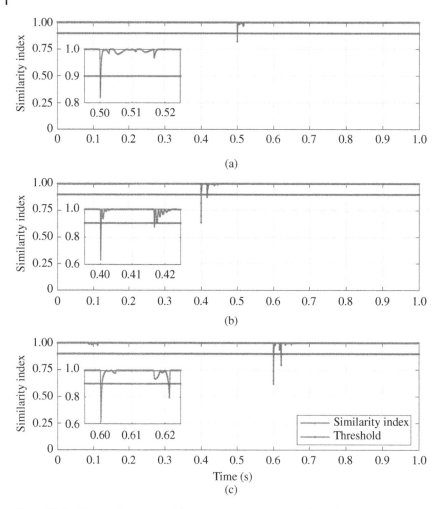

Figure 30.5 The results for event detection, including the similarity index and the adaptive threshold for three example events: (a) a high-impedance fault; (b) a capacitor bank switching; (c) a sub-cycle incipient fault.

detection threshold at this time. However, the similarity index rises to almost 1 after time $t = 0.417$ second. Thus, the *profile* of the similarity index can help characterize the transient component of the event. It starts at time $t = 0.40$ second and lasts for about one cycle till $t = 0.417$ second. This is correct because the capacitor bank switching has a very short transient behavior. We use the information on the similarity index profile to identify the correct duration of the event signature. In this example, we convert the synchronized Lissajous curve from time $t = 0.40$ second to time $t = 0.417$ second to an image. The image will be later used for event classification.

Finally, in Fig. 30.5(c), the profile of the similarity index indicates that an event occurs at time $t = 0.60$ second, which is correct because the incipient fault is a temporary event that occurs for a short period of time. Hence, the results in Fig. 30.5(c) confirm the effectiveness of the proposed event detection method, even for events with very short duration. In this example, we convert the synchronized Lissajous curve from $t = 0.60$ second to $t = 0.617$ second to a synchronized Lissajous image. The image will be later used for event classification.

The above results confirms the effectiveness and the precision of the proposed event detection method.

30.3 Event Classification

Once the power quality event is detected by the proposed method in Section 30.2, the next step in situational awareness is to identify the *type* of the detected events. This section is about the use of synchro-waveform measurements for the propose of event classification. Event classification is defined as what type of events occurs. Thus, we propose a novel method based on *image classification* to categorize each detected event into different classes based on the shape of their one-cycle synchronized Lissajous curves. The one-cycle synchronized Lissajous curve is constructed from the moment that the event is detected and for the duration of one cycle. To the best of our knowledge, no prior study has used any variation of the synchronized Lissajous curves to conduct event classification in this context. Furthermore, all the prior studies are focused on making use of only event classification based on measurements from only one power quality or waveform sensor.

First, we will discuss the factors that affect the shape of the synchronized Lissajous curves and why they make the classification problem a highly challenging task. Second, we will convert the detected synchronized Lissajous curves to images so that they can be classified by using image processing techniques. Third, we will develop an efficient convolution neural network (CNN) to extract features of the synchronized Lissajous images in order to conduct event classification.

30.3.1 Challenging Factors

The shape of the synchronized Lissajous curve depends on not only the type (i.e. the class) of the event, but also other factors such as the angle, the location, and the size of the affected physical components. Therefore, even when we look at *different examples of the exact same class of events*, the shapes of the synchronized Lissajous curves can have considerable differences based on the above various factors. They can make the event classification problem challenging, as we explain next.

30.3.1.1 Impact of the Event Angle
Consider the synchronized Lissajous curves in Fig. 30.6. They both represent *the exact same disturbance*, which is a capacitor bank switching event. However, the firing angle of the switching action is different in these two cases. One switching event occurs near the positive peak of the voltage waveform. The other switching event occurs near the negative peak of the voltage waveform. We can see that the oscillations in the corresponding Lissajous curves start at two different places on the voltage–current plane; making the two curves look different. In fact, one curve is almost the mirror reflection of the other curve. Therefore, we can conclude that the angle of the event can affect the shape of the synchronized Lissajous curve, thereby creating additional challenges and complications in the event classification problem.

30.3.1.2 Impact of the Event Location
Next, consider the synchronized Lissajous curves in Fig. 30.7. They represent *the exact same disturbance*, which is a high-impedance fault with equal fault impedance. However, the location of the

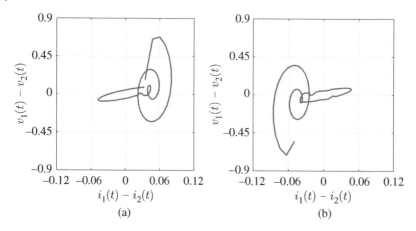

Figure 30.6 The synchronized Lissajous curves during the same event that occurs at two different firing angles: (a) near positive peak; (b) near negative peak.

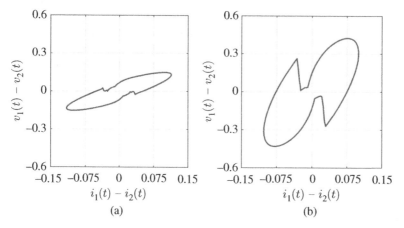

Figure 30.7 The synchronized Lissajous curves during the same event that occurs at two different locations: (a) near WMU 1; (b) near WMU 2.

fault is different in these two cases; one is closer to the substation at the beginning of the feeder; while the other one is closer to the end of the feeder. We can see that the shapes of the two curves are somewhat similar; however, there are major *rotational differences* among these curves. If the fault occurs near the beginning of the feeder, i.e. near WMU 1, then the angle between the voltage difference waveform and the current difference waveform in the synchronized Lissajous curve is *smaller*, see Fig. 30.7(a). However, if the fault occurs near the end of the feeder, i.e. near WMU 2, then the angle between the voltage difference waveform and the current difference waveform in the synchronized Lissajous curve is *larger*, see Fig. 30.7(b). We can conclude that the location of the event can directly affect the shape of the synchronized Lissajous curve, thereby making classification a challenging task.

30.3.1.3 Impact of other Event Parameters

Finally, consider the synchronized Lissajous curves in Fig. 30.8. They show *the exact same disturbance*, which is an incipient fault. However, the impedance of the fault is different in these two cases. One fault has a smaller impedance. The other fault has a larger impedance. We can see that

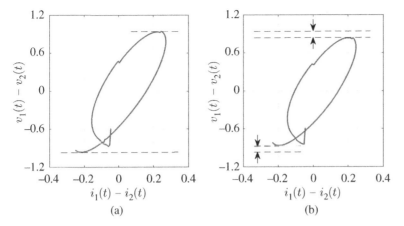

Figure 30.8 The synchronized Lissajous curves during two incipient faults with different fault parameters: (a) lower fault impedance; (b) higher fault impedance.

the shapes of the two curves are almost similar; however, the sizes of the curves are different, which is due to the different fault impedances. If the impedance of the incipient fault is smaller, then the size of the Lissajous curve is larger, see Fig. 30.8(a). Conversely, if the impedance of the incipient fault is larger, then the size of its corresponding Lissajous curve is smaller, see Fig. 30.8(b). We can conclude that the parameters of an event can highly affect not only the shape but also the size of the synchronized Lissajous curve.

30.3.2 Synchronized Lissajous Curve as Image

The challenges in Section 30.3.1 can be addressed if we treat the synchronized Lissajous curves as *images* and subsequently take advantage of the recent advancements in the field of image processing to solve the event classification problem.

There are multiple reasons why it is beneficial to study a synchronized Lissajous curve as an *image*, as opposed to studying the raw synchronized waveform measurements as *time series*. *First*, graphical images can capture the overall *patterns* in the shape of the synchronized Lissajous curves; while such overall patterns are inherently spread over time in the original time series. For example, there are clear similarities between the two synchronized Lissajous curves in Fig. 30.7. It is clear that one image is almost a *squeezed version* of the other image. Therefore, the two Lissajous images belong to the same class of events. However, such similarity would not be clear if we only look at the raw waveform measurements corresponding to these two events. *Second*, the sequential nature of time series is embedded with many important characteristics, which lays outside of a typical time-domain analysis. Therefore, it is difficult to perform classification in time domain using the state-of-the-art sequence classification methods. *Third*, deep machine learning methods have shown particularly promising results in recent years in solving image processing problems. Therefore, if we present the event classification problem based on synchronized Lissajous curves as an image processing problem, then we benefit from powerful image processing tools.

The synchronized Lissajous curves are converted to synchronized Lissajous images by using various readily available conversion functions in MATLAB and/or Python. For example, one option is to use the combination of functions `getframe` and `frame2im` in MATLAB; see [19, 20]. We will verify the importance of treating synchronized Lissajous curves as images through case studies in Section 30.3.5.

30.3.3 Convolutional Neural Networks

Once the synchronized Lissajous curves are converted to images, one can use various advanced image processing methods to classify the events based on their synchronized Lissajous images. In this thesis, we use CNNs to classify the detected Lissajous images into multiple classes of events. CNNs are effective deep machine learning techniques that are widely used in image recognition and speech recognition, among other fields [21, 22].

The structure of CNN includes an input layer, a few hidden layers, and an output layer. The input layer takes as input the synchronized Lissajous images of the detected power quality events. The hidden layers consist of the convolutional, batch normalization, activation, max-pooling, dropout, and the fully-connected layers. The convolutional layer is the key layer to extract features. It includes a series of kernel filters. The batch normalization layer normalizes the input, to speed up the training of the CNN. The activation layer implements non-linearity functions to the CNN model, by using functions such as sigmoid, hyperbolic tangent, or rectified linear unit (ReLU). The max-pooling layer performs down-sampling to summarize the extracted features. The dropout layer randomly assigns zero to the input to prevent over-fitting. The fully-connected layer integrates the features from the previous layers to the softmax activation layer to obtain probabilities of the input. The output layer is the classification layer that determines the label of the input image given the probabilities from the previous layer.

Table 30.1 shows the structure of the proposed CNN for event classification based on Lissajous images. It consists of a four-layer architecture, where each architecture includes multiple layers. Since the size of the input Lissajous images is large, a wide kernel filter is used in the first convolutional layer to extract more features from the Lissajous images. The ReLU is used in the activation layers to speed up learning and improve its performance [23]. Softmax is used in the final activation

Table 30.1 The structure of the proposed CNN model.

Layer	Layer type	Activation
1.1	Convolutional	(120,120,60)
1.2	Batch normalization	(120,120,60)
1.3	ReLU	(120,120,60)
2.1	Convolutional	(120,120,60)
2.2	Batch normalization	(120,120,60)
2.3	ReLU	(120,120,60)
3.4	Max-pooling	(60,60,60)
3.1	Convolutional	(60,60,120)
3.2	Batch normalization	(60,60,120)
3.3	ReLU	(60,60,120)
3.4	Max-pooling	(30,30,120)
3.5	Dropout	(30,30,120)
4.1	Fully-connected	(1,1,3)
4.2	Softmax	(1,1,3)
4.3	Classification	–

layer to get a probability distribution density for the classes. The proposed CNN classification approach is implemented in MATLAB using its available CNN model [24].

It bears mentioning that the size of a synchronized Lissajous image depends on the size of the event. This may affect the results in the classification task. This issue is addressed by *normalizing* each synchronized Lissajous curve with respect to its energy *before* the curves are converted to graphical images.

30.3.4 Event Classification Results

Here we assess the performance of the proposed event classification method by applying it to a database generated via the IEEE 33-bus test system. The one-line diagram of the test system was shown in Fig. 30.4. The information and assumptions of the system are the same as those mentioned in Section 30.2.3. We first generate a database for the synchronized voltage and current waveforms from two WMUs that occur during 120 events. For each event, we capture one second (60 cycles) of voltage and current waveforms at each WMU. Thus, we collect four synchronized waveforms for each event over 60 cycles. The data for each event includes a few cycles before the event and a few cycles after the event. Each event generates $4 \times 60 \times 256 = 61,440$ samples of data. We generate one synchronized Lissajous image for each event. Thus, the number of synchronized Lissajous images is 120 images. The size for input images of the CNN model is 240×240. The database consists of 40 high-impedance faults, 40 capacitor bank switching, and 40 incipient faults with short arcs, where they are labeled in the following three classes of disturbances: Class I for high-impedance faults, Class II for capacitor bank switching, and Class III for incipient faults.

The database is divided into three data sets: training data, validation data, and test data. The training data set includes 70% of the total events which are selected randomly. The validation data set includes 10% of the total events. The test data set includes the remaining 20% of the total events. We use Adam optimization algorithm to train the CNN model, see [25]. The initial learning rate in the training process is set to 1×10^{-4}, which remains constant throughout the training. The maximum number of epochs is set to 100, with the mini-batch size of 32.

The training accuracy of the proposed classification method converges to 100% and the validation accuracy converges to 97%; the figures are not shown here. The small difference between the two accuracies indicates the generalization capability of the proposed classification method to unseen events.

The confusion matrix for the *test* results for the proposed Lissajous-based CNN model is shown in Fig. 30.9(a). The diagonal entries denote the events that are classified *correctly*. The off-diagonal entries denote the events that are classified *incorrectly*. For each class, the accuracy is at least 91.7%. High-impedance faults (Class I) and incipient faults (Class III) are classified better than capacitor bank switching (Class II). The minor shortcoming in the classification of capacitor bank switching is because the image of a capacitor bank switching has some features that also exist in other events. Interestingly, the converse is not true and all the incipient faults are classified correctly. The overall accuracy of the test results is 97.2%, which is close to the accuracy of the training results, i.e. 100%. This indicates the robustness and generalization of the proposed classification method to unseen events.

Next, we use the following various statistical metrics to further evaluate the performance of the proposed classification method: precision, sensitivity (a.k.a. recall), specificity, and F_1 score. Table 30.2 shows each metric for each class. As we can see, the worst-case precision and the worst-case specificity rates of the proposed classification method are 94.4% and 96.8%, respectively, both for Class III (incipient faults). Also, the worse-case sensitivity and the worst-case F_1 score

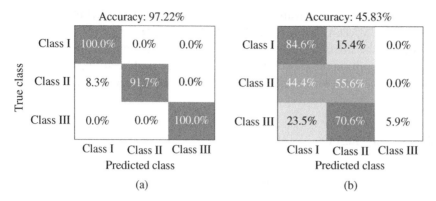

Figure 30.9 Confusion matrix for the test data when we use: (a) the proposed CNN-based image classification method; (b) the competing RNN-based classification method that uses the raw time series measurements.

Table 30.2 Performance metrics of the confusion matrix.

Class	Precision (%)	Sensitivity (%)	Specificity (%)	F_1 Score (%)
I	100.0	92.3	100.0	96.0
II	100.0	100.0	100.0	100.0
III	94.4	100.0	96.8	97.1

rates are 92.3% and 96.0%, respectively, both for Class I (high-impedance faults). These additional metrics further confirm the performance of the proposed event classification method.

Figure 30.10 shows the receiver operating characteristic (ROC) curve of the proposed event classifier at each class. The area under curve (AUC) for each class is marked in the legend box. The ROC curve is a graph that shows the classifier's ability to distinguish different classes for different probability thresholds. It is obtained by plotting the sensitivity versus the 1-specificity at different threshold settings. As we can see in Fig. 30.10, the AUC is at least 0.971, which indicates that the proposed event classifier can almost perfectly distinguish classes from each other. These results reveal the high performance of the proposed event classification model in correctly classifying events.

30.3.5 Classification Based on Images Versus Time Series

As we discussed in Section 30.3, it is highly beneficial to do event classification based on the synchronized Lissajous *images*, as opposed to based on the raw synchronized waveform measurements in time domain. This point is verified here by comparing the proposed CNN classification method that uses synchronized Lissajous images with a recurrent neural network (RNN) classification method that uses the time series of voltage and current waveforms. The latter method is implemented by developing a long short-term memory (LSTM) network to classify the time series of voltage difference waveform in (30.1) and the current difference waveform in (30.2). An LSTM is an RNN that takes time series as input. The hidden layers include two LSTM layers, a dropout layer, a fully connected layer, and a softmax layer. The output layer is the classification layer.

To have a fair comparison, we apply the RNN classification method and our proposed classification method to the *same* training data set and the *same* test data set.

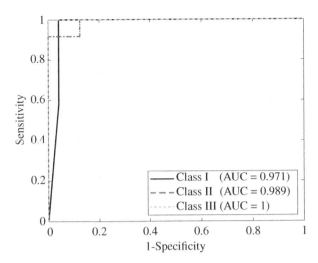

Figure 30.10 The ROC curve of the proposed event classifier at each class. The AUC corresponding to each class is marked inside the legend box. Notice that the ROC curve for Class III appears on the *x*-axis and the *y*-axis.

The confusion matrix for the *test* results of the time series classification model is shown in Fig. 30.9(b). The overall accuracy is 45.83%, which is much smaller than the overall accuracy of our Lissajous-based classification method, i.e. 97.2%.

The above results verify the effectiveness of the proposed classification method based on Lissajous images.

From the above results, there are indeed advantages to conduct event classification by using the synchronized Lissajous curves, as opposed to using the raw time series. The key advantage here is the fact that synchronized Lissajous images can better capture the *fundamental similarities* between the events of the *same type*; that we may sometimes miss if we solely look at the raw measurements as time series. For example, the *same type* of event may create some characteristics in the raw synchronized waveform measurements that can vary depending on the location of the event or the size of the event. Such variations can sometimes make it difficult for the classification algorithms to realize the fact that the events belong to the *same type*, if we examine the raw measurements as time series, see [7]. Thus, when possible, it is recommended to use the synchronized Lissajous curves for event classification.

30.4 Event Location Identification

Once the power quality event is detected by the proposed method in Section 30.2 and classified by the proposed method in Section 30.3, the next step in situational awareness is to identify the *location* of the event. This section is about the use of synchro-waveform measurements for the propose of event location identification. Event location identification is defined as where the event has occurred. Thus, we propose a new method to *identify* the location of transient events, including incipient faults, in power distribution systems, by using synchro-waveform measurements from WMUs. The method is based on conducting a synchronized modal analysis in the system with the focus on the event mode. It consists of three steps. The first step is to characterize the oscillatory modes of the transient components of all the captured synchronized voltage and current waveforms from all WMUs; namely their frequency, damping rate, magnitude, and angle, by conducting a multi-signal modal analysis. The second step is to construct a circuit model for the underlying distribution feeder at the identified dominant mode(s) of the transient event. The final step is to identify the location of the transient event based on certain forward and backward analyses of the constructed circuit model. The proposed method requires installing as few as only two WMUs, one at the beginning of the feeder and one at the end of the feeder.

30.4.1 Modal Analysis of Captured Transient Synchronized Waveform Measurements

The starting point in our proposed location identification methodology is to characterize the *transient component* of the synchronized waveform measurements during an event. Here, we assume that the event *is already detected and classified*, by the methods that we proposed in Sections 30.2 and 30.3 in this chapter.

Figures 30.11 and 30.12 show two examples of WMU measurements that are captured during two different types of transient events. The transient components are marked with boxes with gray shaded areas that start from 0.48 seconds in Figure 30.11 and 0.68 seconds in Figure 30.12. The event in Fig. 30.11 is an *incipient fault*. The event in Fig. 30.12 is a *capacitor bank switching*. In both events, the duration of the transient part is *one cycle or less*.

We propose to characterize the transient component of the event waveforms by conducting *modal analysis*. In this regard, the transient component of the waveforms is characterized as one or more *oscillation modes*. Each oscillation mode itself is characterized based on the following parameters:

- Frequency,
- Damping Rate,
- Magnitude, and
- Phase Angle.

Modal analysis can be done in different ways, such as by using the Prony method [26], matrix pencil method [27], or the methods based on rotational invariance techniques [28].

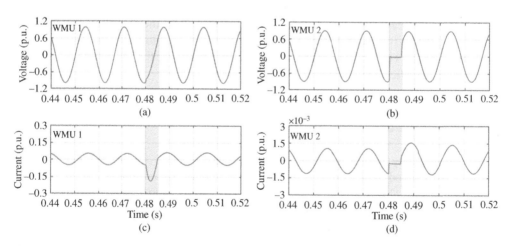

Figure 30.11 Synchronized waveform measurements during an incipient fault: (a)–(c) voltage and current waveforms that are captured by WMU 1; (b)–(d) voltage and current waveforms that are captured by WMU 2. The rectangle with gray shaded area that starts at 0.48 seconds marks the transient component that was the subject of modal analysis.

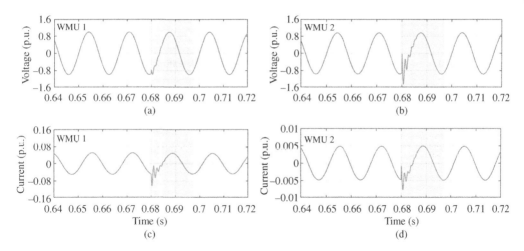

Figure 30.12 Synchronized waveform measurements during a capacitor switching: (a)–(c) voltage and current waveforms that are captured by WMU 1; (b)–(d) voltage and current waveforms that are captured by WMU 2. The rectangle with gray shaded area that starts at 0.68 seconds marks the transient component that was the subject of modal analysis.

30.4.1.1 Single-Signal Versus Multi-Signal Modal Analysis

Regardless of which method is used, modal analysis can be done in two different ways: *single-signal* and *multi-signal*. There is a considerable difference between these two approaches in the context of this chapter, as we explain next.

In *single-signal* modal analysis, each individual waveform is analyzed *independently*; thus, the modes are calculated for each waveform *separately*. For instance, for the cases in Figs. 30.11 and 30.12, where we have two WMUs, we need to do a separate modal analysis for each of the following four signals within the marked boxes with gray shaded areas: voltage waveform at WMU 1, current waveform at WMU 1, voltage waveform at WMU 2, and current waveform at WMU 2. In theory, the frequency should be the same for all the four signals and the damping rate should also be the same for all the four signals; because waveform signals, regardless of where on the circuit they are captured, oscillate at the same frequency and the same damping rate [29].

However, in practice, the results are often slightly different for each signal. This is due to numerical issues, noise in measurements, slight waveform distortions, etc. For example, the fundamental frequency can be obtained as 60.3 Hz from one waveform and 59.9 Hz from another waveform. Such discrepancy can be problematic for the purpose of event location identification that we will discuss in Section 30.4.3.

The above issue can be resolved by using *multi-signal* modal analysis. In this approach, the transient modes are obtained for all waveforms in the same unified estimation analysis. Hence, the frequency is the same for all the four signals. Likewise, the damping rate is the same for all the four signals.

The dominant mode of the incipient fault in Fig. 30.11 is shown in Table 30.3. The dominant modes of the capacitor bank switching event in Fig. 30.12 are shown in Table 30.4. The results in Tables 30.3 and 30.4 are obtained by using the multi-signal Prony method.

In Tables 30.3 and 30.4, the frequency of the dominant mode(s) is the same for all the four waveform signals; and similarly, the damping rate of the dominant mode(s) is the same for all the four waveform signals. The reference for the phase angles is with respect to phase angle of the voltage waveform at WMU 1. Also, notice that, the modal analysis in Table 30.3 includes one dominant

Table 30.3 Dominant mode of the transient event in Fig. 30.11, obtained by using the multi-signal modal analysis.

WMU	Signal	Frequency (Hz)	Damping rate (Hz)	Magnitude (p.u.)	Phase angle (deg.)
1	Voltage			0.96	0.00
	Current	60.00	0.00	0.32	−35.56
2	Voltage			∼ 0.00	−17.94
	Current			∼ 0.00	−32.94

Table 30.4 Dominant modes of the transient event in Fig. 30.12, obtained by using the multi-signal modal analysis.

WMU	Signal	Frequency (Hz)	Damping rate (Hz)	Magnitude (p.u.)	Phase angle (deg.)
1	Voltage			0.98/0.20	0.00/0.00
	Current	60.00/747.72	0.00/−624.30	0.04/0.06	−25.19/82.43
2	Voltage			0.96/0.92	−0.49/−1.07
	Current			0.004/0.004	−25.96/−3.23

*The two most dominant modes are separated with a slash.

mode while the modal analysis in Table 30.4 includes two dominant modes. Next, we discuss the reason for this key difference between the two types of transient events.

30.4.1.2 Selecting the Time Window and the Number of Modes

There are two basic parameters in any modal analysis: the time window and the number of the modes. The choices of these parameters and their required accuracy depend on the type and the duration of the event. For example, the temporary event in Fig. 30.11 has a short duration; therefore, it requires a small window size. As another example, the permanent event in Fig. 30.12 has a much longer duration; therefore, it requires a longer window size; and it is less sensitive to the exact size of the time window for the purpose of the modal analysis. In this section, we obtain the start time of an event by using the event detection method in Section 30.2, see [12, 13], which is proven to accurately obtain the event start time. The event detection method in Section 30.2 also provides us with the end time for an event; although, obtaining the end time of an event is usually more challenging. The window size for the purpose of the modal analysis should be equal or less than the time period between the start time and the end time of the event. For example, if we apply the event detection method in Section 30.2 on the waveforms in Fig. 30.11, the start time of the event is obtained at $t = 480$ ms, and the end time of the event is obtained at $t = 485$ ms. Therefore, time window for modal analysis is set to $485 − 480 = 5$ ms or less, see the lengths of the rectangles with gray shaded areas in Fig. 30.11; to make sure that we do not include the part of the signal that is not related to the event.

In this study, we also use an exhaustive search to further refine the window size and also to select the number of modes in the multi-signal modal analysis. For each event, we seek to select these two parameters such that we minimize the *root mean square error* (RMSE) in modal analysis. This is done by conducting the modal analysis for different time windows that are less than the initial time window that we obtain from the method in Section 30.2 and also for different number of modes. The RMSE is obtained in each case, and the minimum RMSE is identified and the time window and the number of modes are set accordingly.

30.4.1.3 Selecting the Dominant Transient Event Mode(s)

Depending on the nature of the transient event, it may only magnify an *existing mode*; or it may create *new modes*. The former occurred in the case of the incipient fault in Fig. 30.11. The latter occurred in the case of the capacitor switching in Fig. 30.12.

The incipient fault in Fig. 30.11 was due to a momentary arcing in the system. The arc added a new resistance to the circuit; therefore, it did *not* create any new dynamic mode. As a result, the only dominant mode during the transient event in Fig. 30.11 is the fundamental mode, i.e. at 60 Hz, as we saw in Table 30.3.

The situation was different for the capacitor bank switching event in Fig. 30.12. In this case, the event caused a change in the dynamic components of the system; therefore, it created a new dynamic mode of oscillation. As a result, we captured two dominant modes during this transient event. One is the fundamental mode, i.e. at 60 Hz, and the other one is a high-frequency mode, at 748 Hz, as we saw in Table 30.4.

We can use *mode reduction* to decide which dominant mode(s) should be kept for the purpose of our event analysis in Section 30.4.2. One option is to keep the modes with high magnitude. Another option is to check the energy of each mode and keep the modes with high energy.

30.4.1.4 Comparison with Time-Domain Analysis

It is insightful to compare some key aspects of our analysis, which is done in modal phasor domain, versus an analysis that could be done in time domain by using the raw waveform measurements. First, the phasor analysis in this chapter allows us to focus on the dominant event mode of the signals, which makes our analysis more robust to noises, compared to conducting the analysis on the raw time series of the waveform measurements. Second, the phasor representation is easier to work with when it comes to solving the circuit. Note that, our method requires conducting the forward analysis and the backward analysis on the circuit model of the underlying power distribution feeder. If we use time representations; then we would have to deal with solving several differential equations and we would have to also consider an *initial solution*; all of which would unnecessarily complicate the analysis. Third, we use phasor representation only for the exact duration of the event, which ranges from less than a cycle to a few cycles. Thus, we inherently focus on the specific short interval of the *transient* component of the event. Fourth, the proposed method uses the Prony method to capture the dominant event modes in the waveform signals, as opposed to using the fast Fourier transformation. Therefore, although our analysis is done in phasor domain, we do *not* lose the information about the event, unlike in the case of the phasor measurements in PMUs. In fact, we fully capture the transient behavior of the event, even if it is only a short period of time.

30.4.2 Constructing the Feeder Model at the Dominant Transient Modes

Given the dominant modes of the synchronized waveform measurements during the transient event, the next step is to construct the feeder circuit model at those dominant modes. In this regard, consider a power distribution feeder, as in Fig. 30.13, and let us focus on any arbitrary line segment in this feeder, such as the one that is shown in Fig. 30.14(a). Let R and L denote the resistance and inductance of the line segment.

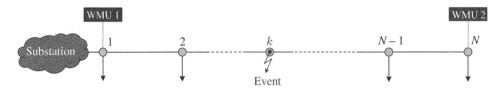

Figure 30.13 A power distribution feeder that is equipped with two WMUs. An event occurs somewhere along the feeder at *unknown* bus *k*.

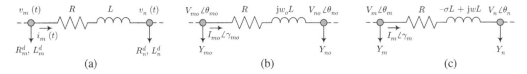

Figure 30.14 Analysis of voltage and current waveforms at a line segment immediately after the transient event occurs: (a) the circuit model in time domain; (b) the circuit model under the *fundamental mode*; (c) the circuit model under the *new transient mode* that might be created by the event.

Suppose a transient event occurs at time $t = 0$ at a bus on the distribution feeder. Suppose the location of the event is *unknown*. The voltage waveform at bus m is denoted by $v_m(t)$; the voltage waveform at bus n is denoted by $v_n(t)$; and the current waveform on the line segment is denoted by $i_m(t)$, where t indicates the timestamp immediately after the event.

As mentioned in Section 30.4.1.3, the transient event may either only magnify the existing fundamental mode; or it may create new dominant modes. Next, we discuss how to model the circuit of the distribution feeder under both circumstances.

30.4.2.1 Case I: Transient Event Does Not Create a New Mode

If the transient event does *not* create any new oscillation mode, e.g. as in Fig. 30.11, then the only dominant mode during the transient event is the fundamental mode, as in Table 30.3.

Let $f_\circ$ and $\omega_\circ = 2\pi f_\circ$ denote the frequency and rotational frequency of the fundamental mode. Also, let $V_{m\circ}$ and $\theta_{m\circ}$ denote the magnitude and phase angle of $v_m(t)$ at the fundamental mode; $V_{n\circ}$ and $\theta_{n\circ}$ denote the magnitude and phase angle of $v_n(t)$ at the fundamental mode; and $I_{m\circ}$ and $\gamma_{m\circ}$ denote the magnitude and phase angle of $i_m(t)$ at the fundamental mode. We can write the voltage difference between buses m and n at the fundamental mode as follows:

$$V_{m\circ}\angle\theta_{m\circ} - V_{n\circ}\angle\theta_{n\circ} = Z_\circ I_{m\circ}\angle\gamma_{m\circ} \qquad (30.15)$$

where

$$Z_\circ = R + \mathrm{j}\omega_\circ L \qquad (30.16)$$

is the impedance of the line at the fundamental mode. The circuit model under the fundamental mode is as in Fig. 30.14(b).

30.4.2.2 Case II: Transient Event Creates a New Mode

If the transient event creates a new oscillation mode, e.g. as in Fig. 30.12, then the dominant modes are not only the fundamental mode but also one or more new modes, as in Table 30.4. Without loss of generality, we assume that there exists only one new dominant mode in the transient event. If the transient event introduces multiple new modes, then we can simply take the dominant mode and the rest of the analysis remains the same.

Let f, $\omega = 2\pi f$, and $-\sigma$ denote the frequency, the rotational frequency, and the damping rate of the new event mode. Also, let V_m and θ_m denote the magnitude and phase angle of $v_m(t)$ at the new event mode; V_n and θ_n denote the magnitude and phase angle of $v_n(t)$ at the new event mode; and I_m and γ_m denote the magnitude and phase angle of $i_m(t)$ at the new event mode. We write the voltage difference between buses m and n at the new event mode as follows:

$$V_m\angle\theta_m - V_n\angle\theta_n = ZI_m\angle\gamma_m, \tag{30.17}$$

where

$$Z = R - \sigma L + j\omega L \tag{30.18}$$

is the impedance of the line at the new event mode. The circuit model under the new event mode is shown in Fig. 30.14(c). Notice the difference between (30.16) and (30.18) and the fact that the damping rate of the new event mode appears as a *resistive* term in (30.18).

30.4.2.3 Load Modeling in Cases I and II

We assume that the active and reactive power loads are given at all buses, either by direct measurements, such as via smart meters; or by using pseudo-measurements, such as via historical data or the ratings of the load transformers. This is a reasonable assumption; because the rating of the load transformers and the substation measurements are always available in practice. Importantly, the proposed method is very robust against errors in pseudo-measurements; as we will verify through case studies in Section 30.4.4.5. Thus, we can estimate the equivalent resistance and inductance of the load at each bus. Let R_m^d and L_m^d denote the resistance and inductance of the load at bus m; and R_n^d and L_n^d denote the resistance and inductance of the load at bus n, as we already marked in Fig. 30.14(a). We can express the admittance of the loads at buses m and n at the fundamental mode in Case I in Section 30.4.2.1 as:

$$\begin{aligned} Y_{m\circ} &= 1/(R_m^d + j\omega_\circ L_m^d), \\ Y_{n\circ} &= 1/(R_n^d + j\omega_\circ L_n^d) \end{aligned} \tag{30.19}$$

Similarly, we can express the admittance of the loads at buses m and n at the new event mode in Case II in Section 30.4.2.2 as:

$$\begin{aligned} Y_m &= 1/(R_m^d - \sigma L_m^d + j\omega L_m^d), \\ Y_n &= 1/(R_n^d - \sigma L_n^d + j\omega L_n^d) \end{aligned} \tag{30.20}$$

Notice the difference between (30.19) and (30.20). The damping rate of the new event mode appears as a *resistive* term in (30.20).

In (30.19) and (30.20), we assume that all loads are constant impedance. However, other types of loads, namely constant current and constant power loads, can also be similarly formulated and integrated into the model using pseudo-measurements. The use of other types of loads is discussed in Appendix B in [5].

30.4.3 Event Location Methodology

Consider the power distribution feeder that we saw in Fig. 30.13. It has N buses. Suppose two WMUs are installed on the distribution feeder, one at the beginning of the feeder at bus 1 and one at the end of the feeder at bus N. Suppose a transient event occurs somewhere along the feeder at *unknown* bus $k \in \{1, \ldots, N\}$.

30.4.3.1 Forward Sweep and Backward Sweep

The starting point in our event location identification method is to conduct a forward sweep and a backward sweep, see [[30], ch. 10], on the constructed circuit model of the distribution feeder.

In forward sweep, we start from the phasor representation of the *dominant mode* that is obtained in WMU 1 at bus 1, and we calculate the nodal voltages at all the buses on the distribution feeder at the dominant mode, all the way forward to WMU 2 at bus N. We denote the results in forward sweep by

$$V_1^f, \ldots, V_{k-1}^f, V_k^f, V_{k+1}^f, \ldots, V_N^f \tag{30.21}$$

In backward sweep, we start from the phasor representation of the *dominant mode* that is obtained in WMU 2 at bus N, and we calculate the nodal voltages at all the buses on the distribution feeder at the dominant mode, all the way back to WMU 1 at bus 1. We denote the results in backward sweep by

$$V_1^b, \ldots, V_{k-1}^b, V_k^b, V_{k+1}^b, \ldots, V_N^b \tag{30.22}$$

Note that, if the transient event does *not* create any new mode, then we use the line impedance in (30.16) and the load admittance in (30.19) to conduct forward sweep and backward sweep. However, if the transient event *does* create any new mode, then we use the line impedance in (30.18) and the load admittance in (30.20) to conduct forward sweep and backward sweep.

30.4.3.2 Minimizing Discrepancy

Let Ψ_i denotes the *discrepancy index* at bus i between the results from the forward sweep in (30.21) and the results from the backward sweep in (30.22):

$$\Psi_i = |V_i^f - V_i^b|, \, \forall \, i = 1, \ldots, N \tag{30.23}$$

where $|.|$ returns the magnitude of a complex number. The location of the transient event is obtained as follows:

$$k^\star = \arg\min_i \Psi_i \tag{30.24}$$

The rational in (30.24) is that the forward sweep and the backward sweep both start from direct measurements at a WMU and they continue to be correct up until we pass the unknown event bus k. At that point, the results of forward sweep and backward sweep both become incorrect. In the forward sweep, $V_1^f, \ldots, V_k^f$ are calculated correctly; while $V_{k+1}^f, \ldots, V_N^f$ are calculated incorrectly. In the backward sweep, $V_1^b, \ldots, V_{k-1}^b$ are calculated incorrectly; while $V_k^b, \ldots, V_N^b$ are calculated correctly. We can conclude that $V_i^f = V_i^b$ for $i = k$, while $V_i^f \neq V_i^b$ for $i \neq k$. Thus, the location of the transient event is obtained as in (30.24).

30.4.3.3 Algorithm

By combining the analysis in Sections 30.4.1, 30.4.2, and 30.4.3, we can develop a three-step algorithm to identify the location of transient events by using WMU measurements, as shown in Algorithm 30.1. In Step I, we extract the characteristics of the transient event from the captured synchronized waveform measurements by doing a multi-signal modal analysis. In Step II, we construct the circuit model of the feeder under the dominant mode(s). In Step III, we conduct a forward sweep and a backward sweep on the constructed circuit model, followed by the discrepancy analysis to identify the location of the event.

Algorithm 30.1: Event Location Identification: Two WMUs

Input: WMU measurements and network data

Output: The location of the transient event

1: // **Step I**:
2: Use multi-signal modal analysis to obtain the dominant mode(s) of the captured wave-forms during the transient event, such as within the gray-shaded boxes in Figs. 30.11 and 30.12.
3: // **Step II**:
4: **if** the event does not create a new mode **then**
5:     Construct the circuit model based on (30.15), (30.16), and (30.19).
6: **else if** the event creates a new mode **then**
7:     Construct the circuit model based on (30.17), (30.18), and (30.20).
8: **end if**
9: // **Step III**:
10: Use forward sweep to obtain the nodal voltages in (30.21).
11: Use backward sweep to obtain the nodal voltages in (30.22).
12: Calculate the voltage discrepancies as in (30.23).
13: Obtain the event bus number by using (30.24).

30.4.3.4 Extension to Arbitrary Number of WMUs

Suppose multiple WMUs are available, one is at the beginning of the feeder, and the rest are at the end of the feeder/laterals, as in Fig. 30.15. Suppose Ω is the set of buses with WMUs. For the WMU at each bus $s \in \Omega \backslash \{1\}$, let us define $\Psi_i^{1,s}$ as the discrepancy index at bus i that is obtained by using (30.23); where we start the forward sweep from the WMU at bus 1 and we start the backward sweep from the WMU at bus s. We define

$$\Psi_i = \sum_{s \in \Omega \backslash \{1\}} \Psi_i^{1,s}, \ \forall \ i = 1, \dots, N \tag{30.25}$$

Accordingly, we identify the location of the transient event at the minimum of the above-combined discrepancy index. The exact procedure is shown in Algorithm 30.2.

30.4.4 Event Location Results

In this section, we assess the performance of the proposed event location identification method by applying it to the IEEE 33-bus test system. The single line diagram of the test system is shown in Fig. 30.15. Five WMUs are installed on this network as marked on the figure. We study different

---

Algorithm 30.2: Event Location Identification: Multiple WMUs

Input: WMU measurements and network data

Output: The location of the transient event

1: **for** the WMU at each bus $s \in \Omega \backslash \{1\}$ **do**
2:     Use Algorithm 30.1 to obtain $\Psi_i^{1,s}$ at each bus i.
3: **end for**
4: Obtain Ψ_i at each bus i using (30.25).
5: Obtain the event bus number using (30.24).

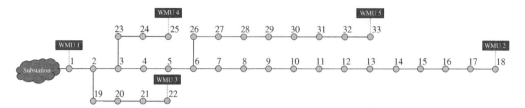

Figure 30.15 The IEEE 33-bus distribution system with five WMUs, where the set of buses with WMUs is $\Omega = \{1,18,22,25,33\}$.

scenarios of transient events, such as sub-cycle incipient faults and multi-cycle incipient faults, and permanent events, such as permanent faults and capacitor bank switching events.

30.4.4.1 Scenario I: Sub-cycle Incipient Fault

Suppose a sub-cycle incipient fault occurs at bus 9 and it lasts for *one quarter of a cycle*. Figure 30.11 in Section 30.4.1 shows the voltage and current waveforms during this event that are captured by WMUs 1 and 2. First, we extract the modes of all the 10 waveforms from all the five WMUs by conducting a multi-signal modal analysis. The results for WMUs 1 and 2 are already shown in Table 30.3. Recall that this event does *not* create any new mode. Next, we construct the circuit model between the WMU at bus 1 and any of the other four WMUs at buses 18, 22, 25, and 33. Finally, we run Algorithm 30.1 for each pair of WMUs; or we run Algorithm 30.2 for all five WMUs.

The results of running Algorithm 30.1 are shown in Fig. 30.16(a)–(d); and the results of running Algorithm 30.2 are shown in Fig. 30.16(e). As shown in Fig. 30.16(a), if the waveform measurements are available only from WMUs 1 and 5, then the discrepancy index is minimized at buses 6 to 18, indicating that the incipient fault occurred somewhere at the downstream of bus 6. As shown in Fig. 30.16(b), if the waveform measurements are available only from WMUs 1 and 4, then the discrepancy index is minimized at buses 3 to 18, and buses 26 to 33, indicating that the fault occurred at one of these buses. As shown in Fig. 30.16(c), if the waveform measurements are available only from WMUs 1 and 3, then the discrepancy index is minimized at buses 2 to 18, and buses 23 to 33, indicating that the fault occurred at one of these buses. As shown in Fig. 30.16(d), if the waveform measurements are available only from WMUs 1 and 2, then the discrepancy index is minimized at bus 9, which is the correct event bus. Finally, as shown in Fig. 30.16(e), if the waveform measurements are available from all the five WMUs, the minimum discrepancy index occurs at bus 9, which is the correct event bus.

From the above cases, we can conclude that the proposed method is able to identify the correct location of the event even if *only two* WMUs are available; as long as the event occurs somewhere *between* those two WMUs. For example, suppose only WMU 1 and WMU 2 are available. In that case, we can *correctly* identify the location of the event if the event occurs anywhere on the main feeder, i.e. at buses 1, 2, 3, …, 17, or 18. However, if the event occurs somewhere on the *first* lateral, i.e. at buses 19, 20, 21, or 22, then we identify bus 2, i.e. the head of the first lateral, as the event bus. This is because we do *not* have any WMU on the first lateral; of course, unless we *do* install WMU 3 at bus 22, which in that case we *can* identify the exact location of the event on the first lateral. Similarly, if the event occurs somewhere on the *second* lateral, i.e. at buses 23, 24, or 25, then we identify bus 3, i.e. the head of the second lateral, as the event bus. This is because we do *not* have any WMU on the second lateral; of course, unless we *do* install WMU 4 at bus 25, which in that case we *can* identify the exact location of the event on the second lateral. Similarly, if the event occurs somewhere on the *third* lateral, i.e. at buses 26, 27, …, 32, or 33, then we identify bus 6, i.e. the head of the third lateral, as the event bus. This is because we do *not* have any WMU on the third

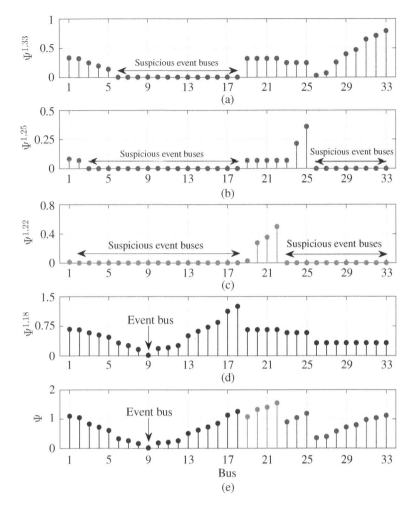

Figure 30.16 Discrepancy index in Scenario I, when the sub-cycle incipient fault occurs at bus 9 using the measurements from: (a) WMUs 1 and 5; (b) WMUs 1 and 4; (c) WMUs 1 and 3; (d) WMUs 1 and 2; (e) WMUs 1 to 5.

lateral; of course, unless we *do* install WMU 5 at bus 33, which in that case we *can* identify the exact location of the event on the third lateral. In summary, the proposed method can work with at least two WMUs; depending on the location of the transient event, certain pairs of WMUs are more suitable to provide the waveform measurements that can lead to correctly identify the location of the event by running Algorithm 30.1. However, since the event bus is *not* known in advance, it is necessary that we use the waveform measurements from all the five WMUs so that we can identify the exact location of the event; whether it occurs on the main feeder or on a lateral. For the rest of this chapter, we focus on identifying the event bus using the waveform measurements from all the five WMUs.

30.4.4.2 Scenario II: Multi-cycle Incipient Fault

Suppose a multi-cycle incipient fault occurs at bus 30 and it lasts for *two cycles*. As in Scenario I, this event does *not* create any new mode. Figure 30.17 shows the results of running Algorithm 30.2

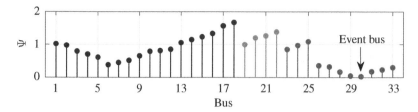

Figure 30.17 Discrepancy index in Scenario II, when the multi-cycle incipient fault occurs at bus 30, based on running Algorithm 30.2 on all five WMUs.

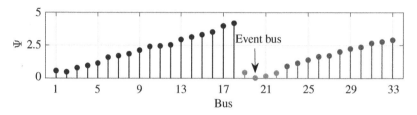

Figure 30.18 Discrepancy index in Scenario III, when the permanent fault occurs at bus 20, based on running Algorithm 30.2 on all five WMUs.

in this scenario based on the waveform measurements from all five WMUs. As we can see, our method is able to correctly identify bus 30 as the location of the incipient fault. This scenario further confirms the accuracy of our method.

30.4.4.3 Scenario III: Permanent Fault

Suppose a permanent symmetric fault occurs at bus 20. We call it permanent because it is *not* self-cleared. It may last until it is cleared by a circuit breaker. As in Scenarios I and II, this permanent fault does *not* create any new mode. The results of running Algorithm 30.2 are shown in Fig. 30.18. The location of the permanent fault is correctly identified at bus 20. The results in this scenario confirm the accuracy of the proposed method even for transient events that lead to permanent events. Of course, our method still focuses only on the transient component of this event; and accordingly, it identifies its location very promptly.

30.4.4.4 Scenario IV: Capacitor Bank Switching Event

Suppose a capacitor bank is switched on at bus 24. Figure 30.12 in Section 30.4.1 shows the voltage and current waveforms during this event that are captured by WMUs 1 and 4. The results of multi-signal modal analysis are already shown in Table 30.4. Unlike in Scenarios I, II, and III, in this scenario, the event not only magnifies the fundamental mode but also creates a new dominant mode, as we saw in Table 30.4. The results of running Algorithm 30.2 are shown in Fig. 30.19. As we can see, the proposed method is able to identify the correct event location.

30.4.4.5 Sensitivity Analysis

Next, we use Monte Carlo simulation to assess the impact of errors in parameters and measurements on the accuracy of the proposed method. The number of random scenarios is 10,000.

(1) **Error in line parameters:** Line inductance and resistance may deviate from their nominal values because of loading, aging, and weather conditions, to name a few. Table 30.5 shows the results for different levels of errors. As we can see, even when the error is at 50%, the proposed

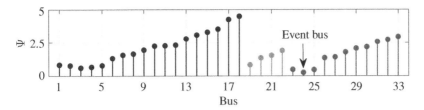

Figure 30.19 Discrepancy index in Scenario IV, when capacitor bank switching occurs at bus 24, based on running Algorithm 30.2 on all five WMUs.

Table 30.5 Impact of error in line parameters.

Error (%)	Correct bus (%)	Neighboring bus (%)	Other bus (%)
25	100.0	0.0	0.0
50	98.9	1.1	0.0
75	93.0	7.0	0.0
100	85.8	14.2	0.0

Table 30.6 Impact of error in pseudo-measurements.

Error (%)	Correct bus (%)	Neighboring bus (%)	Other bus (%)
25	100.0	0.0	0.0
50	100.0	0.0	0.0
75	100.0	0.0	0.0
100	99.8	0.1	0.1

method can identify the correct location for the transient event in 98.9% of the random scenarios. In the remaining 1.1% of the cases, we identify the neighboring bus as the event location. Hence, the robustness of the proposed method is confirmed for errors in line parameters.

(2) **Error in pseudo-measurements:** Table 30.6 shows the location identification accuracy for different levels of errors in pseudo-measurements. Even when the error is at 100%, the proposed method can identify the correct location for the transient even in 99.8% of the random scenarios. In another 0.2% of the cases, we can still identify the neighboring bus. Thus, the robustness of the proposed method is further confirmed.

(3) **Noise and harmonics in waveform measurements:** Table 30.7 shows the results on the accuracy of the proposed event identification method for different levels of harmonics in the system as well as different levels of measurement noise in WMU measurements. The level of harmonics is specified in terms of the *total-harmonic-distortion* (THD) of the current waveforms. The measurement noise level is specified in terms of the *signal-to-noise-ratio* (SNR). As we can see, even when the THD is as high as 3% and the SNR is as low as 20 dB, the proposed method is still able to correctly identify the location of the event in 85.5% of the random scenarios. In another 6.2% of the random scenarios, an immediate neighboring bus of the correct event

Table 30.7 Impact of harmonic distortion and measurement noise on the accuracy of the event location identification method.

THD (%)	SNR (dB)	Correct bus (%)	Neighboring bus (%)	Other bus (%)
1	80	100.0	0.0	0.0
	50	100.0	0.0	0.0
	20	86.8	5.8	7.4
2	80	100.0	0.0	0.0
	50	99.9	0.1	0.0
	20	84.4	7.5	8.1
3	80	100.0	0.0	0.0
	50	99.8	0.2	0.0
	20	85.5	6.2	8.3

bus is identified. The results in Table 30.7 confirm the robustness of the proposed event location identification method even under considerable harmonic and measurement noise levels. It bears mentioning that, identifying the correct location of sub-cycle incipient faults becomes challenging when the levels of noise and harmonics in waveform measurements are high, due to the very short duration of such events.

30.5 Applications

While the focus in this chapter is on the core technical tasks of event detection, event classification, and event location identification in synchro-waveform measurements, the results can ultimately support different real-life applications. Some of these potential applications are discussed as follows.

First, the methodologies that are developed in this chapter can help improve situational awareness with respect to the state of *health* and *safety* of various equipment in power distribution systems. In particular, by detecting and identifying incipient faults, the utility can take remedial actions in a timely manner to prevent catastrophic damages in the future, i.e. to resolve a major future failure while it is still in its early stages. Of course, since incipient faults are usually self-clearing and last for only a very short period of time [5, 31], improving our ability to detect and classify incipient faults can directly benefit the ultimate real-life applications in this area.

Second, detection and identification of specific equipment operations, such as switching on and switching off at capacitor banks, can also help with scrutinizing the operation of certain equipment of interest. This can benefit us with not only identifying any potential malfunctions but also updating the utility models to keep track of the changes in the system due to equipment aging [8]. The latter results can help improve the overall operation of the power distribution system.

Finally, as for the real-life applications of detecting and identifying high-impedance faults, and even some incipient faults, they can be used for instance in wildfire detection and prevention [32–34]. Note that, a high-impedance fault occurs when a line conductor touches a high grounding impedance object, such as during vegetation intrusion or when the power line is down. These circumstances can cause ignition and ultimately lead to wildfire [35]. In fact, many of the most destructive wildfires in California are reported to be caused by power equipment issues. Hence, early detection and identification of high-impedance faults can contribute to improving our ability to detect and prevent wildfires.

30.6 Exercises

1. Consider the following voltage waveform and current waveform during normal operating conditions before an event occurs:

$$v(t) = \sqrt{2}\cos(120\pi t),$$
$$i(t) = 0.1\sqrt{2}\cos(120\pi t - \pi/6).$$

   (a) Obtain the quadratic equation of the corresponding synchronized Lissajous curve as defined in (30.7).
   (b) Verify that the obtained equation always represents an ellipse.
   (c) Obtain the rotational angle of the synchronized Lissajous curve with respect to the x-axis. Noted that the rotational angel is different from the phase angle between the voltage waveform and the current waveform.

2. Consider the following generic waveforms for voltage and current during steady-state conditions after an event occurs:

$$v(t) = \sum_{h=1}^{H} V_h \cos(h\omega t + \theta_h),$$
$$i(t) = \sum_{h=1}^{H} I_h \cos(h\omega t + \gamma_h),$$

   where ω is the fundamental rotational frequency; h is the harmonic order; V_h and θ_h are the magnitude and phase angle of the hth harmonic of the voltage waveform; I_h and γ_h are the magnitude and phase angle of the hth harmonic of the current waveform; and H is the maximum number of harmonic orders.

   (a) Obtain the closed-form area of their corresponding synchronized Lissajous curve in each cycle T as defined in (30.10).
   (b) What is the unit of this area?
   (c) What is the physical meaning associated with this area?

3. Suppose the reporting rate of certain WMUs is 256 samples per cycle, i.e. each cycle takes 256 samples. We use the synchro-waveforms from the WMUs to plot one-cycle synchronized Lissajous curves for the purpose of event detection. The first synchronized Lissajous curve is plotted based on samples 1 to 256. The second synchronized Lissajous curve is plotted based on samples 2 to 257.

   (a) Based on what samples is the 2024th synchronized Lissajous curve plotted?
   (b) Suppose the window of the adaptive threshold is 1028 samples. If we want to obtain the adaptive threshold at sample 2024, i.e. $\mathcal{T}[2024]$, then what similarity indices are needed? Noted that Δt is equal to 1 sample.
   (c) Suppose an event is detected at sample 2024. If we want to obtain the adaptive threshold $\mathcal{T}[2025]$, then what similarity indices are needed?

4. Consider the IEEE 33-bus distribution feeder in Fig. 30.4. Suppose an incipient fault occurs somewhere on the main feeder between WMU 1 and WMU 2. Figure 30.11 that we saw in

Section 30.4.1 shows the synchro-waveforms measured by these two WMUs during this event. Suppose the phasor representations of the voltage and current waveforms at the dominant mode are as in Table 30.3. Assume that the resistance and inductance of each line segment are 0.350 and 0.0004 per unit, respectively. Also, assume that the resistance and inductance of the load at each bus are 1500 and 1.50 per unit, respectively.

(a) Obtain $\Psi_1, \Psi_2, \ldots, \Psi_{18}$ as defined in (30.23).

(b) At what bus number has this event occurred?

References

1 Izadi, M. (2022). Data-driven analysis and applications of time-synchronized waveform measurements in power systems. PhD dissertation. Riverside, CA: University of California.

2 Izadi, M., Mousavi, M.J., and Lim, J. (2023). Power-line event location systems and methods. United States Patent, App. 17/984,245.

3 Cassel, W.R. (1993). Distribution management systems: functions and payback. *IEEE Transactions on Power Systems* 8 (3): 796–801.

4 Mohsenian-Rad, H., Stewart, E., and Cortez, E. (2018). Distribution synchrophasors: pairing big data with analytics to create actionable information. *IEEE Power and Energy Magazine* 16 (3): 26–34.

5 Izadi, M. and Mohsenian-Rad, H. (2021). Synchronous waveform measurements to locate transient events and incipient faults in power distribution networks. *IEEE Transactions on Smart Grid* 12 (5): 4295–4307.

6 Izadi, M., Mousavi, M.J., Lim, J.M., and Mohsenian-Rad, H. (2022). Data-driven event location identification without knowing network parameters using synchronized electric-field and current waveform data. *Proceedings of the IEEE PES General Meeting*, Denver, CO, 1–5.

7 Izadi, M. and Mohsenian-Rad, H. (2021). Characterizing synchronized Lissajous curves to scrutinize power distribution synchro-waveform measurements. *IEEE Transactions on Power Systems* 36 (5): 4880–4883.

8 Izadi, M. and Mohsenian-Rad, H. (2020). Event location identification in distribution networks using waveform measurement units. *Proceedings of the IEEE PES ISGT Europe*, The Hague, Netherlands, 924–928.

9 Bastos, A.F., Santoso, S., Freitas, W., and Xu, W. (2019). Synchrowaveform measurement units and applications. *Proceedings of the IEEE PES General Meeting*, Atlanta, GA, USA, 1–5.

10 SEL. Instruction Manual and Data Sheet. SEL-735 Power Quality and Revenue Meter. https://selinc.com/products/735/ (accessed 22 November 2023).

11 Specifications and Features. Candura PQPro Power Quality Analyzer. https://www.candura.com/products/pqpro.html (accessed 22 November 2023).

12 Izadi, M. and Mohsenian-Rad, H. (2021). A synchronized Lissajous-based approach to achieve situational awareness using synchronized waveform measurements. *Proceedings of the IEEE PES General Meeting*, Washington, DC, 1–5.

13 Izadi, M. and Mohsenian-Rad, H. (2022). A synchronized Lissajous-based method to detect and classify events in synchro-waveform measurements in power distribution networks. *IEEE Transactions on Smart Grid* 13 (3): 2170–2184.

14 Karacor, D., Nazlibilek, S., Sazli, M.H., and Akarsu, E.S. (2014). Discrete Lissajous figures and applications. *IEEE Transactions on Instrumentation and Measurement* 63 (12): 2963–2972.

15 Hong, T. and de León, F. (2015). Lissajous curve methods for the identification of nonlinear circuits: calculation of a physical consistent reactive power. *IEEE Transactions on Circuits and Systems I: Regular Papers* 62 (12): 2874–2885.

16 Abu-Siada, A. and Mir, S. (2019). A new on-line technique to identify fault location within long transmission lines. *Engineering Failure Analysis* 105: 52–64.

17 Leys, C., Ley, C., Klein, O. et al. (2013). Detecting outliers: do not use standard deviation around the mean, use absolute deviation around the median. *Journal of Experimental Social Psychology* 49 (4): 764–766.

18 Manitoba HVDC Research Centre. Ver. 4.2 PSCAD/EMTDC (Software Package). Winnipeg, MB, Canada.

19 MathWorks. Matlab Help Center - Getframe.

20 MathWorks. Matlab Help Center - Frame2im.

21 Simonyan, K. and Zisserman, A. (2015). Very deep convolutional networks for large-scale image recognition. *arXiv preprint arXiv:1409.1556.*

22 Krizhevsky, A., Sutskever, I., and Hinton, G.E. (2012). ImageNet classification with deep convolutional neural networks. In: *Proceedings of the Advances in Neural Information Processing Systems 25 (NIPS 2012)*, 1097–1105.

23 Nair, V. and Hinton, G.E. (2010). Rectified linear units improve restricted Boltzmann machines. *Proceedings of the 27th International Conference on Machine Learning*, Haifa, Israel, 807–814.

24 MathWorks. Matlab Help Center - Convolutional Neural Network.

25 Kingma, D.P. and Ba, J. (2014). Adam: a method for stochastic optimization. *arXiv preprint arXiv:1412.6980.*

26 Hu, Y., Wu, W., and Zhang, B. (2016). A fast method to identify the order of frequency-dependent network equivalents. *IEEE Transactions on Power Apparatus and Systems* 31 (1): 54–62.

27 Crow, M.L. and Singh, A. (2005). The matrix pencil for power system modal extraction. *IEEE Transactions on Power Apparatus and Systems* 20 (1): 501–502.

28 Bollen, M.H.J., Styvaktakis, E., and Gu, I.Y.-H. (2005). Categorization and analysis of power system transients. *IEEE Transactions on Power Delivery* 20 (3): 2298–2306.

29 Trudnowski, D.J., Johnson, J.M., and Hauer, J.F. (1999). Making Prony analysis more accurate using multiple signals. *IEEE Transactions on Power Apparatus and Systems* 14 (1): 226–231.

30 Kersting, W. (2002). *Distribution System Modeling and Analysis.* Boca Raton, FL: CRC Press.

31 Mohsenian-Rad, H. (2022). *Smart Grid Sensors: Principles and Applications.* Cambridge: Cambridge University Press.

32 Izadi, M. Preventing Wildfires Caused by Electrical Power Grid Systems. https://www.youtube.com (accessed 22 November 2023).

33 Mohsenian-Rad, H. (2021). Synchro-waveforms in power distribution with application to wildfire monitoring. *IEEE Power and Energy Society General Meeting Panel Session in Proceedings*, July 2021.

34 Jazebi, S., de León, F., and Nelson, A. (2020). Review of wildfire management techniques—Part I: Causes, prevention, detection, suppression, and data analytics. *IEEE Transactions on Power Delivery* 35 (1): 430–439.

35 Wischkaemper, J.A., Benner, C.L., Russell, B.D., and Manivannan, K.M. (2014). Application of advanced electrical waveform monitoring and analytics for reduction of wildfire risk. *Proceedings of the IEEE PES ISGT*, 1–5, Washington, DC.

31

Traveling Wave Analysis in Microgrids
Soumitri Jena and Peng Zhang

31.1 Introduction

As more inverter-based resources (IBRs) are integrated into the grid, protection schemes that rely on phasor domain signal processing become less effective due to lower fault current and phase angle modulation from the converters. For example, a synchronous generator typically provides fault current six times the rated current, whereas the maximum current capacity of an IBR does not exceed 1.5 times of the full load current. Typically, the IBR does not provide reliable negative and zero-sequence quantities, as synchronous generators do [1]. Therefore, the assumptions made in conventional phasor-based protection design are no longer valid. Moreover, the inherent latency in phasor domain detection is too slow to accurately identify and locate such faults. Distribution system protection schemes also make use of directional components, but the settings are not dynamic. Because of the change in load profiles and bidirectional power flow, these directional protection schemes often mal-operate. The traveling wave-based protection schemes, on the contrary, extract the fault information from the transients in voltage and current signals. The power system is represented by the driving point source in the equivalent network in place of all sources. Therefore, traveling wave methods as an alternative can overcome the challenges in conventional protection and facilitate the widespread deployment of IBRs in microgrids.

This chapter describes a traveling wave-based protection concept to identify and locate faults in the microgrid. Transients in the fault current signal are effectively identified using advanced signal processing techniques. A centralized control unit collects the transient information from the remote terminals and locates the faulted section within few milliseconds into the fault. While sensor and signal processing techniques for traveling wave detection are significant, the algorithms required to locate faults are salable to distribution systems, thus, providing an optimistic roadmap for implementation in real-world microgrids.

31.2 Background Theories

31.2.1 Mathematical Analysis

A fault anywhere in the power system results in abrupt voltage drop in the faulted phases from their pre-fault conditions. Such an abrupt change lunches step-like voltage and current traveling waves(TWs). These waveforms are superimposed on the fundamental frequency waveform [2]. Using superimposed theorem, one can assume the pure-fault circuit as a resistance R_f in series

Microgrids: Theory and Practice, First Edition. Edited by Peng Zhang.
© 2024 The Institute of Electrical and Electronics Engineers, Inc. Published 2024 by John Wiley & Sons, Inc.

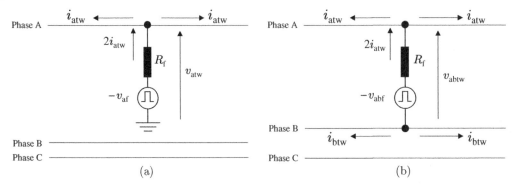

Figure 31.1 Fault-induced TWs: (a) ground: ag and (b) ungrounded: ab fault.

with a step voltage assuming fault point magnitude immediately before fault, but with negative polarity [3].

Figure 31.1(a) and (b) show the pure-fault networks for TW generation in ag (grounded), (ab) ungrounded faults, respectively. It should be pointed out that all other faults can be modeled similarly. For the ag fault shown in Fig. 31.1(a),

$$v_{atw} = -2i_{atw}R_f - v_{af} \tag{31.1}$$

where v_{atw}, and i_{atw} are the voltage and current TWs; R_f is the fault resistance; v_{af} is the pre-fault voltage.

Further,

$$v_{atw} = -2\frac{v_{atw}}{Z_c}R_f - v_{af}; \quad v_{atw} = \frac{-v_{af}Z_c}{2R_f + Z_c} \tag{31.2}$$

where Z_c is the characteristics impedance of the line. The current TW is given by,

$$i_{atw} = \frac{v_{atw}}{Z_c} = \frac{-v_{af}}{2R_f + Z_c} \tag{31.3}$$

Similarly, for the ab fault the voltage and current TWs are,

$$v_{abtw} = \frac{-v_{abf}Z_c}{2R_f + Z_c}; \quad i_{abtw} = \frac{-v_{abf}}{2R_f + Z_c} \tag{31.4}$$

where $v_{abf} = v_{af} - v_{bf}$.

From the analysis presented in this section, it is worthwhile to note that the voltage and current TWs depend on the pre-fault voltage magnitude, fault resistance, and characteristic impedance of the line. The pre-fault voltage $v_f = v_r \sin\theta$; where v_r is the rms voltage and θ is the fault inception angle. A fault at the voltage peak results in the strongest TW whereas fault at a low inception angle produce poor transients and makes TW techniques unfeasible. Similarly, for a higher fault resistance R_f the attenuation in the voltage and current TWs are severe. The characteristics impedance Z_c also influences the magnitude of voltage and current TWs that depend on the line and cable configuration and geometries with ground characteristics. The nature of the TWs is independent of the type of source, which makes them ideal for developing microgrid protection strategies.

31.2.2 Transient Response of Instrument Transformers

The transient responses of current transformers (CT) and capacitor voltage transformers (CVT) are important in TW-based relaying. The CTs provide adequate transient response over a broad

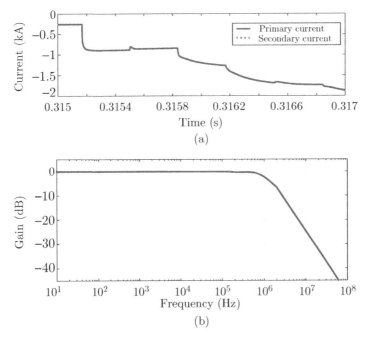

Figure 31.2 (a) Time and (b) frequency response of the CT.

bandwidth (typically greater than 100 kHz), making them suitable for TW extraction [4]. This chapter investigates the time and frequency response of CT and CVT models recommended in [5] for 1 MHz sampling frequency. Figure 31.2(a) depicts the primary and secondary side (scaled to primary) currents in this regard. The secondary waveform is a true reflection of the primary signal. Since the wavefronts are steep, the high-frequency components can be obtained without sacrificing accuracy. The frequency response of the CT (Fig. 31.2(b)) is flat up to 1 MHz. Therefore, the transient response of the CT is appropriate for TW applications.

Unlike CTs, CVTs dampen high-frequency components. The time and frequency domain responses of the CVT used in this work are shown in Fig. 31.3(a) and (b), respectively. As seen, the CVT dampens transients in the secondary waveform. In terms of magnitude, the TW magnitude extracted from the secondary waveform is inaccurate. The frequency response, as shown in Fig. 31.3, suggests that CVT relies on the stray capacitance for extraction of high-frequency components. The issue with the TW-based protection arises in particular in fault conditions such as low inception angle, high fault resistance, or in the case of low bandwidth CVTs. In such cases, the change in voltage signal in the CVT's primary is small, which further attenuates due to the associated poor transient characteristics. On such occasions, the supervisory protection-based incremental quantity or phasor is invoked.

31.3 Challenges for TW Applications in Microgrid

In the future, fault detection, isolation, and service restoration will be critical for advanced microgrid management systems and self-healing capability. The adaptation of the successful fault location (FL) algorithms developed for the transmission system is limited by the topology and

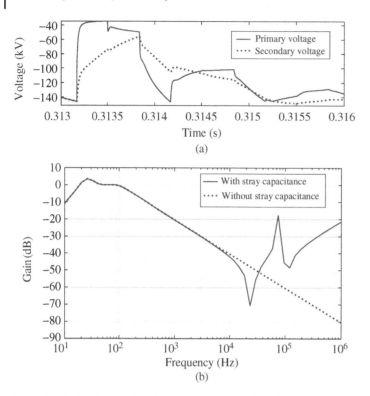

Figure 31.3 (a) Time and (b) frequency response of the CVT.

operating principles of the distribution system (i.e. nonhomogeneous feeders, load taps, laterals, radial operation, and the available measuring equipment). Because laterals are present, the estimated distances could correspond to a variety of fault locations in a microgrid. The fault location techniques for transmission lines are either one-terminal and two-terminal-based methods. Because of the simplifying assumptions and accuracy limitations, the schemes can not be applied to a microgrid architecture. Locating faults in a transmission system using TW conventionally requires few signal processing techniques, the ability to measure with 1 MHz sampling, filter and communication with constant latency. With these settings, the location accuracy is within a tower span of 300 m. However, such accuracy for a microgrid is not desirable as the distribution feeders are of few hundred meters. Because of the presence of laterals and load taps in microgrids, TW signals are difficult to interpret. Moreover, the limited bandwidth of conventional current and voltage transformers directly limits the resolution of fault localization. The current weaknesses of TW-based techniques can be interpreted as follows:

1. Fault-induced TWs propagate along distribution lines in both directions away from the fault point and are reflected at line terminations, junctions, and the fault location. These reflections are observed at relay locations where the spatial context of these multiple reflected waves is not discernible.
2. During a microgrid fault, TWs pass through or be attenuated by various components such as unbalanced short lines, underground cables, taps (points of discontinuity), transformers, changes in conductor size, voltage regulators, and capacitor banks. Therefore, TWs may decay in a short period of time.

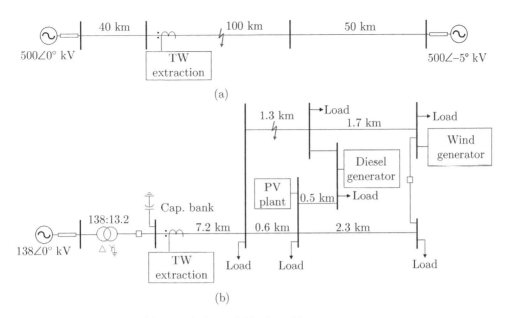

Figure 31.4 Test systems: (a) transmission and (b) microgrid.

31.3.1 Simulations Scenarios

Figure 31.4(a) and (b) depict transmission and microgrid test systems available in the RSCAD resource center [6]. The 500 kV transmission grid has four buses with line lengths of 40, 100, and 50 km. The microgrid has seven buses with very short length lines. The current traveling waves are extracted at the relay location for faults at locations indicated in the figures.

The current traveling waves extracted from the a-phase signal for an ag fault in the transmission and microgrid system are depicted in Fig. 31.5(a) and (b), respectively. The current traveling waves extracted from the transmission system are clearly identifiable. The reflections from the buses and the fault point can be identified in terms of their magnitude, polarity, and time of arrival (ToA). The current traveling waves in the microgrid system, on the contrary, superimpose on each other and cannot be easily identifiable. This superimposition makes it challenging to identify the magnitude, polarity, and ToA of the wavefronts. This suggests the need for advanced signal processing techniques to extract traveling wave information from fault signals in a microgrid.

An immediate observation can be made from the waveforms that while the traveling waves in the transmission system sustain for a longer period (4 ms), the traveling waves in the microgrid die out within 1 ms. This translates to the need for current and voltage sensors with wider bandwidth when compared to the transmission system.

31.4 Proposed Traveling Wave Protection Scheme

A traveling wave protection based on the appearance of periodic traveling waves is developed to overcome the challenges in microgrid applications. Only the ToA and polarity of traveling waves are used to determine the time difference between reflected traveling waves at the relay location, rendering the requirement for an accurate magnitude obsolete. Also, high-bandwidth communication channels are not required.

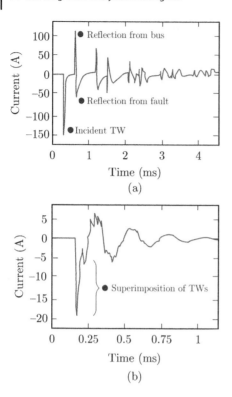

Figure 31.5 Extracted traveling waves: (a) transmission and (b) microgrid.

31.4.1 Periodic Reflected Traveling Waves from IBR

The schematic of a fault in a typical microgrid configuration connected to the grid has been depicted in Fig. 31.6(a). A fault in the microgrid is represented by the pure-fault network, as shown in Fig. 31.6(b).

The pure-fault network is based on the signals that appear only due to the fault (excluding the load voltage and current components) [7]. The network is represented by the driving point source in the equivalent network in place of all sources, i.e. the effects of all other sources are neglected. IBRs with grid following and grid forming control strategies are connected to the grid through a cable. A fault in the cable results in an abrupt voltage drop in the faulted phases from their pre-fault conditions. Such abrupt changes launch step-like voltage-traveling waves. Using the superimposed theorem one can assume the voltage traveling wave $v(0, t)$ as a resistance R_f in series with a step voltage assuming fault point magnitude immediately before fault, but with negative polarity $(-v_f)$. The traveling wave propagates through the cable, represented by an infinitely small distance (dx). $R, L, G,$ and C are the resistance, inductance, capacitance, and conductance, respectively, for unit length. Z_m is the impedance of the synchronous grid. Based on the control strategies, the transient state of IBR-I (grid forming) and IBR-II (grid following) are represented by a constant current source (cs) in parallel with impedance Z_{n1} and a constant voltage source (vs) in series with impedance Z_{n2} [8]. The voltage and current at any distance x from the fault point is the summation of the incident and reflected traveling wave, given in the Laplace domain as,

$$V(x, s) = V^+(s)e^{-\gamma x} + V^-(s)e^{\gamma x}$$
$$I(x, s) = \frac{1}{Z_C}\left(V^+(s)e^{-\gamma x} - V^-(s)e^{\gamma x}\right)$$

(31.5)

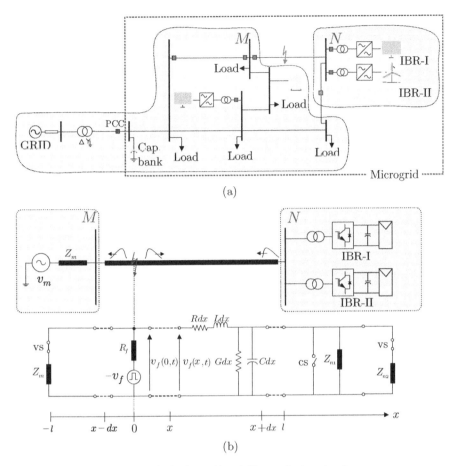

Figure 31.6 (a) Fault in a typical microgrid and (b) pure-fault network.

where

$$\gamma = s\sqrt{LC}; \quad Z_c = \sqrt{\frac{L}{C}}$$

The incident (V^+) and reflected (V^-) waves at the receiving end of the cable ($x = l$),

$$V^-(s) = V^+(s)\Gamma_n e^{-2\gamma l}$$

$$V^+(s) = V(s)\frac{Z_c}{Z_c + Z_n}\left(\frac{1}{1 - \Gamma_n\Gamma_f e^{-2\gamma l}}\right) \tag{31.6}$$

where Z_n is the parallel equivalence of Z_{n1} and Z_{n2}. Γ_n and Γ_f are the reflection coefficients from the IBR terminal and the fault point, respectively. Again, the voltage and current at any point in the cable after the reflection are given by,

$$V(x, s) = V(s)\frac{Z_c}{Z_c + Z_n}\left(\frac{1}{1 - \Gamma_n\Gamma_f e^{-2\gamma \ell}}\right) \tag{31.7}$$

$$\times \left(e^{-\gamma x} + \Gamma_n e^{-2\gamma \ell} e^{\gamma x}\right)$$

$$I(x,s) = \frac{V(s)}{Z_c + Z_n}\left(\frac{1}{1 - \Gamma_n\Gamma_f e^{-2\gamma\ell}}\right)$$
$$\times \left(e^{-\gamma x} - \Gamma_n e^{-2\gamma\ell}e^{\gamma x}\right)$$

(31.8)

Further, the velocity of propagation (u) and propagation time (t) are,

$$u = \frac{1}{\sqrt{LC}}; \quad \tau = \frac{l}{u}$$

(31.9)

Therefore,

$$V(x,s) = V(s)\frac{Z_c}{Z_c + Z_n}\frac{1}{1 - \Gamma_n\Gamma_f e^{-2\tau s}}$$
$$\times \left[e^{-s\frac{x}{v}} + \Gamma_n e^{s\left(\frac{x}{u} - 2\tau\right)}\right]$$

(31.10)

$$I(x,s) = \frac{V(s)}{Z_c + Z_n}\frac{1}{1 - \Gamma_n\Gamma_f e^{-2\tau s}}$$
$$\times \left[e^{-s\frac{x}{v}} - \Gamma_n e^{s\left(\frac{x}{u} - 2\tau\right)}\right]$$

(31.11)

The aforementioned analysis shows that there exist periodic traveling waves between the fault point and both ends of the cable. The frequency of occurrence of these traveling waves depends on the fault distance (x), line length (l), and propagation delay (τ) of the cable. As explained in the following sections, these characteristics can be utilized to identify the faulted cable section and fault location in a microgrid.

Figure 31.7 depicts a schematic for the proposed wide-area-based protection. The line sections incorporate Merging Units (MUs), which take analog current signals from the CT secondary and transform them into time-stamped sampled values (SVs). The traveling wave filtering process

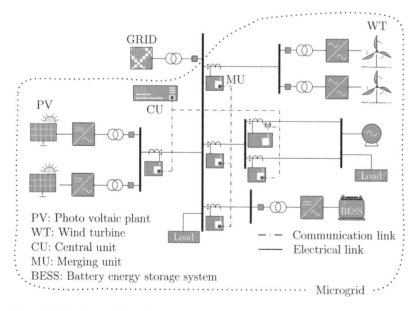

Figure 31.7 Proposed traveling wave-based protection architecture.

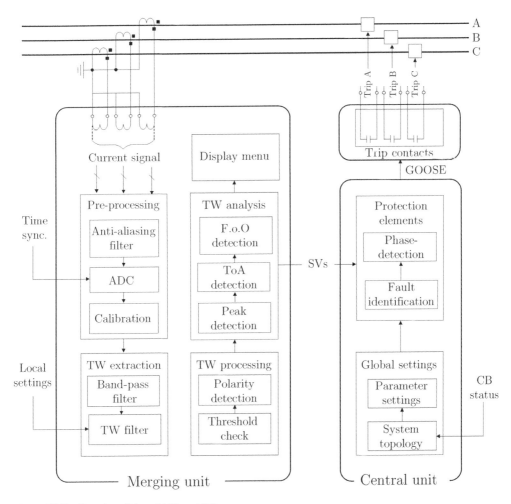

Figure 31.8 Functionalities of MU and CU.

extracts the traveling wave signal and its amplitude and ToA. The frequency of occurrence (F.o.O) of the reflected traveling waves and the traveling wave polarity is communicated with the central unit (CU) using point-to-point ethernet-based communication. In the CU, the F.o.Os and the polarity of traveling waves are utilized to identify the faulted line section and a GOOSE signal is sent to the corresponding circuit breaker (CB) to clear the fault. The detailed functionalities of the MU and CU are showcased in Fig. 31.8 and explained in Section 31.4.2 and 31.4.3.

31.4.2 Merging Unit (MU)

The MUs are placed at one end of each of the cable sections in the microgrid. It acquires the analog signal from the CT secondary and preprocesses for traveling wave extraction.

31.4.2.1 Preprocessing

To avoid signal aliasing, the MU collects current signals from the CT secondary and passes them through an analog anti-aliasing filter (AAF), which limits the spectral bandwidth of the monitored measurements before analog signals are discretized. A cut-off frequency of 600 kHz $(0.8 \times f_N)$

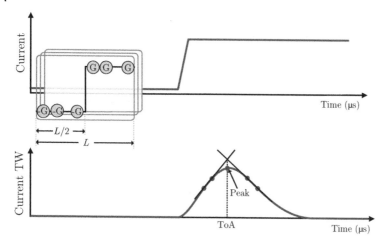

Figure 31.9 Traveling wave extraction and ToA estimation.

has been used, where f_N is the Nyquist sampling frequency requirement. A 16-bit analog-to-digital converter (ADC) samples the analog signal and converts it into digital SVs, which are time-stamped and calibrated through a GPS clock. The time-stamped SVs are utilized for the traveling wave extraction process.

31.4.2.2 Traveling Wave Extraction

A band-pass filter with cut-off frequencies of 1 and 400 kHz rejects both power frequencies and multiple harmonics, as well as high-frequency components such as noise. Figure 31.9 depicts the procedure for the proposed traveling wave extraction and ToA estimation. The filter coefficients (G) are chosen to extract the traveling wave with the same amplitude as a step change. Half of the coefficients are set to G and another half is set to $-G$. The traveling wave samples (i_{TW}) are obtained by convolving the filter coefficients with the raw samples that have been passed through the band-pass filter.

$$i_{TW}(k) = \sum_{h=1}^{L} G(h)i(k - L + h) \tag{31.12}$$

where L is the total number of coefficients.

It must be noted that i_{TW} is zero both before and after the step change when the filter window is fully enclosed. When the coefficients are split into two halves, with one half preceding and the other following a step change, the output of the filter is maximized. The i_{TW} should be equal to A for unitary gain,

$$i_{TW}(k) = \sum_{h=1}^{L/2} - G(h)i(k - L + h)$$
$$+ \sum_{h=(L/2)+1}^{L} G(h)i(k - L + h) \tag{31.13}$$

It can be proved that to obtain unitary gain, the filter coefficients $G = \frac{2}{L}$, where L is the filter window length. The calculation of traveling wave amplitude begins with determining the wavefront's

ToA. The ToA is obtained by interpolating a few samples before and after the peak. After calculating ToA, the wavefront amplitude (I) is calculated as follows:

$$I = G \sum_{n=-M}^{n=M} i_{\text{TW}}(N_{\text{ToA}} - n) \tag{31.14}$$

where G is the scaling factor used to maintain unitary gain; M is the window length with half the number of filter coefficients; i_{TW} is the traveling wave amplitude and N_{ToA} is the sample corresponding to the obtained ToA of the wavefront.

31.4.2.3 Modal Components
To decouple the phase values, we apply Clarke's transformation with reference to phase A.

$$\begin{bmatrix} i_\alpha \\ i_\beta \\ i_0 \end{bmatrix} = \frac{1}{3} \begin{bmatrix} 2 & -1 & -1 \\ 0 & \sqrt{3} & -\sqrt{3} \\ 1 & 1 & 1 \end{bmatrix} \begin{bmatrix} i_a \\ i_b \\ i_c \end{bmatrix} \tag{31.15}$$

where i_α, i_β and i_0 are the alpha, beta, and zero-mode components derived from the instantaneous phase currents.

The alpha-mode components are available for each type of fault and have the highest magnitude among the three; therefore, they are selected in this work.

31.4.2.4 Settings to Avoid Superimposition
To eliminate traveling wave superimposition, the filter window length and gain settings needed to be changed based on the connecting lines. Figure 31.10(a) and (b) show the time-space diagram for a fault at a distance d and the extraction of the incident and reflected traveling waves from MU in the line. The time difference between these two wavefronts is $2(l_1 - d)/v$; where l_1 and v

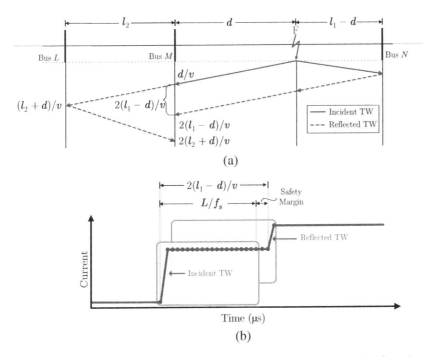

Figure 31.10 (a) Time-space diagram and (b) extraction of the incident and reflected traveling waves.

are the length and propagation velocity of the line. The filter length of the running window is L/f_s microseconds; where D is the number of coefficients and f_s is the sampling frequency in MHz. When the internal fault is near bus M, the reflected traveling wave may arrive within l_1/f_s microseconds into the incident traveling wave. To avoid such a superimposition between these traveling waves,

$$\frac{2(l_1 - d)}{v} > \frac{L}{f_s} \tag{31.16}$$

The scaling factor and the number of filter coefficients are set as per (31.13) and (31.18). The filter window is set assuming the worst condition for waveform superimposition, i.e. faults very close to the relay location ($d = 0$). It may be further noticed that the first reflection may come from the previous bus (bus L). In such a condition,

$$\frac{2l_2}{v} > \frac{L}{f_s} \tag{31.17}$$

Thus, the filter window length for each of the lines in the microgrid is set as per the shortest cable connected to the bus.

$$\frac{2(l_1 - d)}{v} > \frac{L}{f_s} \tag{31.18}$$

The scaling factor and the number of filter coefficients are set as per (31.13) and (31.18). The filter window is set assuming the worst condition for waveform superimposition, i.e. faults very close to the relay location ($d = 0$). It may be further noticed that the first reflection may come from the previous bus (bus L). In such a condition,

$$\frac{2l_2}{v} > \frac{L}{f_s} \tag{31.19}$$

Thus, the filter window length for each of the lines in the microgrid is set as per the shortest cable connected to the bus.

31.4.3 Central Unit (CU)

The CU receives the traveling wave signal, F.o.Os as SVs, and CB status as GOOSE messages from MUs at each of the line sections. The status of the CBs (normally open/normally closed) is used to determine the topology of the microgrid. Each section's length, filter setting, and propagation time are saved as global settings. The protection elements are defined using traveling waves and F.o.O SVs.

31.4.3.1 Estimation of the Frequency of Occurrence (F.o.O)

The proposed protection scheme uses the magnitude and polarity of reflected traveling waves to extract their frequency of occurrence. The first three dominant frequencies are analyzed to locate the fault in the microgrid. The procedure to obtain the frequency of occurrence of the traveling waves is explained in Algorithm 31.1. The magnitude of traveling waves is checked with respect to a predefined threshold ξ to eliminate high-energy events like noise. Thereafter, the peak and ToA of the traveling wave are detected with interpolation of a few samples before and after maximum amplitude. The three dominant frequencies are obtained with zero padding and applying n point fast Fourier transform (FFT), where n is decided based on the length and propagation time of the line. The four highest amplitudes in the frequency bin are saved. Given that the time-domain traveling waves have a DC component as the highest frequency, the next three frequencies ($f_1, f_2,$ and f_3) are checked for the FL criterion. The magnitudes of $f_1, f_2,$ and f_3 are communicated to the CU as time-stamped SVs, along with the traveling wave signal.

Algorithm 31.1: Frequency of Occurrence; $i_{\mathrm{TW}} \rightarrow f_{1,2,3}$

Data: $i_{\mathrm{TW}}(k), \xi$;
Result: f_1, f_2, f_3, ToA;
if $|i_{\mathrm{TW}}(k)| \geq \xi$ **then**
| temp1$(k) \leftarrow i_{\mathrm{TW}}(k)$;
else
| temp1$(k) \leftarrow$ **NULL**;
end
temp2$(k) = $ temp1$(k) - $ temp1$(k-1)$;
if temp2$(k) < 0$ **and** temp2$(k-1) > 0$ **then**
| ToA$(k) \leftarrow k$;
| $i_{\mathrm{peak}}(k) \leftarrow$ temp2(k)
else
| ToA$(k) \leftarrow$ **NULL**;
| $i_{\mathrm{peak}}(k) \leftarrow$ **NULL**;
end
temp3 $=$ fft(i_{peak}, n);
$[dc, f_1, f_2, f_3] = $ maxk(temp3, 4)

31.4.3.2 Fault Location

The time-space diagrams for an internal and an external fault in the microgrid are depicted in Fig. 31.11(a) and (b), respectively. It can be observed that the MU sees the first wavefront from the fault point after $d \cdot \tau$ microseconds into the fault, where τ is the propagation time constant of the line section. The second wavefront from the fault point is observed after $3d \cdot \tau$ microseconds after the fault inception. Therefore, a reflected traveling wave from the fault point is observed at the MU every Δt_1 μs.

$$\Delta t_1 = 3d \cdot \tau - d \cdot \tau = 2d \cdot \tau; \quad f_1 = \frac{1}{\Delta t_1} = \frac{1}{2d \cdot \tau} \tag{31.20}$$

where d is the fault distance from MU expressed in p.u., τ is the propagation time of the line section, and f_1 is the F.o.O of the reflected traveling waves from the fault point.

Further, as observed in Fig. 31.11(a), the MUs experience another periodic reflected traveling waves at constant intervals of Δt_2 μs from the other end after reflection from the fault point.

$$\Delta t_2 = 3(1-d) \cdot \tau - (1-d) \cdot \tau = 2(1-d) \cdot \tau; \quad f_2 = \frac{1}{2(1-d) \cdot \tau} \tag{31.21}$$

where f_2 is the F.o.O of the periodic reflected traveling waves from the other end of the line section from the fault point.

Finally, periodic traveling waves are also observed because of the full-length propagation in the line section.

$$\Delta t = 2\tau; \quad f_3 = \frac{1}{2\tau} \tag{31.22}$$

For a fault outside the line (Fig. 31.11(b)), the MU experiences periodic traveling waves only from the other end of the line ($\Delta t = 1/2\tau$). Therefore, the fault location (FL) criteria for each of the line sections in the microgrid can be derived from (31.20), (31.21), and (31.22) as,

$$FL = \begin{cases} 1, & \text{if } \frac{1}{f_3} = \frac{1}{f_1} + \frac{1}{f_2} \\ 0, & \text{otherwise} \end{cases}$$

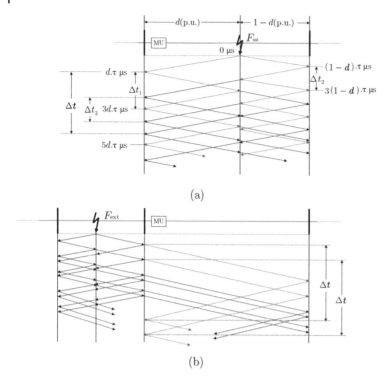

(a)

(b)

Figure 31.11 Time-space diagram and frequency of occurrence of traveling waves: (a) internal fault and (b) external fault.

The FL criteria are checked for each of the line sections in the microgrid when the amplitude of traveling waves is above the predefined threshold (ξ). A margin of $\pm 5\%$ has been incorporated in the *FL* criterion considering the error in ToA estimation, peak detection, and F.o.O estimation processes.

31.5 Performance Analysis

31.5.1 Microgrid Under Study

The efficacy of the proposed scheme has been tested on a real-world microgrid [9] modeled on an RTDS Novacor. The schematic diagram of the microgrid and the cable configurations are depicted in Fig. 31.12. The test system has been simulated in a sub-step environment distributed across three separate cores to facilitate high-frequency sampling (1.5 μs). Two different frequency-dependent underground cable models, as shown in Fig. 31.12(b) and (c), have been utilized. Faults are simulated in 18 different locations. The MUs placed at each of the cable sections, and the traveling wave filter settings for each of the line sections are given in Table 31.1. The filter coefficients and gain settings are decided based on the propagation time of the cable and connected lines.

31.5.2 Extracted Traveling Waves in MUs

To showcase the effectiveness of proposed traveling wave extraction techniques, a single line to ground fault (ag) is applied on cable section C-102 at a distance of 1000 ft from the MU (Case-1).

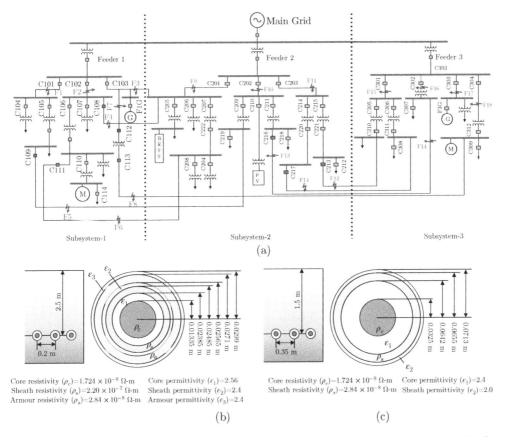

Figure 31.12 Microgrid setup: (a) faults in the test microgrid, (b) cable-1 configuration, and (c) cable-2 configuration.

Figure 31.13 depicts the current traveling waves extracted at MU. An immediate observation can showcase that the wave shape of the currents is clearly distinguishable, and no superimposition of the traveling waves is seen. Further, reflected traveling waves from the fault point at the same intervals (Δ_1) are observed. One can see that the traveling waves attenuate severely during its propagation in the cable. Reflected traveling waves from the connected cable and its remote end are also observed. From the peak of these wavefronts, ToA of the traveling waves is saved. The frequency of occurrence of these periodic traveling waves obtained from the ToA of traveling waves is utilized to determine the fault location.

31.5.3 Fault Location Identification

A fault (ABC-g) in a cable section C-201 at a distance of 2000 ft from the MU is simulated. The traveling waves extracted at the MUs in the cable sections C-201 and C-202 are shown in Fig. 31.14(a) and (b), respectively. It is to be observed that the MU at faulted cable section (C-201) sees reflected traveling waves from the fault point and the other end of the cable at regular intervals. On the contrary, the MU in the connected line (C-202) sees periodic reflected traveling waves from the other end of the cable. The frequency of occurrence of these extracted traveling waves in C-201 and C-202 is depicted in Fig. 31.15(a) and (b), respectively. The dominant frequencies in the traveling waves for the MU in C-201 are 45.1(f_3), 70.9(f_2), and 123.4(f_1) kHz. As these F.o.Os satisfy

Table 31.1 Traveling wave filter settings for MUs.

Fault	Cable section	Filter characteristics	
		Gain	Length
F1	C-101	0.333	6 µs
F2	C-102	0.333	6 µs
F3	C-103	0.333	6 µs
F4	C-108	0.333	6 µs
F5	C-109	0.667	3 µs
F6	C-111	0.667	3 µs
F7	C-112	0.667	3 µs
F8	C-113	0.667	3 µs
F9	C-201	0.333	6 µs
F10	C-202	0.333	6 µs
F11	C-203	0.333	6 µs
F12	C-213	0.333	6 µs
F13	C-216	0.333	6 µs
F14	C-217	0.667	3 µs
F15	C-301	0.667	3 µs
F16	C-302	0.667	3 µs
F17	C-303	0.667	3 µs
F18	C-304	0.667	3 µs

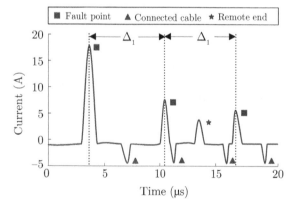

Figure 31.13 Extracted traveling waves for the fault on C-102.

the FL criteria, C-201 has been correctly identified as the faulted line. The dominant frequencies in the connected transmission line C-203 are 145.7, 221.2, and 280.7 kHz. The majority of the reflected traveling waves are from the other end of the cable. Because the *FL* criteria are not met, CU does not recognize it as a faulted line. The *FL* algorithm is only satisfied for C-201 and corresponding circuit breaker is opened. This validates the reliability and selectivity of the proposed scheme.

Figure 31.14 Extracted traveling waves at (a) C-201 and (b) C-202.

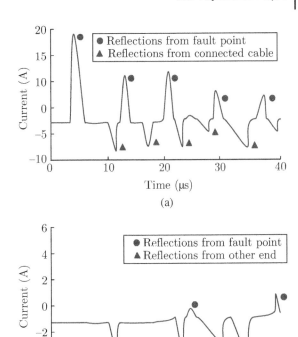

Figure 31.15 Frequency of occurrence (a) C-201 and (b) C-202.

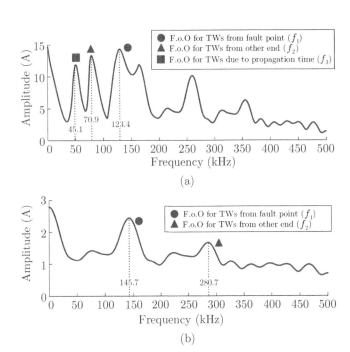

Table 31.2 Fault location identification using the proposed technique.

Fault	Cable	Length	Fault distance	F.o.O: faulted line			F.o.O: connected line		
				f_3 (kHz)	f_2 (kHz)	f_1 (kHz)	f_3 (kHz)	f_2 (kHz)	f_1 (kHz)
F1	C-101	1800 ft	1000 ft	149.1	246.9	380.64	57.77	122.21	145.32
F2	C-102	5500 ft	2000 ft	45.1	70.9	123.4	145.7	221.2	280.7
F3	C-103	1000 ft	500 ft	164.2	250.5	481.2	60.4	122.1	212.3
F4	C-108	1000 ft	400 ft	97.2	205.7	303.7	25.1	124.3	332.3
F5	C-109	2000 ft	1500 ft	101.2	150.3	252.3	21.2	32.4	99.8
F6	C-111	2000 ft	400 ft	118.8	222.1	255.6	92.5	112.1	123.5
F7	C-112	3000 ft	1200 ft	99.3	162.4	261.5	44.6	87.9	97.3
F8	C-113	2000 ft	1000 ft	137.9	200.3	443.5	121.1	132.5	253.6
F9	C-201	5500 ft	1000 ft	74.4	112.3	222.1	23.1	221.1	225.2
F10	C-202	2000 ft	1200 ft	123.8	201.2	322.1	121.1	181.1	221.2
F11	C-203	3000 ft	2000 ft	126.8	201.3	343.2	44.6	131.5	155.6
F12	C-213	1500 ft	1000 ft	157.7	256.7	409.1	56.4	121.4	178.3
F13	C-216	1500 ft	750 ft	140.3	230.2	451.5	57.5	121.3	155.5
F14	C-217	1500 ft	800 ft	151.9	252.1	382.2	181.1	192.1	212.2
F15	C-301	2500 ft	1500 ft	99.7	155.1	279.1	112.3	211.3	323.1
F16	C-302	2000 ft	1200 ft	90.2	121.1	353.3	121.1	155.1	277.2
F17	C-303	2000 ft	1800 ft	98.8	176.5	225.7	134.6	177.5	212.8
F18	C-304	1500 ft	1100 ft	169.0	278.8	430.7	21.8	78.7	156.5

31.5.4 Performance for Faults at Different Locations

To test the performance of the proposed scheme for faults at different locations in the microgrid, a total of 18 fault scenarios are simulated on different cable sections (Fig. 31.12(a)). The fault location in terms of the dominant F.o.O of traveling waves at the MUs is produced in Table 31.2. For a fault anywhere in the microgrid, traveling waves are observed all over the microgrid. The FL criteria are met only for the faulted line in each of the cases. Further, the F.o.Os in the MUs in each of the non-faulted lines (connected line given as an example) depends on the reflections from the other end of the line and do not satisfy the FL criteria. Furthermore, the single-ended fault location scheme correctly identifies the fault distance from the MU. The results from Table 31.2 testify to the reliability and selectivity of the proposed technique.

31.5.5 Performance for Different Fault Resistance

A high resistance fault may result in traveling waves with a low amplitude that go unnoticed. To test the performance of the proposed scheme for different fault resistances, a single line to ground (Ag) fault with three different resistances (0 Ω, 25 Ω, and 50 Ω) is applied on C-301 at a distance of 1500 ft from the MU. The extracted traveling waves and F.o.O of the reflected traveling waves are showcased in Fig. 31.16(a) and (b), respectively. From the results, it is observed that though the magnitude of the traveling waves reduces with an increase in the fault resistance, the dominant frequencies in the waveform remain consistent. As the FL criteria are met for each of the

Figure 31.16 (a) Extracted traveling waves and (b) F.o.O of reflected traveling waves for different faults resistances.

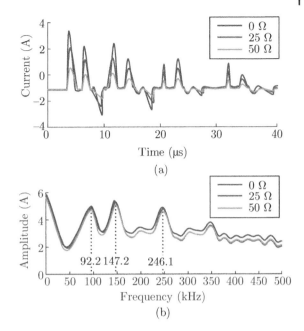

(a)

(b)

fault scenarios, C-301 is identified as the faulty cable section in each of the cases. This shows the effectiveness of the proposed scheme for faults with different resistances.

31.5.6 Performance for Different Fault Inception Angles

Faults with inception angles (with reference to phase-a) in steps of 45° are applied on C-303 at a distance of 1000 ft from the MU. The results for these fault scenarios are depicted in Table 31.3. It can be concluded from the extracted F.o.O of reflected traveling waves that the proposed FL algorithm correctly identifies the faulted cable section (C-303) while restraining the healthy line (C-304). It is worthwhile to know that faults with very small inception angles (close to 0°) do not produce traveling waves with sufficient amplitude. In such cases, the proposed scheme calls

Table 31.3 Performance for different fault inception angles.

FIA	i_{TW} (A)	F.o.O: C-303 (kHz)			F.o.O : C-304 (kHz)		
		f_3	f_2	f_1	f_3	f_2	f_1
0°	0	—	—	—	—	—	—
45°	−1.2	136.4	223.9	347.3	47.2	98.5	102.7
90°	−1.7	135.8	222.9	347.1	47.4	98.4	102.3
135°	−1.2	136.5	221.1	333.6	47.5	98.2	102.2
180°	0	—	—	—	—	—	—
225°	1.2	136.1	221.1	343.4	47.5	98.5	102.8
270°	1.7	135.4	223.5	343.4	47.1	98.5	102.4
315°	1.2	135.4	224.3	346.1	47.3	98.2	102.5
360°	0	—	—	—	—	—	—

for phasor-based supervisory protection while compromising speed. Further, the Appendix can be referred to understand that such a phenomenon is only limited to single-line-to-ground faults at zero crossing and double-line-to-ground faults at the same instantaneous voltage in both phases.

31.5.7 Large-Scale Analysis

The performance of the evaluated protection functions is further analyzed by generating a large number of fault scenarios. To do so, a wide variety of fault scenarios are generated by the system parameters, as shown in Table 31.4. A dataset comprising a total of 216 fault cases is generated by varying the fault location (6), fault type (3), fault resistance (3), and inception angle (4). Also, to analyze the effect of noise, a separate dataset is prepared by adding random white noise with a signal-to-noise ratio of 50 dB per sample.

The scatter plot for comparing the operating times of the conventional over-current element (t_{oc}) versus the proposed fault location technique (t_{FL}) is shown in Fig. 31.17. The operating time of each of the elements is calculated as,

$$t_{op} = t_{ph} + t_{res} + t_{pd} + t_{pe} \tag{31.23}$$

where t_{op} is the operating time; t_{ph} is the time required for phasor computation; t_{res} is the response time in the MU; t_{pd} is the communication propagation delay; and t_{pe} is the program execution time.

Table 31.4 Parameters used for large-scale analysis.

Parameter	Value
Fault location (6)	F1, F5, F10, F12, F15, F17
Fault type (3)	ag, bcg, abcg
Fault resistance (3)	0 Ω, 25 Ω, 50 Ω
Fault inception angle (4)	45°, 90°, 270°, 315°
Total cases	$6 \times 3 \times 3 \times 4 = 216$

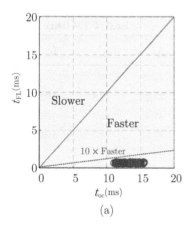
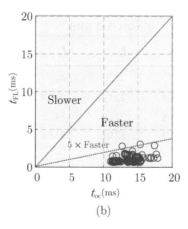

Figure 31.17 Scatter plot for operating time comparison (a) without noise and (b) with noise.

The overcurrent element (OC) uses half-cycle discrete fourier transform (DFT) to obtain the phasors, whereas the proposed fault location (FL) technique does not require phasor computation. The response time is the number of samples between fault inception and decision, multiplied by the time step (1.5 μs for both cases). A fixed propagation delay of 1.5 ms is assumed with ethernet-based communication [IEC-61850], data acquisition, and analog delays. With a 5 MB instruction set, the program execution time is assumed to be 1.25 ms for semiconductor processing speed (0.25 μs = 1/4 GHz).

It can be observed from Fig. 31.17 that the proposed technique is 10 times faster for faults without noise and 5 times faster for noisy signals. Though the proposed technique performed well for signals with noise, one ought to bear in mind that transient-based protection techniques must have supervisory protection when electromagnetic interference is very high.

31.6 Conclusion

This chapter presents a wide-area traveling wave protection scheme for microgrids. The structural difficulties associated with implementing legacy traveling wave protection schemes are overcome by developing an adaptive filtering mechanism and an accurate fault location technique based on periodic reflected traveling waves. The frequency of occurrence of the traveling waves is used to locate the faulted section. The proposed technique, unlike conventional traveling wave-based schemes, does not necessitate high bandwidth communication or complex computations. The use of a current-only technique eliminates the need for voltage-traveling waves. The performance on a real-world microgrid with various fault scenarios demonstrates precise fault location and ultra-high-speed protection capability. Any microgrid architecture can use the centralized protection architecture.

31.7 Exercises

1. A fault-induced wavefront of 20 kV travels in line section-A toward the junction J. At the junction transmission line-A is connected to lines B, C, and cable section-D. The corresponding characteristic impedances (Z_c) are provided in the figure.

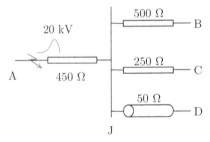

   For the system with given data, calculate the magnitude of reflected voltage and current waves in line-A.
2. For the event mentioned in Question-1, calculate the magnitude of transmitted current through lines-B, C and D.
3. For a 150 km line section, a fault is incepted at 25 km from terminal A. If the velocity of wave propagation is 2.960000×10^8 m/s, using Bewley lattice diagram, find the times of arrival of

first and second traveling waves at terminal A, after fault inception. The time of fault inception recorded is 0.321 seconds.

4. In the system shown below, the length of the line MN is 200 km.

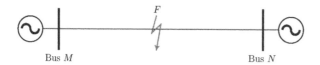

The following traveling waves are recorded at bus M and N :

	First traveling wave	
	I(A)	Time (µs)
Bus M	0.43	205
Bus N	0.29	410

The line parameters are: $L_1 = 8.853 \times 10^{-7}$ H/m; $C_1 = 1.302 \times 10^{-11}$ F/m. Find out the fault location from bus M.

5. In the system shown below, the line length LM and MN are 20 and 50 km. The traveling wave propagation time for lines LM and MN are 200 and 500 µs, respectively. The following traveling waves are recorded at bus M and N.

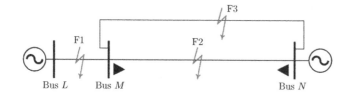

	First traveling wave		Second traveling wave	
	I(A)	Time (µs)	I(A)	Time (µs)
Bus M	0.92	100	0.82	500
Bus N	0.57	400	−0.75	600

Find out the fault location: (A) F1, (B) F2, (C) F3, (D) No-Fault

References

1 Chowdhury, R. and Fischer, N. (2021). Transmission line protection for systems with inverter-based resources – Part I: Problems. *IEEE Transactions on Power Delivery* 36 (4): 2416–2425. https://doi.org/10.1109/TPWRD.2020.3019990.

2 Johns, A.T. and Salman, S.K. (1995). *Digital Protection for Power Systems*, Number 15. IET.

3 Nayak, K., Jena, S., and Pradhan, A.K. (2021). Travelling wave based directional relaying without using voltage transients. *IEEE Transactions on Power Delivery* 36 (5): 3274–3277. https://doi.org/10.1109/TPWRD.2021.3099338.

4 Schweitzer, E., Kasztenny, B., Mynam, M.V. et al. (2016). Defining and measuring the performance of line protective relays. *43rd Annual Western Protective Relay Conference*, Spokane, WA, USA, 1–21.

5 IEEE Power System Relaying and Control Committee (2005). EMTP Reference Models for Transmission Line Relay Testing. https://pes-psrc.org/kb/report/074.pdf.

6 RTDS Technologies Inc. (2022). RSCAD FX: Real-Time Simulation Software Package. Winnipeg, MB-CANADA. https://knowledge.rtds.com/hc/en-us/articles/360046352893-RSCAD-FX-Real-Time-Simulation-Software-Package.

7 Dong, X. (2022). *The Theory of Fault Travel Waves and its Application*. Springer.

8 IEEE Std 1547-2018 (2018). *IEEE Standard for Interconnection and Interoperability of Distributed Energy Resources with Associated Electric Power Systems Interfaces (Revision of IEEE Std 1547-2003)*. https://doi.org/10.1109/IEEESTD.2018.8332112.

9 Salcedo, R., Corbett, E., Smith, C. et al. (2019). Banshee distribution network benchmark and prototyping platform for hardware-in-the-loop integration of microgrid and device controllers. *The Journal of Engineering* 2019 (8): 5365–5373.

32

Neuro-Dynamic State Estimation of Microgrids

Fei Feng, Yifan Zhou, and Peng Zhang

32.1 Background

Dynamic state estimation (DSE) is an indispensable foundation for power system operation, as it provides the most-likely dynamic states of the system to perform online monitoring and control [1, 2]. However, existing DSE methods cannot fulfill the operating requirements of today's networked microgrids (NMs) because the complete and accurate physics model of the whole NMs may not always be attainable to support precise tracking of the fast dynamics of NMs, especially the states of inverter controllers.

Physics-based DSE algorithms, represented by Kalman filter and its variants [3–5], strongly rely on accurate dynamic models of the whole system to estimate the system states [6, 7]. However, in NMs, the complete physical models are often unattainable due to unavailable parameters of distributed inverter controllers, frequently changing control modes and plug-and-play of DERs, data privacy needs, etc. Such complications lead to subsystems with unidentified dynamic models in NMs, which unavoidably make the classical, physics-based DSE algorithms impractical [8].

Recent progresses in learning dynamic model from data shed lights on developing data-driven DSE without requiring explicit physics of the entire system. References [9, 10] apply Koopman operators to establish linear approximations of nonlinear systems by using measurements and hence to track the dynamic states of systems [9, 10]. Reference [11] employs Gaussian processes (GP) to approximate the process and measurement functions of systems, which are then integrated with the unscented Kalman filter (UKF) for DSE [11]. Recently, neural-ordinary-differential-equations (ODE-Net) [12] emerges to become an efficacious paradigm for learning underlying dynamic models of power systems [13], which also ignites new hopes for data-driven DSE because it can best preserve the continuous-time dynamic characteristics. Nevertheless, two fundamental obstacles still hinder the application of existing data-driven approaches to DSE for real-world NMs: (1) data-driven dynamic models learned from limited and noisy measurements may not satisfy the accuracy needs of DSE; and (2) mismatches between data-driven models and real dynamic measurements unavoidably bias the state estimator.

This chapter first briefly introduces physics-based DSE methods. Then, Neuro-DSE algorithm of NMs is presented.

Microgrids: Theory and Practice, First Edition. Edited by Peng Zhang.
© 2024 The Institute of Electrical and Electronics Engineers, Inc. Published 2024 by John Wiley & Sons, Inc.

32.2 Preliminaries of Physics-Based DSE

Given an arbitrary dynamic system governed by its process and measurement functions, DSE targets tracking the system's dynamic states x under noisy measurements y:

$$\begin{cases} x_k = f(x_{k-1}) + w_k \\ y_k = h(x_k) + r_k \end{cases} \tag{32.1}$$

where x_k and y_k denote the state variables and measurement variables at time step k, respectively; $f(\cdot)$ and $h(\cdot)$ denote the discrete-time process and measurement functions, respectively; and w_k and r_k are process and measurement noises, respectively.

Kalman filter is a mainstream algorithm for DSE [3–5]. This section takes the extended Kalman filter (EKF) [3], i.e. a prominent Kalman filter variant, as a representative to introduce the basis of physics-based DSE. EKF consists of two kernel steps, i.e. prediction and correction:

- **Prediction**, which calculates the predicted states $x_{k|k-1}$ at the current step based on the estimated states $x_{k-1|k-1}$ at the previous step:

$$x_{k|k-1} = f(x_{k-1|k-1}) \tag{32.2}$$

- **Correction**, which corrects the predicted states based on the noisy measurements $\tilde{y}_k$ and therefore generates the estimated states $x_{k|k}$ at the current step:

$$x_{k|k} = x_{k|k-1} + K_k \cdot (\tilde{y}_k - h(x_{k|k-1})) \tag{32.3}$$

here, K_k denotes the Kalman gain, which is derived from $f(\cdot)$ and $h(\cdot)$ as follows:

$$\begin{aligned} K_k = {} & (J_k^f \Sigma_{k-1} (J_k^f)^T + W) \cdot (J_k^h)^T \cdot \\ & (J_k^h \cdot (J_k^f \Sigma_{k-1} (J_k^f)^T + W) \cdot (J_k^h)^T + R)^{-1} \end{aligned} \tag{32.4}$$

where J_k^f and J_k^h, respectively, denote the Jacobian matrices of $f(\cdot)$ and $h(\cdot)$ relative to x at kth step; W and R, respectively, denote the noise and measurement covariance matrix; and Σ_k denotes the covariance matrices iteratively calculated by $\Sigma_k = (J_k^f \Sigma_{k-1} (J_k^f)^T + W)(I - (J_k^h)^T \cdot K_k^T)$.

Obviously, conventional Kalman filter (and its variants) relies on physics models to estimate the system states, which is not applicable for NMs DSE with unknown subsystem models. Thus, in the following Section 32.3, we introduce Neuro-DSE, which incorporates a learning-based ODE-Net dynamic model into Kalman filter to enable data-driven DSE.

32.3 Neuro-DSE Algorithm

Figure 32.1 demonstrates the outline of the Neuro-DSE algorithm. The core idea of Neuro-DSE is to establish a data-driven dynamic model of the unidentified subsystems so that the dynamic states of the rest of the NMs can still be estimated via Kalman filter.

As illustrated on the top of Fig. 32.1, without loss of generality, we partition the NMs system into an external subsystem (ExSys), whose dynamic model is unidentified, and an internal subsystem (InSys), whose dynamic model is well defined by its physics natures.[1] Our target is to construct a data-driven dynamic model of ExSys, thereby performing a neural-network-incorporated DSE of InSys.

---

1 Please note that by "well-defined" Insys, we mean the physics model of InSys is fully attainable. However, measurements of InSys can be partial (i.e. not fully observable).

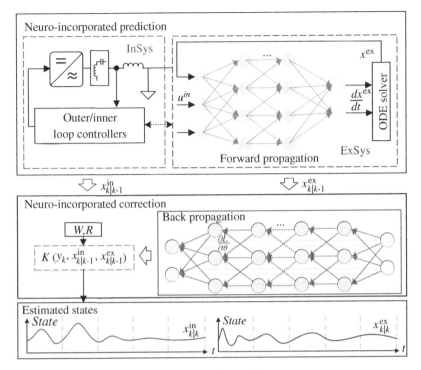

Figure 32.1 Architecture of the Neuro-DSE algorithm.

In the following, we successively establish the ODE-Net-enabled, data-based formulation for ExSys, the physics-based formulation for InSys, and finally the Neuro-DSE algorithm based on the physics-neural-integrated NMs formulation.

32.3.1 ODE-Net-Enabled ExSys Modeling

ODE-Net is capable of learning continuous-time dynamic models from discrete-time measurements, which exhibits superior noise-resilience over conventional deep neural networks (DNNs) [12, 13].

In this section, ODE-Net is employed to formulate Exsys based on available measurements:

$$\frac{d\boldsymbol{x}^{\text{ex}}}{dt} = \boldsymbol{F}(\boldsymbol{x}^{\text{ex}}, \boldsymbol{u}^{\text{in}}) \tag{32.5}$$

In (32.5), $\boldsymbol{x}^{\text{ex}}$ denotes the states measured from Exsys, which can consist of both physics quantities and control signals.[2] $\boldsymbol{u}^{\text{in}}$ represents the measurable state variables of Insys, which reflects the interactions between InSys and ExSys. Function $\boldsymbol{F}$ denotes the state-space model of ExSys governed by the forward propagation of ODE-Net:

$$\boldsymbol{F} = \boldsymbol{f}_l(\,\boldsymbol{f}_{l-1}(\dots\boldsymbol{f}_1(\boldsymbol{x}^{\text{ex}}, \boldsymbol{u}^{\text{in}}, \boldsymbol{\theta}_1)\dots, \boldsymbol{\theta}_{l-1}), \boldsymbol{\theta}_l) \tag{32.6}$$

where $\boldsymbol{f}_l(\cdot)$ and $\boldsymbol{\theta}_l$, respectively, denote the function and the trainable parameters of the lth layer of ODE-Net.

---

2 Specifically, in this work, we assume that, for ExSys, only the boundary dynamic behaviors (i.e. current injections to InSys) and control signals sent to InSys (i.e. secondary control signals of grid-forming inverters in ExSys) can be measured, which represents very limited measurements. However, the method is adaptive to arbitrary measurements from ExSys.

To make ODE-Net best match ExSys' dynamics, the loss function is set as the error between the numerical integration results of (32.5) and the time-series measurements of ExSys:

$$\min_{\theta \in \mathbb{R}} \ L(\theta) = \sum_{k=1}^{n} (x_k^{ex} - \tilde{x}_k^{ex})^2 + \gamma \cdot \theta^2 \tag{32.7a}$$

$$\text{s.t.} \ \boldsymbol{x}_k^{ex} = \tilde{\boldsymbol{x}}_1^{ex} + \int_{t_1}^{t_k} \boldsymbol{F}(\boldsymbol{x}^{ex}, \tilde{\boldsymbol{u}}^{in}, \theta) dt \tag{32.7b}$$

where $\theta = \{\theta_1\} \cup \cdots \cup \{\theta_L\}$ denotes all the trainable parameters of ODE-Net; γ denotes the regularization coefficient; n is the total number of time slides; $\tilde{x}_k^{ex}$ denotes the measurements of ExSys (e.g. current injections to InSys, global control signals sent to InSys) at time point k.

Because (32.7) incorporates numerical integration as constraints, gradient descent based on a continuous backpropagation technique is applied to train the ODE-Net until the loss function converges:

$$\theta - \eta \frac{\partial L}{\partial \theta} \to \theta, \quad \frac{\partial L}{\partial \theta}\bigg|_{t_1} = \int_{t_n}^{t_1} \lambda^T \frac{\partial \boldsymbol{F}}{\partial \theta} \tag{32.8}$$

where η denotes the learning rate; λ denotes the Lagrangian multiplier corresponding to constraints (32.7b).

32.3.2 Physics-Based InSys Modeling

InSys is formulated by its physics natures. To emphasize the participation of inverter-interfaced resources in NMs, three representative grid-forming control strategies are considered in this paper:

- **Droop control**:

$$\begin{cases} \omega = \omega^* - \boldsymbol{m}_p(\boldsymbol{P} - \boldsymbol{P}^*) \\ \boldsymbol{E} = \boldsymbol{E}^* - \boldsymbol{n}_q(\boldsymbol{Q} - \boldsymbol{Q}^*) \end{cases} \tag{32.9}$$

where ω, E, P, and Q represent the angular speeds, voltage magnitudes, active and reactive power outputs of distributed energy resources(DERs), respectively; ω^*, E^*, P^*, and Q^* denote the corresponding nominal values; and $\boldsymbol{m}_p$ and $\boldsymbol{n}_q$ denotes the active/reactive power droop coefficients.
- **Secondary control** based on distributed-averaging [14]:

$$\begin{cases} \dfrac{d\Omega}{dt} = -\alpha(\omega - \omega^*) - A\Omega \\ \dfrac{de}{dt} = -\beta(E - E^*) - BQ \end{cases} \tag{32.10}$$

where Ω and e, respectively, denote the secondary control signals of all DERs corresponding to frequency and voltage regulations; α, β, A, and B denote control parameters [15].
- **Virtual synchronous generator (VSG) control** [16]:

$$\frac{d\omega}{dt} = \frac{1}{2H\omega}(P_{\text{ref}} - P + \frac{1}{\boldsymbol{m}_p}(\omega^* - \omega)) \tag{32.11}$$

where H denotes the inertia constant.

Integrating the dynamic equations of all the DERs (i.e. both grid-forming and grid-following), branches, and power loads leads to the physics-based dynamic model of InSys. Functionally, InSys is formulated as:

$$\frac{d\boldsymbol{x}^{in}}{dt} = \boldsymbol{G}(\boldsymbol{x}^{in}, \boldsymbol{x}^{ex}) \tag{32.12}$$

where x^{ex} denotes ExSys states (see (32.5)); x^{in} denotes InSys states (e.g. state variables of each DER, load, and branch).

32.3.3 Neuro-DSE Algorithm

By integrating and discretizing the ODE-Net-enabled ExSys and the physics-enabled InSys, (32.13) constructs the model basis of Neuro-DSE, which is a discrete-time, physics-neural-integrated NMs model:

$$x_k^{ex} = DF(x_{k-1}^{ex}, u_{k-1}^{in}) + w_k^{ex} \tag{32.13a}$$

$$x_k^{in} = DG(x_{k-1}^{in}, x_{k-1}^{ex}) + w_k^{in} \tag{32.13b}$$

$$y_k = M(x_k^{ex}, x_k^{in}) + r_k \tag{32.13c}$$

where D denotes a discretization operator, which discretizes the neural/physics dynamics $F(\cdot)$ and $G(\cdot)$ presented in (32.5) and (32.12); $M(\cdot)$ denotes the measurement function of NMs; and y_k denotes the measurement variables; w_k^{ex} and w_k^{in} denote the Gaussian processing noises which follow Gaussian noise sequences $\mathcal{N}(0, W)$; r_k is the Gaussian measurement noise following $\mathcal{N}(0, R)$, where W and R are the corresponding covariance matrices.

Without loss of generality, we derive the Neuro-DSE algorithm based on the EKF method. Yet, the algorithm is readily compatible to arbitrary Kalman-type filters. Neuro-DSE is also composed of a predictor and a corrector (see Fig. 32.1). However, because of the incorporation of ODE-Net-based modeling, both the prediction and the correction will be involved with neural network operations:

- **Neuro-incorporated prediction**: The prediction step predicts the InSys and ExSys states based on the estimation at the previous step. While the prediction of x^{in} is trivial, the prediction of x^{ex} involves the forward propagation of ODE-Net according to (32.13a) and (32.6):

$$\begin{aligned} x_{k|k-1}^{ex} = \; & Df_L(f_{L-1}(\dots f_1(x_{k-1|k-1}^{ex}, u_{k-1|k-1}^{in}, \theta_1) \\ & \dots, \theta_{L-1}), \theta_L) \end{aligned} \tag{32.14}$$

where k denotes the current time step.

- **Neuro-incorporated correction**: The correction step generates the estimated states by correcting the predictions:

$$\begin{bmatrix} x_{k|k}^{ex} \\ x_{k|k}^{in} \end{bmatrix} = \begin{bmatrix} x_{k|k-1}^{ex} \\ x_{k|k-1}^{in} \end{bmatrix} + K_k \cdot (\tilde{y}_k - M(x_{k|k-1}^{ex}, x_{k|k-1}^{in})) \tag{32.15}$$

where $\tilde{y}_k$ denotes the noisy measurements of NMs. Specifically, the Kalman gain K_k is given by (32.16), which requires the backward gradients of ODE-Net w.r.t. x^{ex} and u^{in}:

$$\begin{aligned} K_k = \; & (J_k \Sigma_{k-1} J_k^T + W) \cdot (J_k^M)^T \cdot \\ & (J_k^M \cdot (J_k \Sigma_{k-1} J_k^T + W) \cdot (J_k^M)^T + R)^{-1} \end{aligned} \tag{32.16}$$

where $J_k = \begin{bmatrix} \partial(DF)/\partial x^{ex} & \partial(DF)/\partial x^{in} \\ \partial(DG)/\partial x^{ex} & \partial(DG)/\partial x^{in} \end{bmatrix}$ and $J_k^M = \begin{bmatrix} \partial M/\partial x^{ex} & \partial M/\partial x^{in} \end{bmatrix}$, respectively, denote the Jacobian matrices; W and R, respectively, denote the noise and measurement covariance matrix; and Σ_k denotes the covariance matrices iteratively calculated by $\Sigma_k = (J_k \Sigma_{k-1} J_k^T + W)(I - (J_k^M)^T \cdot K_k^T)$.

Consequently, by integrating ODE-Net with the process functions and covariance evolution of Kalman filters, Neuro-DSE enables state estimation of the accessible subsystem of the NMs (i.e. InSys) even without the physics model of the inaccessible subsystems (i.e. ExSys).

32.3.4 Joint Estimation of Dynamic States and Parameters via Neuro-DSE

Besides estimating the states of power systems, estimating parameters that cannot be explicitly known is also essential. This issue is even more critical in NMs, as massive DERs with complicated inverter control are involved and any wrong knowledge of the controller parameters would affect the performance of NMs operations. Therefore, this section extends Neuro-DSE to parameter estimation of NMs.

We take a most representative parameter estimation issue, i.e. inertia estimation, as an example to illustrate the method. The core idea of inertia estimation via Neuro-DSE is to treat the inertia constant H as an additional state of the NMs system so that x^{ex}, x^{in}, and H can be jointly tracked by Kalman filter.

To achieve this target, an augmented physics-neural-integrated process function for the joint Neuro-DSE algorithm can be formulated as:

$$x_k^{ex} = DF(x_{k-1}^{ex}, u_{k-1}^{in}) + w_k^{ex} \tag{32.17a}$$

$$x_k^{in} = DG(x_{k-1}^{in}, x_{k-1}^{ex}, H_{k-1}) + w_k^{in} \tag{32.17b}$$

$$H_k = H_{k-1} + w_k^{H} \tag{32.17c}$$

where H_k is the state of inertia constant at time point k; w_k^H is the corresponding noise; other variables follow the same definition with (32.13). The measurement function is applied as (32.13c).

The aforementioned augmented process and measurement formulations are compatible to the neural-incorporated prediction/correction steps of Neuro-DSE, except that the process function of H (see (32.17c)) is correspondingly integrated into (32.15) and (32.16).

Consequently, by integrating the inertia parameter H into the devised Neuro-DSE algorithm, a joint estimation for NMs dynamics states and parameters can be accomplished.

32.4 Self-Refined Neuro-DSE

Neuro-DSE relies on measurements to learn dynamic models of unidentified subsystems and perform data-driven DSE. Its efficacy might be jeopardized when the excessively limited and noisy measurements fail to generate a qualified ODE-Net. This section enhances Neuro-DSE by devising a self-refined neuro-dynamic state estimation (Neuro-DSE$^+$). Neuro-DSE$^+$ is able to proactively augment and filter the measurements, and therefore, it provides a more efficacious data-driven DSE, especially under limited measurements.

32.4.1 Self-Refined Training of ODE-Net

ODE-Net plays an important role in Neuro-DSE, as it learns the dynamic model of unidentified subsystems. The core idea of Neuro-DSE$^+$ is to enhance the quality and quantity of the data used for ODE-Net training, therefore, to obtain a sufficiently accurate data-driven model even under noisy and limited measurements.

We reformulate the ODE-Net-enabled dynamic model of ExSys with an augmented input:

$$\frac{dx^{ex}}{dt} = F(x^{ex}, x^{in}) \tag{32.18}$$

Comparing (32.18) with (32.5), an obvious distinction is that (32.18) incorporates the complete set of the InSys states x^{in} (rather than merely the measurable states u^{in}) to enrich the input information into ODE-Net.

However, because of the limited measurement, x^{in} may not be fully accessible, meaning that it cannot be directly used for training as Section 32.3. Therefore, we establish a self-refined training procedure for ODE-Net:

$$\min_{\theta \in \mathbb{R}} L^+(\theta) = \sum_{k=1}^{n} (x_k^{\text{ex}} - x_{k|k}^{\text{ex}})^2 + \gamma \cdot \theta^2 \tag{32.19a}$$

$$\text{s.t. } x_k^{\text{ex}} = \tilde{x}_1^{\text{ex}} + \int_{t_1}^{t_k} F(x^{\text{ex}}, x_{i|i}^{\text{in}}, \theta) dt \tag{32.19b}$$

$$\begin{bmatrix} x_{k|k}^{\text{ex}} \\ x_{k|k}^{\text{in}} \end{bmatrix} = \begin{bmatrix} x_{k|k-1}^{\text{ex}} \\ x_{k|k-1}^{\text{in}} \end{bmatrix} + K_k \cdot (\tilde{y}_k - H(x_{k|k-1}^{\text{ex}}, x_{k|k-1}^{\text{in}})) \tag{32.19c}$$

where $x_{k|k}^{\text{ex}}$ denotes the estimated states of ExSys at time point k; $x_{i|i}^{\text{ex}}$ denotes the estimated states of InSys at time point i (i.e. corresponding to time t_i); and other notations are the same, as defined in (32.15).

As illustrated in Fig. 32.2, a salient feature of (32.19) is that it embeds the Kalman filter process (32.19c) into ODE-Net training, which enables: (1) constructing the loss function between the ODE-Net's predictions and the filtered ExSys states and, therefore, mitigating the impact of noisy measurements; (2) constructing the ExSys dynamics (32.19b) using the full states of InSys and therefore significantly enriching the expressibility of ODE-Net.

Again, the continuous backpropagation is applied for optimizing (32.19). Once the training process converges, the corresponding ODE-Net can be integrated into (32.13) for dynamic state estimation.

32.4.2 Procedure of Neuro-DSE$^+$ Algorithm

Algorithm 32.1 summarizes the Neuro-DSE$^+$ algorithm. Three kernel steps are incorporated:

- **Step 1**: (Pretraining of ODE-Net) Neuro-DSE$^+$ initializes an augmented ODE-Net following (32.18). Neuro-DSE pretrains an ODE-Net based on (32.5) directly using the measurement data and estimates the InSys states accordingly.

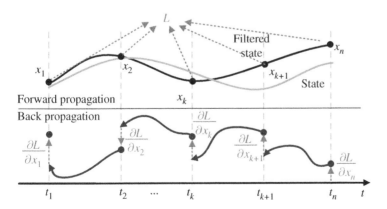

Figure 32.2 Self-refined training of ODE-Net.

Algorithm 32.1: Neuro-DSE$^+$

Initialize: $\theta, \tilde{x}_k^{\text{ex}}, \tilde{u}_k^{\text{in}}, F(\cdot), G(\cdot), M(\cdot), k$;

if *Pre-training* **then**

> Input: $\theta, \tilde{x}_k^{\text{ex}}, \tilde{u}_k^{\text{in}}, F(\cdot), G(\cdot), M(\cdot)$;
>
> Execute $\tilde{x}_1^{\text{ex}} \xrightarrow{F(\cdot),\tilde{u}_k^{\text{in}}} x_k^{\text{ex}}$ Eq. (32.5);
>
> Execute $\min_{\theta \in \mathbb{R}} \sum_{k=1}^n L(x_k^{\text{ex}}, \tilde{x}_k^{\text{ex}})$ Eqs. (32.7) and (32.8);
>
> Estimate and output $x_{k|k}^{\text{ex}}, x_{k|k}^{\text{in}}$ Eqs. (32.6)–(32.16);

else

> **repeat**
>
> > Input: $\theta, \tilde{x}_k^{\text{ex}}, x_{k|k}^{\text{ex}}, x_{k|k}^{\text{in}}, F(\cdot), G(\cdot), M(\cdot)$;
> >
> > Execute $\tilde{x}_1^{\text{ex}} \xrightarrow{F(\cdot),\tilde{x}_{k|k}^{\text{in}}} x_k^{\text{ex}}$ Eq. (32.18);
> >
> > Execute $\min_{\theta \in \mathbb{R}} \sum_{k=1}^n L(x_k^{\text{ex}}, x_{k|k}^{\text{ex}})$ Eqs. (32.19) and (32.8);
> >
> > Output neural function $F(\cdot)$ Eq. (32.18);
> >
> > Estimate $x_{k|k}^{\text{ex}}, x_{k|k}^{\text{in}}$ Eqs. (32.6)–(32.16);
>
> **until** $x_{k|k}^{\text{in}}$ *remain unchanged*;

end

Result: $x_{k|k}^{\text{ex}}, x_{k|k}^{\text{in}}, F(\cdot)$;

- **Step 2**: (Self-refined training of ODE-Net) Neuro-DSE$^+$ performs training based on (32.19) using the estimated InSys states. Once the ODE-Net converges, go to *Step 3*.
- **Step 3**: (InSys states updating) InSys states are reestimated using the up-to-date augmented ODE-Net following the neural-incorporated prediction/correction presented in (32.6) and (32.15). If InSys states remain unchanged, the algorithm terminates, outputting the ODE-Net and the corresponding state estimation results; otherwise, go to *Step 2*.

Neuro-DSE$^+$ filters the noise-contained measurements and augments the unmeasured states to construct the training data for the ODE-Net model. Such a process is particularly beneficial for data-driven DSE under limited and very noisy observations, as it proactively employs the DSE physics of the NMs to refine the measurements and adjust the neural network, rather than merely relying on the observable data.

32.5 Numerical Tests of Neuro-DSE

32.5.1 Test System and Algorithm Settings

The test system is a 4-microgrid NMs (see Fig. 32.3). Five grid-forming inverters are connected to the NMs, which can adopt droop control. Each controller comprises three different parts. The first part is a power controller that adopts droop control or secondary control for power regulation effects. The second and third components of the control system encompass the voltage and current controllers, respectively. These controllers are used to effectively mitigate high-frequency disturbances and ensure adequate damping for the output filter. Detailed control diagrams and system parameters are presented in [13].

We assume microgrid 4 is the ExSys without explicit physics knowledge and will be formulated via a learning-based fashion. The corresponding ODE-Net adopts a two-layer perceptron architecture, with 40 neurons in each layer. Training data for ODE-Net is generated by time-domain simulations under 20% uncertainties of the renewable energy inputs. In this work, only branch current measurements are used for Neuro-DSE, while the internal signals of inverter controllers

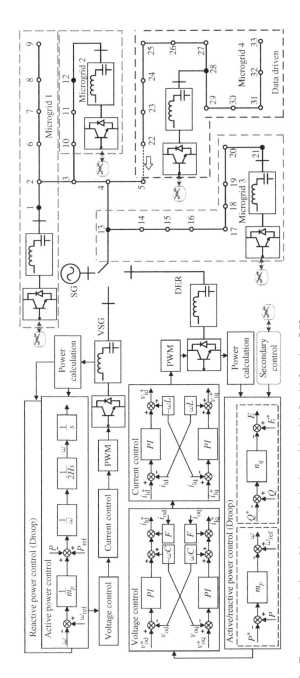

Figure 32.3 Test system: 4-microgrid networked microgrids with 5 grid-forming DERs.

are assumed inaccessible. However, the method is compatible to other types of measurements. The default case is that all the DERs adopt droop controls (including bus 13), and all the branch currents in InSys are measurable. It is assumed that both process and measurement noises of the NMs follow Gaussian distributions $N(\mu, \sigma^2)$, and the default measurement noise is $N(0, e^{-6})$.

32.5.2 Validity of Neuro-DSE

This section validates the effectiveness of the devised Neuro-DSE under various circumstances.

32.5.2.1 Neuro-DSE Under Different Noise Levels
We first study the performance of Neuro-DSE under different noise levels. Two scenarios are considered: (a) measurement noise as $N(0, e^{-6})$ and process noise as $N(0, e^{-6})$; (b) measurement noise increased to $N(0, e^{-4})$ and process noise as $N(0, e^{-6})$.

Figure 32.4 presents the simulation results. The following insights can be obtained:

- Neuro-DSE can track both measurable states (e.g. currents in Fig. 32.4(a-1) and (b-1)) and unmeasurable states (e.g. internal control signals of inverters in Fig. 32.4(a-2) and (b-2)) of NMs.
- Under both large noises (Fig. 32.4(b)) and small noises (Fig. 32.4(a)), the dynamic states estimated from Neuro-DSE are close to the true values.
- Neuro-DSE exhibits powerful compatibility to different Kalman filters. The estimation results from both EKF-based and UKF-based Neuro-DSE are close to the true states under different noise levels as shown in Fig. 32.4.

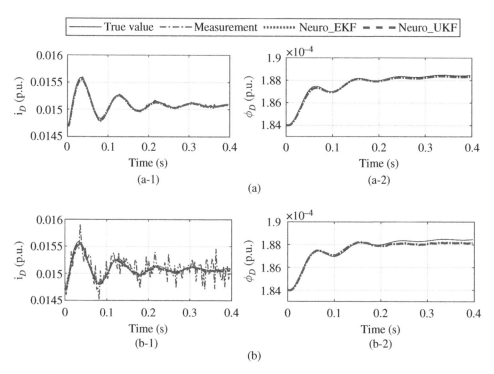

Figure 32.4 Selected states of Neuro-DSE under different noise levels. (a) State trajectories of DER1 at D-axis under $N(0, e^{-6})$ and (b) state trajectories of DER1 at D-axis under $N(0, e^{-4})$ (1: current; 2: voltage control signal).

32.5.2.2 Comparison with Conventional DNN-Based DSE

We then compare Neuro-DSE with conventional DNN-based DSE to illustrate the superiority of the devised method. Two representative DNNs are studied: (1) a residual neural network (ResNet) comprised of 8 hidden layers with double-layer skips and 100 hidden units in each layer; (2) a long short-term memory (LSTM) network with 100 hidden units.

Figures 32.5 and 32.6 clearly illustrate that the devised ODE-Net-based Neuro-DSE outperforms the conventional DNN-based DSE methods. As shown in the figure, ResNet-based DSE shows large differences at the starting stage; LSTM-based DSE tends to have slight biases for the steady-state; and only Neuro-DSE provides accurate estimation during the whole time period.

32.5.2.3 Neuro-DSE Under Different NMs Compositions

Finally, we present Neuro-DSE's powerful universality under different NMs compositions. Besides the droop/secondary-controlled DERs, two additional power sources are studied, i.e. VSG, and synchronous generator (SG). Figure 32.7 presents the simulation results. It can be observed that Neuro-DSE maintains high accuracy for both traditional synchronous generators and inverter-interfaced VSGs, and again exhibits satisfactory estimation performance for tracking the internal controller signals of VSG (see Figure 32.7(b-3) and (b-4)).

32.5.3 Efficacy of Neuro-DSE$^+$

As we introduced, excessively noisy and limited measurements may cause biases of Neuro-DSE, which motivates us to devise a self-refined Neuro-DSE$^+$ to improve the estimation performance.

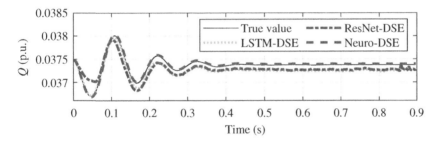

Figure 32.5 Reactive power trajectories of DER1 under different neural networks.

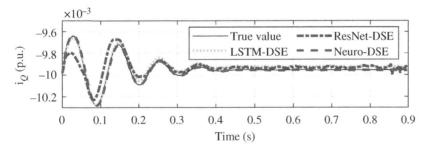

Figure 32.6 Current trajectories of DER1 under different neural networks.

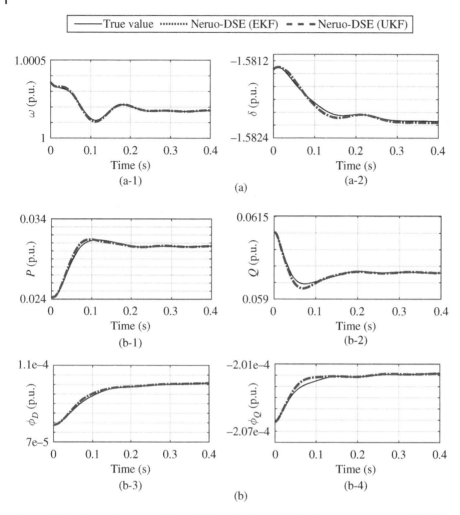

Figure 32.7 Selected states of SG and VSG under Neuro-DSE. (a) State trajectories of SG (1: speed; 2: rotor angle) and (b) state trajectories of VSG (1/2: active/reactive power; 3/4: voltage controller signals at *D/Q* axis).

32.5.3.1 Neuro-DSE$^+$ Under Different Noise Levels

First, we demonstrate the efficacy of Neuro-DSE$^+$ under different noise levels. Figures 32.8 and 32.9 compare the performance of Neuro-DSE and Neuro-DSE$^+$ under two measurement noises $N(0,e^{-6})$ and $N(0,e^{-4})$ to illustrate the superiority of Neuro-DSE$^+$. It can be observed that:

- As shown in Fig. 32.8, Neuro-DSE$^+$ obtains more accurate state estimation results than Neuro-DSE under noisy measurement, evidenced by the fact that the estimated states of Neuro-DSE$^+$ under different noise levels are always closer to the real states than that of Neuro-DSE. For example, in Fig. 32.8(b), when the noise has a distribution of $N(0,e^{-4})$, the estimated **P** from Neuro-DSE has larger deviations to the true value, whereas the result from Neuro-DSE$^+$ remains accurate. This verifies the powerful tracking ability of Neuro-DSE$^+$ under high noise level.

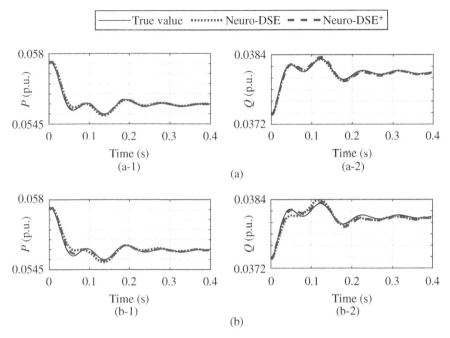

Figure 32.8 Power states of DER1 under different noise levels. (a) State trajectories of DER1 under $N(0, e^{-6})$ and (b) state trajectories of DER1 under $N(0, e^{-4})$ (1/2: active/reactive power).

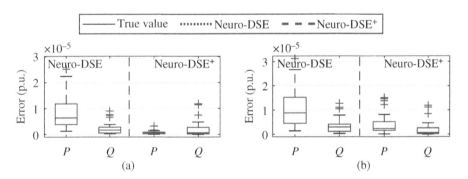

Figure 32.9 Differences of power states of DER1 under different noise levels. (a) Under $N(0, e^{-6})$ and (b) under $N(0, e^{-4})$.

- Figure 32.9 further quantitatively studies the performance of Neuro-DSE$^+$. Box-plots of the estimation error under 40 random noisy scenarios are provided. It is obvious that Neuro-DSE$^+$ outperforms Neuro-DSE in terms of robustness against noises, as the interquartile range of the estimation error of Neuro-DSE$^+$ is significantly smaller than that of Neuro-DSE.

32.5.3.2 Neuro-DSE$^+$ Under Different Measurement Availability

Second, we demonstrate the performance of Neuro-DSE$^+$ under different availability levels of the measurement data. Figure 32.10 presents the DSE results when 100%, 80%, and 70% of branch currents are, respectively, measured. Numerical experiments show that Neuro-DSE$^+$ is capable of providing more accurate estimation results than Neuro-DSE especially under limited measurements. For example, under 70% measurements, the estimated reactive power state Q of Neuro-DSE

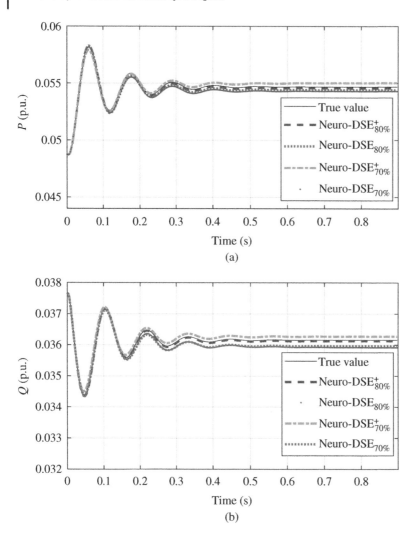

Figure 32.10 State trajectories of DER2 under different measurement levels.

has obvious deviations after 0.25 seconds, while Neuro-DSE+ consistently tracks the true values of the NMs. This is because Neuro-DSE+ automatically supplements the unmeasurable states and uses them as inputs to the ODE-Net model, which significantly enrich the training information.

32.6 Exercises

1. What measurements can we obtain from Supervisory Control and Data Acquisition (SCADA) systems and phasor measurement unit (PMU) systems?
   Please describe the differences between SCADA and PMU.
2. Please describe the Kalman Filter and Extended Kalman Filter.
3. Please describe the differences and the pros/against between the Extended Kalman Filter and the Unscented Kalman Filter.
4. Why is it important to establish neural dynamic state estimation?

5. Give examples of learning-based dynamic state estimation.
6. Describe how neural networks can be integrated into dynamic state estimation. What advantages do they offer over traditional approaches?

References

1 Valverde, G. and Terzija, V. (2011). Unscented Kalman filter for power system dynamic state estimation. *IET Generation, Transmission & Distribution* 5 (1): 29–37.

2 Feng, F., Zhang, P., and Zhou, Y. (2022). Authentic microgrid state estimation. *IEEE Transactions on Power Systems* 37 (2): 1657–1660.

3 Ghahremani, E. and Kamwa, I. (2011). Dynamic state estimation in power system by applying the extended Kalman filter with unknown inputs to phasor measurements. *IEEE Transactions on Power Systems* 26 (4): 2556–2566.

4 Karimipour, H. and Dinavahi, V. (2015). Extended Kalman filter-based parallel dynamic state estimation. *IEEE Transactions on Smart Grid* 6 (3): 1539–1549.

5 Huang, M., Li, W., and Yan, W. (2010). Estimating parameters of synchronous generators using square-root unscented Kalman filter. *Electric Power Systems Research* 80 (9): 1137–1144.

6 Carquex, C., Rosenberg, C., and Bhattacharya, K. (2018). State estimation in power distribution systems based on ensemble Kalman filtering. *IEEE Transactions on Power Systems* 33 (6): 6600–6610.

7 Zhou, N., Meng, D., Huang, Z., and Welch, G. (2015). Dynamic state estimation of a synchronous machine using PMU data: a comparative study. *IEEE Transactions on Smart Grid* 6 (1): 450–460.

8 Revach, G., Shlezinger, N., Ni, X. et al. (2021). KalmanNet: neural network aided Kalman filtering for partially known dynamics. CoRR, abs/2107.10043.

9 Netto, M. and Mili, L. (2018). A robust data-driven Koopman Kalman filter for power systems dynamic state estimation. *IEEE Transactions on Power Systems* 33 (6): 7228–7237.

10 Netto, M., Krishnan, V., Mili, L. et al. (2019). A hybrid framework combining model-based and data-driven methods for hierarchical decentralized robust dynamic state estimation. In: *2019 IEEE Power & Energy Society General Meeting (PESGM)*, 1–5. IEEE.

11 Kumari, D. and Bhattacharyya, S.P. (2017). A data-driven approach to power system dynamic state estimation. *2017 19th International Conference on Intelligent System Application to Power Systems (ISAP)*, 1–6.

12 Chen, R.T.Q., Rubanova, Y., Bettencourt, J., and Duvenaud, D.K. (2018). Neural ordinary differential equations. In: *Advances in Neural Information Processing Systems 31 (NeurIPS 2018)*.

13 Zhou, Y. and Zhang, P. (2022). Neuro-reachability of networked microgrids. *IEEE Transactions on Power Systems* 37 (1): 142–152.

14 Simpson-Porco, J.W., Shafiee, Q., Dörfler, F. et al. (2015). Secondary frequency and voltage control of islanded microgrids via distributed averaging. *IEEE Transactions on Industrial Electronics* 62 (11): 7025–7038.

15 Zhou, Y., Zhang, P., and Yue, M. (2020). Reachable dynamics of networked microgrids with large disturbances. *IEEE Transactions on Power Systems* 36 (3): 2416–2427.

16 Liu, J., Miura, Y., and Ise, T. (2015). Comparison of dynamic characteristics between virtual synchronous generator and droop control in inverter-based distributed generators. *IEEE Transactions on Power Electronics* 31 (5): 3600–3611.

33

Hydrogen-Supported Microgrid toward Low-Carbon Energy Transition

Jianxiao Wang, Guannan He, and Jie Song

33.1 Introduction

Faced with the global problem of carbon emissions and greenhouse effects, many countries have put forward their plans and goals of energy conservation and emission reduction in recent years. The goal of carbon neutrality has been formally put on the agenda of most countries [1]. The combustion of fossil energy is a major source of carbon emissions. To achieve the goal of carbon neutrality, there is an urgent need to reduce dependence on fossil energy, find green substitute energy, and establish a green, safe, and stable energy system.

The proposal of the microgrid provides a promising solution to clean energy transition [2]. Microgrid refers to a small active distribution system composed of distributed energy resources (DERs), energy storage devices, energy conversion devices, loads, monitoring, and protection devices [3]. The concept of microgrid aims to realize the flexible and efficient application of DERs, and to solve the problem of grid integration. There are a large number of DERs, including wind power and photovoltaics in the microgrid. The randomness of the above-mentioned DERs brings challenges to the operation of power grids. It still deserves an in-depth research regarding how to improve the microgrid's ability to accommodate renewable energy while maintaining a reliable operation [4].

With properties of high calorific value and convenient conversion, hydrogen is becoming a new type of secondary energy source attracting global attention [5]. According to the Hydrogen Council, the world is predicted to use hydrogen energy on a large scale starting in 2030 [6]. By 2050, hydrogen energy will account for 18% of the world's end-use energy consumption and contribute 20% to the global carbon dioxide emission reduction. Major developed countries across the world have attached great importance to the development of hydrogen energy [7]. In 2017, Japan issued its basic strategy of hydrogen energy and planned to open the international hydrogen supply chain by 2030 toward a medium- and long-term "hydrogen energy society" by 2050. The Department of Energy (DoE) of the United States issued the "Hydrogen Program Plan" in 2020, to vigorously promote the development of the hydrogen industry. In 2020, the European Commission issued the policy "climate-neutral European Hydrogen strategy," announcing the establishment of the EU hydrogen industry alliance. By converting excess wind power and photovoltaics in the microgrid into hydrogen energy, renewable on-site consumption can be promoted and the pressure on the bulk power grid can be alleviated. Therefore, hydrogen provides a new way for low-carbon energy transition in microgrids.

Microgrids: Theory and Practice, First Edition. Edited by Peng Zhang.
© 2024 The Institute of Electrical and Electronics Engineers, Inc. Published 2024 by John Wiley & Sons, Inc.

33.2 Hydrogen Production in Microgrid Operation

Increasing penetration of uncertain and variable renewable generation in evolving power system leads to fast unexpected changes in net load. California has declared that there will be an increasing demand on ramp-down capacity during sunrise and ramp-up capacity during sunset up to 2030, resulting in "duck curve" on the system net load. Two consequences caused by high penetrations of renewable resources are: (1) increasing ramping requirements posed to online generators, and (2) possibility of ramp capability shortages.

To deal with the inadequacy of the system's ramping capacities, California Independent System Operator (CAISO) and midcontinent independent system operator (MISO) have proposed flexible ramping product (FRP) to improve the dispatch flexibility. FRPs consist of separate products in the upward and downward directions as energy imbalances may be positive or negative. Any real-time schedulable resource has the ability to provide FRPs, while only available from thermal power units currently. Some literature has analyzed the ability of resources such as wind turbines, energy storage and electric vehicles to provide FRPs [8].

In recent years, power-to-hydrogen (P2H) technologies have gained substantial popularity in the role of virtual storage to transform surplus renewable energy to hydrogen that can be utilized in the future [9]. With the rapid development of P2H, such emerging technology has been advocated as a promising solution to enhance the flexibility of renewable energy-dominated microgrid. In practice, P2H can convert the hard-to-distribute wind power into hydrogen during valley hours, and thus store the surplus renewable energy as hydrogen to support fuel cells during peak hours. Therefore, the coordination between P2H and fuel cell will play a critical role in contributing to system operational flexibility in future microgrids [10].

33.2.1 Framework

The framework is designed in Fig. 33.1. We quantify the merits of P2H toward system operational flexibility in a microgrid optimization model, which minimizes the total operation costs with FRP requirements embedded.

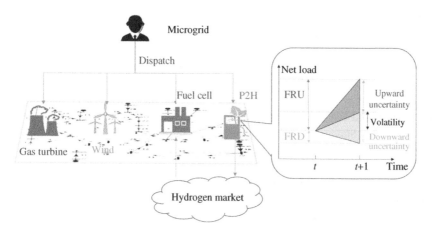

Figure 33.1 Framework of microgrid considering P2H as FRP providers.

33.2.2 Modeling for Power-to-Hydrogen

According to the structure of the electrolysis water and the magnitude of the current and voltage, the power consumption (P^{cons}) and hydrogen production flow (F_{H_2}) are expressed as:

$$P^{cons} = 0.001 \cdot \sum_{i=1}^{N_{stack}} u_{t,i}^{stack} V_{stack} I_c \tag{33.1}$$

$$F_{H_2} - 3600 \cdot \frac{RT_0}{P_0} \cdot \sum_{i=1}^{N_{stack}} u_{t,i}^{stack} \eta_f N_c \frac{I_c}{2\Gamma} \tag{33.2}$$

where $u_{t,i}^{stack}$ is the 0–1 variable characterizing whether the ith electrolyzer stack is operating. $I_c = A_{cell} \cdot i_d$, where A_{cell} is the area of electrode. N_{stack} is the number of electrolyzer stacks. η_f is Faraday efficiency. T_0 and P_0 are standard temperature and pressure, respectively.

Through the above analysis, the dynamic efficiency expression of water electrolysis can be obtained.

$$\eta_{H_2,t} = \frac{HHV_{H_2} \cdot F_{H_2}}{P_t^e} \tag{33.3}$$

where HHV_{H_2} is the high heat value of hydrogen under standard conditions.

In fact, as the power consumption increases, the operating efficiency of the P2G is continuously reduced. In order to improve calculation speed and decrease the difficulty of solving the mixed integer nonlinear model, this subsection proposes a piecewise linearization of operational efficiency.

For the univariate function $\eta(P)$ on the interval $[\overline{p}, p]$, we introduce a number of segmentation points $\alpha_i (i \in 0, 1, \ldots, m)$, making $p = \alpha_0 \le \alpha_1, \ldots \le \alpha_m = \overline{p}$.

The value of the variable p is only located in one of the segmentation intervals, and the corresponding function value is the P2G operation efficiency at the power consumption level. An illustration is shown in Fig. 33.2.

33.2.3 System Model

FRP aims to reserve enough ramping space for the system to simultaneously address the variability and uncertainty of net load. The FRP requirements are composed of two parts: the forecasted net load changes between the current period and the next period and the uncertainty of forecasted

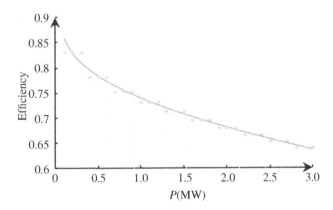

Figure 33.2 Nonlinear efficiency function and piecewise linearization.

net load at next period. FRP includes both upward and downward types. It is called upward ramping products (FRU) and downward ramping products (FRD). The formulas for calculating FRP requirements are:

$$\text{URR}_t^{\text{sys}} = \max\left\{R_{V,t}^{\text{FRU}} + R_{U,t}^{\text{FRU}}, 0\right\} \tag{33.4}$$

$$\text{DRR}_t^{\text{sys}} = \max\left\{R_{V,t}^{\text{FRD}} + R_{U,t}^{\text{FRD}}, 0\right\} \tag{33.5}$$

where $\text{URR}_t^{\text{sys}}$ and $\text{DRR}_t^{\text{sys}}$ are the total FRU and FRD requirements for time t, respectively. $R_{V,t}^{\text{FRU}}$ and $R_{V,t}^{\text{FRD}}$ are the FRU and FRD requirements caused by the system net load variability, respectively. $R_{U,t}^{\text{FRU}}$ and $R_{U,t}^{\text{FRD}}$ are FRU and FRD requirements for the uncertainty of the net load forecast, respectively. The FRP requirement for multiple time periods is shown in Fig. 33.3.

The broken line is the net load forecasting curve, and the difference between the two periods is called the net load volatility. The two double-headed arrows on the far right are the uncertain demand caused by the deviation of the system's net load forecast within a certain confidence interval. The arrows with two colors (light and dark gray) are the FRU (FRD) requirement of the current time period.

The objective of the microgrid model is to minimize operation costs while anticipating the real-time generation and FRP costs.

$$\min \sum_{i=1}^{N_T}\sum_{i=1}^{N_U}\left(C_i^{\text{SU}}y_{i,t} + C_i^{\text{SD}}z_{i,t}\right) + \sum_{s=1}^{N_S}\gamma_s\left[\sum_{i=1}^{N_T}\sum_{i=1}^{N_U}(C_i^G(P_{i,s,t}^G)) + \sum_{i=1}^{N_T}(C_{s,t}^{\text{RU}}S_{s,t}^{\text{UP}} + C_{s,t}^{\text{RD}}S_{s,t}^{\text{DN}})\right] \tag{33.6}$$

where N_T is number of time periods. i is index of units, including the thermal power unit (m is index of thermal power unit) and the gas turbine (g is index of gas turbines). N_U is number of units. N_{TU} is number of thermal power units, N_{GT} is number of gas turbines, $N_U = N_{\text{TU}} + N_{\text{GT}}$. s is the index of scenarios. N_S is number of scenarios. C_i^{SU} and C_i^{SD} are the up and down costs respectively. $y_{i,t}$ and $z_{i,t}$ are startup and shutdown state variable of a unit. r_s is the probability of the scenario s. C_i^G is the power generation cost of unit i. $P_{i,s,t}^G$ is the power generation of unit i at time t under scenario s. $C_{s,t}^R U$ is flexible ramp up surplus price. $C_{s,t}^R D$ is flexible ramp down surplus price. The FRP surplus price is obtained by estimating the probability of power imbalance under a certain amount of FRP by historical data and multiplying it by the value of the load loss. $S_{s,t}^{\text{UP}}$ and $S_{s,t}^{\text{DN}}$ are flexible ramp up and down surplus at time t.

The constraints of the proposed model are listed below:

(1) FRP requirements; (2) available capacity constraints; (3) ramping capability constraints; (4) system balance constraints; (5) unit commitment constraints; (6) power flow constraints; (7) wind

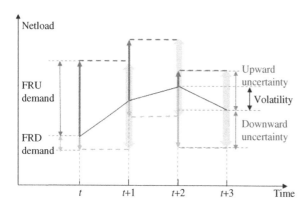

Figure 33.3 FRP requirement for multiple time periods.

power output constraints; (8) P2G operating constraints; (9) gas turbine operating constraints; and (10) hydrogen market contract

33.3 Hydrogen Utilization in Microgrid Operation

Power and transportation industries are the main sources of global greenhouse gas emissions, of which electric power and transportation sectors generally account for 40% and 24%, respectively. In addition, traffic emissions are important causes of air pollution. Due to the emission of particulate matter and NO_x, more than one million people might die prematurely in Europe each year. The direct costs of diseases, production losses, crop yield reductions, and building damage due to air pollution are about 24 billion euros per year, while external costs are estimated to be 330–940 billion euros per year. To cope with the severe challenges caused by resource shortage, climate change and environmental pollution, it is a fundamental way to develop the coordinated revolution of energy and transportation system.

International energy agency (IEA) has pointed out that battery electric vehicles (BEVs) and fuel cell vehicles (FCVs) are the technologies that can provide a sustainable road transportation system with near-zero emissions. Due to lower fuel costs and ready-to-use infrastructure, electric vehicle (EV) has become a key development target for countries all over the world prior to FCV. In 2019, 2.3 million EVs were sold worldwide, which was 9% more than those in 2018. Compared with fossil-fuel vehicles, EVs not only help maintain a clean environment but also reduce the operating costs of transportation. However, the integration of large-scale EVs will significantly enlarge the peak-valley difference of power grids, thus yielding a great burden of power system operation. By 2030, State Grid Corporation of China (SGCC) has expected a 153-TW peak load caused by EV fleet charging.

The concept of energy systems integration has provided a novel pathway to manage the penetration of large-scale EVs. As a promising secondary energy, hydrogen has the advantages of zero emissions and high specific energy relative to most batteries and is becoming an important link between transportation and energy (trans-energy) systems in recent years. Green hydrogen can be generated from renewable energy via power-to-hydrogen (P2H) and be transformed into other chemical products or stored for future use. The relatively low cost of green hydrogen enables FCVs to be a potential development direction.

In this section, an optimal scheduling framework is proposed for trans-energy systems considering the participation of fuel cell hybrid electric vehicles (FCHEVs). During peak load periods, FCHEVs can choose the hydrogen refueling mode instead of the electricity charging mode to reduce the operating burden of the grid. When wind power is rich during night hours, FCHEVs can be charged by electricity to improve the accommodation of wind power, then surplus wind power can be converted into hydrogen for future use. By this means, FCHEVs connect trans-energy systems, thereby promoting the consumption of renewable energy and improving the flexibility of power grids.

33.3.1 Framework

In this subsection, a trans-energy system framework considering FCHEVs' participation is designed to aggregate hydrogen production and transportation, conventional and renewable power plants. The schematic is shown in Fig. 33.4.

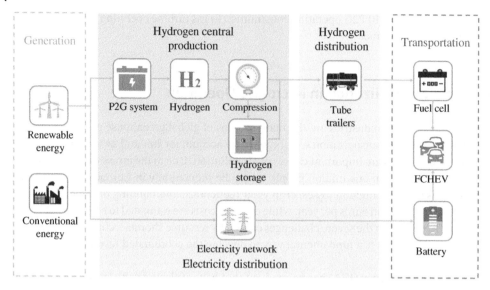

Figure 33.4 Schematic of the trans-energy system connected by FCHEVs.

On the one hand, Fig. 33.4 describes hydrogen supply chains that include hydrogen production through P2H system, compression, transportation, and end-use by FCHEVs. During periods of low load and high incidence of wind power at night, surplus wind power can be converted into storable hydrogen through P2H, thereby facilitating the accommodation of wind power and providing downward flexibility for the power system. The generated hydrogen is then transported to energy consumption terminals for FCHEVs' hydrogen refueling. In addition, large-scale hydrogen storage is one of the low-carbon solutions that can balance the intermittency of wind and solar power generation.

On the other hand, the electricity produced by conventional and renewable energy is transmitted to the charging station through the electricity network to charge the batteries of FCHEVs. FCHEVs can also discharge electricity to the grid, providing upward flexibility during peak load hours.

Here, hydrogen production is considered as proton exchange membrane (PEM) electrolyzers, which has quick response and quick start with a wider dynamic range (20%–150%). Thus, it is more suitable for intermittent power supplies such as wind power. The produced hydrogen needs to be compressed or liquefied to reach sufficient energy density. Currently, highly compressed gas is the most cost-efficient choice for on-board vehicle storage. Excessive compressed hydrogen can be stored in a hydrogen storage tank. In terms of transportation, due to lower infrastructure costs and risks, we select gas tube trailers to transport compressed hydrogen.

FCHEV is a key component that connects transportation system with power grids. By taking advantage of FCHEV's ability to be charged by both hydrogen and electricity, we can fully explore the complementary effects of hydrogen and electricity, thus achieving an in-depth integration of trans-energy systems.

33.3.2 Modeling for FCHEV

33.3.2.1 Techno-Economic Analysis

BEVs, FCVs, and FCHEVs are the mainstreams as being capable of delivering a sustainable road transportation system with near-zero emissions. They have different performances in terms of cost, mileage, and energy supplement as follows:

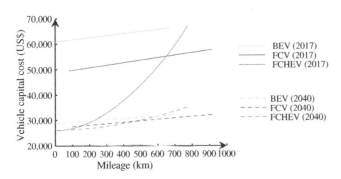

Figure 33.5 Capital cost comparison between BEV, FCV, and FCHEV.

1) **Capital cost**: At present, the investment and operating cost of an FCV are higher than those of a BEV. Figure 33.5 shows the capital cost comparison between BEV, FCV, and FCHEV in 2017 and in 2040. In 2040, the capital costs of the three vehicles tend to be the same.
2) **Mileage and charging time**: "mileage anxiety" is the main obstacle for BEVs, causing 30% of American consumers to be reluctant to pay for it. But this is not a problem for FCVs, as hydrogen has high specific energy relative to batteries, it has longer driving mileage. In terms of charging time, the FCVs can be refueled quickly within 5 minutes while BEVs take several hours.
3) **Energy supplement**: The fuel of FCVs, hydrogen, is very expensive. There are few hydrogen refueling stations currently. BEVs use electric energy as energy source, and the cost is low.

Technically, FCHEVs combine the advantages of BEVs and FCVs, and will have a better performance with longer mileage and shorter charging time. In terms of cost, due to the peak load shifting characteristics of wind power, a large amount of wind power is not fully utilized at night. In practice, wind power curtailment can be used to produce hydrogen via P2G, which can greatly reduce the cost of hydrogen. Thus, the operating cost of FCHEVs is lower.

Therefore, if hydrogen and electricity are used simultaneously as the energy supply for vehicles, it will solve the problems of high fuel cost of FCV and insufficient mileage of BEV. FCHEV can satisfy the above-mentioned needs.

33.3.2.2 FCHEV Mileage Model
FCHEVs use both hydrogen and electricity as energy sources so that they can choose to be charged by electricity or refueled by hydrogen according to dispatching signals. The configuration mainly includes fuel cell and battery as its primary sources, shown in Fig. 33.6. The fuel cell uses a PEM fuel cell, and the battery uses a lithium-ion battery.

To uniformly measure the hydrogen/electricity charging status of FCHEV, mileage is an important indicator which is modeled by the stored electricity and hydrogen mass. The mileage of an FCHEV (R_v) can be expressed as follows:

$$R_v = \frac{E_B}{\varphi_B\left(M_V + \dfrac{E_B k_{m,B}}{\mathrm{SE_{BC}}} + M_{H_2} + M_{FC}P_{FC} + M_{HT}\right)} + \frac{M_{H_2}}{\varphi_H\left(M_V + \dfrac{E_B k_{m,B}}{\mathrm{SE_{BC}}} + M_{H_2} + M_{FC}P_{FC} + M_{HT}\right)} \tag{33.7}$$

where E_B is the battery pack energy, M_{H_2} is the mass of stored hydrogen. φ_B and φ_H are energy consumption efficiency and hydrogen consumption efficiency of FCHEVs, respectively. M_V is the

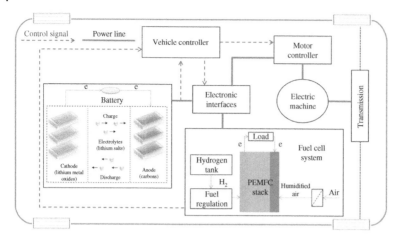

Figure 33.6 Configuration of fuel-cell hybrid electric vehicle.

vehicle mass excluding the battery pack. $k_{m,B}$ is the battery pack mass. SE_{BC} is the specific energy of the battery cell. M_{FC} and M_{HT} are the mass of fuel cell system and hydrogen tank, respectively. P_{FC} is the power of fuel cell system.

The formula (33.7) can be simplified into the following form:

$$R_v = \frac{E_B}{a_1 + b_1 \cdot M_{H_2 c_1 \cdot E_B}} + \frac{M_{H_2}}{a_2 + b_2 \cdot M_{H_2} + c_1 \cdot E_B} \tag{33.8}$$

where a_i, b_i, and c_i are constants. The mileage of FCHEV is sensitive to the battery pack energy and the mass of stored hydrogen. In the denominator, $(b_1 \cdot M_{H_2} + c_1 \cdot E_B)$ and $(b_2 \cdot M_{H_2} + c_2 \cdot E_B)$ account for a small proportion, about $4.85\% \sim 9.23\%$. Thus, the denominator can be set as a constant to simplify the computational complexity. The mileage of FCHEV is determined by the battery's state of charge (SOC) and the mass of stored hydrogen.

$$R_v = SOC^{EV}_{v,s,t} \cdot E^{EV}_{v,\max} \cdot k^E_i + m^{EV}_{v,s,t} \cdot k^H_i \tag{33.9}$$

where k^E_i is the conversion coefficient of battery capacity to mileage and $k^E_i = 1/[\varphi_B(M_v + C_1)]$; k^H_i is the conversion coefficient of hydrogen mass to mileage and $k^h_i = 1/[\varphi_H(M_v + C_2)]$; C_1 and C_2 are constant; $SOC^{EV}_{v,s,t}$ is the SOC of the vth FCHEV at time t; $E^{EV}_{v,\max}$ is the capacity of the battery pack; $m^{EV}_{v,s,t}$ is the mass of stored hydrogen in the vth FCHEV at time t.

33.3.2.3 Charging/Refueling Station Model

In this chapter, the charging/refueling station will dispatch FCHEVs' charging/discharging and refueling according to the peak or valley situations of the net load, thereby improving the system's operational flexibility. FCHEVs will leave the station when the target mileage is met, and the target mileage (R^{dep}_v) is determined by the battery's state of charge (SOC^{EV}_{v,s,T^{dep}_v}) and hydrogen mass (m^{EV}_{v,s,T^{dep}_v}).

33.3.3 Trans-Energy System Scheduling

In this section, an optimal scheduling model for trans-energy systems is proposed considering the participation of FCHEVs. Through the scheduling of FCHEV's charging, discharging, and hydrogen refueling at different times, the strategy for economically optimal operation of FCHEVs in the trans-energy systems is obtained.

33.3.3.1 Optimal Scheduling Model

(1) Objective function

The objective of the scheduling strategy is to minimize the trans-energy systems costs, including the operation cost of power system and hydrogen transportation cost. In Eq. (33.10), the first term is the cost of unit startup and shutdown, and the second term is the generation cost of thermal power units (the generation cost of renewable energy units is taken as zero). Due to the long distance between P2G and the load center, the transportation cost of hydrogen cannot be ignored, as shown in the third term.

$$\min \sum_{t=1}^{N_T} \sum_{i=1}^{N_I} \left(C_i^{SU} y_{i,t} + C_i^{SD} z_{i,t} \right) + \sum_{s=1}^{N_S} \sum_{t=1}^{N_T} \sum_{i=1}^{N_I} \gamma_s C_i^G \left(P_{i,s,t}^G \right) + \sum_{l \to q=1}^{\Omega_Q} \left[C^{HT} \cdot m_{l \to q}^{H_2} \right] d_{l \to q} \tag{33.10}$$

where N_T, N_I, and N_S are the sets of time slots, thermal power units and scenarios, respectively; Ω_Q is the set of paths in the hydrogen transportation network; $y_{i,t}$ and $z_{i,t}$ are start-up and shut-down state variable of unit i; C_i^{SU} and C_i^{SD} are the start/shut cost of the unit i; and r_s is the probability of the scenario s. C_i^G is the power generation cost of unit i; $P_{i,s,t}^G$ is the power generation of thermal power unit i at time t, respectively; C^{HT} is fuel unit cost of hydrogen transportation; $m_{l \to q}^{H_2}$ is hydrogen transportation flow of path $l \to q$; and $d_{l\beta q}$ is the distance between path l and q.

(2) FCHEV charging/refueling station constraints; (3) electricity-hydrogen energy coupling constraints; (4) hydrogen supply chain constraints; and (5) power system constraints.

33.3.3.2 Benefits of FCHEV

The potential benefits of FCHEV are analyzed in Fig. 33.7.

Figure 33.7(a) illustrates the operation situation of BEVs connected to the power system. Compared with BEVs that do not interact with the power system, BEVs' participating in power system dispatching can achieve peak shaving and valley filling. Unfortunately the wind power may not be fully accommodated at night.

In Fig. 33.7(b), an illustration for the scenario of using FCHEVs is provided. When the wind power output is large during valley load periods, the surplus wind power can be converted into storable hydrogen energy via P2G, which facilitates the accommodation of renewable energy. The produced hydrogen can be then refueled by FCHEVs during peak electric load, thereby reducing FCHEVs' demand for charging electricity. Compared with scenario (a), FCHEVs in scenario (b)

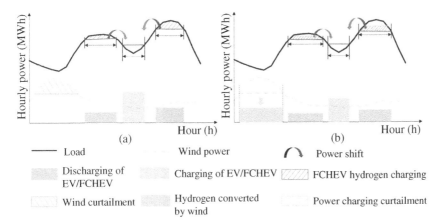

Figure 33.7 Benefit analysis brought by FCHEVs. (a) Operation situation of BEVs connected to the power system. (b) Illustration for the scenario of using FCHEVs.

have a better performance in peak-shaving and valley-filling. With the connection of FCHEV, the transportation and energy systems are integrated, and hydrogen energy and electric energy are multi-energy complementary, thus optimizing the allocation of resources on a larger scale.

33.4 Case Studies

33.4.1 Hydrogen Production in Microgrid Operation

The case is based on a 13-machine system. The system includes 10 thermal power units, 1 wind turbine, a gas turbine and a P2H system. The flexible ramp up/down service price is 247 US$/MW. The forecast error of wind power output and daily load is 10%, and the scheduling period is 24 hours with 1 hour as the interval.

Considering P2H/fuel cell and FRP, the total operating cost of the system within 24 hours is 5.87×10^5 US$, and the wind power utilization rate is 96.3%. Power generation plan is shown in Fig. 33.8.

The FRP demand curve considering the net load volatility and uncertainty is shown in Fig. 33.9. It can be seen that during the peak period of 10–12 hours and 20 hours, the thermal power unit needs to generate to respond users so that it is difficult to reserve enough upward ramping capacity. Gas turbines are committed to assist in providing FRU. Since the user consumes less power during the wind power output increases, the net load drops rapidly at 13–17 hours and 20–24 hours. When there is only thermal power units in the power system, it is difficult to meet the system FRU demand due to the ramping rate limit, resulting in wind power curtailment. P2H starts or stops quickly, which can meet the needs of FRU. As can be seen, P2H meets most of the FRU requirements.

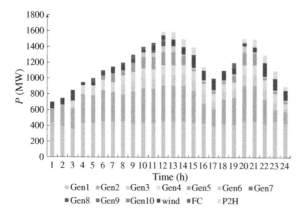

Figure 33.8 Microgrid considering P2G/GT and FRP.

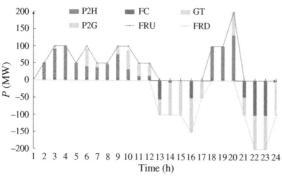

Figure 33.9 FRU/FRD demand curve.

Sufficient FRP capacity will fully address the system's uncertainty and volatility, thereby increasing system flexibility [11].

33.4.2 Hydrogen Utilization in Microgrid Operation

In this subsection, a modified IEEE 30-node electricity network and 16-node transportation network are used to verify the feasibility and effectiveness of the proposed transportation-energy model. The modified IEEE 30-node network has 41 branches and 6 thermal power units, with total installed capacity of 335 MW and wind power installed capacity of 80MW. The four scenarios of wind power and loads are obtained by clustering the actual data of one year, $r_{1-4} = [0.29, 0.40, 0.16, 0.14]$. There are 18 roads in the transportation network. The transportation system and the power system are coupled at five charging/refueling stations. There are 1500 FCHEVs in the trans-energy system (Table 33.1).

To demonstrate the effectiveness and benefits of the proposed framework, three modes in the same system are compared in the case studies as follows:

(i) M1 is the proposed optimal schedule strategy of FCHEV in joint trans-energy systems. As a hub for coupling the energy system and the transportation system, FCHEV can participate in the economic dispatch of the joint system and utilize the hydrogen generated by P2G to absorb excess renewable energy. (ii) In M2, the EVs are interactive with the power system, and can be charged or discharged to alleviate the peak pressure of the power system. (iii) M3 is the traditional case that EVs are connected to the power system as inelastic load, and the randomness of the EVs will increase the operation cost of the power system.

Compared with M2 and M3, M1 can optimize the allocation of resources in a larger range. At the same time, the integrated energy system of electricity and hydrogen also provides flexibility for economic dispatch.

The combination of the power system and the transportation system can expand the spatial and temporal distribution of resources, thereby facilitating the optimal allocation of resources.

Under the conditions of the same number of vehicles and arrival/departure time, the results of the unit commitment(UC) problems in M1, M2, and M3 modes are calculated as follows:

Figure 33.10(a) shows the wind power consumption of the three modes. The wind power utilization rate is the lowest in the M3 mode, and the wind curtailment reaches 414 MWh. The interaction between the EV and the power system in M2 mode has improved wind power consumption to a certain extent, but due to the limited charging time of the EV, the wind power at night cannot be absorbed. Compared with M2, the wind curtailment in M1 is reduced by 75.8%. In M1 mode, night wind power can be consumed through P2G, so the amount of wind curtailment is the least.

Figure 33.10(b) shows the total load in different modes. The total load includes basic load, P2G load and EV charging load. Due to the centralized charging of EVs, the peak-valley difference of the total load in M3 mode reaches 172.4 MW, which is significantly larger than that of M2 and

Table 33.1 Comparison case.

Mode	Vehicle type	Interaction
M1	FCHEV	YES
M2	BEV	YES
M3	BEV	No

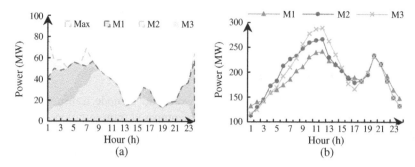

Figure 33.10 Comparison of M1–M3. (a) Wind power consumption and (b) total load.

Table 33.2 Cost comparison in different modes.

Mode	C_{op}(US$)	C_{ss}(US$)	C_t(US$)	Total cost (US$)	$U_{(w)}$(%)
M1	10,887	500	740	12,127	88.23
M2	12,636	900	—	13,536	61.95
M3	12,924	1400	—	14,324	53.93

M1. If the number of vehicles in the system increases, it will cause power transmission congestion. Compared with M2, the peak-valley difference is reduced by 27.7%, which proves that FCHEVs can play a better role in peak load shifting compared with EVs.

where C_{op} is the operation cost; C_{ss} is the start-shut cost; C_t is the transportation cost; U_w is the wind utilization.

Table 33.2 shows the system cost and wind power consumption in one day under different modes. Although the hydrogen transportation cost is taken into account in the M1 mode, the total operating cost is the lowest. On the one hand, because FCHEV can use the hydrogen generated by wind power at night, the utilization rate of wind power is greatly improved, and the operating cost is reduced; on the other hand, the peak-valley difference in M1 mode is small, since the thermal power unit does not need to start up and shut down frequently to meet the load volatility. Thus the start-up/shut-down costs are reduced. Compared with M2, the transportation cost in M1 increases, but due to the increase in wind power utilization, the operating and start-up/shut-down costs are reduced. In this case, the hydrogen energy system reduces the power system cost by about 10.4%.

33.5 Exercises

1. What is the monotonicity of power-to-hydrogen efficiency function with respect to electrolytic power? Please explain why.
2. Please provide two illustrative examples about how hydrogen production and utilization sectors improve electric power grid reliability and economy.
3. What is the relationship between the mileage of a fuel cell hybrid electric vehicle and the mass of stored hydrogen?

33.6 Acknowledgement

This chapter was supported in part by National Key Research and Development Program (2022YFB2405600) and in part by National Natural Science Foundation of China (52277092).

References

1 Yu, Y., Wang, J., Chen, Q. et al. (2023). Decarbonization efforts hindered by China's slow progress on electricity market reforms. *Nature Sustainability* 6: 1006–1015.

2 Dragičević, T., Guerrero, J.M., Vasquez, J.C., and Škrlec, D. (2014). Supervisory control of an adaptive-droop regulated DC microgrid with battery management capability. *IEEE Transactions on Power Electronics* 29 (2): 695–706.

3 Zhang, T., Wang, J., Xia, Q. et al. (2023). Extracting umbrella constraint-based representation of local electricity markets. *IEEE Transactions on Smart Grid* 14 (2): 1632–1641.

4 Zhang, G., Yuan, J., Li, Z. et al. (2020). Forming a reliable hybrid microgrid using electric spring coupled with non-sensitive loads and ess. *IEEE Transactions on Smart Grid* 11 (4): 2867–2879.

5 Li, Q., Qiu, Y., Yang, H. et al. (2020). Stability-constrained two-stage robust optimization for integrated hydrogen hybrid energy system. *CSEE Journal of Power and Energy Systems* 7 (1): 162–171.

6 Wang, J., An, Q., Zhao, Y. et al. (2023). Role of electrolytic hydrogen in smart city decarbonization in China. *Applied Energy* 336: 120699.

7 Pan, G., Gu, W., Lu, Y. et al. (2020). Optimal planning for electricity-hydrogen integrated energy system considering power to hydrogen and heat and seasonal storage. *IEEE Transactions on Sustainable Energy* 11 (4): 2662–2676.

8 Li, X., Wang, L., Yan, N., and Ma, R. (2021). Cooperative dispatch of distributed energy storage in distribution network with PV generation systems. *IEEE Transactions on Applied Superconductivity* 31 (8): 1–4.

9 Wu, Z., Wang, J., Zhou, M. et al. (2023). Incentivizing frequency provision of power-to-hydrogen toward grid resiliency enhancement. *IEEE Transactions on Industrial Informatics* 19 (9): 9370–9381.

10 Wang, T., Li, Q., Qiu, Y. et al. (2022). Power optimization distribution method for fuel cell system cluster comprehensively considering system economy. *IEEE Transactions on Industrial Electronics* 69 (12): 12898–12911.

11 Wu, Z., Chen, L., Wang, J. et al. (2023). Incentivizing the spatiotemporal flexibility of data centers toward power system coordination. *IEEE Transactions on Network Science and Engineering* 10 (3): 1766–1778.

34

Sharing Economy in Microgrid

Jianxiao Wang, Feng Gao, Tiance Zhang, and Qing Xia

34.1   Introduction

Distributed energy resources (DERs) have been dramatically developing across the world in the past decades. However, the increasing penetration of DERs has imposed great challenges to the reliable and economic operation of microgrids [1]. On the one hand, more traditional generation resources should be scheduled to smooth out the fluctuations of DERs. On the other hand, due to the limited controllability, the microgrid cannot fully accommodate DERs, which greatly reduces the utilization of DERs. In China, for example, distributed solar capacity reached 6.06 GW in 2015, but the utilization rate was less than 60%.

With the rapid development of DERs in microgrids, it is challenging to collect the bids from ubiquitous DERs and clear the market in a centralized manner, which may result in considerable market transaction costs [2]. Additionally, distinguished from the generators in a wholesale market, end-use customers pursue a convenient way to schedule their own DERs and may thus be reluctant to frequently bid supply curves. A simple settlement mechanism is needed such that users just share the surplus of DERs and then get paid. To this end, the concept of energy sharing has been advocated as a promising solution to achieve peer-to-peer energy trading among the users in a community [3]. With the central theme of "access over ownership," energy sharing enables electricity users to share the surplus energy from the rooftop solar and batteries with their neighbors and then get paid for the shared energy.

As an emerging business model, sharing economy has gained substantial popularity among transportation and housing sectors [4]. Firms like Uber and Airbnb give individuals economic incentives to provide ridesharing and rent out their houses, which realizes the optimization of resources through sharing excessive goods and services [5]. In recent years, a few studies have focused on the market implementation for sharing energy storage and rooftop photovoltaic (PV) [6]. The symmetric Nash bargaining theory is widely adopted to equally allocate the benefits among market participants. However, as electricity users make different contributions to energy sharing, a well-designed incentive mechanism should identify the values that different users create, and accordingly allocate the benefits.

To fill the aforementioned gaps, an energy sharing scheme is developed, and a novel incentive mechanism is designed for benefit allocation without users bidding on electricity prices [7]. In a distribution grid, an aggregator organizes a number of electricity users to cooperate as a single interest entity. Then the incentive mechanism is implemented to allocate the benefits to users that incentivizes users' participation in energy sharing.

Microgrids: Theory and Practice, First Edition. Edited by Peng Zhang.
© 2024 The Institute of Electrical and Electronics Engineers, Inc. Published 2024 by John Wiley & Sons, Inc.

34.2 Aggregation of Distributed Energy Resources in Energy Markets

34.2.1 Energy Sharing Scheme

In this section, we consider that one aggregator organizes N energy users in a distribution grid in a day-ahead market. Each user has a PV system, an ESS and local load. The energy trading without and with energy sharing are compared.

34.2.1.1 Energy Trading Without Energy Sharing
In this case, energy users are assumed to be price-takers, who purchase electricity from the aggregator at a retail rate, and sell back the surplus power of DERs at a net metering rate (NMR). The following trading events happen in order.

(i) Each user schedules his/her local DERs and determines the net load to minimize individual costs under fixed rates.

(ii) The aggregator collects user' net load information and trades with the connected power grid for energy balance.

(iii) The aggregator charges each user for the net load at a retail rate and pays each user for the net power at an NMR.

34.2.1.2 Energy Trading with Energy Sharing
As electricity is an undifferentiated good, a pool-based energy sharing platform is considered, in which each energy user schedules the amount of shared energy and the aggregator determines the associated payment for each user. The following events happen in order.

(i) All users enroll sharing contracts with the aggregator, which sets up a rule that determines the sharing incentives to users for an amount of shared energy.

(ii) Each energy user schedules his/her DERs and deviates from individual optimum to share DERs.

(iii) The aggregator collects the users' net load and shared energy information and trades with the connected power grid for energy balance.

(iv) The aggregator charges each user for the net load at a retail rate and pays each user for the net power at an NMR and for the shared energy with the sharing incentives.

Instead of only trading with the aggregator, the users can share DERs with each other in the platform. The proposed energy sharing scheme enables the aggregator to organize users to cooperate as a single interest entity and minimize the total costs. In contrast to the case without energy sharing, the proposed scheme achieves Pareto optimality of the aggregator and all users.

34.2.1.3 Decentralized Implementation
The proposed energy sharing scheme aims at maximizing the total benefits of the aggregator and users, which requires detailed information about users' preferences and DERs. However, it is challenging to collect users' private information and schedule energy sharing in a centralized manner. Thus, a decentralized framework is developed to preserve users' privacy. The schematic is shown in Fig. 34.1.

Each user is equipped with an energy management controller (EMC), which controls the hourly load consumption and communicates this load information to the aggregator. The EMC also receives the price signals from the aggregator. Therefore, the bidirectional communication makes

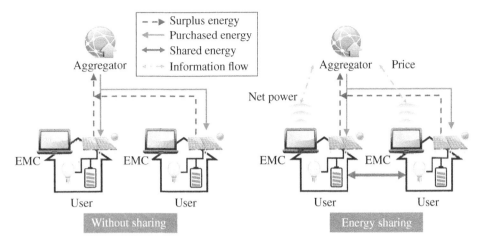

Figure 34.1 Decentralized framework for energy sharing.

the interactions easy between the aggregator and users. Based on the price signal, the EMC of each user optimally schedules local load and DERs. Then the EMCs communicate the net power to the aggregator. After collecting all users' net power, the aggregator updates the price signal and sends it back to the EMCs of users. The proposed energy sharing scheme can be applied in day-ahead and intraday markets.

34.2.2 System Model

34.2.2.1 The Aggregator's Net Cost

To depict the uncertainties in load and solar power, scenario-based stochastic programming is adopted. Thus, the net cost of the aggregator O^A is calculated below:

$$O^A = \sum_{t,s} \gamma_s \left[\lambda_{t,s}^{\text{in}} P_{t,s}^{\text{NL}} - \lambda_{t,s}^{\text{out}} P_{t,s}^{\text{NS}} + \lambda_t^S \sum_i P_{i,t,s}^{\text{NS}} - \lambda_t^R \sum_i P_{i,t,s}^{\text{NL}} \right] \tag{34.1}$$

In (34.1), γ_s is the probability of scenario s. $\lambda_{t,s}^{\text{in}}$ and $\lambda_{t,s}^{\text{out}}$ are the prices for purchasing and selling electricity with the power grid, which are different in practice due to transmission service charges, tax, etc. λ_t^S and λ_t^R are the retail rate and NMR. $P_{t,s}^{\text{NL}}$ and $P_{t,s}^{\text{NS}}$ are users' aggregated net load and surplus power, and $P_{i,t,s}^{\text{NL}}$ and $P_{i,t,s}^{\text{NS}}$ are the net load and surplus power of user i.

34.2.2.2 A User Model Without Energy Sharing

In this section, each user is modeled as an agent with the following objective:

$$\min \sum_{t,s} \gamma_s \left[\lambda_{t,s}^{\text{in}} P_{t,s}^{\text{NL}} - \lambda_{t,s}^{\text{out}} P_{t,s}^{\text{NS}} - U_i(P_{i,t,s}^L) + c_i^{\text{ESS}} (P_{i,t,s,\alpha}^{\text{ESS}} + P_{i,t,s,\beta}^{\text{ESS}}) \right] \tag{34.2}$$

where $U_i(.)$ is user i's utility function. Without loss of generality, we use a quadratic concave utility function. c_i^{ESS} is the operation cost of user i's energy storage system (ESS). The decision variables include user i's net load $P_{i,t,s}^{\text{NL}}$, surplus power $P_{i,t,s}^{\text{NS}}$, hourly load $P_{i,t,s}^L$, solar power $P_{i,t,s}^{\text{PV}}$, the charging and discharging power $P_{i,t,s,\alpha}^{\text{ESS}}$ and $P_{i,t,s,\beta}^{\text{ESS}}$.

In (34.2), let O_i^U denote user i's objective function, which is to minimize the difference between the total costs and his/her utility.

34.2.2.3 The Energy Sharing Model

In the proposed energy sharing scheme, the aggregator organizes all users to cooperate as a single interest entity, and the shared power from users can be optimized. Energy sharing requires users to deviate from individual optimal schedule to accommodate the surplus or demand from their neighbors. Thus, the aggregator should incentivize the users to share DERs by allocating the sharing benefits. The payment from the aggregator to user i is π_i^{ES}. User i's net cost is $O_i^U - \pi_i^{ES}$, and the aggregator's net cost is $O^A + \sum_i \pi_i^{ES}$. The energy sharing model is to minimize the total costs of the aggregator and all users:

$$\min \ O_A + \sum_i \pi_i^{ES} + + \sum_i (O_i^U - \pi_i^{ES}) = O_A + \sum_i O_i^U \tag{34.3}$$

Although the energy sharing model defines the amount of users' shared power, it cannot reveal the payments to users that will incentivize the deviation from individual optimum. Thus, an incentive mechanism is proposed for benefit allocation according to users' contributions.

34.2.3 Profit Sharing Mechanism

As energy sharing requires users to deviate from individual optimal schedule, thus increasing individual costs, an incentive mechanism is needed for benefit allocation so as to incentivize users to participate in energy sharing.

We firstly propose a novel index, termed as sharing contribution rate (SCR), to evaluate users' contributions to energy sharing. Then an asymmetric Nash bargaining (ANB) model considering SCRs is developed for benefit allocation.

34.2.3.1 Sharing Contribution Rate

User i's SCR, denoted by SCR_i, is defined as his/her contributions over the total contributions of all energy users:

$$SCR_i = (1 - \tau^A) \frac{\sum_{t,s} \gamma_s C_{i,t,s}^U}{\sum_{j,t,s} \gamma_s C_{j,t,s}^U}, \quad \forall i \tag{34.4}$$

where $\tau^A \in (0, 1)$ is the aggregator's rate of return, which is predefined and constant in this section; $C_{i,t,s}^U$ is the economic value of the shared DERs.

34.2.3.2 Mechanism Design

The Nash bargaining problem studies how market participants share a surplus that they jointly generate by maximizing the product of market participants' excess utilities. The solution to the ANB model defines the payments to users:

$$\pi_i^{ES*} = SCR_i \cdot \Delta + O_i^{U,1} - O_i^{U,0}, \quad \forall i \tag{34.5}$$

where π_i^{ES*} is the optimal payment to user i, and the total benefit Δ induced by energy sharing is:

$$\Delta = O^{A,1} - O^{A,0} + \sum_i O_i^{U,1} - O_i^{U,0} \geq 0 \tag{34.6}$$

According to (34.5), the optimal payment to a user can be interpreted as two parts. The first is $O_i^{U,1} - O_i^{U,0}$, representing the incremental cost after sharing DERs. The second is $SCR_i\Delta$, indicating a user's share of the total benefits that all market participants jointly generate.

Each market participant's benefit is related to his/her SCR, reflecting the contributions he/she creates. Therefore, all participants can benefit from the energy sharing.

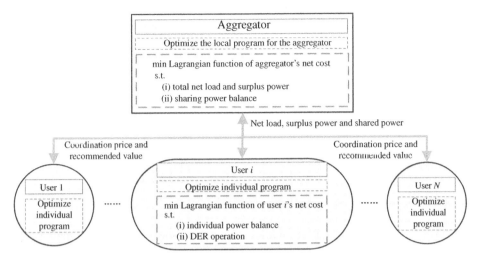

Figure 34.2 Iterative coordination between the aggregator and all users.

34.2.4 Solution Algorithm

In this section, the alternating direction method of multipliers is adopted for decentralized implementation. To decompose the energy sharing model, the following constraints are embedded:

$$\hat{P}^x_{i,s,t} = P^x_{i,s,t} : w^x_{i,s,t}, \quad \forall i, s, t, x \in \{\text{NL}, \text{NS}, \text{ES}\} \tag{34.7}$$

where $\hat{P}^x_{i,s,t}$ is an auxiliary variable, interpreted as user i's net load, net power and shared power recommended by the aggregator, and $w^x_{i,s,t}$ is the Lagrangian multiplier, i.e. the coordination price sent by the aggregator. By relaxing the constraints (34.7), the energy sharing model can be decomposed into a local program for the aggregator and N individual programs for users. Then, the energy sharing model can be iteratively solved by coordinating the aggregator's local program and the users' local programs, which is shown in Fig. 34.2.

Similarly, we can decompose the asymmetric NB model by introducing and relaxing the following constraints:

$$\hat{\pi}^{\text{ES}}_i = pi^{\text{ES}}_i : w^\pi_i, \quad \forall i \tag{34.8}$$

where $\hat{\pi}^{\text{ES}}_i$ is an auxiliary variable, interpreted as the recommended payment to user i, and w^π_i is the Lagrangian multiplier. As the energy sharing model and the asymmetric Nash bargaining (NB) model are convex, the convergence and optimality of the solution algorithm can be guaranteed.

34.3 Aggregation of Distributed Energy Resources in Energy and Capacity Markets

34.3.1 Energy Sharing Scheme

34.3.1.1 Market Framework

In this section, we consider a distribution grid with an aggregator that serves N energy users. An energy user is assumed to be a price-taker in joint energy and capacity markets. Capacity markets have been developed in many regions to address long-term resource adequacy problems, e.g. Midcontinent Independent System Operator (MISO), New York ISO (NYISO) and

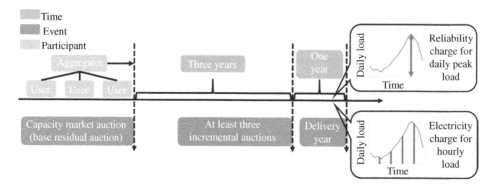

Figure 34.3 PJM's capacity market that coexists with the energy market.

Pennsylvania–New Jersey–Maryland (PJM). Without loss of generality, PJM's capacity market that coexists with the energy market is adopted, in which an auction is held every year. The framework for one complete auction process is shown in Fig. 34.3.

PJM's capacity market has a multi-auction structure designed to procure resources that balance the region's long-term load [8]. A base residual auction (BRA) is held for the delivery year, which is three years in the future. After BRA, at least three incremental auctions are conducted for additional resource commitments to satisfy potential load changes prior to the start of the delivery year. Participation by load serving entities (LSEs) in PJM's capacity market is mandatory.

In this section, we only consider the aggregator's participation in BRA, without the incremental auctions and the realization in the delivery year. In other words, we focus on the decision-making of the aggregator and users before the delivery year considering the uncertainty in renewable generation and prices [9]. Before the delivery year, each user strategically invests in DERs and minimizes annualized individual costs. Then the aggregator implements an energy sharing scheme and an incentive mechanism to encourage users to share DERs. The energy sharing scheme is detailed below.

34.3.1.2 Energy Trading Scheme Comparison
In this section, the energy trading schemes with and without energy sharing are compared, shown in Fig. 34.4.

As one can observe, in the case without energy sharing, users are assumed to be price-takers who only trade with the aggregator. The users purchase electricity from the aggregator at a retail rate, and sell back the surplus energy from DERs at a NMR. Without energy sharing, each user has no information about the demand or surplus of neighbors' DERs. Thus, only individual costs are minimized by scheduling DERs and local load. Then, the aggregator's net cost is calculated as the difference between the price charged by the power grid and the revenues from users. However, users may not receive sufficient and reasonable benefits to justify the investment costs of DERs. For example, the NMR is about 3 cents/kWh for the Pacific Gas and Electric Company (PG&E) in California in 2017, which is much lower than retail rates. Thus, DER planning may be limited without energy sharing.

In the case with energy sharing, a pool-based sharing platform is established, in which users purchase or sell shared energy and the aggregator determines the associated payments to each user. Before the delivery year, all the users enroll sharing contracts with the aggregator, which set up the rule of sharing benefit allocation. Given the sharing contract, each user determines the sizing of DERs anticipating the optimal scheduling for DERs. In the energy sharing platform, the

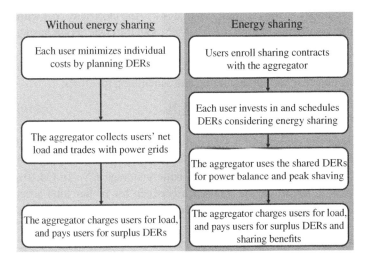

Figure 34.4 Events in cases with and without energy sharing.

aggregator collects information regarding all users' demands and surpluses and helps match DER sharing among users. At the same time, the aggregator takes advantage of shared DERs for peak shaving to reduce the reliability charge. To encourage users to invest in DERs and participate in energy sharing, an incentive mechanism is designed by identifying different users' contributions. The following events occur in order before the delivery year:

(i) Each energy user enrolls in a sharing contract with the aggregator, which sets up the rules for sharing payments to users.

(ii) Each user optimizes the DER capacity and schedules the DERs for energy sharing by deviating from individual optimum.

(iii) To reduce the electricity and reliability charges, the aggregator uses the shared DERs to satisfy local loads and reduce peak loads. Then, the aggregator collects the information on users' hourly net load and power.

(iv) The aggregator charges each user for the net load at retail rates, and pays each user for the net power at NMRs and for the shared DERs with sharing payments.

In this case, the aggregator and users cooperate as a single interest entity to minimize the total costs. The models are formulated in Section 34.3.2.

34.3.2 System Model

In this section, we firstly establish the models for the aggregator and users without energy sharing. Then the energy sharing model is formulated to evaluate the sharing benefits.

34.3.2.1 The Model Without Energy Sharing

The aggregator is responsible for paying the electricity and reliability charges. As the charges are recovered by retail sales to energy users, the aggregator's net cost f^A is expressed as follows:

$$f^A = \sum_s \gamma_s \left(C_s^R + C_s^R - \sum_{i,t} \lambda_t^R P_{i,s,t}^{NL} + \sum_{i,t} \lambda_t^S P_{i,s,t}^{NS} \right) \tag{34.9}$$

In (34.9), the aggregator's net cost consists of the reliability charge is C_s^R, the electricity cost with the power grid is C_s^E, the profits from retail sales to users are $\sum_{i,t} \lambda_t^R P_{i,s,t}^N$ and the costs of purchasing surplus energy from users are $\sum_{i,t} \lambda_t^S P_{i,s,t}^{NS}$.

In the case without energy sharing, each user aims to minimize individual costs. We model each user as an agent with the following objective for optimal sizing and scheduling for DERs:

$$\min_{\mathbf{X}_i} f_i^U = c_i^{PV} Q_i^{PV} + c_i^{ES} Q_i^{ES} + \sum_{s,t} \gamma_s [\lambda_t^R P_{i,s,t}^{NL} - \lambda_t^S P_{i,s,t}^{NS} - U_i(P_{i,s,t}^L)] \tag{34.10}$$

where the decision variables include the capacity of a solar panel Q_i^{PV}, battery storage Q_i^{BS}, the net load $P_{i,s,t}^{NL}$, surplus power $P_{i,s,t}^{NS}$, the electrical load $P_{i,s,t}^L$, the solar power $P_{i,s,t}^{PV}$, the charging and discharging power $P_{i,s,t}^{cha}$ and $P_{i,s,t}^{dis}$, and the stored energy in the battery storage $E_{i,s,t}^{BS}$. In (34.10), the objective of each energy user is to minimize the difference between his or her total costs and utility. User i's utility function $U_i P_{i,s,t}^L$ is a quadratic concave function, related to his or her electric load.

Each user minimizes individual costs and trades only with the aggregator without energy sharing. User i's optimal cost without energy sharing is denoted by $f_i^{U,0}$. The aggregator can then collect all the users' net load, and trade with the power grid. The aggregator's cost is denoted by $f^{A,0}$.

34.3.2.2 The Energy Sharing Model

In the proposed energy sharing scheme, energy users share DERs with the aggregator. The aggregator can then employ the shared power for peak shaving. To share DERs with others, an energy user has to deviate from the optimal individual schedule. Thus, the aggregator should provide incentives to the users who share DERs. The payment from the aggregator to user i is π_i^{ES}, and user i's net cost is $f_i^U - \pi_i^{ES}$. The aggregator's net cost is $f^A + \sum_i \pi_i^{ES}$. The energy sharing model minimizes the total costs of the aggregator and all users:

$$\min_{\mathbf{X}_i, P_{i,s,t}^{ES}} f_i^U = c_i^{PV} Q_i^{PV} + c_i^{ES} Q_i^{ES} + \sum_{s,t} \gamma_s [\lambda_t^R P_{i,s,t}^{NL} - \lambda_t^S P_{i,s,t}^{NS} - U_i(P_{i,s,t}^L)] \tag{34.11}$$

The total benefits brought on by energy sharing are denoted by Δf, known as the cooperative surplus:

$$\Delta f = f^{A,0} - f^{A,1} + \sum_i (f_i^{U,0} - f_i^{U,1}) \geq 0 \tag{34.12}$$

As shown in (34.12), the proposed model cannot reveal the payments to users that incentivize users to deviate from individual optimum for energy sharing. Thus, an incentive mechanism is proposed for benefit allocation.

34.3.3 Profit Sharing Mechanism

A well-designed incentive mechanism should identify different users' contributions and allocate the associated benefits to users. In this section, we propose a sharing contribution rate to quantify different users' contributions to both energy sharing and peak shaving. Based on the indices, an asymmetric NB model is formulated for benefit allocation.

34.3.3.1 Sharing Contribution Rate

An energy user's contributions to energy sharing are twofold. On the one hand, the user deviates from the optimal individual schedule to provide or consume shared power, which helps decrease the electricity costs of the aggregator. On the other hand, the user curtails electrical load or increases DERs during peak hours, which helps reduce the aggregator's reliability charges. Therefore, our proposed SCR considers these two aspects.

We define user i's contribution rate as his or her contributions over the total contributions of all energy users:

$$\text{SCR}_i = \left(1 - \tau^A\right) \frac{\left[\sum_s \left(C_{i,s}^{\text{PS}} + \sum_t C_{i,s,t}^{\text{ES}}\right)\right]}{\sum_j \left[\sum_s \left(C_{j,s}^{\text{PS}} + \sum_t C_{j,s,t}^{\text{ES}}\right)\right]} \tag{34.13}$$

where $\tau^A \in (0,1)$ is the aggregator's rate of return, which is a predefined constant. In practice, τ^A can be regulated by the government or optimally determined in a market-based environment; $C_{i,s,t}^{\text{ES}}$ is user i's contribution of shared DERs; $C_{i,s}^{\text{PS}}$ is user i's contribution to peak shaving. In (34.13), the numerator term represents user i's total contributions to energy sharing and peak shaving, and the denominator is all users' contributions. Note that, each user's total contribution takes a positive value to guarantee that an SCR is nonnegative.

In this section, the users' economic values of energy sharing and peak shaving are naturally chosen to measure their relative contributions. There are alternative characterizations for proportional cost sharing mechanisms, which need an in-depth study in the future.

34.3.3.2 Mechanism Design

Nash bargaining theory studies how market participants share a surplus that they jointly generate by maximizing the product of market participants' excess utilities, which fulfills the properties like those illustrated in Section 34.2.3.

In contrast to the symmetric NB model, the objective of the proposed asymmetric NB model is to maximize the product of market participants' excess utilities with different sharing contribution rates, as shown in (34.14):

$$\min_{\pi_i^{\text{ES}}} \left(f^{A,0} - f^{A,1} - \sum_i \pi_i^{\text{ES}}\right)^{\tau^A} \prod_{i \in \Phi^{\text{ES}}} (f^{U,0} - f^{U,1} + \pi_i^{\text{ES}})^{\text{SCR}_i} \tag{34.14}$$

where the decision variables are $(\pi_i^{\text{ES}}, \forall i)$. Note that in the objective function (34.14), the exponent of each user is his/her SCR, reflecting different users' contributions to energy sharing. However, the exponents in the symmetric NB model are identical for different users. Then, we propose the following theorem:

The optimal payment to user i is:

$$\pi_i^{\text{ES*}} = f^{U,1} - f^{U,0} + \text{SCR}_i \cdot \Delta f, \quad \forall i \tag{34.15}$$

where $\pi_i^{\text{ES*}}$ is the optimal payment to user i, i.e. the solution to the proposed asymmetric NB model. The proposed incentive mechanism satisfies the aforementioned four properties, i.e. (1) Pareto optimality, (2) budget balance, (3) individual rationality, and (4) monotonicity.

As one can observe, the incentive mechanism defines the optimal payment from the aggregator to each user. In (34.15), the optimal payment $\pi_i^{\text{ES*}}$ consists of two parts: The first term $f_i^{U,1} - f_i^{U,0}$ indicates the incremental cost of user i after deviating from individual optimum for energy sharing, and the second term $\text{SCR}_i \cdot \Delta f$ represents the allocation of the total benefits from energy sharing, which is related to user i's contributions to energy sharing.

34.4 Case Studies

34.4.1 Aggregation of DERs in Energy Markets

The quadratic and linear coefficients of users' utility functions are randomly generated from uniform distributions, i.e. $a_i \in U[-0.5, -0.1]$, $b_i \in U[20, 50]$. In the 10-user case, the load and solar data are described in Fig. 34.5.

Users' minimal and maximal loads are set to 0.8 and 1.2 times of the actual loads. Users' daily minimum loads are set to his/her actual daily load demand. The parameters of users' ESSs are listed in Table 34.1. The initial stored energy is randomly generated from $U[5, 30]$ MWh. The ESS's operation cost is estimated as the replacement cost of storage bank over the lifetime throughput. The replacement cost per year is about 13,400 US$/MW, and the yearly charging/discharging hours are 1800 hours.

An energy user's average power curves without and with energy sharing are compared in Fig. 34.6.

In the case without energy sharing, the user has no incentives to arbitrage with the ESS against retail rates. The ESS is only used to store the surplus of solar energy, and discharge to satisfy the night-time load. Due to the low-NMR, the user will minimize his/her net load instead of selling back the surplus power.

However, in the case with energy sharing, the utilization of the ESS is greatly improved for shifting the day-time load to the night hours. As the net load shows, the user consumes electricity at night but supplies DERs at day time. In addition, by comparing the electrical load curves, one can

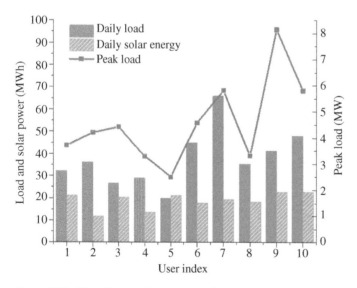

Figure 34.5 Users' load and solar power data.

Table 34.1 Parameters of users' energy storage systems.

$P^{ESS}_{i,t,s,\alpha/\beta}$	η^{ESS}_i	$E^{ESS}_{i,min}$	$E^{ESS}_{i,max}$	c^{ESS}_i
5 MW	95%	5 MWh	30 MWh	3.7 US$/MWh

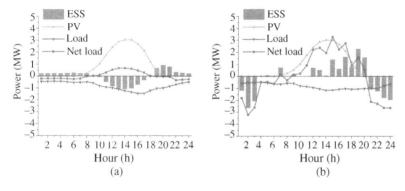

Figure 34.6 A user's power profiles without (a) and with (b) energy sharing.

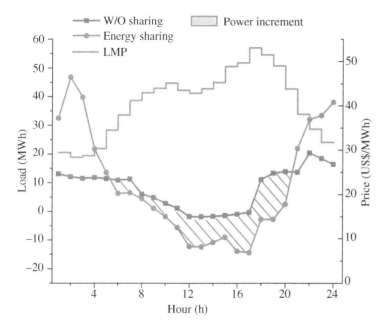

Figure 34.7 Average profiles of the LMPs and aggregated loads.

observe that the load is also shifted from day time to night time after energy sharing. Figure 34.7 shows the average curves of the LMPs and aggregated loads. In the energy sharing scheme, the aggregator organizes users to respond to the LMPs. Compared with the case without sharing, energy sharing can provide additional 132.35 MWh power for the power grid during peak hours from 6:00 to 20:00, thereby contributing to power balance in the power grid.

In symmetric Nash bargaining (SNB), the symmetric Nash bargaining model is used to allocate the sharing benefits. All users have identical weights, without distinguishing the users' contributions to sharing. However, the proposed ANB identifies users' contributions for benefit allocation. The cost savings of users by ANB and SNB are shown in Fig. 34.8.

As one can observe, the cost savings of all users are 0.49 thousand dollars in SNB. However, the cost savings range from 0.44 to 0.53 thousand dollars in ANB. As aforementioned, a user's cost savings are related to his/her SCR, defined as the user's proportion of the economic values of shared

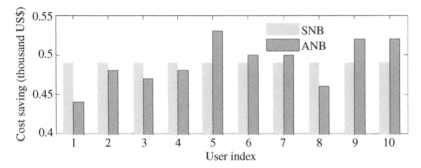

Figure 34.8 Cost savings of users by ANB and SNB.

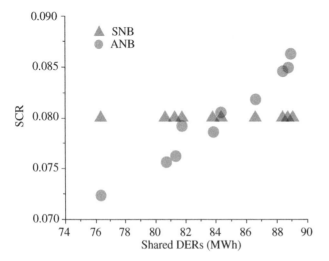

Figure 34.9 Relationship between users' SCRs and shared DERs.

DERs. Figure 34.9 shows the relationship between users' SCRs and shared DERs. Note that a user's shared DERs refers to the total amount of the absolute value.

In ANB, users' SCRs are positively related to the shared DERs, indicating that the more DERs a user shares, the higher level of contributions this user makes. However, all users' SCRs equal 0.08 regardless of the distinct behaviors in SNB. Therefore, the proposed ANB can reveal the contributions of different users and then allocate the benefits.

34.4.2 Aggregation of DERs in Energy and Capacity Markets

The quadratic and linear coefficients of users' utility functions are randomly generated from uniform distributions, i.e. $a_i \in U[-0.5, -0.1]$, $b_i \in U[20, 50]$, $\forall i$. Users' minimal and maximal loads are set to 0.9 and 1.1 times the actual loads for each time slot. Users' daily minimum load requirements are set to his or her actual daily load demand. The daily solar production of a 1-MW solar panel is shown in Fig. 34.10.

The parameters of a 1-MW battery storage system are listed in Table 34.2. The annualized investment costs for a 1-MW solar panel and 1-MW battery storage system are US\$240,000 and US\$80,000, respectively.

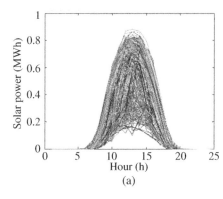
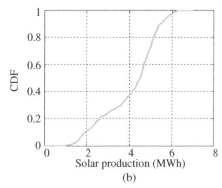

Figure 34.10 Daily solar production of a 1-MW solar panel. (a) Daily solar production scenarios. (b) Cumulative distribution function of solar production.

Table 34.2 Parameters of a 1-MW batter storage system.

Parameter	$P_{i,\max}^{BS}$	$E_{i,\max}^{BS}$	η_i^{BS}
Value	1 MW	2.7 MWh	90%

PG&E summer retail prices are 0.212 US$/kWh from 1:00 to 8:00 and from 22:00 to 24:00, 0.239 US$/kWh from 8:00 to 12:00 and from 18:00 to 22:00, and 0.263 US$/kWh from 12:00 to 18:00. The NMR is 0.03 US$/kWh. The day-ahead energy market prices and the capacity market prices are provided by PJM. The average zonal capacity market price for the 2020–2021 BRA is 120.98 US$/MW-day, and the 2017 yearly LMPs are shown in Fig. 34.11.

The optimal capacities for solar and battery storage in the cases without and with energy sharing are shown in Fig. 34.12.

Compared with the case without energy sharing, the optimal capacities for solar and storage significantly increase after energy sharing. The total solar capacity increases from 23.99 to 91.16 MW, and the total storage capacity increases from 0.89 to 4.17 MW. These results conclude that the proposed energy sharing scheme effectively incentivizes users to invest in DERs. Note that 10 users have different utility levels and electrical load profiles, leading to different investment decisions. For example, the optimal storage capacity of User 6 decreases after energy sharing, indicating that User 6 prefers to consume shared energy by others compared with individual investments.

The aggregator's average net power profiles without and with energy sharing are shown in Fig. 34.13.

With more solar and storage integrated, the aggregator's peak load is reduced by 17.65%, from 20.11 to 16.56 MW. Thus, the annual reliability charges for the aggregator decrease from 0.89 to 0.73 million dollars. In addition, with more solar resources installed, the aggregator can provide the power grid with net power during peak hours from 8:00 to 17:00. The aggregator annually sells 87.75 GWh of electricity back to the power grid.

In SNB, all users are weighted identically, and thus each user's contribution rate in SNB is equal to $(1 - \tau^A)/N = 0.09$. However, the proposed ANB model can identify different users' contributions to energy sharing and peak shaving. The users' SCRs and cost savings in ANB and SNB are compared in Fig. 34.14.

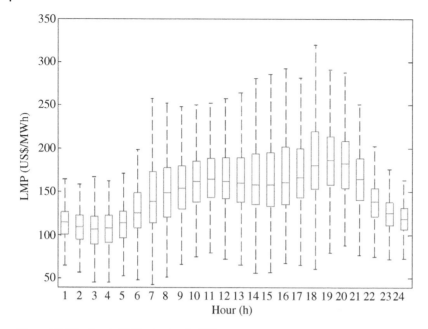

Figure 34.11 Yearly LMPs at a bus in PJM.

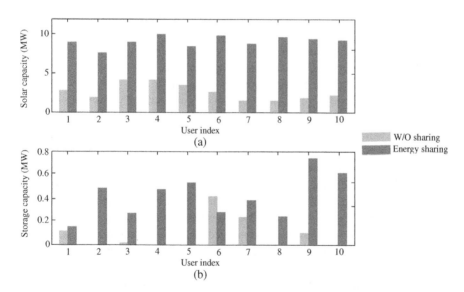

Figure 34.12 Optimal capacities for solar (a) and battery storage (b) without and with energy sharing.

Compared with surplus power (NS), all users can reduce the costs after participating in energy sharing. In SNB, as all users are weighted identically, each user's annual cost savings equal 0.13 million dollars. However, the proposed ANB method can identify the contributions of different users, so the users' cost savings and SCRs are distinguishable. In ANB, users can reduce their costs by 0.09–0.18 million dollars.

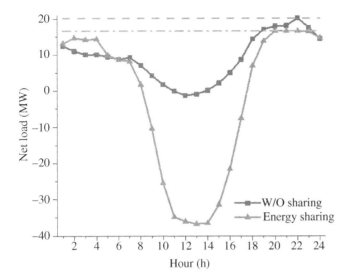

Figure 34.13 Aggregator's average net loads without and with energy sharing.

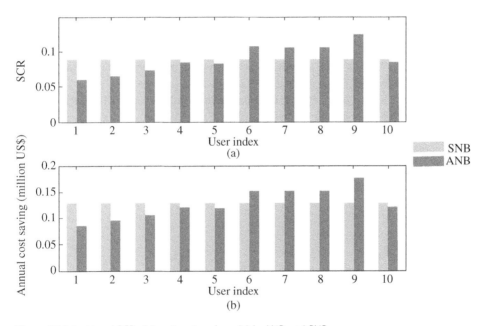

Figure 34.14 Users' SCRs (a) and cost savings (b) in ANB and SNB.

34.5 Exercises

1. Please explain the framework of PJM capacity market. How is the daily reliability charge calculated?
2. Please explain the difference between the energy sharing model and the traditional individually optimal one.
3. Please explain how a sharing contribution rate is determined in energy and energy + capacity markets.

34.6 Acknowledgement

This chapter was supported in part by National Key Research and Development Program (2022YFB2405600) and in part by National Natural Science Foundation of China (52277092).

References

1 Ferreira, P.D.F., Carvalho, P.M.S., Ferreira, L.A.F.M., and Ilic, M.D. (2013). Distributed energy resources integration challenges in low-voltage networks: voltage control limitations and risk of cascading. *IEEE Transactions on Sustainable Energy* 4 (1): 82–88.

2 Yu, Y., Wang, J., Chen, Q. et al. (2023). Decarbonization efforts hindered by China's slow progress on electricity market reforms. *Nature Sustainability* 6: 1006–1015.

3 Peng, Q., Wang, X., Kuang, Y. et al. (2021). Hybrid energy sharing mechanism for integrated energy systems based on the stackelberg game. *CSEE Journal of Power and Energy Systems* 7 (5): 911–921.

4 Wu, Z., Wang, J., Zhong, H. et al. (2023). Sharing economy in local energy markets. *Journal of Modern Power Systems and Clean Energy* 11 (3): 714–726.

5 Tao, Y., Qiu, J., and Lai, S. (2022). A learning and operation planning method for uber energy storage system: order dispatch. *IEEE Transactions on Intelligent Transportation Systems* 23 (12): 23070–23083.

6 Yin, C., Wang, J., Tang, W. et al. (2023). Health-aware energy management strategy toward internet of storage. *IEEE Internet of Things Journal* 10 (9): 7545–7553.

7 Yu, M., Wang, J., Yan, J. et al. (2022). Pricing information in smart grids: a quality-based data valuation paradigm. *IEEE Transactions on Smart Grid* 13 (5): 3735–3747.

8 Chen, H., Pilong, C., Rocha-Garrido, P. et al. (2023). Grid resilience with high renewable penetration: a PJM approach. *IEEE Transactions on Sustainable Energy* 14 (2): 1169–1177.

9 Wang, J., Chen, L., Tan, Z. et al. (2023). Inherent spatiotemporal uncertainty of renewable power in china. *Nature Communications* 14: 5379.

35

Microgrid: A Pathway to Mitigate Greenhouse Impact of Rural Electrification

Jianxiao Wang, Haiwang Zhong, and Jing Dai

35.1 Introduction

China has experienced continuous and dramatic development of the economy and industry over the past three decades [1]. However, as the worlds largest coal consumer and coal-derived electricity producer, one consequence of the resulting massive consumption of fossil fuels is the rise of emerging greenhouse gas and air pollution emissions [2], posing serious threats to global warming and human health [3].

It has become a national strategy to develop a clean-energy society and to preserve the ecological environment. On the one hand, as a promise to the world, China has set an ambitious target to limit the national carbon footprint [4]. In 2015, China agreed on the Paris Agreement and declared that the carbon emissions per GDP in 2030 must decrease by 60–65% of the value in 2005 [5]. On the other hand, a wide variety of domestic actions have been taken to mitigate carbon dioxide and air pollution, e.g. improving energy efficiency [6], facilitating renewable and sustainable energy [7], and enhancing forest carbon sequestration [8].

At the same time, one of the main sources of the air pollutants in Northern China is rural residents' burning raw coal for heating. Due to the relatively low price and high heat value, raw coal has long been a primary heating resource in Northern China in winter. However, without desulfurization and denitrification, the SO_2, NO_x, and other air pollutants from combustion are directly emitted in the atmosphere, thereby resulting in severe environmental pollution [9]. In the Beijing-Tianjin-Hebei region, the annual rural raw coal consumption generally reaches over 40 million tons, contributing to approximately 15% of SO_2, 4% of NO_x, and 23% of particles, respectively. Realistic evidence in Northern China shows that the air quality in winter usually gets much worse than that in summer.

Recent years have witnessed the Chinese government's great efforts to reduce the carbon and pollutant emissions from rural residents. A series of Electric Heating Policies (EHPs) has been issued since 2015, which enforces strict regulations to substitute in place of raw coal in Northern China [10]. For example, in April 2015, the "Action plan for the clean and efficient coal" issued by the National Energy Administration declared that the use of coal with over 16% ash or 1% sulfur content is prohibited. Another policy, "Instructions for substituting electric heating for coal," issued by the National Development and Reform Commission in May 2016, showed a goal of reducing 130 megatons of coal for from 2016 to 2020 in China. Consequently, many provinces such as Hebei and Shanxi have issued regional action plans to popularize electric heating for raw coal abatement.

Microgrids: Theory and Practice, First Edition. Edited by Peng Zhang.
© 2024 The Institute of Electrical and Electronics Engineers, Inc. Published 2024 by John Wiley & Sons, Inc.

Up till now, China's EHP has contributed to significant improvements in air quality and carbon dioxide reductions from rural residential sectors. However, this shift has resulted in a sharp increase in electric loads and may even lead to a higher level of carbon emissions from power generation. Empirical evidence shows that in January 2018, the State Grid Corporation of China (SGCC) encountered a dramatic electric load increase caused by the EHP and required more electricity from coal-fired power plants. Compared with 2017, the largest daily electricity consumption in January 2018 increased by over 15%. Therefore, the conflict between China's EHP and national carbon mitigation has been exposed with the rapid development of electric heating.

A wide variety of existing literature has investigated the environmental impacts of China's residential heating sectors, including the estimation for carbon and pollutant emissions [11], the policy making and analysis for emission control, and the influence on life expectancy and human health. Yet few studies have quantified the greenhouse impacts caused by electric heating in China or explored the emerging incompatibility between China's EHP and carbon mitigation. Therefore, we aim to quantify the extent that China's EHP can contribute to national carbon emissions in this chapter. To quantify CO_2 induced by China's EHP, we propose a theoretical model considering both power generation and rural residential heating sectors. We explore the link between China's EHP and national carbon mitigation, and analyze the key factors leading to the diverse performance of the policy implementation in different regions. To address the incompatibility, we provide policy suggestions for China and other countries with similar situations to facilitate the accommodation of renewable energy and to improve electric heating efficiency in a microgrid [12].

35.2 System Model

35.2.1 Rural Resident Data

We simulate the carbon emissions in four provinces in Northern China, i.e. Hebei, Henan, Shandong, and Shanxi. To assess the emissions caused by rural space heating, the following data are needed: rural resident population (RRP), housing areas, household heating coal consumption, household electric heating load, and rooftop solar power.

According to the Sixth National Census in China, the 2015 rural populations in HB, HN, SD, and SX were 36.14, 50.39, 42.33, and 16.48 million, respectively. The 2015 rural population in 16 provinces (autonomous regions and municipalities) in Northern China was 256.19 million. According to the population target planning in 2020, the rural populations in the HB, HN, SD, and SX will be 33.00, 49.06, 35.88, and 16.60 million, respectively. In 2030, the rural populations in the four provinces are expected to drop to 25.74, 39.10, 26.67, and 9.67 million, respectively.

The housing area of a family is a key factor that determines the heating coal consumption and electric heating load. The housing area data are collected from "Report on Chinese Residential Energy Consumption," published by the National Academy of Development and Strategy, Renmin University of China. The housing areas are divided into eight intervals, i.e. [15,30], (30,50], (50,70], (70,90], (90,120], (120,150], (150,180], and (180,250] m^2, accounting for 1.05%, 3.48%, 7.67%, 13.24%, 24.39%, 16.72%, 14.29%, and 19.16%, respectively.

Given outdoor air temperature, indoor comfort temperature and the housing area as input, a household's heating coal consumption and electric heating load can be simulated by using EnergyPlus, a building energy consumption simulation software developed by Lawrence Berkeley National Laboratory (LBNL) and some other institutions, sponsored by the Department of Energy. The outdoor air temperature data are collected from the weather data set arranged by the

World Meteorological Organization. In HB, we use the weather data from three cities, i.e. Raoyang, Shijiazhuang, and Xingtai. In HN, we use the weather data from seven cities, including Anyang, Lushi, Nanyang, Shangqiu, Xinyang, Zhengzhou, and Zhumadian. In SD, we use the weather data from eight cities, including Chaoyang, Chengshantou, Huimin, Jinan, Juxian, Longkou, Weifang, and Yanzhou. In SX, we use the weather data from seven cities, including Datong, Houma, Jiexiu, Taiyuan, Yuanping, Yuncheng, and Yushe. According to the "Indoor Air Quality Standard" recognized by Chinese government, the indoor air temperature of a rural household is required to reach at least 13–17 °C, which is set as the input of indoor comfort temperature.

In this chapter, we collect hourly residential solar power data in the aforementioned cities in the four provinces from the National Renewable Energy Laboratory (NREL). The average daily generation of a 1-kW solar panel in HB, HN, SD, and SX is 2.07, 2.14, 2.47, and 2.28 kWh, respectively.

35.2.2 Electric Power Data

To assess the thermal coal consumption and associated carbon emissions of China's electric power grids, the following data are needed: thermal generator parameters, renewable power, and electric load.

We collect the parameters of thermal generators in HB, HN, SD, and SX provinces in 2015. In the southern region of HB, there are 97 thermal generators, with the median TCCR equal to 306.11 kg/MWh. The total installed capacity of thermal generators is 28.66 GW, and the median value is 330 MW. In HN, there are 153 thermal generators, with the median TCCR equal to 297.80 kg/MWh. The total installed capacity is 61.49 GW, and the median value is 320 MW. In SD, there are 190 thermal generators, with the median TCCR equal to 314.15 kg/MWh. The total installed capacity is 62.41 GW, and the median value is 320 MW. In SX, there are 190 thermal generators, with the median TCCR equal to 368.00 kg/MWh. The total installed capacity is 56.46 GW, and the median value is 300 MW. Additionally, the installed capacities of wind farms in HB, HN, SD, and SX are 10.22, 0.91, 7.21, and 6.99 GW, respectively. The installed capacities of solar stations in HB, HN, SD, and SX are 2.22, 0.41, 1.33, and 1.11 GW, respectively.

Due to data limitation, we collect the provincial electric loads measured in hours from 01 November 2015 to 31 December 2015 in HB, HN, SD, and SX. The total electric loads during the 2 months in the four provinces are 30.15, 47.44, 67.02, and 27.58 TWh, respectively. In this section, we estimate the carbon emissions during a heating season to be 2.5 times those during the two months because a heating season generally lasts from 1st November to 31st March in the next year.

In 2015, the national installed capacities of thermal, wind and solar generation are 1.01, 0.13, and 0.04 TW, respectively. According to "China Energy & Electricity Outlook" published by State Grid Energy Research Institute (SGERI), such capacities will be 1.19, 0.28, and 0.28 TW in 2020, respectively. In 2030, the capacities are expected to reach 1.53, 0.70, and 0.56 TW, respectively. In addition, the national electric load demands in 2015, 2020, and 2030 are 5.7×10^3, 7.7×10^3 and 11.1×10^3 TWh, respectively.

35.2.3 Rural Heating in Microgrid

In this chapter, household heating coal consumption and electric heating load are simulated by EnergyPlus. When simulating a household's heating coal consumption, we set the heating coil type as gas and transform the gas consumption into coal consumption based on the total amount of heat. The transformation is expressed as follows:

$$Q_i^{\text{HCoal}} = \frac{H_i^{\text{Gas}}}{h^{\text{Coal}}} \tag{35.1}$$

where Q_i^{HCoal} is the heating coal consumption of the ith household; H_i^{Gas} represents the total amount of heat produced by burning gas; and h^{Coal} is the heating value of coal, i.e. 2.93×10^7 J/kg. When we set the heating coil type as electricity, a household's hourly electric heating load can be directly acquired by running EnergyPlus. For electric heaters, we set the efficiency of the electric heating coil as 80%. For air-sourced heat pumps, we change the efficiency from 250% to 400%.

To systematically evaluate the households' heating energy consumption considering different sizes, we conduct sensitivity analyses on the sizes of houses. We scan the length, width, and height of houses from 5 to 20 m, from 3 to 12 m, and from 3 to 5 m, respectively. Then, we categorize the houses with different sizes into eight intervals based on housing areas, i.e. [15,30], (30,50], (50,70], (70,90], (90,120], (120,150], (150,180], and (180,250] m². Let $\overline{Q}_j^{\text{HCoal}}$ and $\overline{P}_j^{\text{Elec}}(t)$ be the average heating coal consumption and hourly electric heating load at time slot t of the households in the jth area interval, respectively. The average household heating coal consumption $\overline{Q}^{\text{HCoal}}$ and the average household hourly electric heating load $\overline{P}^{\text{Elec}}(t)$ in a province can be calculated as follows:

$$\overline{Q}^{\text{HCoal}} = \sum_{j=1}^{8} \gamma_j \overline{Q}^{\text{HCoal}} \tag{35.2}$$

$$\overline{P}^{\text{Elec}}(t) = \sum_{j=1}^{8} \gamma_j \overline{P}_j^{\text{Elec}}(t) \tag{35.3}$$

where $\gamma_j, j = 1, 2, \ldots, 8$ represents the proportion of the jth area interval, i.e. 1.05%, 3.48%, 7.67%, 13.24%, 24.39%, 16.72%, 14.29%, and 19.16%, respectively. Then a province's total heating coal consumption Q^{HCoal} and total hourly electric heating load $P^{\text{Elec}}(t)$ can be obtained by using the following equations:

$$Q^{\text{HCoal}} = \overline{Q}^{\text{HCoal}} \times \rho \times p^{\text{PIR}} \tag{35.4}$$

$$P^{\text{Elec}}(t) = \overline{P}^{\text{Elec}}(t) \times \rho \times p^{\text{PIR}} \tag{35.5}$$

where ρ is provincial RRP; and p^{PIR} represents the policy implementation rate (PIR).

In addition, a household's net electric heating load $P_i^{\text{Net}}(t)$ is calculated as the difference between the load $P_i^{\text{Elec}}(t)$ and rooftop solar power $P_i^{\text{Solar}}(t)$:

$$P_i^{\text{Net}}(t) = P_i^{\text{Elec}}(t) - P_i^{\text{Solar}}(t) \tag{35.6}$$

35.2.4 Electricity Dispatch Model

In this section, a day-ahead unit commitment model is formulated to quantify the thermal coal consumption. The mathematical formulation is shown as follows:

$$\min_{\mathbf{X}} \sum_{i \in \Phi^{\text{CG}}} \sum_{t \in \Phi^{\text{T}}} [C_i^{\text{CG}} P_i^{\text{CG}}(t) + C_i^U Y_i(t) + C_i^D Z_i(t)] \tag{35.7}$$

subject to

$$\sum_{i \in \Phi^{\text{CG}}} P_i^{\text{CG}}(t) + \sum_{j \in \Phi^{\text{RG}}} P_j^{\text{RG}}(t) + \sum_{k \in \Phi^{\text{TL}}} P_i^{\text{TL}}(t) = P^{\text{Orig}}(t) + P^{\text{Net}}(t), \forall t \in \Phi^{\text{T}} \tag{35.8}$$

$$\sum_{i \in \Phi^{\text{CG}}} P_{i,\max}^{\text{CG}} U_i(t) \geq R(t), \forall t \in \Phi^{\text{T}} \tag{35.9}$$

$$P_{i,\min}^{\text{CG}} U_i(t) \leq P_i^{\text{CG}} \leq P_{i,\max}^{\text{CG}} U_i(t), \forall i \in \Phi^{\text{CG}}, \forall t \in \Phi^{\text{T}} \tag{35.10}$$

$$Y_i(t) + Z_i(t) \leq 1, \forall i \in \Phi^{CG}, \forall t \in \Phi^{T} \tag{35.11}$$

$$Y_i(t) - Z_i(t) = U_i(t) - U_i(t - 1), \forall i \in \Phi^{CG}, \forall t \in \Phi^{T} \tag{35.12}$$

$$\sum_{\delta=t}^{t+T_i^{U-1}} U_i(\delta) \geq T_i^{U} Y_i(t), \forall i \in \Phi^{CG}, \forall t \in \Phi^{T} \tag{35.13}$$

$$\sum_{\delta=t}^{t+T_i^{D-1}} [1 - U_i(\delta)] \geq T_i^{D} Z_i(t), \forall i \in \Phi^{CG}, \forall t \in \Phi^{T} \tag{35.14}$$

$$P_i^{CG}(t) \in \mathbb{R}^+ \cup \{0\}, \forall i \in \Phi^{CG}, \forall t \in \Phi^{T} \tag{35.15}$$

$$U_i(t), Y_i(t), Z_i(t) \in \{0, 1\}, \forall i \in \Phi^{CG}, \forall t \in \Phi^{T} \tag{35.16}$$

where the decision variables are denoted by $\mathbf{X}$, including the hourly power of coal-fired generators $P_i^{CG}(t)$, the on/off states of coal-fired generators $U_i(t)$, and the startup/shutdown variables of coal-fired generators $Y_i(t)/Z_i(t)$. As shown in Eq. (35.7), the proposed unit commitment model is aimed at minimizing the generation costs $c_i^{CG} P_i^{CG}(t)$, the startup and shutdown costs, $c_i^{U} Y_i(t)$ and $c_i^{D} Z_i(t)$, of all coal-fired generators over a 24-hour time horizon. Φ^{CG} is the set of coal-fired generators, and Φ^{T} is the set of hours. c_i^{CG} is the cost per MWh of the ith coal-fired generator, and c_i^{U} and c_i^{D} are the startup and shutdown costs per time of the ith coal-fired generator. Equation (35.8) is the balance for power supply and load demand, where $P_j^{RG}(t)$ and $P_k^{TL}(t)$ are the power of the jth renewable generator and the kth interchange tie-line; the system total load $P^{Load}(t)$ consists of two parts, $P^{Orig}(t)$ and $P^{Net}(t)$, i.e. the original system load and the total net electric heating load in Eq. (35.6); Φ^{RG} and Φ^{TL} are the sets of renewable generators and interchange tie-lines, respectively. Constraint (35.9) shows the spinning reserve requirement, where $P_{(i,\max)}^{CG}$ is the installed capacity of the ith coal-fired generator, and $R(t)$ is the reserve requirement at time slot t. Constraint (35.10) shows the lower and upper limits for coal-fired generators' power, where $P_{(i,\min)}^{CG}$ is the minimal power when the ith coal-fired generator is online. Constraints (35.11) and (35.12) show the relationships between $U_i(t)$, $Y_i(t)$, and $Z_i(t)$. Constraints (35.13) and (35.14) are the minimum on/off hours of coal-fired generators, where T_i^{U} and T_i^{D} are the minimum on and off hours of the ith coal-fired generator. In (35.15) and (35.16), the bounds of the decision variables are defined.

By optimizing the unit commitment model in a day-ahead rolling manner, the daily optimal scheduling strategies for coal-fired generators $\mathbf{P}^{CG*}$, $\mathbf{U}^*$, $\mathbf{Y}^*$, and $\mathbf{Z}^*$ can be obtained, which set up the operation plans of generators. Let C_d^{UC} be the optimal objective value of the unit commitment model for the dth day, representing the daily costs of thermal coal. Thus, the total thermal coal consumption during N days can be calculated as follows:

$$Q^{GCoal} = \sum_{d=1}^{N} \left(\frac{C_d^{UC}}{\lambda^{Coal}} \right) \tag{35.17}$$

where Q^{GCoal} is the total thermal coal consumption, and λ^{Coal} is the price of thermal coal.

Note that in the power balance constraint (35.8), the renewable power, the interchange tie-line power and the system load are collected from power grid companies in four provinces. Given a renewable power and electric heating scenario, we can input the renewable power data and net electric heating load data into the unit commitment model accordingly, and obtain the associated thermal coal consumption. For example, for the scenario with additional renewable energy, we proportionally expand the hourly renewable power data. Based on these data, the unit commitment model is optimized and the total thermal coal consumption for this scenario can be obtained.

Due to data limitation, power transmission and distribution networks are not incorporated in the electricity dispatch model.

35.2.5 National Carbon Emission Estimation

In this section, we simulate the heating coal consumption Q^{HCoal} and electric heating load P^{Elec} by EnergyPlus in HB, HN, SD, and SX provinces. Due to data limitation, we estimate the total coal consumption in Northern China by expanding the results of the four provinces.

To estimate the heating coal consumption in Northern China, we firstly calculate the per capita heating coal consumption in the four provinces, which is assumed to equal that in Northern China:

$$\overline{Q}^{HCoal} = \frac{\overline{Q}_{HB}^{HCoal} + \overline{Q}_{HN}^{HCoal} + \overline{Q}_{SD}^{HCoal} + \overline{Q}_{SX}^{HCoal}}{(1 - p^{PIR})(\rho_{HB} + \rho_{HN} + \rho_{SD} + \rho_{SX})} \tag{35.18}$$

where $\overline{Q}_{NC}^{HCoal}$ is the per capita heating coal consumption in Northern China. $\overline{Q}_{HB}^{HCoal}$, $\overline{Q}_{HN}^{HCoal}$, $\overline{Q}_{SD}^{HCoal}$, and $\overline{Q}_{SX}^{HCoal}$ are the provincial heating coal consumption in HB, HN, SD, and SX, respectively. ρ_{HB}, ρ_{HN}, ρ_{SD}, and ρ_{SX} represent the rural population in HB, HN, SD, and SX, respectively. Then the total heating coal consumption in Northern China is estimated as follows:

$$Q_{NC}^{HCoal} = \overline{Q}_{NC}^{HCoal} \times \rho_{NC} \times (1 - p^{PIR}) \tag{35.19}$$

where Q_{NC}^{HCoal} is the heating coal consumption in Northern China. ρ_{NC} represents the rural population in Northern China.

To estimate the thermal coal consumption in Northern China, we firstly calculate the per capita electric heating load in the four provinces, which is assumed to equal that in Northern China:

$$\overline{P}_{NC}^{Elec} = \frac{P_{HB}^{Elec} + P_{HN}^{Elec} + P_{SD}^{Elec} + P_{SX}^{Elec}}{p^{PIR}(\rho_{HB} + \rho_{HN} + \rho_{SD} + \rho_{SX})} \tag{35.20}$$

where $\overline{P}_{NC}^{Elec}$ is the per capita electric heating load in Northern China. P_{HB}^{Elec}, P_{HN}^{Elec}, P_{SD}^{Elec}, and P_{SX}^{Elec} are the provincial electric heating load in HB, HN, SD, and SX, respectively. Then the total electric heating load in another province in Northern China is estimated as follows:

$$P_x^{Elec} = \overline{P}_{NC}^{Elec} \times \rho_x \times p^{PIR} \tag{35.21}$$

where subscript x represents a province in Northern China. In 2015, the rural populations in Heilongjiang, Jilin, Liaoning, Beijing, Tianjin, Inner Mongolia, Shaanxi, Ningxia, Gansu, Xinjiang, and Tibet are 15.70, 12.30, 14.31, 2.93, 2.69, 9.97, 17.48, 2.99, 14.77, 2.92, 12.54, and 2.34 million, respectively. The thermal coal consumption caused by electric heating load in province x is calculated as follows:

$$Q_x^{GCoal} = P_x^{Elec} \times a_x \tag{35.22}$$

where a_x represents the provincial average thermal coal consumption rate (TCCR), measured in kg/MWh. Therefore, the total thermal coal consumption caused by electric heating load in Northern China Q_{NC}^{GCoal} can be estimated as follows:

$$Q_{NC}^{GCoal} = \sum_{x \in \Phi^{NC}} Q_x^{GCoal} \tag{35.23}$$

where Φ^{NC} represents the set of the provinces in Northern China. Given Eqs. (35.18)–(35.23), we can estimate the total amount of heating coal and thermal coal in Northern China.

The heating value of standard coal is $h^{\text{Coal}} = 2.93 \times 10^7$ J/kg, and the net carbon content per energy is $\alpha^{\text{Coal}} = 26.59$ tC/TJ. In this section, the oxidization rate of raw coal is set as $o^{\text{HCoal}} = 83.7\%$, and that of thermal coal is set as $o^{\text{GCoal}} = 99\%$. Therefore, the emission factors of raw coal and thermal coal, denoted by e^{HCoal} and e^{GCoal}, are calculated below:

$$e^{\text{HCoal}} = o^{\text{HCoal}} \times h^{\text{Coal}} \times \alpha^{\text{Coal}} = 2.39 \text{tCO}_2/t \tag{35.24}$$

$$e^{\text{GCoal}} = o^{\text{GCoal}} \times h^{\text{Coal}} \times \alpha^{\text{Coal}} = 2.83 \text{tCO}_2/t \tag{35.25}$$

The carbon emissions from the rural residential heating coal and thermal coal consumption in Northern China are shown as follows:

$$E_{\text{NC}}^{\text{C}} = e^{\text{HCoal}} Q_{\text{NC}}^{\text{HCoal}} + e^{\text{GCoal}} Q_{\text{NC}}^{\text{GCoal}} \tag{35.26}$$

where E_{NC}^{C} is the CO_2 emitted from power generation and rural raw coal for space heating in Northern China.

35.2.6 Cost Analysis for Microgrid Heating

In this section, we analyze the average annualized cost per household in 25 cities in HB, HN, SD, and SX provinces. The price of raw coal is estimated to be 0.76 ¥/kg in 2015. A household's annualized cost consists of two fractions, namely, the investment cost and the electricity bill.

The annualized investment cost is calculated as follows:

$$\overline{C}^{\text{I}} = (C^{\text{I}} - S^{\text{I}}) r / [1 - (1 + r)^{-\text{PP}}] \tag{35.27}$$

where $\overline{C}^{\text{I}}$ and C^{I} are the annualized and one-off investment costs, respectively. S^{I} is the one-off subsidy for electric heating, equaling ¥ 7,400 (about US$ 1000) in Northern China. r is the interest rate, equaling 4.85%, and PP is the payback period, which is assumed to be 10 years.

Note that the one-off investment cost C^{I} varies with different electric heating devices. In Northern China, we estimate that the average electric heating capacity is 8 kW. The one-off investment for an 8-kW EH is ¥ 2000, and that for an 8-kW HP is ¥ 28,000. Additionally, the average rooftop PV capacity of rural residents in China is set to 3 kW, and the one-off investment cost for PV is 5000 ¥/kW.

For the households with EHs or HPs, the electricity bill during the heating season is calculated as follows:

$$\overline{C}^{\text{E}} = \sum_{t \in \Phi^{\text{HS}}} P^{\text{Elec}}(t) \times (\pi^{\text{R}} - S^{\text{E}}) \tag{35.28}$$

where $\overline{C}^{\text{E}}$ is the electricity bill caused by electric heating. π^{R} is the retail tariff, equaling 0.36, 0.38, 0.39, and 0.33 ¥/kWh in HB, HN, SD, and SX, respectively. S^{E} is the discount tariff for electric heating, equal to 0.12 ¥/kWh. Note that each household is settled on an hourly basis in Eq. (35.28), where Φ^{HS} is the set of the hours during the heating season and $P^{\text{Elec}}(t)$ is the hourly electric heating load. For the households with PVEH, the electricity bill is calculated as follows:

$$\overline{C}^{\text{E}} = \sum_{t \in \Phi^{\text{HS}}} P_{+}^{\text{Elec}}(t) \times (\pi^{\text{R}} - S^{\text{E}}) - \sum_{t \in \Phi^{\text{Y}}} P^{\text{Solar}}(t) \times S^{\text{PV}} \tag{35.29}$$

where $P_{+}^{\text{Net}}(t)$ is the net electric heating load after taking the positive value, i.e. $P_{+}^{\text{Net}}(t) = \max\{0, P^{\text{Elec}}(t) - P^{\text{Solar}}(t)\}$. Φ^{Y} is the set of the hours in a year, i.e. 1, 2,..., 8760. S^{PV} represents the subsidy for hourly solar generation, e.g. $S^{\text{PV}} = 0.42$ ¥/kWh for PAP.

35.3 Case Studies

35.3.1 Provincial Carbon Emissions Caused by Electric Heating Policy

We quantify the carbon emissions from power generation and rural residential heating in four provinces in Northern China during the heating season in 2015. According to the data in 11 cities in Northern China, electric heaters (EHs) generally account for 79.24–100% of the electric heating devices among rural residents. Instead, some residents use heat pumps (HPs) and photovoltaic-powered electric heating (PVEH). The average proportions of EHs, HPs, and PVEH are 91.01%, 6.58%, and 2.41%, respectively. As illustrated in Fig. 35.1, the implementation of EHP leads to a significant increase in the provincial carbon emissions. Here, we focus on two major sources of uncertainties, i.e. PIR and electric heating mix (EHM). PIR refers to the population of rural residents using electric heating over that of provincial rural residents. "Plan for Winter Clean Heating in Northern China (2017–2021)," issued by National Development and Reform Commission in 2017, requires 50% of rural residents in Beijing, Tianjin, and 14 other provinces in Northern China to substitute electric heating for raw coal by 2019. We design three comparative cases with the PIR equaling 45%, 50%, and 55%, respectively. EHM refers to the proportions of EHs, HPs, and photovoltaic-powered electric heating (PVEH). The proportions of EHs, HPs, and PVEH are scanned within the intervals [80%, 100%], [0, 20%], and [0, 10%], respectively.

Considering the joint uncertainties, the incremental carbon emissions in Hebei (HB), Henan (HN), Shandong (SD), and Shanxi (SX) reach [9.32, 14.97], [12.32, 21.23], [20.97, 32.16], and [9.21, 14.57] megatons, respectively. Base case is designed with 50% PIR and the average EHM. In the base case, SD releases the largest amount of CO_2 equaling 27.63 megatons, while the other three provinces, i.e. HB, HN, and SX, produce 12.42, 17.20, and 12.15 megatons CO_2, respectively.

We observe that the uncertainty in EHM (the boxplots) may yield a greater impact on the incremental carbon emissions than that in PIR (the bars). Based on the average EHM, the largest deviations of carbon emission induced by PIR uncertainty are 2.46, 3.76, 4.94, and 2.54 megatons in HB, HN, SD, and SX, respectively. However, on the premise of a fixed PIR, the largest deviations

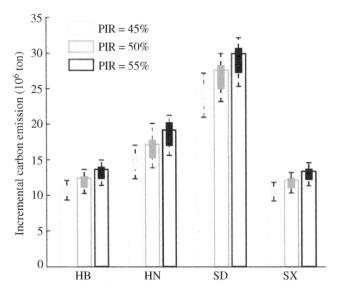

Figure 35.1 Carbon emission estimation in Hebei (HB), Henan (HN), Shandong (SD), and Shanxi (SX) provinces after implementing Electric Heating Policy in 2015.

of carbon emission caused by EHM uncertainty can reach 3.58, 6.22, 6.85, and 3.22 megatons in the four provinces, respectively.

Each three bars for a province represent the incremental carbon emissions with an average electric heating mix when the PIR equals 45%, 50%, and 55%, respectively. Each box represents the incremental carbon emission acquired by scanning the electric heating mix ($n = 12$). The minimum/maximum of each box indicates the minimal/maximal value of incremental carbon emission, and the lower and upper percentiles are 25% and 75%, respectively.

Our results demonstrate that the diversity of provincial carbon emissions mainly comes from three key factors: (1) the climate conditions such as AAT, (2) the RRP using electric heating, and (3) the TCCR. The AAT has a direct impact on household coal consumption and electric heating load (Fig. 35.2(a)). The lower the AAT is, the more heat energy is needed to maintain the indoor

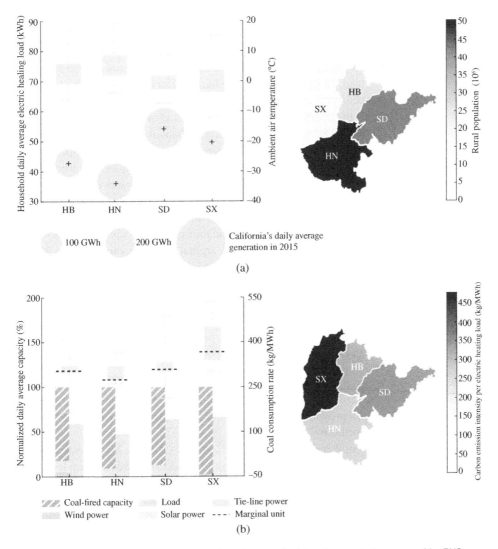

Figure 35.2 Analysis of the key factors influencing provincial carbon emissions caused by EHP. (a) Household daily average electric heating load and rural resident population in Hebei, Henan, Shandong, and Shanxi provinces. (b) Daily average generation capacity factors and carbon emission intensity per electric heating load.

temperature. In Fig. 35.2(a), the median values of hourly AAT in HN and SD in winter are 5.45 and −0.56°C, respectively. As a result, SD has the highest daily average electric heating load of a single household among the four provinces, equaling 54.35 kWh, while the value in HN is the lowest, equaling 36.71 kWh. Figure 35.2(a) also shows that the incremental electric heating loads are positively related to provincial RRP. Such load increment is significant and even comparable to the daily generation of California, USA. The provincial daily average electric heating load in SD is estimated to reach 353.23 GWh, accounting for 65.61% of California's daily average generation in 2015.

Figure 35.2(a) indicates that household daily average electric heating load and RRP in Hebei (HB), Henan (HN), Shandong (SD), and Shanxi (SX) provinces. The left figure shows the relationship between the household daily average electric heating load and hourly ambient air temperature (AAT) in winter. The center of each bubble represents the load of a single household, and the radius represents the provincial daily average electric heating load of all rural residents. Each box shows the distribution of hourly AAT ($n = 1464$). The right figure shows the RRP of four provinces, and the gray area, i.e. the northern region of HB, is excluded from the analysis due to data limitation. Figure 35.2(b) indicates that daily average generation capacity factors and carbon emission intensity per electric heating load. In the left figure, the bars represent the normalized generation capacity and electric load before electric heating, and the total available generation capacity is scaled to 100%. Each box shows the distribution of the TCCRs of coal-fired generators ($n = 97$ for HB, $n = 153$ for HN, and $n = 190$ for SD and SX). The minimum/maximum of each box indicates the minimal/maximal value, and the lower and upper percentiles are 25% and 75%, respectively.

As illustrated in Fig. 35.2(b), the TCCR of the marginal unit directly influences the carbon emission intensity per electric heating load. As China highly relies on coal for electricity generation, the incremental electric heating loads are generally balanced by marginal coal-fired generators on top of the existing generation resources, except for the cases with renewable energy curtailment. In HN, the generation capacity factor is the lowest among the four provinces, equaling 47.78%. The coal-fired generators have relatively low TCCRs, with a marginal value equaling 275.9 kg/MWh. As a result, the carbon emission intensity per electric heating load in HN is only 397.51 kg/MWh. However, as a large power exporting province, the generation capacity factor in SX is the highest, equaling 66.33%. The marginal TCCR in SX is 368 kg/MWh, and thus the carbon emission intensity per electric heating load can reach as high as 635.71 kg/MWh.

35.3.2 National Impacts of Electric Heating Policy

We extend the base case results in HB, HN, SD, and SX to the other provinces in Northern China, considering 50% of the rural residents substituting electric heating in place of raw coal (Fig. 35.3(a)). Our results show three clusters of provincial incremental carbon emissions, which reveals the distribution of rural residents in Northern China.

In the base case, the incremental carbon emission in Northern China in 2015 is estimated to reach 135.60 megatons. Considering the joint uncertainty in PIR and EHM, the emission level may vary from 101.69 to 162.89 megatons (Fig. 35.3(b)). The carbon emissions caused by China's EHP are comparable to the annual total emissions in different countries across the world. For example, such incremental carbon emission approximately accounts for 31.02–49.69% of France's annual emission.

Furthermore, the impacts of EHP on China's carbon mitigation in the future are investigated (Fig. 35.3(c)). In 2020, we estimate that the incremental carbon emission can reach 168.80 megatons in the base case, and may vary from 130.03 to 197.87 megatons due to PIR and EHM uncertainty.

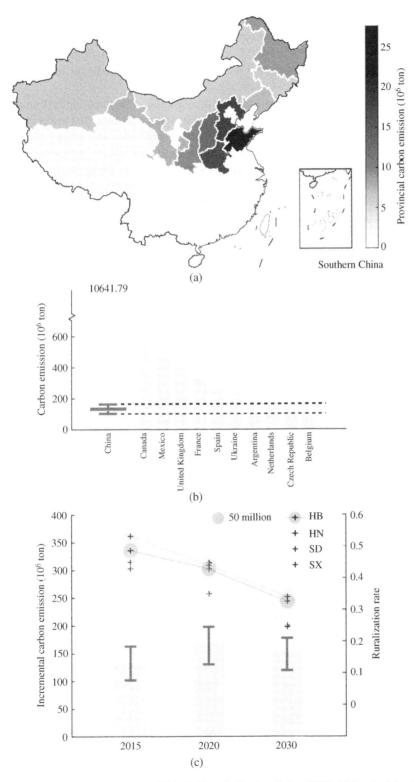

(a)

(b)

(c)

Figure 35.3 National impacts of China's Electric Heating Policy (EHP). (a) Provincial carbon emission increase after implementing EHP among 50% of rural residents. (b) Comparisons of carbon emissions between China and other countries in 2015. (c) Incremental carbon emissions after implementing EHP in Northern China in 2015, 2020, and 2030.

On the other hand, China's urbanization progress will slow down the growth in the carbon emissions caused by EHP. Compared with 2015, the rural population in 2030 is expected to decrease from 48.67% to 32.54% in HB, from 53.15% to 34.00% in HN, from 42.99% to 25.00% in SD and from 44.98% to 24.59% in SX. While the PIR may further increase to about 90% in 2030, the incremental carbon emission caused by EHP will drop to 119.19–177.47 megatons.

Figure 35.3(a) indicates that provincial carbon emission increase after implementing EHP among 50% of rural residents. The gray areas represent the provinces in Southern China that are excluded from the analysis in this section. Figure 35.3(b) indicates that comparisons of carbon emissions between China and other countries in 201,530. The bar represents a country's annual carbon emission in 2015. The box shows the variation of China's incremental carbon emission considering the uncertainty in the PIR and electric heating mix ($n = 36$). The minimum/maximum of the box indicates the minimal/maximal value of national incremental carbon emission, and the lower and upper percentiles are 25% and 75%, respectively. Figure 35.3(c) indicates that incremental carbon emissions after implementing EHP in Northern China in 2015, 2020, and 2030. The bars represent the incremental carbon emissions with an average electric heating mix considering the PIR equaling 50% in 2015, 70% in 2020 and 90% in 2030. The error bar defines the range of incremental carbon emission with the PIR varying from 45% to 55% in 2015, from 65% to 75% in 2020, and from 85% to 95% in 2030 ($n = 2$). The radius of each bubble shows the provincial population in Hebei (HB), Henan (HN), Shandong (SD), and Shanxi (SX), and the center indicates the provincial ruralization rate, i.e. the rural population over the total amount.

35.3.3 Techno-Economic Analysis for Microgrid Heating

Renewable energy curtailment has long been a severe issue in China, yielding an enormous waste of clean energy resources. The national renewable energy curtailment could reach over 80 TWh in 2015, and a lot of wind and solar energy was curtailed in the northern and western provinces in China, including Gansu, Inner Mongolia, Xinjiang, etc. The interconnected ultrahigh-voltage direct/alternating current (UHVDC/AC) transmission systems provide a natural platform for balancing electric heating load with interregional renewable energy. The impacts of matching electric heating load with renewable energy in the four provinces are illustrated in Fig. 35.4(a). With more electric heating loads satisfied by renewable energy, the carbon emissions caused by EHP keep decreasing. To totally offset the incremental carbon emissions, the requirements for annual additional renewable energy in HB, HN, SD, and SX can reach 19.20, 25.21, 36.06, and 12.28 TWh, accounting for 0.60%, 0.71%, 0.70%, and 0.88% of provincial electricity consumption in 2015, respectively. Note that the marginal carbon emission reduction declines due to an increasing curtailment of renewable energy, which is significantly apparent in HB, HN, and SD.

The incremental carbon emissions caused by EHP can be effectively limited by installing distributed photovoltaic (PV) resources (Fig. 35.4(b)). In contrast to the case without PV, the incremental carbon emissions in the four provinces after installing 10-kW PV can be reduced by 5.09, 10.35, 14.71, and 8.89 megatons, respectively. However, as illustrated in Fig. 35.4(b), the accommodation capability for PV declines, leading to more solar energy curtailment and thus a decreasing carbon emission reduction rate. Additionally, we discover that provincial carbon emission reduction per capita per kW is highly related to the level of irradiance. The daily average irradiance in SD is 2.88 kW/m^2, which is stronger than those in HN and HB. Thus, SD has a higher carbon emission reduction rate. However, in spite of a slightly weaker irradiance in SX, the carbon emission reduction rate in SX is higher than that in SD. This is because SX has a greater carbon emission intensity per electric heating load (Fig. 35.2(b)).

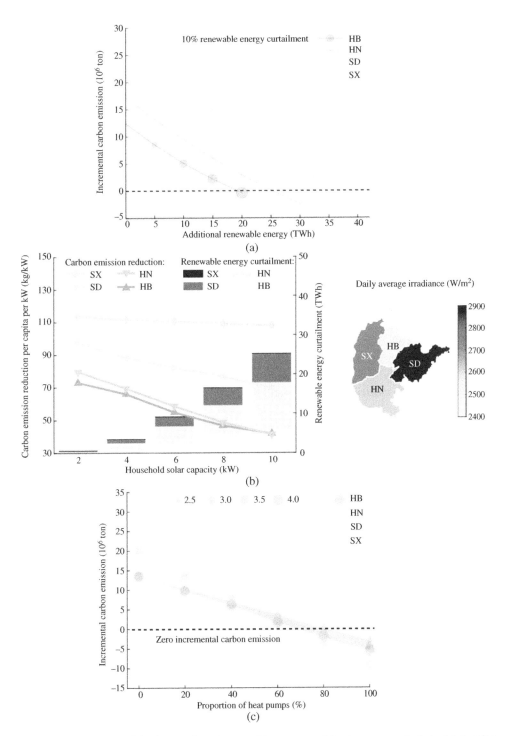

Figure 35.4 Impacts of the integration of renewable energy and the improvement of electric heating devices. (a) Carbon emissions in Hebei, Henan, Shandong, and Shanxi with the integration of renewable energy. (b) Carbon emission reduction capability of distributed solar in Hebei, Henan, Shandong, and Shanxi. (c) Relationship between carbon emission and the proportion of heat pumps used for electric heating.

Figure 35.4(a) indicates that carbon emissions in Hebei (HB), Henan (HN), Shandong (SD), and Shanxi (SX) with the integration of renewable energy. The radius of each bubble represents the curtailment rate for additional renewable energy, and the center indicates the incremental carbon emission caused by electric heating. Figure 35.4(b) indicates that carbon emission reduction capability of distributed PV in the four provinces. The lines represent the carbon emission reduction per capita per kW, and the bars show the curtailment of solar energy. Figure 35.4(c) shows the relationship between carbon emission and the proportion of heat pumps (HPs) used for electric heating. The radius of each bubble represents the coefficient of performance (COP) of an HP, and the shadows are bounded by the lines with COP equaling 2.5 and 4.0.

As illustrated in Fig. 35.4(c), it is an effective solution to reduce carbon emissions by popularizing HPs instead of EHs. This is because the COP of an HP is generally higher than that of an EH. To totally offset the incremental carbon emission, the requirements for HP proportion are estimated to reach [68.33%, 80.26%] in HB, [63.56%, 74.69%] in HN, [70.67%, 82.86%] in SD, and [78.28%, 91.82%] in SX.

Furthermore, we analyze the average annualized cost per household in 25 cities in Northern China (Fig. 35.5). Here, we compare seven cases, including coal, EHs, HPs, rooftop solar with poverty alleviation program (PAP), high-level subsidy, medium-level subsidy, and rooftop solar without subsidy. In China, the solar energy for poverty alleviation program aims to expand over 10 GW distributed PV capacity, benefiting more than 2 million rural households by 2020. The subsidy for PAP is 0.42 ¥/kWh. For other distributed PV systems that are not involved in the PAP, we consider (1) high-level subsidy equaling 0.37 ¥/kWh, implemented before 2019; (2) medium-level subsidy equaling 0.18 ¥/kWh, implemented after 2019; and (3) the case without subsidy. We discover that compared with electric heating, it is still the most cost-saving way for rural residents to

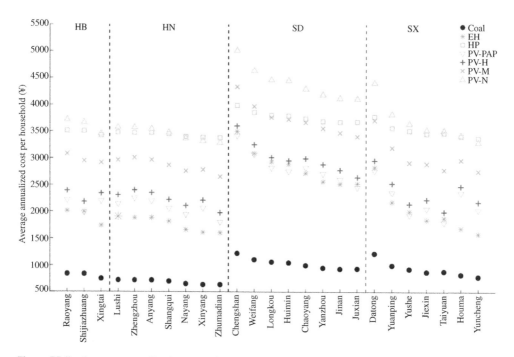

Figure 35.5 Average annualized cost per household in different cities in Northern China.

burn raw coal for space heating in winter. Due to the relatively low price for raw coal, a household only needs to spend ¥ 633.69–1222.40 in winter to consume 333.52–643.37 kg raw coal.

Additionally, electric heating requires rural residents to pay extra money for investment and electricity bills. In most cities, using electric heaters is a cost-efficient solution because of the low capital costs. The average annualized costs for different households vary from ¥ 1583.30 to ¥ 3500.05, which are much less than those spent by using HPs, i.e. ¥ 3377.21–3990.57. This indicates that the current price of an HP is still too high for a residential household. In contrast to an EH, the cost savings from electricity bills cannot even recover the investment for an HP. As illustrated in Fig. 35.5, the subsidy for solar energy plays an important role in popularizing the distributed PV systems among rural residents. The costs in PV-PAP are the least among solar-powered electric heating cases because of the highest subsidies, and even less than those by using EHs in some cities. In PV-PAP and PV-H where the subsidy is high, the average annualized costs vary within ¥ [1792.77, 3413.05] and ¥ [1975.64, 3605.11], respectively. However, Chinese government announced to reduce the subsidy for solar energy to 0.18 ¥/kWh in 2019. The costs in PV-M and PV-N significantly increase to ¥ [2656.17, 4334.98] and ¥ [3300.87, 5026.42], respectively.

"Coal" represents that a rural household burns raw coal for space heating in winter. "EH" and "HP" represent that a rural household uses an 8-kW electric heater and heat pump, respectively. "PV-PAP," "PV-H," "PV-M," and "PV-N" represent that a rural household uses solar-powered electric heating with a 3-kW solar panel and 8-kW electric heater, and the subsidies for solar energy are 0.42, 0.37, 0.18, and 0 ¥/kWh, respectively. Note that "PAP," "H," "M," and "N" are short for poverty alleviation program, high-level subsidy, medium-level subsidy, and no subsidy.

35.4 Discussion

To reduce the pollutant emissions from rural residents, Chinese government has issued Electric Heating Policy to substitute electric heating in place of burning raw coal. However, electric heating can lead to a significant increase in load demands from microgrids. We estimate that in the base case with 50% PIR, the load increase in HB, HN, SD, and SX can reach 24.56, 43.27, 53.87, and 19.11 TWh, respectively. The incremental electric loads require more electricity from coal-fired power plants, thus releasing more carbon emissions. Compared with locally burning raw coal, it is less efficient to use EHs for space heating. The generating efficiency of power plants is generally around 40%, the power loss on transmission and distribution networks is about 6–10%, and the COP of an EH is approximately 80%. This indicates that a large fraction of energy is dissipated along electricity generation, transmission, distribution, and consumption sectors. As a result, substituting 1 kg of raw coal requires 1.89, 1.68, 1.90, and 2.17 kg of thermal coal to satisfy the electric heating load in HB, HN, SD, and SX, respectively (Fig. 35.2(b)). On the other hand, the oxidization rate of thermal coal is much higher than that of raw coal, indicating that thermal coal has a higher carbon emission factor. Therefore, in spite of an effective raw coal reduction among rural residents, China's EHP can lead to significant carbon emissions released from the power sector.

It should be noted that both power transmission and distribution networks are not incorporated in our theoretical model due to data limitation, which may underestimate future curtailment of renewable energy. Therefore, we claim that this chapter provides a conservative estimation for incremental carbon emissions induced by China's EHP. The impact of network congestion on renewable energy curtailment and the associated carbon emissions deserves an in-depth investigation in future work.

Two low-carbon electric heating pathways are suggested for China and other countries with similar situations, i.e. balancing electric heating load with renewable energy, and improving the efficiency of electric heating. Specifically, the carbon emission increase caused by EHP can be effectively offset by integrating the interprovincial renewable energy. In 2018, for example, 85.42 GWh of electric heating load was directly satisfied by wind and solar stations in SX. However, the marginal carbon emission reduction gets low with the increase in renewable energy penetration, which is validated in both cases with province-level renewable energy and distributed PV resources (Fig. 35.4(a) and (b)). On the other hand, to totally offset the carbon emissions induced by EHP, the proportion of HPs is estimated to increase to approximately 60–90% in Northern China (Fig. 35.4(c)).

In the past several years, Chinese government has encouraged rural residents in Northern China to switch to electric heating by offering subsidies. In HB, for example, a household can be subsidized with ¥ 7400 (about US$ 1000) in a one-off scheme to invest in electric heating devices. Additionally, the subsidy for electric heating load can reach 0.12 ¥/kWh, approximately accounting for 20–25% of the retail tariff. In spite of such profitable policy, electric heating is still too expensive for rural residents in Northern China. The average annualized costs per household for using EHs and HPs are estimated to reach ¥ [1583.30, 3500.05] and ¥ [3377.21, 3990.57], respectively, much more than those for burning raw coal, i.e. ¥ [633.69, 1222.40]. In addition, we claim that financial subsidy can yield a great impact on the annualized costs for using PVEH. The costs in PV-PAP can even reach less than those by using EHs in some cities, e.g. Chengshan, Longkou, Huixian (Fig. 35.5). In SD, the costs in PV-PAP are less than those by using EHs because the AATs in these cities are relatively low and high electric heating load is required, leading to extra electricity bills. However, in the southern cities in HN where the AATs are generally high, EHs show a cost-efficient advantage over PV-PAP.

In this section, we summarize three policy suggestions for China and other developing countries. First, the government must explore the potential incompatibility between any new policy and the existing ones. According to our analyses, the underestimation of the greenhouse effect caused by EHP can impede China's carbon mitigation process in the future. Meanwhile, an increasing penetration of electric heating may lead to the shortage of generation capacity and flexible load-following resources, thus threatening the secure and reliable operation of microgrids. Second, the government is suggested to match large-scale renewable generation with electric heating load. Considering the relatively high capital costs for HPs and PV, it is better to accommodate the surplus renewable energy in China's northern and western provinces. Note that the Shanxi government has gained success in organizing the bilateral trading between wind/solar stations and some villages using EHs for heating. However, we also suggest that the carbon reduction performance of developing inter-provincial renewable energy and distributed solar systems should be systematically evaluated in the future considering potential network congestion. Third, we suggest that the government should provide adequate incentives to encourage electric heating among rural residents. In China, the current capital cost for an HP is still too high for rural residential space heating. The one-off subsidy and electricity bill discount are insufficient to popularize HPs as dominant heating devices.

35.5 Exercises

1. Will rural electrification cause an increase in carbon dioxide emissions, and why?
2. Please explain the relationship between the microgrid operation model and the provincial electricity dispatch model.
3. Please propose a potential pathway to mitigate greenhouse impact of rural electrification.

35.6 Acknowledgement

This chapter was supported in part by National Key Research and Development Program (2022YFB2405600) and in part by National Natural Science Foundation of China (52277092).

References

1 Wang, J., Chen, L., Tan, Z. et al. (2023). Inherent spatiotemporal uncertainty of renewable power in China. *Nature Communications* 14: 5379.

2 Liu, J. and Diamond, J.M. (2005). China's environment in a globalizing world. *Nature* 435: 1179–1186.

3 Algarni, A.S., Suryanarayanan, S., Siegel, H.J., and Maciejewski, A.A. (2021). Combined impact of demand response aggregators and carbon taxation on emissions reduction in electric power systems. *IEEE Transactions on Smart Grid* 12 (2): 1825–1827.

4 Wang, J., Zhong, H., Yang, Z. et al. (2020). Exploring the trade-offs between electric heating policy and carbon mitigation in China. *Nature Communications* 11: 6054.

5 Wang, H., Lu, X., Deng, Y. et al. (2019). China's CO_2 peak before 2030 implied from characteristics and growth of cities. *Nature Sustainability* 2 (8): 748–754.

6 Guo, Y., Tian, J., and Chen, L. (2020). Managing energy infrastructure to decarbonize industrial parks in China. *Nature Communications* 11 (1): 981.

7 He, G., Lin, J., Sifuentes, F. et al. (2020). Rapid cost decrease of renewables and storage accelerates the decarbonization of China's power system. *Nature Communications* 11 (1): 2486.

8 Yu, Y., Wang, J., Chen, Q. et al. (2023). Decarbonization efforts hindered by China's slow progress on electricity market reforms. *Nature Sustainability* 6: 1006–1015.

9 Zhi, G., Zhang, Y., Sun, J. et al. (2017). Village energy survey reveals missing rural raw coal in northern China: significance in science and policy. *Environmental Pollution* 223: 705–712.

10 Hao, J., Chen, Q., He, K. et al. (2020). A heat current model for heat transfer/storage systems and its application in integrated analysis and optimization with power systems. *IEEE Transactions on Sustainable Energy* 11 (1): 175–184.

11 Huang, R.-J., Zhang, Y., Bozzetti, C. et al. (2014). High secondary aerosol contribution to particulate pollution during haze events in China. *Nature* 514 (7521): 218–222.

12 Zhang, T., Wang, J., Wang, H. et al. (2022). On the coordination of transmission-distribution grids: a dynamic feasible region method. *IEEE Transactions on Power Systems* 38 (2): 1855–1866.

36

Operations of Microgrids with Meshed Topology Under Uncertainty

Mikhail A. Bragin, Bing Yan, Akash Kumar, Nanpeng Yu, and Peng Zhang

To efficiently operate microgrids with renewable generation, a stochastic modeling through Markov chains is adopted. To prevent fast degradation of expensive equipment due to high fluctuations of voltages, tap changes are penalized. To recover frequency and voltages to nominal values upon the disconnection of microgrids from the main grid, droop controls are used. To avoid approximations leading to errors, an exact AC power flow in rectangular coordinates suitable for meshed topologies is adopted. These considerations will be discussed in the following Section 36.1.

36.1 Self-sufficiency and Sustainability of Microgrids Under Uncertainty

To enable self-sufficiency to hedge against blackouts caused by natural disasters, microgrids typically include generators, combined heat and power, and batteries. With the recent push for clean, green, and renewable energy, microgrids may also include solar panels. Other grid infrastructure may also include electric vehicle (EV) charging stations.

Under normal conditions, microgrids are typically connected to the main grid (e.g. a power distribution system) and may exchange power. The normal operations, therefore, include proactive power generation at the least cost in anticipation of the increase/decrease in customer demand, as well as in anticipation of fluctuations of power generation from renewables. The intermittency of renewables can, however, lead to fluctuations of voltage, which may lead to frequent adjustments of taps within on-load tap changers, thereby leading to fast degradation of expensive equipment and adversely impacting microgrid economic viability. Stochasticity, as well as the discrete nature of the underlying problem, needs to be explicitly captured to design the optimal (or near-optimal) control to ensure a low-cost power supply to local communities.

Under faulty conditions or upon detection of low voltage/frequency on the main grid side, an interconnecting device such as a circuit breaker opens to switch the microgrid to the islanded mode. In this mode, the microgrid's goal is to serve the local loads by using locally available distributed energy resources (DERs). The microgrid's operations are also complicated by the fact that disconnection may result in a drop in voltage and frequency. Therefore, in addition to the considerations described in the previous paragraph, droop controls need to be used to restore voltage and frequency to nominal ranges.

Microgrids: Theory and Practice, First Edition. Edited by Peng Zhang.

To appropriately model droop controls as well as tap changers, both requiring the inclusion of voltages, non-linear AC power flow constraints are necessary. Moreover, because of the high R/X-ratios (the amount of resistance R divided by the amount of reactance X) within distribution systems, the popular DC power flow model is no longer suitable for formulating the power flows within microgrids.

36.1.1 AC Power Flow

36.1.1.1 AC Power Flow for Radial Topologies

The AC Power Flows are known for their non-linearity and non-convexity and the associated AC optimal power flow (AC-OPF) problem is known to be extremely difficult. Several convex relaxations have been developed such as *DistFlow* model [1–4] as well as the *Second-Order Cone Relaxation* (SOCR) [5, 6]. The SOCR technique is summarized below.

Consider a radial (sometimes referred to as "tree") network with $\mathcal{T}$ being a lookahead horizon: $\mathcal{T} = \{1, \dots, T\}$; let $\mathcal{B}$ be a set of buses indexed by b, $\mathcal{I}_b$ be a set of generators at bus b indexed by i, and $\mathcal{L}$ be a sets of lines indexed by l. The following set of constraints captures relationships among active $\left(f_{l,t}^p\right)^2$ and reactive $\left(f_{l,t}^q\right)^2$ power flows squared as well as the current squared $a_{l,t}$ in line l, and the voltage squared $v_{s(l),t}$ at the sending end $s(l)$ of line l as:

$$\left(f_{l,t}^p\right)^2 + \left(f_{l,t}^q\right)^2 \leq v_{s(l),t} \cdot a_{l,t}, l \in \mathcal{L}, t \in \mathcal{T} \tag{36.1}$$

To capture voltage drops across line l between voltages squared at sending $s(l)$ and receiving $r(l)$ buses, the following constraint is used:

$$v_{r(l),t} - v_{s(l),t} = 2 \cdot \left(R_l \cdot f_{l,t}^p + X_l \cdot f_{l,t}^q\right) - a_{l,t} \cdot \left(R_l^2 + X_l^2\right), l \in \mathcal{L}, t \in \mathcal{T} \tag{36.2}$$

where parameters R_l and X_l are the reactance and impedance of line l.

Since power flow at sending and receiving buses of each line l differs due to losses incurred by transmission, the apparent power flow limit $\bar{f}_l$ is enforced for the sending and receiving buses separately within the following two sets of constraints:

$$\left(f_{l,t}^p\right)^2 + \left(f_{l,t}^q\right)^2 \leq \bar{f}_l^2, l \in \mathcal{L}, t \in \mathcal{T} \tag{36.3}$$

$$\left(f_{l,t}^p - a_{l,t} \cdot R_l\right)^2 + \left(f_{l,t}^q - a_{l,t} \cdot X_l\right)^2 \leq \bar{f}_l^2, l \in \mathcal{L}, t \in \mathcal{T} \tag{36.4}$$

The bus voltages squared are limited by $\underline{v_b}^2$ and $\overline{v_b}^2$ as:

$$\underline{v_b}^2 \leq v_{b,t} \leq \overline{v_b}^2, b \in \mathcal{B}, t \in \mathcal{T}. \tag{36.5}$$

With the added load $L_{b,t}^p$ and generation levels $p_{i,t}^p$, the nodal power balance is enforced following [[5], eq. (3)] as:

$$\sum_{l|s(l)=b} f_{l,t}^p - \left(f_{l,t}^p - a_{l,t} \cdot R_l\right)_{l|r(l)=b} - \sum_{i \in \mathcal{I}_b} p_{i,t}^p + L_{b,t}^p + v_{b,t} \cdot G_{l|s(l)=b} = 0, b \in \mathcal{B}, t \in \mathcal{T} \tag{36.6}$$

where $G_l \equiv \frac{R_l}{R_l^2 + X_l^2}$ is conductance of line l. If bus b does not contain generators, then $\sum_{i \in \mathcal{I}_b} p_{i,t}^p = 0$, and if bus b does not contain load, then $L_{b,t}^p = 0$. Because of the radial topology, the summation of power flows is performed with respect to lines whereby b is a "sending" bus because any node can have several children nodes and at most one parent node. Thus, any node is able to receive power from only one parent, although with a loss of $a_{l,t} \cdot G_l$. Nodal power flow for reactive power is similarly introduced:

$$\sum_{l|s(l)=b} f^D_{l,t} - \left(f^q_{l,t} - a_{l,t} \cdot X_l \right)_{l|r(l)=b} - \sum_{i \in \mathcal{I}_b} p^q_{i,t} + L^q_{b,t} + v_{b,t} \cdot B_{l|s(l)=b} = 0, b \in \mathcal{B}, t \in \mathcal{T} \tag{36.7}$$

where $B_l \equiv \frac{-X_l}{R_l^2 + X_l^2}$ is susceptance of line l.

The above "relaxed" formulation is convex, which is amenable for commercial solvers, unlike the original non-convex AC power flow. A notable feature of the formulation is that the voltage squared $v_{b,t}$ and the current squared $v_{b,t}$ are decision variables, and Eq. (36.1) is convex despite the bilinear terms and Eq. (36.5) is linear. For the meshed topologies, however, the above formulation is inexact, and to achieve exactness, the linearity and convexity will not be preserved as will be explained ahead.

36.1.1.2 AC Power Flow for Meshed Topologies

The distribution networks (and microgrids as their part) are not necessarily radial, as acknowledged for the next generation of power distribution systems [7]. The AC power flow in rectangular coordinates is appropriate for both radial and meshed topologies. For every node b, the net active/reactive power generated and transmitted to the node should be equal to the net power consumed and transmitted from node b:

$$\sum_{i \in \mathcal{I}_b} p^p_{i,t} + \sum_{l|r(l)=b} f^p_{l,t} = L^p_{b,t} + \sum_{l|s(l)=b} f^p_{l,t}, b \in \mathcal{B}, t \in \mathcal{T} \tag{36.8}$$

If bus b does not contain generators, then $\sum_{i \in \mathcal{I}_b} p^p_{i,t} = 0$, and if bus b does not contain load, then $L^p_{b,t} = 0$. The nodal power flow balance constraints for reactive power are similarly defined:

$$\sum_{i \in \mathcal{I}_b} p^q_{i,t} + \sum_{l|r(l)=b} f^q_{l,t} = L^q_{b,t} + \sum_{l|s(l)=b} f^q_{l,t}, b \in \mathcal{B}, t \in \mathcal{T} \tag{36.9}$$

Following [16–20], AC power flow is modeled in rectangular coordinates by using complex voltages $v_{b,t} = v^{Re}_{b,t} + j \cdot v^{Im}_{b,t}$. In the complex plane, complex voltages can be represented as row vectors $v_{b,t} = \left(v^{Re}_{b,t}, v^{Im}_{b,t} \right)$ and power flows can be written as:

$$f^p_{l,t} = v_{s(l),t} \cdot \begin{pmatrix} G_{s(l),r(l)} & -B_{s(l),r(l)} \\ B_{s(l),r(l)} & G_{s(l),r(l)} \end{pmatrix} \cdot \left(v_{r(l),t} \right)' \tag{36.10}$$

$$f^q_{l,t} = v_{s(l),t} \cdot \begin{pmatrix} -B_{s(l),r(l)} & -G_{s(l),r(l)} \\ G_{s(l),r(l)} & -B_{s(l),r(l)} \end{pmatrix} \cdot \left(v_{r(l),t} \right)' \tag{36.11}$$

Here $B_{s(l),r(l)}$ is susceptance and $G_{s(l),r(l)}$ is conductance of line $(s(l), r(l))$. Node $s(l)$ denotes the "sending" node of line l, and $r(l)$ denotes the "receiving" node of the line l.

The complex voltage within each node b is subject to the following restrictions:

$$\underline{v_b}^2 \leq (v^{Re}_{b,t})^2 + (v^{Im}_{b,t})^2 \leq \overline{v_b}^2, b \in \mathcal{B}, t \in \mathcal{T} \tag{36.12}$$

Unlike (36.5), inequalities (36.12) are neither linear nor convex.

Power flows in each line l satisfy the following capacity constraints:

$$\left(f^p_{l,t} \right)^2 + \left(f^q_{l,t} \right)^2 \leq \overline{f_l}^2, l \in \mathcal{L}, t \in \mathcal{T} \tag{36.13}$$

Despite the striking similarity to (36.3), the inequality (36.13) is non-convex because of the non-convexity of AC power flows in rectangular coordinates (36.10) and (36.11).

One way to handle the associated non-convexity is by defining a convex hull of the solutions; however, even if the tightest convex relaxation is attained [8], the solution may still be inside of the convex hull rather than at its boundary, which could only be guaranteed for linear programming problems. Another approach to handling non-linearity and non-convexity is through *dynamic linearization* [9] – The linearization around the current operation point, which may change from iteration to iteration. To guarantee feasibility, the so-called "l_1-proximal" terms have been used to gradually penalize the violations of current solutions from previously obtained ones until convergence to a steady-state solution. The latter technique will be explained at length in Section 36.3.2.

36.1.2 Intermittent Renewables

As descried in Chapter 37, deterministic approaches were adopted in several studies on the operation of microgrids (e.g. [10]), where uncertainties were not explicitly captured. To explore the intermittent and uncertain nature, stochastic programming has also been used based on representative scenarios (e.g. [11]), while it is difficult to select an appropriate scenario number while balancing modeling accuracy, computational efficiency, and solution feasibility. To overcome these difficulties, a Markovian approach was developed [12], where renewable states at a particular time period probabilistically capture the past information, thereby reducing the complexity compared to a total enumeration of all possible states; in particular, the complexity is significantly reduced compared to scenario-based methods. In our previous work [13], a Markov-based model was established to integrate intermittent and uncertain PV generation into microgrids.

Following [13], a Markov-based model is adopted for PV generation. Uncertainties due to weather changes follow a Markovian process with N discrete states, with a probability of nth state at time t denoted as $\phi_{n,t}$. Based on historical data, the transition probability to state n from state m is $\pi_{m,n}$.

The probability $\phi_{n,t}$ of a PV state n at time t is a weighted summation of probabilities at time $t-1$:

$$\phi_{n,t} = \sum_{m=1,\dots,N} \pi_{n,m} \cdot \phi_{n,t-1} \tag{36.14}$$

Generally, probabilities of PV levels at subsequent time periods are obtained based on PV levels at previous time periods and the transition matrix consisting of $\{\pi_{m,n}\}$.

36.1.3 Tap Changers

Tap changers [14] are used to reduce voltage deviations due to intermittent renewables by keeping voltage amplitudes within pre-specified limits. However, high levels of renewable penetration, voltage fluctuations, and amplitudes may increase greatly, thereby leading to frequent adjustment of the transformer tap position to preserve power quality and reliability. As a result, the tap-changer transformers may reach the end of their useful lives or experience premature failures. Hawaii utilities, for example, reported that their tap changers, previously maintenance-free for over 40 years, require quarterly maintenance with a prospect of retirement within a mere two years of exploitation, since the taps frequently need to be adjusted over 300 times a day due to voltage fluctuations [15].

Assume that tap changers are associated with the corresponding solar farms. Following [14], and assuming that bus indices are omitted for brevity, tap-changer constraints are:

$$V_{n,t} = \frac{1}{a_{n,t}} V^{\text{in}} - \frac{a_{n,t} Z_{t(a)} S_n}{(V^{\text{in}})^*} \tag{36.15}$$

where V^{in} is voltage input, a is a transformer-turn ratio under no-load conditions as: V^{in}/V_n, $Z_{t(a)}$ is transformer leakage impedance, generally a function of a, but here $Z_{t(a)}$ is assumed to be a complex number, and S is the transformer load. This constraint is non-linear and delineates non-convex regions.

36.2 Microgrid Model: Proactive Operation Optimization Under Uncertainties

Consider a microgrid with a partly connected meshed topology operated by a microgrid system operator (MSO) under an assumption of a balanced single-phase AC power flow. Solar generation is modeled by using N discrete states (denoted by n) with associated probabilities $\phi_{n,t}$ at each time period t and transition probabilities $\pi_{n,m}$ from state n to state m.

36.2.1 Objective

The goal of MSO is to minimize the expected generation cost together with the expected tap-changer cost:

$$\min_{\mathbf{F}, \mathbf{P}, \mathbf{V}, \mathbf{X}} O(\mathbf{P}) = \min_{\mathbf{F}, \mathbf{P}, \mathbf{V}, \mathbf{X}} \left\{ \mathbb{E}(\mathbf{C} \cdot \mathbf{P}') + \mathbb{E}(\mathbf{C}^{\text{tap}} \cdot \mathbf{D}') \right\} \tag{36.16}$$

where $\mathbf{P} = \left\{ p_{i,n,t}^p, p_{i,n,t}^q \right\}$ is a vector consisting of generation levels with the corresponding costs $\mathbf{C} = \left\{ C_{i,t}^p, C_{i,t}^q \right\}$, and $\mathbf{D} = \left\{ d_{b,n,t}^{\text{up}}, d_{b,n,t}^{\text{down}} \right\}$ is a vector of the change of tap position up and down at PV state n with the corresponding tap-changing costs $\mathbf{C}^{\text{tap}}$. The expectation operator is defined as $\mathbb{E}(\mathbf{C} \cdot \mathbf{G}') \equiv \sum_{i,n,t} \phi_{n,t} \cdot C_{i,t}^p \cdot p_{i,n,t}^p + \sum_{i,n,t} \phi_{n,t} \cdot C_{i,t}^q \cdot p_{i,n,t}^q$. The second term in (36.16) is similarly defined. Other decision variables of the problem include $\mathbf{X} = \left\{ x_{i,t} \right\}$ – a vector of binary commitment decision variables, $\mathbf{F} = \left\{ f_{l,n,t}^p, f_{l,n,t}^q \right\}$ – a vector of continuous power flow decision variables and $\mathbf{V} = \left\{ v_{b,n,t}^{\text{Re}}, v_{b,n,t}^{\text{Im}} \right\}$ – a vector of continuous voltage decision variables. The optimization (36.16) is subject to the following constraints.

36.2.1.1 Generation Capacity Constraints
Generation levels satisfy the following generation capacity constraints:

$$\underline{\mathbf{P}} \cdot \mathbf{X} \leq \mathbf{P} \leq \overline{\mathbf{P}} \cdot \mathbf{X} \tag{36.17}$$

where $\underline{\mathbf{P}} = \left\{ \underline{p}_i^p, \underline{p}_i^q \right\}$ and $\overline{\mathbf{P}} = \left\{ \overline{p}_i^p, \overline{p}_i^q \right\}$ are the minimum and maximum generation levels, respectively.

36.2.1.2 Ramp-Rate Constraints
For probable transitions, ramp-rate constraints require that the change of generation levels between two consecutive time periods does not exceed ramp rates $\mathbf{R} = \left\{ r_i^p, r_i^q \right\}$:

$$-\mathbf{R} \leq \mathbf{P}_{n,t} - \mathbf{P}_{m,t-1} \leq \mathbf{R}, \forall \left(\pi_{n,m} \neq 0 \right) \tag{36.18}$$

36.2.1.3 Droop-Control Constraints

It is assumed that generator $\hat{\imath} \in \mathcal{I}_{\hat{b}}$ employs a droop-control strategy and the corresponding droop-control constraints are:

$$f_{n,t} = f_{n,t}^{\text{ref}} - k_f \cdot (p_{\hat{\imath},n,t}^{p,\text{ref}} - p_{\hat{\imath},n,t}^p), \tag{36.19}$$

$$\left| v_{\hat{b},n,t} \right| = \left| v_{\hat{b},n,t}^{\text{ref}} \right| - k_v \cdot (p_{\hat{\imath},n,t}^{q,\text{ref}} - p_{\hat{\imath},n,t}^q), \tag{36.20}$$

where $\left| v_{\hat{b},n,t} \right| \equiv \sqrt{(v_{\hat{b},n,t}) \cdot (v_{\hat{b},n,t})'}$.

36.2.1.4 Tap-Changer Constraints

For mathematical expediency, (36.15) is equivalently written as:

$$a_{n,t}(V^{\text{in}})^* V_n = V^{\text{in}}(V^{\text{in}})^* + (a_{n,t})^2 Z_{t(a)} S_n \tag{36.21}$$

The transformer turn ratio decision variables $a_{n,t}$ are controlled as:

$$a_{n,t} = a_0 + d_{n,t} \cdot \Delta a \tag{36.22}$$

where a_0 is the "rated turn ratio," Δa is the single tap position change, and $\{d_{n,t}\}$ are integer decision variables denoting tap positions as:

$$d_{n,t} = d_{n,t-1} + d_{n,t}^{\text{up}} - d_{n,t}^{\text{down}} \tag{36.23}$$

The above problem belongs to a class of Mixed-Integer Non-Linear Programming problems notable for non-convexities brought by non-linear AC power flows, droop controls, and tap changers as well as by discrete decision variables.

36.3 Solution Methodology

To solve the above problem, the "l_1- proximal" Surrogate Lagrangian Relaxation method [9] is extended to handle tap-changer and droop-control constraints. After linearizing the problem and relaxing nodal flow balance constraints, Lagrangian multipliers are updated based on constraint violations, and the violations are penalized using "absolute-value" penalties; to ensure overall feasibility, "l_1- proximal" terms are used.

36.3.1 Surrogate Absolute-Value Lagrangian Relaxation

36.3.1.1 Relaxed Problem

After relaxing nodal flow balance (36.8) and (36.9), and penalizing their violations, the relaxed problem becomes:

$$\min_{\mathbf{F,P,V,X}} L_c(\mathbf{P}; \Lambda) = \min_{\mathbf{F,P,V,X}} \left\{ O(\mathbf{P}) + \Lambda \cdot \mathbf{R} + c \cdot \|\mathbf{R}\|_1 \right\} \tag{36.24}$$

s.t. (36.10)–(36.13), (36.17)–(36.23),

where $\Lambda_j = \left(\Lambda_j^P, \Lambda_j^Q \right)$ are multipliers relaxing power flow balance constraints (36.8) and (36.9). The vector, $\mathbf{R} = (\mathbf{B}^P, \mathbf{B}^Q)'$ denotes a vector of power flow balance constraint violations. Following [9], multipliers are updated as:

$$\Lambda^k = \Lambda^{k-1} + s^k \cdot \mathbf{R}^k \tag{36.25}$$

where

$$s^k = \alpha^k \cdot s^{k-1} \cdot \frac{\left\| \mathbf{R}^{k-1} \right\|_2}{\left\| \mathbf{R}^k \right\|_2} \tag{36.26}$$

where α^k is defined as

$$\alpha^k = 1 - \frac{1}{M \cdot k^{1 - \frac{1}{k^r}}}, M > 1, r > 0 \tag{36.27}$$

In the beginning, c^k is increased by a constant $\beta > 1$:

$$c^k = c^{k-1} \cdot \beta \tag{36.28}$$

After constraint violations become zero, a feasible solution is obtained, and c^k needs to decrease as:

$$c^k = c^{k-1} \cdot \beta^{-1} \tag{36.29}$$

Subsequently, the c^k is not increased.

36.3.2 Surrogate "l_1-Proximal" Lagrangian Relaxation

Voltage restrictions (36.12), droop control for voltage (36.18), tap-changer constraints (36.21), AC power flows (36.10) and (36.11), as well as line-capacity constraints (36.13) are non-linear. Moreover, the left-hand side constraint of (36.12) as well as constraints (36.10) and (36.11) delineate non-convex regions. Following the work [9], these difficulties are addressed next. The method of [9] is then extended to resolve non-linearity difficulties brought by the newly considered droop control (36.18) and tap-changer constraints (36.21).

In the following, the main linearization principles will be delineated first. Then, the feasibility will be established.

1. **Linearization of cross-product terms within** (36.10) and (36.11): To linearize AC power flow while updating all the voltages, the following formulas are used:

$$\hat{f}^p_{l,t} = \frac{1}{2} v^{k-1}_{s(l),t} \cdot \begin{pmatrix} G_l & -B_l \\ B_l & G_l \end{pmatrix} \cdot v'_{r(l),t} + \frac{1}{2} v_{s(l),t} \cdot \begin{pmatrix} G_l & -B_l \\ B_l & G_l \end{pmatrix} \cdot \left(v^{k-1}_{r(l),t} \right)' \tag{36.30}$$

$$\hat{f}^q_{l,t} = \frac{1}{2} v^{k-1}_{s(l),t} \cdot \begin{pmatrix} -B_l & -G_l \\ G_l & -B_l \end{pmatrix} \cdot v'_{r(l),t} + \frac{1}{2} v_{s(l),t} \cdot \begin{pmatrix} -B_l & -G_l \\ G_l & -B_l \end{pmatrix} \cdot \left(v^{k-1}_{r(l),t} \right)' \tag{36.31}$$

2. **Linearization of voltage restrictions** (36.12): First, Eq. (36.12) is compactly written as $\underline{\mathbf{V}}^2 \leq \mathbf{V} \cdot \mathbf{V}' \leq \overline{\mathbf{V}}^2$. Then, the squared terms within the inequalities are linearized as:

$$\underline{\mathbf{V}}^2 \leq \mathbf{V} \cdot \left(\mathbf{V}^{k-1} \right)' \leq \overline{\mathbf{V}}^2 \tag{36.32}$$

To avoid infeasibility along the iterative procedure, "soft" penalization is introduced through the non-negative penalty variables $\underline{\mathbf{V}}^{\text{pen}} = \{ \underline{v}^{pen}_{b,t} \}$ and $\overline{\mathbf{V}}^{\text{pen}} = \{ \overline{v}^{pen}_{b,t} \}$ as:

$$\underline{\mathbf{V}}^2 - \underline{\mathbf{V}}^{\text{pen}} \leq \mathbf{V} \cdot \left(\mathbf{V}^{k-1} \right)' \leq \overline{\mathbf{V}}^2 + \overline{\mathbf{V}}^{\text{pen}} \tag{36.33}$$

To enforce the feasibility of (36.33), $\underline{\mathbf{V}}^{\text{pen}}$ and $\overline{\mathbf{V}}^{\text{pen}}$ will be penalized.

In a similar vein, Eq. (36.13) is linearized as

$$\hat{\mathbf{F}} \cdot \left(\hat{\mathbf{F}}^{k-1} \right)' \leq \overline{\mathbf{F}}^2 + \overline{\mathbf{F}}^{\text{pen}} \tag{36.34}$$

with subsequent relaxation and penalization of $\overline{\mathbf{F}}^{\text{pen}}$ as explained ahead. Note that here the linearized power flows $\hat{\mathbf{F}}$ from (36.30) and (36.31) are used.

3. **Linearization of tap-changer constraints** (36.21): Tap-changer constraints contain both the cross-products and the squared terms. Following the ideas of above-presented linearization, the linearization is performed as follows:

$$\frac{1}{2}a_{n,t}^{k-1}(V^{\text{in}})^* V_n + \frac{1}{2}a_{n,t}(V^{\text{in}})^* V_n^{k-1} = V^{\text{in}}(V^{\text{in}})^* + a_{n,t}^{k-1}a_{n,t}Z_{t(a)}S_n \tag{36.35}$$

The linearization of line-capacity constraints (36.13) is operationalized in the same way as described in points 2 and 3 above.

4. **Linearization voltage droop-control constraints** (36.20): The first step of linearization of (36.20) is squaring both sides of the equation as such:

$$\left| v_{\hat{b},n,t} \right|^2 = \left| v_{\hat{b},n,t}^{\text{ref}} \right|^2 - 2 \cdot \left| v_{\hat{b},n,t}^{\text{ref}} \right| \cdot \left(k_v \cdot (g_{i,n,t}^{q,\text{ref}} - g_{i,n,t}^q) \right) + k_v^2 \cdot \left(g_{i,n,t}^{q,\text{ref}} - g_{i,n,t}^q \right)^2 \tag{36.36}$$

The left-hand side and the last term of the right-hand side of (36.36) are squared terms, which are linearized as described above. Other terms of (36.36) are either constant or linear and require no further linearization.

To ensure that solutions to the linearized problem satisfy the corresponding original (non-linear) problem terms $\left\| \mathbf{V} - \mathbf{V}^{k-1} \right\|_1$, $\left\| \hat{\mathbf{F}} - \hat{\mathbf{F}}^{k-1} \right\|_1$, and $\left\| \mathbf{A} - \mathbf{A}^{k-1} \right\|_1$, which capture the deviations of voltages $\mathbf{V}$, linearized power flows $\hat{\mathbf{F}}$ and transformer turns ratios $\mathbf{A} \equiv \{a_{n,t}\}$ from previously obtained values are introduced and are penalized by c_p^k. The intention here is to discourage oscillations of solutions while encouraging their approach to common values. To avoid getting trapped at previously obtained values, c_p^k is chosen to be lower than c^k.

36.3.2.1 Linearized Relaxed Problem

The resulting MSO relaxed problem then becomes:

$$\min_{\mathbf{A}, \hat{\mathbf{F}}, \mathbf{P}, \mathbf{V}, \mathbf{X}} \left\{ \begin{array}{c} L_{c^k}(\mathbf{P}; \boldsymbol{\Lambda}^{k-1})+ \\ c_p^k \left(\left\| \mathbf{V} - \mathbf{V}^{k-1} \right\|_1 + \left\| \hat{\mathbf{F}} - \hat{\mathbf{F}}^{k-1} \right\|_1 + \left\| \mathbf{A} - \mathbf{A}^{k-1} \right\|_1 \right)+ \\ \left(\underline{\boldsymbol{\Lambda}}^V \right)^k \cdot \left(\underline{\mathbf{V}}^2 - \mathbf{V} \cdot (\mathbf{V}^{k-1})' \right)+ \\ \left(\overline{\boldsymbol{\Lambda}}^V \right)^k \cdot \left(\mathbf{V} \cdot (\mathbf{V}^{k-1})' - \overline{\mathbf{V}}^2 \right)+ \\ \left(\overline{\boldsymbol{\Lambda}}^F \right)^k \cdot \left(\hat{\mathbf{F}} \cdot (\hat{\mathbf{F}}^{k-1})' - \overline{\mathbf{F}}^2 \right)+ \\ c^k \cdot \underline{\mathbf{V}}^{\text{pen}} + c^k \cdot \overline{\mathbf{V}}^{\text{pen}} + c^k \cdot \overline{\mathbf{F}}^{\text{pen}} \end{array} \right\}, \tag{36.37}$$

s.t., (36.17)–(36.19), (36.22)–(36.23), (36.30)–(36.31), (36.33)–(36.36)

Multipliers are updated as in (36.25), with the difference that the vector of constraint violations $\mathbf{R}$ is appended by $\left(\underline{\mathbf{V}}^2 - \mathbf{V}^{k-1} \cdot \mathbf{V}' \right)$, $\left(\mathbf{V}^{k-1} \cdot \mathbf{V}' - \overline{\mathbf{V}}^2 \right)$, and $\left(\hat{\mathbf{F}} \cdot (\hat{\mathbf{F}}^{k-1})' - \overline{\mathbf{F}}^2 \right)$, and the vector of multipliers $\boldsymbol{\Lambda}$ is appended by $\underline{\boldsymbol{\Lambda}}^V$, $\overline{\boldsymbol{\Lambda}}^V$, and $\overline{\boldsymbol{\Lambda}}^F$. Multipliers corresponding to inequality constraints cannot be negative and are projected onto a positive orthant. Piece-wise linear l_1-norms within (36.37) are linearized by using standard special-ordered sets [21, 22].

36.3.2.2 Feasibility

As c^k increases, violation levels of relaxed constraints decrease. However, the total lack of constraint violations does not imply feasibility because, for example, linearized power flows do not coincide with original power flows. To ensure that $\hat{f}^p_{l,t} \to f^p_{l,t}$ and $\hat{f}^q_{l,t} \to f^q_{l,t}$, c^k_p increases as in (36.28):

$$c^k_p = c^{k-1}_p \cdot \beta_p. \tag{36.38}$$

The intention here is to force "l_1−proximal" terms to zero to ensure feasibility. The convergence of voltages is demonstrated in the following figures.

As demonstrated in Fig. 36.1, two arcs corresponding to the two inequalities (36.12) delineate the boundaries of the feasible region and voltage values v^{k-1} are outside of the feasible region. Linearized constraints encourage voltages to get inside of the feasible region, however, when far from the feasibility, voltages may be prone to oscillations because of the mixed-integer linear nature of the problem solved: 1. because of linearity, solutions may be at endpoints of linearized feasible region; 2. because of the presence of binary variables, there are natural jumps in solutions; as demonstrated in Fig. 36.1, the updated value v^k is also outside of the feasible region. As penalty coefficients for the l_1-proximal terms increase, the drastic changes in voltages are suppressed and the voltages gradually approach feasible values. In Fig. 36.2, the boxed-in portion of Figure 36.1 of the voltages is zoomed in to demonstrate convergence of voltages.

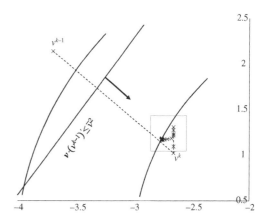

Figure 36.1 Voltage convergence behavior when far from feasibility.

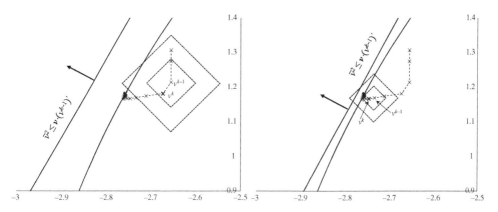

Figure 36.2 Voltage convergence behavior when close to feasibility.

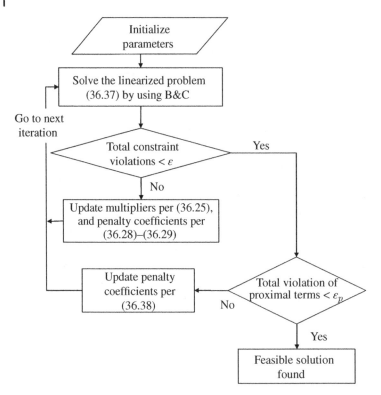

Figure 36.3 Flow chart of the method.

After constraint violations and proximal terms are both zero, the feasible solution is obtained. Forcing the violations exactly to zero may require a considerable number of iterations, so tolerances ϵ and ϵ_p are introduced for constraint violations and proximal terms, respectively. The algorithm is shown in the flow chart in shown in Fig. 36.3.

36.4 Conclusions

Recent blackouts in California, New York, and Britain in 2019, which left millions of customers without power, pushed for an increase in the power system's resilience, stability, and reliability. DERs, such as solar panels and wind turbines, have provided a promising solution. In order to prompt the applications of renewable energy or DERs, microgrid is a promising paradigm. This chapter addresses the issues of sustainability of microgrid operations. Through the consideration of Markov processes to capture stochasticity of renewable generation; through tap-changer constraints penalizing frequent changes of taps, with which the lifespan of the expensive equipment will be greatly extended; through consideration of AC power-flow constraints appropriate for distribution systems; and through droop-control constraints for restoration of frequency and voltage after disconnection from the main grid. The above-mentioned constraints lead to non-convexities and difficulties solving the resulting microgrid operation optimization problems. The methodology based on dynamic linearization and l_1-norm penalization is exact and is amenable to the use of MILP solvers. After linearization, the only source of non-convexities is the presence of discrete variables, which are efficiently handled by MILP solvers. It is demonstrated that voltages approach

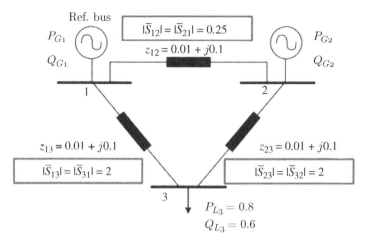

Ref. bus

P_{G1}

Q_{G1}

$|\bar{S}_{12}| = |\bar{S}_{21}| = 0.25$

$z_{12} = 0.01 + j0.1$

P_{G2}

Q_{G2}

1

2

$z_{13} = 0.01 + j0.1$

$|\bar{S}_{13}| = |\bar{S}_{31}| = 2$

$z_{23} = 0.01 + j0.1$

$|\bar{S}_{23}| = |\bar{S}_{32}| = 2$

3

$P_{L3} = 0.8$

$Q_{L3} = 0.6$

Figure 36.4 A 3-bus system having generator and renewable energy.

feasible values, leading to exact satisfaction of AC power flows ensuring the feasibility of operations. As a result of the above, the overall performance of the method leads to convergence. Microgrid operations benefit from the new solution methodology; specifically, the number of tap changes is expected to reduce, thereby ensuring higher sustainability in the presence of voltage fluctuations caused by uncertainties. The main advantage over previous studies is the consideration of key elements that lead to the feasibility of sustainability of microgrid operations; namely, exact AC power flow, droop control, and tap-changer constraints are considered without simplification, and the methodology ensures their exact satisfaction. With an increase in the number of microgrids, which are expected to be interconnected, it is suggested that the methodology is extended to coordinate several microgrids; on the bright side, upon the decomposition inherent in the method developed, the microgrid complexity associated with both the number of discrete variables and the number of stochastic states can be significantly reduced. The present work: 1. has enabled efficient resolution of microgrid subproblems and 2. paves the way for efficient coordination of microgrids. The coupling constraints that connect adjacent microgrids can be dualized by the Surrogate Lagrangian Relaxation to efficiently coordinate microgrids for high resilience and reliability spanning larger geographical areas.

36.5 Exercises

Consider the below 3-bus system in which P_{G1} represents a generator connected on bus-1 and P_{G2} represents photovoltaic (PV) system connected on bus-2 through a step-up transformer. Both the energy sources are connected to each other through $L_{1,2}$ which has $z_{1,2} = 0.01 + j0.1$. Bus-3 has load of $P_{L3} = 0.8$ and $Q_{L3} = 0.6$ and is being powered by two lines $L_{1,3}$ and $L_{2,3}$.

$0 \leqslant P_{G1} \leqslant 3, -2 \leqslant Q_{G1} \leqslant 2$

PV generation at bus b and state n is given as follows:

$$P_{b,n} = \begin{bmatrix} 0 & 0 & 0 \\ 0.1 & 0.2 & 0.3 \\ 0 & 0 & 0 \end{bmatrix}, Q_{b,n} = \begin{bmatrix} 0 & 0 & 0 \\ 0.1 & 0.2 & 0.3 \\ 0 & 0 & 0 \end{bmatrix}$$

Transition Matrix is given as:

$$\pi_{m,n} = \begin{bmatrix} 0.66129 & 0.33871 & 0 \\ 0.215385 & 0.380769 & 0.234615 \\ 0.1 & 0.1 & 0.8 \end{bmatrix}$$

Energy price for generator at each hour is 2 units while energy price for PV at time t is shown below.

$$C_t = \begin{bmatrix} 3 & 3 & 3 & 1 & 1 & 1 & 3 & 3 \end{bmatrix}$$

Based on given data, simulate the above 3-bus system for 8 hours ($t = 1.8$) and determine:

1. Ignoring the tap changer and droop-control costs, find the total cost of energy if PV system and the generator are online all time (8 hours).
2. Ignoring droop control and tap changer, find the total cost if only the generator is online all the time (8 hours) while the PV system is disconnected?
3. Compare the total energy costs obtained from 1 and 2 with the total energy cost when generator is disconnected all the time with no droop control and tap changer.
4. What will be the total energy cost if both the tap changer and droop controls are included in the calculations for $t = 1 : 8$ hours?
5. PV system and generator are operating in synchronization but suddenly generator becomes offline at $t = 5$ and remains offline till $t = 8$. Comment and determine the frequency (Hz) at $t = 5$. Also, find the impact of generator ramp rates on droop-control response and determine the total cost of energy.
6. Plot and compare the power export from each source to the load considering the (4) and (5).

References

1 Yeh, H.G., Gayme, D.F., and Low, S.H. (2012). Adaptive var control for distribution circuits with photovoltaic generators. *IEEE Transactions on Power Apparatus and Systems* 27: 1656–1663.
2 Tan, S., Xu, J., and Panda, S.K. (2013). Optimization of distribution network incorporating distributed generators: an integrated approach. *IEEE Transactions on Power Apparatus and Systems* 28: 2421–2432.
3 Liu, G., Jiang, T., Ollis, T.B. et al. (2019). Distributed energy management for community microgrids considering network operational constraints and building thermal dynamics. *Applied Energy* 239: 83–95.
4 Wang, Z., Chen, H., Wang, J., and Begovic, M. (2014). Inverter-less hybrid voltage/var control for distribution circuits with photovoltaic generators. *IEEE Transactions on Smart Grid* 5: 2718–2728.
5 Farivar, M. and Low, S.H. (2013). Branch flow model: relaxations and convexification—Part I. *IEEE Transactions on Power Apparatus and Systems* 28: 2554–2564.
6 Torbaghan, S.S., Suryanarayana, G., Höschle, H. et al. (2020). Optimal flexibility dispatch problem using second-order cone relaxation of AC power flows. *IEEE Transactions on Power Apparatus and Systems* 35: 98–108.
7 Heydt, G. (2010). The next generation of power distribution systems. *IEEE Transactions on Smart Grid* 1: 225–235.

8 Li, Q. and Vittal, V. (2016). The convex hull of the AC power flow equations in rectangular coordinates. *Proceedings of the 2016 IEEE Power and Energy Society General Meeting (PESGM)*, Boston, MA, USA, 17 July 2016, 1–5.

9 Bragin, M.A. and Dvorkin, Y. (2022). TSO-DSO operational planning coordination through "l_1-Proximal" surrogate Lagrangian relaxation. *IEEE Transactions on Power Systems* 37 (2): 1274–1285. https://doi.org/10.1109/TPWRS.2021.3101220.

10 Guan, X., Xu, Z., and Jia, Q. (2010). Energy-efficient buildings facilitated by microgrid. *IEEE Transactions on Smart Grid* 1: 243–252.

11 Mohammadi, S., Soleymani, S., and Mozafari, B. (2014). Scenario-based stochastic operation management of microgrid including wind, photovoltaic, micro-turbine, fuel cell and energy storage devices. *Electrical Power and Energy Systems* 54: 1–7.

12 Luh, P.B., Yu, Y., Zhang, B. et al. (2014). Grid integration of intermittent wind generation: a Markovian approach. *IEEE Transactions on Smart Grid* 5: 732–741.

13 Yan, B., Luh, P.B., Warner, G., and Zhang, P. (2017). Operation optimization for microgrids with renewables. *IEEE Transactions on Automation Science and Engineering* 14: 573–585.

14 Kasztenny, B., Rosolowski, E., Izykowski, J. et al. (1998). Fuzzy logic controller for on-load transformer tap changer. *IEEE Transactions on Power Delivery* 13: 164–170.

15 Sokugawa, T. and Shawver, M. (2016). Big data for renewable integration at Hawaiian electric utilities. *Proceedings of the DistribuTech Conference & Exhibition*, Orlando, FL, USA, 9–11 February 2016, 9–11.

16 Torres, G.L., Quintana, V.H., and Lambert-Torres, G. (1996). Optimal power flow on rectangular form via an interior point method. *IEEE North American Power Symposium (NAPS)*. https://doi.org/10.13140/RG.2.1.1197.9127.

17 da Costa, V., Martins, N., and Pereira, J.L.R. (1999). Developments in the Newton Raphson power flow formulation based on current injections. *IEEE Transactions on Power Apparatus and Systems* 14: 1320–1326.

18 Zhang, X.P., Petoussis, S.G., and Godfrey, K.R. (2005). Nonlinear interior point optimal power flow method based on a current mismatch formulation. *EE Proceedings - Generation, Transmission and Distribution* 152: 795–805.

19 Bai, X., Wei, H., Fujisawa, K., and Wang, Y. (2008). Semidefinite programming for optimal power flow problems. *International Journal of Electrical Power & Energy Systems* 30: 383–392.

20 Li, Q., Yang, L., and Lin, S. (2015). Coordination strategy for decentralized reactive power optimization based on a probing mechanism. *IEEE Transactions on Power Apparatus and Systems* 30: 555–562.

21 Bragin, M.A., Luh, P.B., Yan, B., and Sun, X. (2019). A scalable solution methodology for mixed-integer linear programming problems arising in automation. *IEEE Transactions on Automation Science and Engineering* 16: 531–541. https://doi.org/10.1109/TASE.2018.2835298.

22 Sun, X., Luh, P.B., Bragin, M.A. et al. (2018). A novel decomposition and coordination approach for large-scale security constrained unit commitment problems with combined cycle units. *IEEE Transactions on Power Apparatus and Systems* 33: 5297–5308. https://doi.org/10.1109/PESGM.2017.8274098.

37

Operation Optimization of Microgrids with Renewables

Bing Yan, Akash Kumar, and Peng Zhang

To reduce energy costs and emissions of microgrids, the daily operation is critical. The problem is to commit and dispatch distributed energy devices with renewable generation to minimize the total energy and emission cost while meeting the forecasted energy demand (both electrical and thermal). Without considering power flows, it is usually formulated as a Mixed-Integer Linear Programming (MILP, with discrete and continuous variables and a linear structure) problem, and is believed to be NP-hard. The problem is challenging because of the intermittent nature of renewables. To efficiently operate microgrids with renewables, the generation uncertainties (take photovoltaic (PV) as an example) are modeled by a Markovian process. For effective coordination among the microgrid, other devices are modeled as Markov processes with states depending on PV states. The entire problem is Markovian, and the resulting combinatorial problem is solved using a powerful MILP method, i.e. branch-and-cut.

37.1 Introduction

With the world's increasing energy demand and growing environmental concerns, efficient utilization of energy is essential for sustainable living, especially renewable energy. As compared with conventional energy systems, microgrids support a flexible and reliable grid by the integration of conventional energy sources and renewables through efficiently incorporating different types of energy sources such as PV panels, wind turbines, and gas turbines (GT) as shown in Fig. 37.1 [1, 2]. Not only do microgrids offer improved efficiencies for electrical and thermal energy but also an isolated operation in emergencies. Moreover, they can transact energy with the main grid, which means that microgrids can sell energy to the grid when energy is in excess and buy from the grid when there is a deficiency. Besides this, microgrids can cater to thermal (i.e. space heating and cooling and domestic hot water (DHW)) demand by integrating natural gas boilers, eclectic chillers, etc., because of their proximity to connected loads. Such a complex network of different energy sources that can satisfy the time-varying loads requires well-coordinated operation to ensure reliability, cost-effectiveness, and sustainability.

In microgrids, different distributed energy devices, such as GTs, PV panels, and natural gas boilers, generate and store different types of energy such as electricity, steam, and hot/chilled water to satisfy time-varying electricity and thermal demand. They should be coordinated through optimized operation to reduce energy costs and greenhouse gas emissions. The microgrid operation

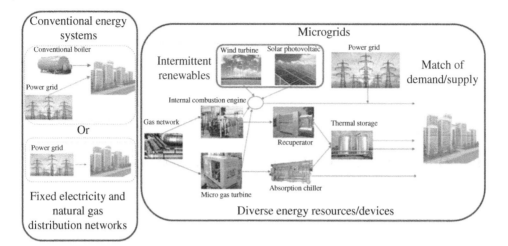

Figure 37.1 Comparison of microgrid and conventional energy systems. Source: Wikimedia Commons, chillersmexico, MICHELE MOSSOP, Fitzer Incorporation.

problem is hierarchical from unit commitment to economic dispatch to optimal power flow. Focusing on the first two, the problem in this chapter is to commit (e.g. on/off and start up) and dispatch (e.g. generation levels) distributed devices to minimize energy and CO_2 emission costs under the grid-connected mode while meeting hourly day-ahead electricity and thermal demand. The islanded mode and the transition between the two modes are not considered since they are not economics.

Optimized microgrid operation, however, is challenging because of the intermittent nature of renewables. In the literature, uncertainties were usually modeled by scenarios in microgrid operation problems as reviewed in Section 37.2. However, it is difficult to select an appropriate number of scenarios to balance modeling accuracy, computational efficiency, and solution feasibility. In this chapter, a mixed-integer linear model is established from the energy and emission point of view in Section 37.3. To avoid the difficulties associated with scenario-based methods, the idea is to model PV generation by a Markovian process with the current state summarizing all the past information. For effective coordination, other devices are modeled as Markov processes correspondingly with states depending on PV states. The entire problem is therefore Markovian. This combinatorial problem is solved using branch-and-cut as described in Section 37.4.

37.2 Existing Work

To model renewables, most of the practical applications adopt the deterministic approach, where intermittent and uncertain renewable generation is represented by its mean value without explicitly considering uncertainties [3]. The problem is formulated in a MILP format and then solved by existing methods, like branch-and-cut. For example in [4], solar irradiation was calculated offline using a deterministic approach.

Without explicitly capturing uncertainties, the solutions obtained by using deterministic approaches are not robust against realizations of renewable generation. On the research side, to model renewable uncertainties, stochastic programming, and robust optimization have been explored. In **stochastic programming**, uncertainties are modeled by representative scenarios [5]. To generate scenarios, renewable generation is typically assumed to follow a certain probability distribution, and each scenario represents a sequenced realization of generation uncertainties over the optimization horizon. In this method, optimization determines a single set of commitment decisions of energy devices to satisfy all the selected scenarios and multiple sets of dispatch decisions for the corresponding scenarios. The objective is to minimize the sum of the commitment cost and the expected dispatch cost. Decomposition methods, such as Benders' decomposition [6–8] and Lagrangian relaxation [9, 10], are commonly used to solve such problems. Since the number of scenarios can be extremely large even with discrete probability distributions, scenario reduction is commonly used [5]. However, it is difficult to determine a proper number of scenarios to balance modeling accuracy, solution feasibility, and computational efficiency.

In **robust optimization**, uncertainties are modeled by a predetermined set [11]. Optimization determines a single set of commitment decisions to be feasible for all possible realizations and a set of dispatch decisions against the worst-case realization. The objective is to minimize the worst-case cost. In robust optimization, Benders' decomposition is usually combined with other methods, such as outer approximation [12, 13] and cutting plane [14–16] to solve the problems. This method gives conservative solutions, and the models involve nonlinear min/max functions which require much computational effort [11].

To overcome the difficulties caused by scenario-based methods, a **Markovian approach** was developed [17]. Wind generation was modeled as a Markov chain, where a state represents the wind generation at a particular time interval, capturing all the past information. Since the number of states increases linearly with that of time intervals, the complexity is significantly reduced compared with scenario-based methods. Testing results in [17, 18] demonstrate the computational efficiency, the effectiveness to accommodate high-level wind penetration, and the ability to capture low-probability high-impact events of this approach. The approach thus represents a new and effective way to address stochastic problems without scenario analysis.

37.3 Mathematical Modeling

The microgrid under consideration involves different distributed energy devices as shown in Fig. 37.2: combined cooling heat and power (CCHP), PV panels, natural gas boilers, electrical chillers, and batteries, chosen among commonly used devices in practical microgrids. The CCHP

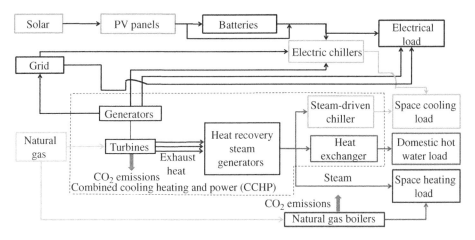

Figure 37.2 A representative microgrid.

system consists of multiple gas turbines and heat recovery steam generators, a steam-driven absorption chiller, and a heat exchanger, as sketched inside dashed lines. Electrical load and electricity required by electric chillers can be satisfied by grid power, CCHP, PV panels, and batteries. The microgrid can also sell extra electricity back to the grid. The electric and steam-driven chillers are used for space cooling, while steam and natural gas boilers for space heating. DHW load can be met by steam through the heat exchanger with sufficient exhaust heat from power generation. From the environmental point of view, combustion of natural gas in CCHP and boilers causes CO_2 emissions.

Consider the daily operation of a microgrid over 24 (T) hours with each hour indexed by t ($1 \leq t \leq T$). For devices, their properties such as cost functions and capacities are assumed known. Energy demand including electricity, space heating/cooling, and DHW is also assumed known at hour t. The operation problem is to decide the device operation strategies such as on/off statuses and generation levels to reduce the total energy and emission cost while meeting the given time-varying demand and satisfying individual device constraints. In this section, for microgrid operation under the grid-connected mode, a MILP model is established from the energy and emission point of view. Section 37.3.1 deals with the modeling of devices, and Section 37.3.2 focuses on PV generation modeling using the Markovian approach. Then battery and other devices are modeled based on the PV states in Section 37.3.3. Section 37.3.4 presents system balance, and the objective function is defined in Section 37.3.5.

37.3.1 Modeling of Devices

The device modeling focuses on ON/OFF statuses and generation levels as in [4, 19–23]. For simplicity, device efficiencies are assumed constant, although they generally depend on generation levels. The reason to choose fixed efficiencies is to maintain problem linearity as often used in the literature for microgrid operation optimization [4, 24]. The modeling of the CCHP, boilers, and chillers is presented as follows, and constraints generally include capacity, energy consumption, and emissions.

37.3.1.1 Modeling of CCHP

For CCHP, GTs are used to meet the electrical load by natural gas, while the fossil fuel combustion causes emissions. Then exhaust heat is recovered by recovery steam generators, and the high-temperature steam could be directly used for space heating or sent to the absorption chiller and heat exchanger for space cooling and DHW, respectively. Constraints for CCHP are presented as follows.

Capacity Constraint of GTs: The generation levels of mth GT $P_m^{\mathrm{GT}}(t)$ (continuous decision) at time t should be within minimum $P_m^{\mathrm{GT,min}}$ and maximum $P_m^{\mathrm{GT,max}}$ if device is ON [ON/OFF binary decision $x_m^{\mathrm{GT}}(t) = 1$], that is

$$P_m^{\mathrm{GT,min}} x_m^{\mathrm{GT}}(t) \leq P_m^{\mathrm{GT}}(t) \leq P_m^{\mathrm{GT,max}} x_m^{\mathrm{GT}}(t) \tag{37.1}$$

For other devices, this constraint is omitted.

Gas Consumption of GTs: The amount of natural gas consumed by mth GT $G_m^{\mathrm{GT}}(t)$ is calculated as:

$$G_m^{\mathrm{GT}}(t) = \frac{P_m^{\mathrm{GT}}(t)}{\eta^{e,\mathrm{GT}} HV^{\mathrm{Gas}}} \tag{37.2}$$

where $\eta^{e,\mathrm{GT}}$ is the gas-to-electric efficiency and HV^{Gas} is the heat value of natural gas.

Carbon Emission of GTs: The amount of CO_2 due to the natural gas combustion in the mth GT $\text{Env}_m^{\text{GT}}(t)$ is given as:

$$\text{Env}_m^{\text{GT}}(t) = G_m^{\text{GT}}(t)\text{HV}^{\text{Gas}}G^{\text{cin}} \tag{37.3}$$

where G^{cin} denotes the carbon intensity of natural gas.

Heat of Exhaust Gas in GTs: The amount of heat contained in the exhaust gas from mth GT $Q_m^{\text{GT}}(t)$ is:

$$Q_m^{\text{GT}}(t) = \frac{P_m^{\text{GT}}(t)\eta^{\text{th,GT}}}{\eta^{e,\text{GT}}} \tag{37.4}$$

where $\eta^{\text{th,GT}}$ is the thermal efficiency of the GT.

Total Steam: Steam generated by all steam generators could be directly used for space heating $Q^{\text{Steam-SH}}(t)$, sent to the absorption chiller for space cooling $Q^{\text{Steam-SC}}(t)$ or sent to the heat exchanger for DHW $Q^{\text{Steam-DHW}}(t)$ (continuous decisions), that is:

$$\sum_m Q_m^{\text{GT}}(t)\eta^{\text{HRSG}} = Q^{\text{Steam-SH}}(t) + Q^{\text{Steam-SC}}(t) + Q^{\text{Steam-DHW}}(t) \tag{37.5}$$

where η^{HRSG} is the energy efficiency of the steam generator.

Heat in the Heat Exchanger: The amount of heat provided by the heat exchanger for DHW $H^{\text{HE-DHW}}(t)$ is:

$$H^{\text{HE-DHW}}(t) = Q^{\text{Steam-DHW}}(t)\eta^{\text{HE}} \tag{37.6}$$

where η^{HE} is efficiency of the heat exchanger.

Cooling in the Absorption Chiller: The amount of cooling provided by the absorption chiller $C^{\text{SChiller}}(t)$ can be modeled as:

$$C^{\text{SChiller}}(t) = Q^{\text{Steam-SC}}(t)\eta^{\text{HR,SChiller}}\text{COP}^{\text{SChiller}} \tag{37.7}$$

where $\eta^{\text{HR,SChiller}}$ denotes the heat recovery efficiency and $\text{COP}^{\text{SChiller}}$ represents the coefficient of performance of the chiller.

37.3.1.2 Modeling of Natural Gas Boilers
Gas Consumption of Natural Gas Boilers: The gas consumption for the nth natural gas boiler $G_n^{\text{Boiler}}(t)$ is modeled as :

$$G_n^{\text{Boiler}}(t) = \frac{H_n^{\text{Boiler}}(t)}{\eta^{\text{Boiler}}\text{HV}^{\text{Gas}}} \tag{37.8}$$

where η^{Boiler} denotes the efficiency of the boiler and $H_n^{\text{Boiler}}(t)$ shows the heat generation level of boiler (continuous decision). The modeling of CO_2 emissions $\text{Env}_n^{\text{boiler}}(t)$ is similar to that of GTs.

37.3.1.3 Modeling of Electric Chillers
Electricity Consumption of Electric Chillers: To provide certain amount of cooling $C_l^{\text{EChiller}}(t)$ (continuous decision), the electricity required by the lth chiller $P_l^{\text{EChiller}}(t)$ is modeled as follows.

$$P_l^{\text{EChiller}}(t) = \frac{C_l^{\text{EChiller}}(t)}{\eta^{\text{EChiller}}\text{COP}^{\text{EChiller}}} \tag{37.9}$$

where η^{EChiller} represents chiller efficiency, and $\text{COP}^{\text{EChiller}}$ denotes chiller's coefficient of performance.

37.3.2 Modeling of Uncertain Renewables

Ideally, the PV generation behaves like a sinusoidal wave with zero values for darkness hours [25, 26], with the amplitude and frequency depending on the PV capacities and locations, and seasons. It can change significantly, depending on weather conditions such as clouds. Therefore, the PV generation depends on the ideal generation and uncertain weather conditions. To avoid the computational complexity caused by scenario-based methods as discussed above in Section 37.2, a Markov-based model is adopted to integrate intermittent and uncertain PV generation into microgrids based on [18]. In the model, weather uncertainties are assumed to be a Markovian process with N sates (as a percentage of the ideal weather conditions), and sate i is denoted as W_i, following related real case studies in [27, 28].

Based on historical data, the probability that the current weather state is j if the previous state was i can be obtained as follows [29],

$$\pi_{i,j} = \frac{\text{observed transitions from state } i \text{ to } j}{\text{occurrences of state } i} \tag{37.10}$$

In this way, the state transition matrix can be established. To solve the problem for a specific region, the historical data should be analyzed to determine the number of states to balance modeling accuracy and computation efficiency. The transition matrix should be also updated by incorporating the latest weather forecast. Because of seasonal behaviors, for each season, a transition matrix is needed.

With the weather conditions modeled about, uncertain PV generation $P_i^{\text{PV}}(t)$ is also a Markovian process as follows:

$$P_i^{\text{PV}}(t) = P^{\text{IPV}}(t)W_iF^{\text{DF}}(t) \tag{37.11}$$

where P^{IPV} is the ideal PV generation, and $F^{\text{DF}}(t)$ is the day-night factor. The probability that the PV generation is $P_i^{\text{PV}}(t)$ at time t, denoted as $\phi_i(t)$, is the sum of the probabilities at time $t-1$ weighted by different transitions:

$$\phi_i(t) = \sum_j \pi_{j,i} \cdot \phi_j(t-1) \tag{37.12}$$

An example of the transition to state i from state j is shown below in Fig. 37.3.

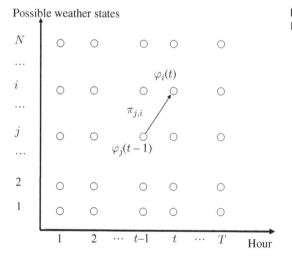

Possible weather states

Figure 37.3 The state transition of a Markovian process.

The probabilities of PV generation levels for future time slots can be obtained based on the initial PV generation state and the transition matrix.

37.3.3 Modeling of Battery and Other Devices Based on the PV States

In this section, the battery is modeled based on the PV states, and the above models for CCHP, boilers, and chillers are extended to incorporate PV states.

37.3.3.1 Modeling of Battery Based on PV States

To capture state dynamics, a simplified battery model is used here, under the assumption that charging/discharging efficiencies are 100%. Battery charge and discharge are extended to depend upon PV states. The state of charge at time t and under PV state i is denoted by $P_i^{\text{Bat}}(t)$. The 1-D equation on the state of change in the literature is extended to 2-D on the state of change and PV states as follows.

$$P_i^{\text{Bat}}(t) = P_j^{\text{Bat}}(t-1) + P_i^{\text{bc}}(t) - P_i^{\text{bd}}(t), \forall j, \forall i \in \{i | \varphi_i(t) \neq 0\} \tag{37.13}$$

where $P_i^{\text{bc}}(t)$ and $P_i^{\text{bd}}(t)$ denote the battery charge and discharge at time t (both continuous decisions), respectively. Moreover, it is to be noted that battery cannot be charged and discharged simultaneously.

37.3.3.2 Modeling of CCHP, Boilers, and Chillers Based on PV States

For an effective coordination, other devices are modeled as Markov processes correspondingly with states depending on the states of PV generation. The generation levels of CCHP, boilers, and electric chillers and the amount of power to/from the grid are therefore modeled to depend on PV states. Consider mth GT in CCHP as an example. For each PV state i, there is a corresponding generation level $P_{m,i}^{\text{GT}}(t)$. The other devices are modeled in a similar way.

37.3.4 Modeling of System Balance

The microgrid under discussion has both electrical and thermal energy sources and these sources must be balanced in terms of supply and demand as discussed below.

37.3.4.1 Electrical Balance

For electrical balance, detailed power flow models are not considered as in [4, 19–23]. The summation of electricity generated by PV panels and CCHP, discharged from batteries and bought from the grid must be equal to the summation of the electricity demand and electricity consumed by electric chillers, sold to the grid, and charged to batteries. This electricity balance constraint must be satisfied at each hour for each PV state where its probability is nonzero, that is

$$P_i^{\text{PV}}(t) + \sum_m P_{m,i}^{\text{GT}}(t) + P_i^{\text{bd}}(t) + P_i^{\text{buy}}(t)$$

$$= P^{\text{dem}}(t) + \sum_l P_{l,i}^{\text{EChiller}}(t) + P_i^{\text{sell}}(t) + P_i^{\text{bc}}(t), \forall i \in \{i | \varphi_i(t) \neq 0\} \tag{37.14}$$

where $P_i^{\text{sell}}(t)$ represents the amount of electricity sold to the grid and $P_i^{\text{buy}}(t)$ represents the amount of electricity bought from the grid (both continuous decisions). The demand $P^{\text{dem}}(t)$ is assumed to be provided.

37.3.4.2 Thermal Balance

For space heating, the summation of heat generated by natural gas boilers and provided by steam equals the demand $H^{\text{dem-SH}}(t)$, that is

$$\sum_n H_{n,i}^{\text{Boiler}}(t) + Q_i^{\text{Steam-SH}}(t) = H^{\text{dem-SH}}(t), \forall i \in \{i | \varphi_i(t) \neq 0\} \tag{37.15}$$

The thermal balance for space cooling and DHW is formulated in a similar way. The entire problem is therefore Markovian.

37.3.5 Objective Function

The objective of the microgrid operation problem is to minimize the total daily cost, i.e. energy and emission costs. The energy cost *Cost* comprises of three terms: buying natural gas from the station, buying electricity from the grid, and selling electricity back to the grid, that is,

$$\text{Cost} = \sum_t \sum_i \varphi_i(t) \left(C^{\text{Gas}} \times \left(\sum_m G_{m,i}^{\text{GT}}(t) + \sum_n G_{n,i}^{\text{boiler}}(t) \right) + C^{\text{Grid,buy}}(t) \right.$$

$$\left. \times P_i^{\text{buy}}(t) - C^{\text{Grid,sell}}(t) \times P_i^{\text{sell}}(t) \right) \Delta t \tag{37.16}$$

where $C^{\text{Grid,buy}}(t)$ and $C^{\text{Grid,sell}}(t)$ represents the unit price of electricity which is bought from grid and the unit price of electricity at which it is sold to grid at time t, respectively; C^{Gas} represents the unit price of natural gas, and Δt is time slot length.

To quantify the cost of CO_2 emissions caused by the natural gas combustion in GTs and gas boilers, carbon Tax *CarbonTax* is considered [30] here, that is,

$$\text{CarbonTax} = P^{\text{CTax}} \sum_t \sum_i \varphi_i(t) \times \left(\sum_m \text{Env}_{m,i}^{\text{GT}}(t) + \sum_n \text{Env}_{n,i}^{\text{Boiler}}(t) \right) \tag{37.17}$$

where P^{CTax} is the carbon tax on CO_2 emissions (\$/kg). Since the carbon tax associated with grid power is already incorporated in the grid price, therefore separate carbon tax for the grid is not considered for this optimization problem.

37.4 Solution Methodology

The problem formulated above is stochastic and linear and involves both discrete and continuous variables. Branch-and-cut, which is powerful for MILP problems, is used. In the method, the integrality requirements on variables are first relaxed, and the relaxed problem can be efficiently solved using a linear programming method. The solution also provides a lower bound to the original MILP problem. If the values of all integer decision variables turn out to be integers, the solution is optimal to the original problem. If not, valid cuts that do not cut off feasible integer solutions are added, trying to obtain the convex hull (the smallest convex set that contains all feasible solutions [31]). Once the convex hull is obtained, the values of all integer decision variables in the solution to the relaxed problem are integers, and this solution is optimal to the original problem. If the convex hull cannot be obtained by cuts, low-efficient branching operations are needed. Optimization stops when computational time reaches the preset stop time or the relative gap (relative difference between the objectives of the optimal relaxed solution and current integer solution) falls below the preset gap [32].

37.5   Exercises

This example is semi-realistic based on the microgrid of Kings Plaza in Brooklyn, NY. Consider a microgrid with PV, CCHP (four GTs, four steam generators, a steam-driven absorption chiller, and a heat exchanger), three electric chillers, and two natural gas boilers. Their sizes and efficiencies are show in Table 37.1.

The thermal efficiency of the GT is assumed the same as its gas-to-electric efficiency. The coefficients of performance of the absorption chiller and electric chillers are both 1.2.

Consider a typical sunny day and the ideal PV generation is given as follows:

$$P^{\text{IPV}}(t)(\text{kW}) = [0\ 0\ 0\ 0\ 0\ 390\ 765\ 1111\ 1414\ 1663\ 1848\ 1962$$
$$2000\ 1962\ 1848\ 1663\ 1414\ 1111\ 765\ 390\ 0\ 0\ 0\ 0]$$

The day-night factor for PV is:

$$F^{\text{DF}}(t) = [0\ 0\ 0\ 0\ 0\ 0.195\ 0.383\ 0.556\ 0.707\ 0.831\ 0.924\ 0.981$$
$$1\ 0.981\ 0.924\ 0.831\ 0.707\ 0.556\ 0.383\ 0.195\ 0\ 0\ 0\ 0]$$

Assume there are ten weather states as shown below, and the initial state is state 1.

$$W_i = \begin{bmatrix} 0 & 0.1 & 0.2 & 0.3 & 0.4 & 0.5 & 0.6 & 0.7 & 0.8 & 0.9 \end{bmatrix}$$

A ten-state transition matrix for PV generation is given as:

$$\pi_{i,j} = \begin{bmatrix} 0.862 & 0.124 & 0.014 & 0 & 0 & 0 & 0 & 0 & 0 & 0 \\ 0.153 & 0.644 & 0.136 & 0.034 & 0.017 & 0 & 0.008 & 0.008 & 0 & 0 \\ 0.041 & 0.219 & 0.467 & 0.178 & 0.041 & 0.027 & 0.027 & 0 & 0 & 0 \\ 0.024 & 0.071 & 0.239 & 0.214 & 0.143 & 0.119 & 0.071 & 0.095 & 0.024 & 0 \\ 0 & 0.059 & 0.147 & 0.088 & 0.235 & 0.354 & 0.088 & 0.029 & 0 & 0 \\ 0 & 0.022 & 0.067 & 0.133 & 0.156 & 0.378 & 0.178 & 0.022 & 0.022 & 0.022 \\ 0 & 0.026 & 0.026 & 0.103 & 0.154 & 0.179 & 0.306 & 0.154 & 0.026 & 0.026 \\ 0 & 0 & 0.029 & 0.057 & 0.057 & 0.029 & 0.2 & 0.342 & 0.257 & 0.029 \\ 0 & 0 & 0 & 0 & 0 & 0.042 & 0.042 & 0.332 & 0.292 & 0.292 \\ 0 & 0 & 0 & 0.067 & 0 & 0 & 0.133 & 0.2 & 0.267 & 0.333 \end{bmatrix}$$

Table 37.1   Device size and efficiencies.

Device	Size (kW)	Minimum generation (kW)	Efficiencies
GT	3200	600	0.403
Steam generator	N/A	N/A	0.8
Absorption chiller	3376	350	0.685
Exchanger	N/A	N/A	0.85
Electric chiller	3325	350	0.7
Boiler	4894	500	0.88

The hourly demand is given as

$$P^{\text{dem}}(t)(\text{kW}) = [1466\ 1466\ 1466\ 1466\ 1083\ 1179\ 1496\ 2406\ 3819$$
$$5489\ 5998\ 6235\ 6450\ 6677\ 6795\ 6815\ 6652\ 4572$$
$$3346\ 3819\ 2936\ 1466\ 1466\ 1466]$$

$$H^{\text{dem-SH}}(t)(\text{kW}) = [0\ 0\ 0\ 0\ 0\ 0\ 0\ 0\ 0\ 0\ 0\ 0$$
$$0\ 0\ 0\ 0\ 0\ 0\ 0\ 0\ 0\ 0\ 0\ 0]$$

$$H^{\text{dem-DHW}}(t)(\text{kW}) = [104\ 100\ 96\ 92\ 92\ 96\ 104\ 108\ 333\ 292\ 333\ 267\ 250$$
$$267\ 333\ 267\ 250\ 292\ 333\ 417\ 425\ 433\ 333\ 208]$$

$$C^{\text{dem-SC}}(t)(\text{kW}) = [1583\ 1250\ 917\ 875\ 833\ 750\ 1167\ 1250\ 1667$$
$$1875\ 2250\ 1875\ 4000\ 4167\ 5000\ 4583\ 4167\ 4167$$
$$4167\ 4250\ 4333\ 2500\ 1250]$$

The time-varying grid price is shown below, where the selling-back price to the grid is set as 75% of the grid price,

$$C^{\text{Grid,buy}}(t)(\$) = [0.1397\ 0.1397\ 0.1397\ 0.1397\ 0.1397\ 0.1397\ 0.1397$$
$$0.1397\ 0.4165\ 0.4165\ 0.4165\ 0.4165\ 0.4165\ 0.4165$$
$$1.1963\ 1.1963\ 1.1963\ 1.1963\ 1.1963\ 0.4165\ 0.4165$$
$$0.4165\ 0.4165\ 0.4165]$$

For natural gas, the heat value is 9.54 kWh/m^3; the price is \$0.27/m^3; and the carbon intensity is 0.202 kg/kWh. The carbon tax is set as \$0.043/kg. Electricity and natural gas can be used with no limits.

Based on the given data, solve the operation problem for 24 hours ($t = 1.24$) and determine:

1. What is the minimum energy cost and mission cost?
2. What are the optimized operation strategies of devices and why?
3. What is the energy cost and mission cost without PV?
4. What is the energy cost and mission cost without the steam-driven absorption chiller?
5. What is the energy cost and mission cost without the natural gas boilers?
6. What is the total energy cost and mission cost if the microgrid is not connected to the grid?
7. Compare the total energy cost and emission obtained from 2 to 6 and analyze the impacts of each device and grid power on the microgrid (both energy and emission).

References

1 Hatziargyriou, N., Asano, H., Iravani, R., and Marnay, C. (2007). Microgrids. *IEEE Power and Energy Magazine* 5 (4): 78–94.

2 Hatziargyriou, N., Asano, H., Iravani, R., and Marnay, C. (2006). Power management strategies for a microgrid with multiple distributed generation units. *IEEE Transactions on Power Systems* 21 (4): 1821–1831.

3 Semero, Y., Zhang, J., and Zheng, D. (2019). Optimal energy management strategy in microgrids with mixed energy resources and energy storage system. *IET Cyber Physical Systems: Theory and Applications* 5 (1): 80–84.

4 Guan, X., Xu, Z., and Jia, Q.-S. (2010). Energy-efficient buildings facilitated by microgrid. *IEEE Transactions on Smart Grid* 1 (3): 243–252.

5 Wang, Y., Tang, L., Yang, Y. et al. (2020). A stochastic-robust coordinated optimization model for CCHP micro-grid considering multi-energy operation and power trading with electricity markets under uncertainties. *Science Direct: Energy* 189 (5): 117273.

6 Nagarajan, A. and Ayyanar, R. (2016). Design and scheduling of microgrids using benders decomposition. *IEEE 43rd Photovoltaic Specialist Conference*, 1843–1847.

7 Zaree, N., Vahidinasab, V., and Estebsari, A. (2018). Energy management strategy of microgrids based on benders decomposition method. *IEEE 2018 International Conference on Environment and Electrical Engineering and* Industrial and Commercial Power Systems Europe, 1–6.

8 Jamalzadeh, R. and Hong, M. (2018). Microgrid optimal power flow using the generalized benders decomposition approach. *IEEE Transactions on Sustainable Energy* 10 (4): 2050–2064.

9 Papadaskalopoulos, D., Pudjianto, D., and Strbac, G. (2014). Decentralized coordination of microgrids with flexible demand and energy storage. *IEEE Transactions on Sustainable Energy* 5 (4): 1406–1414.

10 Wang, L., Zhang, Y., Song, W., and Li, Q. (2021). Stochastic cooperative bidding strategy for multiple microgrids with peer-to-peer energy trading. *IEEE Transactions on Industrial Informatics* 18 (3): 1447–1457.

11 Zhao, Z., Guo, J., Luo, X. et al. (2022). Distributed robust model predictive control-based energy management strategy for islanded multi-microgrids considering uncertainty. *IEEE Transactions on Smart Grid* 13 (3): 2107–2120.

12 Zhao, H., Tanneau, M., and Van Hentenryck, P. (2022). A linear outer approximation of line losses for DC-based optimal power flow problems. *Science Direct Electric Power Systems Research* 212: 108272.

13 Fang, C., Chen, L., Zhang, Y. et al. (2014). Operation of low-carbon-emission microgrid considering wind power generation and compressed air energy storage. *IEEE Proceedings of the 33rd Chinese Control Conference*, 7472–7477.

14 Gao, H., Liu, J., Wang, L., and Liu, Y. (2017). Cutting planes based relaxed optimal power flow in active distribution systems. *Science Direct Electric Power Systems Research* 143: 272–280.

15 Daui, Y., Liu, M., Tang, C., and Wang Z. (2018). Fully distributed dynamic optimal power flow of active distribution networks with multi-microgrids based on cutting plane consensus method and ward equivalent. *IEEE International Conference on Power System Technology*, 1496–1503

16 Lara, J., Oliveras, D., and Cañizares, C. (2018). Robust energy management of isolated microgrids. *IEEE Systems Journal* 13 (1): 680–691.

17 Luh, P.B., Yu, Y., Zhang, B. et al. (2013). Grid integration of intermittent wind generation: a Markovian approach. *IEEE Transactions on Smart Grid* 5 (2): 732–741.

18 Yan, B., Fan, H., Luh, P.B. et al. (2017). Grid integration of wind generation considering remote wind farms: hybrid Markovian and interval unit commitment. *IEEE/CAA Journal of Automatica Sinica* 4 (2): 205–215.

19 Mohammadi, S., Soleymani, S., and Mozafari, B. (2014). Scenario-based stochastic operation management of microgrid including wind, photovoltaic, micro-turbine, fuel cell and energy storage devices. *International Journal of Electrical Power & Energy Systems* 54: 525–535.

20 Chen, Y.H., Lu, S.Y., Chang, Y.R. et al. (2013). Economic analysis and optimal energy management models for microgrid systems: a case study in Taiwan. *Applied Energy* 103: 145–154.

21 Morais, H., Kádár, P., Faria, P. et al. (2010). Optimal scheduling of a renewable micro-grid in an isolated load area using mixed-integer linear programming. *Renewable Energy* 32 (1): 151–156.

22 Olivares, D.E., Lara, J.D., Cañizares, C.A., and Kazerani, M. (2015). Stochastic-predictive energy management system for isolated microgrids. *IEEE Transactions on Smart Grid* 6 (6): 2681–2693.

23 Ross, M., Hidalgo, R., Abbey, C., and Joós, G. (2011). Energy storage system scheduling for an isolated microgrid. *IET Renewable Power Generation* 5 (2): 117–123.

24 Hawkes, A.D. and Leach, M.A. (2009). Modelling high level system design and unit commitment for a microgrid. *Applied Energy* 86 (8-9): 1253–1265.

25 Perez, P.S., Driesen, J., and Belmans, R. (2007). Characterization of the solar power impact in the grid. *Proceedings of International Conference on Clean Electrical Power*, 366–371.

26 Palz, W. (1982). *Photovoltaic Power Generation D*, 45. Reidel Publishing Company.

27 Poggi, P., Notton, G., Muselli, M., and Louche, A. (2000). Stochastic study of hourly total solar radiation in Corsica using a Markov model. *International Journal of Climatology* 20 (14): 1843–1860.

28 Ngoko, B.O., Sugihara, H., and Funaki, T. (2014). Synthetic generation of high temporal resolution solar radiation data using Markov models. *Solar Energy* 103: 160–170.

29 Weber, C., Meibom, P., Barth, R., and Brand, H. (2009). WILMAR: a stochastic programming tool to analyze the large-scale integration of wind energy. In: *Optimization in the Energy Industry. Energy Systems*, Chapter 19 (ed. J. Kallrath, P.M. Pardalos, S. Rebennack, and M. Scheidt), 437–458. Berlin, Heidelberg: Springer-Verlag.

30 Center for Climate and Energy Solutions (2013). Options and Considerations for a Federal Carbon Tax. http://www.c2es.org/publications/optionsconsiderations-federal-carbon-tax (accessed 30 August 2022)

31 Bertsekas, D.P. (2016). *Nonlinear Programming*. Belmont, MA: Athena Scientific.

32 IBM (2022). IBM ILOG CPLEX Optimization Studio CPLEX User's Manual, Sunnyvale, CA, USA.

Index

Microgrids: Theory and Practice, First Edition. Edited by Peng Zhang.
© 2024 The Institute of Electrical and Electronics Engineers, Inc. Published 2024 by John Wiley & Sons, Inc.

IEEE Press Series on Power and Energy Systems

Series Editor: Ganesh Kumar Venayagamoorthy, Clemson University, Clemson, South Carolina, USA.

The mission of the IEEE Press Series on Power and Energy Systems is to publish leading-edge books that cover a broad spectrum of current and forward-looking technologies in the fast-moving area of power and energy systems including smart grid, renewable energy systems, electric vehi- cles and related areas. Our target audience includes power and energy systems professionals from academia, industry and government who are interested in enhancing their knowledge and per- spectives in their areas of interest.

Printed and bound by CPI Group (UK) Ltd, Croydon, CR0 4YY

27/10/2024

14580682-0001